U.S. Customary and Metric Comparisons	To Change From	To	Multiply By
Length			
1 meter =	meters	inches	39.37
39.37 inches	inches	meters	0.0254
1 meter =	meters	feet	3.2808
3.2808 feet	feet	meters	0.3048
1 meter =	meters	yards	1.0936
1.0936 yards	yards	meters	0.9144
1 centimeter =	centimeters	inches	0.3936
0.3937 inch	inches	centimeters	2.54
1 millimeter =	millimeters	inches	0.03937
0.03937 inch	inches	millimeters	25.4
1 kilometer =	kilometers	miles	0.6214
0.6214 mile	miles	kilometers	1.6093
Weight or Mass			
1 gram =	grams	ounces	0.0353
0.0353 ounce	ounces	grams	28.3286
1 kilogram =	kilograms	pounds	2.2046
2.2046 pounds	pounds	kilograms	0.4536
Liquid Capacity			
1 liter =	liters	quarts	1.0567
1.0567 quarts	quarts	liters	0.9463
Capacity or Volume			
1 cubic inch =	cubic inches	cubic centimeters	16.387
16.387 cubic centimeters	cubic centimeters	cubic inches	0.0610
1 cubic inch =	cubic inches	liters	0.01639
0.01639 liters	liters	cubic inches	61.0128
1 cubic foot =	cubic feet	cubic meters	0.0283
0.0283 cubic meter	cubic meters	cubic feet	35.3357
1 teaspoon =	teaspoons	milliliters	4.93
4.93 milliliters	milliliters	teaspoons	0.2028
1 tablespoon =	tablespoons	milliliters	14.97
14.97 milliliters	milliliters	tablespoons	0.0668
1 fluid ounce =	fluid ounces	milliliters	29.57
29.57 milliliters	milliliters	fluid ounces	0.0338
1 cup = 0.24 liters	cups	liters	0.24
	liters	cups	4.1667
1 pint = 0.47 liters	pints	liters	0.47
	liters	pints	2.1277
1 gallon =	gallons	cubic meters	0.00379
0.00379 cubic meters	cubic meters	gallons	263.85

Special Algebra Patterns for Factoring

$a^2 + 2ab + b^2 = (a + b)^2$

$a^2 - b^2 = (a + b)(a - b)$

$a^3 + b^3 = (a + b)(a^2 - ab + b^2)$

$a^3 - b^3 = (a - b)(a^2 + ab + b^2)$

Symbols

$+$	Add
$-$	Subtract
$\times, \cdot, *, (\ )(\ )$	Multiply
$\div, \overline{)\ }, /, -$	Divide
$=$	Equal to
$\approx$	Approximately equal to
$\neq$	Not equal to
$\%$	Percent
$>$	Greater than
$<$	Less than
$\geq$	Greater than or equal to
$\leq$	Less than or equal to
$\sqrt{\ }$	Radical sign or square root
$(\ ), [\ ], \{\ \}, -$	Grouping symbols
$\|\ \|$	Absolute value
$f(x)$	Function notation, read "f of x"
$\overleftrightarrow{AB}$	Line AB
$\overline{AB}$	Line segment AB
$\overrightarrow{AB}$	Ray AB
$\simeq, \cong$	Congruent to
$\sim$	Similar to (geometric figures)
$\angle$	Angle
$\parallel$	Parallel
$\perp$	Perpendicular
$\triangle$	Triangle
$\bigcirc$	Circle
$\llcorner$	Right angle
Δ	Delta, change, used with slope
$\{\ldots | \ldots\}$	Such that, used with set notation
Σ	Summation
x_1	Subscript (1)
$\{\ \}, \phi$	Empty or null set
$\in$	Is an element of
$\cup$	Union (of sets)
$\cap$	Intersection (of sets)
π	Constant—Pi (ratio of diameter to circumference of circle, approximately 3.141592654)
e	Constant—natural exponential; from $\left(1 + \dfrac{1}{n}\right)^n$ where $n \to \infty$, approximately 2.718281828
i	The square root of -1; $\sqrt{-1}$
∞	Infinity
$\therefore$	Therefore
$\exists$	There exists
$\forall$	For every

BASIC MATH, ALGEBRA AND GEOMETRY WITH APPLICATIONS

SECOND EDITION

BY CHERYL CLEAVES AND MARGIE HOBBS

Taken From:

College Mathematics, Seventh Edition
By Cheryl Cleaves and Margie Hobbs

Taken from:

College Mathematics, Seventh Edition
By Cheryl Cleaves and Margie Hobbs
Copyright © 2006 by Prentice-Hall, Inc.
A Pearson Education Company
Upper Saddle River, New Jersey 07458

This special edition published in cooperation with Pearson Custom Publishing.

Printed in the United States of America

10 9 8 7 6 5 4 3 2 1

ISBN 0-536-10595-2

2005360760

RA

Please visit our web site at *www.pearsoncustom.com*

PEARSON CUSTOM PUBLISHING
75 Arlington Street, Suite 300, Boston, MA 02116
A Pearson Education Company

Preface

In *Basic Math, Algebra, and Geometry with Applications,* Second Edition, we have preserved all the features in the previous edition that have made this one of the most appropriate texts on the market for a comprehensive study of mathematics in general education and in career programs. We continue to use real-life situations as a context for applied problems.

Changes in the Second Edition

This edition incorporates many valuable suggestions made by users of the first edition. As a result, we have placed greater emphasis on problem solving, included new material, and rearranged some topics. Content changes include the following:

- Topics have been rearranged to distinguish more clearly between linear and non-linear equations. Graphing linear equations, equations of lines, and systems of linear equations now appear before quadratic equations are introduced.
- Many related formulas have been brought together to form more logical units of study. For example, Heron's formula for finding the area of a triangle has been added. In Chapter 18 both radian and degree measures of angles are introduced and the formulas for finding arc length and the area of a sector are presented for both types of angle measures. Students can better correlate the two notations for angle measure.
- Evaluating and rearranging formulas has been moved to the chapter on linear equations (Chapter 7). This will help students more immediately understand the usefulness of algebra.
- Geometry (perimeter, area, and volume) has been collected in the geometry chapter (Chapter 18). This provides a more organized body of study for students and allows easier reference. It also is a more efficient presentation of the material.
- U.S. customary measures are introduced in Chapter 2, "Review of Fractions." This change allows students to experience authentic applications of fractions earlier.
- Chapter 6, "Interpreting and Analyzing Data," has been renamed "Statistics." Students are more familiar with this terminology. Because these topics can be incorporated at many different points in a program of study, instructors can use this material at any point after Chapter 3.

To help students connect math with their career path and develop collaborative skills, we have added these new features to the book:

- **Focus on Careers:** This feature opens each chapter with an interesting overview of selected careers and their job outlook, so students will become familiar with the type of career information that is provided on the Internet from the U.S. Bureau of Labor Statistics (http://www.bls.gov/oco/).
- **Teamwork Exercises:** These activities provide students the opportunity to develop and refine team interaction skills. They use skills that have been identified by employers as important skills for employees.
- **Career Coding:** Examples, section self-study exercises, and chapter review exercises have been coded using 14 different career categories. This will strengthen the students' ability to make connections between mathematical concepts and career applications. An index of applications by career code follows the table of contents.

AG/H	Agriculture/Horticulture	Agriculture, Horticulture
AUTO	Auto/Diesel	Automobile and Diesel Mechanics
AVIA	Aviation	Aviation, Geographical Information Systems
BUS	Business	Business Administration, Accounting, Personal Finance, Real Estate
CAD/ARC	CAD/Drafting/Architecture	CAD, Drafting, Architecture, Graphic Communication
COMP	Computer Technologies	Computer Tech, Information Systems, Information Technology, Network Technology
CON	Construction Trades	Construction, Carpentry, Electrical, Plumbing, HVAC, Pipe Fitting
ELEC	Electronics	Electronics Technology, Computer Electronics
HELPP	Helping Professions	Criminal Justice, Fire Science, Counseling, Education
HLTH/N	Health and Food	Allied Health, Nursing,
HOSP	Hospitality	Hotel and Restaurant Management
INDTEC	Industrial Technologies	Manufacturing, Industrial, Machine, Engineering Technologies
INDTR	Industrial Trades	Welding, Machine Tools, Industrial Maintenance
TELE	Telecommunication	Telecommunication Technology

- **Cumulative Practice Tests:** To help students assess their understanding of the material as the course progresses, new cumulative tests have been added. Five tests are included after selected chapters to enhance students' assimilation of mathematical concepts.

Our goal is to present a systematic framework for successful learning in mathematics that will strengthen students' *mathematical sense* and give students a greater appreciation for the power of mathematics in everyday life and in the workplace. The new material in this edition has been added to broaden the usefulness of the text. Many of the explanations have been enhanced with carefully constructed visualizations. Exercises have been updated and new ones added.

Commitment to Improving Mathematics Education

The authors continue to be active in the development, revision, and implementation of the standards (*Beyond Crossroads*) of the American Mathematical Association of Two-Year Colleges (AMATYC). We enthusiastically promote the standards and guidelines encouraged by AMATYC, NCTM, MAA, and the SCANS document.

Calculator Usage

Calculator tips appropriate for both scientific and graphing calculators are periodically included. These generic tips guide students to use critical thinking to determine how their calculator operates without referring to a user's manual.

We continue to emphasize the calculator as a tool that *facilitates* learning and understanding. Assessment strategies are included throughout the text and supplementary materials to enable students to test their understanding of a concept independently of their calculator.

Additional Resources

A variety of instructional tools are available with adoption of this text including a printed **Instructor's Resource Manual,** printed **Test Item File, TestGen** computerized test generation software, a **Student Solutions Manual,** and a **Companion Website** at **www.prenhall.com/cleaves.**

The Instructor's Resource Manual, Test Item File, and TestGen computerized test generator are also downloadable from our Instructor Resource Center. Go to **www.prenhall.com,** click the **Instructor Resource Center** link, and then click **Register Today** for an instructor access code. Within 48 hours after registering you will receive a confirming e-mail including an instructor access code. Once you have received your code, go to the site and log on for full instructions on downloading the materials you wish to use.

New! OneKey MyMathTutor in WebCT and Blackboard Distance Learning Courses New for this edition are courses pre-made in WebCT and Blackboard, which contain the following material for instructors and students:

On student site:

* Learning outcomes for each chapter
* Tutorial narrative with additional examples for each learning outcome
* Video tutorial instruction for each learning outcome
* Practice exercises with solutions and explanatory notes for each learning outcome

On instructor site:

* One pre-made quiz per chapter that feeds the gradebook automatically
* Four pre-made sectional exams with gradebook
* Test bank with gradebook
* Instructor's manual with teaching notes and solutions for even-numbered end-of-chapter exercises

Acknowledgments

A project such as this does not come together without help from many people. Our first avenue for input is through students and faculty who use the text. We especially thank our colleagues and students at Southwest Tennessee Community College, University of Memphis, and The University of Mississippi. Their comments and suggestions have been invaluable.

We wish to express thanks to all the people who helped make this edition a reality. In particular, we thank Gary Bauer, Senior Acquisitions Editor, whose belief in our work and support of our ideas have been a major factor in the success of this text. We thank Louise Sette, Prentice Hall production editor. We also thank Ann Imhof of Carlisle Publishers Services.

The teaching of mathematics over time produces a wealth of knowledge about instructional strategies and specific content. We are grateful for the many valuable suggestions received in these areas. We wish to thank the following individuals:

Stan Adamski, Owens Community College, Ohio (OH)

Milton Clark, Florence Darlington Technical College (SC)

Virginia Dewey, York Technical College, South Carolina (SC)

Terry B. Gaalswyk, Western Iowa Technical Community College (IA)

John Gillis, Portsmouth Naval Shipyard Apprentice Program, under the auspices of New Hampshire Community Technical College, (Portsmouth, NJ) and York County Community College, (Wells, ME)

Brent Hamilton, North Iowa Area Community College (IA)

Edwin G. Landauer, Clackamas Community College (OR)

Nicole Muth, Lakeshore Technical College (WI)

Karen Newson, Triangle Tech (PA)

Justin Ostrander, Manhattan Area Technical College, Kansas (KS)

Scott Randby, University of Akron (OH)

Henry Regis, Valencia Community College, Florida (FL)

Behnaz Rouhani, Athens Area Technical Institute (GA)

David C. Shellabarger, Lane Community College (OR)

Joseph Sukta, Moraine Valley Community College (IL)

Jimmie A. Van Alphen, Ozarks Technical Community College (MO)

We appreciate the assistance we received in ensuring the accuracy of the text. These colleagues spent many hours ensuring the accuracy of the text. However we take full responsibility for any misprints or errors that may remain.

Shirley Nehrbass, Ozarks Technical Community College (MO)

Ted Nehrbass, Ozarks Technical Community College (MO)

Marissa Wolfe, Ozarks Technical Community College (MO)

Mary Harrington, The University of Mississippi (MS)

The resources associated with this text have been expanded through the addition of MyMathTutor. We are grateful to the faculty who provided some elements of MyMathTutor.

Ezell Allen, Southwest Tennessee Community College (TN)

Sheldon Dan, Southwest Tennessee Community College (TN)

Lisa Loden, Southwest Tennessee Community College (TN)

Ravi Mehra, Southwest Tennessee Community College (TN)

Bridgett Smith, Southwest Tennessee Community College (TN)

Freddie Wabwire, Southwest Tennessee Community College (TN)

Additionally, we thank Ezell Allen and Lisa Loden for providing the video component of MyMathTutor.

Finally, we thank our familes, especially Charles Cleaves and Allen Hobbs.

Cheryl Cleaves
Margie Hobbs

To the Student

The mathematics you learn from this book will help you advance on your career path. We have given much thought to the best way to teach mathematics and have done extensive research on how students learn. We have provided a wide variety of features and resources so that you can customize your study to your needs and circumstances. The following features are key to helping you learn the mathematics in this text.

Table of Contents. The table of contents is your "roadmap" to this text. Study it carefully to determine how the topics are arranged. This will aid you in relating topics to each other.

Glossary/Index. An extensive glossary/index is an important part of every mathematics book. Use the index to cross-reference topics and to locate other topics that relate to the topic you are studying.

Focus on Careers

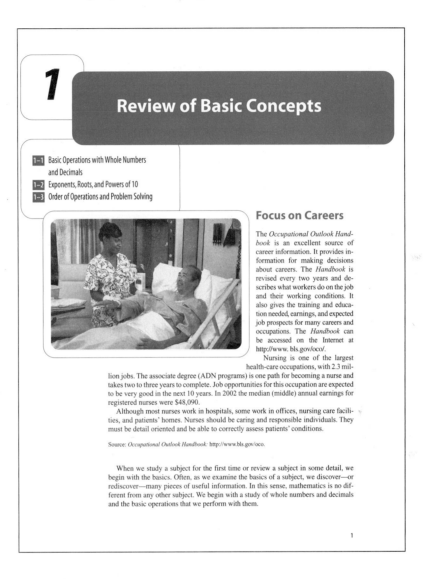

1

Review of Basic Concepts

1–1 Basic Operations with Whole Numbers and Decimals
1–2 Exponents, Roots, and Powers of 10
1–3 Order of Operations and Problem Solving

Focus on Careers

The *Occupational Outlook Handbook* is an excellent source of career information. It provides information for making decisions about careers. The *Handbook* is revised every two years and describes what workers do on the job and their working conditions. It also gives the training and education needed, earnings, and expected job prospects for many careers and occupations. The *Handbook* can be accessed on the Internet at http://www.bls.gov/oco/.

Nursing is one of the largest health-care occupations, with 2.3 million jobs. The associate degree (ADN programs) is one path for becoming a nurse and takes two to three years to complete. Job opportunities for this occupation are expected to be very good in the next 10 years. In 2002 the median (middle) annual earnings for registered nurses were $48,090.

Although most nurses work in hospitals, some work in offices, nursing care facilities, and patients' homes. Nurses should be caring and responsible individuals. They must be detail oriented and be able to correctly assess patients' conditions.

Source: *Occupational Outlook Handbook:* http://www.bls.gov/oco.

When we study a subject for the first time or review a subject in some detail, we begin with the basics. Often, as we examine the basics of a subject, we discover—or rediscover—many pieces of useful information. In this sense, mathematics is no different from any other subject. We begin with a study of whole numbers and decimals and the basic operations that we perform with them.

1

Focus on Careers. Each chapter opens with interesting information about a selected career. The article includes a description of the type of work done in this career, the expected salary, and the job prospects. A good resource for investigating other careers is the website for the U.S. Department of Labor, Bureau of Labor Statistics (http://www.bls.gov/oco/).

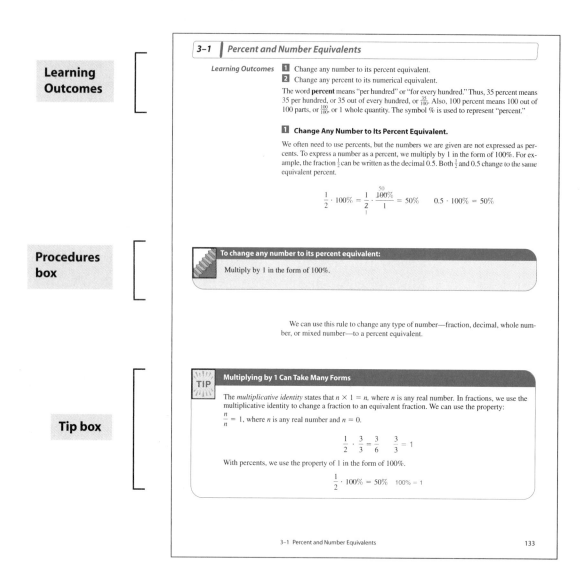

Learning Outcomes column label:

Learning Outcomes

Procedures box column label:

Procedures box

Tip box column label:

Tip box

Contents of the sample page:

Learning Outcomes
1. Change any number to its percent equivalent.
2. Change any percent to its numerical equivalent.

The word **percent** means "per hundred" or "for every hundred." Thus, 35 percent means 35 per hundred, or 35 out of every hundred, or $\frac{35}{100}$. Also, 100 percent means 100 out of 100 parts, or $\frac{100}{100}$, or 1 whole quantity. The symbol % is used to represent "percent."

1 Change Any Number to Its Percent Equivalent.

We often need to use percents, but the numbers we are given are not expressed as percents. To express a number as a percent, we multiply by 1 in the form of 100%. For example, the fraction $\frac{1}{2}$ can be written as the decimal 0.5. Both $\frac{1}{2}$ and 0.5 change to the same equivalent percent.

$$\frac{1}{2} \cdot 100\% = \frac{1}{2} \cdot \frac{\overset{50}{\cancel{100\%}}}{1} = 50\% \qquad 0.5 \cdot 100\% = 50\%$$

To change any number to its percent equivalent:

Multiply by 1 in the form of 100%.

We can use this rule to change any type of number—fraction, decimal, whole number, or mixed number—to a percent equivalent.

TIP

Multiplying by 1 Can Take Many Forms

The *multiplicative identity* states that $n \times 1 = n$, where n is any real number. In fractions, we use the multiplicative identity to change a fraction to an equivalent fraction. We can use the property: $\frac{n}{n} = 1$, where n is any real number and $n = 0$.

$$\frac{1}{2} \cdot \frac{3}{3} = \frac{3}{6} \qquad \frac{3}{3} = 1$$

With percents, we use the property of 1 in the form of 100%.

$$\frac{1}{2} \cdot 100\% = 50\% \qquad 100\% = 1$$

3–1 Percent and Number Equivalents   133

Learning Outcomes.   A learning outcome is what you should be able to do when you master a concept. These outcome statements can guide you through your study plan. Each section begins with a statement of learning outcomes that shows you what you should look for and learn in that section. If you read and think about these outcomes before you begin the section, you will know what to look for as you work through the section. Section Self-Study Exercises are organized by learning outcomes, and the Chapter Review of Key Concepts give procedures to review and a worked example for each learning outcome.

Procedures Boxes.   Each learning outcome has one or more procedures boxes. These boxes provide rules or procedures presented as numbered steps. A procedures box may also present a mathematical property, formula, or fact.

Tip Boxes.   These boxes give helpful hints for doing mathematics, and they draw your attention to important observations and connections that you may have missed in an example.

Six-Step Problem Solving Example with Explanatory Comments

Career coding in examples

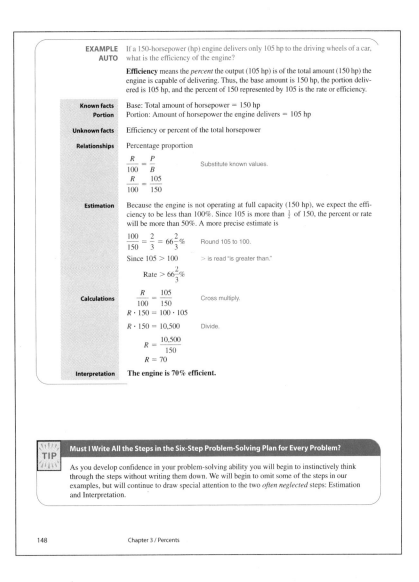

Use of Color in the Text. As you read the text and work through the examples, notice the items shaded with color or gray. These will help you follow the logic of working through the example. Color also highlights important items and boxed features such as the Tips, Learning Outcomes, rules, procedures, and formulas.

Six-Step Approach to Problem Solving. Successful problem solvers use a systematic, logical approach. We use a six-step approach to problem solving. This approach gives you a system for solving a variety of math problems. You will learn how to organize the information given and how to develop a logical plan for solving the problem. You are asked to analyze and compare and to estimate as you solve problems. Estimation helps you decide whether your answer is reasonable. You will learn to interpret the results of your calculations within the problem's context, a skill you will use on the job.

Career-Coded Examples and Exercises. Applied problems focus on a wide variety of careers available as a course of study at your community college, technology center, or university. These careers are grouped into 14 categories, and the examples and exercises are coded to these categories as appropriate. An index of applications is provided after the table of contents for your convenience.

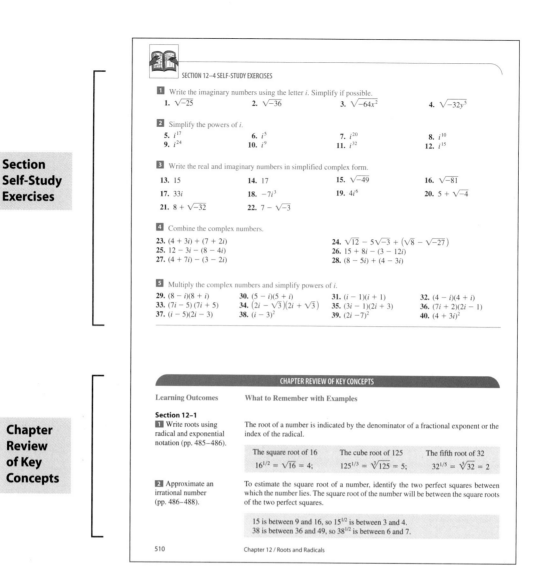

SECTION 12–4 SELF-STUDY EXERCISES

1 Write the imaginary numbers using the letter i. Simplify if possible.

1. $\sqrt{-25}$ 2. $\sqrt{-36}$ 3. $\sqrt{-64x^2}$ 4. $\sqrt{-32y^5}$

2 Simplify the powers of i.

5. i^{17} 6. i^5 7. i^{20} 8. i^{10}

9. i^{24} 10. i^9 11. i^{32} 12. i^{15}

3 Write the real and imaginary numbers in simplified complex form.

13. 15   14. 17   15. $\sqrt{-49}$ 16. $\sqrt{-81}$

17. $33i$ 18. $-7i^3$ 19. $4i^6$ 20. $5 + \sqrt{-4}$

21. $8 + \sqrt{-32}$ 22. $7 - \sqrt{-3}$

4 Combine the complex numbers.

23. $(4 + 3i) + (7 + 2i)$ 24. $\sqrt{12} - 5\sqrt{-3} + (\sqrt{8} - \sqrt{-27})$

25. $12 - 3i - (8 - 4i)$ 26. $15 + 8i - (3 - 12i)$

27. $(4 + 7i) - (3 - 2i)$ 28. $(8 - 5i) + (4 - 3i)$

5 Multiply the complex numbers and simplify powers of i.

29. $(8 - i)(8 + i)$ 30. $(5 - i)(5 + i)$ 31. $(i - 1)(i + 1)$ 32. $(4 - i)(4 + i)$

33. $(7i - 5)(7i + 5)$ 34. $(2i - \sqrt{3})(2i + \sqrt{3})$ 35. $(3i - 1)(2i + 3)$ 36. $(7i + 2)(2i - 1)$

37. $(i - 5)(2i - 3)$ 38. $(i - 3)^2$ 39. $(2i - 7)^2$ 40. $(4 + 3i)^2$

CHAPTER REVIEW OF KEY CONCEPTS

Learning Outcomes

Section 12–1

1 Write roots using radical and exponential notation (pp. 485–486).

2 Approximate an irrational number (pp. 486–488).

What to Remember with Examples

The root of a number is indicated by the denominator of a fractional exponent or the index of the radical.

The square root of 16	The cube root of 125	The fifth root of 32
$16^{1/2} = \sqrt{16} = 4;$	$125^{1/3} = \sqrt[3]{125} = 5;$	$32^{1/5} = \sqrt[5]{32} = 2$

To estimate the square root of a number, identify the two perfect squares between which the number lies. The square root of the number will be between the square roots of the two perfect squares.

15 is between 9 and 16, so $15^{1/2}$ is between 3 and 4.
38 is between 36 and 49, so $38^{1/2}$ is between 6 and 7.

510   Chapter 12 / Roots and Radicals

Using Your Calculator.   Calculators are useful in all levels of mathematics. Some tips introduce easy-to-follow calculator strategies. The tips show you how to analyze the procedure and set up a problem for a calculator solution; a sample series of keystrokes is often included. In addition, the tips help you determine how your type of calculator operates for various mathematical processes.

Section Self-Study Exercises.   These practice sets are keyed to the learning outcomes and appear at the end of each section. Use these exercises to check your understanding of the section. **The answers to every exercise are at the end of the text,** so you can get immediate feedback on whether you understand the concepts.

Chapter Review of Key Concepts.   Each chapter includes a summary in the form of a two-column chart. The first column lists the learning outcomes of the chapter. The second column gives the procedures and examples for each outcome. Page references are included to facilitate your preview or review of the chapter.

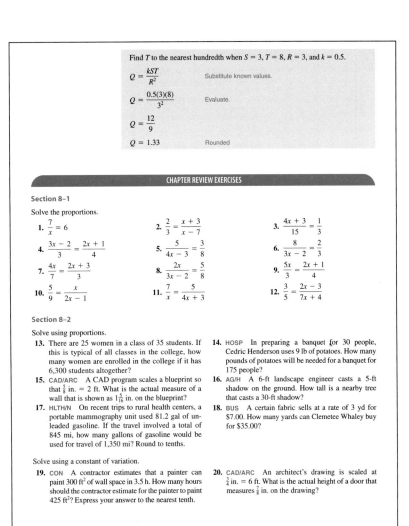

Find T to the nearest hundredth when $S = 3$, $T = 8$, $R = 3$, and $k = 0.5$.

$Q = \dfrac{kST}{R^2}$ Substitute known values.

$Q = \dfrac{0.5(3)(8)}{3^2}$ Evaluate.

$Q = \dfrac{12}{9}$

$Q = 1.33$ Rounded

CHAPTER REVIEW EXERCISES

Section 8–1

Solve the proportions.

1. $\dfrac{7}{x} = 6$

2. $\dfrac{2}{3} = \dfrac{x+3}{x-7}$

3. $\dfrac{4x+3}{15} = \dfrac{1}{3}$

4. $\dfrac{3x-2}{3} = \dfrac{2x+1}{4}$

5. $\dfrac{5}{4x-3} = \dfrac{3}{8}$

6. $\dfrac{8}{3x-2} = \dfrac{2}{3}$

7. $\dfrac{4x}{7} = \dfrac{2x+3}{3}$

8. $\dfrac{2x}{3x-2} = \dfrac{5}{8}$

9. $\dfrac{5x}{3} = \dfrac{2x+1}{4}$

10. $\dfrac{5}{9} = \dfrac{x}{2x-1}$

11. $\dfrac{7}{x} = \dfrac{5}{4x+3}$

12. $\dfrac{3}{5} = \dfrac{2x-3}{7x+4}$

Section 8–2

Solve using proportions.

13. There are 25 women in a class of 35 students. If this is typical of all classes in the college, how many women are enrolled in the college if it has 6,300 students altogether?

14. HOSP In preparing a banquet for 30 people, Cedric Henderson uses 9 lb of potatoes. How many pounds of potatoes will be needed for a banquet for 175 people?

15. CAD/ARC A CAD program scales a blueprint so that $\frac{3}{8}$ in. = 2 ft. What is the actual measure of a wall that is shown as $1\frac{5}{16}$ in. on the blueprint?

16. AG/H A 6-ft landscape engineer casts a 5-ft shadow on the ground. How tall is a nearby tree that casts a 30-ft shadow?

17. HLTH/N On recent trips to rural health centers, a portable mammography unit used 81.2 gal of un-leaded gasoline. If the travel involved a total of 845 mi, how many gallons of gasoline would be used for travel of 1,350 mi? Round to tenths.

18. BUS A certain fabric sells at a rate of 3 yd for $7.00. How many yards can Clemetee Whaley buy for $35.00?

Solve using a constant of variation.

19. CON A contractor estimates that a painter can paint 300 ft² of wall space in 3.5 h. How many hours should the contractor estimate for the painter to paint 425 ft²? Express your answer to the nearest tenth.

20. CAD/ARC An architect's drawing is scaled at $\frac{3}{4}$ in. = 6 ft. What is the actual height of a door that measures $\frac{7}{8}$ in. on the drawing?

Chapter Review Exercises   377

Career Coding in Chapter Review Exercises

Chapter Review Exercises. An extensive set of exercises appears at the end of each chapter so you can review all the learning outcomes presented in the chapter. These exercises, organized by section, may be assigned as homework, or you may want to work them on your own for additional practice. **Answers to the odd-numbered exercises are given at the end of the text,** and worked-out solutions appear in a separate Student Solutions Manual available for purchase. Your instructor has the solutions to the even-numbered exercises in the Instructor's Resource Manual.

Team Problem-Solving Exercises. Employers value an employee's ability to interact with others in a team environment. These exercises will allow you to develop and refine your team-interaction skills.

Concepts Analysis. Too often we focus on the *how to* and overlook the *why* of mathematical concepts. The Concepts Analysis questions further your understanding of a concept and help you see the connections between concepts. Some concepts questions present incorrect solutions to exercises to give you practice in analyzing and correcting errors. Error analysis also reinforces your understanding of concepts. As an added bonus, these exercises strengthen your writing skills. Suggested responses (answers) are found in the Instructor's Resource Manual.

Team Problem Solving Exercises

1. An effective measure of your understanding of a concept is your ability to apply the concept to real-world situations.
   (a) Develop and solve a word problem that can be solved using a direct proportion. Include in your storyline a lawn mower, tanks of gasoline, and acres to be mowed.
   (b) Develop and solve a word problem that can be solved using a direct proportion.

2. Inverse proportions can be used to solve certain types of real-world applications.
   (a) Develop and solve a word problem that can be solved using an inverse proportion. Include in your storyline a belt, pulleys, rpm's and diameters of pulleys.
   (b) Develop and solve a word problem that can be solved using an inverse proportion.

Concept Analysis

1. Illustrate the property of proportions using two equivalent fractions.
3. Explain how to set up a direct proportion.
5. Explain how to set up an inverse proportion.
7. If the constant of variation is positive and y varies directly as x, how will y change when x increases? Use an example to support your answer.
9. Suppose y varies directly as the square of x. How will y change when x is doubled? Illustrate your answer with an example.

2. Explain the difference between a direct proportion and an inverse proportion.
4. Give some examples of situations that are directly proportional.
6. Give some examples of situations that are inversely proportional.
8. If the constant of variation is positive and y varies inversely as x, how will y change when x increases? Use an example to support your answer.
10. Suppose y varies inversely as the square of x. How will y change when x is doubled? Illustrate your answer with an example.

Practice Test

Solve the proportions.

1. $\dfrac{x}{12} = \dfrac{5}{8}$

2. $\dfrac{x}{6} = \dfrac{12}{8}$

3. $\dfrac{640}{24} = \dfrac{x}{360}$

4. $\dfrac{45 \text{ cm}}{18 \text{ cm}} = \dfrac{9 \text{ cm}}{x}$

5. $\dfrac{2\frac{1}{2}}{x} = \dfrac{1\frac{1}{4}}{3\frac{1}{5}}$

6. $\dfrac{3.9}{5.4} = \dfrac{x}{8.1}$

7. A large gear with 300 teeth turns at 40 rpm. Find the rpm of a small gear that has 60 teeth.

Practice Test. The practice test at the end of each chapter lets you check your understanding of the chapter learning outcomes. You should be able to work each problem without referring to any examples in your text or your notes. Take this test before you take the class test to check and verify your understanding of the chapter material. **Answers to the odd-numbered exercises appear at the end of the text,** and their solutions appear in a separate Student Solutions Manual. Your instructor has the solutions to the even-numbered exercises in the Instructor's Resource Manual.

Cumulative Practice Tests. Practice tests for a group of chapters are included after Chapters 3, 6, 10, 15, and 19. These tests will help you prepare for mid-course or end-of-course exams. Periodically reviewing previously learned material will help you retain the concepts for a longer period of time.

Student Solutions Manual. This manual can be purchased at your college bookstore or from online bookstores. It gives you extra *learning insurance* to help you master learning outcomes in the text. The manual contains worked-out solutions to the odd-numbered exercises in the Chapter Review Exercises and the Practice Test for each chapter of the text. Answers to these exercises appear in the back of your text, but using the manual to study the worked-out solutions reinforces your problem-solving skills and your understanding of the concepts.

Companion Website. This website, available at **www.prenhall.com/cleaves**, provides even more practice with the math concepts presented in the form of short quizzes for each section of the text. These quizzes are immediately graded, and you have the opportunity to send the results to your instructor via email.

MyMathTutor in WebCT and Blackboard. Many additional resources are available with MyMathTutor. For each learning outcome there is a tutorial narrative with additional examples and explanatory notes, video tutorial instruction, and practice exercises with solutions and explanatory notes. Also, a self-grading practice test is available for each chapter. Ask your instructor about arranging access to these resources for your class.

We wish you much success in your study of mathematics. Many of the improvements for this book were suggested by students such as yourself. If you have suggestions for improving the presentation, please give them to your instructor or email the authors at **ccleaves@bellsouth.net** or **margiehobbs@bellsouth.net**.

Cheryl Cleaves
Margie Hobbs

Contents

List of Career Applications

Auto/Diesel Technology (AUTO)

Hospitality/Culinary/ Food Technology (HOSP)

Industrial Technology/Manufacturing/ Machine Technology/Engineering Technology (INDTEC)

Industrial Trades/Welding/Machine Tool/Industrial Maintenance (INDTR)

Telecommunications (TELE)

1

Review of Basic Concepts

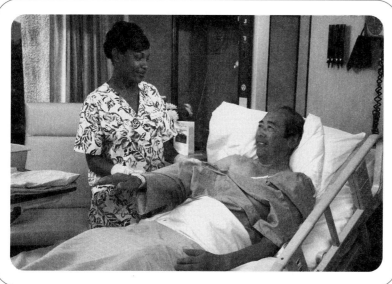

Focus on Careers

The *Occupational Outlook Handbook* is an excellent source of career information. It provides information for making decisions about careers. The *Handbook* is revised every two years and describes what workers do on the job and their working conditions. It also gives the training and education needed, earnings, and expected job prospects for many careers and occupations. The *Handbook* can be accessed on the Internet at http://www. bls.gov/oco/.

Nursing is one of the largest health-care occupations, with 2.3 million jobs. The associate degree (ADN programs) is one path for becoming a nurse and takes two to three years to complete. Job opportunities for this occupation are expected to be very good in the next 10 years. In 2002 the median (middle) annual earnings for registered nurses were $48,090.

Although most nurses work in hospitals, some work in offices, nursing care facilities, and patients' homes. Nurses should be caring and responsible individuals. They must be detail oriented and be able to correctly assess patients' conditions.

Source: *Occupational Outlook Handbook:* http://www.bls.gov/oco.

When we study a subject for the first time or review a subject in some detail, we begin with the basics. Often, as we examine the basics of a subject, we discover—or rediscover—many pieces of useful information. In this sense, mathematics is no different from any other subject. We begin with a study of whole numbers and decimals and the basic operations that we perform with them.

Learning Outcomes

1 Compare whole numbers.

2 Write fractions with power-of-10 denominators as decimal numbers.

3 Compare decimal numbers.

4 Round a whole number or a decimal number to a specified place value.

5 Add and subtract whole numbers and decimals.

6 Multiply and divide whole numbers and decimals.

Our system of numbers, the **decimal-number system,** uses 10 symbols called **digits:** 0, 1, 2, 3, 4, 5, 6, 7, 8, 9. A number can be represented with one or more digits. When a number contains two or more digits, each digit must be in the correct place for the number to have the value we intend it to have. Each place in the system has a specific **place value.**

The numbering system is made up of many different types of numbers. The first two types of numbers that we review are natural numbers and whole numbers. The **natural numbers** begin with the number 1 and continue indefinitely (1, 2, 3, 4, 5, . . . , 101, 102, 103, . . .). Three periods that follow a list of numbers are called an **ellipsis** and mean that the pattern established before the ellipsis continues. The natural numbers are also called **counting numbers.** The set of **whole numbers** includes all the natural numbers and the number 0. Other types of numbers will be introduced as appropriate.

The whole-number place values are arranged in **periods,** or groups of three (Fig. 1–1) reading from right to left. The first period of three is called **units,** the second period of three is called **thousands,** the third period is called **millions,** and the fourth period is called **billions.** Commas are used to separate these periods. The commas make larger numbers easier to read because we can locate specific place values and interpret numbers more easily. Each group of three digits has a hundreds place, a tens place, and a ones place.

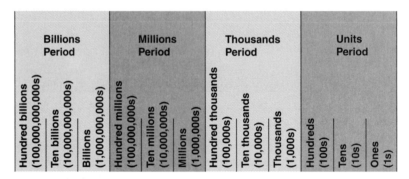

Figure 1–1 Whole-number place values and periods.

In four-digit numbers, the comma separating the units period from the thousands period is optional. Thus, 4,575 and 4575 are both acceptable.

1 Compare Whole Numbers.

Whole numbers can be arranged on a **number line** to show a visual representation of the relationship of numbers by size. The most common arrangement is to begin with zero and place numbers on the line from left to right as they get larger.

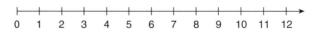

All numbers have a place on the number line and the numbers continue indefinitely without end. A term that is often used to describe this concept is **infinity** and the symbol is ∞.

Whole numbers can be compared by size by determining which of the two numbers is larger or smaller. If two numbers are positioned on a number line, the smaller number is positioned to the left of the larger number. The order relationship can be written in a mathematical statement called an **inequality.** An inequality shows that two numbers are not equal; that is, one is larger than the other. Symbols for showing inequalities are the **less than** symbol $<$ and the **greater than** symbol $>$.

$5 < 7$ Five is less than seven.
$7 > 5$ Seven is greater than five.

To compare whole numbers:

1. Mentally position the numbers on a number line.
2. Select the number that is farther to the left to be the smaller number.
3. Write an inequality using the *less than* symbol.

smaller number $<$ larger number

or

Write an inequality using the *greater than* symbol.

larger number $>$ smaller number

EXAMPLE   Write an inequality comparing the numbers 12 and 19:

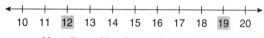

Mentally position the numbers on the line.

12 is the smaller number.   12 is to the *left* of 19.
$12 < 19$ or $19 > 12$ Use appropriate inequality symbol.

Numbers are used to show *how many* and to show *order*. **Cardinal numbers** show *how many* and **ordinal numbers** show *order* or position (such as first, second, third, fourth, etc.). For example, in the statement "three students are doing a presentation," three is a cardinal number (showing how many). In the statement "Margaret is the third tallest student in the class," third is an ordinal number (showing order).

2 Write Fractions with Power-of-10 Denominators as Decimal Numbers.

A notation for writing numbers that are parts of a whole number is called **fraction notation.** In fraction notation, we write one number over another number.

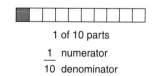

1 of 10 parts

$\dfrac{1}{10}$ numerator
denominator

The bottom number, the **denominator,** represents the number of parts that a whole unit contains. The top number, the **numerator,** represents the number of parts being considered.

A special type of fraction is called a **decimal fraction.**

A decimal fraction is a fraction whose denominator is 10 or some power of 10, such as 100 or 1,000. Often the terms decimal fraction, **decimal number,** and **decimal** are used interchangeably. In fraction notation, 3 out of 10 parts is written as $\frac{3}{10}$. In **decimal notation,** the denominator 10 is not written but is implied by position on the place-value chart (Fig. 1–2). A **decimal point** (.) separates whole-number amounts on the left and fractional parts on the right. The fraction $\frac{3}{10}$ can be written in decimal notation as 0.3.

3 out of 10

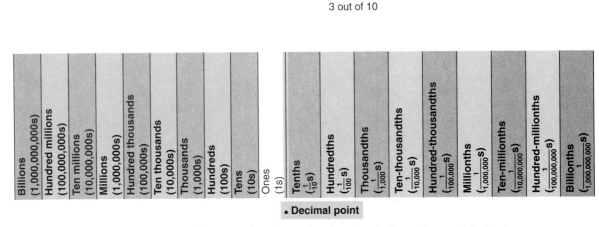

Figure 1–2 Place-value chart for whole numbers and decimals.

To extend the place-value chart to include parts of whole amounts, we place a decimal point (.) after the ones place. The place on the right of the ones place is called the *tenths* place. A decimal point is placed between the ones place and the tenths place to distinguish between whole amounts and fractional amounts.

Informal Use of the Word *Point*

Informally, the decimal point is sometimes read as "point." Thus, 3.6 is read "three and six tenths" or "three point six." The decimal 0.0162 can be read as "one hundred sixty-two ten-thousandths," or "point zero one six two," or "zero point zero one six two." This informal process is often used in verbal communication to ensure that numbers are not miscommunicated.

Unwritten Decimal Points

When we write whole numbers, we usually omit the decimal point; the decimal point is understood to be at the end of the whole number. Therefore, any whole number, such as 32, can be written without a decimal (32) or with a decimal (32.).

Fractions like $\frac{1}{10}$ and $\frac{75}{100}$ have denominators that are powers of 10. Any fraction whose denominator is 10, 100, 1,000, 10,000, and so on, can be written as a decimal number without making any calculations.

Chapter 1 / Review of Basic Concepts

To write a fraction that has a denominator of 10, 100, 1,000, 10,000, and so on, as a decimal:

1. Use the denominator to find the number of decimal places.

   Write $\frac{17}{1,000}$ as a decimal.

   10 → 1 place
   100 → 2 places
   1,000 → 3 places
   10,000 → 4 places

   0.___   Three decimal places are needed.

2. Place the numerator so that the last digit is in the farthest place on the right.

   0._17

3. Fill in any blank spaces with zeros.

   0.017

EXAMPLE   Write $\frac{3}{10}$, $\frac{25}{100}$, $\frac{425}{100}$, and $\frac{3}{1,000}$ as decimal numbers.

$\frac{3}{10}$ is written **0.3.**   One decimal place

$\frac{25}{100}$ is written **0.25.**   Two decimal places

$\frac{425}{100}$ is written **4.25.**   Two decimal places

$\frac{3}{1,000}$ is written **0.003.**   Three decimal places

---

TIP

Do Ending Zeros Change the Value of a Decimal Number?

When we attach zeros on the *right* end of a decimal number, we do not change the value of the number.

$$0.5 = 0.50 = 0.500 \qquad \frac{5}{10} = \frac{50}{100} = \frac{500}{1,000}$$

See equivalent fractions on page 70.

---

3 Compare Decimal Numbers.

As with whole numbers, we often need to compare decimals by size. To make valid comparisons, we must compare like amounts. Whole numbers compare with whole numbers, tenths compare with tenths, thousandths compare with thousandths, and so on.

To compare decimal numbers:

1. Compare whole-number parts.
2. If the whole-number parts are equal, compare digits place by place, starting at the tenths place and moving to the right.
3. Stop when two digits in the same place are different.
4. The digit that is larger determines the larger decimal number.

> **EXAMPLE** Compare the numbers 32.47 and 32.48 to see which is larger.
>
> 32.4**7**   Look at the whole-number parts. They are the same.
> 32.4**8**   Look at the tenths place for each number. Both numbers have a 4 in the tenths place.
>   Look at the hundredths place. They are different and 8 is larger than 7.
>
> **32.48 is the larger number.**

> **EXAMPLE** Write two inequalities for the numbers 0.4 and 0.07.
>   Since the whole-number parts are the same (0), we compare the digits in the tenths place. 0.4 is larger because 4 is larger than 0.
>
> **0.4 > 0.07**   and   **0.07 < 0.4**

Another procedure for comparing decimals is to affix an appropriate number of zeros on the right to make each number have the same number of decimal places.

$$0.4 \;=\; 0.40 \qquad \frac{4}{10} = \frac{40}{100}$$

Now, compare 0.40 and 0.07. The larger number is 0.40. Then, 0.40 > 0.07.

Common Denominators in Decimals

TIP

The denominator of a decimal fraction is determined by the number of decimal places in the number. Decimal fractions have a common denominator if they have the same number of digits to the right of the decimal point. See common denominators of fractions on page 78.

4   **Round a Whole Number or a Decimal Number to a Specified Place Value.**

Rounding a number means finding the closest **approximate number** to a given number. For example, if 37 is rounded to the nearest ten, is 37 closer to 30 or 40? Locate 37 on the number line.

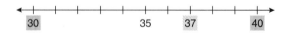

37 is closer to 40 than 30. Thus, 40 is a better approximation to the nearest ten for 37. Another way to say this is that 37 rounded to the nearest ten is 40.

     When rounding a number to a certain place value, we must make sure that we are as accurate as our employer wants us to be. Generally, the size of the number and its use dictate the decimal place to which it should be rounded.

To round a whole or decimal number to a given place value:

1. Locate the digit that occupies the rounding place. Then examine the digit to the immediate right.
2. If the digit to the right of the rounding place is 0, 1, 2, 3, or 4, do not change the digit in the rounding place. If the digit to the right of the rounding place is 5, 6, 7, 8, or 9, add 1 to the digit in the rounding place.
3. Replace all digits to the *right* of the digit in the rounding place with zeros if they are to the left of the decimal point. Drop digits that are to the right of the digit in the rounding place *and also* to the right of the decimal point.

EXAMPLE AG/H Oregon has a land area of 96,187 square miles. What would be a reasonable approximate number for this land area?

96,187 rounds to the following approximate numbers:

96,190 to the nearest ten
96,200 to the nearest hundred
96,000 to the nearest thousand
100,000 to the nearest ten thousand

Deciding to which place to round a number is a judgment depending on what use you will make of the rounded or approximate number.

Both 96,000 and 100,000 are reasonable approximations.

EXAMPLE Round 46.897 to the hundredths place.

46.89**7**   9 is in the hundredths place.
46.8**97**   The next digit to the right is 7, so add 1 to 9. (9 + 1 = 10 and 89 + 1 = 90.)
               Drop the 7 in the thousandths place because it is to the right of the rounding
               place and also to the right of the decimal.

46.90

Nine Plus One Still Equals Ten

When the digit in the rounding place is 9 and must be rounded up, it becomes 10. The 0 replaces the 9 and the 1 is regrouped to the next place to the left.

EXAMPLE Round $293.48 to the nearest dollar.

$29**3**.**4**8   When we round to the nearest dollar, we are rounding to the *ones* place.

$293

EXAMPLE Round $71.8986 to the nearest cent.

$71.8**98**6   One cent is 1 hundredth of a dollar; to round to the nearest cent is to round
                 to the *hundredths* place.

$71.90

Exact Amount Versus Approximate Amount

When an amount has been rounded, it is no longer an exact amount. The rounded amount is now an **approximate amount.** Approximate amounts are used most often in applications.

5 Add and Subtract Whole Numbers and Decimals.

In an addition problem, the numbers being added are called **addends** and the answer is called the **sum** or **total.**

Two useful properties of addition are the **commutative** and **associative** properties. By the *commutative* property, we mean that numbers can be added in any *order.* We can add 7 and 6 in any order and still get the same answer:

$$7 + 6 = 13 \qquad 6 + 7 = 13$$

By the *associative* property, we mean that we can *group* numbers together any way we want when we add and still get the same answer. To add $7 + 4 + 6$, we can group the $4 + 6$ to get 10; then we add the 10 to the 7:

$$7 + (4 + 6) = 7 + 10 = 17$$

Or we can group the $7 + 4$ to get 11; then we add the 6 to the 11:

$$(7 + 4) + 6 = 11 + 6 = 17$$

Addition is a **binary operation;** that is, the rules of addition apply to adding *two* numbers at a time. The associative property of addition shows how addition is extended to more than two numbers.

Using Symbols to Write Properties and Definitions

Many properties and definitions can be written symbolically. Symbolic representation allows a quick recall of the rule or definition.

Some properties are restricted to certain types of numbers and others are appropriate for practically all types of numbers. When writing properties symbolically, we will also note any restrictions and refer to the most inclusive type of numbers that are appropriate. Many properties apply to all *real numbers.* **Real numbers** are introduced later, but for now, whole numbers, fractions, and decimals are included in the set of real numbers.

Commutative Property of Addition: Two numbers may be added in any order and the sum remains the same.

$$a + b = b + a \quad \text{where } a \text{ and } b \text{ are any real numbers.}$$

Associative Property of Addition: Three numbers may be added using different groupings and the sum remains the same.

$$a + (b + c) = (a + b) + c \quad \text{where } a, b, \text{ and } c \text{ are any real numbers.}$$

The associative property of addition also allows other possible groupings and extends to more than three numbers.

$$7 + 4 + 6 = 13 + 4 = 17$$
$$3 + 5 + 7 + 9 = 8 + 16 = 24$$

Adding zero to any number results in the same number. This property is called the **zero property of addition** and zero is called the **additive identity.**

Zero property of addition:

Adding zero to any number results in the same number.

$$n + 0 = n \quad \text{or} \quad 0 + n = n, \text{ where } n \text{ is any real number}$$
$$5 + 0 = 5 \quad \text{or} \quad 0 + 5 = 5$$

When adding numbers of two or more digits, the same place values must be aligned under one another so that all the ones are in the far right column, all the tens in the next column, and so on.

To add numbers of two or more digits:

1. Arrange the numbers in columns so that the ones place values are in the same column.
2. Add the ones column, then the tens column, then the hundreds column, and so on, until all the columns have been added. **Regroup** whenever the sum of a column is more than one digit. This is also called **carrying.**

EXAMPLE   Shipping fees are often charged by the total weight of the shipment. Find the total weight of this order: nails, 250 pounds (lb); tacks, 75 lb; brackets, 12 lb; and screws, 8 lb. Arrange in columns and add.

$$\begin{array}{r} {}^{1\,1} \\ 250 \\ 75 \\ 12 \\ +\ \ 8 \\ \hline 345 \end{array}$$

The sum of the digits in the right column is 15. Record the 5 in the ones column and *regroup* the 1 to the tens column.
Add the tens column. $1 + 5 + 7 + 1 = 14$. Regroup the 1.

The total weight is 345 lb.

When adding, decimal numbers are aligned so that all decimal points fall in the same vertical line. Aligning decimal points has the same effect as using *like* denominators when adding (or subtracting) fractions. Like denominators are discussed in Chapter 2.

To add decimal numbers:

1. Arrange the numbers so that the decimal points are in one vertical line.
2. Add each column.
3. Align the decimal for the sum in the same vertical line.

EXAMPLE Add $42.3 + 17 + 0.36$.

$$
\begin{array}{l}
42.3 \\
17 \\
\underline{0.36}
\end{array}
$$

Note that the decimal in 17 is understood to be at the right end.

$$
\begin{array}{l}
42.30 \\
17.00 \\
\underline{0.36} \\
\mathbf{59.66}
\end{array}
$$

If we prefer, we may write each number so that all have the same number of decimal places by attaching zeros on the right.

Estimating is an increasingly important skill to develop and it relates in part to your *number sense.* Number sense must be developed. We develop and strengthen our number sense by estimating and doing mental calculations. In making calculations it is important to **estimate** and **check** your work.

In estimating or rounding we often refer to a nonzero digit. A **nonzero digit** is a 1, 2, 3, 4, 5, 6, 7, 8, or 9. That is, it is a digit that is not zero.

To estimate the answer to an addition problem:

1. Round each addend to a specific place value or to a number with one nonzero digit.
2. Add the rounded addends.

To check an addition problem:

1. Add the numbers a second time and compare with the first sum.
2. Use a different order or grouping if convenient.

EXAMPLE Elston Home Renovators spent the following amounts on a job: $16,466.15, $23,963.10, and $5,855.20. Estimate the total amount by rounding to thousands. Then find the exact amount and check your answer.

Thousands Place	Estimate	Exact	Check
$16,466.15	$16,000	$16,466.15	$16,466.15
23,963.10	24,000	23,963.10	23,963.10
5,855.20	6,000	5,855.20	5,855.20
	$46,000	**$46,284.45**	**$46,284.45**

The estimate and exact answer are close. The exact answer is reasonable.

Decimal Point and Zeros on the Calculator

The decimal key $\boxed{\cdot}$ is most often located near the number keys on a calculator. This key is pressed when the decimal point appears in the number being entered.

- Does the zero to the left of the decimal have to be entered before the decimal in a number like 0.2? Try it both ways: Add 3 + 0.2 on the calculator.

  Options:　3 $\boxed{+}$ 0 $\boxed{\cdot}$ 2 $\boxed{=}$ ⇒ 3.2　　entering the zero

  　　　　　3 $\boxed{+}$ $\boxed{\cdot}$ 2 $\boxed{=}$ ⇒ 3.2　　not entering the zero

  The symbol ⇒ is used to indicate the calculator result.

- Do zeros that follow a decimal or that are on the right end of a decimal number have to be entered? Try it both ways: Add $3.00 + $1.50 on the calculator.

  Options:　3 $\boxed{\cdot}$ 0 0 $\boxed{+}$ 1 $\boxed{\cdot}$ 50 $\boxed{=}$ ⇒ 4.5　　entering ending zeros

  　　　　　3 $\boxed{+}$ 1 $\boxed{\cdot}$ 5 $\boxed{=}$ ⇒ 4.5　　not entering ending zeros

Ending zeros to the right of the decimal are usually dropped in a calculator display unless the calculator is set to display a specific number of decimal places.

Subtraction is the **inverse operation** of addition. In addition, we add numbers to get their total (such as 5 + 4 = 9), but to solve the subtraction problem 9 − 5 = ? we ask, "What number must be added to 5 to give us 9?" The answer is 4 because 4 added to 5 gives a total of 9. When we subtract two numbers, the answer is called the **difference** or **remainder.** The initial quantity is the **minuend.** The amount being subtracted from the initial quantity is the **subtrahend.**

Subtraction is *not* commutative. 8 − 3 = 5, but 3 − 8 does not equal 5; that is, 3 − 8 ≠ 5. The symbol ≠ is read **is not equal to.**

Subtraction is not associative.

EXAMPLE　Illustrate that subtraction is not associative by showing that 9 − (5 − 1) does not equal (9 − 5) − 1.

$$9 - (5 - 1) = 9 - 4 = \mathbf{5}, \text{ but } (9 - 5) - 1 = 4 - 1 = \mathbf{3}$$

When subtracting two or more numbers, if no grouping symbols are included, perform subtractions from left to right.

EXAMPLE　Subtract 8 − 3 − 1.

$$8 - 3 - 1 = 5 - 1 = \mathbf{4}$$　Subtract from left to right.

Subtracting zero from a number results in the same number:

$$n - 0 = n, \quad \text{where } n \text{ is any real number} \quad 7 - 0 = 7$$

Subtraction and Zeros

Subtracting a number from zero is not the same as subtracting zero from a number; that is, $7 - 0 = 7$, but $0 - 7$ does not equal 7.

To subtract whole numbers that have two or more digits:

1. Arrange the numbers in columns, with the minuend at the top and the subtrahend at the bottom.
2. Make sure the ones digits are in a vertical line on the right.
3. Subtract the ones column first, then the tens column, the hundreds column, and so on.
4. To subtract a larger digit from a smaller digit in a column, **regroup** by subtracting 1 from the digit in the next column to the left. This is the equivalent to *one* group of 10; thus, add 10 to the digit in the given column, then, continue subtracting. The concept of *regrouping* is also referred to as **borrowing.**

EXAMPLE Subtract $9,327 - 3,514$.
Arrange in columns.

$$
\begin{array}{r}
\overset{8\ 13}{} \\
9,3\,27 \\
-\ 3,5\,14 \\
\hline
5,8\,13
\end{array}
$$

Arrange in columns. Subtract each column from the right.

In the hundreds place, 5 is more than 3. Regroup by subtracting 1 group of 10 from 9. $9 - 1 = 8$, $10 + 3 = 13$.

Words and Phrases That Imply Subtraction

These phrases indicate subtraction in applied problems:

how many are left	how many more
how much less	how much larger
how much smaller	

Also, some applied problems require more than one operation. Many problems that require more than one operation involve parts and a total. If we know the total and all the parts but one, we can add all the known parts and subtract the result from the total to find the missing part.

EXAMPLE The Froehlichs left Memphis and drove 356 mi on the first day of their vacation. They drove 426 mi on the second day. If they are traveling to Albuquerque, which is 1,050 mi from Memphis, how many more miles do they have to drive?

The phrase *how many more* indicates subtraction.

$$1{,}050 \text{ mi} = \text{total miles}$$
$$356 \text{ mi} + 426 \text{ mi} + \text{ mi left to drive} = 1{,}050 \text{ mi}$$

To find the miles left to drive, add $356 \text{ mi} + 426 \text{ mi}$ and subtract the result from $1{,}050 \text{ mi}$.

$$356 \text{ mi} + 426 \text{ mi} = \boxed{782 \text{ mi}} \qquad 1{,}050 \text{ mi} - \boxed{782 \text{ mi}} = \mathbf{268 \text{ mi}}$$

1. Arrange the numbers so that the decimal points align vertically.
2. Subtract each column beginning at the right.
3. Interpret blank places as zeros.
4. Place the decimal in the difference in the same vertical line.

EXAMPLE Subtract 7.18 from 15.

Take care to align the decimals properly.

$$\begin{array}{r} 15. \\ -\ 7.18 \\ \hline \end{array}$$ Because 15 is a whole number, its decimal point is placed after the 5.

$$\begin{array}{r} 15.00 \\ -\ 7.18 \\ \hline \mathbf{7.82} \end{array}$$ To subtract, we put zeros in the tenths and hundredths places of 15 and then regroup.

When a worker machines an object using a blueprint as a guide, a certain amount of variation from the blueprint specification is allowed for the machining process. This variation is called the **tolerance**. Thus, if a blueprint calls for a part to be 9.47 in. with a tolerance of $\pm$ 0.05 in. (read **plus or minus** five hundredths), this means that the actual part can be 0.05 in. *more* or 0.05 in. *less* than the specification. To find the *largest* acceptable measure of the object, we add 9.47 in. + 0.05 in. = 9.52 in. To find the *smallest* acceptable size of the object, we subtract 9.47 in. − 0.05 in. = 9.42 in. The dimensions 9.52 in. and 9.42 in. are called the **limit dimensions** of the object. That is, 9.52 in. is the largest acceptable measure, and 9.42 in. is the smallest acceptable measure.

EXAMPLE CON Find the limit dimensions of an object with a blueprint specification of 8.097 in. and a tolerance of $\pm$ 0.005 in. (This is often written 8.097 in. $\pm$ 0.005 in.)

8.097 in. = the blueprint specification for the dimension of an object.
$\pm$ 0.005 in. = tolerance of object's dimension.
Smallest dimension for object = blueprint specification − tolerance.
Largest dimension for object = blueprint specification + tolerance.

Estimation The tolerance of the part is very small—just five thousandths—so the limit dimensions for the part should be only a few thousandths of an inch smaller or larger than the blueprint specification.

8.097 in. − 0.005 in. = 8.092 in.
8.097 in. + 0.005 in. = 8.102 in.

The smallest acceptable dimension for the object is 8.092 in. and the largest acceptable dimension is 8.102 in.

Estimating a subtraction problem is similar to estimating an addition problem. The numbers in the problem are rounded before the subtraction is performed.

To estimate the difference:

1. Round each number to the desired place value or to a number with one nonzero digit.
2. Subtract the rounded numbers.

To check subtraction, we can use the inverse relationship between addition and subtraction. If $9 - 5 = 4$, then $4 + 5$ should equal 9.

To check a subtraction problem:

1. Add the subtrahend and difference.
2. Compare the result of Step 1 with the minuend. If the two numbers are equal, the subtraction is correct.

EXAMPLE Estimate by rounding to hundreds, then find the exact difference, and check.

$$427.45 - 125$$

	Estimate	Exact	Check
427.45	400	427.45	125.00
− 125.00	− 100	− 125.00	+ 302.45
	300	**302.45**	**427.45**

EXAMPLE Find the difference between $53,943.76 and $34,256.45 using a calculator. One option:

53943 $\boxed{\cdot}$ 76 $\boxed{-}$ 34256 $\boxed{\cdot}$ 45 $\boxed{=}$ $\boxed{\text{ENTER}}$ or $\boxed{\text{EXE}}$ may replace $\boxed{=}$.

Calculator display: 19687.31

The difference is $19,687.31.

EXAMPLE Two cuts are made from a 72-in. pipe (see Fig. 1–3). The two lengths cut from the
INDTEC pipe are 28 in. and 15 in. How much of the pipe is left after these cuts are made?

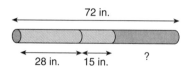

$$
\begin{array}{ll}
72 - 28 - 15 = & \text{Subtract from left to right.} \\
44 - 15 = & 72 - 28 = 44 \\
29 & 44 - 15 = 29
\end{array}
$$

There are 29 in. of pipe left.

Figure 1–3 Lengths cut from pipe.

Often there is more than one way to solve a problem. In the previous example we could have found the sum of the cuts first and then subtracted.

28 in. + 15 in. = 43 in.   Sum of two cuts.
72 in. − 43 in. = 29 in.   Remaining length of pipe.

6 **Multiply and Divide Whole Numbers and Decimals.**

Multiplication of whole numbers is repeated addition. If we have three $10 bills, we have $10 + $10 + $10, or $30. Using multiplication, we see that this is the same as 3 times $10, or $30.

When we multiply two numbers, the first number is called the **multiplicand,** and the number we multiply by is called the **multiplier.** Either number is referred to as a **factor.** The answer or result of multiplication is called the **product.**

2	×	3	=	6
multiplicand		multiplier		product
or factor		or factor		

Various Notations for Multiplication

Besides the familiar × or "times" sign, parentheses ( ), a raised dot (·), and an asterisk (*) are also used to show multiplication. Parentheses are most often used as notation for multiplication.

$$2(3) = 6, \quad (2)(3) = 6, \quad 2 \cdot 3 = 6, \quad 2 * 3 = 6$$

Multiplication is commutative and associative, just like addition. The **commutative property of multiplication** permits two numbers to be multiplied in any order. In symbols, $a(b) = b(a)$ for all real numbers.

$$4(5) = 20, \quad 5(4) = 20$$

When more than two numbers are multiplied, the numbers must be grouped, and the **associative property of multiplication** permits the numbers to be grouped in any way. In symbols, $a(b \cdot c) = (a \cdot b)c$ for all real numbers.

$$\begin{array}{cc} 2(3 \cdot 5) & \text{or} \quad (2 \cdot 3)5 \\ 2(15) & (6)5 \\ 30 & 30 \end{array}$$

EXAMPLE Multiply 3(2)(9)

3(2)(9)	Group any two factors.
(3 · 2)(9)	Multiply grouped factors.
6 (9)	Multiply the factors: 6 and 9.
54	

The product of a number and zero is zero:

$$n(0) = 0, \quad 0(n) = 0, \text{ where } n \text{ is any real number} \quad 4(0) = 0, \quad 0(4) = 0$$

This is called the **zero property of multiplication.**

EXAMPLE Multiply 2(5)(0)(7)

(2)(5)(0)(7)   Use the zero property of multiplication.

0

To multiply factors of two or more digits:

1. Arrange the factors one under the other.
2. Multiply each digit in the multiplicand by each digit in the multiplier. The product of the multiplicand and each digit in the multiplier gives a **partial product.**
   (a) To start, multiply the ones digit in the multiplier by the multiplicand from right to left.
   (b) Align each partial product with its first digit directly under its multiplier digit.
3. Add the partial products.

EXAMPLE Multiply (204)(103).

```
      204
  ×   103
      612      Multiply: 3 × 204 = 612. Align 612 under 3 in the multiplier.
    0 00       Multiply: 0 × 204 = 000. Align 000 under 0 in the multiplier.
   20 4        Multiply: 1 × 204 = 204. Align 204 under 1 in the multiplier.
   21,012      Add the partial products as they are aligned.
```

Partial products 000 and 204 could be combined on a single line.

```
      204
  ×   103
      612
   20 40       0 × 204 = 0. Align under 0 in multiplier.
   21,012      1 × 204 = 204. Align under 1 in multiplier on the same line as the previous
               partial product.
```

Since decimals are fractions, multiplication of decimals causes us to rethink or expand our basic number sense of multiplication. The product of any number and a decimal number less than 1 is less than the original number.

To multiply decimal numbers:

1. Align the numbers as if they were whole numbers and multiply.
2. Count the total number of digits to the right of the decimal in each factor.
3. Place the decimal in the product so that the number of decimal places is the sum of the number of decimal places in the factors.

EXAMPLE Multiply 1.36(0.2).

$$1.36$$
$$\times \quad 0.2$$
$$\textbf{0.272}$$

In multiplication the decimals do *not* have to be in a straight line.

The product has three places to the right of the decimal. Place a zero in the ones place so that the decimal point will not be overlooked.

EXAMPLE Multiply 0.309(0.17).

$$0.309$$
$$\times \quad 0.17$$
$$2163$$
$$309$$
$$\textbf{0.05253}$$

No decimals are placed in the partial products.

We did not have enough digits in the product for five decimal places, so we inserted a zero on the *left*.

EXAMPLE
AG/H The outside diameter of a flower bed is 7.82 meters (m) (see Fig. 1–4). If the brick walk surrounding the bed is 1.56 m thick, find the inside diameter of the flower bed.

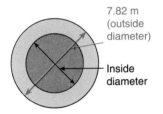

7.82 m (outside diameter)

Inside diameter

The diameter of a circle is the distance across the center of the circle.
7.82 m = outside diameter of the flower bed
1.56 m = width of brick walk surrounding the flower bed
Outside diameter − two widths (one on each end of the inside diameter) of the walk = inside diameter

Figure 1–4

Estimation The outside diameter is nearly 8 m, and the total of the two widths to be subtracted is about 3 m, so the inside diameter should be about 5 m.

$$7.82 - (2 \cdot 1.56) = 7.82 - 3.12 = 4.70$$

The inside diameter of the flower bed is 4.7 m.

The **distributive property of multiplication** means that multiplying a sum or difference by a factor is equivalent to multiplying each term of the sum or difference by the factor.

Distributive property of multiplication:

1. Add or subtract the numbers within the grouping.
2. Multiply the result of Step 1 by the factor outside the grouping.

or

1. Multiply each number inside the grouping by the factor outside the grouping.
2. Add or subtract the products from Step 1.

Symbolically, $a(b + c) = ab + ac$ or $a(b - c) = ab - ac$ for all real numbers.

- Parentheses show multiplication when the distributive property is used: $a(b + c)$ means $a \times (b + c)$.
- The letters represent numbers.
- Letters written together with no operation sign between them imply multiplication: ab means $a \times b$; ac means $a \times c$.

EXAMPLE   Multiply $3(2 + 4)$.

Multiplying first gives: Adding first gives:
$3(2 + 4) =$ $3(2 + 4) =$
$3(2) + 3(4) =$ $3(6) = \mathbf{18}$
$6 + 12 = \mathbf{18}$

EXAMPLE   Multiply $2(6 - 5)$.

Multiplying first gives: Subtracting first gives:
$2(6 - 5) =$ $2(6 - 5) =$
$2(6) - 2(5) =$ $2(1) = \mathbf{2}$
$12 - 10 = \mathbf{2}$

The distributive property is found in many formulas. One example is the formula for the perimeter of a rectangle. The **formula** for finding the perimeter of a rectangle is

$$P = 2(l + w) \qquad \text{or} \qquad P = 2l + 2w$$

EXAMPLE   Find the number of feet of fencing needed to enclose a rectangular pasture that is 1,784.6
AG/H   feet (ft) long and 847.3 ft wide. See Fig. 1–5.

847.3 ft

1,784.6 ft

Figure 1–5
Perimeter of rectangle.

$P = 2(l + w)$ or $P = 2l + 2w$
$P = 2(1{,}784.6 + 847.3)$ $P = 2(1{,}784.6) + 2(847.3)$
$P = 2(2{,}631.9)$ $P = 3{,}569.2 + 1{,}694.6$
$P = 5{,}263.8$ $P = 5{,}263.8$

The amount of fencing needed is 5,263.8 ft.

To estimate the product for a multiplication problem:

1. Round both factors to a chosen or specified place value or to one nonzero digit.
2. Then multiply the rounded numbers.

To check a multiplication problem:

1. Multiply the numbers a second time and check the product.
2. Interchange the factors if convenient.

EXAMPLE AG/H Find the approximate (estimate) and exact costs of 48 flower bulbs if each bulb costs $2.15. Estimate the cost by rounding each factor to a number with one nonzero digit. Then find the exact cost and check your work.

48 = total number of flower bulbs
$2.15 = cost of each bulb

Total cost of flower bulbs = number of bulbs × cost of each bulb

Estimation If 50 bulbs were purchased at $2 each, the total cost would be $100. So the exact cost should be close to $100.

48($2.15) = $103.20   Number of bulbs × Cost of each bulb = Exact cost.

The total cost of 48 flower bulbs is $103.20, which is approximately $100, as estimated.

EXAMPLE AG/H Maintenance Consultants needs to apply fertilizer to a customer's lawn. The lawn is 223.4 ft long and 132.8 ft wide. The fertilizer costs $0.004 per square foot to apply. What is the cost of applying the fertilizer? Use the formula $A = lw$. The area will be expressed in square feet (see Fig. 1–6).

132.8 ft

223.4 ft

Figure 1–6
Area of a rectangle.

Lawn is in the shape of a rectangle.
Length of the lawn = 223.4 ft.
Width of the lawn = 132.8 ft.
Fertilizer costs $0.004 per square foot to apply.
Area of a rectangle is the product of the length times the width ($A = lw$).
Total cost = total number of square feet × cost per square foot.

$A = lw$
$A = 223.4(132.8)$
$A = 29{,}667.52 \text{ ft}^2$ Area of lawn.

$\text{cost} = 29{,}667.52(\$0.004)$
$\text{cost} = \$118.67$ Rounded to the nearest cent

The total cost of applying fertilizer to the lawn is $118.67.

When either or both factors of a multiplication problem end in zeros, a shortcut process such as the one in the following example can be used.

EXAMPLE Multiply 2,600(70).

1. $\begin{array}{r} 2600 \\ \times\ 70 \\ \hline \end{array}$ Separate the ending zeros from the other digits.

2. $\begin{array}{r} 2600 \\ \times\ 70 \\ \hline 182 \end{array}$ Multiply the other digits as if the ending zeros were not there ($26 \times 7 = 182$).

3. $\begin{array}{r} 2600 \\ \times\ 70 \\ \hline \mathbf{182,000} \end{array}$ Attach the ending zeros to the basic product. Note that the number of zeros affixed to the basic product is the same as the sum of the number of zeros at the end of each factor.

This process is sometimes necessary whenever a multiplication problem is too long to fit into a calculator. Many basic calculators have only an eight-digit display window. The problem 26,000,000 times 3,000 would not fit into many basic calculators. This shortcut allows us to work the problem with or without a calculator.

EXAMPLE Multiply 26,000,000(3,000).

$\begin{array}{r} 26,000,000 \\ \times\ 3,000 \\ \hline \mathbf{78,000,000,000} \end{array}$ Separate ending zeros and multiply 26×3.
Attach 9 zeros.

TIP

The Mind Is Often Quicker than the Fingers

Don't use your calculator as a crutch. It is a tool! When multiplying 2,500 times 30, you can multiply 25 times 3 mentally. $25(3) = 75$. Then, attach three zeros to that product.

$$2,500(30) = 75,000$$

This skill does not come automatically. Like playing a musical instrument or mastering a sport, you don't develop skill by watching. You have to practice!

Division is the **inverse operation** of multiplication. Since $4(7) = 28$, then 28 divided by 7 is 4 and 28 divided by 4 is 7. The number being divided is called the **dividend.** The number divided by is the **divisor.** The result is the **quotient.**

Division is *not* commutative. $12 \div 6 = 2$, but $6 \div 12$ does not equal 2; that is, $6 \div 12 \neq 2$.

Division is *not* associative. $(12 \div 6) \div 2 = 2 \div 2 = 1$. But $12 \div (6 \div 2) = 12 \div 3 = 4$. That is, $(12 \div 6) \div 2 \neq 12 \div (6 \div 2)$.

To Write Division Symbolically

1. To use the *divided by* symbol (÷), write the dividend first.

$$28 \div 7 = 4 \longleftarrow \text{quotient}$$

dividend ——↑   ↳—— divisor

2. To use the *long-division symbol* (⟌), write the dividend under the bar.

$$4 \longleftarrow \text{quotient}$$
$$7 \overline{)28}$$

divisor ——↑ ↑—— dividend

3. To use the *division bar* or *slash symbol,* write the dividend on top or first.

——— dividend ———

$$\frac{28}{7} = 4 \longleftarrow \text{quotient} \quad \text{or} \quad 28/7 = 4 \longleftarrow \text{quotient}$$

——— divisor ———

Long division, like long multiplication, involves using one-digit multiplication facts repeatedly to find the quotient.

When the quotient is not a whole number, the quotient may have a *whole-number part* and a **remainder.** When a dividend has more digits than a divisor, parts of the dividend are called **partial dividends,** and the quotient of a partial dividend and the divisor is called a **partial quotient.**

To divide whole numbers:

1. Beginning with its leftmost digit, identify the first group of digits of the dividend that is larger than or equal to the divisor. This group of digits is the first *partial dividend.*

2. For each partial dividend in turn, beginning with the first:

  (a) Divide the partial dividend by the divisor. Write the partial quotient above the rightmost digit of the partial dividend.

  (b) Multiply the partial quotient by the divisor. Write the product below the partial dividend, aligning places.

  (c) Subtract the product from the partial dividend. Write the difference below the product, aligning places. The difference must be less than the divisor.

  (d) Next to the ones place of the difference, write the next digit of the dividend. This is the new partial dividend.

3. When all the digits of the dividend have been used, write the final difference in Step 2c as the remainder (unless the remainder is 0). The whole-number part of the quotient is the number written above the dividend.

EXAMPLE   Divide 881 by 35:

$$35\overline{)881}$$ The first partial dividend is 88.

$$
\begin{array}{r}
2 \\
35\overline{)881} \\
70 \\
\hline
18
\end{array}
$$

The partial quotient for 88 ÷ 35 is 2. Multiply 2(35) = 70. Then subtract 88 − 70 = 18. The difference 18 is less than the divisor 35.

$$
\begin{array}{r}
2 \\
35\overline{)881} \\
70 \\
\hline
181
\end{array}
$$

1 from the dividend is written next to 18 to form the next partial dividend.

$$
\begin{array}{r}
25 \\
35\overline{)881} \\
70 \\
\hline
181 \\
175 \\
\hline
6
\end{array}
$$

The partial quotient for 181 ÷ 35 is 5. The product of 5(35) is 175. The difference of 181 − 175 is 6. The remainder is 6.

881 divided by 35 = **25 R6.**

TIP

Importance of Placing the First Digit Carefully

The correct placement of the first digit in the quotient is critical. If the first digit is out of place, all digits that follow will be out of place, giving the quotient too few or too many digits.

EXAMPLE   Divide $5\overline{)2{,}535}$:

$$
\begin{array}{r}
5 \\
5\overline{)2{,}535} \\
2\,5 \\
\hline
03
\end{array}
$$

5 divides into 25 five times. Write 5 over the last digit of the 25. Subtract and bring down the 3. Note that 3 is less than the divisor, 5.

$$
\begin{array}{r}
507 \\
5\overline{)2{,}535} \\
2\,5 \\
\hline
035 \\
35 \\
\hline
0
\end{array}
$$

5 divides into 3 zero times. Write 0 over the 3 of the dividend. Bring down the next digit, which is 5. Divide 5 into 35: 35 ÷ 5 = 7. Write 7 over the 5 of the dividend. Multiply 7 × 5 = 35. Subtract.

Special properties with division, zero and one.

The quotient of zero divided by a nonzero number is zero:

$$0 \div n = 0, \qquad n\overline{)0,} \qquad \frac{0}{n} = 0 \qquad \text{where } n \text{ is a nonzero real number}$$

$$0 \div 5 = 0, \qquad 5\overline{)0,} \qquad \frac{0}{5} = 0$$

The quotient of a number divided by zero is **undefined** or **indeterminate**:

$$n \div 0 \text{ is } undefined, \qquad 0\overline{\smash{\big)}\,n}^{\ undefined}, \qquad \frac{n}{0} \text{ is undefined}$$
$$12 \div 0 \text{ is } undefined$$
$$0 \div 0 \text{ is } indeterminate$$

Dividing any nonzero number by itself yields 1:

$$n \div n = 1, \qquad \text{if } n \text{ is not equal to zero}; \qquad 12 \div 12 = 1$$

Dividing any number by 1 yields the same number:

$$n \div 1 = n, \qquad 5 \div 1 = 5$$

Since division is *not* associative, if there are no grouping symbols, we divide from left to right.

EXAMPLE   Divide $16 \div 4 \div 2$.

$$16 \div 4 \div 2 = 4 \div 2 = 2 \qquad \text{Divide from left to right.}$$

To divide decimal numbers written in long-division form:

1. Move the decimal in the divisor so that it is on the right side of all digits. (By moving the decimal, you are multiplying by 10, 100, 1,000, and so on.)
2. Move the decimal in the dividend to the right as many places as the decimal was moved in the divisor. Attach zeros if necessary. (This is multiplying the dividend by the same number as was used in Step 1.)
3. Write the decimal point in the answer directly above the new position of the decimal in the dividend. (Do this *before* dividing.)
4. Divide as you would in whole numbers.

EXAMPLE   Divide $4.8 \div 6$.

$$6\overline{\smash{\big)}\,4.8}^{\,.}$$

Insert a decimal point above the decimal point in the dividend.

When the divisor is a whole number, the decimal is understood to be to the right of 6 and is not moved. The decimal is placed in the quotient directly above the decimal in the dividend.

$$6\overline{\smash{\big)}\,4.8}^{\,0.8}$$

Divide.

EXAMPLE Divide 3.12 ÷ 1.2.

$$1.2\overline{)3.12}$$ Move the decimal in the divisor and dividend.

$$12\overline{)31.2}$$ Write the decimal in the quotient, then divide.

$$
\begin{array}{r}
2.6 \\
12\overline{)31.2} \\
\underline{24} \\
7\,2 \\
\underline{7\,2} \\
\end{array}
$$

 To round a quotient to a place value:

1. Divide to one place past the desired rounding place.
2. Attach zeros to the dividend after the decimal if necessary to carry out the division.
3. Round the quotient to the place specified.

EXAMPLE Divide and round the quotient to the nearest tenth.

$$3.2\overline{)15.27}$$

$$
\begin{array}{r}
4.77 \\
3.2\overline{)15.2\,70} \\
\underline{12\,8} \\
2\,4\,7 \\
\underline{2\,2\,4} \\
2\,30 \\
\underline{2\,24} \\
6 \\
\end{array}
$$

Since we are rounding to the nearest tenth, divide to the hundredths.

4.77 rounds to 4.8.

Estimating division is similar to estimating other operations. The numbers are rounded *before* the calculation is made.

 To estimate division of whole numbers:

1. Round the divisor and dividend to one nonzero digit.
2. Find the first digit of the quotient.
3. Attach a zero in the quotient for each remaining digit in the dividend.

Because division and multiplication are inverse operations, division is checked by multiplication.

To check division:

1. Multiply the divisor by the quotient.
2. Add any remainder to the product in Step 1.
3. The result of Step 2 should equal the dividend.

EXAMPLE   Estimate, find the exact answer, and check $913 \div 22$.

Estimate:
$$\begin{array}{r} \mathbf{40} \\ 20\overline{)900} \end{array}$$ 20 divides into 90 four whole times. Attach a zero after 4.

Exact:
$$\begin{array}{r} \mathbf{41\ R11} \\ 22\overline{)913} \\ \underline{88} \\ 33 \\ \underline{22} \\ 11 \end{array}$$ or $$\begin{array}{r} \mathbf{41.5} \\ 22\overline{)913.0} \\ \underline{88} \\ 33 \\ \underline{22} \\ 11\ 0 \\ \underline{11\ 0} \end{array}$$

Check:
$$\begin{array}{r} 41 \\ \times\ 22 \\ \hline 82 \\ 82 \\ \hline 902 \end{array} \quad \begin{array}{r} 902 \\ +\ 11 \\ \hline 913 \end{array} \quad \text{or} \quad \begin{array}{r} 41.5 \\ 2\ 2 \\ \hline 83\ 0 \\ 830 \\ \hline 913.0 \end{array}$$

The answer checks.

We use averages to make comparisons, such as when we compare the average mileage different cars get per gallon of gasoline. There are several types of averages.

In most courses students take, their numerical grade is determined by a process called **numerical averaging.** This average is also called the **arithmetic average,** or **mean.** If the grades are 92, 87, 76, 88, 95, and 96, we can find the average grade by adding the grades and dividing by the number of grades. Because we have six grades, we divide the sum by 6.

$$\frac{92 + 87 + 76 + 88 + 95 + 96}{6} = \frac{534}{6} = 89$$

To find the average of a group of numbers or like measures:

1. Add the numbers or like measures.
2. Divide the sum by the number of addends.

EXAMPLE
AUTO A car involved in an energy efficiency study had the following miles per gallon (mpg or $\frac{mi}{gal}$) listings for five tanks of gasoline: 21.7, 22.4, 26.9, 23.7, and 22.6 $\frac{mi}{gal}$. Find the average miles per gallon for the five tanks of gasoline. Round to the nearest tenth.

Estimate: The low value is 21.7 and the high value is 26.9. The average will be between the two values.

Exact:
$$\frac{21.7 + 22.4 + 26.9 + 23.7 + 22.6}{5} = \frac{117.3}{5} = 23.46$$
$$= 23.5 \frac{mi}{gal} \qquad \text{Rounded}$$

The approximate average is 23.5 $\frac{mi}{gal}$.

SECTION 1–1 SELF-STUDY EXERCISES

1 Write two inequalities to compare each pair of numbers.

1. 6 and 8

2. 42 and 32

3. 196 and 148

4. 2,802 and 2,517

5. 7,809 and 8,902

6. 44,000 and 42,999

7. BUS   A house that sold for $183,500 four years ago has just sold for $198,500. Write two inequalities to compare the housing prices.

8. AG/H   Touliatas Nursery sold 786 flats of annual bedding plants and 583 flats of perennial bedding plants. Write two inequalities to compare the number of plants of each type.

9. HOSP   The Orlando Renaissance Resort sold 758 rooms for a horticulture convention and 893 rooms for a motorcycle trade show. Write two inequalities to compare the number of rooms sold for each of the events.

10. COMP   The first day a spam filter was installed at Phoenix College, 5,982 spam e-mails were blocked. On the second day 2,807 spam e-mails were blocked. Write two inequalities to compare the number of blocked spam e-mails.

2 Write the fractions as decimal numbers.

11. $\frac{5}{10}$

12. $\frac{23}{100}$

13. $\frac{7}{100}$

14. $\frac{683}{100}$

15. $\frac{79}{1,000}$

16. $\frac{468}{1,000}$

3 Compare the number pairs and identify the larger number.

17. 3.72, 3.68

18. 7.08, 7.06

19. 0.23, 0.3

Arrange the numbers in order from smallest to largest.

20. 1.9, 1.87, 1.92

21. 72.1, 72.07, 73

22. INDTEC   Two micrometer readings are recorded as 0.837 in. and 0.81 in. Which is larger?

23. INDTEC   A micrometer reading for a part is 3.85 in. The specifications call for a dimension of 3.8 in. Which is larger, the micrometer reading or the specification?

24. INDTR   A washer has an inside diameter of 0.33 in. Will it fit a bolt that has a diameter of 0.325 in.?

25. INDTEC   Aluminum sheeting can be purchased in thicknesses of 0.04 in. or 0.035 in. Which sheeting is thicker?

26. CON If No. 14 copper wire has a diameter of 0.064 in., and No. 10 wire has a diameter of 0.09 in., which has the larger diameter?

27. HLTH/N A nurse recorded the weights of two patients as 64.8 kilograms and 72.3 kilograms. Which weight is greater?

Write two inequalities for each pair of numbers.

28. 4.2 and 3.8

29. 1.68 and 1.6

30. INDTEC A 100-watt bulb that burns continuously for two minutes uses 0.003 kilowatt-hours of electricity, and an 800-watt toaster uses 0.026 kilowatt-hours for a piece of toast. Which uses the greater number of kilowatt-hours?

31. To change centimeters to inches, multiply by 0.394, and to change kilometers to miles, multiply by 0.621. Which factor is larger?

4 Round to the place value indicated.

32. Nearest hundred: 468

33. Nearest ten thousand: 429,207

34. Nearest billion: 82,629,426,021

35. Nearest ten million: 297,384,726

36. CON A micrometer measure is listed as 0.7835 in. Round this measure to the nearest thousandth.

37. AUTO To the nearest tenth, what is the current of a 2.836-amp (A) motor?

38. HOSP If round steak costs $2.78 per pound, what is the cost per pound to the nearest dollar?

39. CON The average response times (in seconds) for drivers braking when they first see a road hazard were measured as follows: driver A, 0.0275; driver B, 0.0264; driver C, 0.0234; driver D, 0.0284; and driver E, 0.0379. Round each response time to hundredths to identify the drivers whose response times were most similar.

5 Write in columns and add.

40. 4,582 + 86,724 + 482 + 5,826

41. 6,017 + 893 + 15 + 82

42. 17 + 5,804 + 23,907 + 405

43. 4.2 + 3.6 + 7.9

44. 12.8 + 13.52 + 7.86

45. 83.37 + 42 + 1.6 + 3

46. CON A 0.103-in.-thick pipe has an inside diameter of 2.871 in. Find the outside diameter of the pipe. (Hint: The thickness of the pipe is on both sides of the inside diameter.)

47. ELEC The total current in amps in a parallel circuit is found by adding the individual currents. If a circuit has individual currents of 3.98 A, 2.805 A, and 8.718 A, find the total current.

48. BUS A part-time hourly worker earned $25.97 on Monday, $7.48 on Tuesday, $5.88 on Wednesday, $65.45 on Thursday, and $76.47 on Friday. Find the total week's wages.

49. CON A four-sided residential lot that measures 100.8 ft, 87.3 ft, 104.7 ft, and 98.6 ft is to be fenced. How many feet of fencing are required?

50. HLTH/N A patient's normal body temperature registered at 98.2° and his temperature rose 2.7°. What was his increased body temperature?

51. BUS Your investment portfolio totals $25,915.53 at the beginning of the year and it increases by $2,418.48 over the one-year period. What is the value of your portfolio at the end of the year?

Estimate the sum for Exercises 52–53 by rounding to thousands and Exercises 54–56 by rounding to tens.

52.	**53.**	**54.**	**55.**	**56.**
24,003	52,843	0.935	34.07	24.381
5,874	17,497	12.4	15.962	1.1
319,467	13,052	152.07	5.81	17.92
+ 52,855	+ 821	+ 18	+ 0.523	+ 38

57. CON Palmer Associates provided the following prices for items needed to build a sidewalk: concrete, $2,583.45; wire, $43.25; frame material, $18.90; labor, $798. Estimate the cost by rounding each amount to the nearest ten. Find the exact total.

58. BUS Antonio's expenses for one semester are: food, $1,500; lodging, $1,285; books, $288; supplies, $130; transportation, $162. Estimate his expenses by rounding each amount to the nearest hundred. Calculate the exact amount.

59. CON A hardware store filled an order for nails: 25 lb, $2\frac{1}{2}$-in. common; 16 lb, 4-in. common; 12 lb, 2-in. siding; 24 lb, $2\frac{1}{2}$-in. floor brads; 48 lb, 2-in. roofing; and 34 lb, $2\frac{1}{2}$-in. finish. Find the total weight.

60. AG/H If four containers have a capacity of 12 gal, 27 gal, 55 gal, and 21 gal, can 100 gal of fuel be stored in these containers? (Find the total capacity of the containers first.)

61. BUS A printer has three printing jobs that require the following numbers of sheets of paper: 185, 83, and 211. Will one ream of paper (500 sheets) be enough to finish the three jobs?

62. CON How many feet of fencing are needed to enclose the area shown in Fig. 1–7?

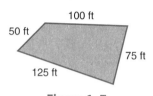

Figure 1–7

Subtract.

63. $3,672 - 2,652$

64. $946 - 831$

65. $53,867 - 831$

66. CON If a mason orders 75 bags of cement for a job and uses only 53, how many bags are left?

67. BUS An inventory sheet shows that 468 outlet boxes were in stock on March 1. Sales during March were 127. How many outlet boxes were left at the end of the month?

Subtract.

68. Subtract 24.38 from 316.2.

69. Subtract 13.5 from 21.

70. Subtract 67.2 from 378.

71. Find the difference between 42 and 37.6.

72. HTH/N Nurse Jennings recorded a temperature of 103.6°F for his patient. If the patient's normal body temperature was 98.6°F, how many degrees of fever did the patient have?

73. INDTR One box of rivets weighs 52.6 lb and another box weighs 37.5 lb. How much more does the first box of rivets weigh?

74. INDTR According to a blueprint, the length of an object is 12.09 in. If the tolerance is ±0.01 in., what are the limit dimensions of the object?

75. CON Two lengths of copper tubing measure 63.6 cm and 3.77 cm. What is the difference in their lengths?

76. CAD/ARC Find the limit dimensions of an object whose blueprint dimension is 4.195 in. ± 0.006 in.

77. CON A bricklayer laid 1,283 bricks on one day. A second bricklayer laid 1,097 bricks. How many more bricks did the first bricklayer lay?

78. INDTR A stockroom has 285.8 in. of bar stock and 173.5 in. of round stock. How many inches of bar stock remain after the object in Fig. 1–8 is made from this stock?

79. CON In Exercise 78, how many inches of round stock remain after the object in Fig. 1–8 is made?

80. CON Find the missing dimension in Fig. 1–9.

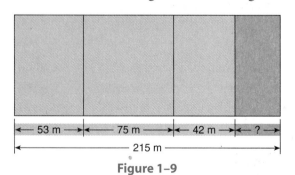

Figure 1–9

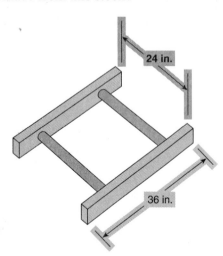

Figure 1–8

6 Multiply the following.

81. (3)(7)(9)(2)

82.
$$\begin{array}{r} 90{,}000 \\ \times\ 7{,}000 \\ \hline \end{array}$$

83.
$$\begin{array}{r} 503 \\ \times\ 204 \\ \hline \end{array}$$

84. BUS Helene Wright counted 8 unopened boxes of washers. Each box contained 512 washers. What total number of washers will be shown on the inventory sheet?

85. AUTO Bill Weppner repaired 6 automobiles a week over a period of 14 weeks. How many automobiles did Bill repair during this time period?

86. HELPP Each officer in the Public Safety office wrote, on average, 25 tickets a week. If there are 7 officers, how many tickets were written over a 4-week period?

87. BUS Jossie Moore is planning to sell candy bars in her Smart Shop. She receives 12 boxes, and each box contains 24 candy bars. If Jossie sells the bars for $1 each, how much money will she get for all the bars?

88.
$$\begin{array}{r} 37.7 \\ \times\ 1.5 \\ \hline \end{array}$$

89.
$$\begin{array}{r} 9.27 \\ \times\ 0.35 \\ \hline \end{array}$$

90.
$$\begin{array}{r} 0.215 \\ \times\ 0.27 \\ \hline \end{array}$$

91.
$$\begin{array}{r} 0.271 \\ \times\ 0.32 \\ \hline \end{array}$$

92. 73.806(2.305)

93. 1.9067 · 0.2013

94. 8.2037 · 0.602

95. 42(0.73)

96. CON A plastic pipe has an inside diameter of 4.75 in. Find the outside diameter if the pipe wall is 0.25 in. thick.

97. CON Electrical switches cost $5.70 wholesale. If the retail price is $7.99 each, how much would an electrician save over retail by buying 12 switches at the wholesale price?

98. CON How much No. 24 electrical wire is needed to make eight pieces each 18.9 in. long?

99. BUS A retailer purchases 15 cases of potato chips at $8.67 per case. If the chips are sold at $12.95 per case, how much profit does the retailer make?

Perform the long-division problems.

100. 5)215

101. 37)1,739

102. CON In a building where 46 outlets are installed, 1,472 ft of cable are used. What is the number of feet of cable used per outlet?

103. INDTR Twelve water tanks are constructed in a welding shop at a total contract price of $14,940. What is the price per tank?

104. INDTEC Five equally spaced holes are drilled in a piece of $\frac{1}{4}$-in. flat metal stock that is 28 in. long. The centers of the first and last holes are 2 in. from the end (see Fig. 1–10). What is the distance between the centers of any two adjacent holes? (*Caution:* How many equal center-to-center distances are there?)

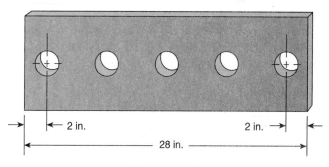

Figure 1–10

Divide.

105. 6.3)68.67

106. 0.23)0.0437

107. 8 ÷ 0.32

108. CON A 7.3-ft-long pipe weighs 43.8 lb. What is the weight of 1 ft of pipe?

109. CON A room requires 770.5 ft² of wallpaper, including waste. How many whole single rolls are needed for the job if a roll covers 33.5 ft² of surface?

110. CON The feed per revolution of a drill is 0.012 in. A hole 7.2 in. deep will require how many revolutions of the drill?

111. INDTR A piece of channel iron 5.6 ft long is cut into eight pieces. Assuming that there is no waste, what is the length of each piece?

Divide and round the quotient to the place indicated.

112. Nearest hundredth: $25\overline{)3.897}$

113. Nearest whole number: $4.1\overline{)34.86}$

114. **INDTR** If 12 lathes cost $6,895, find the cost of each lathe to the nearest dollar.

115. **ELEC** If 12 electrolytic capacitors cost $23.75, find the cost of one capacitor to the nearest cent.

116. **CON** Three bricklayers laid 3,210 bricks on a job in one day. What was the average number of bricks laid by each bricklayer?

117. **CON** ·A developer divides a tract of land into 14 equally valued parcels. If the tract is valued at $147,000, what is the value of each parcel?

118. **INDTR** A shipment of 150 machine parts costs $15,737.50. If this includes a $25 shipping charge, find the cost of each part.

119. **INDTR** If a shop manager earns $23,400 annually, what is the monthly salary?

Find the average, then round to the same place value used in the exercises.

120. Test scores: 86, 73, 95, 85

121. **AG/H** Weight of crates of oranges: 515 lb, 468 lb, 435 lb, 396 lb

122. **BUS** Monthly income: $873.46, $598.21, $293.85, $546.83, $695.83, $429.86, $955.34, $846.95, $1,025.73, $1,152.89, $957.64, $807.25

123. **AG/H** Average rainfall: 1.25 in., 0.54 in., 0.78 in., 2.35 in., 4.15 in., 1.09 in.

124. **CON** Amperes of current: 3.0 A, 2.5 A, 3.5 A, 4.0 A, 4.5 A

125. **AG/H** Crates of strawberries: 347, 623, 491, 387, 519

1–2 | Exponents, Roots, and Powers of 10

Learning Outcomes

1 Simplify expressions that contain exponents.

2 Square numbers and find the square roots of numbers.

3 Use powers of 10 to multiply and divide.

1 Simplify Expressions That Contain Exponents.

The product of repeated factors can be written in shorter form using natural-number exponents. Exponents that are not natural numbers will have a different interpretation.

In the example, $4(4)(4) = 4^3$, the 4 is called the **base** and is the repeated factor. The 3 is the **exponent** and indicates the number of times the factor is used in multiplication. The expression 4^3 is written in **exponential notation** and has a value of 64. The number 64 is written in **standard notation.** Both the exponential notation and the standard notation are called the **power.** The expression 4^3 is read four **cubed** or four to the third power or four raised to the third power. In expressions where 2 is an exponent, such as 4^2, the expression is usually read as four **squared;** however, it may also be read as four to the second power.

To change from natural-number exponential notation to standard notation:

1. Use the base as a factor as many times as indicated by the exponent.
2. Perform the multiplication.

EXAMPLE Identify the base and exponent of the expressions, and write in standard notation.

(a) 5^3　　(b) 1.5^2

(a) 5^3　　**5 is the base; 3 is the exponent.**
$5^3 = 5(5)(5) = 125$　　　Standard notation

(b) 1.5^2　　**1.5 is the base; 2 is the exponent.**
$1.5^2 = 1.5(1.5) = 2.25$　　　Standard notation

Any number with an exponent of 1 is the number itself:

$a^1 = a,$　　for any base a where a is a real number　　$8^1 = 8$　　$2.3^1 = 2.3$

2　Square Numbers and Find the Square Roots of Numbers.

The result of using a natural number as a factor 2 times is a **square** number or a **perfect square.** In the expression $7^2 = 49$, the 49 is a perfect square. This terminology evolved from a common formula for finding the area of a square. $A = s^2$, where A represents the area and s represents the length of one side. When we square 3, we write $3^2 = 3(3) = 9$. We say 9 is the square of 3.

EXAMPLE Find the square.

(a) 2　　(b) 7　　(c) 3.2

(a) 2;　　$2^2 = 2 \times 2 = 4$
(b) 7;　　$7^2 = 7 \times 7 = 49$
(c) 3.2;　　$3.2^2 = 3.2 \times 3.2 = 10.24$

The inverse operation of squaring is taking the square root of a number. The **principal square root** of a perfect square is the number that was used as a factor twice to equal that perfect square. The principal square root of 9 is 3 because 3^2 or $3(3) = 9$.

The **radical sign** $\sqrt{}$ indicates that the square root is to be taken of the number under the bar. This bar serves as a grouping symbol just like parentheses. The number under the bar is called the **radicand.** The entire expression is called a **radical expression.**

radical sign ——┐┌—— bar
$\sqrt{25} = 5 \leftarrow$ principal square root
radicand ——┘

To find the square root of a perfect square by estimation:

1. Select a trial estimate of the square root.
2. Square the estimate.
3. If the square of the estimate is less than the original number, adjust the estimate to a larger number. If the square of the estimate is more than the original number, adjust the estimate to a smaller number.
4. Square the adjusted estimate from Step 3.
5. Continue the adjusting process until the square of the trial estimate is the original number.

EXAMPLE   Find $\sqrt{256}$.

Select 15 as the estimated square root: 15^2 or $15(15) = 225$. The number 225 is less than 256, so the square root of 256 is larger than 15. We adjust the estimate to 17: 17^2 or $17(17) = 289$. The number 289 is more than 256, so the square root of 256 must be smaller than 17. Now adjust the estimate to 16.

Because $16^2 = 256$, 16 is the square root of 256.

The squares of the numbers 1 through 10 are 1, 4, 9, 16, 25, 36, 49, 64, 81, and 100. These are the only natural numbers from 1 through 100 that are perfect squares.

3 Use Powers of 10 to Multiply and Divide.

A **nonzero digit** is any digit except zero. That is, 1, 2, 3, 4, 5, 6, 7, 8, and 9 are nonzero digits. **Powers of 10** are numbers whose only nonzero digit is 1. Thus, 10, 100, 1,000, and so on are powers of 10 because each value can be written in exponential form with a base of 10.

Compare the number of zeros in standard notation with the exponent in exponential notation.

One million	$1,000,000 = 10^6$	6 zeros
One hundred thousand	$100,000 = 10^5$	5 zeros
Ten thousand	$10,000 = 10^4$	4 zeros
One thousand	$1,000 = 10^3$	3 zeros
One hundred	$100 = 10^2$	2 zeros
Ten	$10 = 10^1$	1 zero
One	$1 = 10^0$	0 zeros

The exponents in powers of 10 indicate the number of zeros used in standard notation. Zero exponents will be discussed in Chapter 4.

EXAMPLE   Express as powers of 10.

(a) 10,000,000   (b) 100,000,000   (c) 100,000,000,000

(a) $10,000,000$ $= 10^7$
(b) $100,000,000$ $= 10^8$
(c) $100,000,000,000$ $= 10^{11}$

EXAMPLE Express in standard notation.

(a) 10^5 (b) 10^{13}

(a) $10^5 = \textbf{100,000}$
(b) $10^{13} = \textbf{10,000,000,000,000}$

In some applications, powers of 10 are used to simplify multiplication and division problems. Compare the examples to find the pattern for multiplying a number by a power of 10.

$5(100) = 500$ $5(10^2) = 500$ Decimal point moved two places to the right and two zeros attached.

$27(1,000) = 27,000$ $27(10^3) = 27,000$ Decimal point moved three places to the right and three zeros attached.

To multiply a number by a power of 10 written in standard notation:

1. Move the decimal point in the number to the *right* as many places as the number of zeros in 10, 100, 1,000, and so on.
2. Attach zeros on the right if necessary.

EXAMPLE Multiply 237(100).

$237(100) = 237.00(100) = \textbf{23,700}$ Attach two zeros to the *right* of the 7 in 237, and insert the appropriate comma.

EXAMPLE Multiply 36.2(1,000).

$36.2(1,000) = 36.200(1,000) = \textbf{36,200}$ Move the decimal point three places to the *right*. Two zeros need to be attached.

Examine these divisions.

$32 \div 10 = 3.2$ Decimal point moved one place to the *left*.
$78.9 \div 100 = 0.789$ Decimal point moved two places to the *left*.
$52,900 \div 1,000 = 52.9$ Decimal point moved three places to the *left*.

To divide a number by a power of 10 written in standard notation:

1. Move the decimal to the *left* as many places as the divisor has zeros.
2. Attach zeros to the left if necessary.
3. You may drop zeros to the right of the decimal point if they follow the last nonzero digit of the quotient.

Compare this rule with the rule for multiplying decimal numbers by 10, 100, 1,000, and so on. For multiplication, the decimal shifts to the right; for division, the decimal shifts to the left.

EXAMPLE If 100 lb of floor cleaner costs $63, what is the cost of 1 lb?
Divide $63 by 100.

$63 \div 100 = 0\overset{\frown}{63}. \div 100 = \$0.63$ Decimal is after 3 in 63. Move decimal two places to the left.

The cost of 1 lb is $0.63.

Some scientific and graphing calculators have specific power keys, such as keys for squares and cubes. They also have a "general power" key that can be used for all powers. Even though the "general power" key can be used to square and cube numbers, the "square" and "cube" keys require fewer keystrokes.

TIP

Labels on Calculator Keys Are Not Universal

To show calculator steps, we often use a box to identify a common label for a function. The exact label will vary with the specific calculator model. Some calculators also provide many functions above the keys or on menus instead of on specific keys. Even though we use a key notation in this text to show a calculator function, this function may actually appear above a key or on a menu. Check your calculator manual for exact location and labeling of functions.

Common Labels for Power Keys: $\boxed{x^2}$ $\boxed{x^3}$ $\boxed{\wedge}$ $\boxed{x^y}$

Common Labels for Root Keys: $\boxed{\sqrt{}}$ $\boxed{\sqrt{x}}$ $\boxed{\sqrt[x]{}}$ $\boxed{x^{1/y}}$

EXAMPLE Use the calculator to find 35^2.

35 $\boxed{x^2}$ $\boxed{=}$ On some calculators pressing the equal key may not be required.

1,225

EXAMPLE Use the calculator to find 2.3^5.

2 $\boxed{\cdot}$ 3 $\boxed{\wedge}$ 5 $\boxed{=}$ $\boxed{\wedge}$ and $\boxed{x^y}$ are the most common labels for the general power key.

64.36343

EXAMPLE **(a)** Evaluate $\sqrt{529}$.   **(b)** Evaluate $\sqrt{10.89}$.   **(c)** Evaluate $\sqrt{5}$ to the nearest hundredth.

$\boxed{\sqrt{}}$ 529 $\boxed{=}$ $\boxed{\sqrt{}}$ 10 $\boxed{\cdot}$ 89 $\boxed{=}$ $\boxed{\sqrt{}}$ 5 $\boxed{=}$ $\Rightarrow$ 2.236067977

23   **3.3**   **2.24**   Rounded

Some calculators require that the radicand be entered *before* pressing the square root key. Test your calculator with an example you can do mentally. Other calculators will open a parenthesis when the square root key is pressed.

Parentheses are automatically closed when the equal or enter key is pressed. Get to know your calculator!

SECTION 1–2 SELF-STUDY EXERCISES

1 Give the base and exponent of the expressions.

1. 4^3 **2.** 9^4 **3.** 2.7^9 **4.** 15^2

Simplify the exponential expressions.

5. 10^3 **6.** 2^4 **7.** 3.4^2 **8.** 15^1
9. 8^1 **10.** 9^2

Express with an exponent of 1.

11. 8   **12.** 14.5   **13.** 12   **14.** 23

15. Explain how an exponent of 1 relates to the base.   **16.** Explain what an exponent of 3 means.

2 Write in standard notation.

17. 8^2 **18.** 18^2 **19.** 1.4^2 **20.** 13^2
21. 100^2 **22.** 121^2 **23.** 114^2 **24.** 3.8^2

Square the numbers.

25. 8   **26.** 12   **27.** 18   **28.** 101

Perform the operations.

29. $\sqrt{25}$ **30.** $\sqrt{49}$ **31.** $\sqrt{81}$ **32.** $\sqrt{36}$
33. $\sqrt{196}$ **34.** $\sqrt{225}$ **35.** $\sqrt{121}$ **36.** $\sqrt{144}$

37. What does it mean to square a number?   **38.** What does it mean to take the square root of a number?

3 Multiply the whole numbers using powers of 10.

39. 10×10^2 **40.** 12×10^5 **41.** $2 * 10^4$ **42.** $102 * 100$

Divide the whole numbers using powers of 10.

43. $250 \div 10$ **44.** $210 \div 10$ **45.** $\dfrac{300}{10^2}$ **46.** $2{,}500 \div 10$

47. What does the exponent of a power of 10 mean when you are multiplying?   **48.** What does the exponent of a power of 10 mean when you are dividing?

Use a calculator to evaluate.

49. 15^2 **50.** 7^3 **51.** 5^7 **52.** 12^4
53. $\sqrt{324}$ **54.** $\sqrt{784}$ **55.** $\sqrt{1,089}$ **56.** $\sqrt{196}$

Round to the nearest whole number.

57. $\sqrt{39}$ **58.** $\sqrt{72}$ **59.** $\sqrt{210}$ **60.** $\sqrt{1,095}$

Learning Outcomes

1 Apply the order of operations to a series of operations.

2 Evaluate a formula.

3 Solve applied problems using problem-solving strategies.

1 Apply the Order of Operations to a Series of Operations.

Whenever several mathematical operations are performed, the proper **order of operations** must be followed.

To apply the order of operations:

1. **Parentheses (grouping symbols):** Perform operations within parentheses (or other grouping symbols), beginning with the innermost set of parentheses; or, apply the distributive property.
2. **Exponents and roots:** Evaluate exponential operations and find square roots in order from left to right.
3. **Multiply and divide** in order from left to right.
4. **Add and subtract** in order from left to right.

To summarize, use the following key words:

Parentheses (grouping), Exponents (and roots), Multiplication and Division, Addition and Subtraction

Other grouping symbols are **brackets** [ ], **braces** { }, and a **bar.** The bar can combine with other symbols like the radical sign and be used as a grouping symbol, $\sqrt{4 + 5} = \sqrt{9} = 3$.

A Memory Aid for the Order of Operations

To remember the order of operations, use the sentence, "Please Excuse My Dear Aunt Sally."

*   Parentheses (grouping)   •   Exponents (roots)   •   Multiply/Divide   •   Add/Subtract

EXAMPLE   Evaluate $3(2 + 3)$.

Parentheses may show both a grouping and multiplication. When a number is multiplied by a sum or a difference, the distributive property can also be applied. Using the distributive property:

$$3(2 + 3) = 3(2) + 3(3) \qquad \text{Distribute.}$$
$$= 6 + 9 \qquad \text{Add.}$$
$$= \mathbf{15}$$

Using the order of operations:

$$3(2 + 3) = 3(5) \qquad \text{Do operation in parentheses first.}$$
$$= \mathbf{15} \qquad \text{Multiply.}$$

EXAMPLE   Simplify $4^2 - 5(2) \div (4 + 6)$ by performing the operations in the correct order.

$4^2 - 5(2) \div \boxed{(4 + 6)}$	Do operation within parentheses first: $4 + 6 = 10$.	P
$4^2 - 5(2) \div \boxed{10}$	Evaluate exponential notation: $4^2 = 16$.	E
$16 - \boxed{5(2)} \div 10$	Multiply: $5(2) = 10$.	M D
$16 - \boxed{10} \div 10$	Divide: $10 \div 10 = 1$.	M D
$16 - 1$	Subtract last.	A S
$16 - 1 = \mathbf{15}$		

TIP

Parentheses Indicate Multiplication, the Distributive Property, or a Grouping

Parentheses can indicate multiplication or an operation that should be done first. If the parentheses contain an operation, they indicate a grouping. Otherwise, they indicate multiplication. The expression $5(2)$ indicates multiplication, while $(4 + 6)$ indicates a grouping. The expression $3(2 + 3)$ is an example of the distributive property.

Many calculators have parentheses keys $\boxed{(}$ $\boxed{)}$. These keys are used for all types of grouping symbols.

EXAMPLE   Evaluate $5 \cdot \sqrt{16} - 5 + [15 - (3 \cdot 2)]$.

$5 \cdot \sqrt{16} - 5 + [15 - \boxed{(3 \cdot 2)}]$	Work innermost grouping: $3 \cdot 2$.	P
$5 \cdot \sqrt{16} - 5 + [15 - \boxed{6}\,]$	Work remaining grouping: $15 - 6$.	P
$5 \cdot \boxed{\sqrt{16}} - 5 + 9$	Find square root: $\sqrt{16} = 4$.	E
$5 \cdot \boxed{4} - 5 + 9$	Multiply: $5 \cdot 4$.	M D
$\boxed{20 - 5} + 9$	Add and subtract from left to right.	A S
$\boxed{15} + 9 = \mathbf{24}$		

EXAMPLE   Evaluate $3.2^2 + \sqrt{21 - 5}(2)$.

$3.2^2 + \sqrt{\boxed{21 - 5}}(2)$	Do operation within grouping first; the bar of the radical symbol is a grouping symbol: $21 - 5 = 16$.
$\boxed{3.2^2} + \sqrt{16}(2)$	Evaluate exponent and square root from left to right: $3.2^2 = 10.24$; $\sqrt{16} = 4$.
$\boxed{10.24} + \boxed{4(2)}$	Then multiply: $4(2) = 8$.
$10.24 + \boxed{8}$	Add last.
$10.24 + 8 = \mathbf{18.24}$	

A division bar serves as a grouping symbol in the same way parentheses do.

EXAMPLE Makesha Lee took six tests and scored 87, 92, 76, 85, 95, and 89. Find Makesha's average score.

$$\frac{87 + 92 + 76 + 85 + 95 + 89}{6} =$$ The division bar serves to group the addends. Perform operations that are grouped first.

$$\frac{524}{6} =$$ Perform division.

$$87.333333333$$ Round.

The average score is 87.

Develop Your Calculator Proficiency

Scientific and graphing calculators perform calculations according to the order of operations. Be sure you understand how your calculator requires steps to be entered. This is called calculator **syntax.** Sometimes you can determine the proper syntax by investigation. At other times you must refer to the owner's manual. Always investigate with calculations for which you already know the result.

Rework the examples in this section using a calculator. The given keystrokes are appropriate for many calculators. Other options may also work. *Test your calculator!*

$3(2+3)$ $\quad$ 3 $(\!($ 2 $\boxplus$ 3 $)\!)$ $\boxed{=}$ $\Rightarrow$ 15 $\quad$ Parentheses are required. If open parentheses are not closed, they are automatically closed at the equal sign.

$4^2 - 5(2) \div (4 + 6)$

$\quad$ 4 $\boxed{x^2}$ $\boxminus$ 5 $\boxtimes$ 2 $\boxdiv$ $(\!($ 4 $\boxplus$ 6 $)\!)$ $\boxed{=}$ $\Rightarrow$ 15

$5\sqrt{16} - 5 + [15 - (3 \cdot 2)]$

$\quad$ 5 $\boxed{\sqrt{}}$ 16 $)\!)$ $\boxminus$ 5 $\boxplus$ $(\!($ 15 $\boxminus$ $(\!($ 3 $\boxtimes$ 2 $)\!)$ $)\!)$ $\boxed{=}$ $\Rightarrow$ 24 $\quad$ A parenthesis automatically opened after $\boxed{\sqrt{}}$.

Many calculators do not require a multiplication sign before a parenthesis or other grouping symbol. The radical is also a grouping symbol, and many calculators automatically open a parenthesis. You should close the parentheses appropriately. Otherwise, all open parentheses are automatically closed when $\boxed{=}$ or $\boxed{\text{ENTER}}$ is pressed.

$3.2^2 + \sqrt{21 - 5(2)}$

$\quad$ 3 $\boxed{\cdot}$ 2 $\boxed{x^2}$ $\boxplus$ $\boxed{\sqrt{}}$ 21 $\boxminus$ 5 $)\!)$ $(\!($ 2 $)\!)$ $\boxed{=}$ $\Rightarrow$ 18.24 $\quad$ A parenthesis automatically opened after $\boxed{\sqrt{}}$.

$$\frac{87 + 92 + 76 + 85 + 95 + 89}{6}$$

$\quad$ $(\!($ 87 $\boxplus$ 92 $\boxplus$ 76 $\boxplus$ 85 $\boxplus$ 95 $\boxplus$ 89 $)\!)$ $\boxdiv$ 6 $\boxed{=}$ $\Rightarrow$ 87.333333333

2 Evaluate a Formula.

Formulas are procedures that have been used so frequently to solve certain types of problems that they have become the accepted means of solving these problems. Most formulas are expressed with one or more letter terms rather than words, and these procedures are written as symbolic equations, such as $P = 2(l + w)$, the formula for the perimeter of a rectangle. Electronic spreadsheets and calculator and computer programs are developed through formulas.

The most common use of formulas is for finding missing values. If we know values for all but one letter of a formula, we can find the missing value. To **evaluate** a formula is to substitute known values for the appropriate letters of the formula and perform the indicated operations to find the missing value.

To evaluate a formula:

1. Write the formula.
2. Rewrite the formula substituting known values for letters of the formula.
3. Perform the indicated operations, applying the order of operations.
4. Interpret the solution within the context of the formula.

Some of the first formulas that we use are the formulas for basic geometric shapes called polygons. A **polygon** is a plane or flat, closed figure described by straight-line segments and angles. Polygons have different numbers of sides and different properties. Some common polygons are the parallelogram, rectangle, and square (Fig. 1–11).

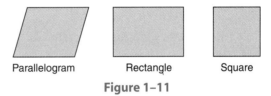

Parallelogram   Rectangle   Square

Figure 1–11

The **base** of any polygon is the horizontal side or a side that would be horizontal if the polygon's orientation is modified.

An **adjacent side** of any polygon is the side that has an end point in common with the base.

A **parallelogram** is a four-sided polygon with opposite sides that are **parallel** (never meet or intersect).

A **rectangle** is a parallelogram with angles that are all **right angles** (square corners).

A **square** is a parallelogram with all sides of equal length and with all right angles. A **square** can also be described as a rectangle with all sides of equal length.

The **perimeter** is the total length of the sides of a plane figure. As we saw earlier, some common applications for perimeter are finding the amount of trim molding for a room, determining the amount of fencing for a yard, finding the amount of edging for a flower bed, and so on.

A general procedure for finding the perimeter of any shape is to add the lengths of the sides. However, shortcuts based on the properties of the shape are given as formulas.

Perimeter

Parallelogram	$P = 2b + 2s$ or $P = 2(b + s)$		b is the base s is an adjacent side
Rectangle	$P = 2l + 2w$ or $P = 2(l + w)$		l is length w is width
Square	$P = 4s$		s is length of a side

Put Cheat Sheet

The perimeter of a parallelogram, like the perimeter of a rectangle, is the sum of its four sides. The sides of the parallelogram are called **base** and **adjacent side** (instead of length and width). Notice the locations of the base and adjacent side in Fig. 1–12.

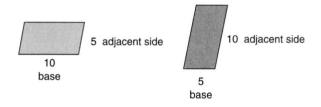

5 adjacent side   10 adjacent side

10
base

5
base

Figure 1–12

EXAMPLE Find the perimeter of a parallelogram with a base of 16 in. and an adjacent side of 8 in. (Fig. 1–13).

Visualize the parallelogram.

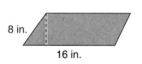

8 in.

16 in.

Figure 1–13

$P_{\text{parallelogram}} = 2(b + s)$ Select the perimeter of parallelogram formula.
$P_{\text{parallelogram}} = 2(16 \text{ in.} + 8 \text{ in.})$ Substitute: $b = 16$, $s = 8$.
$P_{\text{parallelogram}} = 2(24 \text{ in.})$ Evaluate.

$P_{\text{parallelogram}} = \textbf{48 in.}$

EXAMPLE BUS A shop that makes custom picture frames has an order for a frame that has outside measurements of 42 in. by 30 in. (Fig. 1–14). How many inches of picture frame molding are needed for the job?

42 in.

30 in.

Figure 1–14

$P_{\text{rectangle}} = 2(l + w)$ The figure is a rectangle. Select the appropriate formula.
$P_{\text{rectangle}} = 2(42 \text{ in.} + 30 \text{ in.})$ Substitute and evaluate the formula.
$P_{\text{rectangle}} = 2(72 \text{ in.})$
$P_{\text{rectangle}} = 144 \text{ in.}$

When analyzing the dimensions, inches are added to inches, so the result is written in inches. Then the inches are multiplied by a number and the final result is inches.

The frame requires 144 in. of molding.

Chapter 1 / Review of Basic Concepts

In some applications, we need to decrease the total perimeter to account for doorways and other openings in the perimeter.

EXAMPLE AG/H

A chain-link fence is to be installed around a yard measuring 25 m by 30 m (Fig. 1–15). A gate 3 m wide will be installed over the driveway. The gate comes preassembled from the manufacturer. How much fencing does the installer need to put up the fence?

Figure 1–15

$P_{rectangle} = 2(l + w) - 3$ Select the perimeter of a rectangle formula. Subtract the length of the gate.

$P_{rectangle} = 2(30 \text{ m} + 25 \text{ m}) - 3 \text{ m}$ Substitute and perform operations.

$P_{rectangle} = 2(55 \text{ m}) - 3 \text{ m}$ Multiply.

$P_{rectangle} = 110 \text{ m} - 3 \text{ m}$ Subtract.

$P_{rectangle} = 107 \text{ m}$ Perimeter is a linear measure.

The job requires 107 m of fencing.

TIP

Using Subscripts

Subscripted words, letters, or numbers are a handy way to provide additional information. Because we will examine the formulas for the perimeter of several different shapes, we sometimes use subscripts to distinguish among them. For instance, the perimeter of a square, rectangle, or parallelogram can be indicated as P_{square}, $P_{rectangle}$, or $P_{parallelogram}$, respectively.

EXAMPLE CON

How much aluminum edge molding is needed to surround a stainless steel kitchen sink that measures 40 cm on each side (Fig. 1–16)?

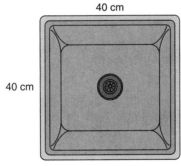

40 cm

40 cm

Figure 1–16

$P_{square} = 4s$ The figure is a square. Select the appropriate formula.

$P_{square} = 4(40 \text{ cm})$ Substitute and evaluate the formula.

$P_{square} = 160 \text{ cm}$

The sink requires 160 cm of molding.

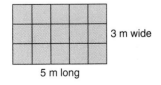

3 m wide

5 m long

Figure 1–17

The **area** of a polygon is the amount of surface of a plane figure. Area is expressed in square units. For example, if a rectangle is 3 m wide and 5 m long, there are 15 m^2 in the area (Fig. 1–17).

$$\text{area} = 15 \text{ m}^2$$

TIP

Writing Square Units

The exponent 2 following a unit of measure indicates *square measure* or area. Thus, 15 m² = 15 square meters, 23 ft² = 23 square feet, 120 cm² = 120 square centimeters, and so on. This is a shortcut way to express a measure that has *already* been "squared."

When do you need the area of a polygon? Some common applications for area are finding the amount of carpeting needed to cover a floor, the amount of paint needed to paint a surface, the amount of fertilizer needed to treat a lawn, and so on.

Area

Rectangle	$A = lw$		l is length w is width
Square	$A = s^2$		s is length of a side
Parallelogram	$A = bh$		b is base h is height

EXAMPLE
CON

A carpet installer is carpeting a room measuring 16 ft by 20 ft. Projecting out from one wall is a fireplace whose hearth measures 3 ft by 6 ft (Fig. 1–18). How many square feet of carpet does the installer require for the job? How much is wasted?

First, compute the area of the room without considering the area of the hearth. This is the amount of carpet needed. Second, compute the area of the fireplace hearth. This is the amount wasted.

Let A_{room} = the area of the room and A_{hearth} = the area of the hearth.

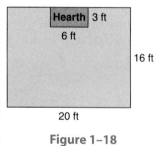

Figure 1–18

$A_{room} = lw$	Select the appropriate formula and substitute.
$A_{room} = 20 \text{ ft}(16 \text{ ft})$	Multiply.
$A_{room} = 320 \text{ ft}^2$	Area is a square measure (ft × ft = ft²).
$A_{hearth} = lw$	Select the appropriate formula and substitute.
$A_{hearth} = 3 \text{ ft}(6 \text{ ft})$	Multiply.
$A_{hearth} = 18 \text{ ft}^2$	Area is a square measure (ft × ft = ft²).

This job requires 320 ft² of carpet. Of this, 18 ft² is waste from the hearth.

The preceding example brings up some interesting questions. Is carpet purchased in square feet? If carpet can only be purchased in 12-ft or 15-ft widths, will there be additional waste? Is it more practical to purchase extra carpet that will be wasted or to spend extra labor costs to install the carpet with several seams? You may want to investigate common practices in the industry.

EXAMPLE
CON

Asphalt roofing shingles are sold in bundles. The number of bundles needed to cover a square (100 ft²) depends on the overlap when the shingles are installed. If the overlap allows 4 in. of each shingle to be exposed, then four bundles are needed per square. With a 5-in. exposure, 3.2 bundles are needed per square. Figure the number of bundles for a 4-in. exposure and for a 5-in. exposure for a roof measuring 30 ft × 20 ft.

$$A_{\text{roof}} = 30 \text{ ft}(20 \text{ ft})$$ 	Dimension analysis: ft(ft) = ft²
$$A_{\text{roof}} = 600 \text{ ft}^2$$
$$600 \text{ ft}^2 \div 100 \text{ ft}^2 = 6 \text{ squares}$$ 	Find the number of squares (100 ft²) by dividing by 100 ft².

$$6(4) = 24 \text{ bundles}$$ 	Find the number of bundles for a 4-in. exposure.
$$6(3.2) = 19.2 \text{ or } 20 \text{ bundles}$$ 	Find the number of bundles for a 5-in. exposure.

Therefore, 24 bundles of shingles are required for a 4-in. exposure, and 20 bundles (from 19.2 bundles) are required for a 5-in. exposure.

EXAMPLE
CON

Lynn Fly, a vinyl floor installer, discovers that one of the squares in the flooring pattern is damaged. He decides to cut out the damaged square and replace it. The damaged square measures 20 cm on each side. What is the area of the square to be replaced?

$$A_{\text{square}} = s^2$$ 	Select the appropriate formula.
$$A_{\text{square}} = (20 \text{ cm})^2$$ 	Substitute measurement of one side and square it.
$$A_{\text{square}} = 400 \text{ cm}^2$$ 	Dimension analysis: cm(cm) = cm²

The vinyl square measures 400 cm².

Dimension Analysis

When we evaluate formulas, the measuring units are sometimes omitted from the written steps. However, it is very important to use the correct unit in your calculations so that the unit in the solution is correct. Compare examples involving perimeter and area.

Example (molding for sink, p. 41)

$$P_{\text{square}} = 4s$$
$$P_{\text{square}} = 4(40 \text{ cm})$$ 	The measuring unit is centimeters; the measure is multiplied by a number.
$$P_{\text{square}} = 160 \text{ cm}$$ 	The measuring unit in the solution is centimeters.

Perimeter is always a linear measure, which means the measuring unit should be to the first power.

Example (vinyl flooring in the preceding example)
$$A_{\text{square}} = s^2$$
$$A_{\text{square}} = (20 \text{ cm})^2$$ 	The measuring unit is centimeters and a measure is squared or a measure is multiplied by a measure.

$$A_{\text{square}} = 400 \text{ cm}^2$$ 	The measuring unit in the solution is square centimeters.

Area is always a square measure, which means that measures of area are to the second power or have an exponent of 2.

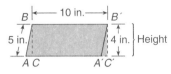

Figure 1–19

Since a rectangle is a parallelogram, the areas of both types of polygons are related.

The **height** of a parallelogram is the perpendicular distance between two parallel sides. Height is also called **altitude.**

If triangle ABC ($\triangle ABC$) in Fig. 1–19 were transposed to the right side of the parallelogram ($\triangle A'B'C'$), we would have a rectangle. Triangles are presented in detail in Chapters 18 and 19.

The base of the original parallelogram is the same as the length of the newly formed rectangle. Similarly, the height of the parallelogram is the same as the width of the rectangle. Thus, the area of the rectangle and parallelogram is 10 in. (4 in.) or 40 in².

EXAMPLE Find the area of a parallelogram with a base of 16 in., an adjacent side of 8 in., and a height of 7 in. (Fig. 1–20).

Visualize the parallelogram.

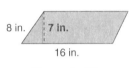

Figure 1–20

$A_{\text{parallelogram}} = bh$ Substitute.

$A_{\text{parallelogram}} = 16 \text{ in. } (7 \text{ in.})$ Multiply.

$A_{\text{parallelogram}} = \mathbf{112 \ in^2}$

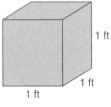

Figure 1–21

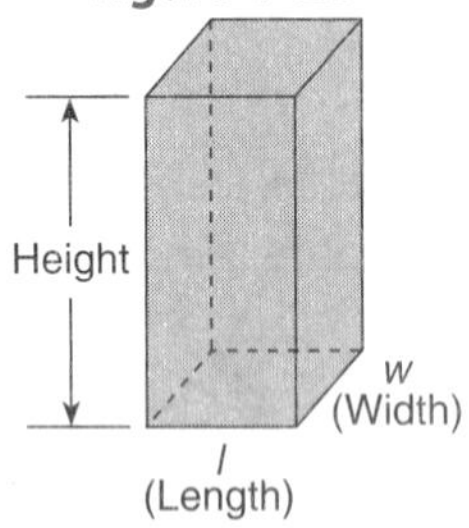

Figure 1–22

Right Rectangular prism

The *volume* of an object, such as a container, is used to estimate how many containers can be loaded into a given-size storage area or shipped in a container of certain dimensions.

The **volume** of a three-dimensional geometric figure is the amount of space it occupies, measured in terms of three dimensions (length, width, and height).

If we have a rectangular box measuring 1 ft long, 1 ft wide, and 1 ft high (Fig. 1–21), it will be a cube representing 1 cubic foot (ft³). Its volume is calculated by multiplying length $\times$ width $\times$ height, or 1 ft $\times$ 1 ft $\times$ 1 ft = 1 ft³. We indicate a **cubic measure** with an exponent 3 after the unit of measure, meaning that the measure is **cubed.**

From this concept of 1 ft³ or 1 cubic foot comes the formula for the volume of a rectangular box as length $\times$ width $\times$ height or $V = lwh$. The mathematical term for a box is a **prism.** Specifically, if the sides and bases all make square corners and the faces and bases are rectangles or squares, we refer to the prism as a **right rectangular prism** (Fig. 1–22). If the length, width, and height of a right rectangular prism are equal, the prism is a **cube.** Prisms will be discussed in more detail in Chapter 18.

Formula for the volume of a right rectangular prism or a cube:

$V_{\text{prism}} = lwh$, where l is the length, w is the width, and h is the height.

$V_{\text{cube}} = s^3$, where s is the length of one side.

EXAMPLE Find the volume of the cube in Figure 1–23 if each side is 7.5 cm. Round to the nearest tenth.

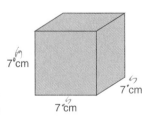

Figure 1–23

$V_{\text{cube}} = s^3$ Select the appropriate formula and substitute 7.5 cm for s.

$V_{\text{cube}} = (7.5 \text{ cm})^3$ Evaluate.

$V_{\text{cube}} = 421.875 \text{ cm}^3$ Round.

The volume of the cube is 421.9 cm³.

3 Solve Applied Problems Using Problem-Solving Strategies.

Problem solving is an important skill in the workplace and in everyday life. In developing good problem-solving skills, it is helpful to use a systematic problem-solving plan. A plan gives you a framework for approaching problems.

You will encounter many different plans for solving problems. They help you organize the problem details so that you can find an appropriate procedure for solving the problem. Our plan, like others, is very structured. As you develop confidence and skill in solving problems, you will probably use less structured and more intuitive approaches.

Six-Step Problem-Solving Plan

1. **Unknown Facts.** What facts are missing from the problem? What are you trying to find?
2. **Known Facts.** What relevant facts are known or given? What facts must you bring to the problem from your own background?
3. **Relationships.** How are the known facts and the unknown facts related? What formulas or definitions can you use to establish a model for solving the problem?
4. **Estimation.** What are some of the characteristics of a reasonable solution? For instance, should the answer be more than a certain amount or less than a certain amount?
5. **Calculations.** Perform the operations identified in the relationships.
6. **Interpretation.** What do the results of the calculations represent within the context of the problem? Is the answer reasonable? Have all unknown facts been found? Do the unknown facts check in the context of the problem?

The 7th Inning buys baseball cards from eight different vendors. In November, the company purchased 8,832 boxes of cards. If an equal number of boxes was purchased from each vendor, how many boxes of cards were supplied by each vendor?

Number of boxes of cards supplied by each vendor

8,832 = total number of boxes purchased
   8 = total number of vendors

An equal number of boxes was purchased from each vendor.

total boxes purchased ÷ number of vendors = boxes purchased from each vendor

If 8,000 boxes were purchased in equal amounts from eight vendors, then 1,000 were purchased from each vendor. Since more than 8,000 boxes were purchased, then more than 1,000 were purchased from each vendor.

Calculations
$$\frac{1,104}{8\overline{)8,832}}$$

Interpretation **Each vendor supplied 1,104 boxes of cards.**

In identifying the relationships of a problem, it is often helpful if you look for key words or phrases that give you clues about the mathematical operations involved in the relationships (Table 1–1). Key words give clues as to whether one quantity is added to, subtracted from, or multiplied or divided by another quantity. For example, if a problem tells you that Carol's salary in 2006 exceeds her 2005 salary by $2,500, you know that you should add $2,500 to her 2005 salary to find her 2006 salary.

Table 1–1   Key Words and What They Generally Imply in Word Problems

Addition	Subtraction	Multiplication	Division	Equality
The sum of	Less than	Times	Divide(s)	Equals
Plus/total	Decreased by	Multiplied by	Divided by	Is/was/are
Increased by	Subtracted from	Of	Divided into	Is equal to
More/more than	Difference between	The product of	Half of (divided by 2)	The result is
Added to	Diminished by	Twice (2 times)	Third of (divided by 3)	What is left
Exceeds	Take away	Double (2 times)	Per	What remains
Expands	Reduced by	Triple (3 times)	How big is each part?	The same as
Greater than	Less/minus	Half of ($\frac{1}{2}$ times)	How many parts can be made from	Gives/giving
Gain/profit	Loss	Third of ($\frac{1}{3}$ times)		Makes
Longer	Lower			Leaves
Older	Shrinks			
Heavier	Smaller than			
Wider	Younger			
Taller	Slower			
Larger than	Shorter			

More than one relationship may be needed to find the unknown facts. When this is the case, you usually need to read the problem several times to find all the relationships and plan your solution strategy.

EXAMPLE BUS   Carlee Anne McAnally needs to ship 78 crystal vases. With standard packing to prevent damage, 5 vases fit in each available box. How many boxes are required to pack the vases?

Unknown fact   Number of boxes required to pack the vases

Known facts
total vases to be shipped = 78
number of vases per box = 5

Relationships
total boxes needed = total number of vases ÷ number per box
total boxes needed = 78 ÷ 5

Estimation
70 ÷ 5 = 14   Round down the dividend.
80 ÷ 5 = 16   Round up the dividend.

Since 78 is between 70 and 80, the number of boxes needed is between 14 and 16.

Calculation   78 ÷ 5 = 15 R3

Interpretation   **16 boxes are needed;** 15 boxes will contain 5 vases each, and 1 box will contain 3 vases. The box with 3 vases will need extra packing.

Using Guessing and Checking to Solve Problems

An effective strategy for solving problems involves guessing. Make a guess that you think may be reasonable, and check to see if the answer is correct. If your guess is not correct, decide if it is too high or too low. Make another guess based on what you learned from your first guess. Continue until you find the correct answer.

Let's try guessing in the previous example. We found that we could pack 70 vases in 14 boxes and 80 vases in 16 boxes. Since we need to pack 78 vases, how many vases can we pack with 15 boxes? 15(5) = 75. Still not enough. Therefore, we will need 16 boxes, but the last box will not be full.

You can probably think of other ways to solve this problem. Some plans will be more efficient than others, but you develop your problem-solving skills by pursuing a variety of strategies.

SECTION 1–3 SELF-STUDY EXERCISES

1 Use the order of operations to evaluate each problem.

1. $5^2 + 4 - 3$
2. $4^2 + 6 - 4$
3. $4(3) - 9 \div 3$
4. $5 \cdot 2.9 - 4 \div 2$
5. $25 \div 5 \cdot 4.8$
6. $64 \div 4(2)$
7. $48 \div 8 \cdot 3$
8. $15 - 2 \cdot 3$
9. $4^2 - (4)(3) + 6$
10. $17 - 4 \cdot 2$
11. $6 \times \sqrt{36} - 2 \times 3$
12. $3 \cdot \sqrt{81} - (3)(4)$
13. $4^2 \cdot 3^2 + (4 + 2)(2)$
14. $2^2 \cdot 5^2 + (2 + 1)(3)$
15. $54 - 3^3 - \dfrac{8}{2}$
16. $156 - 2^3 - \dfrac{9}{3}$
17. $3 - 2 + 3 \cdot 3 - \sqrt{9}$
18. $2 - 1 + (4)(4) - \sqrt{4}$
19. $2^4 \times (7 - 2) \cdot 2$
20. $3^4 \times (9 - 3) \cdot 3$
21. $124 - 8 \cdot 7 + 12$
22. $5 \times 12 \div 6$
23. $72 \div 9 \cdot 3$
24. $32 - 2.05 \cdot 4^2 \div 2$
25. $5.2^2 - 3 \cdot 2^2 \div 6$
26. $4 + 15 \div 3 - \dfrac{0}{7}$
27. $3 + 8 \div 4 - \dfrac{0}{5}$

Use a calculator to perform the operations.

28. $4^3 + 14 - 8$
29. $2^3 + 12 - 7$
30. $2 \cdot \sqrt{16} + (8 - \sqrt{25})$
31. $4 \cdot \sqrt{49} + (9 - \sqrt{64})$
32. $3(2^2 + 1) - 30 \div 3$
33. $4(3^2 + 2) - 60 \div 12$
34. $25 - 5^2 \div (7 - 2)$
35. $36 - 6^2 \div (8 - 2)$
36. $2(3.1^2 + 2) - \sqrt{7.29}$
37. $6 \times 9^2 - \dfrac{12}{4}$
38. $7 \times 8^2 - \dfrac{10}{2}$
★39. $3^4 - 2 \times 4.6 \div \left(\dfrac{10}{2}\right)$

Find the perimeter of the rectangular crop fields in Figs. 1–24 and 1–25.

40. AG/H

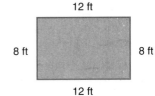

12 ft

8 ft   8 ft

12 ft

Figure 1–24

41. AG/H

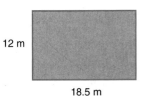

12 m

18.5 m

Figure 1–25

Find the number of square feet in each rectangular garden illustrated in Figs. 1–26 and 1–27. Check your work.

42. CON

8.7 ft

14.2 ft

Figure 1–26

43. CON

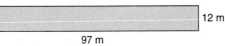

12 m

97 m

Figure 1–27

44. Find the perimeter and area of the parallelogram in Fig. 1–28.

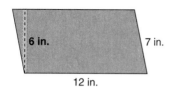

6 in.   7 in.

12 in.

Figure 1–28

45. Find the perimeter and area of the parallelogram in Fig. 1–29.

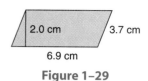

2.0 cm   3.7 cm

6.9 cm

Figure 1–29

Solve the problems.

46. CON   An illuminated sign in the main entrance of a hospital is a parallelogram with a base of 48 in. and an adjacent side of 30 in. How many feet of aluminum molding are needed to frame the sign?

48. CON   A contemporary building has a window in the shape of a parallelogram with a base of 50 in. and an adjacent side of 30 in. How many inches of trim are needed to surround the window?

50. Find the perimeter and area of the rectangle in Fig. 1–30.

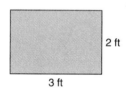

2 ft

3 ft

Figure 1–30

47. AUTO   A customized van has a window cut in each side in the shape of a parallelogram with a base of 20 in. and an adjacent side of 11 in. How many inches of trim are needed to surround the two windows?

49. CON   A table for a reading lab has a top in the shape of a parallelogram with a base of 36 in. and an adjacent side of 18 in. How many inches of edge trim are needed to surround the tabletop?

51. Find the perimeter and area of the rectangle in Fig. 1–31.

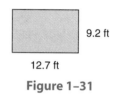

9.2 ft

12.7 ft

Figure 1–31

52. CON   A rectangular parking lot is 340 ft by 125 ft. Find the perimeter of the parking lot.

53. CON   A room is 15 ft by 12 ft. How many feet of chair rail are needed for the room? Disregard openings.

54. CON   How many feet of quarter-round molding are needed to finish around the baseboard after sheet vinyl flooring is installed if the room is 16 ft by 18 ft and there are three 3-ft-wide doorways?

55. CON   The swimming pool in Fig. 1–32 measures 32 ft by 18 ft. How much fencing is needed, including material for a gate, if the fence is to be built 7 ft from each side of the pool?

56. Find the perimeter and area of Fig. 1–33.

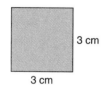

3 cm

3 cm

Figure 1–33

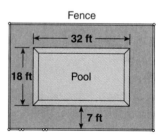

Fence

32 ft

18 ft   Pool

7 ft

Figure 1–32

57. Find the perimeter and area of Fig. 1–34.

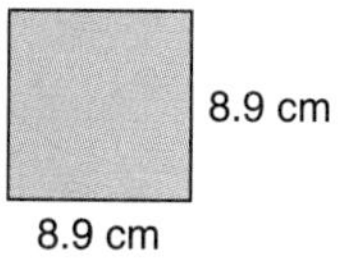

8.9 cm

8.9 cm

Figure 1–34

58. CON The square parking lot of a doctor's office is to have curbs built on all four sides. If the lot is 150 ft on each side, how many feet of curb are needed? Allow 10 ft for a driveway into the parking lot.

60. Find the volume of a cube that is 4.7 ft on each side.

62. Find the volume of a prism if the length is 18 cm, the width is 17 cm, and the height is 21 cm.

59. CON A border of 4-in. × 4-in. wall tiles surrounds the floor of a shower stall that is 48 in. × 48 in. How many tiles are needed for this border? Disregard spaces for grout (connecting material between the tiles).

61. Find the volume of a cube that is 12.8 cm on each side.

63. HELPP A container of fire retardant is a prism that has a length of 2.9 m, a width of 1.6 m, and a height of 3.6 m. Find the volume.

3

64. BUS If you have 348 packages of Halloween candy to rebox for shipment to a discount store and you can pack 12 packages in each box, how many boxes will you need?

66. BUS In a recent year, 21,960 people visited Bio Fach, Germany's biggest ecologically sound consumer goods trade fair. This figure was up from the 18,090 the previous year and 16,300 two years ago. What was the increase in visitors to Bio Fach over the two-year period?

65. BUS If American Communications Network (ACN) has an annual payroll of $5,602,186 for its 214 employees, what is the average salary of an ACN employee?

67. BUS The On-The-Square Card and Gift Shop buys cards from six vendors. In November, the company purchased 6,480 boxes of cards. If the shop purchased an equal number of boxes from each vendor, how many boxes of cards did each vendor supply?

CHAPTER REVIEW OF KEY CONCEPTS

Learning Outcomes

Key Concepts and Examples

Section 1–1

1 Compare whole numbers (pp. 2–3).

Mentally position the numbers on a number line. The leftmost number is the smaller number. Write an inequality using either the *less than* < or *greater than* > symbol.

> Write two inequalities comparing 5 and 9.
> 5 < 9   or   9 > 5

2 Write fractions with power-of-10 denominators as decimal numbers (pp. 3–5).

Write the numerator as the decimal number with the same number of decimal digits as there are zeros in the denominator.

> Express $\dfrac{53}{1,000}$ as a decimal: 0.053.

3 Compare decimal numbers (pp. 5–6).

Compare decimal numbers by comparing the digits in the same place beginning from the left of each number.

> Which decimal is larger, 0.23 or 0.225? Both numbers have the same digit, 2, in the tenths place. 0.23 is larger because it has a 3 in the hundredths place, while 0.225 has a 2 in that place.

4 Round a whole number or a decimal number to a specified place value (pp. 4–8).

Whole number: 1. Locate the rounding place. **2.** Examine the digit to the right. **3.** Round down if the digit is less than 5. **4.** Round up if the digit is 5 or more.

Decimal number: Round as in whole numbers; however, digits after the decimal on the right side of the digit in the rounding place are dropped rather than replaced with zeros.

> Round 3,624 to the tens place: 2 is in the tens place, 4 is the digit to the right; 4 is less than 5, so round down. The rounded value is 3,620.
> Round 5.847 to the nearest tenth: 8 is in the tenths place; 4 is less than 5, so leave 8 as is and drop the 4 and 7. The rounded value is 5.8.

5 Add and subtract whole numbers and decimals (pp. 8–14).

Addition is a binary operation that is commutative and associative.

Zero property of addition (additive identity): $0 + n = n + 0 = n$ where n is any real number.

Add whole numbers:
Arrange addends in columns of like places. Add each column beginning with the ones place. Regroup when necessary.

> $5 + 7 = 7 + 5 = 12$ Commutative property of addition
> $(3 + 2) + 6 = 3 + (2 + 6) = 11$ Associative property of addition
> $0 + 3 = 3 + 0 = 3$ Zero property of addition
>
> 4,824   Addend
> + 745   Addend
> 5,569   Sum

Add decimal numbers:
Add decimal numbers by arranging the addends so that the decimal points are aligned.

> Add: $43.35 + 3.7 + 0.462$
>
> 43.35   Addend
> 3.7   Addend
> + 0.462   Addend
> 47.512   Sum

Estimate and check addition:
Estimate by rounding the addends before finding the sum. Check addition by adding a second time.

> Estimate the sum by rounding to the nearest hundred: $483 + 723$; $500 + 700 = 1,200$. Exact sum: 1,206.

Subtraction is a binary operation that is *not* commutative or associative. Addition and subtraction are inverse operations.

Zero property of subtraction: $n - 0 = n$, where n is any real number.

Subtract whole numbers:
Arrange numbers in columns of like places. Subtract each column, beginning with the ones place. Regroup when necessary.

> If $5 + 4 = 9$, then $9 - 5 = 4$ or $9 - 4 = 5$.   Inverse operations
> $7 - 0 = 7$ Zero property of subtraction
> 4,227   Regroup in tens column: $12 - 4 = 8$
> − 745   Regroup in hundreds column: $11 - 7 = 4$
> 3,482

Subtract decimal numbers:

Subtract decimal numbers by arranging the minuend and subtrahend so that the decimal points are aligned.

> Subtract: $53.824 - 4.0423$
>
> | | |
> |---|---|
> | 53.824 | Minuend |
> | $-\ 4.0423$ | Subtrahend |
> | 49.7817 | Difference |

Estimate and check subtraction:

Estimate by rounding the minuend and subtrahend before finding the difference. Check subtraction by adding the difference and the subtrahend. The result should equal the minuend.

> Estimate the difference by rounding to the nearest hundred: $783 - 423$
> $800 - 400 = 400$. Exact difference: $783 - 423 = 360$.
> Check: $360 + 423 = 783$.

6 Multiply and divide whole numbers and decimals (pp. 15–26).

Multiplication is a binary operation that is commutative and associative.

Zero property of multiplication: $n(0) = 0(n) = 0$, where n is any real number.

Multiply whole-number factors:

Arrange numbers in columns of like place values. Multiply the multiplicand by each digit in the multiplier. Add the partial products.

> $5(7) = 7(5) = 35$ Commutative property of multiplication
> $(3 \cdot 2) \cdot 6 = 3 \cdot (2 \cdot 6) = 36$ Associative property of multiplication
> $5(0) = 0(5) = 0$ Zero property of multiplication
>
> | | |
> |---|---|
> | 259 | Multiplicand |
> | $\times\quad 23$ | Multiplier |
> | 777 | Partial product |
> | 5 18 | Partial product |
> | 5,957 | Product |

Multiply decimal factors:

Multiply the decimal numbers as in whole numbers, and count the number of decimal digits in both factors. Place the decimal in the product to the left of the same number of digits, counting from the right.

> Multiply: $3.25(0.53)$
>
> | | |
> |---|---|
> | 3.25 | Two decimal places |
> | $\times\ 0.53$ | Two decimal places |
> | 9 75 | |
> | 1 62 5 | |
> | 1.72 25 | Four decimal places |

Apply the distributive property:

$a(b + c) = ab + ac$
$a(b - c) = ab - ac$

> $3(5 + 6) = 3(5) + 3(6)$
> $3(11) \quad = 15 + 18$
> $\qquad 33 = 33$

Estimate and check multiplication:

Round the factors, then multiply. Check multiplication by multiplying a second time.

> Estimate the product by rounding to one nonzero digit: 483(72);
> 500(70) = 35,000. Exact product: 34,776.

Use symbols to indicate division:

The division a divided by b can be written as

$$a \div b \qquad b\overline{)a} \qquad \frac{a}{b} \qquad a/b$$

The divisor, b, cannot be zero.

> Write 12 divided by 4 in four ways.
>
> $$12 \div 4 \qquad 4\overline{)12} \qquad \frac{12}{4} \qquad 12/4$$

Divide whole numbers:

Align the numbers properly in long division.

> $$
> \begin{array}{r}
> 20 \text{ R}15 \\
> 23\overline{)475} \\
> \underline{46} \\
> 15 \\
> \underline{0} \\
> 15
> \end{array}
> $$

Divide decimal numbers:

When dividing by a decimal number, move the decimal in the divisor to the right end; move the decimal in the dividend the same number of places to the right. Place the decimal in the quotient.

Divide one place past the specified place value. Attach zeros in the dividend if necessary. Round to the specified place.

> Divide:
>
> $$
> \begin{array}{r}
> 3.8 \\
> 2.1\overline{)7.9\,8} \\
> \underline{6\,3} \\
> 1\,6\,8 \\
> \underline{1\,6\,8}
> \end{array}
> $$
>
> Divide and round to tenths:
>
> $$
> \begin{array}{r}
> 1.70 \approx 1.7 \\
> 15\overline{)25.60} \\
> \underline{15} \\
> 10\,6 \\
> \underline{10\,5} \\
> 10
> \end{array}
> $$

Estimate and check division:

To estimate division, round the dividend and divisor before finding the quotient. Find the first digit of the quotient, and add a zero for each remaining digit in the dividend.

To check division, multiply the quotient by the divisor and add the remainder. The result should equal the dividend.

> Estimate 2,934 ÷ 42.
>
> $$
> \begin{array}{r}
> 70 \\
> 40\overline{)3,000}
> \end{array}
> \qquad
> \begin{array}{r}
> 69 \\
> 42\overline{)2,934} \\
> \underline{2\,52} \\
> 414 \\
> \underline{378} \\
> 36
> \end{array}
> $$
>
> Exact quotient = 69 R36
> Check:
> 69(42) = 2,898
> 2,898 + 36 = 2,934

Find the numerical average:

Find the sum of the values and divide the sum by the number of values.

> Find the average of 74, 65, and 85 to the nearest whole number.
>
> $74 + 65 + 85 = 224$ Find the sum of the values.
> $224 \div 3 = 74.6$ or 75 (rounded)   Divide by the number of values.

Section 1–2

1 Simplify expressions that contain exponents (pp. 30–31).

To simplify an exponential expression, use the base as a factor the number of times indicated by the exponent.

$a^1 = a$, for any base a

> $5^3 = 5(5)(5) = 125$ Five is used as a factor 3 times.
> $7^1 = 7$ Any number raised to the first power is the number.

2 Square numbers and find the square roots of numbers (p. 31).

Squaring and finding square roots are inverse operations.

> $7^2 = 49$
> $\sqrt{49} = 7$

3 Use powers of 10 to multiply and divide (pp. 32–35).

To multiply a decimal number by a power of 10, move the decimal to the *right* as many digits as the power of 10 has zeros. Attach zeros if necessary.

To divide a decimal by a power of 10, move the decimal to the *left*. Note that this process is just the opposite of that for multiplication.

> Multiply:
> $27(10^3) = 27,000$
> $18(100) = 1,800$
> $23.52(1,000) = 23,520$
>
> Divide:
> $3.52 \div 10 = 0.352$
> $400 \div 10^2 = 4$
> $23,000 \div 10^3 = 23$

Section 1–3

1 Apply the order of operations to a series of operations (pp. 36–38).

Order of operations:

1. Parentheses or groupings, innermost first.
2. Exponential operations and roots from left to right.
3. Multiplication and division from left to right.
4. Addition and subtraction from left to right.

> Simplify.
> $3^2 + 5(6 - 4) \div 2$ Work inside parentheses.
> $3^2 + 5(2) \div 2$ Raise to power.
> $9 + 5(2) \div 2$ Multiply.
> $9 + 10 \div 2$ Divide.
> $9 + 5$ Add.
> 14

2 Evaluate a formula (pp. 39–44).	1. Write the formula. 2. Rewrite the formula substituting known values for letters of the formula. 3. Perform the indicated operations, applying the order of operations. 4. Interpret the solution within the context of the formula.

Evaluate the formula $P = 2(l + w)$ for $l = 17$ cm and $w = 11$ cm.

$P = 2(l + w)$ Substitute known values.
$P = 2(17 \text{ cm} + 11 \text{ cm})$ Add numbers in grouping.
$P = 2(28 \text{ cm})$ Multiply.
$P = 56 \text{ cm}$

3 Solve applied problems using problem-solving strategies (pp. 45–47).

Six-Step Problem-Solving Plan

A shipment of textbooks to a college bookstore is sent in two boxes. One box weighs 9 lb, and the second box weighs 14 lb more than the first box. What is the total weight of the boxes?

Unknown facts	Weight of second box Total weight of the two boxes together
Known facts	First box weighs 9 lb. Second box weighs 14 lb more than first box.
Relationships	weight of first box $= 9$ weight of second box $= 9 + 14$ total weight $= 9 + 9 + 14$
Estimation	Since the second box weighs more than twice as much as the first, the two boxes together weigh approximately 3(9) or 27 lb.
Calculation	$9 + 9 + 14 = 32$
Interpretation	The two boxes weigh 32 lb together.

CHAPTER REVIEW EXERCISES

Section 1–1

Write as decimal numbers.

1. (a) $\dfrac{3}{10}$ (b) $\dfrac{15}{100}$ (c) $\dfrac{4}{100}$

2. (a) $\dfrac{75}{1,000}$ (b) $\dfrac{21}{10}$ (c) $\dfrac{652}{100,000}$

3. Round to the indicated place.
   (a) Nearest hundred: 468
   (b) Nearest ten thousand: 49,238
   (c) Nearest tenth: 41.378
   (d) Nearest hundredth: 6.8957
   (e) Nearest ten-thousandth: 23.46097

4. Round to the indicated place.
   (a) Nearest ten: 98
   (b) Nearest ten: 94
   (c) Nearest thousand: 25,786
   (d) Nearest hundredth: 0.0736
   (e) Nearest whole number: 7.93
   (f) Nearest whole number: 1.876

5. Which of these decimal numbers is larger: 4.783 or 4.79?

6. Which of these decimal numbers is smaller: 0.83 or 0.825?

7. Write these decimal numbers in order of size from smallest to largest: 0.021, 0.0216, 0.02.

8. Two measurements of an object are recorded. If the measures are 4.831 in. and 4.820 in., which is larger?

9. The decimal equivalent of $\frac{7}{8}$ is 0.875. The decimal equivalent of $\frac{6}{7}$ is approximately 0.857. Which fraction is larger?

10. Two parts are machined from the same stock. They measure 1.023 in. and 1.03 in. after machining. Which part has been machined more; that is, which part is now smaller?

11. Add.
   (a) $8 + 5 + 3 + 6 + 2 + 4$
   (b) $7 + 4 + 3 + 2 + 5 + 4$

12. (a) $6.2 + 32.7 + 46.82 + 0.29 + 4.237$
   (b) $86.3 + 9.2 + 70.02 + 3 + 2.7$

13. An air conditioner uses 10.4 kW (kilowatts), a stove uses 15.3 kW, a washer uses 2.9 kW, and a dryer uses 6.3 kW. What is the total number of kilowatts used?

14. A do-it-yourself project requires $57.32 for concrete, $74.26 for fence posts, and $174.85 for fence boards. Estimate the cost by rounding to numbers with one nonzero digit, then find the exact cost.

15. Subtract.
   (a) $21.34 - 16.73$
   (b) $15.934 - 12.807$
   (c) $284.73 - 79.831$
   (d) $13,342 - 1,202$

16. Estimate by rounding to hundreds, then find the exact answer.
   (a) $\$12,346.87 - \$4,468.63$
   (b) $3,495 - 3,090$
   (c) $6,767 - 478$
   (d) $293.86 - 148$

17. A blueprint calls for the length of a part to be 8.296 in. with a tolerance of ± 0.005 in. What are the limit dimensions of the part?

18. For a moving sale, a family sold a sofa for $75 and a table for $25. If a newspaper ad for the sale cost $12.75, how much did the family clear on the two items sold?

Use Fig. 1–35 for Exercises 19–22.

19. Find the length of A if $D = 4.237$ in., $B = 1.861$ in., and $C = 1.946$ in.

20. What is the dimension of A if $E = 4.86$ in., and $B = 1.972$ in.?

21. Give the limit dimensions of D if $D = 8.935$ in. with a ± 0.005-in. tolerance.

22. What dimension should be listed for C if D measures 3.7 in. and E measures 1.6 in.?

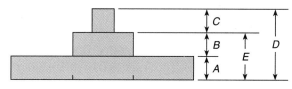

Figure 1–35

Multiply.

23. $(2)(6)(7)$ **24.** $6 \cdot 3 \cdot 2 \cdot 4$ **25.** $(305)(45)$ **26.** $236 \cdot 244$ **27.** $56,002 \cdot 7,040$

28. A college bookstore sold 327 American history textbooks for $39 each. How much did the bookstore receive for the books?

29. A mail-order supplier sells computer keyboards for $67 each. How much would a business pay for 21 keyboards?

30. An automotive tire dealer ran a special on heavy-duty, deluxe whitewall truck tires. If the dealer sold 105 tires for $112 each, how much did the dealer take in on the sale?

31. If a wholesaler ordered 144 computers for $305 each, how much did the dealer pay for the order?

32. A luxury car dealer pays a sound system installer $33.25 per hour. How much is the sound system installer paid for 37 hours of work? Estimate by rounding to a number with one nonzero digit, then find the exact answer.

33. A worker is offered a job that pays $365 per week. If the worker takes the job for 36 weeks, how much will the worker earn? Estimate by rounding to tens, then find the exact answer.

34. Find the area of a field 234.6 ft by 123.2 ft. Estimate the area by rounding to one nonzero digit, then find the exact answer. Check your answer. Express the area in square feet ($A = lw$).

35. A parcel of land measures 1,940.7 ft by 620.4 ft. Estimate the area by rounding to hundreds, then find the exact area. Express the area in square feet ($A = lw$).

36. A piecework employee averages 178.6 pieces per day. If the employee earns $0.28 per item, how much is earned in 5 days?

37. If a steel tape expands 0.00014 in. for each inch when heated, how much will a tape 864 in. long expand?

Divide.

38. $29.25 \div 0.36$ **39.** $325 \div 25$ **40.** $364.8 \div 6$ **41.** $30,126 \div 15$ **42.** $10,160 \div 20$

43. A group of 27 volunteers is seeking contributions to send a first-grade class to the circus. There are 632 envelopes for the collection to be divided equally among the 27 volunteers. How many will each receive? How many will be left over?

44. A school marching band is in a formation of 7 rows, each with the same number of students. If the band has 56 members, how many are in each row?

45. Find the average measure for 42.34 ft, 38.97 ft, 51.95 ft, and 61.88 ft. Round to the nearest hundredth.

46. Five light fixtures cost $74.98, $23.72, $51.27, $125.36, and $85.93. Find the average cost of the fixtures to the nearest cent.

Section 1–2

47. Give the base and exponent of each expression, then simplify.
  (a) 7^3 **(b)** 2.3^4 **(c)** 8^4

48. Give the base and exponent of each expression, then simplify.
  (a) 5^6 **(b)** 1.2^2 **(c)** 10^6

49. Evaluate:
  (a) 1^2 **(b)** 125^2 **(c)** 5.6^2 **(d)** 21^2

50. Find the square root.
  (a) $\sqrt{2500}$ **(b)** $\sqrt{1.44}$
  (c) $\sqrt{289}$ **(d)** $\sqrt{81}$

51. Express as powers of 10.
  (a) 10   **(b)** 1,000
  (c) 10,000   **(d)** 100,000

53. Divide by using powers of 10.
  (a) $700 \div 100$ **(b)** $40.56 \div 1,000$
  (c) $60.5 \div 100$ **(d)** $23,079 \div 10,000$

52. Multiply by using powers of 10.
  (a) 3×100 **(b)** $75 \times 10,000$
  (c) $2.2 \times 1,000$ **(d)** 5×100
  (e) 40.6×10

Section 1–3

Evaluate.

54. $2 + 3 \cdot 3 \div 3$

55. $4^2 \cdot (12 - 7) - 8 + 3$

56. $18 \div 6 - 3$

57. $5 + 21 \div 3 \cdot 7$

58. $82 + 4 \div 2 \times 5$

59. $21 + 7 \cdot 2 - 5 \cdot 4$

60. $15 - 6 \cdot 2 + 3$

61. $18 - 5 \cdot 2 + 7$

62. $24 \div 4 - 18 \div 6$

63. $5 - 2 \cdot 2 + 12$

64. $26 + 8 \div 2 - 3 \cdot 3$

65. $3.1 \cdot 4 \cdot \sqrt{16} - 6^2$

66. $\sqrt{12.25} \cdot (4 - 2) + 8$

67. $5.2^3 - \sqrt{81} \cdot (2 + 1)$

68. $2^4 \div 2 - \sqrt{10 - 1}$

69. $4 + 5 - 2 \cdot 3$

70. $4 + \dfrac{8.6}{2}(2)$

71. $12 \div 4(6)$

72. $8^2 - (3 - 1.5)(5.2)$

73. $27 \div 3 (14 - 8) - 2 + 3^2$

74. $5.13 \div (6.2 - 4.3) + 8.6$

75. If you have 584 packages of hard candy to rebox for shipment to a discount store and you can pack 12 packages in each box, how many boxes will you need?

76. If European Internet Service (EIS) has an annual payroll of $7,460,174,000 for its 194,582 employees, what is the average salary of an EIS employee?

77. Cottage House Antique Mall has 88 booths that rent for $1.40 per square foot. If 48 of the booths have 100 ft^2 and 40 booths have 110 ft^2, what is the total rental from the booths?

78. Find the average of 42, 68, 72, and 96.

79. Judy Ackerman purchases 5 pairs of shoes for $43, $68, $72, $59, and $21. What is the average cost of each pair of shoes?

80. Makisha Brown records the high temperatures for the week of August 13 to be 78°, 72°, 86°, 88°, 90°, 85°, and 82°. What is the average daily temperature?

Find the perimeter of Figs. 1–36 through 1–41.

81.

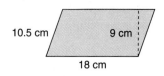

10.5 cm 9 cm 18 cm

Figure 1–36

82.

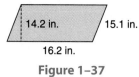

14.2 in. 15.1 in. 16.2 in.

Figure 1–37

83.

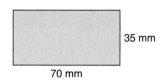

35 mm 70 mm

Figure 1–38

84.

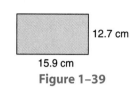

12.7 cm 15.9 cm

Figure 1–39

85.

7.2 m 7.2 m

Figure 1–40

86.

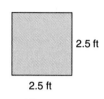

2.5 ft 2.5 ft

Figure 1–41

87. The Tennessee Highway Department has signs in the form of a parallelogram. One set of parallel sides each measures 15 ft and one set of parallel sides each measures 18 ft. Find the perimeter of the sign.

88. Antique tiles were often made in the form of a square that is 6 in. on each side. What is the perimeter of a tile?

89. A rectangular tablecloth measures 84 in. by 60 in. What length of lace is required to trim the edges of the cloth?

90. Find the number of feet of roll fencing needed to fence a square storage area measuring 15.5 ft on a side. A preassembled gate 4 ft wide will be installed.

Find the area of Figs. 1–42 through 1–47.

91.

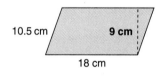

10.5 cm 9 cm 18 cm

Figure 1–42

92.

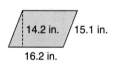

14.2 in. 15.1 in. 16.2 in.

Figure 1–43

93.

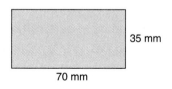

35 mm 70 mm

Figure 1–44

94.

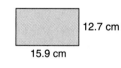

12.7 cm 15.9 cm

Figure 1–45

95.

7.2 m 7.2 m

Figure 1–46

96.

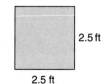

2.5 ft 2.5 ft

Figure 1–47

97. A parking lot for a new hospital, in the shape of a parallelogram, measures 275 ft by 150 ft and has a height of 120 ft. How many square feet need to be paved?

98. A hall wall with no windows or doors measures 25 ft long by 8 ft high. Find the number of square feet to be covered if paneling is installed on the two walls.

99. An office 18 ft by 16.5 ft is to be carpeted. How many square yards of carpeting are needed? Square yards $(yd^2) = ft^2 \div 9$.

100. Vincent Ores, a contractor, is to brick the storefront of a landscape service that has a doorway measuring 7 ft by 6 ft. How many bricks are needed if the storefront is 20 ft by 12 ft and the bricks cover at the rate of 6 per square foot using $\frac{1}{2}$-in. mortar joints?

101. A roof measuring 16 ft by 20 ft is to be covered with asphalt roofing cement. How much would the project cost if the asphalt roofing cement spreads at the rate of 150 square feet per gallon and costs $4.75 per gallon? The cement is purchased by the gallon only.

TEAM PROBLEM-SOLVING EXERCISES

1. Your team is to design a rectangular playground that has 2,160 yd^2 of space.
  (a) Examine different options for the length and width of the playground if one piece of equipment requires at least 15 yd of length. Consider only whole-number options.
  (b) Give three practical options for the dimensions of the playground and explain why you selected each of these options.
  (c) Of all possible rectangular designs, which design requires the least amount of fencing? Consider only whole-number options.

2. An important practice in minimizing calculation errors when using a calculator is to anticipate some characteristics of your answer.
  (a) Find the squares of five different decimal numbers when each has one decimal place.
  (b) Find the squares of five different decimal numbers when each has two decimal places.
  (c) Find the squares of five different decimal numbers when each has three decimal places.
  (d) Examine the patterns established with the squares in parts a, b, and c. Write a statement describing each of these patterns.

CONCEPTS ANALYSIS

1. Addition and subtraction are inverse operations. Write the following addition problem as a subtraction problem, and find the value of the number represented by the letter n: $1.2 + n = 1.7$.

2. Multiplication and division are inverse operations. Write the following multiplication problem as a division problem, and find the value of the number represented by the letter n: $5 \times n = 4.5$.

3. Squaring and finding square roots are inverse operations. Write the following square root as a squaring problem, and find the value of the number represented by the letter n: $\sqrt{n} = 6$.

4. Give an example that shows subtraction is not associative.

5. Give an example that shows division is not commutative.

6. Give the steps in the order of operations.

Find and explain the mistake, then rework each problem correctly.

7. $2.5 + 4.9$
$$\begin{array}{r} 2.5 \\ +\ 4.9 \\ \hline 6.14 \end{array}$$

8. $2 + 5(4) =$
  $7(4) = 28$

9. $\sqrt{9} = 81$

10. How are the procedures for adding whole numbers and adding decimals related?

11. Without making any calculations, do you think 0.004 is a perfect square? Why or why not?

13. Can you find at least one exception to the generalization that a perfect square decimal has an even number of decimal places? Illustrate your answer.

12. Without making any calculations, do you think 0.008 is a perfect cube? Why or why not?

PRACTICE TEST

1. Which number is smaller, 5.09 or 5.1?

3. Round 48.3284 to the nearest tenth.

Perform the indicated operations.

5. $37 + 158 + 764 + 48$

7. $\$13{,}207(702)$

9. $3^2 + 5^3$

11. $3 \cdot 6^2 - 4 \div 2$

13. If a man has a bill for $165 and his paycheck is $475, estimate how much of his paycheck is left after paying the bill by rounding to tens. Find the exact amount left.

15. A softball coach paid $126 for 9 pizzas for a party after a successful season. Estimate the cost of each pizza using numbers with one nonzero digit. Find the exact cost per pizza.

17. Find the product: $42.73 \times 1{,}000$.

19. Round answer to the nearest tenth: $7.2\overline{\smash{)}83.41}$.

21. Find the average of these test scores: 82, 95, 76, 84, 72, and 91. Round to the nearest whole number.

23. Estimate the difference by rounding each number to the nearest tenth: $0.87 - 0.328$.

25. Heating oil costs $1.75 per gallon. What is the cost of 10,000 gal?

2. Round 4.018 to the nearest hundredth.

4. Round $4.834 to the nearest cent.

6. $\$61{,}532 - \$47{,}245$

8. $\$25{,}600 \div 12$

10. 46×10^3

12. $5^3 - (3 + 2) \times \sqrt{9}$

14. A corporation buys 45 DVD players for $335 each. Use numbers with one nonzero digit to estimate the total cost. Find the exact cost.

16. A mathematics professor promises to give 2 extra points for each set of exercises a student works. If one student works 17 sets of exercises, how many points should be given?

18. Divide: $25\overline{\smash{)}27.75}$.

20. Divide: $52.38 \div 10{,}000$.

22. Estimate the sum by rounding to the nearest whole number: $3.85 + 7.46$.

24. The blueprint specification for a machined part calls for its thickness to be 1.485 in. with a tolerance of ± 0.010 in. Find the limit dimensions of the part.

26. A construction job requires 16 pieces of steel, each 7.96 ft long. What length of steel is needed?

Find the perimeter and area of Figs. 1–48 through 1–50.

27.

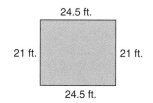

24.5 ft.

21 ft.   21 ft.

24.5 ft.

Figure 1–48

28.

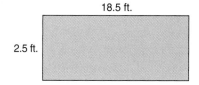

18.5 ft.

2.5 ft.

Figure 1–49

29.

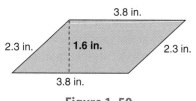

3.8 in.

2.3 in.   **1.6 in.**   2.3 in.

3.8 in.

Figure 1–50

Review of Fractions

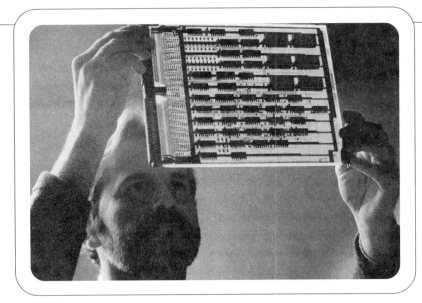

Focus on Careers

Forty-two percent of all engineering technicians are electrical and electronics engineering technicians. An associate degree or extensive job training provides the best opportunities for engineering technicians.

Engineering technicians may assist engineers and scientists to solve technical problems in manufacturing, construction, and maintenance. Some technicians inspect products and processes to ensure quality control. Others work in product design, development, or production.

Nearly a half-million jobs are available in this career, and median earnings range from $36,850 to $51,650, depending on the type of technician. The median annual earnings of electrical and electronics engineering technicians were $42,950 in 2002. Federal government jobs in this career paid the highest annual earnings, and jobs in the telecommunication industry paid the second highest.

Certification for engineering technicians is available at various levels from the National Institute for Certification in Engineering Technologies (NICET). In addition to passing a written examination, job-related experience and a supervisory evaluation and recommendation are required for certification.

Source: *Occupational Outlook Handbook:* http://www.bls.gov/oco.

The language of mathematics is important in the study of mathematics. Mathematical symbols help us describe situations in shortcut fashion, and terminology helps us understand mathematical concepts. In this chapter we study how fractions relate to whole numbers and decimals, and how we add, subtract, multiply, and divide fractions.

Learning Outcomes

1 Find multiples of a natural number.

2 Find all factor pairs of a natural number.

3 Determine the prime factorization of composite numbers.

4 Find the least common multiple and greatest common factor of two or more numbers.

In Chapter 1, we examined a special type of fraction called a *decimal fraction*. A **fraction** is a number that can be expressed as the quotient of two whole numbers.

1 Find Multiples of a Natural Number.

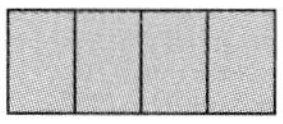

$$1 = \frac{4}{4}$$

Figure 2-1

Symbolically, a **common fraction** is written as $\frac{a}{b}$ or a/b, where a and b are whole numbers and b cannot equal zero ($b \neq 0$). The symbol $\neq$ is read "is not equal to." If one unit is divided into four parts, we can write the fraction $\frac{4}{4}$ to represent this single unit. Fig. 2-1 illustrates a unit divided into four equal parts. The fraction $\frac{4}{4}$ is an example of a *common fraction*. The bottom number, the **denominator**, indicates the number of *equal parts* that makes up one whole unit. The top number, the **numerator**, tells *how many of these parts* are being considered. Figs. 2-2 and 2-3 illustrate the fractions $\frac{1}{4}$ and $\frac{3}{7}$.

Shaded Portion $= \frac{1}{4}$

Figure 2-2

The unit is the *standard* amount when writing fractions. Thus, $\frac{4}{4}$ and $\frac{3}{3}$ each represent one unit. Fractions that represent less than one unit (less than 1), for example $\frac{1}{4}$ or $\frac{3}{7}$, are called **proper fractions.** Fractions that represent one or more units, for example $\frac{7}{5}$, are called **improper fractions.** Figs. 2-4 and 2-5 illustrate the improper fractions $\frac{7}{5}$ and $\frac{10}{5}$.

Shaded Portion $= \frac{3}{7}$

Figure 2-3

Fractions and division are related. In the fraction $\frac{10}{5}$ (two units), $10 \div 5 = 2$. This relationship to division is important to our understanding of fractions.

The numerator and denominator of a fraction are most often separated by a horizontal bar, although sometimes a slash is used. This horizontal bar or slash is the **fraction line,** and it also serves as a division symbol.

When using fraction terminology or notation to describe division, *the numerator is divided by the denominator in all cases.*

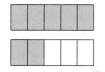

Shaded Portion $= \frac{7}{5}$

Figure 2-4

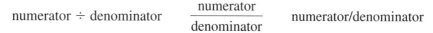

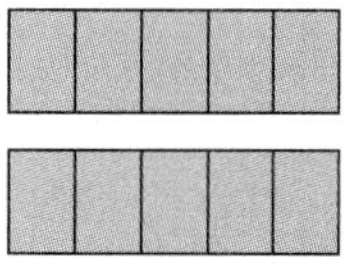

$$\text{numerator} \div \text{denominator} \qquad \frac{\text{numerator}}{\text{denominator}} \qquad \text{numerator/denominator}$$

An improper fraction can be written as a whole number when the denominator divides evenly into the numerator ($\frac{10}{5} = 2$). When the fraction is more than one unit and the denominator cannot divide evenly into the numerator, the improper fraction can be written as a combination of a whole number and a fractional part, such as $\frac{7}{5} = 1\frac{2}{5}$.

Shaded Portion $= \frac{10}{5}$

Figure 2-5

A **mixed number** consists of both a whole number and a fraction. The whole number and fraction are added together. *Example:* $1\frac{2}{5}$ means 1 whole units and $\frac{2}{5}$ of another unit or $1\frac{2}{5} = 1 + \frac{2}{5}$.

Fractions indicate division. Multiplication and division are inverse operations. Let's look at some relationships involving multiplication and division.

If we count by 3s, such as 3, 6, 9, 12, 15, 18, we obtain natural numbers that are *multiples* of 3. Each is the product of 3 and a natural number; that is,

$$3 = 3 \times 1, \qquad 6 = 3 \times 2, \qquad 9 = 3 \times 3$$
$$12 = 3 \times 4, \qquad 15 = 3 \times 5, \qquad 18 = 3 \times 6$$

A **multiple** of a natural number is the product of that number and a natural number.

EXAMPLE   Show that 2, 4, 6, 8, and 10 are multiples of 2 by writing each as the product of 2 and a natural number.

$$2 = 2 \times 1, \quad 4 = 2 \times 2, \quad 6 = 2 \times 3, \quad 8 = 2 \times 4, \quad 10 = 2 \times 5$$

Natural numbers that are multiples of 2 are **even numbers.** Natural numbers that are not multiples of 2 are **odd numbers.**

EXAMPLE   Find five multiples of 16.

$$1 \times 16 = 16, \quad 2 \times 16 = 32, \quad 3 \times 16 = 48, \quad 4 \times 16 = 64, \quad 5 \times 16 = 80$$

16, 32, 48, 64, and 80 are multiples of 16.

We say that a number is **divisible** by another number if the quotient has no remainder or if the dividend is a multiple of the divisor.

We want to be able to determine divisibility by **inspection.** This means that we can examine the number being divided (dividend) and decide if it is divisible by a divisor without actually having to perform the division.

EXAMPLE   Is 35 divisible by 7?

35 is divisible by 7 if $35 \div 7$ has no remainder or if 35 is a multiple of 7.

$$35 \div 7 = 5 \quad \text{or} \quad 35 = 5 \times 7$$

Yes, 35 is divisible by 7.

Rules or tests can help us decide by inspection if certain numbers are divisible by other numbers.

Tests for divisibility:

A number is divisible by
1. 2 if the last digit is an even number (0, 2, 4, 6, or 8).
2. 3 if the sum of its digits is divisible by 3.
3. 4 if the last two digits form a number that is divisible by 4.
4. 5 if the last digit is 0 or 5.
5. 6 if the number is divisible by *both* 2 *and* 3.
6. 7 if the division has no remainder.
7. 8 if the last three digits form a number divisible by 8.
8. 9 if the sum of its digits is divisible by 9.
9. 10 if the last digit is 0.

EXAMPLE Use the tests for divisibility to identify which number in each pair is divisible by the given divisor.

Numbers	Divisor	Answer
874 or 873	2	**874;** the last digit is an even digit (4).
275 or 270	2	**270;** the last digit is an even digit (0).
427 or 423	3	**423;** the sum of the digits is divisible by 3: 4 + 2 + 3 = 9.
5,912 or 5,913	4	**5,912;** the last two digits form a number divisible by 4: 12 ÷ 4 = 3.
80 or 82	5	**80;** the last digit is 0.
56 or 65	5	**65;** the last digit is 5.
804 or 802	6	**804;** the last digit is even and the sum of the digits is divisible by 3: 8 + 0 + 4 = 12.
58 or 56	7	**56;** it divides by 7 with no remainder.
3,160 or 3,162	8	**3,160;** the last three digits form a number divisible by 8: 160 ÷ 8 = 20.
477 or 475	9	**477;** the sum of the digits is divisible by 9: 4 + 7 + 7 = 18. 18 ÷ 9 = 2.
182 or 180	10	**180;** the last digit is 0.

2 **Find All Factor Pairs of a Natural Number.**

Any natural number can be expressed as the product of two natural numbers. These two natural numbers are called a **factor pair** of the number.

Every natural number greater than 1 has at least one factor pair, the number itself and 1.

1 and 3 form a factor pair for 3: $1 \times 3 = 3$.

1 and 5 form a factor pair for 5: $1 \times 5 = 5$.

Many natural numbers have more than one factor pair. For example, list all factor pairs of 12. Start with the pair 1×12. Every number has a factor pair of the number and 1. Next, examine each number from 2 to 11.

2	12 is divisible by 2	12 ÷ 2 = 6	2 × 6 is a factor pair of 12.
3	12 is divisible by 3	12 ÷ 3 = 4	3 × 4 is a factor pair of 12.
4	12 is divisible by 4	12 ÷ 4 = 3	4 × 3 is a factor pair of 12.
5	12 is not divisible by 5		
6	12 is divisible by 6	12 ÷ 6 = 2	6 × 2 is a factor pair of 12.
7	12 is not divisible by 7		
8	12 is not divisible by 8		
9	12 is not divisible by 9		
10	12 is not divisible by 10		
11	12 is not divisible by 11		

Are the factor pairs 2×6 and 6×2 different pairs? No. Both pairs have the same numbers, and since multiplication is commutative, they count as one pair. Similarly, 3×4 and 4×3 are the same factor pair. Is it necessary to examine every number less than 12? No. Once we get the first repeat, 4×3, we can assume we have found all the factor pairs of the number.

The factor pairs of 12 are 1×12, 2×6, and 3×4.

1. Write the factor pair of 1 and the given natural number.
2. Check to see if the number is divisible by 2. If so, write the factor pair of 2 and the quotient of the given number and 2.
3. Check the next number for divisibility; if the given number is divisible by the next number, write the factor pair.
4. Continue Step 3 until you reach a number that already has been found in a previous factor pair.

EXAMPLE List all the factor pairs of 18.

1×18
2×9 18 is divisible by 2: $18 \div 2 = 9$.
3×6 18 is divisible by 3: $18 \div 3 = 6$.
4 or 5   18 is not divisible by 4 or 5.
6   18 is divisible by 6; 6 was found in the factor pair 3×6, so we stop.

The factor pairs of 18 are 1 and 18, 2 and 9, and 3 and 6.

Once we have listed all factor pairs of a number, we can list all factors of a number. From the factor pairs, we list every different factor that appears in any factor pair. The factors of 12 are 1, 2, 3, 4, 6, and 12. The factors of 18 are 1, 2, 3, 6, 9, and 18.

EXAMPLE List all the factor pairs of 48; then write each distinct factor in order from smallest to largest.

1×48
2×24 48 is divisible by 2: $48 \div 2 = 24$.
3×16 48 is divisible by 3: $48 \div 3 = 16$.
4×12 48 is divisible by 4: $48 \div 4 = 12$.
5   48 is not divisible by 5.
6×8 48 is divisible by 6: $48 \div 6 = 8$.
7   48 is not divisible by 7.
8   48 is divisible by 8; 8 was found in the factor pair 6×8, so we stop.

Factor pairs: 1 and 48, 2 and 24, 3 and 16, 4 and 12, 6 and 8.
Factors: 1, 2, 3, 4, 6, 8, 12, 16, 24, 48

---

3   Determine the Prime Factorization of Composite Numbers.

When all the factors of a natural number are listed, some numbers have no other factor than the number itself and 1. These numbers form a special set of numbers called prime numbers. A **prime number** is a whole number *greater than 1* that has factors only of the number itself and 1. Note that *1 is not a prime number.*

EXAMPLE Identify the prime numbers by examining the factors of the numbers.

(a) 8　(b) 1　(c) 3　(d) 9　(e) 7

(a) **8 is *not* a prime number** because its factors are 1, 2, 4, and 8.
(b) **1 is *not* a prime number** because a prime must be greater than 1.
(c) **3 is a prime number** because it has only factors 1 and 3.
(d) **9 is *not* a prime number** because its factors are 1, 3, and 9.
(e) **7 is a prime number** because it has only factors 1 and 7.

A **composite number** is a whole number greater than 1 that is not a prime number. In the preceding example, 8 and 9 are composite numbers. A composite number has at least one factor other than the number itself and 1.

EXAMPLE Identify the composite numbers by examining the factors of the numbers.

(a) 4　(b) 10　(c) 13　(d) 12　(e) 5

(a) **4 is a composite number** because its factors are 1, 2, and 4.
(b) **10 is a composite number** because its factors are 1, 2, 5, and 10.
(c) **13 is *not* a composite number** because its only factors are 1 and 13. It is a prime number.
(d) **12 is a composite number** because its factors are 1, 2, 3, 4, 6, and 12.
(e) **5 is *not* a composite number** because its only factors are 1 and 5. It is a prime number.

Prime Numbers Less Than 50

We can find all the prime numbers that are 50 or less using an ancient technique developed by the mathematician Eratosthenes. This technique is called **sieve of Eratosthenes.**

Step 1. List the numbers from 1 through 50.

Step 2. Eliminate numbers that are not prime, using the systematic process:
(a) 1 is not prime. Eliminate 1.
(b) 2 is prime. Eliminate all multiples of 2.
(c) 3 is prime. Eliminate all multiples of 3.
(d) 4 has already been eliminated.
(e) 5 is prime. Eliminate all multiples of 5.
(f) 6 has already been eliminated.
(g) 7 is prime. Eliminate all multiples of 7.

Step 3. Circle remaining numbers as prime numbers.

1̸	②	③	4̸	⑤	6̸	⑦	8̸	9̸	1̸0̸
⑪	1̸2̸	⑬	1̸4̸	1̸5̸	1̸6̸	⑰	1̸8̸	⑲	2̸0̸
2̸1̸	2̸2̸	㉓	2̸4̸	2̸5̸	2̸6̸	2̸7̸	2̸8̸	㉙	3̸0̸
㉛	3̸2̸	3̸3̸	3̸4̸	3̸5̸	3̸6̸	㊲	3̸8̸	3̸9̸	4̸0̸
㊶	4̸2̸	㊸	4̸4̸	4̸5̸	4̸6̸	㊼	4̸8̸	4̸9̸	5̸0̸

All numbers not eliminated are prime. Why? The numbers 8, 9, and 10 have already been eliminated as multiples of 2, 3, and 5, respectively. Multiples of 11 that are less than 50 have already been eliminated: $11 \times 2 = 22$, $11 \times 3 = 33$, $11 \times 4 = 44$. The product $11 \times 5 = 55$ is greater than 50. Similarly, all other composite numbers have been eliminated.

A composite number can be expressed as a product of prime numbers. **Prime factorization** refers to writing a composite number as the product of *only* prime numbers. These factors are called **prime factors.**

To find the prime factors of a composite number:

1. Test each prime number to see if the composite number is divisible by the prime.
2. Make a factor pair using the first prime number that passes the test in Step 1.
3. Carry forward the prime factors and test the remaining factors by repeating Steps 1 and 2.

EXAMPLE Find the prime factorization of 30.

$$30 = 2(15)$$
first prime

30 is divisible by 2. Factor 30 into a factor pair using its smallest prime factor, 2.

$$30 = 2(3)(5)$$
last primes

Carry the prime factor 2 forward. Factor the composite number 15 using its smallest prime factor, 3. Because 5 is also prime, the factoring is complete.

The prime factorization of 30 is 2(3)(5).

EXAMPLE Find the prime factorization of 16.

$$16 = 2(8)$$
first prime

Factor 16 into two factors using its smallest prime factor.

$$16 = 2(2)(4)$$
second prime

Factor 8 into two factors using its smallest prime factor.

$$16 = 2(2)(2)(2)$$
last two primes

Factor 4 into two factors using its smallest prime factor.

The prime factorization of 16 is 2(2)(2)(2). We can write this expression in exponential notation as 2^4.

Chapter 2 / Review of Fractions

EXAMPLE Find the prime factorization of 210.

$$210 = 2(105)$$

first prime ⟶↑

Factor 210 into two factors using its smallest prime factor.

$$210 = 2(3)(35)$$

second prime ⟶↑

Factor 105 into two factors using its smallest prime factor.

$$210 = 2(3)(5)(7)$$

last two primes ⟶↑↑

Factor 35 into two factors using its smallest prime factor.

The prime factorization of 210 is 2(3)(5)(7).

4 Find the Least Common Multiple and Greatest Common Factor of Two or More Numbers.

The **least common multiple (LCM)** of two or more natural numbers is the smallest number that is a multiple of each number. The LCM is divisible by each number.

To find the least common multiple of 3 and 5, examine the multiples of each number.

multiples of 3: 3, 6, 9, 12, 15, 18, 21, 24, 27, 30, 33, 36, 39, . . .

multiples of 5: 5, 10, 15, 20, 25, 30, 35, 40, . . .

The common multiples in these lists that are less than 40 are 15 and 30; 15 is the *least common multiple* of 3 and 5.

Prime factorization can also be used to find the least common multiple of two or more numbers.

To find the least common multiple of two or more natural numbers using the prime factorization of the numbers:

1. List the prime factorization of each number using exponential notation.
2. List the prime factorization of the least common multiple by including the prime factors appearing in *each* number. If a prime factor appears in more than one number, use the factor with the *largest* exponent.
3. Write the resulting expression in standard notation.

The smallest Num. That each divide into evenly.
LCM - list any number thats in another list with Greatest Exponet!

EXAMPLE Find the least common multiple of 12 and 40 by prime factorization.

$$12 = 2(2)(3) = 2^2(3)$$ Prime factorization of 12.

$$40 = 2(2)(2)(5) = 2^3(5)$$ Prime factorization of 40.

$$LCM = 2^3(3)(5)$$ Prime factorization of LCM.

$$\textbf{LCM} = \textbf{120}$$ LCM in standard notation.

2–1 Multiples and Factors

67

The **greatest common factor (GCF)** of two or more numbers is the largest factor common to each number. Each number is divisible by the GCF.

Let's take the numbers 30 and 42. The prime factors are

$$30 = 2(3)(5) \qquad 42 = 2(3)(7)$$

The *common* prime factors of both 30 and 42 are 2 and 3, which represent the composite factor 6. The *greatest* common factor is the product of the common prime factors, $2(3) = 6$.

GCF = The largest number that divides into each given number evenly

To find the greatest common factor (GCF) of two or more natural numbers:

1. List the prime factorization of each number using exponential notation when appropriate.
2. List the prime factorization of the greatest common factor by including each prime factor appearing in *every* number. If a prime factor appears more than one time in any number (that is, the exponent is greater than 1), use the factor with the *smallest* exponent. If there are no common prime factors, the GCF is 1.
3. Write the resulting expression in standard notation.

List factors that are common to both with smallest exponent.

EXAMPLE   Find the greatest common factor of 15, 30, and 45.

$15 = 3(5) \quad = 3(5)$	Prime factorization of 15.
$30 = 2(3)(5) = 2(3)(5)$	Prime factorization of 30.
$45 = 3(3)(5) = 3^2(5)$	Prime factorization of 45.
$\text{GCF} = 3(5)$	Common prime factors.
$\textbf{GCF} = \textbf{15}$	GCF in standard notation.

EXAMPLE   Find the greatest common factor of 10, 12, and 13.

$10 = 2(5) \quad = 2(5)$	Prime factorization of 10.
$12 = 2(2)(3) = 2^2(3)$	Prime factorization of 12.
$13 = 13 \quad = 13$	Prime factorization of 13.
$\textbf{GCF} = \textbf{1}$	No common prime factors.

LCM Versus GCF

The least common multiple (LCM) and greatest common factor (GCF) are easily confused. Because multiples of a number are the products of the number and any natural number, multiples will be as large as or larger than the original number. The LCM is the *smallest* of the "same or larger" common multiples of the original numbers.

Because factors of a number are the same as or smaller than the given number, the GCF is the *largest* of the "same or smaller" common factors of the original numbers.

SECTION 2–1 SELF-STUDY EXERCISES

1 Show that each number is a multiple of the first number by writing each as the product of the first number and a natural number.

1. 5, 10, 15, 20, 25, 30   **2.** 6, 12, 18, 24, 30, 36   **3.** 8, 16, 24, 32, 40, 48
4. 9, 18, 27, 36, 45, 54   **5.** 10, 20, 30, 40, 50, 60   **6.** 30, 60, 90, 120, 150, 180

Write five multiples of the given number.

7. 6   **8.** 12   **9.** 13
10. 3   **11.** 50   **12.** 4

Use the tests for divisibility to determine which number is divisible by the given number. Explain.

13. 2,434 by 6   **14.** 230 by 5   **15.** 2,434 by 4
16. 1,221 by 3   **17.** 756 by 7   **18.** 920 by 8
19. 621 by 3   **20.** 426 by 6   **21.** 1,232 by 2

2 List all the factor pairs for each number.

22. 4   **23.** 8   **24.** 12   **25.** 15   **26.** 16
27. 20   **28.** 24   **29.** 30   **30.** 36   **31.** 38

List all factors of each number.

32. 40   **33.** 46   **34.** 52   **35.** 64   **36.** 72
37. 81   **38.** 85   **39.** 92   **40.** 96   **41.** 98

3 Identify the prime numbers and the composite numbers by examining the factors of each number.

42. 2   **43.** 6   **44.** 9   **45.** 11   **46.** 14   **47.** 15
48. 16   **49.** 51   **50.** 52   **51.** 53   **52.** 66   **53.** 67

Find the prime factorization.

54. 12   **55.** 18   **56.** 20   **57.** 24   **58.** 25   **59.** 27
60. 29   **61.** 30   **62.** 35   **63.** 40   **64.** 47   **65.** 49
66. 50   **67.** 52   **68.** 65   **69.** 75   **70.** 100   **71.** 105
72. 108   **73.** 115   **74.** 121   **75.** 144   **76.** 156   **77.** 157

Write the prime factorization using exponential notation.

78. 72   **79.** 112   **80.** 124   **81.** 164   **82.** 568   **83.** 900

4 Find the least common multiple.

84. 2 and 3   **85.** 5 and 6   **86.** 7 and 8
87. 3 and 4   **88.** 18 and 24   **89.** 10 and 12
90. 12 and 24   **91.** 9 and 18   **92.** 4, 8, and 12
93. 20, 25, and 35   **94.** 3, 9, and 27   **95.** 2, 8, and 16
96. 6, 15, and 18   **97.** 20, 24, and 30   **98.** 12, 18, and 20
99. 6, 10, and 15   **100.** 8, 12, and 32   **101.** 8, 12, and 18
102. 10, 15, and 20   **103.** 30, 50, and 60   **104.** 6, 11, and 33
105. 8, 13, and 39   **106.** 2, 7, and 14   **107.** 12, 16, and 18

Find the greatest common factor.

108. 2 and 3   **109.** 5 and 6   **110.** 7 and 8
111. 18 and 24   **112.** 10 and 15   **113.** 12 and 24
114. 9 and 18   **115.** 40 and 55   **116.** 2, 8, and 16
117. 18, 30, and 36   **118.** 20, 25, and 35   **119.** 6, 15, and 18
120. 12, 18, and 20   **121.** 6, 10, and 12   **122.** 8, 12, and 18

2–2 | Equivalent Fractions and Decimals

Learning Outcomes

1 Write equivalent fractions with different denominators.
2 Write improper fractions as whole numbers or mixed numbers.
3 Write whole numbers or mixed numbers as improper fractions.
4 Write decimals as fractions and fractions as decimals.
5 Compare fractions, mixed numbers, and decimals.

1 Write Equivalent Fractions with Different Denominators.

There are many different ways to express the same value in fractional form. For example, the whole number 1 can be written as $\frac{1}{1}, \frac{4}{4}, \frac{7}{7}, \frac{15}{15}$, and so on.

Fractions are equivalent if they represent the same value. Compare the illustrations in Fig. 2–6, where line a is one whole unit divided into only 1 part ($\frac{1}{1}$). Line b is one unit divided into 2 parts ($\frac{2}{2}$). Lines c and d are divided into 4 parts ($\frac{4}{4}$), line e is divided into 8 parts ($\frac{8}{8}$), and line f is divided into 16 parts ($\frac{16}{16}$).

Again, look at lines d, e, and f. Line d is divided into 4 parts and 3 of them are shaded. Line e is divided into 8 parts and 6 of them are shaded. Line f is divided into 16 parts and 12 of them are shaded.

Look at the shaded portions of lines d, e, and f. They are the same length, even though they are divided into a different number of parts. Thus, $\frac{3}{4}$, $\frac{6}{8}$, and $\frac{12}{16}$ are **equivalent fractions** since they represent the same shaded part of one whole unit. That is,

$$\frac{3}{4} = \frac{6}{8} = \frac{12}{16}$$

Equivalent fractions are in the same "family of fractions." The first member of the "family" is the fraction in **lowest terms;** that is, no natural number divides evenly into *both* the numerator and the denominator except the number 1. Other "family members" are found by multiplying both the numerator and the denominator by the same natural number.

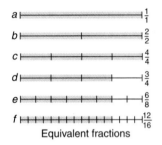
Equivalent fractions

Figure 2–6

EXAMPLE Find five fractions that are equivalent to $\frac{1}{2}$.

$$\frac{1(2)}{2(2)} \quad \text{or} \quad \frac{1}{2}\left(\frac{2}{2}\right) = \frac{2}{4} \qquad \frac{1(5)}{2(5)} \quad \text{or} \quad \frac{1}{2}\left(\frac{5}{5}\right) = \frac{5}{10}$$

$$\frac{1(3)}{2(3)} \quad \text{or} \quad \frac{1}{2}\left(\frac{3}{3}\right) = \frac{3}{6} \qquad \frac{1(6)}{2(6)} \quad \text{or} \quad \frac{1}{2}\left(\frac{6}{6}\right) = \frac{6}{12}$$

$$\frac{1(4)}{2(4)} \quad \text{or} \quad \frac{1}{2}\left(\frac{4}{4}\right) = \frac{4}{8}$$

Other equivalent fractions can be found.

     Chapter 2 / Review of Fractions

TIP | **Multiplication, Division, and 1**

In the preceding example, $\frac{1}{2}$ is multiplied by a fraction whose value is 1, and 1 times any number does not change the value of that number. Written symbolically,

$$\frac{n}{n} = 1 \text{ when } n \text{ is a real number and } n \neq 0 \qquad \text{and} \qquad 1(n) = n \text{ where } n \text{ is any real number}$$

Fractions in the same family can be generated by multiplying the fraction by 1 in the form of $\frac{2}{2}, \frac{3}{3}, \frac{4}{4}$, and so on.

The concept presented in the preceding example and tip is referred to as the **fundamental principle of fractions.** If the numerator and denominator of a fraction are multiplied by the same nonzero number, the value of the fraction remains unchanged.

To change a fraction to an equivalent fraction with a specified larger denominator:

1. Divide the specified larger denominator by the original denominator.
2. Multiply the original numerator and denominator by the quotient found in Step 1. That is, multiply by 1 in the form of $\frac{n}{n}$ when n is a real number and $n \neq 0$.

EXAMPLE Change $\frac{5}{8}$ to an equivalent fraction whose denominator is 32.

$$\frac{5}{8} = \frac{?}{32} \qquad \text{32 ÷ 8 = 4. Apply the fundamental principle of fractions.}$$

$$\frac{5}{8}\left(\frac{4}{4}\right) = \frac{20}{32}$$

Because each fraction has an unlimited number of equivalent fractions, we usually work with fractions in *lowest terms.* When we find an equivalent fraction with smaller numbers and there are no common factors in the numerator and denominator, we have **reduced to lowest terms.**

To change a fraction to an equivalent fraction with a smaller denominator or to reduce a fraction to lowest terms:

1. Find a common factor greater than 1 for the numerator and denominator.
2. Divide both the numerator and the denominator by this common factor.
3. Continue until the fraction is in lowest terms or has the desired smaller denominator.

Note: To reduce to lowest terms in the fewest steps, find the greatest common factor (GCF) in Step 1.

EXAMPLE   Reduce $\frac{8}{10}$ to lowest terms.

Prime factors of 8: **2**(2)(2) or 2^3
Prime factors of 10: **2**(5)
The greatest common factor (GCF) is **2**.

$$\frac{8 \div 2}{10 \div 2} \quad \text{or} \quad \frac{8}{10} \div \frac{2}{2} = \frac{4}{5}$$

TIP

Reducing and the Properties of 1

To reduce the fraction $\frac{8}{10}$, we divide by the whole number 1 in the form of $\frac{2}{2}$. That is, $\frac{8}{10} \div \frac{2}{2} = \frac{4}{5}$. A nonzero number divided by itself is 1, and to divide a number by 1 does not change the value of the number. Symbolically,

$$\frac{n}{n} = 1 \text{ when } n \text{ is a real number and } n \neq 0 \quad \text{and} \quad n \div 1 = n \quad \text{or} \quad \frac{n}{1} = n \text{ when } n \text{ is a real number.}$$

EXAMPLE   Reduce $\frac{18}{24}$ to lowest terms.

Prime factors of 18: 2(3)(3)   or $2 \cdot 3^2$
Prime factors of 24: 2(2)(2)(3)   or $2^3 \cdot 3$
The GCF is 2(3) or 6.

$$\frac{18 \div 6}{24 \div 6} \quad \text{or} \quad \frac{18}{24} \div \frac{6}{6} = \frac{3}{4}$$

TIP

Do You Have to Use the GCF to Reduce to Lowest Terms?

A fraction can be reduced to lowest terms in the fewest steps by using the *greatest common factor*; however, it still can be reduced to lowest terms using any common factor; this just takes a few more steps.

$$\frac{18 \div 2}{24 \div 2} = \frac{9}{12} \quad \text{Reduce with common factor 2.}$$

$$\frac{9 \div 3}{12 \div 3} = \frac{3}{4} \quad \text{Reduce with common factor 3.}$$

The **U.S. customary rule** is divided into inches and parts of an inch and is used to measure length. Each inch is subdivided into fractional parts, usually 8, 16, 32, or 64.

The rule in Fig. 2–7 shows each inch divided into 16 equal parts, so each part is $\frac{1}{16}$ in.; that is, the first mark from the left edge represents $\frac{1}{16}$ in. The left end of the rule represents zero (0).

Chapter 2 / Review of Fractions

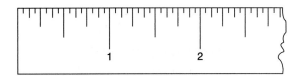

Figure 2–7 U.S. customary rule.

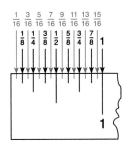

Figure 2–8 1 in.

Fig. 2–8 labels division marks for the parts of 1 in. The second mark from the left edge represents $\frac{2}{16}$ or $\frac{1}{8}$ in. This mark is slightly longer than the first mark.

The fourth mark from the left is labeled $\frac{1}{4}$; that is, $\frac{4}{16} = \frac{1}{4}$. In each case, fractions are always reduced to lowest terms. Notice that the $\frac{1}{4}$ mark is slightly longer than the $\frac{1}{8}$ mark.

The division marks are different lengths to make the rule easier to read. The shortest marks represent fractions that, in lowest terms, are sixteenths ($\frac{1}{16}$, $\frac{3}{16}$, $\frac{5}{16}$, $\frac{7}{16}$, $\frac{9}{16}$, $\frac{11}{16}$, $\frac{13}{16}$, $\frac{15}{16}$). The fractions that reduce to eighths are slightly longer than the sixteenths marks ($\frac{1}{8}$, $\frac{3}{8}$, $\frac{5}{8}$, $\frac{7}{8}$). The marks representing fractions that reduce to fourths are slightly longer than the eighths ($\frac{1}{4}$, $\frac{3}{4}$). The mark for one-half ($\frac{1}{2}$) is longer than the fourths, and the inch mark is the longest.

EXAMPLE Measure line segment *AB* (Fig. 2–9).

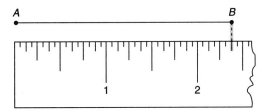

Figure 2–9 Line segment *AB*.

Align point *A* with zero. Line segment *AB* goes past the 2-in. mark but not up to the 3-in. mark. Therefore, the measure of *AB* will be a mixed number between 2 and 3. Point *B* is $\frac{3}{8}$ in. past 2.

AB is $2\frac{3}{8}$ in.

Judge to the Closest Mark

A line segment may not always align exactly with a division mark. Use eye judgment to decide which mark is closer to the end of the line segment.

EXAMPLE Measure line segment *CD* (Fig. 2–10).

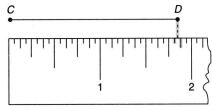

Figure 2–10

Point *D* aligns between $1\frac{13}{16}$ and $1\frac{7}{8}$. Measurements are always approximations; using our best eye judgment, point *D* seems to be halfway between $1\frac{13}{16}$ and $1\frac{7}{8}$.

We say *CD* is $1\frac{13}{16}$ in. or $1\frac{7}{8}$ in. *to the nearest sixteenth of an inch.*

In practice, measurements are considered acceptable if they are within a desired *tolerance*. In the preceding example, the smallest division is $\frac{1}{16}$, so an appropriate tolerance would be plus or minus one-half of one-sixteenth, or $\pm\frac{1}{32}$ ($\frac{1}{2}$ of $\frac{1}{16} = \frac{1}{32}$). That is, the acceptable measure can be $\frac{1}{32}$ more than or $\frac{1}{32}$ less than the ideal measure.

If the ideal measure is halfway between $1\frac{13}{16}$ and $1\frac{7}{8}$ (or equal to $1\frac{27}{32}$) and the tolerance is $\frac{1}{32}$ in., the interval of acceptable values is from $1\frac{27}{32} - \frac{1}{32}$ to $1\frac{27}{32} + \frac{1}{32}$. The acceptable interval is from $1\frac{26}{32}$ to $1\frac{28}{32}$, or $1\frac{13}{16}$ to $1\frac{7}{8}$.

2 Write Improper Fractions as Whole Numbers or Mixed Numbers.

Earlier we noted that an improper fraction is a fraction whose value is equal to or greater than one unit, such as $\frac{6}{3}$, $\frac{15}{7}$, or $\frac{5}{5}$. These fractions can be changed to equivalent whole numbers or mixed numbers.

To write an improper fraction as a whole or mixed number:

1. Perform the division indicated (numerator ÷ denominator).
2. Express any remainder of the division as a fraction or decimal equivalent.

EXAMPLE Write $\frac{15}{3}$ as a whole or mixed number.

$$\frac{15}{3} \text{ means } 15 \div 3 \quad \text{or} \quad 3\overline{)15} \begin{array}{r} 5 \\ \underline{15} \\ 0 \end{array} \quad \text{Then, } \frac{15}{3} = \mathbf{5}.$$

Divide the numerator by the denominator.

If there is no remainder from the division, the improper fraction converts to a whole number.

TIP

Writing a Whole or Mixed Number Is Different from Reducing

Do not confuse converting an improper fraction to a whole or mixed number with reducing a fraction to lowest terms. An improper fraction is in lowest terms if its numerator and denominator have no common factor other than 1. Therefore, the improper fraction $\frac{10}{7}$ is written in lowest terms. The improper fraction $\frac{10}{4}$ is not in lowest terms. It will reduce to $\frac{5}{2}$, which is in lowest terms.

When writing an improper fraction as a whole or mixed number, we must make sure that the fraction is in lowest terms. We can reduce to lowest terms either before dividing or after dividing.

EXAMPLE Write $\frac{28}{8}$ as a mixed number.

$$\frac{28}{8} = \frac{7}{2} \qquad 2\overline{)7} = 3\frac{1}{2}$$
$$\underline{6}$$
$$1$$

Fraction is reduced before dividing.

or

$$\frac{28}{8} \qquad 8\overline{)28}\;{}^{3} = 3\frac{4}{8} = 3\frac{1}{2}$$
$$\underline{24}$$
$$4$$

Fraction is reduced after dividing.

3 **Write Whole Numbers or Mixed Numbers as Improper Fractions.**

Some computation processes require mixed numbers to be expressed as equivalent improper fractions.

To write a mixed number as an improper fraction:

1. Multiply the denominator of the fractional part by the whole number.
2. Add the numerator of the fractional part to the product; this sum becomes the numerator of the improper fraction.
3. The denominator of the improper fraction is the same as the denominator of the fractional part of the mixed number.

EXAMPLE Change $6\frac{2}{3}$ to an improper fraction.

$$6\frac{2}{3} = \frac{(3 \cdot 6) + 2}{3} = \frac{20}{3}$$

Multiply the denominator times the whole number and add the numerator.

Whole numbers can also be written as improper fractions by writing the whole number as a fraction with a denominator of 1.

EXAMPLE Change 8 to fifths.

$$8 = \frac{8}{1}$$

Write 8 as a fraction with a denominator of 1.

$$\frac{8(5)}{1(5)} = \frac{40}{5}$$

Multiply by 1 in the form $\frac{5}{5}$.

4 Write Decimals as Fractions and Fractions as Decimals.

When a decimal is written as a fraction, the number of digits that follow the decimal point determines the denominator of the fraction.

To write a decimal number as a fraction or mixed number in lowest terms:

1. Write the digits without the decimal point and leading zeros as the numerator.
2. Write the denominator as a power of 10 with as many zeros as there are places after the decimal point.
3. Reduce and, if the fraction is improper, convert to a mixed number. See Outcome 2 of this section.

EXAMPLE   Write as a fraction 0.4 and 0.075.

$$0.4 = \frac{4}{10} = \frac{2}{5}$$ Tenths indicates a denominator of 10. Reduce to lowest terms.

$$0.075 = \frac{75}{1,000} = \frac{3}{40}$$ Thousandths indicates a denominator of 1,000. Reduce to lowest terms.

Writing decimal numbers as fractions was introduced in greater detail in Chapter 1, Section 1, Outcome 2.

With increased calculator and computer use, we find it convenient to work with decimals.

The bar separating the numerator and denominator of a fraction indicates division: $\frac{2}{5}$ also means $2 \div 5$ or $5\overline{)2}$.

If we show the decimal after the 2 and attach a zero in the tenths place, we can divide.

$$5\overline{)2.0}^{\,0.4}$$

Therefore, $\frac{2}{5} = 0.4$.

To convert a fraction to a decimal number:

1. Place a decimal point after the numerator.
2. Divide the numerator by the denominator using long division.
3. Attach zeros after the decimal point in the dividend as needed for division.

EXAMPLE   Change $\frac{7}{8}$ to a decimal.

$$7 \div 8 \quad \text{or} \quad 8\overline{)7.000}^{\,0.875}$$

$$\begin{array}{r} 6\,4 \\ \hline 60 \\ 56 \\ \hline 40 \\ 40 \\ \hline 0 \end{array}$$

Attach zeros and divide until the division terminates; that is, it has no remainder.

When we change some fractions to decimals, the quotient does not terminate.

EXAMPLE Write $\frac{1}{3}$ and $\frac{4}{11}$ as decimals to the nearest thousandth.

$$1 \div 3 \quad \text{or} \quad \begin{array}{r} 0.3333 \\ 3\overline{)1.0000} \\ \underline{9} \\ 10 \\ \underline{9} \\ 10 \\ \underline{9} \\ 10 \\ \underline{9} \\ 1 \end{array} = \mathbf{0.333} \quad \text{Rounded.}$$

$$4 \div 11 \quad \text{or} \quad \begin{array}{r} 0.3636 \\ 11\overline{)4.0000} \\ \underline{33} \\ 70 \\ \underline{66} \\ 40 \\ \underline{33} \\ 70 \\ \underline{66} \\ 4 \end{array} = \mathbf{0.364} \quad \text{Rounded.}$$

When fractions are changed into decimals that do not terminate, the decimals are called **repeating decimals.** Repeating decimals can be written with a line over the digits that repeat or an ellipsis at the end to indicate that they repeat. The decimal equivalent can be rounded to any desirable place.

$$\frac{1}{3} = 0.\overline{3} \quad \text{or} \quad 0.333333\ldots \qquad \frac{4}{11} = 0.\overline{36} \quad \text{or} \quad 0.3636\ldots$$

Another option for expressing the decimal equivalent is to carry the division to a specified number of places and to write the remainder as a fraction. In the preceding example the mixed decimal equivalents to the hundredths place would be $0.33\frac{1}{3}$ and $0.36\frac{4}{11}$. The rounded decimal equivalents are **approximate equivalents,** and the repeating decimals and mixed decimals are **exact equivalents.**

To write a mixed number as a decimal number:

1. The whole-number part remains the same.
2. Write only the fraction part as a decimal by dividing the numerator by the denominator.

EXAMPLE Change $3\frac{2}{5}$ to a mixed decimal.

$$2 \div 5 \quad \text{or} \quad \begin{array}{r} 0.4 \\ 5\overline{)2.0} \end{array} \quad \text{Write the fraction part as an equivalent decimal.}$$

Then, $3\frac{2}{5} = \mathbf{3.4}$.

5 Compare Fractions, Mixed Numbers, and Decimals.

Like fractions are fractions that have the same denominator. When fractions are not like fractions, they can be changed to equivalent fractions with like denominators. We call these like denominators **common denominators.**

The **least common denominator (LCD)** for two or more fractions is the *least common multiple* (LCM) of the denominators.

The least common denominator often can be found by inspection. By inspection, we mean examine each denominator and intuitively (or mentally) determine the LCD. For fractions with larger denominators, you may need to use the procedure for finding the LCM discussed in Outcome 1 of this section.

EXAMPLE Find the least common denominator for the fractions $\frac{5}{12}, \frac{4}{15}, \frac{3}{8}$.

$$
\begin{array}{lll}
12 = & 15 = & 8 = \\
2\,(6) = & 3(5) = & 2\,(4) = \\
2\,(2)(3) = & & 2\,(2)(2) = \\
2^2(3) & & 2^3
\end{array}
$$

Find the prime factorization of the denominators.

$$
\begin{aligned}
\text{LCM or LCD} &= 2^3(3)(5) \\
&= 8(3)(5) \\
&= \mathbf{120}
\end{aligned}
$$

Alternative Procedure for Finding the LCM or LCD

We can also find the LCM or LCD of several fractions by dividing duplicated factors and then multiplying.

1. Arrange the denominators horizontally.
2. Divide by any prime factor that divides evenly into *at least two* denominators.
3. The LCM or LCD is the product of the primes and remaining factors.

Look at the denominators from the preceding example.

$$
\begin{array}{l|lll}
2 & 8 & 12 & 15 \\
2 & 4 & 6 & 15 \\
3 & 2 & 3 & 15 \\
\hline
 & 2 & 1 & 5
\end{array}
$$

Divide by the prime factor 2.
Divide by the prime factor 2.
Divide by the prime factor 3.

Primes: 2, 2, 3

Remaining factors: 2, 1, 5

$$
\begin{aligned}
\text{LCM} &= 2 \cdot 2 \cdot 3 \cdot 2 \cdot 1 \cdot 5 \\
&= 120
\end{aligned}
$$

To compare fractions, the denominators must be the same. To compare $\frac{3}{7}$ and $\frac{5}{7}$, which have the same denominators, we compare the numerators, and $\frac{3}{7}$ is smaller than $\frac{5}{7}$.

To compare fractions:

1. Find the least common denominator (LCD).
2. Change each fraction to an equivalent fraction with the least common denominator (LCD) as its denominator.
3. Compare the numerators.

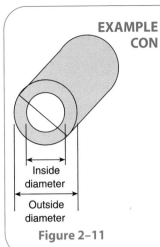

EXAMPLE
CON

Is it possible to have a pipe with an outside diameter of $\frac{5}{8}$ in. and an inside diameter of $\frac{21}{32}$ in. (see Fig. 2–11)?

To answer this question, we need to compare the two fractions $\frac{5}{8}$ and $\frac{21}{32}$.

$$\frac{5}{8} = \frac{5 \cdot 4}{8 \cdot 4} = \frac{20}{32}$$
Change $\frac{5}{8}$ to 32nds.

Is $\frac{20}{32}$, which is equivalent to $\frac{5}{8}$, larger than $\frac{21}{32}$? No, so **$\frac{5}{8}$ in. cannot be the outside diameter of a pipe with an inside diameter of $\frac{21}{32}$ in.**

Inside
diameter

Outside
diameter

Figure 2–11

EXAMPLE
CON

Two drill bits have diameters of $\frac{3}{8}$ in. and $\frac{5}{16}$ in., respectively. Which drill bit makes the larger hole?

$$\frac{3}{8} = \frac{6}{16} \qquad \frac{5}{16} = \frac{5}{16}$$
The least common denominator is 16. Change $\frac{3}{8}$ to 16ths.

Compare the numerators: $\frac{6}{16}$ is larger than $\frac{5}{16}$ because 6 is larger than 5, so $\frac{3}{8}$, which is equivalent to $\frac{6}{16}$, is larger than $\frac{5}{16}$.

The drill bit with a $\frac{3}{8}$-in. diameter will drill the larger hole.

TIP

Comparing Fractions Using Decimal Equivalents

We can compare fractions by changing them to decimal equivalents. Review comparing decimals in Chapter 1, Section 1, Outcome 3.

To compare a fraction and a decimal we change one number so that both numbers are either in fraction form or decimal form. Unless otherwise specified, you may use either form.

EXAMPLE
HELPP

The specifications of a document camera state the length is 15.8 in. A carrying case is $15\frac{7}{8}$ in. long on the inside. Will the camera fit inside the case?

Change $15\frac{7}{8}$ to a decimal. Change the fractional part of the mixed number to an equivalent decimal. $7 \div 8 = 0.875$

$$15\frac{7}{8} = 15.875$$ Compare decimals.

$$15.875 > 15.8$$

The case measurement, 15.875 in., is greater than 15.8 in., so the camera will fit into the case.

1

1. Find five fractions that are equivalent to $\frac{4}{5}$.

2. Find five fractions that are equivalent to $\frac{7}{10}$.

3. Find a fraction that is equivalent to $\frac{3}{4}$ and has a denominator of 24.

Find the equivalent fractions with the indicated denominators.

4. $\frac{3}{8} = \frac{?}{16}$

5. $\frac{4}{7} = \frac{?}{21}$

6. $\frac{9}{11} = \frac{?}{44}$

7. $\frac{1}{3} = \frac{?}{15}$

8. $\frac{5}{6} = \frac{?}{24}$

9. $\frac{7}{8} = \frac{?}{24}$

10. $\frac{2}{5} = \frac{?}{30}$

Reduce the fractions to lowest terms.

11. $\frac{4}{8}$

12. $\frac{6}{10}$

13. $\frac{12}{16}$

14. $\frac{10}{32}$

15. $\frac{16}{32}$

16. $\frac{28}{32}$

17. $\frac{20}{64}$

18. $\frac{2}{10}$

19. $\frac{8}{40}$

20. $\frac{12}{50}$

21. $\frac{10}{16}$

22. $\frac{4}{16}$

23. $\frac{24}{32}$

24. $\frac{12}{64}$

25. $\frac{14}{64}$

Measure line segments 26–35 in Fig. 2–12 to the nearest sixteenth of an inch (tolerance $= \pm\frac{1}{32}$ in.).

26.
27.
28.
29.
30.
31.
32.
33.
34.
35.

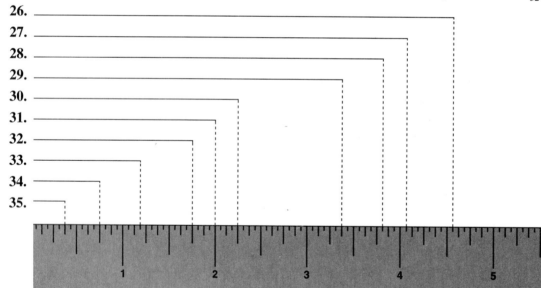

Figure 2–12

2 Write the improper fractions as whole or mixed numbers.

36. $\frac{12}{5}$

37. $\frac{10}{7}$

38. $\frac{12}{12}$

39. $\frac{32}{7}$

40. $\frac{24}{6}$

41. $\frac{15}{7}$

42. $\frac{23}{9}$

43. $\frac{47}{5}$

44. $\frac{86}{9}$

45. $\frac{38}{21}$

46. $\frac{57}{15}$

47. $\frac{64}{4}$

48. $\frac{72}{10}$

49. $\frac{19}{2}$

50. $\frac{36}{4}$

3 Write the mixed numbers as improper fractions.

51. $2\frac{1}{3}$ **52.** $3\frac{1}{8}$ **53.** $1\frac{7}{8}$ **54.** $6\frac{5}{12}$ **55.** $9\frac{5}{8}$

56. $3\frac{7}{8}$ **57.** $7\frac{5}{12}$ **58.** $6\frac{7}{16}$ **59.** $8\frac{1}{32}$ **60.** $1\frac{5}{64}$

61. $7\frac{3}{10}$ **62.** $8\frac{2}{3}$ **63.** $33\frac{1}{3}$ **64.** $66\frac{2}{3}$ **65.** $12\frac{1}{2}$

Change the whole numbers to an equivalent fraction with the given denominator.

66. $5 = \frac{?}{3}$ **67.** $9 = \frac{?}{2}$ **68.** $7 = \frac{?}{8}$ **69.** $8 = \frac{?}{4}$ **70.** $3 = \frac{?}{16}$

4 Change the decimals to their fraction or mixed-number equivalents, and reduce answers to lowest terms.

71. 0.5   **72.** 0.1   **73.** 0.2   **74.** 0.7
75. 0.25   **76.** 0.025   **77.** 3.9   **78.** 4.8
79. 0.378   **80.** 0.875   **81.** 0.375   **82.** 0.625

83. A measure of 0.75 in. represents what fractional part of an inch?

84. What fraction represents 0.1875 ft.?

85. INDTEC   The length of a screw is 2.375 in. Represent this length as a mixed number.

86. An instrument weighs 0.83 lb. Write this as a fraction of a pound.

87. INDTEC   Some sheet metal is 0.3125 in. thick. What is the thickness expressed as a fraction?

88. COMP   A predrilled PC board is 3.125 in. long. Write this length as a mixed number.

Change to decimal numbers.

89. $\frac{3}{5}$ **90.** $\frac{3}{10}$ **91.** $\frac{7}{8}$ **92.** $\frac{3}{8}$ **93.** $\frac{9}{20}$

94. $\frac{49}{50}$ **95.** $\frac{21}{100}$ **96.** $3\frac{7}{8}$ **97.** $1\frac{7}{16}$ **98.** $4\frac{9}{16}$

Write these fractions and mixed numbers as decimals rounded to the nearest hundredth.

99. $\frac{1}{6}$ **100.** $\frac{4}{9}$ **101.** $1\frac{3}{7}$ **102.** $3\frac{5}{11}$

103. $\frac{2}{3}$ **104.** $\frac{3}{11}$ **105.** $\frac{7}{9}$ **106.** $\frac{5}{13}$

Write as decimals rounded to the hundredths place.

107. $\frac{5}{6}$ **108.** $\frac{7}{12}$ **109.** $2\frac{3}{8}$ **110.** $5\frac{4}{7}$

111. TELE   An aerial map shows a building measuring $2\frac{3}{64}$ in. on one side. What is the measure of the side of the building in decimal numbers?

112. BUS   The property tax rate is $45 per $1,000 of the assessed value. Express the tax rate as a decimal.

113. CON   A plan allows a gap of $\frac{1}{8}$ in. between vinyl flooring and the wall for expansion. What is the gap measure in decimal notation?

114. AVIA   A B-767-200 aircraft has a wing span of 47.6 m. What is the wing span written as a mixed number?

5 Find the least common denominator.

115. $\frac{5}{8}, \frac{4}{9}$ **116.** $\frac{3}{10}, \frac{4}{15}$ **117.** $\frac{9}{10}, \frac{4}{25}$ **118.** $\frac{7}{12}, \frac{9}{16}, \frac{5}{8}$ **119.** $\frac{2}{3}, \frac{5}{12}, \frac{7}{8}$

Show which fraction is larger.

120. $\dfrac{2}{3}, \dfrac{3}{5}$ **121.** $\dfrac{5}{12}, \dfrac{7}{16}$ **122.** $\dfrac{8}{9}, \dfrac{7}{8}$ **123.** $\dfrac{5}{8}, \dfrac{11}{16}$ **124.** $\dfrac{15}{32}, \dfrac{29}{64}$

125. $\dfrac{7}{12}, \dfrac{9}{16}$ **126.** $\dfrac{3}{8}, \dfrac{4}{5}$ **127.** $\dfrac{7}{11}, \dfrac{9}{10}$ **128.** $\dfrac{4}{15}, \dfrac{3}{16}$ **129.** $\dfrac{1}{2}, \dfrac{7}{16}$

Show which common fraction or decimal is smaller.

130. $\dfrac{3}{8}, 0.37$ **131.** $\dfrac{4}{5}, 0.82$ **132.** $\dfrac{5}{12}, 0.42$ **133.** $\dfrac{1}{2}, 0.65$ **134.** $\dfrac{3}{4}, 0.34$

135. INDTEC   Is the thickness of a $\frac{3}{16}$-in. sheet of metal greater than the length of a $\frac{15}{64}$-in. sheet metal screw?

136. INDTEC   Will a pipe with a $\frac{5}{16}$-in. outside diameter fit inside a pipe with a $\frac{3}{8}$-in. inside diameter?

137. INDTEC   A hollow-wall fastener has a grip range up to $\frac{7}{16}$ in. Is it long enough to fasten a thin sheet metal strip to a plywood wall if the combined thickness of the wall is $\frac{3}{8}$ in.?

138. CON   A range top is $29\frac{1}{8}$ in. long by $19\frac{1}{2}$ in. wide. Is it smaller than an existing opening $29\frac{3}{16}$ in. long by $19\frac{9}{16}$ in. wide?

139. INDTEC   A wrench is marked $\frac{5}{8}$ at one end and $\frac{19}{32}$ at the other. Which end is larger?

140. INDTEC   Charles Bryant has a wrench marked $\frac{25}{32}$, but it is too large. Would a $\frac{7}{8}$-in. wrench be smaller?

141. INDTEC   For a do-it-yourself project, Brenda Jinkins needs to cut a piece of sheet metal slightly longer than the $10\frac{21}{32}$ in. required in the plans and to trim it down to size. Brenda cuts the piece $10\frac{3}{4}$ in. Is it too short?

142. INDTEC   A plastic anchor for a No. 6 × 1-in. screw requires that at least a $\frac{3}{16}$-in. diameter hole be drilled. Will a $\frac{1}{4}$-in. drill bit be large enough?

143. INDTEC   The plastic anchor in Exercise 142 requires a minimum hole depth of $\frac{7}{8}$ in. Is a $\frac{3}{4}$-in. hole deep enough?

144. INDTEC   Is a $\frac{3}{8}$-in. wrench too large or too small for a $\frac{1}{2}$-in. bolt?

2–3 | *Adding and Subtracting Fractions and Mixed Numbers*

Learning Outcomes
   1   Add fractions and mixed numbers.
   2   Subtract fractions and mixed numbers.

1   Add Fractions and Mixed Numbers.

Adding fractions requires that all fractions being added have the same denominator. Trying to add unlike fractions is like trying to add unlike objects or measures. Before adding unlike fractions, we change the fractions to equivalent fractions with a common denominator.

> **To add fractions:**
>
> 1. If the denominators are not the same, find the least common denominator.
> 2. Change each fraction not already expressed in terms of the common denominator to an equivalent fraction with the common denominator.
> 3. Add the numerators only.
> 4. The common denominator is the denominator of the sum.
> 5. Reduce the sum to lowest terms and change improper fractions to whole or mixed numbers.

EXAMPLE Find the sum of $\frac{3}{8} + \frac{1}{8}$.

Because the denominators are the same, start with Step 3 of the addition procedure.

$$\frac{3}{8} + \frac{1}{8} = \frac{4}{8} = \mathbf{\frac{1}{2}} \qquad \text{Add numerators and reduce.}$$

EXAMPLE Add $\frac{5}{32} + \frac{3}{16} + \frac{7}{8}$.

The least common denominator may be found by inspection. Both 8 and 16 divide evenly into 32, so we use 32 as the common denominator.

$$\frac{5}{32} = \frac{5}{32}, \qquad \frac{3}{16} = \frac{6}{32}, \qquad \frac{7}{8} = \frac{28}{32} \qquad \begin{array}{l}\text{Change each fraction to an equivalent fraction} \\ \text{whose denominator is 32.}\end{array}$$

$$\frac{5}{32} + \frac{6}{32} + \frac{28}{32} = \frac{39}{32} \qquad \text{Add the numerators.}$$

$$\frac{39}{32} = 1\frac{7}{32} \qquad \text{Change to a mixed number.}$$

EXAMPLE CON A plumber uses a $\frac{9}{16}$-in.-diameter copper tube wrapped with $\frac{5}{8}$-in. insulation (Fig. 2–13). What size hole must he bore in the stud (wall support) to install the insulated pipe?

To find the total diameter of the pipe and insulation, add $\frac{9}{16} + \frac{5}{8} + \frac{5}{8}$. The thickness of the insulation is added twice because it counts in the total diameter of the pipe and insulation two times.

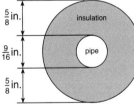

$\frac{5}{8}$ in.

$\frac{9}{16}$ in.

$\frac{5}{8}$ in.

Figure 2–13

$$\frac{9}{16} = \frac{9}{16} \qquad \text{The LCD is 16. Change each fraction to 16ths.}$$

$$\frac{5}{8} = \frac{10}{16}$$

$$+\; \frac{5}{8} = \frac{10}{16} \qquad \text{Add numerators. Keep the common denominator.}$$

$$\frac{29}{16} = 1\frac{13}{16} \qquad \text{Change to a mixed number.}$$

The total diameter is $1\frac{13}{16}$ in., and the diameter of the hole must be at least this large.

To add mixed numbers:

1. Add the whole-number parts.
2. Add the fractional parts and reduce to lowest terms.
3. Change improper fractions to whole or mixed numbers.
4. Add whole-number parts.

EXAMPLE Add $5\frac{2}{3} + 7\frac{3}{8} + 4\frac{1}{2}$.

$$5\frac{2}{3} = 5\frac{16}{24}$$ The LCD is 24. Change fractions to equivalent fractions.

$$7\frac{3}{8} = 7\frac{9}{24}$$ Add whole numbers.

$$+\ 4\frac{1}{2} = 4\frac{12}{24}$$ Add fractional parts.

$$16\frac{37}{24}$$ $\frac{37}{24} = 1\frac{13}{24}$. Change improper fraction to a mixed number.

$$16 + 1\frac{13}{24} = 17\frac{13}{24}$$ Add whole-number parts.

Writing Whole Numbers in Mixed-Number Form

When you add mixed numbers and whole numbers, think of the whole number as a mixed number with zero as the numerator in the fraction.

$$16 \quad = 16\frac{0}{24}$$
$$+\ 1\frac{13}{24} \quad = \quad 1\frac{13}{24}$$
$$\overline{\qquad\qquad 17\frac{13}{24},}$$

EXAMPLE
CAD/ARC Find the largest acceptable measurement of a part if the blueprint calls for the part to be 2 in. long and the tolerance is $\pm\frac{1}{8}$ in.

$$2 + \frac{1}{8} = 2\frac{1}{8}$$ Review tolerance in Chapter 1, Section 1, Outcome 5.

The largest acceptable measure is $2\frac{1}{8}$ in.

2 Subtract Fractions and Mixed Numbers.

The steps for subtracting fractions and mixed numbers are very similar to the steps for adding fractions and mixed numbers.

To subtract fractions:

1. If the denominators are not the same, find the least common denominator.
2. Change each fraction not expressed in terms of the common denominator to an equivalent fraction having the common denominator.
3. Subtract the numerators.
4. The common denominator will be the denominator of the difference.
5. Reduce the difference to lowest terms.

EXAMPLE Subtract $\dfrac{3}{8} - \dfrac{7}{32}$.

$$\dfrac{3}{8} = \dfrac{12}{32}$$ Change $\dfrac{3}{8}$ to an equivalent fraction with a denominator of 32.

$$-\dfrac{7}{32} = \dfrac{7}{32}$$ Subtract numerators and keep the common denominator.

$$\dfrac{5}{32}$$

To subtract mixed numbers:

1. If the fractional parts of the mixed numbers do not have the same denominator, change them to equivalent fractions with a common denominator.
2. When the fraction in the minuend is larger than the fraction in the subtrahend, go to Step 6.
3. When the fraction in the subtrahend is larger than the fraction in the minuend, regroup (borrow) by taking one whole number from the whole-number part of the minuend. This makes the whole number 1 less.
4. Change the whole number that was borrowed to an improper fraction with the common denominator. For example, $1 = \dfrac{3}{3}, 1 = \dfrac{8}{8}, 1 = \dfrac{n}{n}$, where n is the common denominator.
5. Add the borrowed fraction $\left(\dfrac{n}{n}\right)$ to the fraction already in the minuend.
6. Subtract the fractional parts and the whole-number parts.
7. Reduce the difference to lowest terms.

EXAMPLE Subtract $15\dfrac{7}{8} - 4\dfrac{1}{2}$.

$$15\dfrac{7}{8} = 15\dfrac{7}{8}$$ LCD is 8.

$$-4\dfrac{1}{2} = 4\dfrac{4}{8}$$ Subtract like fractions.
Subtract whole numbers.

$$11\dfrac{3}{8}$$

2–3 Adding and Subtracting Fractions and Mixed Numbers

EXAMPLE   Subtract $15\frac{3}{4}$ from $18\frac{1}{2}$.

Study this!!

$$18\frac{1}{2} = 18\frac{2}{4} = 17\frac{4}{4} + \frac{2}{4} = 17\frac{6}{4}$$

LCD = 4. Change $\frac{1}{2}$ to $\frac{2}{4}$.

Regroup. $18 - 1 = 17$, $1 = \frac{4}{4}$, $\frac{4}{4} + \frac{2}{4} = \frac{6}{4}$

$$-15\frac{3}{4} = 15\frac{3}{4} = 15\frac{3}{4} \qquad = 15\frac{3}{4}$$

Subtract fractions.
Subtract whole numbers.

$$2\frac{3}{4}$$

EXAMPLE
HOSP $127\frac{1}{2}$ lb of sugar is used from an inventory of $433\frac{3}{8}$ lb. How many pounds of sugar remain in inventory?

$$433\frac{3}{8} = 433\frac{3}{8} = 432\frac{11}{8}$$

$433 - 1 = 432$, $1 = \frac{8}{8}$, $\frac{8}{8} + \frac{3}{8} = \frac{11}{8}$

$$-127\frac{1}{2} = 127\frac{4}{8} = 127\frac{4}{8}$$

Subtract fractions. Subtract whole numbers.

$$305\frac{7}{8}$$

$305\frac{7}{8}$ lb of sugar remain in inventory.

EXAMPLE   Subtract 27 from $45\frac{1}{3}$.

$$45\frac{1}{3} = 45\frac{1}{3}$$

$$-27 \quad = 27\frac{0}{3}$$

Write 27 as a mixed number. Subtract.

$$18\frac{1}{3}$$

EXAMPLE
TELE   How many feet of coaxial cable are left on a 100-ft roll if $27\frac{1}{4}$ ft are used from the roll?

$$100 \quad = 100\frac{0}{4} = 99\frac{4}{4}$$

Write 100 as a mixed number. Regroup.

$$-27\frac{1}{4} = \quad 27\frac{1}{4} = 27\frac{1}{4}$$

Subtract.

$$72\frac{3}{4}$$

$72\frac{3}{4}$ ft of cable are left on the roll.

EXAMPLE
INDTR

Three lengths measuring $5\frac{1}{4}$ in., $7\frac{3}{8}$ in., and $6\frac{1}{2}$ in. are cut from a 64-in. bar of angle iron. If $\frac{3}{16}$ in. is wasted on each cut, how many inches of angle iron remain?

Visualize the problem by making a sketch (see Fig. 2–14). Then find the total amount of angle iron used. This includes the three lengths and the waste for three cuts.

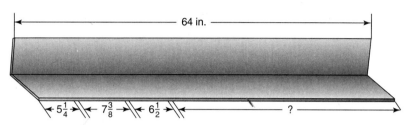

Figure 2–14

Total removed and wasted = 3 lengths + 3 cuts

Three Lengths   Three Cuts

$$5\frac{1}{4} + 7\frac{3}{8} + 6\frac{1}{2} + \frac{3}{16} + \frac{3}{16} + \frac{3}{16} =$$

Change fractions to equivalent fractions with LCD of 16 and add

$5 + 7 + 6 = 18$
$4 + 6 + 8 + 3 + 3 + 3 = 27.$
Regroup mixed number.

$$5\frac{4}{16} + 7\frac{6}{16} + 6\frac{8}{16} + \frac{3}{16} + \frac{3}{16} + \frac{3}{16} = 18\frac{27}{16}$$

$$18\frac{27}{16} = 18 + \frac{27}{16} = 18 + 1\frac{11}{16} = 19\frac{11}{16}$$

Total amount removed and wasted. Regroup and subtract.

$$\text{amount of angle iron remaining} = \frac{\text{beginning}}{\text{length}} - \frac{\text{total iron removed}}{\text{and wasted}}$$

$$64 - 19\frac{11}{16} =$$

Total remaining.

$$63\frac{16}{16} - 19\frac{11}{16} = 44\frac{5}{16}$$

$44\frac{5}{16}$ in. of angle iron remain.

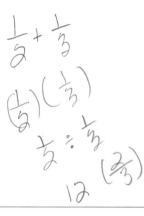

SECTION 2–3 SELF-STUDY EXERCISES

1 Add; reduce answers to lowest terms and write any improper fractions as whole or mixed numbers.

1. $\dfrac{5}{16} + \dfrac{1}{16}$

2. $\dfrac{1}{2} + \dfrac{1}{8} + \dfrac{3}{4}$

3. $\dfrac{1}{8} + \dfrac{1}{2}$

4. $\dfrac{3}{8} + \dfrac{5}{32} + \dfrac{1}{4}$

5. $\dfrac{5}{16} + \dfrac{1}{4}$

6. $\dfrac{15}{16} + \dfrac{1}{2}$

7. $\dfrac{3}{32} + \dfrac{5}{64}$

8. $\dfrac{7}{8} + \dfrac{3}{5}$

9. $\dfrac{3}{4} + \dfrac{8}{9}$

10. $\dfrac{7}{8} + \dfrac{5}{24}$

11. $\dfrac{3}{5} + \dfrac{4}{5}$

12. $\dfrac{5}{7} + \dfrac{4}{21}$

13. **CON** What is the thickness of a countertop made of $\frac{7}{8}$-in. plywood and $\frac{1}{16}$-in. Formica?

14. **INDTR** Three pieces of steel are joined together. What is the total thickness if the pieces are $\frac{1}{2}$ in., $\frac{7}{16}$ in., and $\frac{29}{32}$ in.?

15. **BUS**   Three books are placed side by side. They are $\frac{5}{16}$ in., $\frac{7}{8}$ in., and $\frac{3}{4}$ in. wide. What is the total width of the books if they are polywrapped in one package?

16. **AUTO**   Find the outside diameter of a hose (Fig. 2–15) whose wall is $\frac{1}{2}$ in. thick if its inside diameter is $\frac{7}{8}$ in.

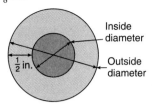

Figure 2–15

17. **INDTEC**   What length bolt is needed to fasten two pieces of metal each $\frac{7}{16}$ in. thick if a $\frac{1}{8}$-in. lock washer and a $\frac{1}{4}$-in. nut are used?

Add; reduce answers to lowest terms and write any improper fractions as whole or mixed numbers.

18. $2\frac{3}{5} + 4\frac{1}{5}$

19. $1\frac{5}{8} + 2\frac{1}{2}$

20. $3\frac{3}{4} + 7\frac{3}{16} + 5\frac{7}{8}$

21. $\frac{1}{6} + \frac{7}{9} + \frac{2}{3}$

22. $2\frac{1}{4} + 2\frac{9}{16}$

23. $1\frac{5}{16} + 4\frac{7}{32}$

24. $3\frac{1}{4} + 1\frac{7}{16}$

25. $4\frac{1}{2} + 9$

26. $518\frac{7}{12} + 483\frac{5}{18}$

27. $291\frac{8}{15} + 78\frac{31}{40}$

28. $309\frac{11}{18} + 805\frac{13}{30}$

29. $78\frac{17}{20} + 46\frac{9}{32}$

30. **CON**   The studs (interior supports) of an outside wall are $5\frac{3}{4}$ in. thick. The inside wallboard is $\frac{7}{8}$ in. thick, and the outside covering is $2\frac{3}{16}$ in. thick. What is the total thickness of the wall?

31. **CAD/ARC**   A blueprint calls for a piece of bar stock $3\frac{7}{16}$ in. long. If a tolerance of $\pm\frac{1}{16}$ in. is allowed, what is the longest acceptable measurement for the bar stock?

32. **HOSP**   If $4\frac{3}{8}$ gal of water are used to dilute $7\frac{1}{4}$ gal of juice, how many gallons are in the mixture?

33. **CON**   How much bar stock is needed to make bars of the following lengths: $10\frac{1}{4}$ in., $8\frac{7}{16}$ in., $5\frac{15}{32}$ in.? Disregard waste.

34. **HLTH/N**   Three pieces of bandage material each measuring $7\frac{5}{8}$ in. are needed to complete a job. How much bandage material is needed?

35. **HOSP**   If $5\frac{1}{8}$ cups of water are mixed with $\frac{3}{4}$ cup of Kool Aid, how many cups are in the mixture?

2   Subtract; reduce when necessary.

36. $\frac{9}{16} - \frac{3}{8}$

37. $\frac{7}{16} - \frac{3}{8}$

38. $\frac{5}{8} - \frac{1}{2}$

39. $\frac{5}{32} - \frac{1}{64}$

40. $23\frac{3}{16} - 5\frac{7}{16}$

41. $9\frac{1}{4} - 4\frac{5}{16}$

42. $9\frac{1}{32} - 3\frac{3}{8}$

43. $14\frac{1}{7} - 12\frac{3}{7}$

Simplify.

44. $2\frac{5}{8} + 7\frac{1}{4} - 6\frac{1}{12}$

45. $21\frac{5}{12} - 7\frac{5}{16} + 4\frac{1}{4}$

46. $555\frac{7}{25} - 388\frac{8}{35}$

47. $843 - 115\frac{7}{8} + 32\frac{3}{14}$

48. **INDTR**   A length of bar stock $16\frac{3}{8}$ in. long is cut so that a piece only $7\frac{9}{16}$ in. long remains. What is the length of the cutoff piece? Disregard waste.

49. **AVIA**   A Delta Airlines freight container is 60.4 in. wide. A freight forwarder has three boxes that have widths of $12\frac{3}{16}$ in., $32\frac{3}{8}$ in., and $15\frac{3}{4}$ in. Will these three boxes fit in the container?

50. **AG/H**   A flower bed includes $7\frac{7}{8}$ in. of base fill. If the bed is to be 18 in. thick, how thick must the topsoil be?

51. **CAD/ARC**   Find the missing length in Fig. 2–16.

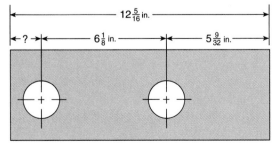

Figure 2–16

52. INDTR A casting is machined so that $22\frac{1}{5}$ lb of metal remain. If the casting weighed $25\frac{3}{10}$ lb, how many pounds were removed by machine?

53. AUTO A bolt $2\frac{5}{8}$ in. long fastens two pieces of metal 1 in. and $1\frac{7}{32}$ in. thick. If a $\frac{3}{32}$-in.-thick lock washer and a $\frac{1}{8}$-in.-thick washer are used, what thickness is the nut if it is even with the end of the bolt? The measure of a bolt length does not include the bolt head.

2–4 | *Multiplying and Dividing Fractions and Mixed Numbers*

Learning Outcomes

1 Multiply fractions and mixed numbers.

2 Raise a fraction to a power.

3 Divide fractions and mixed numbers.

4 Perform calculations involving fractions with a calculator.

When multiplying a fraction by a fraction, we are finding *a part of a part*.

1 Multiply Fractions and Mixed Numbers.

In the multiplication, $\frac{1}{2} \times \frac{1}{2}$ is $\frac{1}{2}$ of $\frac{1}{2}$ (Fig. 2–17). The word "of" is the clue that we must multiply to find the part we are looking for.

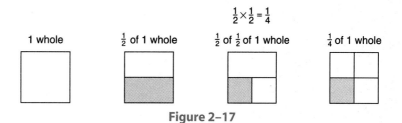

$$\frac{1}{2} \times \frac{1}{2} = \frac{1}{4}$$

1 whole $\frac{1}{2}$ of 1 whole $\frac{1}{2}$ of $\frac{1}{2}$ of 1 whole $\frac{1}{4}$ of 1 whole

Figure 2–17

Adding or subtracting fractions and mixed numbers requires a common denominator. In multiplying fractions, we do *not* change fractions to equivalent fractions with a common denominator. Look at two more examples of taking a part of a part (Figs. 2–18 and 2–19).

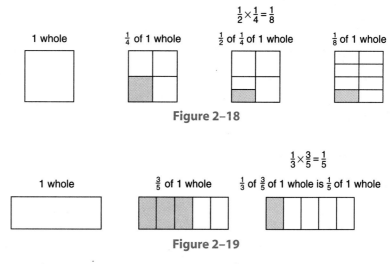

$$\frac{1}{2} \times \frac{1}{4} = \frac{1}{8}$$

1 whole $\frac{1}{4}$ of 1 whole $\frac{1}{2}$ of $\frac{1}{4}$ of 1 whole $\frac{1}{8}$ of 1 whole

Figure 2–18

$$\frac{1}{3} \times \frac{3}{5} = \frac{1}{5}$$

1 whole $\frac{3}{5}$ of 1 whole $\frac{1}{3}$ of $\frac{3}{5}$ of 1 whole is $\frac{1}{5}$ of 1 whole

Figure 2–19

1. Multiply the numerators of the fractions to get the numerator of the product.
2. Multiply the denominators to get the denominator of the product.
3. Reduce the product to lowest terms.

EXAMPLE   Find $\frac{1}{2}$ of $\frac{1}{4}$.

$$\frac{1}{2}\left(\frac{1}{4}\right) = \frac{1}{8}$$ Multiply numerators.
Multiply denominators.

EXAMPLE   Find $\frac{1}{3}$ of $\frac{3}{5}$.

$$\frac{1}{3}\left(\frac{3}{5}\right) = \frac{3}{15} = \frac{1}{5}$$ Multiply numerators.
Multiply denominators. Reduce.

Reduce, or Cancel, Before Multiplying

In the preceding example, reducing is possible. When multiplying fractions, we can reduce common factors before multiplying them.

$$\frac{1}{3}\left(\frac{3}{5}\right) = \frac{1(\overset{1}{\cancel{3}})}{\underset{1}{\cancel{3}}(5)} = \frac{1}{5} \qquad \frac{1(3)}{3(5)} = \frac{1(3)}{5(3)} = \frac{1}{5}\left(\frac{3}{3}\right) = \frac{1}{5}(1) = \frac{1}{5}$$

In this example a numerator and a denominator both have a common factor of 3, so the common factor can be reduced before multiplying. Reducing applies the principles $\frac{n}{n} = 1$ and $1(n) = n$. This process is also referred to as **canceling**.

EXAMPLE   Multiply $\frac{2}{3}\left(\frac{5}{9}\right)\left(\frac{1}{6}\right)$.

$$\frac{\overset{1}{\cancel{2}}}{3}\left(\frac{5}{9}\right)\left(\frac{1}{\underset{3}{\cancel{6}}}\right) = \frac{5}{81}$$ 2 is a common factor of both a numerator and a denominator.
Reduce before multiplying.

Chapter 2 / Review of Fractions

TIP | **Reduce from Any Numerator to Any Denominator**

Common factors that are reduced can be diagonal to each other, one above the other, or separated by another fraction, but one factor *must* be in the numerator and the other in the denominator.

$$\frac{2}{\overset{1}{\cancel{3}}}\left(\frac{\overset{1}{\cancel{3}}}{5}\right) = \frac{2}{5} \qquad \frac{\overset{3}{\cancel{6}}}{8}\left(\frac{3}{5}\right) = \frac{9}{20} \qquad \frac{\overset{1}{\cancel{2}}}{7}\left(\frac{1}{3}\right)\left(\frac{5}{\underset{4}{\cancel{8}}}\right) = \frac{5}{84}$$

EXAMPLE Multiply $\dfrac{2}{5}\left(\dfrac{10}{21}\right)\left(\dfrac{6}{12}\right)$.

$$\frac{\overset{1}{\cancel{2}}}{\cancel{5}}\left(\frac{\overset{2}{\cancel{10}}}{21}\right)\left(\frac{\overset{1}{\cancel{6}}}{\underset{\underset{1}{2}}{\cancel{12}}}\right) = \frac{\mathbf{2}}{\mathbf{21}}$$

5 and 10 are diagonal to each other.
6 is above 12.
2 and 2 are separated by another fraction.

Other patterns of reducing could also have been used.

When multiplying mixed numbers or combinations of whole numbers, fractions, and mixed numbers, we first change each whole number or mixed number to an improper fraction. Then we proceed as in multiplying fractions.

To multiply mixed numbers, fractions, and whole numbers:

1. Change each mixed number or whole number to an improper fraction.
2. Reduce as much as possible.
3. Multiply numerators.
4. Multiply denominators.
5. Write the answer as a whole or mixed number if possible.

EXAMPLE Multiply $2\dfrac{1}{2}\left(5\dfrac{1}{3}\right)$.

$$2\frac{1}{2}\left(5\frac{1}{3}\right) = \qquad \text{Change each mixed number to an improper fraction.}$$

$$\frac{5}{2}\left(\frac{16}{3}\right) = \qquad \text{Reduce.}$$

$$\frac{5}{2}\left(\frac{\overset{8}{\cancel{16}}}{3}\right) = $$
$$\underset{1}{}$$

$$\frac{5}{1}\left(\frac{8}{3}\right) = \qquad \text{Multiply numerators. Multiply denominators.}$$

$$\frac{40}{3} = \qquad \text{Change the product to a mixed number.}$$

$$\mathbf{13\frac{1}{3}}$$

EXAMPLE AG/H Bedding plants are to be planted $3\frac{5}{8}$ in. apart and $3\frac{5}{8}$ in. from the end of the planter. What length planter is needed for 9 plants (Fig. 2–20)?

$$3\frac{5}{8}(10) = \frac{29}{8}\left(\frac{\overset{5}{\cancel{10}}}{1}\right) = \frac{145}{4} = 36\frac{1}{4}$$ 10 equal spaces of $3\frac{5}{8}$ in. are required.

$3\frac{5}{8}$ $3\frac{5}{8}$ $3\frac{5}{8}$

Figure 2–20

A $36\frac{1}{4}$ in. planter will be needed.

EXAMPLE CON Bricks that are $2\frac{1}{4}$ in. thick form a brick wall with $\frac{3}{8}$-in. mortar joints (Fig. 2–21). What is the height of the wall above the foundation after nine courses?

Use the Six-Step Problem-Solving Plan.

Unknown facts   Height of the wall after nine courses of brick have been laid

Known facts $2\frac{1}{4}$ in.   thickness of each brick

$\frac{3}{8}$ in.   thickness of each mortar joint

9   number of courses (or rows) of brick and mortar joints

Relationships   Height of wall = thickness of each mortar joint × number of mortar joints + thickness of each brick × number of rows of brick

Estimation   Each brick is a little more than 2 in. thick, so the wall should be at least 2 × 9 or 18 in. high. Since the mortar joint is not quite $\frac{1}{2}$ in. and the fractional portion of the brick's thickness is less than $\frac{1}{2}$ in., the combined thickness of the brick and mortar joint must be less than 3 in. So the total height of the wall should be less than 3 × 9 or 27 in. Thus, we estimate the wall height to be between 18 and 27 in.

Calculation

$$\left(9 \cdot 2\frac{1}{4}\right) + \left(9 \cdot \frac{3}{8}\right)$$ Write as improper fractions.

$$\left(\frac{9}{1} \cdot \frac{9}{4}\right) + \left(\frac{9}{1} \cdot \frac{3}{8}\right)$$ Multiply.

$$\frac{81}{4} + \frac{27}{8}$$ Change to mixed numbers.

Figure 2–21

Foundation

$$20\frac{1}{4} + 3\frac{3}{8}$$ Write fractions with a common denominator.

$$20\frac{2}{8} + 3\frac{3}{8} = 23\frac{5}{8}$$ Add.

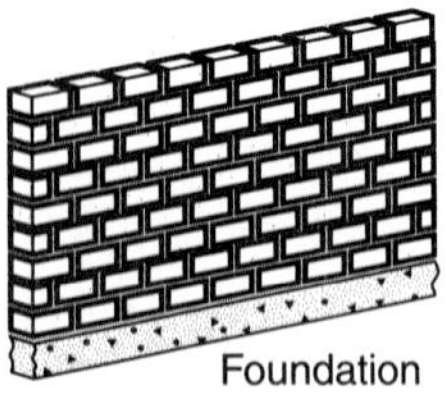

Interpretation   **The wall will be $23\frac{5}{8}$ in. high.**

2 **Raise a Fraction to a Power.**

To raise a fraction or quotient to a power:

1. Raise the numerator to the power.
2. Raise the denominator to the power. The denominator cannot be zero.

Symbolically,

$$\left(\frac{a}{b}\right)^n = \frac{a^n}{b^n} \qquad b \neq 0 \text{ and } a, b, \text{ and } n \text{ are real numbers}$$

EXAMPLE Raise the fractions to the indicated power.

(a) $\left(\dfrac{2}{3}\right)^2$ (b) $\left(\dfrac{1}{2}\right)^3$

(a) $\left(\dfrac{2}{3}\right)^2 = \dfrac{2^2}{3^2} = \dfrac{4}{9}$ Raise the numerator to the power.
                                   Raise the denominator to the power.

(b) $\left(\dfrac{1}{2}\right)^3 = \dfrac{1^3}{2^3} = \dfrac{1}{8}$ $(1)(1)(1) = 1; (2)(2)(2) = 8$

3 Divide Fractions and Mixed Numbers.

If we compare $12 \div 3$ and $\frac{1}{3} \times 12$, we find that both answers are 4. That is, $12 \div 3 = \frac{1}{3}(12)$ or $12(\frac{1}{3})$. Not only is there a relationship between multiplication and division, there is also a relationship between numbers like 3 and $\frac{1}{3}$. Pairs of numbers like $\frac{1}{3}$ and 3 are called *reciprocals*.

Two numbers are **reciprocals** if their product is 1. Thus, $\frac{1}{3}$ and 3 are reciprocals because $\frac{1}{3}(3) = 1$, and $\frac{2}{3}$ and $\frac{3}{2}$ are reciprocals because $\frac{2}{3}(\frac{3}{2}) = 1$. The **multiplicative inverse** of a number is its reciprocal. A number times its multiplicative inverse is 1, the **multiplicative identity.**

 ## To find the reciprocal of a number:

1. Write the number in fractional form.
2. Interchange the numerator and denominator so that the numerator is the denominator and the denominator is the numerator.

 ## Reciprocals and Inverting

Interchanging the numerator and denominator of a fraction is commonly called **inverting** the fraction.

EXAMPLE Find the reciprocal of $\frac{4}{7}$, $\frac{1}{5}$, 3, $2\frac{1}{2}$, 0.8, 1, and 0.

The reciprocal of $\frac{4}{7}$ is $\frac{7}{4}$ or $\mathbf{1\frac{3}{4}}$.   Interchange the numerator and denominator.

The reciprocal of $\frac{1}{5}$ is $\frac{5}{1}$ or **5.**

The reciprocal of 3 is $\frac{1}{3}$.   Write 3 as an improper fraction. $3 = \frac{3}{1}$.

The reciprocal of $2\frac{1}{2}$ is $\frac{2}{5}$.   Write $2\frac{1}{2}$ as an improper fraction. $2\frac{1}{2} = \frac{5}{2}$.

The reciprocal of 0.8 is $\frac{5}{4}$ or **1.25.**   Write 0.8 as a common fraction. $0.8 = \frac{8}{10} = \frac{4}{5}$.

The reciprocal of 1 is **1.**   Write 1 as an improper fraction. $1 = \frac{1}{1}$.

0 has no reciprocal. $0 = \frac{0}{1}$ and $\frac{1}{0}$ is undefined.

Let's review the terminology of division.

$$15 \quad \div \quad 3 \quad = \quad 5$$
$$\text{dividend} \quad \text{divisor} \quad \text{quotient}$$

To identify the divisor, remember that the symbol $\div$ is always read "divided by."

To divide fractions:

1. Change the division to an equivalent multiplication by replacing the divisor with its reciprocal and replacing the division notation with multiplication notation.
2. Perform the resulting multiplication.

EXAMPLE Find $\dfrac{5}{8} \div \dfrac{2}{3}$.

$$\frac{5}{8} \div \frac{2}{3} = \frac{5}{8} \cdot \frac{3}{2} = \frac{15}{16}$$ Change division to an equivalent multiplication.

Put Rules into Your Own Words

Some common phrases for stating the division of fractions rule are:

- Invert the divisor and multiply.
- Invert the second number and multiply.
- Invert the number after the division sign and multiply.

A rule in your own words is often easier for you to remember. Be sure the words guide you to an appropriate process.

EXAMPLE CON   An auger bit advances $\frac{1}{16}$ in. for each turn (see Fig. 2–22). How many turns are needed to drill a hole $\frac{5}{8}$ in. deep? ($\frac{5}{8}$ in. can be divided into how many $\frac{1}{16}$-in. parts?)

$$\frac{5}{8} \div \frac{1}{16} = \frac{5}{8} \cdot \frac{\overset{2}{16}}{\underset{1}{1}} = 10 \qquad \text{Multiply by the reciprocal of the divisor.}$$

Ten turns are needed.

Figure 2–22

To divide mixed numbers and whole numbers, we first write the mixed numbers or whole numbers as improper fractions and then follow the rule for dividing fractions.

To divide mixed numbers, fractions, and whole numbers:

1. Change each mixed number or whole number to an improper fraction.
2. Convert to an equivalent multiplication problem using the reciprocal of the divisor.
3. Multiply according to the rule for multiplying fractions.

EXAMPLE   Find $2\frac{1}{2} \div 3\frac{1}{3}$.

$$2\frac{1}{2} \div 3\frac{1}{3} = \qquad \text{Change each mixed number to an improper fraction.}$$

$$\frac{5}{2} \div \frac{10}{3} = \qquad \text{Change division to equivalent multiplication.}$$

$$\frac{\overset{1}{5}}{2} \cdot \frac{3}{\underset{2}{10}} = \qquad \text{Reduce and multiply.}$$

$$\frac{3}{4}$$

EXAMPLE   Find $5\frac{3}{8} \div 3$.

$$5\frac{3}{8} \div 3 = \qquad \text{Change mixed number and whole number to improper fractions.}$$

$$\frac{43}{8} \div \frac{3}{1} = \qquad \text{Change division to equivalent multiplication.}$$

$$\frac{43}{8} \cdot \frac{1}{3} = \qquad \text{Multiply.}$$

$$\frac{43}{24} = 1\frac{19}{24}$$

EXAMPLE BUS A developer subdivides $5\frac{1}{4}$ acres into lots; each lot is $\frac{7}{10}$ of an acre. How many lots are made?

$$5\frac{1}{4} \div \frac{7}{10} = \frac{21}{4} \div \frac{7}{10} = \frac{\overset{3}{\cancel{21}}}{\underset{2}{\cancel{4}}} \cdot \frac{\overset{5}{\cancel{10}}}{\underset{1}{\cancel{7}}} = \frac{15}{2} = 7\frac{1}{2}$$

Seven lots are made and each is $\frac{7}{10}$ of an acre. The $\frac{1}{2}$ lot is left over or combined with one of the other lots.

EXAMPLE HLTH/N How many Velcro fasteners, each requiring $1\frac{3}{4}$ in. of Velcro, can be made from a roll containing 100 in. of Velcro? See Fig. 2–23.

$1\frac{3}{4}$ in.

$$100 \div 1\frac{3}{4} = \frac{100}{1} \div \frac{7}{4} = \frac{100}{1} \cdot \frac{4}{7} = \frac{400}{7} = 57\frac{1}{7}$$

Figure 2–23

Fifty-seven fasteners can be cut to the desired length. The extra $\frac{1}{7}$ of the desired length is considered waste.

EXAMPLE INDTEC A piece of trophy column stock that is $21\frac{1}{2}$ in. long is cut into four equal trophy columns (Fig. 2–24). If $\frac{1}{16}$ in. is wasted on each cut, find the length of each piece:

Use the Six-Step Problem-Solving Strategy.

Unknown facts	Length of cuts to be made.

Figure 2–24

Known facts	4	Note the number of pieces needed.
	3	Note the number of cuts to be made (4 − 1 = 3).
	$\frac{1}{16}$ in.	Note the waste for each cut.
	$21\frac{1}{2}$ in.	Note the length of trophy column stock.

Relationships Length of each piece = (total length − 3 cuts × amount wasted for each cut) ÷ 4 pieces of stock needed

Estimation If the stock is 20 in. long and cut into 4 pieces and if waste is disregarded, each piece will be 5 in.

Calculation Three cuts are to be made and each cut wastes $\frac{1}{16}$ in. Find the amount wasted.

$$\frac{1}{16} \cdot 3 = \boxed{\frac{3}{16}}$$ Total waste.

$$21\frac{1}{2} - \boxed{\frac{3}{16}}$$ Subtract to find the amount of stock that will be left to divide equally into four trophy columns.

$$21\frac{1}{2} = 21\frac{8}{16}$$ Align vertically and use common denominators.

$$-\frac{3}{16} = \frac{3}{16}$$

$$\overline{}$$

$$21\frac{5}{16} \text{ in.}$$ Amount of stock left to be divided.

Chapter 2 / Review of Fractions

Interpretation Find the length of each trophy column.

$$21\frac{5}{16} \div 4 \; = \; \frac{341}{16} \div \frac{4}{1} \; = \; \frac{341}{16} \times \frac{1}{4} \; = \; \frac{341}{64} \; = \; 5\frac{21}{64}$$

Each trophy column is $5\frac{21}{64}$ in. long.

A **complex fraction** is a fraction in which either the numerator or the denominator or both contain a fraction or a mixed number. Fractions indicate division as we read from top to bottom. The large fraction line is read as "divided by." For example,

$$\frac{2\frac{1}{2}}{7} \quad \text{is read} \quad \text{``} 2\frac{1}{2} \text{ divided by } 7\text{''}$$

$$\frac{4}{1\frac{1}{2}} \quad \text{is read} \quad \text{``} 4 \text{ divided by } 1\frac{1}{2}\text{''}$$

If we think of a fraction as division, then a complex fraction is another way of writing division.

To simplify a complex fraction:

1. Rewrite the fraction with the divided by symbol ÷.
2. Perform the indicated division.

EXAMPLE Simplify $\dfrac{6\frac{3}{8}}{4\frac{1}{2}}$.

$\dfrac{6\frac{3}{8}}{4\frac{1}{2}} =$ Rewrite using the ÷ symbol.

$6\frac{3}{8} \div 4\frac{1}{2} =$ Perform the indicated division. Write mixed numbers as improper fractions.

$\dfrac{51}{8} \div \dfrac{9}{2} =$ Change division to multiplication.

$\dfrac{\overset{17}{\cancel{51}}}{\underset{4}{\cancel{8}}} \cdot \dfrac{\overset{1}{\cancel{2}}}{\underset{3}{\cancel{9}}} =$ Reduce and multiply.

$\dfrac{17}{12} = \mathbf{1\frac{5}{12}}$ Write as a mixed number.

We sometimes use complex fractions to change mixed decimals to fractions.

EXAMPLE Write $0.33\frac{1}{3}$ as a fraction.

$$0.33\frac{1}{3} = \frac{33\frac{1}{3}}{100} = 33\frac{1}{3} \div 100$$

Count places for digits only; that is, do not count the fraction $\frac{1}{3}$ as a place.

$$33\frac{1}{3} \div 100 = \frac{100}{3} \div \frac{100}{1} = \frac{\cancel{100}}{3} \cdot \frac{1}{\cancel{100}} = \frac{1}{3}$$

Change division to multiplication. Reduce and multiply.

What Place Does $\frac{1}{3}$ Hold If the Decimal Is $0.33\frac{1}{3}$?

When changing decimals to fractions, we divide by the place value of the last digit. Is $\frac{1}{3}$ in $0.33\frac{1}{3}$ in the hundredth's or thousandth's place? Hundredth's. The fraction attaches to the last digit.

$0.12\frac{1}{2}$ is read "twelve and one-half hundredths" $0.008\frac{1}{3}$ is read "eight and one-third thousandths"

---

4 **Perform Calculations Involving Fractions with a Calculator.**

Some calculators have a special key for entering fractions. The most common label for a fraction key is $\boxed{a\frac{b}{c}}$. Other calculators have a menu choice that allows the result of a calculation to be displayed in fraction form. Investigate various options and limitations of your calculator or refer to the user's manual if necessary.

To use a calculator fraction key to perform calculations:

To enter a fraction using the $\boxed{a\frac{b}{c}}$ key:

1. Enter the numerator,
2. Press the fraction key.
3. Enter the denominator.

To enter a fraction using division and the fraction display menu choice:

1. Use parentheses.
2. Enter the numerator divided by the denominator. Press $\boxed{\text{ENTER}}$.
3. Display the result in fraction form ($\boxed{\text{MATH}}$, $\boxed{\triangleright\text{Frac}}$, $\boxed{\text{ENTER}}$).

To enter a mixed number using the $\boxed{a\frac{b}{c}}$ key:

1. Enter the whole-number portion of the mixed number.
2. Press the fraction key.
3. Enter the numerator.
4. Press the fraction key.
5. Enter the denominator.

To enter a mixed number using division and the fraction display menu choice:

1. Use parentheses.
2. Enter the whole-number portion of the mixed number followed by the addition operation.
3. Enter the fraction portion of the mixed number as numerator divided by the denominator. Press ENTER.
4. Display the result in fraction form (MATH, ▷Frac, ENTER). The display for a value greater than 1 will be an improper fraction in lowest terms.

EXAMPLE Perform the calculations using a calculator.

(a) $\left(\dfrac{3}{4}\right)\left(\dfrac{5}{6}\right)$

(b) $\dfrac{\dfrac{7}{8} + \dfrac{1}{4}}{3\dfrac{1}{2}}$

(a) $\left(\dfrac{3}{4}\right)\left(\dfrac{5}{6}\right)$

Using the fraction key:

$$3 \boxed{a\frac{b}{c}} 4 \boxed{\times} 5 \boxed{a\frac{b}{c}} 6 \boxed{=} \Rightarrow \dfrac{5}{8}$$

Using the fraction menu choice:

$$\boxed{(}\; 3 \boxed{\div} 4 \boxed{)}\;\; \boxed{(}\; 5 \boxed{\div} 6 \boxed{)}\; \boxed{\text{ENTER}}\; \boxed{\text{MATH}}\; \boxed{\triangleright\text{Frac}}\; \boxed{\text{ENTER}} \Rightarrow \dfrac{5}{8}$$

(b) $\dfrac{\dfrac{7}{8} + \dfrac{1}{4}}{3\dfrac{1}{2}}$

Using the fraction key:

$$\boxed{(}\; 7 \boxed{a\frac{b}{c}} 8 \boxed{+} 1 \boxed{a\frac{b}{c}} 4 \boxed{)}\; \boxed{\div}\; \boxed{(}\; 3 \boxed{a\frac{b}{c}} 1 \boxed{a\frac{b}{c}} 2 \boxed{)}\; \boxed{=} \Rightarrow \dfrac{9}{28}$$

Using the fraction menu choice:

$$\boxed{(}\; 7 \boxed{\div} 8 \boxed{+} 1 \boxed{\div} 4 \boxed{)}\; \boxed{\div}\; \boxed{(}\; 3 \boxed{+} 1 \boxed{\div} 2 \boxed{)}\; \boxed{\text{ENTER}}\; \boxed{\text{MATH}}\; \boxed{\triangleright\text{Frac}}\; \boxed{\text{ENTER}} \Rightarrow \dfrac{9}{28}$$

SECTION 2–4 SELF-STUDY EXERCISES

1 Multiply and reduce answers to lowest terms. Write improper fractions as whole or mixed numbers.

1. $\dfrac{3}{4}\left(\dfrac{1}{8}\right)$

2. $\dfrac{1}{2}\left(\dfrac{7}{16}\right)$

3. $\dfrac{5}{8}\left(\dfrac{7}{10}\right)$

4. $\dfrac{2}{3}\left(\dfrac{7}{8}\right)$

5. $\dfrac{1}{2}\left(\dfrac{3}{4}\right)\left(\dfrac{8}{9}\right)$

6. $\dfrac{3}{8}\left(\dfrac{5}{6}\right)\left(\dfrac{1}{2}\right)$

7. $7\left(3\dfrac{1}{8}\right)$

8. $\dfrac{3}{5}(125)$

9. $2\dfrac{3}{4}\cdot 1\dfrac{1}{2}$

10. $9\dfrac{1}{2}\cdot 3\dfrac{4}{5}$

11. $\dfrac{1}{5}\cdot 7\dfrac{5}{8}$

12. $\dfrac{2}{3}\cdot 3\dfrac{1}{4}$

13. **AUTO** A fuel tank that holds 75 liters (L) of fuel is $\frac{1}{4}$ full. How many liters of fuel are in the tank?

14. **CON** If steps are 12 risers high and each riser is $7\frac{1}{2}$ in. high, what is the total rise of the steps?

15. **INDTEC** An alloy, which is a substance composed of two or more metals, is $\frac{11}{16}$ copper, $\frac{7}{32}$ tin, and $\frac{3}{32}$ zinc. How many kilograms of each metal are needed to make 384 kg of alloy?

odd 1–39 only

2 Raise the fractions to the indicated power.

16. $\left(\dfrac{1}{4}\right)^2$

17. $\left(\dfrac{3}{5}\right)^3$

18. $\left(\dfrac{1}{3}\right)^3$

19. $\left(\dfrac{7}{9}\right)^2$

3 Give the reciprocal.

20. $\dfrac{5}{8}$

21. $2\dfrac{1}{5}$

22. 8

23. 0.9

24. 1.8

Divide and reduce answers to lowest terms. Convert improper fractions to whole or mixed numbers.

25. $\dfrac{1}{2}\div\dfrac{7}{12}$

26. $\dfrac{4}{5}\div\dfrac{8}{9}$

27. $\dfrac{11}{32}\div\dfrac{3}{8}$

28. $\dfrac{3}{4}\div\dfrac{3}{8}$

29. $\dfrac{7}{8}\div\dfrac{3}{16}$

30. $10\div\dfrac{3}{4}$

31. $3\dfrac{1}{8}\div\dfrac{1}{4}$

32. $2\dfrac{1}{2}\div 4$

33. $1\dfrac{1}{7}\div\dfrac{2}{7}$

34. $3\dfrac{3}{4}\div 1\dfrac{1}{2}$

35. **CON** A truck will hold 21 yd^3 (cubic yards) of gravel. If an earth mover has a shovel capacity of $1\frac{3}{4}$ yd^3, how many shovelfuls are needed to fill the truck?

36. **CON** Three shelves of equal length are cut from a 72-in. board. If $\frac{1}{8}$ in. is wasted on each cut, what is the maximum length of each shelf? (Two cuts are made to divide the entire board into three equal lengths.)

37. **CAD/ARC** If $\frac{1}{8}$ in. represents 1 ft on a drawing, find the dimensions of a room that measures $2\frac{1}{2}$ in. by $1\frac{7}{8}$ in. on the drawing. (How many $\frac{1}{8}$s are there in $2\frac{1}{2}$; how many $\frac{1}{8}$s are there in $1\frac{7}{8}$?)

38. **INDTEC** A segment of I-beam is $10\frac{1}{2}$ ft long. Into how many whole $2\frac{1}{4}$-ft pieces can it be divided? Disregard waste.

39. **CON** How many $17\frac{5}{8}$-in. strips of quarter-round molding can be cut from a piece $132\frac{3}{4}$ in. long? Disregard waste.

40. **INDTEC** How many $9\frac{1}{4}$-in. drinking straws can be cut from a $216\frac{1}{2}$-in. length of stock? How much stock is left over?

4 Perform the calculations using a calculator.

41. $5\dfrac{1}{2} + 8\dfrac{7}{15} + 3\dfrac{5}{24}$ **42.** $124\dfrac{8}{35} - 42\dfrac{6}{49}$ **43.** $4\dfrac{5}{12}\left(7\dfrac{7}{15}\right)$ **44.** $1\dfrac{7}{9}\left(3\dfrac{5}{16}\right)$

45. $5\dfrac{5}{6} \div 1\dfrac{1}{14}$ **46.** $8\dfrac{1}{3} \div 2\dfrac{7}{9}$ **47.** $\dfrac{\dfrac{7}{10} + \dfrac{3}{5}}{1\dfrac{11}{15}}$ **48.** $\dfrac{6\dfrac{1}{4}}{\dfrac{5}{8} + \dfrac{7}{8}}$

49. $\dfrac{\dfrac{5}{12} + \dfrac{3}{8}}{\dfrac{4}{19} + \dfrac{1}{38}}$ **50.** $\dfrac{\dfrac{7}{8} - \dfrac{3}{4}}{\dfrac{1}{12} + \dfrac{1}{9}}$

2–5 | The U.S. Customary System of Measurement

Learning Outcomes

1 Convert one unit of measure to another using unit ratios.

2 Convert one unit of measure to another using conversion factors.

3 Add and subtract U.S. customary measures.

4 Multiply and divide U.S. customary measures.

5 Change one U.S. customary rate measure to another.

The **U.S. customary system of measurement** evolved from the English or British system of measurement, and many of the original units in this system are now obsolete. In our discussion of the U.S. customary system of measurement, we will include only selected measuring units.

1 Convert One Unit of Measure to Another Using Unit Ratios.

Length

Four basic units in the U.S. customary system are commonly used to measure length. They are the inch, the foot, the yard, and the mile. Table 2–1 gives the relationships among these measurements of length.

Table 2–1   U.S. Customary Units of Length or Distance

12 inches (in.)[a] = 1 foot (ft)[b]	36 inches (in.) = 1 yard (yd)
3 feet (ft) = 1 yard (yd)	5,280 feet (ft) = 1 mile (mi)

[a]The symbol ″ means inches (8″ = 8 in.) or seconds (60″ = 60 seconds).
[b]The symbol ′ means feet (3′ = 3 ft) or minutes (60′ = 60 minutes).

Weight or Mass

The terms *weight* and *mass* are commonly used interchangeably. In technical, engineering, and scientific applications, the *mass* of an object is the quantity of material that makes up the object. The *weight* is a measure of the earth's gravitational pull on the object. Three commonly used measuring units for weight or mass in the U.S. customary system are the ounce, pound, and ton. Table 2–2 gives the relationships among these measures of weight or mass.

Table 2–2	U.S. Customary Units of Weight or Mass
	16 ounces (oz) = 1 pound (lb)
	2,000 pounds (lb) = 1 ton (T)

Capacity or Volume

The U.S. customary system includes units for both liquid and dry capacity measures; however, the dry capacity measures are seldom used. It is more common to express dry measures in terms of weight than in terms of capacity.

Common U.S. customary units of measure for capacity or volume are the ounce, cup, pint, quart, and gallon. Table 2–3 gives the relationships among the liquid measures for capacity or volume. In the U.S. customary system the term *ounce* represents both weight and liquid capacity. The measures are different and have no common relationship. The context of the problem will suggest whether the unit for weight or capacity is meant.

Table 2–3	U.S. Customary Units of Liquid Capacity or Volume
3 teaspoons (t) = 1 tablespoon (T)	2 tablespoons (T) = 1 ounce (oz)
8 ounces (oz) = 1 cup (c)	4 cups (c) = 1 quart (qt)
2 cups (c) = 1 pint (pt)	4 quarts (qt) = 1 gallon (gal)
2 pints (pt) = 1 quart (qt)	

Using the relationship between two units of measure, we can form a ratio in two different ways that *has a value of 1*. We call this type of ratio a **unit ratio.**

A **ratio** is a fraction. A unit ratio, then, is a fraction with one unit of measure in the numerator and a different, but equivalent, unit of measure in the denominator. Some examples of unit ratios are

$$\frac{12 \text{ in.}}{1 \text{ ft}}, \quad \frac{1 \text{ ft}}{12 \text{ in.}}, \quad \frac{3 \text{ ft}}{1 \text{ yd}}, \quad \frac{1 \text{ mi}}{5,280 \text{ ft}}$$

In each unit ratio, the value of the numerator equals the value of the denominator. A ratio with the numerator and denominator equal has a value of 1. When we convert from one unit of measure to another, we use a unit ratio that contains the original unit and the new unit.

EXAMPLE Write two unit ratios that relate the pair of measures.

(a) ounces and pounds   (b) cups and pints

(a) The relationship between ounces and pounds is 1 lb contains 16 oz. The unit ratios involving ounces and pounds are

$$\frac{1 \text{ lb}}{16 \text{ oz}} \quad \text{and} \quad \frac{16 \text{ oz}}{1 \text{ lb}}$$

(b) The relationship between cups and pints is 1 pint contains 2 cups. The unit ratios involving cups and pints are

$$\frac{1 \text{ pt}}{2 \text{ c}} \quad \text{and} \quad \frac{2 \text{ c}}{1 \text{ pt}}$$

Chapter 2 / Review of Fractions

Unit ratios help us convert from one unit of measure to another.

To change from one U.S. customary unit of measure to another using unit ratios:

1. Set up the original amount as a fraction with the original unit of measure in the numerator.
2. Multiply this fraction by a unit ratio with the original unit in the denominator and the new unit in the numerator.
3. Reduce like units of measure and all numbers wherever possible.

EXAMPLE   Find the number of inches in 5 ft.

Multiply 5 ft by a unit ratio that contains both inches and feet.
Because 5 ft is a whole number, we write it with 1 as the denominator.

$$\frac{5 \text{ ft}}{1} \left(\underline{\hspace{1cm}} \right)$$

Place the original unit with the 5 in the *numerator* of the first fraction.

$$\frac{5 \text{ ft}}{1} \left(\frac{}{\text{ft}} \right)$$

We are changing *from* feet, so we place ft in the *denominator* of the unit ratio, which is shown in parentheses. This allows us to reduce the units later.

$$\frac{5 \text{ ft}}{1} \left(\frac{\text{in.}}{\text{ft}} \right)$$

To change *to* inches, place inches in the *numerator* of the unit ratio.

$$\frac{5 \text{ ft}}{1} \left(\frac{12 \text{ in.}}{1 \text{ ft}} \right) = 60 \text{ in.}$$

Place in the unit ratio the numerical values that make these two units of measure equivalent (1 ft = 12 in.). Complete the calculation, reducing wherever possible.

5 ft = **60 in.**

EXAMPLE   How many pints are in 4.5 quarts?

$$\frac{4.5 \text{ qt}}{1} \left(\frac{2 \text{ pt}}{1 \text{ qt}} \right) = 4.5 \,(2 \text{ pt}) = 9 \text{ pt}$$ From quart (denominator) to pint (numerator).

4.5 qt = **9 pt**

When working with units of measure, it is very important to include the measuring unit in our analysis. Measurements are also referred to as **dimensions,** and the systematic examination of the appropriate measuring units of a solution is referred to as **dimension analysis.**

Sometimes it is necessary to convert a U.S. customary unit to a unit that is *not* the next larger or smaller unit of measure.

TIP   **Changing to Any Larger or Smaller Unit**

To change from a U.S. customary unit to one other than the next larger or smaller unit, proceed as before, but multiply the original amount by as many unit ratios as needed to attain the new U.S. customary unit.

For instance, to change from yards to inches: yards → feet → inches.
To change from gallons to ounces: gallons → quarts → pints → cups → ounces.

EXAMPLE   How many inches are in $2\frac{1}{3}$ yd?

$$2\frac{1}{3} \text{ yd} = \frac{7}{3} \text{ yd}$$

Write $2\frac{1}{3}$ as an improper fraction.

$$\frac{7 \text{ yd}}{3}\left(\frac{\text{ft}}{\text{yd}}\right)\left(\frac{\text{in.}}{\text{ft}}\right)$$

Multiply the improper fraction by two unit ratios. To change from yards to inches, first change from yards to feet, and then from feet to inches. Place the original unit in the *numerator* of the improper fraction, $\frac{7}{3}$.

$$\frac{7 \text{ yd}}{\overset{}{\underset{1}{3}}}\left(\frac{\overset{1}{3} \text{ ft}}{1 \text{ yd}}\right)\left(\frac{12 \text{ in.}}{1 \text{ ft}}\right) = 84 \text{ in.}$$

Insert the appropriate numerical values for each unit ratio (3 ft = 1 yd; 12 in. = 1 ft), and multiply, reducing wherever possible.

$$2\frac{1}{3} \text{ yd} = \textbf{84 in.}$$

Alternative method

If we use the relationship for inches and yards, 36 in. = 1 yd, we need only one unit ratio for the calculation. That is, we convert $2\frac{1}{3}$ yd ($\frac{7}{3}$ yd) to inches as follows:

$$\frac{7 \text{ yd}}{3}\left(\frac{36 \text{ in.}}{1 \text{ yd}}\right)$$

Set up the unit ratio using 36 in. = 1 yd.

$$\frac{7 \text{ yd}}{\underset{1}{3}}\left(\frac{\overset{12}{36 \text{ in.}}}{1 \text{ yd}}\right) = 84 \text{ in.}$$

Reduce units and numbers; then multiply.

$$2\frac{1}{3} \text{ yd} = \textbf{84 in.}$$

TIP

Focus on One Thing at a Time

Sometimes the steps in a multistepped problem can be overwhelming. It is often helpful to focus on one aspect of the problem at a time. In the example changing yards to inches, focus first on just the units of measure or dimensions.

$$\frac{\text{yd}}{1}\left(\frac{\text{ft}}{\text{yd}}\right)\left(\frac{\text{in.}}{\text{ft}}\right)$$

Reduce as appropriate. Yards reduce to 1. Feet reduce to 1. The only measuring unit left is inches. Therefore, the result will be in inches.

Next, focus on the numbers.

$$\frac{7 \text{ yd}}{3}\left(\frac{3 \text{ ft}}{1 \text{ yd}}\right)\left(\frac{12 \text{ in.}}{1 \text{ ft}}\right) \rightarrow \frac{7}{3}\left(\frac{3}{1}\right)\left(\frac{12}{1}\right) = 84$$

Putting the number and unit together, we have 84 in.

---

2 Convert One Unit of Measure to Another Using Conversion Factors.

Unit ratios can be used to develop conversion factors. With conversion factors, you *always* multiply to change from one measuring unit to another.

To develop a conversion factor for converting from one measure to another:

1. Write a unit ratio that changes the given unit to the new unit.
2. Change the fraction (or ratio) to its decimal equivalent by dividing the numerator by the denominator.

EXAMPLE Develop two conversion factors relating pounds and ounces.

Pounds to Ounces

$$\frac{\text{pounds}}{1}\left(\frac{\text{ounces}}{\text{pounds}}\right)$$

$$\frac{\text{pounds}}{1}\left(\frac{16\ \text{ounces}}{1\ \text{pound}}\right)$$

$$\frac{16}{1} = 16$$

pounds × 16 = ounces

Ounces to Pounds

$$\frac{\text{ounces}}{1}\left(\frac{\text{pounds}}{\text{ounces}}\right)$$

$$\frac{\text{ounces}}{1}\left(\frac{1\ \text{pound}}{16\ \text{ounces}}\right)$$

$$\frac{1}{16} \text{ or } 1 \div 16 = 0.0625$$

ounces × 0.0625 = pounds

To change from one U.S. customary unit of measure to another using conversion factors:

1. Select the appropriate conversion factor.
2. Multiply the original measure by the conversion factor.

Table 2–4   U.S. Customary Conversion Factors

	TO CHANGE		
	From	**To**	**Multiply By**
Length or Distance			
12 inches (in.) = 1 foot (ft)	feet	inches	12
	inches	feet	0.0833333
3 feet (ft) = 1 yard (yd)	yards	feet	3
	feet	yards	0.3333333
36 inches (in.) = 1 yard (yd)	yards	inches	36
	inches	yards	0.0277778
5,280 feet (ft) = 1 mile (mi)	miles	feet	5,280
	feet	miles	0.0001894
Weight or Mass			
16 ounces (oz) = 1 pound (lb)	pounds	ounces	16
	ounces	pounds	0.0625
2,000 pounds (lb) = 1 ton (T)	tons	pounds	2,000
	pounds	tons	0.0005
Liquid Capacity or Volume			
8 ounces (oz) = 1 cup (c)	cups	ounces	8
	ounces	cups	0.125
2 cups (c) = 1 pint (pt)	pints	cups	2
	cups	pints	0.5
2 pints (pt) = 1 quart (qt)	quarts	pints	2
	pints	quarts	0.5
4 quarts (qt) = 1 gallon (gal)	gallons	quarts	4
	quarts	gallons	0.25

Additional conversion factors are found on the inside covers of the text.

EXAMPLE   Use a conversion factor from Table 2–4 to convert 56 ounces to pounds.

ounces × 0.0625 = pounds   Conversion factor for ounces to pounds.
56(0.0625) = 3.5 pounds   Multiply.

56 ounces is 3.5 pounds.

Estimation and Dimension Analysis

When estimating unit conversions, first see if the new unit is larger or smaller than the original unit.

Larger to Smaller

Each larger unit can be divided into smaller units. Larger-to-smaller conversions mean *more* smaller units. *More* implies multiplication by a number greater than one.

   To convert a U.S. customary unit to a desired *smaller* unit: Multiply by a conversion factor that is *greater than one.*

2 yd = ____ ft   The smaller unit is feet: 3 ft = 1 yd. Use the conversion factor 3.

2 yd(3 ft) = 6 ft   Multiply number of yards by 3 ft.

2 yd = 6 ft   Dimension analysis: $\dfrac{2\ \text{yd}}{1}\left(\dfrac{3\ \text{ft}}{1\ \text{yd}}\right) = 6\ \text{ft}$

The key word clues are:

larger to smaller unit → obtain more units → multiply by number greater than one

Smaller to Larger

Several small units combine to make one large unit. Thus, smaller-to-larger conversions mean *fewer* large units. *Fewer* implies multiplication by a number less than one.

   To convert a U.S. customary unit to a *larger* unit: *Multiply* by a conversion factor that is *less than one.*

12 ft = ____ yd   The larger unit is yards: 1 yd = 3 ft. Use the conversion factor 0.3333333.

12 ft × 0.3333333 = 4   Divide by 3 ft to get yards.

12 ft = 3.99999996 yd   Dimension analysis: $\dfrac{12\ \text{ft}}{1}\left(\dfrac{1\ \text{yd}}{3\ \text{ft}}\right) = 4\ \text{yd}$

12 ft = 4 yd   Rounded

Using key word clues:

smaller to larger unit → obtain fewer units → multiply by number less than one

Estimation can catch errors in setting up the problem, but it will not likely catch calculation errors.

Measures that use two or more units are called **mixed measures.** A mixed measure is in **standard notation** if the number associated with each unit of measure is smaller than the number required to convert to the next larger unit. The number in the largest unit of measure given may or may not be converted as desired.

To express mixed measures in standard notation:

1. Start with the smallest unit of measure and determine if there are enough units to make one or more of the next larger unit.
2. Regroup to make as many of the larger units as possible.
3. Combine like units.
4. Repeat the process with each given measuring unit.

EXAMPLE Express (a) 8 lb 20 oz and (b) 1 gal 5 qt in standard notation.

(a) 8 lb 20 oz   20 oz = 1 lb 4 oz

      8 lb 20 oz = 8 lb + 1 lb 4 oz   = **9 lb 4 oz**   Standard notation.

(b) 1 gal 5 qt   5 qt = 1 gal 1 qt

      1 gal 5 qt = 1 gal + 1 gal 1 qt   = **2 gal 1 qt**   Standard notation.

EXAMPLE Express 2 yd 4 ft 16 in. in standard notation.

    2 yd 4 ft 16 in. = 2 yd 4 ft + 1 ft 4 in.   Regroup inches.

                                                16 in. = 1 ft 4 in.

                 = 2 yd 5 ft 4 in. = 2 yd + 1 yd 2 ft 4 in.   Regroup feet.

                                                 5 ft = 1 yd 2 ft

                 = **3 yd 2 ft 4 in.**   Standard notation.

Standard Conventions

We can say that 3 ft 15 in. is 4 ft 3 in. in standard notation. Why not change 4 ft 3 in. to 1 yd 1 ft 3 in.? There may be situations when 1 yd 1 ft 3 in. is the desirable form; however, in general, we keep the same units of measure in standard notation as used in the original measure.

---

3 Add and Subtract U.S. Customary Measures.

We can add U.S. customary measures *only* when their units are the same. Measures with the same units are **like measures.** Measures with different units are **unlike measures.**

To add unlike U.S. customary measures:

1. Convert all measures to measures with a common U.S. customary unit.
2. Add. The unit of measure of the sum is the common unit.

EXAMPLE Add 3 ft + 2 in.

Because 2 in. is a fraction of a foot, we can avoid working with fractions by converting the larger unit (feet) to the smaller unit (inches).

$$3 \text{ ft} = \frac{3 \text{ ft}}{1}\left(\frac{12 \text{ in.}}{1 \text{ ft}}\right) = 36 \text{ in.}$$ Convert ft to in., and add like measures.

36 in. + 2 in. = **38 in.**

add mixed U.S. customary measures:

1. Align the measures vertically so the common units are written in the same vertical column.
2. Add common units of measure.
3. Express the sum in standard notation.

EXAMPLE   Add and write the answer in standard form: 6 lb 7 oz and 3 lb 13 oz.

$$
\begin{array}{r}
6 \text{ lb } \ \ 7 \text{ oz} \\
+ \ 3 \text{ lb } 13 \text{ oz} \\
\hline
9 \text{ lb } 20 \text{ oz}
\end{array}
$$

Write in standard notation, 20 oz = 1 lb 4 oz.

Thus, 9 lb 20 oz = 10 lb 4 oz.

To subtract unlike U.S. customary units:

1. Convert both measures to measures with a common U.S. customary unit.
2. Subtract. The unit of measure of the difference is the common unit.

EXAMPLE   Subtract 15 in. from 2 ft.

Changing 15 in. to feet gives us a mixed number, so it is more convenient to convert 2 ft to inches.

2 ft = 24 in.   Convert ft to in. and subtract.
24 in. − 15 in. = **9 in.**

To subtract mixed U.S. customary measures:

1. Align the measures vertically so that the common units are written in the same vertical column.
2. Subtract common units of measure beginning with the smallest unit of measure.

EXAMPLE   Subtract 5 ft 3 in. from 7 ft 4 in.

$$
\begin{array}{r}
7 \text{ ft } 4 \text{ in.} \\
- \ 5 \text{ ft } 3 \text{ in.} \\
\hline
2 \text{ ft } 1 \text{ in.}
\end{array}
$$

Align like measures in a vertical line, and then subtract.

2 ft 1 in.

When we subtract mixed measures, we use our knowledge of regrouping to subtra[ct]
a larger unit from a smaller one.

To subtract a larger U.S. customary unit from a smaller unit in mixed measures:

1. Align the common measures in vertical columns.
2. Regroup by subtracting one unit from the next larger unit of measure in the minuend, convert to the equivalent smaller unit, and add it to the smaller unit.
3. Subtract common measures beginning with the smallest unit of measure.

EXAMPLE   Subtract 3 lb 12 oz from 7 lb 8 oz.

$$\begin{array}{r} 7 \text{ lb } 8 \text{ oz} \\ -3 \text{ lb } 12 \text{ oz} \\ \hline \end{array}$$

We always begin subtraction with the smallest unit, which should be on the *right*.
12 oz cannot be subtracted from 8 oz.

7 lb 8 oz = 6 lb 16 oz + 8 oz =   6 lb 24 oz   Rewrite 7 lb as 6 lb 16 oz. Add
3 lb 12 oz =   −3 lb 12 oz   16 oz to 8 oz to get 24 oz.
                                        3 lb 12 oz

---

4 Multiply and Divide U.S. Customary Measures.

Suppose a tank of weed killer contains 21 gal 3 qt. What is the total amount of weed killer in eight tanks? To find the total amount of weed killer, we multiply 21 gal 3 qt by 8.

To multiply a U.S. customary measure by a number:

1. Multiply the numbers associated with each unit of measure by the given number.
2. Write the resulting measure in standard notation.

EXAMPLE   A tank holds 21 gal 3 qt of weed killer. How much do eight containers hold?
AG/H

$$\begin{array}{r} 21 \text{ gal } 3 \text{ qt} \\ \times \phantom{21 \text{ gal } 3} 8 \\ \hline 168 \text{ gal } 24 \text{ qt} \end{array}$$ Multiply each unit of measure by 8.

Write in standard notation, 24 qt = 6 gal.

168 gal 24 qt = 168 gal + 6 gal = 174 gal in standard notation

The eight containers hold 174 gal.

---

1. Multiply the numbers associated with each like unit of measure.
2. The unit of measure of the product is a square measure or an area.

EXAMPLE A desktop is 2 ft × 3 ft (Fig. 2–25). What is the number of square feet in the surface?

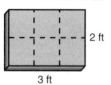
2 ft
3 ft

Figure 2–25

2 ft (3 ft) = **6 ft^2**   Multiply numbers. Product is ft^2.

For a more detailed discussion of area see Chapter 1, Section 3, Outcome 2.
We are frequently required to divide measures by a number.

To divide a U.S. customary measure by a number that divides evenly into each measure:

1. Divide the numbers associated with each unit of measure by the given number.
2. Write the resulting measure in standard notation.

EXAMPLE How much milk is needed for a half-recipe if the original recipe calls for 2 gal 2 qt?

$2\overline{)2\text{ gal }2\text{ qt}}$ Divide each measure by 2.

$$\frac{1\text{ gal }1\text{ qt}}{2\overline{)2\text{ gal }2\text{ qt}}}$$

The half-recipe requires 1 gal 1 qt of milk.

If a given number does not divide evenly into each measure, there will be a remainder.

To divide U.S. customary measures by a number that does not divide evenly into each measure:

1. Set up the problem and proceed as in long division.
2. When a remainder occurs after subtraction, convert the remainder to the same unit used in the next smaller measure, and add it to the quantity in the next smaller measure.
3. Divide the given number into this next smaller unit.
4. If a remainder occurs when the smallest unit is divided, express the remainder as a fractional part of the smallest unit.

EXAMPLE Divide 5 gal 3 qt 1 pt by 3.

$$
\begin{array}{c}
\phantom{3\overline{)}}\ 1\ \text{gal}\quad\ 3\ \text{qt}\quad 1\tfrac{2}{3}\,\text{pt}\\
3\overline{)\ 5\ \text{gal}\quad\ 3\ \text{qt}\quad 1\ \text{pt}}
\end{array}
$$

1 gal 3 qt 1⅔ pt	
3)5 gal 3 qt 1 pt	5 gal ÷ 3 = 1 gal, remainder 2 gal
3 gal	2 gal = 8 qt
2 gal = 8 qt	3 qt + 8 qt = 11 qt
11 qt	11 qt ÷ 3 = 3 qt, remainder 2 qt
9 qt	2 qt = 4 pt
2 qt = 4 pt	1 pt + 4 pt = 5 pt
5 pt	5 pt ÷ 3 = 1 pt, remainder 2 pt
3 pt	
2 pt	Write the final remainder, 2, as a fraction $\tfrac{2}{3}$ pints, and add to 1 pint to get $1\tfrac{2}{3}$ pints.

Thus, 1 gal 3 qt $1\tfrac{2}{3}$ pt is the solution.

If tubing is manufactured in lengths of 8 ft 4 in. and a part is 10 in. long, how many parts can be cut from the length of tubing if we do not account for waste? To solve such a problem, we express both measures in the same unit, just as in adding and subtracting measures. We generally convert to the *smallest* unit used in the example.

To divide a U.S. customary measure by a U.S. customary measure:

1. Convert both measures to the same unit if they are different.
2. Write the division as a fraction, including the common unit in the numerator and the denominator.
3. Reduce the units and divide the numbers.

EXAMPLE If tubing is manufactured in lengths of 8 ft 4 in. and a part is 10 in. long, how many parts can be cut from the length of tubing if we do not account for waste? Divide 8 ft 4 in. by 10 in.

$$8\ \text{ft} = \frac{8\ \text{ft}}{1}\left(\frac{12\ \text{in.}}{1\ \text{ft}}\right) = 96\ \text{in.}\qquad \text{Convert 8 ft to inches. Then add to 4 in.}$$

$$8\ \text{ft}\ 4\ \text{in.} = 96\ \text{in.} + 4\ \text{in.} = 100\ \text{in.}\qquad \text{Write the mixed measure as a measure with one measuring unit.}$$

$$100\ \text{in.} \div 10\ \text{in.}\qquad \text{Divide the total length by the length of each part.}$$

If we write this division in fraction form, we can see more easily that the common units reduce. In other words, the answer will be a number (not a measure) telling *how many equal parts* can be cut from the tubing.

$$\frac{100\ \text{in.}}{10\ \text{in.}} = 10$$

Therefore, 10 parts of equal length can be cut from the tubing.

5 Change One U.S. Customary Rate Measure to Another.

A **rate measure** is a ratio of two different kinds of measures. A rate measure is often referred to as a **rate**. Some examples of rates are 55 miles per hour and 20 cents per mile. In each of these rates, the word **per** means *divided by*.

The rate 55 miles per hour means 55 miles ÷ 1 hour or $\frac{55\text{ mi}}{1\text{ h}}$. The rate 20 cents per mile means 20 cents ÷ 1 mile or $\frac{20\text{ cents}}{1\text{ mi}}$.

Many rate measures involve measures of time. The units we use to measure time are universally accepted. The basic units of time are the year, month, week, day, hour, minute, and second. Table 2–5 gives the relationships among the units for time. These units are often used when working with rates.

Table 2–5   Units of Time

1 year (yr) = 12 months (mo)	1 minute (min) = 60 seconds (s)[c]
1 year (yr) = 365 days (da)	1 millisecond (ms) = $\dfrac{1}{1,000}$ s
1 week (wk) = 7 days (da)	
1 day (da) = 24 hours (h)[a]	
1 hour (h) = 60 minutes (min)[b]	1 nanosecond (ns) = $\dfrac{1}{1,000,000,000}$ s

[a] The abbreviation for hour can also be hr.
[b] The symbol ′ means feet (3′ = 3 ft) or minutes (60′ = 60 minutes).
[c] The symbol ″ means inches (8″ = 8 in.) or seconds (60″ = 60 seconds). The abbreviation for second can also be sec.

To convert one U.S. customary rate measure to another:

1. Compare the units of both numerators and both denominators to determine which units will change.
2. Multiply each unit that changes by a unit ratio containing the new unit so that the unit to be changed will reduce.

EXAMPLE   Change $8\dfrac{\text{pt}}{\text{min}}$ to $\dfrac{\text{qt}}{\text{min}}$.

Estimation   Pints to quarts is *smaller* to *larger,* so there will be fewer quarts.

Examine the rates:

 numerators—pints change to quarts
 denominators—no change

$$\frac{\text{pt}}{\text{min}}\left(\frac{\text{qt}}{\text{pt}}\right)=\frac{\text{qt}}{\text{min}}$$ Develop a unit ratio with pints in the denominator.

$$\frac{\overset{4}{8\ \text{pt}}}{\text{min}}\left(\frac{1\ \text{qt}}{\underset{1}{2\ \text{pt}}}\right)=\frac{4\ \text{qt}}{\text{min}}$$ Insert numbers in the unit ratio and reduce. Multiply.

Interpretation   Thus, $8\dfrac{\text{pt}}{\text{min}}$ equals $4\dfrac{\text{qt}}{\text{min}}$.

A separate unit ratio is used for each unit in the rate that changes. For example, when both the numerator and the denominator of a rate measure change, we multiply by at least two unit ratios to make the conversion.

EXAMPLE Change 60 miles per hour to feet per second.

$$60 \text{ miles per hour} = 60 \frac{\text{mi}}{\text{h}}$$
Write rate as a fraction.

$$\text{feet per second} = \frac{\text{ft}}{\text{s}}$$

Numerators—miles change to feet
Denominators—hours change to seconds
Examine the changes in the measures.

$$60 \frac{\text{mi}}{\text{h}} \underbrace{\left(\frac{5,280 \text{ ft}}{1 \text{ mi}} \right)}_{\substack{\text{(miles} \\ \text{to} \\ \text{feet)}}} \underbrace{\left(\frac{1 \text{ h}}{60 \text{ min}} \right)}_{\substack{\text{(hours} \\ \text{to} \\ \text{minutes)}}} \underbrace{\left(\frac{1 \text{ min}}{60 \text{ s}} \right)}_{\substack{\text{(minutes} \\ \text{to} \\ \text{seconds)}}}$$

Develop appropriate unit ratios and reduce.

$$\frac{5,280 \text{ ft}}{60 \text{ s}} = 88 \frac{\text{ft}}{\text{s}}$$
Divide.

At 60 mi/h you are traveling 88 ft/s.

SECTION 2–5 SELF-STUDY EXERCISES

1 Write two unit ratios that relate the *given* pair of measures.

1. pints and quarts   **2.** feet and miles   **3.** inches and feet   **4.** feet and yards

Use unit ratios to convert the units.

5. 4 ft = _____ in.
6. 7 yd = _____ ft
7. $2\frac{1}{2}$ mi = _____ yd
8. Find the number of yards in 28 ft.
9. Find the number of pounds in $36\frac{4}{5}$ oz.
10. How many ounces are in 45.8 lb?
11. How many quarts are in 5 gal?
12. How many pints are in $6\frac{1}{2}$ qt?

2 Develop two conversion factors for the pairs of units.

13. feet and yards   **14.** quarts and gallons

Use conversion factors to convert the units of measure.

15. 460 oz is equivalent to how many pints?
16. How many pints are in 46 qt?
17. How many pounds are in 580 oz?
18. If a cabinet is $12\frac{3}{4}$ ft long, how many inches is this?
19. How many feet are in 1.5 mi?
20. How many quarts are in 7 gal?

Express the measures in standard notation.

21. 2 ft 20 in.
22. 1 mi 6,375 ft
23. 2 lb $19\frac{1}{2}$ oz
24. 1 gal 5 qt
25. 2 ft 10 in.
26. 5 lb 25 oz
27. 3 gal 5 qt 48 oz
28. 6 qt 20 oz

3 Add and write the answer in standard notation.

29. 5 oz + 2 lb

30. 4 ft + 7 in.

31. 8 lb 2 oz
$+$ 7 lb 9 oz

32. 5 ft 45 in.
$+$ 7 ft 30 in.

33. 5 qt 1 pt
$+$ 2 qt $1\frac{1}{2}$ pt

34. 8 gal 3 qt
$+$ 5 gal 2 qt

35. 7 yd 2 ft
$+$ 1 yd 2 ft

36. 4 yd 2 ft 7 in.
$+$ 2 yd 1 ft 10 in.

Solve the problems. When necessary, express the answers in standard notation.

37. CON   A plumber has a 2-ft length of copper tubing and a 7-in. length of copper tubing. What is the total in inches?

38. COMP   How many feet are two computer cables together if one is 6 ft and the other is 60 in.?

39. HOSP   A mixture for hamburgers contains 2 lb 8 oz of ground round steak and 3 lb 7 oz of regular ground beef. How much does the hamburger mixture weigh?

40. HLTH/N   According to hospital maternity records, one infant twin weighed 6 lb 1 oz and the other weighed 5 lb 15 oz at birth. What was their total weight?

Subtract and write the answer in standard form.

41. 2 ft − 18 in.

42. 2 qt − 3 pt

43. 3 yd − 7 ft

44. 6 lb − 18 oz

45. 3 lb 12 oz
$-$ 2 lb 6 oz

46. 12 lb 7 oz
$-$ 5 lb 12 oz

47. 2 ft 30 in.
$-$ 1 ft 40 in.

48. 5 gal 3 qt 1 pt
$-$ 1 gal 3 qt $1\frac{1}{2}$ pt

49. AUTO   A mechanic has a length of hose 5 ft long. What is its length after 10 in. is cut off?

50. CON   A cabinetmaker cut 9 in. from a 3-ft shelf in a medical lab. How long was the shelf after it was cut?

51. COMP   If a computer sorts a list of names in 1 min 30 s and a faster computer does the same job in 45 s, how much time is saved by using the faster computer?

52. AUTO   A car with a 2.0-L engine accelerates a certain distance in 1 min 27 s. A car with a 5.0-L engine accelerates the same distance in 48 s. How much faster does the car with the 5.0-L engine accelerate?

53. COMP   A package containing a laser printer weighs 74 lb 3 oz. The container and packing material weigh 4 lb 12 oz. How much does the laser printer weigh?

54. AG/H   To weigh a baby pig, Stacey held the pig while standing on a scale. If the scale showed 132 lb 6 oz and Stacey weighs 115 lb 8 oz, how much does the pig weigh?

4 Multiply and write the answers for mixed measures in standard notation.

55. 12 mi
$\times$ 5

56. 18 gal
$\times$ 6

57. 7 lb 3 oz
$\times$ 8

58. 7 ft 3 in.
$\times$ 8

59. HOSP   Tuna is packed in 1-lb 8-oz cans. If a case contains 24 cans, how much does a case weigh?

60. AUTO   A car used 1 qt 1 pt of oil each month for 5 months. Find the total amount of oil used.

Multiply.

61. 5 in. × 7 in.

62. 12 ft × 9 ft

63. 15 yd × 12 yd

64. 4 mi × 27 mi

65. **CON** A room is to be covered with square linoleum tiles that are 1 ft by 1 ft. If the room is 18 ft by 21 ft, how many tiles (square feet) are needed?

66. **AG/H** A horticulturist stores a stock solution of fertilizer in two tanks, each with a capacity of 23 gal 9 oz. How much liquid fertilizer is needed to fill both tanks?

67. **ELEC** Latonya has three containers, each containing 1 qt 3 pt of photographic solution. How much total photographic solution is in all three containers?

68. **AG/H** A package of grass seed weighs 1 lb 4 oz. How much would five packages weigh?

Divide.

69. 3 days 6 h ÷ 2

70. 20 yd 2 ft 6 in. ÷ 2

71. 4 yd 1 ft 9 in. ÷ 3

72. **HELPP** Sixty feet of crime-scene tape are required to complete eight jobs. If each job requires an equal amount of tape, find the amount of tape required for one job. (Express your answer in feet and inches.)

73. **AG/H** A vat holding 10 gal 2 qt of defoliant is emptied equally into three tanks. How many gallons and quarts are in each tank?

74. **CON** How many pieces of $\frac{1}{2}$-in. OD (outside diameter) plastic pipe 8 in. long can be cut from a piece 72 in. long?

75. **ELEC** A roll of No. 14 electrical cable 150 ft long is divided into 30 equal sections. How long is each section?

76. **AG/H** A greenhouse attendant has a container with 6 gal 2 qt 10 oz of potassium nitrate solution that will be divided equally into two smaller containers. How much solution will be stored in each smaller container?

77. **HOSP** For a catering order, Mr. Sonnier prepared 96 lb 12 oz of boiled crawfish. He brought the crawfish to the site in four containers containing equal amounts. How much did the crawfish in each container weigh?

Divide.

78. 36 ft ÷ 12 ft

79. 6 lb 12 oz ÷ 6 oz

80. 2 ft 6 in. ÷ 10 in.

81. **INDTR** How many 6-in. pieces can be cut from 48 in. of pipe?

82. **HOSP** How many 2-lb boxes can be filled from 18 lb of candy?

83. **HOSP** How many 15-oz cans are in a case if the case weighs 22 lb 8 oz?

84. **BUS** How many $8 tickets can be purchased for $72?

5 Change to the indicated rate measure.

85. $\dfrac{45 \text{ lb}}{\text{h}} = \underline{\quad} \dfrac{\text{lb}}{\text{min}}$

86. $\dfrac{3 \text{ mi}}{\text{h}} = \underline{\quad} \dfrac{\text{ft}}{\text{h}}$

87. $\dfrac{144 \text{ lb}}{\text{min}} = \underline{\quad} \dfrac{\text{oz}}{\text{min}}$

88. $\dfrac{30 \text{ gal}}{\text{min}} = \underline{\quad} \dfrac{\text{qt}}{\text{s}}$

89. **INDTEC** A pump that can pump 45 $\frac{\text{gal}}{\text{h}}$ can pump how many quarts per minute?

90. **INDTEC** A pump can dispose of sludge at the rate of 3,200 $\frac{\text{lb}}{\text{h}}$. How many pounds can be disposed of per minute? Round to the nearest tenth.

Learning Outcomes	What to Remember with Examples

Section 2–1

The *numerator* (top number) of a fraction is the number of parts of the whole amount we are considering. The *denominator* (bottom number) of a fraction is the number of parts a whole amount has been divided into. *Proper fractions* are less than 1. *Improper fractions* are equal to or larger than 1. A *mixed number* consists of a whole number and a common fraction written together and indicates addition of the whole number and fraction. A *decimal fraction* is a fraction whose denominator is 10 or a power of 10 and it can be written in decimal notation with the place value representing the denominator of the fraction.

1 Find multiples of a natural number (pp. 61–63).

To find a multiple of a natural number, multiply the number by any natural number.

> Find the first five multiples of 7:
>
> $$7 \times 1 = 7, \quad 7 \times 2 = 14, \quad 7 \times 3 = 21, \quad 7 \times 4 = 28, \quad 7 \times 5 = 35$$
>
> The first five multiples of 7 are 7, 14, 21, 28, and 35.

2 Find all factor pairs of a natural number (pp. 63–64).

1. Write the factor pair of 1 and the number.   **2.** Check to see if the number is divisible by 2. If so, write the factor pair of 2 and the quotient of the beginning natural number and 2.   **3.** Check each natural number in turn for divisibility; and, if divisible by the natural number, write the factor pair.   **4.** Continue Step 3 until you reach a number that has already been found as a quotient in a previous factor pair.

> Find all the factors of 24: 1×24, 2×12, 3×8, 4×6; 5 is not a factor, 6×4 is a repeat. Factors are 1, 2, 3, 4, 6, 8, 12, 24.

3 Determine the prime factorization of composite numbers (pp. 64–67).

1. Test each prime number to see if the composite number is divisible by the prime number. **2.** Make a factor pair using the first prime number that passes the test in Step 1. **3.** Carry forward the prime factors and test the remaining factors by repeating Steps 1 and 2.

> Find the prime factorization of 28:
>
> $$28 = 2 \times 14$$
> $$= 2 \times 2 \times 7 \text{ or } 2^2 \times 7$$

4 Find the least common multiple and greatest common factor of two or more numbers (pp. 67–68).

To find the least common multiple (LCM): **1.** List the prime factorization of each number using exponential notation.   **2.** List the factors of the LCM by including each prime factor appearing in any of the numbers the greatest number of times that it appears in any factor.   **3.** Write the result in standard notation.

> Find the LCM for 12, 15, and 30:
>
> $$12 = 2^2 \times 3 \qquad 15 = 3 \times 5 \qquad 30 = 2 \times 3 \times 5$$
> $$\text{LCM} = 2^2 \times 3 \times 5 \text{ or } 60$$

To find the greatest common factor (GCF): **1.** List the prime factorization of each number in exponential notation.   **2.** The GCF includes factors common to each number.   **3.** Write the result in standard notation.

> Find the GCF of 12, 15, 30:
>
> $$12 = 2^2 \times 3 \qquad 15 = 3 \times 5 \qquad 30 = 2 \times 3 \times 5$$
> $$\text{GCF} = 3$$

Section 2–2

1 Write equivalent fractions with different denominators. (pp. 70–74)

To change a fraction to an equivalent fraction with a larger denominator: **1.** Divide the larger denominator by the original denominator. **2.** Multiply the original numerator and denominator by the quotient found in Step 1.

> Change $\frac{5}{9}$ to an equivalent fraction that has a denominator of 36.
>
> $$\frac{5}{9}\left(\frac{4}{4}\right) = \frac{20}{36} \qquad 36 \div 9 = 4$$

To reduce a fraction to lowest terms: Divide both the numerator and the denominator by the greatest common factor (GCF).

> Reduce $\frac{12}{16}$.
>
> $$\frac{12}{16} = \frac{12}{16} \div \frac{4}{4} = \frac{3}{4} \qquad \text{GCF is 4.}$$

To read the U.S. customary rule: Align the rule along the object (Fig. 2–26). Count the number of whole and fractional inches ($\frac{1}{16}$'s, $\frac{1}{8}$'s, $\frac{1}{4}$'s, and so on) to determine the approximate length. Use eye judgment to estimate closeness of the object to a mark on the rule.

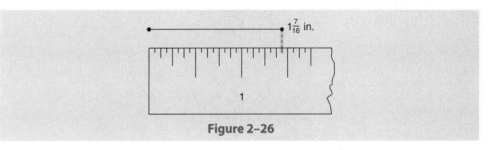

$1\frac{7}{16}$ in.

1

Figure 2–26

2 Write improper fractions as whole numbers or mixed numbers (pp. 74–75).

To convert an improper fraction to a whole or mixed number: **1.** Divide the numerator by the denominator. **2.** Write any remainder as a fraction with the original denominator as its denominator.

> Convert $\frac{18}{6}$ and $\frac{15}{4}$ to whole or mixed numbers.
>
> $$\frac{18}{6} = 3 \qquad \frac{15}{4} = 3\frac{3}{4}$$

3 Write whole numbers or mixed numbers as improper fractions (p. 75).

To convert a whole number to an improper fraction: **1.** Write the whole number as the numerator. **2.** Write 1 as the denominator.

To convert a mixed number to an improper fraction: **1.** Multiply the whole number by the denominator. **2.** Add the numerator to the result of Step 1. **3.** Place the sum from Step 2 over the original denominator.

> Change 7 and $4\frac{7}{8}$ to improper fractions.
>
> $$7 = \frac{7}{1} \qquad 4\frac{7}{8} = \frac{(8 \cdot 4) + 7}{8} = \frac{39}{8}$$

4 Write decimals as fractions and fractions as decimals (pp. 76–78).

To change a decimal to a fraction (or mixed number) in lowest terms: **1.** Write the number without the decimal as the numerator. **2.** Write the denominator as a power of 10 with the same number of zeros as the decimal has decimal places. **3.** Reduce or convert to a mixed number.

Chapter Review of Key Concepts

Write 0.23 as a fraction:

$$\frac{23}{100}$$ Two decimal digits so the denominator is 100.

To change a fraction to a decimal: Divide the numerator by the denominator by placing a decimal after the last digit of the numerator and attaching zeros as needed.

Change $\frac{5}{8}$ to a decimal:

$$
\begin{array}{r}
0.625 \\
8\overline{)5.000} \quad \text{Divide by the denominator.} \\
\underline{4\,8} \\
20 \\
\underline{16} \\
40 \\
\underline{40}
\end{array}
$$

To find the least common denominator (LCD), use the process for finding the least common multiple: **1.** List the prime factorization of each number using exponential notation. **2.** List the factors of the LCM by including each prime factor appearing in any of the numbers the greatest number of times that it appears in any factor. **3.** Write the result in standard notation.

Find the least common denominator for $\frac{7}{18}$ and $\frac{5}{24}$.

$$18 = 2 \cdot 3 \cdot 3 = 2 \cdot 3^2$$
$$24 = 2 \cdot 2 \cdot 2 \cdot 3 = 2^3 \cdot 3$$
$$\text{LCM} = \text{LCD} = 2^3 \cdot 3^2 = 2 \cdot 2 \cdot 2 \cdot 3 \cdot 3 = 72$$

The smallest number that can be divided evenly by both 18 and 24 is 72.

5 Compare fractions, mixed numbers, and decimals (pp. 78–79).

To compare fractions: **1.** Write the fractions as equivalent fractions with common denominators. **2.** Compare the numerators. The larger numerator indicates the larger fraction.

To compare mixed numbers: **1.** Compare the whole-number parts, if different. **2.** If the whole-number parts are equal, write the fractions with common denominators. **3.** Compare the numerators.

Which fraction is smaller, $\frac{2}{5}$ or $\frac{5}{12}$?

$$\frac{2}{5} = \frac{24}{60} \qquad \text{LCD} = 60$$

$$\frac{5}{12} = \frac{25}{60} \qquad \text{Since } \tfrac{24}{60} \text{ is smaller than } \tfrac{25}{60}, \tfrac{2}{5} \text{ is smaller than } \tfrac{5}{12}.$$

Section 2–3

1 Add fractions and mixed numbers (pp. 82–84).

To add fractions: **1.** Find the common denominator. **2.** Change each fraction to an equivalent fraction with the common denominator. **3.** Add the numerators and place the sum over the common denominator. **4.** Reduce the sum if possible. **5.** Convert improper fractions to whole or mixed numbers if desired.

Add:

$$\frac{1}{7} + \frac{3}{7} + \frac{2}{7} = \frac{6}{7} \qquad \frac{5}{8} + \frac{3}{4} = \frac{5}{8} + \frac{6}{8} = \frac{11}{8} = 1\frac{3}{8}$$

Chapter 2 / Review of Fractions

To add mixed numbers: **1.** Change each fraction to an equivalent fraction with the LCD. **2.** Place the sum of the numerators over the LCD. **3.** Add the whole-number parts. **4.** Write the improper fraction from Step 2 as a whole or mixed number; add the result to the whole number from Step 3. **5.** Reduce if necessary.

Add $4\frac{3}{4} + 5\frac{2}{8} + 1\frac{1}{2}$.

$$4\frac{3}{4} = 4\frac{6}{8}$$

Change each fraction to an equivalent fraction with the LCD.

$$5\frac{2}{8} = 5\frac{2}{8}$$

$$1\frac{1}{2} = 1\frac{4}{8}$$

Add fractions.
Add whole numbers.

$$10\frac{12}{8}$$

Simplify. $\frac{12}{8} = 1\frac{4}{8}$

$$10 + 1\frac{4}{8} = 11\frac{4}{8} = 11\frac{1}{2}$$

2 Subtract fractions and mixed numbers (pp. 84–87).

To subtract fractions: **1.** Change each fraction to an equivalent fraction that has the LCD as its denominator. **2.** Subtract the numerators. **3.** Place the difference over the LCD. **4.** Reduce if possible.

Subtract $\frac{5}{8} - \frac{7}{16}$.

$$\frac{5}{8} = \frac{10}{16}$$

Convert each fraction to an equivalent fraction with a common denominator.

$$-\frac{7}{16} = \frac{7}{16}$$

Subtract the numerators.

$$\frac{3}{16}$$

To subtract mixed numbers: **1.** Write fractions as equivalent fractions with common denominators. **2.** Regroup if the fraction in the minuend is smaller than the fraction in the subtrahend. **3.** Subtract the fractions. **4.** Subtract the whole numbers. **5.** Reduce if necessary.

Subtract $5\frac{3}{8} - 3\frac{9}{16}$.

$$5\frac{3}{8} = 5\frac{6}{16} = 4\frac{22}{16}$$

Convert fraction to 16ths. Regroup.

$$-3\frac{9}{16} = 3\frac{9}{16} = 3\frac{9}{16}$$

Subtract fractions. Subtract whole numbers.

$$1\frac{13}{16}$$

Section 2–4

1 Multiply fractions and mixed numbers (pp. 89–92).

To multiply fractions: **1.** Reduce any numerator and denominator that have a common factor. **2.** Multiply the numerators to get the numerator of the product. **3.** Multiply the denominators to get the denominator of the product. **4.** Be sure the product is reduced.

Multiply $\frac{4}{5}\left(\frac{7}{10}\right)\left(\frac{15}{35}\right)$.

$$\frac{\overset{2}{\cancel{4}}}{\underset{1}{\cancel{5}}}\left(\frac{\overset{1}{\cancel{7}}}{\underset{5}{\cancel{10}}}\right)\left(\frac{\overset{3}{\cancel{15}}}{\underset{5}{\cancel{35}}}\right) = \frac{6}{25} \qquad \text{Reduce and multiply.}$$

To multiply fractions, whole numbers, and mixed numbers: **1.** Write whole numbers as fractions with denominators of 1. **2.** Write mixed numbers as improper fractions. **3.** Reduce as much as possible. **4.** Multiply the numerators to get the numerator of the product. **5.** Multiply the denominators to get the denominator of the product. **6.** Write the product as a whole number, mixed number, or fraction in lowest terms.

Multiply $4\left(3\frac{1}{5}\right)\left(\frac{2}{7}\right)$.

$$4\left(3\frac{1}{5}\right)\left(\frac{2}{7}\right) = \frac{4}{1}\left(\frac{16}{5}\right)\frac{2}{7} = \frac{128}{35} = 3\frac{23}{35}$$

2 Raise a fraction to a power (p. 93).

To raise a fraction or quotient to a power: **1.** Raise the numerator to the power. **2.** Raise the denominator to the power.

Raise $\left(\frac{3}{5}\right)^3$ to the indicated power.

$$\left(\frac{3}{5}\right)^3 = \frac{3^3}{5^3} = \frac{3 \cdot 3 \cdot 3}{5 \cdot 5 \cdot 5} = \frac{27}{125}$$

3 Divide fractions and mixed numbers (pp. 93–98).

To write the reciprocal of a number: **1.** Express the number as a fraction. **2.** Interchange the numerator and denominator.

Find the reciprocal of $\frac{3}{5}$, 6, $2\frac{3}{4}$, and 0.2.
The reciprocal of $\frac{3}{5}$ is $\frac{5}{3}$; the reciprocal of 6 is $\frac{1}{6}$; the reciprocal of $2\frac{3}{4}$ or $\frac{11}{4}$ is $\frac{4}{11}$.

The reciprocal of 0.2 or $\dfrac{2}{10}$ is $\dfrac{10}{2}$ or 5.

To divide fractions: **1.** Replace the divisor with its reciprocal. **2.** Change the division to multiplication. **3.** Multiply.

Divide $\frac{4}{5} \div \frac{8}{9}$.

$$\frac{4}{5} \div \frac{8}{9} = \frac{\overset{1}{\cancel{4}}}{5}\left(\frac{9}{\underset{2}{\cancel{8}}}\right) = \frac{9}{10} \qquad \text{Change to an equivalent multiplication and multiply.}$$

To divide mixed numbers: **1.** Write each mixed number as an improper fraction. **2.** Change the division to an equivalent multiplication by using the reciprocal of the divisor. **3.** Reduce if possible. **4.** Multiply.

Divide $4\frac{2}{3} \div 1\frac{1}{6}$.

$$4\frac{2}{3} \div 1\frac{1}{6} = \frac{14}{3} \div \frac{7}{6} = \frac{\overset{2}{\cancel{14}}}{\underset{1}{\cancel{3}}} \times \frac{\overset{2}{\cancel{6}}}{\underset{1}{\cancel{7}}} = \frac{4}{1} = 4$$

4 Perform calculations involving fractions with a calculator (pp. 98–99).

The most common label for a fraction key is $\boxed{a\frac{b}{c}}$. Other calculators have a menu choice ($\boxed{\text{MATH}}$, $\boxed{\triangleright\text{Frac}}$) that will allow the result of a calculation to be displayed in fraction form. Investigate options and limitations of your calculator.

Add $\dfrac{3}{4} + \dfrac{5}{6}$.

Using the fraction key:

$3\ \boxed{a\frac{b}{c}}\ 4\ \boxed{+}\ 5\ \boxed{a\frac{b}{c}}\ 6\ \boxed{=}\ \Rightarrow\ 1\dfrac{7}{12}$

Using the fraction menu choice:

$\boxed{(}\ 3\ \boxed{\div}\ 4\ \boxed{)}\ \boxed{+}\ \boxed{(}\ 5\ \boxed{\div}\ 6\ \boxed{)}\ \boxed{\text{ENTER}}\ \boxed{\text{MATH}}\ \boxed{\triangleright\text{Frac}}\ \boxed{\text{ENTER}}\ \Rightarrow\ \dfrac{19}{12}$

$\dfrac{19}{12} = 1\dfrac{7}{12}$

Section 2–5

1 Convert one unit of measure to another using unit ratios (pp. 101–104).

To write two equivalent measures as a unit ratio, write one measure in the numerator and its equivalent measure in the denominator. The ratio has a value of 1.

Write two unit ratios for the equivalent measures: 1 yd = 36 in.

$$\frac{1\text{ yd}}{36\text{ in.}} \quad \text{or} \quad \frac{36\text{ in.}}{1\text{ yd}}$$

To convert from one U.S. customary unit of measure to another using unit ratios:
1. Write the original measure in the numerator of a fraction with 1 in the denominator.
2. Multiply by a unit ratio with the original unit of measure in the denominator and the new unit in the numerator.

Change 5 ft to inches.

$$\frac{5\text{ ft}}{1} \cdot \frac{12\text{ in.}}{1\text{ ft}} = 60\text{ in.}$$

Change 285 ft to yards.

$$\frac{285\text{ ft}}{1} \cdot \frac{1\text{ yd}}{3\text{ ft}} = 95\text{ yd}$$

Change $3\frac{1}{2}$ qt to cups.

$$3\frac{1}{2}\text{ qt} = \frac{7}{2}\text{ qt}$$

$$\frac{7\text{ qt}}{2} \cdot \frac{2\text{ pt}}{1\text{ qt}} \cdot \frac{2\text{ c}}{1\text{ pt}} = 14\text{ c}$$

More than one unit ratio may be needed.

2 Convert one unit of measure to another using conversion factors (pp. 104–107).

When we use conversion factors to convert units, we always multiply. Two equivalent measures have two conversion factors.

To develop a conversion factor: **1.** Write a unit ratio that changes the given unit to the new unit. **2.** Change the fraction (or ratio) to its decimal equivalent by dividing the numerator by the denominator.

Develop two conversion factors relating pints and quarts. 1 qt = 2 pt

Pints to Quarts   **Quarts to Pints**

$\dfrac{pt}{1}\left(\dfrac{qt}{pt}\right)$ Write appropriate labels. $\dfrac{qt}{1}\left(\dfrac{pt}{qt}\right)$

$\dfrac{pt}{1}\left(\dfrac{1\ qt}{2\ pt}\right)$ Insert numbers to form a unit ratio. $\dfrac{qt}{1}\left(\dfrac{2\ pt}{1\ qt}\right)$

$\dfrac{1}{2} = 0.5$ Convert fraction to whole number, decimal, or mixed decimal equivalent. $\dfrac{2}{1} = 2$

$pt \times 0.5 = qt$ $qt \times 2 = pt$

To convert from one U.S. customary unit of measure to another using conversion factors: **1.** Select the appropriate conversion factor. **2.** Multiply the original measure by the conversion factor.

> Change 9 pt to quarts.
>
> $9\ pt\ (0.5) = 4.5\ qt$ Conversion factor is 0.5.

Standard notation means that each measure is converted to the next larger unit of measure when possible.

> $2\ ft\ 16\ in. = 2\ ft + 1\ ft\ 4\ in. = 3\ ft\ 4\ in.$ 16 in. = 1 ft 4 in.

3 Add and subtract U.S. customary measures (pp. 107–109).

To add or subtract U.S. customary measures: **1.** Convert each measure to a measure with a common U.S. customary unit. **2.** Add or subtract.

> Add 5 lb and 7 oz.   Convert 5 lb to oz.
>
> $\dfrac{5\ lb}{1} \times \dfrac{16\ oz}{1\ lb} = 80\ oz$ Multiply by unit ratio $\dfrac{16\ oz}{1\ lb}$.
>
> $80\ oz + 7\ oz = 87\ oz$ Combine like measures.

To add or subtract mixed U.S. customary measures: **1.** Align like measures in columns. **2.** Add or subtract, carrying or regrouping as necessary.

> Subtract 2 ft 8 in. from 7 ft.
>
> $\begin{array}{rl} 7\ ft = & 6\ ft\ 12\ in. \\ - & 2\ ft\ \ \ 8\ in. \\ \hline & 4\ ft\ \ \ 4\ in. \end{array}$ Rewrite 7 ft as 6 ft 12 in.
> Subtract.

4 Multiply and divide U.S. customary measures (pp. 109–111).

To multiply a measure by a number: **1.** Multiply the numbers associated with each unit of measure by the given number. **2.** Express the answer in standard notation.

> Multiply 2 gal 3 qt by 5.
>
> $\begin{array}{r} 2\ gal\ \ 3\ qt \\ \times\ \ \ \ \ \ \ \ \ 5 \\ \hline 10\ gal\ 15\ qt \\ 13\ gal\ \ \ 3\ qt \end{array}$ 15 qt = 3 gal 3 qt

To multiply a length measure by a like length measure: **1.** Multiply the numbers associated with each like unit. **2.** The product will be a square unit of measure.

Multiply 5 yd by 12 yd.

$$5 \text{ yd } (12 \text{ yd}) = 60 \text{ yd}^2$$

To divide a measure by a number: **1.** Divide the largest measure by the number. **2.** Convert any remainder to the next smaller unit. **3.** Repeat Steps 1 and 2 until no other units are left. **4.** Express any remainder as a fraction of the smallest unit. **5.** The quotient will be a measure.

Divide 7 lb 5 oz by 5.

$$
\begin{array}{r}
1 \text{ lb} \quad 7\frac{2}{5} \text{ oz} \\
\hline
5\overline{)7 \text{ lb} \quad\quad 5 \text{ oz}} \\
\underline{5 \text{ lb}} \\
2 \text{ lb} = \quad \underline{32 \text{ oz}} \\
37 \text{ oz} \\
\underline{35 \text{ oz}} \\
2
\end{array}
$$

To divide a measure by a measure: **1.** Convert the measures to the same unit. **2.** Divide. The quotient is a number that indicates how many.

How many 5-oz glasses of juice can be poured from 1 gallon of juice?

$$\frac{1 \text{ gal}}{1} \left(\frac{4 \text{ qt}}{1 \text{ gal}} \right) \left(\frac{2 \text{ pt}}{1 \text{ qt}} \right) \left(\frac{2 \text{ c}}{1 \text{ pt}} \right) \left(\frac{8 \text{ oz}}{1 \text{ c}} \right) = 128 \text{ oz}$$ Change 1 gal to ounces.

$$\frac{128 \text{ oz}}{5 \text{ oz}} = 25\frac{3}{5} \text{ glasses}$$

5 Change one U.S. customary rate measure to another (pp. 112–113).

1. Compare the units of both numerators and both denominators to determine which units will change. **2.** Multiply by the unit ratio (or ratios) containing the new unit so that the unit to be changed will reduce.

Change $10 \dfrac{\text{mi}}{\text{hr}}$ to $\dfrac{\text{ft}}{\text{hr}}$.

$$\frac{10 \text{ mi}}{1 \text{ hr}} \text{ to } \frac{\text{ft}}{\text{hr}}$$ Miles change to feet. Hours stay the same.

$$\frac{10 \text{ mi}}{1 \text{ hr}} \left(\frac{5{,}280 \text{ ft}}{1 \text{ mi}} \right) = 52{,}800 \frac{\text{ft}}{\text{hr}}$$

Section 2–1

Write five multiples of each number.

1. 12	**2.** 31	**3.** 10	**4.** 9	**5.** 21
6. 14	**7.** 7	**8.** 11	**9.** 8	**10.** 15

Is the number divisible by the given number? Explain.

11. 153 by 3	**12.** 8,234 by 4	**13.** 8,726 by 6	**14.** 5,986 by 5
15. 240 by 10	**16.** 5,845 by 5	**17.** 63,539 by 9	**18.** 52,428 by 8

List all factor pairs for each number, then write the factors in order from smallest to largest.

19. 48	**20.** 50	**21.** 51	**22.** 63	**23.** 74

Identify each number as *prime* or *composite*. Explain.

24. 17	**25.** 18	**26.** 20	**27.** 21	**28.** 29

Find the prime factorization of each number. Write in factored form and then in exponential notation.

29. 42	**30.** 48	**31.** 98	**32.** 120

Find the least common multiple.

33. 18 and 40	**34** 12 and 18	**35.** 12, 18, and 30	**36.** 6, 10, and 12

Find the greatest common factor.

37. 10 and 12	**38.** 12 and 18	**39.** 12, 18, and 30	**40.** 4, 9, and 16

Section 2–2

Find the equivalent fractions using the indicated denominators.

41. $\dfrac{5}{8} = \dfrac{?}{24}$	**42.** $\dfrac{3}{7} = \dfrac{?}{35}$	**43.** $\dfrac{5}{12} = \dfrac{?}{60}$	**44.** $\dfrac{4}{5} = \dfrac{?}{40}$
45. $\dfrac{2}{3} = \dfrac{?}{15}$	**46.** $\dfrac{4}{9} = \dfrac{?}{18}$	**47.** $\dfrac{3}{4} = \dfrac{?}{32}$	**48.** $\dfrac{1}{6} = \dfrac{?}{30}$
49. $\dfrac{1}{5} = \dfrac{?}{55}$	**50.** $\dfrac{7}{8} = \dfrac{?}{64}$	**51.** $\dfrac{4}{5} = \dfrac{?}{20}$	**52.** $\dfrac{1}{6} = \dfrac{?}{15}$

Reduce to lowest terms.

53. $\dfrac{6}{12}$	**54.** $\dfrac{8}{12}$	**55.** $\dfrac{4}{32}$	**56.** $\dfrac{26}{64}$
57. $\dfrac{2}{8}$	**58.** $\dfrac{8}{32}$	**59.** $\dfrac{34}{64}$	**60.** $\dfrac{16}{64}$
61. $\dfrac{12}{32}$	**62.** $\dfrac{45}{90}$	**63.** $\dfrac{6}{8}$	**64.** $\dfrac{75}{100}$

Measure line segments 65–74 in Fig. 2–27 (tolerance $= \pm\frac{1}{32}$ in.).

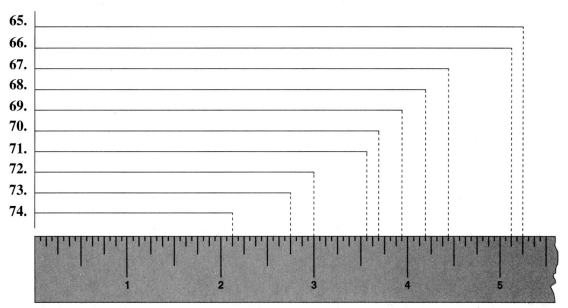

Figure 2–27

Use a U.S. customary rule to measure each line segment.

75.
A —————————— B

76.
C —————————— D

77.
E —————————— F

78.
G —————————— H

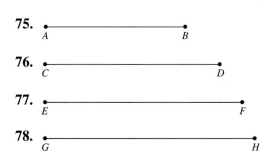

Change the decimals to fractions in lowest terms.

79. 0.7	**80.** 0.83	**81.** 0.95	**82.** 0.25
83. 0.872	**84.** 0.081	**85.** 0.02	**86.** 0.005

Change the fractions to decimals. If necessary, round to the nearest thousandth.

87. $\dfrac{1}{5}$ **88.** $\dfrac{1}{10}$ **89.** $\dfrac{5}{8}$ **90.** $\dfrac{3}{7}$ **91.** $\dfrac{9}{11}$

Write the improper fractions as whole or mixed numbers.

92. $\dfrac{35}{7}$ **93.** $\dfrac{18}{5}$ **94.** $\dfrac{27}{6}$ **95.** $\dfrac{39}{8}$ **96.** $\dfrac{21}{15}$

97. $\dfrac{43}{8}$ **98.** $\dfrac{22}{7}$ **99.** $\dfrac{175}{2}$ **100.** $\dfrac{135}{3}$ **101.** $\dfrac{18}{12}$

Write the whole or mixed numbers as improper fractions.

102. $6\frac{1}{2}$ **103.** 8 **104.** $10\frac{1}{2}$ **105.** $7\frac{1}{8}$ **106.** $5\frac{7}{12}$ **107.** $9\frac{3}{16}$

108. $7\frac{8}{17}$ **109.** $4\frac{3}{5}$ **110.** $9\frac{1}{9}$ **111.** 12 **112.** $16\frac{2}{3}$ **113.** $5\frac{1}{3}$

Change the whole number to an equivalent fraction using the indicated denominator.

114. $5 = \dfrac{?}{3}$ **115.** $2 = \dfrac{?}{10}$ **116.** $6 = \dfrac{?}{4}$ **117.** $11 = \dfrac{?}{3}$ **118.** $7 = \dfrac{?}{5}$

Find the least common denominator.

119. $\dfrac{1}{4}, \dfrac{1}{3}, \dfrac{1}{5}$ **120.** $\dfrac{7}{8}, \dfrac{2}{3}$ **121.** $\dfrac{3}{4}, \dfrac{1}{16}$

122. $\dfrac{1}{12}, \dfrac{3}{4}$ **123.** $\dfrac{5}{12}, \dfrac{3}{10}, \dfrac{13}{15}$ **124.** $\dfrac{1}{12}, \dfrac{3}{8}, \dfrac{15}{16}$

125. INDTR Is a $\frac{5}{8}$-in. wrench larger or smaller than a $\frac{9}{16}$-in. wrench?

126. INDTR An alloy contains $\frac{2}{3}$ metal A, and the same quantity of another alloy contains $\frac{3}{5}$ metal A. Which alloy contains more of metal A?

127. INDTR Is a $\frac{3}{8}$-in.-thick piece of plasterboard thicker than a $\frac{1}{2}$-in.-thick piece?

128. INDTR A $\frac{9}{16}$-in. tube must pass through an opening in a wall. Is a $\frac{3}{4}$-in.-diameter hole large enough?

129. INDTR Is a $\frac{19}{32}$-in. wrench larger or smaller than a $\frac{7}{8}$-in. bolt head?

130. HOSP A cook top that has a width of $22\frac{1}{4}$ in. needs to fit inside an opening of $22\frac{5}{16}$ in. wide. Will the opening need to be made larger?

Which fraction is smaller?

131. $\dfrac{5}{8}, \dfrac{3}{8}$ **132.** $\dfrac{3}{7}, \dfrac{2}{7}$ **133.** $\dfrac{3}{8}, \dfrac{4}{8}$ **134.** $\dfrac{5}{9}, \dfrac{4}{9}$ **135.** $\dfrac{1}{4}, \dfrac{3}{16}$

136. $\dfrac{5}{8}, \dfrac{11}{16}$ **137.** $\dfrac{7}{8}, \dfrac{27}{32}$ **138.** $\dfrac{7}{64}, \dfrac{1}{4}$ **139.** $\dfrac{1}{2}, \dfrac{9}{19}$

Which common or decimal fraction is larger?

140. $0.127, \dfrac{1}{8}$ **141.** $0.26, \dfrac{3}{4}$ **142.** $0.08335, \dfrac{1}{6}$ **143.** $0.272, \dfrac{3}{11}$

Section 2–3

Add; reduce sums to lowest terms and convert improper fractions to mixed numbers or whole numbers.

144. $\dfrac{1}{8} + \dfrac{5}{16}$ **145.** $\dfrac{3}{16} + \dfrac{9}{64}$ **146.** $\dfrac{3}{14} + \dfrac{5}{7}$

147. $\dfrac{3}{5} + \dfrac{5}{6}$ **148.** $3\frac{7}{8} + 7 + 5\frac{1}{2}$ **149.** $2\frac{7}{16} + 6\frac{5}{32}$

150. $9\frac{7}{8} + 5\frac{3}{4}$ **151.** $3\frac{7}{8} + 5\frac{3}{16} + 1\frac{7}{32}$ **152.** $2\frac{1}{4} + 3\frac{7}{8}$

153. HELPP A forest fire advanced $7\frac{5}{8}$ mi in one day and $10\frac{7}{16}$ mi the next. How far did the fire advance?

154. CON Find the total thickness of a wall if the outside covering is $3\frac{7}{8}$ in. thick, the studs (interior supports) are $3\frac{7}{8}$ in., and the inside covering is $\frac{5}{16}$-in. paneling.

155. INDTEC If $7\frac{5}{16}$ in. of a piece of square bar stock is turned (machined) so that it is cylindrical and $5\frac{9}{32}$ in. remains square, what is the total length of the original bar stock?

156. AVIA Two carts of airline food weigh $27\frac{1}{2}$ lb and $20\frac{3}{4}$ lb. What is the total weight of the two carts?

157. INDTEC Three metal rods measuring $3\frac{1}{8}$ in., $5\frac{3}{32}$ in., and $7\frac{9}{16}$ in. are welded together end to end. How long is the welded rod?

158. CON In Fig. 2–28, what is the length of side A?

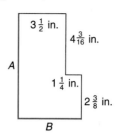

Figure 2–28

159. CON In Fig. 2–28, what is the length of side B?

160. CON A hollow-wall fastener has a grip range up to $\frac{3}{4}$ in. Is it long enough to fasten three sheets of metal $\frac{5}{16}$ in. thick, $\frac{3}{8}$ in. thick, and $\frac{1}{16}$ in. thick?

161. CON Fig. 2–29 shows $\frac{1}{2}$-in. copper tubing wrapped in insulation. What is the distance across the tubing and insulation?

162. CON In Exercise 161, what would be the overall distance across the tubing and insulation if $\frac{3}{8}$-in.-ID (inside diameter) tubing were used?

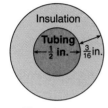

Figure 2–29

Subtract; reduce to lowest terms when necessary.

163. $\dfrac{5}{9} - \dfrac{2}{9}$

164. $\dfrac{11}{32} - \dfrac{5}{64}$

165. $3\dfrac{5}{8} - 2$

166. $7 - 4\dfrac{3}{8}$

167. $8\dfrac{7}{8} - 2\dfrac{29}{32}$

168. $7 - 2\dfrac{9}{16}$

169. $12\dfrac{11}{16} - 5$

170. $48\dfrac{5}{12} - 12\dfrac{11}{15}$

171. $122\dfrac{1}{2} - 87\dfrac{3}{4}$

172. INDTR Pins of $2\frac{3}{8}$ in. and $3\frac{7}{16}$ in. are cut from a drill rod 12 in. long. If $\frac{1}{16}$ in. of waste is allowed for each cut, how many inches of drill rod are left?

173. CON A bolt 2 in. long fastens a piece of $\frac{7}{8}$-in.-thick wood to a piece of metal. If a $\frac{3}{16}$-in.-thick lock washer, a $\frac{1}{16}$-in. washer, and a $\frac{7}{16}$-in.-thick nut are used, what is the thickness of the metal if the nut is even with the end of the bolt after tightening?

174. TELE Four lengths measuring $6\frac{1}{4}$ in., $9\frac{3}{16}$ in., $7\frac{1}{8}$ in., and $5\frac{9}{32}$ in. are cut from 48 in. of cable. How much cable remains? Disregard waste.

175. INDTR A piece of tapered stock has a diameter of $2\frac{5}{16}$ in. at one end and a diameter of $\frac{55}{64}$ in. at the other end. What is the difference in the diameters?

Section 2–4

Multiply and reduce answers to lowest terms. Convert improper fractions to whole or mixed numbers.

176. $\dfrac{3}{5} \times \dfrac{10}{21}$

177. $\dfrac{1}{3} \times \dfrac{7}{8}$

178. $\dfrac{2}{5} \times \dfrac{7}{10}$

179. $\dfrac{7}{9} \times \dfrac{3}{8}$

180. $\dfrac{2}{3} \times \dfrac{5}{8} \times \dfrac{3}{16}$

181. $\dfrac{15}{16} \times \dfrac{4}{5} \times \dfrac{2}{3}$

182. $5 \times \dfrac{3}{4}$

183. $\dfrac{7}{16} \times 18$

184. $\dfrac{3}{16} \times 184$

185. $1\dfrac{1}{2} \times \dfrac{4}{5}$

186. $3\dfrac{1}{3} \times 4\dfrac{1}{2}$

187. $1\dfrac{3}{4} \times 1\dfrac{1}{7}$

188. CON In a concrete mixture, $\frac{4}{7}$ of the total volume is sand. How much sand is needed for 135 cubic yards (yd^3) of concrete?

189. CON Concrete blocks are 8 in. high. If a $\frac{3}{8}$-in. mortar joint is used, how high will a wall of 12 courses of concrete blocks be? (*Hint*: There are 12 rows of mortar joints.)

190. INDTEC An adjusting screw will move $\frac{3}{64}$ in. for each full turn. How far will it move in four full turns?

191. HOSP A chef is making a dessert that is $\frac{3}{4}$ the original recipe. How much flour should be used if the original recipe calls for $3\frac{2}{3}$ cups of flour?

192. INDTEC If an alloy is $\frac{3}{5}$ copper and $\frac{2}{5}$ zinc, how many pounds of each metal are in a casting weighing $112\frac{1}{2}$ lb?

193. CON A water pipe has an outside diameter of $18\frac{3}{4}$ cm. What is the width of eight pipes that have the same diameter?

Raise the fractions to the indicated power.

194. $\left(\dfrac{1}{5}\right)^2$

195. $\left(\dfrac{3}{4}\right)^2$

196. $\left(\dfrac{5}{6}\right)^2$

197. $\left(\dfrac{4}{9}\right)^2$

198. $\left(\dfrac{1}{7}\right)^3$

199. $\left(\dfrac{1}{2}\right)^3$

200. $\left(\dfrac{7}{8}\right)^2$

201. $\left(\dfrac{9}{10}\right)^2$

Give the reciprocal.

202. $\dfrac{7}{8}$

203. 4

204. $2\dfrac{3}{5}$

205. 0.7

206. 1.8

Divide and reduce answers to lowest terms. Convert improper fractions to whole or mixed numbers.

207. $\dfrac{7}{8} \div \dfrac{3}{4}$

208. $\dfrac{4}{9} \div \dfrac{5}{16}$

209. $\dfrac{7}{8} \div \dfrac{3}{32}$

210. $8 \div \dfrac{2}{3}$

211. $18 \div \dfrac{3}{4}$

212. $35 \div \dfrac{5}{16}$

213. $5\dfrac{1}{10} \div 2\dfrac{11}{20}$

214. $27\dfrac{2}{3} \div \dfrac{2}{3}$

215. $7\dfrac{1}{5} \div 12$

216. CAD/ARC On a house plan, $\frac{1}{4}$ in. represents 1 ft. Find the dimensions of a porch that measures $4\frac{1}{8}$ in. by $6\frac{1}{2}$ in. on the plan. (How many $\frac{1}{4}$s are there in $4\frac{1}{8}$; how many $\frac{1}{4}$s are there in $6\frac{1}{2}$?)

217. CON A pipe that is 12 in. long is cut into four equal parts. If $\frac{3}{16}$ in. is wasted per cut, what is the maximum length of each pipe? (It takes three cuts to divide the entire length into four equal parts.)

218. CON A stack of $\frac{5}{8}$-in. plywood is $21\frac{7}{8}$ in. high. How many sheets of plywood are in the stack?

219. CON A rod $1\frac{1}{8}$ yd long is cut into six equal pieces. What is the length of each piece? Disregard waste.

220. AG/H If $7\frac{1}{2}$ gal of liquid are distributed equally among five containers, what is the number of gallons per container?

221. BUS Fabric that is $22\frac{1}{2}$ yd long is cut into lengths of $\frac{5}{8}$ yd. How many equal lengths can be made?

Use a calculator to perform the operations.

222. $\dfrac{\frac{5}{8}}{2\frac{1}{8}}$

223. $\dfrac{\frac{1}{3}}{6}$

224. $\dfrac{4}{\frac{4}{5}}$

225. $\dfrac{8}{1\frac{1}{2}}$

226. $\dfrac{3\frac{1}{4}}{5}$

227. $\dfrac{2\frac{1}{5}}{8\frac{4}{5}}$

228. $\dfrac{16\frac{2}{3}}{3\frac{1}{3}}$

229. $\dfrac{12\frac{1}{2}}{100}$

230. $\dfrac{37\frac{1}{2}}{100}$

Section 2–5

Write two unit ratios that relate the given pair of measures.

231. Feet and inches

232. Hours and days

233. Pounds and tons

234. Yards and miles

Using unit ratios or conversion factors, convert the given measures to the new units.

235. How many ounces are in 5 lb?

236. HELPP A fire suit weighing $57\frac{3}{5}$ lb weighs how many ounces?

237. Find the number of pounds in 680 oz.

238. HOSP A can of fruit weighs 22.4 oz. How many pounds is this?

239. AG/H How many feet of wire are needed to fence a property line $1\frac{1}{4}$ mi long?

240. AVIA The tail height of a B737-200 aircraft is 36 ft 6 in. How many inches is the tail height?

Express the measures in standard notation.

241. 1 ft 19 in.

242. 1 mi 5,375 ft

243. 12 lb $17\frac{1}{2}$ oz

244. 2 gal 7 qt

Add or subtract. Write answers in standard notation.

245.
$$\begin{array}{r} 5 \text{ gal } 3 \text{ qt} \\ + 2 \text{ gal } 3 \text{ qt} \end{array}$$

246.
$$\begin{array}{r} 7 \text{ ft } 9 \text{ in.} \\ - 4 \text{ ft } 6 \text{ in.} \end{array}$$

247.
$$\begin{array}{r} 4 \text{ lb } 9 \text{ oz} \\ - 3 \text{ lb } 11 \text{ oz} \end{array}$$

248. AVIA Two packages to be sent air express weigh 5 lb 4 oz each. What is the shipping weight of the two packages?

249. AUTO A water hose purchased for an RV was 2 ft long. What was its length after 7 in. were cut off?

Multiply and write answers for mixed measures in standard notation.

250.
$$\begin{array}{r} 8 \text{ lb } 3 \text{ oz} \\ \times \quad 9 \end{array}$$

251.
$$\begin{array}{r} 9 \text{ in.} \\ \times 7 \text{ in.} \end{array}$$

252.
$$\begin{array}{r} 10 \text{ gal } 3 \text{ qt} \\ \times \quad 7 \end{array}$$

Divide.

253. 20 yd 2 ft 6 in. ÷ 2

254. 5 gal 3 qt 2 pt ÷ 6

255. HELPP If 18 lb of flame retardant are divided equally into four boxes, express the weight of the contents of each box in pounds and ounces.

256. CON If 32 equal lengths of pipe are needed for a job and each length is to be 2 ft 8 in., how many feet of pipe are needed for the job?

257. 14 ft ÷ 4 ft

258. 2 mi 120 ft ÷ 15 ft

259. $5\,\dfrac{\text{mi}}{\text{min}} = \underline{\hspace{2cm}}\,\dfrac{\text{mi}}{\text{h}}$

260. $2{,}520\,\dfrac{\text{gal}}{\text{h}} = \underline{\hspace{2cm}}\,\dfrac{\text{qt}}{\text{h}}$

261. HOSP How many quarts of milk are needed for a recipe that calls for 3 pt of milk?

262. HOSP How many $\frac{1}{2}$-oz servings of jelly can be made from a $1\frac{1}{2}$-lb container of jelly?

TEAM PROBLEM-SOLVING EXERCISES

Your team is preparing a report that is to be printed on both sides of the paper. It is customary to put odd-numbered pages on the front and even-numbered pages on the back. A new chapter is started on the front of a sheet of paper, even if this creates a preceding blank page. Assign page numbers to the document based on these guidelines.

Chap. 1, 15 pages Chap. 2, 17 pages
Chap. 3, 24 pages Chap. 4, 15 pages

1. What page number will start Chapter 2?
2. How many pages are in the document?
3. How many blank pages are in the document?

1. What two operations require a common denominator?

2. Explain how to find the reciprocal of a fraction.

3. What steps must be followed to find the reciprocal of a mixed number?

4. What number can be written as any fraction that has the same numerator and denominator? Explain why.

5. What operation requires the use of the reciprocal of a fraction?

6. Name the operation that has each of the following for an answer: sum, difference, product, quotient.

7. What operation must be used to solve an applied problem if the total and one of the two parts are given?

8. What does the denominator of a fraction indicate?

9. What does the numerator of a fraction indicate?

10. What kind of fraction has a value less than 1?

Find, explain, and correct the mistakes in these problems.

11. $\dfrac{5}{8} + \dfrac{1}{8} = \dfrac{6}{16} = \dfrac{3}{8}$

12. $\begin{array}{r} 12 \\ -5\dfrac{3}{4} \\ \hline 7\dfrac{3}{4} \end{array}$

13. $\dfrac{3}{5} \times 2\dfrac{1}{5} = 2\dfrac{3}{25}$

14. $\dfrac{5}{8} \div 4 = \dfrac{5}{8} \times \dfrac{4}{1} = \dfrac{5}{2} = 2\dfrac{1}{2}$

15. $12\dfrac{3}{4} = 12\dfrac{6}{8} = 11\dfrac{16}{8}$

$\dfrac{-4\dfrac{7}{8}}{} = \dfrac{-4\dfrac{7}{8}}{} = -4\dfrac{7}{8}$

$7\dfrac{9}{8} = 7 + 1\dfrac{1}{8} = 8\dfrac{1}{8}$

Represent as fractions.

1. 3 out of 4 people in a survey

2. $7 \div 9$

Write as whole or mixed numbers.

3. $\dfrac{9}{3}$

4. $\dfrac{14}{9}$

Write as improper fractions.

5. $4\dfrac{6}{7}$

6. $3\dfrac{1}{10}$

Write the prime factors in exponential notation.

7. 96

8. 132

9. 62

Write the least common multiple (LCM).

10. 48 and 64

11. 36, 45, and 54

Write the greatest common factor (GCF).

12. 15 and 35

13. 12, 18, and 36

14. Reduce $\frac{18}{24}$ to lowest terms.

15. Find the measure of the line segment AB in Fig. 2–30.

16. Write $\frac{7}{20}$ as a decimal.

17. Write $\frac{21}{2}$ as a decimal.

18. Write $5\frac{3}{8}$ as an improper fraction.

19. Which fraction is smaller, $\frac{4}{5}$ or $\frac{7}{10}$?

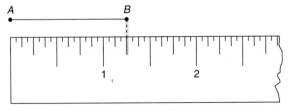

Figure 2–30

Perform the indicated operations.

20. $\dfrac{7}{12} + \dfrac{5}{6}$

21. $3\dfrac{4}{15} + 4\dfrac{3}{10}$

22. $\dfrac{5}{6}\left(\dfrac{3}{10}\right)$

23. $2\dfrac{2}{9}\left(1\dfrac{3}{4}\right)$

24. $\dfrac{7}{8} - \dfrac{1}{4}$

25. $6\dfrac{1}{4} - 2\dfrac{3}{4}$

26. $\dfrac{5}{12} \div \dfrac{5}{6}$

27. $7\dfrac{1}{2} \div \dfrac{5}{9}$

28. $5\dfrac{2}{3} \div 1\dfrac{1}{9}$

Solve the problems.

29. Two of the seven security employees at the local community college received safety awards from the governor. Represent the part of the total number of employees who received an award as a fraction.

30. A candy-store owner mixes $1\frac{1}{2}$ lb of caramels, $\frac{3}{4}$ lb of chocolates, and $\frac{1}{2}$ lb of candy corn. What is the total weight of the mixed candy?

31. A homemaker has $5\frac{1}{2}$ cups of sugar on hand to make a batch of cookies requiring $1\frac{2}{3}$ cups of sugar. How much sugar is left?

32. If $6\frac{1}{4}$ ft of wire is needed to make one electrical extension cord, how many extension cords can be made from $68\frac{3}{4}$ ft of wire?

33. A costume maker figures one costume requires $2\frac{2}{3}$ yd of red satin. How many yards of red satin are needed to make three costumes?

34. Will a $\frac{5}{8}$-in.-wide drill bit make a hole wide enough to allow a $\frac{1}{2}$-in. (outside diameter) copper tube to pass through?

Percents

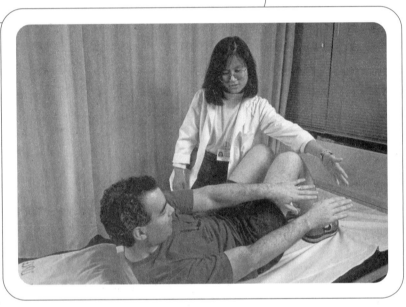

Focus on Careers

Physical therapist assistants (PTAs) assist physical therapists in working with patients who have suffered injuries or debilitating disease. Through physical therapy, patients improve mobility, relieve pain, and prevent permanent physical disabilities. Physical therapy assistants perform treatment procedures including exercises, massages, paraffin baths, and traction. PTAs record patient responses to treatment and discuss treatment outcomes with the physical therapist.

Physical therapist assistant jobs are expected to grow much faster than the average through the year 2012 because the number of individuals with disabilities or debilitating illnesses such as heart disease is expected to increase as the population ages. Persons entering this career usually have an associate degree. Some states require licensure and stipulate specific educational requirements.

Almost three-fourths of all jobs are in hospitals or offices of physical therapists. In 2002, physical therapist assistants had median annual earnings of $36,080, and 25% earned more than $42,780. The median annual pay at general medical offices and surgical hospitals was $35,750 in 2002, and offices of other health practitioners paid a median annual salary of $35,750.

Source: *Occupational Outlook Handbook:* http://www.bls.gov/oco/.

Learning Outcomes **1** Change any number to its percent equivalent.
2 Change any percent to its numerical equivalent.

The word **percent** means "per hundred" or "for every hundred." Thus, 35 percent means 35 per hundred, or 35 out of every hundred, or $\frac{35}{100}$. Also, 100 percent means 100 out of 100 parts, or $\frac{100}{100}$, or 1 whole quantity. The symbol % is used to represent "percent."

1 Change Any Number to Its Percent Equivalent.

We often need to use percents, but the numbers we are given are not expressed as percents. To express a number as a percent, we multiply by 1 in the form of 100%. For example, the fraction $\frac{1}{2}$ can be written as the decimal 0.5. Both $\frac{1}{2}$ and 0.5 change to the same equivalent percent.

$$\frac{1}{2} \cdot 100\% = \frac{1}{\underset{1}{2}} \cdot \frac{\overset{50}{\cancel{100\%}}}{1} = 50\% \qquad 0.5 \cdot 100\% = 50\%$$

To change any number to its percent equivalent:

Multiply by 1 in the form of 100%.

We can use this rule to change any type of number—fraction, decimal, whole number, or mixed number—to a percent equivalent.

TIP

Multiplying by 1 Can Take Many Forms

The *multiplicative identity* states that $n \times 1 = n$, where n is any real number. In fractions, we use the multiplicative identity to change a fraction to an equivalent fraction. We can use the property:

$\frac{n}{n} = 1$, where n is any real number and $n = 0$.

$$\frac{1}{2} \cdot \frac{3}{3} = \frac{3}{6} \qquad \frac{3}{3} = 1$$

With percents, we use the property of 1 in the form of 100%.

$$\frac{1}{2} \cdot 100\% = 50\% \qquad 100\% = 1$$

EXAMPLE Change the fractions to percent equivalents: $\frac{1}{4}$, $\frac{1}{3}$, $\frac{3}{8}$, and $\frac{1}{200}$.

$$\frac{1}{4} \cdot 100\% = \frac{1}{\underset{1}{4}} \cdot \frac{\overset{25}{\cancel{100\%}}}{1} = \mathbf{25\%}$$

$$\frac{1}{3} \cdot 100\% = \frac{1}{3} \cdot \frac{100\%}{1} = \frac{100\%}{3} = \mathbf{33\frac{1}{3}\%}$$

$$\frac{3}{8} \cdot 100\% = \frac{3}{\underset{2}{8}} \cdot \frac{\overset{25}{\cancel{100\%}}}{1} = \frac{75\%}{2} = \mathbf{37\frac{1}{2}\%}$$

$$\frac{1}{200} \cdot 100\% = \frac{1}{\underset{2}{200}} \cdot \frac{\overset{1}{\cancel{100\%}}}{1} = \mathbf{\frac{1}{2}\%}$$ $\frac{1}{2}\%$ means $\frac{1}{2}$ of every hundredth, or $\frac{1}{2}$ of 1%.

EXAMPLE Change the decimals 0.3 and 0.006 to percent equivalents.

$$0.3 \cdot 100\% = 0\underset{\smile}{30}.\% = \mathbf{30\%}$$ Use the shortcut procedure to multiply by 100: Move the decimal point two places to the right.

$$0.006 \cdot 100\% = 000.6\% = \mathbf{0.6\%}$$ 0.6% means 0.6 of every hundredth, or 0.6 of 1%.

EXAMPLE Change the whole numbers 1 and 7 to their percent equivalents.

$$1 \cdot 100\% = \mathbf{100\%}$$ 100 out of 100 or all of 1 quantity.

$$7 \cdot 100\% = \mathbf{700\%}$$ 7 whole quantities, or 7 times a quantity.

EXAMPLE Change the mixed numbers and decimals to their percent equivalents: $1\frac{1}{4}$, $3\frac{2}{3}$, and 5.3.

$$1\frac{1}{4} \cdot 100\% = \frac{5}{\underset{1}{4}} \cdot \frac{\overset{25}{\cancel{100\%}}}{1} = \mathbf{125\%}$$ Write $1\frac{1}{4}$ as an improper fraction and multiply.

$$3\frac{2}{3} \cdot 100\% = \frac{11}{3} \cdot \frac{100\%}{1} = \frac{1,100\%}{3} = \mathbf{366\frac{2}{3}\%}$$ Write $3\frac{2}{3}$ as an improper fraction and multiply.

$$5.3 \cdot 100\% = 5\underset{\smile}{30}.\% = \mathbf{530\%}$$ Multiply by 100 by moving the decimal two places to the right.

2 Change Any Percent to Its Numerical Equivalent.

Percents are a convenient way to express the ratio of any quantity to 100. They are excellent time-savers when we make comparisons or state problems on the job. In making calculations we first convert the percents to fraction-, decimal-, whole-, or mixed-number equivalents.

To change a percent to a numerical equivalent:

Divide by 1 in the form of 100%.

Dividing by 1 Can Take Many Forms

To reduce fractions, we use the property $\dfrac{n}{n} = 1$, where n is any nonzero real number.

With percents we apply this property using 1 in the form of 100%.

The numerical equivalent of a percent can be expressed in fraction or decimal form.

EXAMPLE Change the percents to their fraction and decimal equivalents: 75%, 38%, and 5%.

	Fraction equivalent	**Decimal equivalent**	
75%	$75\% \div 100\%$	$75\% \div 100\%$	For fractions, change division to an equivalent multiplication. Reduce.
	$\dfrac{\overset{3}{\cancel{75\%}}}{1} \cdot \dfrac{1}{\underset{4}{\cancel{100\%}}} = \dfrac{3}{4}$	$0.75 = \mathbf{0.75}$	For decimals, use the shortcut procedure for dividing by 100%: Move the decimal point two places to the left.
38%	$38\% \div 100\%$	$38\% \div 100\%$	
	$\dfrac{\overset{19}{\cancel{38\%}}}{1} \cdot \dfrac{1}{\underset{50}{\cancel{100\%}}} = \dfrac{19}{50}$	$0.38 = \mathbf{0.38}$	$\dfrac{\%}{\%}$ reduces to 1.
5%	$5\% \div 100\%$	$5\% \div 100\%$	
	$\dfrac{\overset{1}{\cancel{5\%}}}{1} \cdot \dfrac{1}{\underset{20}{\cancel{100\%}}} = \dfrac{1}{20}$	$0.05 = \mathbf{0.05}$	

As with fractions, we see that it is convenient to change division to an equivalent multiplication. Is *dividing* by 100% the same as *multiplying* by $\dfrac{1}{100\%}$? Yes.

$$\text{a percent} \div 100\% = \text{a percent} \div \frac{100\%}{1} = \text{a percent} \cdot \frac{1}{100\%}$$

Some percents change more conveniently to a fraction equivalent, and others change more conveniently to a decimal equivalent. In solving problems, we normally change the percent to the most convenient equivalent for the problem. In the following examples, both fraction and decimal equivalents are given, and you can judge for yourself when fraction equivalents are more convenient than decimal equivalents, and vice versa.

EXAMPLE Change the percents to their fraction and decimal equivalents: $33\frac{1}{3}\%$, $37\frac{1}{2}\%$.

Fractional equivalent   **Decimal equivalent**

$$33\frac{1}{3}\% \div 100\% = \frac{\overset{1}{\cancel{100\%}}}{3} \cdot \frac{1}{\underset{1}{\cancel{100\%}}} = \frac{1}{3} \qquad 33\frac{1}{3}\% = \mathbf{0.33\frac{1}{3}} \text{ or } \mathbf{0.33 \text{ (rounded)}}$$

A decimal point separates whole quantities from fraction parts. Therefore, there is an *unwritten* decimal between 33 and $\frac{1}{3}$. Because $\frac{1}{3}$ does not change to a terminating decimal equivalent, using the decimal equivalent of $33\frac{1}{3}\%$ will create more extensive calculations. Using a rounded decimal equivalent changes the result from an exact to an approximate amount.

Fractional equivalent   **Decimal equivalent**

$$37\frac{1}{2}\% \div 100\% = \frac{\overset{3}{\cancel{75}}}{2}\% \cdot \frac{1}{\underset{4}{\cancel{100\%}}} = \frac{3}{8} \qquad 37\frac{1}{2}\% = 37.5\% = \mathbf{0.375}$$

Since no rounding is necessary, the decimal equivalent is an exact amount.

EXAMPLE Change 5.25% to its fractional and decimal equivalents.

First, write the percent in fraction form.

Fractional equivalent   **Decimal equivalent**

$$5.25\% = 5\frac{25}{100}\% = 5\frac{1}{4}\% \qquad 5.25\% = 5.25\% \div 100\% = \mathbf{0.0525}$$

$$5\frac{1}{4}\% = 5\frac{1}{4}\% \div 100\%$$

$$= \frac{21}{4}\% \cdot \frac{1}{100\%}$$

$$= \frac{21}{400}$$

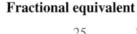

TIP **What Happens to the % (Percent) Sign?**

In multiplying fractions and measurements, we reduced or canceled common factors and units from a numerator to a denominator. Percent signs can also reduce.

$$\frac{\%}{1} \cdot \frac{1}{\%} = 1$$

EXAMPLE Change the percents to their decimal equivalents: $\frac{1}{2}\%$ and 0.25%.

$$\frac{1}{2}\% = 0.5\% = 0.5\% \div 100\% = \mathbf{0.005}$$

$$0.25\% = 0.25\% \div 100\% = \mathbf{0.0025}$$

When quantities are 100% or more, the numerical equivalents will be whole numbers, mixed numbers, mixed decimals, or rounded decimals that are equal to or more than the whole number 1. When solving problems, use the more convenient equivalent.

EXAMPLE Change the percents to their whole-number equivalents or to their rounded decimal equivalents: 700%, 375%, $233\frac{1}{3}\%$, and $462\frac{1}{2}\%$. Round decimals to the nearest thousandth when appropriate.

$$700\% \div 100\% = \mathbf{7}$$

$$375\% \div 100\% = \mathbf{3.75}$$

$$233\frac{1}{3}\% \div 100\% = \frac{\overset{7}{\cancel{700\%}}}{3} \cdot \frac{1}{\underset{1}{\cancel{100\%}}} = \frac{7}{3} = \mathbf{2.333} \text{ (rounded)}$$

$$462\frac{1}{2}\% = 462.5\% = 462.5\% \div 100\% = \mathbf{4.625}$$

TIP **Do Part of the Calculation on the Calculator and the Other Part Mentally**

When you need to find the percent equivalent of a fraction, why not change the fraction to a decimal equivalent by dividing? Then, change the decimal to a percent mentally by multiplying by 100%.

$$\frac{4}{7} = 4 \div 7 = 0.571428571$$

Then, $\frac{4}{7} \approx 57.14\%$ · Mentally move the decimal two places to the right and round to the nearest hundredth of a percent.

1. **ELEC** If $\frac{2}{5}$ of the electricians in a city are self-employed, what percent are self-employed?

2. **CON** If $\frac{7}{10}$ of the bricklayers in a city are male, what percent are male?

Change the numbers to their percent equivalents. Round to the nearest tenth of a percent if necessary.

3. $\frac{5}{8}$ 4. $\frac{7}{9}$ 5. $\frac{7}{1000}$ 6. $\frac{1}{350}$ 7. 0.2

8. 0.14   9. 0.007   10. 0.0125   11. 5   12. 8

13. $1\frac{1}{3}$ 14. $3\frac{1}{2}$ 15. $4\frac{3}{10}$ 16. $2\frac{1}{5}$ 17. 3.05

18. 7.2   19. 15.1   20. 36.25

2 Change to both fractional and decimal equivalents. Round to the nearest ten-thousandth if necessary.

21. 36%   22. 45%   23. 20%

24. 75%   25. $6\frac{1}{4}\%$ 26. 62.5%

27. $66\frac{2}{3}\%$ 28. 0.6%   29. $\frac{1}{5}\%$

30. 0.05%   31. $8\frac{1}{3}\%$ 32. 18.75%

Change to equivalent whole numbers.

33. 800%   34. 400%

Change to both mixed-number and mixed-decimal equivalents.

35. 250%   36. 425%   37. 176%

38. 380%   39. $137\frac{1}{2}\%$ 40. 387.5%

Change to mixed-number equivalents.

41. $166\frac{2}{3}\%$ 42. $316\frac{2}{3}\%$

Change to mixed-decimal equivalents.

43. 115.3%   44. $212\frac{1}{2}\%$ 45. $106\frac{1}{4}\%$

Fill in the missing percents, fractions, or decimals *from memory*.

Percent	Fraction	Decimal	Percent	Fraction	Decimal	Percent	Fraction	Decimal
46. 10%	____	____	47. ____	$\frac{1}{4}$	____	48. ____	____	0.2
49. ____	$\frac{1}{3}$	____	50. 50%	____	____	51. ____	$\frac{4}{5}$	____
52. ____	____	0.75	53. $66\frac{2}{3}\%$	____	____	54. ____	____	1
55. ____	$\frac{3}{10}$	____	56. 40%	____	____	57. ____	____	0.7

Percent	Fraction	Decimal		Percent	Fraction	Decimal		Percent	Fraction	Decimal
58. ___	$\frac{9}{10}$	___	**59.** 60%	___	___		**60.** ___	$\frac{1}{1}$	___
61. ___	___	0.25	**62.** ___	$\frac{2}{3}$	___		**63.** 70%	___	___
64. ___	___	0.5	**65.** ___	$\frac{3}{5}$	___		**66.** ___	___	0.1
67. 20%	___	___	**68.** ___	___	$0.33\frac{1}{3}$		**69.** ___	___	$0.66\frac{2}{3}$
70. ___	$\frac{3}{4}$	___	**71.** 30%	___	___		**72.** ___	___	0.9
73. ___	$\frac{1}{5}$	___	**74.** ___	___	0.6		**75.** 100%	___	___

3-2 | Percentage Problems

Learning Outcomes

1 Identify the portion, base, and rate in percent problems.
2 Solve percent problems using the percentage formula.
3 Solve percent problems using the percentage proportion.
4 Solve business and consumer problems involving percents.

1 Identify the Portion, Base, and Rate in Percent Problems.

All problems involving percents have three basic elements: the rate, the base, and the portion. Knowing what each element is and how all three are related helps us solve problems with percents.

The **rate R** is the percent; the **base B** is the original or total amount; the **portion P** is part of the base. In the statement 50% of 80 is 40, the rate is 50%, the base is 80, and the portion is 40.

Identify the Rate, Base, and Portion

The following descriptions may help you recognize the rate, base, or portion more quickly:

Rate is usually written as a percent, but it may be a decimal or fraction.

Base is the total amount, original amount, or entire amount. It is the amount that the *portion* is a part of. In a sentence the base is often closely associated with the preposition *of*.

Portion can refer to the part, partial amount, amount of increase or decrease, or amount of change. It is a portion of the *base*. In a sentence the portion is often closely associated with a form of the verb *is*. The portion can also be referred to as the **percentage.** We choose to use the term *portion* to minimize the confusion between the terms *percent* and *percentage*.

A common "memory jogger" for finding the percent: $\text{Rate} = \dfrac{is}{of}$.

EXAMPLE Identify the given and missing elements for

(a) 20% of 75 is what number?
(b) What percent of 50 is 30?
(c) Eight is 10% of what number?

        R   *B*   *P*

(a)   20% of 75 is what number?   Use the identifying key words for rate (*percent* or %),
     percent   total   portion   base (*total, original,* associated with the word *of*),
               R   *B*   *P*   and portion (*part,* associated with the word *is*).

(b) What percent of 50 is 30?
            percent   total portion

     P   *R*   *B*

(c) Eight is 10% of what number?
    portion   percent   total

2 Solve Percent Problems Using the Percentage Formula.

The percentage formula, *Portion* = *Rate* × *Base*, can be written as $P = R \times B$. The letters or words represent numbers. When the numbers are put in place of the letters, the formula guides you through the calculations.

Three percentage formulas:

$$\text{Portion} = \text{Rate} \times \text{Base} \qquad P = R \cdot B \qquad \text{For finding the portion.}$$

$$\text{Base} = \frac{\text{Portion}}{\text{Rate}} \qquad B = \frac{P}{R} \qquad \text{For finding the base.}$$

$$\text{Rate} = \frac{\text{Portion}}{\text{Base}}(100\%) \qquad R = \frac{P}{B}(100\%) \qquad \text{For finding the rate.}$$

The rate should be changed from a percent to its numerical equivalent when it is used in the formulas for finding the portion and the base. Multiplying by 100% in the formula for finding the rate causes the rate to be expressed as a percent.

Circles can help us visualize these formulas. The shaded part of the circle in Fig. 3–1 represents the missing amount. The unshaded parts represent the known amounts. If the unshaded parts are *side by side, multiply* their corresponding numbers to find the missing number.

If the unshaded parts are *one on top of the other, divide* the corresponding numbers to find the missing number.

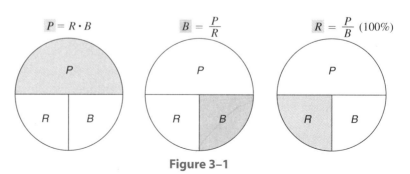

Figure 3–1

1. Identify and classify the two known values and the one missing value.
2. Choose the appropriate percentage formula for finding the missing value.
3. Substitute the known values into the formula. For the rate, use the decimal or fractional equivalent of the percent.
4. Perform the calculation indicated by the formula.
5. Interpret the result. If finding the rate, the formula converts decimal or fractional equivalents of the rate to a percent.

EXAMPLE Solve the problems: (a) 20% of 400 is what number? (b) 20% of what number is 80? (c) 80 is what percent of 400?

(a) 20% of 400 is what number?

$20\% = \text{rate}$	Identify known values and missing value.
$400 = \text{base}$	
Portion is missing.	Choose the appropriate formula.
$P = R \cdot B$	Substitute values using the decimal equivalent of 20%.
$P = 0.2 \cdot 400$	Perform calculation.
$P = \mathbf{80}$	Interpret result.
20% of 400 is 80.	

(b) 20% of what number is 80?

$20\% = \text{rate}$	Identify known values and missing value.
$80 = \text{portion}$	
Base is missing.	Choose the appropriate formula.
$B = \dfrac{P}{R}$	Substitute values. Change the percent to a decimal.
$B = \dfrac{80}{0.2}$	Perform calculation.
$B = \mathbf{400}$	Interpret result.
20% of 400 is 80.	

(c) 80 is what percent of 400?

$80 = \text{portion}$	Identify known values and missing value.
$400 = \text{base}$	
Rate is missing.	Choose the appropriate formula.
$R = \dfrac{P}{B}(100\%)$	Substitute values.
$R = \dfrac{80}{400}(100\%)$	Perform calculation.
$R = \mathbf{20\%}$	Interpret result.

80 is 20% of 400.

The next example shows how to use the fractional equivalent of a mixed-number percent.

EXAMPLE $33\frac{1}{3}\%$ of 282 is what number?

The rate is $33\frac{1}{3}\%$. The key word *of* tells us that 282 is the base. The portion is missing.

$P = RB$ Substitute $\frac{1}{3}$ for $33\frac{1}{3}\%$. $33\frac{1}{3}\% = \frac{33\frac{1}{3}\%}{100\%} = \frac{1}{3}$.

$P = \frac{1}{3}(282)$ Multiply.

$P = \frac{282}{3}$ Divide.

$P = 94$

$33\frac{1}{3}\%$ of 282 is 94.

How Many Digits Do You Use When Using a Calculator?

Even with a calculator, you may prefer to use the fractional equivalent because the decimal equivalent of $33\frac{1}{3}\%$ is a repeating decimal and requires rounding. Examine the effect of rounding to various places.

$$0.3 \cdot 282 = 84.6$$
$$0.33 \cdot 282 = 93.06$$
$$0.333 \cdot 282 = 93.906$$
$$0.3333 \cdot 282 = 93.9906$$
$$0.333333333 \cdot 282 = 93.99999991$$

Because the desired degree of accuracy varies depending on the problem, it is advisable to find the exact answer on the calculator using the fractional equivalent of the percent. A close approximate answer can be found by using the full calculator value for nonterminating decimal equivalents.

It is advisable to anticipate the approximate size of an answer before making any calculations. This approximation helps you discover mistakes, especially when using a calculator.

Estimating a Percent of a Number

Estimate a percent by comparing it to a percent that you can calculate mentally.

To find

1% of a number	*Multiply by 0.01*	Move decimal two places to the left.
2% of a number	$2 \cdot 1\%$ of a number	
3% of a number	$3 \cdot 1\%$ of a number	
10% of a number	*Multiply by 0.1*	Move decimal one place to the left.
5% of a number	$\frac{1}{2}$ of 10% of a number	
20% of a number	$2 \cdot 10\%$ of a number	
30% of a number	$3 \cdot 10\%$ of a number	

50% of a number	$\frac{1}{2}$ of a number	Divide by 2.
25% of a number	$\frac{1}{4}$ of a number	Divide by 4.
75% of a number	$3 \cdot 25\%$ of a number	
$33\frac{1}{3}\%$ of a number	$\frac{1}{3}$ of a number	Divide by 3.
$66\frac{2}{3}\%$ of a number	$2 \cdot 33\frac{1}{3}\%$ of a number	

EXAMPLE   Find 10% of $51.00.

10% of $51.00 = **$5.10.**   Mentally, move the decimal point in the number one place to the left.

EXAMPLE BUS   Estimate a 15% tip on a restaurant bill of $48.18.

10% of $48.00 = $4.80   Find 10% of the rounded total bill.
                             $48.18 rounded to $48.00.

5% of $48.00 = $\frac{1}{2}$ of $4.80 = $2.40   Find 5% of the rounded total bill.

15% = 10% + 5%

$4.80 + $2.40 =   Add the amounts for 10% and 5%.

$7.20   15% tip.

EXAMPLE BUS   A taxi fare is $24.00. Find the amount of a 20% tip.

10% of $24.00 = $2.40   Find 10% of the fare.
2 · $2.40 = $4.80   Double the 10% amount.

A 20% tip is $4.80.

1% of a number can be found mentally. When using a percent that is less than 1%, the portion is less than 1% of the number. Using 1% of the number as an estimate is very useful in checking the decimal placement.

EXAMPLE $\frac{1}{4}\%$ of 875 is what number?

$\frac{1}{4}\%$ is less than 1% and 1% of 875 is 8.75. Therefore, $\frac{1}{4}\%$ of 875 is less than 8.75. Since $\frac{1}{4}\%$ and 0.25% are equivalent, either can be used to solve this problem. We are given the rate and the base, and need to find the portion.

$P = RB$ Substitute known values for the rate and base.
$P = 0.25\%(875)$ Change rate to the decimal equivalent.
$P = 0.0025(875)$ Multiply.
$P = 2.1875$

$\frac{1}{4}\%$ of 875 is 2.1875.

100% of a number is 1 times the number, or the number itself. When using percents that are larger than 100%, the portion will be more than the original number.

EXAMPLE 325% of 86 is what number?

100% of 86 is 86. 325% is more than 3 times 100%. $3 \cdot 86 = 258$. So, 325% of 86 is more than 258. The rate is 325%, the base is 86, and the portion is missing.

$P = RB$ Substitute known values.
$P = 325\%(86)$ 325% = 3.25
$P = 3.25(86)$ Multiply.
$P = 279.5$

325% of 86 is 279.5.

In the following example the rate is more than 100% so we expect the base to be less than the portion.

EXAMPLE 398.18 is 215% of what number?

We are looking for the base, as indicated by the key word *of*. We are given the portion and the rate. Because the rate is more than 100%, we expect the base to be smaller than the portion.

$B = \dfrac{P}{R}$ Substitute values.

$B = \dfrac{398.18}{215\%}$ 215% = 2.15

$B = \dfrac{398.18}{2.15}$ Divide.

$B = 185.2$

398.18 is 215% of 185.2.

3 Solve Percent Problems Using the Percentage Proportion.

A practical method for solving percent problems involves solving the percentage proportion. The equivalent fractions $\frac{5}{10}$ and $\frac{1}{2}$ can be written as the proportion $\frac{5}{10} = \frac{1}{2}$. Each fraction is also called a **ratio**, and two ratios that are equal or equivalent form a **proportion**.

To solve a proportion with a missing element, we use the property that the *cross products* in a proportion are equal. In a proportion, the **cross products** are the product of the numerator of the first fraction times the denominator of the second, and the product of the denominator of the first fraction times the numerator of the second. In the proportion $\dfrac{a}{b} = \dfrac{c}{d}$, the cross products are $a \cdot d$ and $b \cdot c$.

Property of proportions:

The cross products in a proportion are equal. Symbolically, if $\dfrac{a}{b} = \dfrac{c}{d}$, then $a \cdot d = b \cdot c$, provided that b and d are not equal to zero. Also, if $a \cdot d = b \cdot c$, then $\dfrac{a}{b} = \dfrac{c}{d}$.

This property can also help us find a missing element when three of the four elements of a proportion are known.

EXAMPLE   Find the value for a in $\dfrac{a}{12} = \dfrac{3}{9}$.

$$\frac{a}{12} = \frac{3}{9}$$

$a \cdot 9 = 12 \cdot 3$ Find the cross products.

$a \cdot 9 = 36$ The product and one factor are known, so divide the product by the known factor to find the missing factor.

$$a = \frac{36}{9}$$

$a = 4$

A formula that can be used to find any missing element of a percentage problem is the **percentage proportion.** The relationships among the rate, portion, base, and the standard unit of 100 are represented by two fractions that are equal to each other.

Percentage proportion formula:

$\dfrac{R}{100} = \dfrac{P}{B}$ where $\dfrac{R}{100}$ = fractional form of rate or percent, P = portion or part, B = base or total

In a percent problem, if we know any two of the elements, we can find the third element using the property of proportions and cross multiplication. This process is called **solving the proportion.**

To solve the percentage proportion:

1. Cross multiply to find the cross products.
2. Divide the product of the two known factors by the factor with the letter.

Note: An algebraic explanation for solving a proportion is presented in Chapter 8.

EXAMPLE What is 20% of 75?

The rate is 20%, the base is 75, and the portion is missing.

$$\frac{R}{100} = \frac{P}{B}$$ Set up the proportion and substitute the known elements.

$$\frac{20}{100} = \frac{P}{75}$$ Cross multiply.

$$20 \times 75 = 100 \cdot P$$

$$1{,}500 = 100 \cdot P$$ Divide to find P.

$$\frac{1{,}500}{100} = P$$

$$15 = P$$

The portion is 15.

Multiply, Then Divide

Solving percentage proportions always involves one multiplication and one division. One cross product will be two numbers—this is the multiplication. The other cross product will be one number and one letter—this is the division. We divide the cross product with two numbers by the one number in the other cross product.

     Cross product with two numbers is $20 \cdot 75$.
Cross product with one number is $100 \cdot P$. Divide by 100.

$$20 \cdot 75 \div 100 = P$$

$$1{,}500 \div 100 = P$$

$$15 = P$$

To solve proportions with a calculator, we follow a continuous series of steps.

$$20 \; \boxed{\times} \; 75 \; \boxed{\div} \; 100 \; \boxed{=} \; \Rightarrow \; 15$$

Advantages of Using the Percentage Proportion

Using percentage proportions gives us several advantages:

- Only one formula is needed to solve problems for portion, rate, and base.
- A standardized approach to percentage problems works in all cases.
- The standard 100 takes care of having to convert decimal numbers or fractions to percents or to convert percents to decimal or fraction equivalents.
- All problems involve one multiplication and one division step.
- Multiplication by 100 or division by 100 can be done mentally.

20% of what number is 45?

This time we know the rate, 20%, and the portion, 45. We are looking for the base, as indicated by the key word *of*. The base is the original or whole amount. The base is larger than the portion when the rate is less than 100%.

Option 1	Option 2	Option 3	Option 4
$\dfrac{20}{100} = \dfrac{45}{B}$	$\dfrac{1}{5} = \dfrac{45}{B}$	$\dfrac{0.2}{1} = \dfrac{45}{B}$	$B = \dfrac{P}{R}$
$20 \cdot B = 100 \cdot 45$	$B = 5 \cdot 45$	$45 = 0.2 \cdot B$	$B = \dfrac{45}{0.2}$
$20 \cdot B = 4{,}500$	$B = 225$	$\dfrac{45}{0.2} = B$	$B = 225$
$B = \dfrac{4{,}500}{20}$		$225 = B$	
$B = 225$			

20% of 225 is 45.

When solving applied problems, our first task is to identify the two given parts and determine which part is missing.

EXAMPLE
INDTR
If a type of solder contains 55% tin, how many pounds of tin are needed to make 10 lb of solder?

First, let's be sure we understand the word *solder*. **Solder** is a mixture of metals. In this problem, the *total amount* or the *base* is the 10 lb of solder, and it is made of tin and other metals.

Known facts
55% of 10 lb of solder is tin.
Rate: percent of tin = 55%
Base: amount of solder = 10 lb

Unknown fact
Portion or number of pounds of tin = P

Relationships
Percentage proportion

$$\frac{R}{100} = \frac{P}{B}$$ Substitute known values.

$$\frac{55}{100} = \frac{P}{10 \text{ lb}}$$

Estimation
To estimate, 55% is more than $\frac{1}{2}$. $\frac{1}{2}$ of 10 = 5. So the amount of tin should be more than 5 lb.

Calculations

$$\frac{55}{100} = \frac{P}{10}$$ Solve for P. Cross multiply.

$$55 \cdot 10 = 100 \cdot P$$

$$550 = 100 \cdot P$$ Divide by 100.

$$\frac{550}{100} = P$$

$$5\frac{1}{2} = P$$

Interpretation
Thus, $5\frac{1}{2}$ lb of tin are needed to make 10 lb of solder.
To check the reasonableness of the answer, $5\frac{1}{2}$ lb is a little more than 5 lb.

EXAMPLE AUTO	If a 150-horsepower (hp) engine delivers only 105 hp to the driving wheels of a car, what is the efficiency of the engine?

Efficiency means the *percent* the output (105 hp) is of the total amount (150 hp) the engine is capable of delivering. Thus, the base amount is 150 hp, the portion delivered is 105 hp, and the percent of 150 represented by 105 is the rate or efficiency.

Known facts
Portion

Base: Total amount of horsepower = 150 hp
Portion: Amount of horsepower the engine delivers = 105 hp

Unknown facts

Efficiency or percent of the total horsepower

Relationships

Percentage proportion

$$\frac{R}{100} = \frac{P}{B}$$ Substitute known values.

$$\frac{R}{100} = \frac{105}{150}$$

Estimation

Because the engine is not operating at full capacity (150 hp), we expect the efficiency to be less than 100%. Since 105 is more than $\frac{1}{2}$ of 150, the percent or rate will be more than 50%. A more precise estimate is

$$\frac{100}{150} = \frac{2}{3} = 66\frac{2}{3}\%$$ Round 105 to 100.

Since $105 > 100$ > is read "is greater than."

$$\text{Rate} > 66\frac{2}{3}\%$$

Calculations

$$\frac{R}{100} = \frac{105}{150}$$ Cross multiply.

$$R \cdot 150 = 100 \cdot 105$$

$$R \cdot 150 = 10{,}500$$ Divide.

$$R = \frac{10{,}500}{150}$$

$$R = 70$$

Interpretation

The engine is 70% efficient.

TIP

Must I Write All the Steps in the Six-Step Problem-Solving Plan for Every Problem?

As you develop confidence in your problem-solving ability you will begin to instinctively think through the steps without writing them down. We will begin to omit some of the steps in our examples, but will continue to draw special attention to the two *often neglected* steps: Estimation and Interpretation.

The effective value of current or voltage in an ac circuit is 71.3% of the maximum voltage. If a voltmeter shows a voltage of 110 volts (V) as the effective value in a circuit, what is the maximum voltage?

71.3% of the maximum voltage is 110 V. The maximum voltage is the *base*, and the amount of voltage shown on the voltmeter (110 V) is the *portion*.

Estimation

The rate is less than 100%. Therefore, the base is larger than the portion. $B > 110$.

$$\frac{71.3}{100} = \frac{110}{B}$$ Substitute known values. Cross multiply.

$71.3 \cdot B = 100 \cdot 110$

$71.3 \cdot B = 11{,}000$ Divide.

$$B = \frac{11{,}000}{71.3}$$

$B = 154.2776999$

or

$B = 154$ V   To the nearest volt

Interpretation

The maximum voltage is 154 V.

4 Solve Business and Consumer Problems Involving Percents.

No matter what career you pursue, you will encounter business and consumer applications of percents. Spreadsheets are commonly used to solve these applications, and spreadsheets use formulas. Generally, using the appropriate version of the percentage formula is the most efficient method for working with spreadsheets. If you are solving an application using a calculator, either the percentage formula or percentage proportion can be used.

A 5% sales tax is levied on an order of building supplies costing $127.32. What is the amount of sales tax to be paid? What is the total bill?

To find the amount of sales tax to be paid, we need to find 5% of $127.32. The amount of sales tax is the portion, and the cost of the supplies is the base.

Estimation

Tax: 10% of $127.32 = $12.73. $\frac{1}{2}$ of 10% = 5%. $\frac{1}{2}$ of $12.73 > $6.
5% would be greater than $6.
Total bill would be greater than $127 + $6 or $133.

$P = R \cdot B$ *P* is the amount of sales tax.
5% is the percent or sales tax rate.
Substitute known values.

$P = 5\% \cdot \$127.32$ Change 5% to a decimal equivalent.

$P = 0.05(\$127.32)$ Multiply.

$P = \$6.366$ Sales tax

Money amounts are usually rounded to the nearest cent. **Thus, the sales tax is $6.37**.
   To find the total amount to be paid, we add the cost of the supplies and the sales tax.

$\$127.32 + \$6.37 = \mathbf{\$133.69}$ Total bill = Cost of supplies + Sales tax.

Interpretation

The amount of sales tax to be paid is $6.37 and the total bill is $133.69. This amount is very close to our estimation of $133 for the total bill we pay.

EXAMPLE
BUS

If the rate of social security tax is 6.2% of the first $84,900 of earnings in a given year, how much social security tax is withheld on a weekly paycheck of $425 if the total earnings for the year is less than $84,900?

The amount of pay before any deductions are made is called the *gross pay*, and it is the base. The amount of social security tax is the portion. The rate is 6.2%.

Estimation

10% of $425 = $42.50

5% of $425 = $\frac{1}{2}$ of $42.50 = $21.25   6.2% is between 10% and 5%.

6.2% of $425 < $42.50   < is read "is less than."

6.2% of $425 > $21.25   > is read "is greater than."

$P = R \cdot B$ *P* is the amount of social security deduction.

$P = 6.2\% \cdot \$425$ *R* is 6.2%, *B* = $425.
        Substitute known values.
$P = 0.062(\$425)$ Change 6.2% to its decimal equivalent.

$P = \$26.35$ Multiply.

Interpretation   **$26.35 is withheld for social security tax.**

Salaries of salespeople are sometimes based on the amount of sales made. We call this selling on **commission;** such a salary is usually a certain percent of the sales on which the commision is to be paid.

EXAMPLE
AUTO

An automotive parts salesperson earns a salary of $325 per week and 8% commission on all sales over $2,500 per week. The sales during one week were $4,875. What is the salesperson's total pay for that week?

Estimation

To estimate, round the sales to $5,000.

$5,000 - $2,500 = $2,500 eligible for commission.   Round 8% to 10%.
    10% of $2,500 = $250   Estimated commission.
        $250 + $325 = $575   Estimated total pay.

Commission rate and sales are both rounded up so the estimate is more than the actual amount; therefore, the commission is less than $250 and the total salary is less than $575.

First, we need to determine the amount of sales on which the salesperson will earn a commission.

$4,875 - 2,500 = $2,375   Sales that earn a commission.

The salesperson receives an 8% commission on $2,375 (base).

$P = R \cdot B$ *P* is the amount of the commission
        $2,375 is the base on which the commission is earned
$P = 8\%(\$2,375)$ Change 8% to its decimal equivalent.

$P = 0.08(\$2,375)$ Multiply.

$P = \$190$ Commission

The salesperson will receive a $325 base salary and $190 in commission.

$325 + 190 = $515

Interpretation   **The total pay for that week is $515.**

Interest is the amount charged for borrowing or lending money, or the amount of money earned when money is saved or invested. The **amount of interest** is the *portion,* the total amount invested or borrowed is the **principal** or *base,* and the **percent of interest** is the *rate.*

The rate of interest is expressed as a percent per time period. For example, the rate of interest on a loan may be 8% per year, or **per annum.** The rate of interest or finance charge on a charge account may be $1\frac{1}{2}$% per month.

There are many different ways to figure interest. Most banks or loan institutions use compound interest; some institutions use the exact time of a loan; others use an approximate time of a loan, such as 30 days per month or 360 days per year. However, the truth-in-lending law requires that all businesses equate their interest rate to an annual simple interest rate known as the **annual percentage rate (APR)** or **annual percentage yield (APY).** This allows consumers to compare rates of various institutions and to understand exactly what rate they are earning or are being charged.

Simple interest for one time period can be found using the formula $I = PRT$, which is similar to the percentage formula. The time period for the rate should match the unit of time for T.

EXAMPLE **BUS**	A credit-card company charges a finance charge (interest) of $1\frac{1}{2}$% per month on the average daily balance of the account. If the average daily balance on an account is \$157.48, what is the finance charge for the month?

Estimation

1% of \$157.48 = \$1.57
Since $1\frac{1}{2}$% > 1%, the finance charge > \$1.57

$P = R \cdot B$	P is the amount of finance charge for 1 month.
$P = 1.5\%(\$157.48)$	$1\frac{1}{2}$ is the percent of finance charge for 1 month. $1\frac{1}{2} = 1.5$.
$P = 0.015(\$157.48)$	Multiply.
$P = \$2.3622$	
$P = \$2.36$	To the nearest cent

Interpretation **The finance charge is \$2.36.**

EXAMPLE **BUS**	\$1,250 is invested for 6 months at an interest rate of $8\frac{1}{2}$% per year. Find the simple interest earned. Use the simple interest formula $I = PRT$.

Estimation

10% of \$1,250 = \$125	Estimated interest for one year.
6 months = $\frac{1}{2}$ year; $\frac{1}{2}$ of \$125 = \$62.50	Estimated interest for six months.

Since $8\frac{1}{2}$% is less than 10%, the interest < \$62.50

Use the simple interest formula $I = PRT$.

$I = PRT$	P is the principal or base of \$1,250
$I = (\$1,250)(8\frac{1}{2}\%)(\frac{1}{2})$	$8\frac{1}{2}$% is the rate R for 1 year. The time T is 6 months, or $\frac{1}{2}$ year
$I = (\$1,250)(0.085)(0.5)$	Change $8\frac{1}{2}$% and $\frac{1}{2}$ to decimal equivalents.
$I = \$53.125$	Multiply.
$I = \$53.13$	To the nearest cent

Interpretation **The interest earned is \$53.13.**

Identify the Rate, Base, and Percentage in Applied Problems

Applied problems that involve percents are solved using the percentage formula or percentage proportion. For each type of application it is still important to identify the rate, base, and portion.

	Rate	Base	Portion
Sales tax	Sales-tax rate	Purchase price Marked price	Amount of sales tax
Discount	Discount rate	Original price	Amount of discount
Commission	Commission rate	Commissable sales	Amount of commission
Interest	Interest rate per year	Total investment Total loan Principal	Amount of interest per year

SECTION 3–2 SELF-STUDY EXERCISES

1 Identify the given and missing elements as R (rate), B (base), and P (portion). Do not solve.

1. 40% of 18 is what number?

2. HLTH/N   A mail-order pharmacy offers Allegra® at 66% of the retail price of $35.99 for a month's supply. What is the mail-order price?

3. HLTH/N   Premarin® tablets have a discounted price of $12.21, which is 27% of the retail price. What is the retail price?

4. HLTH/N   A nurse spent 83% of a 12-h shift caring for patients. How many hours were devoted to patient care?

5. What percent of 10 is 2?

6. 2 is 20% of what number?

7. Three books is what percent of four books?

8. What percent of 25 students is 5 female students?

9. 6 of 15 motorists is what percent?

10. How many nurses is 20% of the 15 nurses on duty?

11. 35% of how many pieces of sod is 70 pieces?

12. What percent of a total bill of $45 is $3.15?

2 Solve using the percentage formula.

13. 30% of 18 is what number?

14. 42% of 600 is what number?

15. 1.8% of 100 is what number?

16. What is $33\frac{1}{3}$% of 78?

17. What percent of 18 is 9?

18. What percent of 15.6 is 52?

19. What percent of 88 is 77?

20. 150 is what percent of 100?

21. 125% of what number is 80?

22. 0.01% of what number is 7?

3 Solve using the percentage proportion.

23. 20% of 375 is what number?

24. 75% of 84 is what number?

25. $66\frac{2}{3}$% of 309 is what number?

26. 34.5% of 336 is what number?

27. $\frac{3}{4}$% of 90 is what number?

28. 0.2% of 470 is what number?

29. What number is 134% of 115?

30. Find 275% of 84.

31. 400% of 231 is what number?

32. $37\frac{1}{2}$% of 920 is what number?

33. What percent of 348 is 87?

34. What percent of 350 is 105?

35. 72 is what percent of 216?

36. 28 is what percent of 85 (to the nearest tenth percent)?

37. 37.8 is what percent of 240?

38. What percent of 175 is 28?

39. 32 is what percent of 4,000?

40. What percent is 2 out of 300?

41. What percent of 125 is 625?

42. 173.55 is what percent of 156?

43. 50% of what number is 36?

44. 60% of what number is 30?

45. $12\frac{1}{2}$% of what number is 43?

46. 15.87 is 34.5% of what number?

47. $\frac{2}{3}$% of what number is $2\frac{2}{5}$?

48. 0.3% of what number is 0.825?

49. 150% of what number is $112\frac{1}{2}$?

50. 43% of what number is 107.5?

51. 92 is 500% of what number?

52. $133\frac{1}{3}$% of what number is 348?

53. Find the percentage if the base is 75 and the rate is 5%.

54. Find the percentage if the base is 25 and the rate is 2.5%.

55. What is the rate if the base is 10.5 and the percentage is 7?

56. Find the rate when the base is 80 and the percentage is 30.

57. If the percentage is 4.75 and the rate is $33\frac{1}{3}$%, find the base.

58. Find the base when the rate is 15% and the percentage is 52.5.

59. If the rate is $12\frac{1}{2}$% and the base is 75, find the percentage.

60. If the percentage is 11 and the rate is 5%, find the base.

61. Find the rate when the percentage is 15 and the base is 75.

62. What is the base if the percentage is 35 and the rate is 17.5%?

4

63. **INDTEC**   Cast iron contains 4.25% carbon. How much carbon is contained in a 25-lb bar of cast iron?

64. **AG/H**   3,645 rolls of landscape fabric are manufactured during one day. After being inspected, 121 of these rolls are rejected as imperfect. What percent of the rolls is rejected? (Round to the nearest whole percent.)

65. **AUTO**   An engine operating at 82% efficiency transmits 164 hp. What is the engine's maximum capacity (base) in horsepower?

66. **AG/H**   If wrought iron contains 0.07% carbon, how much carbon is in a 30-lb bar of wrought iron?

67. **ELEC**   The voltage of a generator is 120 V. If 6 V are lost in a supply line, what is the rate of voltage loss?

68. **AG/H**   A landscape contractor figures it costs $\frac{1}{2}$% of the total cost of a job to make a bid. What would be the cost of making a bid on a $115,000 job?

69. **INDTEC**   A certain ore yields an average of 67% iron. How much ore is needed to obtain 804 lb of iron?

70. **CON**   A contractor makes a profit of $12,350 on a $115,750 job. What is the percent of profit? (Round to the nearest whole percent.)

71. **HLTH/N**   385 defective alcohol swabs were produced during a day. If 4% of the alcohol swabs produced were defective, how many alcohol swabs were produced in all?

72. **INDTR**   In a welding shop, 104,000 welds are made. If 97% of them are acceptable, how many are acceptable?

73. Estimate a 15% tip on a restaurant bill of $31.15.

74. Estimate a 15% tip for a taxi fare of $43.00.

75. $\frac{1}{8}$% of 320 is what number?

76. $\frac{1}{10}$% of 400 is what number?

77. **BUS**   Find the monthly income a prospective buyer must have to qualify for a home that has a monthly cost (including taxes and insurance) of $1,200 if the mortgage company requires the monthly cost to be no more than 28% of the buyer's gross monthly income.

78. **BUS**   Chloe Duke earns $4,872 monthly and wants to purchase a home, but the mortgage firm requires the monthly payment to be no more than 28% of her gross monthly income. What is the most the monthly mortgage payment can be, including taxes and insurance?

79. **BUS**   Find the sales tax and the total bill on an order of office supplies costing $75.83 if the tax rate is 6%. Round to the nearest cent.

80. **AG/H**   Materials to landscape a property total $785.84. What is the total bill if the sales tax rate is $5\frac{1}{2}$%? Round to the nearest cent.

81. **BUS**   If the rate of social security tax is 6.2%, find the tax on gross earnings of $375.80. Round to the nearest cent.

82. **BUS**   An employee's gross earnings for a pay period are $895.65. The net pay for this salary is $675.23. What percent of the gross pay are the total deductions? Round to the nearest whole percent.

83. **BUS**   An employee has a net salary of $576.89 and a gross salary of $745.60. What percent of the gross salary is the total of the deductions? Round to the nearest whole percent.

84. **BUS**   Madison Duke earns $2,892 in one pay period. Medicare tax is 1.45% of the earnings. How much is deducted for Medicare tax?

85. **BUS** A real estate salesperson earns 4% commission. What is the commission on a property that sold for $295,800?

86. **INDTEC** A manufacturer gives a 2% discount to customers paying cash. If a parts store pays cash for an order totaling $875.84, what amount is saved? Calculate to the nearest cent.

87. **HLTH/N** Find the cash price for an order of hospital supplies totaling $3,985.57 if a 3% discount is offered for cash orders. Calculate to the nearest cent.

88. **BUS** What commission is earned by a salesperson who sells $18,890 in merchandise if a 5% commission is paid on all sales?

89. **TELE** A telecommunications representative is paid a salary of $140 per week and 7% commission on all sales over $3,200 per week. The sales for a recent week were $7,412. What is the representative's salary for that week?

90. **COMP** A computer store manager is paid a salary of $2,153 monthly plus a bonus of 1% of the net earnings of the business. Find the total salary for a month when the net earnings of the business are $105,275.

91. **ELEC** An electrician purchases $650 worth of electrical materials. A finance charge of $1\frac{1}{2}\%$ per month is added to the bill. What is the finance charge for 1 month?

92. **BUS** If $10,000 is invested for 3 months at 6% per year, how much interest is earned?

93. **HLTH/N** A nurse paid $10.24 in monthly interest on a credit-card account that had an average daily balance of $584.87. Find the monthly rate of interest. Round to the nearest hundredth of a percent.

94. **BUS** Find the interest on a loan of $2,450 at 7% per year for 1 year.

95. **BUS** Find the interest on a loan of $5,840 at 10% per year for 2 years, 6 months.

96. **BUS** Find the annual interest rate if a deposit of $5,000 earns $125 in 1 year.

97. **HLTH/N** The adult human skeleton consists of 206 bones. The face has 14 bones. What percent of human bones are found in the face? Round to the nearest tenth percent.

98. **HLTH/N** Each lower limb of the adult human skeleton has 31 bones. What percent of the 206 bones are found in both lower limbs? Round to the nearest tenth percent.

99. **HLTH/N** Approximately 60% of an adult man's body is water. A male that weighs 185 lb has approximately how many pounds of water?

100. **HLTH/N** The average American consumes about $3\frac{1}{2}$ lb of sodium each year. If salt is 40% sodium, how many pounds of salt are consumed, on average, each year by an individual?

101. **CON** Home heat loss through poor-fitting doors and windows can account for 15% of a homeowner's energy cost. A January energy bill is $219.42. Find the cost of unnecessary heat loss.

102. **INDTEC** Alumina makes up 45% of high grade aluminum ore. If 48,000 lb of high-grade aluminum ore are mined, how many pounds of alumina can be obtained?

103. **INDTEC** How many pounds of pure sulfur are contained in 3,400 lb of coal that has 5% pure sulfur content?

104. **INDTEC** ACME Products estimates a product costs $42.80 to manufacture. How much did it cost to make the product if a 12% cost overrun occurred?

3–3 | Increase and Decrease

Learning Outcomes

1 Find the amount of increase or decrease.

2 Find the new amount directly in increase or decrease problems.

3 Find the rate or base in increase or decrease problems.

1 Find the Amount of Increase or Decrease.

Percents are often used in problems dealing with increases or decreases.

TIP

Relating Increases and Decreases to Percents

Increases and decreases are applications of the percentage formula.

	Rate	**Base**	**Portion**
Increase	Rate of increase	Original amount	Amount of increase
Decrease	Rate of decrease	Original amount	Amount of decrease

New amount for increase   =   Original amount   +   Amount of increase
New amount for decrease   =   Original amount   −   Amount of decrease

---

EXAMPLE
HLTH/N

Medical assistants are to receive a 9% increase in wages per hour. If they were making $19.25 an hour, what is the amount of increase per hour (to the nearest cent)? Also, what is the new wage per hour?

The original wage per hour is the base, and we want to find the amount of increase (portion).

Estimation

10% of $19.25 is $1.92. The increase will be less than $1.92. The new wage per hour will be approximately $21.

$P = RB$ *P* represents amount of increase, 9% is the rate of increase, and *B* is the original wage per hour.

$P = 9\%(\$19.25)$ 9% = 0.09
$P = 0.09(19.25)$ Multiply.
$P = 1.7325$ $1.73 to the nearest cent is the amount of increase.

Interpretation

The medical assistants will receive a $1.73 per-hour increase in wages.

$19.25 + $1.73 = $20.98   New amount = original amount + amount of increase.

Their new hourly wage will be $20.98.

EXAMPLE
INDTEC

Molten iron shrinks 1.2% while cooling. What is the cooled length of a piece of iron if it is cast in a 24-cm pattern?

First, we find the amount of shrinkage (amount of decrease, portion). The original amount, 24 cm, is the base.

Estimation

The cooled piece will be less than 24 cm.

$P = RB$ *P* is the amount of decrease, 1.2% is the rate of decrease, and the base is the original amount, or 24 cm.

$P = 1.2\%(24)$ 1.2% = 0.012
$P = 0.012(24)$ Multiply.
$P = 0.288$ Amount of shrinkage (decrease)

Interpretation

The amount of shrinkage is 0.288 cm, so the length of the cooled piece (new amount) is

24 − 0.288 = 23.712 cm   New amount = original amount − amount of decrease.

2 Find the New Amount Directly in Increase or Decrease Problems.

When knowing the amount of increase is not necessary, the new amount can be found directly.

New Amounts and New Rates

	Rate	Base	Portion
New amount for increase	New rate *100% + rate of increase*	Original amount	New amount *original + increase*
New amount for decrease	New rate *100% − rate of decrease*	Original amount	New amount *original − decrease*

EXAMPLE INDTR

A 3% error is acceptable for a machine part to be usable. If the part is intended to be 57 cm long, what is the range of measures that is acceptable for this part?

The machine part can be ±3% from the ideal length of 57 cm. The range of acceptable measures is found by calculating the smallest acceptable measure and the largest acceptable measure. Recall that the symbol ± is read *plus or minus*. It means that the measure can be more (+) or less (−) than the designated amount. In this case, the part can be 3% longer than or shorter than 57 cm and still be usable. The smallest acceptable value is 97% of the ideal length (100% − 3%).

$P = RB$ P is the smallest acceptable amount because 97% is the *smallest acceptable percent*. The base is 57 cm.

$P = 97\%(57)$ 97% = 0.97

$P = 0.97(57)$ Multiply.

$P = 55.29$

Interpretation

The smallest acceptable measure is 55.29 cm.

The largest acceptable value is 103% of the ideal length (100% + 3%).

$P = RB$ P is the largest acceptable amount because 103% is the *largest acceptable percent*. The base is 57 cm.

$P = 103\%(57)$ 103% = 1.03

$P = 1.03(57)$ Multiply.

$P = 58.71$

Interpretation

The largest acceptable value is 58.71 cm. The range of acceptable measures is from 55.29 cm to 58.71 cm.

It is helpful in developing our number sense with percents to think of percents in pairs. 100% is one whole quantity. A percent that is less than 100% represents part of a quantity. For example, 30% represents part of one quantity. Then, 70% is the rest of the quantity if 30% is removed. This percent, 70%, is the complement of 30%. The **complement of a percent** is the difference between 100% and a given percent.

To find the complement of a percent:

Subtract the percent from 100%.

EXAMPLE Find the complement of **(a)** 25%, **(b)** 80%, and **(c)** 36%.

(a) $100\% - 25\% = \mathbf{75\%}$ Subtract from 100%.
(b) $100\% - 80\% = \mathbf{20\%}$ Subtract from 100%.
(c) $100\% - 36\% = \mathbf{64\%}$ Subtract from 100%.

To estimate the amount you pay for a sale item:

1. Round the original price and the percent to numbers you can work with mentally.
2. Find the complement of the rounded percent.
3. Relate the complement to 10% by dividing it by 10%.
4. Find 10% of the rounded original price.
5. Multiply the results from Steps 3 and 4.

EXAMPLE
BUS

Estimate the amount you pay on a $49.99 item advertised at 70% off.

$49.99 rounds to $50.
$100\% - 70\% = 30\%$ Complement of 70% (percent you pay).
$30\% \div 10\% = 3$ Relate to 10%.
10% of $50 = $5   Move decimal one place to the left.
$3 \cdot \$5 = \mathbf{\$15}$ Three times 10% of $50.

EXAMPLE
BUS

Estimate the amount you pay on a $28 item advertised at 18% off.

18% rounds to 20%. $28 rounds to $30.
$100\% - 20\% = 80\%$ Complement of 20% (percent you pay).
$80\% \div 10\% = 8$ Relate to 10%.
10% of $30 = $3   Move decimal one place to the left.
$8 \cdot \$3 = \mathbf{\$24}$ Eight times 10% of $30.

---

3 Find the Rate or Base in Increase or Decrease Problems.

Many kinds of increase or decrease problems involve finding either the rate or the base.

The rate is the **percent of change** or the **percent of increase or decrease.** The base is still the *original amount.*

EXAMPLE
AUTO A worn brake lining is measured to be $\frac{3}{32}$ in. thick. If the original thickness was $\frac{1}{4}$ in., what is the percent of wear?

First, the amount of wear is $\frac{1}{4} - \frac{3}{32}$.

$$\frac{8}{32} - \frac{3}{32} = \frac{5}{32}$$ Find the common denominator. Subtract.

The amount of wear (decrease) is the portion. The base is the original amount.

$$R = \frac{P}{B} \cdot 100\%$$ R is the percent of wear.
P is the amount of wear, or $\frac{5}{32}$ in. B is $\frac{1}{4}$ in.

$$R = \frac{\frac{5}{32}}{\frac{1}{4}} \cdot 100\%$$ Divide.

$$R = \frac{5}{32} \div \frac{1}{4} \cdot 100\%$$ Change to an equivalent multiplication.

$$R = \frac{5}{\overset{}{\underset{8}{32}}} \cdot \frac{\overset{1}{4}}{1} \cdot 100\%$$ Multiply.

$$R = \frac{5}{8} \cdot 100\%$$ Change to an equivalent percent.

$$R = \frac{5}{8} \cdot 100\%$$

$$R = \frac{5}{\underset{2}{8}} \cdot \frac{\overset{25}{100\%}}{1}$$ Reduce and multiply.

$$R = 62\frac{1}{2}\%$$

Interpretation **The percent of wear is $62\frac{1}{2}\%$.**

SECTION 3–3 SELF-STUDY EXERCISES

1 Solve.

1. Find the amount of increase if 432 is increased by 25%.

2. If 78 is increased by 40%, what is the new amount?

3. Find the amount of decrease if 68 is decreased by 15%.

4. If 135 is decreased by 75%, what is the new amount?

5. **HOSP** Jobs in the food preparation and serving occupation totaled 2,206,000 in 2000. According to the Bureau of Labor Statistics, jobs in this occupation are scheduled to grow by 30.5% by 2010. How many new jobs will there be in 2010? Find the total number of jobs expected in this occupation in 2010.

6. **INDTR** The Bureau of Labor Statistics notes that the occupation of customer service representative will be a fast-growing occupation from 2000 to 2010. In 2000 the number of jobs was 1,946,000 and an increase of 32.4% is expected. Find the number of new jobs expected and the total number of jobs available in 2010.

2 Solve.

7. Find the complement of 40%.

8. Find the complement of 18%.

9. Find the complement of 86.3%.

10. Find the complement of $33\frac{1}{3}$%.

11. Find the complement of 0.09%.

12. Find the complement of 1%.

13. **INDTR** Steel rods shrink 10% when cooled from furnace temperature to room temperature. If a tie rod is 30 in. long at furnace temperature, how long is the cooled tie rod?

14. **CON** A contractor needs 1,650 board feet of 1-in. × 8-in. common boards to subfloor a house. If she needs 10% extra flooring to allow for waste when the boards are laid square, how much flooring should she order?

15. **CON** If 17% extra flooring is needed to allow for waste when boards are laid diagonally, how much flooring should be ordered to cover 2,045 board feet of floor? Answer to the nearest whole board foot.

16. **CON** A construction company requires 25,400 bricks for a job. If it allows 2% more bricks for breakage, how many bricks must it order?

17. **CON** When making an estimate on a job, a contractor wants to make a 10% profit. If all the estimated costs are $15,275, what is the total bid of cost and profit for the job?

18. **CON** Rock must be removed from a highway right-of-way. If 976 cubic yards (yd³) of unblasted rock are to be removed, how many cubic yards is this after blasting? Blasting causes a 40% swell in volume.

19. **BUS** Ciara Walker was earning $49,860 and received a 7% raise. Find her new annual earnings directly.

20. **BUS** LaTreas Walker received a 3% salary increase on her weekly earnings of $1,982. What are her new weekly earnings?

21. **BUS** David Dawson earned $4,290 but paid 6.2% in social security taxes and 1.45% of his earnings in Medicare taxes. What was his net earnings after paying social security and Medicare taxes?

22. **BUS** Megan Anders purchased a swimsuit that was priced at $84.00 before a 30% discount. What was the sale price?

3 Solve.

23. **CON** The cost of a pound of nails increased from $2.36 to $2.53. What is the percent of increase to the nearest whole-number percent?

24. **AG/H** A landscape contractor estimated that materials, shrubs, saplings, and labor for a job would cost $5,385. An estimate 1 year later for the same job was $7,808, due to inflation. Find the percent of increase due to inflation to the nearest whole number.

25. **AG/H** A chicken farmer bought 2,575 baby chicks. Of this number, 2,060 lived to maturity. What percent loss was experienced by the chicken farmer?

26. **ELEC** An electrician recorded costs of $1,297 for a job. If he received $1,232 for the job, what was the percent of money lost on the job? Round to the nearest tenth of a percent.

27. **AUTO** An engine that has a 4% loss of power has an output of 336 hp. What is the input (base) horsepower of the engine?

28. **AG/H** A contractor ordered 10.5 yd³ of sand for a job which included a 5% allowance for waste. How much sand was needed for the job?

29. **CON** A floor that would normally require 2,580 board feet is to be laid diagonally. What percent waste allowance was allowed for flooring laid diagonally, if 3,019 board feet were ordered? Round to the nearest whole percent.

30. **BUS** A shop manager records a 14% loss on rivets for waste. If the shop ordered 28.5 lb of rivets to compensate for loss due to waste, how many pounds of rivets were needed?

31. **INDTR** Steel bars shrink 10% when cooled from furnace temperature to room temperature. If a cooled steel bar is 36 in. long, how long was it when it was formed?

32. **CON** The cost of No. 1 pine studs increased from $3.85 each to $4.62 each. Find the percent of increase.

33. **INDTEC** A 141-hp output is required for an engine. If there is a 6% loss of power, what amount of input horsepower (or base) is needed?

34. **HLTH/N** A month's supply of Prevacid® retails for $147.09 and is discounted to $129.06. Find the percent savings for the discounted drug to the nearest whole percent.

Learning Outcomes

What to Remember with Examples

Section 4–1

1 Change any number to its percent equivalent (pp. 133–134).

Multiply a number by 1 in the form of 100% to change a number to a percent. For a shortcut, move the decimal two places to the right.

Change $\frac{1}{2}$, 1.2, and 7 to percents.

$$\frac{1}{2} \cdot 100\% = \frac{100\%}{2} = 50\% \qquad 1.2 \cdot 100\% = 120\% \qquad 7 \cdot 100\% = 700\%$$

2 Change any percent to its numerical equivalent (pp. 135–137).

Divide by 1 in the form of 100% to change a percent to a number. Reduce if possible. For a shortcut, move the decimal two places to the left.

Change 7% to a fraction.

$$7\% \div 100\% = \frac{7}{100}$$

Change $1\frac{1}{4}\%$ to a fraction.

$$1\frac{1}{4}\% \div 100\% = \frac{5\%}{4} \div 100\% = \frac{5\%}{4} \cdot \frac{1}{100\%} = \frac{5}{400} = \frac{1}{80}$$

Convert 3.5% to a decimal.

$$3.5\% \div 100\% = 0.035 = 0.035$$

Convert 245% to a mixed number.

$$245\% \div 100\% = \frac{245\%}{100\%} = 2\frac{45}{100} = 2\frac{9}{20}$$

Convert 124.5% to a decimal.

$$124.5\% \div 100\% = 1.245 = 1.245$$

Convert $100\frac{1}{4}\%$ to a decimal.

$$100\frac{1}{4}\% = 100.25\%$$

$$100.25\% \div 100\% = 1.0025$$

Section 3–2

1 Identify the portion, base, and rate in percent problems (pp. 139–140).

Use the key words for rate (*percent* or %), base (*total, original*, associated with the word *of*), and portion (*part*, associated with the word *is*).

         R   *B* *P*
(a) What percent of 10 is 5?

   R   *B*   *P*
(b) 25% of what number is 3?

   P   *R*   *B*
(c) 3 is 20% of what number?

2 Solve percent problems using the percentage formula (pp. 140–144).

Use key words to identify the known elements and the missing element. Then select the appropriate percentage formula.

$$P = R \cdot B \qquad R = \frac{P}{B}(100\%) \qquad B = \frac{P}{R}$$

What amount is 5% of $200? *Of* identifies $200 as the base.
The percent or rate is 5%. The missing element is the portion.

$P = RB$	Select the appropriate percentage formula and substitute given amounts.
$P = 5\%(\$200)$	Change 5% to a fraction or decimal equivalent by dividing by 100%.
$P = 0.05(200)$	Multiply.
$P = \$10$	Portion.

What percent of 6 is 2?
Is suggests 2 is the portion. *Of* identifies 6 as the base. The rate is missing.

$R = \dfrac{P}{B}(100\%)$	Select the appropriate percentage formula and substitute given amounts.
$R = \dfrac{2}{6}(100\%)$	Reduce fraction or change to decimal equivalent by dividing.
$R = \dfrac{1}{3}(100\%)$	Change to a percent by multiplying by 100%.
$R = 33\dfrac{1}{3}\%$	Rate $\left(\dfrac{1}{3} \times \dfrac{100\%}{1} = \dfrac{100\%}{3} = 33\dfrac{1}{3}\%\right)$.

12 is 24% of what number?
The rate is 24%. 12 is the part or portion and is suggested by *is*. *Of* identifies "what number" as the missing base.

$B = \dfrac{P}{R}$	Select the appropriate percentage formula and substitute given amounts.
$B = \dfrac{12}{24\%}$	Change 24% to a decimal equivalent by dividing by 100%.
$B = \dfrac{12}{0.24}$	Divide.
$B = 50$	Base.

3 Solve percent problems using the percentage proportion (pp. 144–149).

To solve a proportion, multiply diagonally across the equal sign to find the cross products. Then divide the cross product of the two known factors by the number factor of the other cross product.

$\dfrac{3}{y} = \dfrac{2}{5}$	Multiply cross products diagonally across the equal sign.
$3 \times 5 = 2 \times y$	
$15 = 2 \times y$	Divide by 2 (the number factor of the other cross product).
$\dfrac{15}{2} = y$	Change to mixed-number or mixed-decimal equivalent.
$7\dfrac{1}{2} = y \qquad$ or $\qquad y = 7.5$	

Chapter Review of Key Concepts

What amount is 15% of $40? *Of* identifies $40 as the base.
The percent or rate is 15%. The missing element is the portion.

$$\frac{R}{100} = \frac{P}{B}$$ Identify the rate, base, and portion and substitute given amounts into the percentage proportion.

$$\frac{15}{100} = \frac{P}{\$40}$$ Multiply cross products.

$$15 \times \$40 = P \times 100$$

$$\$600 = P \times 100$$ Divide by 100.

$$\frac{\$600}{100} = P$$

$$\$6 = P$$ Portion.

4 Solve business and consumer problems involving percents. (pp. 149–152).

Business and consumer applications are often solved using the percentage formula and an electronic spreadsheet or calculator.

If the sales tax rate is 7%, find the tax and total bill on a purchase of $30.

$P = RB$	Estimation:
$P = 7\%(\$30)$	10% of $30 = $3
$P = 0.07(\$30)$	Sales tax < $3
$P = \$2.1$	Total bill < $30 1 $3 or < $33.
$P = \$2.10$	Interpretation:
$\$30 + \$2.10 = \$32.10$	Sales tax is $2.10 and total cost is $32.10.

Section 3–3

1 Find the amount of increase or decrease (pp. 154–155).

The original amount is the base. The new amount is the original amount plus the increase or the original amount minus the decrease. Subtract the new amount and the original amount to find the increase or decrease.

Julio made $15.25 an hour but took a 20% pay cut. What was the new hourly pay?

	Estimation:
$P = 20\%(\$15.25)$	10% of $15 is $1.50.
$P = 0.2(15.25)$	20% of $15 is $3.00. $15 − $3 = $12
$P = 3.05$	Hourly pay cut (decrease)
$\$15.25 − \$3.05 = \$12.20$	Original amount − decrease = new amount

2 Find the new amount directly in increase or decrease problems (pp. 156–157).

Add the percent of increase to 100% or subtract the percent of decrease from 100%. Use this new percent in the percentage formula.

A project requires 5 lb of galvanized nails. If 15% of the nails will be wasted, how many pounds must be purchased?

$P = 115\%(5)$	$B = 5$ lb, $R = 100\% + 15\% = 115\%$
	P is missing.
$P = 1.15(5)$	Estimation:
	10% of 5 = 0.5.

More than 0.5 or $\frac{1}{2}$ lb of nails must be added for waste.

Order > 5.5 lb.
Interpretation:

$P = 5.75$ lb   5.75 lb must be ordered.

3 Find the rate or base in increase or decrease problems (pp. 157–158).

Subtract the original amount and the new amount to find the amount of increase or decrease. Then use the percentage formula or percentage proportion, with R or B as the missing element.

A 5-in. power edger blade now measures $4\frac{3}{4}$ in. What is the percent of wear?

$$R = \left(\dfrac{\frac{1}{4}}{5}\right)(100\%)$$

$4\frac{4}{4} - 4\frac{3}{4} = \frac{1}{4}$. The portion is $\frac{1}{4}$. The base or original amount is 5 in. R is missing.

$$R = \left(\frac{1}{4} \div 5\right)(100\%)$$

Estimation:

10% of 5 in. = 0.5 or $\frac{1}{2}$ in.

$$R = \frac{1}{20}(100\%)$$

Wear $= \frac{1}{4}$ in.

Wear $< \frac{1}{2}$ in.

$$R = \frac{1}{20}(100\%)$$

Percent of wear $< 10\%$.

Interpretation:

$$R = 5\%$$

The percent of wear is 5%.

A PC has 20% of its hard drive storage capacity filled. If the PC now has 5.12 G of storage capacity available, what was the original storage capacity?

Current storage capacity available is 100% − 20% = 80% of the original storage capacity. 5.12 G is the portion. The original storage capacity is the base.

$$B = \frac{5.12}{80\%}$$

$$B = \frac{5.12}{0.8}$$

$$B = 6.4 \text{ G}$$

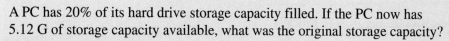

CHAPTER REVIEW EXERCISES

Section 3–1

Change to percent equivalents.

1. $\dfrac{72}{100}$ **2.** $\dfrac{9}{100}$ **3.** $\dfrac{23}{100}$ **4.** $\dfrac{87}{100}$

5. 0.7 **6.** 0.35 **7.** 0.83 **8.** $4\frac{1}{3}$

9. $3\frac{1}{5}$ **10.** $2\frac{1}{2}$ **11.** 125 **12.** $\dfrac{5}{6}$

13. 17.3 **14.** 18

Change to fraction equivalents.

15. 72% **16.** 40% **17.** $12\frac{1}{2}\%$ **18.** $16\frac{2}{3}\%$ **19.** $\frac{2}{3}\%$

20. $\frac{3}{5}\%$ **21.** 275% **22.** 124% **23.** $112\frac{1}{2}\%$ **24.** $183\frac{1}{3}\%$

Change to decimal equivalents.

25. 227.2% **26.** 73.8% **27.** 340% **28.** 92%

29. 83% **30.** 52.7% **31.** $62\frac{1}{2}\%$ **32.** $37\frac{1}{2}\%$

Section 3–2

Find the value of the letter in each proportion.

33. $\dfrac{1}{2} = \dfrac{a}{9}$ **34.** $\dfrac{x}{7} = \dfrac{3}{4}$ **35.** $\dfrac{7}{16} = \dfrac{21}{y}$ **36.** $\dfrac{3}{a} = \dfrac{2}{5}$

Identify the given and missing elements as R (rate), B (base), and P (portion). Do not solve.

37. 5% of 180 is what number?
38. $15 is what percent of $120?
39. 45% of how many dollars is $36?
40. What percent of 10 syringes is 2 syringes?
41. 6 is what percent of 25 sacks of grass seed?
42. How many landscape contractors is 15% of 40 landscape contractors?
43. 18% of 150 pieces of sod is how many?
44. What percent of a total of 28 students is 8 students?

Solve using the percentage formula.

45. 5% of 480 is what number?
46. $62\frac{1}{2}\%$ of 120 is what number?
47. $\frac{1}{4}\%$ of 175 is what number?
48. $233\frac{1}{3}\%$ of 576 is what number?
49. 39 is what percent of 65?
50. What percent of 118 is 42.48?
51. What percent of 65 is 162.5?
52. 80% of what number is 116?

Solve using the percentage proportion.

53. 24% of what number is 19.92?
54. 7.56 is $6\frac{3}{4}\%$ of what number?
55. 260% of what number is 395.2?
56. 3 is 0.375% of what number?
57. 38.25 is what percent of 250?
58. 83% of 163 is what number?
59. What percent of 26 is 130?
60. 4.75% of 348.2 is what number?
61. $10\frac{1}{3}\%$ of what number is 8.68?
62. INDTEC Specifications for bronze call for 80% copper. How much copper is needed to make 300 lb of bronze?

63. INDTEC Zinc makes up 84 lb of an alloy that weighs 224 lb. What percent of the alloy is zinc?

64. CON When a subfloor using common 1-in. × 8-in. boards is laid diagonally, 17% is allowed for waste. How many board feet will be wasted out of 1,250 board feet?

65. INDTEC Of the 2,374 pieces produced by a particular machine, 27 were defective. What percent (to the nearest hundredth of a percent) were defective?

66. ELEC The voltage loss in a line is 2.5 V. If this is 2% of the generator voltage, what is the generator voltage?

67. BUS If a family spends 28% of its income on food, how much of a $950 paycheck goes for food?

68. If a freshman class of 1,125 college students is made up of 8% international students, how many international students are in the class?

69. A city prosecuted 1,475 individuals with traffic citations. If 36,875 individuals received traffic citations, what percent was prosecuted?

70. INDTEC A survey studied 600 people for their views on nuclear power plants near their towns. Of these, 75 people said that they approved of nuclear power plants near their towns. What percent approved?

71. In one college, 67 students made the dean's list. If this was 33.5% of the student body, what was the total number of students in the college?

72. **BUS** It is estimated that only 19% of the licensed big game hunters on a state wildlife management area are successful. If 95 big game hunters were successful, what was the total number of big game hunters in a particular hunting season?

73. **BUS** Denise Knighton's gross pay is $3,296. She paid 6.2% in social security taxes. How much social security tax did she pay?

Solve.

74. **CON** Discontinued paneling is sale priced at 25% off the regular price. If the regular price is $12.50 per sheet, what is the sale price per sheet rounded to the nearest cent?

75. **BUS** A parts distributor is paid a weekly salary of $250 plus an 8% commission on all sales over $2,500 per week. What is the distributor's salary for a given week if the sales for that week totaled $4,873?

76. **BUS** An employee's gross earnings for a month is $1,750. If the net pay is $1,237, what percent of the gross pay is the total of the deductions? Round to the nearest whole percent.

77. **CON** An order of lumber totals $348.25. If a 5% sales tax is added to the bill, what is the total bill? Round to the nearest cent.

78. **CON** A builder purchases a concrete mixer for $785. The builder does not know the sales tax rate, but the total bill is $828.18. Find the sales tax rate. Round to the nearest tenth of a percent.

79. **BUS** An employer must match employees' social security contributions. If the weekly payroll is $27,542 and if the social security rate is 6.2%, what are the employer's contributions? (All employees' year-to-date salaries are under the maximum salary subject to the social security tax.)

80. **CON** A contractor gets a 2% discount on a monthly sand and gravel bill of $1,655.75 if the bill is paid within 10 days of the statement date. How much is saved by paying the bill within the 10-day period?

81. **BUS** A businessperson is charged a $4.96 monthly finance charge on a bill of $283.15. What is the monthly interest rate on the account? Round to the nearest hundredth of a percent.

82. **BUS** Find the interest on a loan of $3,200 if the annual interest rate is 6% and the loan is for 9 months.

83. **BUS** A property owner sells a house for $127,500 and pays off all outstanding mortgages. If she has $52,475 cash left from this transaction and the money is invested at 5% interest per year for 18 months, how much interest does she earn?

84. **BUS** The sales tax in one city is 8.75% of the purchase price. How much is the sales tax on a purchase of $78.56?

85. **BUS** A small town charges 3.25% of the purchase price for sales tax. What is the sales tax on a purchase of $27.45?

86. **BUS** Interest for 1 year on a loan of $2,400 is $192. What is the interest rate?

87. **BUS** A $2,000 certificate of deposit earns $170 interest in 1 year. What is the interest rate?

88. **BUS** If a loan for $500 at 8.75% interest per annum (year) is paid in 3 months, how much is the interest?

89. **BUS** What is the interest on a business loan for $6,500 at 9.75% interest per annum (year) paid in 7 months?

90. **BUS** A salesperson earns a commission of 15% of the total monthly sales. If the salesperson earns $2,145, how much were the total sales?

91. **BUS** A salesclerk in a store is paid a salary plus 3% commission. If the salesclerk earns $10.65 commission for a weekend, how much were the sales?

Section 3–3

92. Find the complement of 23%.

93. Find the complement of 12.9%.

94. Find the complement of 7%.

95. Find the complement of 21.5%.

96. Find the complement of $18\frac{1}{2}$%.

97. Find the complement of $32\frac{2}{5}$%.

98. Find the complement of 2%.

99. Find the complement of 92%.

100. **INDTR** A steel beam expands 0.01% of its length when exposed to the sun. If a beam measures 49.995 ft after being exposed to the sun, what is its cooled length?

101. **CON** A contractor ordered 800 board feet of lumber for a job that required 750 board feet. What percent, to the nearest whole number, of the required lumber was ordered for waste?

102. BUS A brickmason received a 12% increase in wages, amounting to $35.40. Find the amount of wages received before the increase. Find the amount of wages received after the increase.

104. INDTR According to specifications, a machined part may vary from its specified measure by $\pm 0.4\%$ and still be usable. If the specified measure of the part is 75 in. long, what is the range of measures acceptable for the part?

106. AG/H A mixture of uncut loam and clay soil will have a 20% earth swell when it is excavated. If 300 yd^3 of uncut earth are to be removed, how many cubic yards will have to be hauled away if the earth swell is taken into account?

108. BUS A book that did sell for $18.50 now sells for 20% more. How much does the book now sell for?

110. BUS A paperback dictionary originally sold for $4. It now sells for $1 more. What is the percent of the increase?

112. AG/H When earth is dug, it usually increases in volume or expands by 20%. How much earth will a contractor have to haul away if 150 ft^3 are dug?

114. INDTR A 20-in. bar of iron measures 20.025 in. when it is heated. What is the percent of increase?

116. INDTR A casting weighed 130 ounces (oz) when first made. After it dried, it weighed 127.4 oz. What was the percent of weight loss caused by drying?

118. BUS Workers took a 10% pay cut to help their company stay open during economic hard times. What is the reduced annual salary of a worker who originally earned $35,000?

120. The Bureau of Labor Statistics notes there were 5,083,000 jobs requiring an associate degree in 2000 and that number is expected to increase to 6,710,000 by 2010. What is the percent of increase?

122. HOSP According to the Bureau of Labor Statistics the residential care industry is predicted to be the second highest growth industry from 2000 to 2010. The number of jobs is expected to increase from 806,000 to 1,318,000. What is the expected percent of increase?

124. BUS A homeowner with an annual family income of $35,500 spends in a year $6,900 for a home mortgage, $950 for property taxes, $380 for homeowner's insurance, $2,400 for utilities, and $200 for maintenance and repair. To the nearest tenth, what percent of the homeowner's annual income is spent for housing?

126. INDTR A 78-lb alloy of tin and silver contains 69.3 lb of tin. Find the percent of silver in the alloy to the nearest tenth of a percent.

103. INDTR A lathe costing $600 was sold for $516. What was the percent of decrease in the price of the lathe?

105. INDTR A wet casting weighing 145 kg has a 2% weight loss in the drying process. How much does the dried casting weigh?

107. CAD/ARC A blueprint specification for a part lists its overall length as 62.5 cm. If a tolerance of $\pm 0.8\%$ is allowed, find the limit dimensions for the part's length.

109. COMP A computer disk that once sold for $2.25 now sells for 25% less. How much does the computer disk sell for?

111. COMP A laptop computer originally priced at $2,400 now sells for $300 more. What is the percent of the increase?

113. INDTR A shipping carton is rated to hold 50 lb. A larger carton that holds 15% more weight will be used. How much weight will the larger carton hold?

115. BUS An 18-in. pearl necklace is exchanged for a 24-in. one. What is the percent of increase?

117. HELPP A dieter went from 168 lb to 160 lb in 1 week. To the nearest tenth of a percent, what was the percent of weight loss?

119. AUTO A motorist trades an older car with 350 hp for a new car with 17.4% less horsepower. What is the horsepower of the new car to the nearest whole number?

121. COMP The Bureau of Labor Statistics predicts that the computer and data processing services industry will have the fastest growth from 2000 to 2010. The industry had 2,095 thousand jobs in 2000 and expects to have 3,900 thousand jobs in 2010. What is the percent increase?

123. AUTO A motorist with an annual income of $18,250 spends each year $3,420 on automobile financing, $652 on gasoline, $625 on insurance, and $150 on maintenance and repair. To the nearest tenth, what percent of the motorist's annual income is used for automotive transportation?

125. There are 25 women in a class of 35 students. Find the percent of men in the class to the nearest tenth of a percent.

Your team wants to analyze the percent of income that is spent on housing and transportation. Assume you have a family annual income of $35,500.

1. In a year $6,900 is spent for a home mortgage, $950 for property taxes, $380 for homeowner's insurance, $2,400 for utilities, and $200 for maintenance and repair. What percent, to the nearest tenth percent of the family's annual income, is spent for housing?

2. In a year $3,420 is spent on automobile financing, $1,152 on gasoline, $625 on insurance, and $150 on maintenance and repair. What percent, to the nearest tenth percent of the family's annual income, is spent for transportation?

1. Under what conditions are two fractions proportional?

2. Solving a proportion with one missing term requires two computations. In the proportion, $\frac{R}{100} = \frac{65}{26}$, what two computations do you need to perform to find the value of R?

3. Give some clues for determining if a value in a percent problem represents a rate.

4. Give some clues for determining if a value in a percent problem represents a portion.

5. Give some clues for determining if a value in a percent problem represents a base.

6. Is the amount of sales tax required on a purchase in your state determined by a percent? What is the sales tax rate in your state?

7. If the total bill, including sales tax, on a purchase is 107% of the original amount, what is the sales tax rate? Explain.

8. If a dress is marked 25% off the original price, what percent of the original price does the buyer pay? Explain.

9. If a quantity increases 50%, is the new amount twice the original amount? Explain.

10. If the cost of an item decreases 50%, is the new amount half the original amount? Explain.

Find and explain any mistakes in the following, then rework the incorrect problems.

11. $\dfrac{R}{100} = \dfrac{4}{5}$

$\dfrac{R}{25} = \dfrac{1}{5}$

$5 \cdot R = 25 \cdot 1$

$5 \cdot R = 25$

$R = 25 \div 5$

$R = 5\%$

12. $3\% = 0.3$

13. What percent of 25 is 75?

$\dfrac{R}{100} = \dfrac{25}{75}$

$75 \cdot R = 100 \cdot 25$

$75 \cdot R = 2,500$

$R = 2,500 \div 75$

$R = 33\dfrac{1}{3}\%$

14. 26 is 0.5% of what number?

$\dfrac{0.5}{100} = \dfrac{26}{B}$

$0.5 \cdot 26 = 100 \cdot B$

$13 = 100 \cdot B$

$\dfrac{13}{100} = B$

$0.13 = B$

15. If the cost of a $15 shirt increases 10%, what is the new cost of the shirt?

$\dfrac{10}{100} = \dfrac{P}{15}$

$100 \cdot P = 10 \cdot 15$

$100 \cdot P = 150$

$P = \dfrac{150}{100}$

$P = 1.5$

The shirt increased $1.50 in price.

1. Write $\frac{4}{5}$ as a percent.
2. Write 85% as a fraction.
3. Write 0.3% as a decimal.

Identify the given and missing elements as R (rate), B (base), and P (percent).

4. $10 is what percent of $35?
5. 40% of 10 X-ray technicians is how many?
6. What percent of 24 syringes is 8?
7. 9 is what percent of 27 dogwood trees?
8. How many books is 30% of 40 books?
9. 12% of 50 grass plugs is how many?

Solve.

10. 10% of 150 is what number?
11. What number is $6\frac{1}{4}$% of 144?
12. What percent of 275 is 33?
13. 45.75 is 15% of what number?
14. 55 is what percent of 11?
15. 250% of what number is 287.5?
16. What percent of 360 is 1.2?
17. 245% of what number is 164.4? Round to the nearest hundredth.
18. 5.4% of 57 is what number?
19. Find the complement of 88%.
20. Find the complement of 13%.
21. $15,000 is invested at 14% per year for 3 months. How much interest is earned on the investment?

22. A casting measuring 48 cm when poured shrinks to 47.4 cm when cooled. What is the percent of decrease?

23. An electronic parts salesperson earned $175 in commission. If the commission is 7% of sales, how much did the salesperson sell?

24. Electronic parts increased 15% in cost during a certain period, amounting to an increase of $65.15 on one order. How much would the order have cost before the increase? Round to the nearest cent.

25. In 2005, an area vocational school had an enrollment of 325 men and 123 women. In 2006, there were 149 women. What was the percent of increase of women students? Round to the nearest hundredth percent.

26. The payroll for a hobby shop for 1 week is $1,500. If federal income and social security taxes average 28%, how much is withheld from the $1,500?

27. During one period, a bakery rejected 372 items as unfit for sale. In the following period, the bakery rejected only 323 items, a decrease in unfit bakery items. What was the percent of decrease? Round to the nearest hundredth percent.

28. Materials to landscape a new home cost $643.75. What is the amount of tax if the rate is 6%? Round to the nearest cent.

29. A casting weighed 36.6 kg. After milling, it weighs 34.7 kg. Find the percent of weight loss to the nearest whole percent.

30. After soil was excavated for a project, it swelled 15%. If 275 yd^3 were excavated, how many cubic yards of soil were there after excavation?

31. The total bill for machinist supplies was $873.92 before a discount of 12%. How much was the discount? Round to the nearest cent.

32. A business paid a $5.58 finance charge on a monthly balance of $318.76. What was the monthly rate of interest? Round to the nearest hundredth.

1. Evaluate: $12 + 5(4) \div 2 - 8$
2. Evaluate: 6.7^2
3. Find the perimeter of a rectangle that has a width of 24 cm and length of 40 cm.
4. Find the area of the rectangle in Exercise 3.
5. Round 42.8196 to the nearest hundredth.
6. Evaluate: $5^3 - \sqrt{49}(3 - 1)$
7. Write in factored form and using exponential notation the prime factorization of 420.
8. Evaluate: $37 - (2 + 8 \cdot 2 - 10)$
9. $42.8 + 23.06 + 15.9$
10. $196.7 - 84.92$
11. $52.06(8.723)$

Evaluate and simplify.

12. $5\dfrac{3}{8} + 4 + 12\dfrac{1}{8}$
13. $3\dfrac{7}{8} + 9\dfrac{3}{4} + 5\dfrac{1}{2}$
14. $15\dfrac{3}{5} - 9\dfrac{3}{8}$

15. $4\dfrac{1}{2}\left(\dfrac{4}{9}\right)$
16. $\dfrac{3}{5} \div \dfrac{7}{10}$
17. $1\dfrac{3}{4} \div 1\dfrac{5}{12}$

18. 12^3
19. 3.5^2
20. 123^2
21. $\sqrt{1.96}$
22. $\sqrt{361}$
23. $\sqrt{3600}$
24. How many pints are in 268 qt?
25. A shipping container has a width of 125 in. How many feet is this?
26. An LD-8 airplane shipping container has a maximum net weight of 5,100 lb. How many tons can it hold?
27. The length of an MD-88 aircraft is 147 ft 11 in. Express the length in inches.
28. 28% of 15 is what number?
29. What percent of 24 is 20? Round to the nearest percent.
30. 30% of what number is 12?
31. A salary of $42,000 is increased to $45,000. What is the percent of increase? Round to the nearest hundredth percent.
32. Find the complement of 62%.
33. Find the complement of 14%.
34. To calculate the percent of protein in pet food on a *dry matter basis* (DM basis), divide the percent of protein shown on the Guaranteed Analysis label by the complement of the percent of moisture. PMI Nutrition® dry dog food has 21.0% protein and 14% moisture on its guaranteed analysis label. What is the DM basis percent of protein?
35. Alpo Prime Cuts® canned dog food has 8.0% protein and 82.0% moisture on its guaranteed analysis label. What is the DM basis percent of protein in this dog food?

4

Measurement

Focus on Careers

Job prospects for brickmasons, blockmasons, and stonemasons are expected to be excellent over the next few years. Local contractors, trade associations, or local union-management communities sponsor apprenticeships that usually require three years of on-the-job training. Additionally, a minimum of 144 hours of classroom instruction in each of the three years is required. Subjects such as blueprint reading, mathematics, layout work, and sketching are required.

More than one out of four masons are self-employed, and many specialize in contracting to work on small jobs such as patios, walkways, and fireplaces. Some masons become supervisors for masonry contractors and others become owners of businesses employing many workers.

Those choosing this career usually work outdoors and are exposed to all kinds of weather. They stand, kneel, and bend for long periods of time and often have to lift heavy materials. The work may be hazardous; injuries include falls from scaffolds and injuries from tools. Proper safety equipment and practices do minimize these risks.

In 2002, there were 165,000 brickmasons, blockmasons, and stonemasons. The median hourly earnings for brickmasons and blockmasons were $20.11. The middle 50% earned between $15.36 and $25.32 per hour. Stonemasons had median hourly earnings of $16.36. Apprentices and helpers usually start at a lower wage rate. Pay increases as apprentices gain experience and learn new skills.

Source: *Occupational Outlook Handbook,* 2004–2005 Edition, U.S. Department of Labor, Bureau of Labor Statistics.

Learning Outcomes

1 Identify uses of metric measures of length, mass, weight, and capacity.

2 Convert from one metric unit of measure to another.

3 Make calculations with metric measures.

The **metric system** is an international system of measurement that uses standard units and power-of-10 prefixes to indicate other units of measure.

In the metric system, or the **International System of Units (SI),** a *standard unit* represents each type of measurement. A **standard unit** is the unit that assumes the position of the ones place on the place-value chart and is the word to which the prefixes attach. In most cases the standard unit is also the base unit for the type of measure. The **base unit** is the unit that is used most often in practice. The **meter** is used for length or distance, the **gram** is used for mass or weight, and the **liter** is used for capacity or volume. A prefix is affixed to the standard unit to indicate a measure greater than the standard unit or less than the standard unit.

1 **Identify Uses of Metric Measures of Length, Mass, Weight, and Capacity.**

The most common prefixes for units *smaller* than the standard unit are

$$\textbf{deci-}\ \frac{1}{10}\ \text{of} \qquad \textbf{centi-}\ \frac{1}{100}\ \text{of} \qquad \textbf{milli-}\ \frac{1}{1,000}\ \text{of}$$

The most common prefixes for units *larger* than the standard unit are

$$\textbf{deka-}\ 10\ \text{times} \qquad \textbf{hecto-}\ 100\ \text{times} \qquad \textbf{kilo-}\ 1,000\ \text{times}$$

Thousands (1,000)	Hundreds (100)	Tens (10)	Units or ones (1)	Tenths ($\frac{1}{10}$)	Hundredths ($\frac{1}{100}$)	Thousandths ($\frac{1}{1,000}$)
			STANDARD UNIT			
Kilo-	Hecto-	Deka-		Deci-	Centi-	Milli-

• Decimal point

Figure 4–1 Place-Value Metric-Value Comparison

We can compare the decimal place-value chart with the prefixes (Fig. 4–1). The standard unit (whether meter, gram, or liter) corresponds to the *ones* place. All the places to the left are multiples of the standard unit. That is, the value of *deka-* (some sources use *deca-*) is 10 times the standard unit; the value of *hecto-* is 100 times the standard unit; the value of *kilo-* is 1,000 times the standard unit; and so on. All the places to the right of the standard unit are subdivisions of the standard unit. That is, the value of *deci-* is $\frac{1}{10}$ of the standard unit; the value of *centi-* is $\frac{1}{100}$ of the standard unit; the value of *milli-* is $\frac{1}{1,000}$ of the standard unit; and so on.

EXAMPLE Give the value of the metric unit using the standard unit (gram, liter, or meter).

(a) Kilogram (kg) = 1,000 times 1 gram or **1,000 g**

(b) Deciliter (dL) = $\frac{1}{10}$ of 1 liter or **0.1 L**

(c) Hectometer (hm) = 100 times 1 meter or **100 m**

(d) Dekaliter (dkL) = 10 times 1 liter or **10 L**

(e) Milliliter (mL) = $\frac{1}{1,000}$ of 1 liter or **0.001 L**

(f) Centigram (cg) = $\frac{1}{100}$ of 1 gram or **0.01 g**

Other metric prefixes are used for very large and very small amounts. For measurements smaller than one-thousandth of a unit or larger than one thousand times a unit, prefixes that align with periods on a place-value chart are commonly used.

Powers of 10 for decimal places are written with negative exponents. The relationship between the exponent and the value of the number is discussed more completely in Chapter 5.

Metric Prefixes

Prefix	Relationship to Standard Unit*
atto-(a)	quintillionth part ($\times 0.000000000000000001$ or 10^{-18})
femto-(f)	quadrillionth part ($\times 0.000000000000001$ or 10^{-15})
pico-(p)	trillionth part ($\times 0.000000000001$ or 10^{-12})
nano-(n)	billionth of ($\times 0.000000001$ or 10^{-9})
micro-(µ)	millionth of ($\times 0.000001$ or 10^{-6})
milli-(m)	thousandth of ($\times 0.001$ or 10^{-3})
centi-(c)	hundredth of ($\times 0.01$ or 10^{-2})
deci-(d)	tenth of ($\times 0.1$ or 10^{-1})
deka-/deca-(dk)	ten times ($\times 10$ or 10^{1})
hecto-(h)	hundred times ($\times 100$ or 10^{2})
kilo-(k)	thousand times ($\times 1,000$ or 10^{3})
mega-(M)	million times ($\times 1,000,000$ or 10^{6})
giga-(G)	billion times ($\times 1,000,000,000$ or 10^{9})
tera-(T)	trillion times ($\times 1,000,000,000,000$ or 10^{12})
peta-(P)	quadrillion times ($\times 1,000,000,000,000,000$ or 10^{15})
exa-(E)	quintillion times ($\times 1,000,000,000,000,000,000$ or 10^{18})

* See Chapter 5, Section 5 for powers of 10.

All metric units of length, weight, and volume are expressed either as a standard unit or as a standard unit with a prefix. We examine some common metric units to develop an intuitive sense of their size.

Length

Meter: The **meter** is the standard unit for measuring length. Both *meter* and *metre* are acceptable spellings for this unit of measure. A meter is about 39.37 in. It is 3.37 in. longer than a yard (Fig. 4–2). We use the meter to measure lengths and distances like room dimensions, land dimensions, lengths of poles, heights of mountains, and heights of buildings. The abbreviation for meter is m.

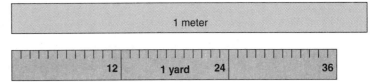

1 meter

1 meter – slightly longer than a yard
(36 inches = 1 yard) (39.37 inches = 1 meter)

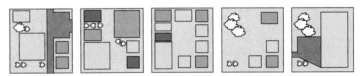

Figure 4–2 Measured in meters.

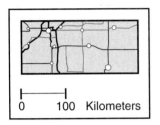

Figure 4–3

Kilometer: A **kilometer** is 1,000 m and is used for longer distances. The abbreviation for kilometer is km. The prefix *kilo* means 1,000. We measure the distance from one city to another, one country to another, or one landmark to another in kilometers (Fig. 4–3). Driving at a speed of 55 mi (or 90 km) per hour, we would travel 1 km in about 40 seconds(s). An average walking speed is 1 km (approximately 5 city blocks) in about 10 min (Fig. 4–4).

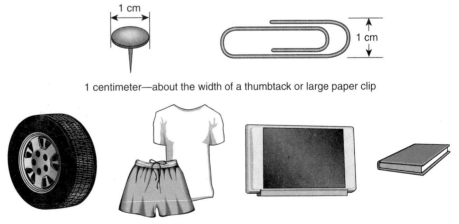

Figure 4–4 Measured in kilometers.

Centimeter: To measure objects less than 1 m long, we commonly use the **centimeter** (cm). The prefix *centi* means "$\frac{1}{100}$ of," and a centimeter is one-hundredth of a meter. A centimeter is about the width of a thumbtack head, somewhat less than $\frac{1}{2}$ in. (Fig. 4–5). We use centimeters to measure medium-sized objects such as tires, clothing, textbooks, and television pictures.

1 cm

1 cm

1 centimeter—about the width of a thumbtack or large paper clip

Figure 4–5 Measured in centimeters.

Millimeter: Many objects are too small to be measured in centimeters, so we use a **millimeter** (mm), which is "$\frac{1}{1,000}$ of" a meter. It is about the thickness of a plastic credit card or a dime (Fig. 4–6). Certain film sizes, bolt and nut sizes, the length of insects, and similar items are measured in millimeters.

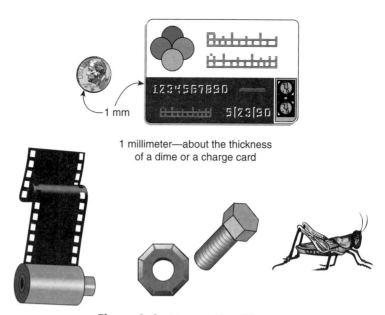

1 millimeter—about the thickness
of a dime or a charge card

Figure 4–6 Measured in millimeters.

Other units of length and their abbreviations are decimeter, dm; dekameter, dkm; and hectometer, hm.

Weight or Mass

Mass and weight are often used interchangeably, but in technical or scientific terms they are different. The weight of an object is a measure of the earth's gravitational pull on the object. As an illustration, an object may have a specific weight and mass on Earth. As an object moves away from Earth, as out in space, the mass remains constant (the same), whereas the weight decreases. When the object is in a "weightless" state, it floats freely in space.

Gram: A metric unit for measuring mass in the metric system is the **gram.** A gram is the mass of 1 cubic centimeter (cm^3) of water at its maximum density. The metric unit for measuring weight is the Newton (N); however, in common usage the gram is used in comparing metric and U.S customary units of weight. A cubic centimeter is a cube whose edges are each 1 centimeter long. It is a little smaller than a sugar cube. The abbreviation for gram is g. We use grams to measure small or light objects such as paper clips, cubes of sugar, coins, and bars of soap (Fig. 4–7).

1 gram—about the weight of two paper clips

Figure 4–7 Measured in grams.

Figure 4–8

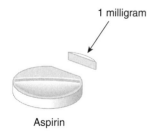

1 milligram

Aspirin

Figure 4–9

Kilogram: A **kilogram** (kg) is 1,000 grams. Since a cube 10 cm on each edge will be 1,000 cm^3, the weight of water required to fill this cube is 1,000 grams or 1 kilogram. A kilogram is approximately 2.2 lb. The kilogram is used to measure the weight of people, books, meat, grain, automobiles, and so on (Fig. 4–8). The kilogram is probably the most commonly used metric measure of mass. It is sometimes referred to as the base metric unit for mass.

Milligram: The gram is used to measure small objects, and the **milligram** ($\frac{1}{1,000}$ of a gram) is used to measure *very* small objects. Milligrams are too small for ordinary uses; however, pharmacists and manufacturers use milligrams (mg) to measure small amounts of drugs, vitamins, and medications (Fig. 4–9).

Chapter 4 / Measurement

1 liter—the volume of a large, individual sized soft drink

Figure 4–10

Other units and their abbreviations are decigram, dg; centigram, cg; dekagram, dkg; and hectogram, hg.

Capacity or Volume

Liter: A **liter** (L) is the volume of a cube 10 cm on each edge. It is the standard metric unit of capacity. Like the meter, it may be spelled *liter* or *litre;* we use the spelling *liter.* A cube 10 cm on each edge filled with water weighs approximately 1 kg, so 1 L of water weighs about 1 kg. One liter is just a little larger than a liquid quart. Soft drinks are sold in both 1-L and 2-L bottles, gasoline is sold by the liter at some service stations, and numerous other products are sold in liter containers (Fig. 4–10).

Milliliter: A liter is 1,000 cm³, so $\frac{1}{1,000}$ of a liter, or a **milliliter,** has the same volume as a cubic centimeter. Most liquid medicine is labeled and sold in milliliters (mL) or cubic centimeters (cc or cm³). Medicines, perfumes, and other very small quantities are measured in milliliters (Fig. 4–11).

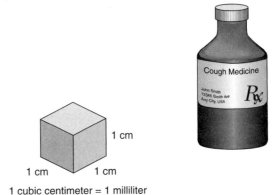

1 cubic centimeter = 1 milliliter

Figure 4–11 Measured in milliliters.

Other units and their abbreviations are deciliter, dL; centiliter, cL; dekaliter, dkL; hectoliter, hL; and kiloliter, kL.

Other standard units in the metric system exist, but those discussed here are the ones most commonly used.

EXAMPLE Choose the most reasonable metric measure.

1. Distance from Jackson, Mississippi, to New Orleans, Louisiana
   (a) 322 m   (b) 322 km   (c) 322 cm   (d) 322 mm
2. Weight of an adult woman
   (a) 56 g   (b) 56 mg   (c) 56 kg   (d) 56 dkg
3. Bottle of eye drops
   (a) 30 dL   (b) 30 dkL   (c) 30 L   (d) 30 mL
4. Weight of an aspirin
   (a) 352 mg   (b) 352 dg   (c) 352 g   (d) 352 kg

1. **(b)** 322 km (about 200 mi)
2. **(c)** 56 kg (about 120 lb)
3. **(d)** 30 mL (about 1 oz)
4. **(a)** 352 mg (one regular-strength aspirin)

2 Convert from One Metric Unit of Measure to Another.

To understand how we change one metric unit to another, let's arrange the units into a place-value chart like the one used for decimals (Fig. 4–12). The units are arranged from left to right and from largest to smallest.

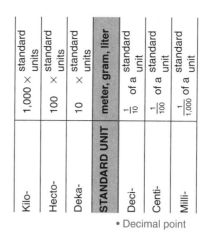

Figure 4–12 Metric-value chart.

Kilo-	Hecto-	Deka-	STANDARD UNIT	Deci-	Centi-	Milli-
1,000 × standard units	100 × standard units	10 × standard units	meter, gram, liter	$\frac{1}{10}$ of a unit	$\frac{1}{100}$ of a unit	$\frac{1}{1,000}$ of a unit

• Decimal point

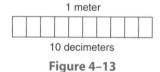

1 meter

10 decimeters

Figure 4–13

As we move from any place on the chart one place to the *right*, the metric unit changes to the next smaller unit. The larger unit is broken down into 10 smaller units, so we are *multiplying* the larger unit by 10 when we move one place to the right (Fig. 4–13).

To change from one metric unit to *any smaller* metric unit:

1. Mentally position the measure on the metric-value chart so that the decimal immediately follows the original measuring unit.
2. Move the decimal to the right so that it *follows* the new measuring unit. (Attach zeros if necessary.)

EXAMPLE   43 dkm = _____ cm?

Place 43 dkm on the metric-value chart so that the last digit is in the dekameters place; that is, the understood decimal that follows the 3 will be *after* the dekameters place (see Fig. 4–14). To change to centimeters, move the decimal point so that it follows the centimeters place. Fill in the empty places with zeros (see Fig. 4–15).

Note the shortcut to multiplication by powers of 10: Move the decimal point one place to the right for each time 10 is used as a factor.

43 dkm = 43,000 cm.

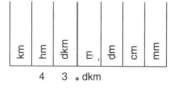

Figure 4–14

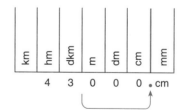

Figure 4–15

EXAMPLE   2.5 dg = _____ mg? (Note the decimal location.)

Place the number 2.5 on the chart so that the decimal point follows the decigrams place (see Fig. 4–16). (Note the decimal after the decigrams place.) To change to milligrams, shift the decimal two places to the right so that it follows the milligram place (see Fig. 4–17). (Note the decimal after the milligrams place.)

2.5 dg = 250 mg.

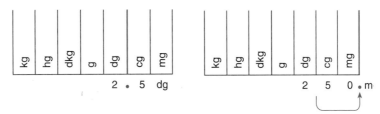

Figure 4–16   **Figure 4–17**

As we move one place to the *left* on the metric-value chart, the metric unit changes to the next larger unit. The smaller units are combined into one larger unit 10 times larger than each smaller unit, so we are *dividing* the smaller unit by 10 when we move one place to the left.

 To change from one metric unit to *any larger* metric unit:

1. Mentally position the measure on the metric-value chart so that the decimal immediately follows the original unit.
2. Move the decimal to the left so that it *follows* the new unit. (Attach zeros if necessary.)

EXAMPLE   3,495 L = _____ kL?

Place the number on the chart so the digit 5 is in the liters place; that is, the decimal point follows the liters place (see Fig. 4–18). (Note the understood decimal point after the liters place.) To change to kiloliters, move the decimal *three* places to the left so the decimal follows the kiloliters place (see Fig. 4–19).

Thus, 3,495 L = 3.495 kL.

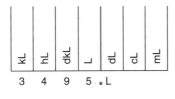   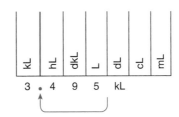

Figure 4–18   **Figure 4–19**

EXAMPLE   2.78 cm = _____ dkm?

Place the number on the chart so that the decimal follows the centimeters place (see Fig. 4–20). (Note the decimal position after the centimeters place.) Move the decimal three places to the left so that it follows the dekameters place (see Fig. 4–21). (Note the decimal position after the dekameters place.)

Therefore, 2.78 cm = 0.00278 dkm.

2 . 7 8 cm

0 . 0 0 2 7 8 dkm

Figure 4–20

Figure 4–21

How Far and Which Way?

To determine the movement of the decimal point when changing from one metric unit to another, answer these questions:

1. *How far* is it from the original unit to the new unit (how many places)?
2. *Which way* is the movement on the chart (left or right)?

Change 28.392 cm to m.

How far is it from cm to m (Fig. 4–22)? **Two places**

Which way? **Left**

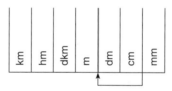

Figure 4–22

Move the decimal in the original measure *two places* to the *left*.

$$28.392 \text{ cm} = 0.28392 \text{ m}$$

Conversion factors can also be used to convert from one metric measure to another. Some of the most commonly used conversion factors are given on the inside covers of this text.

3 Make Calculations with Metric Measures.

We can add or subtract only *like* or *common* measures. For instance, 5 cm and 3 cm are *like* measures. In contrast, 17 kg and 4 hg are *unlike* measures.

To add or subtract *like* or *common* metric measures:

1. Add or subtract the numerical values.   7 cm + 4 cm = 11 cm
2. Give the answer in the common unit of measure.   15 dkg − 3 dkg = 12 dkg

To add or subtract *unlike* metric measures:

1. Change the measures to a common unit of measure.
2. Add or subtract the numerical values.
3. Give the answer the common unit of measure.

EXAMPLE Add 9 mL + 2 cL.

9 mL + 2 cL   Change cL to mL; that is, 2 cL = 20 mL.
9 mL + 20 mL = **29 mL**

Alternative solution:

 9 mL + 2 cL   Change mL to cL. 9 mL = 0.9 cL.
0.9 cL + 2 cL =

  0.9 cL   Note alignment of decimals.
  2   cL   An understood decimal follows the addend 2.
  ——
  2.9 cL   2.9 cL = 29 mL

29 mL or 2.9 cL

EXAMPLE Subtract: 14 km − 34 hm.

  14.0 km   Change hm to km; that is, 34 hm = 3.4 km.
 − 3.4 km   Caution: Notice alignment of decimals.
  ——
  10.6 km

The difference is 10.6 km, or 106 hm.

Incompatible Measures

Can 5 g be added to 2 cm? Can a common unit be found for centimeters and grams? No, grams measure mass and centimeters measure length (Fig. 4–23), so there is no common unit. Thus, we cannot add 5 g + 2 cm.

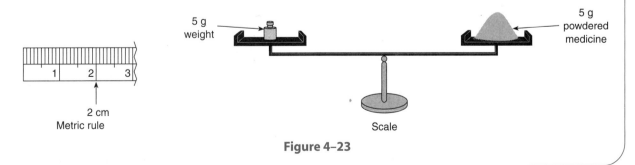

Figure 4–23

Metric measures can be multiplied by a number. If we want to find the total length of three pieces of landscape timber that are each 2.7 m long, we multiply 2.7 m by 3.

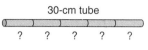

To multiply a metric measure by a number:

1. Multiply the numerical values.
2. Give the answer the same unit as the original measure.

$2.7 \text{ m} \times 3$
$= 8.1 \text{ m}$

30-cm tube

? ? ? ? ?

Figure 4–24

Suppose we have a tube that is 30 cm long and we need it cut into five equal parts (Fig. 4–24). How long is each part? To solve this problem, we divide 30 cm by 5.

$$\frac{30 \text{ cm}}{5} = 6 \text{ cm}$$

To divide a metric measure by a number:

1. Divide the numerical values.
2. Give the quotient the same unit as the original measure.

$30 \text{ cm} \div 5$
$= 6 \text{ cm}$

We can divide all measures by a number because we are dividing a quantity into parts. In some situations, we need to divide a measure by a measure. Suppose we have to administer a 250-mg dose of ascorbic acid in 100-mg tablets. How many tablets are needed per dose?

To divide a metric measure by a like measure:

1. If unlike measures are used, first change them to like measures.
2. Divide the numerical values.
3. Give the quotient as a number that tells how many parts there are.

$\dfrac{250 \text{ mg}}{100 \text{ mg}}$

$= 2.5 \text{ tablets}$

The answer, 2.5, is labeled tablets rather than mg because we are looking for *how many* tablets are needed.

EXAMPLE   Solve.

(a) Four micrometers weighing 752 g each will fit into a shipping carton. If the shipping carton and filler weigh 217 g, what is the total weight of the shipment?
(b) A 5-m-long board is to be cut into four equal parts. How long is each part?
(c) How many 25-g packages of seed can be made from 2 kg of seed?

(a) Total weight = carton and filler + 4 micrometers
   = 217 g + (4 × 752 g)   Multiply.
   = 217 g + 3,008 g   Add.
   = 3,225 g, or 3.225 kg

The total weight is 3,225 g, or 3.225 kg.

(b) $\dfrac{5 \text{ m}}{4} = 1.25 \text{ m}$ Divide length in meters by number of parts. Result is expressed in meters.

Each part is 1.25 m long.

(c) $\dfrac{2 \text{ kg}}{25 \text{ g}} = \dfrac{2,000 \text{ g}}{25 \text{ g}}$ Convert kg to g; then divide. Reduce units.

   $= 80$ Result expresses the number of packages.

80 packages of seed can be made.

1 Give the value of the metric units in standard units.

1. **(a)** kilometer (km)   **(b)** dekaliter (dkL)
   (c) decigram (dg)   **(d)** millimeter (mm)
   (e) hectogram (hg)   **(f)** centiliter (cL)

Choose the most reasonable metric measure.

2. Height of a 6-year-old pediatric patient
   (a) 1.5 km   **(b)** 1.5 m
   (c) 1.5 cm   **(d)** 1.5 mm

3. Diameter of a dime
   (a) 1.5 m   **(b)** 1.5 cm
   (c) 1.5 mm   **(d)** 1.5 km

4. Distance from Los Angeles to San Francisco
   (a) 800 m   **(b)** 800 mm
   (c) 800 km   **(d)** 800 cm

5. Length of a pencil
   (a) 20 km   **(b)** 20 cm
   (c) 20 m   **(d)** 20 mm

6. Overnight accumulation of snowfall
   (a) 8 cm   **(b)** 8 m
   (c) 8 km   **(d)** 8 dkm

7. Width of a home DVD case
   (a) 12 m   **(b)** 12 km
   (c) 12 cm   **(d)** 12 mm

8. Weight of the average male adult
   (a) 70 mg   **(b)** 70 kg
   (c) 70 g   **(d)** 7 dg

9. Weight of a can of tuna
   (a) 184 mg   **(b)** 184 kg
   (c) 184 g   **(d)** 184 cg

10. Weight of a teaspoonful of sugar
    (a) 2 g   **(b)** 2 kg
    (c) 2 mg   **(d)** 2 hg

11. Weight of a vitamin C tablet
    (a) 250 g   **(b)** 250 mg
    (c) 250 kg   **(d)** 250 dkg

12. Weight of a dinner plate
    (a) 350 kg   **(b)** 350 mg
    (c) 350 g   **(d)** 350 cg

13. Weight of a bag of dog food
    (a) 25 g   **(b)** 25 kg
    (c) 25 mg   **(d)** 25 dg

14. Volume of a tank of pesticide
    (a) 60 L   **(b)** 60 mL

15. Dose of cough syrup
    (a) 100 L   **(b)** 100 mL

16. Glass of juice
    (a) 0.12 mL   **(b)** 0.12 L

2 Change to the measure indicated. When using the metric-value chart, place the decimal immediately *after* the measuring unit.

17. 4 m = _____ dm
18. 7 g = _____ dg
19. 58 km = _____ hm
20. 8 hL = _____ dkL
21. 0.25 km = _____ hm
22. 21 dkL = _____ L
23. 8.5 cm = _____ mm
24. 14.2 dg = _____ cg
25. How many milliliters are there in 15.3 cL?
26. How many meters are there in 46 dkm?
27. How many dekagrams are there in 7.5 hg?
28. How many millimeters are there in 16 cm?
29. 4L = _____ cL
30. 8 m = _____ mm
31. 58 km = _____ m
32. 8 hg = _____ g
33. 0.25 km = _____ m
34. 21 dkg = _____ dg
35. 10.25 hm = _____ cm
36. 8.33 L = _____ mL
37. 2 km = _____ mm
38. 0.7 g = _____ cg
39. Change 2.36 hL to liters.
40. Change 0.467 dkm to centimeters.
41. Change 3.8 kg to decigrams.
42. Change 13 dkm to centimeters.
43. 28 m = _____ dkm
44. 238 hL = _____ kL
45. 101 mg = _____ cg
46. 60 hm = _____ km
47. 29 dkL = _____ hL
48. 192.5 g = _____ dkg
49. 17 cm = _____ dm
50. How many decimeters are in 4,389 cm?
51. How many grams are in 47 dg?
52. How many deciliters are in 2.25 cL?

53. 2,743 mm = _____ m   **54.** 385 g = _____ kg   **55.** 15 dkm = _____ km

56. 8 cL = _____ L   **57.** 296,484 m = _____ hm   **58.** 29.83 dg = _____ dkg

59. 0.3 cm = _____ dkm   **60.** 40 dL = _____ kL   **61.** 2,857 mg = _____ kg

62. 15,285 m = _____ km   **63.** Change 297 cm to hectometers.   **64.** Change 0.03 mL to liters.

3 Add or subtract as indicated.

65. 3 m + 8 m   **66.** 7 hL + 5 hL   **67.** 15 cg − 9 cg   **68.** 2 dm + 4 cm

69. 5 cL + 9 mL   **70.** 4 m + 2 L   **71.** 14 kL − 39 hL   **72.** 1 g − 45 cg

73. 3 cg − 5 mL   **74.** 7 km + 2 m

75. HLTH/N A patient absorbs 175 mL of fluid through an IV. If the IV bag has 825 mL left, how much fluid was in the bag to begin with?

76. HLTH/N 653 dkL of orange juice concentrate is removed from a vat containing 8 kL of the concentrate. How much concentrate remains in the vat?

Multiply.

77. 43 m × 12   **78.** 3.4 m × 12   **79.** 50.32 dm × 3

80. CAD/ARC A plot of ground is divided into seven plots, each with road frontage of 138.5 m. What is the total road frontage of the plot of ground?

81. AVIA Earth consists of a series of relatively thin plates which are in constant motion. These plates move at different velocities. The Australian plate moves about 60 mm per year. How many millimeters will the plate move in 78 years?

Divide.

82. 48 g ÷ 3

84. $\dfrac{54 \text{ cL}}{6}$

83. 39 m ÷ 3

85. INDTEC A block of silver weighing 978 g is cut into six equal pieces. How much does each piece weigh?

86. INDTEC Two pieces of steel each 12 m long are cut into a total of 60 equal pieces. How long is each piece?

87. 2.5 cg ÷ 0.5 cg

88. 3 m ÷ 10 cm

89. HLTH/N How many 250-mL prescriptions can be made from a container of 4 L of decongestant?

90. AG/H How many 500-g containers are needed to hold 40 kg of grass seed?

91. Add 4.6 cL + 5.28 dL of photographic developer.

92. Add 3 m + 2 dkm of fabric for draperies.

93. Subtract 19.8 km − 32.3 hm of paved highway.

94. Subtract 13 kL − 39 hL of stored liquid.

95. Multiply 0.25 cL of cologne by 5.

96. Multiply a 35-mm film size by 2.

97. INDTEC A length of satin fabric 30 dm long is cut into 4 equal pieces. How long is each piece?

98. HLTH/N An IV bag holding 250 mL of an antibiotic is calibrated (marked off) into five equal sections. How many milliliters are represented by each section?

99. INDTEC How many containers of jelly can be made from 8,500 L of jelly if each container holds 4 dL of jelly?

100. HELPP How many 2-kg vials of fire retardant (HC1) can be obtained from 38 kg of fire retardant?

Learning Outcomes

1 Convert from one unit of time to another.
2 Make calculations with measures of time.
3 Convert between Fahrenheit temperatures and Celsius temperatures.
4 Examine other useful measures.

As we examined rate measures in Chapter 2, Section 5, Outcome 5, we introduced the units of time. We convert from one unit of time to another just as we convert other compatible units.

1 Convert from One Unit of Time to Another.

The relationships among various units of time are given in Chapter 2, Section 5, Outcome 5. We can use unit ratios or conversion factors to convert from one unit of time to another.

EXAMPLE Convert 3 h to minutes.

Estimation

Hours to minutes → larger to smaller → more minutes

Use a unit ratio to make the conversion:

$$3 \text{ h} \left(\frac{\text{min}}{\text{h}} \right) \qquad \text{Use an appropriate unit ratio.}$$

$$3 \text{ h} \left(\frac{60 \text{ min}}{1 \text{ h}} \right) = 3 \times 60 \text{ min} = 180 \text{ min}$$

Interpretation **There are 180 min in 3 h.**

Or use a conversion factor to make the conversion:

1 h = 60 min   Conversion factor = 60.
3 h × 60 min per hour = 180 min   Multiply hours by 60 min.

---

2 Make Calculations with Measures of Time.

Suppose we have a 1-h meeting and five items to be covered on our agenda. If we are to devote the same amount of time to each item, how much time should we give to each item?

$$1 \text{ h} \div 5 = \frac{1}{5} \text{ h}$$

How long is $\frac{1}{5}$ h? Is this the usual way to express an amount of time? No. When less than 1 h is involved, we often change to minutes.

$$\frac{1}{5} \text{ h} \left(\frac{\overset{12}{60} \text{ min}}{1 \text{ h}} \right) = 12 \text{ min}$$

A technician can assemble one precut picture frame in 7 min. How long will it take to assemble 25 frames?

Estimation

If the frame could be assembled in 6 min, then the technician could assemble 10 per hour (6 min × 10 = 60 min = 1 h). Thus, it would take 2 h to assemble 20 frames and $2\frac{1}{2}$ h to assemble 25 frames. Since it actually takes 7 min per frame, it will take more than $2\frac{1}{2}$ h.

$$\frac{7 \text{ min}}{\text{frame}} \times 25 \text{ frames} = 175 \text{ min}$$

$$175 \text{ min} \times \frac{1 \text{ h}}{60 \text{ min}} = 2.91\overline{6} \text{ h} \qquad \text{Convert min to h using a unit ratio.}$$

If we want to know how many hours and minutes, 2 h = 120 min:

$$175 \text{ min} - 120 \text{ min} = 55 \text{ min}$$

Alternative method for converting a part of an hour to minutes:

$$0.91\overline{6} \text{ h} \times \frac{60 \text{ min}}{1 \text{ h}} = 54.99\overline{9}, \text{ or } 55, \text{ min} \qquad \text{To calculate minutes from the decimal part of an hour, use the decimal portion of 2.91}\overline{6} \text{ h.}$$

Interpretation **It will take 2 h 55 min to assemble 25 frames.**

3 Convert Between Fahrenheit Temperatures and Celsius Temperatures.

The **Kelvin** scale is one scale used to measure temperature in the metric system of measurement. Units on the scale are abbreviated with a capital K (without the symbol ° because these units are called *kelvins*) and are measured from absolute zero, the temperature at which *all* heat is said to be removed from matter. Another metric temperature scale is the **Celsius** scale (abbreviated °C), which has as its zero the freezing point of water. The Kelvin and Celsius scales are related such that absolute zero on the Kelvin scale is the same as −273°C on the Celsius scale. Each unit of change on the Kelvin scale is equal to 1 degree of change on the Celsius scale; that is, the size of a kelvin and a Celsius degree is the same on both scales.

The U.S. customary system temperature scale that starts at absolute zero is called the **Rankine** scale. It is related to the more familiar **Fahrenheit** scale, which places the freezing point of water at 32°. One degree of change on the Rankine scale equals 1 degree of change on the Fahrenheit scale. Absolute zero (the zero for the Rankine scale) corresponds to 460 degrees *below* zero (−460°) on the Fahrenheit scale.

The Celsius and Fahrenheit scales are the most common temperature scales used for reporting air and body temperatures. The formulas for converting temperatures using these two scales are more complicated than the previous ones because 1 degree of change on the Celsius scale does *not* equal 1 degree of change on the Fahrenheit scale (Fig. 4–25).

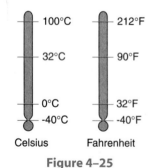

Celsius
- 100°C
- 32°C
- 0°C
- -40°C

Fahrenheit
- 212°F
- 90°F
- 32°F
- -40°F

Figure 4–25

To convert Fahrenheit degrees to Celsius degrees:

Use the formula: $°C = \frac{5}{9}(°F - 32)$, where °C = degrees Celsius and °F = degrees Fahrenheit.

EXAMPLE
HOSP

A chef needs to change 212°F (the boiling point of water) to degrees Celsius.

$$°C = \frac{5}{9}(°F - 32)$$ Substitute 212 for °F in the formula.

$$°C = \frac{5}{9}(212 - 32)$$ Work within grouping; subtract 212 − 32.

$$°C = \frac{5}{9}(180)$$ Multiply. $\frac{5}{\underset{1}{9}} \times \frac{\overset{20}{\cancel{180}}}{1} = 100.$

$$°C = 100$$

212°F = 100°C.

EXAMPLE
HLTH/N

According to Dr. Shotwell, an antifungal powder containing tolnaftate may be stored at a room temperature of 77°F. Change 77°F to degrees Celsius.

Estimation

The Celsius temperature is a smaller value than the Fahrenheit temperature for values near 77°F. From Fig. 4–25 we see that 90°F is approximately 32°C. Then 77°F will be less than 32°C.

$$°C = \frac{5}{9}(°F - 32)$$ Substitute 77 for °F.

$$°C = \frac{5}{9}(77 - 32)$$ Work grouping. 77 − 32 = 45

$$°C = \frac{5}{9}(45)$$ Multiply. $\frac{5}{\underset{1}{9}} \times \frac{\overset{5}{\cancel{45}}}{1}$

$$°C = 25$$

Interpretation   **77°F = 25°C.**   Based on the estimation the solution is reasonable.

To convert Celsius degrees to Fahrenheit degrees:

Use the formula: $°F = \frac{9}{5}°C + 32$, where °F = degrees Fahrenheit and °C = degrees Celsius.

The formula can also be written as °F = 1.8°C + 32.

EXAMPLE

Change 100°C to degrees Fahrenheit.

$$°F = \frac{9}{5}°C + 32$$ Substitute 100 for °C.

$$°F = \frac{9}{5}(100) + 32$$ Multiply first. $\frac{9}{\underset{1}{\cancel{5}}}\left(\frac{\overset{20}{\cancel{100}}}{1}\right) = 180$

$$°F = 180 + 32$$ Add.

$$°F = 212$$

100°C = 212°F.

EXAMPLE
HLTH/N

The label on a dropper bottle of ofloxacin ophthalmic solution warns that the medicine must not be stored at a temperature above 25°C. Change 25°C to degrees Fahrenheit.

Estimation

In Fig. 4–25 we see that the Fahrenheit temperature for values near 25°C is less than 90°F.

$$°F = \frac{9}{5} \,°C + 32 \qquad \text{Substitute 25 for °C.}$$

$$°F = \frac{9}{5} \,(25) + 32 \qquad \text{Multiply. } \frac{9}{\underset{1}{5}}\left(\frac{\overset{5}{25}}{1}\right)$$

$$°F = 45 + 32 \qquad \text{Add.}$$

$$°F = 77$$

Interpretation

25°C = 77°F.   Based on the estimate the solution is reasonable.

5 **Examine Other Useful Measures.**

All SI metric measures use the same system of prefixes for multiples and submultiples. As with the kilogram, the base unit for a type of measure may not always be the standard unit that is used as the stem word for the SI prefixes. Table 4–1 summarizes the SI base units and some other commonly used metric units.

Table 4–1	SI Metric Units	
Base Unit and Abbreviation	**Standard Unit When Different from the Base Unit**	**Type of Measure**
meter (m)		length
kilogram (kg)	gram (g)	mass
liter (L)		volume or capacity
second (s)		time
kelvin (K)		temperature
ampere (A)		electric current
candela (cd)		light intensity
mole (mol)		molecular substance
newton (N)		force
joule (J)		energy
watt (W)		power
square meter (m^2)		area
meter per second (m/s)		speed
cubic meter (m^3)		volume
volt (V)		voltage
ohm (Ω)		resistance
hertz (Hz)		frequency
farad (F)		capacitance
henry (H)		inductance
coulomb (C)		charge

Metric prefixes for very large and small units are given in Section 1, page 172.

EXAMPLE   12 nanoseconds (ns) is what part of a second?

1 ns = 1 billionth of a second, or 0.000000001 s
12 ns = 12 × 0.000000001 = **0.000000012 s**

Using unit ratios:

$$12 \text{ ns} \times \frac{0.000000001 \text{ s}}{1 \text{ ns}} = \textbf{0.000000012 s}$$

EXAMPLE   Give the meaning of the following units.

(a) 115 μA   (b) 3.2 MW

(a) μA means microamperes, microamps, or millionths of an amp.

115 μA = **115 microamps, or 0.000115 A**

(b) MW means megawatts, or 1 million watts.

3.2 MW = **3.2 megawatts or 3,200,000 W**

SECTION 4–2 SELF-STUDY EXERCISES

1 Use ratios or conversion factors to convert the measures of time. Round to hundredths.

1. How many days are in 36 h?

2. Find the number of minutes in 580 s.

3. How many minutes are in 2.5 h?

4. A physical test took 5.3 min to perform. How many seconds is this?

5. If you studied for 3.5 h, how many minutes did you study?

6. A process takes 182 s to perform. How many minutes is this?

7. Convert 7.2 h to minutes.

8. Convert 88 s to minutes.

2 Make the calculations.

9. **INDTEC** A conveyor belt can move 72 lb of rice per minute. How many pounds can be moved on the conveyor belt in 1 hour?

10. **AVIA** Aircraft fuel moves through a pipeline at 40 gal per minute. How many gallons per hour can move through the pipeline?

11. **BUS** A dog kennel uses 30 lb of dog food per day. How many pounds are used in a year?

12. **BUS** A toll station can accommodate on average 28 vehicles per minute. How many vehicles can be accommodated in an hour?

13. **AUTO** A car traveling at the rate of 60 mph (miles per hour) is traveling how many miles per second? Round to the nearest ten-thousandth.

14. **INDTR** A pump that can pump $125 \frac{\text{gal}}{\text{h}}$ can pump how many gallons per minute? Round to the nearest ten-thousandth.

15. **INDTR** A pump can dispose of sludge at the rate of $3,600 \frac{\text{lb}}{\text{h}}$. How many pounds can be disposed of per minute?

16. **HELPP** If water flows through a fire hose at the rate of $96 \frac{\text{gal}}{\text{min}}$, how many gallons will flow per second?

17. **AG/H** Scientists estimate approximately 137 species of life forms become extinct every day. How many to the nearest thousand become extinct each year?

18. **AG/H** Scientists estimate forests are being destroyed on Earth's surface at a rate of 149 acres per minute. Use unit ratios to find the number of acres to the nearest thousand being destroyed each day.

3 Change the Fahrenheit temperatures to Celsius.

19. 95°F **20.** 32°F **21.** 113°F **22.** 41°F **23.** 59°F
24. 50°F **25.** 149°F **26.** 122°F **27.** 176°F **28.** 248°F

4 Change the Celsius temperatures to Fahrenheit.

29. 70°C **30.** 15°C **31.** 45°C **32.** 50°C **33.** 20°C
34. 215°C **35.** 310°C **36.** 410°C **37.** 185°C **38.** 0°C

39. HLTH/N Normal skin temperature in cool weather is 90° to 93°F. What are the Celsius temperatures to the nearest tenth?

40. HLTH/N Persons experiencing mild hypothermia have a core body temperature of 95° to 98.6°F. Express as Celsius temperatures.

41. HLTH/N A patient with a core body temperature of 32.2°C is experiencing severe hypothermia. Express as a Fahrenheit temperature to the nearest whole degree.

42. HLTH/N A person who is wet, improperly dressed, and intoxicated with alcohol can become hypothermic when the air temperature is approximately 20°C. Express as a Fahrenheit temperature.

4 Use the table of metric prefixes on p. 172 and Table 4–1 on p. 186.

43. TELE One hertz is a frequency of one cycle per second. Frequencies of radio and television waves are often measured in kilohertz or megahertz. How many hertz are in 500 MHz?

44. ELEC The henry is a large unit. Inductances in circuits are often measured in millihenrys (mH) or microhenrys (μH). How many henrys are in 420 mH?

45. ELEC The watt (W) is the unit used for measuring electrical power. An electrical device that used 1,400 kW has used how many watts?

46. HLTH/N The average human visual system requires about 50 ms to take in an image. How many different images can be processed per second before they blur?

47. 45 ms is what part of a second?

48. 805 ms is what part of a second?

4–3 | *Metric–U.S. Customary Comparisons*

Learning Outcome **1** Convert between U.S. customary measures and metric measures.

Converting a measure from the U.S. customary system to the metric system, and vice versa, is often necessary because both systems are used in the United States. Most industries use one system exclusively, making conversions from one system to another uncommon. However, an industry that uses the U.S. customary system in the United States often needs to convert to the metric system to market its products in other countries.

1 Convert Between U.S. Customary Measures and Metric Measures.

To convert measures from the U.S. customary system to the metric system, and vice versa, we need only one conversion relationship for each type of measure (length, weight, and capacity). The following equivalency relationships have been rounded to the nearest ten thousandth of a unit:

For length:	1 m = 1.0936 yd
For weight:	1 kg = 2.2046 lb
For capacity:	1 L = 1.0567 liquid qt (liquid measure)

Additional conversion relationships, such as centimeters to inches and kilometers to miles, can be derived from these relationships using unit ratios.

The most common procedure for converting between the U.S. customary and metric systems is to use conversion factors. Conversion factors always involve multiplication, whereas unit ratios may involve multiplication or division. Many conversion factors are given in the table below. Calculators sometimes have conversion factors programmed into the calculator for greater accuracy.

U.S. Customary and Metric Conversion Factors

		From	To	Multiply By
Length				
1 meter = 39.37 inches		meters	inches	39.37
		inches	meters	0.0254
1 meter = 3.2808 feet		meters	feet	3.2808
		feet	meters	0.3048
1 meter = 1.0936 yards		meters	yards	1.0936
		yards	meters	0.9144
1 centimeter = 0.3937 inch		centimeters	inches	0.3937
		inches	centimeters	2.54
1 millimeter = 0.03937 inch		millimeters	inches	0.03937
		inches	millimeters	25.4
1 kilometer = 0.6214 miles		kilometers	miles	0.6214
		miles	kilometers	1.6093
Weight				
1 gram = 0.0353 ounce		grams	ounces	0.0353
		ounces	grams	28.3286
1 kilogram = 2.2046 pounds		kilograms	pounds	2.2046
		pounds	kilograms	0.4536
Liquid Capacity				
1 liter = 1.0567 quarts		liters	quarts	1.0567
		quarts	liters	0.9463

EXAMPLE Change 50 ft to meters.

feet to meters: Conversion factor is 0.3048.

$50 \times 0.3048 = \textbf{15.24 m}$ Multiply.

EXAMPLE How many full cups are in a 2-L soft-drink bottle?

Either conversion factors or unit ratios or a combination of the two can be used.

$$\frac{2\ \text{L}}{1}\left(\frac{1.0567\ \text{qt}}{1\ \text{L}}\right)\left(\frac{2\ \text{pt}}{1\ \text{qt}}\right)\left(\frac{2\ \text{c}}{1\ \text{pt}}\right)$$ Set up unit ratios and multiply.

$= 2(1.0567)(2)(2)\ \text{c} = 8.4536\ \text{c}$ 8 full cups and a portion of a cup left.

There are 8 full cups of soft drink in a 2-L bottle.

Body mass index (BMI) is the standard unit for measuring a person's degree of obesity or emaciation. BMI is body weight in kilograms (kg) divided by height in meters squared, or BMI $= \dfrac{w}{h^2}$.

According to federal guidelines in the U.S., a BMI greater than 25 means that you are overweight. Surveys show that 59% of men and 49% of women have BMIs greater than 25. Extreme obesity is defined as a BMI greater than 40.

To calculate body mass index (BMI):

1. Multiply your weight in pounds by 0.4536 to convert to kilograms.
2. Convert your height to inches.
3. Multiply the inches by 0.0254 to get meters.
4. Square the number found in Step 3.
5. Divide your weight in kilograms by the result from Step 4, and round to the nearest whole number. The result is your BMI.

Study no cheat sheet

EXAMPLE HLTH/N

Alexa May is 5 feet 6 inches and weighs 138 pounds. Find her body mass index.

$138 \times 0.4536 = 62.5968$ Change pounds to kilograms.

$5 \text{ ft } 6 \text{ in.} = 5 \times 12 + 6$ Change feet to inches.

$\qquad\qquad = 60 + 6$

$\qquad\qquad = 66 \text{ in.}$

$66 \times 0.0254 = \boxed{1.6764}$ Change inches to meters.

$\text{BMI} = \dfrac{w}{h^2}$ $w = 62.5968,\ h = 1.6764$

$\text{BMI} = \dfrac{62.5968}{(1.6764)^2}$

$\text{BMI} = \dfrac{62.5968}{2.81031696}$

BMI = 22.27392885

BMI = 22   Rounded to nearest whole number

SECTION 4–3 SELF-STUDY EXERCISES

1 Use unit ratios or conversion factors to change to the units indicated.

1. 9 m to inches	**2.** 120 m to yards	**3.** 42 km to miles
4. 6 L to quarts	**5.** 10 qt to liters	**6.** 27 kg to pounds
7. 50 lb to kilograms	**8.** 7 in. to centimeters	**9.** 18 ft to meters
10. 39 mL to ounces (1 qt = 32 oz)	**11.** $5\frac{3}{4}$ gal to liters (1 gal = 4 qt)	**12.** 7.3 m to feet
13. 12.7 cm to inches	**14.** 12 mm to inches	**15.** 235 km to miles

16. 500 km to miles
17. 28 in. to millimeters
18. 36 in. to centimeters
19. 48 in. to meters
20. 5,280 ft to meters
21. 100 yd to meters
22. 200 mi to kilometers
23. 50 L to quarts
24. 100 qt to liters
25. 24 g to ounces
26. 20 kg to pounds
27. 150 lb to kilograms
28. 45 lb to kilograms
29. 2 oz to grams
30. 0.5 oz to grams
31. 3 kg to pounds

32. ELEC   A spool of wire contains 100 ft of wire. How many meters of wire are on the spool?

33. INTDR   A 60-lb sheet of metal weighs how many kilograms?

34. CAD/ARC   Two cities 150 mi apart are how many kilometers apart?

35. AG/H   A 30-m-wide field is how many yards wide?

36. HOSP   A tourist in Europe traveled 200 km, 60 km, and 120 km by car. How many total miles was this?

37. HLTH/N   A patient in therapy jogged 5 km, 4 km, and 3 km. How many miles did the patient jog?

38. HOSP   A container holds 12 qt. How many liters will the container hold?

39. ELEC   A spool of electrical wire contains 100 m of wire. How many feet of wire are on the spool?

40. HLTH/N   Gary Druckemiller is 6 ft 7 in. and weighs 192 lb. What is his body mass index?

41. HLTH/N   Jo Ella Stearns weighs 121 lb and is 5 ft 8 in. What is her body mass index?

4–4 | Accuracy, Precision, Error, and Measuring Instruments

Learning Outcomes

1. Determine the significant digits of a number.
2. Find the precision and greatest possible error of a measurement.
3. Determine the relative error and the percent error of a measurement.
4. Determine an appropriate approximation of measurement calculations.
5. Read a metric rule.
6. Find the distance and midpoint of a line segment.

1 Determine the Significant Digits of a Number.

Approximate numbers that represent measured values may have varying degrees of accuracy. The **accuracy** of a measurement refers to how close the measured value is to the true or accepted value. **Precision** is the degree to which the measuring process gives consistent results. (See Fig. 4–26.)

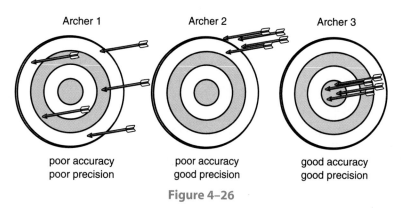

Archer 1   Archer 2   Archer 3

poor accuracy   poor accuracy   good accuracy
poor precision   good precision   good precision

Figure 4–26

One indicator of the degree of precision of a measurement is the number of **significant digits** in a measure. The numbers 2,500, 250, 25, 2.5, 0.25, and 0.025 all have two significant digits. When a number contains no zeros, all the digits are significant. However, zeros are special digits. Sometimes zeros are significant digits, and sometimes they are not significant digits.

To determine the significant digits of a number:

For whole numbers:

1. Start with the leftmost nonzero digit.
2. Count each digit (including zeros) through the rightmost nonzero digit.

For decimals and mixed decimals:

1. Start with the leftmost nonzero digit.
2. Count each digit (including zeros) through the last digit.

Determine the number of significant digits in the following:

200 has 1 significant digit.
28 has 2 significant digits.
1,320 has 3 significant digits.
2,005 has 4 significant digits.

0.07 has 1 significant digit.
2.0 has 2 significant digits.
1.20 has 3 significant digits.
0.01250 has 4 significant digits.

2   Find the Precision and Greatest Possible Error of a Measurement.

The concept of **precision** is associated with approximate numbers. Exact numbers have infinite precision. The precision of a measure is based on the place value or significant digits of a measure. The average of the measurements of 2.3 cm, 4.25 cm, and 5.125 cm can be no more precise than the least precise of the measurements (tenths).

For the calculations $\dfrac{2.3 + 4.25 + 5.125}{3}$, we would round the results to the nearest tenth. The number 3 is an exact number (exactly three measures) and has an infinite precision, so tenths is still the least precise.

$$\frac{2.3 + 4.25 + 5.125}{3} = \frac{11.675}{3} = 3.891666667 = 3.9 \text{ (rounded to tenths)}$$

The **greatest possible error** of a measurement is half of the precision of the measurement.

To find the greatest possible error of a measurement:

Find the greatest possible error of a measurement of $1\dfrac{3}{4}$ ft.

1. Determine the precision of the measurement of the smallest subdivision.

$\text{Precision} = \dfrac{1}{4} \text{ ft}$

2. The greatest possible error is one-half of the precision.

$\dfrac{1}{2} \times \dfrac{1}{4} = \dfrac{1}{8} \text{ ft}$

EXAMPLE Find the greatest possible error of the measurements.

(a) $2\frac{5}{8}$ in. (b) 3.5 cm.

(a) $2\frac{5}{8}$ in.   Precision $= \frac{1}{8}$ in.

$\frac{1}{2} \times \frac{1}{8} = \frac{1}{16}$ **in.**   Greatest possible error is one-half the precision.

(b) 3.5 cm   Precision $= 0.1$ cm.

$0.5(0.1) = $ **0.05 cm.**   Greatest possible error is one-half the precision.

3 Determine the Relative Error and the Percent Error of a Measurement.

Because factories and other industrial concerns are extremely interested in issues related to quality control, most regularly measure parts to ensure that they fall within preestablished tolerance limits for quality. Every measurement taken has some amount of error. There are three ways of reporting the error: as an absolute error, a relative error, and a percent of error. Reporting the error as an absolute error does not give much information, so relative error and percent error are frequently calculated.

- The absolute value of the difference between the observed measurement and the true value is the **absolute error.** That is, the value is positive no matter which value is larger.
- The quotient of the absolute error and the true value is the **relative error.**
- The relative error converted to a percent is the **percent error.**

The concept of absolute value is discussed in detail in Section 5–1 on page 210.

EXAMPLE The blueprint for a part calls for it to be 32.112 mm. A measurement of an actual part
IND/TR is recorded as 32.155 mm. Find the absolute error, relative error, and percent error.

absolute error $= \left| \text{observed value} - \text{true value} \right|$

absolute error $= \left| 32.155 - 32.112 \right|$

absolute error $=$ 0.043 mm

relative error $= \dfrac{\text{absolute error}}{\text{true value}}$

relative error $= \dfrac{0.043}{32.112}$

relative error $=$ 0.0013390633

percent error $=$ relative error $\times 100\%$

percent error $= 0.0013390633 \times 100\%$

percent error $=$ 0.134%   Rounded

4 **Determine an Appropriate Approximation of Measurement Calculations.**

It is often necessary to round the results of calculations. In some instances specific industry standards are appropriate. In other cases, specific directions will be given. However, in many instances, you will determine the appropriate rounding place. When adding or subtracting measurements, we examine the precision of the measurements being added or subtracted.

To round the sum or difference of measurements of different precisions:

1. If necessary, change all measurements to a common unit of measure.
2. Add or subtract the common units of measure.
3. Round the result to have the same precision (place value) as the *least precise* measurement.

EXAMPLE   Add the measurements: 12.5 m, 38 cm, 2.9 m, 43.25 cm.

12.5 m   Change all measurements to meters.
38 cm = 38(0.01 m) = 0.38 m
2.9 m
43.25 cm = 43.25(0.01 m) = 0.4325 m   Add the like measurements.

$$\begin{array}{r} 12.5 \\ 0.38 \\ 2.9 \\ +\ \ 0.4325 \\ \hline 16.2125 \end{array}$$ The least precise measurement is tenths.

16.2125 m rounds to 16.2 m.

When multiplying or dividing measurements, the product or quotient can be no more precise than the measurement with the fewest significant digits.

To round the product or quotient of measurements of different accuracy:

1. Multiply or divide the measurements.
2. Round the product or quotient to have the same number of significant digits as the measurement with the fewest significant digits.

EXAMPLE   Multiply the measurements: (150 m)(105 m)(50 m).

$$(150 \text{ m})(105 \text{ m})(50 \text{ m}) = 787,500 \text{ m}^3 \qquad (\text{m})(\text{m})(\text{m}) = \text{m}^3$$

50 m is the measurement with the fewest significant digits, one.

787,500 m³ rounds to 800,000 m³.

---

5 Read a Metric Rule.

Many standard rulers have both a U.S. customary scale and a metric scale. The metric rule is usually calibrated in centimeters or millimeters. The **metric rule** illustrated in Fig. 4–27 shows centimeters (cm) as the major divisions, represented by the longest lines. Each centimeter is divided into 10 mm, which are the shortest lines. A line slightly longer than the millimeter line divides each centimeter into two equal parts of 5 mm each.

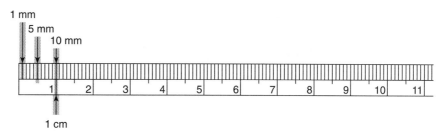

Figure 4–27 The Metric Rule.

The metric rule is read like the U.S. customary rule with the exception that only two measures are indicated, centimeters and millimeters. Other measures are calculated in relation to millimeters or centimeters.

Since 1 mm is $\frac{1}{10}$ cm, we can write metric measures in decimals. For example, a measure of 3 cm and 4 mm is written as either 3.4 cm or 34 mm.

EXAMPLE Find the length of line segment AB (Fig. 4–28) to the nearest millimeter.

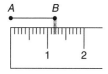

Figure 4–28

The line segment AB extends two marks past 1 cm.

Line segment AB is 12 mm or 1.2 cm to the nearest millimeter.

EXAMPLE Find the length of the line segment CD (Fig. 4–29).

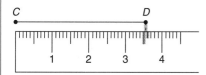

Figure 4–29

The end of line segment CD falls approximately halfway between the 35-mm mark and the 36-mm mark, by eye judgment, measuring about 35.5 mm. Both 35 and 36 mm would be acceptable measures of the line segment CD.

35 mm, 35.5 mm, and 36 mm are acceptable approximations for line segment CD.

---

6 Find the Distance and the Midpoint of a Line Segment.

Two points determine a *line* that is infinite in length. However, the two points also determine a **line segment** that has a certain length. The two points are called the **end points** of the line segment.

The distance between two points on a line is the difference between the coordinates of the points on the line. A **coordinate** is the number that identifies a specific location on a number line or measuring device.

To find the distance between two points on a line:

1. Determine the coordinate for each point.
2. Subtract the value of the coordinate of the leftmost point from the value of the coordinate of the rightmost point.

$$\text{distance} = P_2 - P_1$$

where P_1 and P_2 are points on the number line or measuring device and P_1 is the leftmost point.

EXAMPLE   To find the distance from $2\frac{1}{2}$ to $4\frac{1}{4}$:

Visualize the points on a number line (Fig. 4–30).

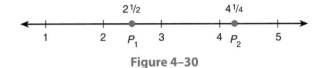

Figure 4–30

$4\dfrac{1}{4} - 2\dfrac{1}{2} =$ Subtract the coordinate of the leftmost point (P_1) from the coordinate of the rightmost point (P_2).

$4\dfrac{1}{4} - 2\dfrac{2}{4} =$ Regroup. $4\dfrac{1}{4} = 3 + \dfrac{4}{4} + \dfrac{1}{4} = 3\dfrac{5}{4}$.

$3\dfrac{5}{4} - 2\dfrac{2}{4} =$ Subtract.

$1\dfrac{3}{4}$ **units**   Distance from $2\dfrac{1}{2}$ to $4\dfrac{1}{4}$.

On a number line or measuring device, the **coordinate of the midpoint** between two points is the average of the coordinates of the points.

To find the midpoint of a line segment between two points on a line:

1. Determine the coordinate for each endpoint.
2. Average the coordinates of the endpoints.

$$\text{Midpoint} = \frac{P_1 + P_2}{2}$$

where P_1 and P_2 are endpoints of the line segment on the number line or measuring device.

EXAMPLE Find the midpoint between two points on a metric rule at 2.8 cm and 5.6 cm.

$$\text{Midpoint} = \frac{P_1 + P_2}{2}$$ Let $P_1 = 2.8$ and $P_2 = 5.6$. Average the two points.

$$\text{Midpoint} = \frac{2.8 + 5.6}{2}$$ Add.

$$\text{Midpoint} = \frac{8.4}{2}$$ Divide.

$$\text{Midpoint} = 4.2 \text{ cm}$$

The midpoint of the points 2.8 and 5.6 is 4.2 cm.

SECTION 4–4 SELF-STUDY EXERCISES

1 Indicate the number of significant digits in each number.

1. 583,000 **2.** 702,500 **3.** 0.0057 **4.** 82.07 **5.** 7.200
6. 860 **7.** 5,080 **8.** 2.0091 **9.** 572,080 **10.** 1,010

2 Find the greatest possible error of each measurement.

11. $7\frac{3}{16}$ in. **12.** 7.2 mm **13.** $5\frac{1}{4}$ in. **14.** 19 oz

15. 8 L **16.** $5\frac{1}{8}$ in. **17.** $12\frac{1}{32}$ in. **18.** $1\frac{5}{8}$ oz

19. 6.2 mi **20.** 7.9 cm **21.** 8.17 dg **22.** 5.3 cg
23. 6.7 cL **24.** 1.4 km **25.** 12.7 km **26.** 125.3 km

3

27. A measurement is observed to be 52.02 cm, against a true measure of 52 cm. Find the absolute error, relative error, and percent error.

28. INDTR   A blueprint for a machine part specifies a measure of 52.2 mm. An actual part has a measurement of 52.4 mm. Find the absolute error, relative error, and percent error.

29. CAD/ARC   The blueprint for a part specifies a measurement of 48.7 cm. An actual part has a measurement of 50.2 cm. Find the absolute error, relative error, and percent error.

30. AUTO   The blueprint for a racing steering wheel specifies a measurement of 14.25 in. An actual wheel has a measurement of 14.54 in. Find the absolute error, relative error, and percent error.

31. INDTEC   A blueprint for the gasoline tank of an industrial lawn mower specifies that it holds 50.05 L of gasoline. Upon close measurement, an actual tank is found to hold 50.25 L of gasoline. Find the absolute error, relative error, and percent error.

32. AUTO   An actual gear is measured to be 15.4 in. in diameter. The blueprint specifies that the diameter of the part be 15 in. Find the absolute error, relative error, and percent error.

4 Add. Give the sum using the appropriate precision.

33. 12.84 m, 13.5 m, 182.3 m, 6.732 m
35. 14.3 cm, 0.7 m, 4.9 m, 23.45 cm

34. 0.93 g, 1.8 g, 42.7 g, 18.932 g
36. 12.78 g, 0.23 kg, 4.752 kg, 0.3466 kg

Give the product or quotient using the appropriate number of significant digits.

37. (202 m)(45 m)(100 m)

38. (125 m)(403 m)(340 m)

39. 1,500 g ÷ 35 g

40. 1,968 km ÷ 30 km

5 Measure line segments 41–50 in Fig. 4–31 to the nearest millimeter. Express answers in millimeters or centimeters.

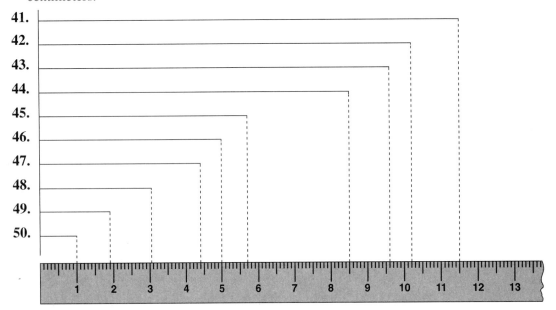

Figure 4–31

Use a metric rule to measure each line segment to the nearest tenth centimeter.

51. A •———————————————————————• B

52. C •————————————————• D

53. E •—————————• F

54. G •—————————————————————• H

6 Find the distance between each pair of points.

55. 5 in. and 8 in.

56. $2\frac{1}{2}$ in. and $5\frac{1}{4}$ in.

57. 8.2 cm and 15.7 cm

58. 4.9 cm and 18.5 cm

59. $12\frac{3}{8}$ in. and $23\frac{5}{8}$ in.

60. $14\frac{7}{8}$ in. and $25\frac{3}{4}$ in.

Find the midpoint between each pair of points.

61. 1.8 cm and 7 cm

62. 3.5 cm and 6.3 cm

63. 5.5 cm and 10.3 cm

64. 3 cm and 5.2 cm

65. $7\frac{1}{2}$ in. and $15\frac{1}{4}$ in.

66. 2 in. and $9\frac{3}{4}$ in.

Learning Outcomes	What to Remember with Examples

Section 4–1

1 Identify uses of metric measures of length, mass, weight, and capacity (pp. 171–175).

Associate small, medium, and large objects with appropriate measuring units of comparable size. Powers-of-10 prefixes are used, such as *kilo* for 1,000.

> Perfume, mL; soda, L; travel, km; racetrack, m; vitamin, mg; potatoes, kg; eye drops, mL

2 Convert from one metric unit of measure to another (pp. 175–178).

Move the decimal point in the original measure to the left or right as many places as necessary to move from the original unit to the new unit on the metric chart of prefixes.

> Change 5.04 cL to liters. Move the decimal two places to the left.
>
> 5.04 cL = 0.0504 L

3 Make calculations with metric measures (pp. 178–180).

To add or subtract metric measures: **1.** Change measures to measures with like units if necessary. **2.** Add or subtract the numerical values. **3.** Give the answer the common unit of measure.

To multiply or divide a metric measure by a number, multiply or divide the numbers and keep the same unit.

To divide a metric measure by a measure: **1.** Change measures to measures with like units if necessary. **2.** Divide the numbers, canceling the units of measure. The answer will be a number.

> 7 km + 34 m =
> 7,000 m + 34 m = 7,034 m
> or
> 7 km + 0.034 km = 7.034 km
>
> 7 mg × 3 = 21 mg
> 5 m ÷ 25 cm = 500 cm ÷ 25 cm
> $\dfrac{500 \text{ cm}}{25 \text{ cm}} = 20$

Section 4–2

1 Convert from one unit of time to another (p. 183).

Use unit ratios or conversion factors to convert one unit of time to another.

> How many days are in 3,420 hours?
>
> 1 day = 24 h
>
> $3{,}420 \text{ h} \times \dfrac{1 \text{ day}}{24 \text{ h}} = 142.5 \text{ days}$

2 Make calculations with measures of time (pp. 183–184).

Measures of time that are the same unit of measure can be added or subtracted by adding or subtracting the quantities and keeping the same unit of measure. A time measure can be multiplied or divided by a number by multiplying or dividing the measure by the number and keeping the same measuring unit.

A printer prints 1 page in 30 s. How long in hours does it take to print 45 pages?

1 min = 60 s; 1 h = 60 min

30 s × 45 = 1,350 s   Seconds to complete job.

$1{,}350 \text{ s} \times \dfrac{1 \text{ min}}{60 \text{ s}} = 22.5 \text{ min}$ Minutes to complete job.

$22.5 \text{ min} \times \dfrac{1 \text{ h}}{60 \text{ min}} = 0.375 \text{ h}$ Portion of hour to complete job.

The printer can print 45 pages in 22.5 min or 0.375 h.

3 Convert between Fahrenheit temperatures and Celsius temperatures (pp. 184–186).

To change degrees Fahrenheit to degrees Celsius use the formula $°C = \dfrac{5}{9}(°F - 32)$.

Change 50°F to °C.

$°C = \dfrac{5}{9}(°F - 32)$ Substitute 50 for °F.

$°C = \dfrac{5}{9}(50 - 32)$ Perform operation in parentheses.

$°C = \dfrac{5}{9}(18)$ Multiply.

$°C = 10$

To change degrees Celsius to degrees Fahrenheit use the formula $°F = \dfrac{9}{5}°C + 32$.

Change 28°C to °F.

$°F = \dfrac{9}{5}°C + 32$ Substitute 28 for °C.

$°F = \dfrac{9}{5}(28) + 32$ Multiply.

$°F = \dfrac{252}{5} + 32$ Divide.

$°F = 50.4 + 32$ Add.

$°F = 82.4$

4 Examine other useful measures (pp. 186–187).

All SI metric measures use the same system of prefixes for multiples and submultiples. Table 4–1 on page 186 summarizes the SI base units and some other commonly used metric units.

What is the meaning of 3 μA?

1 μA = 1 micro ampere
1 μA = 1 millionth of an ampere, or 0.000001 A, or 10^{-6} A
3 μA = 3 × 0.000001 = **0.000003 A, or 3 × 10^{-6} A**

Using a unit ratio:

$$3 \text{ μA} \times \dfrac{0.000001 \text{ A}}{1 \text{ μA}} = 0.000003 \text{ A}$$

Section 4–3

1 Convert between U.S. customary measures and metric measures (pp. 188–190).

1. Set up the original amount as a fraction with the original unit in the numerator and 1 in the denominator. **2.** Multiply by a unit ratio with the original unit in the denominator and the new unit in the numerator.

> How many feet are in 14 m?
>
> $$\frac{14 \text{ m}}{1} \times \frac{3.28 \text{ ft}}{1 \text{ m}} = 45.92 \text{ ft}$$ Use the conversion factor 1 m = 3.28 ft.

Section 4–4

1 Determine the significant digits of a number (pp. 191–192).

For whole numbers: **1.** Start with the leftmost nonzero digit. **2.** Count each digit through the rightmost nonzero digit.

> 258 has three significant digits.
> 1,050 has three significant digits.

For decimal numbers: **1.** Start with the leftmost nonzero digit. **2.** Count each digit through the last digit.

> 0.023 has two significant digits.
> 1.50 has three significant digits.

2 Find the precision and the greatest possible error of a measurement (pp. 192–193).

1. Determine the precision of the measurement of the smallest subdivision. **2.** The greatest possible error is one-half of the precision.

> The precision of $2\frac{5}{8}$ in. is $\frac{1}{8}$ in.
>
> $$\frac{1}{2} \times \frac{1}{8} = \frac{1}{16} \text{ in.}$$
>
> The greatest possible error is $\frac{1}{16}$ in.

3 Determine the relative error and the percent error of a measurement (p. 193).

To find the absolute error, find the absolute value of the difference between the observed measurement and the true value.

To find the relative error, find the quotient of the absolute error and the true value.

To find the percent error, change the relative error to a percent by multiplying by 1 in the form of 100%.

> The actual measure of a widget is 15.963 cm and the blueprint specifies 15.94 cm for the widget.
>
> $$\text{absolute error} = |15.963 - 15.94|$$
> $$= 0.023 \text{ cm}$$
> $$\text{relative error} = \frac{0.023}{15.94} = 0.0014429109$$
> $$\text{percent error} = 0.0014429109 \times 100\% = 0.1443\% \text{ (rounded)}$$

4 Determine an appropriate approximation of measurement calculations (p. 194).

To round the sum or difference of measurements of different precisions: **1.** If necessary, change all measurements to a common unit of measure. **2.** Add or subtract. **3.** Round the result to have the same precision (place value) as the *least precise* measurement.

Chapter Review of Key Concepts

Add the measurements: 34.5 m, 15 cm, 4.6 mm.

Change all measurements to meters.

$$34.5 \text{ m} = 34.5 \text{ m}$$
$$15 \text{ cm} = 15 \times 0.01 \text{ m} = 0.15 \text{ m}$$
$$4.6 \text{ mm} = 4.6 \times 0.01 \text{ m} = 0.046 \text{ m}$$

Add the like measurements.

$$\begin{array}{r} 34.5 \\ 0.15 \\ \underline{0.046} \\ 34.696 \end{array}$$

The least precise measurement is tenths.

34.696 m rounds to 34.7 m.

To round the product or quotient of measurements of different accuracy: **1.** Multiply or divide the measurements. **2.** Round the product or quotient to have the same number of significant digits as the measurement with the fewest significant digits.

Multiply the measurements: (12.3 cm)(5 cm)(0.25 cm).

$$(12.3 \text{ cm})(5 \text{ cm})(0.25 \text{ cm}) = 15.375 \text{ cm}^3 \qquad (\text{cm})(\text{cm})(\text{cm}) = \text{cm}^3$$

5 cm is the measurement with the fewest significant digits, one.

15.375 cm³ rounds to 20 cm³.

5 Read a metric rule (p. 195).

See Fig. 4–32. Align the rule along the object. Count the number of millimeters and/or centimeters (10 mm = 1 cm) to determine the approximate length. Use eye judgment to estimate closeness of the object to a mark on the rule.

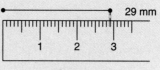

Figure 4–32

6 Find the distance and the midpoint of a line segment (pp. 195–197).

Distance: Subtract the coordinates of each point.
Distance = $P_2 - P_1$

Find the distance between the points 3.5 cm and 6.3 cm on a metric rule.

Distance = 6.3 − 3.5   Let P_1 = 3.5 cm and P_2 = 6.3 cm.
     = 2.8 cm

Midpoint:

Average the coordinates of the end points (P_1 and P_2).

$$\text{Midpoint} = \frac{P_1 + P_2}{2}$$

Find the midpoint between the points 3.5 cm and 6.3 cm on a metric rule.

$$\text{Midpoint} = \frac{3.5 + 6.3}{2} \qquad \text{Let } P_1 = 3.5 \text{ cm and } P_2 = 6.3 \text{ cm.}$$

$$= \frac{9.8}{2}$$

$$= 4.9 \text{ cm}$$

Section 4–1

Give the prefix that relates each number to the standard unit in the metric system.

1. 1,000 times
2. $\frac{1}{10}$ of
3. $\frac{1}{1,000}$ of
4. 10 times
5. $\frac{1}{100}$ of
6. 100 times

Give the value of the prefixes based on a standard measuring unit.

7. dekameter (dkm)
8. hectogram (hg)
9. milligram (mg)
10. centigram (cg)
11. kiloliter (kL)
12. deciliter (dL)

Choose the most reasonable answer.

13. Height of the Washington Monument
    (a) 200 m
    (b) 200 cm
    (c) 200 mm
    (d) 200 km
14. Height of Mt. Rushmore
    (a) 1.6 km
    (b) 1.6 m
    (c) 1.6 cm
    (d) 1.6 mm
15. Weight of an egg
    (a) 50 g
    (b) 50 kg
    (c) 50 mg
16. Weight of a saccharin tablet
    (a) 50 kg
    (b) 50 mg
    (c) 50 g
17. Weight of a man's shoe
    (a) 0.25 g
    (b) 0.25 mg
    (c) 0.25 kg
18. Carton of milk
    (a) 4 L
    (b) 4 mL
19. Bottle of medicine
    (a) 50 L
    (b) 50 mL

Change to the unit indicated.

20. 0.4 dkm = _____ hm
21. 67.1 m = _____ dkm
22. 4 m = _____ dm
23. 2.3 m = _____ mm
24. 5 cm = _____ mm
25. 0.123 hm = _____ mm
26. How many millimeters are in 0.432 km?
27. 23 dkm = _____ mm
28. 42.7 cm = _____ dkm
29. 41,327 dkm = _____ km
30. A board is 1.82 m long. How many centimeters long is the board?
31. 394.5 g = _____ hg
32. 2.7 hg = _____ dg
33. 3,000,974 cg = _____ kg

Perform the operations indicated.

34. 25 mm − 14 mm
35. 12 g + 5 m
36. 17 mg − 8 mL
37. 8 g − 52 cg
38. 43 dkg × 7
39. 6.83 cg × 9
40. $\frac{18 \text{ cm}}{9}$
41. 7.5 kg ÷ 0.5 kg
42. $\frac{8 \text{ hL}}{20 \text{ L}}$
43. 34 hL ÷ 4
44. 2.4 m ÷ 5 cm

45. **BUS** Fabric must be purchased to make seven garments, each requiring 2.7 m of fabric. How much fabric must be purchased?
46. **BUS** Candy weighing 526 g is mixed with candy weighing 342 g. What is the weight of the mixture?
47. **HOSP** A recipe calls for 5 mL of vanilla flavoring and 24 cL of milk. How much liquid is this?
48. **BUS** 20 boxes, each weighing 42 kg, are to be moved. How much weight must be moved?
49. **CON** A metal rod 42 m long is cut into seven equal pieces. How long is each piece?
50. **INDTEC** 32 kg of a chemical are distributed equally among 16 chemistry students. How many kilograms of chemical does each student receive?
51. **HOSP** A serving of punch is 25 cL. How many servings can be obtained from 25 L of punch?
52. **HOSP** How many containers of jelly can be made from 8,548 L of jelly if each container holds 4 dL of jelly?

Section 4–2

Use unit ratios or conversion factors to convert the measures of time.

53. How many days are in 72 h?

54. How many minutes are in 2.4 h?

55. Convert 158 min to hours.

56. HLTH/N   A doctor ordered a surgery patient not to drive for 3 weeks. How many days is this?

57. HLTH/N   A heart surgery patient was held in intensive care for 96 h. How many days is this?

58. HLTH/N   A bone marrow patient remained hospitalized for 72 days. How many weeks is this?

59. BUS   If you have worked for your employer for 39 months, how many years have you worked?

60. There are 96 days remaining in a year. How many weeks remain?

Make the temperature conversions.

61. 15°C = _____ °F

62. 86°F = _____ °C

63. 95°C = _____ °F

64. The freezing point of benzene is 5°C. What is this temperature on the Fahrenheit scale?

65. HLTH/N   The label on a container of the acid-reducer famotidine states that storage above 40°C should be avoided. What is 40°C on the Fahrenheit scale?

66. AUTO   Summer road surface temperatures of 122°F give what reading on the Celsius scale?

67. HOSP   Candy that should reach a cooking temperature of 365°F should reach what temperature using the Celsius scale?

68. AVIA   A jet plane that travels at 250 m/s travels how many kilometers per second?

69. HLTH/N   A hospital patient has a temperature reading of 37.9°C. What is the temperature in degrees Fahrenheit?

Section 4–3

Make the conversions using unit ratios or conversion factors. Round to thousandths.

70. 7 m = ____ inches

71. 215 m = ____ yards

72. 69 km = ____ miles

73. 15 L = ____ quarts

74. 12 qt = ____ liters

75. 32 kg = ____ pounds

76. 10 lb = ____ kilograms

77. 9 in. = ____ centimeters

78. 21 ft = ____ meters

79. 14.8 dkL = ____ quarts

80. $3\frac{1}{2}$ gal = ____ liters

81. How many meters long is 200 ft of pipe?

82. CON   Concrete weighing 90 lb weighs how many kilograms?

83. Two cities 175 mi apart are how many kilometers apart?

84. CON   A room 10 m wide is how many feet wide?

85. HLTH/N   Wanda Williams is 5 ft 9 in. and weighs 142 lb. What is her body mass index?

86. HLTH/N   Ravi Mehra weighs 168 lb and is 5 ft 7 in. What is his body mass index?

Section 4–4

Determine the number of significant digits of each number.

87. 304,243

88. 2,401,000

89. 4.010

90. 0.023

Find the greatest possible error of each measurement.

91. $2\frac{1}{2}$ in.

92. $7\frac{3}{8}$ in.

93. $3\frac{5}{16}$ ft

94. $7\frac{9}{32}$ in.

95. 5.8 cm

96. 12.2 cm

97. 15.3 cm

98. 7.5 oz

99. BUS   A gas pump registers 4.95 gal when the measuring standard of 5.00 gal is used. What is the percent error?

100. HLTH/N   A pharmacy balance reads 2.03 g when a standard 2.00-g weight is used in calibration. What is the percent error?

101. How many significant digits are there in 0.5010?

102. How many significant digits are there in 203.07?

103. What is the greatest possible error of the measurement 4.7 cg?

104. What is the greatest possible error of the measurement $2\frac{5}{32}$ in.?

105. Add and write the sum using appropriate precision:

$$4.2 \text{ m} + 508 \text{ cm} + 31.72 \text{ m} + 5.46 \text{ m}$$

106. Multiply and write the product using the appropriate number of significant digits.

$$(132 \text{ m})(80 \text{ m})(27.5 \text{ m})$$

Measure line segments 107–116 in Fig. 4–33 to the nearest millimeter.

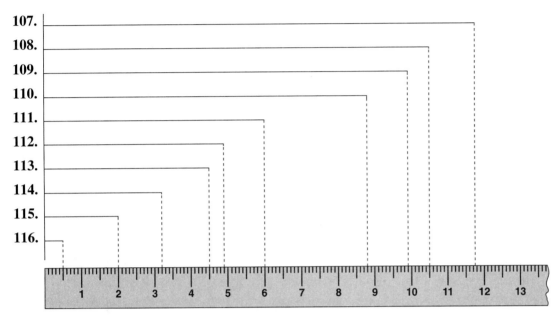

Figure 4–33

Find the distance between each pair of points.

117. $7\frac{5}{8}$ in. and $12\frac{7}{8}$ in.

118. $9\frac{5}{16}$ in. and $11\frac{3}{8}$ in.

119. $3\frac{5}{32}$ in. and 8 in.

120. 9 in. and $16\frac{1}{4}$ in.

121. 17.8 cm and 20.5 cm

122. 31.9 cm and 37 cm

Find the midpoint between each pair of points.

123. $15\frac{3}{8}$ in. and $25\frac{1}{8}$ in.

124. $12\frac{1}{2}$ in. and $15\frac{3}{4}$ in.

125. 2.3 cm and 8.9 cm

126. 1.2 cm and 7 cm

127. 5 cm and 9.8 cm

128. 8 cm and 8.3 cm

TEAM PROBLEM-SOLVING EXERCISES

A **lumen** is the most common measurement of light output, or luminous flux. The lumen rating of a light source is a measure of the total light output of the light source. Light sources are labeled with an output rating in lumens. For example, an R30, 65-W indoor flood lamp may have a rating of 750 lumens. A **watt** is the unit used to measure electrical power. It defines the rate of energy consumption by an electrical device when it is in operation. The energy cost of operating an electrical device is calculated as its wattage time in hours of use.

The measure of electrical energy from which electricity billing is determined is the kilowatt-hour (kWh). For example, a 100-W bulb operated for 1,000 h would consume 100 kWh, (100 W × 1,000 h) = 100,000 Wh, or 100 kWh.

1. A conference room has five R30, 65-W indoor flood lamps. How many lumens are produced if all 5 lamps are lighted?

2. How many kilowatt-hours are consumed if the 5 lamps are lighted for 36 h?

3. At the electrical billing rate of $0.12/kWh, how much will it cost to operate the five lamps over 1,000 h?

1. Give four items that are measured in kilograms.
2. Give four items that are measured in meters.
3. If you were building a house, what units of metric linear measure would you be likely to use?
4. Explain how metric units for length, weight, and capacity are similar.
5. Which measure is longer, a yard or a meter?
6. Which metric measure would you use to dispense a liquid medicine?
7. If your medicine bottle reads 5 mg, is the medicine more likely to be in liquid or capsule form?
8. The kilometer is usually associated with what U.S. customary unit?

Identify the mistake, explain why it is wrong, and correct the mistake.

9. 3.252 dkL = 325.2 kL

10. 418 yd = _____ m?

$$418 \text{ yd} \times \frac{0.9144 \text{ m}}{1 \text{ yd}} = 362 \text{ yd}$$

11. 24 qt = _____ L?

$$24 \text{ qt} \times \frac{1 \text{ L}}{0.9463 \text{ qt}} = 25 \text{ L}$$

12. Compare and contrast the U.S. customary and the metric systems of measurement.

13. Why do you think the metric system has not fully replaced the U.S. customary system in the United States?

Change to the metric unit indicated.

1. 298 m = _____ km
2. 8 dm = _____ mm
3. 5.2 dL of liquid are poured from a container holding 10 L. How many liters of liquid remain in the container?
4. A bar of soap weighs 175 g. How much do 15 bars of soap weigh (in kilograms)?
5. 75 mi = _____ km
6. 25 kg = _____ lb
7. 4 L = _____ pt
8. How many liters of weed killer are contained in a 55-gal drum?
9. Change 48°F to degrees Celsius.
10. Change 70°C to degrees Fahrenheit.
11. Find the number of minutes in 42 h.
12. How many hours are in 5,700 min?
13. A GIS Specialist worked 47 h a week for 5 consecutive weeks. How many hours did she work?
14. A machine was in use for 20 h a day during the 30 days in June. How many hours was the machine producing?
15. The number 840 has how many significant digits?
16. A tool measures 3.5 cm. What is the greatest possible error?
17. A hip replacement part measures 24.75 cm and the blueprint specifies the part is 24.72 cm. What is the relative error and percent error?
18. Find the sum of 32.5 cg, 4.76 cg, 503.2 cg, and 16.3 cg and use an appropriate approximation to report the sum.

Measure line segments 19–20 in Fig. 4–34 to the nearest millimeter.

19.

20.

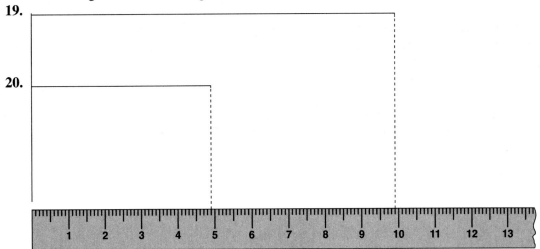

Figure 4–34

21. A 0.243-caliber bullet travels $2{,}450 \frac{\text{ft}}{\text{s}}$ for the first 200 yd. Using conversion factors, change $2{,}450 \frac{\text{ft}}{\text{s}}$ to $\frac{\text{m}}{\text{s}}$.

22. Justin Allen weighs 172 lb and is 5 ft 10 in. Find his body mass index.

23. Find the midpoint of the two measures taken on the metric rule: 3.8 cm and 5.9 cm.

24. Find the distance between the points on a U.S. customary rule: $17\frac{15}{32}$ in. and $23\frac{5}{16}$ in.

5

Signed Numbers and Powers of 10

Focus on Careers

Geographic Information Systems (GIS) specialists work with computer programs and software to create and maintain data or maps that can be combined with geographically referenced data. Different types of data, such as socioeconomic, demographic, political, land use, land cover, environmental, and transportation networks, are related through the use of GIS software.

GIS specialists must have knowledge of geography, written and oral communication, mathematics, critical thinking, information gathering, computer science, and systems evaluation. Most of the GIS specialist's time is spent in front of a computer screen; however, some GIS employees collect data in the field.

Jobs in this field are expected to grow by about 65% over the next 10 years. Hourly wages for jobs in California range from $20.20 to $39.26, with the average being $29.97. GIS specialists usually work 40 hours a week, but longer hours are common.

Many 2-year colleges offer GIS certificate and degree programs. Graduates with a GIS major and significant study in business, sociology, political science, or economics often have an advantage over job seekers who have only a GIS major.

Source: California Employment Development Department.

When a temperature is colder than zero degrees, how can we express this temperature numerically? When the selling price of an item is less than the cost of an item, how can we express the profit made on the sale numerically? When a withdrawal on a bank account exceeds the amount of money in the account, how can we express the account balance numerically? These situations demonstrate a need to express numbers that are less than zero.

Learning Outcomes

1 Compare signed numbers.

2 Add signed numbers with like signs.

3 Add signed numbers with unlike signs.

1 Compare Signed Numbers.

Many physical phenomena have values that are less than zero. Integers are needed to express these values. When the opposite of each natural number is included, the set of whole numbers can be expanded to form the set of **integers.** Fig. 5–1 shows how the set of whole numbers is extended to include all integers. A number line for integers continues indefinitely in *both* directions. Numbers get smaller as we proceed to the left and larger as we proceed to the right.

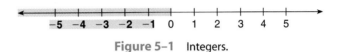

Figure 5–1   Integers.

The **opposite** of a number is the same number of units from zero, but in the opposite direction. Fig. 5–2 shows a number line with numbers and their opposites.

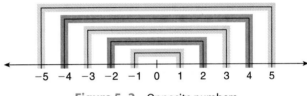

Figure 5–2   Opposite numbers.

The relationship among the three sets of numbers—natural numbers, whole numbers, and integers—is shown in Fig. 5–3.

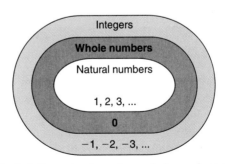

Figure 5–3   Relating natural numbers, whole numbers, and integers.

Are Positive Signs Necessary?

As you can see from the number line in Fig. 5–1, it is not necessary to include the positive sign for positive values. You may sometimes want to include the sign to draw attention to it or to emphasize its direction. Negative signs, however, must always be included.

Integers are just one type of signed numbers. Fractions, mixed numbers, and decimals also have values less than zero. **Signed numbers** are all types of numbers that are either positive or negative or 0. We begin our study of signed numbers with integers and expand to other types of signed numbers later in the chapter.

If a number is positioned to the left of another number on a number line, it represents a smaller value. Larger values are positioned to the right of a number on the number line. As with whole numbers, fractions, and decimals, we use the inequality symbols $<$ and $>$ to compare signed numbers.

EXAMPLE Use the symbols $<$ and $>$ to show the relationship between the two numbers.

(a) 7 ___ 9   (b) 0 ___ -1 (c) -4 ___ -2 (d) -2 ___ 0

(a) $7 < 9$ 7 is smaller—to the left of 9 on the number line.
(b) $0 > -1$ 0 is larger—to the right of -1 on the number line.
(c) $-4 < -2$ -4 is smaller—to the left of -2 on the number line.
(d) $-2 < 0$ -2 is smaller—to the left of 0 on the number line.

The distance between each pair of consecutive integers is the same for all integers located on the number line. This concept of *distance between numbers* on the number line is related to the concept of *absolute value*. The **absolute value** of a number is often described as the number of units or distance the number is from zero. The symbol for absolute value is $|\ |$; $|3|$ is read "the absolute value of 3."

Distance is a physical property, and it cannot have a negative value. The absolute value of a number is always a nonnegative value.

What Does Nonnegative Mean?

Are positive numbers also nonnegative? Yes, every positive number is nonnegative.
Are any nonnegative numbers not positive? Yes, zero is nonnegative *and* nonpositive. Therefore, when we say *nonnegative*, we mean positive or zero. Similarly, *nonpositive* means negative or zero.

Write Facts Symbolically

Positive numbers are represented symbolically as $a > 0$, where a is any real number greater than zero.
Negative numbers are represented by $a < 0$, where a is any real number less than zero.
The symbols $\geq$ which means "is greater than or equal to" and $\leq$ which means "is less than or equal to" are used to represent nonnegative and nonpositive numbers respectively.
Practice reading symbolic statements in words.

$a > 0$	positive numbers	$a \geq 0$	nonnegative numbers (positives and zero)
$a < 0$	negative numbers	$a \leq 0$	nonpositive numbers (negatives and zero)

Then, the *definition of the absolute value* of a number can be written symbolically and read in words.

$|a| = a$, for $a > 0$ The absolute value of a positive number is equal to itself.
$|a| = -a$, for $a < 0$ The absolute value of a negative number is equal to its opposite.
$|0| = 0$ The absolute value of zero is zero.

EXAMPLE Give the absolute value of the following quantities: (a) $|9|$ (b) $|-4|$

(a) $|9| = 9$ 9 is positive. Its absolute value is the number itself.

(b) $|-4| = 4$ -4 is negative. Its absolute value is the opposite of -4.

TIP — Two Components of a Signed Number

Integers and other signed numbers tell us two things. For the integer -8, the negative sign tells us the number is to the left of zero and is referred to as the *directional sign* of the number. The 8 tells us how many units the number is from zero; it is the *absolute value* of the number.

Every number except zero has an opposite. The opposite of a number is also called the **additive inverse** of the number. A number and its additive inverse (opposite) have the same absolute value but different directional signs.

EXAMPLE Find the opposite of each number and show your answer on the number line.

(a) -8 **(b)** 6   **(c)** 0

(a) The opposite of -8 is 8.

(b) The opposite of 6 is -6.

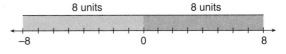

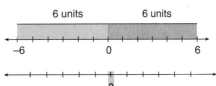

(c) Zero is its own opposite.

2 Add Signed Numbers with Like Signs.

Since our earliest experiences with addition we have added positive numbers. In the addition problem $3 + 2 = 5$, all three numbers are positive. Numbers are understood to be positive when no sign is given. To emphasize that numbers are positive we can write the problem as $+3 + +2 = +5$. We can also write the problem using parentheses to separate the two plus signs, $+3 + (+2) = +5$. The plus symbol serves two purposes in mathematical expressions. It is used as the directional sign for a positive number, and it is used to show addition. In the expression $+3 + (+2) = +5$, the plus sign in front of the parentheses shows the operation of addition. The plus sign inside the parentheses identifies the number as positive.

Adding integers builds on our knowledge of addition of positive numbers.

To add 3 and 2 on the number line we begin at zero and move three units to the right (positive direction). This move takes us to $+3$. From the $+3$ position, we move two units to the right. This second move takes us to the $+5$ position (Fig. 5–4). In other words, $3 + 2 = 5$.

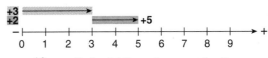

Figure 5–4   Adding using a number line.

To add two negative integers on the number line, $(-3) + (-2)$, we begin at zero and move three units to the left (negative direction). This takes us to -3 on the number line. Then, from -3 we move two units to the left. This second move takes us to -5 (Fig. 5–5). Thus, $-3 + (-2) = -5$.

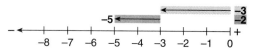

Figure 5–5   Adding two negative numbers using a number line.

When adding the two positive numbers or the two negative numbers, the sum could have been obtained using the following rule.

To add signed numbers with like signs:

1. Add the absolute values of the numbers.
2. Give the sum the common or like sign.

EXAMPLE   Add -12 and -5.

$$-12 + (-5) = -17$$ Add absolute values. Keep common negative sign.

Addition is a **binary operation.** This means that only two numbers are used in the operation. If more than two numbers are added, two are added and then the next number is added to the sum of the first two. Rules developed for addition apply to two numbers at a time.

EXAMPLE   Add $-3 + (-4) + (-1)$.

$$-3 + (-4) + (-1) =$$ Add the first two numbers.

$$-7 + (-1) = -8$$ Add the sum to remaining number.

3   Add Signed Numbers with Unlike Signs.

When adding numbers with unlike signs, we move in the direction indicated by the sign of each number, positive to the right and negative to the left. To add $-4 + 3$, we first move four units to the left of zero (see Fig. 5–6). This takes us to -4. Then, from the -4 position, we move three units to the right. This takes us to -1. Thus, $-4 + 3 = -1$. This process is generalized by a rule using absolute values.

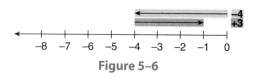

Figure 5–6

To add signed numbers with unlike signs:

1. *Subtract* the smaller absolute value from the larger absolute value.
2. Give the sum the sign of the number with the larger absolute value.

EXAMPLE   Add $7 + (-12)$.

$7 + (-12) =$ Signs are unlike.

$7 + (-12) = -5$ Subtract absolute values: $12 - 7 = 5$. Keep the sign of the -12 (negative), the larger absolute value.

To Add, You . . . Subtract?

Phrases we used in arithmetic for adding integers, such as "find the sum" or "add," can be confusing. In addition of integers, sometimes we add absolute values and sometimes we subtract absolute values. We often use the word "**combine**" to imply the addition of numbers with either like or unlike signs.

EXAMPLE   Combine $-5 + 7$.

$-5 + 7 =$ Signs are unlike.

$-5 + 7 = 2$ Subtract absolute values: $7 - 5 = 2$. Keep the sign of the 7 (positive), the larger absolute value.

When an addition involves several positive numbers and several negative numbers, it is convenient to group all the positive numbers and all the negative numbers and to add each group separately. We then add the sums of each group. We can do this because of the *commutative* and *associative* properties of addition of real numbers. In the next example we combine the commutative and associative properties of addition to add several signed numbers.

EXAMPLE   Add $8 + (-9) + 13 + (-15)$.

$8 + (-9) + 13 + (-15) =$ Signs are unlike. Rearrange and group numbers with like signs.

$[8 + 13] + [-9 + (-15)] =$ The brackets, [ ], show the grouping and separate the operational plus sign from the directional minus sign of -9.

Add groups of numbers with like signs.

$21 + (-24) = -3$ Combine resulting numbers using the rule for adding numbers with unlike signs.

The properties of addition that involve the number zero apply to integers as well as to whole numbers, fractions, and decimals. When zero is added to any real number, the number is not changed. We say that zero is the **additive identity.** The sum of a number and its opposite is zero. We can also write definitions using symbols.

Zero is the *additive identity* because $a + 0 = a$ for all real values of a.
The opposite of a number is the *additive inverse* of the number because $a + (-a) = 0$ for all real values of a.

EXAMPLE   Add: **(a)** $-2 + 0$ **(b)** $5 + (-5)$ **(c)** $0 + (-3) + 4$ **(d)** $5 + (-2) + (-5)$

(a) $-2 + 0 = \mathbf{-2}$	Additive identity.
(b) $5 + (-5) = \mathbf{0}$	Additive inverse.
(c) $0 + (-3) + 4 =$	Additive identity.
$\quad -3 + 4 = \mathbf{1}$	Add numbers with unlike signs.
(d) $\boxed{5} + (-2) + \boxed{(-5)} =$	Apply the commutative and associative properties.
$\boxed{5 + (-5)} + (-2) =$	Additive inverse.
$\boxed{0} + (-2) = \mathbf{-2}$	Additive identity.

SECTION 5–1 SELF-STUDY EXERCISES

1   Use the symbol $>$ or $<$ to show the relationship between each pair of numbers.

1. 5 ___ 8   **2.** -3 ___ -2 **3.** -7 ___ -9 **4.** 0 ___ 5

5. -8 ___ 0   **6.** 12 ___ 7   **7.** -3 ___ 2   **8.** 5 ___ -7

Write the absolute value.

9. $|7|$ **10.** $|-17|$ **11.** $|8|$ **12.** $|-7|$

Write the opposite of each number.

13. 42   **14.** -17 **15.** -78 **16.** 57

2   Add.

17. $8 + 15$ **18.** $-9 + (-7)$ **19.** $-5 + (-8)$

20. $(-12) + (-17)$ **21.** $-24 + (-31)$ **22.** $-78 + (-46)$

23. $7 + 10 + 12$ **24.** $-5 + (-8) + (-7)$ **25.** $12 + 87$

26. $-21 + (-38)$ **27.** $-32 + (-16)$ **28.** $(-58) + (-103)$

29. $-5 + (-12) + (-36)$ **30.** $52 + (+92) + (+88)$ **31.** $-107 + (-502) + (-396)$

3   Add.

32. $5 + (-3)$ **33.** $-3 + 5$ **34.** $8 + (-4)$

35. $-8 + 4$ **36.** $9 + (-3)$ **37.** $3 + (-9)$

38. $-4 + (+2)$ **39.** $-3 + 7$ **40.** $4 + (-6)$

41. $-4 + 6$ **42.** $-18 + 8$ **43.** $32 + (-72)$

44. $21 + (-14)$ **45.** $17 + (-4) + 3 + (-1)$ **46.** $-3 + 2 + (-7)$

47. $47 + (-82) + 2$ **48.** $14 + (-6) + 1$ **49.** $-7 + (-3) + (-1)$

50. $4 + 2 + (-3) + 10$

51. $2 + (-1) + 8$

52. $-2 + 1 + (-8) + 12$

53. $15 + (-15)$

54. $-7 + 7$

55. $92 + (-92)$

56. $(-396) + (396)$

57. $-3 + 0$

58. $0 + (-7)$

59. $6 + (-3) + 5 + (-8)$

60. $12 + (-5) + 8 + (-18)$

61. $4 + (-7) + 6 + (-15)$

62. $18 + (-7) + (-9) + (-16)$

63. $8 + (-4) + (-13) + 22$

64. $18 + (-3) + 5 + (-32)$

65. $12 + (-15) + 73 + (-46)$

66. $(-83) + (-21) + 42 + (-36)$

67. $(-5) + (-18) + 32 + (-7)$

68. $15 + (-32) + 46 + (-19)$

69. **BUS** You open a bank account by depositing $242. You then write checks for $21, $32, and $123. What is your new account balance?

70. **BUS** A stock priced at $42 has the following changes in one week: $-3, +8, -6, -7, +2$. What is the value of the stock at the end of the week?

71. **AVIA** Liquid hydrogen, fuel in the space shuttle, must be kept below $-253°C$. A tank of liquid hydrogen is stored at $-320°C$, but the temperature is raised by $25°C$. What is the new temperature of the hydrogen?

72. The temperature at the South Pole is recorded as $-38°F$ at midnight. The temperature rises $15°F$ by noon. Determine the temperature reading at noon.

5–2 | Subtracting Signed Numbers

Learning Outcomes   **1** Subtract signed numbers.
2 Combine addition and subtraction.

1 Subtract Signed Numbers.

Addition and subtraction are related.
Subtracting a number is the *same as* adding the opposite of the number.

$6 - 9$	Show direction signs of the numbers.
$+6 - (+9)$	Subtract $+9$.
$+6 + (-9)$	Add the opposite of $+9$ or add -9.

Let's write this relationship as a rule.

To subtract signed numbers:

1. Change the subtraction sign to addition.
2. Change the sign of the second number (subtrahend) to its opposite.
3. Apply the appropriate rule for adding signed numbers.

EXAMPLE   Subtract: **(a)** $2 - 6$ **(b)** $-9 - 5$ **(c)** $12 - (-4)$ **(d)** $-7 - (-9)$

(a)	$2 - 6 =$	Subtract positive 6 from positive 2. Write sign of subtrahend.
	$2 - (+6) =$	Change subtraction to addition.
	$2 + (-6) =$	Add the opposite of $+6$, which is -6, to 2.
	$2 + (-6) = \mathbf{-4}$	Apply the rule for adding numbers with unlike signs.
(b)	$-9 - 5 =$	Subtract positive 5 from negative 9. Write sign of subtrahend.
	$-9 - (+5) =$	Change subtraction to addition.
	$-9 + (-5) =$	Add the opposite of $+5$, which is -5, to -9.
	$-9 + (-5) = \mathbf{-14}$	Apply the rule for adding numbers with like signs.

(c) $12 - (-4) =$ Subtract negative 4 from positive 12.

$12 + (+4) =$ Add the opposite of -4, which is $+4$, to 12.

$12 + 4 = \mathbf{16}$ Apply the rule for adding numbers with like signs.

(d) $-7 - (-9) =$ Subtract negative 9 from negative 7.

$-7 + (+9) =$ Add the opposite of -9, which is $+9$, to -7.

$-7 + 9 = \mathbf{2}$ Apply the rule for adding numbers with unlike signs.

TIP — Writing Subtractions as Equivalent Additions

When writing a subtraction as an equivalent addition, you make *two* changes.

1. Change the operation from subtraction to addition.
2. Change the subtrahend to its opposite.

$$2 - (+6) \qquad -9 - (+5) \qquad 12 - (-4) \qquad -7 - (-9)$$
$$2 + (-6) \qquad -9 + (-5) \qquad 12 + (+4) \qquad -7 + (+9)$$
$$\uparrow\ \uparrow \qquad\qquad \uparrow\ \uparrow \qquad\qquad \uparrow\ \uparrow \qquad\qquad \uparrow\ \uparrow$$
$$1.\ \ 2. \qquad\qquad 1.\ \ 2. \qquad\qquad 1.\ \ 2. \qquad\qquad 1.\ \ 2.$$

Subtractions involving zero and opposites are interpreted first as equivalent addition problems. Since zero has no opposite, $+0$ and -0 are still 0.

EXAMPLE Evaluate: **(a)** $8 - 0$ **(b)** $0 - 15$ **(c)** $-32 - (-32)$ **(d)** $-15 - 15$

(a) $8 - 0 =$ Subtract zero from positive 8.

$8 + (0) =$ Add zero to positive 8.

$8 + (0) = \mathbf{8}$ Zero added to any number is that number (additive identity).

(b) $0 - 15 =$ Subtract positive 15 from zero. Write sign of subtrahend.

$0 - (+15) =$ Change subtraction to addition.

$0 + (-15) =$ Add the opposite of $+15$, which is -15, to zero.

$0 + (-15) = \mathbf{-15}$ Any number added to zero (additive identity) is that number.

(c) $-32 - (-32) =$ Subtract negative 32 from negative 32.

$-32 + (+32) =$ Add the opposite of -32, which is $+32$, to -32.

$-32 + 32 = \mathbf{0}$ A number added to its opposite (additive inverse) is zero.

(d) $-15 - (+15) =$ Subtract positive 15 from negative 15.

$-15 + (-15) =$ Add the opposite of $+15$, which is -15, to -15.

$-15 + (-15) = \mathbf{-30}$ Apply the rule for adding numbers with like signs.

---

2 Combine Addition and Subtraction.

When we express the sum and difference of signed numbers with all the appropriate operational and directional signs, we have an expression with both an operational sign and a directional sign between every two numbers. We have four different possibilities when two signs are written together. We can simplify these four possibilities.

Writing mathematical expressions that include every operational and directional sign is cumbersome. In general practice, we omit as many signs as possible. When two signs are written between two numbers, we can write a simplified expression with only one sign.

	Double Signs		**Single Sign**
Like Signs	Plus, Plus: $+3$ $+ (+ 5)$	Add $+3$ and $+5$.	$3 + 5$
	Minus, Minus: $+3$ $- (- 5)$	Change to addition. $+3 + (+5)$	$3 + 5$
Unlike Signs	Plus, Minus: $+3$ $+ (- 5)$	Add $+3$ and -5.	$3 - 5$
	Minus, Plus: $+3$ $- (+ 5)$	Change to addition. $+3 + (-5)$	$3 - 5$

To generalize, *two like signs between signed numbers,* whether both plus or both minus, translate to adding a positive number. Use just one plus sign:

$$+ + \rightarrow + \qquad - - \rightarrow + \qquad \text{like signs} \rightarrow +$$

Two unlike signs between signed numbers, either a plus/minus or a minus/plus, translate to adding a negative number. Use just one minus sign:

$$+ - \rightarrow - \qquad - + \rightarrow - \qquad \text{unlike signs} \rightarrow -$$

To add and subtract more than two signed numbers:

1. Rewrite the problem so that all integers are separated by only one sign.
2. Add the series of signed numbers.

EXAMPLE Evaluate: **(a)** $3 - (-5) - 6$ **(b)** $-8 + 10 - (-7)$

(a) $3 - (-5) - 6 =$ Rewrite with only one sign between signed numbers.

     $3 + 5 - 6 =$ Add numbers with like signs.

        $8 - 6 = \mathbf{2}$ Apply the rule for adding numbers with unlike signs.

(b) $-8 + 10 + (+7) =$ Rewrite with only one sign between signed numbers.

     $-8 + 10 + 7 =$ Add numbers with like signs.

       $-8 + 17 = \mathbf{9}$ Apply the rule for adding numbers with unlike signs.

The key to solving applied problems with signed numbers is to identify which numbers are positive and which are negative.

Positive key words: profits, gains, money in the bank, temperatures above zero, receipts, income, winnings, and so on.

Negative key words: losses, deficits, checks that cleared the bank, temperatures below zero, drops, declines, payments, and so on.

EXAMPLE
BUS

A landscaping business makes a profit of $345 one week, has a loss of $34 the next week, and makes a profit of $235 the third week. What is the total, or net, profit?

The total profit is the sum of the weekly profits and losses.

$345 − $34 + $235   Interpret profits as positive and losses as negative.
profit   loss   profit

345 + 235 − 34   Rearrange and add positives.

580 − 34 = 546   Apply rule for adding numbers with unlike signs.

The total or net profit for   Interpret answer.
the three weeks is $546.

EXAMPLE

In a recent year, 98°F was the highest temperature in Boston, and −2°F was the lowest temperature. What was the temperature range for the city that year? (The range is the difference between the highest and lowest values.)

98 − (−2) =   Subtract −2 from 98.

98 + 2 = 100   Rewrite with only one sign between integers, and apply the appropriate rule for adding integers.

The temperature range for   Interpret answer.
Boston that year was 100°F.

SECTION 5–2 SELF-STUDY EXERCISES

1 Subtract.

1. −3 − 9	**2.** 8 − 2	**3.** 9 − 15
4. (−3) − (−7)	**5.** −11 − 14	**6.** (−6) − (−3)
7. 5 − (−3)	**8.** 8 − 11	**9.** −8 − 1
10. 11 − (−2)	**11.** (−8) − (−7)	**12.** (−15) − (−7)
13. (20) − (−42)	**14.** (−38) − (−27)	**15.** (−42) − (+16)
16. (−21) − (+36)	**17.** (−18) − (+15) − (−18)	**18.** (−12) − (+21) − (+72)
19. 14 − (−21) − 17	**20.** (−14) − (−21) − 24	**21.** −142 − (+46) − (−21)
22. 217 − (−38) − (+172)	**23.** 802 + (196) − (−415)	**24.** 72 − (−23) − (+198)
25. 15 − 0	**26.** 0 − 8	**27.** −12 − 0
28. 0 − (−8)	**29.** 0 − (−7)	**30.** 10 − 0
31. 28 − (−28)	**32.** −46 − 46	**33.** 7 − (−7)
34. −18 − 18	**35.** 5 − 5	**36.** 21 − 21

2 Evaluate.

37. 5 + 8 − 9	**38.** 6 − 4 + 5	**39.** 9 + 2 − 5
40. −1 + 1 − 4	**41.** 5 + 3 − 7	**42.** 7 + 3 − (−4)
43. −8 − 2 − (−7)	**44.** −3 + 4 − 7 − 3	**45.** 2 − 4 − 5 − 6 + 8
46. 8 − 3 + 2 − 1 + 7	**47.** −5 − 3 + 8 − 2 + 4	**48.** 6 − (−3) + 5 − 6 − 9
49. −8 + 2 − 7 + 14	**50.** 3 − 5 + 8 − 11 − 15	**51.** 5 − 2 + 9 − 6 − 12 + 32

52. $18 + 12 - 16 - 32 - 81$ **53.** $-28 + 32 - 17 - 32$ **54.** $78 - 64 + 32 - 78$
55. $52 - 96 - 102 + 86$ **56.** $42 - 17 + 86 - 191$ **57.** $77 - 96 - 102 + 86$
58. $149 - 23 + 6 - 82$ **59.** $54 - 87 + 33 - 21$ **60.** $228 - 21 + 33 - 54$

61. The temperatures for Bowling Green, Kentucky, ranged from 102°F to −5°F. What was the temperature range for the city?

62. New Boston, Texas, registered −8°F as its lowest temperature one year and 99°F as its highest temperature for the same year. What was the temperature range for New Boston?

63. BUS   Computing Solutions records a profit of $28,296 one quarter (three months), a loss of $1,896 for the second quarter, a profit of $52,597 for the third quarter, and a profit of $36,057 for the fourth quarter. What is their net profit for the year?

64. Explain the difference between subtracting zero from a number and subtracting a number from zero.

Use the table showing the U.S. trade balance with Belize for Exercises 65–68.

65. BUS   Express the difference in trade balances in February 2005 and January 2005 with a signed number.

66. BUS   Express the difference in trade balances in April 2005 and March 2005 with a signed number.

67. BUS   Find the difference in trade balances for March 2005 and February 2005.

68. BUS   Use your knowledge of operations with signed numbers to find the ending trade balance on April 30, 2005.

**U.S. Trade with Belize
(in millions of dollars)**

Month	Trade Balance
January 2005	7.5
February 2005	−0.5
March 2005	5.0
April 2005	9.6

Source: U.S. Census Bureau, Foreign Trade Division, Data Dissemination Branch, Washington, DC 20233.

5-3 | *Multiplying and Dividing Signed Numbers*

Learning Outcomes   **1** Multiply signed numbers.
2 Evaluate powers of signed numbers.
3 Divide signed numbers.

Multiplying absolute values of signed numbers is exactly the same as multiplying whole numbers, fractions, and decimals. However, the rules for multiplying signed numbers must also include the assignment of the proper sign to the product.

1 Multiply Signed Numbers.

Just as with whole numbers, fractions, and decimals, multiplication of signed numbers is a binary operation that is commutative and associative; that is, two factors at a time can be multiplied in any order. More than two factors can be grouped in any manner. These examples show how we assign signs to the product of two signed numbers:

Like Signs   **Unlike Signs**

$4(6) = 24,$ $(-3)(-7) = 21$ $-10(2) = -20,$ $8(-2) = -16$

To multiply two signed numbers:

1. Multiply the absolute values of the numbers.
2. If the factors have like signs, the sign of the product is positive: $(+)(+) = +; (-)(-) = +$
3. If the factors have unlike signs, the sign of the product is negative: $(+)(-) = -; (-)(+) = -$

EXAMPLE   Multiply: **(a)** $-12(-2)$ **(b)** $10 \cdot 3$ **(c)** $25(-3)$ **(d)** $-5 * 7$

 (a) $-12(-2) = \mathbf{24}$ Like signs give a positive product.
 (b) $10 \cdot 3 = \mathbf{30}$ Like signs give a positive product.
 (c) $25(-3) = \mathbf{-75}$ Unlike signs give a negative product.
 (d) $-5 * 7 = \mathbf{-35}$ Unlike signs give a negative product.

When the number 1 is multiplied by any real number, the result is the number. The number 1 is the **multiplicative identity** because $a \cdot 1 = 1 \cdot a = a$ for all real values of a.

EXAMPLE   Multiply: **(a)** $4(-2)(6)$ **(b)** $-3(4)(-5)$ **(c)** $-2(-8)(-3)$
     (d) $-2(-3)(-4)(-1)$

 (a) $4(-2)(6) =$ Multiply the first two factors and apply the rule for factors with unlike signs.
   $-8(6) = \mathbf{-48}$ Multiply and apply the rule for factors with unlike signs.

 (b) $-3(4)(-5) =$ Multiply the first two factors and apply the rule for factors with unlike signs.
   $-12(-5) = \mathbf{60}$ Multiply and apply the rule for factors with like signs.

 (c) $-2(-8)(-3) =$ Multiply the first two factors and apply the rule for factors with like signs.
   $16(-3) = \mathbf{-48}$ Multiply and apply the rule for factors with unlike signs.

 (d) $-2(-3)(-4)(-1) =$ Multiply the first two and the last two factors.
   $6(4) = \mathbf{24}$ Multiply and apply the rule for factors with like signs.

If we examine the multiplications in the preceding example more closely, we see that the number of negative factors affects the sign of the answer.

	Number of Negative Factors	Sign of Product
$4(-2)(6)$	1	$-$
$-3(4)(-5)$	2	$+$
$-2(-8)(-3)$	3	$-$
$-2(-3)(-4)(-1)$	4	$+$

To determine the sign of the product when multiplying three or more factors:

1. The sign of the product is *positive* if the number of negative factors is *even.*
2. The sign of the product is *negative* if the number of negative factors is *odd.*

EXAMPLE   Multiply: **(a)** $-2(6)(-1)(-3)$ **(b)** $(2)(-5)(1)(-3)$

(a) $-2(6)(-1)(-3) = \mathbf{-36}$ Multiply absolute values. The odd number of negative factors makes the product negative.

(b) $(2)(-5)(1)(-3) = \mathbf{30}$ Multiply absolute values. The even number of negative factors makes the product positive.

The *zero property of multiplication* extends to integers. The product of zero and any real number is zero: $x \cdot 0 = 0$. If we have two or more factors and one factor is zero, we can immediately write the product as zero without having to work through the steps.

EXAMPLE   Multiply: **(a)** $3(-21)(2)(0)$ **(b)** $-9(-2)(8)(-1)$

(a) $3(-21)(2)(0) = \mathbf{0}$ Zero is a factor.
(b) $-9(-2)(8)(-1) = \mathbf{-144}$ Zero is not a factor.

Applied problems often require multiplication of integers.

EXAMPLE
BUS   In a 3-week period, a technology stock declined approximately 2 points each week. How many points did the stock decline in the 3 weeks?

Because there are equal declines each week, we multiply the amount of weekly decline times the number of weeks: $-2(3) = -6$. Thus, **the stock declined (negative) a total of 6 points over the three-week period.**

2 Evaluate Powers of Signed Numbers.

Raising a number to a natural-number power is an extension of multiplication, so determining the sign of the result is similar to multiplying several integers. Observe the pattern for determining the sign of the result:

Positive Base
$(+4)^2 = (+4)(+4) = +16$
$(+4)^3 = (+4)(+4)(+4) = +64$

Negative Base
$(-4)^2 = (-4)(-4) = +16$
$(-4)^3 = (-4)(-4)(-4) = -64$

To raise signed numbers to a natural-number power, use the following patterns:

1. A positive number raised to any natural-number power is positive.
2. Zero raised to any natural-number power is zero.
3. A negative number raised to an even natural-number power is positive.
4. A negative number raised to an odd natural-number power is negative.

EXAMPLE   Evaluate the powers: **(a)** 4^3 **(b)** 0^8 **(c)** $(-2)^4$ **(d)** $(-3)^5$

(a) $4^3 = 4(4)(4) = \textbf{64}$ Base is positive.
(b) $0^8 = \textbf{0}$ Base is zero.
(c) $(-2)^4 = (-2)(-2)(-2)(-2) = \textbf{16}$ Base is negative, exponent is even.
(d) $(-3)^5 = (-3)(-3)(-3)(-3)(-3) = \textbf{-243}$ Base is negative, exponent is odd.

EXAMPLE
ELEC   Security systems sometimes use a four-digit code to activate the system. How many different codes can be made with four digits? Identify codes that may be impractical.

The number system has 10 digits: 0, 1, 2, 3, 4, 5, 6, 7, 8, and 9, so we can fill each one of the four slots of the code in 10 different ways. We allow the same digit to be used repeatedly. (This is called **sampling with replacement.**) To find the total number of different codes, we multiply $10 \cdot 10 \cdot 10 \cdot 10$. This product can be written as 10^4, or 10,000. **There are 10,000 ways to make a four-digit security code. Some codes, such as 0000, may be impractical.**

Negative Base Versus an Opposite

$(-2)^4$ is not the same expression as -2^4. *To raise a negative number to a power requires parentheses.*
$(-2)^4$ is a negative base raised to a power. -2^4 is the opposite of 2^4.

Negative Base	**Opposite**
$(-2)^4 = (-2)(-2)(-2)(-2) = 16$	$-2^4 = -(2)(2)(2)(2) = -16$

Sometimes the values of two expressions are equal, but the interpretation is different. The expressions $(-3)^5$ and -3^5 give the same result because the exponent is an odd number.

$$(-3)^5 = (-3)(-3)(-3)(-3)(-3) = -243 \qquad -3^5 = -(3)(3)(3)(3)(3) = -243$$

---

3 Divide Signed Numbers.

The rules for determining the sign when dividing signed numbers are similar to the rules for multiplying signed numbers.

To divide two signed numbers:

1. Divide the absolute values of the numbers.
2. If the numbers have like signs, the sign of the quotient is positive: $(+) \div (+) = +$; $(-) \div (-) = +$.
3. If the numbers have unlike signs, the sign of the quotient is negative: $(+) \div (-) = -$; $(-) \div (+) = -$.

EXAMPLE   Divide: (a) $\dfrac{-8}{-2}$ (b) $\dfrac{6}{3}$ (c) $\dfrac{10}{-2}$ (d) $\dfrac{-9}{1}$

(a) $\dfrac{-8}{-2} = \mathbf{4}$ Like signs give a positive quotient.

(b) $\dfrac{6}{3} = \mathbf{2}$ Like signs give a positive quotient.

(c) $\dfrac{10}{-2} = \mathbf{-5}$ Unlike signs give a negative quotient.

(d) $\dfrac{-9}{1} = \mathbf{-9}$ Unlike signs give a negative quotient.

Division with zero works the same for signed numbers as it does for whole numbers. Any real number divided by 0 is not defined because there is not a real number that can be multiplied by 0 to give the original number. We say $\frac{0}{0}$ is indeterminant because *any* real number will multiply by 0 to give 0.

To evaluate division with zero:

Zero divided by any nonzero number is zero.

$$\dfrac{0}{n} = 0; \qquad n \neq 0 \text{ and } n \text{ is any real number} \qquad \dfrac{0}{-5} = 0$$

Division by zero is either undefined or indeterminate.

$\dfrac{n}{0}$ is undefined; $n \neq 0$ and n is any real number $\dfrac{-5}{0}$ is undefined $\dfrac{0}{0}$ is indeterminate

EXAMPLE   Evaluate: (a) $\dfrac{-3}{0}$ (b) $\dfrac{0}{0}$ (c) $\dfrac{0}{-5}$

(a) $\dfrac{-3}{0}$ **is undefined**

(b) $\dfrac{0}{0}$ **is indeterminate**

(c) $\dfrac{0}{-5} = \mathbf{0}$

SECTION 5–3 SELF-STUDY EXERCISES

1 Multiply.

1. $5 \cdot 8$
2. $-4(-3)$
3. $7 * 5$
4. $(-3)(-7)$
5. $-8(-3)$
6. $(-2)(-3)$
7. $5(-3)$
8. $(-2)(5)$
9. $-4 * 8$
10. $-3 \cdot 4$
11. $-7 * 8$
12. $6(-4)$

13. **BUS** Madison Duke had four checks returned for insufficient funds, and her bank charged her a $28 service charge for each check. Use multiplication of signed numbers to show how these transactions affected her checking account balance.

14. **BUS** Chloe Duke made seven withdrawals of $40 each from her checking account. Use signed numbers to show how these transactions affected her checking account balance.

15. $8(-3)(-2)(7)$
16. $5(-2)(3)(2)$
17. $6(1)(-3)(-2)$
18. $4(0)(-12)(3)$
19. $15(-2)(-3)$
20. $5(2)(-3)(0)$
21. $-3(2)(-7)(-1)$
22. $9(-1)(3)(-2)$
23. $(-7)(-5)(-6)$
24. $(-3)(-9)(-12)(-7)$
25. $7(-3)(-10)(12)(-8)$
26. $(7)(8)(-5)(-3)$
27. $(-3)(-8)(-2)(5)$
28. $-8(0)$
29. $5(0)$
30. $0(-12)$
31. $(-15)(0)$
32. $18(0)$
33. $0 \cdot 3$
34. $0(-15)$
35. $0(-17)$
36. $-28 \cdot 0$
37. $46 \cdot 0$
38. $5(1)(-3)(2) + 5(0)$
39. $-8(1)(3)(-7) + 7(0)$
40. $5(0) + 2(-3)(-7)$
41. $4(-2) + 3(0)(-5)$
42. $(-3)(-6)(-8)(0)$
43. $2(-3) - 5(2) + 7(0)(-4)$

44. Review the definitions for additive inverse and additive identity and write a similar definition for multiplicative inverse.

45. Illustrate the commutative property of multiplication using one positive and one negative integer.

2 Evaluate.

46. $(-3)^2$
47. $(-2)^3$
48. $(-5)^2$
49. 0^{10}
50. -2^3
51. $(-8)^3$
52. $(5)^4$
53. 3^4
54. 7^4
55. $(-42)^2$
56. $(5)^1$
57. $(-28)^1$
58. 12^1
59. $(-8)^1$
60. -10^1
61. -8^1
62. $(-3)^3$
63. -11^2
64. $(-11)^2$
65. -5^3

3 Divide.

66. $-15 \div 5$
67. $\dfrac{8}{-2}$
68. $\dfrac{-24}{6}$
69. $-28 \div 7$
70. $\dfrac{-32}{-4}$
71. $\dfrac{-50}{-10}$
72. $\dfrac{36}{-6}$
73. $\dfrac{-48}{-6}$
74. $\dfrac{-25}{-5}$

75. **BUS** You have decreased your house loan balance by $1,800 over the past 6 months. Show this figure as a signed number, and find the average monthly decrease.

76. A temperature of 10°C fell to -8°C in 3 h. Find the average hourly change expressed as a signed number.

77. $\dfrac{12}{0}$
78. $\dfrac{-15}{0}$
79. $\dfrac{0}{+3}$
80. $\dfrac{0}{-12}$
81. $\dfrac{0}{-8}$
82. $\dfrac{-20}{0}$
83. $\dfrac{-100}{0}$
84. $\dfrac{1}{0}$
85. $\dfrac{0}{8}$
86. $\dfrac{-350}{0}$

87. Which two operations have the same rules for handling signs when working with signed numbers?

88. A nor'easter storm blew into Green Bay, Wisconsin, and the temperature changed from 38°F to −22°F between 1 P.M. and 7 P.M. Represent the average hourly change in temperature with a signed number.

89. BUS A business started the year with a net worth of –$4,852. During the first six months of the year, the business recovered and increased its net worth to a value of $15,983. Find the average monthly increase in net worth for the 6-month period.

90. What value or values must be in the numerator and denominator of a division in order for the result to be zero?

5–4 | *Signed Fractions and Decimals*

Learning Outcomes
1 Change a signed fraction to an equivalent signed fraction.
2 Perform basic operations with signed fractions and decimals.
3 Apply the order of operations with signed numbers.

1 Change a Signed Fraction to an Equivalent Signed Fraction.

Fractions and decimals can also have negative values. This extends our types of numbers to include rational numbers. A **rational number** is any type of number that can be written as the quotient of two integers (Fig. 5–7).

A fraction has three signs: the sign of the fraction, the sign of the numerator, and the sign of the denominator. The fraction $\frac{2}{3}$ expressed as a *signed fraction* is $+\frac{+2}{+3}$. When a signed fraction has negative signs, we sometimes change the signed fraction to an equivalent signed fraction.

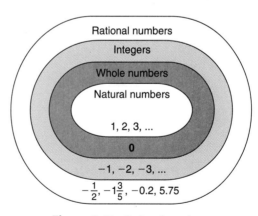

Figure 5–7 Rational numbers.

 To find an equivalent signed fraction:

1. Identify the three signs of the fraction (sign of the fraction, sign of the numerator, and sign of the denominator).
2. Change any two of the three signs to the opposite sign.

EXAMPLE Change $-\dfrac{-2}{-3}$ to three equivalent signed fractions.

$$-\dfrac{-2}{-3} \;=\; +\dfrac{+2}{-3} \;=\; \dfrac{2}{-3}$$ Change the signs of the fraction and the numerator.

$$-\dfrac{-2}{-3} \;=\; +\dfrac{-2}{+3} \;=\; \dfrac{-2}{3}$$ Change the signs of the fraction and the denominator.

$$-\dfrac{-2}{-3} \;=\; -\dfrac{+2}{+3} \;=\; -\dfrac{2}{3}$$ Change the signs of the numerator and the denominator.

Changing the signs of a fraction is a manipulation tool that simplifies our work when we perform basic operations with signed fractions.

Why Would We Want to Change the Signs of a Fraction?

When changing any two signs of a fraction, we can accomplish these desirable outcomes.

- Avoid dealing with negatives.

$$-\dfrac{-3}{+4} = +\dfrac{+3}{+4} \quad \text{or} \quad \dfrac{3}{4}$$

$$+\dfrac{-6}{-7} = +\dfrac{+6}{+7} \quad \text{or} \quad \dfrac{6}{7}$$

$$-\dfrac{+5}{-9} = +\dfrac{+5}{+9} \quad \text{or} \quad \dfrac{5}{9}$$

- Change subtraction to addition.

$$-\dfrac{+5}{+6} = +\dfrac{-5}{+6} \quad \text{or} \quad \dfrac{-5}{6}$$

- Avoid negative denominators.

$$-\dfrac{+3}{-4} = +\dfrac{+3}{+4} \quad \text{or} \quad \dfrac{3}{4}$$

$$+\dfrac{+2}{-5} = +\dfrac{-2}{+5} \quad \text{or} \quad \dfrac{-2}{5}$$

- Deal with fewer negatives.

$$-\dfrac{-5}{-8} = +\dfrac{-5}{+8} \quad \text{or} \quad \dfrac{-5}{8}$$

2 Perform Basic Operations with Signed Fractions and Decimals.

The rules for operating with integers can be extended to apply to signed fractions.

To add and subtract signed fractions:

1. Change the signs of the fractions so the denominators are positive and subtractions are expressed as addition.
2. Apply the appropriate rules for adding signed numbers and for adding fractions.

EXAMPLE   Add $\dfrac{-3}{4} + \dfrac{5}{-8}$.

$\dfrac{-3}{4} + \dfrac{-5}{8} =$ Change the signs of the numerator and denominator in the second fraction so that both denominators are positive.

$\dfrac{-6}{8} + \dfrac{-5}{8} =$ Change to equivalent fractions with a common denominator.

$\dfrac{-11}{8} =$ Add numerators, applying the rule for adding numbers with like signs.

$-1\dfrac{3}{8}$ Change to a mixed number. The sign of the mixed number is determined by the rule for dividing numbers with unlike signs.

EXAMPLE   Subtract $\dfrac{-3}{7} - \dfrac{-5}{7}$.

$\dfrac{-3}{7} + \dfrac{5}{7} =$ Change subtraction to addition by changing the signs of the second fraction and the numerator. Apply the rule for adding numbers with unlike signs.

$\dfrac{2}{7}$

To multiply or divide signed fractions:

1. Write divisions as equivalent multiplications.
2. Apply the appropriate rules for multiplying integers (signed numbers) and for multiplying fractions.

EXAMPLE   Multiply $\left(\dfrac{-4}{5}\right)\left(\dfrac{3}{-7}\right)$.

$\dfrac{-4}{5} \cdot \dfrac{3}{-7} =$ Multiply the numerators and denominators.

$\dfrac{-12}{-35} =$ Apply rule for dividing numbers with like signs to simplify signs.

$\dfrac{12}{35}$

EXAMPLE Simplify $\left(\dfrac{-2}{3}\right)^3$.

$\left(\dfrac{-2}{3}\right)^3 =$ Cube the numerator and cube the denominator.

$\dfrac{(-2)^3}{3^3} =$ $(-2)(-2)(-2) = -8; (3)(3)(3) = 27$

$\dfrac{-8}{27}$ **or** $-\dfrac{8}{27}$ Manipulate signs if desired.

To perform basic operations with signed decimals:

1. Determine the indicated operations.
2. Apply the appropriate rules for signed numbers and for decimals.

EXAMPLE Add -5.32 and -3.24.

$$
\begin{array}{r}
-5.32 \\
-3.24 \\
\hline
\mathbf{-8.56}
\end{array}
$$
     Align decimals and use the rule for adding numbers with like signs.

EXAMPLE Subtract -3.7 from 8.5.

$8.5 - (-3.7) = 8.5 + 3.7 = \mathbf{12.2}$ Change subtraction to an equivalent addition and add numbers with like signs.

EXAMPLE Multiply 3.91 and -7.1.

$$
\begin{array}{r}
3.91 \\
\times \quad -7.1 \\
\hline
391 \\
27\ 37 \\
\hline
\mathbf{-27.761}
\end{array}
$$
     Apply rule for multiplying numbers with unlike signs.

EXAMPLE Divide: $(-1.2) \div (-0.4)$.

$$
-0.4\overline{\smash{)}{-1.2}}^{\,\mathbf{+3.}}
$$
     Shift decimal points and divide. Use rule for dividing numbers with like signs.

The rules of signed numbers apply to all types of real numbers. First apply the appropriate rule for signed numbers, then perform the necessary calculations.

3 **Apply the Order of Operations with Signed Numbers.**

Evaluating operations with signed numbers follows the same order of operations as evaluating with whole numbers.

Perform operations in the following order as they appear from left to right:

1. Parentheses used as groupings and other grouping symbols.   P
2. Exponents (powers and roots).   E
3. Multiplications and divisions.   MD
4. Additions and subtractions.   AS

EXAMPLE   Evaluate $8(4 - 6)$.

Work within the grouping symbols first:

$8(4 - 6) =$ Add or subtract inside parentheses. $4 - 6 = -2$.   **P**

$8(-2) = -16$ Multiply 8 by -2.   **MD**

Or use the distributive principle:

$8(4 - 6) =$ Distribute. Multiply each term in the parentheses by the factor 8.

$8(4) + 8(-6) =$

$32 + -48 = -16$ Add 32 and -48.

The fraction bar as a symbol for division is also a grouping symbol.

EXAMPLE   Evaluate $\dfrac{3 - 5}{2} - \dfrac{9}{2 + 1} + 4(5)$.

$\dfrac{3 - 5}{2} - \dfrac{9}{2 + 1} + 4(5) =$ Perform operations grouped by the fraction bar.

$\dfrac{-2}{2} - \dfrac{9}{3} + 4(5) =$ Divide and multiply.

$-1 - 3 + 20 =$ Add.

$-4 + 20 = 16$ Add.

EXAMPLE Evaluate $-12 \div 3 - (2)(-5)$.

$$-12 \div 3 - (2)(-5) =$$

$$\boxed{-12 \div 3} - \boxed{(2)(-5)} = \qquad \text{Divide and multiply first.}$$

$$-4 - \boxed{(-10)} = \qquad \text{Change to a single sign between the numbers.}$$

$$-4 + 10 = \mathbf{6} \qquad \text{Add integers with unlike signs.}$$

The problem in the preceding example can also be written without parentheses around the 2. The problem is then expressed as

$$-12 \div 3 - 2(-5)$$

We divide and multiply first, but notice how we perform the multiplication.

$$\boxed{-12 \div 3}\ \boxed{-2(-5)} = \qquad \text{Consider the minus sign between the 3 and 2 as the sign}$$
$$\qquad\qquad\qquad\qquad \text{of the 2. } -12 \div 3 = -4 \text{ and } -2(-5) = 10.$$

$$-4\ \boxed{+ 10} = \qquad \text{Add the results of the division and the multiplication.}$$

$$-4 + \boxed{10} = \mathbf{6} \qquad \text{Add integers with unlike signs.}$$

EXAMPLE Evaluate $10 - 3(-2)$.

$$10\ \boxed{-3(-2)} = \qquad \text{Think of the multiplication as } -3 \text{ times } -2.$$

$$10\ \boxed{+6} = \qquad \text{Interpret as addition.}$$

$$10 + 6 = \mathbf{16} \qquad \text{Add integers with like signs.}$$

Only after parentheses and all multiplications and/or divisions are completed do we perform the final additions and/or subtractions from left to right.

EXAMPLE Evaluate $4 + 5(2 - 8)$.

$$4 + 5\ \boxed{(2 - 8)} = \qquad \text{Perform operations in parentheses.}$$

$$4 + \boxed{5\ (-6)} = \qquad \text{Multiply.}$$

$$4 - 30 = \qquad \text{Add integers with unlike signs.}$$

$$4 - 30 = \mathbf{-26}$$

The Order of Operations Is Important

Note what happens if we proceed *out of order* in the preceding example!

Incorrectly Worked

$$4 + 5(2 - 8) \qquad \textit{Incorrectly} \text{ add first instead of last.}$$
$$9(2 - 8) \qquad \text{Perform operation in parentheses second instead of first.}$$
$$9(-6) \qquad \text{Multiply last instead of second.}$$
$$9(-6) = -54 \qquad \text{Incorrect solution.}$$

We get an incorrect answer. *The order of operations must be followed to arrive at a correct solution.*

EXAMPLE   Evaluate $5 + (-2)^3 - 3(4 + 1)$.

$$5 + (-2)^3 - 3\,\boxed{(4 + 1)} =$$ Perform operation in parentheses.

$$5 + \boxed{(-2)^3} - 3\,\boxed{(5)} =$$ Raise to a power.

$$5 + \boxed{(-8)} - \boxed{3(5)} =$$ Multiply.

$$5 + (-8) - \boxed{15} =$$ Change to one sign only between numbers.

$$5 - 8 - 15 =$$ First addition of integers.

$$-3 - 15 =$$ Remaining addition of integers.

$$\mathbf{-18}$$

Since negative numbers are such an integral part of our everyday lives, practically all types of calculators, even the basic calculator, can deal with negative values. However, the notation for negatives and the process for entering negatives varies widely from calculator to calculator. Some of the most common options are discussed.

Some calculators use the subtraction key for both subtracting and entering negative numbers. Others have a special key for entering a **negative sign** that is labeled as a negative sign enclosed in parentheses $\boxed{(-)}$. A common option with basic calculators is the **sign-change key** $\boxed{+/-}$. This key is a **toggle key** and changes the sign of the number in the display to the opposite sign.

EXAMPLE   Use a calculator to evaluate.

(a) $2 - 7$ (b) $(-7)(2)$ (c) $\dfrac{-4}{2} + 14$

The most common options:

(a) $2 - 7$ $2\,\boxed{-}\,7\,\boxed{=}\;\Rightarrow\;\mathbf{-5}$ $\boxed{=}$ may be labeled $\boxed{\text{ENTER}}$ or $\boxed{\text{EXE}}$.

(b) $(-7)(2)$ $\boxed{(-)}\,7\,\boxed{\times}\,2\,\boxed{=}\;\Rightarrow\;\mathbf{-14}$

or

$\boxed{(-)}\,7\,\boxed{(}\,2\,\boxed{)}\,\boxed{=}\;\Rightarrow\;\mathbf{-14}$ Some calculators interpret parentheses with no operational symbol preceding the parentheses as multiplication.

or

$7\,\boxed{+/-}\,\boxed{\times}\,2\,\boxed{=}\;\Rightarrow\;\mathbf{-14}$ The sign-change key is entered after the absolute value of the number.

(c) $\dfrac{-4}{2} + 14$ $\boxed{(-)}\,4\,\boxed{\div}\,2\,\boxed{=}\,\boxed{+}\,14\,\boxed{=}\;\Rightarrow\;\mathbf{12}$ The first $\boxed{=}$ may not be required if your calculator applies the order of operations. A slash $\boxed{/}$ may be used for division on some calculators.

You may impose the appropriate order of operations by using the $\boxed{=}$ key when you perform a series of calculations. This is necessary with a basic calculator when parentheses keys are not available.

When a bar is used as both a grouping and a division symbol on a calculator, we must instruct the calculator to work the grouping first by enclosing the grouping in parentheses.

EXAMPLE Evaluate using a calculator:

$$\frac{5 + 1}{3} + 2$$

$$\frac{5 + 1}{3} + 2 \quad \boxed{(}\,5\,\boxed{+}\,1\,\boxed{)}\,\boxed{\div}\,3\,\boxed{+}\,2\,\boxed{=} \Rightarrow 4 \qquad \text{Using parentheses for grouping.}$$

or

$$5\,\boxed{+}\,1\,\boxed{=}\,\boxed{\div}\,3\,\boxed{+}\,2\,\boxed{=} \Rightarrow 4 \qquad \text{Using equal for grouping.}$$

EXAMPLE Evaluate $-\frac{1}{2} + 4\left(\frac{3}{8} - \frac{5}{8}\right)$.

$$-\frac{1}{2} + 4\left(\frac{3}{8} - \frac{5}{8}\right) = \qquad \text{Perform operation inside grouping. } \frac{3}{8} - \frac{5}{8} = \frac{-2}{8} = -\frac{1}{4}.$$

$$-\frac{1}{2} + 4\left(-\frac{1}{4}\right) = \qquad \text{Multiply. } 4\left(-\frac{1}{4}\right) = -1$$

$$-\frac{1}{2} - 1 = \qquad \text{Change } -1 \text{ to an equivalent fraction with a denominator of 2.}$$

$$-\frac{1}{2} - \frac{2}{2} = \qquad \text{Add like fractions with like signs.}$$

$$-\frac{3}{2} \text{ or } -1\frac{1}{2} \qquad \text{Change to signed mixed number if desired.}$$

EXAMPLE Evaluate $3.2 - 1.7(0.2)^3$.

$$3.2 - 1.7(0.2)^3 = \qquad \text{Raise 0.2 to the third power. } (0.2)(0.2)(0.2) = 0.008$$
$$3.2 - 1.7(0.008) = \qquad \text{Multiply. } -1.7(0.008) = -0.0136$$
$$3.2 - 0.0136 = \qquad \text{Subtract (add opposite).}$$
$$\mathbf{3.1864}$$

SECTION 5–4 SELF-STUDY EXERCISES

1 Change each fraction to three equivalent signed fractions.

1. $+\dfrac{+5}{+8}$ 　　　 **2.** $-\dfrac{3}{4}$ 　　　 **3.** $\dfrac{-2}{-5}$ 　　　 **4.** $-\dfrac{-7}{-8}$ 　　　 **5.** $\dfrac{7}{8}$

2 Perform the operations.

6. $\dfrac{-7}{8} + \dfrac{5}{8}$ 　　　 **7.** $\dfrac{-4}{5} + \left(\dfrac{-3}{10}\right)$ 　　　 **8.** $\dfrac{3}{8} + \left(-\dfrac{7}{16}\right)$

9. $\left(-\dfrac{5}{16}\right) + \dfrac{1}{2}$

10. $\dfrac{1}{3} + \left(-\dfrac{5}{9}\right)$

11. $\left(-3\dfrac{1}{4}\right) + \left(-7\dfrac{3}{8}\right)$

12. $1\dfrac{7}{8} + \left(-4\dfrac{5}{12}\right)$

13. $-7\dfrac{1}{2} + \left(-1\dfrac{3}{4}\right) + \left(5\dfrac{3}{8}\right)$

14. $\dfrac{1}{2} - \left(\dfrac{-3}{5}\right)$

15. $\left(-\dfrac{3}{4}\right) + \left(-\dfrac{5}{8}\right)$

16. $-\dfrac{3}{8} + \left(+\dfrac{5}{16}\right)$

17. $1\dfrac{7}{8} - \left(-\dfrac{5}{6}\right)$

18. $3\dfrac{3}{4} - \left(+4\dfrac{1}{4}\right)$

19. $\dfrac{-3}{5} \times \left(\dfrac{10}{-11}\right)$

20. $\left(-\dfrac{4}{9}\right) \times \left(-\dfrac{5}{6}\right)$

21. $\left(-7\dfrac{1}{3}\right)\left(-4\dfrac{1}{11}\right)$

22. $\dfrac{1}{4} \div (-8)$

23. $-5\left(\dfrac{-2}{-5}\right)$

24. $\left(-\dfrac{2}{7}\right)\left(-\dfrac{14}{15}\right)\left(-\dfrac{3}{8}\right)$

25. $\left(-\dfrac{3}{5}\right)\left(\dfrac{-1}{-2}\right)\left(\dfrac{10}{21}\right)$

26. $-\dfrac{5}{8} \div \dfrac{4}{5}$

Round to hundredths if necessary.

27. $5.823 - 32.12$

28. $-8.32 + 7.21$

29. $-84.23 - 7.21$

30. $34.6(-3.2)$

31. $-7.2(8.2)$

32. $-83.1(-4.1)$

33. $83.2 \div (-3)$

34. $-0.826 \div (-2)$

35. $-3.2 + 7.8$

36. $4.23 - 4.2$

37. $4.6 \div (-2)$

38. $-3.8 + (-1.7)$

3 Evaluate.

39. $3(5 - 8)$

40. $(7 + 8) - (2 - 7)$

41. $3(7) + 9 \div 3$

42. $5 - 3(2)$

43. $-12 + 2(3) - 7$

44. $12 \div 4 \times 7$

45. $15 + 7(-4)$

46. $-8 - 7(-3)$

47. $9 - 3(-2)$

48. $-6(-2) - 4$

49. $-6(-4)$

50. $7(1 - 4) \div 3 + 2$

51. $\dfrac{2 - 8}{3} + 5(-2) \div 2$

52. $\dfrac{5 - 9}{4} + 4(-3) \div 6$

53. $4 \times 8 - 7 \times 3^2 + 18 \div 6$

54. $5(3 - 4) - 7(2 - 5) \div (-3)$

55. $142(3 - 21) + 48(27)$

56. $24 - 3(2 + 7) \div 3 + 12$

57. $\dfrac{12 - 18}{3} + 15 - 81 \div 3$

58. $14 - \dfrac{3 + 7}{5} \div 2 + 5(3)$

59. $2(3 - 12) \div 4 \times 6 - 8$

60. $2(4 - 3) - 7 + 4(2 - 8)$

61. $7 - 21 + 138 - 256$

62. $\dfrac{4}{5} + 3\left(-\dfrac{2}{3} + \dfrac{1}{3}\right)$

63. $\dfrac{-5}{8} + 7\left(\dfrac{1}{8} - \dfrac{7}{8}\right)$

64. $\dfrac{1}{2} - 3\left(2\dfrac{3}{4} - \dfrac{1}{4}\right)$

65. $-\dfrac{3}{5} + \dfrac{1}{2}(2 - 8)$

66. $\left(-\dfrac{2}{3}\right)^2 - \dfrac{1}{2}\left(\dfrac{3}{8}\right)$

67. $-\dfrac{3}{5} + \left(\dfrac{2}{7}\right)^2$

68. $5.2 + 3.8(-4.1)$

69. $1.3 - (2.1)^2(1.2)$

70. $(0.3)^2 + 5.7(-2.1)$

71. $5(8 - 7 + 3) - 2$

72. $7(8 + 2) - 4(7 - 5 - 10)$

73. $128 \div 4(15 - 4 + 3 - 10)$

74. $7(2) - 4(15 - 8 - 5(3))$

5–5 | *Powers of 10*

Learning Outcomes   **1** Multiply and divide by powers of 10.

2 Raise a power of 10 to a power.

1 **Multiply and Divide by Powers of 10.**

We learned that our number system is based on the number 10; that is, each place value is a *power of 10*. To show how each place value is related to the base 10, we need to define two special exponents. A number with an **exponent of zero** is defined to have a value of 1. On the place-value chart, the ones place is shown to be 10^0. The decimal places have the power of 10 in the denominator of the fraction representing the place value. That is, the place value is the *reciprocal* of a power of 10 with a positive exponent. To illustrate these decimal places as powers of 10, we use **negative exponents.** Look at the place-value chart in Fig. 5–8 to see how our number system relates to powers of 10.

Reciprocals and Negative Exponents

Any nonzero number raised to the zero power is equal to 1.

For any real number n, $n^0 = 1$ if $n \neq 0$.

An expression with a *negative exponent* can be written as an equivalent expression with a positive exponent.

For any real number n, $n^{-1} = \dfrac{1}{n}$ $\dfrac{1}{n^{-1}} = n$ if $n \neq 0$.

A nonzero number times its reciprocal equals 1.

For any real number n, $n \cdot \dfrac{1}{n} = 1$ $n^1 \cdot n^{-1} = n^0 = 1$ if $n \neq 0$.

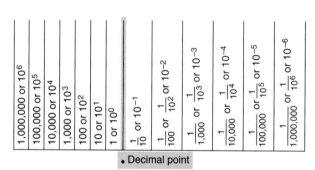

• Decimal point

Figure 5–8 Base-10 place-value chart.

Whole-Number Part		**Fractional or Decimal Part**	
Millions	$1,000,000 = 10^6$	$\dfrac{1}{10} = 0.1 = 10^{-1}$	Tenths
Hundred thousands	$100,000 = 10^5$	$\dfrac{1}{100} = 0.01 = 10^{-2}$	Hundredths
Ten thousands	$10,000 = 10^4$	$\dfrac{1}{1,000} = 0.001 = 10^{-3}$	Thousandths
Thousands	$1,000 = 10^3$	$\dfrac{1}{10,000} = 0.0001 = 10^{-4}$	Ten-thousandths
Hundreds	$100 = 10^2$	$\dfrac{1}{100,000} = 0.00001 = 10^{-5}$	Hundred-thousandths
Tens	$10 = 10^1$	$\dfrac{1}{1,000,000} = 0.000001 = 10^{-6}$	Millionths
Ones	$1 = 10^0$		

What Does the Exponent in a Power of 10 Tell Us?

- When the exponent of 10 is positive, the exponent is the same as the number of zeros in the equivalent whole number in standard notation.
- When the exponent of 10 is negative, the exponent is the same as the number of zeros in the denominator of the equivalent fraction.
- When the exponent of 10 is negative, the exponent is the same as the number of decimal digits in the decimal equivalent.

To multiply a number by a power of 10:

1. If the exponent is positive, shift the decimal point to the *right* the number of places indicated by the *positive* exponent. Attach zeros as necessary.
2. If the exponent is negative, shift the decimal point to the *left* the number of places indicated by the *negative* exponent. Insert zeros as necessary.

EXAMPLE Perform the multiplications using powers of 10.

(a) 275(10)   **(b)** 0.18(100)   **(c)** 2.4(1,000)   **(d)** 43(0.1)

(a) $275(10) = 275(10^1) = \textbf{2,750}$ The exponent is $+1$, so move the decimal one place to the right. Attach one zero.

(b) $0.18(100) = 0.18(10^2) = \textbf{18}$ The exponent is $+2$, so move the decimal two places to the right.

(c) $2.4(1,000) = 2.4(10^3) = \textbf{2,400}$ The exponent is $+3$, so move the decimal three places to the right. Attach two zeros.

(d) $43(0.1) = 43(10^{-1}) = \textbf{4.3}$ The exponent is -1, so move the decimal one place to the left.

To divide by a power of 10:

1. Change the division to an equivalent multiplication.
2. Use the rule for multiplying by a power of 10.

EXAMPLE Perform the divisions using powers of 10.

(a) $3.14 \div 10$ **(b)** $0.48 \div 100$ **(c)** $20.1 \div 1,000$

(a) $3.14 \div 10 = 3.14(\frac{1}{10})$ Change the division to an equivalent multiplication.

$= 3.14(10^{-1})$ Express the fraction $\frac{1}{10}$ as a power of 10.

$= \textbf{0.314}$ Because the exponent of 10 is -1, move the decimal one place to the *left*.

(b) $0.48 \div 100 = 0.48(\frac{1}{100}) = 0.48(10^{-2}) = \textbf{0.0048}$ Move the decimal two places to the left. Insert two zeros.

(c) $20.1 \div 1,000 = 20.1(\frac{1}{1,000}) = 20.1(10^{-3}) = \textbf{0.0201}$ Move decimal three places to the left. Insert one zero.

To multiply or divide a power of 10 by another power of 10, we can use special applications of the *laws of exponents*. These laws are introduced in Chapter 9, but we will look at these special cases now.

To multiply powers of 10 written in exponential notation:

1. The product will be a power of 10.
2. Add the exponents for the power-of-10 factors.

Symbolically, $10^a \cdot 10^b = 10^{a+b}$, where a and b are integers.

To divide powers of 10 written in exponential notation:

1. The quotient will be a power of 10.
2. Subtract the exponents for the power-of-10 factors (numerator minus demoninator).
3. Powers of 10 with negative exponents can be written as reciprocal expressions with positive exponents, if desired.

Symbolically, $\dfrac{10^a}{10^b} = 10^{a-b}$, where a and b are integers; $10^{-a} = \dfrac{1}{10^a}$.

EXAMPLE   Multiply or divide using the laws of exponents.

(a) $10^5(10^2)$ 　　(b) $10^{-1}(10^2)$ 　　(c) $10^0(10^3)$ 　　(d) $\dfrac{10}{10^3}$

(e) $\dfrac{10^5}{10^4}$ 　　(f) $\dfrac{10^2}{10^2}$ 　　(g) $\dfrac{10^{-2}}{10^3}$

(a) $10^5(10^2) = 10^{5+2} = \mathbf{10^7}$ 　　　Add exponents.
(b) $10^{-1}(10^2) = 10^{(-1+2)} = \mathbf{10^1}$ **or 10** 　　Add exponents.
(c) $10^0(10^3) = 10^{0+3} = \mathbf{10^3}$ 　　　Add exponents.

(d) $\dfrac{10}{10^3} = 10^{1-3} = \mathbf{10^{-2}}$ **or** $\dfrac{1}{\mathbf{10^2}}$ 　　Subtract exponents.

(e) $\dfrac{10^5}{10^4} = 10^{5-4} = \mathbf{10^1}$ **or 10** 　　　Subtract exponents.

(f) $\dfrac{10^2}{10^2} = 10^{2-2} = \mathbf{10^0}$ **or 1** 　　　Subtract exponents.

(g) $\dfrac{10^{-2}}{10^3} = 10^{-2-3} = \mathbf{10^{-5}}$ **or** $\dfrac{1}{\mathbf{10^5}}$ 　Subtract exponents.

2   Raise a Power of 10 to a Power.

Recall that raising a number to a natural-number power is an extension of multiplication. Powers of 10 can be raised to a power by applying this same concept and thus establishing another law of exponents.

> **To raise a power of 10 to a power:**
>
> 1. The result will be a power of 10.
> 2. The exponent will be the product of the original exponent and the power.
>
> Symbolically, $(10^a)^b = 10^{ab}$.

EXAMPLE Raise the powers of 10 to the indicated power.

$$\text{(a) } (10^3)^2 \qquad \text{(b) } (10^{-3})^4$$

(a) $(10^3)^2 = 10^{3 \cdot 2} = \mathbf{10^6}$
(b) $(10^{-3})^4 = 10^{-3(4)} = \mathbf{10^{-12}}$

SECTION 5–5 SELF-STUDY EXERCISES

1 Perform the operations.

1. 453×100
2. 0.27×100
3. $5.82 \times 1{,}000$
4. 8.97×0.1
5. 523×0.01
6. 8.06×0.001
7. $5.73 \div 10$
8. $0.293 \div 100$
9. $45.7 \div 1{,}000$
10. $85.79 \div \dfrac{1}{10}$
11. $43.7 \div \dfrac{1}{100}$
12. $8.37 \div \dfrac{1}{1{,}000}$

Multiply or divide as indicated.

13. 0.37×10^2
14. 1.82×10^3
15. 5.6×10^{-1}
16. 142×10^{-2}
17. 78×10^4
18. 62×10^0
19. $4.6 \div 10^4$
20. $6.1 \div 10$
21. $7.2 \div 10^1$
22. $42 \div 10^0$
23. $10^7(10^5)$
24. $10^{-3}(10^4)$
25. $10^0(10^2)$
26. $\dfrac{10^7}{10^3}$
27. $\dfrac{10^5}{10^5}$
28. $\dfrac{10^3}{10^5}$
29. $\dfrac{10^{-5}}{10^7}$
30. $\dfrac{10}{10^7}$
31. $10^4(10^6)$
32. $10^{-3}(10^{-4})$
33. $10^0(10^{-3})$
34. $10^{-3}(10^4)$
35. $10(10^2)$
36. $\dfrac{10^4}{10^2}$
37. $\dfrac{10}{10^4}$
38. $\dfrac{10^4}{10^4}$
39. $\dfrac{10^{-2}}{10^3}$
40. $\dfrac{10^0}{10^1}$

2 Raise the powers of 10 to the indicated exponents.

41. $(10^2)^3$
42. $(10^3)^4$
43. $(10^{-2})^2$
44. $(10^{-5})^{-2}$
45. $(10^4)^{-2}$

5–6 | *Scientific Notation*

Learning Outcomes
1 Change a number from scientific notation to ordinary notation.
2 Change a number from ordinary notation to scientific notation.
3 Multiply and divide numbers in scientific notation.
4 Raise a number in scientific notation to a power.
5 Change among engineering, scientitic, and ordinary notations.

Sure to cover

1 **Change a Number from Scientific Notation to Ordinary Notation.**

A number is expressed in **scientific notation** if it is the product of two factors. The absolute value of the first factor is a number greater than or equal to 1 but less than 10. The second factor is a power of 10.

A number that is written strictly according to place value is an **ordinary number.**

Characteristics of Scientific Notation

- Numbers between 0 and 1 and between -1 and 0 require negative exponents when written in scientific notation.
- The first factor in scientific notation always has an absolute value greater than or equal to one but less than 10. There is only one digit to the left of the decimal and that digit cannot be a zero.
- The use of the times sign ($\times$) for multiplication is the most common representation for scientific notation.
- The second factor is a power of 10.

EXAMPLE Which of the terms is expressed in scientific notation?

(a) 4.7×10^2 (b) 0.2×10^{-1} (c) -3.4×5^2 (d) 2.7×10^0
(e) 34×10^4 (f) 8×10^{-6} (g) $-2.8 \div 10^3$

(a) 4.7 is more than 1 and less than 10. 10^2 is a power of 10. Multiplication is indicated. **Thus, the term is in scientific notation.**
(b) Even though 10^{-1} is a power of 10, **this term is not in scientific notation** because the first factor (0.2) is less than 1.
(c) The absolute value of -3.4 is more than 1 and less than 10, but 5^2 is not a power of 10. **Thus, this term is not in scientific notation.**
(d) 2.7 is more than 1 and less than 10. 10^0 is a power of 10. **Thus, this term is in scientific notation.**
(e) 34 is greater than 10. Even though 10^4 is a power of 10, **this term is not in scientific notation** because the first factor (34) is 10 or more.
(f) 8 is more than 1 and less than 10. 10^{-6} is a power of 10. **Thus, this term is in scientific notation.**
(g) The absolute value of -2.8 is more than 1 and less than 10, but division rather than multiplication is the indicated operation. **This term is not in scientific notation.**

To change from a number written in scientific notation to an ordinary number:

1. Perform the indicated multiplication by moving the decimal point in the first factor the appropriate number of places. Affix or insert zeros as necessary.
2. Omit the power-of-10 factor.

When we multiply by a power of 10, the exponent of 10 tells us how many places and in which direction to move the decimal.

EXAMPLE   Change to ordinary numbers.

(a) 3.6×10^4 (b) 2.8×10^{-2} (c) 1.1×10^0 (d) -6.9×10^{-5}
(e) -9.7×10^6

(a) $3.6 \times 10^4 = 36000. = 36,000$ Move the decimal four places to the right.

(b) $2.8 \times 10^{-2} = 0.028 = 0.028$ Move the decimal two places to the left.

(c) $1.1 \times 10^0 = 1.1$ Move the decimal zero places.

(d) $-6.9 \times 10^{-5} = -0.000069 = -0.000069$ Move the decimal five places to the left.

(e) $-9.7 \times 10^6 = 9700000. = -9,700,000$ Move the decimal six places to the right.

Do examples 238, 239, 240. Write down defen.

2 Change a Number from Ordinary Notation to Scientific Notation.

To express an ordinary number in scientific notation, we reverse the procedures we used before. Shifting the decimal point changes the value of a number. The power-of-10 factor is used to offset or balance this change so the original value of the number is maintained.

Change a number written in ordinary notation to scientific notation:

1. Insert a caret (∧) in the proper place to indicate where the decimal should be positioned in the ordinary number so that the absolute value of the number is valued at 1 or between 1 and 10.
2. Determine the number of places and in which direction the decimal shifts *from* the new position (caret) *to* the old position (decimal point). This number is the exponent of the power of 10.

A Balancing Act: Why Count from the New to the Old?

Moving the decimal in the ordinary number changes the value of the number unless you balance the effect of the move in the power-of-10 factor. When a decimal is moved in the first factor, the value is changed. To offset this change, an opposite change must be made in the power-of-10 factor. Counting from the new to the old position indicates the proper number of places and the direction (positive or negative) for balancing with the power-of-10 factor.

Remember the word **NO.** Count from **N**ew to **O**ld.

$3,800 = 3.8 \times 10^3$ $3_\wedge800. \times 10^3$ N → O = +3.
$0.0045 = 4.5 \times 10^{-3}$ $0.004_\wedge5 \times 10^{-3}$ N → O = -3.

EXAMPLE Express in scientific notation.

(a) 285   (b) 0.007   (c) 9.1   (d) 85,000   (e) 0.00074

(a) $285 \rightarrow 2_\wedge 85 = \mathbf{2.85 \times 10^2}$

The unwritten decimal is after the 5. Place the caret between 2 and 8 so the number 2.85 is between 1 and 10. Count *from* the caret *to* the decimal to determine the exponent of 10. A move two places to the right represents the exponent $+2$.

(b) $0.007 \rightarrow 0.\,007\,_\wedge = \mathbf{7 \times 10^{-3}}$

7 is between 1 and 10. Count *from* the caret *to* the decimal. A move three places to the left represents the exponent -3.

(c) $9.1 = \mathbf{9.1 \times 10^0}$

9.1 is already between 1 and 10, so the decimal does not move; that is, the decimal moves zero places.

(d) $85,000 \rightarrow 8_\wedge 5000 = \mathbf{8.5 \times 10^4}$

From the caret *to* the decimal is four places to the right.

(e) $0.00074 \rightarrow 0.\,0007\,_\wedge 4 = \mathbf{7.4 \times 10^{-4}}$

From the caret *to* the decimal is four places to the left.

TIP

Between 1 and 10: One Digit to the Left of the Decimal and that Digit Cannot Be Zero

The expression "between 1 and 10" means any number that is more than 1 but less than 10. The first factor in scientific notation must be equal to 1 *or* between 1 and 10. That means **there will be one and only one digit to the left of the decimal point and that digit cannot be zero.**

Occasionally a number is written in power-of-10 notation but not in scientific notation because the first factor is not equal to 1 or is not between 1 and 10. Write the first factor in scientific notation and multiply the power-of-10 factors.

EXAMPLE Express in scientific notation.

(a) 37×10^5 (b) 0.03×10^3

(a) $3_\wedge 7 \times 10^5 = 3.7 \times 10^1 \times 10^5$ Write the first factor in scientific notation and
$\qquad\qquad\quad = \mathbf{3.7 \times 10^6}$ multiply the power-of-10 factors.

(b) $0.03_\wedge \times 10^3 = 3 \times 10^{-2} \times 10^3$ Write the first factor in scientific notation and
$\qquad\qquad\quad = \mathbf{3 \times 10^1}$ multiply the power-of-10 factors.

---

3 Multiply and Divide Numbers in Scientific Notation.

We can multiply or divide numbers expressed in scientific notation without first having to convert them to ordinary numbers.

Multiply numbers in scientific notation:

1. Multiply the first factors using the rules of signed numbers.
2. Multiply the power-of-10 factors using the laws of exponents.
3. Examine the first factor of the product (Step 1) to see if its value is equal to 1 or between 1 and 10.
   (a) If so, write the results of Steps 1 and 2.
   (b) If not, write the first factor in scientific notation and multiply power-of-10 factors.

EXAMPLE Multiply and express the product in scientific notation.

(a) $(4 \times 10^2)(2 \times 10^3)$ (b) $(3.7 \times 10^3)(2.5 \times 10^{-1})$ (c) $(8.4 \times 10^{-2})(5.2 \times 10^{-3})$

(a) $(4 \times 10^2)(2 \times 10^3) = \mathbf{8 \times 10^5}$

The product is in scientific notation.

(b) $(3.7 \times 10^3)(2.5 \times 10^{-1}) = \mathbf{9.25 \times 10^2}$

The product is in scientific notation.

(c) $(8.4 \times 10^{-2})(5.2 \times 10^{-3}) = 43.68 \times 10^{-5}$

43.68×10^{-5} is not in scientific notation.

$43.68 \to 4{\scriptstyle\wedge}3.68 \quad$ or $\quad 4.368 \times 10^1 \qquad$ Write first factor in scientific notation.

$4.368 \times 10^1 \times 10^{-5} = 4.368 \times 10^{1-5} \qquad$ Multiply the powers-of-10 factors.

$= \mathbf{4.368 \times 10^{-4}}$

Division involving numbers in scientific notation is similar to multiplication.

To divide numbers in scientific notation:

1. Divide the first factors using the rules of signed numbers.
2. Divide the power-of-10 factors using the laws of exponents.
3. Examine the first factor of the quotient (Step 1) to see if its value is equal to 1 or between 1 and 10.
   (a) If so, write the results of Steps 1 and 2.
   (b) If not, write the first factor in scientific notation and multiply the power-of-10 factors.

EXAMPLE Divide and express the quotient in scientific notation.

(a) $\dfrac{3 \times 10^5}{2 \times 10^2}$ (b) $\dfrac{1.44 \times 10^{-3}}{6 \times 10^{-5}}$ (c) $\dfrac{9.6 \times 10^{29}}{3.2 \times 10^{111}}$ (d) $\dfrac{1.25 \times 10^3}{5}$

(a) $\dfrac{3 \times 10^5}{2 \times 10^2} = \dfrac{3}{2} \times 10^{5-2} = \mathbf{1.5 \times 10^3}$

The first factor is usually written in decimal notation.

(b) $\dfrac{1.44 \times 10^{-3}}{6 \times 10^{-5}} = \dfrac{1.44}{6} \times 10^{-3-(-5)} = 0.24 \times 10^{-3+5} = 0.24 \times 10^2$

0.24 is less than 1, so adjustments are necessary.

$0.24 \to 0.2{\scriptstyle\wedge}4 = 2.4 \times 10^{-1} \qquad$ Write the first factor in scientific notation.

Then $2.4 \times 10^{-1} \times 10^2 = \mathbf{2.4 \times 10^1}. \qquad$ Multiply the powers-of-10 factors.

(c) $\dfrac{9.6 \times 10^{29}}{3.2 \times 10^{111}} = \dfrac{9.6}{3.2} \times 10^{29-111} = \mathbf{3 \times 10^{-82}}$

(d) $\dfrac{1.25 \times 10^3}{5}$ 5 is the same as 5×10^0.

$\dfrac{1.25 \times 10^3}{5 \times 10^0} = \dfrac{1.25}{5} \times 10^{3-0} = 0.25 \times 10^3$ 0.25 is less than 1.

$0.25 \rightarrow 0.2_{\wedge}5 = 2.5 \times 10^{-1}$ Write the first factor in scientific notation and multiply the power-of-10 factors.

Then, $2.5 \times 10^{-1} \times 10^3 = 2.5 \times 10^{-1+3} = \mathbf{2.5 \times 10^2}$.

TIP   Mental Adjustment of Exponents

Once we understand the concept of balancing the effect of moving a decimal by adjusting the exponent of the power-of-10 factor, we can perform this adjustment mentally.

$$43.68 \times 10^{-5} = 4_{\wedge}3.68 \times 10^{-5} \qquad \text{N} \rightarrow \text{O} = +1$$
$$= 4.368 \times 10^{-5+1} \qquad \text{Adjust mentally.}$$
$$= 4.368 \times 10^{-4}$$
$$0.25 \times 10^3 = 0.2_{\wedge}5 \times 10^3 \qquad \text{N} \rightarrow \text{O} = -1$$
$$= 2.5 \times 10^{3-1} \qquad \text{Adjust mentally.}$$
$$= 2.5 \times 10^2$$

TIP   Scientific Notation and the Calculator

Power-of-10 Key

The power-of-10 key, labeled $\boxed{10^x}$, $\boxed{\text{EXP}}$, or $\boxed{\text{EE}}$ on most calculators, is a shortcut key for entering the following keys:

$$\boxed{\times} \; 10 \; \boxed{x^y}$$

The shortcut key is used only for power-of-10 factors, and only the *exponent* of 10 is entered. If you enter $\times 10$, then $\boxed{\text{EXP}}$, your answer will have one extra factor of 10.

Look at $\dfrac{(3 \times 10^5)}{(2 \times 10^2)}$ on the calculator. *Steps will vary on various calculators.*

$\boxed{(}\; 3 \;\boxed{10^x}\; 5 \;\boxed{)} \;\boxed{\div}\; \boxed{(}\; 2 \;\boxed{10^x}\; 2 \;\boxed{)} \;\boxed{=} \Rightarrow 1{,}500$ or $3 \;\boxed{\text{EXP}}\; 5 \;\boxed{\div}\; 2 \;\boxed{\text{EXP}}\; 2 \;\boxed{=} \Rightarrow 1{,}500$

This result may be expressed in scientific notation if desired: $1{,}500 = 1.5 \times 10^3$

The internal program of a calculator has predetermined how the output of a calculation is displayed. For example, even if you would like an answer displayed in scientific notation, it may fall within the guidelines for display as an ordinary number. You must make the conversion to scientific notation yourself. The reverse may also be true.

Experiment using examples for which you already know the answer to determine the limitations and requirements of your calculator. Some calculators require the parentheses and some do not.

EXAMPLE
AVIA

A star is 4.2 light-years from Earth. If 1 light-year is 5.87×10^{12} mi, how many miles from Earth is the star?

To solve this problem, multiply the number of light years times the distance of 1 light year.

Estimation

By rounding we estimate that the star will be approximately 4 times 6×10^{12} mi away.

$$4(6 \times 10^{12}) = 24 \times 10^{12} = 2.4 \times 10^{13}$$

$$4.2(5.87 \times 10^{12}) = \qquad\qquad 4.2 = 4.2 \times 10^0$$

$$24.654 \times 10^{12} = \qquad\qquad \text{Perform scientific notation adjustment.}$$

$$2.4654 \times 10^{12+1} = \qquad\qquad N \to O = +1$$

$$2.4654 \times 10^{13}$$

$$\text{or} \qquad 2.5 \times 10^{13} \qquad\qquad \text{Round the first factor to tenths.}$$

Interpretation **The star is 2.5×10^{13} mi from Earth.**

EXAMPLE
ELEC

An angstrom (Å) is 1×10^{-7} mm. What is the length in millimeters of 14.82 Å?

Solve the problem by multiplying the number of angstrom units times the length of 1 Å.

Estimation

14.82 angstrom units are more than 10 times longer than 1×10^{-7} mm, or more than 1×10^{-6}.

$$14.82(1 \times 10^{-7}) =$$

$$14.82 \times 10^{-7} = \qquad\qquad \text{Adjust the first factor and exponent.}$$

$$1.482 \times 10^{-7+1} = \qquad\qquad N \to O = +1$$

$$1.482 \times 10^{-6} \text{ mm}$$

Interpretation **The length of 14.82 Å is 1.482×10^{-6} mm.**

EXAMPLE
ELEC

One coulomb (C) is approximately 6.28×10^{18} electrons. How many coulombs do 2.512×10^{21} electrons represent?

Divide the total number of electrons by the number of electrons in one coulomb.

$$\frac{2.512 \times 10^{21}}{6.28 \times 10^{18}}$$

Estimation

By rounding the first factors we have

$$\frac{3 \times 10^{21}}{6 \times 10^{18}} = 0.5 \times 10^{21-18} = 0.5 \times 10^3 = 500\,\text{C}$$

$$\frac{2.512 \times 10^{21}}{6.28 \times 10^{18}} = \qquad \text{Divide coefficients, subtract exponents.}$$

$$0.4 \times 10^3 = \qquad \text{Adjust the first factor.}$$

$$4 \times 10^{3-1} = \qquad N \to O = -1$$

$$4 \times 10^2 \text{ or} \qquad \text{Write as an ordinary number.}$$

$$400\,\text{C}$$

Interpretation **2.512×10^{21} electrons represent 400 C.**

4 **Raise a Number in Scientific Notation to a Power.**

Two more of the laws of exponents are applied when raising a number in scientific notation to a power. These laws address raising more than one factor to a power and raising a power to a power. Again, we will examine these laws in Chapter 9; but, for now we will examine the laws as they apply to scientific notation.

To raise a number in scientific notation to a power:

1. Raise the first factor to the power.
2. Raise the power of 10 to the power by multiplying exponents.
3. Adjust the first factor and power of 10 so that the first factor is greater than or equal to 1 or less than 10.

Symbolically, $(n \times 10^a)^b = n^b \times 10^{ab}$, n is any real number, and $n \neq 0$. Adjust so that $n \geq 1$ and $n < 10$.

EXAMPLE   Raise the following to the indicated powers: **(a)** $(2 \times 10^3)^2$ **(b)** $(4.2 \times 10^{-4})^3$

(a) $(2 \times 10^3)^2 = 2^2 \times 10^{3(2)}$ Square each factor.
$\qquad\qquad\qquad = \mathbf{4 \times 10^6}$

(b) $(4.2 \times 10^{-4})^3 = 4.2^3 \times 10^{-4(3)}$ Cube each factor. Evaluate.
$\qquad\qquad\qquad\quad = 74.088 \times 10^{-12}$ Adjust to scientific notation.
$\qquad\qquad\qquad\quad = 7.4088 \times 10^{-12+1}$
$\qquad\qquad\qquad\quad = \mathbf{7.4088 \times 10^{-11}}$

---

5 **Change Among Engineering, Scientific, and Ordinary Notations.**

In Chapter 4 we examined some metric prefixes for very large and very small units of measure. In this chapter we examined power-of-10 exponents and scientific notation. We now examine power-of-10 exponents that are multiples of 3.

Metric Prefixes

Prefix	Abbreviation	Meaning	Power of 10
Yotta-	Y	one septillion times	$\times 10^{24}$
Zetta-	X	one sextillion times	$\times 10^{21}$
Exa-	E	one quintillion times	$\times 10^{18}$
Peta-	P	one quadrillion times	$\times 10^{15}$
Tera-	T	one trillion times	$\times 10^{12}$
Giga-	G	one billion times	$\times 10^{9}$
Mega-	M	one million times	$\times 10^{6}$
kilo	k	one thousand times	$\times 10^{3}$
Standard Unit			$\mathbf{\times 10^{0}}$
milli-	m	one thousandth of	$\times 10^{-3}$
micro-	μ	one millionth of	$\times 10^{-6}$
nano-	n	one billionth of	$\times 10^{-9}$
pico-	p	one trillionth of	$\times 10^{-12}$
femto-	f	one quadrillionth of	$\times 10^{-15}$
atto-	a	one quintillionth of	$\times 10^{-18}$
zepto-	z	one sextillionth of	$\times 10^{-21}$
yocto-	y	one septillionth of	$\times 10^{-24}$

These prefixes correlate with a modification of scientific notation called **engineering notation**. Engineering notation, like scientific notation, has a first factor and a power-of-10 factor. The first factor is greater than or equal to 1 and less than 1,000. The power-of-10 factor is a multiple of three.

To change an ordinary number to engineering notation:

1. Indicate with a caret ($_\wedge$) where the decimal should be positioned.
   (a) If the number is greater than or equal to one and less than 1,000, the decimal will not shift and the power-of-10 factor will be 10^0.
   (b) If the number is greater than or equal to 1,000, insert commas as appropriate to separate the place-value periods, and place the caret (the *new* position of the decimal) at the leftmost comma.
   (c) If there are no nonzero digits to the left of the decimal, count from the decimal to the right in groups of three places until you have at least one, but no more than three, significant digits to the left of the caret (*new* position of the decimal).

2. Determine the exponent of the power-of-10 factor and its sign by counting the number of places from the *new* position of the decimal to the *old* position. The resulting exponent will be a multiple of 3.

EXAMPLE Change to engineering notation:

(a) 2,400,000   (b) 2,400   (c) 24
(d) 0.24   (e) 0.024   (f) 0.0000024

(a) 2,400,000 $2_\wedge 400,000$ Place the caret at the leftmost comma.
                   2.4×10^6 N → O = +6

(b) 2,400 $2_\wedge 400$ Place the caret at the leftmost comma.
                   2.4×10^3 N → O = +3

(c) 24 $24_\wedge$ Place the caret at the decimal point. The power-of-10 exponent can be only 0 or a multiple of 3.
                   24×10^0 N → O = 0

(d) 0.24 $0.240_\wedge$ Place the caret after the third decimal place value. There can be no more than 3 significant digits on the left of the caret.
                   240×10^{-3} N → O = −3

(e) 0.024 $0.024_\wedge$ Place the caret after the third decimal value. There can be no more than 3 significant digits on the left of the caret.
                   24×10^{-3} N → O = −3

(f) 0.0000024 $0.000002_\wedge 4$ Place the caret after the sixth decimal value. There can be no more than 3 significant digits on the left of the caret.
                   2.4×10^{-6} N → O = −6

Measures that are expressed in metric units can be written in engineering notation by showing the power-of-10 factor with the standard unit or by translating the power-of-10 factor to the appropriate metric prefix. When using metric prefixes, a prefix replaces the power-of-10 factor. If the measure has a metric unit other than a standard unit and a first factor that is *less than* 1 or *greater than* 1,000, the metric unit will need to be adjusted. The table of metric prefixes with powers of 10 is on p. 172.

EXAMPLE Write the engineering notation using metric prefixes.

(a) 4,382,000 Ω (b) 0.00001 Å   (c) 5,870 μW   (d) 3,500 MHz

(a) 4,382,000 Ω = 4.382×10^6 Ω 10^6 = M See p. 172.
     = 4.382 MΩ

(b) 0.00001 Å = 10×10^{-6} Å 10^{-6} = μ
     = 10 μÅ

(c) 5,870 μW = 5.87×10^3 μW $\mu = 10^{-6}$
     = $5.87 \times 10^3 \times 10^{-6}$ W   Multiply power-of-10 factors.
     = 5.87×10^{-3} W 10^{-3} = m
     = 5.87 mW

(d) 3,500 MHz = 3.5×10^3 MHz   M = 10^6
     = $3.5 \times 10^3 \times 10^6$ Hz   Multiply power-of-10 factors.
     = 3.5×10^9 Hz 10^9 = G
     = 3.5 GHz

SECTION 5–6 SELF-STUDY EXERCISES

Study this Chapter

1 Write as ordinary numbers.

1. 4.3×10^2	**2.** 6.5×10^{-3}	**3.** 2.2×10^0	**4.** 7.3×10
5. 9.3×10^{-2}	**6.** 8.3×10^4	**7.** 5.8×10^{-3}	**8.** 8×10^4
9. 6.732×10^0	**10.** 5.89×10^{-3}	**11.** 7.83×10	**12.** 1.59×10^3
13. 3.97×10^5	**14.** 4.723×10^{-4}	**15.** 9.91×10^{-6}	**16.** 1.03×10^6

2 Express in scientific notation.

17. 392	**18.** 0.02	**19.** 7.03	**20.** 42,000
21. 0.081	**22.** 0.0021	**23.** 23.92	**24.** 0.101
25. 1.002	**26.** 721	**27.** 42×10^4	**28.** 32.6×10^3
29. 0.213×10^2	**30.** 0.0062×10^{-3}	**31.** $56,000 \times 10^{-3}$	**32.** 0.197×10^{-5}
33. 745×10^{-1}	**34.** 18×10^3	**35.** 0.701×10^2	**36.** $72,500 \times 10^{-5}$

3 Perform the indicated operations. Express answers in scientific notation.

37. $(6.7 \times 10^4)(3.2 \times 10^2)$ **38.** $(1.6 \times 10^{-1})(3.5 \times 10^4)$ **39.** $(5.0 \times 10^{-3})(4.72 \times 10^0)$

40. $(8.6 \times 10^{-3})(5.5 \times 10^{-1})$ **41.** $\dfrac{3.15 \times 10^5}{4.5 \times 10^2}$ **42.** $\dfrac{4.68 \times 10^3}{7.2 \times 10^7}$

43. $\dfrac{4.55 \times 10^{-1}}{6.5 \times 10^{-4}}$ **44.** $\dfrac{7.84 \times 10^{-2}}{9.8 \times 10^0}$ **45.** $\dfrac{1.96 \times 10^{-3}}{8.0 \times 10^{-5}}$

46. A star is 5.5 light years from Earth. If one light-year is 5.87×10^{12} miles, how many miles from Earth is the star?

47. An angstrom (Å) is 1×10^{-7} mm. How many angstroms are in 4.2×10^{-5} mm?

4 Perform the indicated operation and express the result in scientific notation. Round appropriately.

48. $(8.3 \times 10^2)^3$
49. $(7 \times 10^{-2})^4$
50. $(1.93 \times 10^{-4})^2$

51. $(7.2 \times 10^2)^2(3.1 \times 10^1)^2$
52. $\dfrac{(8.2 \times 10^{-3})^2}{(5.73 \times 10^4)^3}$
53. $63{,}000 \times 7{,}000$

54. $(2{,}500)(61{,}000)(300)$
55. $\dfrac{42{,}000}{60{,}000}$
56. $(52{,}000)^2$

57. $(0.00013)^{-2}$
58. $(17{,}000)^2$
59. $(83{,}000)^2(5{,}200)^2$

60. $\dfrac{(0.0071)^3}{(0.02)^4}$
61. $\left(\dfrac{210 \times 300}{510}\right)^2$
62. $\left(\dfrac{5{,}000 \times 820}{4{,}000}\right)^3$

63. $\left(\dfrac{16.7 \times 5{,}200}{0.16 \times 820}\right)^3$
64. $\left(\dfrac{0.0061 \times 2{,}300}{0.23 \times 46{,}000}\right)^3$

5 Write each number in engineering notation.

65. 3,800,000
66. 5,600
67. 78
68. 52,000
69. 80,000,000
70. 5,830,000
71. 1,736,500,000
72. 41,980,000
73. 0.78
74. 0.33
75. 0.0000011
76. 0.000000008
77. 0.0009832
78. 0.0000719
79. 0.0120307
80. 0.000675

Change each unit to engineering notation using metric prefixes.

81. 428,000 Ω
82. 5,700,000 V
83. 3,520,000,000 W
84. 79,000,000 Hz
85. 0.000081 s
86. 0.0973 Å
87. 0.00000000541 s
88. 0.89 s
89. 0.00058 MΩ
90. 0.00077 μs
91. 2,980,000 ps
92. 7,810,000 mW
93. 42,300 kV
94. 1,572,000 kW
95. 5,096,000 Ω
96. 0.000000008 Å
97. 182,000 Hz
98. 1,600 Ω
99. 5,200,000 V
100. 97,000,000 W

CHAPTER REVIEW OF KEY CONCEPTS

Learning Outcomes

What to Remember with Examples

Section 5–1

1 Compare signed numbers (pp. 209–211).

Positive numbers are to the right of zero and negative numbers are to the left of zero on the number line.

> Arrange from smallest to largest: 5, −3, 0, 8, −5
>
> $$-5, -3, 0, 5, 8$$

The "greater than" symbol is $>$.
The "less than" symbol is $<$.

> Use $>$ or $<$ to make a true statement: $5 \; ? \; -3; \; 5 > -3$ or $-3 < 5$

The absolute value of a number is its *distance* from zero without regard to direction.

> Evaluate the following absolute values: $|-3|, |5|$
>
> $$|-3| = 3, \qquad |5| = 5$$

Opposites are numbers that have the same absolute value but opposite signs.

> Give the opposite of the following: 8, -4, -2, $+4$
> $$-8, \quad 4, \quad 2, \quad -4$$

2 Add signed numbers with like signs (pp. 211–212).

To add signed numbers with like signs: **1.** Add the absolute values. **2.** Give the sum the common or like sign.

> Add: $13 + 7 = 20$ $\qquad -35 + (-13) = -48$

3 Add signed numbers with unlike signs (pp. 212–214).

To add signed numbers with unlike signs: **1.** Subtract the smaller absolute value from the larger absolute value. **2.** Give the sum the sign of the number with the larger absolute value.

> Add: $-12 + 7$
>
> Subtract absolute values: $12 - 7 = 5$. Give the 5 the negative sign because -12 has the larger absolute value.
> $$-12 + 7 = -5$$

When zero is added to a signed number, the result is unchanged: $0 + a = a + 0 = a$.

> Add: $-42 + 0 = -42$ $\qquad 0 + 38 = 38$

Section 5–2

1 Subtract signed numbers (pp. 215–216).

To subtract signed numbers: **1.** Change subtraction to addition. **2.** Change the sign of the second number, the subtrahend. **3.** Use the appropriate rule for adding signed numbers.

> Subtract: $-32 - (-28)$
> $$-32 - (-28) = -32 + (+28)$$
> $$= -4$$

To subtract with zero: **1.** Change the subtraction to addition. **2.** Use the appropriate rule for addition.

> $$-8 - 0 = -8 + (0) = -8 \qquad 0 - (-4) = 0 + (+4) = 4$$

Subtracting an opposite from a number is the same as adding a number to itself.

> $$8 - (-8) = 8 + 8 = 16 \qquad -12 - 12 = -24$$

2 Combine addition and subtraction (pp. 216–218).

1. Change all subtractions to additions. **2.** Add the series of signed numbers from left to right.

> Simplify: $5 - 3 + 2 - (-4)$
> $$5 + (-3) + 2 + 4 =$$
> $$2 + 2 + 4 = 4 + 4 = 8$$

Chapter 5 / Signed Numbers and Powers of 10

Section 5–3

1 Multiply signed numbers (pp. 219–221).

To multiply signed numbers with like signs: **1.** Multiply the absolute values. **2.** Make the sign of the product positive.

$$-6(-7) = +42 \qquad 8 \cdot 6 = 48$$

To multiply signed numbers with unlike signs: **1.** Multiply the absolute values. **2.** Make the sign of the product negative.

$$-7(2) = -14 \qquad 8(-3) = -24$$

To multiply several signed numbers: **1.** The sign of the product is *positive* if the number of negative factors is even. **2.** The sign of the product is *negative* if the number of negative factors is odd.

$$5(-2)(3) = -10(3) = -30$$

Any number (including signed numbers) multiplied by zero results in zero: $a \times 0 = 0 \times a = 0$

$$0 \times 3 = 0 \qquad -5 \times 0 = 0 \qquad 0(-7) = 0$$

2 Evaluate powers of signed numbers (pp. 221–222).

A positive number raised to a natural-number power results in a positive number. Zero raised to any natural-number power is zero. To raise a negative number to a power requires parentheses. A negative number raised to an even natural-number power results in a positive number. A negative number raised to an odd natural-number power is a negative number.

$$(3)^3 = 27 \qquad (-2)^4 = 16 \qquad -2^4 = -16$$
$$0^5 = 0 \qquad (-2)^3 = -8 \qquad -2^3 = -8$$

3 Divide signed numbers (pp. 222–223).

To divide signed numbers: **1.** Divide the absolute values of the dividend and the divisor. **2.** If the dividend and divisor have like signs, the sign of the quotient is positive. **3.** If the dividend and divisor have unlike signs, the sign of the quotient is negative.

$$-12 \div (-4) = 3 \qquad 15 \div (-3) = -5$$

Zero divided by any nonzero signed number is zero. $0 \div a = 0$ (if a is not equal to zero). A nonzero number *cannot* be divided by zero. $a \div 0$ is undefined; $0 \div 0$ is indeterminate.

$$0 \div 12 = 0 \qquad -7 \div 0 \text{ is undefined.}$$
$$0 \div 0 \text{ is indeterminate.}$$

Section 5–4

1 Change a signed fraction to an equivalent signed fraction (pp. 225–226).

If any two of the three signs of a signed fraction are changed, the fraction's value does not change.

Write three equivalent signed fractions for $-\frac{7}{8}$.

$$-\frac{7}{8} = -\frac{+7}{+8} \text{ Equivalent fractions are } +\frac{+7}{-8} \text{ or } -\frac{-7}{-8} \text{ or } +\frac{-7}{+8}$$

Chapter Review of Key Concepts

2 Perform basic operations with signed fractions and decimals (pp. 226–229).

To add or subtract signed fractions: **1.** Write equivalent fractions that have positive integers as denominators and have common denominators. **2.** Add or subtract the numerators using the rules for adding signed numbers.

To multiply or divide signed fractions: **1.** Multiply or divide the fractions. **2.** Apply the rules for multiplying or dividing signed numbers.

Add $\dfrac{-5}{8} + \dfrac{7}{8}$

$$\dfrac{-5}{8} + \dfrac{7}{8} = \dfrac{2}{8} = \dfrac{1}{4}$$

To perform basic operations with signed decimals: **1.** Determine the indicated operations. **2.** Apply the appropriate rules for signed numbers and for decimals.

Add $4.37 + (-2.91)$

$$4.37 - 2.91 = 1.46 \qquad \text{Add decimals with unlike signs.}$$

3 Apply the order of operations with signed numbers (pp. 229–232).

Expressions are evaluated in the following order from left to right: **1.** Parentheses. **2.** Exponents (powers and roots). **3.** Multiplication and division. **4.** Addition and subtraction.

Evaluate: $5 - 4(7 + 2)^2 - 15 \div 5$

$5 - 4(7 + 2)^2 - 15 \div 5 =$	Inside parentheses.
$5 - 4(9)^2 - 15 \div 5 =$	Raise to power.
$5 - 4(81) - 15 \div 5 =$	Multiply.
$5 - 324 - 15 \div 5 =$	Divide.
$5 - 324 - 3 =$	Add signed numbers.
$-319 - 3 =$	
-322	

Evaluate $-\dfrac{1}{4} - \dfrac{2}{3}\left(-\dfrac{1}{2}\right)$

$-\dfrac{1}{4} - \dfrac{2}{3}\left(-\dfrac{1}{2}\right) =$ Multiply using rules for like signs and multiplying fractions.

$-\dfrac{1}{4} + \dfrac{1}{3} =$ Change to equivalent fractions with the LCD.

$-\dfrac{3}{12} + \dfrac{4}{12} =$ Use the rule for adding numbers with unlike signs.

$\dfrac{1}{12}$

Evaluate $(-0.3)^2 - 2.8$

$(-0.3)^2 - 2.8 =$	Square signed decimal.
$0.09 - 2.8 =$	Add decimals with unlike signs.
-2.71	

Section 5–5

1 Multiply and divide by powers of 10 (pp. 233–236).

To multiply by powers of 10, add the exponents and keep the base of 10. To divide by powers of 10, subtract the exponents and keep the base of 10.

Multiply. $10^6(10^7) = 10^{6+7} = 10^{13}$ Divide. $10^3 \div 10^5 = 10^{3-5} = 10^{-2}$

2 Raise a power of 10 to a power (pp. 236–237).	**1.** The result will be a power of 10. **2.** The exponent will be the product of the original exponent and the power. Symbolically, $(10^a)^b = 10^{ab}$.

> Raise the powers of 10 to the indicated exponents.
>
> $(10^4)^2$ $(10^{-3})^2$
>
> $(10^4)^2 = 10^{4\cdot2} = 10^8$ $(10^{-3})^2 = 10^{-3\cdot2} = 10^{-6}$

Section 5–6

1 Change a number from scientific notation to ordinary notation (pp. 238–239).	To change a number from scientific notation to ordinary notation, perform the indicated multiplication by moving the decimal point in the first factor the appropriate number of places. Insert zeros as necessary. Omit the power-of-10 factor.

> Write 3.27×10^{-4} as an ordinary number.
>
> $00003.27 = 0.000327.$ Shift the decimal 4 places to the *left*.

2 Change a number from ordinary notation to scientific notation (pp. 239–240).	To change a number written in ordinary notation to scientific notation, insert a caret in the proper place to indicate where the decimal should be positioned so that the absolute value of the number is valued at 1 or between 1 and 10. Then determine how many places and in which direction the decimal shifts from the new position (caret) to the old position (decimal point). This number is the exponent of the power of 10.

> Write 54,000 in scientific notation.
>
> $5_\wedge 4000. = 5.4 \times 10^4$ From New to Old is 4 places to the right so the exponent is +4.

3 Multiply and divide numbers in scientific notation (pp. 240–243).	To multiply numbers in scientific notation, multiply the first factors, then multiply the powers of 10 by adding exponents. Next, examine the first factor of the product to see if its absolute value is equal to 1 or is between 1 and 10. If the absolute value of the factor is 1 or is between 1 and 10, the process is complete. If the absolute value of the factor is not 1 or not between 1 and 10, shift the decimal so that the factor is equal to 1 or is between 1 and 10, and adjust the exponent of the power of 10 accordingly.

> Multiply.
>
> $(4.5 \times 10^{89})(7.5 \times 10^{36}) =$ Multiply first factors then power-of-10 factors.
> $33.75 \times 10^{125} =$ Write first factor in scientific notation.
> $3_\wedge 375 \times 10^1 \times 10^{125} =$ Multiply powers of 10.
> 3.375×10^{126}

To divide numbers in scientific notation, use steps similar to multiplication, but apply the rule for the division of signed numbers and subtract exponents.

> Divide.
>
> $(3 \times 10^{-3}) \div (4 \times 10^2) =$ Divide first factors, then power-of-10 factors.
> $0.75 \times 10^{-5} =$ Write first factor in scientific notation.
> $07_\wedge 5 \times 10^{-1} \times 10^{-5} =$ Multiply powers of 10.
> 7.5×10^{-6}

Chapter Review of Key Concepts

4 Raise a number in scientific notation to a power (p. 244).

1. Raise the first factor to the power. **2.** Raise the power of 10 to the power by multiplying exponents. **3.** Adjust the first factor and power of 10 so that the first factor is greater than or equal to 1 or less than 10.

Symbolically, $(n \times 10^a)^b = n^b \times 10^{ab}$. Adjust so that $n \geq 1$ and $n < 10$.

Raise the following to the indicated powers:

(a) $(3 \times 10^4)^2$ (b) $(2.3 \times 10^{-3})^3$

(a) $(3 \times 10^4)^2 = 3^2 \times 10^{4 \cdot 2} = 9 \times 10^8$ Square each factor.
(b) $(2.3 \times 10^{-3})^3 = 2.3^3 \times 10^{-3 \cdot 3} = 12.167 \times 10^{-9}$ Cube each factor.

$= 1.2167 \times 10^{-9+1}$ Adjust the first factor and the power-of-10 factor.

$= 1.2167 \times 10^{-8}$

5 Change among engineering, scientific, and ordinary notations (pp. 244–246).

To change an ordinary number to engineering notation.
1. Indicate with a caret where the decimal should be positioned. (a) If the number is greater than or equal to 1 and less than 1,000, the decimal will not shift and the power-of-10 factor will be 10^0. (b) If the number is greater than or equal to 1,000, insert commas as appropriate to separate the place-value periods and place the caret (the *new* position of the decimal) at the leftmost comma. (c) If there are no nonzero digits to the left of the decimal, count from the decimal to the right in groups of three places until you have at least one, but no more than three, significant digits to the left of the caret (*new* position of the decimal). **2.** Determine the exponent of the power-of-10 factor and its sign by counting the number of places from the *new* position of the decimal to the *old* position. The resulting exponent will be a multiple of 3.

Write 82,000,000 Hz in engineering notation.

$$82{,}000{,}000 \text{ Hz} = 82 \times 10^6 \text{ Hz}$$

Write 0.00017 μs in engineering notation.

$$0.00017 \text{ μs} = 170 \times 10^{-6} \text{ μs}$$

CHAPTER REVIEW EXERCISES

Section 5–1

Give the value of each number:

1. $|5|$ **2.** $|-8|$ **3.** $|+7|$ **4.** $|-52|$

Give the opposite of each number:

5. -12 **6.** 8   **7.** -2 **8.** -13 **9.** 87

Add.

10. $-3 + (-8)$ **11.** $(-15) + 8$ **12.** $7 + (-11)$ **13.** $-25 + 0 + 12 + 7$

14. BUS A publicly traded company has a profit of $256,872 for one year and a loss of $38,956 for the following year. What is the net profit over the two-year period?

15. A football team gained and lost the following yardage during a series of plays beginning with first down: $+4, -5, +9$. What is the net yardage for the three plays?

16. AVIA Because of stormy weather, a pilot flying at 35,000 ft descends 8,000 ft. What is his new altitude?

17. BUS Agnes opens a checking account by depositing $500. She then writes checks for $42, $18, and $21. What is her balance after depositing another $150?

Section 5–2

Evaluate.

18. $8 - 5$

19. $-9 - 4$

20. $-7 - (-2)$

21. $11 - (-3)$

22. $12 + 3 + (-8) - 5$

23. $-6 + 3 - 5 - 7$

24. Temperatures in northern Canada ranged as high as $37°F$ one summer. That same year the lowest temperature was $-28°F$. What was the range of temperatures for the year?

25. What is the difference (or range) in temperatures of $43°F$ above zero and $27°F$ below zero?

26. What is the difference in temperatures of $47°$ below zero and $28°$ below zero?

27. HLTH/N Two successive recordings for a surgery patient's temperature were $103.2°F$ and $97.8°F$. Express the temperature change with a signed number.

28. CON A 10-ft-long fence post is placed in a hole that is 3 ft deep. How much of the post is above the ground?

29. If the temperature changes from $-22°C$ to $14°C$, what is the change?

Use the table showing Federal Air Transportation Revenue by Source for Exercises 30–33.

30. AVIA Express the difference in passenger ticket revenue in 2003 and 2002 with a signed number.

31. AVIA Express the difference in passenger ticket revenue in 2001 and 2000 with a signed number.

32. AVIA Find the difference in international departure tax revenue for 2002 and 2001.

33. AVIA Find the difference in total revenue from these two taxes for 2002 and 2001.

Federal Air Transportation Revenue by Source
(current dollars in millions)

Year	Passenger Ticket	International Departure Tax	Total of these Two Taxes
2000	5,103	1,349	6,452
2001	4,805	1,336	6,141
2002	4,726	1,282	6,008
2003	4,655	1,426	6,081

Source: U.S. Department of Transportation, Bureau of Transportation Statistics.

Section 5–3

Evaluate.

34. $-3(-7)$

35. $7(-2)$

36. $-7(3)$

37. $2(3)(-7)(0)$

38. $5(-2)(-1)(-3)$

39. $4(3)(-2)(7)$

40. $(-3)^2$

41. $(7)^3$

42. $(-4)^3$

43. -4^2

44. -2^3

45. 5^2

46. BUS A stock dropped $2.00 for each of seven straight weeks. What was the total change in the stock value?

47. On one winter day, the temperature dropped $2°$ each hour for 5 h. What was the total drop in temperature?

48. If the temperature in Exercise 47 was $8°$ originally, what was the temperature at the end of the 5-h period?

49. BUS An article states, "XYZ stock has dropped 4 points each week for the past 5 weeks." Was the stock price higher or lower 5 weeks ago than it is today? How much higher or lower? Use negative numbers to express drops in prices. Use a signed number to express the amount the stock price changed.

Divide.

50. $-8 \div (-4)$

51. $12 \div 3$

52. $\dfrac{14}{-7}$

53. $\dfrac{-20}{-5}$

54. $\dfrac{16}{-4}$

55. $\dfrac{-51}{-3}$

56. $\dfrac{0}{-8}$

57. $\dfrac{-7}{0}$

58. $\dfrac{-51}{3}$

59. $\dfrac{51}{-17}$

60. BUS A company records the following gains and losses in net profit for a six-month period: $22,973; −$12,357; −$2,791; $32,872; $18,930; and $2,093. Find the net gain or loss for the six-month period.

Section 5–4

Use the order of operations to evaluate the following. Verify the results with a calculator.

61. $7(3 + 5)$

62. $-2(3 - 1)$

63. $\dfrac{15 - 7}{8}$

64. $-20 \div 4 - 3(-2)$

65. $4 + (-3)^4 - 2(5 + 1)$

66. $(-3)^3 + 1 - 8$

Perform the indicated operations.

67. $\dfrac{-4}{5} \div -\dfrac{7}{15}$

68. $3.23 + (-4.61)$

69. $-12.4 \div 0.2$

70. $\dfrac{-11}{12} - \left(\dfrac{-7}{8}\right)$

71. $-\dfrac{7}{8} + \left(-\dfrac{5}{12}\right)$

72. $1\dfrac{3}{5} \div \left(-7\dfrac{5}{8}\right)$

73. $-2\dfrac{5}{8} \times 4\dfrac{1}{2}$

74. $\dfrac{3}{4} - \left(-\dfrac{1}{2}\right)^2$

75. $0.2 - 3.1(-7.6)$

76. $-0.7 - (-7.2 + 5)$

Section 5–5

Perform the indicated operations. Express as ordinary numbers.

77. $10^5 \cdot 10^7$

78. $10^{-2} \cdot 10^8$

79. $10^7 \cdot 10^{-10}$

80. 4.2×10^5

81. $8.73 \div 10^{-3}$

82. $5.6 \div 10^{-2}$

Write as ordinary numbers.

83. 3.75×10^5

84. 4.23×10^4

85. 3.87×10^{-5}

86. 7.37×10^{-9}

Write in scientific notation.

87. 52,000

88. 4,500

89. 0.00017

90. 3,800,000

91. 0.000000008

Perform the indicated operation and write the result in scientific notation.

92. $(4.2 \times 10^5)(3.9 \times 10^{-2})$

93. $\dfrac{1.25 \times 10^3}{3.7 \times 10^{-8}}$

Raise to the indicated power and express in scientific notation. Round appropriately.

94. $(5.2 \times 10^3)^4$

95. $(8.3 \times 10^{-2})^3$

96. $(4.5 \times 10^{-2})^{-2}$

97. The United States population is approximately 250 million. Write the number in scientific notation.

98. One coulomb (C) is approximately 6.28×10^{18} electrons. How many coulombs do 4.87×10^{15} electrons represent?

Write each number in engineering notation.

99. 0.00000092

100. 0.004

101. 8,400,000

102. 5,900

103. 41

104. 31,000

105. 17,000,000

106. 129,000,000

107. 3,084,000,000

108. 0.0982

109. 0.0000018

110. 0.000989

111. 0.007

112. 0.12

113. 0.00000000035

114. 5,200,000,000

115. 0.00049 s

116. 0.00052 Å

117. 0.588 Å

118. 10.53 s

Change each unit to engineering notation using metric prefixes.

119. 246.7 V

120. 5,082 W

121. 42,000 mW

122. 7,800 kΩ

123. 5,729 μW

124. 25,000 ps

125. 4,800 GHz

126. 4,000 ns

1. Your team is examining the patterns that could be used for automobile license plates. Determine how many license plates can be made using the following patterns. Assume that no plate will be discarded from these patterns.
   (a) 3 digits followed by 3 letters and both digits and letters can be repeated
   (b) the same option from part (a) with either the group of letters or numbers coming first
   (c) 6 letters and all letters can be repeated

2. The national debt at one point was approximately $7.5 trillion. Assuming that a dollar bill is approximately 0.2 mm thick, answer the following questions.
   (a) Suppose the national debt is represented with dollar bills stacked on top of each other. How many kilometers high will the stack reach?
   (b) The distance from Earth to the moon is 380,000 km. How many times will the stack of bills represented by the national debt reach from Earth to the moon?
   (c) If there are approximately 300 million people in the United States, how much would it take per person to pay off the national debt?

CONCEPTS ANALYSIS

1. What two operations for integers use similar rules for handling the signs? Explain the rules for these operations.

2. Explain what is meant by "the absolute value of a number." Give an example.

3. What operation with 0 is not defined?

4. Describe the process of adding two integers that have different signs.

5. Write a statement using the symbol for "is greater than."

6. Describe the correct order for operations with integers in words.

7. Explain how to find the sign of a power if the base is a negative integer. Give an example for an even exponent and for an odd exponent.

8. Give an example of multiplying two negative integers, and give the product.

9. Draw a number line that shows positive and negative integers and zero, and place the following integers on the number line: $-3, 8, -2, 0, 3, 5$.

10. Find and correct the mistakes in the following problem.
$$(-8)^2 - 3(2)$$
$$16 - 3(2)$$
$$13(2)$$
$$26$$

PRACTICE TEST

Use the symbols $>$ and $<$ to write the following as *true* statements.

1. -8 is less than 0

2. 2 is less than 3

3. -5 is more than -10

Answer the questions.

4. What is the value of $|-12|$?

5. What is the opposite of 8?

Perform the operations.

6. $-8 - 2$

7. $-3 + 7$

8. $\dfrac{8}{-2}$

9. $2(6)(-4)$

10. $-\dfrac{2}{5} + \dfrac{1}{10}$

11. $-1.3 - 2.4 + 5.8$

12. $-7 + (-3)$

13. $(-8)(3)(0)(-1)$

14. $-3 + 5 + 0 + 2 + (-5)$

15. $\dfrac{-7}{0}$

16. $4(-13)$

17. $\dfrac{4}{-2}$

18. $2 + 10(2) + 6(7)$

19. $2(3 - 9) \div 2^2 + 7$

20. $5(2 - 3) + 16 \div 4$

21. $(10^3)^2$

22. $\dfrac{10^{-5}}{10^3}$

Write as ordinary numbers.

23. 42×10^3

24. 0.83×10^2

25. 5.9×10^{-2}

Write in scientific notation.

26. 5.2301

27. 0.021

28. 52.3×10^2

29. 783×10^{-5}

Perform the indicated operations. Express the answers in scientific notation.

30. $(5.9 \times 10^5)(3.1 \times 10^4)$

31. $\dfrac{5.25 \times 10^4}{1.5 \times 10^2}$

32. A star is 3.4 light-years from Earth. If 1 light-year is 5.87×10^{12} mi, how many miles from Earth is the star?

33. The total resistance (in ohms) of a dc series circuit equals the total voltage divided by the total amperage. If the total voltage is 3×10^3 V and the total amperage is 2×10^{-3} A, find the total resistance (in ohms) expressed as an ordinary number.

34. In Grand Rapids, Minnesota, the temperature ranged from 42°F at 12 noon to -14°F at 7 P.M. Represent the change in temperature with a signed number.

35. Temperatures around the world may range from 135°F in Seville, Spain, to -40°F in Fairbanks, Alaska. What is the range (difference) of temperatures?

Change to engineering notation using metric prefixes.

36. 0.000001 Ms

37. 0.0047 GΩ

38. 0.0000829 W

39. 0.000000023 fs

6

Statistics

Focus on Careers

Careers as teachers and teacher's assistants can be very rewarding. Teacher's assistants often work part time while continuing to work toward teacher certification. Teacher's assistants provide instructional and clerical support for classroom teachers. They also provide instructional reinforcement by working with individuals or small groups.

Teacher's assistants responsible for instruction require more training than those who provide nonteaching duties. Teacher's assistants in Title 1 schools are required to have at least 2 years of college, hold a 2-year or higher degree, or pass rigorous state and local assessments. Many community colleges offer associate degree programs that prepare graduates to work as teacher's assistants. Students can meet most of the course requirements in the first 2 years of teacher-preparation programs at these community colleges and then matriculate to 4-year colleges for the remainder of their certification requirements.

Employment for teacher's assistants and for teachers is expected to grow faster than the average through 2012. Opportunities will be best for persons with at least 2 years of formal education beyond high school, and persons who speak other languages should be in high demand in school systems where large numbers of students speak a language other than English in the home.

The middle 50% of teacher assistants earned between $14,880 and $23,600 in 2002, and the highest 10% earned more than $29,050.

Source: *Occupational Outlook Handbook,* 2004–2005 Edition, U.S. Department of Labor, Bureau of Labor Statistics.

A **graph** shows information visually. Graphs may show how our tax dollars are divided among various government services, trace the fluctuations in a patient's temperature, or illustrate regional planting seasons. Other graphs may show equations, inequalities, and their solutions. Tables, on the other hand, usually list data. For example, income tax tables list taxes due on different incomes.

6-1 | Reading Circle, Bar, and Line Graphs

Learning Outcomes

1 Read circle graphs.
2 Read bar graphs.
3 Read line graphs.

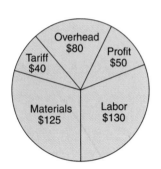

Figure 6–1 Distribution of wholesale price for a $425 color television.

Graphs give us useful information at a glance if we interpret them properly. Three common graphs used to represent data are the circle graph, the bar graph, and the line graph.

1 Read Circle Graphs.

A **circle graph** uses a circle to show pictorially how a whole quantity is divided into parts.

The complete circle represents one whole quantity. The circle is divided into parts so that the sum of all the parts equals the whole quantity. These parts can be expressed as fractions, decimals, or percents. Fig. 6–1 is a circle graph.

When we "read" a graph, we examine the information on the graph.

To read a circle, bar, or line graph:

1. Examine the title of the graph to find out what information is shown.
2. Examine the parts to see how they relate to one another and to the whole.
3. Examine the labels for each part of the graph and any explanatory remarks that may be given.
4. Use the given parts to calculate additional amounts or percents.

EXAMPLE
BUS

Use Fig. 6–1 to answer these questions.

(a) What percent of the wholesale price is the cost of labor?
(b) What percent of the wholesale price is the cost of materials?
(c) What would the wholesale price be if no tariff (tax) was paid on imported parts?

(a) $R = \dfrac{P}{B}(100\%)$ Use the percentage formula to find R.

$R = \dfrac{\$130}{\$425}(100\%)$ R is the percent of the wholesale price ($425) that is attributed to labor cost. The labor cost is $130.

$R = 0.3058823529(100\%)$ Round to tenths.

$R = \mathbf{30.6\%}$ **(labor)**

(b) $R = \dfrac{P}{B}(100\%)$ Use the percentage formula.

$R = \dfrac{\$125}{\$425}(100\%)$ *R* is the percent of the wholesale price ($425) that is attributed to materials cost. The materials cost is $125.

$R = 0.2941176471(100\%)$ Round to tenths.

$R = 29.4\%$ **(materials)**

(c) Price – tariff = $425 − $40 = $385 (cost without tariff)

2 Read Bar Graphs.

Different types of graphs allow us to access different types of information. A **bar graph** uses two or more bars to compare two or more amounts.

The bar lengths represent the amounts being compared. Bars can be drawn either horizontally or vertically.

The **axis,** or **reference line,** that runs along the length of the bars is a scale of the amounts being compared; in Fig. 6–2 this line is horizontal. The other reference line (vertical in this case) labels the bars. Fig. 6–2 is a horizontal bar graph.

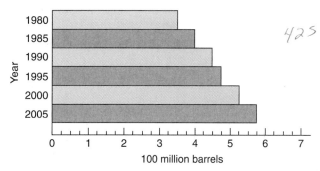

425

Figure 6–2 Company oil production.

EXAMPLE
BUS

Use Fig. 6–2 to answer these questions.

(a) How many more 100 million barrels of oil are indicated for the company in 2005 than in 1985?

(b) Judging from the graph, should company oil production in 2010 be more or less than in 2005?

(c) How many 100 million barrels of oil did the company produce in 1985?

(a) 2005 production – 1985 production = 5.75 − 4.00 = 1.75 hundred million barrels.

(b) More, because the trend has been toward greater production

(c) Four hundred million barrels in 1985

3 Read Line Graphs.

Line graphs are encountered in industrial reports, handbooks, and the like. A **line graph** uses one or more lines to show changes in data.

The horizontal axis on a line graph usually represents periods of time or specific times. The vertical axis represents numerical amounts. Line graphs show trends in data and high and low values at a glance. Fig. 6–3 is a line graph.

EXAMPLE
HLTH/N

Use Fig. 6–3 to answer these questions regarding a patient's temperature.

(a) On what date and time of day did the patient's temperature first drop to within 0.2° of normal (98.6°F)?

(b) On which post-op (post-operative) days did the patient's temperature remain within 0.2° of normal?

(c) What was the highest temperature recorded for the patient?

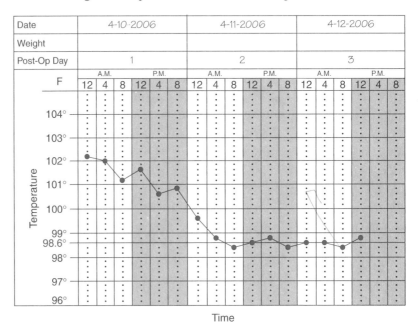

Figure 6–3 Patient's graphic temperature chart.

(a) **4-11-2006 at 4 A.M.** (Each "dot" is 0.2°, so the temperature was 98.8°F.)
(b) **Post-op days 2 and 3** (beginning at 4 A.M. on day 2)
(c) **102.2 degrees** (recorded at 12 A.M. on 4-10-2006)

SECTION 6–1 SELF-STUDY EXERCISES

1 Use Fig. 6–4 to answer Exercises 1–3.

1. What percent of the gross salary goes into retirement?
2. What percent of the take-home pay is federal income tax? Round to tenths.
3. What percent of the gross pay is the take-home pay? Round to tenths.

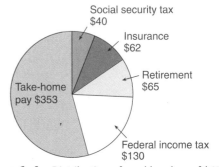

Figure 6–4 Distribution of weekly salary of $650.

Use Fig. 6–5 to answer Exercises 4–6. Round to tenths.

4. What percent of the day is spent working?
5. What percent of the day is spent sleeping?
6. The amount of time spent studying is what percent of the time spent in class?

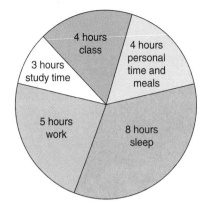

Figure 6–5 Distribution of a student's typical day.

Use Fig. 6–6 to answer Exercises 7–10. Round to two significant digits.

7. **AG/H** How many bushels of wheat are used for livestock feed?
8. **AG/H** How many bushels of wheat are exported annually?
9. **AG/H** If 1 bushel of wheat yields 42 lb of flour after milling, how many pounds of flour are consumed in the U.S. each year?
10. **AG/H** If U.S. wheat farmers harvested 48,653,000 acres of wheat, how many whole bushels per acre were averaged?

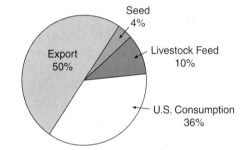

Figure 6–6 Distribution by percent of annual wheat production of 2.3 billion bushels.

Use Fig. 6–7 to answer Exercises 11–12.

11. **HLTH/N** What type of hospital discharge accounted for the greatest percentage of discharges?
12. **HLTH/N** Hospital discharges attributed to long-term care and home-health care accounted for what percentage of discharges?

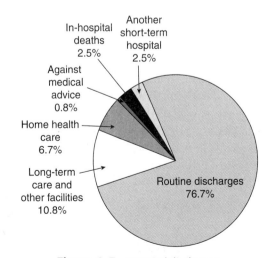

Figure 6–7 Hospital discharges.
Source: http://www.ahrq.gov/data/hcup/factkt1/index.html, 2005.

Use Fig. 6–8 to answer Exercises 13–15.

13. **BUS** What expenditure is expected to be the same next year as this year?
14. **BUS** What two expenditures are expected to increase next year?
15. **BUS** What two expenditures are expected to decrease next year?

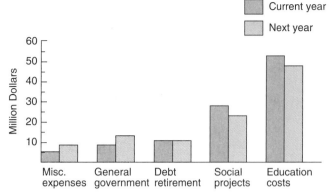

Figure 6–8 Distribution of local tax dollars.

Use Fig. 6–9 to answer Exercises 16–20. Round to one significant digit.

16. **INDTEC** How many barrels of oil were produced in 1995?
17. **INDTEC** Find the percent increase in world oil consumption from 1995 to 2000.
18. **INDTEC** By how many gallons did world oil consumption increase from 1975 to 2000?
19. **INDTEC** What was the percent of increase in world oil consumption between 1975 and 1990?
20. **INDTEC** Examine the graph to guess which 5-year period(s) had the greatest percentage increase.

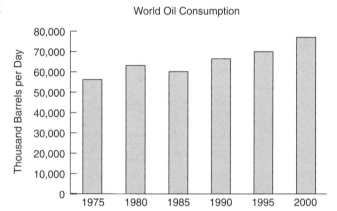

Figure 6–9 World oil consumption. *Source: Energy Information Administration, Department of Energy, U.S. Government, 2005.*

Use Fig. 6–10 to answer Exercises 21–24.

21. What year had the largest daily number of barrels of oil produced on shore?
22. What was the total number of barrels of oil produced daily in 1990? In 2000?
23. By what percent did daily offshore oil production increase from 1960 to 2000? Round to tenths.
24. In what year was the total daily oil production at its lowest level for the 5 decades reported?

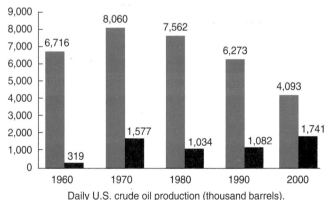

Figure 6–10 Daily U.S. crude oil production (thousand barrels). *Source: Energy Information Administration, U.S. Government, 2005.*

3 Use Fig. 6–11 to answer Exercises 25–28.

25. ELEC How many amperes of current are produced by 50 V when the resistance is 10 Ω?

26. ELEC How many volts are needed to produce 2 A of current when the resistance is 25 Ω?

27. ELEC Approximately how many volts are required to produce a current of 3.5 A when the resistance is 10 ohms?

28. ELEC Find the resistance when 100 V is needed to produce a current of 4 A.

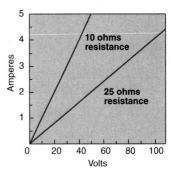

Figure 6–11   Amperage produced by voltage across two resistances.

Use Fig. 6–12 to answer Exercises 29–33.

29. AUTO What year was the motor gasoline supply 6,580,000 barrels of oil?

30. AUTO Which petroleum product consistently had the greatest supply?

31. AUTO For which 10-year time period was the increase in supply of motor gasoline the greatest?

32. AVIA What time period saw the greatest percentage of increase in jet fuel?

33. AVIA In 2000 the supply of jet fuel was what percent of the motor gasoline supply? Round to tenths.

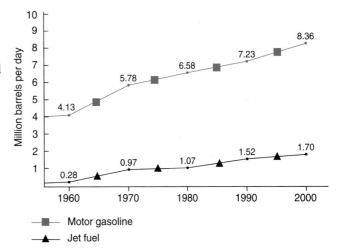

Figure 6–12   Petroleum products supplied by type in the United States. *Source: Energy Information Administration, U.S. Government.*

Use Fig. 6–13 to answer Exercises 34–35.

34. AG/H Which type of cattle consistently had the largest inventory from 1990–2000?

35. AG/H For which type of cattle were inventories consistently increasing?

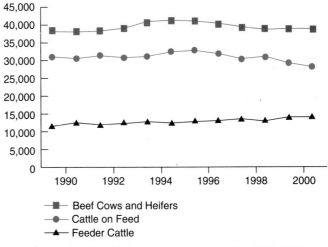

Figure 6–13   U.S. January cattle inventories, 1990–2001. *Source: IMPLAN Model for the United States, 2005.*

Use Fig. 6–14 to answer Exercises 36–38.

36. AG/H Approximately how many metric tons of U.S. beef and variety meats were exported in 2000?

37. AG/H Approximately what was the percent increase in the export tons in the 20-year period shown on the graph?

38. AG/H What 5-year period showed the most dramatic increase in value of U.S. beef and variety meat exports, including hides?

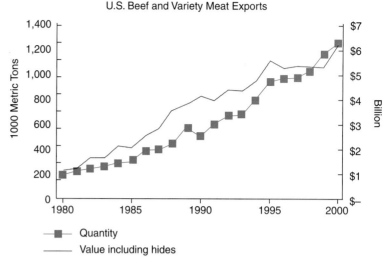

Figure 6–14 U.S. beef and variety meat exports. *Source: IM PLAN model for the United States, 2005.*

Use Fig. 6–15 to answer Exercises 39–42.

39. INDTEC What trend is evident in world production of silver?

40. INDTEC What trend is evident in silver mine production?

41. INDTEC By what percent did world silver production increase from 1950 to 2000?

42. INDTEC In what 10-year period did mine production dip below 1000 metric tons?

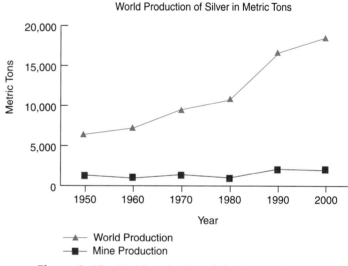

Figure 6–15 World production of silver in metric tons. *Source: Kenneth E. Porter and Henry E. Hilliard, 2004.*

6–2 | *Averages and Frequency Distributions*

Learning Outcomes

1 Find the arithmetic mean.

2 Find the median and the mode.

3 Make and interpret a frequency distribution.

4 Find the mean of grouped data.

1 Find the Arithmetic Mean.

In this age of information explosion we have massive amounts of data available to us. **Data** are facts or information from which conclusions may be drawn. To use these data effectively in the decision-making process, we need to examine the data and summarize key trends and characteristics. This summary is generally in the form of statistical measurements. A **statistical measurement** or **statistic** is a standardized, meaningful measure of a set of data that reveals a certain feature or characteristic of the data. Such statistics are called **descriptive statistics.**

An **average** is an approximate number that is a central value of a set of data. The most common average is the arithmetic mean. The **arithmetic mean,** also called **statistical mean,** is the sum of the quantities in the data set divided by the number of quantities.

Symbols can be used to express the procedures for statistical measures. In symbols, the formula for the mean for a set of data is

$$\bar{x} = \frac{\Sigma x_i}{n}$$

The formula is read as "the mean $\bar{x}$ (read "x-bar") equals the sum Σ of each value of x_i (read "x sub i") divided by the number of values n. The Greek capital letter **sigma, Σ,** is a summation symbol and indicates the addition of a set of values. The notation x_i identifies each data value with a subscript: x_1 is the first value, x_2 is the second value, and x_n is the nth value.

To find the arithmetic mean:

Find the mean of 22, 31, and 37.

$x_1 = 22,\ x_2 = 31,\ x_3 = 37$

1. Add the quantities.

Σx_i

$\Sigma x_i = 22 + 31 + 37$

$\Sigma x_i = 90$

2. Divide the sum by the number of quantities.

$\bar{x} = \dfrac{\Sigma x_i}{n} = \dfrac{90}{3}$

$\bar{x} = 30$

Symbolically, $\bar{x} = \dfrac{\Sigma x_i}{n}$, where $\bar{x}$ = arithmetic mean; Σ means sum;

x_i represents each x value; n = number of values.

EXAMPLE
HLTH/N

Find the mean of each set of quantities.

Mean: find the average

(a) Pulse rates: 68, 84, 76, 72, 80

There are five pulse rates, so we find their sum and divide by 5:

$$\bar{x} = \frac{\Sigma x_i}{n} = \frac{68 + 84 + 76 + 72 + 80}{5} = \frac{380}{5} = 76$$

The mean pulse rate is 76.

(b) Pounds: 21, 33, 12.5, 35.2 (to the nearest whole number)

There are four weights, so we find their sum and divide by 4.

$$\bar{x} = \frac{\Sigma x_i}{n} = \frac{21 + 33 + 12.5 + 35.2}{4} = \frac{101.7}{4} = 25.425$$

The mean weight is 25 lb (to the nearest pound).

A graphing calculator has many features to facilitate finding statistical measures. The first step is to build a list containing the data to be measured.

To build a list using a TI-83 or a TI-84 calculator:

Clear any previous list that you no longer care to keep.

[STAT] [4:ClrList] [LIST] [1:L₁] [ENTER].

Enter your new list as List 1.

[STAT] [1:Edit...] Enter each data amount, followed by [ENTER]. After the last data amount is entered, press [QUIT].

Remember if a feature is labeled above a key, an access key [2ND] is pressed first. On the TI-83 or TI-84, the [QUIT] and numbered lists (L₁, L₂, L₃, . . . , L₆) features are labeled above a key.

Using the data from Part (a) of the previous example:

[STAT] [1:Edit...] 68 [ENTER] 84 [ENTER] 76 [ENTER] 72 [ENTER] 80 [ENTER] [QUIT].

To find the mean:

[LIST] [MATH] [3:mean] [LIST] [1:L₁] [ENTER] [ENTER] ⇒ 76

To access the [MATH] menu from the [LIST], use the right-arrow keys.

EXAMPLE AUTO An automobile used 41 gal of regular gasoline on a trip of 876 mi. What was the average miles per gallon ($\frac{mi}{gal}$) to the nearest gallon?

The number of miles traveled on the 41 gal of gasoline is 876 and represents the total miles. Divide the total miles by 41.

$$\frac{876 \text{ mi}}{41 \text{ gal}} = 21.36585366$$

The car averaged 21 $\frac{mi}{gal}$ on the trip to the nearest gallon.

EXAMPLE AG/H Table 6–1 lists the final grades of a horticulture student. Find the student's QPA (quality point average) to the nearest hundredth based on a 4-point system.

To find the QPA for a term, the quality points for the letter grade of each course are multiplied by the credit hours of each course to obtain the total quality points for each course. This total is divided by the total credit hours earned. The quality points awarded are A = 4, B = 3, and C = 2.

Table 6–1   Quality Points for Final Grades

Subject	Grade	Credit Hours		Quality Points per Hour		Total Points
Algebra 101	B	3	×	3	=	9
Spray chemicals 102	A	3	×	4	=	12
Landscape 301	A	3	×	4	=	12
English 101	C	4	×	2	=	8
		13 hr				41 points

$$\frac{41}{13} = 3.153846154 \text{ or } 3.15 \qquad \text{To nearest hundredth.}$$

Thus, the student's QPA for the term is 3.15.

EXAMPLE   A community college student has the following grades in Physics 101: 73, 84, 80, 62, and 70. What grade is needed on the last test for the student to get a C on the final grade or a 75 average?

One way to find the needed grade is to assume that each grade is 75 for the 6 tests:

$$\frac{75 + 75 + 75 + 75 + 75 + 75}{6} = \frac{450}{6} = 75 \qquad \text{The sum of the six grades is 450.}$$

Then a total of 450 points are needed on six tests to have an average of 75.

$$73 + 84 + 80 + 62 + 70 = \boxed{369} \qquad \text{Add the five test grades.}$$

$$450 - \boxed{369} = \boxed{81} \qquad \text{Find the difference between the sum of the five grades and 450.}$$

The student must earn a score of 81 or higher on the last test to have an average of 75 or higher.

$$\text{Check:} \qquad \frac{73 + 84 + 80 + 62 + 70 + \boxed{81}}{6} = \frac{450}{6} = 75$$

2   Find the Median and the Mode.

Besides the arithmetic mean, we also use the *median* and the *mode* to describe groups of data. These are called measures of *central tendency.* **Measures of central tendency** are descriptive statistics that determine the center of a data set from a different perspective.

The **median** is the middle value when the data values are arranged in order of size. The median is frequently used for data relating to annual incomes, housing price ranges, etc.

To find the median:

1. Arrange the values in order of size, either smaller to larger or larger to smaller.
2. If the number of values is odd, the median is the middle value.
3. If the number of values is even, the median is the average of the two middle values.

EXAMPLE HLTN/N A TPR chart shows a patient's temperature, pulse rate, and respiration rate. The following pulse rates were recorded on a TPR chart: 68, 88, 76, 64, 72. What is the patient's median pulse rate?

In descending order of size: 88
76
72   ← median or middle value in an odd number of values
68
64

The median pulse rate is 72.

EXAMPLE The following temperatures were recorded: 56°, 48°, 66°, and 62°. What is the median temperature?

The number of temperatures is even, so we find the average of the two middle values.

In ascending order of size   48
56
62   $\left. \begin{array}{c} \\ \\ \end{array} \right\}$ $\dfrac{56 + 62}{2} = \dfrac{118}{2} = 59$
66

The median temperature is 59°.

Find the Median Using a Graphing Calculator

A median feature is included on many graphing calculators.

Examine the calculator steps to find the median for the previous example.

Build a list:

$\boxed{\text{STAT}}$ $\boxed{\text{4:ClrList}}$ $\boxed{\text{STAT}}$ $\boxed{\text{1:L}_1}$ $\boxed{\text{ENTER}}$.   Clear previous list in L_1.

Enter your new list as List 1.

$\boxed{\text{STAT}}$ $\boxed{\text{1:Edit...}}$ 48 $\boxed{\text{ENTER}}$ 56 $\boxed{\text{ENTER}}$ 62 $\boxed{\text{ENTER}}$ 66 $\boxed{\text{ENTER}}$ $\boxed{\text{QUIT}}$

Find the median:

$\boxed{\text{LIST}}$ $\boxed{\text{MATH}}$ $\boxed{\text{4:median(}}$ $\boxed{\text{STAT}}$ $\boxed{\text{1:L}_1}$ $\boxed{\text{ENTER}}$ ⇒ 59

The **mode** is the value that occurs most frequently in the data set.

1. Identify the value or values that occur with the greatest frequency as the mode.
2. If no value occurs more than another value, there is *no mode* for the data set.
3. If more than one value occurs with the greatest frequency, the modes of the data set are the values that have the greatest frequency.

EXAMPLE
BUS

The hourly pay rates at a local fast-food restaurant are as follows: cooks, $8.50; servers, $7.15; bussers, $7.15; dishwashers, $7.25; managers, $10.50. Find the mode.

Identify the value or values that appear most often.

The hourly pay rate of $7.15 occurs more than any other rate. It is the mode.

EXAMPLE

The daily work shifts at a mall clothing store are 4, 6, and 8 hours. Find the mode.

No shift occurs more frequently than another, so there is no mode.

3 **Make and Interpret a Frequency Distribution.**

For a class of 25 students the instructor records the following grades:

76 91 71 83 97 87 77 88 93 77 93 81 63
79 74 77 76 97 87 89 68 90 84 88 91

It is difficult to make sense of all these numbers as they appear here. But the instructor can arrange the scores into several smaller groups, called **class intervals.** The word *class* means a special category.

These scores can be grouped into class intervals of 5, such as 60–64, 65–69, 70–74, 75–79, 80–84, 85–89, 90–94, and 95–99. Each class interval has an odd number of scores. The *middle score* of each interval is a **class midpoint.**

The instructor can now *tally* the number of scores that fall into each class interval to get a **class frequency,** the number of scores in each class interval.

A compilation of class intervals, midpoints, tallies, and class frequencies is called a **grouped frequency distribution.**

EXAMPLE

Examine the grouped frequency distribution in Table 6–2, and answer the following questions.

Table 6–2   Frequency Distribution of 25 Scores

Class Interval	Midpoint	Tally	Class Frequency
60–64	62	/	1
65–69	67	/	1
70–74	72	//	2
75–79	77	//// /	6
80–84	82	///	3
85–89	87	////	5
90–94	92	////	5
95–99	97	//	2

(a) How many students scored 70 or above?

$$2 + 6 + 3 + 5 + 5 + 2 = 23$$

Add the frequencies for class intervals with scores 70 or higher.

23 students scored 70 or above.

(b) How many students made As (90 or higher)?

$$5 + 2 = 7$$

Add the frequencies for class intervals 90–94 and 95–99.

7 students made As (90 or higher).

(c) What percent of the total grades were As (90s)?

$$\frac{7\ \text{As}}{25\ \text{total}} = \frac{7}{25} = 0.28 = \mathbf{28\%\ As}$$

The portion or part is 7 and the base or total is 25.

(d) Were the students prepared for the test?

The relatively high number of 90s (7) compared to the relatively low number of 60s (2) suggests that **in general, most students were prepared for the test.**

(e) What is the ratio of As (90s) to Fs (60s)?

$$\frac{7\ \text{As}}{2\ \text{Fs}} = \frac{7}{2}$$

The ratio is $\dfrac{7}{2}$.

EXAMPLE Students in a history class reported their credit-hour loads as shown. Make a grouped frequency distribution of their credit hours. Credit hours carried: 3, 12, 15, 3, 6, 6, 12, 9, 12, 9, 6, 3, 12, 18, 6, 9.

To establish a class interval with an easy-to-find midpoint, use an odd number of points in the interval. Here, an interval of 5 is used; that is, 0–4 contains five possibilities: 0, 1, 2, 3, and 4. The middle number is the midpoint, 2. Make a tally mark for each time the credit hours of a student falls in the interval. Then count the tally marks to get the class frequency (Table 6–3).

Table 6–3	Frequency Distribution of Credit-Hour Loads		
Class Interval	**Midpoint**	**Tally**	**Class Frequency**
0–4	2	///	3
5–9	7	⌿⌿⌿ //	7
10–14	12	////	4
15–19	17	//	2

4 **Find the Mean of Grouped Data.**

When data are grouped, it may be desirable to find the mean of the grouped data. To do this we extend our frequency distribution.

To find the mean of grouped data:

1. Make a frequency distribution.
2. Find the products (xf) of the midpoint (x) of the interval and the frequency (f) for each interval for all intervals.
3. Find the sum of the frequencies (Σf).
4. Find the sum of the products (Σxf).
5. Divide the sum of the products by the sum of the frequencies.

Symbolically, $\text{mean}_{\text{grouped}} = \dfrac{\Sigma xf}{\Sigma f}$.

EXAMPLE Find the grouped mean to the nearest whole number of the data in the frequency distribution in Table 6–4.

Table 6–4 Frequency Distribution of Credit-Hour Loads

Class Interval	Midpoint x	Frequency f	Product xf
0–4	2	3	6
5–9	7	7	49
10–14	12	4	48
15–19	17	2	34
Total		16	137

$$\Sigma f = 16 \qquad \text{Add the frequencies.}$$

$$\Sigma xf = 137 \qquad \text{Add the products.}$$

$$\text{mean}_{\text{grouped}} = \frac{\Sigma xf}{\Sigma f} \qquad \text{Substitute.}$$

$$\text{mean}_{\text{grouped}} = \frac{137}{16} \qquad \text{Divide.}$$

$$\text{mean}_{\text{grouped}} = 8.5625 \qquad \text{Round.}$$

$$\text{mean}_{\text{grouped}} = 9$$

Is the Mean of Grouped Data Exact?

No. The mean of grouped data is based on the assumption that all the data in an interval have a mean that is exactly equal to the midpoint of the interval. Because this is usually not the case, the mean of grouped data is a reasonable approximation.

1 Find the mean of the given values. Round to hundredths if necessary.

1. 12, 14, 16, 18, 20
2. 13, 15, 17, 19, 21
3. 68, 54, 73, 69
4. 85, 68, 77, 65
5. 37.6, 29.8
6. 65.3, 67.9
7. 32°F, 41°F, 54°F
8. 10°C, 13°C, 15°C
9. $27, $32, $65, $29, $21
10. $32, $43, $22, $63, $36
11. 11 in., 17 in., 16 in.
12. 9 in., 7 in., 8 in.
13. Respiration rates: 16, 24, 20
14. Pulse rates: 68, 84, 76

15. A baseball player batted 276 home runs over a 16-year period. What was the average number of home runs per year to the nearest tenth?

16. Noel Womack scored 87, 96, 86, 92, and 93 in English 101. What must he score on the last test to earn an average score of 92?

17. **AUTO** An automobile used 32 gal of regular gasoline on a 786-mi trip. What was the average miles per gallon to the nearest tenth?

18. **AUTO** A pickup truck used 25 gal of regular gasoline on a 256-mi trip. What was the average miles per gallon to the nearest tenth?

19. **CON** U.S. cement export data for the last 10 years in thousand tons:

746, 625, 633, 759, 803,
791, 743, 694, 738, 834

Find the mean number of tons exported for the period.

20. **CON** World cement production data for the last 20 years in million tons:

916.6; 941.1; 959.4; 1,008; 1,053; 1,118; 1,042; 1,043; 1,185; 1,123; 1,291; 1,370; 1,445; 1,493; 1,547; 1,540; 1,600; 1,650; 1,730; 1,800

Find the mean number of tons of cement produced in the world. Report two significant digits.

2 Find the median for each data set.

21. 32, 56, 21, 44, 87
22. 78, 23, 56, 43, 38
23. 12, 21, 14, 18, 15, 16
24. 21, 33, 18, 32, 19, 44
25. $22, $35, $45, $30, $29
26. $66, $54, $76, $55, $69

27. **BUS** The following hourly pay rates are used at fast-food restaurants: cooks, $8.15; servers, $8.25; bussers, $8.15; dishwashers, $8.25; managers, $11.25. Find the median pay rate.

28. **BUS** The following hourly pay rates are used at a locally owned store: clerks, $8.45; bookkeepers, $9.25; operators, $8.15; assistant managers, $10.95. Find the median pay rate.

29. **CON** Find the median number of tons of concrete exported for the 10-year period in Exercise 19. Report three significant digits.

30. **CON** Find the median number of tons of concrete produced in the world for the data in Exercise 20.

Find the mode for each data set.

31. 2, 4, 6, 2, 8, 2
32. 5, 12, 5, 5, 20
33. 21, 32, 67, 34, 23, 22
34. 32, 45, 41, 23, 56, 77
35. $56, $67, $32, $78, $67, $20, $67, $56
36. $32, $87, $67, $32, $32, $87, $77, $22

37. **BUS** These weekend work shifts are in effect at a mall clothing store: 4 hours in A.M., 6 hours in P.M., 4 hours in P.M. Find the mode for the number of hours.

38. **BUS** These special prices are in effect at a fast-food restaurant: $1.75, hamburgers; $1.97, hot ham sandwiches; $2.38, chicken fillet sandwiches; $1.97, roast beef sandwiches. Find the mode.

39. **CON** Find the mode for the data in Exercise 19.

40. **CON** Find the mode for the data in Exercise 20.

3 Use Table 6–5 to answer these questions. The frequency distribution shows the ages of 25 college students in a landscaping class.

41. How many students are 22 or younger?

42. How many students are older than 34?

43. What is the ratio of the number of students 38–40 to the number of students 17–19?

44. What is the ratio of the smallest class frequency to the largest class frequency?

45. What percent of the total class are students age 17–19?

46. What percent of the total class are students age 20–22?

Table 6–5	Frequency Distribution of 25 Ages		
Class Interval	Midpoint	Tally	Class Frequency
38–40	39	/	1
35–37	36	/	1
32–34	33	//	2
29–31	30	///	3
26–28	27	//	2
23–25	24	### /	6
20–22	21	### //	7
17–19	18	///	3

47. What two age groups make up the smallest number of students in the class?

48. What two age groups make up the largest number of students in the class?

49. How many students are over age 28?

50. How many students are under age 26?

Use the given hourly pay rates (rounded to the nearest whole dollar) for 33 support employees in a private college to complete a frequency distribution using the format shown in Table 6–6.

Table 6–6	Pay Rates of 33 Support Employees		
	Class Interval	Midpoint	Tally	Class Frequency
51.	$14–16	_____	____	_____
52.	$11–13	_____	____	_____
53.	$8–10	_____	____	_____
54.	$5–7	_____	____	_____

$6 $6 $10 $7 $6 $6 $6
$6 $7 $7 $8 $8 $6 $6
$11 $10 $7 $11 $8 $16 $6
$6 $9 $6 $7 $9 $6 $6
$12 $13 $7 $15 $5

Use the given 40 test scores of two physics classes to complete a frequency distribution using the format in Table 6–7.

Table 6–7	Test Scores of 40 Physics Students		
	Class Interval	Midpoint	Tally	Class Frequency
55.	91–95	_____	____	_____
56.	86–90	_____	____	_____
57.	81–85	_____	____	_____
58.	76–80	_____	____	_____
59.	71–75	_____	____	_____
60.	66–70	_____	____	_____
61.	61–65	_____	____	_____
62.	56–60	_____	____	_____

57 91 76 89 82 59 72 88
76 84 67 59 77 66 56 76
77 84 85 79 69 88 75 58
85 65 67 66 93 83 69 81
80 64 78 76 72 90 79 90

63. Students recorded the number of hours they studied each week as: 3, 15, 18, 0, 2, 9, 12, 16, 7, 8, 5, 10, 14, 9, 7, 3, 4, 14, 17, 16, 6, 4, 8, 11, 14, 13, 10, 15, 16, 5.

Create a grouped frequency distribution with four classes beginning with the class 0–4. Show the class interval, midpoint, tally, and class frequency for each class interval.

64. The number of animals at the city animal shelter varies daily. Use the data to make a frequency distribution with five classes beginning with the class 11–20. Show the class interval, midpoint, tally, and class frequency for each class interval: 22, 31, 32, 27, 29, 16, 12, 18, 21, 30, 46, 52, 43, 51, 42, 26, 42, 17, 19, 25.

65. Use the data in Exercise 19 to create a grouped frequency distribution with three classes beginning with the class 601–700 thousand tons. Show the class interval, midpoint, tally, and class frequency for each class interval.

66. Use the data in Exercise 20 to create a grouped frequency distribution with five classes beginning with the class 751–1,000 million tons. Show the class interval, midpoint, tally, and class frequency for each class interval.

4 Round the grouped mean.

67. Find the grouped mean for the data in Table 6–5. Round to the nearest whole number.

68. Find the grouped mean for the data in Table 6–6. Round to the nearest dollar.

69. Find the grouped mean for the data in Table 6–7. Round to the nearest whole number.

70. Find the grouped mean for the data in Exercise 63. Round to the nearest whole number.

6–3 | Range and Standard Deviation

Learning Outcomes

1 Find the range.
2 Find the standard deviation.

1 Find the Range.

The mean, the median, and the mode are *measures of central tendency*. Other statistical measures are **measures of variation** or **dispersion**. The variation or dispersion of a set of data may also be referred to as the **spread**. One of these measures of dispersion is the **range**. The range is the difference between the highest value and the lowest value in a set of data.

To find the range:

1. Find the highest and lowest values.
2. Find the difference between the highest and lowest values.

EXAMPLE BUS Find the range for the data described in the example for fast-food restaurant hourly pay rates on page 269.

The high value is $10.50. The low value is $7.15.

$$\text{range} = \$10.50 - \$7.15 = \mathbf{\$3.35}$$

Use More Than One Statistical Measure

A common mistake when making conclusions or inferences from statistical measures is to examine only one statistic, such as the range. To obtain a complete picture of the data requires looking at more than one statistic.

2 Find the Standard Deviation.

Although the range gives us some information about dispersion, it does not tell us whether the highest or lowest values are typical values or extreme **outliers.** We can get a clearer picture of the data set by examining how much each value *differs* or *deviates* from the mean.

The **deviation from the mean** of a data value is the difference between the value and the mean.

To find the deviations from the mean:

Data set: 38, 43, 45, 44.

1. Find the mean of a set of data.

$$\bar{x} = \frac{\text{sum of data values}}{\text{number of values}} = \frac{\Sigma x_i}{n}$$

$$\frac{38 + 43 + 45 + 44}{4} = \frac{170}{4} = 42.5$$

2. Find the amount that each data value deviates or is different from the mean.

deviation from the mean =
data value − mean = $x_i - \bar{x}$

$38 - 42.5 = -4.5$ (below the mean)
$43 - 42.5 = 0.5$ (above the mean)

$45 - 42.5 = 2.5$ (above the mean)
$44 - 42.5 = 1.5$ (above the mean)

For values smaller than the mean, the difference is represented by a *negative* number indicating the value is *below* or less than the mean. For values larger than the mean, the difference is represented by a positive number indicating the value is *above* or greater than the mean. *The absolute value of the sum of the deviations below the mean should equal the sum of the deviations above the mean.* In the example in the box, only one value is below the mean, and its deviation is −4.5. Three values are above the mean, and the sum of these deviations is $0.5 + 2.5 + 1.5 = 4.5$. We say that *the sum of all deviations from the mean is zero.* This is true for all sets of data.

We have not gained any statistical insight or new information by analyzing the sum of the deviations from the mean or even by analyzing the average of the deviations.

$$\text{average deviation} = \frac{\text{sum of deviations}}{\text{number of values}} = \frac{0}{n} = 0$$

EXAMPLE Find the deviations from the mean for the set of data 45, 63, 87, and 91.

$$\bar{x} = \frac{\Sigma x_i}{n} = \frac{45 + 63 + 87 + 91}{4} = \frac{286}{4} = 71.5 \qquad \text{Mean.}$$

To find the deviation from the mean, subtract the mean, $\bar{x}$, from each value of x. We arrange these values in a table.

Values x_i	Deviations $x_i - \bar{x}$
45	$45 - 71.5 = $ **−26.5**
63	$63 - 71.5 = $ **−8.5**
87	$87 - 71.5 = $ **15.5**
91	$91 - 71.5 = $ **19.5**
Sum 286	0

As we might expect, the sum of the deviations in the example equals zero because the sum of the negative deviations ($-26.5 + -8.5 = -35$) equals the sum of the positive deviations ($15.5 + 19.5 = 35$). To compensate for this situation, mathematicians employ a statistical measure called the **standard deviation,** which uses the square of each deviation from the mean. The square of a negative value is always positive. The sum of the squared deviations is divided by 1 less than the number of values, and the result is called the sample **variance:**

$$\text{sample variance} = v = \frac{\Sigma(x_i - \bar{x})^2}{n - 1}$$

The square root of the variance is the standard deviation. Various formulas exist for finding the standard deviation of a set of values, but we examine only one formula. Several calculations are necessary and are best organized in a table.

$$\text{sample standard deviation} = s = \sqrt{\frac{\Sigma(x_i - \bar{x})^2}{n - 1}}$$

To find the standard deviation of a set of sample data:

1. Find the mean, $\bar{x}$.
2. Find the deviation of each value from the mean: $(x_i - \bar{x})$
3. Square each deviation: $(x_i - \bar{x})^2$
4. Find the sum of the squared deviations: $\Sigma(x_i - \bar{x})^2$
5. Divide the sum of the squared deviations by *one less than* the number of values in the data set. The quotient is called the *sample variance:*

$$v = \frac{\Sigma(x_i - \bar{x})^2}{n - 1}.$$

6. Find the sample standard deviation by taking the square root of the sample variance:

$$s = \sqrt{\frac{\Sigma(x_i - \bar{x})^2}{n - 1}}.$$

EXAMPLE Find the sample standard deviation for the values 45, 63, 87, and 91.

From the previous example the mean is 71.5 and the number of values is 4.

$$\bar{x} = 71.5, n = 4$$

Values x_i	Deviations from the mean $x_i - \bar{x}$	Squares of the deviations from the mean $(x_i - \bar{x})^2$
45	$45 - 71.5 = -26.5$	$(-26.5)^2 = 702.25$
63	$63 - 71.5 = -8.5$	$(-8.5)^2 = 72.25$
87	$87 - 71.5 = 15.5$	$(15.5)^2 = 240.25$
91	$91 - 71.5 = 19.5$	$(19.5)^2 = 380.25$
Sum of values 286	Sum of deviations 0	Sum of squared deviations 1,395

$$v = \frac{\Sigma(x_i - \bar{x})^2}{n - 1} \qquad \frac{\text{Sum of squared deviations}}{n - 1}.$$

$$v = \frac{1,395}{3} = 465$$

$$s = \sqrt{v} \qquad \text{Square root of variance.}$$

$$s = \sqrt{465} = 21.56385865 \; or \; \textbf{21.6 sample standard deviation (rounded).}$$

Find the Standard Deviation Using a Graphing Calculator

A standard deviation feature is included on many graphing calculators.

Examine the calculator steps to find the median on page 268.

Build a list:

[STAT] [4:ClrList] [LIST] [1:L₁] [ENTER].   Clear previous list in L₁.

Enter your new list as List 1.

[STAT] [1:Edit...] 45 [ENTER] 63 [ENTER] 87 [ENTER] 91 [ENTER] [QUIT]

Find the standard deviation:

[LIST] [MATH] [7:stdDev(] [LIST] [1:L₁] [ENTER] ⇒ 21.56385865.

A small standard deviation indicates that the mean is a typical value in the data set. A large standard deviation indicates that the mean is not typical, and other statistical measures should be examined to better understand the characteristics of the data set. Let's examine the various statistics for the data set on a number line (Fig. 6–16).

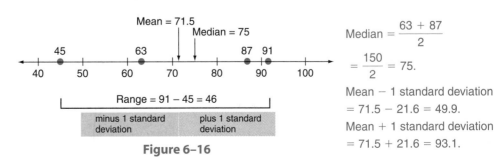

Figure 6–16

$$\text{Median} = \frac{63 + 87}{2}$$

$$= \frac{150}{2} = 75.$$

Mean − 1 standard deviation
$= 71.5 - 21.6 = 49.9.$
Mean + 1 standard deviation
$= 71.5 + 21.6 = 93.1.$

We can confirm visually that the dispersion of the data is broad and the mean is not a typical value in the data set.

Another interpretation of the standard deviation is in its relationship to the **normal distribution**. If we assume that data are normally distributed, we make speculations about where data are located in relation to the population mean of the data set. Procedures for determining whether data are normally distributed are presented in advanced studies of statistics.

The graph of a normal distribution is a bell-shaped curve, as in Fig. 6–17. The curve is *symmetrical*; that is, if folded at the highest point of the curve, the two halves would match. The mean of the data set is at the highest point or fold line. Then, half the data (50%) is to the left or *below* the mean and half the data (50%) is to the right or *above* the mean. Other characteristics of the normal distribution are:

68.3% of the data are within **1** standard deviation of the mean.
95.4% of the data are within **2** standard deviations of the mean.
99.7% of the data are within **3** standard deviations of the mean.

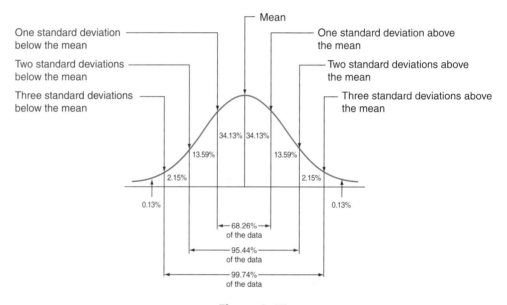

Figure 6–17

To solve applied problems involving the mean, standard deviation, and a normal distribution of a population:

1. Locate the mean and the desired values on the normal curve.

$$\frac{\text{value} - \text{population mean}}{\text{population standard deviation}} = \text{number of standard deviations from (above or below) the mean}$$

2. Highlight the desired regions of the normal curve based on the conditions of the problem.
3. Add the percents associated with the highlighted regions in Step 2.

EXAMPLE
AUTO

An Auto Zone Duralast Gold automobile battery has a population mean life of 46 months with a population standard deviation of 4 months. In an order of 100 batteries, what percent do you expect to last at least 54 months?

$$\frac{\text{value} - \text{population mean}}{\text{population standard deviation}} = \text{Number of standard deviations } from \text{ (above or below) the mean}$$

$$\frac{54 - 46}{4} = \frac{8}{4} = 2 \qquad \text{Standard deviations above the mean.}$$

Highlight all regions that are below the mean and two standard deviations above the mean (Fig. 6–18).

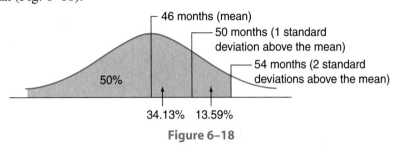

Figure 6–18

$50\% + 34.13\% + 13.59\% = 97.72\%$ Add the percents associated with the highlighted regions.

97.72% of the batteries should last less than 54 months.

EXAMPLE
AUTO

Using the facts of the previous example, how many batteries do you expect to last at least 54 months? Round to the nearest battery.

Locate regions on the normal curve that are more than 2 standard deviations above the mean (Fig. 6–19).

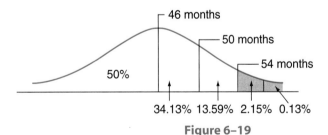

Figure 6–19

$2.15\% + 0.13\% = 2.28\%$ Add the percents associated with the shaded regions.

Alternative method for finding the percent:

$50\% + 34.13\% + 13.59\% = 97.72\%$ Add the percents associated with the unshaded regions.

$100\% - 97.72\% = 2.28\%$ Complement of 97.72%

$2.28\% (100) \text{ batteries} = 0.0228(100) = 2.28 \text{ batteries}$ Percent of 100 batteries

2 batteries (rounded) should last at least 54 months.

1 Find the range for each data set.

1. 22, 36, 41, 41, 17

2. 28, 33, 36, 13, 28

3. 10, 23, 12, 17, 13, 16

4. 23, 23, 18, 32, 29, 14

5. $25, $15, $25, $40, $19

6. $36, $44, $26, $52, $19

7. 23°F, 37°F, 29°F, 54°F, 46°F, 71°F, 67°F

Table 6–8	**Production in Tons of Sweet Cherries by State**								
State	1997	1998	1999	2000	2001	2002	2003	2004	2005*
CA	49,200	15,200	79,500	47,000	53,300	55,500	62,000	73,000	45,000
ID	1,600	2,200	7,900	3,000	1,400	1,700	2,900	3,100	2,200
MI	27,000	35,000	27,000	19,600	23,000	2,700	13,000	24,700	27,000
MT	1,100	2,050	720	1,100	2,020	2,350	1,920	2,360	1,300
NY	650	700	1,505	900	1,100	350	600	900	950
OR	50,000	55,000	50,000	37,000	40,000	31,000	38,000	43,000	29,000
PA	500	550	800	500	580	355	340	400	
UT	720	2,700	1,150	2,400	700	400	2,200	1,600	1,100
WA	95,000	98,000	67,000	95,000	106,000	87,000	116,000	133,000	120,000

*Forecast for 2005
Source: National Agricultural Statistics Service (NASS), Agricultural Statistics Board, U.S. Department of Agriculture (USDA), 2005.

Use Table 6–8 to answer Exercises 8–13.

8. AG/H   Find the range for cherry production in 2003.

9. AG/H   Find the range for cherry production in states in 2004.

10. AG/H   Find the range for cherry production in states in 2002.

11. AG/H   Find the range for cherry production for Washington state across the years 1997–2005.

12. AG/H   Find the range for cherry production for California state across the years 1997–2005.

13. AG/H   Find the range for cherry production for New York state across the years 1997–2005.

2 Find the standard deviation for each data set. Round to the nearest hundredth.

14. 12, 14, 16, 18, 20

15. 68, 54, 73, 69

16. 32°F, 41°F, 54°F

17. $27, $32, $65, $29, $21

18. Respiration rates: 16, 24, 20

19. Pulse rates: 68, 84, 76

Use Table 6–8 to answer Exercises 20–22.

20. AG/H   Find the standard deviation for cherry production for Washington state for the years reported. Round to the nearest ton.

21. AG/H   Find the standard deviation for cherry production for California for the years reported. Round to the nearest ton.

22. AG/H   Find the standard deviation for cherry production for New York state for the years reported. Round to the nearest ton.

23. HLTH/N   The population mean length of a hospital stay for surgery is 5.8 days and the standard deviation is 1.9 days. What percentage of patients are hospitalized for 3.9 days or less?

24. HLTH/N   In a sample of 200 surgery patients, how many are expected to stay 9.7 or more days if the mean length of stay is 5.8 days and the standard deviation is 3.9 days?

25. HLTH/N   Pediatricians work an average of 50 h per week. The standard deviation is 16 hours. What percentage of pediatricians work less than 18 h per week?

26. HLTH/N In a sample of 80 pediatricians, how many work less than 18 h per week if the mean is 50 h per week and the standard deviation is 16 h?

27. HLTH/N Research has documented that the mean brain weight of people with Alzheimer's disease is 1,076.8 g and the standard deviation is 105.8. What percent of patients have a brain weight greater than 1,288.4 g?

28. HLTH/N In a sample of 500 Alzheimer's patients, how many will have a brain weight less than 865.2 g?

6–4 | *Counting Techniques and Simple Probabilities*

Learning Outcomes
1 Count the number of ways objects in a set can be arranged.
2 Determine the probability of an event occurring if an activity is repeated over and over.

1 Count the Number of Ways Objects in a Set Can Be Arranged.

A **set** is a well-defined group of objects or **elements.** The numbers 2, 4, 6, 8, and 10 can be a set of even numbers between 1 and 12. Women, men, and children can be a set of people. A, B, and C can be a set of the first three capital letters in the alphabet.

   Counting, in this section, means determining all the possible ways the elements in a set can be arranged. One way to count is to *list* all possible arrangements and then count the number of arrangements. In a single arrangement, all elements are included and no element appears more than one time. We refer to this process as arranging **without replacement** or **without repetition.**

EXAMPLE List and count the ways the elements in the set *A, B,* and *C* can be arranged.

A is first, 2 choices	*B* is first, 2 choices	*C* is first, 2 choices
A BC	*B AC*	*C AB*
A CB	*B CA*	*C BA*

Therefore, *A, B,* and *C* can be arranged in six ways. Each of these arrangements can also be called a set.

If more than three elements are in the set, the procedure becomes more challenging. It may be helpful to use a **tree diagram,** which allows each new set of possibilities to branch out from a previous possibility.

To make a tree diagram of possible arrangements of items in a set:

1. List the choices for putting an item in the first slot.
2. From each choice, list the remaining choices for the second slot (first branch).
3. From the end of each branch, make a branch to the remaining choices for the third slot.
4. Continue the process until there are no remaining choices.
5. To list an arrangement, start with the first slot and follow a branch to its end and record the choices.
6. To count the total number of possible arrangements, count the number of branch ends in the last slot.

EXAMPLE Make a tree diagram and count the number of ways the elements in the set containing letters *W*, *X*, *Y*, and *Z* can be arranged without repetition.

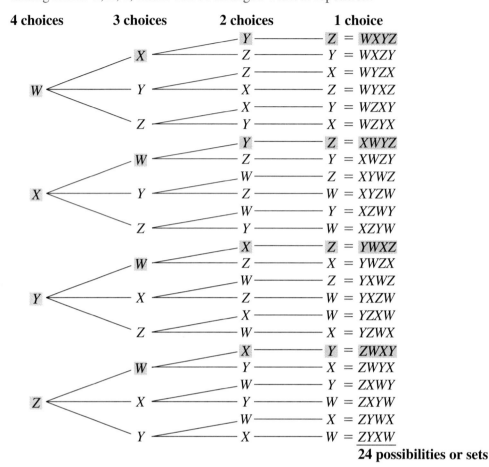

24 possibilities or sets

As evident in the previous example, the greater the number of elements in a set, the greater the complexity and the time required to list all possible arrangements. We can use logic and common sense to obtain a count of the possible arrangements of a set of elements.

There are four possibilities for the first letter: *W, X, Y,* or *Z*. For each of the four possible first letters, we have three choices for second letters. Now, we have 4 × 3 or 12 possibilities. For each of the 12 possibilities two choices remain for the third letter: 12 × 2 = 24. Then, for each of the 24 three-letter combinations, only one choice is left: 24 × 1 = 24. So, we have a total of 24 possibilities.

Another way to visualize this is to think of drawing letters from a bag or bowl (Fig. 6–20).

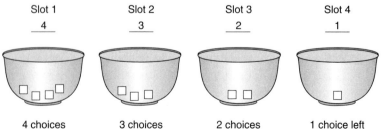

Figure 6–20

By multiplying the number of possible choices for each position, we can determine the total number of possibilities without listing them: $4 \cdot 3 \cdot 2 \cdot 1 = 24$.

To determine the number of choices for arranging a specified number of items without repetition:

1. Determine the number of slots to be filled.
2. Determine the number of choices for each slot.
3. Multiply the numbers from Step 2.

Next, we will look at a situation that allows repetition.

EXAMPLE   A coin is tossed three times. With each toss, the coin falls heads up or tails up. How many possible outcomes of heads and tails are there with three tosses of the coin?

There are three tosses. Each toss has only two possibilities, heads or tails; that is,

1st toss	2nd toss	3rd toss
2 possibilities	2 possibilities	2 possibilities

By multiplying the number of possible outcomes for each toss, we get $2 \cdot 2 \cdot 2 = 8$.

So there are 8 possible outcomes.

Check: 1st toss   2nd toss   3rd toss

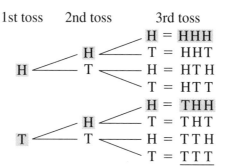

This tree diagram lists the possible outcomes of three tosses.

8 possibilities or sets

To Repeat or Not Repeat

In some situations, as in arranging *W, X, Y,* and *Z,* once a choice or selection is made, that choice cannot be repeated. In these situations the number of choices decreases with each selection.

Number of possible arrangements of *W, X, Y, Z* = $4 \cdot 3 \cdot 2 \cdot 1 = 24$

In other situations, as in tossing a coin, every coin toss can result in a head or a tail. The result of any coin toss can repeat the result of a previous toss.

Number of possible results of three coin tosses = $2 \cdot 2 \cdot 2 = 2^3 = 8$

When counting the number of possible outcomes, *first determine if repeats are allowed or not.*

EXAMPLE Henry has three ties: a red tie, a blue tie, and a green tie. He also has three shirts: a white shirt, a pink shirt, and a yellow shirt. How many sets of shirts and ties are possible?

If we start with the shirts, there are three possibilities (white, pink, and yellow). For each shirt, there are three possible ties (red, blue, and green). **So we have 3 · 3 = 9 possible outcomes.**

EXAMPLE Given the digits 1, 2, 3, 4, 5, and 6, how many three-digit numbers can be made without repeating a digit?

The numbers are to contain three digits, so there are three positions to fill. We have six digits to work with. The first digit can be one of six. For each of these six digits, there are five possible second digits. For each of the five second digits, there are four possible third digits.

Positions: 1st   2nd   3rd
Possibilities: 6   5   4

By multiplying the possibilities for each position, we get 6 · 5 · 4 = **120 possible outcomes or ways to make a three-digit number.**

2 Determine the Probability of an Event Occurring if an Activity Is Repeated Over and Over.

Probability means the chance of an event occurring if an activity is repeated over and over. The probability of an event occurring is expressed as a ratio or a percent.

Weather forecasters use percents when they forecast a 60% chance of rain or a 20% chance of snow. This text will use ratios like $\frac{3}{5}$ for 3 chances out of 5 or $\frac{2}{3}$ for 2 chances out of 3. The decimal equivalent of a ratio can also be used to express a probability: $\frac{3}{5} = 0.6$.

When a coin is tossed, two outcomes are possible, heads or tails. But only one side will be up. The probability of tossing heads is 1 out of 2, $\frac{1}{2}$, 0.5, or 50%.

To express the probability of an event occurring successfully:

1. Determine the total number of elements in the set (total possible outcomes).
2. Determine the number of elements that are defined as successful.
3. Make a ratio with the number of choices that are defined as successful divided by the total number of elements in the set.

$$\frac{\text{number of successful possibilities}}{\text{number of total possibilities}}$$

4. Express the ratio in lowest terms. The ratio can be expressed as a decimal or percent if desired.

The probability of an event occurring ranges from 0 for an impossible event to 1 for a certain event.

EXAMPLE

When a die is rolled, what is the probability that a 3 will appear on the top face (Fig. 6–21)?

Figure 6–21

A die has six sides where dots represent 1–6. Each side is an element in the set of six sides, so the set has a total of six elements.
Only one element is a 3.

The probability of rolling a 3 is $\frac{1}{6}$.

EXAMPLE
BUS

A holiday gift shopper wrapped eight men's ties in separate boxes. There were two solid-color ties and six striped ties. If the gift boxes were given at random to eight men, what is the probability of a man receiving a solid-color tie?

eight ties   Total possibilities
two solid-color ties   Successful possibilities

$$\text{probability} = \frac{\text{successful possibilities}}{\text{total possibilities}} = \frac{2}{8} = \frac{1}{4}$$

The probability of receiving a solid-color tie is $\frac{1}{4}$ or 0.25 or 25%.

EXAMPLE
HLTH/N

A practical nurse has a box of 144 syringes in individual sterile packets. Three of the syringes have torn packets. What is the probability that the first syringe picked will have a torn packet? If the packet for the first syringe selected is torn, what is the probability of picking a second syringe with a torn packet?

On the first pick, the probability of picking one syringe with a torn packet is $\frac{3}{144}$, which reduces to $\frac{1}{48}$. We assume that the first pick was a syringe with a torn packet, so this leaves a total of 143 syringes and now only 2 have torn packets. **On the second pick, the probability of picking a syringe with a torn packet is $\frac{2}{143}$.**

SECTION 6–4 SELF-STUDY EXERCISES

1 Complete the exercises using counting techniques.

1. List and count all the possible arrangements for Keaton, Brienne, and Renee to be seated in three adjacent seats at a basketball game.

2. List and count all the possible outcomes for arranging books A, B, C, and D on a shelf.

3. **BUS** Count all the possible outcomes for Jim Riddle to arrange a T-shirt, a sport shirt, a dress shirt, and a sweater on a shelf for display.

4. **HLTH/N** How many outcomes are possible for arranging containers of cotton balls, gauze pads, swabs, tongue depressors, and adhesive tape in a row on a shelf in a doctor's examining room?

5. **AG/H** A landscaping process involves five steps. The steps can be arranged in any order. The landscaping company efficiency officer wants to determine the most efficient order for the five steps. How many arrangements of steps are possible?

6. **AG/H** A farmer has three plots of land and intends to plant corn, beans, tomatoes, rice, or cotton. How many ways can he arrange his plantings without repeating any crop?

7. A drawing will be held to award door prizes. If the names of 24 people (no repeated names) are in the pool for the drawing, what is the probability that David's name will be pulled at random for the first prize? If his name is pulled and not replaced, what is the probability that Gaynell's name will be pulled next?

8. A box of greeting cards contains 20 friendship cards, 10 get-well cards, and 10 congratulations cards. What is the probability of picking a get-well card at random?

9. Mimi tosses 21 pennies, 16 nickels, and 11 dimes into a container. What is the probability of reaching in and picking a dime?

10. A jar holds 10 lock washers and 15 flat washers. What is the probability of drawing a lock washer at random?

11. A TV quiz program puts all questions in a box. If the box contains five hard questions, five average questions, and five easy questions, what is the probability of being asked an easy question?

12. When a single die is rolled, what is the probability that a 1 will appear?

CHAPTER REVIEW OF KEY CONCEPTS

Learning Outcomes

Section 6–1

1 Read circle graphs (pp. 258–259).

What to Remember with Examples

A circle graph compares parts to a whole.

If the total monthly revenue at a used car dealership is $75,000, what is the revenue from trucks? (See Fig. 6–22.)

35% of 75,000 =
0.35 × 75,000 = $26,250

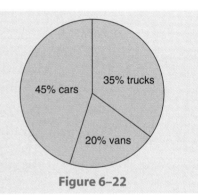

45% cars   35% trucks   20% vans

Figure 6–22

2 Read bar graphs (p. 259).

Bar graphs are used to compare values to each other.

What is the ratio of men's salaries to women's salaries in the pants department? (See Fig. 6–23.)

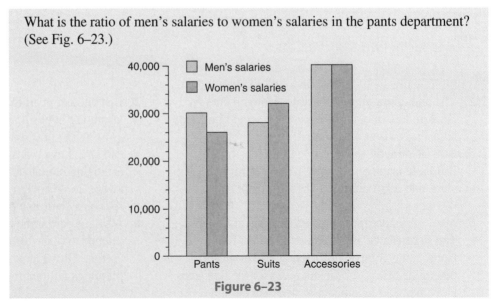

Figure 6–23

Men's salary in the pants department = $30,000
Women's salary in the pants department = $26,000
Ratio of men's salaries to women's salaries = $\frac{30}{26} = \frac{15}{13}$

3 Read line graphs
(pp. 259–260).

A line graph shows how an item changes with time.

Use the line graph of the average prices for textbooks to find the year when textbooks averaged $67 per book. (See Fig. 6–24.)

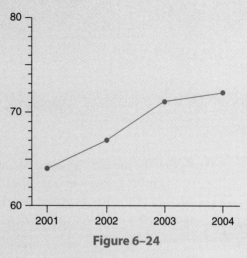

Figure 6–24

In 2002, the average price of textbooks was $67.

Section 6–2

1 Find the arithmetic mean or arithmetic average (pp. 264–267).

To find the mean, add the values; divide by the number of values.

Find the arithmetic mean of the test scores: 76, 86, 93, 87, 68, 76, 88

$$76 + 86 + 93 + 87 + 68 + 76 + 88 = 574$$

$$574 \div 7 = 82$$

The mean is 82.

2 Find the median and the mode (pp. 267–269).

The median of an odd number of values is the middle value when the values are arranged in order of size. For an even number of values, average the two middle values.

The mode is the value that occurs most frequently. A set of values may have no mode or more than one mode.

Find the median and mode of the set of test scores:

68, 76, 76, 86, 87, 88, 93   Already arranged in order.
     ↑ ↑

The median is 86.   Middle value.
The mode is 76.   Most frequent value.

3 Make and interpret a frequency distribution (pp. 269–270).

To make a frequency distribution, determine the appropriate interval for classifying the data. Tally the data.

Make a frequency distribution with the following data, indicating leave days for State College employees. (See Table 6–9.)

2	2	4	4	4	5	5	6	6	8
8	8	9	12	12	12	14	15	20	20

Table 6–9   Annual Leave Days of 20 State College Employees

Class Interval	Midpoint	Tally	Class Frequency
16–20	18	//	2
11–15	13	###	5
6–10	8	### /	6
1–5	3	### //	7

4 Find the mean of grouped data (pp. 270–271).

To find the mean of grouped data: **1.** Make a frequency distribution. **2.** Find the product (xf) of the midpoint (x) of the interval and the frequency (f) for the interval. **3.** Find the sum of the frequencies (Σf). **4.** Find the sum of the products (Σxf). **5.** Divide the sum of the products by the sum of the frequencies.

Symbolically, $\text{mean}_{\text{grouped}} = \dfrac{\Sigma xf}{\Sigma f}$.

Find the mean of the grouped data in the frequency distribution in Table 6–10.

Table 6–10   Annual Leave Days of 20 State College Employees

Class Interval	Midpoint x	Frequency f	Product xf
16–20	18	2	36
11–15	13	5	65
6–10	8	6	48
1–5	3	7	21
Total		20	170

$$\Sigma f = 20 \qquad \text{Add the frequencies.}$$

$$\Sigma xf = 170 \qquad \text{Add the products.}$$

$$\text{mean}_{\text{grouped}} = \frac{\Sigma xf}{\Sigma f} \qquad \text{Substitute.}$$

$$\text{mean}_{\text{grouped}} = \frac{170}{20} \qquad \text{Divide.}$$

$$\text{mean}_{\text{grouped}} = 8.5$$

Section 6–3

1 Find the range (pp. 274–275).

The range for a set of data is the difference between the largest value and the smallest value.

Chapter 6 / Statistics

Find the range of the set of test scores:

$$68, 76, 76, 86, 87, 88, 93$$

range $= 93 - 68 = 25$

---

2 Find the standard deviation (pp. 275–279).

To find the standard deviation from the mean: **1.** Find the mean, $\bar{x}$. **2.** Find the deviation of each value from the mean: $(x_i - \bar{x})$. **3.** Square each deviation: $(x_i - \bar{x})^2$. **4.** Find the sum of the squared deviations: $\Sigma(x_i - \bar{x})^2$. **5.** Divide the sum of the squared deviations by one less than the number of values in the data set. This is called the sample variance, $v = \dfrac{\Sigma(x_i - \bar{x})^2}{n - 1}$. **6.** Find the sample standard deviation by taking the square root of the variance.

$$s = \sqrt{\frac{\Sigma(x_i - \bar{x})^2}{n - 1}}$$

Find the standard deviation of the test scores: 68, 76, 76, 86, 87, 88, 93

$$\bar{x} = \frac{68 + 76 + 76 + 86 + 87 + 88 + 93}{7} = \frac{574}{7} = 82$$

$x_i - \bar{x}$	$(x_i - \bar{x})^2$
$68 - 82 = -14$	196
$76 - 82 = -6$	36
$76 - 82 = -6$	36
$86 - 82 = 4$	16
$87 - 82 = 5$	25
$88 - 82 = 6$	36
$93 - 82 = 11$	121

$$\Sigma(x_i - \bar{x})^2 = 466$$

$$v = \frac{\Sigma(x_i - \bar{x})^2}{n - 1} = \frac{466}{6} = 77.66666667$$

$$s = \sqrt{\Sigma\frac{(x_i - \bar{x})^2}{n - 1}} = \sqrt{77.66666667} = 8.812869378 = 8.8 \quad \text{Round.}$$

Section 6–4

1 Count the number of ways objects in a set can be arranged (pp. 281–284).

Multiply the number of choices for each position in the arrangement.

Renee Smith's closet has two new blazers (navy and red) and four new skirts (gray, black, tan, and brown). How many outfits can she make from the new clothes?

$$1 \text{ blazer} + 1 \text{ skirt} = 1 \text{ outfit}$$

$$\begin{pmatrix} \text{blazer} \\ \text{choices} \end{pmatrix} \cdot \begin{pmatrix} \text{skirt} \\ \text{choices} \end{pmatrix} = \begin{pmatrix} \text{possible} \\ \text{outfits} \end{pmatrix}$$

$$2 \quad \cdot \quad 4 \quad = \quad 8$$

2 Determine the probability of an event occurring if an activity is repeated over and over (pp. 284–285).

The probability of an event occurring is the ratio of the number of possible successful outcomes to the number of possible outcomes.

> A box in a doctor's office contains thirty $\frac{1}{2}$-in. adhesive strips and seventy $\frac{3}{4}$-in. adhesive strips. What is the probability of picking a $\frac{3}{4}$-in. adhesive strip at random?
>
> $$\frac{70}{100} = \frac{7}{10}$$ Total items from which to select = 30 + 70 = 100
>
> The probability of picking a $\frac{3}{4}$-in. adhesive strip is $\frac{7}{10}$ or 0.7 or 70%.

CHAPTER REVIEW EXERCISES

Section 6–1

Use Fig. 6–25 to answer Exercises 1–4.

1. In what year(s) did women use more sick days than men?

2. In what year(s) did men use about five sick days?

3. In what year(s) did men use more sick days than women?

4. What was the greatest number of sick days for men?

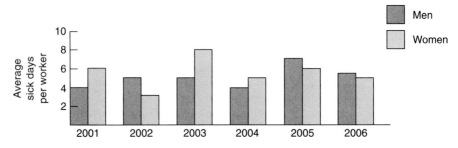

Figure 6–25   Comparison of sick days for men and women.

Use Fig. 6–26 to answer Exercises 5–7.

5. **CON**   What percent of the total cost is the cost of the lot? Round to the nearest tenth.

6. **CON**   What percent of the total cost is the cost of the house? Round to the nearest tenth.

7. **CON**   The cost of the lot and landscaping represents what percent of the total cost? Round to the nearest tenth.

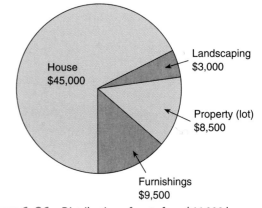

Figure 6–26   Distribution of costs for a $66,000 home.

Use Fig. 6–27 to answer Exercises 8–9.

8. **HLTH/N** What was the patient's highest pulse rate? The highest respiration rate?

9. **HLTH/N** On what date and at what time were the patient's pulse rate and respiration rate at their highest level?

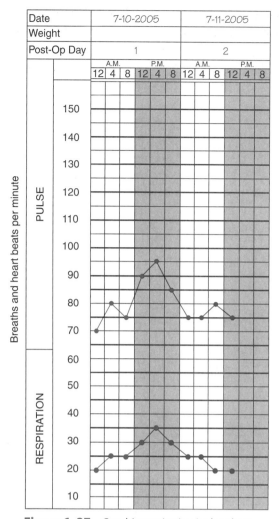

Figure 6–27 Graphic respiration/pulse chart.

Use Fig. 6–28 to answer Exercises 10–13.

10. **AG/H** The cranberry crop forecast for Oregon is 27% above the 2002 actual production. What was the actual production in 2002? Round to the nearest thousand barrels.

11. **AG/H** What is the total U.S. production in barrels of cranberries forecast for 2004?

12. **AG/H** Which state produces the greatest percentage of cranberries in the United States?

13. **AG/H** Massachusetts produces what percentage of U.S. cranberries? Round to tenths.

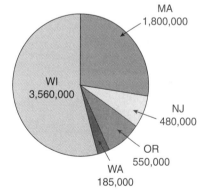

Figure 6–28 U.S. cranberry 2004 forecasted production by state in barrels. *Source: University of Kentucky Cooperative Extension Service, 2005.*

Use Fig. 6–29 to answer Exercises 14–17.

14. **AG/H** Use the graph to discuss trends in bearing acres used in U.S. orange production from 1994 to 2004.
15. **AG/H** In what year did the number of bearing acres used in U.S. orange production peak?
16. **AG/H** For which years was the number of bearing acres for U.S. orange production greater than 800,000 acres?
17. **AG/H** Identify the year that had the greatest increase in bearing acres for U.S. orange production.

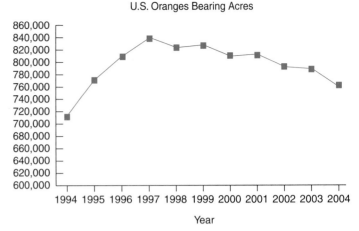

Figure 6–29 U.S. oranges: Bearing acres. *Source: Agricultural Statistics Board NASS, USDA, 2005.*

Use Fig. 6–30 to answer Exercises 18–21.

18. **AG/H** In what year was utilized production of U.S. oranges the greatest?
19. **AG/H** What was the percent increase in orange production from 2003 to 2004? Round to the nearest whole percent.
20. **AG/H** What was the percent of decrease in orange production from 1998 to 1999? Round to the nearest whole percent.
21. **AG/H** In what year were 11,545,000 tons of oranges produced in the United States?

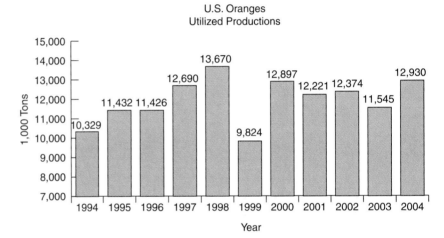

Figure 6–30 U.S. oranges: Utilized production. *Source: Agricultural Statistics Board NASS, USDA, 2005.*

Section 6–2

Solve the problems.

22. Jim Smith made 176 baskets over a 16-game period. What was the average number of baskets per game to the nearest tenth?
23. **BUS** Lee Vance sold 63 new cars over a 5-month period. What was the average number of cars sold per month to the nearest tenth?
24. **AUTO** An automobile used 22 gal of regular gasoline on a trip of 358 mi. What was the average miles per gallon to the nearest tenth?
25. **AUTO** A delivery truck used 21 gal of regular gasoline on a trip of 289 mi. What was the average miles per gallon to the nearest tenth?
26. **BUS** These hourly pay rates are used at fast-food restaurants: cooks, $8.25; servers, $8.95; bussers, $7.20; dishwashers, $7.20; managers, $10.25. Find the median pay rate.
27. **BUS** These hourly pay rates are used at a locally owned store: office assistants, $9.85; bookkeepers, $10.20; cashiers, $9.45; assistant managers, $11.90. Find the median pay rate.
28. **BUS** These weekend work shifts are in effect in a mall clothing store: 3 hours in A.M., 6 hours in P.M., 3 hours in P.M. Find the mode.
29. **BUS** These special prices are in effect at a fast-food restaurant: $1.85, hamburgers; $1.98, hot ham sandwiches; $2.28, chicken sandwiches; $1.85, roast beef sandwiches. Find the mode.

30. Natalie Bradley, a student at a technical college, earned the final grades listed in Table 6–11 for the past term. Find Natalie's QPA (quality point average) to the nearest hundredth.

Table 6–11 **Grade Distribution**

Subject	Grade	Hours	Quality Points per Hour
Electronics 101	A	4	4
Circuits 201	A	4	4
Algebra 101	B	4	3

31. Tami Murphy earned the final grades listed in Table 6–12 for the past term. Find Tami's QPA (quality point average) to the nearest hundredth.

Table 6–12 **Grade Distribution**

Subject	Grade	Hours	Quality Points per Hour
English 201	A	3	4
History 202	B	3	3
Philosophy 101	A	3	4

32. Janice Van Dyke has these grades in Algebra 102: 98, 82, 87, 72, and 82. What minimum grade does she need on the last test to have a B or 85 average?

33. Sarah Smith has the following grades in American History: 99, 93, 91, 88, and 86. What grade does she need on the last test to get an A or 90 average?

A regional horticultural association is composed of 54 members of several clubs. Use the format shown in Table 6–13 and make a frequency distribution of the members' ages: 17, 18, 20, 21, 21, 24, 24, 29, 29, 29, 31, 31, 33, 33, 34, 35, 35, 38, 38, 38, 39, 41, 42, 43, 43, 43, 43, 45, 45, 47, 47, 48, 48, 48, 49, 50, 51, 51, 52, 56, 56, 58, 58, 60, 60, 62, 64, 64, 65, 66, 68, 70, 71, 71.

Table 6–13 **Ages of Club Members of Regional Horticultural Association**

	Class Interval	Midpoint	Tally	Class Frequency
34.	66–75	_____	____	_____
35.	56–65	_____	____	_____
36.	46–55	_____	____	_____
37.	36–45	_____	____	_____
38.	26–35	_____	____	_____
39.	16–25	_____	____	_____

Answer Exercises 40–47 based on the information in Table 6–13.

40. In what age group is the least number of members?

41. In what age group is the greatest number of members?

42. How many members are under age 36?

43. How many members are over age 55?

44. What is the ratio of the number of members aged 66–75 to members aged 16–25?

45. What is the ratio of the number of members aged 66–75 to the number of members aged 46–55?

46. What is the ratio of the number of members aged 46–55 to the members aged 16–25?

47. What percent (to the nearest tenth of a percent) of the total number of members are the members aged 46–55?

48. Use seven class intervals beginning with 66–70 to make a frequency distribution of these scores on an English language test: 68, 70, 74, 77, 78, 82, 82, 84, 86, 86, 86, 88, 89, 90, 90, 93.

49. Use three class intervals beginning with 20–24 as the first interval to make a frequency distribution of these miles per gallon reported in one week by customers to an automotive rental company for six-cylinder cars: 22, 22, 23, 23, 23, 24, 24, 25, 26, 26, 27, 29, 29, 30, 30, 31, 31.

50. Find the grouped mean for the data in Table 6–13. Round to tenths.

51. Find the grouped mean for the data in Exercise 49. Round to tenths.

Section 6–3

52. Find the range of the test scores: 67, 87, 76, 89, 70, 69, 82. Round to hundredths.

53. Find the range for Marcus Johnson's test scores of 92, 83, 39, 98, 88, 90.

54. Find the standard deviation of the test scores in Exercise 52. Round to hundredths.

55. Find the standard deviation of the test scores in Exercise 53. Round to hundredths.

56. **HLTH/N** The heights of 25-year-old men are normally distributed with a mean of 1.72 m and a standard deviation of 0.27 m. What percent of men have heights between 1.45 m and 2.26 m?

57. **HLTH/N** In a sample of 1,000 25-year-old men, how many will be taller than 1.45 m? The population mean is 1.72 m and standard deviation is 0.27 m.

58. **HLTH/N** Juice sold in 12-oz containers has a mean of 11.9 oz and a standard deviation of 0.2 oz. What percent of containers hold 11.7 oz to 12.1 oz?

59. **HLTH/N** Refer to Exercise 58 to determine how many containers in a batch of 3,000 containers of juice are filled to 12.1 oz or more?

Section 6–4

60. For her first professional job, Martha Deskin, a recent college graduate, purchased four pairs of shoes, three business suits, and five blouses. How many outfits are possible?

61. **HLTH/N** There are three magazines on horses and ten magazines on fashion in Dr. Nelson Company's waiting room. If Shirley Riddle sends her toddler to get two magazines, what is the probability that he will randomly pick a fashion magazine on the first draw? If he is successful, what is the probability of his picking a fashion magazine on the second draw?

62. Two coins are tossed three times. Count the number of possible outcomes of heads and tails with three tosses of the two coins.

63. If four coins are tossed, what is the number of possible outcomes of heads and tails? *Hint:* HHHT and HTHH are different outcomes.

64. Cy Pipkin is puzzled over a true-false question on a test and does not know the answer. What is the probability that he will pick the right answer by chance?

TEAM PROBLEM-SOLVING EXERCISES

1. Statistical reports are an important part of a statistical study. Conduct a study on a topic of interest to the team and prepare a statistical report that includes the following components:
  (a) Question or questions that you are attempting to answer
  (b) Description of the process for collecting data
  (c) Summary of the data with appropriate statistics
  (d) Conclusion

2. Most statistical reports include graphical representations of the data collected. Computer software such as Excel™ allows you to prepare graphs or charts from the data entered into a spreadsheet.
  (a) Prepare spreadsheets with appropriate formulas to make the necessary calculations for the data in Exercise 1.
  (b) Prepare charts to illustrate the data collected in Exercise 1.

1. What type of information does a circle graph show?

2. Describe a situation where it would be appropriate to organize the data in a circle graph.

3. What type of information does a bar graph show?

4. Describe a situation where it would be appropriate to organize the data in a bar graph.

5. What type of information does a line graph show?

6. Describe a situation where it would be appropriate to organize the data in a line graph.

7. Explain why the grouped mean may not be the same as the mean for the data set.

8. Explain the differences among the three types of averages (measures of central tendency): the mean, the median, and the mode.

9. Use an example to show how using a tree diagram for counting gives the same result as multiplying the number of choices for each position in the set.

10. If there are five red marbles and seven blue marbles in a bag, is the probability of drawing a red marble $\frac{5}{7}$? Explain your answer.

1. A _____ graph shows how different data values relate to each other.

2. A _____ graph shows how a whole quantity is related to its parts.

3. A _____ graph shows how an item or items change over time.

Answer Questions 4 and 5 from the line graphs in Fig. 6–31.

4. How many degrees warmer was it indoors at midnight than outdoors?

5. What was the change in outdoor temperature between 11:00 A.M. and noon?

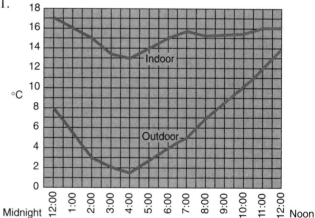

Figure 6–31 Indoor-outdoor temperature.

Answer the questions about the manufacturing costs for the electronic game as shown in the circle graph in Fig. 6–32.

6. What percent of the total cost is materials?

7. What would the profit be if there were no tariff on imported parts?

8. How much are overhead and materials together?

9. What percent of the total cost is the profit?

10. What is the ratio of labor to total cost?

11. What is the ratio of the tariff to total cost?

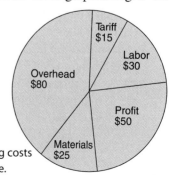

Figure 6–32 Manufacturing costs of a $200 electronic game.

Use the bar graph in Fig. 6–33 to answer Questions 12–15 about the academic-year starting salaries of women and men college professors in various academic departments of a college.

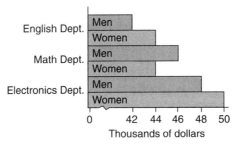

Figure 6–33   Salaries of women and men college professors.

12. In what department(s) do men make more than women?

13. In what departments do women make more than men?

14. What percent of women's salaries are men's salaries in the English Department (to the nearest whole percent)?

15. What percent of men's salaries are women's salaries in the Electronics Department (to the nearest whole percent)?

Use the frequency distribution shown in Table 6–14 to answer Questions 16–18. The distribution shows the number of correct answers on a 30-question test in a science class.

16. How many students scored more than 25 correct?

17. What is the ratio of students scoring 6–10 correct to those scoring 21–25 correct?

18. Find the grouped mean for the data in Table 6–14.

19. Find the range, mean, median, and mode for these test scores: 81, 78, 69, 75, 81, 93, 68. Round to tenths.

20. Find the standard deviation of the data: 77, 87, 77, 89, 70, 69, 82.

21. Find the standard deviation of the data in Exercise 19. Round to hundredths.

Table 6–14	Frequency Distribution of Correct Answers		
Class Interval	**Midpoint**	**Tally**	**Class Frequency**
26–30	28	///	3
21–25	23	////	4
16–20	18	✝✝✝ //	7
11–15	13	///	3
6–10	8	//	2
1–5	3	/	1

22. List and count the ways the elements in the set *L, M, N,* and *O* can be arranged.

23. Rayford has two ties: a red tie and a blue tie. He also has three shirts: a white shirt, a green shirt, and a yellow shirt. How many combinations of shirts and ties are possible?

24. Millie has in her makeup bag three green eye shadows, four white eye shadows, and two black eye shadows. What is the probability of her pulling out a black eye shadow?

25. An envelope contains the names of two men and three women to be interviewed for a promotion at Washington's Landscape Service. The interviewer draws names to determine the order of the interviews. What is the probability of drawing a woman's name first?

26. If a small boy has one red marble and three yellow marbles in his pocket, what is the probability of his pulling out the red one on the first try?

27. **HLTH/N**   The average hospital cost of an angioplasty/ stint surgery for heart attack is $57,417 with a standard deviation of $3,584. What percent of hospital stays cost between $53,833 and $57,417 if the data are normally distrubuted?

28. **HLTH/N**   Use the information in Exercise 27 to find in 562 angioplasty/stint procedures how many cost between $57,417 and $61,001? Round to the nearest whole number.

Change to the indicated unit.

1. 14.3 dm = _____ m
2. 12 m = _____ cm
3. 30,002 dg = _____ kg
4. 0.159 dkm = _____ dm
5. 86°F = _____ °C
6. 45°C = _____ °F
7. What is the greatest possible error of the measurement $3\frac{1}{4}$ cm?
8. How many significant digits are in 40.240?
9. How many significant digits are in 0.00356?
10. A scale calibrated to 10 grams actually reads 10.05 grams. What is the percent error?

11. Add and write the sum using appropriate precision:
5.7 cm + 3.05 cm + 21.46 cm

Perform the indicated operations and simplify.

12. $5 + (-2) + (-8)$
13. $-3 - 8 - 5 - (-2)$
14. $(-8)(2)(-4)$
15. $-24 \div 4$
16. $(-2)^3 - 5 + (-6) - 2^2$
17. $-5\frac{1}{2} + 4\frac{1}{4}$

18. Write 58,000 in scientific notation.
19. Write 5.3×10^{-2} as an ordinary number.
20. Write 4.7×10^4 as an ordinary number.
21. Write 0.0035 in scientific notation.
22. Write 42,300,000 in engineering notation.
23. Write 0.0001457 in engineering notation.

Simplify and write the result in scientific notation.

24. $(5.1 \times 10^2)(4.1 \times 10^3)$
25. $\dfrac{4.6 \times 10^4}{2.3 \times 10^{-2}}$
26. $(4 \times 10^2)^3$

Use Figure 6–34 to answer questions 27–29.

27. How many cwts (100 lbs) of long grain rice are produced by Arkansas?
28. Which state produces the most medium grain rice?
29. Which two states produce only long grain rice?

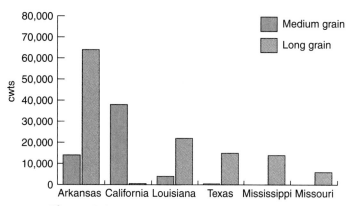

Figure 6–34 Top Six Rice Producing States in U.S.

Source: http://www.sagevfoods.com/MainPages/Rice101/Production.htm

Use Figure 6–35 to answer questions 30–32.

30. If 560 million metric tons of rice were grown worldwide, how many metric tons were produced by India?
31. Which country produces more rice than any other?
32. How many metric tons of the world rice production are produced by the USA?

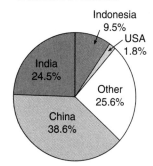

Figure 6–35 Source: http://www.sagevfoods.com/MainPages/Rice101/Production.htm

The data represent the number of miles driven each day for a cross country moving van. 583, 417, 456, 498, 523, 456.

33. Find the mean, median, and mode for the data set.

34. Find the range and standard deviation for the data set.

35. A container has 30 assorted coins. Seventeen are quarters, 8 are nickels, 3 are pennies, and 2 are dimes. What is the probability that one coin selected from the container is a nickel is each coin is equally likely to be selected?

7

Linear Equations

Focus on Careers

The hospitality industry provided more than 1.8 million jobs in 2002. Some of these were self-employed workers who run bed-and-breakfast inns and provide other services. Most employment is available in densely populated cities or resort areas and most employers have more than 100 employees.

There are many careers in this service sector. Hotel manager, lodging manager, food service manager, sales manager, purchasing manager, and executive chef are just a few careers found in the hospitality industry.

Skills and experience needed by workers vary widely and almost all workers undergo on-the-job training. About 200 community colleges offer programs in hotel and restaurant management. Graduates of these programs, especially those with good communication skills, have a better-than-average opportunity for entry and advancement in these careers.

There is a wide range of pay in the industry. Nonsupervisory workers earned $10.01 an hour on average, but hourly earnings for restaurant cooks and maintenance and repair workers were $10.48 on average. Salaries of managers are usually higher, and managers may earn bonuses ranging up to 20% of their basic salaries.

Employment in this industry is expected to increase by about 17% for the 2002–2012 period compared with a 16% growth rate projected for all industries combined.

Source: *Occupational Outlook Handbook,* 2004–2005 Edition, U.S. Department of Labor, Bureau of Labor Statistics.

Learning Outcomes
1 Identify equations, terms, factors, constants, variables, and coefficients.
2 Write verbal interpretations of symbolic statements.
3 Translate verbal statements into symbolic statements using variables.
4 Simplify variable expressions.

1 **Identify Equations, Terms, Factors, Constants, Variables, and Coefficients.**

Before we work with equations to solve problems, we need to understand the basic concepts and terminology of equations. In mathematics, we use the symbol "=" to show that quantities are "equal to" each other. We write the statement "5 is equal to 2 plus 3" as $5 = 2 + 3$. This symbolic statement is called an *equation*.

An **equation** is a symbolic statement that two expressions or quantities are equal in value. An equation may be true or false.

To verify that an equation is true:

1. Find the value of the expression on each side of the equal sign.
2. Compare the values for each side.
   (a) If the values for each side are equal, the equation is true.
   (b) If the values for each side are not equal, the equation is not true.

EXAMPLE   Verify that the statements are true equations.

(a) $3(8) = 24$ (b) $12 - 3 = 2 + 7$

(a) $3(8) = 24$ 3 times 8 is 24.
         $24 = 24$ The equation is true.

(b) $12 - 3 = 2 + 7$ 12 minus 3 is 9. 2 plus 7 is 9.
            $9 = 9$ The equation is true.

Some types of equations contain a *missing* or *unknown* number. We **solve** an equation by finding the missing number that makes the equation *true*. We use letters such as x, a, or z to represent the missing number in the equation. These letters are called **variables.** In each equation, the letter has a certain but unknown value that depends on the other numbers and relationships in the equation.

EXAMPLE   Find the value of the variable that makes the equation true.

(a) $x = 3 + 8$ (b) $\dfrac{18}{3} = n$ (c) $y = 3(4) - 5$

(a) $x = 3 + 8$ 8 added to 3 is 11.

      $x = \mathbf{11}$

(b) $\dfrac{18}{3} = n$ 18 divided by 3 is 6.

      $\mathbf{6 = n}$

(c) $y = 3(4) - 5$ Multiply first. 3 times 4 is 12.

      $y = 12 - 5$ Add 12 and -5. 12 plus $-5 = 7$.

      $y = \mathbf{7}$

Because both sides of an equation are equivalent, either side can come first in the equation. The equation, $x = 3 + 8$ means the same as $3 + 8 = x$. Also $n = 6$ is equivalent to $6 = n$. The unknown, or variable, may appear on the left or the right side of an equation.

Symmetric property of equality:

If the sides of an equation are interchanged, equality is maintained.

Symbolically, if $a = b$, then $b = a$, where a and b are any real numbers.

In algebra we carefully distinguish between terms and factors. **Factors** are numbers or variables that are multiplied. **Terms** are algebraic expressions that are single quantities or products or quotients of quantities.

A term can be a single letter; a single number; the product of numbers and letters; the product of numbers, letters, and/or groupings; or the quotient of numbers, letters, and/or groupings. In a division, the fraction line implies that the numerator and/or the denominator is grouped. *Terms are separated by plus & minus signs*

Multiplication Notation Conventions

When a term is the product of letters alone (such as ab or xyz) or the product of a number and one or more letters (such as $3x$ or $2ab$), parentheses or other symbols of multiplication are usually omitted. Thus, ab means a times b, and $3x$ means 3 times x.

To identify terms:

1. Separate an expression into terms by identifying addition or subtraction signs that are not within a grouping.
2. The sign of each term is the sign that precedes the term.

$$3a + b \qquad 3(a + b) \qquad \frac{3}{a + b}$$

     two terms   one term   one term

EXAMPLE   Identify the terms in each expression by highlighting each term.

(a) $3x + 5$ (b) $2ab + 4a - b + 2(a + b)$

(c) $\dfrac{2a + 1}{3}$ (d) $\dfrac{2a}{3} + 1$

(a) $3x$ + 5

Terms are separated by "+" and "−" signs that are not within a grouping. The expression has two terms.

(b) $2ab$ + $4a$ − b + $2(a + b)$

The plus sign within the grouping does not separate terms. The expression has four terms.

(c) $\dfrac{2a + 1}{3}$

The fraction bar is a grouping symbol. The numerator is a grouping, $2a + 1$. The expression has one term.

(d) $\dfrac{2a}{3}$ + 1

The expression has two terms.

A term that contains only numbers is a **number term** or **constant.** A term that contains only one letter, several letters used as factors, or a combination of letters and numbers used as factors is a **letter term** or **variable term.**

The numerical factor of a term is the **numerical coefficient.** Unless otherwise specified, we will use **coefficient** to mean the *numerical coefficient.* The numerical factor should be written *in front* of the variable.

To identify the numerical coefficient of a term:

1. Since a term contains only factors, the coefficient is the numerical factor or the product of all numerical factors.
2. If the term is a fraction (an indicated division),
   (a) Write the numerical factor in the denominator as an equivalent multiplication.
   (b) Write the product of all numerical factors.
   (c) The numerical coefficient is the product from Step 2b.

EXAMPLE   Identify the numerical coefficient in each term.

(a) $2x$ (b) $-3ab$ (c) $4(x + 3)$ (d) $-\dfrac{n}{3}$ (e) $\dfrac{2b}{5}$

(a) The coefficient of $2x$ is **2.**
(b) The coefficient of $-3ab$ is **−3.**
(c) The coefficient of $4(x + 3)$ is **4.**

(d) $-\dfrac{n}{3}$ is the same as $-\dfrac{1}{3}n$, so the coefficient is $-\dfrac{1}{3}.$

(e) $\dfrac{2b}{5} = \dfrac{1}{5}(2b) = \dfrac{1}{5}(2)b = \dfrac{2}{5}b.$ The coefficient is $\dfrac{2}{5}.$

TIP   **Coefficient of 1 or −1**

When a variable term has no written numerical coefficient, the coefficient is understood to be 1, so $x = 1x$. Similarly, the numerical coefficient of $-x$ is -1, so $-x = -1x$.

2 Write Verbal Interpretations of Symbolic Statements.

In symbolic statements we commonly use letters to represent missing values. That is, we write $5 + x = 8$ or $5 + y = 8$ or $5 + a = 8$. The choice of letters is not usually important. The position of the letter and the other conditions of the statement are important. We can phrase a mathematical statement several ways: $5 + x = 8$ can be stated

"What number added to 5 gives 8?"

or

"5 plus a number has a result of 8. What is the missing number?"

To write verbal interpretations of symbolic statements:

1. Locate the variable or variable terms.
2. Examine the operations that link the factors and terms.
3. Write a verbal statement or statements that describe all the conditions of the symbolic statement.

EXAMPLE State the equations in words.

(a) $x - 7 = 4$ (b) $\dfrac{x}{5} = 3$ (c) $2x + 3 = 15$ (d) $2(x + 3) = 14$

Each equation can be stated several ways. One choice is given for each equation.

(a) **When 7 is subtracted from a number, the result is 4.**
(b) **A number divided by 5 is 3.**
(c) **15 is the result when 3 is added to 2 times a number.**
(d) **If the sum of a number and 3 is doubled, the result is 14.**

3 Translate Verbal Statements into Symbolic Statements Using Variables.

Symbolic representations of written statements or real-life situations allow us to determine the value of missing amounts more systematically.

To translate verbal statements into symbolic statements:

1. Assign a letter to represent the missing number.
2. Identify key words or phrases that imply or suggest specific operations.
3. Translate words into symbols.

EXAMPLE   Translate the statements into symbols.

(a) The sum of 12, 23, and a third number is 52.
The third number is missing.
Let x represent the third number.

$$12 + 23 + x = 52$$ Translate the entire statement. *Sum* indicates addition. *Is* translates to *equals*.

(b) When a number is subtracted from 45, the result is 17.
The number being subtracted (the subtrahend or second number) is missing.
Let x represent the missing number.

$$45 - x = 17$$ 45 is the minuend or first number, and *the result* translates to *equals*.

Key words can be examined to identify operations.

Addition: the sum of, plus, increased by, more than, added to, exceeds, longer, total, heavier, older, wider, taller, gain, greater than, more, expands
Subtraction: less than, decreased by, subtracted from, the difference between, diminished by, take away, reduced by, less, minus, shrinks, younger, lower, shorter, narrower, slower, loss
Multiplication: times, multiply, of, the product of, multiplied by
Division: divide, divided by, divided into, how big is each part, how many parts can be made from

Some words that imply multiplication or division may indicate a specific number in the multiplication or division. Examples include twice (2 times), double (2 times), triple (3 times), and half of (1/2 times or divided by 2).

EXAMPLE   Translate the statement into symbols.

How many shelves that are each 3 feet long can be made from a board that is 12 feet long?

The missing number is the number of shelves that can be made from one 12-ft board. Let x represent the number of shelves. The number of shelves, x, times the length of each shelf, 3, equals 12.

$$3x = 12$$

EXAMPLE   Explain the difference between the two statements.

Study!!

4 times the difference between a number and 8 is 24.
The difference between 4 times a number and 8 is 24.

4 times the difference between a number and 8 is 24. This statement is written symbolically as $4(x - 8) = 24$. It shows that the difference is taken first and the result is multiplied by 4.
    The difference between 4 times a number and 8 is 24. This statement is written symbolically as $4x - 8 = 24$. A number is multiplied by 4 and then 8 is subtracted from the result.

4 **Simplify Variable Expressions.**

Terms are **like terms** if they are numbers or if they are variable terms with exactly the same letter factors.

The terms $4y$ and $2y$ are like terms because both contain the same letter (y). Similarly, 3 and 1 are like terms because both are numbers. We *cannot* combine 3 and $4y$ because they are unlike terms (constant and variable term), and we *cannot* combine $4y$ and $2x$ because they are also unlike terms (different variables).

To simplify variable expressions:

1. Change subtraction to addition if appropriate.
2. Add the constants using the appropriate rule for adding signed numbers. The sum is a signed number.
3. Add the numerical coefficients of the like variables using the appropriate rule for adding signed numbers. The sum has the same variable or variables as the like terms being added.

EXAMPLE Simplify the expressions by combining like terms.

(a) $5a + 2a - a$ (b) $3x + 5y + 8 - 2x + y - 12$

(a) $5a + 2a - a = 6a$ All terms are like terms. Add coefficients of a:
                               $5 + 2 - 1 = 6$.

(b) $3x + 5y + 8 - 2x + y - 12$ Add like terms.

  $= x + 6y - 4$ $3x - 2x = x$
                               $5y + y = 6y$
                               $8 - 12 = -4$

The **distributive principle** can be extended to include multiplying numbers and variables. When an expression has an instance of the distributive principle, first remove the grouping symbol by multiplying by the factor or factors in front of the grouping.

EXAMPLE Simplify by applying the distributive principle.

(a) $3(5x - 2)$ (b) $-7(x + y - 3z)$ (c) $-(-3x + 4)$

(a) $3(5x - 2) = 3(5x) - 3(2)$ Apply the distributive principle
    $= 15x - 6$ by multiplying $(5x - 2)$ by 3.

(b) $-7(x + y - 3z) = -7(x) -7(y) -7(-3z)$ Apply the distributive principle by
    $= -7x - 7y + 21z$ multiplying each term of the
                                               grouping by -7.

(c) $-(-3x + 4) = -1(-3x) -1(+4)$ Apply the distributive principle by
    $= 3x - 4$ multiplying each term of the
                                       grouping by -1.

1 Verify that the statements are true.

1. $5(-3) = -15$

2. $17 - 6 = 11$

3. $12 - 5(2) = -7 + 9$

4. $8 - 3(5) = 2(-1 - 3) + 1$

5. $7(3 - 8) = 4(-8) - 3$

6. $4(-7) + 3(-2) = -5(8 + 2) - 4$

Find the value of the variable that makes the equation true.

7. $n = 4 + 7$

8. $m = 8 - 2$

9. $5 - 9 = y$

10. $\dfrac{12}{2} = x$

11. $p = 2(5) - 1$

12. $b = 6 - 3(5)$

Identify the terms in each expression by drawing a box around each term.

13. $7 + c$

14. $4a - 7$

15. $3x - 2(x + 3)$

16. $\dfrac{a}{3}$

17. $7xy + 3x - 4 + 2(x + y)$

18. $14x + 3$

19. $\dfrac{7}{(a + 5)}$

20. $\dfrac{4x}{7} + 5$

21. Write an algebraic expression that contains three terms.

Identify the numerical coefficient of each term.

22. $5x$

23. $-4xy$

24. $\dfrac{n}{5}$

25. $\dfrac{2a}{7}$

26. $6(x + y)$

27. $\dfrac{-4}{5x}$

28. $7(3x - 2y)$

29. $-(x - 3)$

30. Write an expression that has one term and a numerical coefficient of -15.

2 State the equations in words.

31. $x + 4 = 7$

32. $x - 5 = 2$

33. $3x = 15$

34. $3x + 1 = 7$

3 Write the statements in symbols.

35. 5 more than a number is 12.

36. A number divided by 6 is 9.

37. 4 times the difference of a number and 3 is 12.

38. 3 less than 4 times a number is 12.

39. The sum of 12, 7, and a third number is 17.

40. 7 more than twice a number is 21.

41. **INDTR** If the temperature rises $15°$, it will be $48°$. What is the temperature?

42. **HLTH/N** If 15 mL of water are added to a medicine, there are 45 mL in all. What is the volume of the medicine?

43. **INDTEC** How many 5-ft pieces of I-beam can be cut from a piece that is 45 ft long?

44. **CON** A piece of oak flooring that is 18 ft long has an unacceptable flaw that requires 3 ft to be trimmed from one end. How many 5-ft boards can be cut from the remaining length of flooring?

4 Simplify by combining like terms.

45. $3a - 7a + a$

46. $-8x + y - 3y$

47. $5x - 3y + 2x + y$

48. $-4a + b + 9 + a - b - 3$

49. $2x + 8 - x + 4 - x - 2$

50. $3a + 5b + 8c + 1 + b$

Simplify by applying the distributive principle and combining like terms if appropriate.

51. $5(2x - 4)$
52. $4 + 3(2x + 3)$ study
53. $3 - (4x - 2)$
54. $5 - 2(6a + 1)$
55. $-5(x - y + 2z)$
56. $5 - (2a - 3b + 7)$
57. $8(x + 2y) - 3(7x - 3y + 5)$
58. $-3(2x + 5y - 6) - (5y - 2)$

7–2 | Solving Linear Equations

Learning Outcomes

1 Solve linear equations using the addition axiom.
2 Solve linear equations using the multiplication axiom.
3 Solve linear equations with like terms on the same side of the equation.
4 Solve linear equations with like terms on opposite sides of the equation.

A **linear equation in one variable** is an equation in which the same letter is used in all variable terms and the exponent of the variable is 1.

1 Solve Linear Equations Using the Addition Axiom.

An equation is **solved** when the letter or variable is alone on one side of the equal sign. That is, the coefficient of the variable is +1 and no other terms are on the same side as the variable. The number on the side opposite the variable is called the **solution** or **root** of the equation. The solution is the number that makes the equation true.

The underlying principle for solving equations is the **basic principle of equality.**

Basic principle of equality:

To preserve equality, if an operation is performed on one side of an equation, the *same* operation must be performed on the other side.

We can illustrate this principle by visualizing a balanced scale. If we have a scale with 1 oz on one side and 1 oz on the other side, the scale is balanced. If we increase or decrease the weight on one side, we need to do the same on the other side or one side would be heavier than the other and the equality of both sides would be lost (see Fig. 7–1).

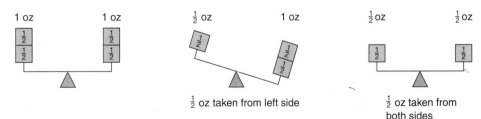

Figure 7–1

If $\frac{1}{2}$ oz is taken from both sides, the scale is still balanced. The purpose in solving an equation is to **isolate** the variable on one side of the equation. Terms can be moved from one side of an equation to the other using the addition axiom. The **addition**

axiom states that the same quantity can be added to both sides of an equation without changing the equality of the two sides. Symbolically, if $a = b$, then $a + c = b + c$ for all real numbers a, b, and c.

Since subtraction is the same as adding the opposite, the addition axiom applies to both addition and subtraction.

To solve a linear equation using the addition axiom:

1. Locate the variable in the equation.
2. Identify the constant that is associated with the variable by addition (or subtraction).
3. Add the opposite of the constant to both sides of the equation.

EXAMPLE Solve the equation $x + 4 = 8$.

$$x + 4 = 8$$ 4 is added to x. Add -4, the opposite of 4, to both sides (addition axiom).

$$x + 4 - 4 = \boxed{8 - 4}$$ $4 - 4 = 0$, $8 - 4 = 4$

$$x + 0 = \boxed{4}$$ $x + 0 = x$

$$x = \mathbf{4}$$

Adding Opposites and Zero

A number plus its opposite equals zero.

$$n + (-n) = 0, \text{ where } n \text{ is any real number}$$

Zero plus a number or variable leaves the number or variable unchanged.

$$n + 0 = n, \text{ where } n \text{ is any real number}$$

These two properties and the addition axiom allow us to mentally move a term to the other side of the equation as its opposite.

$x + 3 = 8$	becomes	$x = 8 - 3$	Mentally add -3 to both sides.
$x - 2 = 4$	becomes	$x = 4 + 2$	Mentally add $+2$ to both sides.
$5 + x = 2$	becomes	$x = 2 - 5$	Mentally add -5 to both sides.
$10 = x - 1$	becomes	$10 + 1 = x$	Mentally add $+1$ to both sides.

EXAMPLE Solve the equations.

(a) $x + 3 = 8$ **(b)** $x - 2 = 4$ **(c)** $5 + x = 2$ **(d)** $10 = x - 1$

(a) $x + 3 = 8$ To isolate x mentally add -3 to both sides.

$\quad\quad x = 8 - 3$ Combine like terms.

$\quad\quad x = \mathbf{5}$

(b) $x - 2 = 4$ To isolate x mentally add $+2$ to both sides.

$\quad\quad x = 4 + 2$ Combine like terms.

$\quad\quad x = \mathbf{6}$

(c) $5 + x = 2$ To isolate x mentally add -5 to both sides.

$\qquad x = 2 - 5$ Combine like terms.

$\qquad \mathbf{x = -3}$

(d) $\qquad 10 = x - 1$ To isolate x mentally add $+1$ to both sides.

$\qquad 10 + 1 = x$ Combine like terms.

$\qquad \mathbf{11 = x}$ Apply the symmetric property of equality.

$\qquad$ or

$\qquad \mathbf{x = 11}$

TIP   Use Simple Cases to Learn and Understand a Process or Procedure

Many equations can be solved mentally, or by using basic arithmetic and our knowledge of the relationships among the operations of addition, subtraction, multiplication, and division. In the example of $x + 3 = 8$, we can determine mentally that the missing number is 5. Our goal is to learn a procedure for solving simple equations, so that we can apply the same procedure to more complex equations.

---

2   **Solve Linear Equations Using the Multiplication Axiom.**

The basic principle of equality can be extended to multiplication and division. The **multiplication axiom** states that both sides of an equation may be multiplied or divided by the same *nonzero* quantity without changing the equality of the two sides. Symbolically, if $a = b$ and $c \neq 0$, then $ac = bc$. Similarly, if $a = b$ and $c \neq 0$, then $\dfrac{a}{c} = \dfrac{b}{c}$, for all real numbers a, b, and c.

To solve a linear equation using the multiplication axiom:

1. Multiply both sides of the equation by the reciprocal of the coefficient of the variable term; or
2. Divide both sides of the equation by the numerical coefficient of the variable term.

EXAMPLE   Solve the equation $4x = 48$.

Multiplication and division are inverse operations, so dividing by a number is the same as multiplying by its multiplicative inverse—its reciprocal.

$\qquad 4x = 48$ Multiply by the reciprocal of the coefficient of the variable term.

$\qquad \dfrac{1}{4}(4x) = \dfrac{1}{4}(48)$

$\qquad x = 12$

$\qquad 4x = 48$ Divide by the coefficient of the variable term.

$\qquad \dfrac{4x}{4} = \dfrac{48}{4}$

$\qquad x = 12$

EXAMPLE   Solve the equation $45 = 0.5x$.

$$\frac{45}{0.5} = \frac{0.5x}{0.5}$$ Divide both sides by the coefficient of the variable term.

$$90 = x$$ Apply the symmetric property of equality.

or

$$x = 90$$

On Which Side Should the Variable Be?

Multiplication and division are effective regardless of which side of the equation contains the variable term. Once the equation is solved, however, many people prefer to put the variable on the left ($x = 90$ instead of $90 = x$). Either way is correct.

When the equation contains only whole numbers or decimals as coefficients or constant terms, it is generally more convenient to divide by the coefficient of the variable term than to multiply by the reciprocal of the coefficient of that variable.

When the equation contains a fraction, it is usually preferable to multiply by the reciprocal of the coefficient of the variable.

EXAMPLE   Solve the equation $\frac{1}{4}n = 7$.

Using multiplication: $\frac{1}{4}n = 7$ Multiply both sides by the reciprocal of the coefficient of the variable term.

$$\left(\frac{4}{1}\right)\frac{1}{4}n = 7\left(\frac{4}{1}\right)$$ The reciprocal of $\frac{1}{4}$ is $\frac{4}{1}$.

$$n = 28$$

Using division: $\frac{1}{4}n = 7$ Divide both sides by the coefficient of the variable term.

$$\frac{\frac{1}{4}n}{\frac{1}{4}} = \frac{7}{\frac{1}{4}}$$ $7 \div \frac{1}{4} = \frac{7}{1} \cdot \frac{4}{1} = 28$

$$n = 28$$

An equation is *solved* when the coefficient of the letter term is $+1$. If the coefficient is -1, we apply the multiplication axiom to complete the solution.

> **EXAMPLE** Solve the equation $-n = 25$.
>
> The coefficient of n is -1, so the equation is not solved. The coefficient of n must be $+1$ for the equation to be solved.
>
> $$\frac{-1n}{-1} = \frac{25}{-1}$$ Divide by the coefficient of the variable term.
>
> $$n = -25$$

ask about

When an equation has been solved, the *solution,* or *root,* can be checked to verify it is the solution.

To verify the solution of an equation:

1. Substitute the solution in place of the variable in the original equation.
2. Perform all indicated operations on each side of the equation.
3. If the solution is correct, the value of the left side of the equation should equal the value of the right side.

> **EXAMPLE** In the example $4x = 48$ the solution was found to be 12. Check this root.
>
> $$4x = 48$$ Substitute 12 for x.
> $$4(12) = 48$$ Perform the multiplication.
> $$48 = 48$$ The root is verified if the final equation is true.

3 Solve Linear Equations with Like Terms on the Same Side of the Equation.

We can combine only *like* terms and only if they are on the *same side* of the equation. That means constant terms can be added only to other constant terms. Variable terms can be added only to other like variable terms. By **combine** we mean add like terms according to the appropriate signed number rule.

To solve linear equations with like terms on the same side of the equation:

1. Combine (add or subtract) like terms that are on the same side of the equal sign.
2. *Multiply* both sides of the equation by the reciprocal of the coefficient of the variable or *divide* both sides of the equation by the coefficient of the variable.

> **EXAMPLE** Solve the equation $4x - x = 7 + 8$.
>
> $$4x - x = 7 + 8$$ The coefficient of x is -1.
>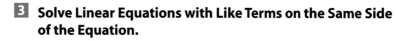
> $$4x - x = \boxed{7 + 8}$$ Combine like terms on each side of the equal sign.
> $$3x = \boxed{15}$$ Divide both sides of the equation by 3, the coefficient of x.

$$\frac{3x}{3} = \frac{15}{3}$$

$$x = 5$$

4 Solve Linear Equations with Like Terms on Opposite Sides of the Equation.

Like terms do not always appear on the same side of the equal sign. In such cases, the like terms must be manipulated so that they appear on the same side of the equal sign *before* they can be combined. The basic strategy for solving an equation is to *isolate* the variable terms on one side of the equation.

> **To solve linear equations with like terms on opposite sides of the equation:**
>
> 1. Identify the like terms and determine which term or terms should be moved to create an equation with like terms on each side.
> 2. Add to both sides of the equation the opposite of each term that is to be moved (addition axiom).
> 3. Combine like terms on each side of the equation.
> 4. Solve the resulting equation by using the multiplication axiom.

EXAMPLE Solve the equation $2x + 4 = 8$.

$2x \boxed{+ 4} = 8$	4 and 8 are like terms on opposite sides of the equal sign. Mentally add -4, the opposite of 4, to both sides (addition axiom).
$2x = 8 \boxed{- 4}$	Combine like terms. $8 - 4 = 4$
$2x = 4$	Divide both sides by the coefficient of the variable term, $2x$ (multiplication axiom).
$\dfrac{2x}{2} = \dfrac{4}{2}$	
$x = 2$	

EXAMPLE Solve the equation $9 - 4x = 8x$.

Manipulate the terms in this equation so that both variable terms are isolated on one side of the equation and the number term is on the other side.

$9 \boxed{- 4x} = 8x$	$-4x$ and $8x$ are like terms on opposite sides of the equal sign. Mentally add $4x$, the opposite of $-4x$, to both sides (addition axiom).
$9 = 8x \boxed{+ 4x}$	$-4x + 4x = 0$, $8x + 4x = 12x$.
$9 = 12x$	Divide both sides by the coefficient of x, which is 12 (multiplication axiom).
$\dfrac{9}{12} = \dfrac{12x}{12}$	
$\dfrac{9}{12} = x$	Reduce to lowest terms.
$\dfrac{3}{4} = x$ or $x = \dfrac{3}{4}$	Apply the symmetric property of equality if desired.

Manipulating terms by applying the addition axiom can be described as **sorting** the terms so that like terms are on the same side of the equation.

EXAMPLE Solve the equation $4x - 3 = 3x - 3 + x$.

$4x - 3 = 3x - 3 + x$	Combine like terms. $3x + x = 4x$
$4x - 3 = 4x - 3$	Sort terms (addition axiom). Add $-4x$ to both sides. Add $+3$ to both sides.
$4x - 4x = -3 + 3$	Combine like terms. $4x - 4x = 0$; $-3 + 3 = 0$
$0 = 0$	Since $0 = 0$ is a *true* statement, the original equation is true for any real number value of x.

The solution is all real numbers.

Equations like $4x - 3 = 3x - 3 + 2$ that have many solutions are called **identities.**

EXAMPLE Solve the equation $3x - 7 = x + 2x - 11$.

$3x - 7 = x + 2x - 11$	Combine like terms. $x + 2x = 3x$
$3x - 7 = 3x - 11$	Sort terms (addition axiom). Add $-3x$ to both sides. Add $+7$ to both sides.
$3x - 3x = -11 + 7$	Combine like terms. $3x - 3x = 0$. $-11 + 7 = -4$.
$0 = -4$	Zero and -4 are not equal. Therefore, there is no solution of the equation.

No solution

When the variable terms add to zero, we examine the resulting statement. A true statement indicates the solution set is the *set of all real numbers.* A false statement indicates the equation has *no solution.*

TIP

Align Equal Signs! One Equal Sign per Line!

Organizing the steps of the solution to an equation reduces errors.

- Arrange steps under each other and align the equal signs in a vertical line.
  This helps you determine if a term has moved from one side to the other.
  As a term is eliminated from one side, move it to the other side as its opposite.
- Have only one equal sign per line.

Each line should be one complete equation or statement.

Good Form	**Poor Form**
$x = \dfrac{2}{4}$	$x = \dfrac{2}{4} = \dfrac{1}{2}$
$x = \dfrac{1}{2}$	

An equation can be solved two ways: (1) by isolating the variable terms on the left and (2) by isolating the variable terms on the right. When we choose the side for the variable terms, the constants must be on the opposite side.

EXAMPLE Solve the equation $9x + 2 = 6x - 10$ two ways.

Variable Terms Isolated on Left		**Variable Terms Isolated on Right**
$9x + 2 = 6x - 10$	Sort.	$9x + 2 = 6x - 10$
$9x - 6x = -10 - 2$	Combine like terms.	$2 + 10 = 6x - 9x$
$3x = -12$	Divide.	$12 = -3x$
$\dfrac{3x}{3} = \dfrac{-12}{3}$		$\dfrac{12}{-3} = \dfrac{-3x}{-3}$
$x = -4$		$x = -4$

EXAMPLE Solve the equation $2x - 3 + 5x = 8 - 6x$.

$2x - 3 + 5x = 8 - 6x$ Combine the like terms $2x$ and $5x$ on the left side.

$7x - 3 = 8 - 6x$ Sort to collect like terms on the same side.

$7x + 6x = 8 + 3$ Combine the like terms on each side of the equation.

$13x = 11$ Divide by the coefficient of x.

$$\frac{13x}{13} = \frac{11}{13}$$

$$x = \frac{11}{13}$$

SECTION 7–2 SELF-STUDY EXERCISES

1 Solve the equations. Use the addition axiom.

1. $x - 3 = 5$	**2.** $x - 7 = 8$	**3.** $x - 9 = -12$	**4.** $x - 11 = -19$
5. $x + 5 = 9$	**6.** $x + 8 = -3$	**7.** $x + 11 = -15$	**8.** $x + 14 = 17$
9. $12 = x + 7$	**10.** $-15 = x - 8$	**11.** $14 = x - 2$	**12.** $18 = x + 9$
13. $7 = 5 - x$	**14.** $5 = 9 - x$	**15.** $17 = 11 + x$	**16.** $21 - x = 12$

2 Solve the equations. Use the multiplication axiom.

17. $7x = 56$	**18.** $2x = 18$	**19.** $15 = 3x$	**20.** $36 = 1.2n$
21. $-3a = 27$	**22.** $-18 = 9b$	**23.** $-7c = -49$	**24.** $-72 = 8x$
25. $5x = 32$	**26.** $36 = 8y$	**27.** $-6n = 15$	**28.** $-28 = -3b$

29. $-x = 7$ **30.** $-2 = -y$ **31.** $\dfrac{1}{3}x = 5$ **32.** $4 = \dfrac{1}{2}y$

33. $\dfrac{3}{5}x = 2$ **34.** $-\dfrac{3}{4}y = 5$ **35.** $-\dfrac{1}{2}x = -6$ **36.** $21 = \dfrac{3}{8}n$

3 Solve.

37. $3x + 7x = 60$ **38.** $42 = 8m - 2m$ **39.** $5a - 6a = 3$
40. $3m - 9m = 3$ **41.** $y + 3y = 32$ **42.** $0 = 2x - x$
43. $8y - 6y = -14$ **44.** $5y - y = -16$ **45.** $-6x - 12x = 36$

Do Tonight.

4 Solve.

46. $-36 = 9x + 18$ **47.** $b + 6 = 5$ **48.** $1 = x - 7$
49. $5t - 18 = 12$ **50.** $10 - 2x = 4$ **51.** $2y + 7 = 17$
52. $3x + 7 = x$ **53.** $3a - 8 = 7a$ **54.** $10x + 18 = 8x + 18 + 2x$
55. $4x + 7 = 8 + 4x$ **56.** $4x = 5x + 8$ **57.** $2t + 6 = t + 13$
58. $12 + 5x = 6 - x$ **59.** $4y - 8 = 2y + 14$ **60.** $8 - 7y = y + 24$

61. $\dfrac{2x}{3} = 18$ **62.** $\dfrac{y + 1}{2} = 7$ **63.** $\dfrac{8 - R}{76} = 1$

64. $P = \dfrac{1}{2} + \dfrac{1}{3}$ **65.** $x + \dfrac{1}{7}x = 16$ **66.** $\dfrac{2}{5} - x = \dfrac{1}{2}x + \dfrac{4}{5}$

67. $\dfrac{3}{7}m - \dfrac{1}{2} = \dfrac{2}{3}$ **68.** $\dfrac{1}{4}s = \dfrac{1}{4} + \dfrac{1}{10} + \dfrac{1}{20}$ *Do Tonight*   **69.** $m = 2 + \dfrac{1}{4}m$

70. $\dfrac{7}{9} + 3 = \dfrac{1}{2}T$ **71.** $2.3x = 4.6$ **72.** $4.5x - 5.1 = 3.9$

73. $0.8R = 0.6$ (round to nearest tenth)   **74.** $0.33x + 0.25x = 3.5$ (round to nearest hundredth)
75. $0.04x = 0.08 - x$ (round to nearest hundredth)   **76.** $0.47 = R + 0.4R$ (round to nearest hundredth)

7–3 | *Applying the Distributive Property in Solving Equations*

Learning Outcome   **1** Solve linear equations that contain parentheses.

When an equation contains an addition or subtraction in parentheses and that quantity in parentheses is multiplied by another factor, we have an example of the *distributive property*. (See Chapter 1, Section 1, Outcome 6.)

1 **Solve Linear Equations That Contain Parentheses.**

To solve linear equations that contain parentheses:

1. Apply the distributive property to *remove parentheses*.
2. *Combine like terms* on each side of the equation.
3. *Sort terms* to collect the variable terms on one side and constants on the other (addition axiom).
4. *Combine like terms* on each side of the equation.
5. *Multiply* by the reciprocal of the coefficient of the variable term or *divide* by the coefficient of the variable term (multiplication axiom).

EXAMPLE   Solve the equation $28 = 7x - 3(x - 4)$.

$$28 = 7x - 3(x - 4)$$ Each term in the parentheses is multiplied by -3 using the distributive property. $-3(x) = -3x$. $-3(-4) = 12$

$$28 = \boxed{7x - 3x} + 12$$ Combine like terms on the right.

$$28 = \boxed{4x} + \boxed{12}$$ Sort terms.

$$28 \boxed{-12} = 4x$$ Combine like terms on the left.

$$16 = 4x$$ Divide.

$$\frac{16}{4} = \frac{4x}{4}$$

$$x = 4$$

As equations get more involved, the importance of checking becomes more apparent. To check the root 4, we substitute 4 for x in the equation.

$$28 = 7x - 3(x - 4)$$ Substitute 4 for x.
$$28 = 7(4) - 3(4 - 4)$$ Simplify the grouping $4 - 4 = 0$.
$$28 = 7(4) - 3(0)$$ Multiply.
$$28 = 28 - 0$$ Subtract.
$$\mathbf{28 = 28}$$ Solution checks.

EXAMPLE   Solve the equation $6 - (x + 3) = 2x$ for x.

Since $(x + 3)$ is in parentheses, it is a grouping and we handle it as one term. The sign of the term is negative and the numerical coefficient is understood to be -1.

The first step in solving this equation is to apply the distributive property by multiplying $(x + 3)$ by its understood coefficient, -1.

$$6 - (x + 3) = 2x$$ -1 is the understood coefficient of $(x + 3)$.

$$6 - 1(x + 3) = 2x$$ Distribute.

$$6 - x - 3 = 2x$$ Combine. $6 - 3 = 3$

$$3 - x = 2x$$ Sort.

$$3 = 2x + x$$ Combine. $2x + x = 3x$

$$3 = 3x$$ Divide.

$$\frac{3}{3} = \frac{3x}{3}$$

$$x = 1$$

The usefulness of solving linear equations is in solving applied problems. We will continue to use the Six-Step Problem-Solving Plan as our guide for solving applied problems.

EXAMPLE **AG/H**	A horticulturist marks off a 32-m-wide rectangular nursery plot with 158 m of fencing. Because the plants must be properly spaced, she needs to know the length of the plot. Find the length in meters.
Unknown facts	The length of the nursery plot in meters.
Known facts	The nursery plot is rectangular. The width is 32 m. The fencing gives us the perimeter of 158 m.
Relationships	Using the formula for the perimeter of a rectangle, we know that the perimeter is twice the sum of the length and width, or $P = 2(l + w)$.
Estimation	Two widths ($2 \cdot 32$ m) will use 64 m of fencing. There will be less than 100 m of fencing for the two lengths. The length will be less than 50 m.

Calculations

$$P = 2(l + w) \qquad \text{Substitute in the formula.}$$
$$158 = 2(l + 32) \qquad \text{Distribute.}$$
$$158 = 2(l) + 2(32) \qquad \text{Multiply.}$$
$$158 = 2l + 64 \qquad \text{Sort.}$$
$$158 - 64 = 2l \qquad \text{Combine.}$$
$$94 = 2l \qquad \text{Divide.}$$
$$\frac{94}{2} = \frac{2l}{2}$$
$$47 = l$$

Interpretation **Thus, the length of the nursery plot is 47 m.**

Check:
$$P = 2(l + w) \qquad \text{Substitute.}$$
$$158 = 2(47 + 32) \qquad \text{Distribute.}$$
$$158 = 94 + 64 \qquad \text{Combine.}$$
$$158 = 158$$

EXAMPLE **INDTEC**	A machine part weighs 2.7 kg and is to be shipped in a carton weighing x kg. If the total weight of three packaged machine parts is 9.3 kg, how much does each carton weigh? (See Fig. 7–2.)

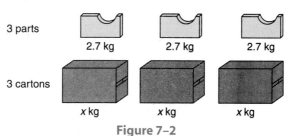

3 parts

2.7 kg   2.7 kg   2.7 kg

3 cartons

x kg x kg x kg

Figure 7–2

Unknown facts	The weight of each empty carton.
Known facts	There are three parts and three cartons. Each part weighs 2.7 kg. The three packaged parts (in cartons) weigh a total of 9.3 kg.
Relationships	Let x equal the weight of one empty carton. Each packaged part weighs the sum of the part (2.7 kg) plus the weight of a carton (x kg). The sum of the three packaged parts equals 9.3 kg; that is, $3(2.7 + x) = 9.3$.

7–3 Applying the Distributive Property in Solving Equations   317

Estimation	Three packaged parts weigh approximately 9 kg, or 3 kg per packaged carton. Since one part weighs 2.7 kg, then the carton weighs about 0.3 kg.

Calculations

$$3(2.7 + x) = 9.3 \qquad \text{Distribute.}$$

$$3(2.7) + 3(x) = 9.3 \qquad \text{Make calculations.}$$

$$8.1 + 3x = 9.3 \qquad \text{Sort.}$$

$$3x = 9.3 - 8.1 \qquad \text{Combine.}$$

$$3x = 1.2 \qquad \text{Divide.}$$

$$\frac{3x}{3} = \frac{1.2}{3}$$

$$x = 0.4$$

Interpretation **Each carton weighs 0.4 kg.**

Check:

$$3(2.7 + \boxed{x}) = 9.3 \qquad \text{Substitute.}$$

$$3(2.7 + \boxed{0.4}) = 9.3 \qquad \text{Make calculations.}$$

$$3(3.1) = 9.3$$

$$9.3 = 9.3$$

Additional Problem-Solving Strategies

- Read the problem carefully. Read it several times and read it phrase by phrase.
- Understand all the words in the problem.
- Analyze the problem:
  What are you asked to find?
  What facts are given?
  What facts are implied?
- Visualize the problem.
- State the conditions or relationships of the problem "symbolically."
- Examine the options.
- Develop a *plan* for solving the problem.
- Write your *plan* symbolically; that is, write an equation.
- Anticipate the characteristics of a reasonable solution.
- Solve the equation.
- Verify your answer with the conditions of the problem.

There are many different problem-solving plans. Additional strategies can be considered.

SECTION 7–3 SELF-STUDY EXERCISES

1 Solve.

1. $4(x - 5) = 8$
2. $3(y - 7) = -15$
3. $6(x + 3) = -12$
4. $5(x + 7) = 25$
5. $-3(x + 8) = -39$
6. $-2(x - 7) = 26$
7. $3(2x - 1) = 3$
8. $4(x - 1) = 28$
9. $-4(-3x + 1) = 8$

10. $-9(2x - 3) = 27$

11. $7(x - 3) = 7x + 21$

12. $5x + 15 = 5(x + 3)$

13. $8(x - 5) = 8x + 12$

14. $6x + 7 = 3(2x + 4)$

15. $2(3x - 5) = 7x - 12$

16. $5(3x - 7) + 2 = 7x + 7$

17. $2x - 3 + 6x = 4(3x - 2) + 3$

18. $7(x - 1) = 2(x + 3) + 4(x - 7)$

19. $3(x - 4) - 2(3x - 1) = 17$

20. $6x - 4 + 2x = -(x - 5)$

21. $8x - (2x - 9) = -3$

22. $6 - 2(3x - 1) = 14$

23. $12 - 5(2x - 3) = -5$

24. $8 - 3(x - 2) = x - 6$

Write the statements as equations and solve.

25. The sum of x and 4 equals 12. Find x.

26. 4 less than 2 times a number is 6. Find the number.

27. INDTEC   Three parts totaling 27 lb are packaged for shipping. Two parts weigh the same. The third part weighs 3 lb less than each of the two equal parts. Find the weight of each part.

28. HLTH/N   How many gallons of water must be added to 24 gal of pure disinfectant to make 60 gal of diluted disinfectant?

29. INDTEC   A wet casting weighing 4.03 kg weighs 3.97 kg after drying. Write and solve an algebraic equation to find the weight loss due to drying.

30. CON   A plumber needs 3 times as much perforated pipe as solid pipe to lay a drain line 400 ft long. How much of each type of pipe is needed?

31. ELEC   An engineering student purchased a new circuits text, a used Spanish text, and a graphing calculator. She remembered that the calculator cost twice as much as the Spanish book and that the circuits text cost $70.00. The total before tax was $235.00. What was the cost of the calculator and the Spanish text?

32. CON   Mary Jefferson purchased a home on a square lot 150 ft on each side. She wants to enclose the entire lot with a cedar fence. One estimate for the job was for $14.00 per linear foot. How much will the fence cost?

33. AG/H   Jake Drewrey, owner of Jake's Landscape Service, knows that one of his fertilizer tanks holds twice as many gallons of liquid fertilizer as a second tank. The two tanks together hold 325 gal. If both tanks are filled to capacity, how many gallons of fertilizer does each tank hold?

34. AG/H   Natalie Bradley hired a stone mason to build a rectangular flower bed at one end of her patio. She needs enough mulch to cover 60 ft^2, the area of the flower bed. If the flower bed has a length of 10 ft, how wide is the bed?

7-4 | Solving Linear Equations with Fractions and Decimals by Clearing the Denominators

Learning Outcomes

1 Solve fractional equations by clearing the denominators.

2 Solve applied problems involving rate, time, and work.

3 Solve decimal equations by clearing the decimals.

1 Solve Fractional Equations by Clearing the Denominators.

Thus far, we have used mostly integers in equations. However, real-world situations require us to deal with various fraction and decimal quantities when we solve equations.

The techniques we used in Sections 7–2 and 7–3 will also help us solve equations with fractions and decimals. Other techniques minimize the calculations with fractions and decimals and allow more steps to be performed mentally. One of these techniques involves *clearing* the equation of all denominators in the first step. The resulting equation, which contains no fractions, is then solved using the procedures we learned earlier.

This process, called **clearing the fractions,** is another application of the multiplication axiom.

Apply the multiplication axiom by multiplying the entire equation by the fraction's denominator.

EXAMPLE Solve $\dfrac{4d}{3} = 12$.

$$\dfrac{4d}{3} = 12 \qquad \text{Multiply both sides by the denominator 3. Reduce where possible.}$$

$$(\overset{1}{\cancel{3}})\dfrac{4d}{\underset{1}{\cancel{3}}} = (3)(12)$$

$$4d = 36 \qquad \text{Divide.}$$

$$\dfrac{4d}{4} = \dfrac{36}{4}$$

$$d = 9$$

Check: $\qquad \dfrac{4(\overset{3}{\cancel{9}})}{\underset{1}{\cancel{3}}} = 12$

$$12 = 12$$

EXAMPLE Solve $2x + \dfrac{3}{4} = 1$ by clearing the fraction.

$$2x + \dfrac{3}{4} = 1 \qquad \text{Multiply each term by the denominator 4.}$$

$$4\,(2x) + 4\left(\dfrac{3}{4}\right) = 4\,(1) \qquad \text{Reduce and multiply.}$$

$$4(2x) + \overset{1}{\cancel{4}}\left(\dfrac{3}{\underset{1}{\cancel{4}}}\right) = 4(1)$$

$$8x + 3 = 4 \qquad \text{Sort.}$$

$$8x = 4 - 3 \qquad \text{Combine.}$$

$$8x = 1 \qquad \text{Divide.}$$

$$\dfrac{8x}{8} = \dfrac{1}{8}$$

$$x = \dfrac{1}{8}$$

Check: $2x + \dfrac{3}{4} = 1$ Substitute $\dfrac{1}{8}$ in place of x.

$\overset{1}{2}\left(\dfrac{1}{\underset{4}{8}}\right) + \dfrac{3}{4} = 1$ Multiply.

$\dfrac{1}{4} + \dfrac{3}{4} = 1$ Add.

$1 = 1$

One advantage of clearing fractions is that the process can be applied to equations in which the variable is in the *denominator* of the fraction. Let's use this procedure to solve $\dfrac{10}{x} = 2$. Note that the term $\dfrac{10}{x}$ is *not* the product of 10 and x. Therefore, the coefficient of x is *not* 10.

EXAMPLE   Solve $\dfrac{10}{x} = 2$.

$\dfrac{10}{x} = 2$ Multiply both sides by denominator x. Reduce where possible.

$(\cancel{x})\overset{1}{\dfrac{10}{\cancel{x}}} = (\cancel{x})2$

$10 = 2x$ Divide.

$\dfrac{10}{2} = \dfrac{2x}{2}$

$\mathbf{5 = x}$

Check: $\dfrac{\overset{2}{\cancel{10}}}{\underset{1}{\cancel{5}}} = 2$

$2 = 2$

Check for Extraneous Roots

When you solve equations with a variable in the fraction's denominator, you may get a "root" that does not make a true statement when substituted in the original equation. Solutions or roots that do not make a true statement in the original equation are called **extraneous roots.** When solving an equation with the variable in the denominator, you *must* check the root to see if it makes a true statement in the original equation.

One situation that produces an extraneous root is a value that causes the denominator of any fraction to be zero. Such values are called **excluded values.**

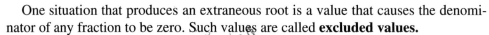 

Ask to explain excluded value

To identify an excluded value that causes a denominator of zero:

1. Write an equation for each fraction that has a variable in the denominator by setting the denominator equal to zero.
2. Solve each resulting equation.
3. The solution for each equation from Step 1 is an excluded value.

EXAMPLE Solve $\dfrac{0}{x} = 5$.

Excluded value:

$$x = 0$$

Set the denominator containing a variable equal to zero. When the denominator is 0, the fraction $\dfrac{0}{x}$ is indeterminate.

Solve the equation:

$$\frac{0}{x} = 5$$ Multiply both sides by x.

$$x\left(\frac{0}{x}\right) = 5(x)$$

$$\frac{0}{5} = \frac{5x}{5}$$

$$\boxed{0} = x$$ Possible solution.

Check: $\dfrac{\boxed{0}}{0} \neq 5$ Does not check. Zero is the excluded value.

The equation $\dfrac{0}{x} = 5$ has no solution.

Equations in which the fraction contains more than one term in its numerator or denominator can also be cleared of fractions.

EXAMPLE Solve $\dfrac{12}{Q + 6} = 1$. Identify any excluded values.

Excluded value:

$$Q + 6 = 0$$ Set denominator equal to zero and solve for Q.

$$Q = 0 - 6$$

$$Q = -6$$ Excluded value.

$$\frac{12}{Q + 6} = 1$$ Multiply both sides of the equation by the denominator, the quantity $Q + 6$. Reduce where possible.

Chapter 7 / Linear Equations

$$(Q + 6)\frac{\overset{1}{12}}{\underset{1}{Q + 6}} = (Q + 6)\,1 \qquad \text{Distribute.}$$

$$12 = Q + 6 \qquad \text{Sort.}$$

$$12 - 6 = Q \qquad \text{Combine like terms.}$$

$$\boxed{6} = Q \qquad \text{Because the numerical coefficient of } Q \text{ is already 1, the equation is solved.}$$

Check: $\dfrac{12}{6 + 6} = 1$ \qquad Substitute 6 for Q and evaluate.

$$\frac{12}{12} = 1$$

$$1 = 1$$

The solution of the equation is $Q = 6$.

When an equation has more than one fractional term, we expand our process.

To solve an equation by clearing all fractions:

1. Multiply each term of the *entire* equation by the least common multiple (LCM) of the denominators of the equation.
2. Apply the distributive property to remove parentheses.
3. Combine like terms on each side of the equation.
4. Sort terms to collect the variable terms on one side and constants on the other (addition axiom).
5. Combine like terms on each side of the equation.
6. Solve the resulting equation by multiplying by the reciprocal of the coefficient of the variable term or dividing by the coefficient of the variable term (multiplying axiom).

EXAMPLE Solve $-\dfrac{1}{4}x = 9 - \dfrac{2}{3}x$ by clearing all fractions first.

There are no excluded values since there are no variables in a denominator.

$$-\frac{1}{4}x = 9 - \frac{2}{3}x \qquad \begin{array}{l}\text{Multiply each term by the LCM of the}\\\text{denominators. LCM} = 12 \text{ or } 4(3).\end{array}$$

$$(4)(3)\left(-\frac{1}{4}x\right) = (4)(3)(9) - (4)(3)\left(\frac{2}{3}x\right) \qquad \text{Reduce.}$$

$$(\overset{1}{4})(3)\left(-\frac{1}{4}x\right) = (4)(3)(9) - (4)(\overset{1}{3})\left(\frac{2}{3}x\right) \qquad \text{Multiply the remaining factors.}$$

$$-3x = 108 - 8x \qquad \begin{array}{l}\text{This equation contains no fractions.}\\\text{Sort terms.}\end{array}$$

$$-3x + 8x = 108 \qquad \text{Combine like terms.}$$

$$5x = 108 \qquad \text{Divide by the coefficient of } x.$$

$$\frac{5x}{5} = \frac{108}{5}$$

$$x = \frac{108}{5}$$

Check using a calculator:

$$-\frac{1}{4}x = 9 - \frac{2}{3}x \qquad \text{Substitute } \frac{108}{5} \text{ for } x.$$

$$-\frac{1}{4}\left(\frac{108}{5}\right) = 9 - \frac{2}{3}\left(\frac{108}{5}\right)$$

Left Side: $((-) 1 \div 4) \times (108 \div 5) = \Rightarrow -5.4$

Right Side: $9 - (2 \div 3) \times (108 \div 5) = \Rightarrow -5.4$

Solutions of equations are not always whole numbers. Such solutions are usually written as *proper* or *improper fractions* in lowest terms. This representation is the *exact* solution. Solutions of applied problems are usually written in decimal or mixed-number form. If the decimal form is rounded, the rounded decimal is the approximate solution.

EXAMPLE
ELEC

Find the total resistance in a parallel dc circuit with three branches rated at 4 Ω, 10 Ω, and 20 Ω, respectively. Solve $\frac{1}{R} = \frac{1}{4} + \frac{1}{10} + \frac{1}{20}$ by clearing fractions first.

$R = 0$ is the excluded value.

$$\frac{1}{R} = \frac{1}{4} + \frac{1}{10} + \frac{1}{20} \qquad \begin{array}{l}\text{The LCM is } 20R \text{ because } 20R \text{ is} \\ \text{evenly divisible by } R, 4, 10, \text{ and } 20.\end{array}$$

$$20R\left(\frac{1}{R}\right) = \overset{5}{20R}\left(\frac{1}{\underset{1}{4}}\right) + \overset{2}{20R}\left(\frac{1}{\underset{1}{10}}\right) + \overset{1}{20R}\left(\frac{1}{\underset{1}{20}}\right) \qquad \begin{array}{l}\text{Multiply each term in the } \textit{entire} \\ \text{equation by } 20R \text{ and reduce.}\end{array}$$

$$20 = 5R + 2R + R \qquad \text{Combine like terms.}$$

$$20 = 8R \qquad \text{Divide by the coefficient of } R.$$

$$\frac{20}{8} = \frac{8R}{8} \qquad \text{Reduce.}$$

$$\frac{5}{2} = R \qquad \frac{5}{2} = 2.5$$

Interpretation   **The resistance is 2.5 Ω.**

Check:

$$\frac{1}{R} = \frac{1}{4} + \frac{1}{10} + \frac{1}{20}$$

Substitute $\frac{5}{2}$ for R.

$$\frac{1}{\frac{5}{2}} = \frac{1}{4} + \frac{1}{10} + \frac{1}{20}$$

Perform calculations on each side of the equation.

$$\frac{1}{\frac{5}{2}} = \frac{5}{20} + \frac{2}{20} + \frac{1}{20}$$

$$1 \cdot \frac{2}{5} = \frac{8}{20}$$

$$\frac{2}{5} = \frac{2}{5}$$

Fractions versus Decimals

Sometimes applied problems that require fractions in their equations require that their solutions be expressed as decimal numbers. In these cases, we perform the division indicated by the fraction.

The equation in the preceding example is derived from the formula for finding total resistance in a parallel dc circuit with three branches rated at 4, 10, and 20 Ω. Ohms are expressed in decimal numbers, so in an application the solution should be converted to a decimal equivalent.

$$R = \frac{5}{2} \,\Omega \qquad \text{or} \qquad 2.5 \,\Omega$$

2 Solve Applied Problems Involving Rate, Time, and Work.

A **rate measure** often involves a unit of time. If car A travels 50 mi in 1 h, then car A's **rate of work** (travel) is 50 mi per 1 h, or $\frac{50 \text{ mi}}{1 \text{ h}}$ expressed as a fraction. If car A travels for 3 h, then the **amount of work** is $\frac{50 \text{ mi}}{\text{h}} \times 3 \text{ h} = 150 \text{ mi}$. Review rate measures in Chapter 2, Section 5, Outcome 5.

To find the amount of work produced by one individual or machine:

1. Identify the rate of work and the time worked.
2. Use the formula for amount of work.

Formula for amount of work:

$$\text{amount of work completed} = \text{rate of work} \times \text{time worked}$$

EXAMPLE CON	A carpenter can install 1 door in 3 h. Find the number of doors the carpenter can install in 40 h.
Known facts	Rate of work = 1 door per 3 h, $\frac{1}{3}$ door per hour, or $\frac{1 \text{ door}}{3 \text{ h}}$
	Time worked = 40 h
Unknown facts	W = amount of work or number of doors installed
Relationships	Amount of work = rate of work × time worked
Estimation	It would take 30 h to install 10 doors. More than 10 doors can be installed in 40 h.
Calculations	$W = \dfrac{1 \text{ door}}{3 \text{ h}} \times 40 \text{ h}$ Multiply. Reduce dimensions.
	$W = \dfrac{40}{3}$ doors
	$W = 13\frac{1}{3}$ doors
Interpretation	**Thus, 13 doors can be installed in 40 h.**

If two workers or machines do a job together, we can find the amount of work done by each worker or machine. Combined, the amounts equal 1 total job.

 To find the amount of work each individual or machine produces when working together:

1. Identify the rate of work for each individual or machine. If unknown, assign a letter to represent the unknown.
2. Identify the time worked for each individual or machine. If unknown, assign a letter to represent the unknown. *Note:* Only one letter should be used and other unknowns should be written in relationship to the one letter.
3. Use the formula for completing one job.

Formula for completing one job when A and B are working together:

$$\left(\begin{array}{c}\text{A's}\\\text{amount of}\\\text{work}\end{array}\right) + \left(\begin{array}{c}\text{B's}\\\text{amount of}\\\text{work}\end{array}\right) = 1 \text{ completed job}$$

or

$$\left(\begin{array}{c}\text{A's}\\\text{rate of} \times \text{time}\\\text{work} \qquad \text{worked}\end{array}\right) + \left(\begin{array}{c}\text{B's}\\\text{rate of} \times \text{time}\\\text{work} \qquad \text{worked}\end{array}\right) = 1 \text{ completed job}$$

EXAMPLE
AG/H
Pipe 1 fills a tank in 6 min and pipe 2 fills the same tank in 8 min (Fig. 7–3). How long does it take for both pipes together to fill the tank?

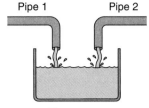

Figure 7–3

Known facts	Pipe 1 fills the tank at a rate of 1 tank per 6 min, $\frac{1}{6}$ tank per minute, or $\frac{1 \text{ tank}}{6 \text{ min}}$.
	Pipe 2 fills the tank at a rate of 1 tank per 8 min, $\frac{1}{8}$ tank per minute, or $\frac{1 \text{ tank}}{8 \text{ min}}$.
Unknown facts	T = time (in minutes) for both pipes together to fill the tank.
Relationships	Amount of work of pipe 1 $= \frac{1 \text{ tank}}{6 \text{ min}} (T)$. Amount of work of pipe 2 $= \frac{1 \text{ tank}}{8 \text{ min}} (T)$.
	Amount of work together = pipe 1's work + pipe 2's work.
Estimation	Both pipes together should fill the tank more quickly than the faster rate, or in less than 6 min.

Calculations

$$\frac{1 \text{ tank}}{6 \text{ min}}(T \text{ min}) + \frac{1 \text{ tank}}{8 \text{ min}}(T \text{ min}) = 1 \text{ tank} \qquad \text{The LCM is } \boxed{24}.$$

$$(24)\left(\frac{1}{6}T\right) + (24)\left(\frac{1}{8}T\right) = (24)(1) \qquad \text{Clear fractions.}$$

$$(\overset{4}{\cancel{24}})\left(\frac{1}{\underset{1}{\cancel{6}}}T\right) + (\overset{3}{\cancel{24}})\left(\frac{1}{\underset{1}{\cancel{8}}}T\right) = (24)(1) \qquad \text{Reduce and multiply.}$$

$$4T + 3T = 24 \qquad \text{Combine.}$$

$$7T = 24 \qquad \text{Divide.}$$

$$\frac{7T}{7} = \frac{24}{7}$$

$$T = \frac{24}{7}\left(\text{or } 3\frac{3}{7}\right) \text{ min}$$

Interpretation **Both pipes together fill the tank in $3\frac{3}{7}$ min.**

TIP

Interpretation of Improper Fractions as Mixed Numbers, Decimal Equivalents, or Mixed Measurements

In applied problems, it is desirable to change improper fractions like $\frac{24}{7}$ into mixed numbers or decimal equivalents. The problem statement generally dictates the interpretation.

In the preceding example, $\frac{24}{7}$ min can be interpreted as $3\frac{3}{7}$ min or 3.4 min (rounded). In some instances, you may need to change $\frac{3}{7}$ min to seconds ($\frac{3}{7}$ min $\times \frac{60 \text{ s}}{1 \text{ min}} = 25.7$ s). Then, $3\frac{3}{7}$ min becomes 3 min 26 s (rounded).

Two pipes, one a faucet and the other a drain, have opposite functions. Pipe 1 fills the tank. Pipe 2 is a drain and empties the tank. In this case, we subtract the work done by the drain from the work done by the faucet. This combined action, if the faucet fills at a faster rate than the drain empties, results in a full tank.

Formula for completing one job when A and B are working in opposition:

$$\left(\begin{array}{c} \text{A's} \\ \text{amount of} \\ \text{work} \end{array}\right) - \left(\begin{array}{c} \text{B's} \\ \text{amount of} \\ \text{work} \end{array}\right) = 1 \text{ completed job}$$

or

$$\left(\begin{array}{c} \text{A's} \\ \text{rate of} \\ \text{work} \end{array} \times \begin{array}{c} \text{time} \\ \text{worked} \end{array}\right) - \left(\begin{array}{c} \text{B's} \\ \text{rate of} \\ \text{work} \end{array} \times \begin{array}{c} \text{time} \\ \text{worked} \end{array}\right) = 1 \text{ completed job}$$

based on A's amount of work being greater than B's amount of work.

EXAMPLE AG/H

A faucet fills a tank in 6 min. A drain empties the tank in 8 min (Fig. 7–4). If both the faucet and drain are open, in how many minutes will the tank start to overflow if the faucet is not turned off?

Known facts

Faucet's rate of work = 1 tank filled per 6 min, $\frac{1}{6}$ tank per min, or $\dfrac{1 \text{ tank}}{6 \text{ min}}$.

Drain's rate of work = 1 tank emptied per 8 min, $\frac{1}{8}$ tank per min, or $\dfrac{1 \text{ tank}}{8 \text{ min}}$.

Unknown facts

T = time (min) until tank overflows with faucet and drain both working.

Relationships

amount of work of faucet = $\dfrac{1 \text{ tank}}{6 \text{ min}}(T)$

amount of work of drain = $\dfrac{1 \text{ tank}}{8 \text{ min}}(T)$

amount of work when both faucet and drain are open = faucet's work − drain's work

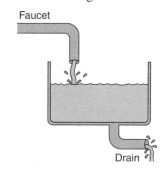

Faucet

Drain

Figure 7–4

Estimation

With both the faucet and drain open, it should take longer to fill the tank than if the faucet was open and the drain closed. It should take longer than 6 min.

Calculations

$$\dfrac{1 \text{ tank}}{6 \text{ min}}(T \text{ min}) - \dfrac{1 \text{ tank}}{8 \text{ min}}(T \text{ min}) = 1 \text{ tank} \qquad \text{Analyze dimensions.}$$

$$\frac{1}{6}T - \frac{1}{8}T = 1 \qquad \text{Clear fractions. LCM = 24}$$

$$(24)\left(\frac{1}{6}T\right) - (24)\left(\frac{1}{8}T\right) = (24)(1) \qquad \text{Reduce.}$$

$$(\overset{4}{24})\left(\frac{1}{\underset{1}{6}}T\right) - (\overset{3}{24})\left(\frac{1}{\underset{1}{8}}T\right) = (24)(1) \qquad \text{Multiply remaining factors.}$$

$$4T - 3T = 24 \qquad \text{Combine like terms.}$$

$$T = 24 \text{ min}$$

Interpretation

With the faucet and drain both open, the tank will be full in 24 min.

Chapter 7 / Linear Equations

3 Solve Decimal Equations by Clearing the Decimals.

Since a decimal is a type of fraction, we can also solve equations containing decimals by clearing decimals. The place value of the last digit of the decimal determines the denominator of the fraction.

To solve a decimal equation by clearing decimals:

1. Multiply each term of the *entire* equation by the least common denominator (LCD) of the fractional amounts represented by the decimals.
2. Follow the same steps used in solving a linear equation.

LCD for Decimals

Digits to the right of the decimal point represent fractions whose denominators are determined by the place value. To find the LCD for all the decimal numbers in an equation, find the decimal with the most digits after the decimal point. Use its denominator to clear the decimals.

This procedure allows you to avoid dividing by a decimal, which can be a common source of error when making calculations by hand.

EXAMPLE Solve $0.38 + 1.1y = 0.6$ by first clearing the equation of decimals.

$$0.38 + 1.1y = 0.6 \qquad \text{The LCD is } \boxed{100}.$$

$$\boxed{100}(0.38) + \boxed{100}(1.1y) = \boxed{100}(0.6) \qquad \text{Multiply by 100.}$$

$$38 + 110y = 60 \qquad \text{Sort.}$$

$$110y = 60 - 38 \qquad \text{Combine.}$$

$$\frac{110y}{110} = \frac{22}{110} \qquad \text{Divide.}$$

$$\mathbf{y = 0.2}$$

The interest formula resembles the percentage formula, $P = RB$, but it includes the time period over which the money was borrowed or invested. If we know three of the four elements, we can find the fourth.

Formula for simple interest:

Cheat sheet

$$I = PRT$$

where $I = $ **interest,** $P = $ **principal,** $R = $ **rate** or percent in decimal form, and $T = $ **time.** The rate in decimal form. time-years

For comparison with the percentage formula, interest is the part or portion and the principal is the base.

EXAMPLE BUS

A $1,000 investment is made for $2\frac{1}{2}$ years at 6.25%. Find the amount of interest.

When we use the interest formula, we change the percent to a decimal equivalent. When no time period is given for the rate, we assume the rate to be per year.

$$I = PRT$$ Substitute the values in the formula.

principal   rate   time

$$I = \$1,000(0.0625)(2.5 \text{ years})$$ 6.25% = 0.0625; $2\frac{1}{2}$ years = 2.5 years.

$$I = \$156.25$$

The interest for $2\frac{1}{2}$ years is \$156.25.

TIP

Clearing Decimals Versus Using a Calculator

Working the preceding example by clearing decimals illustrates that in some cases the method is *not* the most efficient way to solve an equation with decimals. The numbers produced may be large and cumbersome.

$$I = \$1,000 \cdot 0.0625 \cdot 2.5$$ LCM is 10,000.

$$10,000I = \overset{1}{\cancel{10,000}}\left(\$1,\cancel{000} \cdot \frac{\overset{100}{\cancel{625}}}{\cancel{10,000}} \cdot \frac{\overset{25}{\cancel{25}}}{\cancel{10}}\right)$$ Multiply both sides of the equation by 10,000. Reduce.

$$10,000I = 100(625)(25)$$ Multiply.

$$10,000I = 1,562,500$$ Divide.

$$I = \frac{1,562,500}{10,000}$$

$$I = \$156.25$$

A calculator gives the solution more efficiently if we proceed using the decimals.

EXAMPLE ELEC

The formula for voltage V is wattage W divided by amperage A: $V = \frac{W}{A}$. Find the voltage to the nearest hundredth needed for a circuit of 1,280 W with a current of 12.23 A.

Estimation $1,200 \div 12 = 100$

$$V = \frac{W}{A}$$ Substitute values.

$$V = \frac{1,280}{12.23}$$ Divide.

$$V = 104.6606705, \quad \text{or} \quad 104.66 \text{ V}$$

Interpretation   **The voltage is 104.66 V.**

Chapter 7 / Linear Equations

1 Solve the equations. Identify excluded values as appropriate.

1. $\frac{2}{7}x = 8$

2. $-7 = \frac{21}{33}p$

3. $\frac{1}{3}r = \frac{6}{7}$

4. $0 = -\frac{2}{5}c$

5. $-\frac{5}{3}m = 9$

6. $-9m = \frac{5}{3}$

7. $\frac{5}{8}t = 1$

8. $-\frac{5}{7}p = -\frac{11}{21}$

9. $10 = -\frac{1}{35}t$

10. $\frac{5}{12}z = 20$

11. $\frac{2x}{3} = 18$

12. $\frac{7}{Q} = 21$

13. $\frac{y+1}{2} = 7$

14. $\frac{7}{p-4} = -8$

15. $0 = \frac{x}{4}$

16. $-\frac{8}{P} = -72$

17. $\frac{P}{-8} = -72$

18. $-8 = \frac{4B}{B-6}$

19. $\frac{3P}{7} = 12$

20. $\frac{8-R}{76} = 1$

21. $\frac{2}{9}c + \frac{1}{3}c = \frac{3}{7}$

22. $-\frac{1}{4}x = 9 - \frac{2}{3}x$

23. $\frac{2}{7}y + \frac{3}{8} = \frac{1}{7}y + \frac{5}{3}$

24. $\frac{1}{3}x + \frac{1}{2}x = \frac{20}{3}$

25. $\frac{7}{R} - \frac{2}{R} = -1$

26. $S = \frac{1}{15} + \frac{1}{5} + \frac{1}{30}$

27. $18 - \frac{1}{4}x = \frac{1}{2}$

28. $\frac{1}{7}H - \frac{1}{3}H = 0$

29. $\frac{7}{16}h + \frac{1}{9} = \frac{1}{3}$

30. $x + \frac{1}{4}x = 8$

31. $3y + 9 = \frac{1}{4}y$

32. $18 = \frac{4}{3x} - \frac{3}{2x}$

33. $\frac{2}{7}p + 1 = \frac{1}{3}p$

34. $S = \frac{1}{10} + \frac{1}{25} + \frac{1}{50}$

35. $-x + \frac{1}{7} = \frac{1}{2}x$

36. $0 = 1 + \frac{2}{9}c - c$

2 Set up an equation and solve.

37. INDTEC   A kiln fires 8 large vases in 2 h. How many vases can be fired in 20 h?

38. INDTR   A printing press produces 1 day's newspaper in 4 h. A higher-speed press does 1 day's newspaper in 2 h. How much time does it take both presses to produce 1 day's newspaper?

39. INDTEC   One machine packs 1 day's salmon catch in 8 h. A second machine packs 1 day's catch in 5 h. How much time does it require for 1 day's catch to be packed if both machines are used?

40. CON   A painter can paint a house in 6 days. Another painter takes 8 days to paint the same house. If they work together, how much time will it take them to paint the house?

41. INDTEC   One bottling machine can fill 400 bottles of water in 1 h and another machine can fill 400 bottles of water in $1\frac{1}{2}$ h. If both machines are working, how much time does it take to fill 400 bottles of water?

42. AG/H   A tank has two pipes entering. Pipe 1 alone fills the tank in 4 min and pipe 2 takes 12 min to fill the tank. How much time does it take to fill the tank if both pipes are operating at the same time?

43. AG/H   A tank has two pipes entering. Pipe 1 alone fills the tank in 30 min and together the pipes take 10 min to fill the tank. How much time does pipe 2 need to fill the tank alone?

44. AG/H   A tank has two pipes entering it and one leaving it. Pipe 1 fills the tank in 3 min. Pipe 2 takes 7 min to fill the same tank. Pipe 3, however, empties the tank in 21 min. How much time does it take to fill the tank with all three pipes operating at the same time?

For Exercises 45–48, see example on page 324. Round to hundredths.

45. **ELEC**   Find the total resistance in a parallel dc circuit with two branches rated at 12 Ω and 30 Ω, respectively.

46. **ELEC**   A parallel dc circuit has three branches rated at 12 Ω, 15 Ω, and 20 Ω. Find the total resistance.

47. **ELEC**   Three branches of a parallel dc circuit are rated at 16 Ω, 24 Ω, and 32 Ω. Find the total resistance.

48. **ELEC**   A parallel dc circuit has a total resistance of 3 Ω. One branch alone produces 18 Ω resistance. What is the resistance of the other branch?

3   Solve the equations.

49. $2.3x = 4.6$
50. $0.8R = 0.6$ (round to nearest tenth)
51. $0.33x + 0.25x = 3.5$ (round to nearest hundredth)
52. $0.3a = 4.8$
53. $1.5p = 7$ (round to nearest tenth)
54. $0.04x = 0.08 - x$ (round to nearest hundredth)
55. $0.4p = 0.014$
56. $0.47 = R + 0.4R$ (round to nearest hundredth)
57. $2.3 = 5.6 + y$
58. $4.3 = 0.3x - 7.34$
59. $2x + 3.7 = 10.3$
60. $0.16 + 2.3x = -0.3$
61. $1.5x + 2.1 = 3$
62. $3.82 - 2.5y = 1$
63. $0.15p = 2.4$

Solve the problems using decimal equations.

64. **AUTO**   If the formula for force is force = pressure × area, how many pounds of force are produced by a pressure of 35 pounds per square inch on a piston whose surface area is 2.5 in^2? Express the pounds of force as a decimal number.

65. **INDTR**   The circumference of a circle equals π times the diameter. If a steel rod has a diameter of 1.5 in., what is the circumference of the rod to the nearest hundredth? (Circumference is the distance around a circle.) Use the calculator π key.

66. **AUTO**   The distance formula is distance = rate × time. If a trucker drove 682.5 mi at 55 mi per h (mi/h), how long did she drive? (Answer to the nearest whole number.)

67. **AUTO**   The distance formula is distance = rate × time. If a tractor-trailer rig is driven 422.5 mi at 65 mi/h on interstate highways, how long does the trip take?

68. **ELEC**   Electrical resistance in ohms (Ω) is voltage V divided by amperage A. Find the resistance to the nearest tenth for a motor with a voltage of 12.4 V requiring 1.5 A.

69. **BUS**   Rita earned $84.75 working 7.5 h at a college bookstore. What was her hourly wage?

70. **ELEC**   The formula for electrical power is $V = \frac{W}{A}$: voltage (V) equals wattage (W) divided by amperage (A). Find the voltage needed for a circuit of 500 W with a current of 3.2 A.

71. **BUS**   Find the interest paid on a loan of $2,400 for 1 year at an interest rate of 11%.

72. **BUS**   Find the interest paid on a loan of $800 at $8\frac{1}{2}$% interest for 2 years.

73. **BUS**   Find the total amount of money (maturity value) that the borrower will pay back on a loan of $1,400 at $12\frac{1}{2}$% simple interest for 11 years.

74. **BUS**   Find the rate of interest on an investment of $2,500 made by Nurse Honda for a period of 2 years if she received $612.50 in interest.

75. **AG/H**   Maddy Brown needed start-up money for her landscape service. She borrowed $12,000 for 30 months and paid $360 interest on the loan. What interest rate did she pay?

Learning Outcomes   **1** Evaluate formulas.
2 Rearrange formulas to solve for a specified variable.

The most common use of formulas is for finding missing values. If we know values for all but one variable of a formula, we can find the unknown value.

1 Evaluate Formulas.

To **evaluate** a formula is to substitute known values for some variables and perform the indicated operations to find the value of the variable in question. In Chapter 1, Section 3, Outcome 2, we evaluated formulas when the missing value was already isolated on the left. We can now evaluate formulas when the missing value is not isolated.

To evaluate a formula:

1. Write the formula.
2. Rewrite the formula substituting known values for variables of the formula.
3. Solve the equation from Step 2 for the missing variable.
4. Interpret the solution within the context of the formula.

EXAMPLE   Solve the formula $P = 2(l + w)$ for w if $P = 12$ ft and $l = 4$ ft.

$$P = 2(l + w)$$ Perimeter of a rectangle = 2 times the sum of the length and the width. Substitute values.

$$12 = 2(4 + w)$$ Apply the distributive property.

$$12 = 8 + 2w$$ Isolate the term with the variable (addition axiom).

$$12 - 8 = 2w$$ Combine like terms.

$$4 = 2w$$ Divide.

$$\frac{4}{2} = \frac{2w}{2}$$

$$2 = w$$ Interpret solution.

The width is 2 ft.

The interest formula may be used to illustrate a variable that is one of several factors.

EXAMPLE
BUS

Evaluate the formula $I = PRT$ for principal (P) if interest $(I) = \$94.50$, rate $(R) = 21\%$, and time $(T) = \frac{1}{2}$ year.

For convenience in using a calculator, convert $\frac{1}{2}$ year to 0.5 year. In this formula, the rate should be expressed as a decimal equivalent, $21\% = 0.21$.

$I =$	PRT	Substitute values and solve for P.
$94.50 =$	$P(0.21)(0.5)$	Multiply 0.21 and 0.5.
$94.50 =$	$0.105\,P$	Divide.

$$\frac{94.50}{0.105} = \frac{0.105\,P}{0.105}$$

$900 = P$ \hspace{1cm} Interpret solution.

The principal is \$900.

EXAMPLE
AUTO

Evaluate the formula $E = \dfrac{I - P}{I}$ to the nearest thousandth if $I = 24{,}000$ calories (cal) and $P = 8{,}600$ cal. Round to thousandths.

$E = \dfrac{I - P}{I}$ \hspace{1cm} Engine efficiency = difference between heat input and heat output divided by heat input.

$E = \dfrac{24{,}000 - 8{,}600}{24{,}000}$ \hspace{1cm} Substitute given values. Perform calculations in numerator grouping.

$E = \dfrac{15{,}400}{24{,}000}$ \hspace{1cm} Divide.

$E = 0.6416666667$ \hspace{1cm} Round.

The engine efficiency is 0.642 (64.2% efficient).

EXAMPLE
ELEC

Evaluate the formula $R_T = \dfrac{R_1 R_2}{R_1 + R_2}$ if $R_1 = 10\ \Omega$ and $R_2 = 6\ \Omega$.

$R_T = \dfrac{R_1 R_2}{R_1 + R_2}$ \hspace{1cm} Total resistance = product of first resistance and second resistance divided by sum of first and second resistances.

$R_T = \dfrac{10(6)}{10 + 6}$ \hspace{1cm} Substitute given values. Perform calculations in numerator and denominator groupings.

$R_T = \dfrac{60}{16}$ \hspace{1cm} Divide.

$R_T = 3.75$ \hspace{1cm} Dimension analysis: $\dfrac{\text{ohms (\cancel{ohms})}}{\cancel{ohms}} = \text{ohms}$

The total resistance in the circuit is 3.75 Ω.

The next example has many steps and it is important to apply the order of operations. The formula is used to find the length of a belt connecting two pulleys.

EXAMPLE INDTEC

Evaluate $L = 2C + 1.57(D + d) + \dfrac{D + d}{4C}$ if $C = 24$ in., $D = 16$ in., and $d = 4$ in.

Round to hundredths. See Fig. 7–5.

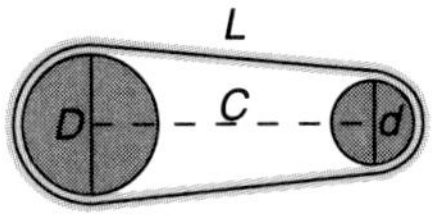

L = length of belt joining two pulleys
C = distance between centers of pulleys
D = diameter of large pulley
d = diameter of small pulley

Figure 7–5

$L = 2C + 1.57(D + d) + \dfrac{D + d}{4C}$ Substitute the given values.

$L = 2(24) + 1.57(16 + 4) + \dfrac{16 + 4}{4(24)}$ Work groupings in parentheses, numerator, and denominator.

$L = 2(24) + 1.57(20) + \dfrac{20}{96}$ Work multiplications and division.

$L = 48 + 31.4 + 0.2083333333$ Add.

$L = 79.60833333$ Round.

The pulley belt is 79.61 in. long.

2 Rearrange Formulas to Solve for a Specified Variable.

Formula rearrangement generally refers to isolating a letter term other than the one already isolated in the formula. Solving formulas in this manner shortens our work when doing repeated formula evaluations. After we solve the formula for the desired variable, we rewrite the formula with the variable on the left side for convenience and for use in electronic spreadsheets.

The following formulas require applying the addition axiom.

EXAMPLE BUS

Solve the markup formula, $M = S - C$, for S (selling price).

$M = S - C$ Markup = selling price − cost. Isolate S.

$M + C = S$ The coefficient of S is positive, so the formula is solved.

$S = M + C$ Rewrite S on the left for convenience.

TIP

Where Is the Variable or Unknown in a Formula?

It may help to think of the one letter we are solving for as the unknown or variable, and to think of the other letters as if they were *coefficients* or *constants* in an ordinary equation.

1. Determine which variable of the formula will be isolated (solved for).
2. Highlight or mentally locate all instances of the variable to be isolated.
3. Treat all other variables of the formula as you would treat numbers in an equation, and perform normal steps for solving an equation.
4. If the isolated variable is on the right side of the equation, interchange the sides so that it appears on the left side.

EXAMPLE
BUS

Solve the formula $M = S - C$ for C (cost).

$M = S - C$	Markup = selling price − cost. Isolate C.
$M - S = -C$	The coefficient of C is negative, so divide both sides by −1.
$\dfrac{M - S}{-1} = \dfrac{-C}{-1}$	Note effect on signs after division by −1.
$-M + S = C$	Write the positive term first.
$S - M = C$	Rewrite with C on the left for convenience.
$C = S - M$	

When a formula contains a term of several factors and we need to solve for one of those factors, we treat the factor being solved for as the variable and the other factors as its coefficient.

EXAMPLE
BUS

Solve for R in the formula $I = PRT$.

$I = PRT$	Interest = Principal × Rate × Time. Since we are solving for R, PT is its coefficient. Divide both sides by the coefficient of the variable.
$\dfrac{I}{PT} = \dfrac{PRT}{PT}$	Reduce.
$\dfrac{I}{PT} = R$	Rewrite with R on the left.
$R = \dfrac{I}{PT}$	

Sometimes formulas contain addition (or subtraction) and multiplication in which we must use the distributive property to solve for a particular letter. In the formula $A = P(1 + ni)$, we must use the distributive property to solve for either n or i because each appears inside the grouping.

EXAMPLE
BUS

Solve for n in the formula for compound amount, $A = P(1 + ni)$.

$A = P(1 + \boxed{n} i)$ Identify the variable to be isolated. Use the distributive property to remove n from parentheses.

$A = P + P\,\boxed{n}\,i$ Use the addition axiom to isolate the term containing n.

$A - P = P\,\boxed{n}\,i$ Divide both sides by the coefficient of n, Pi.

$\dfrac{A - P}{Pi} = \dfrac{P\,\boxed{n}\,i}{Pi}$ Reduce.

$\dfrac{A - P}{Pi} = \boxed{n}$ Rewrite with n on the left.

$\boxed{n} = \dfrac{A - P}{Pi}$

When the formula contains division, we clear the denominator before taking further steps.

EXAMPLE
ELEC

The formula $V = \dfrac{P}{I}$ represents the relationship among the voltage drop (V), the electrical power (P), and the current (I). Rearrange the formula to solve for P.

$V = \dfrac{P}{I}$ Identify the variable to be isolated. Multiply both sides by the denominator I to clear it.

$(I)V = \dfrac{P}{I}(I)$ Reduce.

$IV = \boxed{P}$ Rewrite with P on the left.

$P = IV$

Spreadsheets.

Information is often displayed in a table of rows and columns called a **spreadsheet.** These rows and columns may show the results of calculations, such as totals or percents, in addition to the original data. Many software programs such as Excel help you build an **electronic spreadsheet** and draw appropriate graphs using formulas and equations. The calculations are made automatically once the formulas are defined in the spreadsheet.

In a spreadsheet, key information is placed in a specific location called an **address,** and formulas are developed using these addresses as the variables. Spreadsheet programs are useful because key information can be changed while retaining the basic formulas of the spreadsheet. This process allows one to see quickly how various changes in the key data affect the results.

Let's examine a spreadsheet for the following situation. *The 7th Inning Sports Memorabilia Shop is developing an annual operating budget. The budget categories and the projected amount of expense for each category are shown in the spreadsheet in Fig. 7–6. The spreadsheet can be used to find the total operating budget for the year and the percent of total annual budget for each category in the projected annual budget. Formulas must be developed to calculate this information.*

	A	B	C
1	The 7th Inning Budgeted Operating Expenses		
2			
3	Expense	Budget Amount	Percent of Total Budget
4			
5	Salaries	$42,000.00	
6	Rent	$36,000.00	
7	Depreciation	$9,000.00	
8	Utilities and Phone	$14,500.00	
9	Taxes and Insurance	$14,000.00	
10	Advertising	$3,500.00	
11	Purchases	$120,000.00	
12	Other	$500.00	
13			
14	Total		

Figure 7–6

The spreadsheet program labels columns with letters like A, B, and C, and the rows with numbers. We'll use row 1 for the title of the spreadsheet, row 3 to label the columns of data, and column A to label the rows of data. Each position on the spreadsheet is a **cell,** and the program identifies each cell by its column letter and row number. For example, the amount budgeted for taxes and insurance is in cell B9 and is $14,000.00.

Spreadsheet Addresses

A spreadsheet address is an individual cell that is the intersection of a column and a row. This address is generally given in two parts: column and row.

Address	Interpretation
C 8	Column C and Row 8
J 30	Column J and Row 30
AA 5	Column AA (follows Column Z) and Row 5

Now we develop formulas to make the calculations. Each spreadsheet program gives various shortcuts for writing formulas and formats for giving instructions unique to that program; however, we will write the basic concepts used and add program-specific conventions as appropriate.

EXAMPLE   Write the formulas and make the calculations to complete the Fig. 7–6 spreadsheet.

To find the total budget to be placed in cell B14, we need to add the amounts in cells B5–B12. To do this, we give the addresses of the cells to be added in a formula: $B14 = B5 + B6 + B7 + B8 + B9 + B10 + B11 + B12$.

The percent of the total budget is calculated by dividing the specific amount by the total budget and then multiplying by 100. We will write a formula for each line of data. Most programs use an asterisk (*) to show multiplication, and a forward slash (/) to show division.

$C5 = B5/B14*100$ $C6 = B6/B14*100$ $C7 = B7/B14*100$
$C8 = B8/B14*100$ $C9 = B9/B14*100$ $C10 = B10/B14*100$
$C11 = B11/B14*100$ $C12 = B12/B14*100$

There are two ways to determine the value for cell C14. If we use the percent method, the percentage and the base are the same amount, so the total percent is 100%. To build in a *check* against the spreadsheet formulas, however, it is advisable to find the total percent by adding the calculated percents. It is easy to make a typing error in the formulas or to place the formula in the wrong cell. The total should be 100% or extremely close. There may be a small discrepancy due to the effects of rounding. $C14 = C5 + C6 + C7 + C8 + C9 + C10 + C11 + C12$. The spreadsheet with the completed calculations is shown in Fig. 7–7.

	A	B	C
1	The 7th Inning Budgeted Operating Expenses		
2			
3	Expense	Budget Amount	Percent of Total Budget
4			
5	Salaries	$42,000.00	17.5
6	Rent	$36,000.00	15.0
7	Depreciation	$9,000.00	3.8
8	Utilities and Phone	$14,500.00	6.1
9	Taxes and Insurance	$14,000.00	5.8
10	Advertising	$3,500.00	1.5
11	Purchases	$120,000.00	50.1
12	Other	$500.00	0.2
13			
14	Total	$239,500.00	100.0

Figure 7–7

1 Evaluate the formulas.

1. $D = RT$ if $R = 40$ mi/h and $T = 7$ h
2. $S = C + M$ if $C = \$40$ and $M = \$80$
3. $A = 4\pi r^2$ if $\pi = 3.14$ and $r = 15$ in.
4. $A = bh$ if $b = 12$ m and $h = 9.8$ m
5. $P = a + b + c$ if $a = 50$ cm, $b = 43$ cm, and $c = 45$ cm

Evaluate the interest formula $I = PRT$ using the following values.

6. **BUS** Find the interest if $P = \$800$, $R = 15.5\%$, and $T = 2\frac{1}{2}$ years.

7. **BUS** Find the rate if $I = \$427.50$, $P = \$1,500$, and $T = 2$ years.

8. **BUS** Find the time if $I = \$236.25$, $P = \$750$, and $R = 10.5\%$.

9. **BUS** Find the principal if $I = \$838.50$, $R = 21\frac{1}{2}\%$, and $T = 1\frac{1}{2}$ years.

Evaluate Exercises 10–12 using the percentage formula, $P = RB$.

10. **AG/H** Find the portion if $R = 15\%$ and $B = 600$ lb.

11. Find the rate if $P = 24$ kg and $B = 300$ kg.

12. **BUS** Find the base if $P = \$250$ and $R = 7.4\%$. Round to hundredths.

13. Evaluate the rate formula, $R - \dfrac{P}{B}$ for P if $R = 16\%$ and $B = 85$.

14. Evaluate the base formula, $B = \frac{P}{R}$, for R if $B = \$2,200$ and $P = \$374$.

15. **CAD/ARC** Find the length of a rectangular work area if the perimeter is 180 in. and the width is 24 in. Use the formula $P = 2(l + w)$.

16. Find the area of a circle whose radius is 54.5 cm using the formula $A = \pi r^2$ and using 3.14 for π. Round to the nearest tenth.

17. **BUS** Find the cost (C) if the markup (M) on an item is $\$5.25$ and the selling price (S) is $\$15.75$. Use the formula $M = S - C$.

18. Using the formula for the side of a square, $s = \sqrt{A}$, find the length of a side of a square field whose area is 16 mi^2.

19. **CON** Evaluate the formula for the area of a circle, $A = \pi r^2$, if $r = 7$ in. and $\pi = 3.14$. Round to the nearest tenth.

20. Evaluate the formula for the area of a square, $A = s^2$, if $s = 2.5$ km.

21. **BUS** Using the markup formula, $M = S - C$, find the selling price (S) if the markup (M) is $\$12.75$ and the cost (C) is $\$36$.

22. **ELEC** Use the formula $R_T = \dfrac{R_1 R_2}{R_1 + R_2}$ to find the total resistance (R_T) if one resistance (R_1) is 12 Ω and the second resistance (R_2) is 8 Ω.

23. **AUTO** What is the percent efficiency (E) of an engine if the input (I) is 25,000 calories and the output (P) is 9,600 calories? Use the formula $E = \dfrac{I - P}{I}$.

24. **AUTO** Distance is rate times time, or $D = RT$. Find the rate if the distance traveled is 140 mi and the time traveled is 4 h.

25. **ELEC** The formula for voltage (Ohm's law) is $E = IR$. Find the amperes of current (I) if the voltage (E) is 120 V and the resistance (R) is 80 Ω.

26. **INDTEC** According to Boyle's law, if temperature is constant, the volume of a gas is inversely proportional to the pressure on it. Find the final volume (V_2) of a gas using the formula $\dfrac{V_1}{V_2} = \dfrac{P_2}{P_1}$ if the original volume (V_1) is 15 ft^3, the original pressure (P_1) is 60 lb per square inch (psi), and the final pressure (P_2) is 150 psi.

27. **ELEC** The formula for power (P) in watts (W) is $P = I^2R$. Find the current (I) in amperes if a device draws 63 W and the resistance (R) is 7 Ω.

28. AUTO   Use the formula $H = \dfrac{D^2N}{2.5}$ to find the number of cylinders (N) required in an engine of 3.2 hp (H) if the cylinder diameter (D) is 2 in.

29. INDTEC   The formula for the speed (s) of a driven pulley in revolutions per minute (rpm) is $s = \dfrac{DS}{d}$. Find the speed of a driven pulley with a diameter (d) of 5 in. if the diameter (D) of the driving pulley is 10 in. and its speed (S) is 800 rpm.

30. INDTEC   If the distance (C) between the centers of the pulleys in Exercise 29 is 24 in., find the length (L) to the nearest hundredth of the belt connecting them using the formula $L = 2C + 1.57(D + d) + \dfrac{D + d}{4C}$

31. ELEC   Find the impedance (Z) in ohms using the formula $Z = \sqrt{R^2 + X^2}$ if the reactance (X) is 15 Ω and the resistance (R) is 8 Ω. Round to tenths.

2   Rearrange the formulas.

32. BUS   Solve $I_n = I - S$ for S, where I_n = new inventory, I = current inventory, and S = sales.

33. Solve $S = 2\pi rh$ for r.

34. Solve $y = mx + b$ for b.

35. Solve $V = \pi r^2 h$ for h.

36. Solve $S = C + M$ for C.

37. Solve $\dfrac{R}{100} = \dfrac{P}{B}$ for R.

38. Solve $P = 2(b + s)$ for b.

39. Solve $C = 2\pi r$ for r.

40. Solve $A = lw$ for l.

41. Solve $R = AC - B$ for C.

42. Solve $E = IR$ for R.

43. Solve $D = RT$ for R.

44. Solve $S = P - D$ for D.

45. Solve $C = \pi d$ for d.

46. Solve $C = 2\pi r$ for r.

47. BUS   The formula for finding the amount of a repayment on a loan is $A = I + P$, where A is the amount of the repayment, I is the interest, and P is the principal. Solve the formula for interest.

48. BUS   The formula for finding interest is $I = PRT$, where I represents interest, P represents principal, R represents rate, and T represents time. Rearrange the formula to find the time.

49. Develop formulas to complete the spreadsheet in Fig. 7–8 to show data for the actual expenses for the 7th Inning Sports Memorabilia Shop.

	A	B	C	D	E	F
1	The 7th Inning Budgeted Operating Expenses And Actual Expenses					
2						
3	Expense	Budget Amount	Percent of Total Budget	Actual Expenses	% of Actual Total Expense	% Difference from Budget
4						
5	Salaries	$45,000.00		$42,000.00		
6	Rent	$37,000.00		$36,000.00		
7	Depreciation	$12,000.00		$14,000.00		
8	Utilities and Phone	$13,000.00		$10,862.56		
9	Taxes and Insurance	$15,000.00		$13,583.29		
10	Advertising	$2,000.00		$2,847.83		
11	Purchases	$125,000.00		$132,894.64		
12	Other	$2,000.00		$1,356.35		
13						
14	Total					

Figure 7–8

50. Use the formulas to complete the spreadsheet for the 7th Inning Sports Memorabilia Shop. The percent difference from the budget uses the budget as the base. Negative percents show percents under budget and positive percents show percents over budget.

Learning Outcomes

What to Remember with Examples

Section 7–1

1 Identify equations, terms, factors, constants, variables, and coefficients (pp. 300–303).

An *equation* is a statement that two quantities are equal. A *variable* is a letter that represents an unknown value. A *root* or *solution* of an equation is the value of the variable that makes the equation a true statement.

$x = 5 + 2$ is an *equation*. x is the *variable*. 7 is the *root* or *solution*.

Factors are expressions of multiplication.

$5a$ means 5 *times a*. 5 is a *factor* of $5a$. a is a *factor* of $5a$.

Terms are algebraic expressions that are single quantities or products or quotients of quantities.

In the expression $5a + 3b + 7$, $5a$ is a *term* and $3b$ is a *term*. 7 is a *term*.

Constants are terms that contain only numbers.

In $5a + 3b + 7$, the *constant* term is 7.

Variable terms are terms that have at least one letter.

In $5a + 3b + 7$, $5a$ and $3b$ are *variable* terms.

A *coefficient* is one factor as it relates to the remaining factors of a term.

In $5a + 7$, 5 is the *coefficient* of a and a is the *coefficient* of 5. 7 has no coefficient. The coefficient 5 is also called the *numerical coefficient*.

2 Write verbal interpretations of symbolic statements (p. 303).

Mathematical symbols can be translated into phrases and statements.

$2x - 3 = 5$ can be translated into "3 less than twice a number is five."

3 Translate verbal statements into symbolic statements using variables (pp. 303–304).

Statements can be translated into mathematical symbols.

The statement "A number increased by 13 results in 52" is translated into $x + 13 = 52$.

4 Simplify variable expressions (p. 304).

Combine like terms by adding or subtracting the coefficients of the terms and by using the same letter or letters for the sum or difference.

$5m + 4m = (5 + 4)m = 9m$

Apply the distributive principle by multiplying the factor outside the parentheses by each term inside the parentheses.

$-4(x + 2y - 5) = -4x - 8y + 20$

Distributing -1 changes only the sign of each term inside the parentheses.

$-(2x - 3) = -1(2x - 3) = -2x + 3$

Section 7–2

1 Solve linear equations using the addition axiom (pp. 307–309).

1. Locate the variable in the equation. **2.** Identify the constant that is associated with the variable by addition (or subtraction). **3.** Add the opposite of the constant to both sides.

$$x + 3 = 2 \qquad \text{Add } -3 \text{ to both sides.}$$
$$x + 3 \boxed{-3} = 2 - 3 \qquad \text{Combine like terms.}$$
$$x + 0 = -1 \qquad x + 0 = x$$
$$x = -1$$

$$9 = x - 6 \qquad \text{Add 6 to both sides.}$$
$$9 + 6 = x - 6 + 6 \qquad \text{Combine like terms.}$$
$$15 = x + 0 \qquad x + 0 = x$$
$$15 = x$$

2 Solve linear equations using the multiplication axiom (pp. 309–311).

Multiply both sides of the equation by the reciprocal of the coefficient of the letter term *or* divide both sides of the equation by the coefficient of the letter term.

Solve:

$$\frac{x}{5} = 8 \qquad\qquad -8x = 24 \qquad\qquad \frac{1}{3}x = \frac{4}{9}$$

$$\frac{5}{1}\left(\frac{x}{5}\right) = 8\left(\frac{5}{1}\right) \qquad \frac{-8x}{-8} = \frac{24}{-8} \qquad \left(\frac{\overset{1}{\cancel{3}}}{1}\right)\left(\frac{1}{\cancel{3}}x\right) = \left(\frac{4}{\cancel{9}}\right)\left(\frac{\cancel{3}}{1}\right)$$

$$x = 40 \qquad\qquad x = -3$$

$$x = \frac{4}{3}$$

$$7.5 = 2.5x \qquad\qquad \frac{x}{0.6} = 2.9$$

$$\frac{7.5}{2.5} = \frac{2.5x}{2.5} \qquad 0.6\left(\frac{x}{0.6}\right) = 0.6(2.9)$$

$$3 = x \qquad\qquad x = 1.74$$

To check the solution of an equation, substitute the value of the variable in each place it appears in the equation. Perform operations on both sides of the equation. The two sides of the equation should be equal.

Verify that $x = 3$ is the solution for the equation $2x - 1 = 5$.

$$2(3) - 1 = 5$$
$$6 - 1 = 5$$
$$5 = 5$$

3 Solve linear equations with like terms on the same side of the equation (pp. 311–312).

Combine the like terms on the same side of the equation. Solve the remaining equation using the multiplication axiom.

Solve $3x - 5x = 12$ for x.

$$3x - 5x = 12$$
$$-2x = 12$$
$$\frac{-2x}{-2} = \frac{12}{-2}$$
$$x = -6$$

Chapter Review of Key Concepts

343

| **4** Solve linear equations with like terms on opposite sides of the equation (pp. 312–314). | Use the addition axiom to move variable terms to one side of the equation and constants to the other. Combine like terms. Solve the remaining equation using the multiplication axiom. |

Solve $x - 5 = 7$ for x.

$$x - 5 = 7$$
$$x - 5 + 5 = 7 + 5$$
$$x = 12$$

Solve $7.2 = x - 3.5$ for x.

$$7.2 = x - 3.5$$
$$7.2 + 3.5 = x$$
$$10.7 = x$$

Solve $x - \dfrac{3}{8} = \dfrac{5}{8}$ for x.

$$x - \dfrac{3}{8} = \dfrac{5}{8}$$
$$x = \dfrac{5}{8} + \dfrac{3}{8}$$
$$x = \dfrac{8}{8}$$
$$x = 1$$

Solve $3x - 5 = 5x + 7$ for x.

$$3x - 5 = 5x + 7$$
$$3x - 5x = 7 + 5$$
$$-2x = 12$$
$$\dfrac{-2x}{-2} = \dfrac{12}{-2}$$
$$x = -6$$

Section 7–3

| **1** Solve linear equations that contain parentheses (pp. 315–318). | Remove parentheses using the distributive property. Continue solving the equation. |

Solve $3(2x - 5) = 8x + 7$ for x.

$3(2x - 5) = 8x + 7$	Distribute.
$6x - 15 = 8x + 7$	Sort terms (addition axiom).
$6x - 8x = 7 + 15$	Combine like terms.
$-2x = 22$	Divide by coefficient of x (multiplication axiom).
$\dfrac{-2x}{-2} = \dfrac{22}{-2}$	
$x = -11$	

Six-Step Problem-Solving Plan

A shipment of college textbooks is sent in two boxes weighing 37 lb total. One box weighs 9 lb more than the other. What does each box weigh?

Unknown facts

Weight of each box.

Known facts

Total weight of two boxes is 37 lb.
One box weighs 9 lb more than the other.

Relationships

x = weight of one box $x + x + 9 = 37$
$x + 9$ = weight of other box

Estimation

If the boxes were the same weight, each would weigh $18\frac{1}{2}$ lb.
Thus, one box will be less than and one box will be more than $18\frac{1}{2}$ lb.

Calculations

$$x + x + 9 = 37 \qquad \text{Combine like terms.}$$

$$2x + 9 = 37 \qquad \text{Sort terms (addition axiom).}$$

$$2x = 37 - 9 \qquad \text{Combine like terms.}$$

$$2x = 28 \qquad \text{Divide (multiplication axiom).}$$

$$\frac{2x}{2} = \frac{28}{2}$$

$$x = 14 \text{ lb} \qquad \text{Weight of one box}$$

$$x + 9 = 23 \text{ lb} \qquad \text{Weight of the other box}$$

Interpretation

The two boxes weigh 14 and 23 lb.

Section 7–4

1 Solve fractional equations by clearing the denominators (pp. 319–325).

To clear the equation of denominators: Multiply the *entire* equation by the least common multiple (LCM) of the denominators of the equation.

$$\frac{3}{a} + \frac{1}{3} = \frac{2}{3a} \qquad \text{LCM} = 3a. \text{ Excluded value} = 0.$$

$$(3a)\left(\frac{3}{a}\right) + (3a)\left(\frac{1}{3}\right) = (3a)\left(\frac{2}{3a}\right) \qquad \text{Multiply by } 3a.$$

$$(\overset{3}{\cancel{3a}})\left(\frac{3}{\cancel{a}}\right) + (\overset{a}{\cancel{3a}})\left(\frac{1}{\cancel{3}}\right) = (\overset{1}{\cancel{3a}})\left(\frac{2}{\cancel{3a}}\right) \qquad \text{Reduce.}$$

$$9 + a = 2$$

$$a = 2 - 9$$

$$a = -7$$

If a variable is in a denominator, check the root to see if it makes a true statement and is not an *extraneous root.*

Check the example above:

$$\frac{3}{-7} + \frac{1}{3} = \frac{2}{3(-7)} \qquad \text{Substitute } -7 \text{ for } a.$$

$$\frac{3}{-7} + \frac{1}{3} = \frac{2}{-21} \qquad \text{LCM} = 21.$$

$$-\frac{9}{21} + \frac{7}{21} = \frac{2}{-21}$$

$$-\frac{2}{21} = -\frac{2}{21} \qquad \text{The root makes a true statement.}$$

2 Solve applied problems involving rate, time, and work (pp. 325–328).

Amount of work: Rate $\times$ time = amount of work

A *rate* is a ratio of two measures such as 1 card per 15 min or $\dfrac{1 \text{ card}}{15 \text{ min}}$.

Lashonda can install a PC video card in 15 min. How many cards can she install in $1\frac{1}{2}$ hours?

amount of work (W) = rate $\times$ time

$$W = \frac{1 \text{ card}}{15 \text{ min}} (90 \text{ min}) \qquad 1\frac{1}{2}\text{ h} = 90 \text{ min}$$
$$W = 6 \text{ cards}$$

Completing one job when working together:
(A's rate $\times$ time) + (B's rate $\times$ time) = 1 job

Let T = time. Let 1 = 1 completed job.

Galenda assembles a product in 10 min and Marcus assembles the product in 15 min. How long will it take them together to assemble one product?
Galenda's rate $\times$ time: $\frac{1}{10}T$. Marcus's rate $\times$ time: $\frac{1}{15}T$.

$$\frac{1}{10}T + \frac{1}{15}T = 1$$

Estimation: less than 10 min but more than half of 10, or 5 min for both to do one job.
Galenda's rate: $\frac{1}{10}$ job per min.

$$(30)\left(\frac{1}{10}T\right) + (30)\left(\frac{1}{15}T\right) = (30)(1)$$

Marcus's rate: $\frac{1}{15}$ job per min.
Time in minutes: T. LCM = 30.
Multiply by 30.

$$\overset{3}{(\cancel{30})}\left(\frac{1}{\underset{1}{\cancel{10}}}T\right) + \overset{2}{(\cancel{30})}\left(\frac{1}{\underset{1}{\cancel{15}}}T\right) = (30)(1)$$

Reduce.

$$3T + 2T = 30$$
$$5T = 30 \qquad\qquad \text{Combine like terms.}$$
$$\frac{5T}{5} = \frac{30}{5} \qquad\qquad \text{Divide.}$$
$$T = 6 \text{ min}$$

Completing one job when working in opposition:
(A's rate $\times$ time) $-$ (B's rate $\times$ time) = 1 job

Solve as above but *subtract* the two amounts of work.

An inlet valve fills a vat in 2 h. A drain valve empties the vat in 5 h. With both valves open, how long does it take for the vat to fill?
Inlet valve's rate $\times$ time: $\frac{1}{2}T$. Drain valve's rate $\times$ time: $\frac{1}{5}T$.

$$\frac{1}{2}T - \frac{1}{5}T = 1 \qquad\qquad \text{Estimation: more than 2 h.}$$

$$(10)\left(\frac{1}{2}T\right) - (10)\left(\frac{1}{5}T\right) = (10)1 \qquad\qquad \begin{array}{l}\text{LCM} = 10.\\ \text{Multiply by 10.}\end{array}$$

$$\overset{5}{(\cancel{10})}\left(\frac{1}{\underset{1}{\cancel{2}}}T\right) - \overset{2}{(\cancel{10})}\left(\frac{1}{\underset{1}{\cancel{5}}}T\right) = (10)1 \qquad\qquad \text{Reduce.}$$

$$5T - 2T = 10$$
$$3T = 10$$
$$\frac{3T}{3} = \frac{10}{3}$$
$$T = 3\frac{1}{3}\text{ h}$$

Chapter 7 / Linear Equations

3 Solve decimal equations by clearing the decimals (pp. 329–330).

To clear the equation of decimals: Multiply the *entire* equation by the least common denominator (LCD) of the fractional amounts represented by the decimals. (The place value of the decimal amount with the most places after the decimal point will be the LCD.)

$$3.5x + 2.75 = 10 \qquad \text{Place value of the LCM} = 100.$$
$$(100)(3.5x) + (100)(2.75) = (100)(10) \qquad \text{Multiply by 100.}$$
$$350x + 275 = 1{,}000 \qquad \text{Sort.}$$
$$350x = 1{,}000 - 275 \qquad \text{Combine like terms.}$$
$$350x = 725 \qquad \text{Divide.}$$
$$\frac{350x}{350} = \frac{725}{350}$$
$$x = 2.071428571$$
$$x = 2.07 \quad \text{(rounded)}$$

Section 7–5

1 Evaluate formulas (pp. 333–335).

Substitute values. Solve using rules for solving equations and/or the order of operations.

Find the length (l) if the area (A) is 8 ft^2 and the width (w) is 2 ft.

$$A = lw \qquad \text{Substitute values.}$$
$$8 = (l)(2) \qquad \text{Divide.}$$
$$l = 4 \text{ ft}$$

If the perimeter (P) of a square is 12 in., find the length of a side.

$$P = 4s \qquad \text{Substitute value for } P.$$
$$12 = 4s \qquad \text{Divide.}$$
$$3 = s \qquad \text{Interchange sides.}$$
$$s = 3 \text{ in.}$$

2 Rearrange formulas to solve for a specified variable (pp. 335–339).

Isolate the desired variable so that it appears on the left. This can make evaluation or using spreadsheets more efficient. Apply appropriate rules for solving equations.

Solve the formula $S = \dfrac{R + P}{2}$ for R.

$$(2)S = \frac{R + P}{2}(2) \qquad \text{Clear the denominator.}$$
$$2S = R + P \qquad \text{Isolate } R \text{ (sort terms).}$$
$$2S - P = R \qquad \text{Interchange sides so the variable appears on the left.}$$
$$R = 2S - P$$

CHAPTER REVIEW EXERCISES

Section 7–1

Verify that the statements are true equations.

1. $5 + 7 = 18 - 6$

2. $9(8) = 12 + 3(4) + 7^2 - 1$

3. $8 - 9 + 3^2 = -2^2 + 12$

4. $7 + 3 = 8 \cdot 1 - 5 + 4 + 3$

Find the value of the variable that makes the equation true.

5. $x = 5 + 9$

7. $y = \dfrac{48}{-6}$

6. $8 - 5 = x$

8. $7 + 3(5 - 2) = x$

Identify the terms in the expressions by drawing a box around each term.

9. $15x - \dfrac{3a}{7} + \dfrac{(x - 7)}{5}$

10. $5x - 8 + \dfrac{3}{y}$

State the equations in words.

11. $x + 5 = 2$

13. $\dfrac{x}{8} = 7$

12. $x - 7 = 11$

14. $3(x + 7) = -3$

Write the following statements in symbols.

15. 7 more than twice a number is 11.

16. A certain stock listed on the New York Stock Exchange closed at 42.375, a decrease of 3.125 points from the opening price.

17. Twice the sum of a number and 8 is 40.

18. The print shop used 31 cases of copy paper during one month. End-of-the-month inventory indicated 172 cases on hand. Write an equation to find the number of cases on hand at the beginning of the month. Then, solve the equation.

Simplify.

19. $3a + 2a$

22. $-(4y - 7)$

25. $11 - 2(x - 5)$

28. $3(x - 2) - 2(5x - 7y + 1)$

20. $7a - 3b + 5 + 7a - 9b$

23. $-3(a + 2) - 5$

26. $-8(5x - 3y - 15)$

29. $15a + 3(5a - 7b + 4)$

21. $3(2y - 4) + y$

24. $5 - (a - 3)$

27. $3 - (2x - 4y - 8)$

30. $12 - (x + y - 5)$

Section 7–2

Solve.

31. $x - 5 = 8$

34. $x - 15 = -7$

37. $x + 7 = 10$

40. $3 + y = -5$

43. $3x + 4 = 19$

46. $5 = 3x - 7$

49. $5x - 12 = 9x$

52. $4x = -28$

55. $3 = \dfrac{1}{5}x$

58. $0.6 = -a$

61. $42 = -\dfrac{6}{7}x$

64. $5y - 7y = 14$

67. $21 = x + 2x$

70. $36 = 9a - 5a$

73. $-12 = -8 - 2x$

76. $10 + 4x = 5 - x$

79. $7x - 5 + 2x = 3 - 4x + 12$

32. $y + 7 = 3$

35. $x - 5 = 14$

38. $x - 5 = 3$

41. $1 = a - 4$

44. $4x - 3 = 9$

47. $-7 = 6x - 31$

50. $7x = 8x + 4$

53. $-15 = 2b$

56. $-\dfrac{2}{7}x = 8$

59. $-\dfrac{3}{8}x = -24$

62. $-\dfrac{5}{8}x = -10$

65. $2b - 7b = 10$

68. $8x - 3x = 6 + 9$

71. $20 - 4 = 2x - 6x$

74. $12x + 27 = 3x$

77. $7 - 4y = y + 22$

80. $2x - 3 + 15 = 7x - 8 - 6x$

33. $x - 8 = -10$

36. $-2 = 8 - x$

39. $x - 3 = -4$

42. $t + 7 = 12$

45. $15 - 3x = -6$

48. $4 = 7 - 4x$

51. $3x = 21$

54. $-5 = -m$

57. $-7y = -49$

60. $\dfrac{1}{2}x = -5$

63. $\dfrac{1}{7}x = 12$

66. $0 = 4t - t$

69. $4x + x = 25$

72. $13 - 27 = 3x - 10x$

75. $3x + 9 = 10 + 3x + 1$

78. $7x - 1 = 4x + 17 + 7x$

81. $4y + 8 = 3y - 4$

82. $y - 5 = 6y + 30$
83. $8 - 2y = 15 - 3y$
84. $6 - 7x = 15 - x$
85. $5x - 12 = 2x + 15$
86. $18x - 21 = 15x + 33$
87. $3x - 5x + 2 = 6x - 5 + 12x$
88. $\dfrac{R}{7} - 6 = -R$
89. $0 = \dfrac{8}{9}c + \dfrac{1}{4}$
90. $0.9R = 0.3$ (round to tenths)
91. $0.86 = R + 0.4R$ (round to hundredths)
92. $0.04y = 0.02 - y$ (round to hundredths)
93. $18 = 6(2 - y)$
94. $4(6 + x) = 36$
95. $7x - 3(x - 8) = 28$
96. $3(x + 2) - 5 = 2x + 7$
97. $5(3 - 2x) = -5$
98. $3(2x + 1) = -3$

Section 7–3

Solve the equations.

99. $3x = 3(9 + 2x)$
100. $4a = 8 - (a + 7)$
101. $5x = 7 + (x + 5)$
102. $3(x + 2) - 5 = 2x + 7$
103. $4(3 - x) = 2x$
104. $3(2x - 4) = 4x - 6$
105. $-2(4 - 2x) = -16 + 2x$
106. $-16 = -2(-2x + 4)$
107. $8 = 6 - 2(3x - 1)$
108. $4x - (x + 3) = 3x - 3$
109. $3(x - 1) = 18 - 2(x + 3)$
110. $-(x - 1) = 2(x + 7)$
111. $-(2x + 1) = -7$
112. $2 + 3(x - 4) = 2x - 5$
113. $7 = 3 + 4(x + 2)$
114. $7(x + 2) = -6 + 2x$
115. $3(4x + 3) = 3 - 4(x - 1)$
116. $3(2 - x) - 1 = 4(3 - x)$

Write the statements as equations and solve.

117. The difference between x and 6 is 8. Find x.

118. Twice a number increased by 5 is 17. Find the number.

119. 5 times the sum of x and 6 is 42 more than x. Find x.

120. How many gallons of water must be added to 46 gal of pure alcohol to make 100 gal of alcohol solution?

121. **BUS** If one technician works 3 hr less than another and their total hours worked are 51, how many hours has each technician worked?

122. **CON** The shorter side of an L-shaped carpenter's square is 6 in. shorter than the longer side. If the total length of the carpenter's square is 24 in., what is the measure of each side?

123. **AG/H** Ms. Galendez's backyard is a rectangle whose length is twice the width. If the perimeter of the yard is 720 ft, what are the dimensions of her yard?

124. **BUS** Brubakers catered 32 chicken dinners for $409. This included a $25 delivery charge. Find the cost of each dinner excluding the delivery charge.

Section 7–4

Solve the equations.

125. $m + \dfrac{1}{4} = \dfrac{3}{4}$
126. $\dfrac{3}{5}y = 12$
127. $p = \dfrac{1}{2} + \dfrac{1}{3}$
128. $x + \dfrac{1}{7}x = 16$
129. $\dfrac{2}{5} - x = \dfrac{1}{2}x + \dfrac{4}{5}$
130. $\dfrac{R}{7} - 6 = -R$
131. $\dfrac{3}{7}m - \dfrac{1}{2} = \dfrac{2}{3}$
132. $\dfrac{1}{4}S = \dfrac{1}{4} + \dfrac{1}{10} + \dfrac{1}{20}$
133. $m = 2 + \dfrac{1}{4}m$

Solve the equations. Identify excluded values.

134. $\dfrac{1}{R} = \dfrac{1}{10} + \dfrac{1}{3} + \dfrac{1}{6}$
135. $\dfrac{2}{P} = \dfrac{1}{2} + \dfrac{1}{4} - \dfrac{5}{12}$
136. $\dfrac{3}{x} + 4 = \dfrac{1}{5} - 7$

Set up an equation with fractions and solve.

137. **AG/H** Melissa can complete a landscape project in 3 h. Henry can complete the same project in 7 h. How long would it take Melissa and Henry working together to complete the landscape project?

138. **COMP** One optical scanner reads a stack of sheets in 20 min. A second scanner reads the same stack in 12 min. How long does it take for both scanners together to process the one stack of sheets?

139. ELEC An apprentice electrician can install 5 light fixtures in 2 h. How many light fixtures can be installed in 10 h?

140. CON A brick mason can erect a retaining wall in 6 h. The brick mason's apprentice can do the same job in 10 h. How much time does it take both of them working together to erect the retaining wall?

141. ELEC A parallel dc circuit has 3 branches rated at $2\,\Omega$, $6\,\Omega$, and $12\,\Omega$. Find the total resistance to the nearest hundredth. See example on p. 324.

142. ELEC Find the total resistance of a parallel circuit with 2 branches rated at $12\,\Omega$ and $16\,\Omega$. Round to hundredths.

Solve the equations. Round to tenths if necessary.

143. $2.3x - 4.1 = 0.5$

144. $0.22 + 1.6x = -0.9$

145. $0.3x - 2.15 = 0.8x + 3.75$

146. AUTO The distance formula is distance = rate $\times$ time. If a portable MRI unit traveled 350.8 mi to and from a rural hospital at 50 mi/h, how long to the nearest hour did the trip to and from the hospital take?

147. ELEC Electrical resistance in ohms (Ω) is voltage (V) divided by amperage (A). Find the resistance of a small motor with a voltage of 8.5 V requiring 0.5 A.

Section 7–5

Evaluate the interest formula, $I = PRT$, using the given values.

148. Find the rate if $I = \$2,484$, $P = \$4,600$, and $T = 3$ years.

149. BUS Find the time if $I = \$387.50$, $P = \$1,550$, and $R = 12.5\%$.

150. Evaluate the formula for the area of a square, $A = s^2$, if $s = 3.25$ km.

151. ELEC Ohm's law is $E = IR$. Find the amperes of current (I) if the voltage (E) is 220 V and the resistance (R) is $80\,\Omega$.

152. ELEC Use the formula $R_t = \dfrac{R_1 R_2}{R_1 + R_2}$ to find the total resistance (R_t) if one resistance (R_1) is $10\,\Omega$ and the second resistance (R_2) is $9\,\Omega$. Round to tenths.

153. ELEC The formula for power (P) in watts (W) is $P = I^2 R$. Find the current (I) in amperes if a device draws 392 W and the resistance (R) is $8\,\Omega$.

154. AUTO Use the formula $E = \dfrac{I - P}{I}$ to find the percent efficiency (E) of an engine if the input (I) is 22,600 cal and the output (P) is 5,600 cal. Round to the nearest tenth of a percent.

Solve the formulas for the indicated variable.

155. $V = lwh$ for w

156. $s = c + m$ for c

157. $s = r - d$ for r

158. $s = r - d$ for d

159. $v = v_0 - 32t$ for t

160. $V = \frac{1}{3}Bh$ for h

161. BUS The formula for finding the sale price on an item is $S = P - D$, where S is the sale price, P is the original price, and D is the discount. Solve the formula for the original price.

162. BUS The formula for finding tax is $T = RM$, where T represents tax, R represents the tax rate, and M represents the marked price. Rearrange the formula to find the marked price.

1. A formula is an equation that gives a model for solving a certain type of application. Devise formulas for the following relationships.
   (a) An electrical power company computes the monthly charges by multiplying the kilowatts of power used times the cost per kilowatt and adds to that a fixed monthly fee.
   (b) A store calculates the ending balance on a charge account by multiplying the interest rate times the previous unpaid balance and then adding the previous balance, the interest, and purchases and subtracting payments.
   (c) Profit on the sale of a particular item is the product of the number of items sold and the difference between the selling price of the item and its cost to the seller.

2. Formula rearrangement is a way of devising variations of formulas.
   (a) Explain the usefulness of formula rearrangement.
   (b) Select a formula that has at least three variables. Find a variation of the formula for each variable of the formula.
   (c) Describe at least two occasions when it is desirable to rearrange a formula.

1. Describe like terms and explain how they are combined. Give an example and combine the terms.
3. If you solve an equation and get $7 = x$, but want to write your root as $x = 7$, what mathematical property allows you to do so?

5. Write the following in the proper sequence for solving equations. An item can be used more than once.
   (a) Arrange the equation using the addition axiom so that number terms are on one side and letter terms are on the other.
   (b) Divide both sides by the coefficient of the letter term or multiply both sides by the reciprocal of the coefficient of the letter term (multiplication axiom).
   (c) Apply the distributive property to remove any parentheses.
   (d) Combine like terms that are on the same side of the equal sign.
7. Write an example of an equation that requires the use of the addition axiom. Solve the equation. If several steps are required, identify the step that applies the addition axiom.
9. Write an example of an equation that requires the distributive property. Solve the equation. If several steps are required, identify the step that applies the distributive property.

2. Explain the difference between *factors* and *terms*. Give an example of each.
4. Write a paragraph explaining how the natural numbers, integers, rational numbers, and real numbers are related. Diagram the relationship of these numbers, and give examples of each type of number.
6. Give an example of an equation that requires the use of the multiplication axiom. Solve the equation. If the solution requires several steps, identify the step that applies the multiplication axiom.

8. Solve the equation $\frac{3}{4}x = \frac{5}{8}$ in two different ways. Why are the two ways equivalent? What property of equality was used?

10. Explain the estimation step we use to solve applied problems. Why is this step important in the problem-solving process?

Identify and explain the first mistake encountered in each problem. Then work the problem correctly.

11. $7(x - 3) + 2 = 4 + 2x$
$7x - 3 + 2 = 4 + 2x$
$7x - 1 = 4 + 2x$
$5x - 1 = 4$
$5x = 5$
$x = 1$

12. $5 - 3(x + 2) = 3(x + 1)$
$2(x + 2) = 3(x + 1)$
$2x + 4 = 3x + 3$
$-x = -1$
$x = 1$

13. $5x - (3x - 6) = 18$
$5x - 3x - 6 = 18$
$2x - 6 = 18$
$2x = 18 + 6$
$2x = 24$
$x = 12$

PRACTICE TEST

Solve. Round to hundredths if necessary.

1. $x + 5 = 19$

2. $-8y = 72$

3. $\dfrac{x}{2} = 5$

4. $5x + 2x = 49$

5. $5 - 2x = 3x - 10$

6. $5x + 3 - 7x = 2x + 4x - 11$

7. $3(x + 4) = 18$

8. $3x + 2 = 4(x - 1) - 1$

9. $\dfrac{8}{y + 2} = -7$

10. $\dfrac{4}{5}z + z = 8$

11. $5x + \dfrac{3}{5} = 2$

12. $3x + 2 = \dfrac{2}{3}$

13. $\dfrac{3}{5}x + \dfrac{1}{10}x = \dfrac{1}{3}$

14. $\dfrac{1}{x} = \dfrac{1}{3} + \dfrac{5}{6}$

Solve the equations. Round to hundredths when necessary.

15. $1.3x = 8.02$

16. $4.5y + 1.1 = 3.6$

17. $0.18x = 300 - x$

18. $7.9 = 0.5x - 8.35$

19. $0.23 + 7.1x = -0.8$

Solve the problems involving fractions and decimal numbers.

20. A pipe fills 1 tank in 4 h. If a second pipe empties 1 tank in 6 h, how long does it take for the tank to fill with both pipes operating?

21. Ohm's law is $E = IR$. Find the amperes of current (I) if the voltage (E) is 110 V and the resistance (R) is 50 Ω?

22. The formula for the volume (V) of a solid rectangular figure is $V = lwh$ (length $\times$ width $\times$ height). If the volume of a mailing container is 7.5 m³, its length is 1.5 m, and its width is 0.5 m, what is its height?

23. The electrical resistance of a wire is found from the formula $R = \dfrac{PL}{A}$. Rearrange the formula to find the length L of the wire.

24. A pipe fills 1 tank in 8 h. A second pipe fills the tank in 12 h. How long does it take to fill the tank with both pipes filling the tank?

25. Engine displacement d is found using the formula $d = \pi r^2 sn$. Solve to find s.

26. Use the formula $R_t = \dfrac{R_1 R_2}{R_1 + R_2}$ to find the total resistance (R_t) if one resistance (R_1) is 9 Ω and the second resistance (R_2) is 8 Ω. Round to tenths.

27. If the efficiency (E) of an engine is 70% and the input (I) is 40,000 cal, find the output (P) in calories. Use the formula $E = \dfrac{I - P}{I}$.

8

Ratio and Proportion

Focus on Careers

Machinists use tools such as lathes and milling machines to produce precision metal parts. Machinists produce large quantities of the same part and also produce small batches of one-of-a-kind parts. They must be able to read blueprints, work with machines, and know the properties of metals.

Machinists learn the required knowledge and skills at community or technical colleges or vocational schools, and some learn through a combination of apprenticeship programs and schooling. Machinists should be mechanically inclined, have good problem-solving abilities, and be able to do highly accurate work. In some highly precise parts, tolerances may reach 0.00001 in.! Mathematics, blueprint reading, metalworking, and drafting courses are musts for machinists.

Job opportunities for machinists are expected to continue to be excellent because the number of workers entering the field is expected to be less than the number of job openings that arise from employment growth and from replacing machinists who retire.

The median hourly wage for machinists in 2002 was $15.66, but the top 10% of machinists earned more than $23.17 per hour. Metalworking machinery manufacturing jobs paid the highest hourly wage, and motor vehicle parts manufacturing paid slightly lower wages.

Source: *Occupational Outlook Handbook,* 2004–2005 Edition, U.S. Department of Labor, Bureau of Labor Statistics.

Learning Outcome **1** Solve equations that are proportions.

A common type of equation that contains fractions is a *proportion.* In a proportion, *each side* of the equation is a fraction or ratio. In Chapter 3, Section 2, Outcome 3, we used the percentage proportion to solve percentage problems.

1 Solve Equations That Are Proportions.

An equation in the form of a proportion can be used to solve many types of applied problems. As with the percentage proportion, we use the property of proportions to solve equations in the form of a proportion.

Property of proportions:

The cross products in a proportion are equal. Symbolically, if $\dfrac{a}{b} = \dfrac{c}{d}$, then $a \cdot d = b \cdot c$, provided that b and d are not equal to zero. Also, if $a \cdot d = b \cdot c$, then $\dfrac{a}{b} = \dfrac{c}{d}$, a, b, c and d represent any real number provided that b and d are not zero.

EXAMPLE Solve the proportions.

(a) $\dfrac{x}{4} = \dfrac{9}{6}$ (b) $\dfrac{4x}{5} = \dfrac{17}{20}$ (c) $\dfrac{x-2}{x+8} = \dfrac{3}{5}$ (d) $\dfrac{3}{x} = 7$

(a) $\dfrac{x}{4} = \dfrac{9}{6}$ Cross multiply. $x(6) = \boxed{6x}$; $4(9) = \boxed{36}$

$6x = \boxed{36}$ Solve for x.

$\dfrac{6x}{6} = \dfrac{36}{6}$

$x = 6$

(b) $\dfrac{4x}{5} = \dfrac{17}{20}$ Cross multiply. $4x(20) = \boxed{80x}$; $5(17) = \boxed{85}$

$80x = \boxed{85}$ Divide by coefficient of x.

$\dfrac{80x}{80} = \dfrac{85}{80}$

$x = \dfrac{85}{80}$ Simplify.

$x = \dfrac{17}{16}$

(c) $\dfrac{x-2}{x+8} = \dfrac{3}{5}$ Cross multiply.

$5(x-2) = 3(x+8)$ Distribute.

$5x - 10 = 3x + 24$ Sort.

$5x - 3x = 24 + 10$ Combine terms.

$2x = 34$ Solve for x.

$\dfrac{2x}{2} = \dfrac{34}{2}$

$\mathbf{x = 17}$

(d) $\dfrac{3}{x} = 7$ Write 7 as a fraction. $7 = \dfrac{7}{1}$

$\dfrac{3}{x} = \dfrac{7}{1}$ Cross multiply.

$3 = 7x$ Solve for x.

$\dfrac{3}{7} = \dfrac{7x}{7}$

$\dfrac{\mathbf{3}}{\mathbf{7}} = \mathbf{x}$

$5(x-2) = 3(x+8)$
$5x - 10 = 3x + 24$

TIP Cross Products and the Multiplication Axiom

The property of cross products is a shortcut application of the multiplication axiom for b and d not equal to zero.

$\dfrac{a}{b} = \dfrac{c}{d}$ Multiply both sides of the equation by both denominators.

$b(d)\left(\dfrac{a}{b}\right) = b(d)\left(\dfrac{c}{d}\right)$ Simplify.

$\cancel{b}(d)\left(\dfrac{a}{\cancel{b}}\right) = b(\cancel{d})\left(\dfrac{c}{\cancel{d}}\right)$

$da = bc$ Apply the commutative property of multiplication.

$ad = bc$ Same as property of cross products

8–1 Ratio and Proportion

1 Solve the proportions. Round to four significant digits if necessary.

1. $\dfrac{x}{5} = \dfrac{9}{15}$

2. $\dfrac{3x}{16} = \dfrac{3}{8}$

3. $\dfrac{x-1}{x+6} = \dfrac{4}{5}$

4. $\dfrac{5}{x} = 8$

5. $\dfrac{2}{7} = \dfrac{x-4}{x+3}$

6. $\dfrac{2x+1}{8} = \dfrac{3}{7}$

7. $\dfrac{3x-2}{3} = \dfrac{2x+1}{3}$

8. $\dfrac{5}{2x-2} = \dfrac{1}{8}$

9. $\dfrac{8}{3x+2} = \dfrac{8}{14}$

10. $\dfrac{2x}{8} = \dfrac{3x+1}{7}$

11. $\dfrac{-5}{x-2} = \dfrac{5}{x}$

12. $\dfrac{7}{x-6} = \dfrac{-3}{x}$

13. $\dfrac{4}{8} = \dfrac{x}{60}$

14. $\dfrac{5.2}{16} = \dfrac{x}{8.3}$

15. $\dfrac{5.2}{340} = \dfrac{12.8}{x}$

16. $\dfrac{3x+1}{4} = \dfrac{5x}{8}$

17. $\dfrac{7.35}{4} = \dfrac{5.21}{x}$

18. $\dfrac{9.12}{8} = \dfrac{x}{1.03}$

19. $\dfrac{2x}{7} = \dfrac{5.3}{6.7}$

20. $\dfrac{7x}{5} = \dfrac{91}{20}$

21. $\dfrac{0.107}{4x} = \dfrac{0.04}{321}$

22. $\dfrac{0.7x}{2.3} = \dfrac{5.6}{12.8}$

23. $\dfrac{0.04x}{5.2} = \dfrac{3.16}{14.08}$

24. $\dfrac{10^4}{x} = \dfrac{10^7}{10^5}$

25. $\dfrac{10^8}{10^5} = \dfrac{x}{10^9}$

26. $\dfrac{2x-3}{7} = \dfrac{3}{28}$

27. $\dfrac{32}{21} = \dfrac{8}{5x-1}$

28. $\dfrac{4}{3x-5} = \dfrac{18}{12}$

29. $\dfrac{12 \text{ in.}}{15 \text{ in.}} = \dfrac{8 \text{ in.}}{x}$

30. $\dfrac{15 \text{ ft}}{38 \text{ ft}} = \dfrac{x}{57 \text{ ft}}$

31. $\dfrac{12 \text{ k}\Omega}{8 \text{ k}\Omega} = \dfrac{x}{6 \text{ k}\Omega}$

32. $\dfrac{x}{15 \text{ W}} = \dfrac{60 \text{ W}}{3.5 \text{ W}}$

33. $\dfrac{150 \text{ V}}{0.6 \text{ V}} = \dfrac{65 \text{ V}}{x}$

34. $\dfrac{12 \text{ mA}}{8.2 \text{ mA}} = \dfrac{x}{0.5 \text{ mA}}$

35. $\dfrac{5.12}{14.87} = \dfrac{x}{3.91}$

36. $\dfrac{21.25}{3.2x} = \dfrac{212.5}{32}$

37. $\dfrac{\frac{3}{5}}{\frac{5}{8}} = \dfrac{\frac{4}{5}}{x}$

38. $\dfrac{x}{\frac{5}{9}} = \dfrac{\frac{3}{10}}{\frac{4}{5}}$

39. $\dfrac{2\frac{1}{4}}{x} = \dfrac{\frac{7}{10}}{\frac{4}{9}}$

40. $\dfrac{4\frac{3}{8}}{\frac{5}{8}} = \dfrac{x}{\frac{4}{5}}$

41. $\dfrac{50 \text{ mi}}{1 \text{ h}} = \dfrac{400 \text{ mi}}{x \text{ h}}$

42. $\dfrac{2{,}500 \text{ mi}}{3.8 \text{ h}} = \dfrac{500 \text{ mi}}{x \text{ h}}$

43. $\dfrac{\$245{,}000}{2{,}500 \text{ ft}^2} = \dfrac{\$x}{1 \text{ ft}^2}$

44. $\dfrac{\$495{,}000}{5{,}200 \text{ ft}^2} = \dfrac{\$x}{1 \text{ ft}^2}$

45. $\dfrac{\$4.12}{20 \text{ oz}} = \dfrac{\$x}{1 \text{ oz}}$

46. $\dfrac{15 \text{ gal}}{40 \text{ acres}} = \dfrac{x \text{ gal}}{1 \text{ acres}}$

47. $\dfrac{350 \text{ mg}}{7.8 \text{ cm}^3} = \dfrac{x \text{ mg}}{1 \text{ cm}^3}$

48. $\dfrac{350 \text{ mg}}{60 \text{ cm}^3} = \dfrac{x \text{ mg}}{1 \text{ cm}^3}$

49. $\dfrac{5 \text{ gal}}{20 \text{ acres}} = \dfrac{x \text{ gal}}{1 \text{ acre}}$

50. $\dfrac{44 \text{ gal}}{1 \text{ min}} = \dfrac{x \text{ gal}}{3 \text{ min}}$

51. $\dfrac{1{,}500 \text{ mi}}{2.7 \text{ h}} = \dfrac{x \text{ mi}}{1 \text{ h}}$

52. $\dfrac{400 \text{ mg}}{20 \text{ cm}^3} = \dfrac{x \text{ mg}}{1 \text{ cm}^3}$

53. $\dfrac{1{,}000 \text{ mg}}{120 \text{ cm}^3} = \dfrac{x \text{ mg}}{1 \text{ cm}^3}$

54. $\dfrac{4{,}500 \text{ mg}}{560 \text{ cm}^3} = \dfrac{x \text{ mg}}{1 \text{ cm}^3}$

8–2 | *Direct and Joint Variation*

Learning Outcomes

1 Solve problems of direct variation using proportions.

2 Solve problems of direct variation using a constant of variation.

3 Solve problems of combined variation using a constant of variation.

Apples	Cost
4	$1
8	$2
12	$3
16	$4

1 **Solve Problems of Direct Variation Using Proportions.**

Many problems in the workplace can be solved using proportions. The details of the problem can be grouped into two pairs of data that can be *directly* related.

A **direct variation** is one in which the quantities being compared are directly related, so that as one quantity increases (or decreases), the other quantity also increases (or decreases).

Data are often arranged in tables like the one on the left so these relationships can be examined. If 4 apples cost $1, set up a table to examine related costs of other amounts of apples.

In a pair of data, one item is identified as the **independent variable,** such as the *number of apples* purchased. The other item depends on the first item and it is identified as the **dependent variable,** such as the *total cost* of the apples.

To set up a direct variation:

1. Establish two pairs of related data.
2. Write one pair of data in the numerators of two ratios.
3. Write the other pair of data in the denominators of two ratios.
4. Form a proportion using the two ratios.

EXAMPLE
BUS

(a) Find the cost of 10 apples if 4 apples cost $1. (b) How many apples can be purchased for $10?

(a) Pair 1: 4 apples cost $1 .

Pair 2: 10 apples cost c dollars .

Estimation

8 apples would cost $2. Therefore, 10 apples will cost more than $2.

$$\frac{4 \text{ apples}}{10 \text{ apples}} = \frac{\$1}{\$c}$$
Pair 1 is the numerator of each ratio.

Pair 2 is the denominator of each ratio. Cross multiply.

$$4c = 10$$
Divide.

$$\frac{4c}{4} = \frac{10}{4}$$

$$c = 2.50$$
To the nearest cent.

Interpretation

10 apples cost $2.50.

(b) Pair 1: 4 apples cost $1 .

   Pair 2: a apples cost $10 .

Estimation

If 10 apples cost $2.50, 4 times as many apples can be bought for $10. That is, 40 apples can be bought.

$$\frac{4 \text{ apples}}{a \text{ apples}} = \frac{\$1}{\$10} \qquad \text{Pair 1}$$
$$\text{Pair 2 Cross multiply.}$$

$$40 = a$$

Interpretation

40 apples can be bought for $10.

Directly related data pairs can also be set up by making each pair a ratio. In the preceding example, we would write the ratio as

$$\text{Pair 1} \qquad \frac{4 \text{ apples}}{\$1} = \frac{10 \text{ apples}}{\$c} \qquad \text{Pair 2}$$

EXAMPLE AUTO

A truck travels 102 mi on 6 gal of gasoline. How far will it travel on 30 gal of gasoline?

Known facts

Pair 1: 102 mi uses 6 gal of gasoline.

Unknown facts

Pair 2: m mi uses 30 gal of gasoline.

Estimation

30 gal ÷ 6 gal = 5. Then, approximately 5 times 100 miles or 500 miles can be driven on 30 gal.

Dimension Analysis

Calculations

$$\frac{102 \text{ mi}}{m \text{ mi}} = \frac{6 \text{ gal}}{30 \text{ gal}} \qquad \frac{\text{distance}_1}{\text{distance}_2} = \frac{\text{gasoline}_1}{\text{gasoline}_2} \quad \begin{matrix}\text{Pair 1}\\\text{Pair 2}\end{matrix}$$

$$\frac{102}{m} = \frac{6}{30} \qquad \text{Cross multiply. } \frac{\text{mi}}{\text{mi}} = \frac{\text{gal}}{\text{gal}}$$

$$102(30) = 6m \qquad \text{Divide. mi(gal)} = \text{gal(mi)}$$

$$\frac{3,060}{6} = \frac{6m}{6} \qquad \text{Reduce. } \frac{\text{mi}(\text{gal})}{\text{gal}} = \frac{\text{gal}(\text{mi})}{\text{gal}}$$

$$510 = m \qquad m \text{ is expressed in miles.}$$

Interpretation

The truck will travel 510 mi on 30 gal of gasoline.

TIP

Analyze Dimensions

Even though we sometimes remove the written dimensions from an equation, we should analyze the dimensions to be sure we use the correct units in the solution.

EXAMPLE
INDTEC

If a metal rod tapers 1 in. for every 24 in. of length, what is the amount of taper of a 30-in. piece of rod? (See Fig. 8–1.)

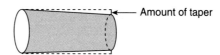
← Amount of taper

Figure 8–1

Known facts	Pair 1: 1-in. taper for 24-in. length
Unknown facts	Pair 2: x-in. taper for 30-in. length
Estimation	A 30-in. rod will taper more than 1 in.
Calculations	$\dfrac{\text{1-in. taper}}{x\text{-in. taper}} = \dfrac{\text{24-in. length}}{\text{30-in. length}}$ Pair 1 Pair 2

$$\frac{1}{x} = \frac{24}{30} \qquad \text{Cross multiply.}$$

$$1(30) = 24x$$

$$30 = 24x \qquad \text{Divide.}$$

$$\frac{30}{24} = \frac{24x}{24}$$

$$\frac{5}{4} = x \qquad \text{or} \qquad 1\frac{1}{4} = x$$

Interpretation   **The amount of taper for a 30-in. length of rod is $1\frac{1}{4}$ in.**

---

2   Solve Problems of Direct Variation Using a Constant of Variation.

Another way to use directly related data is to identify a data pair in which both values are known then use that relationship to find a conversion factor. For example, if 4 apples cost $1, how much does 1 apple cost?

$$\$1 \div 4 \text{ apples} = \$0.25 \text{ cost per apple}$$

A conversion factor is used to multiply by the number of items. If each apple costs $0.25, then 10 apples cost 10 × $0.25, or $2.50. A conversion factor is also called a **constant of direct variation.** The direct variation formula is $y = kx$, where x is the **independent variable,** y is the **dependent variable,** and k is the **constant of direct variation.** Another way to express this is $k = \dfrac{y}{x}$.

To find the constant of direct variation:

1. Identify a pair of related data in which both values are known.
2. Write the data pair as a ratio. The units in the denominator of the ratio should match the units of the independent variable. $k = \dfrac{y}{x}$
3. Leave the ratio as a fraction or change it to a decimal equivalent.

EXAMPLE
BUS

A 15-oz box of cereal costs $2.29. Find the constant of direct variation to the nearest ten-thousandth of a dollar for the cost per ounce. (This is also referred to as the **unit cost**.)

$$k = \frac{y}{x}$$
$$\begin{array}{c}\text{Unit cost}\\ \text{or}\\ \text{Constant of direct variation}\end{array} = \frac{\text{cost}}{\text{oz}} = \frac{\$}{\text{oz}} \qquad \frac{\text{dependent variable}}{\text{independent variable}}$$

$$k = \frac{\$2.29}{15 \text{ oz}} \qquad \text{Change to a decimal equivalent. Round to the nearest ten-thousandth.}$$

$$k = \$0.1527/\text{oz} \qquad \text{Cost per ounce.}$$

The unit cost or constant of direct variation is $0.1527 \dfrac{\$}{\text{oz}}$.

EXAMPLE
BUS

A 15-oz box of cereal costs $2.29 or a 36-oz box of cereal costs $4.89. Which is the better buy? Round the unit costs to the nearest ten-thousandth of a dollar.

In the preceding example we found the cost per ounce (unit cost) of the 15-oz box of cereal to be $0.1527.

Find the unit cost of the 36-oz box of cereal.

$$k = \frac{\$4.89}{36} = \$0.1358/\text{oz} \qquad \text{Ratio of } \frac{\$}{\text{oz}}$$

Compare the unit costs.

Unit cost of 15-oz box = $0.1527 per oz; unit cost of 36-oz box = $0.1358 per oz

The 36-oz box of cereal costs less per ounce and is the better buy.

In the preceding example, we make a judgment solely on the basis of the mathematical facts. In reality, other factors are considered. Do you have the money to buy the larger box? Will you be able to use the larger amount of cereal before it gets stale?

To solve problems of direct variation using a constant of variation:

1. Find the constant of direct variation using known values for x and y and the formula $k = \dfrac{y}{x}$.

2. Evaluate the formula of direct variation

$$y = kx,$$

where x is the independent variable, y is the dependent variable, and k is the constant of direct variation.

EXAMPLE An architect's drawing is scaled at 0.25 in. = 5 ft. Find the actual measurements for the following measures shown on the drawing.

(a) 0.75 in.   **(b)** 1.25 in.

Determine the scaled measure per foot of actual measurement (constant of direct variation). The actual measure is the independent variable and the scaled measure is the dependent variable.

$$\text{constant of direct variation} = \frac{\text{dependent variable}}{\text{independent variable}} \qquad k = \frac{y}{x}. \text{ Substitute known values.}$$

$$k = \frac{5 \text{ ft}}{0.25 \text{ in.}} \qquad\qquad \text{Divide.}$$

$$k = 20\frac{\text{ft}}{\text{in.}}$$

Use the direct variation formula to find the respective values of the dependent variables.

(a) $y = kx$ Substitute $k = 20$ and $x = 0.75$.

$$y = 20\,\frac{\text{ft}}{\text{in.}}\,(0.75 \text{ in.}) \qquad \text{Multiply and analyze the dimensions.}$$

$$y = \mathbf{15\ ft}$$

(b) $y = kx$ Substitute $k = 20$ and $x = 1.25$.

$$y = 20\,\frac{\text{ft}}{\text{in.}}\,(1.25 \text{ in.}) \qquad \text{Multiply.}$$

$$y = \mathbf{25\ ft}$$

TIP

Which Method for Direct Variation Is Preferred?

The proportion method is the most versatile for a variety of situations. Using the constant of direct variation is useful in computer programs and electronic spreadsheets.

---

3 **Solve Problems of Joint Variation Using a Constant of Variation.**

One quantity may vary directly as a product of two or more other quantities. This variation is called **joint variation.**

To solve problems of joint variation using a constant of variation:

1. Symbolically represent the relationships among the variables.
2. Find the constant of joint variation by substituting known values into the equation from Step 1 and solving for k.
3. Evaluate the equation from Step 1 using the constant of joint variation from Step 2 and other known values.

An example of an equation of joint variation is $y = kxz$. It can be read as "y varies jointly as x and z" or "y is jointly proportional to x and z." A third statement is "$y = kxz$ for some constant k."

EXAMPLE Find w if w varies jointly as x and y and $x = 140$, $y = 6$, and the constant of joint variation is 0.5.

Write the equation using the joint variation model.

$w = kxy$ Substitute known values and solve for w.
$w = 0.5(140)(6)$ Evaluate.
$w = \mathbf{420}$

EXAMPLE The *simple interest* for an account is jointly proportional to the time and principal. If the quarterly interest for an account balance of $7,000 is $87.50, find the interest on a balance of $8,500 for 9 months.

In the joint variation equation, let y represent the interest, x represent the time, and z represent the principal.

$\text{interest} = k(\text{time})(\text{principal})$ Find k. Substitute known values for interest, time, and principal.

Quarterly means 3 months or $\dfrac{3}{12}$ year, or 0.25 year.

$\$87.50 = k(0.25)(\$7,000)$ Solve for k.

$\dfrac{\$87.50}{(0.25)\$7,000} = k$ Evaluate.

$k = 0.05$ Constant of joint variation.

Use the constant of joint variation to find the interest.

$\text{interest} = k(\text{time})(\text{principal})$ Find the interest. Substitute known values for k, the time, and the principal. 9 months = 0.75 year.

$\text{interest} = (0.05)(0.75)(\$8,500)$ Evaluate.

$\text{interest} = \$318.75$

In the preceding example note the similarity between the joint variation model and the simple interest formula—Interest = principal(rate)(time) or $I = PRT$. The constant of joint variation is represented by the rate R in the simple interest formula.

1 Solve using proportions.

1. **BUS** If 7 cans of dog food sell for $4.13, how much will 10 cans sell for?

2. **BUS** If 6 cans of coffee sell for $22.24, how much will 20 cans sell for?

3. **AUTO** A mechanic took 7 h to tune up 9 fuel-injected engines. At this rate, how many fuel-injected engines can be tuned up in 37.5 h? Round to the nearest whole number.

4. **BUS** A costume maker took 9 h to make 4 headpieces for a Mardi Gras ball. At this rate, how many complete headpieces can be made in 35 h?

5. How far can a family travel in 5 days if it travels at the rate of 855 mi in 3 days?

6. **AUTO** How far can a tractor-trailer rig travel in 8 days if it travels at the rate of 1,680 mi in 4 days?

7. **AG/H** How much crystallized insecticide does 275 acres of farmland need if the insecticide treats 50 acres per 100 lb?

8. **AG/H** How much fertilizer does 2,625 ft² of lawn need if the fertilizer treats 1,575 ft² per gallon? Express the answer to the nearest tenth of a gallon.

9. **BUS** A can of tomatoes contains 793 g and costs $1.49. Find the constant of direct variation for the cost per gram. Round to the nearest ten-thousandth.

10. **BUS** A can of tomatoes contains 411 g and costs $0.89. Find the constant of direct variation for the cost per gram. Round to the nearest ten-thousandth.

11. From Exercises 9 and 10 compare the unit cost of the small can of tomatoes with the unit cost of the large can to determine which size can is more economical.

12. **INDTEC** A bottling machine can fill 800 bottles in 2.5 h. Find the constant of direct variation for the number of bottles filled per hour.

13. **INDTEC** Two gears have a ratio of 8 to 2. If the larger gear has 48 teeth, how many teeth does the smaller gear have?

14. **INDTEC** Two gears have a ratio of 9 to 4. If the larger gear has 72 teeth, how many teeth does the smaller gear have?

15. **INDTR** The diameter of the larger of two pulleys connected with a belt is 36 cm. If the ratio is 9:1, what is the diameter of the smaller pulley?

16. **INDTR** The diameter of the smaller of two pulleys connected with a belt is 48 cm. If the ratio is 8:1, what is the diameter of the larger pulley?

17. **CON** The slope of a roof is the ratio of the rise to the run $\left(\dfrac{\text{rise}}{\text{run}}\right)$ (Fig. 8–2). A roof has a slope of $\dfrac{3}{12}$. What is the rise for this roof if it has a run of 28 ft?

18. **CON** A roof has a slope of $\dfrac{1}{6}$. What is the rise for this roof if it has a run of 36 ft?

19. **CON** The *pitch* of a roof is the ratio of the rise of a roof to its span (Fig. 8–3). A roof has a pitch of 1:4. If the span is 40 ft, what is the rise?

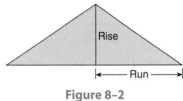

Figure 8–2

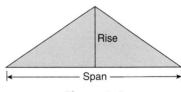

Figure 8–3

20. **CON** A roof has a pitch of 1:3. If the span is 42 ft, what is the rise?

21. **HLTH/N** A person who weighs 142 lb should be given how many milligrams of medication if the dosage is 25 mg for every 10 lb?

22. **HLTH/N** A person who weighs 185 lb should be given how many milligrams of medication if the dosage is 15 mg for every 10 lb?

23. **HLTH/N** The pediatric dosage for chlorpromazine hydrochloride is 0.25 mg/lb. What is the dosage for a child that weighs 40 lb?

Solve using a constant of variation.

24. **HLTH/N** Chlorpromazine injection strength contains 50 mg per 2 mL. A patient is prescribed 0.5 g. How many milliliters should be administered (1 g = 1,000 mg)?

25. **CAD/ARC** A blueprint has a scale of $\frac{1}{2}$ in. = 1 ft. On a blueprint a wall is drawn $7\frac{1}{2}$ in. long. What is the actual measure of the wall?

26. **CAD/ARC** A blueprint has a scale of $\frac{3}{4}$ in. = 1 ft. On a blueprint a wall is drawn $8\frac{3}{4}$ in. long. What is the actual measure of the wall?

27. **CAD/ARC** A building that is 120 ft long would be shown as what length on a blueprint if the scale is $\frac{1}{4}$ in. = 1 ft?

28. **CAD/ARC** A building that is 320 ft long would be shown as what length on a blueprint if the scale is $\frac{1}{8}$ in. = 1 ft?

29. **ELEC** The length of a wire is proportional to its resistance. The resistance of 200 ft of a certain wire is 0.0062 Ω. What is the resistance of 750 ft of the same wire?

30. **ELEC** The length of a wire is proportional to its resistance. The resistance of 100 ft of a certain wire is 0.048 Ω. What is the resistance of 1,200 ft of the same wire?

31. **AVIA** Use the map in Fig. 8–4 to find the air distance to the nearest mile from Upper Sandusky to Lima.

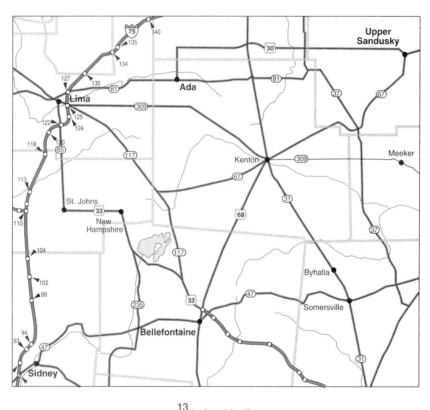

$\frac{13}{16}$ inch = 10 miles

Figure 8–4

32. **AVIA** Use the map in Fig. 8–4 to find the air distance to the nearest mile from Lima to Bellefontaine.

33. **AVIA** Use the map in Fig. 8–4 to find the air distance to the nearest mile from Sidney to Bellefontaine.

34. **INDTR** A machinist is to make the metal plate represented in Fig. 8–5. What is the overall length of the actual plate?

35. **INDTR** What is the overall width of the actual plate represented in Fig. 8–5?

36. **INDTR** What is the diameter in inches of the semicircle in Fig. 8–5? The diameter is the distance across the center of a circle.

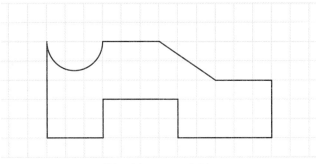

Side of 1 square = 2.5 in.

Figure 8–5

Use square-ruled paper to make a line drawing of objects described in Exercises 37–41.

37. A square 10 cm on a side, using a scale of 1 square = 1 cm

38. A rectangle 10 ft by 7.5 ft, using a scale of 1 square = 2.5 ft

39. A rectangle 15 m by 12 m, using a scale of 1 square = 3 m

40. A circle with a 24-in. diameter, using a scale of 1 square = 4 in.

41. A rectangle 4.5 cm by 3.5 cm, using a scale of 1 square = 0.5 cm

42. **HLTH/N** A doctor ordered streptomycin 250 mg IM for a patient. The dosage available for use contains 1 g per 4 mL. How many milliliters should be injected (1 g = 1,000 mg)?

43. **HLTH/N** A patient is prescribed 50 mg of Librium® IM. The medication is available as 0.4 mg per 2 mL. How many milliliters should be injected?

44. **HLTH/N** A patient is prescribed an injection of 0.2 mg of Atropine® IM. The drug is available as 0.5 mg per milliliter. How many milliliters should be injected?

45. **AG/H** Instructions for mixing a chemical pesticide state that 6 oz of chemical should be mixed with 64 oz of water. How many ounces of chemical should be used with 160 oz of water?

46. **ELEC** An electrical transformer shown in Fig. 8–6 is constructed by winding a number of turns of wire into a primary. Another set of turns is constructed by winding to form one or more secondaries. The ratio of the primary voltage (E_p) to the secondary voltage (E_s) is the same as the ratio of the number of primary turns (T_p) to the number of secondary turns (E_s).

$$\frac{E_p}{E_s} = \frac{T_p}{T_s}$$

Find the secondary voltage if the transformer has a 360-turn primary and a 30-turn secondary and 150 V are applied.

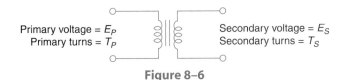

Primary voltage = E_P
Primary turns = T_P

Secondary voltage = E_S
Secondary turns = T_S

Figure 8–6

3 Solve problems of joint variation using a constant of variation.

47. If x varies jointly with y and z, find x when $y = 18$ and $z = 20$ if the constant of variation is $k = 3$.

48. If I varies jointly with R and T, find I when $R = 0.05$, $T = 3$, and the constant of variation is $P = 5,000$.

49. In a relationship, r varies jointly as s and t. If r is 84 when s is 7 and t is 4, find r when s is 12 and t is 9.

50. **BUS** Find the simple interest I on a principal P of $8,000 if the rate R is 6% and the time T is 5 years. I varies jointly as R and T and the principle P is the constant of variation.

51. BUS Using the constant of variation given in Exercise 50, find the interest if the rate is 3.5% and the time is 7 years.

52. ELEC For an appliance the wattage rating P varies jointly as the resistance, R, and the square of the current I. If the wattage rating is 60 W when the current is 0.1 A and the resistance is 500 Ω, find the wattage rating when the current is 0.3 A and the resistance is 100 Ω.

8–3 | Inverse and Combined Variation

Learning Outcomes

1 Solve problems of inverse variation using proportions.
2 Solve problems of inverse variation using a constant of variation.
3 Solve problems of combined variation using a constant of variation.

1 Solve Problems of Inverse Variation Using Proportions.

An **inverse variation** is one in which the quantities being compared are inversely related. That is, as one quantity increases, the other decreases, or as one quantity decreases, the other increases.

For example, as we *increase* pressure on a gas, the gas compresses and so *decreases* in volume. Or, as we *decrease* pressure on the gas, it *increases* in volume as it expands. This relationship is the opposite, or inverse, of direct variation.

If 3 workers frame a house in 2 weeks and the contractor *increases* the number of workers to 6, the framing time *decreases* to 1 week, assuming the workers work at the same rate. In other words, the framing time is *inversely proportional* to the number of workers on the job.

Unlike the directly related ratios in a proportion, inversely related ratios do not allow us the flexibility we had in setting up ratios of unlike measures.

To set up an inverse proportion:

1. Establish two pairs of related data.
2. Arrange one pair as the numerator of one ratio and the denominator of the other ratio.
3. Arrange the other pair so that each ratio contains like measures.
4. Form a proportion using the two ratios.

EXAMPLE INDTEC If 5 machines take 12 days to complete a job, how long will it take for 8 machines to do the job?

As the number of machines *increases,* the amount of time required to do the job *decreases.* Thus, the quantities are *inversely proportional.*

Pair 1: 5 machines finish in 12 days .

Pair 2: 8 machines finish in *x* days . .

Estimation   We expect more machines to do the job in less than 12 days. Also, we did not double the number of machines, so it will take more than half of the time (6 days).

$$\frac{5 \text{ machines}}{8 \text{ machines}} = \frac{x \text{ days for 8 machines}}{12 \text{ days for 5 machines}}$$ Each ratio uses like measures. The pairs are arranged inversely.

Dimension Analysis

$$\frac{5}{8} = \frac{x}{12}$$ $\dfrac{\text{machines}}{\text{machines}} = \dfrac{\text{days}}{\text{days}}$

$$5(12) = 8x$$ Cross multiply. machines(days) = machines(days)

$$60 = 8x$$ Divide by machines.

$$\frac{60}{8} = x$$ $\dfrac{\text{machines (days)}}{\text{machines}} = \text{days}$

$$x = \frac{15}{2} \quad \text{or} \quad 7\frac{1}{2} \text{ days}$$

Interpretation   **It will take $7\frac{1}{2}$ days for 8 machines to do the job.**

Why Is Estimation So Important?

Suppose we arrange the data so that each pair forms a ratio of unlike measures and then we invert one of the ratios. Look at the value to see if the answer conforms to what we expect the answer to be.

Pair 1   Reciprocal of Pair 2

$$\frac{5 \text{ machines}}{12 \text{ days}} = \frac{x \text{ days}}{8 \text{ machines}}$$

$$12x = 40$$

$$x = 3.33 \text{ days}$$

How does this compare to our estimate? We expected our answer to be less than 12 days but more than 6 days. The answer above, 3.33 days, is not consistent with our estimate.

$$\frac{\text{machines}}{\text{days}} = \frac{\text{days}}{\text{machines}}$$ Analyzing the dimensions we produce an incorrect statement.

machines(machines) = days(days)

These measures are not equal. Estimation and dimension analysis can identify proportions that are set up incorrectly!

The speed and size of gears and pulleys involve *inverse* relationships. Suppose a large gear and a smaller gear are in mesh, or a large pulley is connected by a belt to a smaller pulley. The larger gear or pulley has the *slower* speed, and the smaller gear or pulley has the *faster* speed.

EXAMPLE
INDTEC
A 10-in.-diameter gear is in mesh with a 5-in.-diameter gear (Fig. 8–7). If the larger gear has a speed of 25 rpm, at how many rpm does the smaller gear turn?

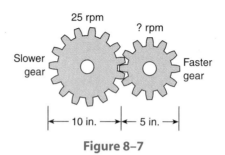

Figure 8–7

Because gears in mesh are *inversely* related, we set up an inverse proportion. Each ratio uses like measures and the ratios are in inverse order.

Pair 1: 25 rpm of larger gear for 10-in. size of larger gear.

Pair 2: *x* rpm of smaller gear for 5-in. size of smaller gear.

Estimation
We expect the speed of the smaller gear to be faster or greater than the 25-rpm speed of the larger gear.

$$\frac{25 \text{ rpm of larger gear}}{x \text{ rpm of smaller gear}} = \frac{5 \text{ in. of smaller gear}}{10 \text{ in. of larger gear}} \quad \begin{matrix} \text{Pair 2} \\ \text{Pair 1} \end{matrix}$$ Set up ratios in inverse order.

$$\frac{25}{x} = \frac{5}{10}$$ Cross multiply.

$$25(10) = 5x$$

$$250 = 5x$$ Divide.

$$\frac{250}{5} = x$$

$$50 = x$$

Interpretation **The smaller gear turns at the faster, or greater, speed of 50 rpm.**

2 Solve Problems of Inverse Variation Using a Constant of Variation.

Inverse variation has a conversion factor similar to the constant of direct variation. The inverse variation formula is $y = \dfrac{k}{x}$, where x is the nonzero independent variable, y is the dependent variable, and k is the **constant of inverse variation.** This formula can also be written as $k = xy$.

To find the constant of inverse variation:

1. Identify a pair of inversely related data in which both values are known.
2. Write the related pair as a product. $k = xy$
3. Multiply to find the constant of inverse variation.

To solve problems of inverse variation using a constant of variation:

1. Find the constant of inverse variation using known values for x and y and the formula $k = xy$.
2. Evaluate the formula of inverse variation

$$y = \frac{k}{x}$$

where x is the independent variable, y is the dependent variable, and k is the constant of inverse variation.

EXAMPLE A landscaper needs a five-person crew to complete a commercial project in 30 h. How many hours will it take to complete the project for the following number of workers?

(a) 3 workers   (b) 8 workers

Determine the number of hours necessary to complete the job per worker (constant of inverse variation). The number of workers is the independent variable and the number of hours to complete the job is the dependent variable.

constant of inverse variation = (independent variable)(dependent variable)

$k = xy$	Substitute known values.
$k = (5 \text{ workers})(30 \text{ h})$	Multiply.
$k = 150 \text{ worker-hours}$	

Use the inverse variation formula to find the respective values of the dependent variables.

(a) $y = \dfrac{k}{x}$ Substitute $k = 150$ and $x = 3$.

     $y = \dfrac{150 \text{ worker-hours}}{3 \text{ workers}}$ Divide.

     $y = 50 \text{ h}$

(b) $y = \dfrac{k}{x}$ Substitute $k = 150$ and $x = 8$.

     $y = \dfrac{150 \text{ worker-hours}}{8 \text{ workers}}$ Divide.

     $y = 18.75 \text{ h}$

TIP | How to Distinguish Between Direct and Inverse Variation

We can distinguish between direct and inverse variation by anticipating cause-and-effect situations.

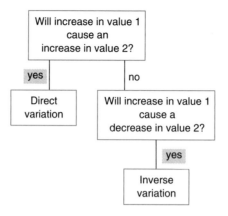

Similarly,

$$\text{decrease causes decrease} \Rightarrow \text{direct variation}$$
$$\text{decrease causes increase} \Rightarrow \text{inverse variation}$$

3 Solve Problems of Combined Variation Using a Constant of Variation.

In certain real-life situations the variables may be related through both direct and inverse variations. Such variations are called **combined variation.**

To solve problems of combined variation using a constant of variation:

1. Symbolically represent the relationships among the variables.
2. Find the constant of combined variation by substituting known values into the equation from Step 1 and solving for k.
3. Evaluate the equation from Step 1 for the established constant of variation (Step 2) and other known values.

An example of an equation of combined variation is $y = \dfrac{kx}{z}$. It can be read as "y varies directly as x and inversely as z" or "y is directly proportional to x and inversely proportional to z."

EXAMPLE z varies directly as the square of x and inversely as y. Find z when $x = 8$, $y = 15$, and the constant of combined variation is 12.

Write an equation of combined variation.

$z = \dfrac{kx^2}{y}$ Substitute known values.

$z = \dfrac{12(8^2)}{15}$ Evaluate.

$z = \dfrac{12(64)}{15}$

$z = \dfrac{768}{15}$

$z = \mathbf{51.2}$

EXAMPLE The number of hours h that it takes p persons to assemble m machines varies directly as the number of machines and inversely as the number of persons. Three persons can assemble 10 machines in 5 h. How many persons would it take to assemble 36 machines in 8 h?

Write an equation of combined variation.

$$h = \frac{km}{p}$$

The number of hours h varies directly with the number of machines m and inversely with the number of persons p.

Find the constant of combined variation.

$$h = \frac{km}{p}$$

Substitute known values and solve for k.

$$5 = \frac{k(10)}{3}$$

Clear fraction.

$$15 = 10k$$

$$\frac{15}{10} = k$$

Reduce and apply symmetric property of equality.

$$k = \frac{3}{2}, \quad \text{or } 1.5$$

Constant of combined variation.

Find p for $h = 8$ and $m = 36$.

$$h = \frac{1.5m}{p}$$

Substitute $h = 8$ and $m = 36$. Solve for p.

$$8 = \frac{1.5(36)}{p}$$

$$8p = 1.5(36)$$

$$8p = 54$$

$$p = \frac{54}{8} \text{ or } 6.75$$

Round up to the next whole number.

Seven persons are required to complete the job in the allotted time.

SECTION 8–3 SELF-STUDY EXERCISES

1 Solve using proportions.

1. INDTR The fan pulley and alternator pulley are connected by a fan belt on an automobile engine. The fan pulley is 225 cm in diameter and the alternator pulley is 125 cm in diameter. If the fan pulley turns at 500 rpm, how many revolutions per minute does the faster alternator pulley turn?

2. INDTR The volume of a certain gas is inversely proportional to the pressure on it. If the gas has a volume of 160 in^3 under a pressure of 20 lb per square inch (psi), what is the volume if the pressure is decreased to 16 psi?

3. **INDTR** A pulley that measures 15 in. across (diameter) turns at 1,600 rpm and drives a larger pulley at the rate of 1,200 rpm. What is the diameter of the larger pulley in this inverse relationship?

4. **CON** Six painters can trim the exterior of all the new brick homes in a subdivision in 9 weeks. The contractor wants to have the homes ready in just 3 weeks and so needs more painters. How many painters are needed for the job if they all work at the same rate.

5. **AG/H** Two groundskeepers take 25 h to prepare a golf course for a tournament. How long would it take five groundskeepers to prepare the golf course?

6. **INDTEC** A gear measures 5 in. across. It turns another gear 2.5 in. across. If the larger gear has a speed of 25 rpm, what is the rpm of the smaller gear?

7. **INDTR** A small pulley 3 in. in diameter turns 250 rpm and drives a larger pulley at 150 rpm. What is the diameter (distance across) of the larger pulley?

8. **CON** Two painters working at the same speed can paint 800 ft^2 of wall space in 6 h. If a third painter paints at the same speed, how long will it take all three to paint the same wall space?

9. **INDTEC** Three machines complete a printing project in 5 h. How many machines are needed to finish the same project in 3 h?

10. **INDTEC** A gear measures 6 in. across. It is in mesh with another gear with a diameter of 3 in. If the larger gear has a speed of 60 rpm, what is the rpm of the smaller gear?

11. **HLTH/N** Nurse Lee prepares dosages for her patients in 30 min. If she gets help from assistants, who also work at her rate, and together they complete the preparation in 6 min, how many helpers did she get?

12. **INDTR** A 4.5-in. pulley turning at 1,000 rpm is belted to a larger pulley turning at 500 rpm. What is the size of the larger pulley?

2

13. **INDTR** A 12-in. pulley turns at 30 rpm. Find the constant of inverse variation.

14. **INDTR** A 16-in. gear makes 60 rpm. Find the constant of inverse variation.

The diameter of two years in mesh is inversely proportional to the number of revolutions per minute each turns. Complete the table for two years in mesh using a constant of variation.

	LARGE GEAR		SMALL GEAR	
	DIAMETER	RPM	DIAMETER	RPM
15.	30 in.	120	15 in.	
16.	24 in.		18 in.	300
17.		70	14 cm	3,500
18.	20 cm	100		500
19.		180	12 cm	540
20.	32 in.	225	12 in.	

The number of teeth of two gears in mesh is inversely proportional to the number of revolutions per minute each turns. Complete the table for two gears in mesh using a constant of variation.

	LARGE GEAR		SMALL GEAR	
	NUMBER OF TEETH	RPM	NUMBER OF TEETH	RPM
21.	60	300	45	
22.	250	28	50	
23.		1.7	60	8.5
24.	300	70		600
25.	448		56	300
26.	84	24		96

27. **INDTEC** A gear with 40 teeth turns 30 rpm on another gear with 120 teeth. At how many rpm does the second gear turn?

28. **INDTEC** A gear turning at 19 rpm with 150 teeth turns a smaller gear with 75 teeth. At how many rpm does the smaller gear turn?

29. **INDTEC** A large gear turns at 40 rpm in mesh with a gear that has 60 teeth and turns at 120 rpm. How many teeth does the larger gear have?

30. **INDTEC** A small gear turns at 150 rpm in mesh with a large gear that has 180 teeth and turns at 90 rpm. How many teeth has the small gear?

3 Solve problems of combined variation using a constant of variation.

31. In a relationship, P varies directly as m and inversely as n and the constant of variation is 18. Find P if $m = 14$ and $n = 28$.

32. If x varies directly as y and inversely as z and $x = 18$ when $y = 12$ and $z = 2$, find x when $y = 75$ and $z = 15$.

33. If x varies directly as y and z and inversely as the square of r, and $x = 16$ when $y = 3$, $z = 12$ and $r = 3$, find x when $y = 5$, $z = 30$, and $r = 10$.

34. J varies directly as the cube of R and inversely with the square of S. If $J = 972$ when $R = 3$ and $S = 6$, find J when $R = 4$ and $S = 8$.

35. A varies directly as the square of B and inversely with C, and $A = 5{,}000$ when $B = 40$ and $C = 8$. Write an equation that relates the variables. Find the constant of variation.

36. Find A in Exercise 35 if $B = 20$ and $C = 2$.

37. A horizontal beam safely supports a load P that varies jointly as the product of the width W of the beam and the square of the depth D and inversely as the length L. A beam that has a width of 5 in., depth of 12 in., and length of 8 in. can safely support 1,260 lb. Determine the safe load in pounds of a beam made from the same material if the beam is 20 in. long.

38. Use the relationship in Exercise 37 to find the load P of a horizontal beam that has the same constant of variation and has a width of 6 in., a depth of 14 in., and a length of 12 in.

CHAPTER REVIEW OF KEY CONCEPTS

Section 8–1

1 Solve equations that are proportions (pp. 354–355).

Property of proportions: If $\dfrac{a}{b} = \dfrac{c}{d}$, then $ad = bc$ ($b, d \neq 0$).

To solve: **1.** Find the cross products. **2.** Divide both sides by the coefficient of the variable.

$$\frac{4}{2} = \frac{x}{6}$$

$(2)(x) = (4)(6)$ — Cross products.

$2x = 24$ — Divide by the coefficient of x.

$$\frac{2x}{2} = \frac{24}{2}$$

$x = 12$

Section 8–2

1 Solve problems of direct variation using proportions (pp. 357–359).

Verify that the problem involves direct proportion: As one amount increases (decreases), the other amount increases (decreases).

Set up a direct proportion: **1.** Establish two pairs of related data. **2.** Write one pair of data in the numerators of two ratios. **3.** Write the other pair of data in the denominators of the two ratios. **4.** Form a proportion using the two ratios.

Kinta makes $72 for 8 h of work in a hospital business office. How many hours must he work to earn $150?

Pairs are directly related: As hours increase, pay increases.

$$\frac{\$72}{\$150} = \frac{8 \text{ h}}{x \text{ h}}$$
Pair 1. $72; 8h.
Pair 2. $150; x h.

Estimation: x h > 8 h because $150 > $72.

$$\frac{72}{150} = \frac{8}{x}$$
Cross multiply.

$$(72)(x) = (8)(150)$$

$$72x = 1{,}200$$
Divide.

$$\frac{72x}{72} = \frac{1{,}200}{72}$$

$$x = 16\frac{2}{3} \text{ h}$$
Kinta must work $16\frac{2}{3}$ h to earn $150.

2 Solve problems of direct variation using a constant of variation (pp. 359–361).

1. Find the constant of direct variation using known values for x and y and the formula $k = \frac{y}{x}$. **2.** Evaluate the formula of direct variation $y = kx$ where x is the independent variable, y is the dependent variable, and k is the constant of direct variation.

A caterer can purchase 4 cartons of Coca Cola® for $10. What is the cost of 5 cartons?

$$\text{Constant of direct variation} = \frac{\text{dependent variable}}{\text{independent variable}}$$

$$k = \frac{y}{x}$$
Substitute known values.

$$k = \frac{\$10}{4 \text{ cartons}}$$
Divide to simplify ratio.

$$k = \frac{\$2.50}{\text{carton}}$$
$2.50 per carton

$$y = kx$$
Substitute k = $2.50 per carton and x = 5 cartons.

$$y = \left(\frac{\$2.50}{\text{carton}}\right)(5 \text{ cartons})$$
Multiply.

$$y = 12.50$$

Five cartons of Coca Cola® cost $12.50.

3 Solve problems of joint variation using a constant of variation (pp. 361–362).

1. Symbolically represent the relationships among the variables. **2.** Find the constant of joint variation by substituting known values into an equation from Step 1 and solving for k. **3.** Evaluate the equation from Step 1 for the established constant of joint variation (Step 2) and other known values.

p varies jointly as q and r. If $p = 42$ when $q = 5$ and $r = 7$, find p when $q = 9$ and $r = 12$.

Find the constant of variation:

$p = kqr$ Substitute known values and solve for k.

$42 = k(5)(7)$

$42 = k(35)$

$\dfrac{42}{35} = k$

$1.2 = k$ Constant of variation

Find p when $q = 9$ and $r = 12$:

$p = kqr$ Substitute known values and solve for p.

$p = 1.2(9)(12)$

$p = 129.6$

Section 8–3

1 Solve problems of inverse variation using proportions (pp. 366–368).

Verify that the problem involves inverse proportion: As one amount increases (decreases), the other amount decreases (increases).

Set up an inverse proportion: **1.** Establish two pairs of related data. **2.** Arrange one pair as the numerator of one ratio and the denominator of the other. **3.** Arrange the other pair so that each ratio contains like measures. **4.** Form a proportion using the two ratios.

Jane can paint a room in 4 h. If she has two helpers who also can paint a room in 4 h, how long will it take all three to paint the same room?

Pairs are inversely related: As the number of painters increases, time decreases.

$\dfrac{1 \text{ painter}}{3 \text{ painters}} = \dfrac{x \text{ h}}{4 \text{ h}}$ Pair 1: 1 painter, 4 h.

                       Pair 2: 3 painters, x h.   Estimation: $x \text{ h} < 4 \text{ h}$ since

$\dfrac{1}{3} = \dfrac{x}{4}$ 1 painter < 3 painters.

$(3)(x) = (1)(4)$

$3x = 4$

$\dfrac{3x}{3} = \dfrac{4}{3}$

$x = 1\dfrac{1}{3}$ h for 3 painters to paint the room.

2 Solve problems of inverse variation using a constant of variation (pp. 368–370).

1. Find the constant of inverse variation using known values for x and y and the formula $k = xy$. **2.** Evaluate the formula of inverse variation $y = \dfrac{k}{x}$ where x is the independent variable, y is the dependent variable, and k is the constant of inverse variation.

The number of teeth on two gears in mesh is inversely proportional with the speed at which the gears turn. If a gear with 15 teeth turns at 600 revolutions per minute $\left(\dfrac{r}{min}\right)$, how many teeth are on a gear in mesh that turns at $750\dfrac{r}{min}$?

constant of inverse variation = (independent variable)(dependent variable).

$$k = xy \qquad \text{Substitute known values.}$$

$$k = \left(600\,\frac{r}{min}\right)(15\text{ teeth}) \qquad \text{Multiply.}$$

$$k = 9{,}000\,\frac{r}{min}\text{-teeth}$$

$$y = \frac{k}{x} \qquad\qquad\qquad\qquad\qquad\qquad \text{Substitute.}$$

$$y = \frac{9{,}000\,\dfrac{r}{min}\text{-teeth}}{750\,\dfrac{r}{min}} \qquad\qquad \begin{array}{l} k = 9{,}000\,\dfrac{r}{min}\text{-teeth and} \\[4pt] x = 750\,\dfrac{r}{min}. \end{array}$$

$$\qquad\qquad\qquad\qquad\qquad\qquad\quad \text{Divide.}$$

$$y = 12\text{ teeth}$$

The gear that turns at $750\,\dfrac{r}{min}$ has 12 teeth.

3 Solve problems of combined variation using a constant of variation (pp. 370–371).

1. Symbolically represent the relationships among the variables. 2. Find the constant of combined variation by substituting known values into an equation from Step 1 and solving for k. 3. Evaluate the equation from Step 1 for the established constant of combined variation (Step 2) and other known values.

Q varies jointly as S and T and inversely as the square of R. Find the constant of combined variation when $S = 5$, $T = 7$, $R = 2$, and $Q = 4.375$.

$$Q = \frac{kST}{R^2} \qquad \begin{array}{l}\text{Symbolic relationships}\\ \text{Substitute known values.}\end{array}$$

$$4.375 = \frac{k(5)(7)}{2^2} \qquad \text{Solve for } k.$$

$$4.375 = \frac{k(35)}{4}$$

$$4.375(4) = 35k$$

$$17.5 = 35k$$

$$\frac{17.5}{35} = k$$

$$0.5 = k \qquad \text{Constant of variation.}$$

Find T to the nearest hundredth when $S = 3$, $T = 8$, $R = 3$, and $k = 0.5$.

$$Q = \frac{kST}{R^2} \qquad \text{Substitute known values.}$$

$$Q = \frac{0.5(3)(8)}{3^2} \qquad \text{Evaluate.}$$

$$Q = \frac{12}{9}$$

$$Q = 1.33 \qquad \text{Rounded}$$

CHAPTER REVIEW EXERCISES

Section 8–1

Solve the proportions.

1. $\dfrac{7}{x} = 6$

2. $\dfrac{2}{3} = \dfrac{x + 3}{x - 7}$

3. $\dfrac{4x + 3}{15} = \dfrac{1}{3}$

4. $\dfrac{3x - 2}{3} = \dfrac{2x + 1}{4}$

5. $\dfrac{5}{4x - 3} = \dfrac{3}{8}$

6. $\dfrac{8}{3x - 2} = \dfrac{2}{3}$

7. $\dfrac{4x}{7} = \dfrac{2x + 3}{3}$

8. $\dfrac{2x}{3x - 2} = \dfrac{5}{8}$

9. $\dfrac{5x}{3} = \dfrac{2x + 1}{4}$

10. $\dfrac{5}{9} = \dfrac{x}{2x - 1}$

11. $\dfrac{7}{x} = \dfrac{5}{4x + 3}$

12. $\dfrac{3}{5} = \dfrac{2x - 3}{7x + 4}$

Section 8–2

Solve using proportions.

13. There are 25 women in a class of 35 students. If this is typical of all classes in the college, how many women are enrolled in the college if it has 6,300 students altogether?

14. **HOSP** In preparing a banquet for 30 people, Cedric Henderson uses 9 lb of potatoes. How many pounds of potatoes will be needed for a banquet for 175 people?

15. **CAD/ARC** A CAD program scales a blueprint so that $\frac{5}{8}$ in. = 2 ft. What is the actual measure of a wall that is shown as $1\frac{5}{16}$ in. on the blueprint?

16. **AG/H** A 6-ft landscape engineer casts a 5-ft shadow on the ground. How tall is a nearby tree that casts a 30-ft shadow?

17. **HLTH/N** On recent trips to rural health centers, a portable mammography unit used 81.2 gal of unleaded gasoline. If the travel involved a total of 845 mi, how many gallons of gasoline would be used for travel of 1,350 mi? Round to tenths.

18. **BUS** A certain fabric sells at a rate of 3 yd for $7.00. How many yards can Clemetee Whaley buy for $35.00?

Solve using a constant of variation.

19. **CON** A contractor estimates that a painter can paint 300 ft^2 of wall space in 3.5 h. How many hours should the contractor estimate for the painter to paint 425 ft^2? Express your answer to the nearest tenth.

20. **CAD/ARC** An architect's drawing is scaled at $\frac{3}{4}$ in. = 6 ft. What is the actual height of a door that measures $\frac{7}{8}$ in. on the drawing?

21. **ELEC**   A wire 825 ft long has a resistance of 1.983 Ω. How long is a wire of the same diameter if the resistance is 3.247 Ω? Round to the nearest whole foot.

22. **BUS**   A coffee company mixes 1.6 lb of chicory with every 3.5 lb of coffee. At this ratio, how many pounds of chicory are needed to mix with 2,500 lb of coffee? Round to the nearest whole number.

23. If b varies jointly with d and e and the constant of variation is 0.8, find b when $d = 12$ and $e = 25$.

24. If p varies jointly as the square of q and the cube of r and $p = 108$ when $q = 3$ and $r = 2$, find p when $q = 4$ and $r = 3$.

25. The area of a triangle A varies jointly as b and h. If $A = 225$ in^2 when $b = 18$ in. and $h = 25$ in., find A when $b = 35$ in. and $h = 40$ in.

Section 8–3

Solve using proportions.

26. **AG/H**   It takes five people 7 days to clear an acre of land of debris left by a tornado; inversely, more people can do the job in less time. How long would it take seven people all working at the same rate?

27. **INDTEC**   If 15 machines can complete a job in 6 weeks, how many machines are needed to complete the job in 4 weeks?

28. **INDTR**   A 10-in. pulley makes 900 revolutions every minute. It drives a larger pulley at 500 rpm. What is the diameter of the larger pulley in this inverse relationship?

29. A car with a speed control device travels 100 mi at 50 mi/h. The trip takes 2 h. If the car traveled at 40 mi/h, how much time would the driver need to reach the same destination?

30. **INDTR**   A pulley whose diameter is 3.5 in. is belted to a pulley whose diameter is 8.5 in. In this inverse relationship, if the smaller, faster pulley turns at the rate of 1,200 rpm, what is the rpm of the slower pulley? Round to the nearest whole number.

31. **INDTR**   A gear with a diameter of 45 cm is in mesh with a gear that has a diameter of 30 cm. If the larger gear turns at 1,000 rpm, how many revolutions per minute does the smaller gear make?

Solve using a constant of variation.

32. **INDTR**   A small pulley with a 6-in. diameter turns 350 rpm and drives a larger pulley at 150 rpm. What is the diameter (distance across) of the larger pulley?

33. **INDTEC**   Three workers take 5 days to assemble a shipment of microwave ovens; inversely, more workers can do the job in less time. How long will it take five workers to do the same job?

34. **INDTEC**   Waylon can install a hard drive in 10 PCs in 6 h. He gets help from assistants who work at his rate and together they complete the installations in 2 h. How many helpers did he get?

35. **INDTR**   A gear is 4 in. across. It turns a smaller gear 2 in. across. If the larger gear has a speed of 30 rpm, what is the rpm of the smaller gear?

36. If x varies directly as the square of y and inversely with z and the constant of variation is 3.2, find x when $y = 14$ and $z = 16$.

37. P varies directly as r and inversely as the square of s. If $P = 175$, $r = 252$, and $s = 6$, find P when $r = 196$ and $s = 7$.

38. The load P that a beam can support varies jointly as the product of the beam width W and the square of the beam depth D and inversely as the beam length L. A beam 8 in. wide, 15 in. deep, and 40 in. long can safely support 1,116 lb. How many pounds can be supported by a beam that is 12 in. wide, 6 in. deep, and 20 in. long?

1. An effective measure of your understanding of a concept is your ability to apply the concept to real-world situations.
   (a) Develop and solve a word problem that can be solved using a direct proportion. Include in your storyline a lawn mower, tanks of gasoline, and acres to be mowed.
   (b) Develop and solve a word problem that can be solved using a direct proportion.

2. Inverse proportions can be used to solve certain types of real-world applications.
   (a) Develop and solve a word problem that can be solved using an inverse proportion. Include in your storyline a belt, pulleys, rpm's and diameters of pulleys.
   (b) Develop and solve a word problem that can be solved using an inverse proportion.

1. Illustrate the property of proportions using two equivalent fractions.
2. Explain the difference between a direct proportion and an inverse proportion.
3. Explain how to set up a direct proportion.
4. Give some examples of situations that are directly proportional.
5. Explain how to set up an inverse proportion.
6. Give some examples of situations that are inversely proportional.
7. If the constant of variation is positive and y varies directly as x, how will y change when x increases? Use an example to support your answer.
8. If the constant of variation is positive and y varies inversely as x, how will y change when x increases? Use an example to support your answer.
9. Suppose y varies directly as the square of x. How will y change when x is doubled? Illustrate your answer with an example.
10. Suppose y varies inversely as the square of x. How will y change when x is doubled? Illustrate your answer with an example.

Solve the proportions.

1. $\dfrac{x}{12} = \dfrac{5}{8}$

2. $\dfrac{x}{6} = \dfrac{12}{8}$

3. $\dfrac{640}{24} = \dfrac{x}{360}$

4. $\dfrac{45 \text{ cm}}{18 \text{ cm}} = \dfrac{9 \text{ cm}}{x}$

5. $\dfrac{2\frac{1}{2}}{x} = \dfrac{1\frac{1}{4}}{3\frac{1}{5}}$

6. $\dfrac{3.9}{5.4} = \dfrac{x}{8.1}$

7. A large gear with 300 teeth turns at 40 rpm. Find the rpm of a small gear that has 60 teeth.

Solve the problems involving fractions, decimal numbers, and proportions.

8. A pipe fills a 120-gal tank in 4 h. How long does it take for the pipe to fill a 420-gal tank?

9. A 9-in. gear is in mesh with a 4-in. gear. If the larger gear makes 75 rpm, how many revolutions per minute does the smaller gear make in this inverse relationship?

10. One employee can wallpaper a room in 12 h. Four employees working at the same rate can wallpaper the same room in how many hours?

11. A force of 32.75 lb exerts pressure on a surface of 24.65 in^2 in a hydraulic system? A force of 117.9 lb exerts the same pressure on what area? Force and area are directly proportional when the pressure is constant.

12. A $34,475 loan for the purchase of an electronically controlled assembly machine in a factory cost the management $2,758 in simple interest. How much would a machine cost if the interest was $1,879? Round to the nearest dollar.

13. If a compact car used 62.5 L of unleaded gasoline to travel 400 mi, how many liters of gasoline would the driver use to travel 350 mi? Round to tenths.

14. If three workers take 8 days to complete a job, how many workers would be needed to finish the same job in only 6 days if each worked at the same rate? (More workers take fewer days.)

15. If voltage is 40 V and amperage is 3.5 A, find the equivalent amperage for a voltage of 100 V. Voltage and amperage are directly proportional. Express the answer as a decimal rounded to tenths.

16. The ratio of men to women in technical and trade occupations is estimated to be 3 to 1, that is, $\frac{3}{1}$. If 56,250 men are employed in such occupations in a certain city, how many employees are women?

17. If an ice maker produces 75 lb of ice in $3\frac{1}{2}$ h, how many pounds of ice would it produce in 5 h? Round to the nearest whole number.

18. If D varies jointly with E and F and $D = 2,100$ when $E = 12$ and $F = 35$, find D when $E = 18$ and $F = 40$.

19. Simple interest I varies jointly as the interest rate r and the time t in years. The simple interest for a loan at 8% for 4 years requires $320. How much interest is required for a loan of the same amount of money if it is made at 9% for 5 years?

20. H varies directly as J and inversely with M. If the constant of variation is $\frac{3}{4}$, find H when $J = 16$ and $M = 24$.

9 Graphing Linear Equations and Functions

Focus on Careers

Have you always wanted to be a firefighter? Every year fires cause thousands of deaths and injuries and destroy billions of dollars worth of property. Firefighters are frequently the first emergency personnel to arrive at the scene of a fire or other emergency. They are required to put out fires, treat injuries, and perform many other important duties.

Firefighting requires a high level of organization and team-work, and firefighters are required to follow directions at the scene of a fire. They may be asked to connect hose lines to hydrants, position ladders, control water hoses that deliver water to the fire, rescue victims, and provide emergency medical assistance.

About 90% of firefighters are employed by municipal (city) or county fire departments. Competition for these jobs is very high, and applicants usually are required to pass written, physical, and medical examinations.

A high degree of skills and abilities is required of firefighters. Some firefighters attend the U.S. National Fire Academy and many colleges and universities offer 2-year and 4-year degrees in fire science.

Employment of firefighters is expected to grow about as fast as average through 2012, but competition for the available jobs is expected to be very high. The median hourly wage of firefighters was $17.42 in 2002. Local governments paid the highest median hourly earnings at $17.92, and state governments paid $13.58 per hour.

Source: *Occupational Outlook Handbook,* 2004–2005 Edition, U.S. Department of Labor, Bureau of Labor Statistics.

Learning Outcomes

1 Locate points on a rectangular coordinate system.

2 Represent an equation in two variables as a function.

3 Make a table of solutions for a linear equation or function.

4 Graph a linear equation or function using a table of solutions.

Linear equations in one variable have at most one solution. The solution can be represented graphically as a point on a number line. A linear equation in two variables has an unlimited number of solutions. The solutions of a linear equation in two variables can be represented graphically as a line on a rectangular coordinate system.

1 Locate Points on a Rectangular Coordinate System.

The number line shows values pictorially. This kind of visual representation is called a **one-dimensional graph.** The one dimension represents the distance and direction that a value is from zero.

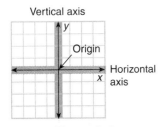

Vertical axis

Origin

Horizontal axis

Figure 9–1

The **rectangular coordinate system** gives a graphical representation of two-dimensional values. In the rectangular coordinate system, two number lines are positioned to form a right angle or square corner (see Fig. 9–1). The number line that runs from left to right is the **horizontal axis** or **x-axis,** and is represented by the letter *x*. The number line that runs from top to bottom is the **vertical axis** or **y-axis,** and is represented by the letter *y*.

Zero on both number lines is located at the point where the two number lines cross. This point is called the **origin.** Points are located by two dimensions, the horizontal distance from the origin and the vertical distance from the *x*-axis.

EXAMPLE Describe the location of the points in Fig. 9–2 by giving the amount of horizontal and vertical movement from the origin and vertical movement from the x-axis.

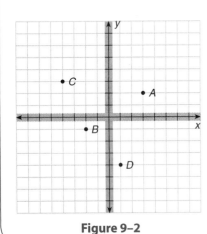

Figure 9–2

Point *A:* horizontal movement, $+3$; vertical movement, $+2$

Point *B:* horizontal movement, -2; vertical movement, -1

Point *C:* horizontal movement, -4; vertical movement, $+3$

Point *D:* horizontal movement, $+1$; vertical movement, -4

The location of points on a rectangular coordinate system can be written in an abbreviated, symbolic form. **Point notation** uses two signed numbers to show the horizontal movement from the origin and the vertical movement from the *x*-axis to a given point. Horizontal movement is represented by the ***x*-coordinate** and is written first. Vertical movement is represented by the ***y*-coordinate** and is written second. These signed numbers are separated with a comma and enclosed in parentheses.

Symbolically, we represent the location of a point as an **ordered pair** of numbers

$$(x, y)$$

where x is the horizontal movement from the origin, indicated by the x-coordinate, and y is the vertical movement from the x-axis, indicated by the y-coordinate.

EXAMPLE Write the points in the preceding example (Fig. 9–2) using point notation.

Point A = (horizontal movement of $+3$, vertical movement of $+2$) or **($+3, +2$)**
Point B = (horizontal movement of -2, vertical movement of -1) or **($-2, -1$)**
Point C = (horizontal movement of -4, vertical movement of $+3$) or **($-4, +3$)**
Point D = (horizontal movement of $+1$, vertical movement of -4) or **($+1, -4$)**

To **plot** a point means to show its location on the rectangular coordinate system.

To plot a point on the rectangular coordinate system:

1. Start at the origin.
2. Count to the left or right the number of units of the first signed number (x-coordinate) in the ordered pair.
3. From the ending point of Step 2, count up or down the number of units from the second signed number (y-coordinate).
4. Place a dot to show the point and write the coordinates beside the point.

EXAMPLE Plot these points: Point A = (3, 1), point B = (−2, 5), point C = (−3, −2), point D = (1, −3).

See Fig. 9–3.

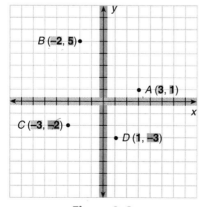

A: **right 3, up 1**
B: **left 2, up 5**
C: **left 3, down 2**
D: **right 1, down 3**

Figure 9–3

Develop Your Spatial Sense

Moving in the wrong direction is a common mistake when plotting points. To develop your spatial sense, consider the signs of the coordinates of points being plotted. Then, visualize in which part of the graph the point will fall. The four regions of the rectangular coordinate systems are called **quadrants.**

Refer to the points $A = (3, 1)$; $B = (-2, 5)$; $C = (-3, -2)$; and $D = (1, -3)$.

Point A (3, 1): move *right* and *up,* as shown in Fig. 9–4. This is quadrant I.

Point B (−2, 5): move *left* and *up,* as shown in in Fig. 9–5. This is quadrant II.

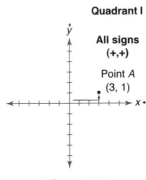

Figure 9–4

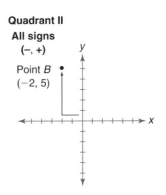

Figure 9–5

Point C (−3, −2): move *left* and *down,* as shown in Fig. 9–6. This is quadrant III.

Point D (1, −3): move *right* and *down,* as shown in Fig. 9–7. This is quadrant IV.

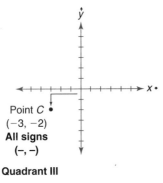

Figure 9–6

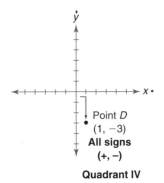

Figure 9–7

A point that shows movement in only one direction is written in point notation by using 0 as the coordinate that represents no movement. Points on the *x*-axis have no vertical movement. Points on the *y*-axis have no horizontal movement.

EXAMPLE Plot the points: $A = (2, 0)$, $B = (0, 2)$, $C = (-2, 0)$, $D = (0, -2)$.

See Fig. 9–8.

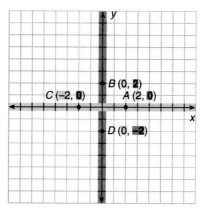

Point $A = (2, 0)$	right 2, then stop
Point $B = (0, 2)$	up 2
Point $C = (-2, 0)$	left 2, then stop
Point $D = (0, -2)$	down 2

Figure 9–8

2 Represent an Equation in Two Variables as a Function.

The equations that we have examined so far have been linear equations in one variable. We will expand our study of linear equations to include equations with two variables. In general, the variable that is the *independent variable* will be represented with the letter *x*. The *dependent variable* will be represented with the letter *y*.

An example of a linear equation in two variables is $y = x + 3$. This type of equation does not have just one solution. It has an unlimited number of solutions. For instance, if *x* is 2, then *y* is 2 + 3, or 5. If *x* is 3, then *y* is 3 + 3, or 6. Each solution of an equation in two variables has a value for each variable. We write these solutions as *ordered pairs* (independent variable, dependent variable).

The mathematical term for a set of ordered pairs is a **relation.** The first component of each ordered pair or the set of values for the independent variable is called the **domain.** The second component of each ordered pair or the set of values for the dependent variable is called the **range.** When each value of the domain corresponds to exactly one value in the range, the relation is called a **function.** A function shows the association between the values from the domain and the range.

A linear equation in two variables is one type of function. In function notation we write the dependent variable *y* as $f(x)$ (read "function of *x*"). This notation more clearly illustrates the association between the independent and dependent variables.

To write a linear equation in two variables using function notation:

1. Solve the equation in two variables for the dependent variable *y*.
2. Rewrite *y* as $f(x)$.

EXAMPLE Rewrite the equations in function notation.

(a) $y = 2x - 3$ (b) $2x + y = 7$

(a) $y = 2x - 3$ Rewrite y as $f(x)$.
    $f(x) = 2x - 3$
(b) $2x + y = 7$ Solve for y.
    $y = -2x + 7$ Rewrite y as $f(x)$.
    $f(x) = -2x + 7$

3 Make a Table of Solutions for a Linear Equation or Function.

Since an equation or function in two variables has an unlimited number of solutions, we can organize several of these solutions in a **table of solutions.** We determine the values of the independent variable that we will examine. That is, we select values from the domain. Then, we find corresponding values of the dependent variable.

To make a table of solutions for an equation or function:

1. Solve the equation for y and write the equation in function notation, if desired.
2. Select appropriate values for the independent variable.
3. Perform the calculations associated with the function for each selected value of the independent variable.
4. Write the results of each calculation as an ordered pair.

EXAMPLE Make a table of solutions for the following equations.

(a) $y = 2x - 3$ (b) $2x + y = 7$

(a) $y = 2x - 3$ Rewrite as $f(x)$.
    $f(x) = 2x - 3$ Select appropriate values for the independent variable x.

The domain of the function is all real numbers. Select values in the domain that are easy to work with and reasonable to graph. Any number of values can be selected but we will select at least three that are not too close together. Let's select -3, 0, and 3.

Table of Solutions

x	$f(x)$	y	(x, y)
-3	$f(x) = 2x - 3$ $f(-3) = 2x - 3$ $f(-3) = 2(-3) - 3$ $f(-3) = -6 - 3$ $f(-3) = -9$	-9	$(-3, -9)$
0	$f(0) = 2x - 3$ $f(0) = 2(0) - 3$ $f(0) = 0 - 3$ $f(0) = -3$	-3	$(0, -3)$
3	$f(3) = 2x - 3$ $f(3) = 2(3) - 3$ $f(3) = 6 - 3$ $f(0) = 3$	3	$(3, 3)$

Chapter 9 / Graphing Linear Equations and Functions

(b) $2x + y = 7$ Solve for y.

$\quad\quad\quad y = 2x + 7$ Rewrite y as $f(x)$.

$\quad\quad f(x) = 2x + 7$

Select appropriate values of x. Let's select -2, 0, and 2. In making our table, we will perform some of the steps mentally.

x	$f(x)$	y	(x, y)
-2	$f(-2) = -2(-2) + 7$	11	$(-2, 11)$
0	$f(0) = -2(0) + 7$	7	$(0, 7)$
2	$f(2) = -2(2) + 7$	3	$(2, 3)$

TIP   Make a Table of Solutions on a Calculator

A graphing calculator can build a table of solutions for a function having two variables if the equation is solved for the dependent variable.

Set the desired characteristics of the table (starting point of the desired domain and the desired increment of the table values of the independent variable):

Access $\boxed{\text{TBLSET}}$ feature. Enter starting value of desired domain after $\boxed{\text{TblStart} =}$.

Enter the desired increment after $\boxed{\triangle \text{Tbl} =}$.

Be sure that both the independent and dependent variables are set to $\boxed{\text{Auto}}$.

Enter the function: $\boxed{Y=}$ $\boxed{\text{CLEAR}}$. Then, enter the right side of the function at $\boxed{\text{Y}_1 =}$. Enter a variable using the $\boxed{\text{X}, \theta, \text{T}, n}$ key.

View the table by pressing $\boxed{\text{TABLE}}$. The arrow keys allow you to move up and down the table.

Use Part (b) of the previous example ($y = -2x + 7$) to build a table of solutions.

$\boxed{\text{TBLSET}}$ $\boxed{\text{TblStart} =}$ -3 $\boxed{\triangle \text{Tbl} =}$ 1 $\boxed{\text{Auto}}$ $\boxed{\text{Auto}}$ $\boxed{\text{QUIT}}$

$\boxed{Y=}$ $\boxed{\text{CLEAR}}$ $\boxed{(-)}$ 2 $\boxed{\text{X}, \theta, \text{T}, n}$ $\boxed{+}$ 7 $\boxed{\text{TABLE}}$. See Fig. 9–9.

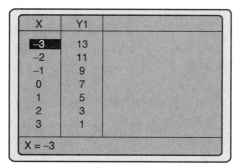

Figure 9–9

4   Graph a Linear Equation or Function Using a Table of Solutions.

The **table of solutions** or **table-of-values** procedure for graphing an equation works for *all* types of equations. Other methods for graphing equations focus on specific properties of each type of equation; these methods are generally less time-consuming and more effecient that the table-of-solutions procedure.

To graph a linear equation or function using a table of solutions:

1. Prepare a table of solutions by evaluating the equation or function using different values (at least three) in the domain of the independent variable.
2. Plot the points from the table of solutions on a rectangular coordinate system.
3. Connect the points with a straight line. Extend the graph beyond the three points and place an arrow on each end as appropriate to indicate that the line extends indefinitely.

EXAMPLE Graph the equations $y = 2x - 3$ and $2x + y = 7$ using the table-of-solutions method.

Use the table of solutions from the examples in Outcome 3.
$y = 2x - 3$ (see Fig. 9–10.)

x	y
-3	-9
0	-3
3	3

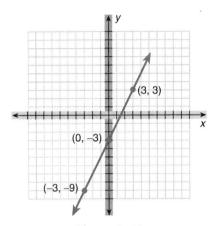

Figure 9–10

$2x + 7 = y$ (see Fig. 9–11.)   Solve the equation for y.
$\qquad y = -2x + 7$ Since the solution, $(-2, 11)$ is off the graph, another point was selected. $f(5) = -2(5) + 7 = -10 + 7 = -3$.

x	y
-2	11
0	7
2	3
5	-3

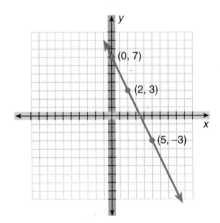

Figure 9–11

A **linear equation** is an equation for which the graph is a straight line. We can *check* each solution of an equation in two variables by substituting the values for the respective variables and performing the operations indicated.

To check a solution of an equation with two variables:

1. Substitute the values from the ordered pair for the respective variables in the equation.
2. Evaluate each side of the equation.
3. If the ordered pair is a solution of the equation, the two sides of the equation will be equal.

EXAMPLE Check the ordered pair $(-2, 16)$ for the equation $2x + y = 12$.

$$2x + y = 12 \qquad \text{Substitute } -2 \text{ for } x \text{ and } 16 \text{ for } y.$$
$$2(-2) + 16 = 12 \qquad \text{Simplify the left side.}$$
$$-4 + 16 = 12$$
$$12 = 12 \qquad \text{True.}$$

The ordered pair makes the equation true. So $(-2, 16)$ is a solution of the equation.

EXAMPLE Check the ordered pair $(3, -2)$ for the equation $y = 2x - 5$.

$$y = 2x - 5 \qquad \text{Substitute 3 for } x \text{ and } -2 \text{ for } y.$$
$$-2 = 2(3) - 5 \qquad \text{Simplify the right side.}$$
$$-2 = 6 - 5$$
$$-2 = 1 \qquad \text{False.}$$

The ordered pair does not make the equation true. So $(3, -2)$ is not a solution.

If we are given the value of one variable for an equation with two variables, we can substitute the given value in the equation and solve the equation for the other variable.

To find a specific solution of an equation with two variables when given the value of one variable:

1. Substitute the given value of the variable in each place that variable occurs in the equation.
2. Solve the equation for the other variable.

EXAMPLE Solve the equation $2x - 4y = 14$, if $x = 3$.

$$2x - 4y = 14 \qquad \text{Substitute 3 for } x.$$
$$2(3) - 4y = 14$$
$$6 - 4y = 14 \qquad \text{Sort terms.}$$
$$-4y = 14 - 6$$
$$-4y = 8 \qquad \text{Divide both sides by } -4.$$
$$\frac{-4y}{-4} = \frac{8}{-4}$$
$$y = -2$$

The solution is $(3, -2)$; $x = 3$ and $y = -2$.

The **standard form of a linear equation in two variables** is $ax + by = c$ for all real numbers a, b, and c. Both a and b are not zero; however, either a or b may be zero.

When a or b is zero, the variable for which zero is the coefficient is eliminated from the equation.

$ax + by = 15$	Let $a = 5$ and $b = 0$.
$5x + 0(y) = 15$	$0(y) = 0$.
$5x + 0 = 15$	Linear equation in one variable.
$5x = 15$	Divide both sides by 5.
$\dfrac{5x}{5} = \dfrac{15}{5}$	
$x = 3$	

An equation like $x = 3$ is a linear equation, and we think of the coefficient of the missing variable term as zero:

$$x = 3 \quad \text{is considered as} \quad x + 0y = 3 \quad \text{in standard form}$$

This means that x is 3 for any value of y.

EXAMPLE For the equation $x = 6$, complete the given ordered pairs: $(\ , 2); (\ , -3); (\ , 1)$.

Equation	**Ordered Pairs**			
$x = 6$	$(\ , 2)$	$(\ , -3)$	$(\ , 1)$	x is 6 for *any* value of y.
	$(6, 2)$	$(6, -3)$	$(6, 1)$	

Numerous application problems can be solved with equations in two variables.

EXAMPLE BUS A small business photocopied a report that included 12 black-and-white pages and 25 color pages. The cost was $23.70. Letting b equal the cost for a black-and-white page and c equal the cost for a color page, write an equation with two variables.

Cost of black-and-white copies:	$12b$ (12 times cost per page)
Cost of color copies:	$25c$ (25 times cost per page)
Total cost:	cost of black-and-white plus cost of color $= \$23.70$

Equation: $12b + 25c = \$23.70$

An equation containing two variables shows the relationship between the two unknown amounts. If a value is known for either amount, the other amount can be found by solving the equation for the missing amount.

EXAMPLE Solve the equation from the preceding example to find the cost for each black-and-white copy if color copies cost $0.90 each.

$$12b + 25c = 23.70$$ Substitute 0.90 for c.
$$12b + 25(0.90) = 23.70$$
$$12b + 22.50 = 23.70$$ Sort terms.
$$12b = 23.70 - 22.50$$
$$12b = 1.20$$ Divide both sides by 12.
$$\frac{12b}{12} = \frac{1.20}{12}$$
$$b = 0.10$$

Each black-and-white copy costs $0.10.

SECTION 9–1 SELF-STUDY EXERCISES

1 Locate the points in Fig. 9–12 by giving the amount of horizontal and vertical movement from the origin.

1. R **2.** S **3.** T **4.** U **9.** Write the coordinates for the points A, B, C, and
5. V **6.** W **7.** X **8.** Y D on the graph in Fig. 9–13.

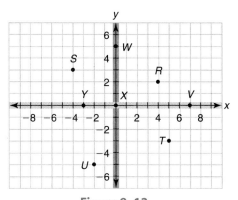

Figure 9–12

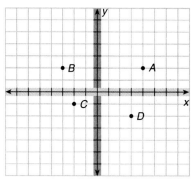

Figure 9–13

Draw a rectangular coordinate system, plot the points, and give the quadrant for each point in Exercises 10–15.

10. $A = (-7, 4)$ **11.** $B = (0, -10)$ **12.** $C = (-2, 0)$
13. $D = (-7, -7)$ **14.** $E = (3, 0)$ **15.** $F = (4, -3)$

16. What common property describes the coordinates of the points on the x-axis?

17. What common property describes the coordinates of the points on the y-axis?

18. Describe the signs of the coordinates of points that lie in the upper-left quarter of a graph.

19. Describe the signs of the coordinates of points that lie in the lower-right quarter of a graph.

2 Rewrite the equation in function notation.

20. $y = 5x - 7$ **21.** $y = -3x + 2$ **22.** $2y = 8x - 10$
23. $4y = -7x - 3$ **24.** $3x + y = 12$ **25.** $6x - 2y = 8$
26. $3x + 2y = 9$ **27.** $x - 7y = 14$ **28.** $5x - 3y = 7$

3 Make a table of solutions to obtain at least three sets of coordinates that satisfy each equation.

29. $y = 5x - 7$ **30.** $y = -3x + 2$ **31.** $2y = 8x - 10$

32. $4y = -7x - 3$ **33.** $3x + y = 12$ **34.** $6x - 2y = 8$

35. Make a table of at least three solutions for the equation $y = 3x - 7$.

36. Make a table of three solutions for the equation $y = -2x + 3$.

Prepare a table of three solutions for the equations and functions.

37. $y = -4x + 5$ **38.** $y = \dfrac{1}{2}x + 3$ **39.** $y = -\dfrac{2}{3}x - 1$

40. $y = 5x + 2$ **41.** $f(x) = -\dfrac{1}{2}x + 2$ **42.** $y = \dfrac{1}{3}x + 2$

4 Use the table of solutions prepared in Exercises 35–42 to graph the following solutions.

43. $y = 3x - 7$ (See Exercise 35.)   **44.** $y = -2x + 3$ (See Exercise 36.)

45. $y = -4x + 5$ (See Exercise 37.)   **46.** $y = \dfrac{1}{2}x + 3$ (See Exercise 38.)

47. $y = -\dfrac{2}{3}x - 1$ (See Exercise 39.)   **48.** $y = 5x + 2$ (See Exercise 40.)

49. $y = -\dfrac{1}{2}x + 2$ (See Exercise 41.)   **50.** $y = \dfrac{1}{3}x + 2$ (See Exercise 42.)

Determine which of the ordered pairs are solutions for the equation $2x - 5y = 9$.

51. $(4, 5)$ **52.** $(17, 5)$ **53.** $(12, 3)$

54. $(6, 1)$ **55.** $(7, 1)$ **56.** $(4.5, 0)$

Determine which of the ordered pairs are solutions for the equation $3y = 2 - 4x$.

57. $(2, -2)$ **58.** $(-2, 2)$ **59.** $(5, -6)$

60. $(-6, 5)$ **61.** $(0, \frac{2}{3})$ **62.** $(6, -7\frac{1}{3})$

Write the specific solution for each equation in ordered pair form.

63. $x - y = 3$, if $x = 8$ **64.** $x + y = 7$, if $x = 2$ **65.** $x + 2y = -4$, if $y = 1$

66. $x - 3y = 12$, if $y = 3$ **67.** $5x - y = 10$, if $x = 5$ **68.** $4x + y = 8$, if $x = 2$

Complete the ordered pairs for each equation.

69. $x = 1$: $(\ , 3)$; $(\ , -2)$; $(\ , 0)$; $(\ , 1)$ **70.** $y = 4$: $(-1, \ )$; $(3, \ )$; $(0, \ )$; $(2, \ )$

Solve using equations with two variables.

71. BUS Anne Richards purchased three shirts at one price and five belts at another price. Let s = the price of each shirt and b = the price of each belt. Each shirt cost $22. How much did each belt cost if the total purchase price was $126?

72. AUTO Allen and Cecile Bell paid $13 for 4 spark plugs and 5 quarts of motor oil. Let p = the cost of each plug and q = the cost of each quart of oil. If the oil cost $1 per quart, how much did each spark plug cost?

73. BUS Lola Bowers charged $160 for four pairs of pants and two shirts. If the shirts cost $12 each, how much did each pair of pants cost? Let p = the cost of each pair of pants and s = the cost of each shirt.

74. AUTO Coach Barton paid $15,000 for an automobile. The purchase price included three extended-warranty payments of $400 each. How much did the car cost without the extended warranty? Let c = the cost of the car and w = the cost of each warranty payment.

Learning Outcomes

1 Graph linear equations using intercepts.
2 Graph linear equations using the slope and y-intercept.
3 Graph linear equations using a graphing calculator.
4 Solve a linear equation in one variable using a graph.

We have graphed linear equations using a table of solutions. Now let's examine other graphing procedures and specific properties of linear equations in two variables.

1 Graph Linear Equations Using Intercepts.

Two important points on the graph of an equation are the points where the graph crosses the x- and y-axis. These points are called **intercepts.** The **x-intercept** is the point on the x-axis through which the line of the equation passes; that is, the y-value is zero $(x, 0)$. The **y-intercept** is the point on the y-axis through which the line of the equation passes; that is, the x-value is zero $(0, y)$.

To find the intercepts of a linear equation:

1. Find the x-intercept; let $y = 0$ and solve for x.
2. Find the y-intercept; let $x = 0$ and solve for y.

EXAMPLE   Find the intercepts of the equation $3x - y = 5$.

$3x - y = 5$	For the x-intercept, let $y = 0$.
$3x - 0 = 5$	Solve for x.
$3x = 5$	
$x = \dfrac{5}{3}$	A mixed number or decimal is easier to plot than an improper fraction.
$x = 1\dfrac{2}{3}, \quad$ or $\quad 1.67$	$1\frac{2}{3}$ is the exact value and 1.67 is an approximate value.

The x-intercept is $(1\frac{2}{3}, 0)$.

$3x - y = 5$	For the y-intercept, let $x = 0$.
$3(0) - y = 5$	Solve for y.
$-y = 5$	
$y = -5$	

The y-intercept is $(0, -5)$.

To graph linear equations by the intercepts method:

1. Find the x- and y-intercepts.
2. Plot the intercepts on a rectangular coordinate system.
3. Draw the line through the two points and extend it beyond each point.
4. Check by examining one additional solution of the equation.

EXAMPLE Graph the equation $3x - y = 5$ by using the intercepts of each axis.

Plot the two intercepts found in the preceding example, $(1\frac{2}{3}, 0)$ and $(0, -5)$. Draw the line connecting these points and extending beyond. See Fig. 9–14. We can check by finding the coordinates of one other point using the table-of-solutions method and plotting the point. If the point is on the line, the graph is correct. Find y when $x = 2$:

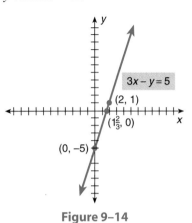

$$3x - y = 5 \qquad \text{Substitute 2 for } x.$$
$$3(2) - y = 5 \qquad \text{Solve for } y.$$
$$6 - 5 = y$$
$$1 = y \qquad \text{The point (2, 1) is on the graph.}$$

Figure 9–14

Graphing Equations When Both Intercepts Are (0, 0)

If both intercepts are $(0, 0)$, the intercepts coincide at the origin and form only *one point*. An additional point must be found by the table-of-solutions method to have two distinct points for drawing the line. A third point is still useful to check your work.

EXAMPLE Graph $y = 7x$ using the intercepts method.

Find the intercepts.

$$y = 7x \qquad \text{Let } x = 0.$$
$$y = 7(0)$$
$$y = 0 \qquad \text{Coordinates of the } y\text{-intercept are (0, 0).}$$
$$y = 7x \qquad \text{Let } y = 0.$$
$$0 = 7x$$
$$\frac{0}{7} = \frac{7x}{7}$$
$$0 = x \qquad \text{Coordinates of the } x\text{-intercept are (0, 0).}$$

Find the coordinates of another point.

$$y = 7x \qquad \text{Let } x = 1.$$
$$y = 7(1)$$
$$y = 7$$

Additional point $= (1, 7)$

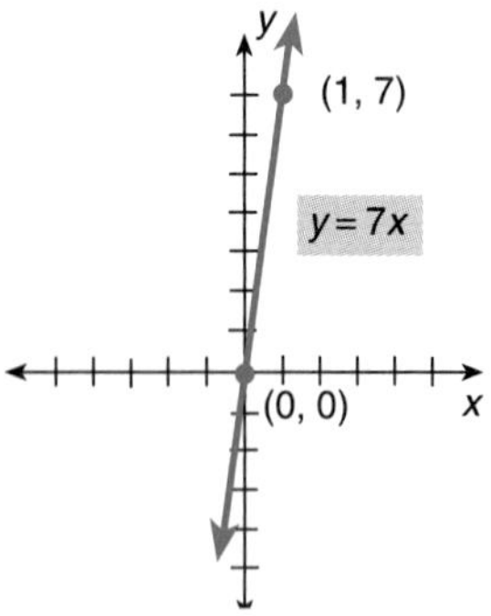

Plot the points $(0, 0)$ and $(1, 7)$. Draw the line through the two points and extend it beyond each point, as shown in Fig. 9–15.

Figure 9–15

2 Graph Linear Equations Using the Slope and *y*-Intercept.

We can identify characteristics of the graph of an equation by inspection. To see these characteristics, we solve the equation for *y*. The equation will be in the form *y* = *mx* + *b*, where *m* is the coefficient of *x* and *b* is a constant.

First, let's examine the characteristics identified by the constant. Let the coefficient of *x* equal 1 and look at the graphs of the equations in Fig. 9–16.

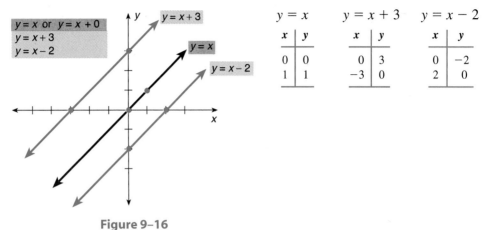

$y = x$		$y = x + 3$		$y = x - 2$	
x	*y*	*x*	*y*	*x*	*y*
0	0	0	3	0	−2
1	1	−3	0	2	0

Figure 9–16

The three graphs have the same slope (steepness or slant), but they have different *y*-intercepts. That is, they cross the *y*-axis at different points.

$$y = x + 0 \quad \text{crosses the } y\text{-axis at } 0; y\text{-intercept} = (0, 0)$$

$$y = x + 3 \quad \text{crosses the } y\text{-axis at } +3; y\text{-intercept} = (0, 3)$$

$$y = x - 2 \quad \text{crosses the } y\text{-axis at } -2; y\text{-intercept} = (0, -2)$$

The *y*-coordinate of the point where the graph crosses the *y*-axis is the same as the constant in the equation. The constant (*b*) in an equation in the form *y* = *mx* + *b* is the *y*-coordinate of the *y*-intercept and can be written (0, *b*).

Let's examine equations that have graphs with a different slant or steepness but the same *y*-intercept. The *x*- and *y*-intercepts in each of the three equations in Fig. 9–17 are (0, 0). We will find one additional point for each equation.

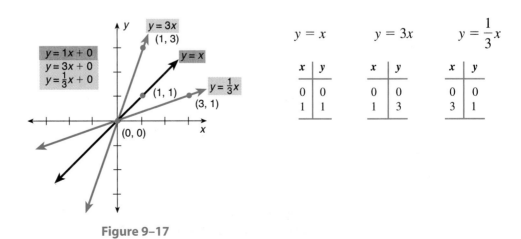

$y = x$		$y = 3x$		$y = \dfrac{1}{3}x$	
x	*y*	*x*	*y*	*x*	*y*
0	0	0	0	0	0
1	1	1	3	3	1

Figure 9–17

The three graphs have the same y-intercepts, but they have different slopes (steepness or slants). We define this slope or steepness so that a numerical value identifies it. The **slope** of a line is the ratio of the vertical change to the horizontal change.

$$\text{slope} = \frac{\text{vertical change}}{\text{horizontal change}}$$

A slope of 1 (written as $\frac{1}{1}$) means a change of 1 vertical unit for every 1 horizontal unit of change. A slope of 3 (written as the ratio $\frac{3}{1}$) means a change of 3 vertical units for every 1 horizontal unit. A slope of $\frac{1}{3}$ means a change of 1 vertical unit for every 3 horizontal units (see Fig. 9–18).

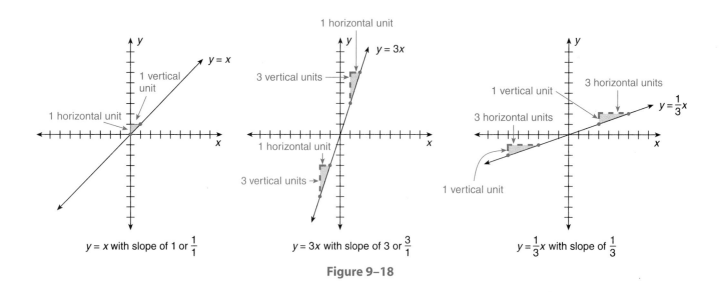

Figure 9–18

Notice that the slope of the line is the same as the fractional form of the coefficient of the x-term in the equation. Look again at the equations in Figs. 9–16 and 9–17.

From Fig. 9–16	**Slope-Intercept Form**	**From Fig. 9–17**	**Slope-Intercept Form**
$y = x$	$y = 1x + 0$	$y = x$	$y = 1x + 0$
$y = x + 3$	$y = 1x + 3$	$y = 3x$	$y = 3x + 0$
$y = x - 2$	$y = 1x - 2$	$y = \dfrac{1}{3}x$	$y = \dfrac{1}{3}x + 0$

In each case, the equation is solved for y. This form of equation is called the **slope-intercept form of a linear equation,** or $y = mx + b$, where m is the slope and b is the y-coordinate of the y-intercept. When equations are written in this form, the slope and y-intercept can be identified by inspection.

Slope-intercept form of an equation:

$$y = mx + b$$

where m and b are real numbers and $m =$ slope and $b = y$-coordinate of the y-intercept

Chapter 9 / Graphing Linear Equations and Functions

To determine the slope and *y*-intercept of a linear equation:

1. Solve the equation for *y* and write it in the form $y = mx + b$.
2. The *slope* is *m* which is the coefficient of *x*.
3. The *y*-coordinate of the *y*-intercept is *b*. The *y*-intercept is (0, *b*).

EXAMPLE   Write the equations in slope-intercept form and identify the slope and *y*-intercept.

(a) $2x + y = 4$ (b) $5x - y = -2$ (c) $3x + 4y = -12$

(a) $2x + y = 4$ Solve for *y*.

$$y = -2x + 4$$ Slope-intercept form.

$$\text{slope} = -2 \text{ or } \frac{-2}{1} \text{ or } \frac{2}{-1}$$ Coefficient of *x*.

y-coordinate of y-intercept = 4   Constant.

y-intercept = 4 or (0, 4)

(b) $5x - y = -2$ Solve for *y*. Divide by −1.

$$\frac{-y}{-1} = \frac{-5x - 2}{-1}$$ Write the right side as separate terms and simplify.

$$y = 5x + 2$$ Slope-intercept form.

$$\text{slope} = 5 \text{ or } \frac{5}{1} \text{ or } \frac{-5}{-1}$$ Coefficient of *x*.

y-coordinate of y-intercept = 2   Constant.

y-intercept = (0, 2)

(c) $3x + 4y = -12$ Solve for *y*. Sort terms to isolate 4*y*. Divide by 4.

$$\frac{4y}{4} = \frac{-3x - 12}{4}$$ Write the right side as separate terms and reduce.

$$y = -\frac{3}{4}x - 3$$ Slope-intercept form.

$$\text{slope} = -\frac{3}{4} \text{ or } \frac{-3}{4} \text{ or } \frac{3}{-4}$$ Coefficient of *x*.

y-coordinate of y-intercept = −3 Constant.

y-intercept = −3 or (0, −3)

An equation written in the slope-intercept form can be graphed using just the slope and *y*-intercept.

To graph a linear equation in the form *y* = *mx* + *b* using the slope and *y*-intercept method:

1. Locate the *y*-intercept on the *y*-axis.
2. Using the slope in fractional form, determine the amount of vertical and horizontal movement indicated.
3. From the *y*-intercept, locate additional points on the graph of the equation by counting the indicated vertical and horizontal movement.
4. Draw the line connecting the points and extending beyond the points.

EXAMPLE Graph the equations using the slope and *y*-intercept method.

(a) $2x + y = 4$ (b) $5x - y = -2$ (c) $3x + 4y = -12$

(a) $2x + y = 4$ Solve for *y*.

$$y = -2x + 4$$

y-intercept $= (0, 4)$ Locate this point on the *y*-axis.

slope $= \dfrac{-2}{1}$ or $\dfrac{2}{-1}$ Write the slope in fractional form as two equivalent fractions with a positive sign of the fraction.

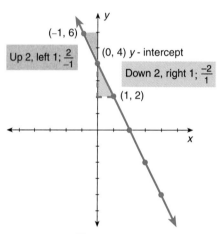

Figure 9–19

$\dfrac{-2}{1}$ indicates vertical movement of -2 and horizontal movement of $+1$ from the *y*-intercept.

$\dfrac{2}{-1}$ indicates vertical movement of $+2$ and horizontal movement of -1. See Fig. 9–19.

(b) $5x - y = -2$ Solve for *y*.

$$y = 5x + 2$$ Slope – intercept form

y-intercept $= (0, 2)$ Locate this point on the *y*-axis.

slope $= \dfrac{5}{1}$ or $\dfrac{-5}{-1}$

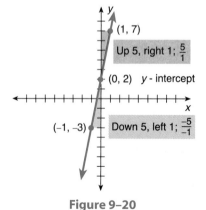

Figure 9–20

$\dfrac{5}{1}$ indicates vertical movement of $+5$ and horizontal movement of $+1$ from the *y*-intercept.

$\dfrac{-5}{-1}$ indicates vertical movement of -5 and horizontal movement of -1. See Fig. 9–20.

(c) $3x + 4y = -12$ Solve for y.

$$y = \frac{3}{4}x - \frac{12}{4}$$ Slope – intercept form

$$y = -\frac{3}{4}x - 3$$

y-intercept $= (0, -3)$ Locate this point on the y-axis.

$$\text{slope} = \frac{-3}{4} \text{ or } \frac{3}{-4}$$

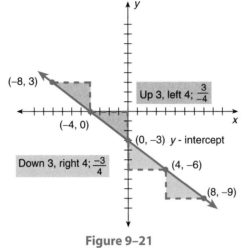

$\frac{-3}{4}$ indicates vertical movement of -3 and horizontal movement of $+4$ from the y-intercept.
$\frac{3}{-4}$ indicates vertical movement of $+3$ and horizontal movement of -4. See Fig. 9–21.

Figure 9–21

3 Graph Linear Equations Using a Graphing Calculator.

A graphing calculator can be used to graph linear equations written in the form $y = mx + b$. Also, computer software can be used to graph linear equations. These tools free us from the tedious task of graphing equations and allow us to focus on the patterns and properties of graphs.

Every model of graphing calculator and every brand of computer software may have a different set of keystrokes for graphing equations. You will need to refer to the owner's manual to adapt to a particular model or brand. However, we will illustrate the usefulness of these tools by showing a few features common to most calculators.

Feature	Purpose
Y =	Screen for entering equation that is solved for y
Window	Setting the range (high and low values of each axis) for the viewing window
Zoom	Zooming to get a closer or wider view or a view to "fit" the full range of y-values for the selected x-values
Trace	Tracing to locate specific points and determine their coordinates
Graph	Displays the graph of all equations that have been entered
Calc	Menu for shortcuts to specific calculations
Table	Screen for displaying a table of solutions

EXAMPLE   Graph $y = 5x + 2$ using a graphing calculator. Find the y-value for $x = -2$.

Perform the appropriate function on the graphing calculator of your choice.

Clear graphing screen.	Erase previous equations from the Y = screen.
Set or initialize range.	Define the viewing range of the graph window. One option is to select the standard option from the Zoom screen. That means the range will be reset at a predetermined, factory-set domain and range.
Enter equation.	Equation must be solved for y. On the Y = screen enter the equation using the appropriate key for the variable. This key is often labeled $\boxed{\text{X, θ, T, } n}$.
View graph.	Use the Graph key. Determine a different viewing window by setting the values on the Window screen or using the Zoom feature.
Determine key points on the graph.	Using the trace function, move the cursor with the left or right arrow keys. The x- and y-coordinates of the point at the cursor will appear at the bottom of the screen.

Find the corresponding y-value for a given x-value by using the *value* function from the CALC screen. Enter $\boxed{(-)}$ 2 for x at the prompt.

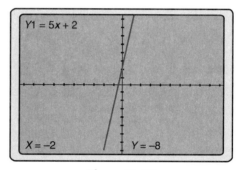

Figure 9–22

When $x = -2, y = -8$.

4 Solve a Linear Equation in One Variable Using a Graph.

A linear equation written in slope-intercept form is also referred to as a *function* of x, the independent variable. An equation in one variable, such as $2x + 5 = 3$, can be written in the form of a function by first rewriting all terms on either side of the equation.

$$2x + 5 = 3 \qquad \text{or} \qquad 2x + 5 = 3$$
$$2x + 5 - 3 = 0 \qquad\qquad\qquad 0 = 3 - 2x - 5$$
$$2x + 2 = 0 \qquad\qquad\qquad\quad 0 = -2x - 2$$

Then, write the nonzero side of each equation as a function of x.

$$y = 2x + 2 \qquad \text{or} \qquad y = -2x - 2$$

The equation can also be written as $f(x) = 2x + 2$ or $f(x) = -2x - 2$.

To write an equation in one variable as a function:

1. Rewrite the equation so that all terms are on one side of the equation.
2. Write the equation in function notation using $y =$ the nonzero side of the equation from Step 1.

The solution of the original equation in one variable is the value of x that makes the value of either function zero.

If an equation in one variable is written in function notation, the solution of the equation is shown on the graph at the point where the graph crosses the x-axis. In Fig. 9–23, both functions cross the x-axis at $(-1, 0)$ and the solution of the equation $2x + 5 = 3$ is $x = -1$. Thus, we can solve equations in one variable by using the graphing function on a calculator or computer.

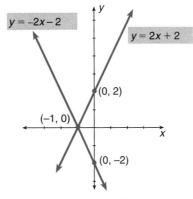

Figure 9–23

To solve an equation in one variable using a graph:

1. Write the equation as a function.
2. Graph the function using a calculator or computer.
3. The solution is the x-intercept. This solution is also referred to as the *zero* ($y = 0$) of the function.

TIP

Find the Zero of a Function Using a Calculator

On most graphing calculators there is a feature to find the zero point or points of a function in one variable. On one popular model, the TI-84 Plus Silver Edition™, the feature is a selection on the CALC screen.

The option for finding the zero point requires several steps that are illustrated in the next example.

EXAMPLE   Solve the equation $2x + 5 = 3$ using the TI-84 Plus Silver Edition™.

$$2x + 5 = 3$$
$$2x + 5 - 3 = 0$$
$$2x + 2 = 0$$
$$y = 2x + 2 \qquad \text{Function form of the equation.}$$

$\boxed{Y=}$	Enter right side of equation: 2 $\boxed{\text{X, θ, T, } n}$ + 2
CALC	$\boxed{\text{CALC}}$.
Zero	Option 2. Press $\boxed{\text{ENTER}}$.
Left bound?	Move cursor to any point to the left of the *x*-intercept. Press $\boxed{\text{ENTER}}$.
Right bound?	Move cursor to any point to the right of the *x*-intercept. Press $\boxed{\text{ENTER}}$.
Guess?	Move the cursor close to the *x*-intercept. Press $\boxed{\text{ENTER}}$.
$\boldsymbol{x = -1}$	Read solution from *x* value at the bottom left of the screen.

SECTION 9–2 SELF-STUDY EXERCISES

1 Graph the equations using the intercepts procedure.

1. $x + y = 5$

2. $x + 3y = 5$

3. $\dfrac{1}{2}y = 4 + x$

4. $y = 3x - 1$

5. $5x = y + 2$

6. $x = -2y + 3$

2 Determine the slope and *y*-intercept by inspection.

7. $y = 4x + 3$

8. $y = -5x + 6$

9. $y = -\dfrac{7}{8}x - 3$

10. $y = 3$

Rewrite the following equations in slope-intercept form and determine the slope and *y*-intercept.

11. $4x - 2y = 10$

12. $2y = 5$

13. $\dfrac{1}{2}y + x = 3$

14. $2.1y - 4.2x = 10.5$

15. $2x = 8$

16. $x - 5 = 4$

Graph the equations using the slope-intercept procedure.

17. $y = 2x - 3$

18. $y = -\dfrac{1}{2}x - 2$

19. $y = -\dfrac{3}{5}x$

20. $x - 2y = 3$

21. $2x + y = 1$

22. $y = \dfrac{3}{4}x + 2$

3 Graph the equations using a graphing calculator and find the value of *y* as indicated. Use appropriate viewing window.

23. $y = 0.3x + 0.5$

24. Find *y* when $x = -0.4$ in $y = 0.3x + 0.5$.

25. $y = -2.6x - 1.7$

26. Find *y* when $x = 1.3$ in $y = -2.6x - 1.7$.

27. $y = 212x + 757$

28. Find *y* when $x = 5$ in $y = 212x + 757$.

29. $y = 48x - 200$

30. Find *y* when $x = 7$ in $y = 48x - 200$.

31. **BUS** The cost of printing a magazine is $5,000 to typeset and prepare copy for printing and $3 per copy to print. This is expressed as an equation $y = 3x + 5{,}000$, where x is the number of copies printed and y is the total cost of printing x copies. Find the cost of printing 10,000 copies. Find the cost of printing 30,000 copies.

32. **BUS** Income is expressed by the equation $y = \$8.30x$, where x is the number of hours per week an employee works. Find the income for an employee who works 40 hours in a week. Find the annual income if a person works for an entire year (52 weeks).

33. **HLTH/N** Some researchers say a person's maximum heart rate in beats per minute can be found by subtracting age from 220. This can be expressed as an equation $y = 220 - x$. What is the maximum heart rate for a person 25 years old? What is the maximum heart rate for a person 65 years old?

34. **BUS** An inventory shows 196 jigsaw puzzles in stock. These puzzles can be produced at a rate of 15 per hour. The number of puzzles on hand can be expressed as an equation $y = 15x + 196$, where x represents the number of hours of production and y represents the total inventory. How many puzzles are in inventory if they have been produced for 60 h?

4 Solve the following equations using a graphing calculator.

35. $3x + 2 = 7$

36. $5x - 4 = 2$

37. $4 = 6x - 1$

38. $3x + 7 = 4x + 6$

39. $3x - 1 = 4(x + 2)$

40. $2x - 8 = 4(x + 2)$

9–3 | Slope

Learning Outcomes

1 Calculate the slope of a line, given two points on the line.

2 Determine the slope of a horizontal or vertical line.

The concept of the rate of change is important in real-world applications. This concept is often referred to as *slope* and is used when we consider the slant of a roof or the grade of a roadway. We use the rate of change in applications, like changes in temperature, in the quality of a product, or in sales.

1 **Calculate the Slope of a Line, Given Two Points on the Line.**

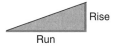

Run

Figure 9–24

The *slope* of a straight line is the rate of change of the line or the ratio of the vertical rise of a line to the horizontal run of the line (see Fig. 9–24).

$$\text{slope} = \frac{\text{rise}}{\text{run}} \quad \text{or} \quad \frac{\text{vertical change}}{\text{horizontal change}}$$

To find the slope of a line from two given points on the line:

1. Designate either point as point 1 with coordinates (x_1, y_1). Designate the other point as point 2 with coordinates (x_2, y_2).
2. Calculate the change (*difference*) in the y-coordinates to find the vertical *rise* $(y_2 - y_1)$ and the change in the x-coordinates to find the horizontal *run* $(x_2 - x_1)$.
3. Write a ratio of the rise to the run and reduce the ratio to lowest terms.

Symbolically, if P_1 and P_2 are any two points on the line,

$$\text{slope} = \frac{\Delta y}{\Delta x} = \frac{y_2 - y_1}{x_2 - x_1}$$

where $P_1 = (x_1, y_1)$ and $P_2 = (x_2, y_2)$.

The Greek capital letter delta (Δ) is used to indicate a change and to write the slope definition symbolically.

EXAMPLE Find the slope of a line if the points $(2, -1)$ and $(5, 3)$ are on the line (see Fig. 9–25).

Identify $(2, -1)$ as point 1 (P_1) and $(5, 3)$ as point 2 (P_2). The coordinates of P_1 are (x_1, y_1) and the coordinates of P_2 are (x_2, y_2). Write the slope definition symbolically and substitute known values.

$$\text{change in } y = \Delta y = \text{rise} = y_2 - y_1 = 3 - (-1)$$
$$\text{change in } x = \Delta x = \text{run} = x_2 - x_1 = 5 - 2$$
$$\text{slope} = \frac{\Delta y}{\Delta x} = \frac{\text{rise}}{\text{run}} = \frac{y_2 - y_1}{x_2 - x_1} = \frac{3 - (-1)}{5 - 2} = \frac{4}{3}$$

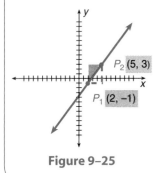

Figure 9–25

The slope of the line through points $(2, -1)$ and $(5, 3)$ is $\dfrac{4}{3}$.

Any line has an infinite number of points. *Any* two points on the line can be used to find the slope. Also, the designation of P_1 and P_2 is not critical. Let's find the slope of the line in the preceding example by designating P_1 as $(5, 3)$ and P_2 as $(2, -1)$.

$$\Delta y = \text{rise} = y_2 - y_1 = -1 - 3 = -4$$
$$\Delta y = \text{run} = x_2 - x_1 = 2 - 5 = -3$$
$$\frac{\Delta y}{\Delta x} = \frac{\text{rise}}{\text{run}} = \frac{-4}{-3} = \frac{4}{3}$$

EXAMPLE Find the slope of the line passing through the pair of points.

$(-5, -1)$ and $(3, -2)$

Visualize the line on a coordinate system (Fig. 9–26).

Let $P_1 = (-5, -1)$ and $P_2 = (3, -2)$.

$$\text{slope} = \frac{\Delta y}{\Delta x} = \frac{y_2 - y_1}{x_2 - x_1} = \frac{-2 - (-1)}{3 - (-5)} = \frac{-2 + 1}{3 + 5} = \frac{-1}{8} = -\frac{1}{8}$$

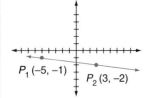

Figure 9–26

The slope is $-\dfrac{1}{8}$.

Chapter 9 / Graphing Linear Equations and Functions

Examine the slopes found in the two previous examples and the corresponding figures (Fig. 9–25 and Fig. 9–26). The first example has a positive slope of $\frac{4}{3}$. The graph of the line in Fig. 9–25 *rises* from left to right. The second example has a negative slope of $-\frac{1}{8}$. The graph of the line in Fig. 9–26 *falls* from left to right.

Lines with positive or negative slope:

- A line that has a positive slope will *rise* as the line moves from left to right.
- A line that has a negative slope will *fall* as the line moves from left to right.

EXAMPLE Table 9–1 lists the annual cost in dollars of tuition and fees at public 2-year colleges for selected years. Find the rate of change in tuition and fees from the 1998–1999 academic year to the 1999–2000 academic year. Also, find the rate of change from 1999–2000 to 2004–2005.

Find the rate of change of tuition and fees from 1998–1999 to 1999–2000.

Table 9–1 **Tuition and Fees at 2-Year Public Colleges Current Doillars**	
Academic Year	Tuition and Fees
94–95	$1,310
95–96	$1,330
96–97	$1,465
97–98	$1,567
98–99	$1,554
99–00	$1,649
00–01	$1,642
01–02	$1,608
02–03	$1,674
03–04	$1,909
04–05	$2,076

Source: *Annual Survey of Colleges*, The College Board, New York, NY.

(1998, 1,554) and (1999, 1,649) Write the data as ordered pairs. Use the initial year of the academic year as the reference.

$$m = \frac{y_2 - y_1}{x_2 - x_1}$$

Use the slope formula to find the rate of change.

$$\text{rate of change} = \frac{1{,}649 - 1{,}554}{1999 - 1998}$$

$$\text{rate of change} = \frac{95}{1} = \textbf{\$95 per year}$$

Find the rate of change from 1999–2000 to 2004–2005.

(1999, 1,649) and (2004, 2,076)

$$m = \frac{y_2 - y_1}{x_2 - x_1}$$

$$\text{rate of change} = \frac{2{,}076 - 1{,}649}{2004 - 1999}$$

$$\text{rate of change} = \frac{427}{5} = \textbf{\$85.40 per year}$$

The rate of change in tuition for one year from 1998–1999 to 1999–2000 was $95. But the rate of change from 1999–2000 to 2004–2005 was $85.40 per year. This rate represents an average change per year over the 5-year period.

2 Determine the Slope of a Horizontal or Vertical Line.

Let's examine the slope of a horizontal line and a vertical line (Fig. 9–27 and 9–28). First look at the horizontal line that passes through the points $(3, 2)$ and $(-3, 2)$.

EXAMPLE Find the slope of the horizontal line that passes through $(3, 2)$ and $(-3, 2)$.

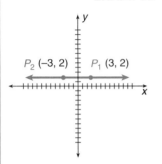

Let $P_1 = (3, 2)$ and $P_2 = (-3, 2)$. See Fig. 9–27.

$$\text{slope} = \frac{\Delta y}{\Delta x} = \frac{y_2 - y_1}{x_2 - x_1} = \frac{2 - 2}{-3 - 3} = \frac{0}{-6} = \mathbf{0}$$

The horizontal line has a slope of 0.

Figure 9–27

Slope of horizontal lines:

The slope of a horizontal line is zero.

- Two points are on the same horizontal line if their y-coordinates are equal.
- A horizontal line has no rise or no change in y-values.

Next, let's look at the slope of the vertical line that passes through the points $(3, 2)$ and $(3, -2)$.

EXAMPLE Find the slope of the vertical line that passes through the points $(3, 2)$ and $(3, -2)$.

Let $P_1 = (3, 2)$ and $P_2 = (3, -2)$. See Fig. 9–28.

$$\text{slope} = \frac{\Delta y}{\Delta x} = \frac{y_2 - y_1}{x_2 - x_1} = \frac{-2 - 2}{3 - 3} = \frac{-4}{0}$$

Division by zero is undefined; **therefore, the slope of the vertical line passing through $(3, 2)$ and $(3, -2)$ is undefined.**

The vertical line has undefined slope.

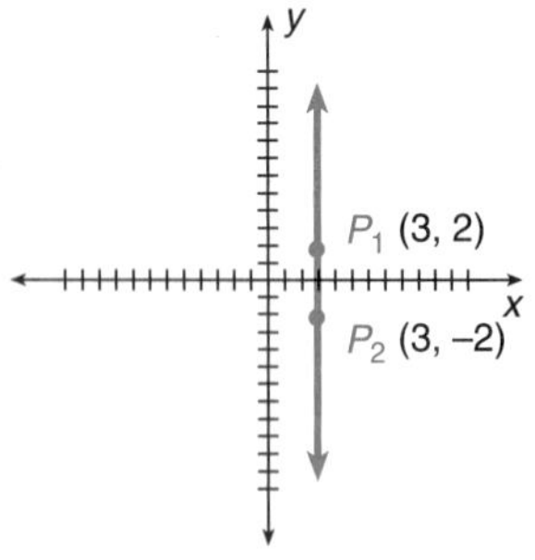

Figure 9–28

Chapter 9 / Graphing Linear Equations and Functions

Slope of vertical lines:

The slope of a vertical line is undefined.

- Two points are on the same vertical line if their *x*-coordinates are equal.
- A vertical line has no run or no change in *x*-values.

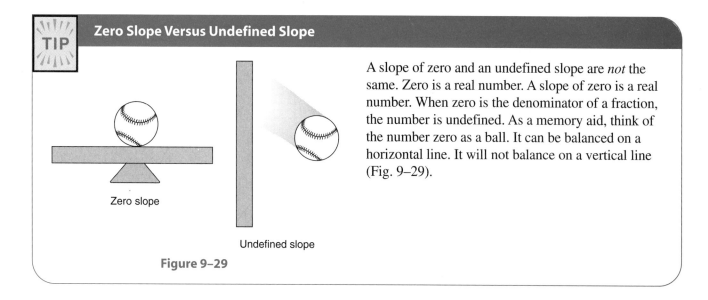

TIP

Zero Slope Versus Undefined Slope

A slope of zero and an undefined slope are *not* the same. Zero is a real number. A slope of zero is a real number. When zero is the denominator of a fraction, the number is undefined. As a memory aid, think of the number zero as a ball. It can be balanced on a horizontal line. It will not balance on a vertical line (Fig. 9–29).

Zero slope

Undefined slope

Figure 9–29

To identify a horizontal or vertical line:

1. Examine the *y*-coordinates of two points on the line. If the *y*-coordinates are the same, the line is horizontal and the slope is 0.
2. Examine the *x*-coordinates of two points on the line. If the *x*-coordinates are the same, the line is vertical and the slope is undefined.

SECTION 9–3 SELF-STUDY EXERCISES

1 Find the slope of the line passing through the pairs of points.

1. $(-3, 3)$ and $(1, 5)$
2. $(4, -1)$ and $(1, 4)$
3. $(3, 1)$ and $(5, 7)$
4. $(-1, -1)$ and $(3, 3)$
5. $(4, 5)$ and $(-4, -1)$
6. $(7, 2)$ and $(-3, 2)$
7. $(4, -5)$ and $(0, 0)$
8. $(2, -2)$ and $(5, -5)$
9. $(-5, 2)$ and $(-5, -4)$
10. $(4, -4)$ and $(1, 5)$
11. $(3, 5)$ and $(-4, 5)$
12. $(-7, 3)$ and $(-7, 5)$

13. BUS Dee Wallace's salary was $62,000 in 2005 and increased to $64,800 in 2007. What was the annual rate of change in his salary?

14. BUS If 100 backpacks cost $2,000 to produce and 800 backpacks cost $9,000 to produce, find the unit cost (slope) of producing the backpacks.

15. AVIA An airplane takes off from the ground (altitude is 0 ft and time is 0 min) and after 3 min it has an altitude of 6,000 ft. Find the rate of change in feet per minute.

Use Table 9–2 for Exercises 16–22. Use the initial year of the academic year as the reference.

16. **BUS** Find the rate of change in tuition and fees at public 4-year colleges from 1993 to 2003.
17. **BUS** Find the rate of change in tuition and fees at public 4-year colleges from 1998 to 2003.
18. Find the rate of change in tuition and fees at public 4-year colleges from 02–03 to 03–04.
19. Find the rate of change in tuition and fees at public 2-year colleges from 93–94 to 03–04.
20. Find the rate of change in tuition and fees at public 2-year colleges from 98–99 to 02–03.
21. The rate of change in tuition and fees for the 10-year period 93–94 to 03–04 is not the same as the rate of change from 98–99 to 02–03 for public 4-year colleges. Will the data graph to form a straight line? Explain.
22. Compare the rate of change in tuition and fees at public 4-year colleges with the rate of change in tuition and fees at public 2-year colleges for the 10-year period 94–95 to 04–05.

Table 9–2 Average Published Tuition and Fee Charges, 1992–93 to 2004–05 (Enrollment-Weighted)

Acacemic Year	Public Four-Year	Public Two-Year
92–93	3,102	1,483
93–94	3,285	1,613
94–95	3,407	1,650
95–96	3,447	1,631
96–97	5,547	1,747
97–98	3,644	1,835
98–99	3,742	1,791
99–00	3,766	1,847
00–01	3,796	1,777
01–02	4,004	1,710
02–03	4,263	1,741
03–04	4,729	1,943
04–05	5,132	2,076

2 Sketch the line passing through the pair of points and identify the line as having a positive slope, a negative slope, a zero slope, or an undefined slope.

23. $(-3, 3)$ and $(1, 5)$
24. $(4, -1)$ and $(4, 4)$
25. $(3, 7)$ and $(5, 7)$
26. $(-1, -1)$ and $(3, 3)$
27. $(4, 5)$ and $(-4, -1)$
28. $(7, 2)$ and $(-3, 2)$
29. $(4, -5)$ and $(0, 0)$
30. $(5, -2)$ and $(5, -5)$
31. $(-5, 2)$ and $(-5, -4)$
32. $(4, -4)$ and $(1, 5)$
33. $(3, 5)$ and $(-4, 5)$
34. $(-7, 3)$ and $(-7, 5)$

9–4 | *Linear Equation of a Line*

Learning Outcomes

1 Find the equation of a line, given the slope and one point.
2 Find the equation of a line, given two points on the line.
3 Find the equation of a line, given the slope and *y*-intercept.
4 Find the equation of a line, given a point on the line and an equation of a line parallel to that line.
5 Find the equation of a line, given a point on the line and an equation of a line perpendicular to that line.

1 Find the Equation of a Line, Given the Slope and One Point.

It is often desirable to know the equation of the line. The equation of the line can serve as a model for finding any point on the line and for solving various applications.

To find the equation of a line if the slope and one point on the line are known:

1. Use the following version of the **point-slope form of a linear equation** of a straight line.

$$y - y_1 = m(x - x_1)$$

where x_1 and y_1 are coordinates of the known point, m is the slope of the line, and x and y are the variables of the equation.

2. Substitute known values for x_1, y_1, and m.

3. Rearrange the equation to be in standard form $(ax + by = c)$ or slope-intercept form $(y = mx + b)$.

Standard form of an equation:

$$ax + by = c$$

The characteristics of an equation in standard form are

- Both variable terms are on the left.
- The leading term has the x variable and is positive.
- The equation contains no fractions.

EXAMPLE   Find the equation of the line passing through the point $(3, -2)$ with a slope of $\frac{2}{3}$. Write the equation in slope-intercept form.

$$y - y_1 = m(x - x_1)$$ Substitute into the point-slope form. $m = \frac{2}{3}$; $x_1 = 3$; $y_1 = -2$

$$y - (-2) = \frac{2}{3}(x - 3)$$ Simplify left side.

$$y + 2 = \frac{2}{3}(x - 3)$$ Distribute $\dfrac{2}{\underset{1}{3}} \cdot \dfrac{\overset{-1}{-3}}{1} = -2$.

$$y + 2 = \frac{2}{3}x - 2$$ Sort terms.

$$y = \frac{2}{3}x - 2 - 2$$ Combine like terms.

$$y = \frac{2}{3}x - 4$$ Solved for y.

EXAMPLE Write $y = \frac{2}{3}x - 4$ in standard form.

$$y = \frac{2}{3}x - 4$$
Rearrange with variable terms on the left with the *x*-variable as the first term.

$$-\frac{2}{3}x + y = -4$$
Clear the fraction by multiplying by the denominator 3.

$$3\left(-\frac{2}{3}x + y\right) = 3(-4)$$
Distribute.

$$-2x + 3y = -12$$
Multiply by -1 to make the leading term positive.

$$\mathbf{2x - 3y = 12}$$
Standard form.

2 Find the Equation of a Line, Given Two Points on the Line.

To find the equation of a line when two points on the line are known, we use both the slope and the point-slope formulas.

To find the equation of a line if two points on the line are known:

1. Use the slope formula to find the slope, given two points.

$$m = \frac{y_2 - y_1}{x_2 - x_1}$$

2. Use the point-slope form of an equation of a straight line, the calculated slope from Step 1, and the coordinates of either one of the given points.

$$(y - y_1) = m(x - x_1)$$

3. Write the equation in slope-intercept or standard form.

EXAMPLE Find the equation of the line that passes through the points $(0, 8)$ and $(5, 0)$. Write the equation in slope-intercept form.

$$\text{slope} = \frac{\Delta y}{\Delta x} = \frac{y_2 - y_1}{x_2 - x_1} = \frac{0 - 8}{5 - 0} = -\frac{8}{5}$$
Substitute known values into the slope formula. Let $P_1 = (0, 8)$ and $P_2 = (5, 0)$.

$$y - y_1 = m(x - x_1)$$
Substitute the slope and values for one point into the point-slope form. Use P_1.

$$y - 8 = -\frac{8}{5}(x - 0)$$
Simplify.

$$y - 8 = -\frac{8}{5}x$$
Solve for *y*.

$$\mathbf{y = -\frac{8}{5}x + 8}$$

EXAMPLE Find the equation of the line that passes through the points $(-3, 4)$ and $(7, 4)$.

We let $P_1 = (-3, \boxed{4})$ and $P_2 = (7, \boxed{4})$.

$$\text{slope} = \frac{\Delta y}{\Delta x} = \frac{y_2 - y_1}{x_2 - x_1} = \frac{4 - 4}{7 - (-3)} = \frac{0}{7 + 3} = \frac{0}{10} = 0 \qquad \text{Substitute values.}$$

Use the point-slope form of an equation and P_1.

$y - y_1 = m(x - x_1)$ Substitute values.

$y - 4 = 0[x - (-3)]$ Simplify.

$y - 4 = 0$ Solve for y.

$\mathbf{y = 4}$

The situation in the preceding example is generalized in the equation of a horizontal line.

Equation of a horizontal line:

The equation of a line with a slope of zero (a horizontal line) is

$$y = k$$

where k is the common y-coordinate for all points on the line.

There is one special situation in which the point-slope form of an equation *cannot* be used to determine the equation of a line, that is, when the slope is undefined.

Equation of a vertical line:

The equation of a line with a slope that is undefined (a vertical line) is

$$x = k$$

where k is the common x-coordinate for all points on the line.

EXAMPLE Find the equation of the line that passes through the points $(5, 3)$ and $(5, 7)$.

Since the x-coordinate is the same in both points, the slope of the line is undefined. The equation of the line is $x = k$, where k is the common x-coordinate. In this example, $k = 5$.

The equation of the line through $(5, 3)$ and $(5, 7)$ is $x = 5$.

3 Find the Equation of a Line, Given the Slope and *y*-Intercept.

So far we have written the equation of a line when we know either two points on the line or the slope and at least one point on the line. If the one point is the *y*-intercept $(0, b)$, then we use the slope-intercept form of an equation, $y = mx + b$, and we can write the equation by inspection.

To find the equation of a line if the slope and *y*-intercept are known:

1. Use the slope-intercept form of an equation of a straight line:

$$y = mx + b$$

   where m = slope and b = the *y*-coordinate of the *y*-intercept.
2. Substitute known values for m and b.
3. Write the equation in standard form or slope-intercept form.

EXAMPLE Write the equation for a line with a slope of -3 and a *y*-intercept of 5.

Slope = $m = -3$;   *y*-intercept = $b = 5$

$y = mx + b$ Substitute values.

$y = -3x + 5$ Slope-intercept form.

The necessary facts for writing the equation of a line are often obtained from the graph of the equation.

EXAMPLE Fig. 9–30 shows the cost of producing picture frames.

(a) Write an equation that represents the graph.
(b) Use the equation to find the cost of producing 20 picture frames.

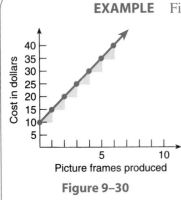

Figure 9–30

(a) The *y*-intercept represents the fixed cost of producing picture frames. $b = \$10$. The slope m is 5 units vertical change for every 1 unit horizontal change, which is $\frac{5}{1} = 5$. Notice the different scales used for the *x*- and *y*-axis.

$$y = 5x + 10$$ Equation of the line.

This type of equation is a **cost function.**

(b) Use the equation of the line found in part (a) to find y when $x = 20$.

$y = 5x + 10$ Substitute $x = 20$.
$y = 5(20) + 10$ Evaluate.
$y = 100 + 10$
$y = \$110$ Interpret.

The cost of producing 20 frames is \$110.

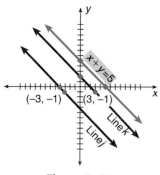

Figure 9–31

4 Find the Equation of a Line, Given a Point on the Line and an Equation of a Line Parallel to That Line.

Parallel lines are two or more lines that are the same distance apart everywhere. They have no points in common. See Fig. 9–31.

If we write $x + y = 5$ in slope-intercept form, we have $y = -x + 5$ and the slope is -1. *Any* equation that has a slope of -1 either is parallel to or coincides with the line formed by the equation $x + y = 5$. **Coincides** means one line lies on top of the other; that is, they are the same line.

Although parallel lines have the same slopes, they have different x- and y-intercepts.

Slope of parallel lines:

The slopes of parallel lines are equal.

EXAMPLE   Write the equations for lines a and b in Fig. 9–31.

$\quad\quad x + y = 5$ Solve the given equation for y to determine the slope of all parallel lines.
$\quad\quad y = -x + 5$ Slope $= -1$.

The y-intercepts for lines j and k can be determined from the graph.

Line j: $y = -x - 4$ Substitute $m = -1$ and $b = -4$
Line k: $y = -x + 2$ Substitute $m = -1$ and $b = 2$

To write the equation of a line that is parallel to a given line when at least one point on the parallel line is known:

1. Solve the equation of the given line for y.
2. Determine the slope m from the equation in Step 1.
3. The slope of the parallel line is the same as the slope of the given line.
4. Use the point-slope form of a straight line $y - y_1 = m(x - x_1)$ and substitute values for m, x_1, and y_1.
5. Rearrange the equation to be in standard form ($ax + by = c$) or slope-intercept form ($y = mx + b$).

EXAMPLE   Write the equation of a line that is parallel to $2y = 3x + 8$ and passes through the point (2, 1). Write the new equation in slope-intercept form and in standard form.

$\quad\quad 2y = 3x + 8$ Write the original equation in slope-intercept form.

$\quad\quad \dfrac{2y}{2} = \dfrac{3x + 8}{2}$

$\quad\quad y = \dfrac{3}{2}x + 4$ Identify the slope.

9–4 Linear Equation of a Line   413

The slope of $2y = 3x + 8$ is $\frac{3}{2}$.

$y - y_1 = m(x - x_1)$ Substitute. The slope of a parallel line is equal to the slope of the given line. $m = \frac{3}{2}$; $x_1 = 2$; $y_1 = 1$

$y - 1 = \frac{3}{2}(x - 2)$ Distribute.

$y - 1 = \frac{3}{2}x - 3$ Solve for y.

$y = \frac{3}{2}x - 3 + 1$ Combine like terms.

$y = \frac{3}{2}x - 2$ Slope-intercept form.

To write the equation in standard form, begin by clearing fractions.

$2y = 2\left(\frac{3}{2}x - 2\right)$ To clear fractions, multiply by 2.

$2y = 3x - 4$ Move variable term $3x$ to the left side of the equation.

$-3x + 2y = -4$ Multiply each term by -1 so that the coefficient of x is positive.

$3x - 2y = 4$ Standard form.

5   Find the Equation of a Line, Given a Point on the Line and an Equation of a Line Perpendicular to That Line.

Perpendicular lines are two lines that intersect to form right angles (90° angles). Another term for *perpendicular* is **normal**. In Fig. 9–32, the perpendicular (or normal) line that passes through the given point is indicated by the symbol ⌐, which means that a right angle is formed by the lines.

To explore the relationship of the slopes of perpendicular lines, examine the graphs and equations in Fig. 9–32. The slopes of the lines on the left are 2 and $-\frac{1}{2}$, whereas the slopes of the two lines on the right are $\frac{2}{3}$ and $-\frac{3}{2}$.

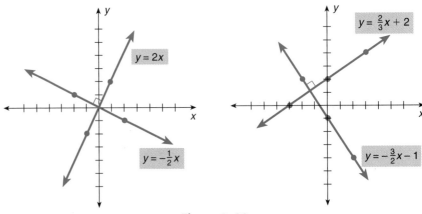

Figure 9–32

The slope of *any* line that is perpendicular to a given line is the **negative reciprocal** of the slope of the given line.

Negative Reciprocals

The negative reciprocal of a given number is not necessarily a negative value. It is the *opposite* of the given number.

To find the negative reciprocal of a number:

1. Interchange the numerator and denominator to form the reciprocal.
2. Give the reciprocal the sign opposite the sign of the original number.

The negative reciprocal of -5 is $+\dfrac{1}{5}$. The negative reciprocal of $-\dfrac{3}{4}$ is $+\dfrac{4}{3}$.

The negative reciprocal of $\dfrac{4}{5}$ is $-\dfrac{5}{4}$. The negative reciprocal of 3 is $-\dfrac{1}{3}$.

To find the equation of a line that is perpendicular (normal) to a given line and passes through a given point:

1. Determine the slope of the *given* line.
2. Find the *negative reciprocal* of this slope. The negative reciprocal is the slope of the perpendicular line.
3. Write the equation for the perpendicular line by substituting the coordinates of the *given* point for x_1 and y_1 and the slope of the *perpendicular* line for m into the point-slope form of the equation $y - y_1 = m(x - x_1)$.
4. Write the equation in slope-intercept or standard form.

Words as Subscripts

In the next example, we clarify the notation with subscripts. The phrase *slope of a given line* is notated as "slope$_{\text{given}}$." The *slope of the perpendicular line* is notated as "slope$_{\text{perpendicular}}$."

EXAMPLE   Find the equation of the line that is perpendicular to $4x + y = -3$ and passes through $(0, -3)$.

$$4x + y = -3 \qquad \text{Given equation. Solve for } y.$$

$$y = -4x - 3 \qquad \text{Identify the slope.}$$

$$\text{slope}_{\text{given}} = -4$$

$$\text{slope}_{\text{perpendicular}} = +\frac{1}{4} \qquad \text{Negative reciprocal of } -4.$$

9–4 Linear Equation of a Line

$$y - y_1 = m(x - x_1) \qquad \text{Point-slope form. Substitute } m = \frac{1}{4},\ P_1 = (0, -3).$$

$$y - (-3) = \frac{1}{4}(x - 0) \qquad \text{Simplify.}$$

$$y + 3 = \frac{1}{4}x \qquad \text{Solve for } y.$$

$$y = \frac{1}{4}x - 3 \qquad \text{Slope-intercept form. For standard form clear the fraction and rearrange.}$$

or

$$4y = x - 12 \qquad \text{Move the } x\text{-variable term to the left and make the leading coefficient positive.}$$

$$x - 4y = 12 \qquad \text{Standard form.}$$

EXAMPLE Which of the equations represents a line that is perpendicular to $2x - 3y = 1$ passing through $(2, -1)$?

(a) $y = \dfrac{2}{3}x - \dfrac{1}{3}$ $\qquad$ (b) $y = \dfrac{2}{3}x - \dfrac{7}{3}$

(c) $y = -\dfrac{3}{2}x - \dfrac{1}{3}$ $\qquad$ (d) $y = -\dfrac{3}{2}x + 2$

$$2x - 3y = 1 \qquad \text{Rewrite the given equation in slope-intercept form, i.e., solve for } y.$$

$$-3y = -2x + 1$$

$$\frac{-3y}{-3} = \frac{-2x}{-3} + \frac{1}{-3}$$

$$y = \frac{2}{3}x - \frac{1}{3} \qquad \text{Slope-intercept form of given equation.}$$

$$\text{slope}_{\text{given}} = \frac{2}{3} \qquad \text{From } y = \frac{2}{3}x - \frac{1}{3}$$

$$\text{slope}_{\text{perpendicular}} = -\frac{3}{2} \qquad \text{Negative reciprocal of given slope.}$$

The choice is now limited to (c) or (d) since (a) and (b) have slopes of $\dfrac{2}{3}$.

$$y - y_1 = m(x - x_1) \qquad \text{Substitute } m = -\frac{3}{2},\ x_1 = 2,\ y_1 = -1.$$

$$y - (-1) = -\frac{3}{2}(x - 2) \qquad \text{Distribute and simplify.}$$

$$y + 1 = -\frac{3}{2}x + 3 \qquad \text{Solve for } y.$$

$$y = -\frac{3}{2}x + 3 - 1$$

$$y = -\frac{3}{2}x + 2$$

The correct equation is $y = -\dfrac{3}{2}x + 2$, or choice (d).

SECTION 9–4 SELF-STUDY EXERCISES

1 Find the equation of a line passing through the given point with the given slope. Solve the equation for y when necessary.

1. $(-8, 3)$, $m = \dfrac{2}{3}$ **2.** $(4, 1)$, $m = -\dfrac{1}{2}$ **3.** $(-3, -5)$, $m = 2$ **4.** $(0, -1)$, $m = 1$

2 Find the equation of a line passing through the given pairs of points. Solve the equation for y.

5. $(4, 6)$ and $(7, 1)$ **6.** $(-1, 6)$ and $(-1, 4)$ **7.** $(-1, -3)$ and $(3, -3)$
8. $(-4, 4)$ and $(-4, -2)$ **9.** $(-2, -4)$ and $(5, 10)$ **10.** $(-4, 0)$ and $(6, 0)$

11. BUS If 50 DVDs cost \$2,185 to produce and 2,000 DVDs cost \$11,935 to produce, write a cost function for producing DVDs.

12. BUS The first 5,000 leather bags can be produced at a cost of \$160,000, and 50,000 bags can be produced at a cost of \$1,150,000. Write a cost function for producing these bags.

3 Write the equation of the line with the given slope and y-intercept.

13. $m = \dfrac{1}{4}$, $b = 7$ **14.** $m = -8$, $b = -4$ **15.** $m = -2$, $b = 3$ **16.** $m = \dfrac{3}{5}$, $b = -2$

17. $m = 1$, $b = 0$ **18.** $m = 5$, $b = -\dfrac{1}{5}$ **19.** $m = 2$, $b = -2$ **20.** $m = -\dfrac{3}{4}$, $b = 0$

21. BUS A local business rents computer time for a \$3 setup charge and \$0.20 for every minute the computer is used. Write an equation to represent the cost y of using the computer for x minutes. How much does 45 min of computer time cost?

22. BUS The cost of producing fine china plates is \$8 per plate plus a one-time equipment charge of \$12,000. Write an equation that represents the total cost y of producing x plates. What is the total cost of producing 8,000 plates?

Identify the slope and y-intercept to write the equations of the lines that are graphed in Figs. 9–33 through 9–35.

23.

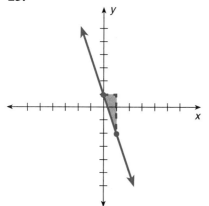

Figure 9–33

24.

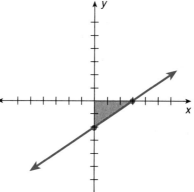

Figure 9–34

25.

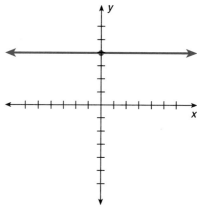

Figure 9–35

Solve Exercises 26 and 27 using the information given in Fig. 9–36, which shows the cost of producing widgets in dollars.

26. **INDTEC** Write an equation that represents the cost of producing widgets shown in the graph.
27. **INDTEC** Use the equation to find the cost of producing 10 widgets.
28. **AUTO** A snowplow has a maximum speed of 40 mi/h per hour on a dry, flat road surface. Its speed decreases 0.9 mi/h per hour for every inch of snow on the highway. Write an equation that represents the speed y of the snowplow when moving x inches of snow. What is the maximum speed of the plow in moving 10 in. of snow?

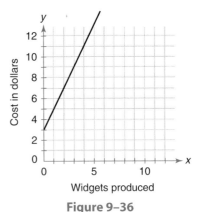

Figure 9–36

4 Write the new equations in standard form. Verify your answers using a graphing calculator.

29. Find the equation of the line that is parallel to the line $x + y = 6$ and passes through the point $(2, -3)$.
30. Find the equation of the line that is parallel to the line $2x + y = 5$ and passes through the point $(1, 7)$.
31. Find the equation of the line that is parallel to the line $3y = x - 2$ and passes through the point $(4, 0)$.
32. Find the equation of the line that is parallel to the line $3x - y = -2$ and passes through the point $(-3, -2)$.
33. Find the equation of the line that is parallel to the line $x + 3y = 7$ and passes through the point $(4, 1)$.
34. Find the equation of the line that is parallel to the line $3x - y = 4$ and passes through the point $(0, 3)$.
35. Find the equation of the line that is parallel to the line $2x + 3y = 5$ and passes through the point $(1, 1)$.
36. Find the equation of the line that is parallel to the line $3x + 2y = 1$ and passes through the point $(2, 0)$.
37. Find the equation of the line that is parallel to the line $2x - 5y = 0$ and passes through the point $(3, -1)$.
38. Find the equation of the line that is parallel to the line $-3x + 4y = -1$ and passes through the point $(\frac{1}{2}, 0)$.
39. **BUS** The cost of producing tires on an assembly line includes a fixed cost of $45,000 and a variable cost of $12 per tire. This can be written as a cost function $y = 12x + 45{,}000$, where x is the number of tires produced and y is the total cost of x tires. Research shows that 10,000 tires can be produced on a newly installed assembly line at a total cost of $140,000 with the same variable cost per tire as the original assembly line. Write an equation that represents the total cost of producing tires on the new assembly line.
40. **BUS** A car rental agency uses the function $y = 0.5x + 25$, where x is the number of miles driven and y is the total cost of car rental. Another car rental agency has the same variable cost of $0.50 per mile and quotes the total cost of driving 500 mi as $295. Write an equation to represent the second agency's rental charges.

5 Write the new equations in standard form. Verify your results using a graphing calculator. Be sure the range is set so the vertical and horizontal increments are equal.

41. Find the equation of the line that is perpendicular to the line $x + y = 6$ and passes through the point $(2, 3)$.
42. Find the equation of the line that is perpendicular to the line $2x + y = 5$ and passes through the point $(1, 7)$.
43. Find the equation of the line that is perpendicular to the line $3y = x - 2$ and passes through the point $(4, 0)$.
44. Find the equation of the line that is perpendicular to the line $3x - y = 2$ and passes through the point $(-3, -2)$.
45. Find the equation of the line that is perpendicular to the line $x + 3y = 7$ and passes through the point $(4, 1)$.
46. Find the equation of the line that is perpendicular to the line $x + 2y = 7$ and passes through the point $(-2, 3)$.

47. Find the equation of the line that is perpendicular to the line $2x + 3y = 4$ and passes through the point $(3, -1)$.

48. Find the equation of the line that is perpendicular to the line $4x + y = 1$ and passes through the point $(0, 0)$.

49. Find the equation of the line that is perpendicular to the line $2x + 2y = 3$ and passes through the point $(\frac{1}{2}, 2)$.

50. Find the equation of the line that is perpendicular to the line $5x - y = 6$ and passes through the point $(5, -\frac{1}{5})$.

CHAPTER REVIEW OF KEY CONCEPTS

Learning Outcomes	What to Remember with Examples

Section 9–1

1 Locate points on a rectangular coordinate system (pp. 382–385).

To plot a point on the rectangular coordinate system: **1.** Start at the origin. **2.** Count to the left or right (horizontally) the number of units of the first signed number. **3.** Start at the ending point found in Step 2, and count up or down (vertically) the number of units of the second signed number.

Draw a coordinate system and locate the following points: $A(-3, -2)$, $B(4, -1)$, $C(0, -3)$, $D(2, 3)$. See Fig. 9–37.

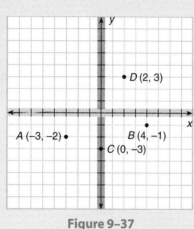

Figure 9–37

2 Represent an equation in two variables as a function (pp. 385–386).

1. Solve the equation in two variables for the dependent variable y. **2.** Rewrite y as $f(x)$.

Rewrite $3x - y = -2$ as a function.

$$3x - y = -2 \qquad \text{Solve for } y.$$
$$-y = -3x - 2 \qquad \text{Multiply entire equation by } -1.$$
$$y = 3x + 2 \qquad \text{Rewrite } y \text{ as } f(x).$$
$$f(x) = 3x + 2$$

3 Make a table of solutions for a linear equation or function (pp. 386–387).

1. Write the equation in function notation. **2.** Select appropriate values for the independent variable, **3.** Perform the calculations associated with the function for each selected value of the independent variable. **4.** Write the results as an ordered pair.

Make a table of solutions for the equation $y = -x + 3$. The domain is all real numbers.

When $x = -2$
$$y = -(-2) + 3$$
$$y = 2 + 3$$
$$y = 5$$

When $x = 0$
$$y = -0 + 3$$
$$y = 3$$

When $x = 2$
$$y = -2 + 3$$
$$y = 1$$

x	y
-2	5
0	3
2	1

Each solution is an ordered pair that can be written in point notation: $(-2, 5)$, $(0, 3)$, and $(2, 1)$. The range is all real numbers.

4 Graph a linear equation or function using a table of solutions (pp. 387–391).

1. Prepare a table of solutions. **2.** Plot the points on a rectangular coordinate system. **3.** Connect the points with a straight line.

Prepare a table of values and graph $y = -x + 3$ (Fig. 9–38).

x	y
-2	5
0	3
2	1

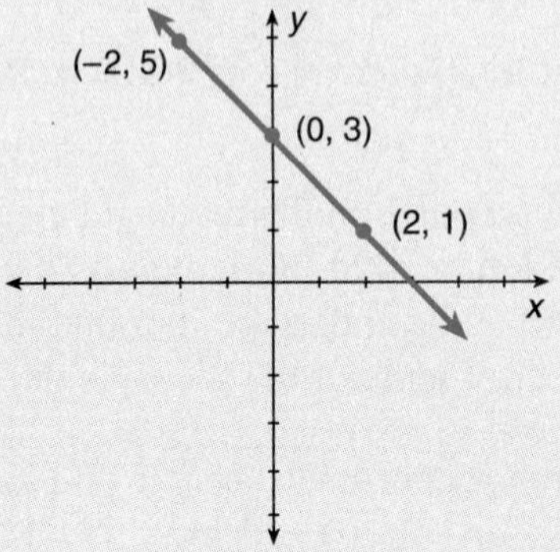

Figure 9–38

To check the solution of an equation in two variables: **1.** Substitute the value for x in each place it occurs and substitute the value for y in each place it occurs. **2.** Simplify each side of the equation using the order of operations. **3.** The solution makes a true statement; that is, both sides of the equation will equal the same number.

A solution for the equation $3x - y = 1$ is $(1, 2)$. Check to verify the solution.

$3x - y = 1$	Substitute values.
$3(1) - 2 = 1$	Evaluate.
$3 - 2 = 1$	
$1 = 1$	True. Solution checks.

To evaluate an equation when the value of one variable is given: **1.** Substitute the given value of the variable in each place that variable occurs in the equation. **2.** Solve the equation for the other variable.

Evaluate the equation $4x - y = 6$ for x, if $y = 14$.

$4x - y = 6$	Substitute 14 for y.
$4x - 14 = 6$	Solve for x.
$4x = 20$	
$x = 5$	

Solution: $(5, 14)$

Section 9–2

1 Graph linear equations using intercepts (pp. 393–394).

1. Find the x-intercept by letting $y = 0$ and solving for x. **2.** Find the y-intercept by letting $x = 0$ and solving for y. **3.** Plot the two points and graph the equation. **4.** Find a third point on the graph as a check point.

Graph the equation $y = 2x - 1$ by using the intercepts method.

x-intercept: $0 = 2x - 1$ Substitute $y = 0$. y-intercept: $y = 2(0) - 1$ Substitute $x = 0$.

$\qquad\qquad 1 = 2x$ Solve for x. $y = 0 - 1$ Solve for y.

$\qquad\qquad \dfrac{1}{2} = x$ $y = -1$

$\qquad\qquad\qquad\qquad\qquad\qquad\qquad\qquad\qquad\qquad\quad\; (0, -1)$

$\left(\dfrac{1}{2}, 0\right)$

Plot the two points: then draw the graph (Fig. 9–39).

Check point: For $x = 3$
$y = 2x - 1$
$y = 2(3) - 1$
$y = 6 - 1$
$y = 5$
$(3, 5)$

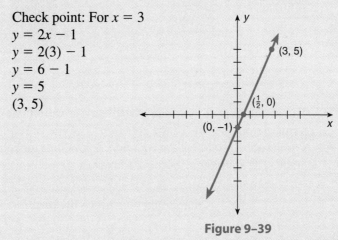

Figure 9–39

2 Graph linear equations using the slope and y-intercept (pp. 395–399).

1. Write the equation in the form $y = mx + b$. **2.** The slope is m. In fraction form the numerator of m is the vertical movement and the denominator of m is the horizontal movement. **3.** The y-coordinate of the y-intercept is b.

To graph using the slope-intercept method: **1.** Locate the y-intercept or b by counting vertically along the y-axis. **2.** From this point, count the slope (vertical, then horizontal) and locate the second point. **3.** Draw the line through the two points.

Use the slope-intercept method to graph $y = 2x - 1$. The y-coordinate of the y-intercept is -1, so count down 1 from the origin. The coordinates of this point are $(0, -1)$. From this point, move $+2$ vertically and $+1$ horizontally. The coordinates of the second point are $(1, 1)$. Connect the two points (Fig. 9–40).

y-intercept $= (0, -1)$

$$m = 2 = \frac{2}{1} \text{ or } \frac{-2}{-1}$$

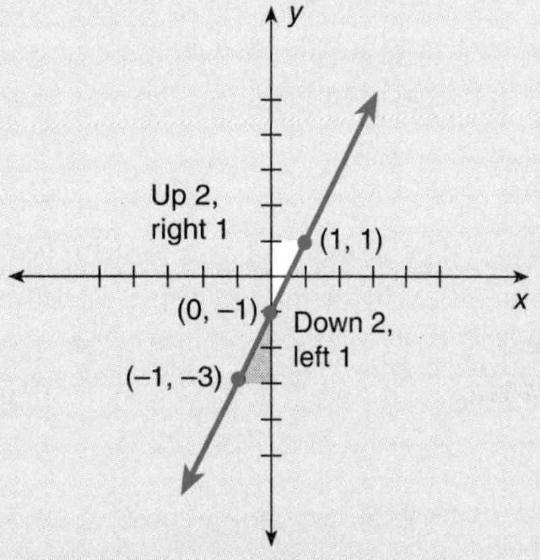

Up 2, right 1 $(1, 1)$

$(0, -1)$ Down 2, left 1

$(-1, -3)$

Figure 9–40

3 Graph linear equations using a graphing calculator (pp. 399–400).

To graph a linear equation using a graphing calculator: **1.** Solve the equation for y. **2.** Press the function key $\boxed{Y=}$; then enter the right side of the equation. **3.** Press the graph key to show the graph on the screen. Adjust the view with the Window or Zoom key if appropriate.

Graph the equation $y = 2x - 1$.
Using a TI-84 Plus Silver Edition™:

$\boxed{Y=}$ $\boxed{\text{CLEAR}}$ 2 $\boxed{\text{X, θ, T, } n}$ $\boxed{-}$ 1
$\boxed{\text{GRAPH}}$
(See Fig. 9–41.)

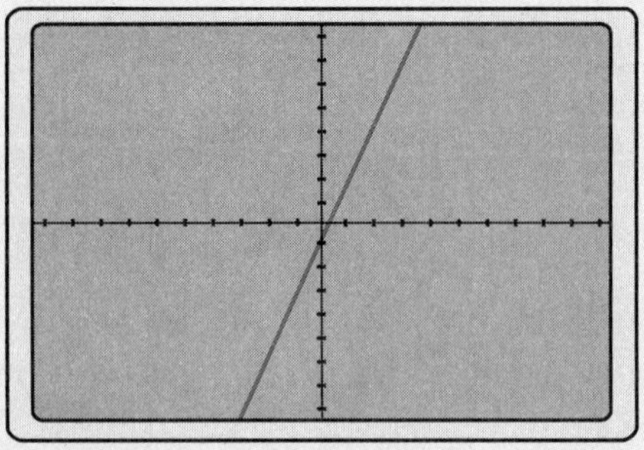

Figure 9–41

4 Solve a linear equation in one variable using a graph (pp. 400–402).

To write an equation in one variable as a function: **1.** Rewrite the equation so that all terms are on one side of the equation. **2.** Write the equation in function notation using $y = $ the nonzero side of the equation. **3.** Evaluate the function for at least two values and graph. The solution is the x-coordinate of the point where the line crosses the x-axis.

Write $2x + 3 = 6$ as a function, graph the function, and find the solution from the graph.

$$2x + 3 = 6$$
$$2x + 3 - 6 = 0$$
$$2x - 3 = 0$$

$$y = 2x - 3 \qquad \text{At } x = 0$$
$$y = 2(0) - 3$$
$$y = -3$$

$$y = 2(1) - 3 \qquad \text{At } x = 1$$
$$y = -1$$

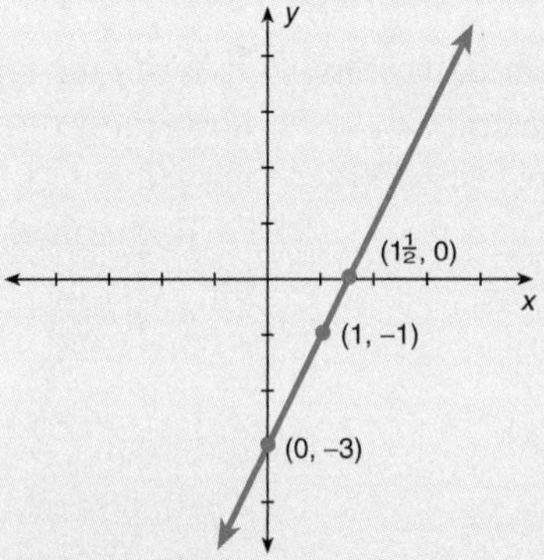

Figure 9–42

From the CALC menu, choose the zero option.
Solution: $x = 1.5$ or $1\frac{1}{2}$, the x-coordinate of the point where the line crosses the x-axis (Fig. 9–42).

Section 9–3

1 Calculate the slope of a line, given two points on the line (pp. 403–405).

The slope of a line joining two points is the difference of the y-coordinates divided by the difference of the x-coordinates, that is, $\frac{\text{rise}}{\text{run}}$.

$$m = \frac{\text{rise}}{\text{run}} = \frac{\Delta y}{\Delta x} = \frac{y_2 - y_1}{x_2 - x_1}$$

Find the slope of the line passing through the points $(3, 1)$ and $(-2, 5)$.

$$\frac{y_2 - y_1}{x_2 - x_1} \qquad \frac{5 - 1}{-2 - 3} = \frac{4}{-5} \quad \text{or} \quad -\frac{4}{5} \qquad p_1 = (3, 1), p_2 = (-2, 5)$$

Chapter 9 / Graphing Linear Equations and Functions

2 Determine the slope of a horizontal or vertical line (pp. 406–407).

The slope of a horizontal line is zero, and all points on the same horizontal line have the same y-coordinate. The slope of a vertical line is not defined, and all points on the same vertical line have the same x-coordinate.

> Examine the coordinate pairs and indicate which points lie on the same horizontal line and which lie on the same vertical line: $A\,(4, -3)$, $B\,(2, 5)$, $C\,(4, 5)$, $D\,(2, -3)$.
>
> Points A and D and points B and C lie on the same horizontal lines, respectively.
>
> Points A and C and points B and D lie on the same vertical lines, respectively.

Section 9–4

1 Find the equation of a line, given the slope and one point (pp. 408–410).

Use the point-slope form of an equation to find the equation when given the slope and coordinates of a point on a line.

$$y - y_1 = m(x - x_1)$$

> A line has slope 2 and passes through the point $(3, 4)$. Find the equation of the line.
>
> | | |
> |---|---|
> | $y - y_1 = m(x - x_1)$ | Substitute values. $m = 2$, $x_1 = 3$, $y_1 = 4$. |
> | $y - 4 = 2(x - 3)$ | Distribute. |
> | $y - 4 = 2x - 6$ | Solve for y. $-6 + 4 = -2$ |
> | $y = 2x - 2$ | |

2 Find the equation of a line, given two points on the line (pp. 410–411).

Find the slope, then use the point-slope form of the equation to find the equation when given coordinates of two points on a line.

> Find the equation of a line that passes through the two points $(3, 1)$ and $(-3, 2)$. First, find the slope.
>
> | | |
> |---|---|
> | $m = \dfrac{2 - 1}{-3 - 3} = \dfrac{1}{-6}$ or $-\dfrac{1}{6}$ | Find the slope. Substitute values from the two points. |
> | $y - y_1 = m(x - x_1)$ | Substitute $m = -\dfrac{1}{6}$, $x_1 = 3$, and $y_1 = 1$. |
> | $y - 1 = -\dfrac{1}{6}(x - 3)$ | Distribute. |
> | $y - 1 = -\dfrac{1}{6}x + \dfrac{1}{2}$ | Solve for y. $\dfrac{1}{2} + 1 = \dfrac{1}{2} + \dfrac{2}{2} = \dfrac{3}{2}$ |
> | $y = -\dfrac{1}{6}x + \dfrac{3}{2}$ | Slope-intercept form. |
> | or | |
> | $6y = -1x + 9$ | Clear fractions and rearrange. |
> | $x + 6y = 9$ | Standard form. |

3 Find the equation of a line, given the slope and y-intercept (p. 412).

Use the slope-intercept form of the equation, $y = mx + b$, and substitute values for m and b, the slope and y-intercept, respectively, to find the equation of a line.

Write the equation of a line that has slope $\frac{2}{3}$ and the y-intercept $(0, -1)$. Use the slope-intercept form of the equation:

$y = mx + b$ Substitute values, $m = \dfrac{2}{3}$, $b = -1$.

$y = \dfrac{2}{3}x - 1$

4 Find the equation of a line, given a point on the line and an equation of a line parallel to that line (pp. 413–414).

The slopes of parallel lines are equal; that is, lines that have the same slope are parallel lines. Equations that have equal slopes have graphs that are parallel lines.

Determine which two of the three given equations have graphs that are parallel lines:

(a) $y = 3x - 5$ (b) $3x - 2y = 10$ (c) $6x - 2y = 8$

Write each of the three equations in slope-intercept form and compare the slopes (coefficients of x).

(a) $y = 3x - 5$ (b) $y = \dfrac{3}{2}x - 5$ (c) $y = 3x - 4$

(a) and (c) have the same slope and their graphs are parallel.

To find the equation of a line parallel to a given line and passing through a given point, use the point-slope form of the linear equation. Substitute the value of the slope of the given equation and the coordinates of the given point into the point-slope form of an equation.

Find the equation of a line that passes through the point $(2, 3)$ and is parallel to the line that has equation $y = 4x - 1$. Write the new equation in slope-intercept form.
Use the slope of the given equation, 4, for the slope of the new equation.

$y - y_1 = m(x - x_1)$ Substitute $m = 4$, $x_1 = 2$, and $y_1 = 3$.
$y - 3 = 4(x - 2)$ Distribute.
$y - 3 = 4x - 8$ Solve for y. $-8 + 3 = -5$
$y = 4x - 5$ Slope-intercept form.

5 Find the equation of a line, given a point on the line and an equation of a line perpendicular to that line (pp. 414–416).

The slopes of perpendicular lines are negative reciprocals. Equations with slopes that are negative reciprocals have graphs that are perpendicular lines.

Determine which two of the three given equations have graphs that are perpendicular lines:

(a) $y = 3x - 5$ (b) $3x + 9y = 10$ (c) $9x + 3y = 12$

Write each of the three equations in slope-intercept form and compare the slopes (coefficients of x).

(a) $y = 3x - 5$ (b) $y = -\dfrac{1}{3}x + \dfrac{10}{9}$ (c) $y = -3x + 4$

(a) and (b) have slopes that are negative reciprocals and their graphs are perpendicular lines. Note that (a) and (c) are not perpendicular. Their slopes are opposites but *not* reciprocals. The slopes of (b) and (c) are reciprocals but *not* opposites.

Chapter 9 / Graphing Linear Equations and Functions

To write the equation of a line perpendicular to a given line, substitute the negative reciprocal of the slope of the given equation and the coordinates of the given point into the point-slope form of an equation.

Find the equation of a line that passes through the point $(-1, 2)$ and is perpendicular to the line represented by the equation $y = \frac{1}{3}x - 5$. Write the equation in standard form.

The slope for the new equation is the negative reciprocal of $\frac{1}{3}$, which is -3.

$y - y_1 = m(x - x_1)$	Substitute $m = -3$, $x_1 = -1$, and $y_1 = 2$.
$y - 2 = -3[x - (-1)]$	Simplify and distribute.
$y - 2 = -3x - 3$	Rearrange with variables on left.
$3x + y = -3 + 2$	Combine like terms.
$3x + y = -1$	Standard form.

CHAPTER REVIEW EXERCISES

Section 9–1

Draw a rectangular coordinate system and locate the points.

1. $A = (5, -2)$

2. $B = (-8, -3)$

3. $C = (0, -4)$

4. $D = (3, 7)$

5. $E = (-3, 2)$

6. $F = (-3, 0)$

7. What are the coordinates of the origin?

8. Which of the four quarters of the coordinate system is used to plot points with coordinates that are both negative?

9. Write the coordinates for points A through E on the graph in Fig. 9–43.

10. Write the coordinates of a point that lies on the y-axis and is four units below the x-axis.

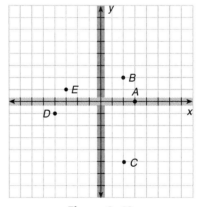

Figure 9–43

Represent the solutions of the equations in a table of solutions and on a graph.

11. $y = 2x - 3$

12. $y = -4x + 1$

13. $y = 3x$

14. $y = 4x$

15. $y = -3x$

16. $y = -4x$

17. $y = 2x + 1$

18. $y = 2x + 5$

19. $y = 4x - 2$

20. $y = -5x + 1$

21. $y = \frac{1}{2}x - 2$

22. $y = -\frac{1}{4}x + 1$

Which of the coordinate pairs are solutions for the equation $2x - 3y = 12$?

23. $(-2, -3)$

24. $(1, -3)$

25. $(3, -2)$

26. $(0, 4)$

27. $(6, 0)$

28. $(3, 4)$

29. In the equation $y = 3x - 1$, find y when $x = -2$.

30. In the equation $2x + y = 8$, find x when $y = 10$.

31. Find y in the equation $x - 3y = 5$ when $x = 8$.

32. Find y if $x = 7$ in the equation $3x - 4y = -2$.

Chapter Review Exercises

Section 9–2

Graph the equations using the intercepts procedure.

33. $x = -4y - 1$

34. $x + y = -4$

35. $3x - y = 1$

36. $x = -4y$

37. $\frac{1}{2}x + \frac{1}{3}y = 1$

38. $2x + 3y = 6$

Graph using the slope and y-intercept procedure.

39. $y = 5x - 2$

40. $y = -x$

41. $y = -3x - 1$

42. $y = \frac{1}{2}x + 3$

43. $x - y = 4$

44. $2y + 4 = -3$

45. $x - 2y = -1$

46. $x + y = 5$

47. $y - 2x = -2$

48. $y + 3x = 4$

Graph the equations using a graphing calculator. Use appropriate viewing window.

49. $y = 0.5x - 3$

50. $y = 2.1x + 0.5$

51. $y = 20x - 15$

52. $y = -30x + 10$

53. The function $y = 5x + 8,000$ is used to express the cost of manufacturing a widget, where $8,000 is the fixed cost of constructing molds and $5 is the cost of materials to make each widget. What is the cost of manufacturing 10,000 widgets?

54. A salesperson earns a salary of $200 weekly plus $12.50 commission for each photo sitting. Weekly income is expressed by the function $y = 12.50x + 200$, where x represents the number of sittings in a week. Find the weekly income if 28 sittings are sold.

Write each equation as a function and find the solution graphically.

55. $x + 2 = 8$

56. $3x - 7 = 1$

57. $2x + 1 = 5x + 7$

58. $-14 = 2(x - 7)$

59. $5(x + 2) = 3(x + 4)$

60. $2x + 1 = 7x + 2 - 8$

Determine the slope and y-intercept of the given equations by inspection.

61. $y = 3x + \dfrac{1}{4}$

62. $y = \dfrac{2}{3}x - \dfrac{3}{5}$

63. $y = -5x + 4$

64. $y = 7$

65. $x = 8$

66. $y = \dfrac{1}{3}x - \dfrac{5}{8}$

67. $y = \dfrac{x}{8} - 5$

68. $y = -\dfrac{x}{5} + 2$

Write the given equations in slope-intercept form and determine the slope and y-intercept.

69. $2x + y = 8$

70. $4x + y = 5$

71. $3x - 2y = 6$

72. $5x - 3y = 15$

73. $\dfrac{3}{5}x - y = 4$

74. $2.2y - 6.6x = 4.4$

75. $3y = 5$

76. $3x - 6y = 12$

Section 9–3

Find the slope of the line passing through the given pairs of points.

77. $(-2, 2)$ and $(1, 3)$

78. $(3, -1)$ and $(1, 3)$

79. $(3, 2)$ and $(5, 6)$

80. $(-1, -1)$ and $(2, 2)$

81. $(4, 3)$ and $(-4, -2)$

82. $(6, 2)$ and $(-3, 2)$

83. $(3, -4)$ and $(0, 0)$

84. $(1, -1)$ and $(5, -5)$

85. $(-4, 1)$ and $(-4, 3)$

86. $(4, -4)$ and $(1, 3)$

87. $(5, 0)$ and $(-2, 4)$

88. $(-2, 1)$ and $(0, 3)$

89. $(-4, -8)$ and $(-2, -1)$

90. $(3, 3)$ and $(3, 0)$

91. $(5, -3)$ and $(-1, -3)$

92. $(-5, -1)$ and $(-7, -3)$

93. $(-7, 0)$ and $(-7, 5)$

94. $(3, 5)$ and $(2, 5)$

95. Write the coordinates of two points that lie on the same horizontal line.

96. Write the coordinates of two points that lie on the same vertical line.

Use Table 9–2 on page 408 for Exercises 97–102. Use the initial year of the academic year as the reference.

97. What is the rate of change in tuition and fees at public 2-year colleges from 1998 to 1999?

98. What is the rate of change in tuition and fees at public 2-year colleges from 2002 to 2003?

99. What is the rate of change in tuition and fees at public 2-year colleges from 1993 to 1994?

100. What is the rate of change in tuition and fees at public 2-year colleges from 1992 to 2003?

101. Explain why the rate of change found in Exercises 98, 99, and 100 are different.

102. If the data in Table 9–2 formed a straight line when graphed, what would you expect to find for the rate of change for the data in Exercises 98, 99, and 100?

Section 9–4

Find the equation of a line passing through the given point with the given slope. Solve the equation for y if necessary.

103. $(-6, 2)$, $m = \dfrac{1}{3}$

104. $(3, 2)$, $m = -\dfrac{2}{5}$

105. $(4, 0)$, $m = \dfrac{3}{4}$

106. $(0, -2)$, $m = 2$

107. $(2, 3)$, $m = 4$

108. $(6, 0)$, $m = -1$

Find the equation of a line passing through the given pairs of points. Solve the equation for y if necessary.

109. $(-5, 2)$ and $(6, 1)$

110. $(1, 4)$ and $(-1, 3)$

111. $(-1, -3)$ and $(3, 4)$

112. $(-3, 0)$ and $(4, 0)$

113. $(-2, -3)$ and $(3, 6)$

114. $(2, -4)$ and $(3, -4)$

115. $(5, 2)$ and $(6, 3)$

116. $(4, 6)$ and $(1, -1)$

117. $(-1, -2)$ and $(-3, -4)$

118. $(4, 0)$ and $(4, -3)$

119. $(5, -2)$ and $(3, -2)$

120. $(5, 4)$ and $(0, 4)$

121. A salesperson sells 80 items and earns \$3,800 in one month, and in another month 120 items are sold resulting in \$4,200 earnings. If we assume this is a linear function, write a function that expresses the salesperson's monthly salary plus commission, S, as a function of the number of items sold, x. Salary is the y-intercept.

122. Simple interest is a linear function. Write an equation to express interest earned if \$500 is earned in 2 years and \$2,000 is earned in 8 years.

Write the equations using the given slope and y-intercept.

123. $m = 3$, $b = -2$

124. Slope $= \dfrac{3}{5}$; y-intercept $= -7$

Identify the slope and y-intercept to write the equations using information from the graphs in Figs. 9–44 and 9–45.

125.

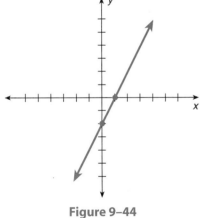

Figure 9–44

126.

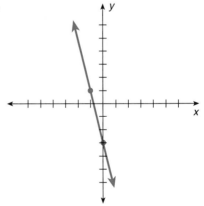

Figure 9–45

Write equations in standard form. Graph both equations for each exercise on a graphing calculator to verify parallelism.

127. Find the equation of the line that is parallel to the line $x + y = 4$ and passes through the point $(2, 5)$.

128. Find the equation of the line that is parallel to the line $3x + y = 6$ and passes through the point $(1, 0)$.

129. Find the equation of the line that is parallel to the line $2y = x - 3$ and passes through the point $(2, -3)$.

130. Find the equation of the line that is parallel to the line $4x - y = -1$ and passes through the point $(0, -2)$.

131. Find the equation of the line that is parallel to the line $x - 3y = 5$ and passes through the point $(5, -5)$.

132. Find the equation of the line that is parallel to the line $3x - 2y = 2$ and passes through the point $(0, -3)$.

133. Find the equation of the line that is parallel to the line $x + 3y = 6$ and passes through the point $(-4, -2)$.

134. Find the equation of the line that is parallel to the line $4x + 3y = 1$ and passes through the point $(3, -\frac{1}{2})$.

135. Find the equation of the line that is parallel to the line $3x - 4y = 0$ and passes through the point $(\frac{1}{3}, 2)$.

136. Find the equation of the line that is parallel to the line $-2x + 3y = 2$ and passes through the point $(-1, -1)$.

Write the equations in standard form. Use a graphing calculator to verify that the two lines in each exercise are perpendicular.

137. Find the equation of the line that is perpendicular to the line $x + y = 4$ and passes through the point $(-3, 1)$.

138. Find the equation of the line that is normal to the line $3x + y = 6$ and passes through the point $(1, 4)$.

139. Find the equation of the line that is perpendicular to the line $x + 2y = 5$ and passes through the point $(-2, 0)$.

140. Find the equation of the line that is normal to the line $2x + 2y = 4$ and passes through the point $(0, 0)$.

141. Find the equation of the line that is normal to the line $5x + y = 8$ and passes through the point $(-1, 2)$.

142. Find the equation of the line that is perpendicular to the line $3y = x - 4$ and passes through the point $(2, 3)$.

143. Find the equation of the line that is perpendicular to the line $5x - y = 10$ and passes through the point $(\frac{1}{2}, 3)$.

144. Find the equation of the line that is normal to the line $x - 3y = 6$ and passes through the point $(-2, 4)$.

145. Find the equation of the line that is perpendicular to the line $4x - y = 8$ and passes through the point $(4, -\frac{1}{2})$.

146. Find the equation of the line that is perpendicular to the line $4y = 2x + 1$ and passes through the point $(3, -1)$.

TEAM PROBLEM-SOLVING EXERCISES

1. A 1-gal can of indoor house paint is advertised to cover 400 ft^2 of wall surface.
  (a) Make a table to show the amount of wall surface area that can be covered by 1, 2, 3, . . . , 10 gal of paint.
  (b) Graph these data and extend the line beyond the data to 15 gal of paint.
  (c) Use the graph to decide how many 1-gal cans of paint it would take to cover 5,500 ft^2 of wall surface.
  (d) The paint being used for this job can be purchased for $19.05 a gallon. Find the cost of the paint for the 5,500 ft^2 of wall surface.
  (e) The sales tax rate is 8.25%. Calculate the total cost of the paint.

2. Linda Kodama is introducing a new lipstick in Brightglow's product line. As product development manager, she estimates the cost of the new lipstick to be $4.53 per item, plus an additional cost of $5,000 for product development.
  (a) Make a table of the estimated cost for producing 0 lipsticks, 100 lipsticks, 1,000 lipsticks, and 2,000 lipsticks.
  (b) Represent these costs as ordered pairs (number of lipsticks, cost).
  (c) Plot the ordered pairs and graph a line that fits the ordered pairs.
  (d) Write an equation that represents the cost of the new lipstick as a function of the number of lipsticks produced.
  (e) Linda projects the selling price of the new lipstick to be $8.99. How many lipsticks must the company sell to recover the cost of producing the new items?

CONCEPTS ANALYSIS

1. Describe the graph of a line with slope that is positive.

2. Describe the graph of a line with slope that is negative.

3. Describe the graph of a line with slope that is a fraction between 0 and 1.

4. Describe the graph of a line with slope that is a number > 1.

5. Describe the graph of a line with slope that is a fraction between -1 and 0.

6. Describe the graph of a line with slope that is a number < -1.

7. What is the slope of a horizontal line? Why?

8. What is the slope of a vertical line? Why?

9. If the standard form of an equation of a horizontal line is $y = k$, what does k represent?

10. If the standard form of an equation of a vertical line is $x = k$, what does k represent?

11. If an equation is solved for y, what information about the graph can be read from the equation?

12. How do the slopes of parallel lines compare?

13. How do the slopes of perpendicular lines compare?

14. Give two basic characteristics of an equation which graphs into a straight line.

15. List the steps to find the x-intercept of a line if you are given two points on the line.

16. What do we mean when we say a point is represented by an ordered pair of numbers?

17. What does the graph of an equation represent?

18. Explain the procedure for plotting the point $(5, -2)$ on a grid representing the rectangular coordinate system.

19. Discuss the similarities and differences in the table-of-solutions method, the intercepts method, and the slope and y-intercept method for graphing a linear equation.

20. Why is it generally helpful to solve an equation for y before using the table-of-solutions method for graphing?

PRACTICE TEST

Make a table of values to represent the solutions to the following equations and show the solutions on a graph.

1. $y = \dfrac{1}{2}x$

2. $y = \dfrac{1}{2}x + 1$

3. $y = 2x - 4$

4. $y = 5 - x$

Write as functions and find the solution graphically.

5. $3x + 2 = 5$

6. $2(x + 1) = 3x$

Write the specific solution in ordered-pair form for each equation.

7. $x + y = 7$, if $x = 2$

8. $2x - y = 1$, if $y = -3$

9. A cake bakery estimates the profit function to be $y = 3x - 800$, where x is the number of cakes produced in a month and $800 is cost of overhead such as utilities. Find the profit if 8,000 cakes are produced.

10. A cost function is known to be $y = 10x + 250$, where x is the number of units produced and $250 is the fixed cost. Find the total cost of producing 5,000 units.

Graph using the intercepts method.

11. $x + y = -5$

12. $2x - 3y = 6$

13. $x + 2y = 8$

Graph using the slope-intercept method.

14. $y = -3x + 1$

15. $2x + y = -3$

16. $x + 2y = 1$

Find the slope of the line passing through the given pairs of points.

17. $(-3, 6)$ and $(3, 2)$

18. $(0, 4)$ and $(-1, 6)$

Write the equations in slope-intercept form and determine the slope and y-intercept.

19. $-2x + y = 34$

20. $x - y = 4$

21. $x = 4y$

22. $2y - x = 3$

Find the equation of the line passing through the given point with the given slope. Solve the equation for y.

23. $(3, -5)$, $m = \dfrac{2}{3}$

24. $(5, 1)$, $m = -2$

Find the equation of the line passing through the given pairs of points. Solve the equation for y.

25. $(1, 3)$ and $(4, 5)$ **26.** $(-1, 1)$ and $(4, -4)$ **27.** $(5, 2)$ and $(-1, 2)$

28. Write the equation in slope-intercept form of the line that has a slope $= -2$ and y-intercept $= -3$.

29. Write the equation of the line shown in Fig. 9–46 in slope-intercept form.

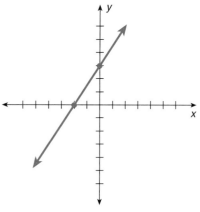

Figure 9–46

Write the equations in standard form.

30. Find the equation of the line that is parallel to $y - 2x = 3$ and passes through the point $(2, 5)$.

31. Find the equation of the line that is parallel to $2x + y = 4$ and passes through the point $(4, -3)$.

32. Find the equation of the line perpendicular to $y - 2x = 3$ and passing through the point $(2, 5)$.

33. Find the equation of the line perpendicular to $2x + y = 4$ and passing through the point $(4, -3)$.

Systems of Linear Equations

10–1 Solving Systems of Linear Equations Graphically

10–2 Solving Systems of Linear Equations Using the Addition Method

10–3 Solving Systems of Linear Equations Using the Substitution Method

10–4 Problem Solving Using Systems of Linear Equations

Focus on Careers

Careers in telecommunications include voice, video, and Internet communication services. Those persons who have up-to-date technical skills will have the best job opportunities. Maintaining up-to-date technical skills in this environment is a must.

Fiber-optic cables to transmit high-speed, high-capacity communications and radio towers to provide wireless telecommunications services are just two of the areas where highly trained technicians find employment. Wireless telephone communication is a primary sector of this industry. Replacement of landlines with cellular phone service will become increasingly common as new technology is deployed.

Jobs in the telecommunications industry are steady, year-round jobs and overtime is sometimes required. More than 60% of those employed worked for companies with 100 or more employees. Most jobs are found in cities.

Although some jobs in this industry require a high school education, most companies choose persons that have training at 2-year or 4-year colleges. Computer literacy and keyboarding skills are important.

Average weekly earnings in the telecommunications industry were $761 in 2002. This is significantly higher than the average weekly earnings of $506 for all of private industry.

Source: *Occupational Outlook Handbook,* 2004–2005 Edition. U.S. Department of Labor, Bureau of Labor Statistics.

Many real-world situations involve problems in which several conditions or constraints have to be considered. These conditions can be written in separate equations that form a system of equations. The solution of the system will be the value or values that satisfy all conditions.

431

Learning Outcome **1** Solve a system of linear equations by graphing.

1 Solve a System of Linear Equations by Graphing.

A **system of two linear equations,** each having two variables, is solved when we find the one ordered pair of solutions that satisfies *both* equations. One method of solving systems of two equations is to graph each equation and find the intersection of these graphs. The point where the two graphs intersect represents the ordered pair of solutions that the two graphs have in common.

To solve a system of two linear equations with two variables by graphing:

1. Graph each equation on the same pair of axes.
2. The solution will be the common point.

EXAMPLE CON A board is 20 ft long. It needs to be cut so that one piece is 2 ft longer than the other. What should be the length of each piece?

Write two equations to describe all the conditions of the problem.

Since the board is not cut into equal pieces, we let the letter *l* represent the *longer* piece and the letter *s* represent the *shorter* piece.

Condition 1: The total length of the board is 20 ft. Thus, the two pieces (*l* and *s*) total 20 ft: $l + s = 20$.

Condition 2: One piece is 2 ft longer than the other. Thus, the shorter piece plus 2 ft equals the longer piece: $s + 2 = l$.

The two equations become a *system of equations.*

$l + s = 20$ Condition 1.
$s + 2 = l$ Condition 2.

Graph each equation on the same set of axes and examine the intersection of the graphs. Let *s* be the independent variable and *l*, the dependent variable.

Condition 1	**Condition 2**
$l + s = 20$	$s + 2 = l$
or	or
$l = 20 - s$ Domain: [0, 20]	$l = s + 2$ Domain: [0, 20]

Make a table of solutions for each equation.

s	l		s	l	
8	12	Choose values that are near	0	2	Choose only zero and positive
10	10	the middle of the domain.	5	7	values.
12	8		10	12	

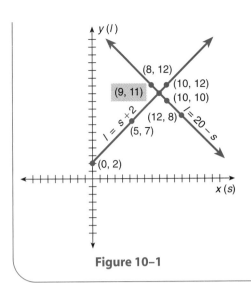

The point of intersection (Fig. 10–1) is (9, 11), which means that $s = 9$ and $l = 11$. **The shorter length is 9 ft and the longer length is 11 ft.**

Check to see if the solution ($s = 9, l = 11$) satisfies both equations.

$$l + s = 20 \qquad s + 2 = l$$
$$11 + 9 = 20 \qquad 9 + 2 = 11$$
$$20 = 20 \qquad 11 = 11$$

The ordered pair checks in both equations.

Figure 10–1

Line relationships:

When graphing two straight lines, three possibilities can occur.

1. The two lines can intersect at **just one point.** This means the system is **independent** and one pair of values satisfies both equations.
2. The two lines **do not intersect** at all. This means the system is **inconsistent** and no pair of values satisfies both equations.
3. The two lines **coincide** or fall exactly in the same place. This means the system is **dependent** and the equations are identical or are multiples, and any pair of values that satisfies one equation satisfies both equations.

To solve a system of equations using a graphing calculator:

1. Enter both equations in the $\boxed{Y=}$ screen.
2. Graph both equations on the same screen.
3. Use the Trace feature or the CALC/Intersect feature to determine the approximate coordinates of the intersection of the graphs.
4. Use the Zoom or Box (Window) feature to get a closer view and thus a more accurate approximation of the intersection of the graphs.

SECTION 10–1 SELF-STUDY EXERCISES

1 Solve the systems of equations by graphing.

1. $x + y = 12$
   $x - y = 2$

2. $2x + y = 9$
   $3x - y = 6$

3. $2x - y = 5$
   $4x - 2y = 8$

4. $x - y = 9$
   $3x - 3y = 27$

5. $3x + 2y = 8$
   $x + y = 2$

6. $y = x - 2$
   $y = -2x + 7$

7. $y = -3$
   $y = x - 7$

8. $y = 5x$
   $y = -2x - 7$

9. $y = -2x$
   $y = x - 3$

10. $x - y = 3$
$x + 2y = 6$

11. $5x + y = -2$
$x - y = 8$

12. $2x - y = -3$
$x + 2y = 6$

13. $4x + y = -1$
$x + y = -4$

14. $x + y = 3$
$2y = 3x - 4$

15. $y = 2x - 5$
$y = x + 2$

16. $x + 2y = 6$
$x + y = 2$

17. $2x - y = 7$
$x + y = 5$

18. $3x + y = 5$
$x - y = 7$

19. $2x - y = 6$
$5x + y = 8$

20. $5x + y = 0$
$x = y$

21. $-x + y = -2$
$2x + y = 7$

10–2 | Solving Systems of Linear Equations Using the Addition Method

Learning Outcomes

1 Use the addition method to solve a system of linear equations that contains opposite variable terms.

2 Use the addition method to solve a system of linear equations that does not contain opposite variable terms.

3 Apply the addition method to a system of linear equations with no solution or with many solutions.

We can see that solving systems of equations graphically can be tedious and time-consuming. Also, many solutions to systems of equations are not whole numbers. Graphically, it is difficult to plot fractions or to read fractional intersection points. Therefore, we need a convenient algebraic procedure for solving systems of equations. We examine the two methods most commonly used to solve a system of equations with two variables, but there are other methods.

1 **Use the Addition Method to Solve a System of Linear Equations That Contains Opposite Variable Terms.**

The **addition method** incorporates the concept that *equals added to equals give equals.* Thus, when we add two equations, the result is still an equation. However, the addition method is effective *only* if one variable is *eliminated* in the addition process. This method is also called the **elimination method.**

To solve a system of linear equations that contains opposite variable terms by the addition (elimination) method:

1. Write each equation in standard form ($ax + by = c$).
2. Add the two equations.
3. Solve the equation from Step 2. The result is one part of the ordered pair solution.
4. Substitute the solution from Step 3 in either equation and solve for the remaining variable. This completes the solution ordered pair.
5. Check the solution in one or both original equations.

EXAMPLE   Solve the systems of equations by the addition method:

(a) $x + y = 7$ **(b)** $3x + 8y = 7$
$ x - y = 5$ $ -3x - 3y = 8$

(a) $x + y = 7$ Equation 1. Add the equations to eliminate the *y*-variable.

$\underline{x - y = 5}$ Equation 2

$2x = 12$ Solve for *x*.

$ x = 6$ Substitute 6 in place of *x* in *either* of the given equations.

$x + y = 7$ Equation 1. Substitute *x* = 6.

$6 + y = 7$ Solve for *y*.

$ y = 7 - 6$ Simplify.

$ y = 1$

The solution is $x = 6$ and $y = 1$, or (6, 1).

Check in the *other* equation:

$x - y = 5$ Equation 2. Substitute values from the solution ordered-pair.

$6 - 1 = 5$ Simplify.

$ 5 = 5$ The solution checks.

Checking in both equations helps ensure that no errors have been made in the addition process or in the substitution process.

(b) $3x + 8y = 7$ Equation 1. Add the equations to eliminate the *x*-variable.

$-3x - 3y = 8$ Equation 2

$ 5y = 15$ Solve for *y*.

$ y = 3$ Substitute 3 for *y* in Equation 1.

$3x + 8y = 7$ Equation 1. Substitute *y* = 3.

$3x + 8(3) = 7$ Solve for *x*.

$3x + 24 = 7$

$3x = 7 - 24$

$3x = -17$

$x = \dfrac{-17}{3}$

The solution is $(-\frac{17}{3}, 3)$.

Check in equation 2:

$-3x - 3y = 8$ Equation 2. Substitute $x = -\dfrac{17}{3}$ and $y = 3$.

$-3\left(\dfrac{-17}{3} \right) - 3(3) = 8$ Multiply.

$17 - 9 = 8$ Subtract.

$8 = 8$ The solution checks.

2 Use the Addition Method to Solve a System of Linear Equations That Does Not Contain Opposite Variable Terms.

In the addition method one unknown must be eliminated; that is, the terms must add to 0. Thus, if neither pair of variable terms in a system of equations is opposite terms that will add to 0, we multiply one or both of the equations by numbers that *cause* the terms of one variable to add to 0. This is an application of the multiplication axiom (see Chapter 7, Section 2, Outcome 2).

To solve a system of linear equations using the addition method (elimination):

1. Write each equation in standard form ($ax + by = c$).
2. If necessary, multiply one or both equations by numbers that cause the terms of one variable to add to zero.
3. Add the two equations to eliminate a variable.
4. Solve the equation from Step 3 for the remaining variable.
5. Substitute the solution from Step 4 in either equation and solve for the remaining variable.
6. Check the solution in one or both original equations.

EXAMPLE Solve the system of equations: $2x + y = 7$ and $x + y = 3$.

There are several possibilities for eliminating one variable by applying the multiplication axiom. We will examine three choices.

Choice 1.
$$\begin{aligned} -2x - y &= -7 \\ x + y &= 3 \\ \hline -x \phantom{{}-y} &= -4 \\ x &= 4 \end{aligned}$$

Multiply the first equation by -1 and add the equations to eliminate the y-terms.

Solve for x.

Choice 2.
$$\begin{aligned} 2x + y &= 7 \\ -x - y &= -3 \\ \hline x \phantom{{}-y} &= 4 \end{aligned}$$

Multiply the second equation by -1 and add the equations to eliminate the y-terms.

Choice 3.
$$\begin{aligned} 2x + y &= 7 \\ -2x - 2y &= -6 \\ \hline -y &= 1 \\ y &= -1 \end{aligned}$$

Multiply the second equation by -2 and add the equations to eliminate the x-terms.

For choice 2, substitute $x = 4$:

$$2x + y = 7$$
$$2(4) + y = 7$$
$$8 + y = 7$$
$$y = 7 - 8$$
$$y = -1$$

For choice 3, substitute $y = -1$:

$$2x + \boxed{y} = 7$$
$$2x + (\boxed{-1}) = 7$$
$$2x - 1 = 7$$
$$2x = 7 + 1$$
$$2x = 8$$
$$x = 4$$

The solution is $(4, -1)$.

Check the solution in the first original equation:	Check the solution in the second original equation:

$$2x + y = 7$$
$$2(4) + (-1) = 7$$
$$8 + (-1) = 7$$
$$7 = 7$$

$$x + y = 3$$
$$4 + (-1) = 3$$
$$3 = 3$$

When an original equation is altered, it is important to check your solution in *both* original equations. This enables you to identify mistakes such as forgetting to multiply *each* term in the equation by a number.

EXAMPLE Solve the system of equations: $2x + 3y = 1$ and $3x + 4y = 2$.

In this system, no integer can be multiplied by just one equation to eliminate a letter. Therefore, we need to multiply each equation by some number. There are several possibilities; we examine just two.

Choice 1.

$-3(2x + 3y) = -3(1)$	Multiply Equation 1 by -3.
$2(3x + 4y) = 2(2)$	Multiply Equation 2 by $+2$.
$-6x - 9y = -3$	Add equations to eliminate the x-variable.
$\underline{6x + 8y = 4}$	
$-y = 1$	Solve for y.
$y = -1$	y-value of solution
$2x + 3y = 1$	Substitute $y = -1$.
$2x + 3(-1) = 1$	Solve for x.
$2x - 3 = 1$	
$2x = 1 + 3$	
$2x = 4$	
$x = 2$	x-value of solution

The solution is $(2, -1)$.

Choice 2.

$4(2x + 3y) = 4(1)$	Multiply Equation 1 by 4.
$-3(3x + 4y) = -3(2)$	Multiply Equation 2 by -3.
$8x + 12y = 4$	Add equations to eliminate the y-variable.
$\underline{-9x - 12y = -6}$	
$-x = -2$	Solve for x.
$x = 2$	x-value of solution
$2x + 3y = 1$	Substitute $x = 2$.
$2(2) + 3y = 1$	Solve for y.
$3y = 1 - 4$	
$3y = -3$	
$y = -1$	y-value of solution

The solution is $(2, -1)$.

Check the solution in the original equations:

$2x + 3y = 1$	$3x + 4y = 2$	Substitute $x = 2$ and $y = -1$.
$2(2) + 3(-1) = 1$	$3(2) + 4(-1) = 2$	
$4 + (-3) = 1$	$6 + (-4) = 2$	
$1 = 1$	$2 = 2$	Solution checks.

3 Apply the Addition Method to a System of Linear Equations with No Solution or with Many Solutions.

Sometimes a system of equations has no solution; the graphs of the two equations do not intersect. There are also instances when a system of equations has all solutions in common; the graphs of the two equations coincide.

EXAMPLE Solve the system $x + y = 7$ and $x + y = 5$.

$$x + y = 7 \qquad \text{Multiply Equation 2 by } -1.$$
$$-1(x + y) = -1(5)$$

$$x + y = 7 \qquad \text{Add the equations to eliminate a variable.}$$
$$\underline{-x - y = -5}$$
$$0 = 2 \qquad \text{Both variables are eliminated.}$$

Notice, both variables are eliminated and the resulting equation, $0 = 2$, is *false*.

There are no solutions to this system.

If both variables in a system of equations are eliminated and the resulting statement is false, the equations are **inconsistent** and have no solution. The graphs of the equations are parallel lines.

EXAMPLE Solve the system $2x - y = 7$ and $4x - 2y = 14$.

$$-2(2x - y) = -2(7) \qquad \text{Multiply Equation 1 by } -2.$$
$$-4x + 2y = -14 \qquad \text{Add the equations to eliminate a variable.}$$
$$\underline{4x - 2y = 14}$$
$$0 = 0 \qquad \text{Both variables are eliminated.}$$

Both variables are eliminated; however, the result is a *true* statement ($0 = 0$). In this situation, **all solutions of one equation are also solutions of the other equation.** For example, $x = 4$ and $y = 1$ is a solution of both equations.

If both variables in a system of equations are eliminated and the resulting statement is true, then the equations are **dependent** and have many solutions. The graphs of the equations coincide.

SECTION 10–2 SELF-STUDY EXERCISES

1 Solve the systems of equations using the addition method.

1. $a - 2b = 7$
$3a + 2b = 13$

2. $3m + 4n = 8$
$2m - 4n = 12$

3. $x - 4y = 5$
$-x - 3y = 2$

4. $a - b = 6$
$2a + b = 3$

5. $x + 2y = 5$
$3x - 2y = 3$

6. $x - 5y = 7$
$2x + 5y = 5$

7. $5x + 2y = -3$
$-5x - 4y = -7$

8. $-8x - 3y = 7$
$8x + 5y = 3$

9. $x + 3y = 5$
$-x + 3y = 13$

2 Solve the systems of equations using the addition method.

10. $3x + y = 9$
$x + y = 3$

11. $7x + 2y = 17$
$y = 3x + 2$

12. $a + 6b = 18$
$4a - 3b = 0$

13. $3x + y = -1$
$4x - 2y = -8$

14. $a = 6y$
$2a - y = 11$

15. $5x - 3y = -4$
$2x - 3y = -7$

16. $2x - 3y = 10$
$3x + 2y = 2$

17. $4x - 3y = -19$
$-2x - 4y = 4$

18. $2x + 5y = 0$
$-3x - 2y = -11$

19. $x + 2y = 6$
$2x + 5y = 15$

20. $5x - y = -8$
$x - 2y = -7$

21. $-x + 2y = -7$
$x - 6y = -5$

3 Solve the systems of equations using the addition method.

22. $x + y = 8$
$x + y = 3$

23. $x + y = 9$
$x + y = 3$

24. $3x - 2y = 6$
$9x - 6y = 18$

25. $2a + 4b = 10$
$a + 2b = 5$

26. $3a - 2b = 14$
$3a = 2b + 2$

27. $5x - 7y = 8$
$10x - 14y = 16$

28. $-3x + 2y = 3$
$9x - 6y = -8$

29. $2x + 5y = 3$
$16x + 40y = 38$

30. $-x - 2y = 3$
$4x + 8y = 15$

10–3 | *Solving Systems of Linear Equations Using the Substitution Method*

Learning Outcome   **1** Use the substitution method to solve a system of linear equations.

1 Use the Substitution Method to Solve a System of Linear Equations.

Another method for solving systems of equations is by substitution. Recall that in formula rearrangement (Chapter 7, Section 5, Outcome 2), whenever more than one variable is used in an equation or formula, we can rearrange the equation to solve for a particular variable. In the **substitution method** for solving systems of equations, we solve one equation for one variable and then substitute the equivalent expression in place of the variable in the other equation.

To solve a system of equations by substitution:

1. Rearrange either equation to isolate one variable.
2. Substitute the equivalent expression from Step 1 into the *other* equation and solve for the remaining variable.
3. Substitute the value of the variable found in Step 2 into the equation from Step 1 to find the remaining value of the solution.
4. Check the solution in both original equations.

EXAMPLE Solve the system of equations $x + y = 15$ and $y = 2x$ using the substitution method.

$$x + y = 15 \qquad \text{Equation 1}$$
$$\boxed{y = 2x} \qquad \text{Equation 2 is already solved for } y.$$
$$x + \boxed{y} = 15 \qquad \text{Substitute } 2x \text{ for } y \text{ in Equation 1 and solve.}$$
$$x + \boxed{2x} = 15 \qquad \text{Solve for } x.$$
$$3x = 15$$
$$\boxed{x = 5} \qquad x\text{-value of solution}$$
$$y = 2\,\boxed{x} \qquad \text{Substitute the solution for } x \text{ in Equation 2 to find } y.$$
$$y = 2(\,\boxed{5}\,) \qquad \text{Simplify.}$$
$$\boxed{y = 10} \qquad y\text{-value of solution}$$

The solution is (5, 10).

Check:

$\boxed{x} + \boxed{y} = 15$	$\boxed{y} = 2\,\boxed{x}$	Substitute $x = 5$ and $y = 10$ in both equations.
$\boxed{5} + \boxed{10} = 15$	$\boxed{10} = 2(\,\boxed{5}\,)$	Simplify.
$15 = 15$	$10 = 10$	The solution checks in both equations.

EXAMPLE Solve the following system of equations using the substitution method.

$$2x - 3y = -14$$
$$x + 5y = 19$$

Either equation can be solved for either unknown. In this example, the x-term in the second equation has a coefficient of 1, so the simplest choice would be to solve the second equation for x.

Step 1	**Step 2**	**Step 3**
$x + 5y = 19$	$2\,\boxed{x} - 3y = -14$	$x = 19 - 5\,\boxed{y}$
$\boxed{x = 19 - 5y}$	$2(\,\boxed{19 - 5y}\,) - 3y = -14$	$x = 19 - 5(\,\boxed{4}\,)$
	$38 - 10y - 3y = -14$	$x = 19 - 20$
	$38 - 13y = -14$	$\boxed{x = -1}$
	$-13y = -14 - 38$	
	$-13y = -52$	
	$\dfrac{-13y}{-13} = \dfrac{-52}{-13}$	
	$\boxed{y = 4}$	

The solution is $(-1, 4)$.

Check the solution $x = -1$, $y = 4$ in both original equations:

Step 4

$2\,\boxed{x} - 3\,\boxed{y} = -14$	$\boxed{x} + 5\,\boxed{y} = 19$
$2(\,\boxed{-1}\,) - 3(\,\boxed{4}\,) = -14$	$\boxed{-1} + 5(\,\boxed{4}\,) = 19$
$-2 - 12 = -14$	$-1 + 20 = 19$
$-14 = -14$	$19 = 19$

Long Problems Don't Have to Be Difficult

Sometimes we let ourselves become overwhelmed by the mere length of a problem. Look at the previous example. Each step of the solution involves skills we have previously used many times. Here are some tips to help you manage longer problems.

- Get a global or overall understanding of the problem you are solving.
- Make a prediction or estimate of the solution if appropriate.
- Get a global or overall understanding of the process you are using to solve the problem.
- List in your own words (as briefly as possible) the steps of the process.
- Focus on one step at a time.
- Examine the solution to see if it matches your prediction or estimate. Check if appropriate.

SECTION 10–3 SELF-STUDY EXERCISES

1 Solve the systems of equations using the substitution method.

1. $2a + 2b = 60$
$a = 10 + b$

2. $7r + c = 42$
$3r - 8 = c$

3. $x - 35 = -2y$
$3x - 2y = 17$

4. $x + y = 12$
$x = 2 + y$

5. $2p + 3k = 2$
$2p - 3k = 0$

6. $x + 2y = 5$
$x = 3y$

7. $x - 3 = 2y$
$x = 3y - 2$

8. $a = 3x - 1$
$x = a + 5$

9. $5x + 2y = 7$
$x = 2y - 1$

10. $x + y = 7$
$x = y - 5$

11. $x - 2y = 6$
$4x + 3y = 35$

12. $y = 3x - 2$
$y = x + 6$

13. $2x - 3y = 0$
$3x - y = 7$

14. $2x - 3y = -3$
$4x + 2y = 18$

15. $6x - 2y = 3$
$3x + 4y = 9$

16. $6x + 2y = 3$
$3x - 4y = -1$

17. $8x - 6y = -2$
$4x + 9y = 7$

18. $4x + 6y = 1$
$8x - 4y = -6$

19. $x - y = -4$
$5x + 9y = 8$

20. $x - 6y = 8$
$x + 3y = -1$

21. $x + 3y = 1$
$2x + 4y = -4$

10–4 | *Problem Solving Using Systems of Linear Equations*

Learning Outcome   **1** Use a system of linear equations to solve application problems.

1 Use a System of Linear Equations to Solve Application Problems.

Many job-related problems can be solved by setting up and solving systems of equations. The addition method or the substitution method can be used to solve application problems.

Two dry cells connected in series have a total internal resistance of 0.09 Ω. The difference between the internal resistances of the individual dry cells is 0.03 Ω. How much is each internal resistance?

Known facts

The total internal resistance of the two dry cells is 0.09 Ω.
The difference in the internal resistances of the two dry cells is 0.03 Ω.

Unknown facts

What is the resistance of dry cell 1 (r_1)?
What is the resistance of dry cell 2 (r_2)?

Relationships

$r_1 + r_2 = 0.09$ Equation 1
$r_1 - r_2 = 0.03$ Equation 2

Estimation

If resistances were the same, they would each be 0.045 Ω. Because they are not the same, one will be more than 0.045 Ω and one will be less than 0.045 Ω.

Calculations

$$r_1 + r_2 = 0.09$$ Add the equations to eliminate r_2.
$$r_1 - r_2 = 0.03$$
$$2r_1 \quad = 0.12$$ Solve for r_1.
$$r_1 = \frac{0.12}{2}$$
$$r_1 = 0.06$$ r_1-value of solution
$$r_1 + r_2 = 0.09$$ Substitute for r_1 in Equation 1.
$$0.06 + r_2 = 0.09$$ Solve for r_2.
$$r_2 = 0.09 - 0.06$$
$$r_2 = 0.03$$ r_2-value of solution

Interpretation

The larger internal resistance is 0.06 Ω and the smaller internal resistance is 0.03 Ω.

A tank holds a solution that is 10% herbicide. Another tank holds a solution that is 50% herbicide. If a farmer wants to mix the two solutions to get 200 gal of a solution that is 25% herbicide, how many gallons of each solution should be mixed?

Known facts

There are two strengths of herbicide, 10% and 50%.
200 gal of 25% herbicide are needed.

Unknown facts

How many gallons of 10% herbicide (h) are needed?
How many gallons of 50% herbicide (H) are needed?

Relationships

200 gal of the new herbicide are needed: $h + H = 200$
Amount of pure herbicide in h gal of 10% herbicide: $0.1h$
Amount of pure herbicide in H gal of 50% herbicide: $0.5H$
Amount of pure herbicide in 200 gal of 25% herbicide: $0.25(200)$

$$h + H = 200$$ Equation 1 (total gallons)
$$0.1h + 0.5H = 0.25(200)$$ Equation 2 (gallons of pure herbicide)

Estimation

If equal amounts of herbicide were needed, we would need 100 gal of each solution. However, since the desired solution strength is not exactly halfway between the two original herbicide strengths, we will need unequal amounts of herbicide. One amount will be less than 100 gal and the other will be more than 100 gal.

Solve by the substitution method.

$$h + H = 200$$ Solve Equation 1 for h.

$$h = 200 - H$$ Equivalent expression for h

$$0.1\,h + 0.5H = 0.25(200)$$ Substitute $(200 - H)$ for h in Equation 2.

$$0.1(200 - H) + 0.5H = 0.25(200)$$ Solve for H.

$$20 - 0.1H + 0.5H = 50$$

$$20 + 0.4H = 50$$

$$0.4H = 50 - 20$$

$$0.4H = 30$$

$$H = \frac{30}{0.4}$$

$$H = 75 \text{ gal}$$ H-value of solution

$$h + H = 200$$ Substitute 75 for H in Equation 1.

$$h + 75 = 200$$ Solve for h.

$$h = 200 - 75$$

$$h = 125 \text{ gal}$$ h-value of solution

Interpretation

The farmer must mix 75 gal of the 50% herbicide and 125 gal of the 10% herbicide to make 200 gal of a 25% herbicide.

EXAMPLE BUS

Rosita has $5,500 to invest and for tax purposes wants to earn exactly $500 interest for 1 year. She wants to invest part at 10% and the remainder at 5%. How much must she invest at each interest rate to earn exactly $500 interest in 1 year?

Let x = the amount invested at 10%. Let y = the amount invested at 5%. Interest for 1 year = rate $\times$ amount invested. Convert percents to decimals. Using these relationships, we derive a system of equations.

Known facts

Total of $5,500 to be invested
$500 interest to be earned in one year

Unknown facts

How much should be invested at 10%?
How much should be invested at 5%?

Relationships

Amount invested at 10%: x
Interest earned at 10%: $0.1x$
Amount invested at 5%: y
Interest earned at 5%: $0.05y$

$$x + y = 5,500$$ Equation 1 (total investment)
$$0.1x + 0.05y = 500$$ Equation 2 (total interest in 1 year)

Estimation

If the total amount were invested at 10%, the interest (in 1 year) would be $550 ($0.1 \times \$5,500$). Since we want $500 in interest, most of the money will need to be invested at 10%.

Calculations   Solve by the substitution method.

$$x + y = 5{,}500$$ Solve Equation 1 for x.

$$x = 5{,}500 - y$$ Equivalent expression for x

$$0.1\,x + 0.05y = 500$$ Substitute $(5{,}500 - y)$ for x in Equation 2.

$$0.1(\,5{,}500 - y\,) + 0.05y = 500$$ Solve for x.

$$550 - 0.1y + 0.05y = 500$$ $-0.1y + 0.05y = -0.05y$

$$550 - 0.05y = 500$$

$$-0.05y = 500 - 550$$

$$-0.05y = -50$$

$$y = \frac{-50}{-0.05}$$

$$y = \$1{,}000$$ Amount invested at 5%

$$x + y = \$5{,}500$$ Substitute $\$1{,}000$ for y in Equation 1.

$$x + \$1{,}000 = \$5{,}500$$

$$x = \$5{,}500 - \$1{,}000$$

$$x = \$4{,}500$$ Amount invested at 10%

Interpretation   **Rosita must invest \$4,500 at 10% and \$1,000 at 5% for 1 year to earn \$500 interest.**

SECTION 10–4 SELF-STUDY EXERCISES

1   Solve the problems using systems of equations with two unknowns.

1. **CON**   Two lengths of board total 48 in. If one board is 17 in. shorter than the other, find the length of each board.

2. **BUS**   A broker invested \$35,000 in two different stocks. One earned dividends at 4% and the other at 5%. If a \$1,570 dividend was earned on both stocks together, how much was invested in each? (*Reminder:* Change 4% and 5% to decimals.)

3. **BUS**   A department store buyer ordered 12 shirts and 8 hats for \$380 one month and 24 shirts and 10 hats for \$664 the following month. What was the cost of each shirt and each hat?

4. **AUTO**   A mechanic makes \$105 on each 8-cylinder engine tune-up and \$85 on each 4-cylinder engine tune-up. If the mechanic did 10 tune-ups and made a total of \$990, how many 8-cylinder jobs and how many 4-cylinder jobs were completed?

5. **ELEC**   30 resistors and 15 capacitors cost \$12. And 10 resistors and 20 capacitors cost \$8.50. How much does each capacitor and resistor cost?

6. **AVIA**   A private airplane flew 420 mi in 3 h with the wind. The return trip against the wind took 3.5 h. Find the rate of the plane in calm air and the rate of the wind.

7. **BUS**   In 1 year, Dee Wallace earned \$660 in interest on two investments totaling \$8,000. If he received 7% and 9% rates of return, how much did he invest at each rate?

8. A motorboat went 40 mi with the current in 3 h. The return trip against the current took 4 h. How fast was the current? What would have been the speed of the boat in calm water? Round to the nearest hundredth mile per hour.

9. **BUS** A visitor to south Louisiana purchased 3 lb of dark-roast pure coffee and 4 lb of coffee with chicory for $27.30 in a local supermarket. Another visitor at the same store purchased 2 lb of coffee with chicory and 5 lb of dark-roast pure coffee for $28. How much did each coffee cost per pound?

10. **AG/H** A lawn-care technician wants to spread a 200-lb seed mixture that is 50% bluegrass. If the technician has on hand a mixture that is 75% bluegrass and a mixture that is 10% bluegrass, how many pounds of each mixture are needed to make 200 lb of the 50% mixture? Round to the nearest whole pound.

11. **BUS** A college bookstore received a partial shipment of 50 scientific calculators and 25 graphing calculators at a total cost of $2,200. Later the bookstore received the balance of the calculators: 25 scientific and 50 graphing at a cost of $3,800. Find the cost of each calculator.

12. **BUS** A consumer received two 1-yr loans totaling $10,000 at interest rates of 10% and 15%. If the consumer paid $1,300 interest, how much money was borrowed at each rate?

13. **HOSP** For the first performance at the Overton Park Shell, 40 reserved seats and 80 general admission seats were sold for $2,000. For the second performance, 50 reserved seats and 90 general admission seats were sold for $2,350. What was the cost for a reserved seat and for a general admission seat?

14. A photographer has a container with a solution of 75% developer and a container with a solution of 25% developer. If she wants to mix the solutions to get 8 pt of solution with 50% developer, how many pints of each solution does she need to mix?

15. **ELEC** Two resistances have a sum of 21 Ω. Their difference is 13 Ω. Write two equations using R_1 as the first resistance and R_2 as the second resistance. Find each resistance.

16. **AG/H** The total weight of a fertilizer composed of nitrogen and potassium is 480 lb. The fertilizer has three times as much nitrogen as potassium. How many pounds of each chemical are in the fertilizer?

17. **AG/H** A plant nursery purchased holly shrubs that cost $4 each and nandinas that cost $5 each. The total cost was $260 for 60 shrubs. How many of each type of shrub did the nursery purchase?

18. **CON** A mortar mix contains five times as much sand as water. The total volume is 12 ft³. How much of each ingredient is in the mix?

CHAPTER REVIEW OF KEY CONCEPTS

Learning Outcomes

What to Remember with Examples

Section 10–1

1 Solve a system of linear equations by graphing (pp. 432–433).

Graph each equation. The intersection of the two lines is the solution to the system. (The table-of-solutions, intercepts, or slope-intercept method may also be used to graph each equation.)

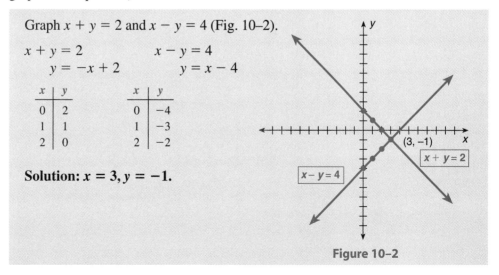

Graph $x + y = 2$ and $x - y = 4$ (Fig. 10–2).

$$x + y = 2 \qquad x - y = 4$$
$$y = -x + 2 \qquad y = x - 4$$

x	y
0	2
1	1
2	0

x	y
0	−4
1	−3
2	−2

Solution: $x = 3$, $y = -1$.

Figure 10–2

Section 10–2

1 Use the addition method to solve a system of linear equations that contains opposite variable terms (pp. 434–435).

To solve a system of equations with two opposite variable terms, add the equations so that the opposite terms add to zero. Solve the new equation and substitute the root in an original equation to find the value of the remaining variable.

Solve $x - y = 6$ and $x + y = 4$ ($-y$ and $+y$ are opposites).

$$x - y = 6 \qquad \text{Equation 1. Add the equations to eliminate } y.$$
$$\underline{x + y = 4} \qquad \text{Equation 2}$$
$$2x = 10 \qquad \text{Solve for } x.$$

$$\frac{2x}{2} = \frac{10}{2}$$

$$x = 5 \qquad x\text{-value of solution}$$

$$x - y = 6 \qquad \text{Equation 1. Substitute } x = 5.$$

$$5 - y = 6 \qquad \text{Solve for } y.$$
$$-y = 6 - 5$$
$$-y = 1$$
$$y = -1 \qquad y\text{-value of solution}$$

Check:
$$x - y = 6$$
$$5 - (-1) = 6$$
$$5 + 1 = 6$$
$$6 = 6$$

$$x + y = 4$$
$$5 + (-1) = 4$$
$$4 = 4$$

Solution checks.

The solution is (5, −1).

2 Use the addition method to solve a system of linear equations that does not contain opposite variable terms (pp. 436–437).

To solve a system of equations that does not contain opposite variables: **1.** Multiply one or both equations by signed numbers that cause the terms of one variable to add to zero. **2.** Add the equations. **3.** Solve the equation from Step 2. **4.** Substitute the root in an original equation and solve for the remaining variable. **5.** Check solutions.

Solve $2x - y = 6$ and $4x + 2y = 4$.

$$2(2x - y) = 2(6) \qquad \text{Multiply Equation 1 by 2. Add equations to eliminate } y.$$
$$4x - 2y = 12$$
$$\underline{4x + 2y = 4}$$
$$8x = 16 \qquad \text{Solve for } x.$$

$$\frac{8x}{8} = \frac{16}{8}$$

$$x = 2 \qquad x\text{-value of solution}$$
$$4x + 2y = 4 \qquad \text{Equation 2. Substitute } x = 2.$$
$$4(2) + 2y = 4$$
$$8 + 2y = 4 \qquad \text{Solve for } y.$$
$$2y = 4 - 8$$
$$2y = -4$$

$$\frac{2y}{2} = \frac{-4}{2}$$

$$y = -2 \qquad y\text{-value of solution.}$$

Check:
$$2x - y = 6$$
$$2(2) - (-2) = 6$$
$$4 + 2 = 6$$
$$6 = 6$$

$$4x + 2y = 4$$
$$4(2) + 2(-2) = 4$$
$$8 - 4 = 4$$
$$4 = 4$$

Solution checks.

Solution: (2, −2)

3 Apply the addition method to a system of linear equations with no solution or with many solutions (p. 438).

Apply the addition rule. But note that *both* variables are eliminated. If the resulting statement is false, there is no solution. The system is *inconsistent*. If the resulting statement is true, there are many solutions. The system is *dependent*.

Solve $a + b = 2$ and $a + b = 4$.

$$-1(a + b) = -1(2)$$ Multiply the first equation by -1. Add to eliminate variables.
$$-a - b = -2$$
$$\underline{a + b = 4}$$
$$0 = 2$$ False. No solution. Inconsistent system.

Section 10–3

1 Use the substitution method to solve a system of linear equations (pp. 439–441).

To solve a system of equations by substitution: **1.** Rearrange either equation to isolate one of the variables. **2.** Substitute the equivalent expression from Step 1 in the *other* equation and solve. **3.** Substitute the root from Step 2 in one of the original equations and solve for the other variable. **4.** Check.

Solve $2x + y = 6$ and $2x - 2y = 8$.

$$y = 6 - 2x$$ Isolate y in the first equation.
$$2x - 2y = 8$$ Substitute $6 - 2x$ for y in second equation.
$$2x - 2(6 - 2x) = 8$$
$$2x - 12 + 4x = 8$$ Solve for x.
$$6x - 12 = 8$$
$$6x = 8 + 12$$
$$6x = 20$$
$$\frac{6x}{6} = \frac{20}{6}$$
$$x = \frac{10}{3}$$ x-value of solution

Check: $2x + y = 6 \qquad\qquad 2x - 2y = 8$

$$y = 6 - 2\left(\frac{10}{3}\right) \qquad 2\left(\frac{10}{3}\right) + \left(\frac{-2}{3}\right) = 6 \qquad 2\left(\frac{10}{3}\right) - 2\left(-\frac{2}{3}\right) = 8$$

$$y = \frac{18}{3} - \frac{20}{3} \qquad\qquad \frac{20}{3} + \frac{-2}{3} = 6 \qquad\qquad \frac{20}{3} + \frac{4}{3} = 8$$

$$y = -\frac{2}{3} \quad \begin{array}{l}\text{y-value}\\\text{of solution}\end{array} \qquad \frac{18}{3} = 6 \qquad\qquad \frac{24}{3} = 8$$

$$\qquad\qquad\qquad\qquad 6 = 6 \qquad\qquad 8 = 8$$

Solution: $\left(\dfrac{10}{3}, -\dfrac{2}{3}\right)$ Solution checks.

Section 10–4

1 Use a system of linear equations to solve application problems (pp. 441–444).

Let two variables represent the unknown values. Use numbers and the variables to represent the conditions of the problem. Set up a system of equations and solve.

A taxidermist bought two pairs of glass deer eyes and five pairs of glass duck eyes for $15. She later purchased three pairs of deer eyes and five pairs of duck eyes for $20. Find the price per pair of each type of glass eye.

Let x = the price of a pair of deer eyes. Let y = the price of a pair of duck eyes. Price = the number of pairs of eyes times price of eye type:

$$2x + 5y = 15 \qquad \text{2 pairs of deer eyes plus 5 pairs of duck eyes cost \$15.}$$
$$3x + 5y = 20 \qquad \text{3 pairs of deer eyes plus 5 pairs of duck eyes cost \$20.}$$
$$-1(2x + 5y) = -1(15) \qquad \text{Multiply first equation by } -1. \text{ Add equations to eliminate } y.$$
$$\begin{aligned}-2x - 5y &= -15\\ \underline{3x + 5y} &= \underline{20}\\ x &= 5 \qquad \text{x-value of solution}\end{aligned}$$
$$2x + 5y = 15 \qquad \text{Use first equation to find } y. \text{ Substitute } x = 5.$$
$$2(5) + 5y = 15$$
$$10 + 5y = 15 \qquad \text{Solve for } y.$$
$$5y = 15 - 10$$
$$5y = 5$$
$$\frac{5y}{5} = \frac{5}{5}$$
$$y = 1 \qquad \text{y-value of solution}$$

The deer eyes cost $5 a pair and the duck eyes cost $1 a pair.

CHAPTER REVIEW EXERCISES

Section 10–1

Solve the systems of equations by graphing.

1. $y = 5x - 1$
   $y = 3x + 3$
2. $y = -x + 5$
   $y = 3x - 7$
3. $y = 5x + 1$
   $y = 3x - 1$
4. $2x - y = 5$
   $4x - 2y = 2$
5. $x + y = 8$
   $x - y = 2$
6. $3x + 2y = 13$
   $x - 2y = 7$
7. $2x + 2y = 10$
   $3x + 3y = 15$
8. $x + y = 1$
   $3x - 4y = 10$

Section 10–2

Solve the systems of equations using the addition method.

9. $3x + y = 9$
   $2x - y = 6$
10. $7x - y = 4$
    $8x - y = 3$
11. $5x - 3y = 8$
    $-5x + 2y = -7$
12. $2a + 3b = 8$
    $a - b = 4$
13. $Q = 2P + 8$
    $2Q + 3P = 2$
14. $4j + k = 3$
    $8j + 2k = 6$
15. $r = 2y + 6$
    $2r + y = 2$
16. $3a + 3b = 3$
    $2a - 2b = 6$
17. $c = 2y$
    $2c + 3y = 21$
18. $2x + 4y = 9$
    $x + 2y = 3$
19. $3R - 2S = 7$
    $-14 = -6R + 4S$
20. $x - 3 = -y$
    $2y = 9 - x$
21. $c = 2 + 3d$
    $3c - 14 = d$
22. $Q - 10 = T$
    $T = 2 - 2Q$
23. $x - 18 = -6y$
    $4x - 0 = 3y$

24. $R + S = 3$
$S - 9 = -3R$

25. $3a - 2b = 6$
$6a - 12 = b$

26. $a - b = 2$
$a + b = 12$

27. $x + 2y = 7$
$x - y = 1$

28. $2c + 3b = 2$
$2c - 3b = 0$

29. $x + 2r = 5.5$
$2x = 1.5r$

30. $7x + 2y = 6$
$4y = 12 - x$

Section 10–3

Solve the systems of equations using either the addition or the substitution method.

31. $a + 7b = 32$
$3a - b = 8$

32. $x + y = 1$
$4x + 3y = 0$

33. $c - d = 2$
$c = 12 - d$

34. $3a + 4b = 0$
$a + 3b = 5$

35. $7x - 4 = -4y$
$3x + y = 6$

36. $5Q - 4R = -1$
$R + 3Q = -38$

37. $a = 2b + 11$
$3a + 11 = -5b$

38. $y = 5 - 2x$
$3x - 2y = 4$

39. $c = 2q$
$2c + q = 2$

40. $3x + 2y = 10$
$y = 6 - x$

41. $4x - 2.5y = 2$
$2x - 1.5y = -10$

42. $2a - c = 4$
$a = 2 + c$

43. $4d - 7 = -c$
$3c - 6 = -6d$

44. $x = 10 - y$
$5x + 2y = 11$

45. $3.5a + 2b = 2$
$0.5b = 3 - 1.5a$

46. $c + d = 12$
$c - d = 2$

47. $x + 4y = 20$
$4x + 5y = 58$

48. $a + 5y = 7$
$a + 4y = 8$

49. $3a + 1 = -2b$
$4b + 23 = 15a$

50. $6y + 0 = 5p$
$4y - 3p = 38$

Section 10–4

Solve the problems using systems of equations with two unknowns.

51. ELEC Three electricians and four apprentices earned a total of $365 on one job. At the same rate of pay, one electrician and two apprentices earned a total of $145. How much pay did each apprentice and electrician receive?

52. AG/H Six bushels of bran and 2 bushels of corn weigh 182 lb. If 2 bushels of bran and 4 bushels of corn weigh 154 lb, how much do 1 bushel of bran and 1 bushel of corn each weigh?

53. CON A painter paid $22.50 for 2 qt of white shellac and 5 qt of thinner. If 3 qt of shellac and 2 qt of thinner cost the painter's helper $14.50, what is the cost of each quart of shellac and thinner?

54. ELEC A main current of electricity is the sum of two smaller currents whose difference is 0.8 A. What are the two smaller currents if the main current is 10 A?

55. The sum of two angles is 175°. Their difference is 63°. What is the measure of each angle?

56. HOSP A tour boat traveled 20 mi in 2 h with the current. The return trip took 3 h. Find the rate of the boat in calm water and the rate of the current.

57. BUS In 1 year, Sholanda Brown earned $280 on two investments totaling $5,000. If she received 5% and 6% rates of return, how much did she invest at each rate?

58. AVIA A plane flew 300 km against the wind in 4 h. The return trip with the wind took 3 h. How fast was the wind? What would have been the speed of the plane in calm air?

59. HOSP A restaurant purchased 30 lb of Colombian coffee and 10 lb of blended coffee for $190. In a second purchase, the same restaurant paid $120 for 20 lb of Colombian coffee and 5 lb of blended coffee. How much did each coffee cost per lb?

60. AG/H A rancher wants to spread a 300-lb grass seed mixture that is 50% tall fescue. If the rancher has on hand a seed mixture that is 80% tall fescue and a mixture that is 20% tall fescue, how many pounds of each mixture are needed to make 300 lb of the 50% mixture?

61. AUTO An automotive service station purchased 25 maps of Ohio and 8 maps of Alaska at a total cost of $65.55. Later the station purchased 20 maps of Ohio and 5 maps of Alaska for $49.50. Find the cost of each map.

63. BUS At the first of the month, store buyer Selena Henson placed a $12,525 order for 20 name-brand suits and 35 suits with generic labels. At the end of the month, she placed a $15,725 order for 30 name-brand suits and 35 generic-label suits. How much did she pay for each type of suit?

65. BUS Jorge makes 5% commission on telephone sales and 6% commission on showroom sales. If his sales totaled $40,000 and his commission was $2,250, how much did he sell by telephone? How much did he sell on the showroom floor?

62. BUS Emily Harrington made two 1-year investments totaling $7,000 at interest rates of 4% and 7%. If she received $415 in return, how much money was invested at each rate?

64. BUS A taxidermist has a container with a solution of 10% tanning chemical and a container with a solution of 50% tanning chemical. If the taxidermist wants to mix the solutions to get 10 gal of solution with 25% tanning chemical, how many gallons of each solution should be mixed?

66. BUS Sing-Fong has 60 coins in nickels and quarters. The total value of the coins is $12. How many coins of each type does she have?

TEAM PROBLEM-SOLVING EXERCISES

1. A true measure of your understanding of a concept is the ability to apply this understanding to real-world situations.

   (a) Write a story line for a mixture problem that can be solved using the system

$$0.5x + 0.3y = 8$$
$$x + y = 20$$

   (b) Solve the system and interpret the results within the context of your story line.

2. Select a career or business situation of interest to your team.

   (a) Write a story line for a mixture problem that can be solved using the system

$$3x + 5y = 28$$
$$x - y = 4$$

   (b) Solve the system and interpret the results within the context of your story line.

CONCEPTS ANALYSIS

1. What does the graphical solution of a system of two equations with two unknowns represent?

3. What can be said about the solution of a system of two equations if the graphs of the two coincide?

5. Identify the first error that occurs in the solution. Solve the system correctly.

$$
\begin{array}{ll}
x + 2y = 3 & x + 2y = 3 \\
2x - 3y = -1 & x + 2(5) = 3 \\
2x + 4y = 6 & x + 10 = 3 \\
2x - 3y = -1 & x = -7 \\
\quad\quad y = 5 & \text{Solution: } (-7, 5)
\end{array}
$$

7. Explain the substitution method of solving a system of two equations with two unknowns.

2. Describe the solution of a system of two equations if the graphs of the two equations are parallel?

4. Describe the solution of a system of two equations if their graphs intersect in exactly one point?

6. Explain the addition or elimination method of solving a system of two equations with two unknowns.

8. What does it mean when solving a system of two equations with two unknowns if the system produces a true statement like $0 = 0$?

9. What does it mean when solving a system of two equations with two unknowns if the system produces a statement like $3 = 0$?

10. Can a system of two linear equations in two unknowns have exactly two ordered pairs as solutions? Explain your answer.

PRACTICE TEST

Solve the systems of equations graphically.

1. $2a + b = 10$
$a - b = 5$

2. $x + y = 5$
$2x - y = 4$

3. $3x + 4y = 6$
$x + y = 5$

4. $a - 3b = 7$
$a - 5 = b$

5. $2c - 3d = 6$
$c - 12 = 3d$

Solve the systems of equations using the addition method.

6. $x + y = 6$
$x - y = 2$

7. $p + 2m = 0$
$2p = -m$

8. $6p + 5t = -16$
$3p - 3 = 3t$

9. $3x + y = 5$
$2x - y = 0$

10. $7c - 2b = -2$
$c - 4b = -4$

Solve the systems of equations with two variables using the substitution method.

11. $4x + 3y = 14$
$x - y = 0$

12. $a + 2y = 6$
$a + 3y = 3$

13. $7p + r = -6$
$3p + r = 6$

Solve the systems of equations with two variables using either the addition or the substitution method.

14. $4x + 4 = -4y$
$6 + y = -6x$

15. $38 + d = -3a$
$5a + 1 = 4d$

Solve the problems using systems of equations with two variables.

16. Two lengths of stereo speaker wire total 32.5 ft. One length is 2.9 ft longer than the other. How long is each length of speaker wire?

17. Two currents add to 35 A and their difference is 5 A. How many amperes are in each current?

18. Six packages of common nails and four packages of finishing nails weigh 6.5 lb. If two packages of common nails and three packages of finishing nails weigh 3.0 lb, how much does one package of each kind of nail weigh?

19. The length of a piece of sheet metal is $1\frac{1}{2}$ times the width. The difference between the length and the width is 17 in. Find the length and the width.

20. A mixture of fieldstone is needed for a construction job and will cost $954 for the 27 tons of stone. The stone is of two types, one costing $38 per ton and one costing $32 per ton. How many tons of each are required?

21. A broker invested $25,000 in two different stocks. One stock earned dividends at 3.5% and the other at 4%. If a dividend of $900 was earned on both stocks together, how much was invested in each stock?

22. A mason purchased 2-in. cold-rolled channels and $\frac{3}{4}$-in. cold-rolled channels whose total weight was 820 lb. The difference in weight between the heavier 2-in. and lighter $\frac{3}{4}$-in. channels was 280 lb. How many pounds of each type of channel did the mason purchase?

23. A total capacitance in parallel is the sum of two capacitances. If the capacitances total 0.00027 farad (F) and the difference between the two capacitances is 0.00016 F, what is the value of each capacitance in the system?

Simplify.

1. $4 - 3(2x - 5)$

2. $5x - (3x - 2)$

Solve.

3. $7x - 3(x - 8) = 28$

4. $18 - 6(2 - y) = 24$

5. $\dfrac{5}{12}x - \dfrac{3}{4} = \dfrac{1}{9} - \dfrac{2}{3}x$

6. $4x - 3.2 + x = 3.3 - 2.4x$

7. Find the impedance (Z) in ohms using the formula $Z = \sqrt{R^2 + X^2}$ if the resistance (R) is $4\ \Omega$ and the reactance (X) is $7\ \Omega$. Round to tenths.

Solve.

8. $\dfrac{3x}{8} = \dfrac{3}{4}$

9. $\dfrac{2x}{7} = \dfrac{3}{5}$

10. $\dfrac{2}{12} = \dfrac{7}{4x}$

11. A large gear has 400 teeth and turns at 30 rpm. Find the number of teeth of a small gear in mesh that turns at 75 rpm.

12. A blueprint has a scale of $\dfrac{1}{4}$ in. $= 1$ ft. What is the actual measurement of a wall that is five inches on the blueprint?

13. Graph using a table of solutions: $y = -2x + 4$

14. Graph using the intercepts procedure: $2x - y = 6$

15. Graph using the slope and y-intercept procedure:

$y - \dfrac{3}{4}x = 2$

Determine the slope and y-intercept.

16. $y = \dfrac{3}{4}x + 2$

17. $3x - 2y = 10$

18. Find the slope of the line passing through the points $(5, 2)$ and $(-2, 3)$.

Find the equation of the line passing through the given point with the given slope. Write the equation in slope y-intercept form.

19. $(5, -4),\ m = -\dfrac{2}{3}$

20. $(-1, -5),\ m = 3$

Find the equation of the line passing through the given pairs of points. Solve the equation for y.

21. $(3, 2)$ and $(5, 6)$

22. $(-2, 1)$ and $(4, -2)$

Write the equation with the given slope and y-intercept. Solve the equation for y.

23. $m = 4,\ b = 3$

24. slope $= -2$, y-intercept $= \dfrac{1}{3}$

25. Write the equation in standard form of the line that is parallel to the line $x - y = 3$ and contains the point $(5, 4)$.

26. Write the equation in standard form of the line that is perpendicular to the line $x - 2y = 6$ and passes through the point $(3, 0)$.

Solve the system of equations using the addition method.

27. $2x - y = -1$
$\quad\ x + y = 4$

28. $3x - 2y = -1$
$\quad 2x + 4y = 10$

29. Two resistances have a sum of $32\ \Omega$ and a difference of $8\ \Omega$. Find each resistance.

30. The total liquid in a herbicide mixture is 20 L. If the mixture has 4 times as much water as herbicide, how many liters of each substance are in the mixture?

11

Powers and Polynomials

Focus on Careers

If you have always liked to take apart and fix things, you may want to consider a career as a diesel service technician or mechanic. The nation's trucks and buses are powered by diesel engines because the diesel engine is more powerful and more durable than comparable gasoline-burning engines. Diesel technicians repair and maintain these trucks and buses. They also maintain locomotive engines and other heavy equipment such as bulldozers, cranes, road graders, fork lifts, farm tractors, and other farm equipment.

Most community colleges and trade and vocational schools offer programs in diesel repair. Employers prefer to hire graduates of formal training programs when they are hiring entry-level diesel mechanics. These training programs offer the most current diesel technology and instruction, and graduates also acquire increased skills needed to interpret technical manuals and communicate with coworkers and customers.

The National Institute for Automotive Service Excellence offers certification as master truck technicians or in specific areas of truck repair. Certifications give job seekers a clear advantage over those without certification.

The middle 50% of diesel engine specialists earned between $13.13 and $24.54 per hour in 2002 and the highest 10% earned more than $24.61 per hour! Most service technicians work a standard 40-h week and usually earn overtime for additional hours and earn at a higher rate for evening, night, or weekend hours.

Source: *Occupational Outlook Handbook,* 2004–2005 Edition. U.S. Department of Labor, Bureau of Labor Statistics.

Learning Outcomes
1 Multiply powers with like bases.
2 Divide powers with like bases.
3 Find a power of a power.

We have seen in various equations and formulas that a variable can represent all types of numbers: natural numbers, whole numbers, integers, fractions, decimals, and signed numbers. When raising a variable to a power, we must consider all the types of numbers the variable can be.

The **laws of exponents** have evolved from the patterns that formed when interpreting powers as repeated multiplication. These laws allow us to shorten our processes and to take advantage of the accessibility of technology. Be sure to notice for what circumstances a law is applicable. For example, many of the laws apply to factors with *like* bases.

1 Multiply Powers with Like Bases.

To multiply 2^3 by 2^2 using repeated multiplication, we have $2(2)(2)$ times $2(2)$ or $2(2)(2)(2)(2)$. Another way of writing this is $2^3 \times 2^2 = 2^5$. We can multiply x^4 and x^2. Even though the value of x is not known, $x^4(x^2)$ is $x(x)(x)(x)$ times $x(x)$ or $x(x)(x)(x)(x)(x)$.

$$x^4(x^2) = x^6$$

The product contains $4 + 2$ or 6 factors of x. The following law is a shortcut for using repeated multiplication.

To multiply powers that have like bases:

1. Verify that the bases are the same. Use this base as the base of the product.
2. Add the exponents for the exponent of the product.

This rule can be stated symbolically as

$$a^m(a^n) = a^{m+n} \quad \text{where } a, m, \text{ and } n \text{ are real numbers}$$

EXAMPLE Write the products.

(a) $y^4(y^3)$ **(b)** $a(a^2)$ **(c)** $b(b)$ **(d)** $x(x^3)(x^2)$ **(e)** $x^2(y^3)$ **(f)** $5x^3(2x^4)$

(a) $y^4(y^3) = y^{4+3} = \mathbf{y^7}$ The bases are the same, so add the exponents.
(b) $a(a^2) = a^{1+2} = \mathbf{a^3}$ When the exponent for a base is not written, it is 1.
(c) $b(b) = b^{1+1} = \mathbf{b^2}$
(d) $x(x^3)(x^2) = x^{1+3+2} = \mathbf{x^6}$
(e) $x^2(y^3) = \mathbf{x^2 y^3}$ Bases are unlike.
(f) $5x^3(2x^4) = \mathbf{10x^7}$ Multiply coefficients. Add exponents of like bases.

The previous law applies *only* to expressions with *like* bases. Thus, $x^2(y^3)$ can be written only as x^2y^3. No other simplification can be made.

2 Divide Powers with Like Bases.

When reducing fractions, we can reduce to a factor of 1 any factors common to both the numerator and denominator. We use this concept when dividing powers that have like bases.

EXAMPLE Reduce or simplify the fractions; that is, perform the division indicated by each fraction.

(a) $\dfrac{x^5}{x^2}$ **(b)** $\dfrac{a^2}{a}$ **(c)** $\dfrac{b^7}{c^5}$ **(d)** $\dfrac{m^3}{m^3}$ **(e)** $\dfrac{y^3}{y^4}$

(a) $\dfrac{x^5}{x^2} = \dfrac{x(x)(x)(x)(x)}{x(x)} = \boxed{x^3}$ The quotient contains $5 - 2 = 3$ factors of x, or x^3.

(b) $\dfrac{a^2}{a} = \dfrac{a(a)}{a} = \boxed{a}$ **(or a^1)**   The quotient contains $2 - 1 = 1$ factor of a.

(c) $\dfrac{b^7}{c^5} = \dfrac{b^7}{c^5}$ Bases are unlike, so no simplification can be made.

(d) $\dfrac{m^3}{m^3} = \dfrac{(m)(m)(m)}{(m)(m)(m)} = \boxed{1}$ All factors of m reduce to 1.

(e) $\dfrac{y^3}{y^4} = \dfrac{(y)(y)(y)}{(y)(y)(y)(y)} = \boxed{\dfrac{1}{y}}$ The denominator has more factors of y than the numerator.

In the previous example examine the exponents in the division and the exponent of the quotient. In parts a and b, the exponent of the quotient is the difference of the exponents of the dividend and the divisor.

(a) $\dfrac{x^5}{x^2} = x^{5-2} = x^3$ (b) $\dfrac{a^2}{a} = a^{2-1} = a^1 = a$

The results in parts d and e illustrate that the interpretation of exponents goes beyond just repeated multiplication; a new notation for 1 and for reciprocals is needed.

(d) $\dfrac{m^3}{m^3} = m^{3-3} = m^0 = 1$ Any nonzero base raised to the zero power equals 1.

(e) $\dfrac{y^3}{y^4} = y^{3-4} = y^{-1} = \dfrac{1}{y^1} = \dfrac{1}{y}$ An expression with a negative exponent equals the reciprocal of the expression written with a positive exponent having the same absolute value.

Reciprocals and Negative Exponents

Any nonzero number raised to the zero power is equal to 1.

$$n^0 = 1 \qquad \text{where } n \text{ is a real number and } n \neq 0$$

An expression with a *negative exponent* can be written as an equivalent expression with a positive exponent.

$$n^{-1} = \frac{1}{n} \qquad \frac{1}{n^{-1}} = n \qquad \text{where } n \text{ is a real number and } n \neq 0$$

A nonzero number times its reciprocal equals 1.

$$n \cdot \frac{1}{n} = 1 \qquad n^1 \cdot n^{-1} = n^0 = 1 \qquad \text{where } n \text{ is a real number and } n \neq 0$$

To divide powers that have like bases:

1. Verify that the bases are the same. Use this base as the base of the quotient.
2. Subtract the exponents for the exponent of the quotient.

This rule can be stated symbolically as

$$\frac{a^m}{a^n} = a^{m-n} \qquad \text{where } a, m, \text{ and } n \text{ are real numbers except that } a \neq 0$$

Because expressions with positive integral exponents are evaluated by using repeated multiplication, it is preferable to rewrite expressions with negative exponents as equivalent expressions with positive exponents. This manipulation is accomplished by applying the definition of negative exponents and exponents of zero.

EXAMPLE Write the quotients using positive exponents.

(a) $\dfrac{x^5}{x^8}$ (b) $\dfrac{a}{a^4}$ (c) $\dfrac{y^{-3}}{y^2}$ (d) $\dfrac{x^3}{x^{-5}}$ (e) $\dfrac{12x^7}{8x^5}$

(a) $\dfrac{x^5}{x^8} = x^{5-8} = x^{-3} = \dfrac{1}{x^3}$ (b) $\dfrac{a}{a^4} = a^{1-4} = a^{-3} = \dfrac{1}{a^3}$

(c) $\dfrac{y^{-3}}{y^2} = y^{-3-2} = y^{-5} = \dfrac{1}{y^5}$ (d) $\dfrac{x^3}{x^{-5}} = x^{3-(-5)} = x^{3+5} = x^8$

(e) $\dfrac{12x^7}{8x^5} = \dfrac{3x^{7-5}}{2} = \dfrac{3x^2}{2} \text{ or } \dfrac{3}{2}x^2$

The inverse relationship of multiplication and division allows us flexibility in applying the laws of exponents.

Look at Part (a) of the previous example.

$$\frac{x^5}{x^8} \quad \text{is} \quad x^5 \div x^8 \quad \text{or} \quad x^5 \cdot \frac{1}{x^8} \quad \text{or} \quad x^5 \cdot x^{-8}$$

The multiplication law of exponents can be used.

$$x^5 \cdot x^{-8} = x^{5+(-8)} = x^{-3}$$

A *factor* can be moved from a numerator to a denominator (or vice versa) by changing the sign of its exponent.

$$x^{-2} = \frac{x^{-2}}{1} = \frac{1}{x^2}, \qquad \frac{1}{x^{-3}} = \frac{x^3}{1} = x^3$$

This property *does not apply* to a *term* that is part of a numerator or denominator that has two or more terms.

$$\frac{x^{-2} + 1}{3} \quad \textit{does not equal} \quad \frac{1}{3x^2}.$$

In other words, *only factors* of the entire numerator or denominator can be moved. If more than one term is in the numerator, each term is divided by the denominator.

$$\frac{x^{-2} + 1}{3} = \frac{x^{-2}}{3} + \frac{1}{3} = \frac{1}{3x^2} + \frac{1}{3}$$

EXAMPLE   Simplify the expressions and make all exponents positive.

    (a) $\dfrac{a^2 b^{-3}}{ab^{-1}}$ **(b)** $\dfrac{xy^{-1}}{xy^2}$ **(c)** $\dfrac{x^3 + y^2}{xy}$

There is more than one way to simplify the expressions.

(a)

Option 1

$$\frac{a^2 b^{-3}}{ab^{-1}} = a^{2-1} b^{-3-(-1)}$$

Apply the division law of exponents. $2 - 1 = 1$; $-3 - (-1) = -3 + 1 = -2$. Be careful with the signs.

$$= ab^{-2}$$

Make all exponents positive.

$$= \frac{a}{b^2}$$

Option 2

$$\frac{a^2 \, b^{-3}}{ab^{-1}} = \frac{a^2 \, a^{-1} \, b^{-3} \, b^1}{1}$$

Apply the property of negative exponents to write all factors in the numerator. Apply the multiplication law of exponents.

$$= ab^{-2} = \frac{a}{b^2}$$

Make all exponents positive.

(b) $\dfrac{xy^{-1}}{xy^2} = x^{1-1}y^{-1-(2)}$

Apply the division law of exponents. $x^0 = 1$.

$= y^{-3}$

Make exponent positive.

$= \dfrac{1}{y^3}$

(c) $\dfrac{x^3 + y^2}{xy} = \dfrac{x^3}{xy} + \dfrac{y^2}{xy}$

The fraction is already simplified. It can be written as two separate fractions. Separate into two terms and apply division law to like bases.

$= \dfrac{x^{3-1}}{y} + \dfrac{y^{2-1}}{x}$

$= \dfrac{x^2}{y} + \dfrac{y}{x}$

---

3 Find a Power of a Power.

In the term $(2^3)^2$, we have a power raised to a power. 2^3 is the base of the expression and 2 is the exponent.

$$(2^3)^2 \quad \text{is} \quad (2^3)(2^3) = 2^{3+3} = 2^6 \quad \text{or} \quad 64$$

Let's look at some examples of raising numerical and variable powers to a power.

EXAMPLE Find the powers of powers by first writing each expression as repeated multiplication and then applying the multiplication law of exponents.

(a) $(3^2)^3$ **(b)** $(x^3)^4$ **(c)** $(a^2)^2$ **(d)** $(n^3)^5$

(a) $(3^2)^3 = (3^2)(3^2)(3^2) = 3^{2+2+2} = 3^6 = \mathbf{729}$

(b) $(x^3)^4 = (x^3)(x^3)(x^3)(x^3) = x^{3+3+3+3} = \boldsymbol{x^{12}}$

(c) $(a^2)^2 = (a^2)(a^2) = a^{2+2} = \boldsymbol{a^4}$

(d) $(n^3)^5 = (n^3)(n^3)(n^3)(n^3)(n^3) = n^{3+3+3+3+3} = \boldsymbol{n^{15}}$

From the preceding example we can see a pattern developing. In each case, if we multiply the exponents, we get the exponent of the new power.

To raise a power to a power:

1. Multiply exponents.
2. Keep the same base.

$(a^m)^n = a^{mn}$ where a, m, and n are real numbers

Applying this rule to the problems in the preceding example, we have

(a) $(3^2)^3 = 3^{2(3)} = 3^6 = 729$ (b) $(x^3)^4 = x^{3(4)} = x^{12}$

(c) $(a^2)^2 = a^{2(2)} = a^4$ (d) $(n^3)^5 = n^{3(5)} = n^{15}$

Other laws of exponents are extensions or combinations of the three laws we have already examined. These additional laws are useful tools for simplifying expressions containing exponents.

To raise a fraction or quotient to a power:

1. Raise the numerator (dividend) to the power.
2. Raise the denominator (divisor) to the power.

$$\left(\frac{a}{b}\right)^n = \frac{a^n}{b^n} \quad \text{where } a, b, \text{ and } n \text{ are real numbers and } b \neq 0$$

EXAMPLE Raise the fractions to the indicated power.

(a) $\left(\dfrac{2}{3}\right)^2$ **(b)** $\left(\dfrac{-3}{4}\right)^2$ **(c)** $\left(\dfrac{-1}{3}\right)^3$ **(d)** $\left(\dfrac{x}{y^3}\right)^3$ **(e)** $\left(\dfrac{x^2}{y^3}\right)^4$

(a) $\left(\dfrac{2}{3}\right)^2 = \dfrac{2^2}{3^2} = \dfrac{4}{9}$

(b) $\left(\dfrac{-3}{4}\right)^2 = \dfrac{(-3)^2}{4^2} = \dfrac{9}{16}$ $(-3)(-3) = +9$

(c) $\left(\dfrac{-1}{3}\right)^3 = \dfrac{(-1)^3}{3^3} = \dfrac{-1}{27}$ or $-\dfrac{1}{27}$ $(-1)(-1)(-1) = -1$

(d) $\left(\dfrac{x}{y^3}\right)^3 = \dfrac{x^3}{(y^3)^3} = \dfrac{x^3}{y^9}$

(e) $\left(\dfrac{x^2}{y^3}\right)^4 = \dfrac{(x^2)^4}{(y^3)^4} = \dfrac{x^8}{y^{12}}$

To raise a product to a power:

Raise each factor to the indicated power.

$$(ab)^n = a^n b^n \quad \text{where } a, b, \text{ and } n \text{ are real numbers}$$

EXAMPLE Raise the products to the indicated powers.

(a) $(ab)^2$ **(b)** $(a^2b)^3$ **(c)** $(xy^2)^2$ **(d)** $(3x)^2$ **(e)** $(2x^2y)^3$ **(f)** $(-5xy)^2$

(a) $(ab)^2 = a^{1(2)} b^{1(2)} = a^2 b^2$ Unwritten exponents are understood to be 1.

(b) $(a^2b)^3 = a^{2(3)} b^{1(3)} = a^6 b^3$

(c) $(xy^2)^2 = x^{1(2)} y^{2(2)} = x^2 y^4$

(d) $(3x)^2 = 3^{1(2)} x^{1(2)} = 3^2 x^2 = 9x^2$ Evaluate numerical factors.

(e) $(2x^2y)^3 = 2^{1(3)} x^{2(3)} y^{1(3)} = 2^3 x^6 y^3 = 8x^6 y^3$

(f) $(-5xy)^2 = (-5)^{1(2)} x^{1(2)} y^{1(2)} = (-5)^2 x^2 y^2 = 25x^2y^2$

Limitations of the Laws of Exponents

It is very important to understand what the laws of exponents *do not* include.

- The product-raised-to-power law applies to factors, not terms.

$$(a + b)^3 \qquad \textbf{does not equal} \qquad a^3 + b^3$$

- The multiplication-of-powers law applies to *like* bases.

$$a^2(b^3) \qquad \textbf{does not equal} \qquad ab^5 \qquad \text{or} \qquad (ab)^5$$

- An exponent affects only the one factor or grouping immediately to the left.

$$3x^2 \qquad \text{and} \qquad (3x)^2 \qquad \textbf{are not equal}$$

In the term $3x^2$, the numerical coefficient 3 is multiplied times the square of x. In the term $(3x)^2$, $3x$ is squared. Thus, $(3x)^2 = 3^2x^2 = 9x^2$.

- A negative coefficient of a base is not affected by the exponent.

$$-x^3 \qquad \textbf{means} \qquad -(x)(x)(x)$$

If $x = 4$, $-x^3 = -(4)^3$ or $-(64) = -64$. If $x = -4$, $-x^3 = -(-4)^3$ or $-(-64) = 64$.

SECTION 11–1 SELF-STUDY EXERCISES

1 Write the products.

1. $x^3(x^4)$ **2.** $m(m^3)$ **3.** $a(a)$

4. $x^2(x^3)(x^5)$ **5.** $y(y^2)(y^3)$ **6.** $a^2(b)$

7. $3a^5(4b^7)$ **8.** $2x^3(5x^8)$ **9.** $(5x^2yz)(-2xy^2z^3)$

10. $(-8x^2y)(-3xy^3)$ **11.** $\left(\dfrac{3}{5}a^2b\right)\left(\dfrac{10}{21}ab^4\right)$ **12.** $\left(\dfrac{2}{3}a^3b^2\right)\left(\dfrac{9}{16}ab\right)$

13. $(-1.2m^2n^7)(3.5m^{-4}n^3)$ **14.** $(-3a^2)(-4a^3)(-6a)$ **15.** $(32x^3)\left(\dfrac{3}{4}x^4\right)$

16. $(4a^2b)(-3ab)(-2a^{-2}b^2)$

2 Write the quotients with positive exponents.

17. $\dfrac{y^7}{y^2}$ **18.** $\dfrac{x^5}{x}$ **19.** $\dfrac{a^3}{a^4}$ **20.** $\dfrac{b^6}{b^5}$

21. $\dfrac{m^2}{m^2}$ **22.** $\dfrac{x}{x^3}$ **23.** $\dfrac{y^5}{y}$ **24.** $\dfrac{n^2}{n^{-5}}$

25. $\dfrac{x^{-4}}{x^6}$ **26.** $\dfrac{8n^2}{4n^3}$ **27.** $\dfrac{18x^7}{12x^0}$ **28.** $\dfrac{x^{-3}}{x^2}$

29. $\dfrac{x^2y^{-2}}{xy^4}$ **30.** $\dfrac{ab^2}{a^{-2}b^4}$ **31.** $\dfrac{x^2y^4}{xy^2}$ **32.** $\dfrac{a^3b^2}{a^2b^3}$

33. $\dfrac{21a^3b^{-4}}{7a^2b}$ **34.** $\dfrac{39r^2s^7}{26rs^{-5}}$ **35.** $\dfrac{12x^3y^5}{18x^7y^3}$ **36.** $\dfrac{51m^2n^5}{34m^{-4}n^2}$

3 Simplify the expressions. Make all exponents positive.

37. $(4^2)^3$ **38.** $(x^4)^2$ **39.** $(y^7)^0$ **40.** $(x^{10})^4$

41. $(a^7)^3$ **42.** $\left(-\dfrac{1}{2}\right)^3$ **43.** $\left(-\dfrac{2}{7}\right)^2$ **44.** $\left(\dfrac{a}{b}\right)^4$

45. $(2m^2n)^3$ **46.** $\left(\dfrac{x^2}{y}\right)^3$ **47.** $(x^2y^4)^3$ **48.** $(-2a)^2$

49. $(x^2y)^3$ **50.** $(-3ab^2)^3$ **51.** $(-7x^2)^5$ **52.** $(4x^2y^8)^2$

53. $-7x^3(x^{-5})^4$ **54.** $-x^4$ **55.** $(-x)^4$ **56.** $-6(x^2)^4$

57. $-x^3y(x^2)^5$ **58.** $-3x^9(x^{-2})^3$ **59.** $-x^2y^{-7}$ **60.** $(-3a^2bc)^3$

11–2 | Polynomials

Learning Outcomes

1 Identify polynomials, monomials, binomials, and trinomials.
2 Identify the degree of terms and polynomials.
3 Arrange polynomials in descending order.

1 Identify Polynomials, Monomials, Binomials, and Trinomials.

A **polynomial** is an algebraic expression in which the exponents of the variables are nonnegative integers and there are no variables in a denominator.

EXAMPLE Identify which expressions are polynomials. If an expression is not a polynomial, explain why.

(a) $5x^2 + 3x + 2$ (b) $5x - \dfrac{3}{x}$ (c) 9 (d) $-\dfrac{1}{2}x + 3x^{-2}$

(a) $5x^2 + 3x + 2$ **is a polynomial.**

(b) $5x - \dfrac{3}{x}$ **is not a polynomial** because the term $-\dfrac{3}{x}$ has a variable in the denominator.

(c) **9 is a polynomial** because it is equivalent to $9x^0$ and the exponent, zero, is a nonnegative integer.

(d) $-\dfrac{1}{2}x + 3x^{-2}$ **is not a polynomial** because $3x^{-2}$ has a negative exponent.

Some polynomials have special names depending on the number of terms contained in the polynomial.

A **monomial** is a polynomial containing one term. A term may have more than one factor.

$$3,\ -2x,\ 5ab,\ 7xy^2,\ \dfrac{3a^2}{4}\ \text{are monomials.}$$

A **binomial** is a polynomial containing two terms.

$$x + 3,\ 2x^2 - 5x,\ x + \dfrac{y}{4}\ \text{are binomials.}$$

A **trinomial** is a polynomial containing three terms.

$$a + b + c,\ x^2 - 3x + 4,\ x + \dfrac{2a}{7} - 5\ \text{are trinomials.}$$

To identify polynomials, monomials, binomials, and trinomials:

1. Write all variables in the numerator if necessary.
2. Expressions that have a negative exponent in any term are not polynomials.
3. Identify the expression based on the number of terms it contains.

EXAMPLE   Identify which of the expressions are polynomials; then state whether each polynomial is a monomial, binomial, or trinomial.

(a) $x^2y - 1$ (b) $4(x - 2)$

(c) $\dfrac{2x + 5}{2y}$ (d) $3x^2 - x + 1$

(a) $x^2y - 1$ **Binomial**

(b) $4(x - 2)$ **Monomial**

(c) $\dfrac{2x + 5}{2y}$ This one-term expression is **not a monomial because it is not a polynomial** (there is a variable in the denominator).

(d) $3x^2 - x + 1$ **Trinomial**

2 Identify the Degree of Terms and Polynomials.

The **degree of a term** that has only one variable with a nonnegative exponent is the same as the exponent of the variable.

$$3x^4, \quad \text{fourth degree} \qquad -5x, \quad \text{first degree}$$

The degree of a *constant* is 0. A variable to the zero power is implied.

$$5 = 5x^0, \quad \text{degree zero}$$

If a term has more than one variable, the degree of the term is the *sum* of the exponents of all the variable factors.

$$2xy = 2x^1y^1, \quad \text{second degree} \qquad -2ab^2 = -2a^1b^2, \quad \text{third degree}$$

To identify the degree of a term:

1. Exponents of variable factors must be integers greater than zero.
2. For a term that is a constant, the degree is zero.
3. For a term that has only one variable factor, the degree is the same as the exponent of the variable.
4. For a term that has more than one variable factor, the degree is the sum of the exponents of the variables.

EXAMPLE Identify the degree of each term in the polynomial $5x^3 + 2x^2 - 3x + 3$.

$5x^3$,   degree 3 (or third degree)
$2x^2$,   degree 2 (or second degree)
$-3x$,   degree 1 (or first degree)
3,   degree 0

Special names are associated with terms of degree 0, 1, 2, and 3. A **constant term** has degree 0. A **linear term** has degree 1. A **quadratic term** has degree 2. A **cubic term** has degree 3.

The **degree of a polynomial** that has only one variable and only positive integral exponents is the degree of the term with the largest exponent.

A **linear polynomial** has degree 1. A **quadratic polynomial** has degree 2. A **cubic polynomial** has degree 3.

$5x^3 - 2$ has a degree of 3 and is a cubic polynomial.

$x + 7$ has a degree of 1 and is a linear polynomial.

$7x^2 - 4x + 5$ has a degree of 2 and is a quadratic polynomial.

To identify the degree of a polynomial:

1. Identify the degree of each term of the polynomial.
2. Compare the degrees of each term of the polynomial and select the greatest degree as the degree of the polynomial.

EXAMPLE Identify the degree of the polynomials.

(a) $5x^4 + 2x - 1$ (b) $3x^3 - 4x^2 + x - 5$ (c) 7 (d) $x - \dfrac{1}{2}$

(a) $5x^4 + 2x - 1$ has a **degree of 4.**
(b) $3x^3 - 4x^2 + x - 5$ has a **degree of 3 and is a cubic polynomial.**
(c) 7 has a **degree of 0 and is a constant.**
(d) $x - \dfrac{1}{2}$ has a **degree of 1 and is a linear polynomial.**

3   **Arrange Polynomials in Descending Order.**

The terms of a polynomial in one variable are customarily arranged in order based on the degree of each term of the polynomial. The terms can be arranged beginning with the term with the highest degree (*descending order*) or beginning with the term with the lowest degree (*ascending order*).

Most Common Arrangement of Polynomials

Polynomials are most often arranged in **descending order** so that the degree of the polynomial is the degree of the first term.

The first term of a polynomial arranged in descending order is called the **leading term** of the polynomial. The coefficient of the leading term of a polynomial is called the **leading coefficient.**

To arrange polynomials in descending order of a variable:

1. Identify the variable on which the terms of the polynomial will be arranged if the polynomial has more than one variable.
2. Compare the degrees of the selected variable for each term.
3. List the term with highest degree of the specified variable first.
4. Continue to list the terms of the polynomial in descending order of the selected variable.

EXAMPLE   Arrange each polynomial in descending order and identify the degree, the leading term, and the leading coefficient of the polynomial.

(a) $5x + 3x^3 - 7 + 6x^2$ (b) $x^4 - 2x + 3$ (c) $4 + x$ (d) $x^2 + 5$

(a) $5x + 3x^3 - 7 + 6x^2 = \mathbf{3x^3 + 6x^2 + 5x - 7.}$
   Third degree, leading term is $3x^3$, leading coefficient is 3.
(b) $\mathbf{x^4 - 2x + 3}$ is already in descending order.
   Fourth degree, leading term is x^4, leading coefficient is 1.
(c) $4 + x = \mathbf{x + 4.}$
   First degree or linear polynomial, leading term is x, leading coefficient is 1.
(d) $\mathbf{x^2 + 5}$ is already in descending order.
   Second degree or quadratic polynomial, leading term is x^2, leading coefficient is 1.

A polynomial arranged in descending order can be written with every successive degree represented. Missing terms have a coefficient of 0.

$$x^4 - 2x + 3 \text{ is the same as } x^4 + 0x^3 + 0x^2 - 2x^1 + 3x^0.$$

SECTION 11–2 SELF-STUDY EXERCISES

1   Identify each of the following expressions as a monomial, binomial, or trinomial.

1. $2x^3y - 7x$	**2.** $5xy^2 + 8y$	**3.** $3xy$	**4.** $7ab$
5. $5(3x - y)$	**6.** $(x - 5)(2x + 4)$	**7.** $5x^2 - 8x + 3$	**8.** $7y^2 + 5y - 1$
9. $\dfrac{4x - 1}{5}$	**10.** $\dfrac{x}{6} - 5$	**11.** $4(x^2 - 2) + x^3$	**12.** $3x^2 - 7(x - 8)$

Identify the degree of each term in each polynomial.

13. $6x$

14. $8x^2$

15. $6x^2 - 8x + 12$

16. $7x^3 - 8x + 12$

17. $x - 12$

18. $3x^2 - 8$

19. 15

20. 21

21. $2x - \dfrac{1}{4}$

22. $8x^2 + \dfrac{5}{6}$

23. $5x^2y^3 + 7xy$

24. $8r^4s - 5r^3s^5$

Identify the degree of each polynomial.

25. $5x^2 + 8x - 14$

26. $x^3 - 8x^2 + 5$

27. $9 - x^3 + x^6$

28. $12 - 15x^2 - 7x^5$

29. $2x - \dfrac{4}{5}x^2$

30. $\dfrac{7}{8} - x$

31. $5x^3y + 3x^2y^2 + 5xy^5$

32. $7x^2y - 4xy^5$

3 Arrange each polynomial in descending powers of x, and identify the degree, leading term, and leading coefficient of the polynomial.

33. $5x - 3x^2$

34. $7 - x^3$

35. $4x - 8 + 9x^2$

36. $5x^2 + 8 - 3x$

37. $7x^3 - x + 8x^2 - 12$

38. $7 - 15x^4 + 12x$

39. $-7x + 8x^6 - 7x^3$

40. $15 - 14x^8 + x$

41. $12x + 8x^4 + 15x^3$

42. $5x^3 - 7x^4 + 3x - 8$

43. $3x - 5x^4 + 2x^2 - 5$

44. $-3 + 7x - 8x^4 - x^3$

11-3 | *Basic Operations with Polynomials*

Learning Outcomes

1 Add and subtract polynomials.

2 Multiply polynomials.

3 Use the FOIL method to multiply two binominals.

4 Multiply polyniminals that result in special products.

5 Divide polynomials.

1 Add and Subtract Polynomials.

Now that variables with exponents have been introduced, we can broaden our concept of **like terms.** For variable terms to be like terms, the variables as well as the exponents of the variables must be exactly the same; that is, $2x^2$ and $-4x^2$ are like terms because the x's are both squared. But $2x^2$ and $4x$ are not like terms because the x's do not have the same exponents.

EXAMPLE Which pairs of terms are like terms?

(a) $3a^2b$ and $-\frac{2}{3}a^2b$ **(b)** $-8xy^2$ and $7x^2y$ **(c)** $10ab^2$ and $(-2ab)^2$

(d) $5x^4y^3$ and $-2y^3x^4$ **(e)** $2x^2$ and $3x^3$

(a) $3a^2b$ **and** $-\frac{2}{3}a^2b$ **are like terms.** All variables and their corresponding exponents are the same.

(b) $-8xy^2$ **and** $7x^2y$ **are** *not* **like terms.** In $-8xy^2$ the exponent of x is 1, and in $7x^2y$ the exponent of x is 2. Also, the exponents of y are not the same.

(c) $10ab^2$ **and** $(-2ab)^2$ **are** *not* **like terms.** In the term $10ab^2$, only the b is squared. In $(-2ab)^2$, the entire term is squared, and the result of the squaring is $4a^2b^2$.

(d) $5x^4y^3$ and $-2y^3x^4$ are like terms. Both x factors have an exponent of 4, and both y factors have an exponent of 3. Because multiplication is commutative, the order of factors does not matter.

(e) $2x^2$ and $3x^3$ are *not* like terms. The exponents of x are not the same.

Algebraic expressions containing several terms are simplified as much as possible by combining like terms.

To combine like terms:

1. Combine the coefficients of the like terms using the rules for adding or subtracting signed numbers.
2. The variable factors and their exponents do not change in the sum or difference.

What Does *Simplify* Mean?

The instructions to "simplify the expression" are vague but often used in mathematics exercises. In general, to simplify an expression means to write the expression using fewer terms or reduced or with lower coefficients or exponents. When the instructions say "simplify," the intent is for you to examine the expression and see which laws or operations allow you to rewrite the expression in a simpler form.

EXAMPLE Simplify the algebraic expressions by combining like terms.

(a) $5x^3 + 2x^3$ (b) $x^5 - 4x^5$ (c) $a^3 + 4a^2 + 3a^3 - 6a^2$
(d) $3m + 5n - m$ (e) $y + y - 5y^2 + y^3$

(a) $5\,x^3 + 2\,x^3 = (5 + 2)\,x^3 = \mathbf{7\,x^3}$ $5x^3$ and $2x^3$ are like terms; add the coefficients 5 and 2. The sum has the same variable factor and exponent as the like terms.

(b) $x^5 - 4\,x^5 = (1 - 4)\,x^5 = \mathbf{-3\,x^5}$ x^5 and $-4x^5$ are like terms. $1 - 4 = -3$. The difference has the same variable factor and exponent as the like terms.

(c) $a^3 + 4a^2 + 3a^3 - 6a^2 = \mathbf{4a^3 - 2a^2}$ a^3 and $3a^3$ are like terms. $4a^2$ and $-6a^2$ are like terms. Combine coefficients mentally.

(d) $3m + 5n - m = \mathbf{2m} + \mathbf{5n}$ $3m$ and $-m$ are like terms.

(e) $y + y - 5y^2 + y^3 = \mathbf{2y} - 5y^2 + y^3$ y and y are like terms. Coefficients are 1.

When an algebraic expression contains a grouping preceded immediately by a negative sign, we subtract the entire grouping. Another interpretation is that the grouping is multiplied by -1, the implied coefficient of the grouping. Either interpretation causes *each* sign within the grouping to be changed to its opposite and at the same time removes the parentheses. If the grouping is preceded by a positive sign or an unexpressed positive sign, parentheses are removed without changing signs, as if each term in the grouping were multiplied by $+1$, the implied coefficient.

EXAMPLE Simplify the expressions.

(a) $y^2 + 2y - (3y^2 + 5y)$
(b) $(m^3 - 3m^2 - 5m + 4) - (4m^3 - 2m^2 - 5m + 2)$

(a) $y^2 + 2y - (3y^2 + 5y) =$

$y^2 + 2y - 1(3y^2 + 5y) =$ Distribute the implied coefficient of -1.

$y^2 + 2y - 3y^2 - 5y =$ Combine like terms.
$$-2y^2 - 3y$$

(b) $(m^3 - 3m^2 - 5m + 4) - (4m^3 - 2m^2 - 5m + 2) =$ Distribute the implied coefficient of -1.

$(m^3 - 3m^2 - 5m + 4) - 1(4m^3 - 2m^2 - 5m + 2) =$ Combine like terms.

$m^3 - 3m^2 - 5m + 4 - 4m^3 + 2m^2 + 5m - 2 =$
$$-3m^3 - m^2 + 2$$

2 Multiply Polynomials.

Simplifying algebraic expressions may also involve multiplication.

To multiply by a monomial:

1. Multiply the coefficients using the rules for signed numbers.
2. Multiply the letter factors using the laws of exponents for factors with like bases.
3. Distribute if multiplying a polynomial by a monomial.

EXAMPLE Multiply.

(a) $(4x)(3x^2)$ (b) $(-6y^2)(2y^3)$ (c) $(-a)(-3a)$

(a) $(4x)(3x^2) = 4(3)(x^{1+2}) = \mathbf{12\,x^3}$ Multiply coefficients.
Add exponents.

(b) $(-6y^2)(2y^3) = -6(2)(y^{2+3}) = \mathbf{-12\,y^5}$ Multiply coefficients.
Add exponents.

(c) $(-a)(-3a) = -1(-3)(a^{1+1}) = \mathbf{3\,a^2}$ Multiply coefficients. Add exponents.
The coefficient of $-a$ is -1. The exponent of $-a$ is 1.

If factors are to be multiplied times more than one term, apply the distributive property.

EXAMPLE Perform the multiplications.

$\quad$ **(a)** $2x(x^2 - 4x)$ $\qquad$ **(b)** $-2y^3(2y^2 + 5y - 6)$ $\qquad$ **(c)** $4a(3a^3 - 2a^2 - a)$

$\quad$ **(a)** $2x\,(x^2 - 4x) = 2x^3 - 8x^2$ $\qquad\qquad$ Distribute.

$\quad$ **(b)** $-2y^3\,(2y^2 + 5y - 6) = -4y^5 - 10y^4 + 12y^3$ $\qquad$ Distribute.

$\quad$ **(c)** $4a\,(3a^3 - 2a^2 - a) = 12a^4 - 8a^3 - 4a^2$ $\qquad$ Distribute.

We multiply two binomials by applying the distributive property more than once. According to the distributive property, each term of the first factor is multiplied by each term of the second factor. This means we are required to use the distributive property more than once.

EXAMPLE Multiply $(x + 4)(x + 2)$.

$$(x + 4)\,(x + 2) = x\,(x + 2) + 4\,(x + 2) \qquad \text{Apply the distributive property.}$$
$$= x^2 + 2x + 4x + 8 \qquad \text{Combine like terms.}$$
$$= x^2 + 6x + 8$$

To multiply two polynomials:

1. Use the distributive property to multiply each term of the first polynomial times the entire second polynomial.
2. Combine like terms.

Symbolically,

$$(a + b)\,(c + d) = a\,(c + d) + b\,(c + d) = ac + ad + bc + bd$$
$$(a + b + c)\,(d + e + f) = a\,(d + e + f) + b\,(d + e + f) + c\,(d + e + f)$$
$$= ad + ae + af + bd + be + bf + cd + ce + cf$$

EXAMPLE Multiply $(2x^2 + 3x - 2)(3x^2 - 5x + 6)$.

$$(2x^2 + 3x - 2)\,(3x^2 - 5x + 6) =$$
$$2x^2\,(3x^2 - 5x + 6) + 3x\,(3x^2 - 5x + 6) - 2\,(3x^2 - 5x + 6) = \quad \text{Distribute.}$$
$$6x^4 - 10x^3 + 12x^2 + 9x^3 - 15x^2 + 18x - 6x^2 + 10x - 12 = \quad \text{Combine like terms.}$$

$$6x^4 - x^3 - 9x^2 + 28x - 12$$

Use Long Multiplication to Multiply Polynomials

The preceding problem can be organized using a procedure similar to the long-multiplication procedure in arithmetic.

1. Multiply each term in the multiplier times the entire multiplicand.
2. Align partial products so that like terms are in columns.
3. Combine like terms.

$$
\begin{array}{r}
3x^2 - 5x + 6 \\
2x^2 + 3x - 2 \\
\hline
-6x^2 + 10x - 12 \\
9x^3 - 15x^2 + 18x \\
6x^4 - 10x^3 + 12x^2 \\
\hline
6x^4 - x^3 - 9x^2 + 28x - 12
\end{array}
$$

Multiply by -2.
Multiply by $3x$.
Multiply by $2x^2$.
Combine like terms.

Since multiplication is commutative, the polynomials can be interchanged.

$$
\begin{array}{r}
2x^2 + 3x - 2 \\
3x^2 - 5x + 6 \\
\hline
+12x^2 + 18x - 12 \\
-10x^3 - 15x^2 + 10x \\
6x^4 + 9x^3 - 6x^2 \\
\hline
6x^4 - x^3 - 9x^2 + 28x - 12
\end{array}
$$

Multiply by 6.
Multiply by $-5x$.
Multiply by $3x^2$.
Combine like terms.

3 Use the FOIL Method to Multiply Two Binomials.

When multiplying two binomials, we can guide ourselves through the repeated applications of the distributive property by using the acronym FOIL.

To multiply two binomials using the FOIL method:

1. Write the product of the *first* term of each factor.
2. Write the product of the two *outer* terms of the factors.
3. Write the product of the two *inner* terms of the factors.
4. Write the product of the *last* term of each factor.
5. Combine like terms.

$$(a + b)(c + d) = ac + ad + bc + bd$$

First, Last, Inner, Outer, First, Last, Outer, Inner

EXAMPLE Use the FOIL method to multiply $(2x - 3)(x + 1)$.

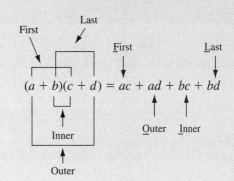

$$
\begin{aligned}
(2x - 3)(x + 1) &= 2x^2 + 2x - 3x - 3 \\
&= 2x^2 - x - 3
\end{aligned}
$$

4 Multiply Polynomials That Result in Special Products.

When the factors being multiplied have certain characteristics or conditions, we can anticipate the product without going through all the steps. Identifying and using these special characteristics allows us to do many multiplications mentally.

When we multiply the sum of two terms by the difference of the same two terms, we make some special observations about the product.

$$(x + 3)(x - 3) = x^2 \underbrace{- 3x + 3x}_{0} - 9 = x^2 - 9$$

F   O   I   L

Examine the final product. Notice that the product has only *two* terms and is a *difference.* The sum of the outer and inner products is zero ($-3x + 3x = 0$). The terms of the product are *perfect squares.*

Compare the product with its factors. The first term of the product is the *square of the first term* in either of its factors. The second term in the product is the *square of the second term* in either of its factors.

We call pairs of factors like $(a + b)(a - b)$ **the sum and difference of the same two terms,** or **conjugate pairs.** The product, $a^2 - b^2$, is called **the difference of two perfect squares.**

To mentally multiply the sum and difference of the same two terms (conjugate pairs):

1. Square the first term of either binomial.
2. Insert a minus sign.
3. Square the second term of either binomial.

Symbolically, $(a + b)(a - b) = a^2 - b^2$

EXAMPLE   Find the products mentally.

(a) $(x + 2)(x - 2)$ **(b)** $(2a + y)(2a - y)$

Since each of these is the sum and difference of the same two terms, the product is the difference of two perfect squares.

(a) $(\,x\, + \,2\,)(\,x\, - \,2\,) = x^2 - 4$ Square *x*. Square 2.

(b) $(\,2a\, + \,y\,)(\,2a\, - \,y\,) = 4a^2 - y^2$ Square 2*a*. Square *y*.

When we multiply the same two binomials, we make some special observations. Because $x \cdot x = x^2$, $(x + 3)(x + 3) = (x + 3)^2$. The quantity $(x + 3)^2$ is called a **binomial square** or the **square of a binomial.**

EXAMPLE Find the products of the binomial squares using the FOIL method.

(a) $(x + 2)^2$ (b) $(2x - 5)^2$ (c) $(3a - 2b)^2$

To find the products, write each binomial square as two factors; then use the FOIL method to multiply.

(a) $(x + 2)^2 = (x + 2)(x + 2)$

             $= x^2 + \boxed{4x} + 4$ Outer + Inner: $2x + 2x = 4x$

(b) $(2x - 5)^2 = (2x - 5)(2x - 5)$

             $= 4x^2 \boxed{-20x} + 25$ Outer + Inner: $-10x - 10x = -20x$

(c) $(3a - 2b)^2 = (3a - 2b)(3a - 2b)$

             $= 9a^2 \boxed{-12ab} + 4b^2$ Outer + Inner: $-6ab - 6ab = -12ab$

Notice that each product is a trinomial (three terms). These are special trinomials called **perfect-square trinomials.**

To mentally square a binomial:

1. Square the first term of the binomial for the first term of the perfect-square trinomial.
2. Double the product of the two terms of the binomial for the middle term of the trinomial.
3. Square the second term of the binomial for the third term of the trinomial.

Symbolically, $(a + b)^2 \quad = \quad a^2 + 2ab + b^2$

                         $(a - b)^2 \quad = \quad a^2 - 2ab + b^2$

        binomial square   perfect-square trinomial

EXAMPLE Square the binomials mentally.

(a) $(x + 2)^2$ (b) $(2x - 3)^2$ (c) $(5x + 1)^2$

	Square First Term	Double Product of Terms	Square Second Term
(a) $(x + 2)^2 =$	$(x)^2$	$2 \cdot (2x)$	$(2)^2$
	x^2	$+4x$	$+4$
(b) $(2x - 3)^2 =$	$(2x)^2$	$2 \cdot (-6x)$	$(-3)^2$
	$4x^2$	$-12x$	$+9$
(c) $(5x + 1)^2 =$	$(5x)^2$	$2 \cdot (5x)$	$(1)^2$
	$25x^2$	$+10x$	$+1$

Many patterns emerge when we multiply polynomials. However, it is not practical to develop a special process for every pattern. We will examine only the most common special products.

Let's look at the special products of the following two polynomials:

$$(a + b)(a^2 - ab + b^2)$$

$$(a - b)(a^2 + ab + b^2)$$

Recognizing Patterns

Special products are products that fit a specific pattern. If the factors fit the pattern, then the product is predictable and can be done mentally. Look carefully at the pattern developed by the special products:

$$(a + b)(a^2 - ab + b^2) \quad \text{and} \quad (a - b)(a^2 + ab + b^2)$$

- The first and last terms of the trinomial are the squares of the two terms of the binomial.
- The absolute value of the middle term of the trinomial is the product of the two terms of the binomial.
- When the binomial is a *sum* $(+)$, the middle term of the trinomial is *negative* $(-)$.
- When the binomial is a *difference* $(-)$, the middle term of the trinomial is *positive* $(+)$.

Now, let's multiply the binomial and trinomial by an extension of the FOIL method or repeated applications of the distributive property.

EXAMPLE Multiply $(a + b)(a^2 - ab + b^2)$.

Multiply each term in the trinomial by each term in the binomial.

First   Middle   Last

$$(a + b)(a^2 - ab + b^2) = a^3 \begin{array}{l} - a^2b + ab^2 \\ + a^2b - ab^2 \end{array} + b^3 \qquad \begin{array}{l} a(a^2 - ab + b^2) \\ b(a^2 - ab + b^2) \end{array}$$

$$\underline{ a^3 + b^3}$$

The four middle terms add to zero.

Thus, $(a + b)(a^2 - ab + b^2) = a^3 + b^3$, the **sum of two perfect cubes.**

EXAMPLE Multiply $(a - b)(a^2 + ab + b^2)$.

Multiply each term in the trinomial by each term in the binomial.

First   Middle   Last

$$(a - b)(a^2 + ab + b^2) = a^3 \begin{array}{l} + a^2b + ab^2 \\ - a^2b - ab^2 \end{array} - b^3 \qquad \begin{array}{l} a(a^2 + ab + b^2) \\ -b(a^2 + ab + b^2) \end{array}$$

$$\underline{ a^3 - b^3}$$

The four middle terms add to zero.

Thus, $(a - b)(a^2 + ab + b^2) = a^3 - b^3$, the **difference of two perfect cubes.**

To mentally multiply a binomial and a trinomial of the types $(a + b)(a^2 - ab + b^2)$ and $(a - b)(a^2 + ab + b^2)$:

1. Cube the first term of the binomial.
2. If the binomial is a sum, insert a plus sign. If it is a difference, insert a minus sign.
3. Cube the second term of the binomial.

Symbolically,

$$(a + b)(a^2 - ab + b^2) = a^3 + b^3$$

$$(a - b)(a^2 + ab + b^2) = a^3 - b^3$$

TIP

Using Symbolic Patterns

The pattern $(a + b)(a^2 - ab + b^2)$ is almost impossible to describe in sentence form. However, we can identify the pattern at a glance when it is written symbolically.

When we apply this pattern to different polynomials, we substitute for the letters a and b in the pattern.

$$\text{Multiply } (2x + 3)(4x^2 - 6x + 9).$$

Do these factors match the pattern? Let $2x$ represent a and 3 represent b.

$a^2 = (2x)^2 = 4x^2$	Matches pattern.
$b^2 = 3^2 = 9$	Matches pattern.
$ab = (2x)(3) = 6x$	Matches pattern.
Sign of binomial is positive. Middle sign of trinomial is negative.	Matches pattern.

Now, apply the pattern of the product.

$$(a + b)(a^2 - ab + b^2) = a^3 + b^3$$
$$a^3 = (2x)^3 = 8x^3$$
$$b^3 = 3^3 = 27$$

Then, $(2x + 3)(4x^2 - 6x + 9) = 8x^3 + 27$.

EXAMPLE   Mentally multiply $(m + p)(m^2 - mp + p^2)$.

$(a + b)(a^2 - ab + b^2) = a^3 + b^3$ Use the pattern and substitute m for a and p for b.

$(m + p)(m^2 - mp + p^2) = \mathbf{m^3 + p^3}$

EXAMPLE   Mentally multiply $(4c - d)(16c^2 + 4d + d^2)$.

$(a - b)(a^2 + ab + b^2) = a^3 - b^3$ Substitute $4c$ for a and d for b.

$(4c - d)(16c^2 + 4cd + d^2) = \mathbf{64c^3 - d^3}$ $4^3 = 64$

5 Divide Polynomials.

Simplifying algebraic expressions may also involve division.

To divide a monomial by a monomial:

1. Divide the coefficients using the rules for signed numbers.
2. Divide the variable factors using the laws of exponents for factors with like bases.

EXAMPLE   Divide. Express answers with positive exponents.

(a) $\dfrac{2x^4}{x}$ (b) $\dfrac{-6y^5}{2y^3}$ (c) $\dfrac{-4x}{4x^2}$ (d) $\dfrac{-5x^4}{15x^2}$ (e) $\dfrac{3x}{12x^3}$

(a) $\dfrac{2x^4}{x} = \dfrac{2}{1}(x^{4-1}) = \mathbf{2\,x^3}$

Divide coefficients. Subtract exponents. The coefficient of x in the denominator is 1.

(b) $\dfrac{-6y^5}{2y^3} = \dfrac{-6}{2}(y^{5-3}) = \mathbf{-3\,y^2}$

Divide coefficients. Subtract exponents.

(c) $\dfrac{-4x}{4x^2} = \dfrac{-4}{4}(x^{1-2}) = \mathbf{-1\,x^{-1}}$ or $-\dfrac{\mathbf{1}}{\mathbf{x}}$

Divide coefficients. Subtract exponents. Write factors with negative exponents as equivalent positive exponents.

This can also be written as $\dfrac{1}{-x}$ or $\dfrac{-1}{x}$.

(d) $\dfrac{-5x^4}{15x^2} = \dfrac{-5}{15}(x^{4-2}) = \mathbf{-\dfrac{1}{3}x^2}$ or $-\dfrac{\mathbf{x^2}}{\mathbf{3}}$

Reduce coefficients. Subtract exponents.

$-\dfrac{1}{3}x^2$ is the same as $-\dfrac{1}{3}\left(\dfrac{x^2}{1}\right)$ or $-\dfrac{x^2}{3}$.

(e) $\dfrac{3x}{12x^3} = \dfrac{3}{12}(x^{1-3}) = \dfrac{1}{4}x^{-2} = \dfrac{\mathbf{1}}{\mathbf{4\,x^2}}$

Reduce coefficients. Subtract exponents. Make exponent positive.

Since $\dfrac{1}{4}x^{-2} = \dfrac{1}{4}\left(\dfrac{1}{x^2}\right)$, this can be written as $\dfrac{1}{4x^2}$.

When a polynomial is divided by a monomial, *each* term in the dividend (numerator) is divided by the divisor (denominator).

EXAMPLE   Perform the divisions.

(a) $\dfrac{18a^4 + 15a^3 - 9a^2 - 12a}{3a}$ (b) $\dfrac{3x^3 - x^2}{x^3}$ (c) $\dfrac{6x^3 + 2x^2}{2x^2}$

(a) $\dfrac{18a^4 + 15a^3 - 9a^2 - 12a}{3a} = \dfrac{18a^4}{3a} + \dfrac{15a^3}{3a} - \dfrac{9a^2}{3a} - \dfrac{12a}{3a}$ Write as separate terms.

$= \mathbf{6a^3 + 5a^2 - 3a - 4}$ Simplify each term.

(b) $\dfrac{3x^3 - x^2}{x^3} = \dfrac{3x^3}{x^3} - \dfrac{x^2}{x^3} = 3 - x^{-1}$ **or** $3 - \dfrac{1}{x}$

Write as separate terms and simplify each term.

(c) $\dfrac{6x^3 + 2x^2}{2x^2} = \dfrac{6x^3}{2x^2} + \dfrac{2x^2}{2x^2} = 3x + 1$

Write as separate terms and simplify each term.

TIP — Why Can't We Reduce Terms?

You reduce *factors*, but not *terms*. Look at part c of the previous example, $\dfrac{6x^3 + 2x^2}{2x^2}$. A common *mistake* is to reduce the terms.

$$\dfrac{6x^3 + 2x^2}{2x^2} = 6x^3 \quad \text{Incorrect!}$$

Why is this not correct? To check a division, multiply the quotient by the divisor (denominator). The result will be the dividend (numerator).

Does $6x^3(2x^2) = 6x^3 + 2x^2$? NO!

Expressions that have several operations can be simplified.

EXAMPLE Simplify and write all exponents as positive exponents.

(a) $\dfrac{5x^5}{10x^2} + 3x(2x^2)$ **(b)** $\dfrac{3ab^2 - a^2b + 4}{ab}$ **(c)** $3x(4xy^2)^2$

(a) $\dfrac{5x^5}{10x^2} + 3x(2x^2) =$ Simplify each term.

$\dfrac{1}{2}x^3 + 6x^3 =$ Combine terms. $\dfrac{1}{2} + 6 = \dfrac{1}{2} + \dfrac{12}{2} = \dfrac{13}{2}$

$\dfrac{13}{2}x^3$ or $\dfrac{13x^3}{2}$

(b) $\dfrac{3ab^2 - a^2b + 4}{ab} =$ Write or mentally visualize as separate terms.

$\dfrac{3ab^2}{ab} - \dfrac{a^2b}{ab} + \dfrac{4}{ab} =$ Simplify each term.

$3b - a + \dfrac{4}{ab}$

(c) $3x(4xy^2)^2 =$ Follow the order of operations. Raise to a power first.

$3x(16x^2y^4) =$ Multiply.

$48x^3y^4$

A polynomial can be divided by a polynomial by using a long-division procedure. If the remainder is 0, both the divisor and the quotient are factors of the dividend.

To divide a polynomial by a polynomial using long division:

1. Divide the first term of the dividend by the first term of the divisor. The partial quotient is placed above the first term of the dividend.
2. Multiply the partial quotient times the divisor and align the product under like terms of the dividend.
3. Subtract (change subtrahend to opposite and use addition rules).
4. Bring down the next term of the dividend and repeat Steps 1–3.
5. Repeat Steps 1–4 until all terms of the dividend have been brought down. The result of the last subtraction is the remainder.
6. Write the remainder as a fraction with the remainder as the numerator and the divisor as the denominator.

EXAMPLE Divide $2x^3 - 9x^2 + 7x - 12$ by $x - 4$.

$$
\begin{array}{r}
2x^2 - x + 3 \\
x - 4 \overline{\smash{)}2x^3 - 9x^2 + 7x - 12} \\
\underline{2x^3 - 8x^2} \\
-x^2 + 7x \\
\underline{-x^2 + 4x} \\
3x - 12 \\
\underline{3x - 12} \\
0
\end{array}
$$

Divide: $\dfrac{2x^3}{x} = 2x^2$

Multiply: $2x^2(x - 4) = 2x^3 - 8x^2$

Subtract: $2x^3 - 2x^3 = 0; -9x^2 - (-8x^2) = -x^2$

Divide: $\dfrac{-x^2}{x} = -x$. Multiply: $-x(x - 4) = -x^2 + 4x$

Subtract.

Divide: $\dfrac{3x}{x} = 3$. Multiply: $3(x - 4) = 3x - 12$

Subtract.

Remainder = 0.

The quotient is $2x^2 - x + 3$.

EXAMPLE Divide $x^3 + 1$ by $x - 1$.

$$
\begin{array}{r}
x^2 + x + 1 + \dfrac{2}{x - 1} \\
x - 1 \overline{\smash{)}x^3 + 0x^2 + 0x + 1} \\
\underline{x^3 - x^2} \\
x^2 + 0x \\
\underline{x^2 - x} \\
x + 1 \\
\underline{x - 1} \\
2
\end{array}
$$

Represent missing powers of x with terms having a coefficient of 0.

Subtract: $0x^2 - (-x^2) = x^2$.

Subtract: $0x - (-x) = x$.

Subtract: $1 - (-1) = 2$.

The quotient is $x^2 + x + 1 + \dfrac{2}{x - 1}$.

Chapter 11 / Powers and Polynomials

1 Simplify.

1. $3a^2 + 4a^2$
2. $5x^3 - 2x^3$
3. $b^2 + 3a^2 + 2b^2 - 5a^2$
4. $3a - 2b - a$
5. $x - 3x - 2x^2 - 3x^2$
6. $3a^2 - 2a^2 + 4a^2$
7. $x^2 + 3y - (2x^2 + 5y)$
8. $4m^2 - 2n^2 - (2m^2 - 3n^2)$
9. $7a + 3b + 8c + 2a - (b - 2c)$
10. $5x + 3y - (7x - 2z)$
11. $(4x^2 - 3) - (3x^2 + 2) - (x^2 - 1)$
12. $5x - (3x + 7) - (-x - 8)$

2 Multiply.

13. $7x(2x^2)$
14. $(-2m)(-m^2)$
15. $(-3m)(7m)$
16. $(-y^3)(2y^3)$
17. $4x^2(2x - 7)$
18. $-3ab(2a^2b - 5ab^2)$
19. $7x(5x^3 - 3x^2 - 7)$
20. $2xy(3x^2y - 5xy^2)$
21. $3x(x - 6)$
22. $4x(3x^2 - 7x + 8)$
23. $-4x(2x - 3)$
24. $2x^2(5 + 2x)$

Multiply.

25. $(x + 7)(x + 3)$
26. $(x + 8)(x + 5)$
27. $(2x - 1)(x + 2)$
28. $(3x + 7)(x - 5)$
29. $(x + 5)(x^2 + 2x - 1)$
30. $(x + 7)(x^2 - 3x + 2)$
31. $(x - 7)(x^2 - 5x + 2)$
32. $(x - 3)(x^2 - 8x + 1)$
33. $(2x - 5)(x^2 - x + 1)$
34. $(3x - 2)(x^2 + x - 3)$
35. $(2x + 7)(3x^2 - 5x + 2)$
36. $(5x - 3)(4x^2 - 2x - 3)$
37. $(x - 2)(x^2 + 2x + 4)$
38. $(2x - 3)(4x^2 + 6x + 9)$
39. $(x + 3)(x^2 - 3x + 9)$
40. $(3x + 2)(9x^2 - 6x + 4)$
41. $(x + 5)(x^2 - 5x + 25)$
42. $(11x - 3)(121x^2 + 33x + 9)$

Use the FOIL method to find the products. Practice combining the outer and inner products mentally.

43. $(a + 3)(a + 8)$
44. $(x - 4)(x + 5)$
45. $(y - 2)(y - 9)$
46. $(y - 7)(y - 3)$
47. $(2a + 3)(a + 4)$
48. $(3a - 5)(a + 1)$
49. $(3a - 2b)(a - 2b)$
50. $(5x - y)(x - 5y)$
51. $(3x - 4)(2x - 3)$
52. $(a - b)(2a - 5b)$
53. $(7 - m)(3 - 7m)$
54. $(5 - 2x)(8 - x)$
55. $(x + 7)(x + 4)$
56. $(y - 7)(y - 5)$
57. $(m + 3)(m - 7)$
58. $(3b - 2)(x + 6)$
59. $(4r - 5)(3r + 2)$
60. $(5 - x)(7 - 3x)$
61. $(4 - 2m)(1 - 3m)$
62. $(2 + 3x)(3 + 2x)$
63. $(x + 3)(2x - 5)$
64. $(5x - 7y)(4x + 3y)$
65. $(2a + 3b)(7a - b)$
66. $(5a + 2b)(6a - 5b)$
67. $(9x - 2y)(3x + 4y)$
68. $(5x - 8y)(4x - 3y)$
69. $(7m - 2n)(3m + 5n)$

3 Find the special products using patterns.

70. $(a + 3)(a - 3)$
71. $(2x + 3)(2x - 3)$
72. $(a - y)(a + y)$
73. $(4r + 5)(4r - 5)$
74. $(5x + 2)(5x - 2)$
75. $(7 + m)(7 - m)$
76. $(3y - 5)(3y + 5)$
77. $(8y + 3)(8y - 3)$
78. $(3a - 11b)(3a + 11b)$
79. $(5y - 3)(5y + 3)$
80. $(x - 7)(x + 7)$
81. $(x - 11)(x + 11)$
82. $(2 - 3x)^2$
83. $(3x + 4)^2$
84. $(Q + L)^2$
85. $(a^2 + 1)^2$
86. $(2d - 5)^2$
87. $(3a + 2x)^2$
88. $(3x - 7)^2$
89. $(6 + Q)^2$
90. $(y + 5x)^2$
91. $(4 - 3j)^2$
92. $(3m - 2p)^2$
93. $(m^2 + p^2)^2$
94. $(2a - 7c)^2$
95. $(9 - 13a)^2$
96. $(2x - 5y)^2$

Use the extension of the FOIL method or the special products patterns to find the products.

97. $(x + p)(x^2 - xp + p^2)$
98. $(Q + L)(Q^2 - QL + L^2)$
99. $(3 + a)(9 - 3a + a^2)$
100. $(2x + 5p)(4x^2 - 10xp + 25p^2)$
101. $(3m + 2)(9m^2 - 6m + 4)$
102. $(6 - p)(36 + 6p + p^2)$

103. $(5y - p)(25y^2 + 5yp + p^2)$ **104.** $(x - 2y)(x^2 + 2xy + 4y^2)$ **105.** $(Q - 6)(Q^2 + 6Q + 36)$

106. $(3t - 2)(9t^2 + 6t + 4)$ **107.** $(2x + 3y)(4x^2 - 6xy + 9y^2)$ **108.** $(3x - 4y)(9x^2 + 12xy + 16y^2)$

4 Divide.

109. $\dfrac{6x^4}{3x^2}$ **110.** $\dfrac{-5a^2}{10a}$ **111.** $\dfrac{-7x}{-14x^3}$ **112.** $\dfrac{-9x^5}{12x^2}$

113. $\dfrac{6x^2 - 4x}{2x}$ **114.** $\dfrac{12x^5 - 6x^3 - 3x^2}{3x^2}$ **115.** $\dfrac{7x^4 - x^2}{x^3}$ **116.** $\dfrac{8x^4 + 6x^3}{2x^2}$

117. $\dfrac{6x^4 + 8x^2}{18x^2}$ **118.** $\dfrac{5a^2b^3 - 3ab^2 - 7}{ab}$ **119.** $\dfrac{15a^2b^3 - 3ab^2}{6ab}$ **120.** $\dfrac{18x^2 - 12y^2 - 6xy}{6xy}$

121. $\dfrac{15x^2}{3x} - 2x(7x^3)$ **122.** $\dfrac{6x(2x^3y^2)^3 + 8x}{2x}$ **123.** $\dfrac{-3x - (5x + 8x^3)}{2x}$ **124.** $\dfrac{3x^2 - 5y^2}{15xy} - \dfrac{8x^3}{4x}$

Perform the indicated division and determine if the binomial is a factor of the dividend.

125. $x - 5 \,\overline{)\,x^2 - x - 20}$ **126.** $x - 3 \,\overline{)\,x^2 + 3x - 18}$

127. $x + 1 \,\overline{)\,x^3 - 2x^2 - x + 2}$ **128.** $x + 2 \,\overline{)\,x^3 + 3x^2 - 3x - 10}$

129. $x + 3 \,\overline{)\,3x^3 + 7x^2 - 3x + 8}$ **130.** $x - 5 \,\overline{)\,2x^3 - 3x^2 - 33x + 8}$

131. $x + 1 \,\overline{)\,x^3 + 2x^2 - 5x - 6}$ **132.** $x + 3 \,\overline{)\,x^3 + 2x^2 - 5x - 6}$

133. $x - 3 \,\overline{)\,x^3 - 27}$ **134.** $x + 5 \,\overline{)\,x^3 + 125}$

CHAPTER REVIEW OF KEY CONCEPTS

Learning Outcomes	What to Remember with Examples

Section 11–1

1 Multiply powers with like bases (pp. 454–455).

To multiply powers with like bases, add the exponents and keep the common base as the base of the product. $a^m(a^n) = a^{m+n}$, where a, m, and n are real numbers and $a \neq 0$.

> Multiply. $x^5(x^{-7})$
>
> $$x^{5+(-7)} = x^{-2} = \frac{1}{x^2}$$

2 Divide powers with like bases (pp. 455–458).

To divide powers with like bases, subtract the exponents and keep the common base as the base of the quotient. $\dfrac{a^m}{a^n} = a^{m-n}$, where a, m, and n are real numbers except that $a \neq 0$.

> Divide. $\dfrac{a^7}{a^4}$
>
> $$a^{7-4} = a^3$$

3 Find a power of a power (pp. 458–460).

To find a power of a power, multiply the exponents and keep the same base. $(a^m)^n - a^{mn}$, where a, m, and n are real numbers and $a \neq 0$.

Simplify. $(y^5)^4$

$$y^{5(4)} = y^{20}$$

To raise a fraction or quotient to a power, raise both the numerator and the denominator to the power. $\left(\dfrac{a}{b}\right)^n = \dfrac{a^n}{b^n}$ and $b \neq 0$.

$$\left(\frac{-2}{x^2}\right)^3 = \frac{(-2)^3}{(x^2)^3} = \frac{-8}{x^6}$$

To raise a product to a power, raise each factor to the power. $(ab)^n = a^n b^n$.

$$(5x^2 y)^2 = 5^2 x^4 y^2 = 25x^4 y^2$$

Section 11–2

1 Identify polynomials, monomials, binomials, and trinomials (pp. 461–462).

Polynomials are algebraic expressions in which the exponent of the variable is a nonnegative integer and there are no variables in a denominator. A monomial is a polynomial with a single term. A binomial is a polynomial containing two terms. A trinomial is a polynomial containing three terms.

Give an example of a polynomial, a monomial, a binomial, and a trinomial.

Polynomial: $4x^3 + 6x^2 - x + 3$ Monomial: $8x$

Binomial: $6x + 3$ Trinomial: $8x^2 - 4x - 3$

2 Identify the degree of terms and polynomials (pp. 462–463).

The degree of a term containing one variable is the exponent of the variable. The degree of a term of more than one variable is the sum of the exponents of all the variable factors. The degree of a constant term is zero. The degree of a polynomial that has only one variable is the degree of the term that has the largest exponent.

Identify the degree of each term and the degree of the polynomial.

$$4x^3 \quad + \quad 6x^2 \quad - \quad x \quad + \quad 3 \qquad \text{The polynomial has a degree of 3.}$$

degree 3   degree 2   degree 1   degree 0

3 Arrange polynomials in descending order (pp. 463–464).

To arrange polynomials in descending order, list the term that has the highest degree first, the term that has the next highest degree second, and so on, until all terms have been listed.

Arrange the polynomial in descending order.

$$4 - 2x + 7x^5 - 3x^2 + x^3$$

Descending order:

$$7x^5 + x^3 - 3x^2 - 2x + 4$$

Section 11–3

1 Add and subtract polynomials (pp. 465–467).

Like terms are terms that not only have the same variable factors, but also have the same exponent. To add or subtract like terms, add or subtract the coefficients and keep the variable factors and their exponents exactly the same.

Simplify. $3x^4 + 8x^2 - 7x^2 + 2x^4 = 5x^4 + x^2$

Chapter Review of Key Concepts

2 Multiply polynomials (pp. 467–469).

To multiply expressions containing powers, multiply the coefficients; then add the exponents of like bases. Write with positive exponents.

> Simplify. $(3x^4y^5)(7x^2yz) = 21x^6y^6z$

To multiply two polynomials distribute by multiplying each term of the first factor times each term of the second factor. Combine like terms.

> Multiply $(x^2 + 3x + 1)(x^2 - 3x + 2)$.
> $$x^2(x^2 - 3x + 2) + 3x(x^2 - 3x + 2) + 1(x^2 - 3x + 2) =$$
> $$x^4 - 3x^3 + 2x^2 + 3x^3 - 9x^2 + 6x + x^2 - 3x + 2 =$$
> $$x^4 - 6x^2 + 3x + 2$$

3 Use the FOIL method to multiply two binomials (p. 469).

To multiply two binomials by the FOIL method: Multiply <u>F</u>irst terms, <u>O</u>uter terms, <u>I</u>nner terms, and <u>L</u>ast terms. Combine like terms.

> Multiply $(3a - 2)(a + 3)$.
>
>   F   O   I   L
> $$3a^2 + 9a - 2a - 6 \quad \text{Combine like terms.}$$
> $$3a^2 + 7a - 6$$

4 Multiply polynomials that result in special products (pp. 470–473).

To multiply the sum and difference of the same two terms: **1.** Square the first term. **2.** Insert minus sign. **3.** Square the second term.

> Multiply $(3b - 2)(3b + 2)$.
> $$9b^2 - 4 \qquad (3b)^2 = 9b^2; (2)^2 = 4$$

To square a binomial: **1.** Square the first term. **2.** Double the product of the two terms. **3.** Square the second term.

> Multiply $(2a - 3)^2$.
> $$4a^2 - 12a + 9 \qquad (2a)^2 = 4a^2; 2(2a)(-3) = -12a; 3^2 = 9$$

To multiply a binomial and a trinomial of the types $(a + b)(a^2 - ab + b^2)$ and $(a - b)(a^2 + ab + b^2)$:
1. Cube the first term of the binomial. **2.** If the binomial is a sum, insert a plus sign; if the binomial is a difference, insert a minus sign. **3.** Cube the second term of the binomial.

> Multiply $(8b - 3)(64b^2 + 24b + 9)$.
> $$512b^3 - 27 \qquad (8b)^3 = 512b^3; 3^3 = 27$$
> Multiply $(2x + 5)(4x^2 - 10x + 25)$.
> $$8x^3 + 125 \qquad (2x)^3 = 8x^3; 5^3 = 125$$

5 Divide polynomials (pp. 474–476).

To divide expressions containing powers, divide (or reduce) the coefficients; then subtract the exponents of the like bases. Write with positive exponents. To divide a polynomial by a monomial, devide each term of the polynomial by the monomial.

> Simplify. $\dfrac{10x^7y^4}{5x^8y^2} = 2x^{-1}y^2 = \dfrac{2y^2}{x}$ $\dfrac{3x^2 = 5x}{x} = \dfrac{3x^2}{x} + \dfrac{5x}{x} = 3x + 5$

     Chapter 11 / Powers and Polynomials

To divide a polynomial by a polynomial using long division:

1. Divide the first term of the dividend by the first term of the divisor. The partial quotient is placed above the first term of the dividend. **2.** Multiply the partial quotient times the divisor and align the product under like terms of the dividend. **3.** Subtract (change subtrahend to opposite and use addition rules). **4.** Bring down the next term of the dividend and repeat Steps 1–3. **5.** Repeat Steps 1–4 until all terms of the dividend have been brought down. The result of the last subtraction is the remainder. **6.** Write the remainder as a fraction with the remainder as the numerator and the divisor as the denominator.

Divide $(3x^2 + 13x - 10) \div (x + 5)$

$$
\begin{array}{r}
3x - 2 \\
x + 5 \overline{\smash{\big)}\, 3x^2 + 13x - 10} \\
\underline{3x^2 + 15x} \\
-2x - 10 \\
\underline{-2x - 10} \\
0
\end{array}
$$

CHAPTER REVIEW EXERCISES

Section 11–1

Perform the indicated operations. Write the answers with positive exponents.

1. $x^5 \cdot x^5$

2. $x^2(x^4)$

3. $3x^4 \cdot 7x^5$

4. $5x^3 \cdot 8x$

5. $\dfrac{x^8}{x^5}$

6. $\dfrac{x^3}{x^5}$

7. $\dfrac{21x^4}{3x}$

8. $\dfrac{24y^7}{18y^{10}}$

9. $\dfrac{x^3 y^{-1}}{x^2 y^2}$

10. $\dfrac{x^3 y^2}{x^2 y}$

11. $(x^3)^4$

12. $(-x^3)^3$

13. $(x^{-3})^{-5}$

14. $(x^2 y^5)^2$

15. $(-3x^2)^3$

16. $\dfrac{a^2 b^7}{ab^2}$

17. $\dfrac{xy^3}{xy^5}$

18. $(-x^4)^3$

19. $5x^{-2} \cdot 8x^4$

20. $\dfrac{28x^3 y^{-3}}{7xy^{-4}}$

Section 11–2

Identify the degree of each polynomial.

21. $7m^2 - 8m + 12m^4$

22. $5a^4 - 7a^3 + 12a^2 - 38$

23. $2x^3 y^2 - 15xy^3 + 21y^4$

24. Is the expression $5x^3 - 3x^{-2}$ a polynomial? Why or why not?

Arrange the following polynomials in descending order and identify the degree of each polynomial, the leading term, and the leading coefficient.

25. $5x + 3x^3 - 8 + x^2$

26. $3y^5 - 7y - 8y^4 + 12$

27. $5x^2 - 12x + 2x^4 - 32$

Section 11–3

Simplify the following:

28. $4x^3 + 7x - 3x^3 - 5x$

29. $8x - 2x^4 - (3x^3 + 5x - x^3)$

30. $5x - 3x + (7x^2 - 8x)$

31. $4x^2 - (3y^2 + 7x^2 - 8y^2)$

32. $4x^3(-3x^4)$

33. $-7x^8(-3x^{-2})$

34. $5a(a^2 - 7)$

35. $2x(x^2 + 3x - 5)$

36. $-2y(3y^2 - 7y - 12)$

37. $-2x(x^3 - 7x^2 + 15)$

38. $(y - 7)(y - 5)$

39. $(m + 3)(m - 7)$

40. $(3b - 2)(x + 6)$

41. $(4r - 5)(3r + 2)$

42. $(5 - x)(7 - 3x)$

43. $(4 - 2m)(1 - 3m)$

44. $(2 + 3x)(3 + 2x)$

45. $(x + 3)(2x - 5)$

46. $(5x - 7y)(4x + 3y)$

47. $(2a + 3b)(7a - b)$

48. $(5a + 2b)(6a - 5b)$

49. $(9x - 2y)(3x + 4y)$

50. $(5x - 8y)(4x - 3y)$

51. $(7m - 2n)(3m + 5n)$

52. $(x + 3)(2x^2 + 3x + 2)$

53. $(5x + 3)(x^2 - 3x + 1)$

Find the special products mentally.

54. $(y - 4)(y + 4)$

55. $(6x - 5)(6x + 5)$

56. $(3m + 4)(3m - 4)$

57. $(7y + 11)(7y - 11)$

58. $(5x - 2y)(5x + 2y)$

59. $(8a - 5b)(8a + 5b)$

60. $(12r - 7s)(12r + 7s)$

61. $(\sqrt{8} + 2)(\sqrt{8} - 2)$

62. $(\sqrt{5} + \sqrt{2})(\sqrt{5} - \sqrt{2})$

63. $(3 + i)(3 - i)$

64. $(7 - i)(7 + i)$

65. $(x + 9)^2$

66. $(x - 7)^2$

67. $(x - 3)^2$

68. $(2x - 3)^2$

69. $(4x - 15)^2$

70. $(5 + 3m)^2$

71. $(8 + 7m)^2$

72. $(5x - 13)^2$

73. $(4x - 11)^2$

74. $(K + L)(K^2 - KL + L^2)$

75. $(g - h)(g^2 + gh + h^2)$

76. $(4 + a)(16 - 4a + a^2)$

77. $(2H - 3T)(4H^2 + 6HT + 9T^2)$

78. $(3a - 5)(9a^2 + 15a + 25)$

79. $(6 + i)(36 - 6i + i^2)$

80. $(9y - p)(81y^2 + 9yp + p^2)$

81. $(z + 2t)(z^2 - 2zt + 4t^2)$

82. $(g - 2)(g^2 + 2g + 4)$

83. $(7T + 2)(49T^2 - 14T + 4)$

Find the quotients.

84. $\dfrac{12x^5}{6x^3}$

85. $\dfrac{12x^7}{-18x^4}$

86. $\dfrac{11x^4}{22x^7}$

87. $\dfrac{42x^3y}{-15x^3y^3}$

88. $\dfrac{-8x^3y^5}{20x^4y}$

89. $\dfrac{6x^3 - 12x^2 + 21x}{3x}$

90. $\dfrac{25y^5 - 85y^3 + 70y^2}{-5y}$

91. $\dfrac{4x^5}{8x^2} - 3x^2(2x^4)$

92. $2x(3x^2y)^3 + \dfrac{7x^3}{21x^2}$

93. $\dfrac{16x^3 + 12x^4 - 20x^5}{-4x^2}$

94. $\dfrac{9x^2y^5 - 15xy^3}{3xy^3}$

95. $x - 9 \overline{\smash{\big)}\, x^2 - 11x + 18}$

96. $2x - 3 \overline{\smash{\big)}\, 6x^2 - 13x + 21}$

97. $5x + 2 \overline{\smash{\big)}\, 15x^2 - 4x - 4}$

98. $x + 2 \overline{\smash{\big)}\, x^3 - x^2 - 4x + 4}$

99. $x - 5 \overline{\smash{\big)}\, x^3 - 4x^2 - 10x + 25}$

100. $2x - 1 \overline{\smash{\big)}\, 6x^3 - 7x^2 + 4x - 3}$

101. $2x^2 - x - 1 \overline{\smash{\big)}\, 6x^3 - x^2 - 4x - 1}$

TEAM PROBLEM-SOLVING EXERCISES

1. The symbolic representation of the laws of exponents illustrates many properties and restrictions. Explain these properties and restrictions in words.

   (a) $a^m \cdot a^n = a^{m+n}$

   (b) $\dfrac{a^m}{a^n} = a^{m-n}, a \neq 0$

   (c) $(a^m)^n = a^{mn}$

   (d) $\left(\dfrac{a}{b}\right)^n = \dfrac{a^n}{b^n}, b \neq 0$

   (e) $(ab)^n = a^n b^n$

2. Give an example to illustrate each of the laws of exponents.

   (a) $a^m \cdot a^n = a^{m+n}$

   (b) $\dfrac{a^m}{a^n} = a^{m-n}, a \neq 0$

   (c) $(a^m)^n = a^{mn}$

   (d) $\left(\dfrac{a}{b}\right)^n = \dfrac{a^n}{b^n}, b \neq 0$

   (e) $(ab)^n = a^n b^n$

Find the mistake in the examples. Explain the mistake and correct it.

1. $\dfrac{9x^3 - 12x^2 + 3x}{3x} = 3x^2 - 4x$

2. $x^5(x^3) = x^{15}$

3. $4x^3 + 3x^3 = 7x^6$

4. Explain how the FOIL process for multiplying two binomials is an application of the distributive property.

5. **(a)** How is the distributive property used to multiply a binomial and a trinomial?
(b) How is the distributive property used to multiply two trinomials?

6. Give an example of the product of a binomial and a trinomial to illustrate your answer to Question 5a.

7. Give an example of the product of two trinomials to illustrate your answer to Question 5b.

8. List the properties of the two binomials that result in the product of the difference of two perfect squares.

9. List the properties of a perfect-square trinomial that result from squaring a binomial.

10. Explain the difference in the processes used to divide a polynomial by a monomial and to divide a polynomial by a binomial.

Perform the indicated operations. Write the answers with positive exponents.

1. $(x^4)(x)$

2. $\dfrac{x^0}{x^2}$

3. $\left(\dfrac{4}{7}\right)^2$

4. $\dfrac{24x^2y^{-1}}{16xy^3}$

5. $\dfrac{x^{-7}}{x^3}$

6. $(6a^2b)^2$

7. $\left(\dfrac{x^2}{y}\right)^2$

8. $\dfrac{12x^2}{4x^3}$

9. $4a(3a^2 - 2a + 5)$

10. $\dfrac{60x^3 - 45x^2 - 5x}{5x}$

11. $5x^2 - 3x - (2x + 4x^2)$

12. $4x^3 + 2x - 7 + (3x^3 - 7x - 5)$

13. $-7x^3(-8x^4)$

14. $-(2x - 8) + 12$

15. $4xy - (3x - 2) + 4$

What is the degree of the polynomial?

16. $5x^3 - 4x^3 + x^2$

17. $14x - 3x + 21x$

Arrange each polynomial in decreasing powers of x.

18. $4 - 12x + 15x^3$

19. $6 - 3x^4 - 2x^3$

20. $6x^4 - 2x^5 - 5$

Find the products.

21. $(m - 7)(m + 7)$

22. $(3x - 2)(3x + 2)$

23. $(a + 3)^2$

24. $(2x - 7)^2$

25. $(x - 3)(2x - 5)$

26. $(7x - 3)(2x + 1)$

27. $(x - 2)(x^2 + 2x + 4)$

28. $(3y + 4)(9y^2 - 12y + 16)$

29. $(5a - 3)(25a^2 + 15a + 9)$

30. $(a + 6)(a^2 - 6a + 36)$

Divide using long division.

31. $x - 3 \overline{\smash{\big)}\,2x^2 + x - 21}$

32. $x + 3 \overline{\smash{\big)}\,2x^3 - 3x^2 + 7x - 6}$

12

Roots and Radicals

Focus on Careers

Want a high-paying job? Get an education! An associate or bachelor's degree is the most significant source of education for 10 of the 20 fastest-growing occupations.

People holding only high school diplomas or less have the lowest expected lifetime earnings. As the successive level of education increases, so do the lifetime earnings. Persons with an associate degree are expected to earn 1.6 times as much as a person with some high school but no diploma. And a person with a bachelor's degree can expect to earn 2.1 times as many dollars as a person with some high school but no diploma.

For a number of years, employment in the United States has been shifting to service-providing jobs and that is expected to continue through 2012. Jobs in education and health services are expected to grow faster and add more jobs than any other employment sector. Approximately 25% (over 5 million) of new jobs created in the U.S. economy are expected to be in the health-care, social assistance, or private educational services sector.

Professional and business services are expected to add nearly 5 million new jobs by 2012. In this sector, the fastest-growing industry is expected to be employment services. Management, scientific, and technical consulting services are expected to grow very rapidly because of increased use of new technology and computer software.

Sources: *Occupational Outlook Handbook,* 2004–2005 Edition. U.S. Department of Labor, Bureau of Labor Statistics. "The Big Payoff: Educational Attainment and Synthetic Estimates of Work-Life Earnings," *Current Population Reports,* pp. 23–210, by Jennifer Cheeseman Day and Eric C. Newburger.

Learning Outcomes

1 Write roots using radical and exponential notation.

2 Approximate an irrational number.

3 Write powers and roots using rational exponents and radical notation.

In Chapter 1 we saw that the opposite or inverse operation for raising to powers was finding roots. The *square root* of a number is the number that is used as a factor 2 times to equal the square. The *radical notation* for square root is $\sqrt{}$, $\sqrt{25} = 5$. The *index* of a square root is 2, but is not required in radical notation.

A *cube root* is the number that is used as a factor 3 times to give the cube. The index of a cube root is 3. For cube roots, the index 3 is written in the $\sqrt{}$ portion of the radical sign: $\sqrt[3]{8} = 2$. For *fourth roots*, the index 4 is written in the $\sqrt{}$ portion of the radical sign: $\sqrt[4]{81} = 3$.

The results when natural numbers are squared, cubed, raised to the fourth power, and so on are called *perfect powers*. The principal (or positive) root of a perfect power is a natural number.

Powers of 2	Powers of 3	Powers of 4	Powers of 5
$2^1 = 2$	$3^1 = 3$	$4^1 = 4$	$5^1 = 5$
$2^2 = 4$	$3^2 = 9$	$4^2 = 16$	$5^2 = 25$
$2^3 = 8$	$3^3 = 27$	$4^3 = 64$	$5^3 = 125$
$2^4 = 16$	$3^4 = 81$	$4^4 = 256$	$5^4 = 625$
$2^5 = 32$	$3^5 = 243$	$4^5 = 1,024$	$5^5 = 3,125$
$2^6 = 64$	$3^6 = 729$	$4^6 = 4,096$	$5^6 = 15,625$

Selected roots

$\sqrt[4]{16} = 2$	$\sqrt[5]{243} = 3$	$\sqrt[3]{64} = 4$	$\sqrt[6]{15,625} = 5$

1 **Write Roots Using Radical and Exponential Notation.**

A rational or fractional exponent is another notation used to indicate a root. To indicate a square root, use the exponent $\frac{1}{2}$. To indicate a cube root, use the exponent $\frac{1}{3}$. For a fourth root, use the exponent $\frac{1}{4}$. To generalize, the exponential notation for the root of a number is the number written with the fractional exponent. The index of the root is the denominator of the fractional exponent and 1 is the numerator.

$$\sqrt[n]{a} = a^{\frac{1}{n}}$$ where *n* is a natural number greater than 1 and *a* is a positive number unless otherwise indicated.

To write roots using radical and exponential notation:

1. Identify the index of the root.
2. For radical notation, place the index of the root in the $\sqrt{}$ portion of the radical sign with the radicand under the bar portion of the sign. The index 2 for square roots does not have to be written.
3. For exponential notation, write the radicand as the base and the index of the root as the denominator of a fractional exponent that has a numerator of 1. The decimal equivalent of the fractional exponent can also be used.

TIP | Equivalent Notations for Roots

Just as $\frac{1}{2}$, 0.5, and 50% are three notations for writing equivalent values, radical notation and exponential notation are notations for writing equivalent values for roots.

In Words	Radical Notation	Exponential Notation
Square root of n	$\sqrt{n}$	$n^{1/2}$ or $n^{\frac{1}{2}}$ or $n^{0.5}$
Cube root of n	$\sqrt[3]{n}$	$n^{1/3}$ or $n^{\frac{1}{3}}$ or $n^{0.\overline{3}}$
Fourth root of n	$\sqrt[4]{n}$	$n^{1/4}$ or $n^{\frac{1}{4}}$ or $n^{0.25}$

EXAMPLE Write the roots using radical notation and exponential notation.

(a) Square root of 25   (b) Cube root of 125   (c) Fourth root of 625

(a) Square root of 25 $= \sqrt{25} = 25^{1/2}$ or $25^{0.5}$

(b) Cube root of 125 $= \sqrt[3]{125} = 125^{1/3}$ or $125^{0.\overline{3}}$

(c) Fourth root of 625 $= \sqrt[4]{625} = 625^{1/4}$ or $625^{0.25}$

2 | Approximate an Irrational Number.

Not all numbers are perfect powers. The root of a nonperfect power is an **irrational number.**

To find the two whole numbers that are closest to the value of an irrational number expressed as a radical:

1. Make a list of perfect powers that go beyond the given radicand.
2. Identify the two perfect powers that the radicand is between.
3. Find the principal roots of the two perfect powers from Step 2.
4. The root of the radicand is between the roots found in Step 3.

EXAMPLE Find two whole numbers that are closest to (a) $\sqrt{75}$ and (b) $\sqrt[3]{120}$.

(a) 1, 4, 9, 16, 25, 36, 49, 64, 81   List perfect squares. 75 is between 64 and 81.

$\sqrt{64} = 8$ $\sqrt{81} = 9$ Square roots of 64 and 81

$\sqrt{75}$ **is between 8 and 9.**

(b) 1, 8, 27, 64, 125   List perfect cubes. 120 is between 64 and 125.

$\sqrt[3]{64} = 4$ $\sqrt[3]{125} = 5$ Cube roots of 64 and 125

$\sqrt[3]{120}$ **is between 4 and 5.**

Real numbers include both rational and irrational numbers (Fig. 12–1).

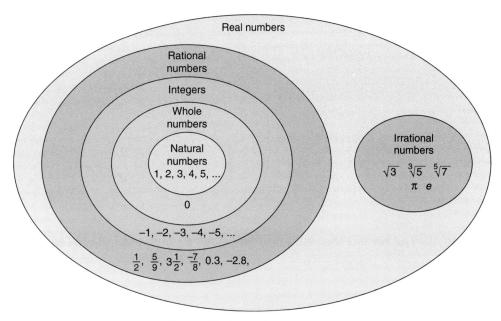

Figure 12–1 Real number system.

Irrational numbers are ordered and have a position on the number line similar to the ordering of rational numbers.

Because of the availability of scientific and graphing calculators, radical notation, which once was the most popular notation for roots, is now being replaced by the rational exponent notation for roots.

Various Options for Roots on Your Calculator

Scientific and graphing calculators use a variety of symbols and processes to find roots.

Special root keys:

Some calculators have special function keys or menu options for the most common roots: square roots $\boxed{\sqrt{}}$ and cube roots $\boxed{\sqrt[3]{}}$. To use these functions, select the special root option, enter the base or radicand, and then press the $\boxed{=}$ or $\boxed{\text{ENTER}}$ key.

To find roots in general, other calculator functions are necessary.

General root key:

Use the general root key, $\boxed{x^{1/y}}$, on your calculator, to find the fourth root of 16. Enter the base or radicand, press the general root key, enter only the index of the root (denominator of exponent), and then press $\boxed{=}$. Some calculators have a general root key in radical notation, $\boxed{\sqrt[y]{x}}$. Some typical sequences of keys are

16 $\boxed{x^{1/y}}$ 4 $\boxed{=}$ $\Rightarrow$ 2

16 $\boxed{\sqrt[y]{x}}$ 4 $\boxed{=}$ $\Rightarrow$ 2   Enter index second.

4 $\boxed{\sqrt[y]{x}}$ 16 $\boxed{=}$ $\Rightarrow$ 2   Enter index first.

12–1 Irrational Numbers and Real Numbers   487

General power key:

Using the general power key, $\boxed{\wedge}$ or $\boxed{x^y}$, on a calculator, find the fourth root of 16.

Exponent as indicated division: $16 \boxed{\wedge} \boxed{(}\; 1 \boxed{\div} 4\boxed{)} \;\boxed{=} \Rightarrow 2$

$$16^{1/4} = 2$$

Exponent as fraction: $16 \boxed{\wedge} 1 \boxed{a\frac{b}{c}} 4 \boxed{=} \Rightarrow 2$

Some calculators with a fraction key require the fraction be placed in parentheses, others do not.

$$16^{1/4} = 2$$

Exponent as decimal equivalent: $16 \boxed{\wedge} \boxed{\cdot} 25 \boxed{=} \Rightarrow 2$ $\dfrac{1}{4} = 0.25$

$$16^{0.25} = 2$$

Many calculators will automatically open a parenthesis when using a root function. You will close the parenthesis after the radicand has been entered or it will automatically close when $\boxed{=}$ or $\boxed{\text{ENTER}}$ is pressed.

Test various sequences on your calculator to determine what options you have for finding roots.

EXAMPLE   Use calculator values to find the approximate positions of the square roots on the number line (Fig. 12–2).

(a) $\sqrt{2}$ (b) $\sqrt{3}$ (c) $\sqrt{5}$ (d) $\sqrt{6}$

(a) $\sqrt{2} \approx 1.414$ **(b)** $\sqrt{3} \approx 1.732$

(c) $\sqrt{5} \approx 2.236$ **(d)** $\sqrt{6} \approx 2.449$

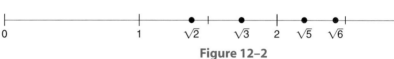

Figure 12–2

---

3 Write Powers and Roots Using Rational Exponents and Radical Notation.

Powers and roots are inverse operations for nonnegative values. In the order of operations, powers and roots (Exponents in "Please Excuse My Dear Aunt Sally") have the same priority. The order in which the operations are performed does not matter.

	Rational Exponent Notation	**Radical Notation**
For $x \geq 0$ and n a natural number greater than 1:	$(x^{1/n})^n = x$	$\left(\sqrt[n]{x}\right)^n = x$
	$(x^n)^{1/n} = x$	$\sqrt[n]{x^n} = x$

EXAMPLE   Write both the rational exponent and radical notations.

(a) The fourth root of the square of x.
(b) The square of the fourth root of x.
(c) The square root of the cube of x.
(d) The cube of the square root of x.

Rational Exponent Notation	Radical Notation
(a) $(x^2)^{1/4}$	$\sqrt[4]{x^2}$
(b) $(x^{1/4})^2$	$(\sqrt[4]{x})^2$
(c) $(x^3)^{1/2}$	$\sqrt{x^3}$
(d) $(x^{1/2})^3$	$(\sqrt{x})^3$

To write powers and roots using rational exponents and radical notation:

Rational exponent to radical notation:

1. Write a decimal exponent as an equivalent fractional exponent.
2. Write the numerator of a rational exponent as the power.
3. Write the denominator of the rational exponent as the index of the root.

$$\text{Symbolically, } x^{\text{power/root}} = \sqrt[\text{root}]{x^{\text{power}}} \quad \text{or} \quad \left(\sqrt[\text{root}]{x}\right)^{\text{power}}$$

Radical to rational exponent:

1. Write the power as the numerator of the rational exponent.
2. Write the index of the root as the denominator of the rational exponent.
3. Write the fraction in lowest terms or as a decimal equivalent.

$$\text{Symbolically, } \quad \sqrt[\text{root}]{x^{\text{power}}} = x^{\text{power/root}} \quad \text{or} \quad x^{\frac{\text{power}}{\text{root}}}$$

EXAMPLE Write $x^{2/3}$ in radical form.

$x^{2/3}$ Write the numerator of the rational exponent as the power.
    Write the denominator of the rational exponent as the index of the root.

$\sqrt[3]{x^2}$

EXAMPLE Write $\sqrt[4]{x^3}$ in rational exponent notation.

$\sqrt[4]{x^3}$ Write the power of the radicand as the numerator of the fractional exponent.
    Write the index as the denominator of the fractional exponent.

$x^{3/4}$ **or** $x^{0.75}$

Interpreting an exponent depends on the type of number used as the exponent. Let's summarize the different types of numbers we have used as exponents.

Different Types of Numbers Used as Exponents Indicate Different Interpretations

TIP

Exponent Type	Means
Natural number exponents greater than 1	Repeated multiplication

$$4^3 = 4 \times 4 \times 4 = 64$$

Exponent of 1 — Expression equals the base

$$7^1 = 7$$

Exponent of 0 when base is not equal to 0 — Expression equals 1

$$5^0 = 1$$

Negative integral exponents — Reciprocal of power

$$3^{-2} = \frac{1}{3^2} = \frac{1}{9}$$

Fractional or decimal exponents — Roots

$$9^{1/2} = 9^{0.5} = \sqrt{9} = 3$$

$$8^{1/3} = 8^{0.\overline{3}} = \sqrt[3]{8} = 2$$

SECTION 12–1 SELF-STUDY EXERCISES

1   Write the roots in radical notation and exponential notation.

1. Square root of 36   **2.** Cube root of 27   **3.** Fourth root of 81
4. Fifth root of 32   **5.** Square root of 81   **6.** Cube root of 1,331

2   Find the two whole numbers that are closest to each root.

7. $\sqrt{40}$ **8.** $\sqrt{68}$ **9.** $\sqrt{12}$ **10.** $\sqrt{120}$
11. $\sqrt[3]{12}$ **12.** $\sqrt[3]{36}$ **13.** $\sqrt[3]{50}$ **14.** $\sqrt[3]{65}$

Use calculator values to find the approximate position. Then position each root on a number line. Label the points.

15. $\sqrt{25}$ **16.** $\sqrt{17}$ **17.** $\sqrt{20}$ **18.** $\sqrt{10}$
19. $\sqrt{18}$ **20.** $\sqrt{32}$ **21.** $\sqrt{7}$ **22.** $\sqrt{40}$

3   Write each expression in rational exponent notation and radical notation.

23. Square root of x^3 **24.** Cube root of the square of x **25.** Fourth power of cube root of x
26. Square of cube root of x **27.** Cube of fourth root of x **28.** Fifth power of cube root of x

Write in radical notation.

29. $x^{3/5}$ **30.** $x^{2/5}$ **31.** $x^{3/4}$ **32.** $x^{5/8}$
33. $y^{1/5}$ **34.** $y^{7/8}$ **35.** $y^{3/8}$ **36.** $y^{4/5}$

Write in rational exponent notation in simplest form.

37. $\sqrt[3]{x^2}$ **38.** $\sqrt[4]{x^2}$ **39.** $\sqrt[5]{x^2}$ **40.** $\sqrt[5]{x^3}$
41. $\sqrt[6]{x^5}$ **42.** $\sqrt[6]{x^3}$ **43.** $\sqrt[6]{x^4}$ **44.** $\sqrt[7]{x^4}$

Learning Outcomes **1** Find the square root of variables.

2 Simplify square-root radicals using rational exponents and the laws of exponents.

3 Simplify square-root radical expressions containing perfect-square factors.

1 Find the Square Root of Variables.

The square root of a variable is best understood by using the rational exponent notation and the laws of exponents. We will continue to consider only real numbers, and we will assume that the variables represent positive values.

$$\sqrt{x^2} = (x^2)^{1/2} = x^1 = x$$
$$\sqrt{x^4} = (x^4)^{1/2} = x^2$$
$$\sqrt{x^6} = (x^6)^{1/2} = x^3$$

TIP

Perfect-Square Variable Factors

A positive variable with an even-number exponent is a perfect square. To find the square root of a variable factor, take $\frac{1}{2}$ of the exponent.

EXAMPLE   Find the square root of the positive variables.

(a) $\sqrt{x^8}$ (b) $-\sqrt{x^{12}}$ (c) $\pm\sqrt{x^{18}}$ (d) $\sqrt{\dfrac{a^2}{b^4}}$

(a) $\sqrt{x^8} = (x^8)^{1/2} = \boldsymbol{x^4}$ **(b)** $-\sqrt{x^{12}} = -(x^{12})^{1/2} = \boldsymbol{-x^6}$

(c) $\pm\sqrt{x^{18}} = \pm(x^{18})^{1/2} = \boldsymbol{\pm x^9}$ **(d)** $\sqrt{\dfrac{a^2}{b^4}} = \dfrac{(a^2)^{1/2}}{(b^4)^{1/2}} = \dfrac{\boldsymbol{a}}{\boldsymbol{b^2}}$

TIP

Principal and Negative Square Roots

There are two square roots for a positive radicand. It is customary to indicate whether the positive (principal) square root, the negative square root, or both square roots are to be considered by putting the appropriate sign in front of the radicand or its coefficient.

$$\sqrt{4} = 2 \qquad -\sqrt{4} = -2 \qquad \pm\sqrt{4} = \pm 2$$

2 Simplify Square-Root Radicals Using Rational Exponents and the Laws of Exponents.

Even though the calculator makes the pencil-and-paper method of taking roots outmoded, it is still helpful to perform some mental manipulations on expressions containing powers and roots before evaluating these expressions using the calculator. One common manipulation is to simplify rational exponents and radical expressions by applying the laws of exponents.

To simplify radicals using rational exponents and the laws of exponents:

1. Convert the radicals to equivalent expressions using rational exponents.
2. Apply the laws of exponents and the arithmetic of fractions.
3. Convert simplified expressions back to radical notation if desired.

EXAMPLE Convert the radical expressions to equivalent expressions using rational exponents and simplify if appropriate.

(a) $\sqrt[3]{x}$ (b) $\sqrt[5]{2y}$ (c) $\left(\sqrt{ab}\right)^3$ (d) $\sqrt[4]{16b^8}$ (e) $\left(\sqrt[3]{27xy^5}\right)^4$

(a) $\sqrt[3]{x} = x^{1/3}$ **(b)** $\sqrt[5]{2y} = (2y)^{1/5}$ or $2^{1/5}y^{1/5}$

(c) $\left(\sqrt{ab}\right)^3 = (ab)^{3/2}$ or $a^{3/2}b^{3/2}$

(d) $\sqrt[4]{16b^8} =$ 16 is a perfect fourth power. $2^4 = 16$

   $(2^4b^8)^{1/4} =$ Raise each factor to the $\dfrac{1}{4}$ power.

   $(2^4)^{1/4}(b^8)^{1/4} =$ $(2^4)^{1/4} = 2^1 = 2;\ (b^8)^{1/4} = b^2$

   $\mathbf{2b^2}$

(e) $\left(\sqrt[3]{27xy^5}\right)^4 =$ 27 is a perfect cube. $3^3 = 27$

   $(3^3xy^5)^{4/3} =$ Raise each factor to the $\dfrac{4}{3}$ power.

   $(3^3)^{4/3}x^{4/3}(y^5)^{4/3} =$ $(3^3)^{4/3} = 3^4 = 81;\ (y^5)^{4/3} = y^{20/3}$

   $\mathbf{81\,x^{4/3}y^{20/3}}$

Simplifying Coefficients

TIP

When coefficients of variable terms have rational exponents, the numerical equivalent can be determined if desired. Usually, if the coefficient is a perfect power of the indicated root, we will evaluate the coefficient. Otherwise, we may leave the coefficient with the rational exponent.

$$8^{2/3} = (8^{1/3})^2 = ((2^3)^{1/3})^2 = 2^2 = 4 \qquad 5^{2/3} \text{ does not simplify}$$

The laws of exponents are applied to rational exponents of factors having *like bases*. The following example illustrates the laws of exponents applied to rational exponents.

EXAMPLE Perform the following operations and simplify. Express answers with positive exponents in lowest terms.

(a) $(x^{3/2})(x^{1/2})$ (b) $(3a^{1/2}b^3)^2$ (c) $\dfrac{x^{1/2}}{x^{1/3}}$ (d) $\dfrac{10a^3}{2a^{1/2}}$

Chapter 12 / Roots and Radicals

(a) $(x^{3/2})(x^{1/2}) = x^{3/2+1/2} = x^{4/2} = \boldsymbol{x^2}$ Add exponents: $\frac{3}{2} + \frac{1}{2} = \frac{4}{2} - 2$

(b) $(3a^{1/2}b^3)^2 = 3^2ab^6 = \boldsymbol{9ab^6}$ Multiply exponents: $\frac{1}{2} \cdot 2 = 1; 3 \cdot 2 = 6$

(c) $\dfrac{x^{1/2}}{x^{1/3}} = x^{1/2-1/3} = \boldsymbol{x^{1/6}}$ Subtract exponents: $\frac{1}{2} - \frac{1}{3} = \frac{3}{6} - \frac{2}{6} = \frac{1}{6}$

(d) $\dfrac{10a^3}{2a^{1/2}} = 5a^{3-1/2} = \boldsymbol{5a^{5/2}}$ Reduce coefficients. Subtract exponents:
$3 - \frac{1}{2} = \frac{3}{1} - \frac{1}{2} = \frac{6}{2} - \frac{1}{2} = \frac{5}{2}$

EXAMPLE   Evaluate the expressions (a) and (d) from the preceding example for $x = 2$ and $a = 3$ both before and after simplifying.

(a) $(x^{3/2})(x^{1/2}) =$

Before simplifying: $(x^{3/2})(x^{1/2}) = (2^{3/2})(2^{1/2})$ Substitute $x = 2$.

$2\ \boxed{\wedge}\ \boxed{(}\ 3\ \boxed{\div}\ 2\ \boxed{)}\ \boxed{\times}$ Enter into the calculator using the general power key $\boxed{\wedge}$ or $\boxed{x^y}$.

$2\ \boxed{\wedge}\ \boxed{(}\ 1\ \boxed{\div}\ 2\ \boxed{)}\ \boxed{=}\ \Rightarrow \boldsymbol{4}$

After simplifying: $(x^{3/2})(x^{1/2}) = x^{3/2+1/2}$ Add exponents.

$\qquad\qquad\qquad = x^2$ Substitute $x = 2$.

$\qquad\qquad\quad 2^2 = \boldsymbol{4}$ Evaluate.

(d) $\dfrac{10a^3}{2a^{1/2}} =$

Before simplifying: $\dfrac{10a^3}{2a^{1/2}} = \dfrac{10 \cdot 3^3}{2 \cdot 3^{1/2}}$ Substitute $a = 3$.

$\boxed{(}\ 10\ \boxed{\times}\ 3\ \boxed{\wedge}\ 3\ \boxed{)}\ \boxed{\div}$ Both the numerator and denominators are groupings.
$\boxed{(}\ 2\ \boxed{\times}\ 3\ \boxed{\wedge}\ \boxed{(}\ 1\ \boxed{\div}\ 2\boxed{)}\ \boxed{)}\ \boxed{=}$
$\Rightarrow \boldsymbol{77.94228634}$

After simplifying: $\dfrac{10\,a^3}{2a^{1/2}} = \dfrac{10}{2} \cdot \dfrac{a^3}{a^{1/2}}$ Reduce coefficients. Subtract exponents.

$\qquad\qquad\qquad = 5a^{3-\frac{1}{2}}$ $3 - \dfrac{1}{2} = \dfrac{6}{2} - \dfrac{1}{2} = \dfrac{5}{2}$

$\qquad\qquad\qquad = 5a^{5/2} \text{ or } 5a^{2.5}$ Substitute $a = 3$.

$\qquad\qquad\qquad = 5 \cdot 3^{2.5}$ Evaluate.

$5\ \boxed{\times}\ 3\ \boxed{\wedge}\ 2.5\ \boxed{=}\ \Rightarrow$

$\qquad \boldsymbol{77.94228634}$

Is it really worth the effort to simplify an expression before evaluating? Yes, in most cases. Long sequences of calculator steps are very tedious to enter.

As we saw in part (d) of the preceding example, *decimal exponents* are often desirable in finding roots.

EXAMPLE Perform the operations. Express the answers with positive exponents.

(a) $a^{2.3}(a^3)$ (b) $(5a^{3.5}b^{0.5})^2$

(a) $a^{2.3}(a^3) = a^{2.3+3} = a^{5.3}$

What does $a^{5.3}$ mean? If written as an improper fraction, $a^{5.3} = a^{5\frac{3}{10}} = a^{53/10}$.

This means we take the tenth root of a to the 53rd power or $\sqrt[10]{a^{53}}$.

(b) $(5a^{3.5}b^{0.5})^2 = 5^2 a^{3.5(2)} b^{0.5(2)}$
$$= 25a^7 b$$

3 Simplify Square-Root Radical Expressions Containing Perfect-Square Factors.

Some radicands are perfect squares, whereas others are not perfect squares. In the latter case, it may be useful to simplify the radicand. This is usually done by first writing the radicand as appropriate factors.

To **factor an algebraic expression** is to write it as the indicated product of two or more factors, that is, as a multiplication.

Some radicands that are not perfect squares can be *factored* into a perfect square times other factors. *We simplify radicals by taking the square root of all the perfect-square factors.*

To simplify square-root radicals containing perfect-square factors:

1. If the radicand is a perfect square, express it as a square root without the radical sign.
2. If the radicand is *not* a perfect square, factor the radicand so that one factor is the largest possible perfect square factor. The square roots of the perfect-square factors appear *outside* the radical and the other factors stay *inside* (under) the radical sign.

If the radicand is *not* a perfect square and *cannot* be factored into one or more perfect-square factors, it is in simplest radical form.

EXAMPLE Simplify the radicals.

(a) $\sqrt{32}$ (b) $\sqrt{y^7}$ (c) $\sqrt{18x^5}$ (d) $\sqrt{75xy^3z^5}$ (e) $\sqrt{7x}$

When factoring coefficients, some perfect squares that can be used are 4, 9, 16, 25, 36, 49, 64, and 81.

(a) $\sqrt{32} = \sqrt{16 \cdot 2}$
<blockquote>What is the largest perfect-square factor of 32? 4 is a factor of 32, but 16 is also a factor of 32. Use the largest perfect-square factor.</blockquote>

$$= \sqrt{16}\,(\sqrt{2}) = 4\sqrt{2}$$

(b) $\sqrt{y^7} = \sqrt{y^6 \cdot y^1} = \sqrt{y^6}\,(\sqrt{y})$
<blockquote>The largest perfect-square factor of y^7 is y^{7-1}, or y^6. For variables, perfect square factors have even-number exponents.</blockquote>

$$= y^3\sqrt{y}$$

(c) $\sqrt{18x^5}$

Factor radicand into as many perfect-square factors as possible.

$$= \sqrt{9 \cdot 2 \cdot x^4 \cdot x^1}$$

Write each factor as a separate radical.

$$= \sqrt{9}(\sqrt{2})(\sqrt{x^4})(\sqrt{x})$$

Take the square root of the two perfect squares.

$$= 3(\sqrt{2})(x^2)\sqrt{x}$$

Multiply coefficients. Multiply radicands.

$$= 3x^2\sqrt{2x}$$

In the previous example we showed each step in the simplifying process. However, we customarily do most of these steps mentally.

(d) $\sqrt{75xy^3z^5} = \sqrt{25 \cdot 3 \cdot x \cdot y^2 \cdot y \cdot z^4 \cdot z}$

Write the square roots of the perfect-square factors outside the radical sign. The remaining factors are written under the radical sign.

$$= 5yz^2\sqrt{3xyz}$$

(e) $\sqrt{7x} = \sqrt{7x}$

7 and x contain no perfect-square factors. The radical is in simplest form.

TIP

Finding Perfect-Square Factors

- Natural-number perfect squares: 1, 4, 9, 16, 25, 36, 49, 64, 81, 100, 121, 144, ...
- 1 is a factor of any number: $8 = 8 \cdot 1$. To factor using the perfect square 1 does not simplify a radicand.
- Any variable with an exponent greater than 1 is a perfect square or has a perfect-square factor.
- Perfect-square variables have even numbers as exponents:

$$x^2, x^4, x^6, x^8, x^{10}, \ldots$$

- Variables with a perfect-square factor:

$$x^3 = x^2 \cdot x^1, \qquad x^5 = x^4 \cdot x^1$$
$$x^7 = x^6 \cdot x^1, \qquad x^9 = x^8 \cdot x^1$$

- A convenient way to keep track of perfect-square factors is to circle them. The square roots of circled factors are written outside the radical sign. The uncircled factors stay in the radicand as is.

$$\sqrt{75ab^4c^3} = \sqrt{(25) \cdot 3 \cdot a \cdot (b^4) \cdot (c^2) \cdot c^1}$$
$$= 5b^2c\sqrt{3ac}$$

EXAMPLE Simplify.

(a) $\sqrt{7x^2}$ (b) $\sqrt{9a}$ (c) $\sqrt{32m^5n^6}$

(a) $\sqrt{7(x^2)} = x\sqrt{7}$ **(b)** $\sqrt{(9)a} = 3\sqrt{a}$

(c) $\sqrt{32m^5n^6} = \sqrt{(16) \cdot 2 \cdot (m^4) \cdot m \cdot (n^6)} = 4m^2n^3\sqrt{2m}$

1 Find the square root of the positive variables.

1. $\sqrt{x^{10}}$

2. $\sqrt{x^6}$

3. $\pm\sqrt{x^{16}}$

4. $\sqrt{\dfrac{x^{14}}{y^{24}}}$

5. $\sqrt{a^6b^{10}}$

6. $-\sqrt{a^2b^4c^{12}}$

7. $\sqrt{\dfrac{x^2y^4}{z^{10}}}$

8. $\sqrt{\dfrac{a^4}{b^{10}c^{12}}}$

2 Convert the radical expressions to equivalent expressions using rational exponents, and simplify if appropriate.

9. $\sqrt[3]{y}$

10. $\sqrt[3]{5x}$

11. $\left(\sqrt{xy}\right)^5$

12. $\left(\sqrt{r}\right)^7$

13. $\sqrt[4]{81x^8}$

14. $\sqrt[4]{16r^{12}}$

15. $\left(\sqrt[3]{8x^3y^6}\right)^4$

16. $\left(\sqrt[4]{16xy^5}\right)^5$

17. $\sqrt[4]{x}$

18. $\sqrt[5]{7x}$

19. $\left(\sqrt[3]{xy}\right)^4$

20. $\sqrt[4]{81b^{20}}$

21. $\left(\sqrt[3]{8xy^7}\right)^2$

Perform the operations. Express answers with positive exponents in lowest terms.

22. $x^{5/2}\cdot x^{3/2}$

23. $y^{4/3}\cdot y^{5/3}$

24. $(4a^{1/3}b^2)^3$

25. $(2x^{2/3}y^{5/6})^3$

26. $\dfrac{x^{1/3}}{x^{2/3}}$

27. $\dfrac{x^{4/5}}{x^{1/5}}$

28. $\dfrac{y^{1/3}}{y^{1/2}}$

29. $\dfrac{b^{2/3}}{b^{1/2}}$

30. $\dfrac{12x^4}{3x^{1/2}}$

31. $\dfrac{20x^3}{4x^{2/3}}$

32. Evaluate $(x^{1/3})(x^{2/3})$ if $x=2$.

33. Evaluate $\dfrac{12x^2}{2x^{1/3}}$ if $x=8$.

Perform the operations.

34. $a^{1.2}(a^4)$

35. $(7a^{1.7}\cdot a^{2.3})^2$

36. $\dfrac{x^{3.7}}{x^{1.7}}$

37. $\dfrac{12x^{2.9}}{4x^{0.9}}$

38. $2x^{2.1}(3x^{3.9})$

3 Simplify.

39. $\sqrt{24}$

40. $\sqrt{98}$

41. $\sqrt{48}$

42. $\sqrt{x^9}$

43. $\sqrt{y^{15}}$

44. $\sqrt{12x^3}$

45. $\sqrt{56a^5}$

46. $\sqrt{72a^3x^4}$

47. $\sqrt{44x^5y^2z^7}$

48. Create a square-root radical with the product of a constant factor that is not a perfect square but has a perfect-square factor and a variable factor with an odd exponent greater than 1. Then simplify.

Learning Outcomes

1 Add or subtract square-root radicals.

2 Multiply or divide square-root radicals.

1 **Add or Subtract Square-Root Radicals.**

As you recall, when adding or subtracting measures or algebraic terms, we only add or subtract like quantities. Similarly, only *like* radicals are added or subtracted.

Like radicals are radical expressions with radicands that are identical and have the same order or index.

To add or subtract like square-root radicals:

1. Add or subtract the coefficients of like radicals.
2. Use the common radical as a factor in the solution.

Symbolically, $a\sqrt{b} + c\sqrt{b} = (a + c)\sqrt{b}, b > 0$

EXAMPLE   Add or subtract the radicals when possible.

(a) $3\sqrt{7} + 2\sqrt{7}$ (b) $4\sqrt{2} + \sqrt{2}$ (c) $5\sqrt{3} + 7\sqrt{5} + 2\sqrt{3} - 4\sqrt{5}$

(d) $3 + \sqrt{3}$ (e) $3\sqrt{11} - 3\sqrt{11}$ (f) $2\sqrt{2} - \sqrt{3}$

(a) $3\sqrt{7} + 2\sqrt{7} = 5\sqrt{7}$ Radicals are like. Add coefficients.

(b) $4\sqrt{2} + \sqrt{2} = 5\sqrt{2}$ When no coefficient is written in front of a radical, the coefficient is 1.

(c) $5\sqrt{3} + 7\sqrt{5} + 2\sqrt{3} - 4\sqrt{5} =$ Combine like radicals by adding appropriate coefficients.

$7\sqrt{3} + 3\sqrt{5}$

(d) $3 + \sqrt{3}$ No addition can be performed. These are not like radicals.

(e) $3\sqrt{11} - 3\sqrt{11} = 0\sqrt{11} = 0$ Zero times any number is zero.

(f) $2\sqrt{2} - \sqrt{3}$ The terms cannot be combined. These are not like radicals.

We add or subtract square-root radical expressions only if they have *like* radicands. However, when radicals are not in simplest form, we simplify the radical expressions. If we obtain like radicands after simplifying, then we add or subtract.

EXAMPLE   Add or subtract the radical expressions when possible.

(a) $12\sqrt{5} + 3\sqrt{20}$ (b) $\sqrt{3} - \sqrt{27}$ (c) $6\sqrt{3} + 2\sqrt{8}$

(a) $12\sqrt{5} + 3\sqrt{20}$ $\sqrt{20}$ can be simplified. $\sqrt{20} = \sqrt{4 \cdot 5}$

$12\sqrt{5} + 3\sqrt{4 \cdot 5}$ $\sqrt{4} = 2$. The 2 becomes a coefficient. 3 and 2 are both coefficients of $\sqrt{5}$.

$12\sqrt{5} + 3 \cdot 2\sqrt{5}$ Multiply 3 and 2.

$12\sqrt{5} + 6\sqrt{5}$ Now we have like radicands. Add coefficients.

$\mathbf{18\sqrt{5}}$ Keep like radicand.

(b) $\sqrt{3} - \sqrt{27}$ $\sqrt{27}$ can be simplified. $\sqrt{27} = \sqrt{9 \cdot 3}$

$\sqrt{3} - \sqrt{9 \cdot 3}$ $\sqrt{9} = 3$. The 3 becomes a coefficient.

$\sqrt{3} - 3\sqrt{3}$ Now we have like radicals. Add coefficients. $1 - 3 = -2$

$\mathbf{-2\sqrt{3}}$ Keep like radicand.

(c) $6\sqrt{3} + 2\sqrt{8}$ $\sqrt{8}$ can be simplified. $\sqrt{8} = \sqrt{4 \cdot 2}$

$6\sqrt{3} + 2\sqrt{4 \cdot 2}$ $\sqrt{4} = 2$. The 2 becomes a coefficient.

$6\sqrt{3} + 2 \cdot 2\sqrt{2}$ $2 \cdot 2\sqrt{2} = 4\sqrt{2}$

$\mathbf{6\sqrt{3} + 4\sqrt{2}}$ Terms cannot be combined. Radicals are still unlike radicals.

---

2 **Multiply or Divide Square-Root Radicals.**

When multiplying two square-root radicals, the expressions under the radical signs (radicands) are multiplied together. Numbers in front of the radical signs are coefficients of the radicals and are multiplied separately.

To multiply square-root radicals:

1. Multiply coefficients to give the coefficient of the product.
2. Multiply radicands to give the radicand of the product.
3. Simplify if possible.

Symbolically, $a\sqrt{b} \cdot c\sqrt{d} = ac\sqrt{bd}$.

Express Procedures in Your Own Words

The precise details of a procedure can sometimes be overwhelming. The procedure written symbolically often guides you through the process. However, it is desirable to describe the procedure in your own words.

For example, the procedure for multiplying square-root radicals could be casually phrased in various ways.

Outside times outside $\Rightarrow$ Stays outside

Inside times inside $\Rightarrow$ Stays inside

$$(2\sqrt{3})(5\sqrt{2}) = 10\sqrt{6}$$

Another example might be used with the procedure for simplifying perfect-square radicands. The radical symbol defines a confined area. "Perfect" factors get out; "imperfect factors" stay in. As perfect factors come out, they are different (square root is taken).

$$\sqrt{9x^2y} =$$ 9 is "perfect." It comes out as 3.

x^2 is "perfect." It comes out as x.

y is "imperfect." It stays in as y.

$$3x\sqrt{y}$$

EXAMPLE   Multiply the following radicals.

(a) $\sqrt{3} \cdot \sqrt{5}$ (b) $\sqrt{\dfrac{7}{8}} \cdot \sqrt{\dfrac{2}{3}}$ (c) $3\sqrt{2} \cdot 4\sqrt{3}$ (d) $\sqrt{3} \cdot \sqrt{12}$

(a) $\sqrt{3} \cdot \sqrt{5} = \sqrt{15}$ $\sqrt{15}$ will not simplify.

(b) $\sqrt{\dfrac{7}{8}} \cdot \sqrt{\dfrac{2}{3}} = \sqrt{\dfrac{14}{24}} = \sqrt{\dfrac{7}{12}}$ Reduce.

(c) $3\sqrt{2} \cdot 4\sqrt{3} = 12\sqrt{6}$ Multiply coefficients 3 and 4. Then multiply radicands 2 and 3.

(d) $\sqrt{3} \cdot \sqrt{12} = \sqrt{36} = 6$ 36 is a perfect square. Take the square root of 36.

The distributive property can be used to multiply radical expressions.

EXAMPLE   Multiply by distributing.

(a) $5(\sqrt{7} - 2)$

(b) $\sqrt{3}(\sqrt{11} - 5)$

(c) $\sqrt{2}(\sqrt{5} - \sqrt{7})$

(a) $5(\sqrt{7} - 2) = 5\sqrt{7} - 5 \cdot 2$ Distribute. Multiply $5 \cdot 2$.

$= 5\sqrt{7} - 10$

(b) $\sqrt{3}(\sqrt{11} - 5) = \sqrt{3} \cdot \sqrt{11} - 5\sqrt{3}$ Distribute. Multiply $\sqrt{3} \cdot \sqrt{11}$.

$= \sqrt{33} - 5\sqrt{3}$

(c) $\sqrt{2}(\sqrt{5} - \sqrt{7}) = \sqrt{2} \cdot \sqrt{5} - \sqrt{2} \cdot \sqrt{7}$ Distribute. Multiply radicals.

$= \sqrt{10} - \sqrt{14}$

The FOIL process works for all types of numbers and expressions.

EXAMPLE Find the product $(2 + \sqrt{3})(5 + 2\sqrt{3})$.

$$(2 + \sqrt{3})(5 + 2\sqrt{3})$$ Multiply applying the FOIL process.

$\underline{\textbf{F}} \qquad \underline{\textbf{O}} \qquad \underline{\textbf{I}} \qquad \underline{\textbf{L}}$

$2 \cdot 5 + 2 \cdot 2\sqrt{3} + \sqrt{3} \cdot 5 + \sqrt{3} \cdot 2\sqrt{3} =$ $\sqrt{3} \cdot 2\sqrt{3} = 2\sqrt{9} = 6$

$10 + 4\sqrt{3} + 5\sqrt{3} + 6 =$ Combine like terms.

$\mathbf{16 + 9\sqrt{3}}$

Factor pairs like $(a + b)(a - b)$ are called *the sum and difference of the same two terms,* or *conjugate pairs.* Their product, $a^2 - b^2$, is called *the difference of two perfect squares.*

To multiply the sum and difference of conjugate pairs:

1. Square the first term of either binomial.
2. Insert a minus sign.
3. Square the second term of either binomial.

Symbolically, $(a + b)(a - b) = a^2 - b^2$, where a and b are real numbers.

The product of conjugate pairs is useful when expressions have irrational or imaginary terms or factors.

EXAMPLE Multiply (a) $(3 + \sqrt{5})(3 - \sqrt{5})$, (b) $(2\sqrt{3} + 4)(2\sqrt{3} - 4)$
(c) $(\sqrt{2} + \sqrt{3})(\sqrt{2} - \sqrt{3})$.

(a) $(3 + \sqrt{5})(3 - \sqrt{5})$ These are conjugate pairs.

$(a + b)(a - b) = a^2 - b^2$ Apply the special product pattern.

$(3 + \sqrt{5})(3 - \sqrt{5}) = 9 - 5$ $(\sqrt{5})^2 = 5$. Combine like terms.

$= \mathbf{4}$

(b) $(2\sqrt{3} + 4)(2\sqrt{3} - 4)$ These are conjugate pairs.

$(a + b)(a - b) = a^2 - b^2$ Apply the special product pattern.

$(2\sqrt{3} + 4)(2\sqrt{3} - 4) = 12 - 16$ $(2\sqrt{3})^2 = 4 \cdot 3 = 12$

$= \mathbf{-4}$

(c) $(\sqrt{2} + \sqrt{3})(\sqrt{2} - \sqrt{3})$ These are conjugate pairs.

$(a + b)(a - b) = a^2 - b^2$ Apply the special product pattern.

$(\sqrt{2} + \sqrt{3})(\sqrt{2} - \sqrt{3}) = 2 - 3$ $(\sqrt{2})^2 = 2; (\sqrt{3})^2 = 3$

$= \mathbf{-1}$

EXAMPLE Multiply $(3 + \sqrt{5})^2$.

$$(3 + \sqrt{5})^2 = \qquad \text{Apply the squaring process or FOIL.}$$

$$9 + 2 \cdot 3\sqrt{5} + (\sqrt{5})^2 = \qquad \text{Simplify.}$$

$$9 + 6\sqrt{5} + 5 = \qquad \text{Combine like terms.}$$

$$\mathbf{14 + 6\sqrt{5}}$$

When dividing square-root radicals, we follow a similar procedure as with multiplication. Coefficients and radicands are divided (or reduced) separately.

To divide square-root radicals:

1. Divide coefficients to give the coefficient of the quotient.
2. Divide radicands to give the radicand of the quotient.
3. Simplify if possible.

$$\frac{a\sqrt{b}}{c\sqrt{d}} = \frac{a}{c}\sqrt{\frac{b}{d}} \qquad c \text{ and } d \neq 0$$

EXAMPLE Divide the following radicals.

(a) $\dfrac{\sqrt{12}}{\sqrt{4}}$ (b) $\dfrac{\sqrt{\frac{2}{3}}}{\sqrt{\frac{7}{4}}}$ (c) $\dfrac{3\sqrt{6}}{6}$ (d) $\dfrac{5\sqrt{20}}{\sqrt{10}}$ (e) $\dfrac{8x^3\sqrt{9x^2}}{2x^2\sqrt{16x^3}}$

(a) $\dfrac{\sqrt{12}}{\sqrt{4}} = \sqrt{\dfrac{12}{4}} = \sqrt{3}$ Divide or reduce.

(b) $\dfrac{\sqrt{\frac{2}{3}}}{\sqrt{\frac{7}{4}}} = \sqrt{\dfrac{\frac{2}{3}}{\frac{7}{4}}} = \sqrt{\dfrac{2}{3}\left(\dfrac{4}{7}\right)} =$ Multiply $\frac{2}{3}$ by the reciprocal of $\frac{7}{4}$. ($\frac{2}{3} \div \frac{7}{4} = \frac{2}{3} \cdot \frac{4}{7}$)

$\sqrt{\dfrac{8}{21}} = \sqrt{\dfrac{4 \cdot 2}{21}} = \dfrac{2\sqrt{2}}{\sqrt{21}}$ Factor and take the square root of perfect-square factors.

(c) $\dfrac{3\sqrt{6}}{6}$ The coefficient 3 and the denominator 6 can be divided (or reduced) because they are both outside the radical.

$\dfrac{\overset{1}{\cancel{3}}\sqrt{6}}{\underset{2}{\cancel{6}}} = \dfrac{\sqrt{6}}{2}$ A coefficient of 1 does not have to be written in front of the radical.

(d) $\dfrac{5\sqrt{20}}{\sqrt{10}} = 5\sqrt{2}$

The 20 and 10 are divided because they are both square-root radicands.

(e) $\dfrac{8x^3\sqrt{9x^2}}{2x^2\sqrt{16x^3}} =$

Reduce the factors outside the radical.

$\dfrac{4x\sqrt{9x^2}}{\sqrt{16x^3}} =$

Reduce the factors inside the radical.

$\dfrac{4x\sqrt{9}}{\sqrt{16x}} =$

Remove perfect-square factors from all radicands.

$\dfrac{4x(3)}{4\sqrt{x}} =$

Reduce the resulting coefficients.

$\dfrac{3x}{\sqrt{x}} =$

Fractions with radical factors in the denominator are often rewritten in an equivalent form. Before the common use of calculators, this was a popular manipulation. Dividing by a rational number, and most often a whole number, was easier and more accurate than dividing by a rounded decimal approximation of the irrational number. Even now, finding common denominators and other procedures are easier if all denominators contain only rational numbers. Thus, $\frac{1}{\sqrt{3}}$ and $\sqrt{\frac{2}{5}}$ are generally rewritten so that the denominator is a rational number. This procedure is called **rationalizing the denominator.**

To rationalize a denominator:

1. If the denominator has an irrational factor, multiply the denominator by another irrational factor so that the resulting radicand is a perfect square.
2. To preserve the value of the fraction, multiply the numerator by the same radical factor used in step 1. Thus, in steps 1 and 2 together, we have multiplied by an equivalent of 1.
3. Simplify all radicals and reduce the resulting fraction, if possible.

EXAMPLE   Rationalize all denominators. Simplify the answers if possible.

(a) $\dfrac{5}{\sqrt{7}}$ (b) $\dfrac{2}{\sqrt{x}}$ (c) $\dfrac{4}{\sqrt{8}}$ (d) $\dfrac{5\sqrt{2}}{x^2\sqrt{3x}}$ (e) $\sqrt{\dfrac{2}{3}}$

(a) $\dfrac{5}{\sqrt{7}} = \dfrac{5}{\sqrt{7}} \cdot \dfrac{\sqrt{7}}{\sqrt{7}} = \dfrac{5\sqrt{7}}{7}$

$\sqrt{7} \cdot \sqrt{7} = (\sqrt{7})^2 = 7$. Denominator now has no irrational factor.

(b) $\dfrac{2}{\sqrt{x}} = \dfrac{2}{\sqrt{x}} \cdot \dfrac{\sqrt{x}}{\sqrt{x}} = \dfrac{2\sqrt{x}}{x}$

$\sqrt{x} \cdot \sqrt{x} = (\sqrt{x})^2 = x$. Denominator now has no irrational factor.

(c) $\dfrac{4}{\sqrt{8}} = \dfrac{4}{\sqrt{4 \cdot 2}} = \dfrac{4}{2\sqrt{2}} = \dfrac{2}{\sqrt{2}}$

Simplify perfect-square factors in radicands and reduce.

$$= \dfrac{2}{\sqrt{2}} \cdot \dfrac{\sqrt{2}}{\sqrt{2}} = \dfrac{2\sqrt{2}}{2} = \sqrt{2}$$

Rationalize the denominator by multiplying by $\dfrac{\sqrt{2}}{\sqrt{2}}$.

Reduce coefficients.

(d) $\dfrac{5\sqrt{2}}{x^2\sqrt{3x}} = \dfrac{5\sqrt{2}}{x^2\sqrt{3x}} \cdot \dfrac{\sqrt{3x}}{\sqrt{3x}} = \dfrac{5\sqrt{6x}}{x^2 \cdot 3x} = \dfrac{5\sqrt{6x}}{3x^3}$

(e) $\sqrt{\dfrac{2}{3}} = \dfrac{\sqrt{2}}{\sqrt{3}} \cdot \dfrac{\sqrt{3}}{\sqrt{3}} = \dfrac{\sqrt{6}}{3}$

Separate fractional radicand into a fraction with a radical factor in both the numerator and denominator.

Are calculator values less accurate if the denominator is not rationalized? Examine the next example.

EXAMPLE Use a calculator to compare the approximate value of the expressions:

$$\dfrac{1}{\sqrt{2}} \quad \text{and} \quad \dfrac{\sqrt{2}}{2}$$

$$\dfrac{1}{\sqrt{2}} = 1 \div \boxed{\sqrt{}}\ 2\ \boxed{=} \Rightarrow 0.7071067812$$

$$\dfrac{\sqrt{2}}{2} = \boxed{\sqrt{}}\ 2 \div 2\ \boxed{=} \Rightarrow 0.7071067812$$

The approximate values of the radical expressions are equivalent.

Exact Values Versus Approximate Values

When calculators are used to convert irrational numbers to decimal notation, they become *approximate values.* That is, a decimal notation is a rounded value. In applications we are most often interested in approximate numbers. *Exact values* in simplest form are useful in minimizing the calculator steps when finding approximate values.

Radical expressions are in simplest form if:

1. There are no perfect-square factors in any radicand.
2. There are no fractional radicands.
3. The denominator of a radical expression is rational (contains no radicals).

A logical sequence for simplifying radical expressions follows.

To simplify a radical expression:

1. Reduce radicands and coefficients whenever possible.
2. Simplify expressions by removing perfect-square factors from all radicands.
3. Reduce radicands and coefficients whenever possible.
4. Rationalize denominators that contain radical factors.
5. Reduce radicands and coefficients whenever possible.

EXAMPLE Perform the operation and simplify if possible.

(a) $\sqrt{y^3} \cdot \sqrt{8y^2}$ (b) $\sqrt{\dfrac{1}{3}} \cdot \sqrt{\dfrac{8}{3x}}$

(a) $\sqrt{y^3} \cdot \sqrt{8y^2} = \sqrt{8y^5} = \sqrt{4 \cdot 2 \cdot y^4 \cdot y} = 2y^2\sqrt{2y}$

(b) $\sqrt{\dfrac{1}{3}} \cdot \sqrt{\dfrac{8}{3x}} = \sqrt{\dfrac{8}{9x}} = \dfrac{\sqrt{8}}{\sqrt{9x}} = \dfrac{\sqrt{4 \cdot 2}}{\sqrt{9 \cdot x}} = \dfrac{2\sqrt{2}}{3\sqrt{x}} \cdot \dfrac{\sqrt{x}}{\sqrt{x}} = \dfrac{2\sqrt{2x}}{3x}$

SECTION 12–3 SELF-STUDY EXERCISES

1 Add or subtract. Simplify radicals where necessary.

1. $5\sqrt{3} + 7\sqrt{3}$

2. $8\sqrt{5} - 12\sqrt{5}$

3. $4\sqrt{7} + 3\sqrt{7} - 5\sqrt{7}$

4. $2\sqrt{3} - 8\sqrt{5} + 7\sqrt{3}$

5. $9\sqrt{11} - 3\sqrt{6} + 4\sqrt{6} - 12\sqrt{11}$

6. $44\sqrt{2} + \sqrt{3} - \sqrt{2} + 5\sqrt{3}$

7. $2\sqrt{3} + 5\sqrt{12}$ **8.** $7\sqrt{5} + 2\sqrt{45}$ **9.** $4\sqrt{63} - \sqrt{7}$ **10.** $3\sqrt{6} - 2\sqrt{54}$

11. $8\sqrt{2} - 3\sqrt{28}$ **12.** $2\sqrt{3} + \sqrt{48}$ **13.** $3\sqrt{5} + 4\sqrt{180}$ **14.** $7\sqrt{98} - 2\sqrt{2}$

15. $6\sqrt{40} - 2\sqrt{90}$ **16.** $\sqrt{12} - \sqrt{27}$ **17.** $3\sqrt{112} + 5\sqrt{7}$ **18.** $2\sqrt{32} - 3\sqrt{8}$

2 Multiply and simplify if possible.

19. $\sqrt{6} \cdot \sqrt{7}$ **20.** $\sqrt{18} \cdot \sqrt{2}$ **21.** $5\sqrt{2} \cdot 3\sqrt{5}$

22. $8\sqrt{3} \cdot 5\sqrt{12}$ **23.** $5\sqrt{3x} \cdot 4\sqrt{5x^2}$ **24.** $2x\sqrt{3x^4} \cdot 7x^2\sqrt{8x}$

25. $7(\sqrt{2} + 5)$ **26.** $4(12 - \sqrt{3})$ **27.** $\sqrt{5}(\sqrt{2} - 3)$

28. $\sqrt{7}(4 + \sqrt{3})$ **29.** $\sqrt{3}(\sqrt{5} + \sqrt{2})$ **30.** $(\sqrt{11} - 5)(\sqrt{2} + \sqrt{3})$

31. $(2\sqrt{3} - 1)(\sqrt{2} + 3)$ **32.** $(\sqrt{5} - 4)(\sqrt{2} - 3)$ **33.** $(\sqrt{7} + \sqrt{2})(\sqrt{7} + 3\sqrt{5})$

34. $(\sqrt{3} - 2)(\sqrt{3} + 2)$ **35.** $(\sqrt{6} - 1)(\sqrt{6} + 1)$ **36.** $(2\sqrt{3} - 5)^2$

Perform the indicated operations and simplify if possible.

37. $\sqrt{x^2} \cdot \sqrt{3x}$

38. $\sqrt{\dfrac{2}{3}} \cdot \sqrt{\dfrac{4}{5y}}$

39. $\sqrt{\dfrac{1}{x}} \cdot \sqrt{\dfrac{8}{7x}}$

40. $\sqrt{\dfrac{4}{9y^2}} \cdot \sqrt{\dfrac{1}{2y}}$

41. $\sqrt{\dfrac{8x}{3}} \cdot \sqrt{\dfrac{2x^2}{3}}$

42. $\sqrt{7} \cdot \sqrt{x^2}$

43. $\sqrt{\dfrac{4x^2}{x^3}} \cdot \sqrt{12x}$

44. $\sqrt{\dfrac{1}{2}} \cdot \sqrt{\dfrac{x^2}{3}}$

45. $\sqrt{\dfrac{y^3}{2}} \cdot \sqrt{\dfrac{2}{7y}}$

46. $\sqrt{8x} \cdot \sqrt{x^2}$

47. $\sqrt{3x} \cdot \sqrt{18x^2}$

48. $\sqrt{12x^3} \cdot \sqrt{9y^4}$

Divide and simplify.

49. $\dfrac{\sqrt{18}}{\sqrt{2}}$

50. $\dfrac{5\sqrt{10}}{10}$

51. $\dfrac{4\sqrt{12}}{2\sqrt{6}}$

52. $\dfrac{15\sqrt{24}}{9\sqrt{2}}$

53. $\dfrac{12x^2\sqrt{8x}}{15x\sqrt{6x^3}}$

54. $\dfrac{6x^4\sqrt{25x^3}}{2x\sqrt{16x^2}}$

Rationalize the denominator and simplify.

55. $\dfrac{5}{\sqrt{3}}$

56. $\dfrac{6}{\sqrt{5}}$

57. $\dfrac{1}{\sqrt{8}}$

58. $\dfrac{\sqrt{5}}{\sqrt{11}}$

59. $\dfrac{\sqrt{8}}{\sqrt{12}}$

60. $\dfrac{5x}{\sqrt{3x}}$

61. $\dfrac{2x^2}{\sqrt{7x^2}}$

62. $\dfrac{4x^5}{\sqrt{12x^3}}$

12–4 | Complex and Imaginary Numbers

Learning Outcomes

1 Write imaginary numbers using the letter i.

2 Raise imaginary numbers to powers.

3 Write real and imaginary numbers in complex form, $a + bi$.

4 Combine complex numbers.

5 Multiply radical expressions and complex numbers.

1 Write Imaginary Numbers Using the Letter *i*.

Taking the square root of a negative number introduces a new type of number. This new type of number is called an *imaginary number*. An **imaginary number** has a factor of $\sqrt{-1}$, which is represented by $i\,(i = \sqrt{-1}\,)$.

Imaginary numbers and real numbers combine to form the set of complex numbers (Fig. 12–3). A **complex number** is a number that can be written in the form $a + bi$, where a and b are real numbers and i is $\sqrt{-1}$.

The square root of a negative number, say, $\sqrt{-16}$, can be simplified as $\sqrt{-1 \cdot 16}$ or $4\sqrt{-1}$. Because the square root of -1 is an *imaginary-number* factor and 4 is a real-number factor, we will use the letter i to represent $\sqrt{-1}$ and rewrite $4\sqrt{-1}$ as $4i$. Similarly, $\sqrt{-4} = \sqrt{-1 \cdot 4} = 2\sqrt{-1} = 2i$.

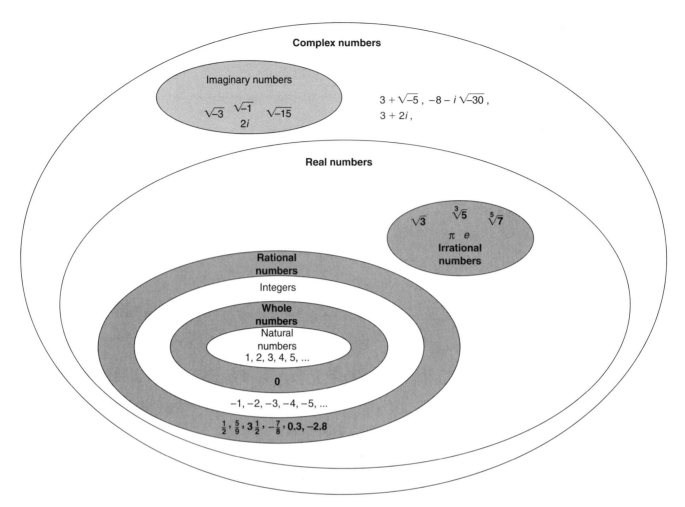

Figure 12–3 Complex-number system.

TIP

Is *i* Different from *j*?

Electronics and other applications of imaginary numbers may use *j* rather than *i* to represent $\sqrt{-1}$. Thus, $3 + 4i$ and $3 + 4j$ represent the same quantity.

EXAMPLE Rewrite the imaginary numbers using the letter *i* for $\sqrt{-1}$.

(a) $\sqrt{-9}$ (b) $\sqrt{-25}$ (c) $\sqrt{-7}$

(a) $\sqrt{-9} = \sqrt{-1 \cdot 9} = 3\sqrt{-1} = \mathbf{3i}$

(b) $\sqrt{-25} = \sqrt{-1 \cdot 25} = 5\sqrt{-1} = \mathbf{5i}$

(c) $\sqrt{-7} = \sqrt{-1 \cdot 7} = \sqrt{7} \cdot \sqrt{-1} = \sqrt{7}\mathbf{i}$ or $\mathbf{i}\sqrt{7}$.

Are Numerical Coefficients Always First?

$\sqrt{7}i$ and $\sqrt{7i}$ do not represent the same amount. In the first term, $\sqrt{7}i$, i is not under the radical symbol. In the second term, $\sqrt{7i}$, i is under the radical symbol. Because it is easy to confuse the two terms, when the coefficient of an imaginary number is an irrational number, we write the i factor first: $\sqrt{7}i = i\sqrt{7}$.

2 Raise Imaginary Numbers to Powers.

Some powers of imaginary numbers are real numbers. Examine the pattern that develops with powers of i.

$i = \sqrt{-1} = i$ $\qquad$ $i^5 = i^4 \cdot i^1 = 1(i) = i$ $\qquad$ $i^9 = i^4 \cdot i^4 \cdot i^1 = 1(1)(i) = i$

$i^2 = (\sqrt{-1})^2 = \boxed{-1}$ $\qquad$ $i^6 = i^4 \cdot i^2 = 1(-1) = \boxed{-1}$ $\qquad$ $i^{10} = i^4 \cdot i^4 \cdot i^2 = 1(1)(-1) = \boxed{-1}$

$i^3 = i^2 \cdot i^1 = -1i = -i$ $\qquad$ $i^7 = i^4 \cdot i^3 = 1(-i) = -i$ $\qquad$ $i^{11} = i^4 \cdot i^4 \cdot i^3 = 1(1)(-i) = -i$

$i^4 = i^2 \cdot i^2 = -1(-1) = \boxed{1}$ $\qquad$ $i^8 = i^4 \cdot i^4 = 1(1) = \boxed{1}$ $\qquad$ $i^{12} = i^4 \cdot i^4 \cdot i^4 = 1(1)(1) = \boxed{1}$

- All powers of i simplify to either i, -1, $-i$, or 1.
- If the exponent of i is a multiple of 4, the result is 1.
- If the exponent of i is even but not a multiple of 4, the result is -1.
- All even powers of i are real numbers.

These and other patterns allow us to develop a shortcut process for simplifying powers of i.

To simplify a power of i.

1. Divide the exponent by 4 and examine the **remainder.**
2. The power of i simplifies as follows, based on the remainder in Step 1.

Remainder of 0 $\Rightarrow$ $i^0 = 1$ $\quad$ Remainder of 2 $\Rightarrow$ $i^2 = -1$

Remainder of 1 $\Rightarrow$ $i^1 = i$ $\quad$ Remainder of 3 $\Rightarrow$ $i^3 = -i$

EXAMPLE Simplify the powers of i.

(a) i^{15} $\qquad$ (b) i^{20} $\qquad$ (c) i^{33} $\qquad$ (d) i^{18}

(a) $i^{15} = -i$ $\qquad$ $15 \div 4 = 3$ R3. Remainder of 3 $\Rightarrow$ $i^3 = -i$.

(b) $i^{20} = 1$ $\qquad$ $20 \div 4 = 5$. Remainder 0 $\Rightarrow$ $i^0 = 1$.

(c) $i^{33} = i$ $\qquad$ $33 \div 4 = 8$ R1. Remainder 1 $\Rightarrow$ $i^1 = i$.

(d) $i^{18} = -1$ $\qquad$ $18 \div 4 = 4$ R2. Remainder 2 $\Rightarrow$ $i^2 = -1$.

3 Write Real and Imaginary Numbers in Complex Form, *a* + *bi*.

A complex number has two parts, a **real part** and an **imaginary part.** In $a + bi$, if $a = 0$, then $a + bi$ is the same as $0 + bi$ or bi, which is an imaginary number. If $b = 0$, then $a + bi$ is the same as $a + 0 \cdot i$ or a, which is a real number. Thus, real numbers and imaginary numbers are also complex numbers.

EXAMPLE Rewrite in the form $a + bi$.

(a) 5 (b) $\sqrt{-36}$ (c) $-3i^2$ (d) $6 - \sqrt{-5}$

(a) $5 = \boxed{\textbf{5 + 0}i}$ $a = 5$, $b = 0$

(b) $\sqrt{-36} = 6i = \boxed{0} + 6i$ $a = \boxed{0}$, $b = 6$

(c) $-3i^2 = -3(-1) = 3 = \boxed{\textbf{3 + 0}i}$ $a = 3$, $b = 0$

(d) $6 - \sqrt{-5} = 6 - \sqrt{(5)(-1)} = \boxed{\textbf{6}} - i\sqrt{\textbf{5}}$ $a = 6$, $b = -\sqrt{5}$

---

4 Combine Complex Numbers.

Complex numbers are combined by adding *like* parts. Real parts are added together and imaginary parts are added together.

To add or subtract complex numbers:

1. Add or subtract real parts for the real part of the answer.
2. Add or subtract imaginary parts for the imaginary part of the answer.

Symbolically, $(a + bi) + (c + di) = (a + c) + (b + d)i$

EXAMPLE Combine the complex numbers.

(a) $(3 + 5i) + (8 - 2i)$ (b) $(-3 + i) - (7 - 4i)$ (c) $-2 + (5 + 6i)$

(a) $(3 + \boxed{5i}) + (8 - \boxed{2i}) = \boxed{11} + \boxed{3i}$ $3 + 8 = 11$; $5i - 2i = 3i$

(b) $(-3 + i) - (7 - 4i) =$ Distribute -1.

$\quad(-3 + \boxed{i}) + (-7 + \boxed{4i}) = \boxed{-10} + \boxed{5i}$ $-3 - 7 = -10$; $i + 4i = 5i$.

(c) $\boxed{-2} + (5 + \boxed{6i}) = \boxed{\textbf{3}} + \boxed{\textbf{6}i}$ $-2 + 5 = 3$.

---

5 Multiply Radical Expressions and Complex Numbers.

Multiplying complex numbers uses procedures similar to multiplying polynomials.
   Use the FOIL process to multiply binomials with complex numbers.

EXAMPLE Find the product $(5 + 3i)(6 - 7i)$

$$(5 + 3i)(6 - 7i)$$ Multiply applying the FOIL process.

$$\underline{F} \quad \underline{O} \quad \underline{I} \quad \underline{L}$$
$$5 \cdot 6 + 5 \cdot (-7i) + 3i \cdot 6 + 3i(-7i) =$$

$$30 - 35i + 18i - 21i^2 =$$ $i^2 = -1; -21i^2 = -21(-1) = 21$

$$\mathbf{51 - 17i}$$

The product of conjugate pairs of complex numbers is useful when expressions have imaginary terms or factors.

EXAMPLE Multiply.

(a) $(7 + 3i)(7 - 3i)$ (b) $\left(5i - \sqrt{2}\right)\left(3i + \sqrt{2}\right)$

(a) $(7 + 3i)(7 - 3i)$ These are conjugate pairs.

$(a + b)(a - b) = a^2 - b^2$ Apply the special product pattern.

$(7 + 3i)(7 - 3i) = 49 - (-9)$ $(3i)^2 = 9i^2 = 9(-1) = -9$

$= \mathbf{58}$

(b) $\left(5i - \sqrt{2}\right)\left(5i + \sqrt{2}\right)$ These are conjugate pairs.

$\left(5i - \sqrt{2}\right)\left(5i + \sqrt{2}\right) = (5i)^2 - \left(\sqrt{2}\right)^2$ $(5x)^2 = 25x^2 = 25(-1) = -25$

$= -25 - 2$

$= \mathbf{-27}$

Squaring a binomial with a complex term uses the same procedures as squaring other binomials.

EXAMPLE Multiply.

(a) $(3 + 5i)^2$ (b) $(5 - 2i)^2$

(a) $(3 + 5i)^2 =$ Apply the squaring process.

$9 + 2 \cdot 3 (5i) + (5i)^2 =$ Simplify: $2 \cdot 3 (5i) = 30i; (5i^2) = -25$

$9 + 30i - 25 =$ Terms.

$\mathbf{-16 + 3i}$

(b) $(5 - 2i)^2 =$ Apply the squaring process.

$25 + 2 \cdot 5(-2i) + (-2i)^2 =$ Simplify.

$25 - 20i + 4i^2 =$ $4i^2 = 4(-1) = -4$

$25 - 20i - 4 =$ Combine like terms.

$\mathbf{21 - 20i}$

SECTION 12–4 SELF-STUDY EXERCISES

1 Write the imaginary numbers using the letter i. Simplify if possible.

1. $\sqrt{-25}$ **2.** $\sqrt{-36}$ **3.** $\sqrt{-64x^2}$ **4.** $\sqrt{-32y^5}$

2 Simplify the powers of i.

5. i^{17} **6.** i^5 **7.** i^{20} **8.** i^{10}
9. i^{24} **10.** i^9 **11.** i^{32} **12.** i^{15}

3 Write the real and imaginary numbers in simplified complex form.

13. 15   **14.** 17   **15.** $\sqrt{-49}$ **16.** $\sqrt{-81}$

17. $33i$ **18.** $-7i^3$ **19.** $4i^6$ **20.** $5 + \sqrt{-4}$

21. $8 + \sqrt{-32}$ **22.** $7 - \sqrt{-3}$

4 Combine the complex numbers.

23. $(4 + 3i) + (7 + 2i)$ **24.** $\sqrt{12} - 5\sqrt{-3} + \left(\sqrt{8} - \sqrt{-27}\right)$
25. $12 - 3i - (8 - 4i)$ **26.** $15 + 8i - (3 - 12i)$
27. $(4 + 7i) - (3 - 2i)$ **28.** $(8 - 5i) + (4 - 3i)$

5 Multiply the complex numbers and simplify powers of i.

29. $(8 - i)(8 + i)$ **30.** $(5 - i)(5 + i)$ **31.** $(i - 1)(i + 1)$ **32.** $(4 - i)(4 + i)$
33. $(7i - 5)(7i + 5)$ **34.** $\left(2i - \sqrt{3}\right)\left(2i + \sqrt{3}\right)$ **35.** $(3i - 1)(2i + 3)$ **36.** $(7i + 2)(2i - 1)$
37. $(i - 5)(2i - 3)$ **38.** $(i - 3)^2$ **39.** $(2i - 7)^2$ **40.** $(4 + 3i)^2$

CHAPTER REVIEW OF KEY CONCEPTS

Learning Outcomes	What to Remember with Examples
Section 12–1	
1 Write roots using radical and exponential notation (pp. 485–486).	The root of a number is indicated by the denominator of a fractional exponent or the index of the radical.

The square root of 16	The cube root of 125	The fifth root of 32
$16^{1/2} = \sqrt{16} = 4;$	$125^{1/3} = \sqrt[3]{125} = 5;$	$32^{1/5} = \sqrt[5]{32} = 2$

2 Approximate an irrational number (pp. 486–488).	To estimate the square root of a number, identify the two perfect squares between which the number lies. The square root of the number will be between the square roots of the two perfect squares.

> 15 is between 9 and 16, so $15^{1/2}$ is between 3 and 4.
> 38 is between 36 and 49, so $38^{1/2}$ is between 6 and 7.

Position irrational numbers on the number line (Fig. 12–4) by approximating the value of the irrational number mentally or with a calculator.

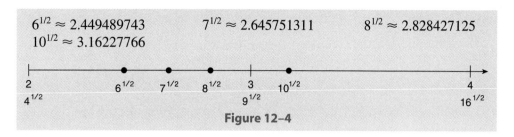

$6^{1/2} \approx 2.449489743$ $7^{1/2} \approx 2.645751311$ $8^{1/2} \approx 2.828427125$
$10^{1/2} \approx 3.16227766$

Figure 12–4

3 Write powers and roots using rational exponents and radical notation (pp. 488–490).

To write a radical expression as an expression with rational exponents, the radicand is the base of the expression. The exponent of the radicand is the numerator of the rational exponent, and the index of the root is the denominator of the rational exponent.

Symbolically, $\sqrt[\text{root}]{x^{\text{power}}} = x^{\text{power}/\text{root}}$.

Write in rational exponent notation.
$$\sqrt[5]{3^2} = 3^{2/5}$$

Write in radical notation:
$$5^{1/2} = \sqrt{5}; \qquad 7^{3/5} = \sqrt[5]{7^3} \text{ or } \left(\sqrt[5]{7}\right)^3$$

Section 12–2

1 Find the square root of variables (p. 491).

Variable factors are perfect squares if the exponent is divisible by 2. To find the square root of a perfect-square variable with an even-number exponent, take one-half of the exponent and keep the same base.

Give the square root of the following: x^6, x^{10}, x^{24}.
$$\sqrt{x^6} = x^3, \qquad \sqrt{x^{10}} = x^5, \qquad \sqrt{x^{24}} = x^{12}$$

2 Simplify square-root radicals using rational exponents and the laws of exponents (pp. 491–494).

All the laws of exponents that were used for integral exponents apply to expressions with rational exponents. The arithmetic of fractions is also applied.

Simplify.
$$x^{1/3} \cdot x^{2/3} = x^{1/3+2/3} = x^{3/3} = x; \qquad \frac{x^{7/8}}{x^{3/4}} = x^{7/8-3/4} = x^{7/8-6/8} = x^{1/8}$$

3 Simplify square-root radical expressions containing perfect-square factors (pp. 494–495).

To simplify square-root radical expressions containing perfect-square factors, factor constants and variables using the largest possible perfect-square factor of the constant and of each variable. The largest perfect square of a variable will be written with the largest possible even-numbered exponent.

Simplify the following radical expressions: $\sqrt{98}, \sqrt{x^{13}}, \sqrt{72y^9}, 5\sqrt{12}$.
$$\sqrt{98} = \sqrt{49(2)} = 7\sqrt{2}; \qquad \sqrt{x^{13}} = \sqrt{x^{12}(x)} = x^6\sqrt{x}$$
$$\sqrt{72y^9} = \sqrt{36 \cdot 2 \cdot y^8 \cdot y} = 6y^4\sqrt{2y}$$
$$5\sqrt{12} = 5\sqrt{4 \cdot 3} = 5 \cdot 2\sqrt{3} = 10\sqrt{3}$$

Section 12–3

1 Add or subtract square-root radicals (pp. 497–498).

To add or subtract square-root radicals, first simplify all radical expressions; then add or subtract the coefficients of like radical terms. Like radical terms are terms that have exactly the same factors in the radicand and have the same index.

Add or subtract: $3\sqrt{5x} + 7\sqrt{5x}$; $2\sqrt{18} - 5\sqrt{8}$.

$$3\sqrt{5x} + 7\sqrt{5x} = 10\sqrt{5x}$$

$$2\sqrt{18} - 5\sqrt{8} = 2\sqrt{9 \cdot 2} - 5\sqrt{4 \cdot 2} = 2 \cdot 3\sqrt{2} - 5 \cdot 2\sqrt{2} =$$

$$6\sqrt{2} - 10\sqrt{2} = -4\sqrt{2}$$

2 Multiply or divide square-root radicals (pp. 498–504).

To multiply radicals that have the same index, multiply the coefficients and write as the coefficient of the product, and multiply the radicands and write as the radicand of the product; then simplify the radical.

Multiply. $5\sqrt{7} \cdot 8\sqrt{14} = 40\sqrt{98} = 40\sqrt{49 \cdot 2} = 40 \cdot 7\sqrt{2} = 280\sqrt{2}$;

$$\sqrt{5}(\sqrt{7} - \sqrt{2}) = \sqrt{5} \cdot \sqrt{7} - \sqrt{5} \cdot \sqrt{2} = \sqrt{35} - \sqrt{10}$$

To divide radicals that have the same index, divide the coefficients for the coefficient of the quotient; then divide the radicands for the radicand of the quotient. Simplify any remaining radical expressions that can be simplified; then simplify any resulting coefficients that can be simplified.

Divide. $$\frac{12\sqrt{75}}{8\sqrt{6}} = \frac{3\sqrt{3 \cdot 25}}{2\sqrt{3 \cdot 2}} = \frac{3\sqrt{25}}{2\sqrt{2}} = \frac{3 \cdot 5}{2\sqrt{2}} = \frac{15}{2\sqrt{2}}$$

To rationalize a denominator of a radical expression, remove perfect-square factors from all radicands. Multiply the denominator by another radical so that the resulting radicand is a perfect square. Then multiply the numerator by the same radical the denominator was multiplied by, simplify the resulting radicals, and reduce the resulting fraction, if possible.

Rationalize the denominator. $$\sqrt{\frac{3}{5}} = \frac{\sqrt{3}}{\sqrt{5}} \cdot \frac{\sqrt{5}}{\sqrt{5}} = \frac{\sqrt{15}}{5}$$

Section 12–4

1 Write imaginary numbers using the letter i (pp. 505–507).

The letter i is used to represent $\sqrt{-1}$. Thus, the square root of negative numbers can be expressed as imaginary numbers and simplified using the letter i. Be careful to distinguish when the negative is *outside* the radical and when it is *under* the radical.

Simplify. $\sqrt{-48} = \sqrt{-1 \cdot 16 \cdot 3} = 4i\sqrt{3}$

2 Raise imaginary numbers to powers (p. 507).

Imaginary numbers can be raised to powers by examining the remainder when the exponent is divided by 4. The remainder can be used as a simplified exponent. Then, $i^0 = 1, i^1 = i, i^2 = -1$, and $i^3 = -i$.

Simplify. $i^{17} = i^{16} \cdot i^1 = i^1 = i$; $i^{42} = i^{40} \cdot i^2 = i^2 = -1$

Chapter 12 / Roots and Radicals

3 Write real and imaginary numbers in complex form, $a + bi$ (p. 508).

A complex number is a number that can be written in the form $a + bi$, where a and b are real numbers and i is $\sqrt{-1}$. Either a or b can be zero. If a is zero, the number is an imaginary number; if b is zero, the number is a real number.

Rewrite as complex numbers: $\sqrt{-81}, 38, \sqrt{9}, \sqrt{-19}$.

$$\sqrt{-81} = 9i = 0 + 9i; \qquad 38 = 38 + 0i; \qquad \sqrt{9} = 3 = 3 + 0i$$
$$\sqrt{-19} = i\sqrt{19} = 0 + i\sqrt{19}$$

4 Combine complex numbers (p. 508).

To combine complex numbers, add the real parts for the real part of the result; then add the imaginary parts for the imaginary part of the result. Be careful with signs when subtracting complex numbers.

Simplify. $(5 + 3i) + (7 - 8i) = (5 + 7) + (3 - 8)i = 12 - 5i;$
$(3 - 8i) - (4 + 6i) = (3 - 4) + (-8 - 6)i = -1 - 14i$

5 Multiply complex numbers (pp. 508–509).

Multiplying radical expressions and complex numbers apply similar procedures to multiplying polynomials.

Multiply $(3 + \sqrt{5})(2 - 3\sqrt{5})$.

$$\underline{F} \qquad \underline{O} \qquad \underline{I} \qquad \underline{L}$$
$$(3 + \sqrt{5})(2 - 3\sqrt{5}) = 3(2) + 3(-3\sqrt{5}) + 2\sqrt{5} + \sqrt{5}(-3\sqrt{5})$$
$$= 6 - 9\sqrt{5} + 2\sqrt{5} - 3(5)$$
$$= -9 - 7\sqrt{5}$$

Multiply $(2 + 5i)(2 - 5i)$.

$(2 + 5i)(2 - 5i) = 4 - 25i^2 = 4 - 25(-1) = 4 + 25 = 29$

CHAPTER REVIEW EXERCISES

Section 12–1

Write the roots in radical notation and exponential notation.

1. Square root of 49
2. Cube root of 125
3. Fourth root of 16
4. Fifth root of 3,125
5. Square root of 121
6. Cube root of 343

Find the two whole numbers that are closest to each root.

7. $\sqrt{38}$
8. $\sqrt{15}$
9. $\sqrt{135}$
10. $\sqrt{75}$
11. $\sqrt[3]{60}$
12. $\sqrt[3]{135}$

Position each square root on a number line. Label the points. Estimate the roots and check estimation with a calculator.

13. $\sqrt{15}$
14. $\sqrt{27}$
15. $\sqrt{5}$
16. $\sqrt{38}$

Write in both rational exponent and radical notations.

17. The square root of the seventh power of x
18. The cube root of the fourth power of x
19. The square of the cube root of x
20. The fifth power of the fifth root of x

Write in rational exponent notation in simplest form.

21. $\sqrt{x}$

22. $\sqrt[3]{x^5}$

23. $\sqrt[5]{x^4}$

24. $\sqrt[4]{9x}$

25. $\left(\sqrt[3]{xy}\right)^4$

26. $\sqrt[3]{64x^{10}}$

27. $\sqrt{7}$

Write in radical notation.

28. $x^{5/8}$

29. $y^{3/5}$

30. $a^{1/4}$

Section 12–2

Find the square roots if all variables represent positive numbers.

31. $\sqrt{y^{12}}$

32. $\sqrt{a^{10}}$

33. $-\sqrt{b^{18}}$

34. $\pm\sqrt{\dfrac{x^4}{y^6}}$

Convert the radicals to equivalent expressions using rational exponents and simplify if appropriate.

35. $\sqrt[3]{x}$

36. $\sqrt[4]{p}$

37. $\sqrt[5]{4y}$

38. $\left(\sqrt{ab}\right)^6$

39. $\sqrt[3]{8b^{12}}$

40. $\left(\sqrt{49x^2y^3}\right)^4$

Perform the operations. Express the answers with positive exponents in lowest terms.

41. $(a^{1/2})(a^{3/2})$

42. $(a^{4/3})(a^{2/3})$

43. $y^{3/4} \cdot y^{1/4}$

44. $y^{5/8} \cdot y^{1/8}$

45. $(3x^{1/4}y^2)^3$

46. $(2x^{3/4}\,y)^2$

47. $(4ax^{1/2})^3$

48. $(x^{1/2})^{1/3}$

49. $\dfrac{x^{3/4}}{x^{1/4}}$

50. $\dfrac{x^{1/6}}{x^{5/6}}$

51. $\dfrac{a^{5/6}}{a^{-1/3}}$

52. $\dfrac{a^{7/10}}{a^{2/5}}$

53. $\dfrac{x^{5/8}}{x^{3/4}}$

54. $\dfrac{y^{1/3}}{y^{5/6}}$

55. $\dfrac{a^3}{a^{1/3}}$

Use a calculator to evaluate the expressions in Exercises 56 to 63 for $a = 2$, $b = 1$, and $x = 3$ both before and after simplifying.

56. $\dfrac{a^2}{a^{3/5}}$

57. $\dfrac{12a^4}{6a^{1/2}}$

58. $\dfrac{27x^3}{9x^{2/3}}$

59. $\dfrac{15a^{3/5}}{10a^5}$

60. $\dfrac{14a^{5/6}}{24a^2}$

61. $a^{2.3}(a^4)$

62. $(3a^{1.2}b^2)^3$

63. $(4a^6b^8)^{1/2}$

Simplify the expressions.

64. $\left(\sqrt{x^5}\right)^2$

65. $\sqrt{x^2}$

66. $\left(\sqrt{8}\right)^2$

67. $\sqrt{9P^3}$

68. $\sqrt{8^2}$

69. $\sqrt{18a^2b}$

70. $\sqrt{12x^2y^3}$

71. $\sqrt{32x^5y^2}$

72. $\sqrt{63x^4y^7}$

73. $\sqrt{75x^{10}y^9}$

74. $\sqrt{125xy^3}$

75. $\sqrt{147xy^8}$

Section 12–3

Add or subtract the radicals.

76. $12\sqrt{11} - 5\sqrt{11}$

77. $5\sqrt{3} - 7\sqrt{3}$

78. $4\sqrt{2} + 3\sqrt{5} - 8\sqrt{2} + 6\sqrt{5}$

79. $3\sqrt{7} - 2\sqrt{28}$

80. $\sqrt{2} - \sqrt{8}$

81. $2\sqrt{6} + 3\sqrt{54}$

82. $3\sqrt{5} - 2\sqrt{45}$

83. $4\sqrt{3} - 8\sqrt{48}$

84. $\sqrt{40} + \sqrt{90}$

85. $5\sqrt{8} - 3\sqrt{50}$

86. $5\sqrt{7} - 4\sqrt{63}$

87. $3\sqrt{2} - 5\sqrt{32}$

Multiply the radicals and simplify.

88. $\sqrt{6} \cdot \sqrt{3}$

89. $2\sqrt{8} \cdot 3\sqrt{6}$

90. $2\sqrt{a} \cdot \sqrt{b}$

91. $5\sqrt{3} \cdot 8\sqrt{7}$

92. $2\sqrt{3} \cdot 5\sqrt{18}$

93. $-8\sqrt{5} \cdot 4\sqrt{30}$

94. $5(\sqrt{3} - 2)$

95. $\sqrt{3}(\sqrt{12} - 5)$

96. $\sqrt{2}(\sqrt{6} - \sqrt{10})$

97. $\sqrt{3}(\sqrt{6} - \sqrt{15})$

98. $(\sqrt{7} - 5)(\sqrt{5} - 3)$

99. $(\sqrt{5} - 8)(\sqrt{5} + 8)$

Divide the radicals and simplify.

100. $\dfrac{4\sqrt{8}}{2}$

101. $\dfrac{3\sqrt{5}}{2\sqrt{20}}$

102. $\dfrac{2\sqrt{90}}{\sqrt{5}}$

103. $\dfrac{6\sqrt{18}}{8\sqrt{12}}$

104. $\dfrac{14\sqrt{56}}{7\sqrt{7}}$

105. $\dfrac{5\sqrt{48}}{20\sqrt{20}}$

106. $\dfrac{\sqrt{9x}}{\sqrt{3x}}$

107. $\dfrac{\sqrt{3y^3}}{\sqrt{y^3}}$

108. $\left(\sqrt{\dfrac{25}{36}}\right)^2$

109. $\left(\sqrt{\dfrac{9}{16}}\right)^2$

110. $\sqrt{\dfrac{9c^4}{25y^6}}$

111. $\sqrt{\dfrac{36x^8}{81y^{10}}}$

Rationalize the denominator and simplify.

112. $\dfrac{1}{\sqrt{8}}$

113. $\dfrac{\sqrt{7}}{\sqrt{12}}$

114. $\dfrac{\sqrt{3}}{\sqrt{7x}}$

115. $\dfrac{\sqrt{3}}{\sqrt{8}}$

116. $\dfrac{\sqrt{7}}{5\sqrt{18}}$

117. $\dfrac{5\sqrt{3}}{\sqrt{24}}$

118. $\dfrac{\sqrt{15}}{5\sqrt{7}}$

119. $\dfrac{2\sqrt{5}}{\sqrt{18}}$

Section 12–4

Write the numbers using the letter i. Simplify if possible.

120. $\sqrt{-144}$

121. $\sqrt{-100}$

122. $-\sqrt{-16x^2}$

123. $\pm\sqrt{-24y^7}$

Simplify the powers of i.

124. i^6

125. i^{14}

126. i^{98}

127. i^{77}

Write as complex numbers in simplified form.

128. 5

129. $15i$

130. $3 + \sqrt{-9}$

131. $-12i^5$

132. $-6i^{11}$

Perform the indicated operation and simplify.

133. $(5 + 3i) + (2 - 7i)$

134. $(4 - i) - (3 - 2i)$

135. $(7 - \sqrt{-9}) + (4 + \sqrt{-16})$

136. $(5i - 3)(2i - 3)$

137. $(4i + 3)(4i - 3)$

138. $(7i - 4)^2$

1. Laws of radicals that pattern the laws of exponents can be used to simplify radicals that are not square-root radicals. For example, $\sqrt[n]{xy} = \sqrt[n]{x} \cdot \sqrt[n]{y}$, for positive values of x and y and a natural number n greater than 1.
   (a) Illustrate this property with a numerical example for a natural-number value of n that is greater than 2.
   (b) Illustrate with a numerical example the property $\sqrt[m]{\sqrt[n]{x}} = \sqrt[n]{\sqrt[m]{x}} = \sqrt[mn]{x}$ for a positive value of x and a natural-number value of n that is greater than 2.

2. Illustrate with a numerical example the following properties of radicals and rational exponents.
   (a) $\sqrt[n]{\dfrac{x}{y}} = \dfrac{\sqrt[n]{x}}{\sqrt[n]{y}}$ for positive values of x and y and a natural-number value of n that is greater than 2.

   (b) $x^{-m/n} = \dfrac{1}{x^{nn}}$ for a positive value of x and values of m and n that are natural numbers greater than 2.

Write the rules in words. Assume that all radicands represent positive values.

1. $a\sqrt{b} \cdot c\sqrt{d} = ac\sqrt{bd}$

2. $\dfrac{a\sqrt{b}}{c\sqrt{d}} = \dfrac{a}{c}\sqrt{\dfrac{b}{d}}; c, d \neq 0$

3. $a\sqrt{b} + c\sqrt{b} = (a + c)\sqrt{b}$

4. $\left(\sqrt{x}\right)^2 = x$ or $\sqrt{x^2} = x$, for positive values of x.

5. List the conditions for a radical expression to be in simplest form.

6. What does it mean to *rationalize* a denominator? What calculations or manipulations with fractions are easier if the denominator is a rational number?

7. Write the property in words.
$$x^{1/n} = \sqrt[n]{x}$$

8. Write the following property in words:
$$x^{m/n} = \sqrt[n]{x^m} \text{ or } \left(\sqrt[n]{x}\right)^m$$

9. Explain how to analyze a power of i to determine if the power simplifies to be i, 1, -1, and $-i$.

10. Write an example of two binomials that are conjugates and contain radicals. Multiply the conjugates to eliminate the radical.

Perform the indicated operations. Simplify if possible. Rationalize the denominators if needed.

1. $2\sqrt{7} \cdot 3\sqrt{2}$

2. $\dfrac{\sqrt{8}}{\sqrt{2}}$

3. $4\sqrt{3} + 2\sqrt{3}$

4. $3\sqrt{8} - 4\sqrt{8}$

5. $\dfrac{4\sqrt{2}}{\sqrt{3}}$

6. $\sqrt{3y^3} \cdot \sqrt{15y^2}$

7. $\dfrac{6\sqrt{8}}{2\sqrt{3}}$

8. $\dfrac{3\sqrt{a}}{\sqrt{b}}$

9. $\dfrac{3\sqrt{5}}{2} \cdot \dfrac{7}{\sqrt{3x}}$

Convert the radical expressions to equivalent expressions using rational exponents and simplify.

10. $\sqrt[6]{x}$

11. $\sqrt[3]{27x^{15}}$

12. $\left(\sqrt{5x^4}\right)^6$

13. $\sqrt[5]{x^{10}y^{15}z^{30}}$

Perform the operations. Express answers with positive exponents in lowest terms.

14. $a^{4/5} \cdot a^{1/5}$

15. $(125x^{1/2}y^6)^{1/3}$

16. $\dfrac{b^{3/4}}{b^{1/4}}$

17. $\dfrac{12x^{3/5}}{6x^{-2/5}}$

18. $\dfrac{r^{-1/5}s^{1/3}}{r^{3/5}s^{-5/3}}$

Simplify.

19. i^{23}

20. i^{88}

Add or subtract.

21. $(5 + 3i) - (8 - 2i)$

22. $(7 - i) + (4 - 3i)$

23. $5\sqrt{3} + 8\sqrt{5} - 7\sqrt{3}$

24. $2\sqrt{24} - 7\sqrt{54} + 3\sqrt{96}$

Multiply and simplify.

25. $\sqrt{7}(\sqrt{5} - 4)$

26. $\sqrt{3}(\sqrt{6} - \sqrt{12})$

27. $(5\sqrt{2} - 3)(5\sqrt{2} + 3)$

28. $(3i - 8)(3i + 8)$

13

Factoring

Focus on Careers

Landscape and grounds mainte-nance workers create and main-tain a pleasant and functional outdoor environment. Landscap-ing workers install and maintain landscaped areas.

Supervisors of landscaping workers prepare cost estimates, schedule work for crews, perform quality checks, suggest and imple-ment changes in work procedures to improve efficiencies, hire em-ployees, and keep employees' time and work-performed records. Pes-ticide handlers mix and apply pes-ticides, herbicides, fungicides, or insecticides.

The Professional Grounds Management Society offers certification to grounds managers who have a combination of 8 years of experience and formal education be-yond high school and who pass an examination.

Most states require certification for workers who apply pesticides. Certification usually includes passing a test on the proper and safe use and disposal of these chem-icals and some formal education beyond high school. Many 2-year colleges have land-scape maintenance programs.

Employment of landscape and grounds maintenance workers is expected to grow faster than the average for all occupations through 2012. Both new construction and renovation of existing landscaping and grounds will provide many jobs.

Supervisors and managers of landscaping and lawn service workers had median hourly earnings of $15.89 in 2002, and tree trimmers and pruners had median hourly earnings of $12.07 in 2002.

Source: *Occupational Outlook Handbook,* 2004–2005 Edition. U.S. Department of Labor, Bureau of Labor Statistics.

Throughout our study of mathematics, we have examined products and factors. To reduce fractions, we looked for factors common to both the numerator and the denominator. In this chapter, we again find it useful to examine products and factors.

13-1 | The Distributive Property and Common Factors

Learning Outcome **1** Factor an expression containing a common factor.

We discussed the distributive property and finding common factors earlier in the text and applied them in different contexts. In this section, rather than use the distributive property to multiply and obtain a product, we start with a product and regenerate the factors that produce the product. In other words, we want to undo the multiplication. Factoring resembles division, which is the inverse operation of multiplication.

1 Factor an Expression Containing a Common Factor.

The multiplication problem $7a(3a + 2)$ is written in **factored form.** It is the indicated product of $7a$ and the grouped quantity $3a + 2$. After the expression is multiplied, we have two terms written as the sum $21a^2 + 14a$. This is the **expanded form.** To rewrite the expression $21a^2 + 14a$ as the indicated product $7a(3a + 2)$ is to **factor** it.

Let's look at a general example of the distributive property:

$$\underset{\text{factored form}}{a(x + y)} \quad = \quad \underset{\text{expanded form}}{ax + ay}$$

Notice that a appears as a factor in both terms in the expanded form. When a factor appears in each of several terms, it is called a **common factor** of the terms. The distributive property in reverse can be used to write the addition as a multiplication. In other words, we can *factor* the expression.

$$ax + ay = a(x + y)$$

To factor an expression containing a common factor:

1. Find the *greatest* factor common to *each* term of the expression.
2. Divide each term by the common factor. Divide mentally if practical.
3. Rewrite the expression as the indicated product of the greatest common factor (GCF) and the quotients in Step 2.

EXAMPLE Write $3a + 3b$ in factored form.

We can use the distributive property to factor the expression.

$$3\,a + 3\,b = \qquad \text{Write 3 as a factor and divide each term by 3.}$$

$$3\left(\frac{3a}{3} + \frac{3b}{3}\right) = \qquad \text{Simplify each fraction in parentheses.}$$

$$\mathbf{3(a + b)} \qquad \text{Factored form}$$

The distributive property also applies if we have more than two terms.

EXAMPLE   Write $3ab + 9a + 12b$ in factored form.

$$3\ ab\ +\ 9a\ +\ 12b$$ 3 is the common factor. Divide.

$$3 \cdot 3 \quad 3 \cdot 4$$

$$3\left(\frac{3ab}{3} + \frac{3 \cdot 3a}{3} + \frac{3 \cdot 4b}{3} \right)$$ Simplify.

$$3(ab + 3a + 4b)$$ Factored form.

When looking for common factors, we always look for *all* common factors.

EXAMPLE   Factor $10a^2 + 6a$ completely.

$$10a^2 + 6a =$$ The GCF is 2a. Write 2a as a factor and divide each term by 2a.

$$2a\left(\frac{10a^2}{2a} + \frac{6a}{2a} \right) =$$ Simplify.

$$2a(5a + 3)$$ Factored form.

EXAMPLE   Factor $2x^2 + 4x^3$ completely.

$$2x^2 + 4x^3 =$$ The GCF is 2x^2. Write 2x^2 as a factor and divide each term by 2x^2.

$$2x^2\left(\frac{2x^2}{2x^2} + \frac{4x^3}{2x^2} \right) =$$ Simplify.

$$2x^2(1 + 2x)$$ Term of 1 must be written.

EXAMPLE   Write $2x + 3y$ in factored form.

$$2x + 3y =$$ The GCF is 1. The expression can be written in factored form only as $1(2x + 3y)$.

$$1(2x + 3y)$$ Factored form.

When an expression can be written in factored form only as 1 times the entire expression, the expression is a **prime polynomial.**

When Is It Necessary to Write a 1?

We have found that it is not always necessary to write the number 1. When is it necessary?

When 1 is a term, it must be written.

$$2x^2 - x = x\left(\frac{2x^2}{x} - \boxed{\frac{x}{x}}\right) = x(2x - \boxed{1})$$

When 1 is a factor, writing the 1 is optional: $1 \cdot n = n$.

$$2a + 2b = 2\left(\frac{2a}{2} + \frac{2b}{2}\right) = 2(1a + 1b) \qquad \text{or} \qquad 2(a + b)$$

When 1 is an exponent, writing the 1 is optional: $n^1 = n$.

$$2x^3 - 5x^2 = x^2\left(\frac{2x^3}{x^2} - \frac{5x^2}{x^2}\right) = x^2(2x^1 - 5x^0) = x^2(2x - 5)$$

Also, recall that $n^0 = 1$ for any real number n, $n \neq 0$.

Sometimes a binomial factor or a grouping is the common factor.

EXAMPLE   Factor $7y(2y - 5) + 3(2y - 5)$.

$$7y\,(2y - 5) + 3\,(2y - 5) = \qquad \text{Common factor is } (2y - 5).$$

$$(2y - 5)\left[\frac{7y\cancel{(2y - 5)}}{\cancel{(2y - 5)}} + \frac{3\cancel{(2y - 5)}}{\cancel{(2y - 5)}}\right] =$$

$$(2y - 5)(7y + 3)$$

If the leading coefficient of a polynomial is negative, it is often helpful to factor a common factor of -1.

EXAMPLE   Factor $-3x^2 + 2x - 5$.

$$-3x^2 + 2x - 5 = \qquad \text{Common factor is } -1.$$

$$-1\left(\frac{-3x^2}{-1} + \frac{2x}{-1} - \frac{5}{-1}\right) =$$

$$-1(3x^2 - 2x + 5)$$

1 Factor completely. Check.

1. $7a + 7b$
2. $12x + 12y$
3. $m^2 + 2m$
4. $5y^3 + 8y^2$
5. $6x^2 + 3x$
6. $12y^3 + 18y^4$
7. $12x^5 - 6x^4$
8. $5x - 15xy$
9. $5y + 3z$
10. $8x - 7y$
11. $5ab + 10a + 20b$
12. $4ax^2 + 6a^2x + 10a^2x^2$
13. $5a - 7ab + 35b$
14. $12a^2 - 15a + 6$
15. $3x^3 - 9x^2 - 6x$
16. $8a^2b + 14ab^3 + 28a^3b^3$
17. $3m^2 - 6m^3 + 12m^4$
18. $12x^2y - 18xy^3 + 24x^2y^2$
19. $15a^2bc + 18a^3b^2c^3 - 21a^4bc^5$
20. $20x^2y^3z - 35x^3y^2z - 40x^2y^2z$
21. $8x^4y^2 - 12x^2y^4 - 4x^2y^2$

Write in factored form so the leading coefficient of the polynomial factor is positive.

22. $-x - 7$
23. $-3x - 8$
24. $-5x + 2$
25. $-12x + 7$
26. $-x^2 + 3x - 8$
27. $-2x^2 - 7x - 11$
28. $-2x^2 + 6x - 8$
29. $-3x^2 - 9x + 15$
30. $-7x^2 - 21x + 14$
31. $-12x^2 + 18x + 6$
32. $5x(x + 3) + 8y(x + 3)$
33. $3x(2x - 1) + 5(2x - 1)$
34. $4y(3y - 5) + 7(3y - 5)$
35. $7a(a - b) + 2b(a - b)$
36. $5m(2m - 3n) - 7n(2m - 3n)$
37. $y(y - 2) - 3(y - 2)$
38. $3x(2x - 7) - 8(2x - 7)$
39. $7y(9y - 2) - 5(9y - 2)$
40. $5\sqrt{7} + 10$
41. $8\sqrt{3} - 12$
42. $3\sqrt{2} - 9\sqrt{3}$

13–2 | *Factoring Special Products*

Learning Outcomes

1 Recognize and factor the difference of two perfect squares.

2 Recognize and factor a perfect-square trinomial.

3 Recognize and factor the sum or difference of two perfect cubes.

To factor any of the special products we used in Chapter 11, we apply the inverse of the process. That is, we start with the product and "work back" to the factors that produce these special products.

To rewrite a special product in factored form, we must recognize the product as a pattern. Once we identify the special product, then we must know the pattern of the product in factored form.

1 **Recognize and Factor the Difference of Two Perfect Squares.**

Before we can factor such a special product, we must be able to recognize an expression as a special product. First, we examine the *sum and difference of the same two terms*. The product is the difference of two perfect squares.

To identify a binomial as the difference of two perfect squares:

1. Verify that the expression is a binomial (two terms).
2. Verify that the absolute value of each term is a perfect square.
3. Verify that the second term is subtracted from the first term.

The pattern is $a^2 - b^2$.

EXAMPLE   Identify the special products that are the difference of two perfect squares.

(a) $x^2 - 9$ (b) $a^2 + 49$ (c) $m^2 - 27$ (d) $3y^2 - 25$
(e) $9x^2 - 4$ (f) $4x^2 - 4x + 1$

(a) Difference of two perfect squares.
(b) Not the difference of two perfect squares; this is a *sum,* not a difference.
(c) Not the difference of two perfect squares; 27 is not a perfect square.
(d) Not the difference of two perfect squares; in $3y^2$, 3 is not a perfect square.
(e) Difference of two perfect squares.
(f) Not the difference of two perfect squares; this is a trinomial, not a binomial.

To factor the difference of two perfect squares:

1. Take the square root of the first term.
2. Take the square root of the second term.
3. Write one factor as the *sum* of the square roots found in Steps 1 and 2, and write the other factor as the *difference* of the square roots from Steps 1 and 2.

Symbolically,

$$a^2 - b^2 = (a + b)(a - b)$$

EXAMPLE   Factor the special products, which are the differences of two perfect squares.

(a) $a^2 - 9$ (b) $x^2 - 36$ (c) $4x^2 - 1$ (d) $-49 + 16m^2$

(a) $a^2 - 9 = (a + 3)(a - 3)$
(b) $x^2 - 36 = (x + 6)(x - 6)$
(c) $4x^2 - 1 = (2x + 1)(2x - 1)$
(d) $-49 + 16m^2 = 16m^2 - 49 = (4m + 7)(4m - 7)$

Order of Factors

TIP

Because multiplication is commutative, the factors given as answers in the preceding example may be expressed in any order, such as $(a + 3)(a - 3)$ or $(a - 3)(a + 3)$, $(x + 6)(x - 6)$ or $(x - 6)(x + 6)$, and so on.

2 Recognize and Factor a Perfect-Square Trinomial.

A trinomial is a **perfect-square trinomial** if the first and last terms are positive perfect squares and the absolute value of the middle term is *twice* the product of the square roots of the first and last terms. We need to be able to distinguish these special products from other expressions before we factor them.

13–2 Factoring Special Products

To identify a perfect-square trinomial:

1. Verify that the expression is a trinomial.
2. Verify that the first and last terms are positive and perfect squares.
3. Mentally take the square root of the first and last terms and multiply the results. Two times this product should equal the absolute value of the middle term of the original trinomial.

The pattern is $a^2 + 2ab + b^2$ or $a^2 - 2ab + b^2$.

EXAMPLE Verify that the trinomials are perfect-square trinomials.

(a) $x^2 + 14x + 49$ (b) $4m^2 - 12m + 9$ (c) $9x^2 + 24xy + 16y^2$

(a) The first and last terms, x^2 and 49, are positive perfect squares. The middle term, $14x$, has an absolute value that is twice the product of the square roots of x^2 and 49. That is, $2(7x) = 14x$.

(b) The first and last terms, $4m^2$ and 9, are positive perfect squares. The middle term, $-12m$, has an absolute value that is twice the product of the square roots of $4m^2$ and 9. That is, $2(2m \cdot 3) = 12m$.

(c) The first and last terms, $9x^2$ and $16y^2$, are positive perfect squares. The middle term, $24xy$, has an absolute value that is twice the product of the square roots of $9x^2$ and $16y^2$. That is, $2(3x \cdot 4y) = 24xy$.

EXAMPLE Explain why the trinomials are *not* perfect-square trinomials.

(a) $x^2 + 2x - 1$ (b) $4x^2 + 6x + 9$
(c) $x^2 - 5x + 4$ (d) $-4x^2 - 4x + 1$

(a) The last term, -1, is negative. This term must be positive in a perfect-square trinomial.

(b) The middle term, $6x$, is not *twice* the product of $2x$ and 3.

(c) The middle term, $-5x$, is not *twice* the product of x and 2.

(d) The first term, $-4x^2$, is negative. It should be positive.

To factor a perfect-square trinomial:

1. Write the square root of the first term.
2. Write the sign of the middle term.
3. Write the square root of the last term.
4. Indicate the square of this binomial quantity.

Symbolically,

$$a^2 + 2ab + b^2 = (a + b)^2 \text{ or } a^2 - 2ab + b^2 = (a - b)^2$$

EXAMPLE Factor the perfect-square trinomials.

(a) $x^2 + 14x + 49$ (b) $4m^2 - 12m + 9$ (c) $9x^2 + 24xy + 16y^2$

	Square root of *first* term	Sign of *middle* term	Square root of *last* term	*Square the quantity*
(a) $x^2 + 14x + 49$	x	$+$	7	$(x + 7)^2$
(b) $4m^2 - 12m + 9$	$2m$	$-$	3	$(2m - 3)^2$
(c) $9x^2 + 24xy + 16y^2$	$3x$	$+$	$4y$	$(3x + 4y)^2$

---

3 Recognize and Factor the Sum or Difference of Two Perfect Cubes.

Before we factor the sum or difference of two perfect cubes, we should be able to recognize an expression that matches the pattern.

To identify the sum or difference of two perfect cubes:

1. Verify that the expression is a binomial.
2. Verify that each term is a perfect cube.

The pattern is $a^3 + b^3$ or $a^3 - b^3$.

EXAMPLE Identify the expressions that are the sum or difference of two perfect cubes.

(a) $a^3 - 27$ (b) $8x^3 + 9$ (c) $8y^3 - 125$ (d) $64 - 8a^3$
(e) $a^3 + 16b^3$ (f) $c^3 + y^3 - 27$ (g) $27x^3 + 8$ (h) $3a^3 - 64$

(a) Difference of two perfect cubes.
(b) Not the sum or difference of two perfect cubes; 9 is not a perfect cube.
(c) Difference of two perfect cubes.
(d) Difference of two perfect cubes.
(e) Not the sum or difference of two perfect cubes; 16 is not a perfect cube.
(f) Not the sum or difference of two perfect cubes; this is a trinomial.
(g) Sum of two perfect cubes.
(h) Not the sum or difference of two perfect cubes; 3 is not a perfect cube.

To factor the sum of two perfect cubes:

1. Write the binomial factor as the *sum* of the cube roots of the two terms.
2. Write the trinomial factor as the square of the first term from Step 1, *minus* the product of the two terms from Step 1, *plus* the square of the second term from Step 1.

Symbolically,

$$a^3 + b^3 = (a + b)(a^2 - ab + b^2)$$

13–2 Factoring Special Products

525

To factor the difference of two perfect cubes:

1. Write the binomial factor as the *difference* of the cube roots of the two terms.
2. Write the trinomial factor as the square of the first term from Step 1, *plus* the product of the two terms from Step 1, *plus* the square of the second term from Step 1.

Symbolically,

$$a^3 - b^3 = (a - b)(a^2 + ab + b^2)$$

Notice, the sign of the third term of the trinomial is positive when factoring either the sum or the difference of two cubes.

EXAMPLE Factor the expressions that are the sum or difference of two perfect cubes.

(a) $27a^3 - 8$ (b) $m^3 + 125n^3$

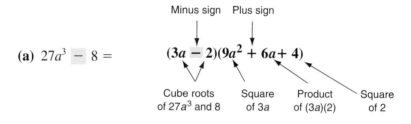

(a) $27a^3 - 8 =$ (3a − 2)(9a² + 6a+ 4)

Minus sign / Plus sign
Cube roots of $27a^3$ and 8 / Square of $3a$ / Product of $(3a)(2)$ / Square of 2

(b) $m^3 + 125n^3 =$ (m + 5n)(m² − 5mn + 25n²)

Plus sign / Minus sign
Cube roots of m^3 and $125n^3$ / Square of m / Product of $m(5n)$ / Square of $5n$

SECTION 13–2 SELF-STUDY EXERCISES

1 Identify the special products that are the difference of two perfect squares.

1. $r^2 - s^2$
2. $d^2 - 4d + 10$
3. $4y^2 - 16$
4. $36m^2 - 9n^2$
5. $9 + 4a^2$
6. $64p^2 - q^2$

Factor the following special products.

7. $y^2 - 49$
8. $16x^2 - 1$
9. $9a^2 - 100$
10. $4m^2 - 81n^2$
11. $9x^2 - 64y^2$
12. $25x^2 - 64$
13. $100 - 49x^2$
14. $4x^2 - 49y^2$
15. $121m^2 - 49n^2$
16. $81x^2 - 169$
17. $-9 + 4a^2$
18. $-16 + 25r^2$
19. $36x^2 - 49y^2$
20. $49 - 144x^2$
21. $16x^2 - 81y^2$

2 Identify the trinomials that are perfect-square trinomials.

22. $4y^2 + 2y + 16$
23. $9m^2 - 24mn + 16n^2$
24. $16a^2 + 8a - 1$
25. $-9r^2 + 12r + 4$
26. $y^2 - 14y + 49$
27. $p^2 + 10p + 25$

Factor the special products.

28. $x^2 + 6x + 9$ **29.** $x^2 + 14x + 49$ **30.** $x^2 - 12x + 36$

31. $x^2 - 16x + 64$ **32.** $4a^2 + 4a + 1$ **33.** $25x^2 - 10x + 1$

34. $9m^2 - 48m + 64$ **35.** $4x^2 - 36x + 81$ **36.** $x^2 - 12xy + 36y^2$

37. $4a^2 - 20ab + 25b^2$ **38.** $y^2 - 10y + 25$ **39.** $9x^2 + 60xy + 100y^2$

40. $-x^2 - 12x - 36$ **41.** $-9x^2 + 6x - 1$ **42.** $-x^2 - 8x - 16$

3   Identify the special products that are the sum or difference of two perfect cubes.

43. $8b^3 - 125$ **44.** $y^2 - 14y + 49$ **45.** $T^3 + 27$

46. $c^3 + 16$ **47.** $125x^3 - 8y^3$ **48.** $64 + a^3$

Factor the special products, which are the sum or difference of two perfect cubes.

49. $m^3 - 8$ **50.** $y^3 - 125$ **51.** $Q^3 + 27$

52. $c^3 + 1$ **53.** $125d^3 - 8p^3$ **54.** $a^3 + 64$

55. $216a^3 - b^3$ **56.** $x^3 + Q^3$ **57.** $8p^3 - 125$

58. $27 - 8y^3$ **59.** $-a^3 - 8$ **60.** $-x^3 - 27$

13–3 | *Factoring General Trinomials*

Learning Outcomes

1   Factor general trinomials whose squared term has a coefficient of 1.

2   Remove common factors after grouping an expression.

3   Factor a general trinomial by grouping.

4   Factor any binomial or trinomial that is not prime.

Many trinomials do not have a common factor and do not match the pattern of a special product. Some will still factor as the product of two binomials. First, let's examine a trinomial whose squared term has a coefficient of 1.

1   **Factor General Trinomials Whose Squared Term Has a Coefficient of 1.**

Trinomials that are not perfect-square trinomials are **general trinomials.**

To factor a trinomial with a squared term that has a coefficient of 1:

1. Ensure the trinomial is arranged in descending powers of one variable.
2. Determine the signs of the second term of each binomial by examining the sign of the *third* term of the trinomial.

    $+ \Rightarrow$ like signs (both $+$ or both $-$, matching the middle sign)

    $- \Rightarrow$ unlike signs (one $+$ and one $-$, with the sign of the larger absolute value of the factor pair matching the sign of the middle term of the trinomial)

3. Write all factor pairs of the coefficient of the third term.
4. Select the factor pair that adds (algebraically) to the coefficient of the middle term of the trinomial. Include appropriate signs.
5. Write the binomial factors of the trinomial with the square root of the first-term variable as the first term of each factor and the two factors from Step 4 as the second terms of the binomials.

EXAMPLE   Factor $x^2 + 6x + 5$.

$x^2 + 6x + 5$ Terms are already in descending order.

$(\ + \ )(\ + \ )$ Last term $+$ and middle term $+$ means signs are alike and both positive.
Only factor pair of 5 is 1(5). The algebraic sum is $+1 + (+5) = +6$.

$(x + 1)(x + 5)$ First term of each factor is x. Second terms are $+1$ and $+5$.

Check by using FOIL to multiply.

EXAMPLE   Factor $x^2 + 4x - 12$.

$x^2 + 4x - 12$ Terms are already in descending order. Last term $-$ means signs of thc bionomials are unlike.

$(\ + \ )(\ - \ )$ Middle term $+$ means the factor with the larger absolute value in the selected factor pair will be $+$.

Factor pairs of 12:

$1 \cdot 12$

$\boxed{2 \cdot 6}$ Algebraic sum of -2 and $+6$ is $+4$.

$3 \cdot 4$

$(x - 2)(x + 6)$ First term of each factor is x. Second terms are -2 and 6.

EXAMPLE   Factor $x^2 - 20 - x$.

$x^2 - x - 20$ Arrange in descending order of x. Last sign $-$ means signs of bionominls are unlike.

Middle sign $-$ means the factor with the larger absolute value in the selected factor pair will be $-$.

Factor pairs of 20:

$1 \cdot 20$

$2 \cdot 10$

$\boxed{4 \cdot 5}$ Algebraic sum of $+4$ and -5 is -1.

$(x + 4)(x - 5)$ First term of each factor is x. Second terms are $+4$ and -5.

---

2 **Remove Common Factors After Grouping an Expression.**

Algebraic expressions that have more than three terms and have no factors common to every term in the expression may have common factors for some groups of terms. In these cases, the expression may be written as groupings.

To remove common factors after grouping an expression:

1. Identify groupings.
2. Factor all common factors from each grouping.
3. Examine each term to see if there are common groupings. If so, factor out the common grouping.

EXAMPLE   Write the expression $2x^2 - 2xb + ax - ay$ as an alternate expression by grouping pairs of terms. Factor common factors from each grouping.

$2x^2 - 2xb + ax - ay$ Group two terms in each grouping.
$(2x^2 - 2xb) + (ax - ay)$ Factor common factors from each grouping.

$2x(x - b) + a(x - y)$ There is no common grouping in the terms.

Groupings Are Not Necessarily Factors

In the preceding example the groupings $(2x^2 - 2xb)$ and $(ax - ay)$ are not factors. They are groupings that are added. To rewrite the expression as $2x(x - b) + a(x - y)$ is not in factored form. The expression has two terms and each term has two or more factors. *An expression in factored form is only one term.*

EXAMPLE   Write the expressions in factored form by using grouping.

(a) $mx + 2m - 4x - 8$ (b) $y^2 + 2xy + 3y + 6x$
(c) $3x^2 - 9x - 7x + 21$ (d) $2x^2 + 8x + 5y - 15$

(a) $mx + 2m - 4x - 8 =$ Group the four-termed expression into two terms.

$(mx + 2m) + (-4x - 8) =$ Factor common factors in each of the two terms.

$m(x + 2) + {}^-4(x + 2) =$ Convert double signs to an equivalent single sign.

$m\,(x + 2)\ -\ 4\,(x + 2) =$ Factor into one term by factoring out the common binomial factor $(x + 2)$.

$(x + 2)(m - 4)$ or $(m - 4)(x + 2)$ Check the result by using the FOIL method.

$(x + 2)(m - 4) = xm - 4x + 2m - 8$ Rearrange terms and factors (commutative properties of multiplication and addition).

$= mx + 2m - 4x - 8$ The factoring checks.

(b) $y^2 + 2xy + 3y + 6x =$ Group into two terms.
$(y^2 + 2xy) + (3y + 6x) =$ Factor the common factor in each term.

$y\,(y + 2)\ +\ 3\,(y + 2) =$ Factor into one term by factoring the common binomial factor.

$(y + 2)\,(y + 3)$ or $(y + 3)(y + 2)$

(c) $3x^2 - 9x - 7x + 21 =$ Group into two terms.
$(3x^2 - 9x) + (-7x + 21) =$ Factor each term.
$3x(x - 3) + {}^-7(x - 3) =$ Convert double signs to a single sign.

$3x\,(x - 3)\ -\ 7\,(x - 3) =$ Factor into one term by factoring the common binomial factor.

$(x - 3)\,(3x - 7)$ or $(3x - 7)(x - 3)$

(d) $2x^2 + 8x + 5y - 15 =$ Group into two terms.
   $(2x^2 + 8x) + (5y - 15) =$ Factor each term.
   $2x(x + 4) + 5(y - 3)$

In this example, the two terms do not have a common factor. Thus, the expression cannot be factored into one term. Even if we rearrange the terms, we will not be able to write the expression as a single term.

$2x^2 + 8x + 5y - 15$ is prime.

TIP Make Leading Coefficient of Binomial Positive

When factoring by grouping, manipulate the signs of the common factor so that the leading coefficient of the binomial is positive.

$$-3x + 9$$

Factor as $-3(x - 3)$, *not* $3(-x + 3)$. Parts a and c of the preceding example illustrate this tip.

3 Factor a General Trinomial by Grouping.

We can now use a systematic method to factor general trinomials with leading integral coefficients that are positive and not equal to 1.

To factor a general trinomial of the form $ax^2 + bx + c$ by grouping:

1. Verify that a is a positive integer or rewrite with -1 as the common factor.
2. Multiply the coefficient of the first term of the trinomial by the coefficient of the last term.
3. Factor the product from Step 2 into a pair of factors:
   (a) whose *sum* is the coefficient of the middle term if the sign of the last term is positive, or
   (b) whose *difference* is the coefficient of the middle term if the sign of the last term is negative.
   If there is no factor pair that meets one of these conditions, the original trinomial is prime.
4. Rewrite the trinomial as a polynomial with four terms by replacing the middle term with two terms that have the coefficients identified in Step 3.
5. Group the polynomial with four terms from Step 4 into two groups of two terms.
6. Factor the common factors from each of the two groups.
7. Factor the common binomial.

EXAMPLE   Factor $6x^2 + 19x + 10$ by grouping.

   $6(10) = 60$ Multiply the coefficients of the first and third terms.

   $60 = (1)60$ List all factor pairs of 60.
      $(2)30$
      $(3)20$
      $(4)15$ Identify the pair that *adds* to 19, since the sign of the last term is positive.
      $(5)12$
      $(6)10$

$$6x^2 + 19x + 10 =$$ Separate $+19x$ into two terms using the coefficients 4 and 15.

$$6x^2 + 4x + 15x + 10 =$$ Group into two terms.

$$(6x^2 + 4x) + (15x + 10) =$$ Factor common factors in each term.

$$2x(3x + 2) + 5(3x + 2) =$$ Factor common binomial.

$$(3x + 2)(2x + 5) \text{ or } (2x + 5)(3x + 2)$$ Check using the FOIL method.

EXAMPLE   Factor $20x^2 - 23x + 6$ by grouping.

$$20(6) = 120$$ Find the product of 20 and 6.

List the factor pairs of 120 and select the pair that has a sum of 23.

$$120 = \begin{array}{l} (1)120 \\ (2)60 \\ (3)40 \\ (4)30 \\ (5)24 \\ (6)20 \\ (8)15 \\ (10)12 \end{array}$$ List all factor pairs of 120.

Identify the pair that *adds* to 23, since the sign of the last term is *positive*.

$$20x^2 - 23x + 6 =$$ Separate $-23x$ into two terms using the coefficients -8 and -15.

$$20x^2 - 8x - 15x + 6 =$$ Group into two terms. Be sure the second grouping keeps the negative sign with $15x$.

$$(20x^2 - 8x) + (-15x + 6) =$$ Factor common factors in each term so that the leading coefficient in each bionomial is positive.

$$4x(5x - 2) - 3(5x - 2) =$$ Factor the common binomial.

$$(5x - 2)(4x - 3)$$ Check using the FOIL method.

If the third term of a trinomial has a negative sign, we use the same procedure but look for two factors whose *difference* is the coefficient of the middle term.

EXAMPLE   Factor $10x^2 + 19x - 15$ by grouping.

$$10(15) = 150$$ Find the product of 10 and 15.

List the factor pairs of 150 and select the pair that has a difference of 19.

$$150 = \begin{array}{l} (1)150 \\ (2)75 \\ (3)50 \\ (5)30 \\ (6)25 \\ \\ (10)15 \end{array}$$ List all factor pairs of 150.

Identify the pair that *subtracts* to 19, since the last term is *negative*.

The factors 25 and 6 have a difference of 19. When we rewrite the trinomial, we write $19x$ as $25x - 6x$.

$$10x^2 + 19x - 15 =$$

Separate $+19x$ into two terms using the coefficients 25 and -6. The signs will be different and the larger coefficient will be positive since the sign of $19x$ is positive.

$$10x^2 + 25x - 6x - 15 =$$

Group into two terms.

$$(10x^2 + 25x) + (-6x - 15) =$$

Factor common factors in each term so that the leading coefficient in each term is positive.

$$5x(2x + 5) - 3(2x + 5) =$$

Factor the common binomial.

$$(2x + 5)(5x - 3)$$

Check using the FOIL method.

TIP | Shortening the Process for Factoring by Grouping

The following is a variation of the factor-by-grouping method. The variation is mathematically sound and employs a strategy using the property of 1 that is often overlooked.

Factor $6x^2 + 7x - 20$.

$6x^2 + 7x - 20$ Multiply the coefficients of the first and third terms: $6(-20) = -120$

$120 =$

1	120
2	60
3	40
4	30
5	24
6	20
8	15
10	12

A negative product means the factor pair has unlike signs. List all factor pairs of 120.

Identify the factor pair with a *difference* of 7.
The larger factor will be positive. **$-8 + 15 = 7$**

Variation in procedure begins here.

$$\frac{(6x \quad)(6x \quad)}{6}$$

Use the coefficient of the first term as the coefficient of the first term in *each* binomial. This gives us an extra factor of 6, and we compensate by *dividing the expression by 6*.

$$\frac{(6x - 8)(6x + 15)}{6} =$$

Use the factors of -120 that have an algebraic sum of 7 as the second term of each binomial. Signs will be unlike.

$$\frac{2(3x - 4)(3)(2x + 5)}{6} =$$

Factor the common factors from each binomial and simplify the numerical factors.

$$\frac{\cancel{6}(3x - 4)(2x + 5)}{\cancel{6}} =$$

The factors of 6 in the numerator and denominator reduce.

$$(3x - 4)(2x + 5)$$

Check using the FOIL method.

Now, let's shorten the process again.

Once we have selected the pair of factors whose sum or difference matches the middle term, *make two fractions using the selected factor pair as the denominators. The leading coefficient of the trinomial will be the numerator of each fraction.*

$$6x^2 + 7x - 20 \qquad 6(-20) = -120$$

The factor pair of -120 that has a sum of $+7$ is -8 and $+15$. Make fractions and reduce each fraction.

$$\frac{6}{-8} = \frac{3}{-4} \qquad \frac{6}{+15} = \frac{2}{+5}$$ The leading coefficient of the trinomial is the numerator of both fractions.

Each reduced fraction gives the coefficients of one of the binomial factors. 3, -4 and 2, $+5$

$$(3x - 4)(2x + 5)$$

Let's try this process with another example.

If the trinomial has a common factor, we must factor the common factor before finding the factor pair and writing the fractions. The common factor will be part of the final answer.

Factor $30x^2 + 8 - 32x$.

$30x^2 - 32x + 8 =$ Arrange in descending powers of x.

$2\,(15x^2 - 16x + 4)$ Factor any common factors.

$15(4) = 60$ Write the factor pairs of 60.

1	60
2	30
3	20
4	15
5	12
6	10

Identify the pair that has a *sum* of 16.
Signs will be alike, both negative. $-6 + (-10) = -16$

Make fractions with the factor pair -6 and -10 as the denominators and the leading coefficient of the factored trinomial, 15, as the numerators.

$$\frac{15}{-6} = \frac{5}{-2} \qquad \frac{15}{-10} = \frac{3}{-2}$$ Reduce each fraction.

Write the factors using the coefficients 5, -2 and 3, -2. Don't forget the common factor.

$2\,(5x - 2)(3x - 2)$

TIP

Practice Moves You from Systematic Processes to Intuitive Processes

Systematic processes help you understand the concepts and build confidence. They also help you develop your mathematical senses such as your number sense and your spatial sense. The more you practice, the more you develop your mathematical senses.

Many students instinctively and automatically move to shortened and mental processes. Many times you can test a few combinations of coefficients and find the correct factors without going through the entire systematic process. With practice, you can write expressions in factored form more readily.

---

4 Factor Any Binomial or Trinomial That Is Not Prime.

Now let's develop a strategy for factoring any binomial or trinomial that can be factored. This strategy allows us to say with confidence that a particular binomial or trinomial will not factor and is prime.

To factor any binomial or trinomial:

Perform the following steps in order.

1. Factor the greatest common factor (if any).
2. Check the binomial or trinomial to see if it is a special product.
   (a) If it is the difference of two perfect squares, use the pattern.
   (b) If it is a perfect-square trinomial, use the pattern.
   (c) If it is the sum or difference of two perfect cubes, use the appropriate pattern.
3. If there is no special product, factor by grouping or by any appropriate process.
4. Examine each factor to see if it can be factored further.
5. Check factoring by multiplying.

The steps for factoring any binomial or trinomial are presented visually in the flowchart (Fig. 13–1).

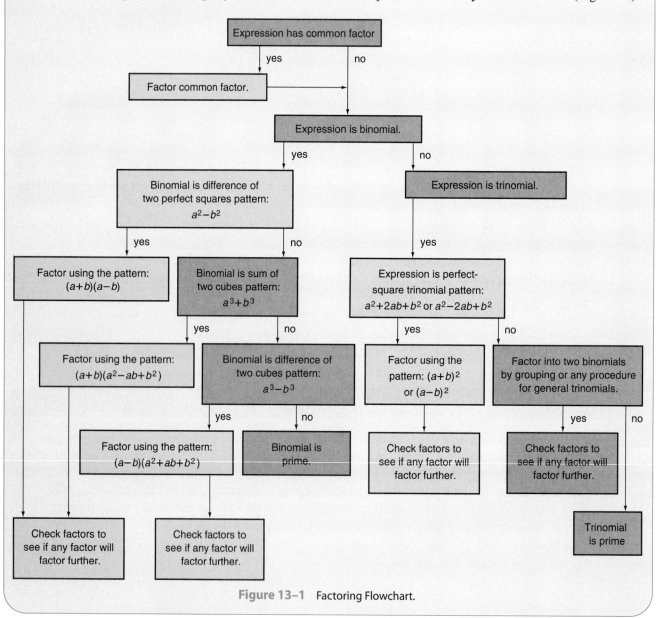

Figure 13–1   Factoring Flowchart.

EXAMPLE Completely factor $4x^3 - 2x^2 - 6x$.

$2x(2x^2 - x - 3)$ $2x$ is the common factor. The trinomial is not a special product.

$\mathbf{2x(2x - 3)(x + 1)}$ Factor the trinomial using any appropriate process but *keep* the $2x$ factor. Check by multiplying.

EXAMPLE Completely factor $12x^2 - 27$.

$3(4x^2 - 9)$ 3 is the common factor. The binomial is the difference of two perfect squares.

$\mathbf{3(2x - 3)(2x + 3)}$ Factor the difference of two perfect squares. Keep the common factor of 3. Check.

EXAMPLE Completely factor $-18x^3 + 24x^2 - 8x$.

$-2x(9x^2 - 12x + 4)$ Look for a common factor: $-2x$. We factor a negative when the leading coefficient is negative. It is preferred that the leading coefficient in the binomial or trinomial be positive. The resulting trinomial is a perfect-square trinomial.

$\mathbf{-2x(3x - 2)^2}$ Factor the perfect-square trinomial.

SECTION 13–3 SELF-STUDY EXERCISES

1 Factor.

1. $x^2 + 7x + 6$	**2.** $x^2 - 7x + 6$	**3.** $x^2 - 5x + 6$	**4.** $x^2 + 5x + 6$
5. $x^2 - 11x + 28$	**6.** $x^2 + 8x + 12$	**7.** $x^2 - 8x + 12$	**8.** $x^2 + 13x + 12$
9. $x^2 - 13x + 12$	**10.** $x^2 + 7x + 12$	**11.** $x^2 - 7x + 12$	**12.** $x^2 - 4x + 3$
13. $x^2 + 8x + 7$	**14.** $x^2 + 7x + 10$	**15.** $x^2 - x - 6$	**16.** $x^2 + x - 6$
17. $x^2 - 5x - 6$	**18.** $x^2 + 5x - 6$	**19.** $x^2 - x - 12$	**20.** $x^2 + x - 12$
21. $x^2 + 4x - 12$	**22.** $x^2 - 4x - 12$	**23.** $x^2 - 11x - 12$	**24.** $x^2 + 11x - 12$
25. $y^2 - 3y - 10$	**26.** $y^2 - y - 20$	**27.** $b^2 + 2b - 3$	**28.** $-14 - 5b + b^2$
29. $12 - 7x + x^2$	**30.** $-30 - x + x^2$	**31.** $11x + 18 + x^2$	**32.** $-9x + x^2 + 18$
33. $-7x - 18 + x^2$	**34.** $-18 + x^2 + 17x$	**35.** $x^2 + 9x + 20$	**36.** $x^2 - 12x + 20$
37. $x^2 - 10x + 16$	**38.** $x^2 - 17x + 16$	**39.** $x^2 - 13x - 14$	**40.** $x^2 - 5x - 14$

2 Factor the polynomials by removing the common factors after grouping.

41. $x^2 + xy + 4x + 4y$	**42.** $6x^2 + 4x - 3xy - 2y$	**43.** $3mx + 5m - 6nx - 10n$
44. $30xy - 35y - 36x + 42$	**45.** $x^2 - 2x + 8x - 16$	**46.** $6x^2 - 2x - 21x + 7$
47. $x^2 - 4x + x - 4$	**48.** $8x^2 - 4x + 6x - 3$	**49.** $x^2 - 5x + 4x - 20$
50. $3x^2 - 6x + 5x - 10$	**51.** $4x^2 + 8x - 3x - 6$	**52.** $4x^2 - 8x + 3x - 6$
53. $4x^2 + 8x + 3x + 6$	**54.** $4x^2 - 8x - 3x + 6$	**55.** $8x^2 + 4x - 6x - 3$

56. $3x^2 + 7x + 2$ **57.** $3x^2 + 14x + 8$ **58.** $6x^2 + 13x + 6$ **59.** $8x^2 + 2x - 3$

60. $6x^2 - 17x + 12$ **61.** $2x^2 - 9x + 10$ **62.** $6x^2 - 13x + 5$ **63.** $8x^2 + 10x + 3$

64. $6x^2 - 11x + 5$ **65.** $8x^2 + 26x + 15$ **66.** $15x^2 - 22x - 5$ **67.** $8x^2 - 10x + 3$

68. $2x^2 - 5x - 7$ **69.** $12x^2 + 8x - 15$ **70.** $10x^2 + x - 3$ **71.** $12x^2 + 5x - 2$

72. $12x^2 - 5x - 2$ **73.** $12x^2 + 11x + 2$ **74.** $12x^2 - 11x + 2$ **75.** $24x^2 + 5x - 1$

76. $24x^2 - 11x + 1$ **77.** $6x^2 + 7xy - 10y^2$ **78.** $6a^2 - 17ab - 14b^2$ **79.** $18x^2 - 3x - 10$

80. $20x^2 - xy - 12y^2$ **81.** $7x^2 + x - 8$ **82.** $2a^2 + 21a + 19$

4 Factor the polynomials completely. Identify common factors first, then identify special cases.

83. $4x - 4$ **84.** $x^2 + x - 56$ **85.** $2x^2 + x - 3$ **86.** $x^2 - 9$

87. $4x^2 - 16$ **88.** $m^2 + 2m - 15$ **89.** $2a^2 + 6a + 4$ **90.** $b^2 + 6b + 9$

91. $16m^2 - 8m + 1$ **92.** $x^2 - 8x + 7$ **93.** $2m^2 + 5m + 2$ **94.** $2m^2 - 5m - 3$

95. $2a^2 - 3a - 5$ **96.** $3x^2 + 10x - 8$ **97.** $6x^2 + x - 15$ **98.** $8x^2 + 10x - 3$

99. $-2x^2 + 6x - 4$ **100.** $x^4 - 16$ **101.** $x^6 - 81$ **102.** $x^6 - 27$

103. $3x^4 - 48$ **104.** $6x^3 - 48$ **105.** $3x^2 - 6x - 24$ **106.** $64a^9 + b^3$

CHAPTER REVIEW OF KEY CONCEPTS

Learning Objectives	What to Remember with Examples

Section 13–1

1 Factor an expression containing a common factor (pp. 519–521).

To factor an expression containing a common factor: **1.** Find the greatest factor common to each term of the expression. **2.** Divide each term by the common factor. **3.** Rewrite the expression as the indicated product of the greatest common factor and the remaining quantity.

> Factor $3ab + 9b^2$.
>
> $$3b\left(\frac{3ab}{3b} + \frac{9b^2}{3b}\right) = 3b(a + 3b)$$

Section 13–2

1 Recognize and factor the difference of two perfect squares (pp. 522–523).

The difference of two perfect squares is a binomial in the form $a^2 - b^2$.

> Identify which expression is the special product, the difference of two squares.
>
> **(a)** $36 - 27A^2$; no, 27 is not a perfect square.
> **(b)** $9c^2 - 4$; yes, both terms are perfect squares. The terms are subtracted.

To factor the difference of two perfect squares: **1.** Take the square root of the first term. **2.** Take the square root of the second term. **3.** Write one factor as the sum and one factor as the difference of the square roots from Steps 1 and 2. Symbolically, $a^2 - b^2 = (a + b)(a - b)$.

> Factor $9c^2 - 4$.
>
> $$(3c + 2)(3c - 2)$$

2 Recognize and factor a perfect-square trinomial (pp. 523–525).

A perfect-square trinomial is a trinomial in the form $a^2 + 2ab + b^2$. The first and last terms are positive perfect squares and the absolute value of the middle term is twice the product of the square roots of the first and last terms.

> Identify which expression is a perfect-square trinomial.
>
> **(a)** $4x^2 + 18x + 9$; not a perfect-square trinomial. The middle term is not twice the product of the square roots of the first and last terms. $(2)(2x)(3) = 12x$
> **(b)** $9a^2 + 12a + 4$; the first and last terms are perfect squares. The middle term is twice the product of the square roots of the first and last terms $(2 \cdot 3a \cdot 2 = 12a)$, so it is a perfect-square trinomial.

To factor a perfect-square trinomial: **1.** Write the square root of the first term. **2.** Write the sign of the middle term. **3.** Write the square root of the last term. **4.** Indicate the square of the quantity.

Symbolically, $a^2 + 2ab + b^2 = (a + b)^2$ and $a^2 - 2ab + b^2 = (a - b)^2$.

> Factor $9a^2 + 12a + 4$.
> $$(3a + 2)^2$$

3 Recognize and factor the sum or difference of two perfect cubes (pp. 525–526).

The sum of two perfect cubes is a binomial in the form $a^3 + b^3$. The difference of two perfect cubes is a binomial in the form $a^3 - b^3$.

> Identify the sum or difference of two perfect cubes:
>
> **(a)** $b^3 - 27$; yes, both terms are perfect cubes; difference of two cubes.
> **(b)** $x^3 - 6$; no, 6 is not a perfect cube.
> **(c)** $8 + y^3$; yes, both terms are perfect cubes; sum of two cubes.

To factor the sum (difference) of two perfect cubes:
1. Write the binomial factor as the sum (difference) of the cube roots of the two terms.
2. Write the trinomial factor as the square of the first term from Step 1, insert a minus sign if factoring a sum (a plus sign if factoring a difference). Write the product of the two terms from Step 1. The third term of the trinomial is the square of the second term from Step 1.

Symbolically, $a^3 + b^3 = (a + b)(a^2 - ab + b^2)$ and $a^3 - b^3 = (a - b)(a^2 + ab + b^2)$.

> Factor $8 + y^3$.
> $$(2 + y)(4 - 2y + y^2)$$
> Factor $b^3 - 27$.
> $$(b - 3)(b^2 + 3b + 9)$$

Section 13–3

1 Factor general trinomials whose squared term has a coefficient of 1 (pp. 527–528).

To factor a general trinomial whose squared term has a coefficient of 1: Factor using the FOIL method in reverse. Use the factors of the third term that will give the desired algebraic sum that produces the middle term and sign.

> Factor $a^2 - 5a + 6$.
>
> $(\quad)(\quad)$ Factors of 6 that add to 5 are 3 and 2.
> $(a \quad 3)(a \quad 2)$ Use terms with like signs that are negative and add to -5.
> $(a - 3)(a - 2)$ These factors of 6, -3 and -2, give $-5a$ as the middle term.

| **2** Remove common factors after grouping an expression (pp. 528–530). | Arrange terms of expression into groups of two terms. Factor common factors from each group. Factor the common binomial factor if there is one. |

Factor $2x^2 - 4x + xy - 2y$ completely.

$(2x^2 - 4x) + (xy - 2y)$	Arrange into groups of two terms. Factor common factors.
$2x(x - 2) + y(x - 2)$	Factor common binomial.
$(x - 2)(2x + y)$	

| **3** Factor a general trinomial by grouping (pp. 530–533). | To factor a general trinomial by grouping: **1.** Verify that the first term is positive or rewrite with -1 as the common factor. **2.** Multiply the coefficients of the first term and last term. **3.** Factor the product from Step 2 into two factors (a) whose sum is the coefficient of the middle term if the last term is positive, or (b) whose difference is the coefficient of the middle term if the last term is negative. **4.** Rewrite the trinomial as four terms so the coefficients of the middle term use the factors from Step 3. **5.** Group the polynomial from Step 4 in two groups of two terms. **6.** Factor the common factors from each grouping. **7.** Factor the common binomial factor. |

Factor $6x^2 - 17x + 12$.

$6 \cdot 12 = 72$	Factor pairs of 72.
$1 \cdot 72$	Find the factor pair that has a sum of 17.
$2 \cdot 36$	
$3 \cdot 24$	
$4 \cdot 18$	
$6 \cdot 12$	
$8 \cdot 9$	Selected pair: $-8 + (-9) = -17$
$6x^2 - 8x - 9x + 12$	Rewrite the middle term. $-17x = -8x + (-9x)$
$(6x^2 - 8x) + (-9x + 12)$	Factor common factor from each binomial.
$2x(3x - 4) - 3(3x - 4)$	Factor the common binomial.
$(3x - 4)(2x - 3)$	

| **4** Factor any binomial or trinomial that is not prime (pp. 533–535). | To factor any binomial or trinomial: **1.** Factor the greatest common factor (if any). **2.** Check for any special products. **3.** If there is no special product, factor by grouping. **4.** Examine each factor to see if it can be factored further. **5.** Check factoring by multiplying. |

Factor $12x^3 + 6x^2 - 18x$ completely.

$6x(2x^2 + x - 3) =$	Factor common factors.
$6x(2x^2 + 3x - 2y - 3) =$	Separate x into two terms.
$6x[(2x^2 + 3x) + (-2x - 3)] =$	Group and factor common factors from each grouping.
$6x[x(2x + 3) - 1(2x + 3)] =$	Factor common grouping.
$6x(2x + 3)(x - 1)$	

(After removing the common factor, procedures for factoring general trinomials can be used to factor the remaining trinomial.)

Section 13–1

Factor by removing the greatest common factor.

1. $5x + 5y$ **2.** $2x + 5x^2$ **3.** $12m^2 - 8n^2$
4. $25x^2y - 10xy^3 + 5xy$ **5.** $2a^3 - 14a^2 - 2a$ **6.** $30a^3 - 18a^2 - 12a$
7. $15x^3 - 5x^2 - 20x$ **8.** $9a^2b^3 - 6a^3b^2$ **9.** $18a^3 - 12a^2$
10. $a^2b + ab^2$ **11.** $-6x^2 - 10x$ **12.** $-12n^2 - 18$
13. $5\sqrt{3} + 15\sqrt{7}$ **14.** $12\sqrt{5} - 21a\sqrt{5}$ **15.** $\sqrt{12} - 10\sqrt{7}$

Section 13–2

Identify the special products that are the difference of two perfect squares.

16. $25m^2 - 4n^2$ **17.** $64 + 4a^2$ **18.** $4f^2 - 9g^2$
19. $H^2 - G^2$ **20.** $s^2 - 4s + 10$ **21.** $64b^2 - 49$

Verify which of the following trinomials are perfect-square trinomials.

22. $16c^2 + 8c - 1$ **23.** $-9x^2 + 12x + 4$ **24.** $4d^2 + 2d + 16$
25. $9t^2 - 24tp + 16p^2$ **26.** $a^2 - 14a + 49$ **27.** $j^2 + 10j + 25$

Identify the special products that are the sum or difference of two perfect cubes.

28. $R^3 + 81$ **29.** $125a^3 - 8b^3$ **30.** $64 + m^3$
31. $8z^3 - 125$ **32.** $d^3 - 12d + 36$ **33.** $64W^3 + 27$

Factor the expressions that are special products. Explain why the other expressions are not special products.

34. $x^2 - 81$ **35.** $25y^2 - 4$ **36.** $100a^2 - 8ab^2$
37. $a^2b^2 + 49$ **38.** $121 - 9m^2$ **39.** $a^2 + 2a + 1$
40. $4x^2 + 12x + 9$ **41.** $16c^2 - 24bc + 9b^2$ **42.** $y^2 - 12y - 4$
43. $n^2 + 169 - 26n$ **44.** $16d^2 - 20d + 25$ **45.** $36a^2 + 84ab + 49b^2$
46. $4x^2 - 25y^2$ **47.** $49 - 14x + x^2$ **48.** $9x^2y^2 - 49z^2$
49. $64 + 25x^2$ **50.** $4x^2 + 12xy + 9y^2$ **51.** $16x^2 + 24x + 9y^2$
52. $36 - x^2$ **53.** $49 - 81y^2$ **54.** $64x^2 - 25y^2$
55. $9x^2 - 100y^2$ **56.** $a^2 - 10a + 25$ **57.** $9x^2 - 6xy + y^2$
58. $25 - 16a^2b^2$ **59.** $9x^2y^2 - z^2$ **60.** $x^2 - y^2$
61. $x^2 + 4x + 4$ **62.** $\frac{1}{4}x^2 - \frac{1}{9}y^2$ **63.** $\frac{4}{25}x^2 - \frac{1}{16}y^2$
64. $27v^3 - 8$ **65.** $T^3 - 8$ **66.** $8r^3 + 27$
67. $d^3 + 729$ **68.** $125c^3 - 216d^3$ **69.** $27K^3 + 64$

Section 13–3

Factor the trinomials.

70. $x^2 + 10x + 21$ **71.** $x^2 + 11x + 24$ **72.** $x^2 + 29x + 28$
73. $x^2 + 13x + 30$ **74.** $x^2 - 13x + 40$ **75.** $x^2 - 9x + 8$
76. $x^2 - 17x - 18$ **77.** $x^2 - 11x - 26$ **78.** $x^2 + x - 30$
79. $x^2 + 5x - 24$ **80.** $6x^2 + 25x + 14$ **81.** $6x^2 + 25x + 4$
82. $4x^2 - 23x + 15$ **83.** $5x^2 - 34x + 24$ **84.** $3x^2 - x - 14$
85. $6x^2 - x - 35$ **86.** $3x^2 + 11x - 4$ **87.** $7x^2 - 13x - 24$

Factor the polynomials. Look for common factors and special cases.

88. $5mn - 25m$

89. $9a^2 - 100$

90. $a^2 - b^2$

91. $2x^2 - 3x - 2$

92. $b^3 - 8b^2 - b$

93. $a^2 - 81$

94. $x^2 - 14x + 13$

95. $y^2 - 14y + 49$

96. $m^2 - 3m + 2$

97. $b^2 + 8b + 15$

98. $x^2 - 13x + 30$

99. $169 - m^2$

100. $5x^2 + 13x + 6$

101. $x^2 - 4x - 32$

102. $x^2 + 9x + 14$

103. $x^2 + 19x - 20$

104. $x^2 + 8x - 20$

105. $2x^2 - 4x - 16$

106. $5x^2 - 20$

107. $2x^3 - 10x^2 - 12x$

108. $25m^2 - 121n^2$

TEAM PROBLEM-SOLVING EXERCISES

1. A standard-sized rectangular swimming pool is 25 ft long and 15 ft wide. This gives a water-surface area of 375 ft². A customer wants to examine some options for varying the size of the pool.

   (a) Write an expression in both factored and expanded form for finding the water-surface area of a pool that is changed by x feet in length and y feet in width.

   (b) Write an expression using one variable in both factored and expanded form for the water-surface area of a pool when the length and width increase by the same amount.

2. Use the expressions written in Exercise 1 to answer the following.

   (a) Will the expressions work for both increasing and decreasing the size of the pool?

   (b) Illustrate your answer to part (a) with numerical examples.

CONCEPTS ANALYSIS

1. List the properties of a binomial that is the difference of two perfect squares.

3. Explain in sentence form how the sign of the third term of a general trinomial affects the signs between the terms of its binomial factors.

5. Write a brief comment explaining each lettered step of the following example.

   Factor $10x^2 - 19x - 12$ using grouping.

   (a)
$$\frac{120}{}$$
1, 120
2, 60
3, 40
4, 30
5, 24*   Factors with difference of 19.
6, 20
8, 15
10, 12
$10x^2 - 19x - 12$

   (b) $10x^2 + 5x - 24x - 12$

   (c) $5x(2x + 1) - 12(2x + 1)$

   (d) $(2x + 1)(5x - 12)$

7. Write the pattern for factoring the difference of two cubes and illustrate it with an example.

9. Write the pattern for factoring the difference of two perfect squares and illustrate it with an example.

2. List the properties of a perfect-square trinomial.

4. What do we mean if we say that a polynomial is prime?

6. Find a mistake in each of the following. Briefly explain the mistake, then work the problem correctly.

   (a) $(3x - 2y)^2 = 9x^2 + 4y^2$

   (b) $5xy^2 - 45x = 5x(y^2 - 9)$

   (c) $2x^2 - 28x + 96 = (2x - 12)(x - 8)$

   (d) $x^2 - 5x - 6 = (x - 2)(x - 3)$

8. What is the value of being able to recognize the special products when we factor algebraic expressions?

10. Write the pattern for factoring a perfect square trinomial and illustrate it with an example.

Factor by removing the greatest common factor.

1. $7x^2 + 8x$

2. $6ax + 15bx$

3. $7a^2b - 14ab$

Factor into special products.

4. $a^2 - 25$

5. $9x^2 - 25$

6. $x^2 + 4x + 4$

7. $x^2 - 18x + 81$

8. $125m^3 - 8n^3$

9. $27r^3 + 64s^3$

Factor into two binomials.

10. $x^2 + 5x - 24$

11. $6x^2 - 5x - 6$

12. $y^2 - y - 6$

13. $a^2 + 16ab + 64b^2$

14. $x^2 - 7x + 10$

15. $b^2 - 3b - 10$

16. $3x^2 + x - 4$

17. $3m^2 - 5m + 2$

18. $3x^2 + 11xy + 3y^2$

Factor completely.

19. $3x^2 - 12$

20. $3x^2 - 3x - 18$

21. $5x^2 - 20$

22. $27a^2 - 48$

23. $3x^2 + 12x + 12$

24. $3x^2 - 15x - 18$

14

Rational Expressions and Equations

14–1 Simplifying Rational Expressions
14–2 Multiplying and Dividing Rational Expressions
14–3 Adding and Subtracting Rational Expressions
14–4 Solving Equations with Rational Expressions

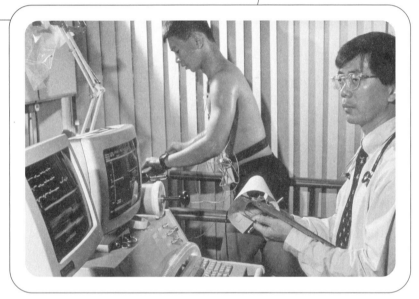

Focus on Careers

Cardiovascular technologists and technicians assist physicians in diagnosing and treating cardiac (heart) and vascular (blood-vessel) illnesses. Some prepare patients for catheterization or angioplasty and monitor blood pressure and heart rate during the procedures. Others perform medical history and use ultrasound to record blood flow, blood pressure, oxygen saturation, etc.

About 75% of jobs in this career are in hospitals. Others are in physician's offices or in diagnostic laboratories.

Cardiovascular technologists normally complete a community college program of study with specialized instruction in cardiovascular or vascular technology. Persons who graduate from a program accredited by the Joint Review Committee on Education in Cardiovascular Technology are eligible for certification from Cardiovascular Credentialing International.

Jobs for this field are expected to grow faster than average for all career paths through 2012. Median earnings for cardiovascular technologists and technicians were $36,430 in 2002, and the highest 10% earned more than $56,080. Employees in physicians' offices generally earned slightly more than employees in general medical and surgical hospitals.

Source: Bureau of Labor Statistics, U.S. Department of Labor, *Occupational Outlook Handbook,* 2004–2005 Edition, Cardiovascular Technologists and Technicians, on the Internet at http://www.bls.gov/oco/ocos100.htm.

Once we become proficient in factoring algebraic expressions, we can use this skill to simplify *rational expressions,* also called *algebraic fractions.*

14–1 | *Simplifying Rational Expressions*

Learning Outcome **1** Simplify or reduce rational expressions.

1 Simplify or Reduce Rational Expressions.

In arithmetic, we learned to reduce fractions by reducing or dividing *factors* that are common to both the numerator and denominator. In our study of the laws of exponents, we also *reduced* or *simplified* algebraic fractions by reducing common factors. This process is accomplished whenever the numerator and denominator of the fraction are written in factored form. In the following example, we review the arithmetic process with fractions and the division laws of exponents. To focus on key aspects of the process, let's write steps that we often do mentally.

EXAMPLE Simplify or reduce the following fractions.

(a) $\dfrac{12}{15}$ (b) $\dfrac{24}{36}$ (c) $\dfrac{2x^2yz^3}{4xy^2z}$ (d) $\dfrac{9a^2b^3}{3ab}$

(a) $\dfrac{12}{15} = \dfrac{(2)(2)(3)}{(3)(5)} =$ Write in factored form using prime factorization.

$\dfrac{(2)(2)(3)}{(3)(5)} = \dfrac{\mathbf{4}}{\mathbf{5}}$ Reduce the common factors and multiply the remaining factors.

(b) $\dfrac{24}{36} = \dfrac{(2)(2)(2)(3)}{(2)(2)(3)(3)} =$ Write in factored form using prime factorization.

$\dfrac{(2)(2)(2)(3)}{(2)(2)(3)(3)} = \dfrac{\mathbf{2}}{\mathbf{3}}$ Reduce the common factors.

(c) $\dfrac{2x^2yz^3}{4xy^2z} = \dfrac{2x^{2-1}y^{1-2}z^{3-1}}{(2)(2)} =$ Write coefficients in factored form and reduce. Apply the laws of exponents to variable factors with like bases.

$\dfrac{xy^{-1}z^2}{2} = \dfrac{\mathbf{xz^2}}{\mathbf{2y}}$ Express all variable factors with positive exponents.

(d) $\dfrac{9a^2b^3}{3ab} = \dfrac{(3)(3)a^{2-1}b^{3-1}}{3}$ Write coefficients in factored form and reduce. Apply the laws of exponents to variable factors with like bases.

$= \mathbf{3ab^2}$

Before continuing, we must understand that we are always reducing common *factors.* This procedure *does not* apply to *addends* or *terms.*

Does $\frac{5}{10}$ reduce to $\frac{3}{8}$? We know from previous experience that $\frac{5}{10}$ reduces to $\frac{1}{2}$, or 0.5. Then what is wrong with the following argument?

Both the numerator and the denominator of the fraction $\frac{5}{10}$ can be rewritten as addends or terms:

Does $\dfrac{5}{10} = \dfrac{2+3}{2+8} = \dfrac{3}{8}$?   Can common addends be reduced? No!

We can verify with our calculator that this is an incorrect statement. Common *addends* cannot be reduced. To correct the process, we reduce *only factors*. Thus, we rewrite $\frac{5}{10}$ in factored form:

$$\frac{5}{10} = \frac{(1)(\cancel{5})}{(2)(\cancel{5})} = \frac{1}{2}$$

If common factors are reduced,

$$\frac{5}{10} = \frac{1}{2}$$

Now let's use our knowledge of factoring *polynomials* to simplify rational expressions. A **rational expression** is an algebraic fraction in which the numerator or denominator or both are polynomials.

First, let's look at some examples that are written in factored form.

EXAMPLE   Simplify the rational expressions that are already in factored form.

(a) $\dfrac{x(x+2)}{(x+2)(x+3)}$ (b) $\dfrac{3ab(a+4)}{6ab(a+3)}$

(c) $\dfrac{2x^2(2x-1)}{x^3(x-1)}$ (d) $\dfrac{(x+3)(x-4)}{(x-3)(x+4)}$

(a) $\dfrac{x(x+2)}{(x+2)(x+3)} = \dfrac{x(\cancel{x+2})}{(\cancel{x+2})(x+3)}$ Reduce common binomial factors.

$$= \dfrac{x}{x+3}$$

The remaining x and $x+3$ cannot be reduced. The x in the numerator is a factor ($x = 1 \cdot x$); however, the x in the denominator is an *addend* or *term*.

(b) $\dfrac{3ab(a+4)}{6ab(a+3)} = \dfrac{\overset{1}{\cancel{3ab}}(a+4)}{\underset{2}{\cancel{6ab}}(a+3)} =$ Reduce common factors:

$1(a+4) = a+4$

$\dfrac{a+4}{2(a+3)}$ or $\dfrac{a+4}{2a+6}$ Reduced rational expressions can be written in *either* factored or expanded form.

Reduce common factor:

(c) $\dfrac{2x^2(2x-1)}{x^3(x-1)} =$ $\dfrac{x^2}{x^3} = x^{2-3} = x^{-1} = \dfrac{1}{x}$

$\dfrac{2(2x-1)}{x(x-1)}$ or $\dfrac{4x-2}{x^2-x}$ Reduced rational expressions can be written in *either* factored or expanded form.

(d) $\dfrac{(x + 3)(x - 4)}{(x - 3)(x + 4)}$ or $\dfrac{x^2 - x - 12}{x^2 + x - 12}$ There are no common factors; therefore, the fraction is already in lowest terms. Reduced rational expressions can be written in *either* factored or expanded form.

To simplify rational expressions:

1. Factor *completely* both the numerator and denominator.
2. Reduce factors common to both the numerator and denominator.
3. Write the simplified expression in either factored or expanded form.

EXAMPLE   Reduce each rational expression to its simplest form.

(a) $\dfrac{x + y}{4x^2 + 4xy}$ (b) $\dfrac{a^2 + b^2}{a^2 - b^2}$ (c) $\dfrac{x^2 - 6x + 9}{x^2 - 9}$

(d) $\dfrac{3x^2 - 12}{6x + 12}$ (e) $\dfrac{a - b}{b - a}$

(a) $\dfrac{x + y}{4x^2 + 4xy} = \dfrac{(x + y)}{4x(x + y)} =$ Factor the common factor in the denominator and recall that the numerator is a grouping. Reduce common binomial factor.

$\dfrac{(x + y)}{4x(x + y)} = \dfrac{1}{4x}$ Numerator of 1 is necessary: $x + y = 1(x + y)$

(b) $\dfrac{a^2 + b^2}{a^2 - b^2} = \dfrac{a^2 + b^2}{(a + b)(a - b)}$ Factor the difference of the squares in the denominator.

The numerator, which is the *sum* of the squares, will not factor. There are no factors common to both the numerator and denominator. Thus, the fraction is in simplest form:

$\dfrac{a^2 + b^2}{(a + b)(a - b)}$ **or** $\dfrac{a^2 + b^2}{a^2 - b^2}$

(c) $\dfrac{x^2 - 6x + 9}{x^2 - 9} = \dfrac{(x - 3)(x - 3)}{(x + 3)(x - 3)}$ Write both the numerator and the denominator in factored form.

$= \dfrac{(x - 3)(x - 3)}{(x + 3)(x - 3)} = \dfrac{x - 3}{x + 3}$ Reduce common binomial factors.

(d) $\dfrac{3x^2 - 12}{6x + 12} = \dfrac{3(x^2 - 4)}{6(x + 2)} = \dfrac{3(x + 2)(x - 2)}{(3)(2)(x + 2)}$ Factor the numerator and the denominator completely.

$= \dfrac{3(x + 2)(x - 2)}{(3)(2)(x + 2)} = \dfrac{x - 2}{2}$ Reduce common numerical and binomial factors.

14–1 Simplifying Rational Expressions   545

(e)

$$\frac{a - b}{b - a} =$$

Write terms in numerator and denominator in same order.

$$\frac{a - b}{-a + b} =$$

Factor the denominator so that the leading coefficient of the binomial is positive.

$$\frac{a - b}{-1(a - b)} =$$

Reduce the common binomial factor.

$$\frac{1}{-1} = -1$$

TIP

Don't Forget 1 or −1

In the preceding example, part (e), we have some very important observations to make.

- When all other factors of a numerator or denominator reduce, a factor of 1 remains.
- Polynomial factors are opposites when every term of one grouping has an opposite in the other grouping.

$(a - b)$ and $(b - a)$ are opposites.   From part e

$(2m - 5)$ and $(5 - 2m)$ are opposites.

- Opposites differ by a factor of −1.

$$b - a = -1(-b + a) = -1(a - b) \quad \text{From part e}$$
$$5 - 2m = -1(-5 + 2m) = -1(2m - 5)$$

- When a numerator and denominator are opposites, the fraction reduces to −1.

$$\frac{a - b}{b - a} = \frac{1(a - b)}{-1(a - b)} = -1 \qquad \frac{2m - 5}{5 - 2m} = \frac{1(2m - 5)}{-1(2m - 5)} = -1$$

- It is helpful in recognizing common factors to rearrange or factor with −1 as a common factor so that leading coefficients within a grouping are positive: $(-x + 5) = -1(x - 5)$ or $(5 - x)$.
- It is helpful in recognizing common factors for the terms in both the numerator and denominator to be arranged in the same order.

$$\frac{x + 5}{5 + x} = \frac{x + 5}{x + 5} \qquad \frac{x - 2}{2 - x} = \frac{x - 2}{-1(x - 2)}$$

SECTION 14–1 SELF-STUDY EXERCISES

1 Simplify the rational expressions.

1. $\dfrac{8}{18}$

2. $\dfrac{9}{24}$

3. $\dfrac{4a^2b^3}{2ab}$

4. $\dfrac{27a^3bc^2}{18a^2b^4c^2}$

5. $\dfrac{x(x + 3)}{(x + 3)(x + 2)}$

6. $\dfrac{3x^2(3x + 2)}{x^3(2x - 1)}$

7. $\dfrac{(x + 2)(x - 5)}{(x + 5)(x - 2)}$

8. $\dfrac{x + y}{2x + 2y}$

9. $\dfrac{a + b}{a^2 - b^2}$

10. $\dfrac{x^2 - 4x + 4}{x^2 - 4}$

11. $\dfrac{4x^2 - 16}{6x + 12}$

12. $\dfrac{3m - 11}{11 - 3m}$

13. $\dfrac{x^2 - x - 6}{3 - x}$

14. $\dfrac{b - a}{2a - 2b}$

15. $\dfrac{3x - 6}{2 - x}$

16. $\dfrac{2x - 3y}{6y - 4x}$

17. $\dfrac{5m - 10n}{2n - m}$

18. $\dfrac{2x - 6}{-x^2 + 5x - 6}$

19. $\dfrac{x^2 - 3x - 10}{-x^2 - 3x - 2}$

20. $\dfrac{-x^2 - 9x - 14}{x^2 + 9x + 14}$

21. $\dfrac{x^2 - 4x - 21}{x^2 - 5x - 14}$

22. $\dfrac{2x^2 - 7x + 3}{2x^2 + 5x - 3}$

23. $\dfrac{3x^2 - 4x - 7}{3x^2 + 5x - 28}$

24. $\dfrac{6x^2 + x - 2}{9x^2 - 4}$

14–2 │ *Multiplying and Dividing Rational Expressions*

Learning Outcomes

1 Multiply and divide rational expressions.

2 Use multiplication and conjugates to rationalize a numerator or denominator of a fraction that has a binomial with an irrational term.

1 Multiply and Divide Rational Expressions.

We can connect our knowledge of arithmetic of multiplying and dividing fractions and the laws of exponents to expand our mathematical experience. We express results in lowest terms or in simplified form.

EXAMPLE Multiply or divide as indicated. Express answers in simplest form.

(a) $\dfrac{5}{8} \cdot \dfrac{16}{25}$

(b) $\dfrac{7}{16} \div 2$

(c) $\dfrac{15xy^2}{xy^3} \cdot \dfrac{(xy)^3}{5x}$

(a) $\dfrac{5}{8} \cdot \dfrac{16}{25} = \dfrac{\overset{1}{\cancel{5}}}{\underset{1}{\cancel{8}}} \cdot \dfrac{\overset{2}{\cancel{16}}}{\underset{5}{\cancel{25}}}$ Reduce factors common to a numerator and denominator. Multiply remaining factors.

$= \dfrac{2}{5}$

If all factors that are common to a numerator and denominator are reduced or canceled before multiplying, the product will be in lowest terms.

(b) $\dfrac{7}{16} \div 2$ Rewrite 2 as a fraction.

$\dfrac{7}{16} \div 2 = \dfrac{7}{16} \div \dfrac{2}{1}$ Rewrite division as multiplication and multiply.

$= \dfrac{7}{16} \cdot \dfrac{1}{2} = \dfrac{7}{32}$

(c) $\dfrac{15xy^2}{xy^3} \cdot \dfrac{(xy)^3}{5x}$ Raise grouping to power.

$= \dfrac{15xy^2}{xy^3} \cdot \dfrac{x^3y^3}{5x} = \dfrac{15x^4y^5}{5x^2y^3}$ Apply the laws of exponents.

$= \dfrac{15x^4y^5}{5x^2y^3}$ Reduce. (Reducing could have occurred before multiplication.)

$= \dfrac{15\left(x^{4-2}\right)\left(y^{5-3}\right)}{5}$

$= 3x^2y^2$

When applying the process of multiplying or dividing fractions to rational expressions, the key fact to remember is that the numerators and denominators should be written in *factored* form whenever possible.

To multiply or divide rational expressions:

1. Convert any division to an equivalent multiplication.
2. Factor completely every numerator and denominator.
3. Reduce factors that are common to a numerator and denominator.
4. Multiply remaining factors.
5. The result can be written in factored or expanded form.

EXAMPLE   Perform the operation and simplify.

(a) $\dfrac{x^2 - 4x - 12}{2x - 12} \cdot \dfrac{x - 4}{x^2 + 4x + 4}$ (b) $\dfrac{4y^2 - 9}{2y^2} \div \dfrac{y^2 - 2y - 15}{4y^2 + 12y}$

(c) $\dfrac{2x - 6}{1 - x} \div \dfrac{x^2 - 2x - 3}{x^2 - 1}$

(a) $\dfrac{x^2 - 4x - 12}{2x - 12} \cdot \dfrac{x - 4}{x^2 + 4x + 4}$ Factor each numerator and denominator.

$= \dfrac{(x + 2)(x - 6)}{2(x - 6)} \cdot \dfrac{x - 4}{(x + 2)(x + 2)}$ Reduce factors common to a numerator and denominator.

$= \dfrac{(x + 2)(x - 6)}{2(x - 6)} \cdot \dfrac{x - 4}{(x + 2)(x + 2)}$ Multiply remaining factors.

$= \dfrac{x - 4}{2(x + 2)}$ or $\dfrac{x - 4}{2x + 4}$ Write in factored or expanded form.

(b) $\dfrac{4y^2 - 9}{2y^2} \div \dfrac{y^2 - 2y - 15}{4y^2 + 12y}$ Convert to multiplication.

$$= \dfrac{4y^2 - 9}{2y^2} \cdot \dfrac{4y^2 + 12y}{y^2 - 2y - 15}$$ Factor each numerator and denominator.

$$= \dfrac{(2y + 3)(2y - 3)}{2y^2} \cdot \dfrac{4y(y + 3)}{(y + 3)(y - 5)}$$ Reduce factors common to a numerator and denominator.

$$= \dfrac{(2y + 3)(2y - 3)}{2\overset{}{\underset{y}{y^2}}} \cdot \dfrac{\overset{2}{4y}(y + 3)}{(y + 3)(y - 5)}$$ Multiply remaining factors.

$$= \dfrac{2(2y + 3)(2y - 3)}{y(y - 5)} \quad \text{or} \quad \dfrac{2(4y^2 - 9)}{y^2 - 5y} \quad \text{or} \quad \dfrac{8y^2 - 18}{y^2 - 5y}$$

(c) $\dfrac{2x - 6}{1 - x} \div \dfrac{x^2 - 2x - 3}{x^2 - 1}$ Convert to multiplication.

$$= \dfrac{2x - 6}{1 - x} \cdot \dfrac{x^2 - 1}{x^2 - 2x - 3}$$ Factor each numerator and denominator. Note the effect of factoring -1 in the denominator of the first fraction so the $x - 1$ will reduce with $x - 1$ in the numerator of the second fraction.

$$= \dfrac{2(x - 3)}{-1(x - 1)} \cdot \dfrac{(x + 1)(x - 1)}{(x + 1)(x - 3)}$$ Reduce factors common to a numerator and denominator.

$$= \dfrac{2(x - 3)}{-1(x - 1)} \cdot \dfrac{(x + 1)(x - 1)}{(x + 1)(x - 3)}$$ Multiply remaining factors.

$$= \dfrac{2}{-1} = -2$$

A **complex rational expression** is a rational expression that has a rational expression in its numerator or its denominator or both. Some examples of complex rational expressions are

$$\dfrac{\dfrac{4xy}{2x}}{5} \qquad \dfrac{\dfrac{x^2 - y^2}{2x}}{\dfrac{x - y}{3x^2}} \qquad \dfrac{1 + \dfrac{1}{x}}{\dfrac{2}{3}} \qquad \dfrac{\dfrac{2}{x} - \dfrac{5}{2x}}{\dfrac{3}{4x} + \dfrac{3}{x}}$$

To simplify a complex rational expression:

1. Rewrite the complex rational expression as a division then convert to an equivalent multiplication of rational expressions.
2. Multiply the rational expressions.

Before we look at examples involving complex rational expressions, the following Tip will eliminate some of our written steps.

Mentally Converting from Complex Form to Division and Then to Multiplication

Examine the symbolic representation for converting a complex expression to multiplication. The numerator is multiplied by the reciprocal of the denominator.

$$\frac{\dfrac{a}{b}}{\dfrac{c}{d}} = \frac{a}{b} \div \frac{c}{d} = \frac{a}{b} \cdot \frac{d}{c} \qquad b, c, \text{ and } d \neq 0$$

When simplifying a complex rational expression, we can eliminate some written steps by making this conversion mentally; for example,

$$\frac{\dfrac{4xy}{2x}}{5} \qquad \text{becomes} \qquad \frac{4xy}{1} \cdot \frac{5}{2x}$$

EXAMPLE Simplify.

(a) $\dfrac{\dfrac{4xy}{2x}}{5}$ (b) $\dfrac{\dfrac{x^2 - y^2}{2x}}{\dfrac{x - y}{3x^2}}$

(a) $\dfrac{\dfrac{4xy}{2x}}{5} = \dfrac{4xy}{1} \cdot \dfrac{5}{2x} =$ Multiply the numerator by the reciprocal of the denominator.

$\dfrac{\overset{2}{\cancel{4}xy}}{1} \cdot \dfrac{5}{\cancel{2}x} =$ Reduce. Multiply remaining factors.

$\dfrac{10y}{1} = \mathbf{10y}$ Simplify.

(b) $\dfrac{\dfrac{x^2 - y^2}{2x}}{\dfrac{x - y}{3x^2}} = \dfrac{x^2 - y^2}{2x} \cdot \dfrac{3x^2}{x - y} =$ Multiply the numerator by the reciprocal of the denominator. Factor the numerator of the first fraction.

$\dfrac{(x + y)(x - y)}{2x} \cdot \dfrac{3x^2}{x - y} =$ Reduce.

$\dfrac{(x + y)(x \cancel{- y})}{2\cancel{x}} \cdot \dfrac{3\overset{x}{\cancel{x^2}}}{\cancel{x - y}}$ Multiply remaining factors.

$= \dfrac{\mathbf{3x(x + y)}}{\mathbf{2}} \quad \text{or} \quad \dfrac{\mathbf{3x^2 + 3xy}}{\mathbf{2}}$ Factored or expanded form.

2 **Use Multiplication and Conjugates to Rationalize a Numerator or Denominator of a Fraction That has a Binomial with an Irrational Term.**

Our understanding of fractions and rational expressions can be used to manipulate fractional expressions that have irrational terms. In an advanced study of mathematics, there are occasions when you may want a numerator or denominator clear of irrational terms.

To use multiplication and conjugates to rationalize a numerator or denominator of a fraction that has a binomial with an irrational term:

1. Determine which part of the fraction (numerator or denominator) is to be cleared of a binomial with an irrational term.
2. Multiply by 1 in the form of $\dfrac{n}{n}$, where n is the conjugate of the binomial with the irrational term.
3. Simplify the resulting expression.

EXAMPLE Rationalize the denominator of the fraction $\dfrac{1 + 2\sqrt{3}}{2 - 5\sqrt{2}}$.

$\dfrac{1 + 2\sqrt{3}}{2 - 5\sqrt{2}}$

The conjugate of $2 - 5\sqrt{2}$ is $2 + 5\sqrt{2}$.

Multiply by 1 in the form of $\dfrac{2 + 5\sqrt{2}}{2 + 5\sqrt{2}}$.

$\dfrac{1 + 2\sqrt{3}}{2 - 5\sqrt{2}} \cdot \dfrac{2 + 5\sqrt{2}}{2 + 5\sqrt{2}}$

FOIL the numerators and use the special product for the sum and difference of two terms for the denominator.

$\dfrac{2 + 1(5\sqrt{2}) + 2(2\sqrt{3}) + 2\sqrt{3}(5\sqrt{2})}{2^2 - (5\sqrt{2})^2}$

Simplify by multiplying coefficients.

$\dfrac{2 + 5\sqrt{2} + 4\sqrt{3} + 10\sqrt{6}}{4 - 25(2)} =$

Simplify the denominator.

$\dfrac{2 + 5\sqrt{2} + 4\sqrt{3} + 10\sqrt{6}}{4 - 50} =$

Combine like terms in the denominator.

$\dfrac{2 + 5\sqrt{2} + 4\sqrt{3} + 10\sqrt{6}}{-46}$

Manipulate the signs of the fraction to make the denominator positive.

or

$-\dfrac{2 + 5\sqrt{2} + 4\sqrt{3} + 10\sqrt{6}}{46}$

EXAMPLE Rationalize the numerator of the fraction $\dfrac{1 + 2\sqrt{3}}{2 - 5\sqrt{2}}$.

$$\frac{1 + 2\sqrt{3}}{2 - 5\sqrt{2}} =$$

The conjugate of $1 + 2\sqrt{3}$ is $2\sqrt{3}$. Multiply by 1 in the form of $\dfrac{1 - 2\sqrt{3}}{1 - 2\sqrt{3}}$.

$$\frac{1 + 2\sqrt{3}}{2 - 5\sqrt{2}} \cdot \frac{1 - 2\sqrt{3}}{1 - 2\sqrt{3}} =$$

Use the special product for the sum and difference of two terms for the numerator and FOIL the denominator.

$$\frac{1^2 - (2\sqrt{3})^2}{2 + 2(-2\sqrt{3}) + 1(-5\sqrt{2}) - 5\sqrt{2}(-2\sqrt{3})} =$$

Simplify.

$$\frac{1 - 4(3)}{2 - 4\sqrt{3} - 5\sqrt{2} + 10\sqrt{6}} =$$

$$\frac{-11}{2 - 4\sqrt{3} - 5\sqrt{2} + 10\sqrt{6}} \quad \text{or} \quad \frac{11}{2 - 4\sqrt{3} - 5\sqrt{2} + 10\sqrt{0}}$$

TIP

Does Rationalizing a Numerator or Denominator of a Fraction Make the Expression Easier to Evaluate?

Probably not. Rationalizing is done whenever you need to manipulate a fraction so that one term has only rational numbers. In more advanced studies of mathematics, there are specific situations when it is desirable to perform this manipulation.

SECTION 14–2 SELF-STUDY EXERCISES

1 Multiply or divide the fractions. Reduce to simplest form.

1. $\dfrac{7}{12} \cdot \dfrac{18}{21}$

2. $\dfrac{3}{8} \cdot \dfrac{4}{15}$

3. $\dfrac{5x^2}{3y} \cdot \dfrac{2x}{3y}$

4. $\dfrac{8a^2}{15ab} \cdot \dfrac{21ab}{24a^2}$

5. $\dfrac{(x + 2)(x - 7)}{(x + 1)(x - 2)} \cdot \dfrac{3(x + 1)}{(x - 7)(x + 2)}$

6. $\dfrac{-4(x - 6)}{(x - 3)(x + 5)} \cdot \dfrac{(3 - x)(x + 5)}{(x - 6)(x - 2)}$

7. $\dfrac{a^2 - b^2}{4} \cdot \dfrac{12}{a + b}$

8. $\dfrac{2u + 2v}{5} \cdot \dfrac{10}{u + v}$

9. $\dfrac{a^2 - 49}{b^2 - 25} \cdot \dfrac{b - 5}{a + 7}$

10. $\dfrac{x^2 + 2x + 1}{5x - 5} \cdot \dfrac{15}{x + 1}$

11. $\dfrac{13r^2}{20a^2} \div \dfrac{39r^2}{5a}$

12. $\dfrac{4}{5} \div \dfrac{8}{15}$

13. $\dfrac{3}{8} \div \dfrac{9}{16}$

14. $\dfrac{5x^2y}{3y} \div \dfrac{10x^3}{9y^3}$

15. $\dfrac{7a^2b}{15b} \div \dfrac{14a}{9b}$

16. $\dfrac{a-b}{4} \div \dfrac{a-b}{2}$

17. $\dfrac{b-a}{7} \div \dfrac{a-b}{14}$

18. $\dfrac{2x-y}{x+y} \div \dfrac{y-2x}{-x-y}$

19. $\dfrac{x}{x^2-4x+4} \div \dfrac{1}{x-2}$

20. $\dfrac{5a^2-5b^2}{a^2b^2} \div \dfrac{a+b}{10ab}$

21. $\dfrac{x^2+4x+3}{x^2-4x-5} \div \dfrac{x+3}{x-5}$

2 Simplify.

22. $\dfrac{\frac{5}{9}}{\frac{3}{5}}$

23. $\dfrac{\frac{7}{8}}{\frac{5}{6}}$

24. $\dfrac{\frac{x-5}{4}}{3x}$

25. $\dfrac{\frac{6}{x-2}}{9x}$

26. $\dfrac{\frac{4x}{8x}}{x+3}$

27. $\dfrac{\frac{7x}{21x}}{x-3}$

28. $\dfrac{\frac{x^2-y^2}{3x+y}}{\frac{y-x}{-3x-y}}$

29. $\dfrac{\frac{x^2-5x+4}{x^2+4x+3}}{\frac{4-x}{x+3}}$

30. $\dfrac{\frac{x^2-36}{5x}}{\frac{x+6}{15x}}$

31. $\dfrac{\frac{x^2-5x}{8}}{\frac{2x-10}{12x}}$

3 Rationalize the denominator of each fraction.

32. $\dfrac{3}{2-\sqrt{5}}$

33. $\dfrac{13}{5-2\sqrt{3}}$

34. $\dfrac{5-\sqrt{2}}{4+\sqrt{2}}$

35. $\dfrac{7+\sqrt{7}}{5+\sqrt{7}}$

36. $\dfrac{2+3\sqrt{5}}{7-2\sqrt{3}}$

37. $\dfrac{5-\sqrt{6}}{5-2\sqrt{6}}$

Rationalize the numerator of each fraction.

38. $\dfrac{2-\sqrt{5}}{7}$

39. $\dfrac{5-\sqrt{11}}{15}$

40. $\dfrac{4+\sqrt{6}}{20}$

41. $\dfrac{3+\sqrt{2}}{7}$

42. $\dfrac{4+3\sqrt{2}}{7}$

43. $\dfrac{5+2\sqrt{3}}{8}$

14–3 | *Adding and Subtracting Rational Expressions*

Learning Outcomes

1 Add and subtract rational expressions.

2 Use addition and subtraction of rational expressions to simplify complex fractions.

1 Add and Subtract Rational Expressions.

In adding and subtracting fractions, we can add or subtract only fractions with like denominators. Whenever we have unlike fractions, we must first find a common denominator for the fractions. We then convert each fraction to an equivalent fraction with the common denominator and add or subtract the numerators. Let's refresh our memory of the procedure for adding and subtracting fractions.

EXAMPLE Add or subtract as indicated.

(a) $\dfrac{3}{8} + \dfrac{1}{8}$ (b) $\dfrac{7}{12} - \dfrac{1}{3}$ (c) $5 - \dfrac{2}{3}$ (d) $\dfrac{5}{2x} + \dfrac{3}{x} + \dfrac{9}{4}$

(a) $\dfrac{3}{8} + \dfrac{1}{8} = \dfrac{4}{8} = \dfrac{1}{2}$

Add the numerators and *keep* the *like* denominator. Reduce.

(b) $\dfrac{7}{12} - \dfrac{1}{3} =$

Select a common denominator and change to equivalent fractions with the common denominator of 12:

$\dfrac{1}{3} = \dfrac{1}{3}\left(\dfrac{4}{4}\right) = \dfrac{4}{12}$

$\dfrac{7}{12} - \dfrac{4}{12} = \dfrac{3}{12} = \dfrac{1}{4}$

Subtract and reduce.

(c) $5 - \dfrac{2}{3} =$

Convert 5 to a fraction with a denominator of 3:

$\dfrac{5}{1}\left(\dfrac{3}{3}\right) = \dfrac{15}{3}$

$\dfrac{15}{3} - \dfrac{2}{3} = \dfrac{13}{3}$ or $4\dfrac{1}{3}$

Subtract. Write improper fraction as a mixed number if desired.

(d) $\dfrac{5}{2x} + \dfrac{3}{x} + \dfrac{9}{4} =$

Convert to equivalent fractions with a common denominator of $4x$.

$\dfrac{5(2)}{2x(2)} + \dfrac{3(4)}{x(4)} + \dfrac{9(x)}{4(x)} =$

Multiply by 1 in the form of $\frac{n}{n}$ to get an equivalent fraction with a denominator of $4x$.

$\dfrac{10}{4x} + \dfrac{12}{4x} + \dfrac{9x}{4x} =$

Add numerators and *keep* the like (common) denominator.

$\dfrac{22 + 9x}{4x}$

Writing Improper Fractions Versus Mixed Numbers

In arithmetic, we often write an improper fraction as a mixed number. In algebra, we use the mixed-number form only when the final result contains only numbers and when we are interpreting the result within the context of an applied problem. A rational expression like the solution in the preceding example, part (d), can be written in an alternative form.

$$\dfrac{22 + 9x}{4x} = \dfrac{22}{4x} + \dfrac{9x}{4x} = \dfrac{11}{2x} + \dfrac{9}{4}$$

Unless the context of the problem requires the step, there's usually no reason to do the extra work.

To add or subtract rational expressions with unlike denominators:

1. Find the *least common denominator (LCD)*.
2. Change *each* fraction to an equivalent fraction with the least common denominator.
3. Add or subtract numerators.
4. Keep the same (common) denominator.
5. Reduce (or simplify) if possible.

EXAMPLE   Add $\dfrac{4}{x+3} + \dfrac{3}{x-3}$.

First, convert to equivalent expressions with a common denominator. The LCD is the product $(x+3)(x-3)$.

$$\frac{4}{x+3} = \frac{4(x-3)}{(x+3)(x-3)} = \frac{4x-12}{(x+3)(x-3)}$$

Multiply the numerator and denominator by $(x-3)$.

$$\frac{3}{x-3} = \frac{3(x+3)}{(x-3)(x+3)} = \frac{3x+9}{(x+3)(x-3)}$$

Multiply the numerator and denominator by $(x+3)$. Order does not matter in multiplication.

Next, use the equivalent expressions to proceed.

$$\frac{4x-12}{(x+3)(x-3)} + \frac{3x+9}{(x+3)(x-3)} =$$

Add numerators.

$$\frac{4x-12+3x+9}{(x+3)(x-3)} =$$

Combine like terms in the numerator.

$$\frac{7x-3}{(x+3)(x-3)} \quad \text{or} \quad \frac{7x-3}{x^2-9}$$

Factored or expanded form.

Some of the steps in the preceding example can be combined into one step, thus requiring fewer *written* steps, but the example illustrates each step that must be performed, either mentally or on paper, to add the rational expressions. Look at the next example.

EXAMPLE   Subtract $\dfrac{x}{x-2} - \dfrac{5}{x+4}$.

Change to equivalent expressions with a common denominator. The LCD is the product $(x-2)(x+4)$.

$$\frac{x}{x-2} = \frac{x(x+4)}{(x-2)(x+4)} = \frac{x^2+4x}{(x-2)(x+4)}$$

Multiply the numerator and denominator by $(x+4)$.

$$\frac{5}{x+4} = \frac{5(x-2)}{(x+4)(x-2)} = \frac{5x-10}{(x-2)(x+4)}$$

Multiply the numerator and denominator by $(x-2)$.

Use equivalent expressions and proceed.

$$\frac{x^2 + 4x}{(x - 2)(x + 4)} - \frac{5x - 10}{(x - 2)(x + 4)} =$$
Subtract numerators.

$$\frac{x^2 + 4x - (5x - 10)}{(x - 2)(x + 4)} =$$
Note: the *entire* numerator following the subtraction sign is subtracted.

$$\frac{x^2 + 4x - 5x + 10}{(x - 2)(x + 4)} =$$
Be careful with the signs when the parentheses are removed.

$$\frac{x^2 - x + 10}{(x - 2)(x + 4)} \quad \text{or} \quad \frac{x^2 - x + 10}{x^2 + 2x - 8}$$
$x^2 - x + 10$ will not factor, so the fraction cannot be reduced.

2 Use Addition and Subtraction of Rational Expressions to Simplify Complex Fractions.

Recall that a complex rational expression may have rational expressions in the numerator or denominator or both. Thus, there may be complex expressions that require the addition or subtraction of rational expressions before conversion to an equivalent multiplication.

To simplify a complex fraction that requires adding or subtracting rational terms:

1. Use addition or subtraction to combine terms in the numerator or denominator or both.
2. Rewrite the complex expression as a multiplication of the reciprocal of the denominator.
3. Proceed using the procedure for multiplying rational expressions.

EXAMPLE   Simplify.

(a) $\dfrac{1 + \dfrac{1}{x}}{\dfrac{2}{3}}$ 　　(b) $\dfrac{\dfrac{2}{x} - \dfrac{5}{2x}}{\dfrac{3}{4x} + \dfrac{3}{x}}$

(a) $\dfrac{1 + \dfrac{1}{x}}{\dfrac{2}{3}} = \dfrac{\dfrac{x}{x} + \dfrac{1}{x}}{\dfrac{2}{3}} =$

Add the terms in the numerator. The common denominator for the numerator is x: $1 = \dfrac{x}{x}$

$$\dfrac{\dfrac{x + 1}{x}}{\dfrac{2}{3}} =$$
Multiply the numerator by the reciprocal of the denominator.

$$\dfrac{x + 1}{x} \cdot \dfrac{3}{2} =$$

$$\dfrac{3(x + 1)}{2x} \quad \text{or} \quad \dfrac{3x + 3}{2x}$$
Factored or expanded form.

(b) $\dfrac{\dfrac{2}{x} - \dfrac{5}{2x}}{\dfrac{3}{4x} + \dfrac{3}{x}} = \dfrac{\dfrac{4}{2x} - \dfrac{5}{2x}}{\dfrac{3}{4x} + \dfrac{12}{4x}} =$

Add or subtract fractions in the numerator and in the denominator. Use the common denominator for the numerator, $2x$, and the denominator, $4x$, and convert to equivalent fractions.

$\dfrac{\dfrac{-1}{2x}}{\dfrac{15}{4x}} =$

Multiply the numerator by the reciprocal of the denominator.

$\dfrac{-1}{2x} \cdot \dfrac{4x}{15} =$

Reduce and multiply.

$\dfrac{-1}{\overset{}{\underset{1}{2x}}} \cdot \dfrac{\overset{2}{4x}}{15} =$

$\dfrac{-2}{15} = -\dfrac{2}{15}$

SECTION 14–3 SELF-STUDY EXERCISES

1 Add or subtract the rational expressions. Reduce to simplest terms.

1. $\dfrac{3}{7} + \dfrac{2}{7}$

2. $\dfrac{3}{8} + \dfrac{7}{16}$

3. $\dfrac{2x}{3} + \dfrac{5x}{6}$

4. $\dfrac{3x}{8} + \dfrac{7x}{12}$

5. $\dfrac{2}{x} + \dfrac{3}{2}$

6. $\dfrac{5}{2x} + \dfrac{6}{x} + \dfrac{2}{3x}$

7. $\dfrac{5}{x + 1} + \dfrac{4}{x - 1}$

8. $\dfrac{6}{x + 3} + \dfrac{4}{x - 1}$

9. $\dfrac{5}{2x + 1} - \dfrac{2}{x - 1}$

10. $\dfrac{7}{x - 6} - \dfrac{6}{x - 5}$

2 Simplify.

11. $\dfrac{2 + \dfrac{1}{x}}{\dfrac{3}{5}}$

12. $\dfrac{3 + \dfrac{2}{x}}{\dfrac{5}{8}}$

13. $\dfrac{\dfrac{3}{x} + \dfrac{5}{3x}}{\dfrac{3}{6x} - \dfrac{2}{x}}$

14. $\dfrac{\dfrac{x}{6} - \dfrac{3x}{9}}{\dfrac{5x}{12} + \dfrac{3x}{4}}$

15. $\dfrac{\dfrac{4}{x} + \dfrac{3}{2x}}{\dfrac{7x}{5} + \dfrac{x}{10}}$

16. $\dfrac{\dfrac{2}{3x} + \dfrac{3}{5}}{\dfrac{1}{5x} + \dfrac{2}{3}}$

17. $\dfrac{\dfrac{7}{x + 2} - \dfrac{5}{x + 2}}{\dfrac{3}{2 - x} + \dfrac{2}{x - 2}}$

18. $\dfrac{\dfrac{5}{x - 3} + \dfrac{2}{3 - x}}{\dfrac{4}{2x - 6} - \dfrac{5}{3x - 9}}$

Learning Outcomes   **1** Exclude certain values as solutions of rational equations.
2 Solve rational equations with variable denominators.

1 Exclude Certain Values as Solutions of Rational Equations.

A rational equation is an equation that contains one or more rational expressions. Some examples of rational equations are

$$\frac{1}{2} + \frac{1}{x} = \frac{1}{3}, \qquad x - \frac{3}{x} = 4, \qquad \frac{x}{x-2} = \frac{1}{x+3}$$

We solved some types of rational equations in Chapter 7 by first clearing the equation of all fractions. An important caution is necessary when a variable is in the denominator of a rational expression. Division by zero is impossible, and any value of the variable that makes the value of any denominator zero is excluded as a possible solution or root of an equation. These values are called **excluded values.**

To find excluded values of rational equations:

1. Set each denominator containing a variable equal to zero and solve for the variable.
2. Each equation in Step 1 produces an excluded value; however, some values may be repeats.

EXAMPLE Determine the value or values that must be excluded as possible solutions.

(a) $\dfrac{1}{x} + \dfrac{1}{2} = 5$

(b) $\dfrac{1}{x} = \dfrac{1}{x-3}$

(c) $\dfrac{1}{x+2} + \dfrac{1}{x-2} = \dfrac{1}{4}$

(d) $\dfrac{3x}{x+2} = 3 - \dfrac{5}{2x}$

(a) $\dfrac{1}{x} + \dfrac{1}{2} = 5$ Set the denominator containing a variable equal to 0.

$$x = 0$$

Excluded value is 0.

(b) $\dfrac{1}{x} = \dfrac{1}{x-3}$ Set each denominator containing a variable equal to 0.

$$x = 0, \qquad x - 3 = 0$$ Solve each equation.

$$x = 3$$

Excluded values are 0 and 3.

(c) $\dfrac{1}{x+2} + \dfrac{1}{x-2} = \dfrac{1}{4}$ Set each denominator containing a variable equal to 0.

$$x + 2 = 0 \qquad x - 2 = 0$$ Solve each equation.

$$x = -2 \qquad x = 2$$

Excluded values are −2 and 2.

(d) $\dfrac{3x}{x+2} = 3 - \dfrac{5}{2x}$ 　　　Set each denominator containing a variable equal to 0.

$$x + 2 = 0 \qquad 2x = 0$$ 　　Solve each equation.

$$x = -2 \qquad x = 0$$

Excluded values are −2 and 0.

Finding excluded values is another way to check whether roots are *extraneous roots*.

2 　Solve Rational Equations with Variable Denominators.

To solve rational equations:

1. Determine the excluded values.
2. Clear the equation of all denominators by multiplying the entire equation by the least common multiple (LCM) of the denominators.
3. Complete the solution using previously learned strategies.
4. Eliminate any excluded values as solutions.
5. Check the solutions.

EXAMPLE 　Solve the rational equations. Check.

(a) $\dfrac{2}{y} + \dfrac{1}{3} = 1$ 　　**(b)** $\dfrac{2}{y-2} = \dfrac{4}{y+1}$ 　　**(c)** $\dfrac{5}{x-2} - \dfrac{3x}{x-2} = -\dfrac{1}{x-2}$

(a) $\dfrac{2}{y} + \dfrac{1}{3} = 1$ 　　　　　　　　Excluded value: $y = 0$

$$3y\left(\dfrac{2}{y}\right) + 3y\left(\dfrac{1}{3}\right) = 3y(1)$$ 　　Multiply by LCM, $3y$, and reduce.

$$(3\cancel{y})\left(\dfrac{2}{\cancel{y}}\right) + (\cancel{3}y)\left(\dfrac{1}{\cancel{3}}\right) = 3y(1)$$ 　　Multiply remaining factors.

$$6 + y = 3y$$ 　　　Sort.

$$6 = 3y - y$$ 　　Combine.

$$6 = 2y$$ 　　　Divide.

$$\dfrac{6}{2} = \dfrac{2y}{2}$$ 　　　Simplify.

$$\mathbf{3 = y}$$ 　　　3 is not an excluded value.

Check: $\dfrac{2}{y} + \dfrac{1}{3} = 1$ 　Substitute 3 for y.

$$\dfrac{2}{3} + \dfrac{1}{3} = 1$$ 　Combine like terms.

$$\dfrac{3}{3} = 1$$ 　Simplify.

$$1 = 1$$ 　Solution checks.

(b) $\dfrac{2}{y-2} = \dfrac{4}{y+1}$

Excluded values:
$y = 2$, $y = -1$

$$(y-2)(y=1)\left(\dfrac{2}{y-2}\right) = (y-2)(y=1)\left(\dfrac{4}{y+1}\right)$$

Multiply by LCM, $(y-2)\,(y+1)$, and reduce.

$$(y-2)(y+1)\dfrac{2}{y-2} = (y-2)(y+1)\dfrac{4}{y+1}$$

Multiply remaining factors.

$$2(y+1) = 4(y-2)$$

Distribute.

$$2y + 2 = 4y - 8$$

Sort.

$$2y - 4y = -8 - 2$$

Combine like terms.

$$-2y = -10$$

Divide.

$$\dfrac{-2y}{-2} = \dfrac{-10}{-2}$$

$$\mathbf{y = 5}$$

5 is not an excluded value.

Check: $\dfrac{2}{y-2} = \dfrac{4}{y+1}$

Substitute 5 for y.

$$\dfrac{2}{5-2} = \dfrac{4}{5+1}$$

Simplify.

$$\dfrac{2}{3} = \dfrac{4}{6}$$

Reduce $\frac{4}{6}$.

$$\dfrac{2}{3} = \dfrac{2}{3}$$

Solution checks.

(c) $\dfrac{5}{x-2} - \dfrac{3x}{x-2} = -\dfrac{1}{x-2}$

Excluded value:
$x = 2$

$$(x-2)\dfrac{5}{x-2} - (x-2)\dfrac{3x}{x-2} = (x-2)\left(-\dfrac{1}{(x-2)}\right)$$

Multiply by LCM, $x-2$, and reduce.

$$(x-2)\dfrac{5}{x-2} - (x-2)\dfrac{3x}{x-2} = (x-2)\left(-\dfrac{1}{(x-2)}\right)$$

$$5 - 3x = -1$$

Sort.

$$-3x = -1 - 5$$

Combine.

$$-3x = -6$$

Divide.

$$\dfrac{-3x}{-3} = \dfrac{-6}{-3}$$

$$x = 2$$

This is an excluded value. The root will not check.

There is no solution.

Check: $\dfrac{5}{x-2} - \dfrac{3x}{x-2} = -\dfrac{1}{x-2}$

Substitute 2 for x.

$$\dfrac{5}{2-2} - \dfrac{3(2)}{2-2} = -\dfrac{1}{2-2}$$

$$\dfrac{5}{0} - \dfrac{6}{0} = -\dfrac{1}{0}$$

Division by zero is impossible. The equation has no solution.

EXAMPLE
HLTH/N

Nurse Vance can prepare dosages for her patients in 30 min. If she gets help from assistants who work at the same rate, together they can complete the preparation in 5 min. How many assistants helped her?

Unknown facts

a = number of assistants

Known facts

30 min = time it took Nurse Vance to complete the task alone
5 min = time it took Nurse Vance and assistants to complete the task
Assistants worked at the same rate as Nurse Vance.

Relationships

$a + 1$ = total number of assistants (Nurse Vance and a assistants)

$\dfrac{30 \text{ min}}{a + 1}$ = amount of time that it would take to complete the task with $a + 1$ assistants

$\dfrac{30 \text{ min}}{a + 1} = 5$ — Both sides of the equation represent the amount of time to complete the task with $a + 1$ assistants.

Estimation

Two people (one additional assistant) could do the job in 15 min. Four people would take 7.5 minutes. Eight people would take less than 4 min. The total number of people will be between 4 and 8 and the number of assistants will be between 3 and 7.

Calculations

$$\frac{30 \text{ min}}{a + 1} = 5$$

Excluded value is -1, which is not appropriate within the context of the problem. Clear the equation of fractions.

$$(a + 1)\left(\frac{30 \text{ min}}{a + 1}\right) = 5(a + 1)$$ Simplify.

$$30 = 5a + 5$$ Solve for a.

$$30 - 5 = 5a$$

$$25 = 5a$$

$$5 = a$$ Number of additional assistants.

Interpretation

Nurse Vance got 5 assistants to help her.

SECTION 14–4 SELF-STUDY EXERCISES

1 Determine the value or values that must be excluded as possible solutions.

1. $\dfrac{5}{x} + \dfrac{3}{5} = 7$

2. $\dfrac{6}{x} + 5 = \dfrac{5}{12}$

3. $\dfrac{9}{x - 5} = \dfrac{7}{x}$

4. $\dfrac{15}{2x} = \dfrac{6}{x + 9}$

5. $\dfrac{7}{x + 8} + \dfrac{2}{x - 8} = \dfrac{5}{x^2 - 64}$

6. $\dfrac{9}{x - 2} + \dfrac{5}{x + 2} = \dfrac{1}{x^2 - 4}$

7. $\dfrac{2x}{4x - 3} - 4 = \dfrac{5}{6x}$

8. $\dfrac{1}{9x} + \dfrac{8x}{5x + 15} = 11$

14–4 Solving Equations with Rational Expressions

2 Solve the equations. Check for extraneous roots.

9. $\dfrac{6}{7} + \dfrac{5}{x} = 1$

10. $\dfrac{4}{x} - \dfrac{3}{4} = \dfrac{1}{20}$

11. $\dfrac{2}{x} + \dfrac{5}{2x} - \dfrac{1}{2} = 4$

12. $\dfrac{5}{2x} - \dfrac{1}{x} = 6$

13. $\dfrac{2}{x-3} + \dfrac{7}{x+3} = \dfrac{12}{x-3}$

14. $\dfrac{10}{x+5} = \dfrac{1}{x-5} - \dfrac{2}{x+5}$

Use rational equations to solve the problems. Check for extraneous roots.

15. **BUS** Shawna can complete an electronic lab project in 3 days. Tom can complete the same project in 4 days. How many days will it take to complete the project if Shawna and Tom work together? (*Hint:* Rate × Time = Amount of work.)

16. **BUS** A group of investors purchased land in Maine for $12,000. More investors joined the group and the cost dropped $500 per person in the original group. How many investors were in the original group if the new group had $1\frac{1}{2}$ times as many investors as the original group?

17. **AUTO** Frank commutes 50 mi on a motorbike to a college, but on the return trip he travels 40 mi/h and makes the trip in $\frac{3}{4}$ h less time. Find Frank's speed going to school. (*Hint:* Time = Distance ÷ Rate.)

18. **INDTR** Pipe 1 fills a tank in 2 min and pipe 2 empties the tank in 6 min. How long does it take to fill the tank if both pipes are open? (*Hint:* Rate × Time = Amount of work.)

19. **BUS** Kim Denley bought $150 worth of medium-roast coffee and, at the same cost per pound, $100 worth of dark-roast coffee. If she bought 25 pounds more of the medium-roast coffee than the dark-roast coffee, how many pounds of each did she buy and what was the cost per pound?

20. **ELEC** Find resistance$_2$ if resistance$_1$ is 10 Ω and resistance$_t$ is 3.75 Ω, using the formula

$$R_t = \dfrac{R_1 R_2}{R_1 + R_2}.$$

CHAPTER REVIEW OF KEY CONCEPTS

Learning Outcomes

What to Remember with Examples

Section 14–1

1 Simplify or reduce rational expressions (pp. 543–546).

If the rational expression is not in factored form, factor completely the numerator and the denominator. Then, reduce factors common to both the numerator and denominator.

Simplify $\dfrac{a+b}{16a^2 + 16ab}$. Factor common factor of 16a in the denominator.

$\dfrac{a+b}{16a(a+b)}$ Reduce the common binomial. Factor.

$\dfrac{\cancel{a+b}}{16a\cancel{(a+b)}} = \dfrac{1}{16a}$

Section 14–2

1 Multiply and divide rational expressions (pp. 547–550).

Write the numerator and denominator in factored form whenever possible. For multiplication, reduce factors common to both the numerator and denominator. Then, multiply remaining factors.

Chapter 14 / Rational Expressions and Equations

Multiply $\dfrac{x}{2x + 8} \cdot \dfrac{2}{x + 1}$. Write each numerator and denominator in factored form.

$\dfrac{x}{2(x + 4)} \cdot \dfrac{2}{x + 1} =$ Reduce.

$\dfrac{(2)(x)}{2(x + 4)(x + 1)} =$ Multiply.

$\dfrac{x}{(x + 4)(x + 1)}$ or $\dfrac{x}{x^2 + 5x + 4}$ Factored or expanded form.

For division, multiply by the reciprocal of the second fraction. Factor, reduce factors common to both a numerator and denominator, then multiply remaining factors.

Divide $\dfrac{x}{x + 2} \div \dfrac{2}{x + 2}$. Multiply by the reciprocal of the divisor. Reduce.

$\dfrac{x}{x + 2} \cdot \dfrac{x + 2}{2} = \dfrac{x}{2}$ Multiply remaining factors.

To simplify complex rational expressions, rewrite as the division of two rational expressions. Next, rewrite the division as a multiplication of the first fraction by the reciprocal of the second fraction. Factor and perform the indicated multiplication.

Simplify.

$\dfrac{\dfrac{x - 3}{5x}}{\dfrac{4x - 12}{15x^2}}$ Write as division.

$\dfrac{x - 3}{5x} \div \dfrac{4x - 12}{15x^2} =$ Rewrite as multiplication, factor, and reduce.

$\dfrac{x - 3}{5x} \cdot \dfrac{\overset{3x}{15x^2}}{4(x - 3)} =$ Multiply remaining factors.

$\dfrac{3x}{4}$

2 Use multiplication and conjugates to rationalize a numerator or denominator of a fraction that has a binomial with an irrational term (pp. 551–552).

1. Determine which term of the fraction (numerator or denominator) is to be cleared of a binomial with an irrational term.
2. Multiply by 1 in the form of $\dfrac{n}{n}$, where n is the conjugate of the binomial with the irrational term.
3. Simplify the resulting expression.

Rationalize the denominator in the expression $\dfrac{5x}{3 + 2\sqrt{5}}$.

$\dfrac{5x}{3 + 2\sqrt{5}} \cdot \dfrac{3 - 2\sqrt{5}}{3 - 2\sqrt{5}} =$ Multiply by 1 in the form of $\dfrac{3 - 2\sqrt{5}}{3 - 2\sqrt{5}}$.

$\dfrac{5x(3 - 2\sqrt{5})}{3^2 - (2\sqrt{5})^2} =$ Distribute and simplify.

Chapter Review of Key Concepts

$$\frac{15x - 10x\sqrt{5}}{9 - 4(5)} = \frac{15x - 10x\sqrt{5}}{9 - 20}$$

$$\frac{15x - 10x\sqrt{5}}{-11} \quad \text{or} \quad -\frac{15x - 10x\sqrt{5}}{11}$$

Section 14–3

1 Add and subtract rational expressions (pp. 553–556).

Add (or subtract) only rational expressions with the same denominator. For rational expressions with unlike denominators:

1. Find the LCD.
2. Change each fraction to an equivalent fraction having the LCD.
3. Add (or subtract) the numerators.
4. Keep the same LCD.
5. Reduce if possible.

Perform the following operations:

(a) $\dfrac{x}{x + 4} + \dfrac{2x}{x + 4} = \dfrac{3x}{x + 4}$ Add numerators. Keep like denominator.

(b) $\dfrac{3}{x + 2} - \dfrac{2}{x + 1}$ LCD $= (x + 2)(x + 1)$. Change each fraction to an equivalent fraction with the LCD.

$$\frac{3}{x + 2} = \frac{3(x + 1)}{(x + 2)(x + 1)}$$ First fraction

$$= \frac{3x + 3}{(x + 2)(x + 1)}$$

$$\frac{2}{x + 1} = \frac{2(x + 2)}{(x + 1)(x + 2)}$$ Second fraction

$$= \frac{2x + 4}{(x + 1)(x + 2)}$$

$$\frac{3x + 3}{(x + 2)(x + 1)} - \frac{2x + 4}{(x + 2)(x + 1)} = \frac{3x + 3 - (2x + 4)}{(x + 1)(x + 2)}$$ Subtract numerators. Keep common denominator.

$$\frac{3x + 3 - 2x - 4}{(x + 1)(x + 2)} = \frac{x - 1}{(x + 1)(x + 2)}$$

2 Use addition and subtraction of rational expressions to simplify complex fractions (pp. 556–557).

1. Use addition or subtraction to combine terms in the numerator or denominator or both.
2. Rewrite the complex fraction as a division then convert to an equivalent multiplication of rational expressions.
3. Use the procedure for multiplying rational expressions.

Simplify $\dfrac{\dfrac{3}{x} + 5}{\dfrac{2}{3x} - \dfrac{1}{x}}$.

$$\frac{\dfrac{3}{x} + \dfrac{5}{1}\left(\dfrac{x}{x}\right)}{\dfrac{2}{3x} - \dfrac{1}{x}\left(\dfrac{3}{3}\right)}$$

Change to equivalent fractions in numerator with LCD of x.

Change to equivalent fractions in denominator with LCD of $3x$.

$$\frac{\dfrac{3}{x} + \dfrac{5x}{x}}{\dfrac{2}{3x} - \dfrac{3}{3x}}$$

Add fractions in the numerator.

Subtract fractions in the denominator.

$$\frac{\dfrac{3 + 5x}{x}}{\dfrac{-1}{3x}}$$

Rewrite as multiplication of reciprocal.

$$\frac{3 + 5x}{x} \cdot \frac{3x}{-1} =$$

Reduce and multiply.

$$\frac{(3 + 5x)3}{-1} =$$

Distribute.

$$\frac{9 + 15x}{-1} = -(9 + 15x) \quad \text{or} \quad -9 - 15x$$

Section 14–4

1 Exclude certain values as solutions of rational equations (pp. 558–559).

To find excluded values, set each denominator with a variable equal to zero. The excluded values obtained cannot be accepted as solutions of the rational equation.

Find the excluded values for the equation:

$$4 = \frac{2}{3 + y} \qquad \text{Set denominator equal to zero.}$$

$$3 + y = 0$$

$$y = -3 \qquad \text{Excluded value.}$$

2 Solve rational equations with variable denominators (pp. 559–561).

To solve rational equations with variable denominators, clear the equation of fractions by multiplying each term of the equation by the LCM of all denominators in the equation. Solve the resulting equation. Check all solutions to determine if any are extraneous roots.

Solve $\dfrac{2}{x - 3} - \dfrac{1}{x + 3} = \dfrac{1}{x^2 - 9}$.

Excluded values are 3 and -3.
LCM is $x^2 - 9$ or $(x - 3)(x + 3)$.

$$(x - 3)(x + 3)\frac{2}{x - 3} - (x - 3)(x + 3)\frac{1}{x + 3} = (x - 3)(x + 3)\frac{1}{(x - 3)(x + 3)}$$

$$2(x + 3) - 1(x - 3) = 1$$

$$2x + 6 - x + 3 = 1$$

$$x + 9 = 1$$

$$x = -8$$

Chapter Review of Key Concepts

Check:
$$\frac{2}{-8-3} - \frac{1}{-8+3} = \frac{1}{(-8)^2 - 9}$$
$$\frac{2}{-11} - \frac{1}{-5} = \frac{1}{64-9}$$
$$\frac{-2}{11} + \frac{1}{5} = \frac{1}{55}$$
$$\frac{-10}{55} + \frac{11}{55} = \frac{1}{55}$$
$$\frac{1}{55} = \frac{1}{55}$$

CHAPTER REVIEW EXERCISES

Section 14–1

Simplify the fractions.

1. $\dfrac{18}{24}$

2. $\dfrac{24}{42}$

3. $\dfrac{5a^2b^3c}{10a^3bc^2}$

4. $\dfrac{3x^2y}{9x^3y^3}$

5. $\dfrac{4xy(x-3)}{8xy(x+3)}$

6. $\dfrac{(x+7)(3-x)}{(x-3)(x-7)}$

7. $\dfrac{(x-4)(x+2)}{(x+2)(4-x)}$

8. $\dfrac{x+1}{3x+3}$

9. $\dfrac{m^2-n^2}{m^2+n^2}$

10. $\dfrac{3x^2+8x-3}{2x^2+5x-3}$

11. $\dfrac{x}{x+xy}$

12. $\dfrac{2x}{4x+6}$

13. $\dfrac{5x+15}{x+3}$

14. $\dfrac{x^3+2x^2-3x}{3x}$

15. $\dfrac{y^2+2y+1}{y+1}$

16. $\dfrac{y-1}{y^2-2y+1}$

17. $\dfrac{2x-6}{x^2+3x-18}$

18. $\dfrac{x^2-4x+4}{x^2-2x}$

19. $\dfrac{3x-9}{x-3}$

20. $\dfrac{4x-12}{x-3}$

Section 14–2

Multiply or divide the fractions. Reduce to simplest terms.

21. $\dfrac{3x^2}{2y} \cdot \dfrac{5x}{6y}$

22. $\dfrac{x^2-y^2}{6} \cdot \dfrac{18}{x-y}$

23. $\dfrac{9}{x+b} \cdot \dfrac{5x+5b}{3}$

24. $\dfrac{81-x^2}{16-f^2} \cdot \dfrac{4-f}{9+x}$

25. $\dfrac{4y^2-4y+1}{6y-6} \cdot \dfrac{24}{2y-1}$

26. $\dfrac{x-3}{x+5} \cdot \dfrac{2x^2+10x}{2x-6}$

27. $\dfrac{5-x}{x-5} \cdot \dfrac{x-1}{1-x}$

28. $\dfrac{3-x}{x-2} \cdot \dfrac{2x-4}{x-3}$

29. $\dfrac{x^2+6x+9}{x^2-4} \cdot \dfrac{x-2}{x+3}$

30. $\dfrac{x^2}{x^2-9} \cdot \dfrac{x^2-5x+6}{x^2-2x}$

31. $\dfrac{2a+b}{8} \div \dfrac{2a+b}{2}$

32. $\dfrac{17a^2}{21y^2} \div \dfrac{34a^2}{68}$

33. $\dfrac{y^2-2y+1}{y} \div \dfrac{1}{y-1}$

34. $\dfrac{x^2y^2}{3x^2-3y^2} \div \dfrac{8xy}{x-y}$

35. $\dfrac{y^2+6y+9}{y^2+4y+4} \div \dfrac{y+3}{y+2}$

36. $\dfrac{2x+2y}{3} \div \dfrac{x^2-y^2}{y-x}$

37. $\dfrac{3x^2+6x}{x} \div \dfrac{2x+4}{x^2}$

38. $\dfrac{x^2-7x}{x^2-3x-28} \div \dfrac{1}{-x-4}$

39. $\dfrac{y^2-16}{y+3} \div \dfrac{y-4}{y^2-9}$

40. $\dfrac{12x+24}{36x-36} \div \dfrac{6x+12}{8x-8}$

Simplify.

41. $\dfrac{\dfrac{5}{x-3}}{4}$

42. $\dfrac{6ab}{\dfrac{3a}{4}}$

43. $\dfrac{\dfrac{x^2-4x}{6x}}{\dfrac{x-4}{8x^2}}$

44. $\dfrac{\dfrac{2}{7}}{\dfrac{3}{4}}$

Rationalize the denominator of each fraction.

45. $\dfrac{12}{6-\sqrt{5}}$

46. $\dfrac{8}{3-\sqrt{2}}$

47. $\dfrac{7+\sqrt{3}}{7-\sqrt{3}}$

48. $\dfrac{4-\sqrt{7}}{3+\sqrt{5}}$

49. $\dfrac{5+\sqrt{2}}{7-3\sqrt{5}}$

50. $\dfrac{4-\sqrt{7}}{8+2\sqrt{11}}$

Rationalize the numerator of each fraction.

51. $\dfrac{8+2\sqrt{3}}{5}$

52. $\dfrac{7-\sqrt{5}}{11}$

53. $\dfrac{4-\sqrt{13}}{12}$

54. $\dfrac{1-3\sqrt{7}}{8}$

55. $\dfrac{5-\sqrt{7}}{16}$

56. $\dfrac{2+\sqrt{15}}{15}$

Section 14–3

Add or subtract the fractions. Reduce to simplest terms.

57. $\dfrac{2}{9}+\dfrac{4}{9}$

58. $\dfrac{2}{3}+\dfrac{5}{12}$

59. $\dfrac{3x}{7}+\dfrac{2x}{14}$

60. $\dfrac{5x}{3}-\dfrac{2x}{4}$

61. $\dfrac{3x}{4}+\dfrac{5x}{6}$

62. $\dfrac{3}{4}+\dfrac{7}{x}$

63. $\dfrac{5}{x}-\dfrac{7}{3}$

64. $\dfrac{3}{x}+\dfrac{5}{7}$

65. $\dfrac{3}{4x}+\dfrac{2}{x}+\dfrac{3}{6x}$

66. $\dfrac{6}{2x+3}+\dfrac{2}{2x-3}$

67. $\dfrac{7}{x-3}+\dfrac{3}{x+2}$

68. $\dfrac{4}{3x+2}-\dfrac{3}{x-2}$

69. $\dfrac{8}{x+3}-\dfrac{2}{x-4}$

70. $\dfrac{3}{x-2}+\dfrac{5}{2-x}$

71. $\dfrac{x}{x-5}-\dfrac{3}{5-x}$

Simplify.

72. $\dfrac{2+\dfrac{3}{x}}{\dfrac{2}{x}}$

73. $\dfrac{\dfrac{5}{x}-\dfrac{3}{4x}}{\dfrac{1}{3x}+\dfrac{2}{x}}$

74. $\dfrac{5-\dfrac{x-2}{x}}{\dfrac{x-4}{2x}-2}$

75. $\dfrac{\dfrac{3x}{6}-\dfrac{5}{x}}{\dfrac{x}{3}+\dfrac{4}{2x}}$

Section 14–4

Determine the value or values that must be excluded as possible solutions of the following.

76. $\dfrac{3}{x}-\dfrac{4}{5}=2$

77. $\dfrac{4}{x}=\dfrac{3}{x-2}$

78. $\dfrac{3}{x-5}-\dfrac{4}{x+5}=\dfrac{1}{25}$

79. $\dfrac{5}{2x-1}-\dfrac{6}{x}=\dfrac{4}{3x}$

Solve the equations. Check for extraneous roots.

80. $\dfrac{3}{x} + \dfrac{2}{3} = 1$

81. $\dfrac{4}{x} = \dfrac{1}{x+5}$

82. $\dfrac{3}{x-4} + \dfrac{1}{x+4} = \dfrac{1}{x^2-16}$

83. $-\dfrac{4x}{x+1} = 3 - \dfrac{4}{x+1}$

Use rational equations to solve the following problems. Check for extraneous or inappropriate roots.

84. INDTEC Fugita can complete the assembly of 50 widgets in 6 days. Ohn can complete the same job in 8 days. How many days will it take to complete the project if Fugita and Ohn work together? (*Hint:* Rate × Time = Amount of work.)

85. Several students chipped in a total of $120 to buy a small refrigerator to use in the dorm. If the group was increased to four students, the cost to the original group dropped $10 per person. How many students were in the original group?

86. Pipe 1 fills a tank in 4 min and pipe 2 fills the same tank in 3 min. What part of the tank will be filled if both pipes are open for 1 min? (*Hint:* Rate × Time = Amount of work.)

87. One machine can do a job in 5 h alone. How many hours would it take a second machine to complete the job alone if both machines together can do the job in 3 h?

TEAM PROBLEM-SOLVING EXERCISES

1. Applications that involve several facts and relationships may vary in the information that is known and unknown. For example, an application of two pipes working together to fill a tank may be interested in filling the tank for only a certain period of time. Solve the following problem by adapting your strategy to the given information.

Pipe 1 fills a tank in 4 min and pipe 2 fills the same tank in 3 min. What portion of the tank will be filled if both pipes are open for 1 min?

2. Use your strategy for two quantities working together and your knowledge of setting up an equation to model the conditions of an application to solve the following.

One machine can do a job in 5 h alone. How many hours would it take a second machine to complete the job alone if both machines together can do the job in 3 hr?

CONCEPTS ANALYSIS

1. How are the properties $\dfrac{n}{n} \times 1$ and $1 \times n = n$ applied in the following example: $\dfrac{4}{6} = \dfrac{2}{3}$?

3. Write a brief comment for each step of the following example.

$$\dfrac{x^2 + 3x + 2}{x^2 - 9} \div \dfrac{2x^2 + 3x - 2}{2x^2 - 7x + 3} =$$

$$\dfrac{x^2 + 3x + 2}{x^2 - 9} \cdot \dfrac{2x^2 - 7x + 3}{2x^2 + 3x - 2} =$$

$$\dfrac{(x+1)(x+2)}{(x+3)(x-3)} \cdot \dfrac{(2x-1)(x-3)}{(2x-1)(x+2)} = \dfrac{x+1}{x+3}$$

2. Why is the following problem incorrect?

$$\dfrac{5}{10} = \dfrac{2+3}{2+8} = \dfrac{3}{8}$$

Find the mistake or mistakes in each problem and briefly explain each mistake. Then, rework the problem correctly.

4. $\dfrac{x}{x+3} = \dfrac{\cancel{x}}{\cancel{x}+3} = \dfrac{1}{3}$

5. $\dfrac{x^2-4}{x^2+4x+4} \div x+2 =$

$\dfrac{(\cancel{x+2})(x-2)}{(\cancel{x+2})(x+2)} \cdot \dfrac{\cancel{x+2}}{1} = x-2$

6. $\dfrac{5}{x+2} + \dfrac{3}{x-3} = \dfrac{8}{(x+2)(x-3)}$

7. $\dfrac{3}{x-1} - \dfrac{5}{x+1} =$

$\dfrac{3(x+1)}{(x-1)(x+1)} - \dfrac{5(x-1)}{(x-1)(x+1)} =$

$\dfrac{3x+3-5x-5}{(x+1)(x-1)} =$

$\dfrac{-2x-2}{(x+1)(x-1)} =$

$\dfrac{-2(x+1)}{(x+1)(x-1)} = \dfrac{-2}{x-1}$

8.

$\dfrac{14}{3x} = \dfrac{12}{7x}$

$\dfrac{\overset{2}{\cancel{14}}}{\cancel{3x}} = \dfrac{\overset{4}{\cancel{12}}}{\cancel{7x}}$

$\dfrac{\overset{1}{}}{} 8 = x^2 \dfrac{\overset{1}{}}{}$

$\pm\sqrt{8} = x$

$\pm2\sqrt{2} = x$

9. How is a division of rational expressions related to a multiplication of rational expressions?

10. In your own words, write a rule for multiplying rational expressions.

PRACTICE TEST

Perform the indicated operations.

1. $\dfrac{x-3}{2x-6}$

2. $\dfrac{x^2-16}{x-4}$

3. $\dfrac{6x^2-11x+4}{2x^2+5x-3}$

4. $\dfrac{x^2-6x+8}{2-x}$

5. $\dfrac{(x-2)(x-4)}{(4-x)(x+2)}$

6. $\dfrac{y^2+x^2}{y^2-x^2}$

7. $\dfrac{6xy}{ab} \cdot \dfrac{a^2b}{2xy^2}$

8. $\dfrac{x^2-y^2}{2x+y} \cdot \dfrac{4x^2+2xy}{y-x}$

9. $\dfrac{x-2y}{x^3-3x^2y} \div \dfrac{x^2-4y^2}{x-3y}$

10. $\dfrac{2a^2-ab-b^2}{6x^2+x-1} \div \dfrac{a^2-b^2}{8x+4}$

11. $\dfrac{2x^2+3x+1}{x} \div \dfrac{x+1}{1}$

12. $\dfrac{4y^2}{2x} \cdot \dfrac{x}{8x}$

13. $\dfrac{1}{x+2} - \dfrac{1}{x-3}$

14. $\dfrac{2}{x+2} + \dfrac{3}{x-1}$

15. $\dfrac{3}{x} + \dfrac{1}{4}$

16. $\dfrac{2}{3y} - \dfrac{7}{y}$

17. $\dfrac{5}{3x - 2} + \dfrac{7}{2 - 3x}$

18. $\dfrac{2x}{3} - \dfrac{5x}{2}$

19. $\dfrac{x - 2y}{x^2 - 4y^2}$

20. $\dfrac{3}{2y} + \dfrac{2}{y} + \dfrac{1}{5y}$

21. $\dfrac{2x}{1 - \dfrac{3}{x}}$

22. $\dfrac{\dfrac{1}{a} + \dfrac{1}{b}}{ab}$

Determine the excluded values of x.

23. $\dfrac{5}{x} = \dfrac{2}{x + 3}$

24. $\dfrac{3}{x - 4} + \dfrac{5}{x + 4} = 6$

Solve the rational equations. Check for extraneous roots.

25. $\dfrac{3x}{x - 2} + 4 = \dfrac{3}{x - 2}$

26. $\dfrac{x}{x + 3} = 5$

27. $\dfrac{5}{x - 2} = \dfrac{-4}{x + 1}$

Use rational equations to solve the problems. Check for extraneous roots.

28. Henry can assemble four computers in 3 h. Lester can assemble four computers in 4 h. How long will it take to assemble four computers if both work together? (*Hint:* Rate × Time = Amount of work.)

29. A medical group purchased land in Colorado for $100,000. When the size of the group doubled, the cost dropped $10,000 per person in the original group. How many persons were in the original group?

30. Cedric Partee drove 300 mi in one day while on a vacation to the mountains, but on the return trip by the same route he drove 10 mi per hour less and the return trip took 1 h longer. Find Partee's speed to and from the mountains. (*Hint:* Time = Distance ÷ Rate.)

Quadratic and Higher-Degree Equations

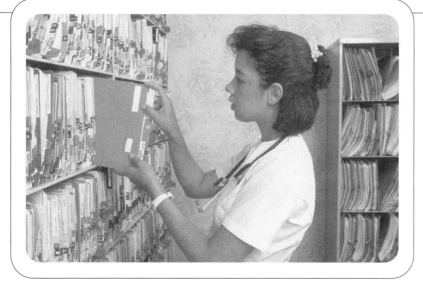

Focus on Careers

Hospitals and doctor's offices are required to maintain accurate records of observations, medical and surgical procedures and treatments, and outcomes of the procedures and treatments. Medical-records and health-information technicians organize these records and evaluate them for accuracy and completeness.

Medical-records technicians have little or no direct contact with patients. They communicate with physicians and other health-care professionals to clarify diagnoses or obtain additional information to complete the patient's medical record.

These careers require an associate degree from a community or junior college and courses include anatomy, physiology, medical terminology, legal aspects of health information, statistics, database management, and computer science. Job prospects for this career are very good, especially in physicians' offices. Job growth is expected to be much faster than average through 2012.

Medical-records technicians usually work a 40-h week and some overtime may be required. Hospitals have medical records technicians on duty 24 h a day, 7 days a week. Because accuracy is essential, these technicians must be very detail-oriented.

Registered health-information technicians must pass a written examination after graduating from a 2-year associate degree program accredited by the Commission on Accreditation of Allied Health Education Programs of the American Medical Association.

In 2002, the highest 10% of medical-records technicians earned in excess of $38,640. The middle 50% of medical-records technicians earned between $19,550 and $30,600 annually. More jobs are found in nursing-care facilities, where the median annual earnings were $25,160.

Source: *Occupational Outlook Handbook,* 2004–2005 Edition. U.S. Department of Labor, Bureau of Labor Statistics.

The equations we have solved so far are called linear equations. In such equations, the variable appears only to the first power, such as x or y. In this chapter, we examine quadratic and higher-degree equations.

15–1 | Solving Quadratic Equations by the Square-Root Method

Learning Outcomes

1 Write quadratic equations in standard form.

2 Identify the coefficients of the quadratic, linear, and constant terms of a quadratic equation.

3 Solve pure quadratic equations ($ax^2 + c = 0$) by the square-root method.

1 Write Quadratic Equations in Standard Form.

There are several methods for solving quadratic equations. Each method has strengths and weaknesses depending on the characteristics of the equation being solved. First, let's examine the basic characteristics of quadratic equations, starting with quadratic equations having only one variable.

A **quadratic equation** is an equation in which at least one variable term is raised to the second power and no variable term has a power more than 2 or less than 0. The **standard form** for a quadratic equation is $ax^2 + bx + c = 0$, where a, b, and c are real numbers and $a > 0$.

EXAMPLE   Arrange the quadratic equations in standard form.

(a) $3x^2 - 3 + 5x = 0$ (b) $7x - 2x^2 - 8 = 0$
(c) $12 = 7x^2 - 4x$ (d) $7x^2 = 3$

(a) $3x^2 - 3 + 5x = 0$ Arrange the terms in descending powers of x.
  $\mathbf{3x^2 + 5x - 3 = 0}$ Standard form

(b) $7x - 2x^2 - 8 = 0$ Arrange the terms in descending powers of x.
  $-2x^2 + 7x - 8 = 0$ Multiply both sides by -1 to make the coefficient of the x^2 term positive.

  $-1(-2x^2 + 7x - 8) = -1(0)$
  $\mathbf{2x^2 - 7x + 8 = 0}$ Standard form

(c) $12 = 7x^2 - 4x$ Rearrange the terms so that one side of the equation equals zero.

  $0 = 7x^2 - 4x - 12$ Interchange sides of equation.
  $\mathbf{7x^2 - 4x - 12 = 0}$

(d) $7x^2 = 3$ Rearrange so that one side equals zero.
  $\mathbf{7x^2 - 3 = 0}$ There is no first-degree variable term.

Chapter 15 / Quadratic and Higher-Degree Equations

A quadratic equation is a second-degree equation, which means that it can have as many as two solutions. As we examine the process for solving quadratic equations, we shall see quadratic equations that have two, one, or no solutions.

2 Identify the Coefficients of the Quadratic, Linear, and Constant Terms of a Quadratic Equation.

Some methods for solving quadratic equations can be used for any type of quadratic equation but are time consuming. Other methods are quicker but apply only to certain types of quadratic equations. In choosing an appropriate method it is helpful to be able to recognize similar and different characteristics of quadratic equations.

The standard form of quadratic equations, $ax^2 + bx + c = 0$, has three types of terms.

1. ax^2 is a **quadratic term;** that is, the degree of the term is 2. In standard form, this term is the **leading term** and a is the **leading coefficient.**
2. bx is a **linear term;** that is, the degree of the term is 1 and the coefficient is b.
3. c is a **numerical term** or **constant term;** that is, the degree of the term is 0. The coefficient is c ($c = cx^0$).

A quadratic equation that has all three types of terms is sometimes referred to as a **complete quadratic equation.**

$$ax^2 + bx + c = 0 \qquad \text{Complete quadratic equation}$$

A quadratic equation that has a quadratic term (ax^2) and a linear term (bx) but no constant term (c) is sometimes referred to as an **incomplete quadratic equation.**

$$ax^2 + bx = 0 \qquad \text{Incomplete quadratic equation}$$

A quadratic equation that has a quadratic term (ax^2) and a constant (c) but no linear term (bx) is sometimes referred to as a **pure quadratic equation.**

$$ax^2 + c = 0 \qquad \text{Pure quadratic equation}$$

It is helpful in planning our strategy for solving a quadratic equation to be able to identify the types of terms in a quadratic equation and to identify the coefficients a, b, and c.

EXAMPLE Write each equation in standard form and identify a, b, and c.

(a) $3x^2 + 5x - 2 = 0$ (b) $5x^2 = 2x - 3$
(c) $3x = 5x^2$ (d) $-6x^2 + 2x = 0$
(e) $5x^2 - 4 = 0$ (f) $x^2 = 9$
(g) $x^2 + 3x = 5x$ (h) $7 - x^2 = 3 + x$

(a) $3x^2 + 5x - 2 = 0$ In standard form
     $a = 3, \quad b = 5, \quad c = -2$

(b) $5x^2 = 2x - 3$ Write in standard form.
     $5x^2 - 2x + 3 = 0$
     $a = 5, \quad b = -2, \quad c = 3$

(c) $3x = 5x^2$ Write in standard form.
     $0 = 5x^2 - 3x$ Interchange sides of equation.
     $5x^2 - 3x = 0$
     $a = 5, \quad b = -3, \quad c = 0$ No constant term means $c = 0$.

(d) $-6x^2 + 2x = 0$ Multiply by -1.

$\quad -1(-6x^2 + 2x) = (-1)0$

$\quad 6x^2 - 2x = 0$ Standard form

$\quad \mathbf{a = 6, \quad b = -2, \quad c = 0}$ No constant term

(e) $5x^2 - 4 = 0$ Standard form

$\quad \mathbf{a = 5, \quad b = 0, \quad c = -4}$ No linear term means $b = 0$.

(f) $x^2 = 9$ Write in standard form.

$\quad x^2 - 9 = 0$

$\quad \mathbf{a = 1, \quad b = 0, \quad c = -9}$ No linear term

(g) $x^2 + 3x = 5x$ Write in standard form.

$\quad x^2 + 3x - 5x = 0$ Combine like terms.

$\quad x^2 - 2x = 0$

$\quad \mathbf{a = 1, \quad b = -2, \quad c = 0}$ No constant term

(h) $7 - x^2 = 3 + x$ Rearrange.

$\quad -x^2 - x + 7 - 3 = 0$ Combine like terms.

$\quad -x^2 - x + 4 = 0$ Multiply by -1.

$\quad -1(-x^2 - x + 4) = (-1)0$

$\quad x^2 + x - 4 = 0$ Standard form

$\quad \mathbf{a = 1, \quad b = 1, \quad c = -4}$

---

3   **Solve Pure Quadratic Equations ($ax^2 + c = 0$) by the Square-Root Method.**

To solve pure quadratic equations, solve for the squared variable and apply the **square-root property of equality.** That is, take the square root of both sides of the equation.

To solve a pure quadratic equation ($ax^2 + c = 0$):

1. Rearrange the equation, if necessary, so that the quadratic term is on one side and the constant is on the other side of the equation.
2. Combine like terms, if appropriate.
3. Rewrite the equation so that the quadratic term has a coefficient is $+1$.
4. Apply the square-root property of equality by taking the square root of both sides.
5. Check each solution in the original equation.

EXAMPLE   Solve $3y^2 = 27$.

$\quad 3y^2 = 27$ Divide both sides by the coefficient of the quadratic term.

$\quad y^2 = 9$ Apply the square-root property of equality.

$\quad \mathbf{y = \pm 3}$ Solutions are $y = 3$ and $y = -3$.

Check $y =$ ⬚3⬚   Check $y =$ ⬚-3⬚

$\quad 3y^2 = 27$ $3y^2 = 27$

$\quad 3(⬚3⬚)^2 = 27$ $3(⬚-3⬚)^2 = 27$

$\quad 3(9) = 27$ $3(9) = 27$

$\quad 27 = 27$ $27 = 27$ Solutions check.

EXAMPLE Solve $4x^2 - 9 = 0$.

$$4x^2 - 9 = 0 \qquad \text{Sort terms.}$$

$$4x^2 = 9 \qquad \text{Divide by 4.}$$

$$x^2 = \frac{9}{4} \qquad \text{Apply the square-root property of equality.}$$

$$x = \pm\frac{3}{2} \qquad \text{Divide if a decimal answer is desired.}$$

$$x = \pm 1.5 \qquad \text{Decimal solutions are } x = 1.5 \text{ and } x = -1.5.$$

Check $x = \boxed{1.5}$

$$4x^2 - 9 = 0$$
$$4(\boxed{1.5})^2 - 9 = 0$$
$$4(2.25) - 9 = 0$$
$$9 - 9 = 0$$
$$0 = 0$$

Check $x = \boxed{-1.5}$

$$4x^2 - 9 = 0$$
$$4(\boxed{-1.5})^2 - 9 = 0$$
$$4(2.25) - 9 = 0$$
$$9 - 9 = 0$$
$$0 = 0 \qquad \text{Solutions check.}$$

In some applications, the fractional answer may be more convenient. In other applications, the decimal answer may be more convenient.

TIP

Roots of Pure Quadratic Equations

In a pure quadratic equation, the value of b is zero: $ax^2 + 0x + c = 0$.

- The equation has real number solutions only if c is negative when the equation is in standard form. Remember, standard form requires that a be positive.
- The two solutions have the same absolute value and are opposites.
- There will be either two solutions or no real solution unless $c = 0$.
- When $c = 0$, the solution is $x = 0$ and it is **double root.**

SECTION 15–1 SELF-STUDY EXERCISES

1 Arrange the quadratic equations in standard form.

1. $5 - 4x + 7x^2 = 0$ **2.** $8x^2 - 3 = 6x$ **3.** $7x^2 = 5$

4. $x^2 = 6x - 8$ **5.** $x^2 - 3x = 6x - 8$ **6.** $8 - x^2 + 6x = 2x$

7. $3x^2 + 5 - 6x = 0$ **8.** $5 = x^2 - 6x$ **9.** $x^2 - 16 = 0$

10. $8x^2 - 7x = 8$ **11.** $8x^2 + 8x - 2 = 8$ **12.** $0.3x^2 - 0.4x = 3$

2 Write each equation in standard form and identify a, b, and c.

13. $5x = x^2$ **14.** $7x - 3x^2 = 5$ **15.** $7x^2 - 4x = 0$

16. $8 + 3x^2 = 5x$ **17.** $5x - 6 = x^2$ **18.** $11x^2 = 8x$

19. $x = x^2$ **20.** $9x^2 - 7x = 12$ **21.** $5 = x^2$

22. $-x^2 - 6x + 3 = 0$ **23.** $0.2x - 5x^2 = 1.4$ **24.** $\frac{2}{3}x^2 - \frac{5}{6}x = \frac{1}{2}$

25. $1.3x^2 - 8 = 0$ **26.** $\sqrt{3}x^2 + \sqrt{5}x - 2 = 0$ **27.** $5 = -3x - 15x^2$

Write equations in standard form.

28. The coefficient of the quadratic term is 8, the coefficient of the linear term is -2, the constant term is -3.

29. The constant is 0, the coefficient of the linear term is 3, and the coefficient of the quadratic term is 1.

30. $a = 5, b = 2, c = -7$

31. $a = 2.5, c = -0.8$

3 Solve the equations. Round to thousandths when necessary.

32. $x^2 = 9$

33. $x^2 - 49 = 0$

34. $9x^2 = 64$

35. $16x^2 - 49 = 0$

36. $0.09y^2 = 0.81$

37. $0.04x^2 = 0.81$

38. $2x^2 = 10$

39. $3x^2 + 4 = 7$

40. $5x^2 - 8 = 12$

41. $7x^2 - 2 = 19$

42. $12x^2 - 27 = 0$

43. $28x^2 = 112$

44. AG/H   The label on a new swimming pool cover shows it is a square and has an area of 529 ft². What is the length of each side of the pool cover?

45. AG/H   A farmer has a field designed in the shape of a circle for efficiency in irrigation. The area of the field is known to be 21,000 yd². What length of irrigation line is required to provide water to the entire field? (*Note:* The line moves from the center of the field in a circle around the field. The area of a circle is $A = \pi r^2$.)

46. Describe the process for solving a pure quadratic equation.

47. What is the relationship between the roots of a pure quadratic equation?

15–2 | *Solving Quadratic Equations by Factoring*

Learning Outcomes

1 Solve incomplete quadratic equations $(ax^2 + bx = 0)$ by factoring.

2 Solve complete quadratic equations $(ax^2 + bx + c = 0)$ by factoring.

1 Solve Incomplete Quadratic Equations $(ax^2 + bx = 0)$ by Factoring.

An incomplete quadratic equation always has a variable common factor, so we can factor the left side of the equation $x^2 + 2x = 0$ to $x(x + 2) = 0$. We have two factors that have a product of 0. The **zero-product property** can be used to solve the equation.

Zero-product property

If $ab = 0$, then $a = b$ or $b = 0$ where a and b are real numbers.

To solve an incomplete quadratic equation ($ax^2 + bx = 0$) by factoring:

1. Use the addition axiom to write the equation in standard form.
2. Factor out all common factors and set each factor containing a variable equal to zero using the zero-product property.
3. Solve for the variable in each equation formed in Step 2.
4. Check each solution in the original equation.

EXAMPLE   Solve $2x^2 = 5x$ for x.

$$2x^2 = 5x$$ Write in standard form.

$$2x^2 - 5x = 0$$ Factor common factor.

$$x\,(2x - 5) = 0$$ Set each factor equal to zero.

$$x = 0 \qquad 2x - 5 = 0$$ Solve each equation.

$$x = \mathbf{0} \qquad\quad 2x = 5$$

$$x = \frac{5}{2}$$

Check $x = 0$ $\qquad\qquad$ Check $x = \dfrac{5}{2}$

$$2x^2 = 5x \qquad\qquad\qquad 2x^2 = 5x$$

$$2(0)^2 = 5(0) \qquad\qquad 2\left(\frac{5}{2}\right)^2 = 5\left(\frac{5}{2}\right)$$

$$2(0) = 5(0) \qquad\qquad\quad \overset{1}{2}\left(\frac{25}{\underset{2}{4}}\right) = \frac{25}{2}$$

$$0 = 0 \qquad\qquad\qquad\quad \frac{25}{2} = \frac{25}{2} \qquad \text{Solutions check.}$$

EXAMPLE   Solve $4x^2 - 8x = 0$ for x.

$$4x^2 - 8x = 0$$ Standard form. Factor common factors.

$$4x\,(x - 2) = 0$$ Set each variable factor equal to zero.

$$4x = 0 \qquad x - 2 = 0$$ Solve each equation.

$$x = \frac{0}{4} \qquad\quad x = 2$$

$$x = \mathbf{0}$$

Check $x = 0$ $\qquad\qquad\qquad$ Check $x = 2$

$$4(0)^2 - 8(0) = 0 \qquad\qquad 4(2)^2 - 8(2) = 0$$

$$4(0) - 8(0) = 0 \qquad\qquad\quad 4(4) - 8(2) = 0$$

$$0 - 0 = 0 \qquad\qquad\qquad\quad 16 - 16 = 0$$

$$0 = 0 \qquad\qquad\qquad\qquad\quad 0 = 0 \qquad \text{Solutions check.}$$

In an incomplete quadratic equation, the value of $c = 0$: $ax^2 + bx + 0 = 0$.

$$x(ax + b) = 0$$

$$x = 0 \qquad ax + b = 0$$

$$ax = -b$$

$$\frac{ax}{a} = \frac{-b}{a}$$

$$x = -\frac{b}{a}$$

- One root is always zero.
- The other root is $-\dfrac{b}{a}$.

2 Solve Complete Quadratic Equations ($ax^2 + bx + c = 0$) by Factoring.

Complete quadratic equations have all three types of terms. If the expression on the left will factor into two binomials, we can apply the zero-product property. If the trinomial will not factor, we must use another method for solving the quadratic equation.

To solve a complete quadratic ($ax^2 + bx + c = 0$) equation by factoring:

1. Arrange the equation in standard form.
2. Factor the trinomial into the product of two binomials, and set each binomial factor equal to zero using the zero-product property.
3. Solve for the variable in each equation formed in Step 2.
4. Check if desired.

EXAMPLE Solve $x^2 + 6x + 5 = 0$ for x.

$$x^2 + 6x + 5 = 0 \qquad \text{Standard form. Factor as the product of two binomials.}$$

$$(x + 5)(x + 1) = 0 \qquad \text{Set each factor equal to 0.}$$

$$x + 5 = 0 \quad x + 1 = 0 \qquad \text{Solve for } x \text{ in each equation.}$$

$$x = -5 \qquad x = -1$$

Check $x = -5$	Check $x = -1$
$x^2 + 6x + 5 = 0$	$x^2 + 6x + 5 = 0$
$(-5)^2 + 6(-5) + 5 = 0$	$(-1)^2 + 6(-1) + 5 = 0$
$25 + (-30) + 5 = 0$	$1 + (-6) + 5 = 0$
$-5 + 5 = 0$	$-5 + 5 = 0$
$0 = 0$	$0 = 0$ Solutions check.

EXAMPLE Solve $6x^2 + 4 = 11x$ for x.

$$6x^2 + 4 = 11x \qquad \text{Write in standard form.}$$

$$6x^2 - 11x + 4 = 0 \qquad \text{Factor trinomial by grouping. Write } -11x \text{ as } -3x - 8x.$$

$$6x^2 - 3x - 8x + 4 = 0 \qquad \text{Group.}$$

$$(6x^2 - 3x) + (-8x + 4) = 0 \qquad \text{Factor common factors.}$$

$$3x(2x - 1) - 4(2x - 1) = 0 \qquad \text{Factor the common grouping.}$$

$$(2x - 1)\,(3x - 4) = 0 \qquad \text{Set each factor equal to 0.}$$

$$2x - 1 = 0 \qquad 3x - 4 = 0 \qquad \text{Solve each equation.}$$

$$2x = 1 \qquad 3x = 4$$

$$x = \frac{1}{2} \qquad x = \frac{4}{3}$$

Check $x = \dfrac{1}{2}$

$$6x^2 + 4 = 11x$$

$$6\left(\frac{1}{2}\right)^2 + 4 = 11\left(\frac{1}{2}\right)$$

$$\overset{3}{\cancel{6}}\left(\frac{1}{\underset{2}{\cancel{4}}}\right) + 4 = \frac{11}{2}$$

$$\frac{3}{2} + \frac{8}{2} = \frac{11}{2} \qquad \left(4 = \frac{8}{2}\right)$$

$$\frac{11}{2} = \frac{11}{2}$$

Check $x = \dfrac{4}{3}$

$$6x^2 + 4 = 11x$$

$$6\left(\frac{4}{3}\right)^2 + 4 = 11\left(\frac{4}{3}\right)$$

$$\overset{2}{\cancel{6}}\left(\frac{16}{\underset{3}{\cancel{9}}}\right) + 4 = \frac{44}{3}$$

$$\frac{32}{3} + \frac{12}{3} = \frac{44}{3} \qquad \left(4 = \frac{12}{3}\right)$$

$$\frac{44}{3} = \frac{44}{3} \qquad \text{Solutions check.}$$

Roots of Complete Quadratic Equations

Because all three types of terms are included in complete quadratic equations, the roots do *not* have the same characteristics as the roots of pure or incomplete quadratic equations.

- Zero is not a root.
- The two roots are not opposites.
- There can be just one distinct root. The equation has two equal roots called a double root.

SECTION 15–2 SELF-STUDY EXERCISES

1 Solve by factoring.

1. $x^2 - 3x = 0$	**2.** $x^2 - 6x = 0$	**3.** $5x^2 - 10x = 0$
4. $2x^2 + x = 0$	**5.** $8x^2 - 4x = 0$	**6.** $5x^2 - 15x = 0$
7. $3x^2 - 7x = 0$	**8.** $y^2 + 4y = 0$	**9.** $3x^2 + 2x = 0$
10. $9x^2 = 12x$	**11.** $6x^2 = 18x$	**12.** $9x^2 = 6x$

13. The square of a number is 8 times the number. Find the number.

14. A square rug has an area 15 times the length of one of the sides. What is the length of a side?

15. How does an incomplete quadratic equation differ from a pure quadratic equation?

16. Will one of the two roots of an incomplete quadratic equation always be zero? Justify your answer.

2 Solve the equations by factoring.

17. $x^2 + 5x + 6 = 0$

18. $x^2 - 6x + 9 = 0$

19. $x^2 - 5x - 14 = 0$

20. $x^2 + 3x - 18 = 0$

21. $x^2 + 7x + 12 = 0$

22. $y^2 - 8y = -15$

23. $a^2 - 13a = 14$

24. $b^2 - 9b = -18$

25. $2x^2 - 7x + 3 = 0$

26. $3x^2 + 13x + 4 = 0$

27. $10x^2 - x - 3 = 0$

28. $6x^2 + 11x + 3 = 0$

29. $2x^2 + 13x + 15 = 0$

30. $3x^2 - 10x + 8 = 0$

31. $6x^2 + 17x - 3 = 0$

32. $3x^2 + 14x + 8 = 0$

33. $2x^2 - 13x + 15 = 0$

34. $6x^2 - 7x + 2 = 0$

35. $8x^2 + 3 = 10x$

36. $5x^2 + 3x = 2$

37. $6x^2 = x + 15$

38. $9x^2 + 18x = -5$

39. $6x^2 + 3 = 11x$

40. $5x^2 + 13x = 6$

41. CON A rectangular hallway is 6 ft longer than its width. The area is 55 ft². What are the length and width of the hallway?

42. AUTO A rectangular metal plate covering a spare tire well that has an opening of 378 in.² is broken and must be reconstructed. The width is 3 in. less than the length. What dimensions should the metalsmith use when making the replacement part?

15–3 | ***Solving Quadratic Equations by Completing the Square or Using the Formula***

Learning Outcomes **1** Solve quadratic equations by completing the square.
2 Solve quadratic equations using the quadratic formula.

1 Solve Quadratic Equations by Completing the Square.

Not all complete quadratic equations can be solved by factoring. One procedure for solving these equations is called **completing the square.**

This method incorporates manipulations that result in a perfect-square trinomial that factors into the square of a binomial. Look again at the relationship between a perfect-square trinomial and the square of a binomial.

$$a^2 + 2ab + b^2 = (a + b)^2$$

To solve a quadratic equation by completing the square:

1. Write the equation in the form $ax^2 + bx = -c$. That is, isolate the terms with variable factors.
2. Divide the equation by a so that the coefficient of x^2 is 1.

$$x^2 + \frac{b}{a}x = -\frac{c}{a}$$

3. Form a perfect-square trinomial on the left by adding $\left(\dfrac{b}{2a}\right)^2$ or $\dfrac{b^2}{4a^2}$ to both sides of the equation.

$$x^2 + \frac{b}{a}x + \frac{b^2}{4a^2} = -\frac{c}{a} + \frac{b^2}{4a^2}$$

4. Factor the perfect-square trinomial into the square of a binomial.

$$\left(x + \frac{b}{2a}\right)^2 = -\frac{c}{a} + \frac{b^2}{4a^2}$$

5. Take the square root of both sides of the equation.

$$x + \frac{b}{2a} = \pm\sqrt{-\frac{c}{a} + \frac{b^2}{4a^2}}$$

6. Rearrange terms under the radical.

$$x + \frac{b}{2a} = \pm\sqrt{\frac{b^2}{4a^2} - \frac{c}{a}}$$

7. Make equivalent fractions using the common denominator for terms under radical

$$x + \frac{b}{2a} = \pm\sqrt{\frac{b^2}{4a^2} - \frac{4ac}{4a^2}}$$

8. Write terms under radical as one fraction.

$$x + \frac{b}{2a} = \pm\sqrt{\frac{b^2 - 4ac}{4a^2}}$$

9. Solve for x.

$$x = \frac{-b \pm \sqrt{b^2 - 4ac}}{2a}$$

EXAMPLE   Solve $x^2 - 5x - 2 = 0$ by completing the square.

$$x^2 - 5x - 2 = 0$$
The coefficient of x^2 is 1. Isolate the terms with variable factors.

$$x^2 - 5x = 2$$
Add $\left(\dfrac{b}{2a}\right)^2$ to both sides ($a = 1$, $b = -5$). $\left(\dfrac{-5}{2(1)}\right)^2$

$$x^2 - 5x + \left(\frac{-5}{2}\right)^2 = 2 + \left(\frac{-5}{2}\right)^2$$
Simplify.

$$x^2 - 5x + \frac{25}{4} = 2 + \frac{25}{4}$$
Combine like terms.

$$x^2 - 5x + \frac{25}{4} = \frac{8}{4} + \frac{25}{4}$$

$$x^2 - 5x + \frac{25}{4} = \frac{33}{4}$$
Write the perfect-square trinomial as the square of a binomial. $\left(x + \dfrac{b}{2a}\right)^2$

$$\left(x - \frac{5}{2}\right)^2 = \frac{33}{4}$$
Take the square root of both sides.

$$x - \frac{5}{2} = \pm\sqrt{\frac{33}{4}}$$
Simplify and solve for x.

$$x = \frac{5}{2} \pm \frac{\sqrt{33}}{2}$$
Identify each root.

$$x = \frac{5}{2} + \frac{\sqrt{33}}{2} \qquad \text{or} \qquad x = \frac{5}{2} - \frac{\sqrt{33}}{2}$$
Exact roots.

$$x \approx 2.5 + 2.872281323 \qquad\qquad x \approx 2.5 - 2.872281323$$

$$\mathbf{x \approx 5.37} \qquad \text{or} \qquad \mathbf{x \approx -0.37}$$
Approximate roots.

EXAMPLE   Solve $3x^2 - 3x - 7 = 0$ by completing the square.

$$3x^2 - 3x - 7 = 0 \qquad \text{Make the leading coefficient 1 by dividing the equation by 3.}$$

$$\frac{3x^2}{3} - \frac{3x}{3} - \frac{7}{3} = 0$$

$$x^2 - x - \frac{7}{3} = 0 \qquad \text{Isolate the terms with variable factors.}$$

$$x^2 - x = \frac{7}{3} \qquad \text{Add } \left(\frac{-1}{2(1)}\right)^2 \text{ to both sides.}$$

$$x^2 - x + \left(\frac{-1}{2}\right)^2 = \frac{7}{3} + \left(\frac{-1}{2}\right)^2 \qquad \text{Simplify.}$$

$$x^2 - x + \frac{1}{4} = \frac{7}{3} + \frac{1}{4} \qquad \frac{7}{3} + \frac{1}{4} = \frac{28}{12} + \frac{3}{12} = \frac{31}{12}$$

$$x^2 - x + \frac{1}{4} = \frac{31}{12} \qquad \text{Write the perfect-square trinomial as the square of the binomial.}$$

$$\left(x - \frac{1}{2}\right)^2 = \frac{31}{12} \qquad \text{Take the square root of both sides.}$$

$$x - \frac{1}{2} = \pm\sqrt{\frac{31}{12}} \qquad \text{Simplify and solve for } x.$$

$$\sqrt{\frac{31}{12}} = \frac{\sqrt{31}}{2\sqrt{3}} \cdot \frac{\sqrt{3}}{\sqrt{3}} = \frac{\sqrt{93}}{6}$$

$$x = \frac{1}{2} \pm \frac{\sqrt{93}}{6} \qquad \text{Identify each root.}$$

$$x = \frac{1}{2} + \frac{\sqrt{93}}{6} \qquad \text{or} \qquad x = \frac{1}{2} - \frac{\sqrt{93}}{6} \qquad \text{Exact solutions.}$$

$$x \approx 0.5 + 1.607275127 \qquad \qquad x \approx 0.5 - 1.607275127$$

$$x \approx \mathbf{2.11} \qquad \text{or} \qquad x \approx \mathbf{-1.11} \qquad \text{Approximate solutions.}$$

TIP

Completing-the-Square Method Works for All Types of Quadratic Equations

The completing-the-square method works for all types of quadratic equations. However, factoring is more efficient for incomplete quadratic equations and complete quadratic equations that factor.

Solve $x^2 - 7x = 0$.

By Factoring

$$x^2 - 7x = 0$$

$$x(x - 7) = 0$$

$$x = 0 \qquad x - 7 = 0$$

$$x = 7$$

By Completing the Square

$$x^2 - 7x + \left(\frac{-7}{2}\right)^2 = 0 + \left(\frac{-7}{2}\right)^2$$

$$x^2 - 7x + \frac{49}{4} = \frac{49}{4}$$

$$\left(x - \frac{7}{2}\right)^2 = \frac{49}{4}$$

$$x - \frac{7}{2} = \pm\frac{7}{2}$$

$$x = \frac{7}{2} \pm \frac{7}{2}$$

$$x = \frac{7}{2} + \frac{7}{2} \qquad x = \frac{7}{2} - \frac{7}{2}$$

$$x = \frac{14}{2} \qquad\qquad x = 0$$

$$x = 7$$

Solve $x^2 + 5x + 6 = 0$.

By Factoring

$$x^2 + 5x + 6 = 0$$

$$(x + 2)(x + 3) = 0$$

$$x + 2 = 0 \qquad x + 3 = 0$$

$$x = -2 \qquad x = -3$$

By Completing the Square

$$x^2 + 5x + 6 = 0$$

$$x^2 + 5x = -6$$

$$x^2 + 5x + \left(\frac{5}{2}\right)^2 = -6 + \left(\frac{5}{2}\right)^2$$

$$x^2 + 5x + \frac{25}{4} = -6 + \frac{25}{4}$$

$$\left(x + \frac{5}{2}\right)^2 = \frac{-24}{4} + \frac{25}{4}$$

$$\left(x + \frac{5}{2}\right)^2 = \frac{1}{4}$$

$$x + \frac{5}{2} = \pm\sqrt{\frac{1}{4}}$$

$$x = -\frac{5}{2} \pm \frac{1}{2}$$

$$x = \frac{-5}{2} + \frac{1}{2} \qquad x = \frac{-5}{2} - \frac{1}{2}$$

$$x = -\frac{4}{2} \qquad\qquad x = -\frac{6}{2}$$

$$x = -2 \qquad\qquad x = -3$$

2 Solve Quadratic Equations Using the Quadratic Formula.

Another method for solving quadratic equations is to use the **quadratic formula.** This formula results from solving the standard quadratic equation, $ax^2 + bx + c = 0$, by completing the square.

Quadratic formula:

$$x = \frac{-b \pm \sqrt{b^2 - 4ac}}{2a}$$

where a, b, and c are real-number coefficients of a quadratic equation in the form $ax^2 + bx + c = 0$.

Quadratic Formula versus Completing the Square

Look again at the completing-the-square method for solving quadratic equations on pages 580–581. The symbolic representation of the method shows how the quadratic formula is derived from this method. Between steps 8 and 9 the equation is solved for x and the fractions are combined.

To evaluate the quadratic formula:

1. Write the equation in standard form.
2. Identify a, b, and c.
3. Substitute numbers for a, b, and c in the quadratic formula.
4. Use the order of operations to simplify the expression under the radical.
5. Simplify the radical.
6. Factor any common factors in the numerator.
7. Simplify by reducing.
8. Write as two distinct solutions.
9. Write as exact solution or approximate solution as desired.

EXAMPLE Use the quadratic formula to solve $x^2 + 5x + 6 = 0$ for x.

$a = 1, \qquad b = 5, \qquad c = 6$ Identify a, b, and c.

$x = \dfrac{-b \pm \sqrt{b^2 - 4ac}}{2a}$ Quadratic formula. Substitute for a, b, and c.

$x = \dfrac{-5 \pm \sqrt{5^2 - 4 \cdot 1 \cdot 6}}{2 \cdot 1}$ Evaluate power and multiply in radicand.

$x = \dfrac{-5 \pm \sqrt{25 - 24}}{2}$ Combine terms in radicand.

$x = \dfrac{-5 \pm \sqrt{1}}{2}$ Evaluate radical.

$x = \dfrac{-5 \pm 1}{2}$

At this point, we separate the solutions into two parts, one using the $+1$ and the other using the -1.

$$x = \dfrac{-5 + 1}{2} \qquad\qquad x = \dfrac{-5 - 1}{2}$$

$$x = \dfrac{-4}{2} \qquad\qquad x = \dfrac{-6}{2}$$

$$x = -2 \qquad\qquad x = -3$$

Check $x = -2$ Check $x = -3$

$$x^2 + 5x + 6 = 0 \qquad\qquad x^2 + 5x + 6 = 0$$

$$(-2)^2 + 5(-2) + 6 = 0 \qquad\qquad (-3)^2 + 5(-3) + 6 = 0$$

$$4 - 10 + 6 = 0 \qquad\qquad 9 - 15 + 6 = 0$$

$$-6 + 6 = 0 \qquad\qquad -6 + 6 = 0$$

$$0 = 0 \qquad\qquad 0 = 0 \qquad \text{Solutions check.}$$

Common Cause for Error

When you use the quadratic formula to solve problems, begin by writing the formula to help you remember it. When you write the formula, be sure to extend the fraction bar beneath the *entire* numerator. This omission is a common cause for errors.

EXAMPLE Use the quadratic formula to solve $3x^2 - 3x - 7 = 0$ for x. Round answers to the nearest hundredth.

$a = 3, \quad b = -3, \quad c = -7$ Identify a, b, and c.

$x = \dfrac{-b \pm \sqrt{b^2 - 4ac}}{2a}$ Quadratic formula. Substitute.
$-(b) = -(-3) = +3$

$x = \dfrac{+3 \pm \sqrt{(-3)^2 - 4(3)(-7)}}{2 \cdot 3}$ Perform calculations in radicand.

$x = \dfrac{+3 \pm \sqrt{9 + 84}}{6}$ Combine terms in the radicand.

$x = \dfrac{+3 \pm \sqrt{93}}{6}$ Exact solutions

Find the approximate solutions:

$x \approx \dfrac{+3 \pm 9.643650761}{6}$ Evaluate the radical. Separate the two solutions.

$x \approx \dfrac{3 + 9.643650761}{6}$ $x \approx \dfrac{3 - 9.643650761}{6}$

$x \approx \dfrac{12.643650761}{6}$ $x \approx \dfrac{-6.643650761}{6}$

$x \approx \mathbf{2.11}$ Rounded $x \approx \mathbf{-1.11}$ Approximate solutions

Notice that we round to hundredths *after* making the final calculation.

Check $x \approx 2.11$ Check $x \approx -1.11$

$3x^2 - 3x - 7 \approx 0$ $3x^2 - 3x - 7 \approx 0$

$3(2.11)^2 - 3(2.11) - 7 \approx 0$ $3(-1.11)^2 - 3(-1.11) - 7 \approx 0$

$3(4.4521) - 6.33 - 7 \approx 0$ $3(1.2321) + 3.33 - 7 \approx 0$

$13.3563 - 6.33 - 7 \approx 0$ $3.6963 + 3.33 - 7 \approx 0$

$0.0263 \approx 0$ $0.0263 \approx 0$ Solutions check.

Rounding Discrepancies and Checking with a Calculator

The symbol $\approx$ means "is approximately equal to." Another symbol for approximations is $\doteq$. If we had checked *before* rounding in the example on the previous page, the checks would have been "closer" to zero.

It is best to use the full calculator value to check. The full calculator value improves the accuracy of the check. Calculate the first root in the previous example.

$$3 \boxed{+} \boxed{\sqrt{}} \; 93 \boxed{=} \boxed{\div} \; 6 \boxed{=} \Rightarrow 2.107275127$$

Check:

If your calculator allows you to insert the answer of your last calculation, use this function in checking. Otherwise, you can store an answer in memory and recall as appropriate.

$$3 \boxed{\text{ANS}} \boxed{x^2} \boxed{-} 3 \boxed{\text{ANS}} \boxed{-} 7 \boxed{=} \Rightarrow 0$$

Calculate the second root. $$3 \boxed{-} \boxed{\sqrt{}} \; 93 \boxed{=} \boxed{\div} \; 6 \boxed{=} \Rightarrow -1.107275127$$

Check: $$3 \boxed{\text{ANS}} \boxed{x^2} \boxed{-} 3 \boxed{\text{ANS}} \boxed{-} 7 \boxed{=} \Rightarrow 0.$$

Even though quadratic equations have two roots, a root may be disregarded in an applied problem because it is not appropriate within the context of the problem.

EXAMPLE   Find the length and width of a rectangular table if the length is 8 in. more than the width and the area is 260 in^2.

Unknown facts   Length and width of rectangle

Known facts   Area = 260 in^2, length = 8 in. more than width

Relationships   Let x = number of inches in the width
$x + 8$ = number of inches in the length
Area = length times width, or $A = lw$

Estimation   The square root of 260 is between 16 and 17 ($16^2 = 256$, $17^2 = 289$); the width should be less than 16 and the length more than 16.

Calculations

$260 = (x + 8)(x)$ Substitute into the area formula and distribute.

$260 = x^2 + 8x$ Write in standard form.

$0 = x^2 + 8x - 260$ or $x^2 + 8x - 260 = 0$

$a = 1, \qquad b = 8, \qquad c = -260$

$x = \dfrac{-b \pm \sqrt{b^2 - 4ac}}{2a}$ Quadratic formula
Substitute.

$x = \dfrac{-8 \pm \sqrt{(8)^2 - 4(1)(-260)}}{2(1)}$ Simplify the radicand.

$x = \dfrac{-8 \pm \sqrt{64 + 1{,}040}}{2}$ Combine terms in radicand.

$x = \dfrac{-8 \pm \sqrt{1{,}104}}{2}$ Simplify radical.

$x = \dfrac{-8 \pm 4\sqrt{69}}{2}$ Factor and reduce.

$x = -4 \pm 2\sqrt{69}$ Exact solutions

Find the approximate solutions:

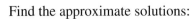

$$x \approx \frac{-8 \pm 33.22649545}{2}$$

Evaluate the radical.
Separate into the two solutions.

$$x \approx \frac{-8 + 33.22649545}{2} \qquad x \approx \frac{-8 - 33.22649545}{2}$$

$$x \approx \frac{25.22649545}{2} \qquad x \approx \frac{-41.22649545}{2}$$

$$x \approx 12.6 \quad \text{width} \qquad x \approx -20.6$$

Disregard negative solution.

$$x + 8 \approx 20.6 \quad \text{length}$$

Interpretation   Measurements are positive, so disregard the negative solution.
The width is approximately 12.6 in. and the length is approximately 20.6 in.

SECTION 15–3 SELF-STUDY EXERCISES

1 Solve by completing the square. Give the exact solutions. Simplify radicals and reduce.

1. $x^2 + 2x - 48 = 0$ **2.** $x^2 - 8x - 9 = 0$ **3.** $x^2 - 10x + 9 = 0$

4. $x^2 - 10x + 24 = 0$ **5.** $x^2 + 2x = 24$ **6.** $x^2 + 16x = -28$

7. $x^2 - 3x - 5 = 0$ **8.** $x^2 - 5x - 2 = 0$ **9.** $x^2 - 5x - 3 = 0$

10. $x^2 - 3x - 1 = 0$ **11.** $2x^2 - 4x - 3 = 0$ **12.** $2x^2 - 2x - 1 = 0$

13. $3x^2 - 6x + 2 = 0$ **14.** $3x^2 + 12x + 8 = 0$ **15.** $4x^2 - 12 = 8x$

16. $7x^2 = 3x + 7$ **17.** $5x^2 - 11x + 2 = 0$ **18.** $3x^2 - x = 7$

2 Solve the quadratic equations by using the quadratic formula.

19. $3x^2 - 7x - 6 = 0$ **20.** $x^2 + x - 12 = 0$ **21.** $5x^2 - 6x = 11$

22. $x^2 - 6x + 9 = 0$ **23.** $x^2 - x - 6 = 0$ **24.** $8x^2 - 2x = 3$

Solve the quadratic equations by using the quadratic formula. Round each approximate answer to the nearest hundredth.

25. $x^2 = -9x + 20$ **26.** $3x^2 + 6x + 1 = 0$ **27.** $\dfrac{2}{x} + \dfrac{5}{2} = x$

28. $\dfrac{2}{x} + 3x = 8$ **29.** $2x^2 + 3x + 3 = 0$ **30.** $x^2 - x + 2 = 0$

Solve the applied problems.

31. CAD/ARC   A rectangular tabletop is 3 cm longer than it is wide. Find the length and width to the nearest hundredth if the area is 47.5 cm². Round to hundredths. (Area = length × width, or $A = lw$.)

32. INDTEC   A rectangular instrument case has an area of 40 in². If the length is 6 in. more than the width, find the dimensions (length and width) of the instrument case.

33. CON   A bricklayer plans to build an arch with a span (s) of 8 m and a radius (r) of 4 m. How high (h) is the arch? (Use the formula
$$h^2 - 2hr + \frac{s^2}{4} = 0.)$$

34. CON   A rectangular patio slab is 180 ft². If the length is 1.5 times the width, find the width and length of the slab to the nearest whole number.

35. AG/H A farmer normally plants a rectangular field 80 ft by 120 ft. This year the government requires the planting area to be decreased by 20%, and the farmer chooses to decrease the width and length of the field by an equal amount. Draw both the original field and the reduced-size field. If x is the amount by which the length and width are decreased, find the length and width of the new field. Round to the nearest whole number.

36. HLTH/N The recommended dosage of a certain type of medicine is determined by the patient's weight. The formula to determine the dosage is given by $D = 0.1w^2 + 5w$, where D is the dosage in milligrams (mg) and w is the patient's body weight in kilograms. Find the weight of a patient for whom the recommended dosage is 1,800 mg. Round to the nearest tenth.

15–4 | *Graphing Quadratic Functions*

Learning Outcomes

1 Graph quadratic functions using the table-of-solutions method.
2 Graph quadratic functions by examining properties.
3 Solve a quadratic equation from a graph of a corresponding quadratic function.
4 Graph quadratic functions using a graphing calculator.
5 Determine the nature of the roots of a quadratic equation by examining the discriminant.

1 Graph Quadratic Functions Using the Table-of-Solutions Method.

The graphs of linear equations are *straight* lines. We will examine the graphs of some equations that have a degree higher than 1, which are not linear equations.

The graph of an equation of a degree higher than 1 is a *curved* line. The curved line can be a parabola, hyperbola, circle, ellipse, or an irregular curved line. We do not define these terms at this time, but Fig. 15–1 illustrates them. The equation for a **parabola** that opens up or down is distinguished from other quadratic equations. The y variable has degree 1. The x variable must have one term with degree 2 and no term with a degree higher than 2. The quadratic equation in two variables for a parabola can be written as the function $y = ax^2 + bx + c$ or $f(x) = ax^2 + bx + c$.

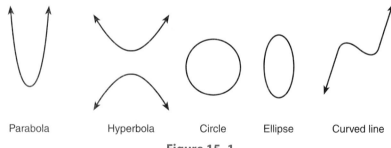

| Parabola | Hyperbola | Circle | Ellipse | Curved line |

Figure 15–1

One method of graphing quadratic functions is to form a table of solutions and plot the points. This method requires *more* points than we customarily use in graphing linear equations.

EXAMPLE Prepare a table of solutions and graph the function $y = x^2 + x - 6$ using integers of x between -3 and 3, inclusive. Determine the domain and range.

$y = x^2 + x - 6$

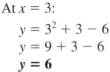

x	y
-3	0
-2	-4
-1	-6
0	-6
1	-4
2	0
3	6

At $x = -3$:

$y = (-3)^2 + (-3) - 6$
$y = 9 - 3 - 6$
$\mathbf{y = 0}$

At $x = -2$:

$y = (-2)^2 + (-2) - 6$
$y = 4 - 2 - 6$
$\mathbf{y = -4}$

At $x = -1$:

$y = (-1)^2 + (-1) - 6$
$y = 1 - 1 - 6$
$\mathbf{y = -6}$

At $x = 0$:

$y = 0^2 + 0 - 6$
$y = 0 + 0 - 6$
$\mathbf{y = -6}$

At $x = 1$:

$y = 1^2 + 1 - 6$
$y = 1 + 1 - 6$
$\mathbf{y = -4}$

At $x = 2$:

$y = 2^2 + 2 - 6$
$y = 4 + 2 - 6$
$\mathbf{y = 0}$

At $x = 3$:

$y = 3^2 + 3 - 6$
$y = 9 + 3 - 6$
$\mathbf{y = 6}$

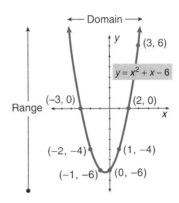

Figure 15–2

Plot the points indicated in the table of solutions and connect the points with a smooth, continuous curve. The curve is symmetrical (Fig. 15–2).

The domain is the set of all real numbers. The range is the set of real numbers greater than or equal to the lowest point of the graph $(-6\frac{1}{4})$.

Finding the value of the lowest point of a parabola is discussed in the learning outcome 3.

2 Graph Quadratic Functions by Examining Properties.

As with linear equations, graphing quadratic equations that form a parabola by the table-of-solutions method can be time-consuming. The graph can be drawn and other important characteristics determined by examining some key properties of the parabola.

A parabola is **symmetrical;** that is, it can be folded in half and the two halves will match. The fold line is called the **axis of symmetry.** For a parabola in the form $y = ax^2 + bx + c$, the equation of the axis of symmetry is $x = -\dfrac{b}{2a}$. The coefficient a does *not* have to be positive.

The point of the graph that crosses the axis of symmetry is the **vertex** of the parabola. Thus, the x-coordinate of the vertex of the parabola is $-\dfrac{b}{2a}$.

> **To graph quadratic function in the form $y = ax^2 + bx + c$:**
>
> 1. Find the axis of symmetry: $x = -\dfrac{b}{2a}$.
> 2. Find the vertex: x-coordinate of vertex $= -\dfrac{b}{2a}$. To find the y-coordinate of the vertex, substitute the value for the x-coordinate into the original function and solve for y.
> 3. Find one or two additional points that are to the right of the axis of symmetry.
> 4. Apply the property of symmetry to find additional points. The symmetrical point will have an x-coordinate that is the same distance from the axis of symmetry on the opposite side. The two points have the same y-coordinate.
> 5. Connect the plotted points with a smooth, continuous curved line.

EXAMPLE Graph the function $y = x^2$ by finding the vertex, the axis of symmetry, and some additional points. Determine the domain and range.

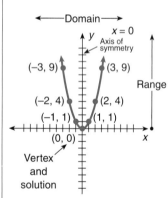

x	y
-3	9
-2	4
-1	1
Vertex 0	0
1	1
2	4
3	9

Figure 15–3

$y = x^2$

$y = x^2 + 0x + 0$ Standard form: $a = 1$, $b = 0$, $c = 0$

$x = -\dfrac{b}{2a}$ Substitute values.

$x = -\dfrac{0}{2(1)}$ Simplify.

$x = 0$ Axis of symmetry and x-coordinate of vertex.

$y = 0^2$ Substitute 0 for x in $y = x^2$.

$y = 0$ y-coordinate of vertex.

Vertex: $(0, 0)$; Axis of symmetry: $x = 0$

Find y for $x = 1$: $y = 1^2$ Find y for $x = 2$: $y = 2^2$ Find y for $x = 3$: $y = 3^2$
$\qquad\qquad\qquad y = 1 \qquad\qquad\qquad\qquad\qquad y = 4 \qquad\qquad\qquad\qquad\qquad y = 9$

Apply the principle of symmetry to complete the table (Fig. 15–3). Plot the points and connect them with a smooth, continuous curve (Fig. 15–4).

Figure 15–4

The domain is the set of all real numbers. The range is the set of real numbers greater than or equal to 0.

3 Solve a Quadratic Equation from a Graph of a Corresponding Quadratic Function.

A quadratic equation in one variable can be written as a function by writing the equation in standard form ($ax^2 + bx + c = 0$) and then writing as a function ($y = ax^2 + bx + c$ or $f(x) = ax^2 + bx + c$). As with linear equations, the real-number solutions of a quadratic equation are at the x-intercepts of the graph of the equation written as a function. The x-coordinate of an x-intercept is also called a **zero** of a function.

Chapter 15 / Quadratic and Higher-Degree Equations

To find the real solutions of a quadratic equation from the graph of a corresponding quadratic function:

1. Write the quadratic equation in standard form and then as a function.
2. Graph the function.
3. Determine the x-coordinate of all x-intercepts of the graph. That is, find the zeros of the function.
4. If there are no x-intercepts, there are no real solutions.

When we graph a quadratic function, we can visually determine the number of real solutions of the corresponding quadratic equation. If the graph has no x-intercept, the quadratic equation has no real solution. When the graph has exactly one x-intercept, the equation has a **double root** (one distinct root or solution). If the graph has two x-intercepts, the corresponding quadratic equation has two real solutions.

EXAMPLE Find the real solutions of the equations by examining the graphs.

(a) $x^2 - 2x - 3 = 0$ **(b)** $-3x^2 = 0$ **(c)** $x^2 + 5 = 0$

Write each equation as a corresponding quadratic function.

     (a) $y = x^2 - 2x - 3$ (b) $y = -3x^2$ (c) $y = x^2 + 5$

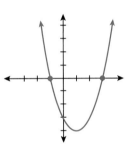 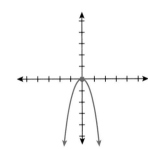 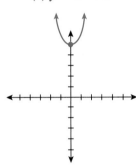

    Figure 15–5   **Figure 15–6**   **Figure 15–7**

   Solutions: $x = -1$ **Solution:** $x = 0$ **No real solutions**
              $x = 3$

The solution (x-coordinates of x-intercepts) of a corresponding quadratic equation can be found algebraically and used to graph a quadratic function.

EXAMPLE Graph the function $y = x^2 + x - 6$ from the example on p. 589 by using the axis of symmetry, the vertex, and the x-intercepts. Determine the domain and range of the function.

Axis of symmetry: $x = -\dfrac{b}{2a}$ Substitute values $a = 1$ and $b = 1$.

                   $x = -\dfrac{1}{2(1)}$ Simplify.

                   $x = -\dfrac{1}{2}$ Also, x-coordinate of vertex.

Vertex: $\left(-\dfrac{1}{2}, y\right)$

Use the x-coordinate, $-\dfrac{1}{2}$, to find the y-coordinate of the vertex.

$y = x^2 + x - 6$ for $x = -\dfrac{1}{2}$ Substitute.

$y = \left(-\dfrac{1}{2}\right)^2 + \left(-\dfrac{1}{2}\right) - 6$ Raise to power.

$y = \dfrac{1}{4} - \dfrac{1}{2} - 6$ Write as equivalent fractions with a common denominator.

$y = \dfrac{1}{4} - \dfrac{2}{4} - \dfrac{24}{4}$ Combine fractions.

$y = -\dfrac{25}{4}$ or $-6\dfrac{1}{4}$ y-coordinate of vertex

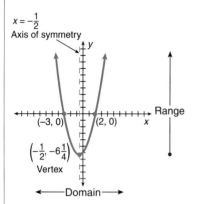

$x = -\dfrac{1}{2}$
Axis of symmetry

$(-3, 0)$ $(2, 0)$ Range

$\left(-\dfrac{1}{2}, -6\dfrac{1}{4}\right)$
Vertex

Domain

Figure 15–8

$\left(-\dfrac{1}{2}, -6\dfrac{1}{4}\right)$

x-intercepts:

$0 = x^2 + x - 6$ Substitute y = 0 and factor.

$0 = (x + 3)(x - 2)$ Set factors equal to 0.

$x + 3 = 0 \qquad x - 2 = 0$ Solve each equation.

$x = -3 \qquad x = 2$ x-coordinates of x-intercepts

$(-3, 0); (2, 0)$ Point notation of x-intercepts

Using the axis of symmetry, vertex, and two x-intercepts, we can get a general idea of the shape of the graph (Fig. 15–8).

The domain is the set of all real numbers. The range is the set of real numbers greater than or equal to $-6\frac{1}{4}$.

EXAMPLE   Graph the function $y = -x^2 + 3$ by using the axis of symmetry, the vertex, and the x-intercepts. Determine the domain and range.

Axis of symmetry: $x = -\dfrac{b}{2a}$ $y = -x^2 + 0x + 3$. Substitute values. $a = -1, b = 0$

$x = -\dfrac{0}{2(-1)}$ Simplify.

$x = 0$ Also, x-coordinate of vertex

Vertex: $(0, y)$ Find y-coordinate of vertex.

$y = -0^2 + 0 + 3$ Substitute x = 0 and solve for y.

$y = 3$ y-coordinate of vertex

$(0, 3)$

x-intercepts:

$0 = -x^2 + 3$ Substitute y = 0 and solve for x.

$x^2 = 3$ Apply the square-root property.

$x = \pm\sqrt{3}$ Exact value of x-intercept

$x \approx \pm 1.7$ Approximate value of x-intercept

$(1.7, 0); (-1.7, 0)$ Point notation of x-intercepts

Plot the points and connect them with a smooth, continuous curve. Two additional points are plotted to give a more complete view of the parabola (Fig. 15–9).

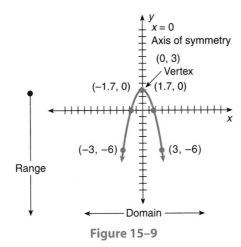

When $x = 3$: $y = -(3)^2 + 3$
$y = -9 + 3$
$y = -6$
$(3, -6)$ is on the graph.
Then, $(-3, -6)$ is on the graph
by the property of symmetry.

Figure 15–9

The domain is the set of all real numbers. The range is the set of real numbers less than or equal to 3.

Tips for Sketching Curves

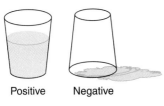

Positive   Negative

When the coefficient of the quadratic term is negative as in the preceding example, the graph of the parabola opens downward. When the coefficient of the quadratic term is positive, the graph of the parabola opens upward. Think of the parabola as a cup or glass: positive holds water; negative spills water (Fig. 15–10).

Figure 15–10

4 Graph Quadratic Functions Using a Graphing Calculator.

Quadratic functions are graphed on graphing calculators using the same procedure as for linear equations. The equation must first be written in standard form and then as a function. Critical values on the graph such as the vertex, x-intercepts or solutions of the corresponding quadratic equation, and the y-intercepts can be determined using calculator functions.

To find the vertex of a quadratic function from the calculator display of the graph:

1. Graph the function on the calculator.
2. Adjust the view of the graph by using the Window or Zoom feature so that the vertex is visible. If the graph turns upward, the vertex is a low point or **minimum.** If the graph turns downward, the vertex is a high point or **maximum** (Fig. 15–11).

Minimum

Maximum

Fig. 15–11

3. Select the appropriate minimum or maximum option from the CALC menu.
4. Move the cursor and press ENTER to a mark point to the left of the vertex (left bound?), to the right of the vertex (right bound?), and near the vertex (guess?).
5. Read the coordinates of the vertex at the bottom of the display screen.

1. Write the equation in standard form and then as a function.
2. Graph the function on the calculator.
3. Adjust the view of the graph by using the Window or Zoom feature so that one or two *x*-intercepts are visible or so that it can be determined that there are no *x*-intercepts.
4. If there are no *x*-intercepts, there are no real solutions.
5. If there are one or two intercepts, select the zero option from the CALC menu.
6. Move the cursor and press ENTER to mark points to the left (left bound?), right (right bound?), and near each *x*-intercept (guess?). Each *x*-intercept is one real solution.
7. Each solution is represented by the *x*-coordinate of each *x*-intercept at the bottom of the display screen.

EXAMPLE   Graph the equation $2x^2 - 5x - 3 = 0$ by writing it as a function and by using a graphing calculator. Find the vertex and the zeros of the function from the graph (Fig. 15–12).

Enter the equation into the calculator.

$\boxed{Y=}$ 2 $\boxed{X, \theta, T, n}$ $\boxed{x^2}$ $\boxed{-}$ 5 $\boxed{X, \theta, T, n}$ $\boxed{-}$ 3 $\boxed{ZOOM}$ $\boxed{6:ZStandard}$

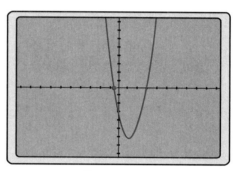

Figure 15–12

Find the vertex: The vertex is a minimum. $\boxed{CALC}$ $\boxed{3:minimum}$

Vertex = (1.25, −6.125)

Mark left, right, and guess. $x \approx 1.25$ (rounded), $y \approx -6.125$

Find the zeros: There are two *x*-intercepts. $\boxed{CALC}$ $\boxed{2:zero}$

Mark left, right, and guess for leftmost solution.

***x*-intercept 1: *x* = −0.5**

$x = -0.5$, $y = 0$

$\boxed{CALC}$ $\boxed{2:zero}$

Mark left, right, and guess for rightmost solution.

***x*-intercept 2: *x* = 3**

$x = 3$, $y = 0$

5   Determine the Nature of the Roots of a Quadratic Equation by Examining the Discriminant.

We have made various observations about the roots of the different types of quadratic equations both from algebraic and graphical methods. Now, we make additional observations about the roots of quadratic equations.

In general, follow these suggestions for solving quadratic equations.

1. Write the equation in standard form: $ax^2 + bx + c = 0$.
2. To solve the equation algebraically:
    a. Identify the type of quadratic equation.
    b. Solve pure quadratic equations by the square-root method.
    c. Solve incomplete quadratic equations by finding common factors.
    d. Solve complete quadratic equations by factoring into two binomials, if possible.
    e. If factoring is not possible or is difficult, use the completing-the-square method or the quadratic formula to solve complete quadratic equations.
3. To solve the equation graphically:
    a. Write the equation as a function.
    b. Graph the function.
    c. Find the x-coordinates of the x-intercepts (zeros) of the function.

The nature of the roots of a quadratic equation can be determined before solving the equation. All types of quadratic equations can be solved using the completing-the-square method or the quadratic formula. Only certain types of quadratic equations can be solved by factoring or applying the square-root property. The radicand of the radical portion of the quadratic formula, $b^2 - 4ac$, is called the **discriminant.** The general characteristics of the roots of a quadratic equation can be determined by examining the discriminant.

Properties of the discriminant, $b^2 - 4ac$:

1. If $b^2 - 4ac \geq 0$, the equation has real-number roots.
    a. If $b^2 - 4ac$ is a perfect square, there are two rational roots.
    b. If $b^2 - 4ac = 0$, there are two equal rational roots, sometimes called a double root.
    c. If $b^2 - 4ac$ is not a perfect square, there are two irrational roots.
2. If $b^2 - 4ac < 0$, the equation has no real roots. The roots are imaginary or complex.

EXAMPLE Examine the discriminant of each equation and determine the nature of the roots. Then solve the equation.

(a) $5x^2 + 3x - 1 = 0$ (b) $3x^2 + 5x = 2$ (c) $4x^2 + 2x + 3 = 0$

(a) $5x^2 + 3x - 1 = 0$ $a = 5, b = 3, c = -1$.

$$b^2 - 4ac = 3^2 - 4(5)(-1)$$ Examine the discriminant.

$$= 9 + 20$$

$$= 29$$ **There will be two irrational roots.** Use the quadratic formula.

$$x = \frac{-b \pm \sqrt{b^2 - 4ac}}{2a}$$ Substitute 29 for the discriminant.

$$x = \frac{-3 \pm \sqrt{29}}{2(5)}$$ Identify each root.

$$x = \frac{-3 + \sqrt{29}}{10} \quad \text{or} \quad x = \frac{-3 - \sqrt{29}}{10}$$ Exact irrational roots

(b)
$$3x^2 + 5x = 2$$

$$3x^2 + 5x - 2 = 0$$

Standard form: $a = 3$, $b = 5$, $c = -2$

$$b^2 - 4ac = 5^2 - 4(3)(-2)$$

Examine the discriminant.

$$= 25 + 24$$

$$= 49$$

The discriminant is a perfect square. **There are two rational roots** and the trinomial will factor.

$$(3x - 1)(x + 2) = 0$$

Factor. Set each factor equal to zero.

$$3x - 1 = 0 \qquad x + 2 = 0$$

Solve each equation.

$$3x = 1$$

$$\mathbf{x = \frac{1}{3}} \qquad \mathbf{x = -2}$$

Exact rational roots

(c) $4x^2 + 2x + 3 = 0$

$a = 4$, $b = 2$, $c = 3$.

$$b^2 - 4ac = 2^2 - 4(4)(3)$$

Examine the discriminant.

$$= 4 - 48$$

$$= \boxed{-44}$$

The discriminant is negative. There are no real roots. **The roots are imaginary or complex.**

$$x = \frac{-2 \pm \sqrt{-44}}{2 \cdot 4}$$

Substitute -44 for the discriminant in the quadratic formula. Simplify the radical.

$$x = \frac{-2 \pm 2i\sqrt{11}}{8}$$

Factor a common factor in the numerator.

$$x = \frac{2(-1 \pm i\sqrt{11})}{8}$$

Reduce.

$$\mathbf{x = \frac{-1 \pm i\sqrt{11}}{4}}$$

The roots are complex.

SECTION 15–4 SELF-STUDY EXERCISES

1 Use a table of solutions to graph the quadratic equations. Determine the domain and range.

1. $y = x^2$

2. $y = 3x^2$

3. $y = \dfrac{1}{3}x^2$

4. $y = -4x^2$

5. $y = -\dfrac{1}{4}x^2$

6. $y = x^2 - 4$

7. $y = x^2 + 4$

8. $y = x^2 - 6x + 9$

9. $y = -x^2 + 6x - 9$

2 Graph the equations using the vertex, x-intercepts, and one other point on the graph. Determine the domain and range.

10. $y = x^2 - 4x + 4$　　　　**11.** $y = x^2 + 4x - 2$　　　　**12.** $y = 2x^2 + 10x + 8$

13. $y = -3x^2 - 6x + 9$　　　**14.** $y = -3x^2 - 6x - 6$　　　**15.** $y = \frac{1}{2}x^2 - 6x + 3$

3 Find all real solutions of the equations by writing and graphing the equations of the corresponding quadratic functions.

16. $x^2 - 5x + 6 = 0$　　　　**17.** $4x^2 = 0$　　　　　　　**18.** $3x^2 = -5$

19. BUS　The revenue R for a leather wallet is based on the price P of the item and is given by the formula $R = -4p^2 + 36p$. What prices generate $0 revenue?

20. TELE　The distance y that an object falls in a vacuum because of gravity in t seconds after it is released is given by the formula $d = 4.9t^2$. How much time is elapsed when $d = 0$?

4 Graph using a graphing calculator. Adjust the window if necessary to show the vertex of the parabola.

21. $y = 3x^2 + 5x - 2$　　**22.** $y = (2x - 3)(x - 1)$　　**23.** $y = 2x^2 - 9x - 5$　　**24.** $y = -2x^2 + 9x + 5$

5 Examine the discriminant of each equation and determine the nature of the roots. Solve each equation. Round to hundredths if necessary.

25. $3x^2 + x - 2 = 0$　　　　**26.** $x^2 - 3x = -1$　　　　**27.** $2x^2 + x = 2$
28. $3x^2 - 2x + 1 = 0$　　　　**29.** $x^2 - 3x - 7 = 0$　　　　**30.** $3x^2 + 5x - 6 = 0$

31. Describe the discriminant of a quadratic equation that has real roots.

32. Describe the discriminant of a quadratic equation that has rational and unequal roots.

| **15–5** | *Solving Higher-Degree Equations by Factoring* |

Learning Outcomes

1 Identify the degree of an equation.
2 Solve higher-degree equations by factoring.
3 Graph higher-degree equations.
4 Distinguish between a function and a relation using the vertical line test.
5 Find the domain and range of a relation from a graph.

Some equations contain terms that have a higher degree than 2.

1 **Identify the Degree of an Equation.**

In Section 15–1, we defined a *quadratic* equation as an equation that has at least one variable term raised to the second power and no variable terms having a power higher than 2. Similarly, a **cubic equation** is an equation that has at least one variable term raised to the third power and no variable terms having a power higher than 3. A quadratic equation can also be referred to as a **second-degree equation,** and a cubic equation as a **third-degree equation.** The **degree of an equation** in one variable is the highest power of any variable term that appears in the equation.

When we solved linear or first-degree equations, we obtained at most one solution for the equation. Quadratic or second-degree equations have at most two solutions. Cubic or third-degree equations have at most three solutions, and fourth-degree equations have at most four solutions.

EXAMPLE State the degree of each of the equations.

(a) $x^2 + 3x + 4 = 0$ (b) $x^3 = 27$ (c) $x^4 + 3x^3 + 2x^2 + x + 4 = 0$

(d) $x^8 = 256$ (e) $3x + 7 = 5x - 3$

(a) $x^2 + 3x + 4 = 0$
   Quadratic or second-degree equation

(b) $x^3 = 27$
   Cubic or third-degree equation

(c) $x^4 + 3x^3 + 2x^2 + x + 4 = 0$
   Fourth-degree equation

(d) $x^8 = 256$
   Eighth-degree equation

(e) $3x + 7 = 5x - 3$
   Linear or first-degree equation

---

2 Solve Higher-Degree Equations by Factoring.

Higher-degree equations that are written in factored form or can be written in factored form are presented here. We solve these equations by using the *zero-product property;* that is, if the product of the factors equals 0, then all factors containing a variable may be equal to 0. (See Section 15–2.)

To solve a higher-degree equation by factoring:

1. Write the equation in standard form.
2. Factor the polynomial side of the equation.
3. Set each factor containing a variable equal to zero.
4. Solve each equation from Step 3.

EXAMPLE Solve the equations for the exact roots by using the zero-product property.

(a) $x(x + 4)(x - 3) = 0$ (b) $(x + 5)(x - 7)(2x - 1) = 0$

(c) $2x^3 - 14x^2 + 20x = 0$ (d) $3x^3 - 15x = 0$

(a) $x\ (x + 4)\ (x - 3) = 0$ Already factored. Set each factor equal to zero.

$x = 0$ $x + 4 = 0$ $x - 3 = 0$ Solve each equation.

$\mathbf{x = 0}$ $\mathbf{x = -4}$ $\mathbf{x = 3}$

(b) $(x + 5)\ (x - 7)\ (2x - 1) = 0$ Already factored. Set each factor equal to zero.

$x + 5 = 0$ $x - 7 = 0$ $2x - 1 = 0$ Solve each equation.

     $\mathbf{x = -5}$ $\mathbf{x = 7}$ $2x = 1$

$$\frac{2x}{2} = \frac{1}{2}$$

$$\mathbf{x = \frac{1}{2}}$$

(c) $2x^3 - 14x^2 + 20x = 0$ Factor the common factors.

     $2x(x^2 - 7x + 10) = 0$ Factor the trinomial.

     $2x \,(x - 5)\,(x - 2) = 0$ Set each factor equal to zero.

     $2x = 0$ $x - 5 = 0$ $x - 2 = 0$ Solve each equation.

     $\dfrac{2x}{2} = \dfrac{0}{2}$ $x = 5$ $x = 2$

     $x = 0$

(d) $3x^3 - 15x = 0$ Factor the common factors.

     $3x(x^2 - 5) = 0$ Factor the difference of two squares, if possible.

     $3x(x^2 - 5) = 0$ Set each factor equal to zero.

     $3x = 0$ $x^2 - 5 = 0$

     $\dfrac{3x}{3} = \dfrac{0}{3}$ $x^2 = 5$ Solve each equation.

     $x = 0$ $x = \pm\sqrt{5}$ The exact roots are $x = 0$, $x = \sqrt{5}$, or $x = -\sqrt{5}$.

---

3 **Graph Higher-Degree Equations.**

Any equation can be graphed by writing it as function and using the table-of-solutions method. It is important to include enough points to get a complete view of the graph.

EXAMPLE   Prepare a table of solutions and graph the function $f(x) = x^3$ using integral values of x between -2 and 2, inclusive. Determine the domain and range.

$$f(-2) = (-2)^3 \qquad f(-1) = (-1)^3 \qquad f(0) = 0^3$$
$$f(-2) = -8 \qquad f(-1) = -1 \qquad f(0) = 0$$

$$f(1) = 1^3 \qquad f(2) = 2^3$$
$$f(1) = 1 \qquad f(2) = 8$$

x	$f(x)$
-2	-8
-1	-1
0	0
1	1
2	8

Plot the points indicated in the table of solutions and connect the points with a smooth, continuous curve (Fig. 15–13).

The domain and range are the set of all real numbers.

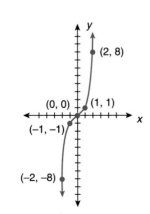

Figure 15–13

EXAMPLE Use a calculator to graph $y = 2x^3 - 1$. Determine the range for a domain from -3 to 3 inclusive.

Enter the equation into the calculator (Fig. 15–14).

$\boxed{Y=}$ $\boxed{\text{CLEAR}}$ 2 $\boxed{X, \theta, T, n}$ $\boxed{\wedge}$ 3 $\boxed{-}$ 1 $\boxed{\text{ZOOM}}$ $\boxed{\text{6:ZStandard}}$

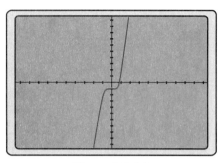

Figure 15–14

Examine a table for the domain from -3 to 3 inclusive (Fig. 15–15).

$\boxed{\text{TBLSET}}$ $\boxed{(-)}$ 3 $\boxed{\text{ENTER}}$ 1 $\boxed{-}$ $\boxed{\text{TABLE}}$

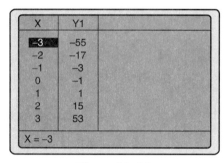

Figure 15–15

The range for a domain from -3 to 3 inclusive is -55 to 53 inclusive.

4 **Distinguish Between a Function and a Relation Using the Vertical Line Test.**

Equations in two variables describe a **relation** between the two variables. This connection between the two variables defines a set of ordered pairs. The equations in two variables that we have examined have been functions. A **function** is a relation that has exactly one value of the dependent variable (*y*-value) for every value of the independent variable (*x*-value). In other words, all functions are also relations, but not all relations are functions. A relation that is not a function will have at least one case where a value of the independent variable (*x*-value) corresponds to *more than one* value of the dependent variable (*y*-value).

To distinguish between a function and a relation using the vertical line test:

1. Graph the relation on a coordinate system.
2. Apply the vertical line test.
   (a) If a vertical line can be drawn so that it intersects a graph at more than one point, then the graph is the graph of a relation that is not a function.
   (b) If no vertical line can be drawn so that it intersects a graph at more than one point, then the graph is the graph of a function.

EXAMPLE Determine which graphs in Fig. 15–16 are graphs of functions.

(a)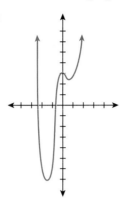

(b)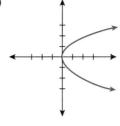

(c)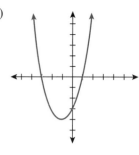

Figure 15–16

(a) **The graph is a graph of a function.** No vertical line can be drawn so that it intersects the graph at more than one point.
(b) **The graph is not a graph of a function.** For all values of x greater than zero, a vertical line will intersect the graph at two points.
(c) **The graph is a graph of a function.** No vertical line can be drawn so that it intersects the graph at more than one point.

5 Find the Domain and Range of a Relation from a Graph.

The **domain** of a relation is the set of all values that are appropriate for the independent variable and is the first component of the ordered pairs of the relation. The **range** of a relation is the set of all resulting values of the dependent variable that correspond to the independent variable and is the second component of the ordered pairs of the relation.

TIP

Relation Versus Function

- For every value of the domain of a relation there is at least one corresponding value of the range.
- For every value of the domain of a function there is exactly one corresponding value of the range.

$x = y^2$

x	y
4	-2
1	-1
0	0
1	1
4	2

Domain: 0, *1, *4

Range: -2, -1, 0, 1, 2

*x-values with two different y-values
$x = y^2$ **is a relation, but not a function.**

$y = x^2$

x	y
-2	4
-1	1
0	0
1	1
2	4

Domain: -2, -1, 0, 1, 2

Range: 0, 1, 4

Each x-value only has one y-value.
Two different x-values can have the same y-value.
$y = x^2$ **is a function.**

1. Graph the relation on a coordinate system.
2. The domain is the set of all the values on the graph for the independent variable (x-values).
3. The range is the set of all the values on the graph for the dependent variable (y-values).

EXAMPLE Determine the domain and range of each relation. (Fig. 15–17)

(a)
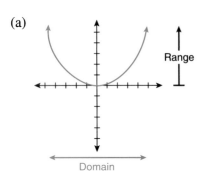

(b)
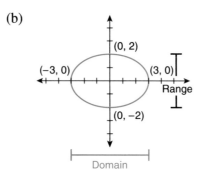

(c)
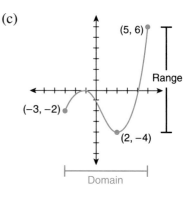

Figure 15–17

(a) **Domain:** the set of all real numbers

   Range: the set of real numbers greater than or equal to zero

(b) **Domain:** the set of real numbers from -3 to 3 inclusive

   Range: the set of real numbers from -2 to 2 inclusive

(c) **Domain:** the set of real numbers from -3 to 5 inclusive

   Range: the set of real numbers from -4 to 6 inclusive

SECTION 15–5 SELF-STUDY EXERCISES

1 State the degree of the equations.

1. $x^2 + 2x - 3 = 0$
4. $2x - 7 - 4x = 3$

2. $3x + 2x = 5$
5. $3x^3 + 2x^4 + 3 = x^2$

3. $x^4 = 42$
6. $x^7 = 128$

2 Find the roots of the equations. Factor if necessary.

7. $x(x - 2)(x + 3) = 0$
10. $x^3 - 7x^2 + 10x = 0$
13. $2x^3 - 18x = 0$

8. $2x(2x - 1)(x + 3) = 0$
11. $3x^3 - 3x^2 = 18x$
14. $12x^3 = 3x$

9. $3x(2x - 5)(3x - 2) = 0$
12. $4x^3 + 10x^2 + 4x = 0$
15. $16x^3 = 9x$

16. 32 times a number is subtracted from twice the cube of the number and the result is zero. How many numbers meet these conditions? What are they?

17. Find all the numbers that satisfy the following conditions: A number cubed is increased by 5 times the number and the result is zero.

18. BUS   A shipping container is a large cardboard box. The volume is found by multiplying the length times the width times the height. The length of the box is 3 ft more than the width, and the height is 7 ft more than the width. The volume is 421 ft^3. Write an equation to find the length, width, and height. Can this problem be solved by the methods found in Section 15–5? If so, solve it. If not, explain why it can't be solved.

3 Graph the equations using the table-of-solutions method. Determine the domain and range.

19. $y = 2x^3$

20. $y = \frac{1}{2}x^3$

21. $y = x^3 - x^2 - 4x + 4$

22. $y = x^3 + 2$

23. $y = -x^3 + 4$

24. $y = x^3 - 3x$

4 Use the vertical line test to determine which are graphs of functions.

25. $y = x^3 - x^2 - 8x + 8$

26. $y = 3x^2 - 2x - 5$

27. $x = y^3 - y^2 - 3y - 3$

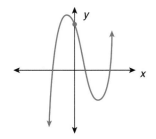

Figure 15–18

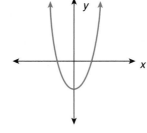

Figure 15–19

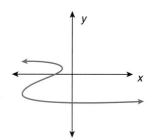

Figure 15–20

28. $x = y^2 - y - 5$

29. $y = x^4 - 3x^3 + x^2$

30. $x = 2y + 3$

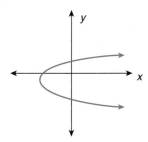

Figure 15–21

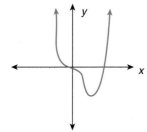

Figure 15–22

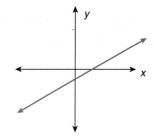

Figure 15–23

5 Find the domain and range of the functions or relations from the graph.

31.

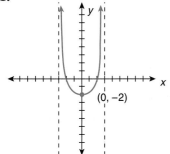

Figure 15–24

32.

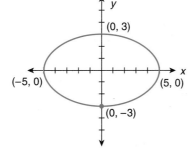

Figure 15–25

33.

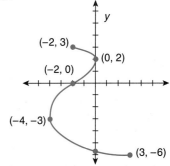

Figure 15–26

34.

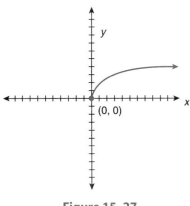

Figure 15–27

35.

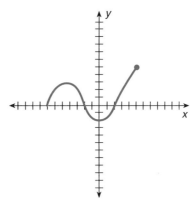

Figure 15–28

36.

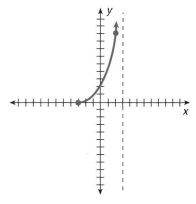

Figure 15–29

CHAPTER REVIEW OF KEY CONCEPTS

Learning Outcomes

What to Remember with Examples

Section 15–1

1 Write quadratic equations in standard form (pp. 572–573).

Write the equation with all terms on the left side of the equation and zero on the right and arrange with the terms in descending powers. The leading coefficient should be positive.

> Write the equation $5x + 3 = 4x^2$ in standard form.
>
> $$5x + 3 = 4x^2 \qquad \text{Rearrange all terms on the left and arrange in descending order.}$$
> $$-4x^2 + 5x + 3 = 0 \qquad \text{Make leading coefficient positive.}$$
> $$-1(-4x^2 + 5x + 3) = -1(0)$$
> $$4x^2 - 5x - 3 = 0$$

2 Identify the coefficients of the quadratic, linear, and constant terms of a quadratic equation (pp. 573–574).

The coefficient of the quadratic term is the coefficient of the squared variable, the coefficient of the linear term is the coefficient of the first-power variable, and the constant term is a single factor.

> Identify the coefficient of the quadratic and linear terms and identify the constant term in the following:
>
> $$5x = 4x^2 - 3 \qquad \text{Write the equation in standard form.}$$
> $$4x^2 - 5x - 3 = 0 \qquad a = 4, b = -5, c = -3$$
>
> The coefficient of the quadratic term is 4. The coefficient of the linear term is -5 and the constant term is -3.

Chapter 15 / Quadratic and Higher-Degree Equations

3 Solve pure quadratic equations ($ax^2 + c = 0$) by the square-root method (pp. 574–575).

To solve pure quadratic equations, solve for the squared variable; then take the square root of both sides of the equation. The two roots have the same absolute value.

Solve $2x^2 - 72 = 0$.

$2x^2 - 72 = 0$	Sort.
$2x^2 = 72$	Divide by 2.
$x^2 = 36$	Take the square root of both sides.
$x = \pm 6$	

Section 15–2

1 Solve incomplete quadratic equations ($ax^2 + bx = 0$) by factoring (pp. 576–578).

Arrange the equation in standard form. Factor the common factor. Then set each of the two factors equal to zero and solve for the variable in each equation. Both roots are rational and one root is always zero.

Solve $5x^2 - 15x = 0$.

$5x^2 - 15x = 0$	Factor common factors.
$5x(x - 3) = 0$	Set each factor equal to 0.
$5x = 0 \qquad x - 3 = 0$	Solve each equation.
$x = 0 \qquad\quad x = 3$	

2 Solve complete quadratic equations ($ax^2 + bx + c = 0$) by factoring (pp. 578–579).

Arrange the quadratic equation in standard form and factor the trinomial. Then set each factor equal to zero and solve for the variable. If the expression factors, the roots are rational.

Solve $2x^2 - 5x - 3 = 0$.

$2x^2 - 5x - 3 = 0$	Factor the trinomial.
$(x - 3)(2x + 1) = 0$	Set each factor equal to 0.
$x - 3 = 0 \qquad 2x + 1 = 0$	Solve each equation.
$x = 3 \qquad\qquad 2x = -1$	
$x = -\dfrac{1}{2}$	

Section 15–3

1 Solve quadratic equations by completing the square (pp. 580–583).

1. Write the equation in the form $ax^2 + bx = c$.
2. Divide each term of the equation by a.
3. Form a perfect-square trinomial on the left by adding $\left(\dfrac{b}{2a}\right)^2$ or $\dfrac{b^2}{4a^2}$ to both sides of the equation.
4. Factor the perfect square trinomial into the square of a binomial. $\left(x + \dfrac{b}{2a}\right)^2$
5. Take the square root of both sides of the equation.
6. Solve for x.

Solve $2x^2 - x + 4 = 0$ by completing the square.

$2x^2 - x + 4 = 0$	Isolate the variable terms.
$2x^2 - x = -4$	Divide equation by 2.
$x^2 - \dfrac{x}{2} = \dfrac{-4}{2}$	Simplify and add $\left(\dfrac{-1}{2}\right)^2$ to both sides.
$x^2 - \dfrac{x}{2} + \left(\dfrac{-1}{2}\right)^2 = -2 + \left(\dfrac{-1}{2}\right)^2$	Simplify.

$$x^2 - \frac{x}{2} + \frac{1}{4} = -2 + \frac{1}{4}$$

Write the left side of the equation as the square of a binomial.

$$\left(x - \frac{1}{2}\right)^2 = \frac{-8}{4} + \frac{1}{4}$$

$$\left(x - \frac{1}{2}\right)^2 = -\frac{7}{4}$$

Take the square root of both sides.

$$x - \frac{1}{2} = \pm\sqrt{-\frac{7}{4}}$$

Solve for x.

$$x = \frac{1}{2} \pm \frac{i\sqrt{7}}{2}$$

Exact roots.

2 Solve quadratic equations using the quadratic formula (pp. 583–587).

In the quadratic formula, $x = \dfrac{-b \pm \sqrt{b^2 - 4ac}}{2a}$, a is the coefficient of the quadratic term, b is the coefficient of the linear term, and c is the constant when the equation is written in standard form. The variable is x.

Solve using the quadratic formula:

$$5x^2 + 7x - 6 = 0$$

Identify a, b, and c.

$$a = 5, \qquad b = 7, \qquad c = -6$$

$$x = \frac{-b \pm \sqrt{b^2 - 4ac}}{2a}$$

Substitute values.

$$x = \frac{-7 \pm \sqrt{7^2 - 4(5)(-6)}}{2(5)}$$

Simplify the radicand.

$$x = \frac{-7 \pm \sqrt{49 + 120}}{10}$$

Add in the radicand.

$$x = \frac{-7 \pm \sqrt{169}}{10}$$

Evaluate the square root.

$$x = \frac{-7 \pm 13}{10}$$

Separate into two cases.

$$x = \frac{-7 + 13}{10} \qquad x = \frac{-7 - 13}{10}$$

Solve each equation.

$$x = \frac{6}{10} \qquad x = -\frac{20}{10}$$

$$x = \frac{3}{5} \qquad x = -2$$

Exact roots

Solving applied problems requires knowledge of other mathematical formulas, for example, $A = lw$ (area of a rectangle).

Find the length and width of a rectangular parking lot if the length is to be 12 m longer than the width and the area is to be 6,205 m^2.

Let x = number of meters in the width
$x + 12$ = number of meters in the length

Estimate: If the parking lot was a square, one side would be $s = \sqrt{A}$ or $s = 78.8$ m. In a rectangle where the length and width are different, the length is more than 78.8 m and the width is less than 78.8 m.

$$lw = A$$
$$x(x + 12) = 6{,}205$$
$$x^2 + 12x = 6{,}205$$
$$x^2 + 12x - 6{,}205 = 0$$

Use a calculator with the formula.

$$x = \frac{-12 \pm \sqrt{12^2 - 4(1)(-6{,}205)}}{2(1)}$$ Substitute values for a, b, and c.

$$x = \frac{-12 \pm \sqrt{144 + 24{,}820}}{2}$$ Simplify radicand.

$$x = \frac{-12 \pm \sqrt{24{,}964}}{2}$$ Evaluate the square root for an approximate value.

$$x = \frac{-12 \pm 158}{2}$$ Separate into two cases.

$$x = \frac{-12 + 158}{2} \qquad x = \frac{-12 - 158}{2}$$ Solve each case.

$$x = \frac{146}{2} = 73 \qquad x = \frac{-170}{2} = -85$$ Disregard the negative root for a length measure.

$$x = 73 \text{ width}$$
$$x + 12 = 85 \text{ length}$$

Section 15–4

1 Graph quadratic functions using the table-of-solutions method (pp. 588–589).

1. Prepare a table of solutions of approximately five ordered pairs. **2.** Plot the points on a rectangular coordinate system. **3.** Connect the points with a smooth, continuous curve.

Graph $y = x^2 - 6x + 8$ using a table of solutions. Start with three values of x, -2, 0, and 2.

at $x = -2$:
$y = (-2)^2 - 6(-2) + 8$
$y = 4 + 12 + 8$
$y = 24$

at $x = 0$:
$y = 0^2 - 6(0) + 8$
$y = 0 - 0 + 8$
$y = 8$

at $x = 2$:
$y = 2^2 - 6(2) + 8$
$y = 4 - 12 + 8$
$y = 0$

Try two more, $x = 4$ and $x = 6$.

at $x = 4$:
$y = 4^2 - 6(4) + 8$
$y = 16 - 24 + 8$
$y = 0$

at $x = 6$:
$y = 6^2 - 6(6) + 8$
$y = 36 - 36 + 8$
$y = 8$

Finally, see what happens between $x = 2$ and $x = 4$ (Fig. 15–30).

at $x = 3$:
$y = 3^2 - 6(3) + 8$
$y = 9 - 18 + 8$
$y = -1$

x	y
-2	24
0	8
2	0
3	-1
4	0
6	8

Figure 15–30

---

2 Graph quadratic functions by examining properties (pp. 589–590).

1. Determine the x-coordinate of the vertex of the parabola by finding the value of $-\dfrac{b}{2a}$.

Use the x-coordinate of the vertex to write an equation $x = -\dfrac{b}{2a}$, which is the axis

of symmetry. **2.** Substitute the value as the x-coordinate into the equation to find the corresponding y-value of the vertex. **3.** Replace y with zero and solve the resulting equation for x. Plot these points if any, which are the solutions for the equation. **4.** Use symmetry to find additional points. **5.** Connect the plotted points and the vertex with a smooth, continuous curved line.

Graph $y = x^2 - 6x + 8$ by examining properties (Fig. 15–31).

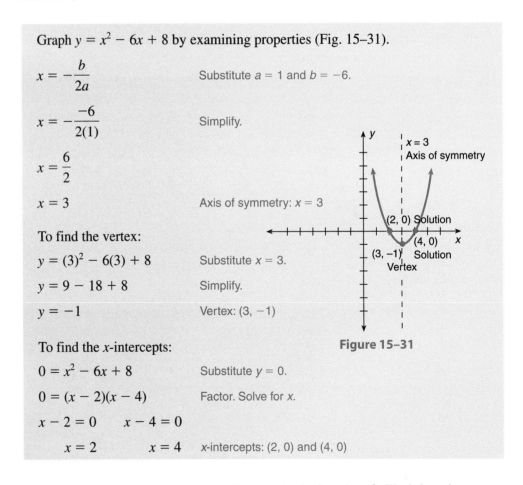

$x = -\dfrac{b}{2a}$ Substitute $a = 1$ and $b = -6$.

$x = -\dfrac{-6}{2(1)}$ Simplify.

$x = \dfrac{6}{2}$

$x = 3$ Axis of symmetry: $x = 3$

To find the vertex:
$y = (3)^2 - 6(3) + 8$ Substitute $x = 3$.

$y = 9 - 18 + 8$ Simplify.

$y = -1$ Vertex: $(3, -1)$

Figure 15–31

To find the x-intercepts:
$0 = x^2 - 6x + 8$ Substitute $y = 0$.

$0 = (x - 2)(x - 4)$ Factor. Solve for x.

$x - 2 = 0 \qquad x - 4 = 0$

$x = 2 \qquad\quad x = 4$ x-intercepts: $(2, 0)$ and $(4, 0)$

---

3 Solve a quadratic equation from a graph of a corresponding quadratic function (pp. 590–593).

1. Write the equation as a corresponding quadratic function. **2.** Find the x-intercepts on the graph. **3.** The solutions of the equation are the x-coordinates of the x-intercepts.

     Chapter 15 / Quadratic and Higher-Degree Equations

Find the solutions for the equation $x^2 - 6x + 8 = 0$ from the graph (Fig. 15–32).

The corresponding quadratic function is $y = x^2 - 6x + 8$. The x-intercepts are (2, 0) and (4, 0). The solutions of $x^2 - 6x + 8 = 0$ are $x = 2$ and $x = 4$.

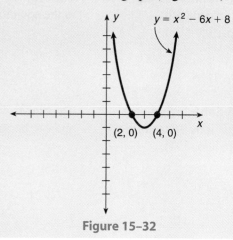

Figure 15–32

4 Graph quadratic functions using a graphing calculator (pp. 593–594).

To graph quadratic equations using a graphing calculator: **1.** Clear the graphing screen. **2.** Enter the equation that has been solved for y and press the graph function.

Graph $y = x^2 - 6x + 8$.

Clear the graphing screen.
Enter equation.
Graph.
Change viewing window if appropriate.

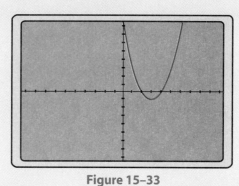

Figure 15–33

5 Determine the nature of the roots of a quadratic equation by examining the discriminant (pp. 594–596).

The radicand of the quadratic formula, $b^2 - 4ac$, is the discriminant of the quadratic equation.

1. If $b^2 - 4ac \geq 0$, the equation has real-number solutions.
  a. If $b^2 - 4ac$ is a perfect square, the two solutions are rational.
  b. If $b^2 - 4ac = 0$, there is one rational solution, a double root.
  c. If $b^2 - 4ac$ is not a perfect square, the two roots are irrational.
2. If $b^2 - 4ac < 0$, the equation has no real-number solutions; that is, the solutions are complex numbers.

Use the discriminant to determine the characteristics of the roots of the equation $3x^2 - 5x + 7 = 0$.

$a = 3, \quad b = -5, \quad c = 7$ Identify a, b, and c.
$(-5)^2 - 4(3)(7) = 25 - 84$ Substitute values into $b^2 - 4ac$ and simplify.
$\qquad\qquad\qquad = -59$

Since -59 is less than zero, the   Interpret result.
roots are not real; they are complex.

Section 15–5

1 Identify the degree of an equation (pp. 595–596).

The degree of an equation in one variable is the highest power of any term that appears in the equation.

> State the degree: $x^3 + 4x = 0$ is a cubic or third-degree equation.
> $x^4 = 81$ is a fourth-degree equation.

2 Solve higher-degree equations by factoring (pp. 596–597).

The higher-degree equations discussed in this section have a common variable factor and can be solved by factoring.

> Solve $x^3 + 2x^2 - 3x = 0$.
>
> | | |
> |---|---|
> | $x^3 + 2x^2 - 3x = 0$ | Factor common factor. |
> | $x(x^2 + 2x - 3) = 0$ | Factor trinomial. |
> | $x(x + 3)(x - 1) = 0$ | Set each factor equal to 0. |
> | $x = 0 \quad x + 3 = 0 \qquad x - 1 = 0$ | Solve each equation. |
> | $x = -3 \qquad\quad x = 1$ | |

3 Graph higher-degree equations (pp. 597–598).

Prepare a table of solutions with enough points to get a complete view of the graph.

> Prepare a table of solutions and graph the function $f(x) = x^3 + 2$ (Fig. 15–34).
>
> | x | y | |
> |-----|-----|---|
> | -2 | -6 | $f(-2) = (-2)^3 + 2 = -8 + 2 = -6$ |
> | -1 | 1 | $f(-1) = (-1)^3 + 2 = -1 + 2 = 1$ |
> | 0 | 2 | $f(0) = 0^3 + 2 = 2$ |
> | 1 | 3 | $f(1) = 1^3 + 2 = 1 + 2 = 3$ |
> | 2 | 10 | $f(2) = 2^3 + 2 = 8 + 2 = 10$ |
>
> $f(x) = x^3 + 2$
>
> **Figure 15–34**

4 Distinguish between a function and a relation using the vertical line test (pp. 598–599).

1. Graph the relation on a coordinate system.
2. Apply the vertical line test.

    a. If a vertical line can be drawn so that it intersects a graph at more than one point, then the graph is the graph of a relation that is not a function.

    b. If no vertical line can be drawn so that it intersects a graph in more than one point, then the graph is the graph of a function.

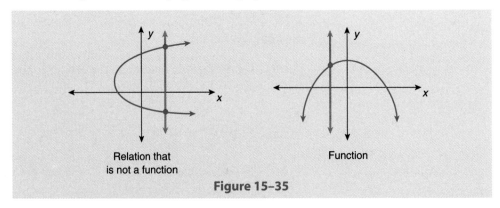

Relation that is not a function

Function

Figure 15–35

5 Find the domain and range of a relation from a graph (pp. 599–600).

1. Graph the relation on a coordinate system.
2. The domain is all the values on the graph for the independent variable (*x*-values).
3. The range is all the values on the graph for the dependent variable (*y*-values).

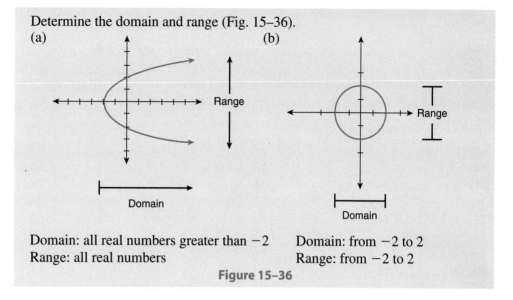

Determine the domain and range (Fig. 15–36).
(a)   (b)

Domain: all real numbers greater than -2 Domain: from -2 to 2
Range: all real numbers   Range: from -2 to 2

Figure 15–36

CHAPTER REVIEW EXERCISES

Section 15–1

Identify the quadratic equations as pure, incomplete, or complete.

1. $x^2 = 49$
2. $x^2 - 5x = 0$
3. $5x^2 - 45 = 0$
4. $3x^2 + 2x - 1 = 0$
5. $8x^2 + 6x = 0$
6. $5x^2 + 2x + 1 = 0$
7. $x^2 - 32 = 0$
8. $x^2 + x = 0$
9. $3x^2 + 6x + 1 = 0$

Write the equations in standard form.

10. $x^2 - 7x = 8$
11. $2x^2 - 5 = 8x$
12. $7x = 2x^2 - 3$
13. $5 + x^2 - 7x = 0$
14. $x - 4 = 3x^2$
15. $3x = 1 - 4x^2$

Solve the equations by the square-root method. Round to thousandths when necessary.

16. $x^2 = 121$
17. $x^2 = 100$
18. $x^2 - 64 = 0$
19. $4x^2 = 9$
20. $64x^2 - 49 = 0$
21. $0.36y^2 = 1.09$
22. $0.16x^2 = 0.64$
23. $5x^2 = 40$
24. $2x^2 - 5 = 3$
25. $6x^2 + 4 = 34$
26. $5x^2 - 6 = 19$
27. $3x^2 = 12$
28. $3x^2 - 4 = 8$
29. $2x^2 = 34$
30. $5x^2 - 9 = 30$
31. $3y^2 - 36 = -8$
32. $2x^2 + 3 = 51$
33. $\frac{1}{2}x^2 = 8$
34. $\frac{2}{3}x^2 = 24$
35. $\frac{1}{4}x^2 - 1 = 15$
36. $\frac{2}{5}x^2 + 2 = 8$

37. A circle has an area of 845 cm². What is the radius of the circle? Use the formula $A = \pi r^2$ and the calculator key for π. Round to the nearest tenth centimeter.

38. The area of an oil painting that is a square is known to be 9,072 cm². What are the inside dimensions of its picture frame?

Section 15–2

Solve the equations by factoring.

39. $x^2 - 5x = 0$
40. $4x^2 = 8x$
41. $6x^2 - 12x = 0$
42. $3x^2 + x = 0$
43. $10x^2 + 5x = 0$
44. $3y^2 = 12y$

45. $y^2 - 7y = 0$
46. $x^2 = 16x$
47. $12x^2 + 8x = 0$
48. $8x^2 - 12x = 0$
49. $x^2 + 3x = 0$
50. $4x^2 - 28x = 0$
51. $5x^2 = 45x$
52. $7x^2 = 28x$
53. $y^2 + 8y = 0$
54. $z^2 - 6z = 0$
55. $3m^2 - 5m = 0$
56. $4n^2 - 3n = 0$
57. $2x^2 = x$
58. $5y^2 = y$

59. 3 times the square of a number is the same as 12 times the number. What is the number?

60. Describe the steps for solving an incomplete quadratic equation. Include a clear description of the nature of the roots.

Solve the equations by factoring.

61. $x^2 - 4x + 3 = 0$
62. $x^2 + 7x + 12 = 0$
63. $x^2 + 3x = 10$
64. $x^2 - 7x + 12 = 0$
65. $x^2 + 7x = -6$
66. $x^2 + 3 = -4x$
67. $x^2 - 6x + 8 = 0$
68. $6y + 7 = y^2$
69. $6y^2 - 5y - 6 = 0$
70. $5y^2 + 23y = 10$
71. $10y^2 - 21y - 10 = 0$
72. $6x^2 - 16x + 8 = 0$
73. $4x^2 + 7x + 3 = 0$
74. $3x^2 = -7x + 6$
75. $12y^2 - 5y - 3 = 0$
76. $x^2 - 3x = 18$
77. $x^2 + 19x = 42$
78. $3x^2 + x - 2 = 0$
79. $3y^2 + y - 2 = 0$
80. $2x^2 - 4x - 6 = 0$
81. $2x^2 - 10x + 12 = 0$
82. $y^2 + 18y + 45 = 0$
83. $x^2 - 3x - 18 = 0$
84. $3x^2 - 9x - 30 = 0$
85. $2y^2 + 22y + 60 = 0$
86. $x^2 + 13x + 12 = 0$
87. $x^2 + 7x - 18 = 0$

88. AG/H An office building in the shape of a rectangle is known to have 47,500 ft^2 of space on the ground floor. The tenant wants to landscape the two longer sides of the building. The tenant also knows that the building is about 60 ft longer than it is wide. How many feet of land along the building need to be landscaped?

89. CAD/ARC Jerri Amour is an architect who is designing a hospital. She knows that a kidney dialysis machine needs a space that is 7 ft longer than it is wide. The total area needed is 228 ft^2 of space. Find the length and width of the space needed for the machine.

Section 15–3

Solve each equation by completing the square. Give the answer as an exact solution.

90. $x^2 + 2x - 3 = 0$
91. $x^2 - 4x + 4 = 0$
92. $x^2 - 6x + 8 = 0$
93. $x^2 - 8x + 12 = 0$
94. $x^2 - 6x + 7 = 0$
95. $x^2 - 8x + 14 = 0$
96. $x^2 - 6x + 4 = 0$
97. $x^2 - 6x + 12 = 0$
98. $x^2 - 2x + 6 = 0$
99. $x^2 - 5x + 4 = 0$
100. $x^2 - 3x - 18 = 0$
101. $x^2 - 3x = 7$

Indicate the values for a, b, and c in the quadratic equations.

102. $5x^2 + x + 6 = 0$
103. $x^2 - 2x = 8$
104. $x^2 - 7x + 12 = 0$
105. $x^2 + 3x = 4$
106. $3x^2 = 2x + 7$
107. $x^2 - 3x = -2$

Solve the quadratic equations by using the quadratic formula.

108. $x^2 - 9x + 20 = 0$
109. $x^2 - 8x - 9 = 0$
110. $x^2 - 5x = -6$
111. $x^2 + 2x = 8$
112. $x^2 - x - 12 = 0$
113. $2x^2 - 3x - 2 = 0$

Solve the quadratic equations by using the quadratic formula. Round each final answer to the nearest hundredth.

114. $3x^2 + 6x + 2 = 0$
115. $2x^2 - 3x - 1 = 0$
116. $5x^2 + 4x - 8 = 0$
117. $3x^2 + 5x + 1 = 0$

118. CON A bricklayer plans to build an arch with a span (s) of 10 m and a radius (r) of 5 m. How high (h) is the arch? (Use the formula $h^2 - 2hr + \dfrac{s^2}{4} = 0$.)

119. CON A rectangular kitchen contains 240 ft^2. If the length is 2 times the width, find the length and width of the room to the nearest whole number. (Area = length $\times$ width, or $A = lw$.)

Bills Owed (Net Expenses) Total
                            (2900.00)

                                    400.00

1). Car Note   -   1000.00
2). Shop License - 600.00
3). Boys   -   500.00
4). Tithes + Offering - 800.00

120. CON What are the dimensions of a rectangular tool storage room if the area is 45.5 m² and the room is 0.5 m longer than it is wide?

121. INDTEC Find the length and width of a rectangular piece of fiberglass if its length is 3 times the width and the area is 591 in². Round to the nearest inch.

Section 15–4

Graph the quadratic functions. Determine the domain and range.

122. $y = x^2 - 1$

123. $y = -x^2 - 1$

124. $y = x^2 + 3x - 10$

125. $y = x^2 - 6x + 8$

126. $y = x^2 - 2x + 1$

127. $y = -x^2 + 2x - 1$

Graph the quadratic functions by using the vertex, x-intercepts, and one other point.

128. $y = x^2 - 4x + 3$

129. $y = x^2 - 2x - 8$

130. $y = x^2 + 12x + 35$

131. $y = x^2 + 8x - 12$

Find all real solutions of the equations by writing and graphing the equations of the corresponding quadratic functions.

132. $x^2 + x - 12 = 0$

133. $x^2 - 4x - 12 = 0$

134. $x^2 - 5x = 14$

135. $x^2 + 8x = -16$

Use a calculator to graph the following. Adjust the window to display the vertex.

136. $y = x^2 - 2x + 1$

137. $y = -2x^2$

138. $y = \frac{1}{2}x^2 - 3$

Use the discriminant of each equation to determine the nature of the roots of the equations.

139. $x^2 + 8x + 16 = 0$

140. $2x^2 - 3x - 5 = 0$

141. $5x^2 - 100 = 0$

142. $3x^2 - 2x + 4 = 0$

143. $2x = 5x^2 - 3$

Section 15–5

State the degree of each equation.

144. $2x + 5x = 15$

145. $3x - 2x^3 + 8 = 0$

146. $16 = x^4$

147. $6 - 3x - 3 = 2x + 4$

148. $y^6 = 729$

149. $5y^8 + 2y^3 - 6 = y^2$

Find the roots of the following equations. Factor if necessary.

150. $x(x + 2)(x - 3) = 0$

151. $2x(3x - 2)(x - 2) = 0$

152. $3x(2x + 1)(x + 4) = 0$

153. $2x^3 + 10x^2 + 12x = 0$

154. $x^3 = 2x^2$

155. $2x^3 + 9x^2 = 5x$

156. $6x^3 + 3x^2 - 18x = 0$

157. $3x^3 - 6x^2 = 0$

158. $3x^3 - x^2 - 2x = 0$

159. $x^3 + 6x^2 + 8x = 0$

160. $x^3 - 8x^2 + 15x = 0$

161. $x^3 - x^2 - 20x = 0$

162. $x^3 + x^2 - 20x = 0$

163. $y^3 - 6y^2 + 7y = 0$

164. $y^3 + 2y^2 + 5y = 0$

165. $x^3 - 3x^2 - 4x = 0$

166. $y^3 + 7y^2 + 12y = 0$

167. $2y^3 + 6y^2 + 4y = 0$

Graph the equations using the table-of-solutions method. Determine the domain and range.

168. $y = x^3 - x^2 - 8x + 1$

169. $y = 5x^3$

Use the vertical line test to determine which are graphs of functions.

170. $x = 6 - 3y$

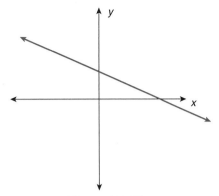

Figure 15–37

171. $x = -y^3 + 2y^2 + 2y + 3$

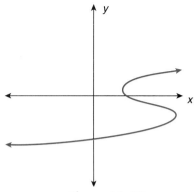

Figure 15–38

172. $y = x^4 - 3x^3 + x$

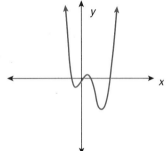

.Figure 15–39

173. $\dfrac{x^2}{25} + \dfrac{y^2}{4} = 1$

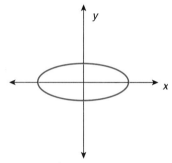

Figure 15–40

Find the domain and range of the relations from the graph.

174.

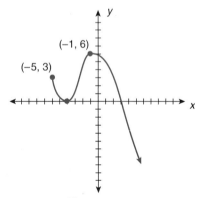

Figure 15–41

175.

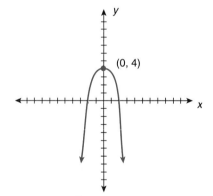

Figure 15–42

176.

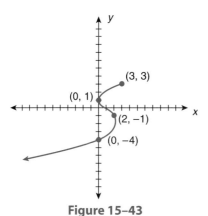

Figure 15–43

177.

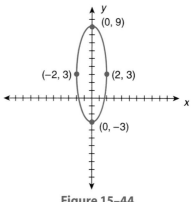

Figure 15–44

TEAM PROBLEM-SOLVING EXERCISES

1. Many objects are designed with dimensions according to the *Golden Ratio*. Objects that have measurements according to this ratio are said to be most pleasing to the eye. The *Golden Rectangle* has dimensions of length (l) and height (h) that satisfy the formula

$$\frac{l + h}{l} = \frac{l}{h}$$

When you cross multiply the formula for the Golden Rectangle, a quadratic equation results.

(a) An artist wants a canvas proportioned according to the Golden Rectangle. If the length is to be 36 in., find the height to the nearest inch.

(b) What should the length of a canvas be if the height is 20 in.?

2. Determine the largest-size wall hanging that has dimensions of the Golden Rectangle that can be placed on a wall that is 32 ft long and 20 ft high. The wall hanging must also be at least 2 ft from the ceiling, floor, and each of the side corners.

CONCEPTS ANALYSIS

1. State the zero-product property and explain how it applies to solving quadratic equations.

3. Use the equation $y = ax^2 + x + 1$ to graph equations when $a = 1, 2, 3, 4, 5$. Describe the changes in the graph caused by changes in a.

5. Under what conditions will a quadratic equation have irrational roots?

2. Under what conditions would the square-root method be used to solve an equation?

4. Under what conditions will a quadratic equation have no real solutions?

6. Under what conditions will a quadratic equation have rational roots?

Find a mistake in each of the following. Correct and briefly explain the mistake.

7. $2x^2 - 2x - 12 = 0$
$2(x^2 - x - 6) = 0$
$2(x + 2)(x - 3) = 0$
$2 = 0 \quad x + 2 = 0 \quad x - 3 = 0$
$\qquad\qquad x = -2 \qquad x = 3$
The roots are 0, -2, and 3.

8. $2x^3 + 5x^2 + 2x = 0$
$$\frac{x(2x^2 + 5x + 2)}{x} = \frac{0}{x}$$
$2x^2 + 5x + 2 = 0$
$(2x + 1)(x + 2) = 0$
$2x + 1 = 0 \qquad x + 2 = 0$
$2x = -1 \qquad\quad x = -2$
$$x = -\frac{1}{2}$$
The roots are $-\frac{1}{2}$ and -2.

9. Under what conditions would the quadratic formula be used to solve a quadratic equation?

10. What is the maximum number of roots the equation $x^4 = 16$ *could* have? How many real roots does the equation have?

11. How is the graph of a quadratic function different from the graph of a linear function?

12. What do *axis of symmetry* and *vertex* refer to on the graph of a quadratic function that represents a parabola?

PRACTICE TEST

Identify the quadratic equations as pure, incomplete, or complete.

1. $3x^2 = 42$
3. $5x^2 = 7x$

2. $7x^2 - 3x + 2 = 0$
4. $4x^2 - 1 = 0$

Solve the quadratic equations using the square-root method. Give the exact roots.

5. $x^2 = 81$

6. $81x^2 - 64 = 0$

Solve the equations by factoring.

7. $3x^2 - 6x = 0$
9. $x^2 - 5x + 6 = 0$
11. $2x^2 + 12 = 11x$

8. $2x^2 + 3x + 1 = 0$
10. $3x^2 - 2x - 1 = 0$
12. $x^2 - 3x - 4 = 0$

Solve the equations using the quadratic formula. Round to the nearest hundredth when necessary.

13. $2x^2 + 3x - 5 = 0$
15. $x^2 - 3x - 5 = 0$

14. $3x^2 - 5x + 4 = 0$
16. $4x^2 - 2x = 3$

Describe the roots of the equations without solving.

17. $x^2 - 8x + 12 = 0$

18. $3x^2 - 2x + 3 = 0$

Graph the quadratic equations by examining properties. Show the axis of symmetry, vertex, and *x*-intercepts.

19. $y = x^2 + 2x + 1$

20. $y = x^2 + 4x + 4$

21. **ELEC** Find the diameter (d) in mils to the nearest hundredth of a copper wire conductor whose resistance (R) is 1.314 Ω and whose length (L) is 3,642.5 ft. (Formula: $R = \dfrac{KL}{d^2}$, where K is 10.4 for copper wire.)

22. Find the radius (r) of a circle whose area (A) is 35.15 cm^2. Round the answer to the nearest hundredth centimeter. (Formula: $A = \pi r^2$.)

23. ELEC What is the current (I) in amps if the resistance (R) of the circuit is 52.29 Ω and 205 watts (W) are used? Round to the nearest hundredth.

$$\left(\text{Formula: } R = \frac{W}{I^2}\right)$$

24. AG/H A square parcel of land has an area of 156.25 m². What is the length of a side? (Use the formula $A = s^2$, where A is the area and s is the length of a side.)

25. ELEC In the formula $E = 0.5\,mv^2$, solve for v if $E = 180$ and $m = 10$ and $v > 0$.

Solve each equation.

26. $x(2x - 5)(x - 3) = 0$

27. $6x^3 + 21x^2 = 45x$

28. $2x^3 - x^2 - 6x = 0$

29. $6x^3 - 18x^2 = 0$

Use the vertical line test to identify the graphs as functions or relations. Determine the domain and range.

30. $y = x^5 + 2x^4 - x^3$

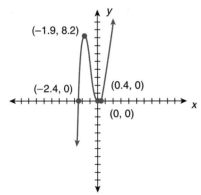

Figure 15–45

31. $x = y^3 - 2y^2 - y$

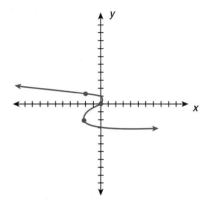

Figure 15–46

Perform the indicated operations.

1. $5x^7 \cdot 8x^3$

2. $\dfrac{x^3y^{-3}}{x^4y^{-5}}$

3. $(x^4)^8$

Arrange in descending order.

4. $15 - 32x^3 + 6x^4 - 8x^2 + 40x$

Simplify.

5. $5x^4 - 3x^2 + 7x^4 + 8x^2$

Multiply.

6. $-8x^2(3x^2 - 4x + 7)$

Divide and simplify.

7. $\dfrac{10x^5 - 20x^6 + 15x^4}{5x^3}$

Factor by removing the greatest common factor.

8. $12x^3 - 18x^2y$

9. $24x^3 - 16x^2 - 8x$

Find the product.

10. $(3x - 2)(x + 1)$

11. $(4x - 5)(3x - 2)$

Find the quotient.

12. $x + 3 \overline{)x^2 - 4x - 21}$

13. $x - 5 \overline{)x^2 - 8x + 15}$

Factor completely.

14. $49y^2 - 36$

15. $x^2 - 16x + 64$

16. $x^2 - 8x + 15$

17. $x^2 + x - 42$

18. $3x^2 - 17x + 10$

19. $6x^2 + 17x + 12$

20. $x^3 - 5x^2 - x$

Solve by the square-root method. Round to thousandths if necessary.

21. $3x^2 - 7 = 20$

22. $4x^2 - 9 = 39$

Solve by factoring.

23. $x^2 - 8x = 0$

24. $x^2 - x - 12 = 0$

25. $2x^2 - 7x + 3 = 0$

26. $4x^2 + 5x = 6$

Solve using the quadratic formula. Round to the nearest hundredth.

27. $4x^2 - 3x - 2 = 0$

28. $x^2 + 6x = 2$

Solve.

29. $x^3 - 3x^2 - 10x = 0$

30. $2x^3 - 5x^2 + 3x = 0$

16

Exponential and Logarithmic Equations

Focus on Careers

Pharmacy technicians work closely with pharmacists to prepare prescribed medication for patients. Pharmacy aides work closely with pharmacy technicians as clerks or cashiers who handle money, stock shelves, and perform other clerical duties. Pharmacy technicians usually perform more technical tasks than pharmacy aides, but in some states the duties and job titles overlap.

Pharmacy technicians receive prescriptions from doctors' offices and verify the prescription information is complete and accurate. They retrieve, count, pour, weigh, measure, and sometimes mix the medication, then prepare and affix labels to the selected container. Registered pharmacists must check each prescription filed by pharmacy technicians before it is given to the patient.

Most pharmacy technician jobs are in retail pharmacies. Working hours may vary, because some pharmacies remain open 24 hours, 7 days a week.

Pharmacy technicians must complete classroom and laboratory work in a variety of areas. Students who complete the training receive an associate degree, a certificate, or a diploma and may be required to pass the National Pharmacy Technician Certification Examination.

Jobs in this career are expected to grow faster than the average for all occupations through 2012 because of the increased pharmaceutical needs of a larger and older population. In 2002 pharmacy technicians earned a median hourly pay of $10.70. The highest 10% of pharmacy technicians were paid more than $15.82 hourly during this same time.

Source: Bureau of Labor Statistics, U.S. Department of Labor, *Occupational Outlook Handbook,* 2004–2005 Edition, Pharmacy Technicians, on the Internet at http://www.bls.gov/oco/ocos252.htm.

Learning Outcomes

1. Evaluate formulas with at least one exponential term.
2. Evaluate formulas that contain a power of the natural exponential, e.
3. Solve exponential equations in the form $b^x = b^y$, where $b > 0$ and $b \neq 1$.
4. Graph an exponential function.

Many scientific, technical, and business phenomena have the property of exponential growth; that is, the growth rate does not remain constant as certain physical properties increase. Instead, the growth rate increases exponentially. For example, notice the difference between $2x$ and 2^x when x increases. If we write the expressions in function notation, we have two different functions of x.

$$f(x) = 2x$$
$$g(x) = 2^x$$

Examine the values in Table 16–1 and the graphical representation of these two functions (Fig. 16–1). 2^x is said to increase *exponentially*. $g(x) = 2^x$ is an *exponential function*. On the other hand, $2x$ increases at a constant rate. $f(x) = 2x$ is a *linear function* since $2x$ is a linear polynomial.

Before the common availability of scientific and graphing calculators, examining and evaluating exponential functions was very tedious and time-consuming. Exponential expressions are found in many formulas. Even with the use of a scientific or graphing calculator, an understanding of exponential expressions is necessary.

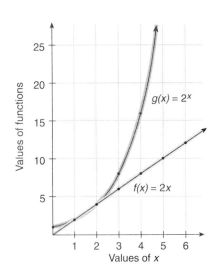

Figure 16–1

Table 16–1   Values of $f(x)$ and $g(x)$

x	$f(x) = 2x$	$g(x) = 2^x$
1	$f(1) = 2(1) = 2$	$g(1) = 2^1 = 2$
2	$f(2) = 2(2) = 4$	$g(2) = 2^2 = 4$
3	$f(3) = 2(3) = 6$	$g(3) = 2^3 = 8$
4	$f(4) = 2(4) = 8$	$g(4) = 2^4 = 16$
5	$f(5) = 2(5) = 10$	$g(5) = 2^5 = 32$
6	$f(6) = 2(6) = 12$	$g(6) = 2^6 = 64$

1   Evaluate Formulas with at Least One Exponential Term.

Many formulas have terms that contain exponents. An **exponential expression** is an expression that contains at least one term that has a variable exponent. A **variable exponent** is an exponent that has at least one variable factor. Exponential expressions can be evaluated on a calculator by using the general power key $\boxed{\wedge}$ or other power keys as appropriate. An **exponential equation** or formula contains at least one term that has a variable exponent.

EXAMPLE   Evaluate using a calculator.

(a) 5^6 (b) 4^{-3} (c) $8^{2.5}$

(a) $5^6 = \mathbf{15,625}$ 5 $\boxed{\wedge}$ 6 $\boxed{\text{ENTER}}$ $\Rightarrow$ 15625

(b) $4^{-3} = \mathbf{0.015625}$ 4 $\boxed{\wedge}$ $\boxed{(-)}$ 3 $\boxed{\text{ENTER}}$ $\Rightarrow$ .015625

(c) $8^{2.5} = \mathbf{181.019336}$ 8 $\boxed{\wedge}$ 2 $\boxed{\cdot}$ 5 $\boxed{\text{ENTER}}$ $\Rightarrow$ 181.019336

A commonly used formula that contains a variable exponent is the formula for calculating the compound amount for compound interest. **Compound interest** is the interest calculated at the end of each period and then added to the principal for the next period. The **compound amount** or the accumulated amount is the combined principal and interest accumulated over a period of time.

Compound amount (accumulated amount):

$$A = P\left(1 + \frac{r}{n}\right)^{nt}$$

where A = accumulated amount
  P = original principal
  t = time in years
  r = rate per year expressed as a decimal equivalent
  n = compounding periods per year

EXAMPLE Using the formula $A = P\left(1 + \dfrac{r}{n}\right)^{nt}$ and a calculator, find the accumulated amount on an investment of \$1,500, invested at an interest rate of 9% for 3 years, if the interest is compounded quarterly.

Estimation We expect to have more than \$1,500.

$A = P\left(1 + \dfrac{r}{n}\right)^{nt}$ $P = \$1,500; r = 9\%$ or $0.09; n =$ quarterly or 4 times a year; $t = 3$ years.

$A = 1,500\left(1 + \dfrac{0.09}{4}\right)^{(4)(3)}$ Simplify exponent and division term in grouping.

$A = 1,500(1 + 0.0225)^{12}$ Combine terms in grouping.

$A = 1,500(1.0225)^{12}$ $1.0225\ \boxed{\wedge}\ 12\ \boxed{=}\ \Rightarrow 1.30604999$

$A = 1,500(1.30604999)$ Multiply.

$A = 1,959.07$ Rounded

Interpretation **The accumulated amount of the \$1,500 investment after 3 years is \$1,959.07 to the nearest cent.**

TIP

Your Mind Is Often Quicker (and More Accurate) Than Your Fingers

Should we do every step on the calculator? The following sequence of keystrokes is one option for performing all the calculations in the preceding example in a continuous sequence.

1500 $\boxed{\times}$ $\boxed{(}$ $\boxed{(}$ 1 $\boxed{+}$ .09 $\boxed{\div}$ 4 $\boxed{)}$ $\boxed{\wedge}$ $\boxed{(}$ $\boxed{(}$ 4 $\boxed{\times}$ 3 $\boxed{)}$ $\boxed{=}$ $\Rightarrow$ 1959.074985

Whenever possible, it is advisable to do some calculations mentally. This can greatly decrease the complexity of the calculator sequences. $4(3) = 12$.

1500 $\boxed{\times}$ $\boxed{(}$ $\boxed{(}$ 1 $\boxed{+}$ .09 $\boxed{\div}$ 4 $\boxed{)}$ $\boxed{\wedge}$ 12 $\boxed{=}$ $\Rightarrow$ 1959.074985

Try each sequence on your calculator.

The accumulated amount is also the **future value,** or **maturity value.** Many business calculators or computer software programs have a future value function (FV).

Interpreting a Formula

In working with formulas it is important to understand what is represented by each letter of the formula. The compound interest formula can also be given in terms of the number of **compounding periods** and the **interest rate per period.**

$$\text{Total compounding periods } (N) = \frac{\text{Compounding periods per year } (n)}{\text{Number of years } (t)} \text{ times}$$

$$N = nt$$

$$\text{Interest rate per period } (R) = \frac{\text{Interest rate per year } (r)}{\text{Compounding periods per year } (n)}$$

$$R = \frac{r}{n}$$

The compound amount or future value formula can also be written as

$$A = P(1 + R)^N \qquad \text{or} \qquad FV = P(1 + R)^N$$

where A or FV = accumulated amount or future value
$\qquad P$ = original principal
$\qquad R$ = interest rate per compounding period
$\qquad N$ = total number of compounding periods

To find the compound interest:

1. Find the accumulated amount or future value using the formula

$$A = P(1 + R)^N$$

where A or FV = accumulated amount or future value
$\qquad P$ = original principal
$\qquad R$ = interest rate per compounding period
$\qquad N$ = total number of compounding periods

2. Subtract the original principal from the accumulated amount.

$$I = A - P$$

Combining both formulas:

$$I = P(1 + R)^N - P$$

or

$$I = P\left((1 + R)^N - 1\right)$$

EXAMPLE Find the compound interest on an investment of $10,000 at 8% annual interest compounded semiannually for 3 years.

$$P = 10{,}000, R = \frac{0.08}{2} = 0.04 \text{ per period}, N = 2 \times 3 = 6 \text{ periods}$$

$I = P(1 + R)^N - P$	Substitute values.
$I = 10{,}000(1 + 0.04)^6 - 10{,}000$	Combine terms inside grouping.
$I = 10{,}000(1.04)^6 - 10{,}000$	Raise to the power.
$I = 10{,}000(1.265319018) - 10{,}000$	Multiply.
$I = 12{,}653.19 - 10{,}000$	Subtract.
$I = \$2{,}653.19$	

The interest is $2,653.19.

The **present value** of an investment is the *lump sum* amount that should be invested now at a given interest rate for a specific period of time to *yield* a specific accumulated amount in the future.

Present value:

$$PV = \frac{FV}{(1 + R)^N}$$

where PV = present value
FV = future value
R = interest rate per period
N = total number of periods

EXAMPLE The 7th Inning Sports Shop needs $20,000 in 10 years to replace engraving equipment. Find the amount the firm must invest at the present if it receives 10% interest compounded annually.

$$FV = 20{,}000, R = \frac{0.10}{1} = 0.1, N = 10(1) = 10$$

$PV = \dfrac{20{,}000}{(1 + 0.1)^{10}}$	Combine terms inside grouping.
$PV = \dfrac{20{,}000}{(1.1)^{10}}$	Raise denominator to the power.
$PV = \dfrac{20{,}000}{2.59374246}$	Divide.
$PV = \$7{,}710.87$	Rounded

$7,710.87 should be invested now.

In advertising or in stating the terms of an investment or loan it is common to equate the compound interest rate to a comparable simple interest rate. This rate is referred to as the **effective rate, annual percentage rate (APR),** and **annual percentage yield (APY).**

EXAMPLE   Find the effective interest rate for a loan of $600 at 10% compounded semiannually.

$E = \left(1 + \dfrac{r}{n}\right)^n - 1$ Substitute values. $r = 0.1$; $n = 2$

$E = \left(1 + \dfrac{0.1}{2}\right)^2 - 1$ Simplify inside grouping.

$E = (1 + 0.05)^2 - 1$

$E = (1.05)^2 - 1$ Raise to the power.

$E = 1.1025 - 1$ Subtract.

$E = 0.1025$

Effective interest rate is 10.25%.

An **annuity** is a fund that accumulates compound interest as periodic payments add to the principal. An **ordinary annuity** has periodic payments that are made at the end of each payment period.

EXAMPLE Find the future value of an ordinary annuity of $6,000 for 5 years at 6% annual interest compounded semiannually.

$$FV = P\left[\frac{(1 + R)^N - 1}{R}\right], \quad P = \$6,000, \quad R = \frac{0.06}{2} = 0.03, \quad N = 5(2) = 10$$

$$FV = 6,000\left[\frac{(1 + 0.03)^{10} - 1}{0.03}\right] \qquad \text{Simplify innermost grouping.}$$

$$FV = 6,000\left[\frac{(1.03)^{10} - 1}{0.03}\right] \qquad \text{Raise to the power.}$$

$$FV = 6,000\left[\frac{1.343916379 - 1}{0.03}\right] \qquad \text{Subtract in numerator.}$$

$$FV = 6,000\left[\frac{0.343916379}{0.03}\right] \qquad \text{Divide in grouping.}$$

$$FV = 6,000(11.46387931) \qquad \text{Multiply.}$$

$$FV = \$68,783.28$$

When you have a specific future goal or target amount that you want to accumulate, a **sinking fund payment** is the amount you would invest in periodic payments to reach this goal. To determine the sinking fund payment, the *known values* are the future goal or amount, the amount of time, and the expected or guaranteed interest rate.

To find the sinking fund payment to produce a specified future value:

Apply the formula

$$P = FV\left[\frac{R}{(1 + R)^N - 1}\right]$$

where P = sinking fund payment
FV = future value or goal
R = interest rate per period
N = total number of periods

EXAMPLE A municipality has established a sinking fund to retire a bond issue of $500,000, which is due in 10 years. The account pays 8% quarterly interest. Find the amount of the quarterly sinking fund payment.

$$P = FV\left[\frac{R}{(1 + R)^N - 1}\right] \qquad \text{Substitute known values.}$$

$$FV = \$500,000, \quad R = \frac{0.08}{4} = 0.02, \quad N = 10(4) = 40$$

$$P = 500,000\left[\frac{0.02}{(1 + 0.02)^{40} - 1}\right]$$ Simplify grouping in denominator.

$$P = 500,000\left[\frac{0.02}{(1.02)^{40} - 1}\right]$$

$$P = 500,000\left[\frac{0.02}{2.208039664 - 1}\right]$$

$$P = 500,000\left[\frac{0.02}{1.208039664}\right]$$ Divide.

$$P = 500,000[0.0165557478]$$ Multiply.

$$P = \$8,277.87$$

Finding the monthly payment to repay a loan is similar to the process for finding the sinking fund payment. The repayment of the loan in equal installments that are applied to the principal and interest over a specified amount of time is called the **amortization of a loan.**

To find the monthly payment for an amortized loan:

Apply the formula

$$M = P\left[\frac{R}{1 - (1 + R)^{-N}}\right]$$

where M = monthly payment
P = principal or initial amount of the loan
R = interest rate per month
N = total number of months

EXAMPLE Find the monthly payment on a 25-year home mortgage of $135,900 at 8%.

$$P = \$135,900, \qquad R = \frac{0.08}{12} = 0.0066666667, \qquad N = 25(12) = 300$$

$$M = P\left[\frac{R}{1 - (1 + R)^{-N}}\right]$$ Substitute values.

$$M = 135,900\left[\frac{0.0066666667}{1 - (1 + 0.0066666667)^{-300}}\right]$$ Simplify denominator.

$$M = 135,900\left[\frac{0.0066666667}{1 - (1.0066666667)^{-300}}\right]$$

$$M = 135,900\left[\frac{0.0066666667}{1 - 0.1362365146}\right]$$

$$M = 135,900\left[\frac{0.0066666667}{0.8637634854}\right]$$ Divide in grouping and multiply.

$$M = \$1,048.90$$

2 Evaluate Formulas That Contain a Power of the Natural Exponential, *e*.

In many applications involving circles, the irrational number π (approximately equal to 3.141592654) is used. Another irrational number, *e*, arises in the discussion of many physical phenomena. Many formulas contain a power of the natural exponential, *e*.

Exponential change is an interesting phenomenon. Let's look at the value of the expression $\left(1 + \dfrac{1}{n}\right)^n$ as *n* gets larger and larger. See Table 16–2.

Table 16–2 Values of $(1 + \frac{1}{n})^n$

n	$(1 + \frac{1}{n})^n$	Result
1	$(1 + \frac{1}{1})^1$	2
2	$(1 + \frac{1}{2})^2$	2.25
3	$(1 + \frac{1}{3})^3$	2.37037037
10	$(1 + \frac{1}{10})^{10}$	2.59374246
100	$(1 + \frac{1}{100})^{100}$	2.704813829
1,000	$(1 + \frac{1}{1,000})^{1,000}$	2.716923932
10,000	$(1 + \frac{1}{10,000})^{10,000}$	2.718145927
100,000	$(1 + \frac{1}{100,000})^{100,000}$	2.718268237
1,000,000	$(1 + \frac{1}{1,000,000})^{1,000,000}$	2.718280469

The value of the expression changes very little as the value of *n* gets larger. We can say that the value approaches a given number. We call the number **e, the natural exponential.** The natural exponential, *e*, like π, is an irrational number and will never terminate nor repeat as more decimal places are examined.

The natural exponential, *e*, like π, is a constant. That is, the value is always the same; it does not vary. The natural exponential, *e*, is the limit that the value of the expression $\left(1 + \dfrac{1}{n}\right)^n$ approaches as *n* gets larger and larger without bound. The value of *e* to nine decimal places is 2.718281828.

To evaluate formulas containing the natural exponential, *e*, we can use a calculator or computer.

TIP

The Natural Exponential, *e*, on Your Calculator

The natural exponential key is generally labeled $\boxed{e^x}$. On most calculators, the exponent is entered after pressing the $\boxed{e^x}$ key. To find $e^{2.3}$, enter the following sequence:

$$\boxed{e^x}\ 2.3\ \boxed{=}\ \Rightarrow\ 9.974182455$$

Always test your calculator. A good test for the $\boxed{e^x}$ key is to find e^0.

$e^0 = 1$ Also, $e^1 = 2.718281828$

EXAMPLE   Evaluate using a calculator.

(a) $e^{2.7}$ (b) $e^{-3.2}$

(a) $e^{2.7} = \mathbf{14.87973172}$ $\boxed{e^x}\ 2\ \boxed{\cdot}\ 7\ \boxed{\text{ENTER}}\ \Rightarrow 14.87973172$

(b) $e^{-3.2} = \mathbf{0.040762204}$ $\boxed{e^x}\ \boxed{(-)}\ 3\ \boxed{\cdot}\ 2\ \boxed{\text{ENTER}}\ \Rightarrow .040762204$

EXAMPLE The formula for the atmospheric pressure (in millimeters of mercury) is $P = 760e^{-0.00013h}$, where h is the height above sea level in meters. Find the atmospheric pressure at 100 m above sea level ($h = 100$).

Estimation Developing your number sense with powers of e comes after much examination of the power of e. For now you will minimize errors by making the calculations in the calculator two times and preferably two different ways. A negative exponent means you will multiply by a number less than 1 and the product will be smaller than the original value.

$P = 760e^{-0.00013h}$	$h = 100$
$P = 760e^{-0.00013(100)}$	Simplify exponent. Multiply -0.00013 times 100 mentally.
$P = 760e^{-0.013}$	$e^{-0.013} = 0.987084135$ (leave in calculator)
$P = 760(0.987084135)$	Multiply by 760.
	One continuous calculator sequence: 760 $\boxed{\times}$ $\boxed{e^x}$ $\boxed{(-)}$ .013 $\boxed{=}$
$P = 750.1839426$	

Interpretation **The atmospheric pressure at 100 m above sea level is 750.18 mm.**

As the number of compounding periods per year increases, the effect of compounding levels off or reaches a limit. Therefore the natural exponential e can be substituted into the compound interest formula to accomplish **continuous compounding.** Continuous compounding can be interpreted as daily compounding, compounding every minute, compounding every second, and so on.

To find the accumulated amount (future value) for continuous compounding:

1. Determine the principal (P), rate per year (r), and the number of years (t).
2. Evaluate the formula

$$A = Pe^{rt}$$

where A = accumulated amount or future value
P = principal
r = rate per year
t = time in years

EXAMPLE Find the compound amount of $5,000 invested at an annual rate of 4% compounded continuously for 5 years.

$P = 5,000, r = 0.04$ (from 4%), $t = 5$	
$A = Pe^{rt}$	Substitute values.
$A = 5,000e^{(0.04)(5)}$	Simplify exponents.
$A = 5,000e^{0.2}$	Raise to the power.
$A = 5,000(1.221402758)$	Multiply.
$A = 6,107.01$	

The accumulated amount is $6,107.01.

3 **Solve Exponential Equations in the Form $b^x = b^y$, where $b > 0$ and $b \neq 1$.**

Some applications of exponents may require finding the value of an unknown exponent. If we can write an equation in a specific form, $b^x = b^y$, where $b > 0$ and $b \neq 1$, we will be able to solve the equation using many of our previously learned skills for solving linear equations. Otherwise, we will need to learn some new strategies involving logarithms, which are introduced in the next section.

To illustrate a property of exponential equations, we look at the equation $2^x = 32$. In the equation, the value of x is the power of 2 that gives a result of 32. We can rewrite 32 as 2^5. Thus, $2^x = 2^5$. When an equation can be written in this form, a special property applies.

To solve an exponential equation in the form of $b^x = b^y$:

Apply the following property and solve for x. If $b^x = b^y$, and $b > 0$ and $b \neq 1$, then $x = y$ where x and y are any real numbers.
Note: This property only applies when bases are *like bases*.

Using Conditional Properties (if . . . then)

Many mathematical properties are phrased using an "if . . . then" format. That means, before the property can be used, the conditions or restrictions must be examined. It also means that the property is not appropriate under other conditions.

What are the conditions of the previous property?

• Bases must be like.
• Bases must be positive.
• Bases cannot equal 1.

To illustrate why the property does not work unless these conditions are met, look at this equation, $1^5 = 1^8$. This is a true statement because $1^5 = 1$ and $1^8 = 1$. But, applying the property "if $b^x = b^y$, then $x = y$," does $5 = 8$? No!

EXAMPLE   Solve the equation $2^x = 32$.

$2^x = 32$ Rewrite 32 as a power of 2.
$2^x = 2^5$ $b = 2$, so $b > 0$, $b \neq 1$
If $2^x = 2^5$, then **$x = 5$.**   Apply property.

Check to see if the solution is appropriate.   Does $2^5 = 32$? Yes, so 5 is the correct
                                                solution.

EXAMPLE Solve the equation, $3^{x+1} = 27$.

$3^{x+1} = 27$ Rewrite 27 as a power of 3.

$3^{x+1} = 3^3$ $b = 3$, so $b > 0$, $b \ne 1$

If $3^{x+1} = 3^3$, then $x + 1 = 3$.   Apply property.

$x + 1 = 3$ Solve for x.

$x = 3 - 1$

$x = 2$

Check to see if the solution, 2, is correct.   Does $3^{2+1} = 27$?

$$3^3 = 27$$

Thus, the solution, $x = 2$, is correct.

It will not always be possible to rewrite an exponential equation as an equation with like bases. In such cases, other methods for solving the exponential equation are used.

4 Graph an Exponential Function.

A table of solutions can be used to graph an exponential equation. As with other non-linear functions, several points should be included in the table.

EXAMPLE Prepare a table of solutions and graph the function $f(x) = 2^x$ using integral values of x between -3 and 3, inclusive. Determine the domain and range.

$f(-3) = 2^{-3}$ $f(-2) = 2^{-2}$ $f(-1) = 2^{-1}$

$f(-3) = \dfrac{1}{2^3}$ $f(-2) = \dfrac{1}{2^2}$ $f(-1) = \dfrac{1}{2^1}$

$f(-3) = \dfrac{1}{8}$ or **0.125** $f(-2) = \dfrac{1}{4}$ or **0.25** $f(-1) = \dfrac{1}{2}$ or **0.5**

$f(0) = 2^0$ $f(1) = 2^1$ $f(2) = 2^2$ $f(3) = 2^3$

$f(0) = 1$ $f(1) = 2$ $f(2) = 4$ $f(3) = 8$

$f(x) = 2^x$

x	$f(x)$
-3	0.125
-2	0.25
-1	0.5
0	1
1	2
2	4
3	8

Plot the points indicated in the table of solutions and connect the points with a smooth, continuous curve (Fig. 16–2).

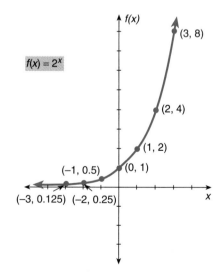

Figure 16–2

The domain is all real numbers. The range is real numbers greater than 0.

The $\boxed{\text{GRAPH}}$ and $\boxed{\text{TABLE}}$ features of a graphing calculator can be used to evaluate an exponential function.

EXAMPLE
HLTH/N

The exponential equation for the breakdown of a particular drug in the human body is $D = D_0(1 + r)^t$, where D is the amount of drug in the bloodstream after a specified amount of time, D_0 is the initial amount of drug in the bloodstream, r is the decimal equivalent of the rate of change, and t is the amount of lapsed time. A certain drug breaks down at the rate of 15% per hour for the first 24 h. (a) Find the amount of drug in the bloodstream after 2 h if the initial amount is 9 mg. (b) After approximately how many hours will the amount of drug in the bloodstream be less than 1 mg?

Known facts

The breakdown of a drug represents a negative rate of change.
Rate of change $= -15\%$
Initial amount of drug in the bloodstream $= 9$ mg

Unknown facts

Amount of drug in the bloodstream after 2 h
How long it will take for the drug level to be less than 1 mg

Relationships

$D = D_0(1 + r)^t$

Estimate

Even though the rate of change is not linear, a linear relationship can approximate the breakdown of the drug.
10% of 9 mg is 0.9 mg; 30% of 9 mg is 2.7 mg.
$D \approx 9 - 2.7 \approx 6.3$ mg

Calculation

(a) Find the amount of drug in the bloodstream after 2 h if the initial amount is 9 mg.

$D = D_0(1 + r)^t$ Substitute known values.
$D = 9(1 - 0.15)^t$ Write as a function of x.
$y = 9(1 - 0.15)^x$

Graph the function for the standard window.

$\boxed{\text{ZOOM}}$ $\boxed{\text{6:ZStandard}}$ $\boxed{Y=}$ 9 $\boxed{(}$ 1 $\boxed{-}$ $\boxed{.}$ 15 $\boxed{)}$ $\boxed{\wedge}$ $\boxed{X, \theta, T, n}$ $\boxed{\text{GRAPH}}$

See Fig. 16–3.

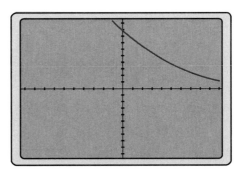

Figure 16–3

16–1 Exponential Expressions, Equations, and Formulas

Evaluate the functions for $x = 2$.

[CALC] [1:value] [ENTER] 2 [ENTER]   See Fig. 16–4.

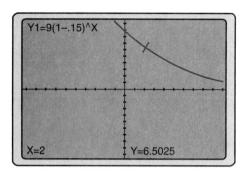

Figure 16–4

Interpretation

The drug level in the bloodstream after 2 hours is 6.5025 mg.

Calculation

(b) After approximately how many hours will the drug level in the bloodstream be less than 1 mg?

[TBLSET] , [TblStart = 0] , [△Tbl = 1] [TABLE]

View [TABLE]. Scroll using arrow keys until Y_1 is less than 1.

See Fig. 16–5.

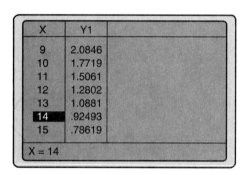

Figure 16–5

Interpretation

The drug level in the bloodstream will be less than 1 mg between 13 and 14 h after taking 9 mg of the drug.

1 Using the compound amount formula, $A = P\left(1 + \dfrac{r}{n}\right)^{nt}$, find the accumulated amount.

1. **BUS** Principal = $1,500, rate = 10%, compounded annually, time = 5 years.

2. **BUS** Principal = $1,750, rate = 8%, compounded quarterly, time = 2 years.

3. **INDTEC** The number of grams of a chemical that will dissolve in a solution is given by the formula $C = 100e^{0.02t}$, where t = temperature in degrees Celsius. Evaluate when
   (a) $t = 10$ (b) $t = 20$
   (c) $t = 25$ (d) $t = 30$

4. **BUS** The compound amount when an investment is compounded continually (every instant) is expressed by the formula $A = Pe^{rt}$, where A = compounded amount, P = principal, t = time in years, and r = rate per year. Find the compound amount when
   (a) Principal = $1,000, interest = 9%, for 2 years
   (b) Principal = $1,500, interest = 10%, for 6 months

Evaluate using a scientific or graphing calculator.

5. 4^3 6. 3^{-5} 7. 5^{10} 8. 8^{-3} 9. $9^{2.5}$ 10. $10^{-\frac{5}{2}}$

2 A formula for electric current is $I = 1.50e^{-200t}$, where t is the time in seconds. Calculate the current for each time. Express answers in scientific notation.

11. **ELEC**   1 sec   12. **ELEC**   1.1 sec   13. **ELEC**   0.5 sec

Evaluate using a scientific or graphing calculator. Round to hundredths.

14. e^2 15. e^{-3} 16. $e^{0.21}$ 17. $e^{-3.5}$

Use the compound interest formula for Exercises 18–27.

18. **BUS** First State Bank loaned Doug Morgan $2,000 for 4 years compounded annually at 8%. How much interest was Doug required to pay on the loan?

19. **BUS** A loan of $8,000 for 2 acres of woodland is compounded quarterly at an annual rate of 12% for 5 years. Find the compound amount and the compound interest.

20. **BUS** Compute the compound amount and the interest on a loan of $10,500 compounded annually for 4 years at 10%.

21. **BUS** Find the future value of an investment of $10,500 if it is invested for 4 years and compounded quarterly at an annual rate of 8%.

22. **BUS** You have $8,000 that you plan to invest in a compound-interest-bearing instrument. Your investment agent advises you that you can invest the $8,000 at 8% compounded quarterly for 3 years or you can invest the $8,000 at $8\frac{1}{4}\%$ compounded annually for 3 years. Which investment should you choose to receive the most interest?

23. **BUS** Find the future value of $50,000 at 6%, compounded semiannually for 10 years.

24. **BUS** Find the compound interest on $2,500 at $6\frac{3}{4}\%$, compounded daily by Leader Financial Bank for 20 days.

25. **BUS** How much compound interest is earned on a deposit of $1,500 at 6.25%, compounded daily for 30 days?

26. **BUS** Ezell Allen has found a short-term investment opportunity. He can invest $8,000 at 8.5% interest for 15 days. How much interest will he earn on this investment if the interest is compounded daily?

27. **BUS** What is the compound interest on $8,000 invested at 8% for 180 days if it is compounded daily?

28. **BUS** Find the effective interest rate for the investment described in Exercise 21. Use the formula for the effective rate.

30. **BUS** Betty Veteto has a loan of $8,500, compounded quarterly for 4 years at 6%. What is the effective interest rate for the loan? Use the formula.

29. **BUS** What is the effective interest rate for a loan of $5,000 at 10%, compounded semiannually for 3 years? Use the effective rate formula.

31. **BUS** What is the effective interest rate for a loan of $20,000 for 3 years if the interest is compounded quarterly at a rate of 12%?

Use the present value formula for Exercises 32–39.

32. **BUS** Compute the amount of money to be set aside today to ensure a future value of $2,500 in 1 year if the interest rate is 11% annually, compounded annually.

34. **BUS** Ronnie Cox has just inherited $27,000. How much of this money should he set aside today to have $21,000 to pay cash for a Ventura Van, which he plans to purchase in 1 year? He can invest at 7.9% annually, compounded annually.

36. **BUS** Joe Brozovich needs $2,000 in 3 years to make the down payment on a new car. How much must he invest today if he receives 8% interest annually, compounded annually?

38. **BUS** Kristen Bieda plans to open a business in 4 years when she retires. How much must she invest today to have $10,000 when she retires if the bank pays 10% annually, compounded quarterly?

33. **BUS** How much should Latonia Shegog set aside now to buy equipment that costs $8,500 in 1 year? The current interest rate is 7.5% annually, compounded annually.

35. **BUS** Shirley Riddle received a $10,000 gift from her mother and plans a minor renovation to her home and an investment for 1 year, at which time she plans to take a trip projected to cost $6,999. The current interest rate is 8.3% annually, compounded annually. How much should be set aside today for her trip?

37. **BUS** Calculate the amount of money that must be invested now at 6% annually, compounded quarterly, to obtain $1,500 in 3 years.

39. **BUS** Charlie Bryant has a child who will be college age in 5 years. How much must he set aside today to have $20,000 for college tuition in 5 years if he gets 8% annually, compounded annually?

Use the future value formula for Exercises 40–51.

40. **BUS** Find the future value of an ordinary annuity of $3,000 annually for 2 years at 9% annual interest. Find the total interest earned.

42. **BUS** Jake Drewrey plans to pay an ordinary annuity of $5,000 annually for 3 years so he can take a year's sabbatical to study for a master's degree in business. The annual rate of interest is 8%. How much will Jake have at the end of 3 years?

44. **BUS** Find the future value of an ordinary annuity of $6,500 semiannually for 7 years at 10% annual interest, compounded semiannually. How much was invested?

46. **BUS** Rosa Kavanaugh established an ordinary annuity of $1,000 annually at 7% annual interest. What is the future value of the annuity after 15 years?

41. **BUS** Len and Sharron Smith are saving money for their daughter Heather to attend college. They set aside an ordinary annuity of $4,000 annually for 2 years at 7% annual interest. How much will Heather have for college when she graduates from high school in 2 years? Find the total interest earned.

43. **BUS** Joe Freeman is planning to establish a small business to provide consulting services in computer networking. He is committed to an ordinary annuity of $3,000 annually at 8.5% annual interest. How much will Joe have to establish the business after 3 years?

45. **BUS** Jimmie Van Alphen pays an ordinary annuity of $2,500 quarterly at 8% annual interest, compounded quarterly, to establish supplemental income for retirement. How much will Jimmie have available at the end of 5 years?

47. **BUS** You invest in an ordinary annuity of $500 annually at 8% annual interest. Find the future value of the annuity at the end of 10 years.

48. BUS You invest in an ordinary annuity of $1,000 annually at 8% annual interest. What is the future value of the annuity at the end of 5 years?

49. BUS Make a chart comparing your results for Exercises 47 and 48. Use these headings: Years, Total Investment, Total Interest. What general conclusion might you draw about effective investment strategy?

50. BUS Find the monthly payment on a home mortgage of $128,600 if the home is financed for 25 years at 7% interest.

51. BUS Find the monthly payment on a home mortgage of $128,600 if the home is financed for 15 years at 7% interest.

3 Solve for x.

52. $3^x = 3^7$

53. $5^x = 5^{-3}$

54. $3^{x+4} = 3^6$

55. $2^{x-3} = 2^7$

56. $6^x = 6^7$

57. $3^x = 27$

58. $2^x = 64$

59. $3^x = \frac{1}{81}$

60. $2^x = \frac{1}{64}$

61. $4^{3x} = 128$

62. $3^{2x} = 243$

63. $4^{3-x} = \frac{1}{16}$

64. $2^{4-x} = \frac{1}{16}$

65. $\left(\frac{1}{3}\right)^x = 27$

66. $\frac{8}{27} = \left(\frac{2}{3}\right)^{5x-12}$

67. $\left(\frac{1}{16}\right)^{2x} = (64)^{6x-10}$

4 Graph the equations using the table-of-solutions method.

68. $y = 3^x$

69. $y = -3^x$

70. $y = 2^x - 3$

71. $y = -2^x - 3$

72. $y = 2(2)^x$

73. $y = 4(2)^x$

74. $y = 6(2)^x$

75. $y = -2(2)^x$

76. $y = -4(2)^x$

Use a graphing calculator for Exercises 77–82.

77. HLTH/N Insulin-delivery systems are designed to release insulin slowly. Even though the rate varies among individuals, the typical pattern of insulin decrease is modeled by the equation $y = 10(0.95)^x$, where x is the number of minutes after insulin enters the bloodstream and y is the number of insulin units in the blood. (a) How many insulin units remain in the bloodstream after 18 min? (b) After approximately how many minutes will the bloodstream contain less than 7.5 units of insulin?

78. HLTH/N Using the equation $y = 10(0.95)^x$, how many minutes will it take for the drug to be reduced to half the original dosage? This length of time is called the *half-life* of the drug.

79. HLTH/N Penicillin, the most famous antibiotic, discovered in 1929, remains in the bloodstream on average according to the formula $y = I(0.6)^x$, where I is the initial dosage, x is the number of hours after the initial injection, and y is the number of milliliters remaining in the bloodstream. Make a table that shows the amount of penicillin in the blood stream up to 5 h later for a 250-mg injection that was administered at 3:00 P.M. How many milliliters remained in the bloodstream at 6:00 P.M.?

80. HLTH/N What is the half-life of penicillin? See Exercise 78 for *half-life*.

81. BUS If you invest $1,000 upon graduating from college and each year the accumulated amount earns on average 5% per year, how much will you have after 1 year? After 2 years? After 10 years? After 30 years? After 50 years? An investment grows according to the exponential equation $y = P(1.05)^x$, where P is the initial amount invested, x is the number of years of the investment, and y is the amount at the end of the time period.

82. Ultrasound is mechanical vibration or sound waves with a typical frequency of between 1.0 and 3.0 MHz (1 MHz = 1 million cycles per second). The intensity of a 3-MHz therapeutic ultrasound can be approximated by the equation $y = 100(0.5)^{0.5x}$, where x represents the depth of penetration in centimeters and y represents the ultrasound intensity at the given depth with 100% intensity at skin surface. What is the ultrasound intensity 6 cm below the skin surface? Use a calculator table to approximate the half-value depth for therapeutic ultrasound with 3-MHz frequency.

16–2 | Logarithmic Expressions, Equations, and Formulas

Learning Outcomes

1 Write exponential equations as equivalent logarithmic equations.
2 Write logarithmic equations as equivalent exponential equations.
3 Evaluate common and natural logarithmic expressions using a calculator.
4 Evaluate logarithms with a base other than 10 or e.
5 Evaluate formulas containing at least one logarithmic term.
6 Graph a logarithmic function.
7 Simplify logarithmic expressions by using the properties of logarithms.

Logarithms were first introduced as a relatively fast way of carrying out lengthy calculations. The calculator and computer have diminished the importance of logarithms as a computational device; however, the importance of logarithms in advanced mathematics, electronics, and theoretical work is more evident than ever. Many formulas use logarithms to express the relationships of physical properties. Logarithms are also used to solve many exponential equations.

1 Write Exponential Equations as Equivalent Logarithmic Equations.

A **logarithmic expression** is an expression that contains at least one term with a logarithm. The three basic components of a power are the base, exponent, and result of exponentiation or power. The word **power** has more than one meaning, so we use the **result of exponentiation** terminology.

$$3^4 = 81 \leftarrow \text{result of exponentiation (power)}$$

with "exponent" labeling the 4 and "base" labeling the 3.

When finding a *power* or the result of exponentiation, you are given the base and exponent. When finding a *root,* you know the base (radicand) and the index of the root (exponent). A third type of calculation involves finding the exponent when the base and the result of exponentiation are given. The exponent in this process is called the *logarithm.* In the logarithmic form $\log_b x = y$, b is the **base,** y is the **exponent** or **logarithm,** and

x is the **result of exponentiation.** This equation is read, "The log of x to the base b is y." Logarithm is written in abbreviated form as "log." In the exponential form $x = b^y$, b is also the base, y is the exponent, and x is the result of the exponentiation.

Algebraic expressions written in logarithmic or exponential form have the same three components: base, exponent, result of exponentiation. Let's examine the mapping from one form to the other.

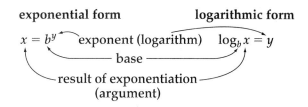

exponential form   **logarithmic form**

$x = b^y$ exponent (logarithm) $\log_b x = y$

base

result of exponentiation
(argument)

To convert an exponential equation to a logarithmic equation:

If $x = b^y$, then $\log_b x = y$, provided that $b > 0$ and $b \neq 1$.

1. The exponent in the exponential form is the dependent variable in the logarithmic form.
2. The base in the exponential form is the base in the logarithmic form.
3. The dependent variable in the exponential form is the result of exponentiation in logarithmic form. This result is sometimes referred to as the **argument.**

EXAMPLE   Write in logarithmic form.

(a) $2^4 = 16$ (b) $3^2 = 9$ (c) $2^{-2} = \dfrac{1}{4}$

(a) $2^4 = 16$ converts to $\mathbf{\log_2 16 = 4}$ Base = 2, exponent = 4, result of exponentiation = 16.

(b) $3^2 = 9$ converts to $\mathbf{\log_3 9 = 2}$ Base = 3, exponent = 2, result of exponentiation = 9.

(c) $2^{-2} = \dfrac{1}{4}$ converts to $\mathbf{\log_2 \dfrac{1}{4} = -2}$ Base = 2, exponent = -2, result of exponentiation = $\frac{1}{4}$.

---

2   **Write Logarithmic Equations as Equivalent Exponential Equations.**

The inverse of the process in the preceding learning outcome changes a logarithmic expression to an exponential expression.

To convert a logarithmic equation to an exponential equation:

$\log_b x = y$ converts to $x = b^y$, provided that $b > 0$ and $b \neq 1$.

1. The dependent variable in the logarithmic form is the exponent in the exponential form.
2. The base in the logarithmic form is the base in the exponential form.
3. The result of the exponentiation in logarithmic form is the dependent variable in exponential form.

EXAMPLE Write in exponential form.

(a) $\log_2 32 = 5$ (b) $\log_3 81 = 4$ (c) $\log_5 \dfrac{1}{25} = -2$ (d) $\log_{10} 0.001 = -3$

(a) $\log_2 32 = 5$ converts to $\mathbf{2^5 = 32}$

Base = 2, exponent = 5, result of exponentiation = 32.

(b) $\log_3 81 = 4$ converts to $\mathbf{3^4 = 81}$

Base = 3, exponent = 4, result of exponentiation = 81.

(c) $\log_5 \dfrac{1}{25} = -2$ converts to $\mathbf{5^{-2} = \dfrac{1}{25}}$

Base = 5, exponent = −2, result of exponentiation = $\frac{1}{25}$.

(d) $\log_{10} 0.001 = -3$ converts to $\mathbf{10^{-3} = 0.001}$

Base = 10, exponent = −3, result of exponentiation = 0.001.

3 Evaluate Common and Natural Logarithmic Expressions Using a Calculator.

When the base of a logarithm is 10, the logarithm is referred to as a **common logarithm.** If the base is omitted in a logarithmic expression, the base is assumed to be 10. Thus, $\log_{10} 1,000 = 3$ is normally written as $\log 1,000 = 3$. On a calculator, expressions containing common logarithms can be evaluated using the $\boxed{\log}$ key.

A logarithm with a base of e is a **natural logarithm** and is abbreviated as ln. Calculators normally have a $\boxed{\ln}$ key. Thus, $\log_e 1 = 0$ is normally written as $\ln 1 = 0$. Expressions containing natural logarithms can be evaluated using the $\boxed{\ln}$ key.

TIP

Finding Logarithms Using Your Calculator

Calculators generally have two logarithm keys, $\boxed{\log}$ for *common logarithms* and $\boxed{\ln}$ for *natural logarithms.*

Evaluate log 2 and ln 2.

Most Calculators:

To find the common log of 2, press the $\boxed{\log}$ key, followed by the result of exponentiation, 2.

$$\boxed{\log}\ 2\ \boxed{=}\ \Rightarrow\ 0.3010299957$$

We can interpret this as $10^{0.3010299957} = 2$.

To find the natural log of 2, press the $\boxed{\ln}$ key, followed by the result of exponentiation, 2.

$$\boxed{\ln}\ 2\ \boxed{=}\ \Rightarrow\ 0.6931471806$$

That is, $e^{0.6931471806} = 2$.

Experiment with your calculator to determine the necessary keystrokes. A good test is to verify that $\log 1 = 0$ and $\ln 1 = 0$.

EXAMPLE Evaluate using a calculator or exponential equations.

(a) $\log 10,000$ (b) $\log 0.000001$ (c) $\log_4 256$ (d) $\log_3 \dfrac{1}{27}$

(a) $\log 10,000 = \mathbf{4}$ $\boxed{\text{log}}\,10000\,\boxed{=}$. Check: $10^4 = 10,000$

(b) $\log 0.000001 = \mathbf{-6}$ Enter $\boxed{\text{log}}\,.000001\,\boxed{=}$. Check: $10^{-6} = 0.000001$

(c) Calculators generally do not have a key for logarithms with a base different from 10 or e. If possible, rewrite the logarithmic equation as an exponential equation with like bases and solve.

$\begin{aligned} \log_4 256 &= x \text{ is } 4^x = 256 && \text{Write 256 as a power of 4. We have } 256 = 4^4. \\ 4^x &= 4^4 && \text{Like bases.} \\ \boxed{x = 4} && \text{Check: } 4^4 = 256 \end{aligned}$

(d) If possible, rewrite the logarithmic equation as an exponential equation with like bases and solve.

$\begin{aligned} \log_3 \dfrac{1}{27} &= x \text{ is } 3^x = \dfrac{1}{27} && \text{Write } \tfrac{1}{27} \text{ as a power of 3. That is, } 27 = 3^3; \tfrac{1}{27} = 3^{-3}. \\ 3^x &= 3^{-3} \\ \boxed{x = -3} && \text{Check: } 3^{-3} = \tfrac{1}{27} \end{aligned}$

EXAMPLE Evaluate using a calculator. Express the answers to the nearest ten-thousandth.

(a) $\ln 5$ (b) $\ln 4.5$ (c) $\ln 948$

(a) $\ln 5 = \mathbf{1.6094}$ **(b)** $\ln 4.5 = \mathbf{1.5041}$ **(c)** $\ln 948 = \mathbf{6.8544}$

---

4 Evaluate Logarithms with a Base Other Than 10 or e.

Many situations require us to find a logarithm with a base other than 10 or e when we cannot convert to an equivalent exponential equation. A conversion formula is used to find the logarithm for a base other than 10 or e using a calculator.

To evaluate a logarithm with a base b other than 10 or e:

$$\log_b a = \frac{\log a}{\log b}$$

A similar process can be used with natural logarithms.

$$\log_b a = \frac{\ln a}{\ln b}$$

To check, evaluate b^x where b is the base of the logarithm and x is the logarithm. The result should equal the argument.

EXAMPLE   Find $\log_7 343$.

$$\log_7 343 = \frac{\log 343}{\log 7} = 3$$ $\boxed{\log}$ 343 $\boxed{\div}$ $\boxed{\log}$ 7 $\boxed{=}$ $\Rightarrow$ 3; Check: $7^3 = 343$

---

5 **Evaluate Formulas Containing at Least One Logarithmic Term.**

Many formulas use common and natural logarithms.

EXAMPLE   The loudness of sound is measured by a unit called a **decibel** (dB). A very faint sound, called the **threshold sound,** is assigned an intensity of I_0. Other sounds have an intensity of I, which is a specified number times the threshold sound ($I = nI_0$).

Then the decibel rate is given by the formula dB $= 10 \log \frac{I}{I_0}$. Find the decibel rating to the nearest decibel for sounds having the following intensities (I).

(a) A whisper, $110I_0$
(b) A speaking voice, $230I_0$
(c) A busy street, $9{,}000{,}000I_0$
(d) Loud music, $875{,}000{,}000{,}000I_0$
(e) A jet plane at takeoff, $109{,}000{,}000{,}000{,}000I_0$

(a) A whisper, $I = 110I_0$ $n = 110$

$$\text{dB} = 10 \log \frac{110I_0}{I_0}$$ $\frac{I_0}{I_0} = 1$

$$\text{dB} = 10 \log 110$$ $\log 110 = 2.041392685$

dB = 20 (rounded)

(b) A speaking voice, $I = 230I_0$ $n = 230$

$$\text{dB} = 10 \log \frac{230I_0}{I_0}$$ $\frac{I_0}{I_0} = 1$

$$\text{dB} = 10 \log 230$$ $\log 230 = 2.361727836$

dB = 24 (rounded)

(c) A busy street, $I = 9{,}000{,}000I_0$ $n = 9{,}000{,}000$

$$\text{dB} = 10 \log \frac{9{,}000{,}000I_0}{I_0}$$ $\frac{I_0}{I_0} = 1$

$$\text{dB} = 10 \log 9{,}000{,}000$$ $\log 9{,}000{,}000 = 6.954242509$

dB = 70 (rounded)

(d) Loud music, $I = 875{,}000{,}000{,}000I_0$ $n = 875{,}000{,}000{,}000$

$$\text{dB} = 10 \log \frac{875{,}000{,}000{,}000I_0}{I_0}$$ $\frac{I_0}{I_0} = 1$

$$\text{dB} = 10 \log 875{,}000{,}000{,}000$$ $875{,}000{,}000{,}000 = 8.75 \times 10^{11}$

$$\text{dB} = 10 \log (8.75 \times 10^{11})$$ 10 $\boxed{\log}$ 8.75 $\boxed{\text{EXP}}$ 11 $\boxed{=}$ $\Rightarrow$ 119.4200805

dB = 119 (rounded)

(e) A jet plane at takeoff, $I = 109{,}000{,}000{,}000{,}000 I_0 = (1.09 \times 10^{14}) I_0$

$$dB = 10 \log \frac{(1.09 \times 10^{14}) I_0}{I_0} \qquad n = 1.09 \times 10^{14}$$

$$dB = 10 \log (1.09 \times 10^{14}) \qquad \frac{I_0}{I_0} = 1$$

$$dB = 10(14.0374265)$$

dB = 140 (rounded)

To find the time for an investment to reach a specified amount if compounded continuously:

$$t = \frac{(\ln A - \ln P)}{r}$$

t = time (in years) of investment
A = accumulated amount
P = initial invested amount
r = annual rate of interest

EXAMPLE   Find the length of time for \$32,750 to reach \$47,650.97 when invested at 5% compounded continuously.

$A = \$47{,}650.97, \qquad P = \$32{,}750, \qquad r = 0.05$

$$t = \frac{\ln A - \ln P}{r} \qquad\qquad \text{Substitute known values.}$$

$$t = \frac{\ln 47{,}650.97 - \ln 32{,}750}{0.05} \qquad\qquad \text{Evaluate numerator.}$$

$$t = \frac{10.77165827 - 10.39665824}{0.05}$$

$$t = \frac{0.3750000246}{0.05} \qquad\qquad \text{Divide.}$$

$$t = \textbf{7.5 years}$$

EXAMPLE   How long does it take an investment of \$10,000 to double if it is invested at 5% annual interest compounded continuously?

$A = \$20{,}000, \qquad P = 10{,}000, \qquad r = 0.05$

$$t = \frac{\ln A - \ln P}{r} \qquad\qquad \text{Substitute known values.}$$

$$t = \frac{\ln 20{,}000 - \ln 10{,}000}{0.05} \qquad\qquad \text{Evaluate numerator.}$$

$$t = \frac{9.903487553 - 9.210340372}{0.05}$$

$$t = \frac{0.6931471806}{0.05} \qquad\qquad \text{Divide.}$$

$$t = \textbf{13.9 years}$$

6 Graph a Logarithmic Function.

Logarithmic functions can be graphed using a table of solutions or a graphing calculator.

EXAMPLE Prepare a table of solutions and graph the function $f(x) = \ln 4x$ using the values of x: 0.125, 0.25, 0.5, 1, 2, 3, 4. Determine the domain and range.

$f(x) = \ln 4x$

x	$f(x)$
0.125	−0.7
0.25	0
0.5	0.7
1	1.4
2	2.1
3	2.5
4	2.8

Using a calculator:

$f(0.125) = \ln 4(0.125)$
$f(0.125) = \ln 0.5$
$\mathbf{f(0.125) = -0.6931471806}$

$f(0.25) = \ln 4(0.25)$
$f(0.25) = \ln 1$
$\mathbf{f(0.25) = 0}$

$f(0.5) = \ln 4(0.5)$
$f(0.5) = \ln 2$
$\mathbf{f(0.5) = 0.6931471806}$

$f(1) = \ln 4(1)$
$f(1) = \ln 4$
$\mathbf{f(1) = 1.386294361}$

$f(2) = \ln 4(2)$
$f(2) = \ln 8$
$\mathbf{f(2) = 2.079441542}$

$f(3) = \ln 4(3)$
$f(3) = \ln 12$
$\mathbf{f(3) = 2.48490665}$

$f(4) = \ln 4(4)$
$f(4) = \ln 16$
$\mathbf{f(4) = 2.772588722}$

Plot the points indicated in the table of solutions and connect the points with a smooth, continuous curve (Fig. 16–6).

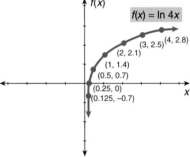

The domain is the set of real numbers greater than 0. The range is the set of all real numbers.

Figure 16–6

EXAMPLE A scale for measuring the intensity of an earthquake is the **Richter scale.** Richter scale rating $= \log \dfrac{I}{I_0}$, where I_0 is the measure of the intensity of a very small (faint) earthquake and I is the measure of an earthquake times I_0.

Find the Richter scale ratings for intensity values of powers of 10 from 10^3 to 10^8. Describe the relationship between the Richter scale rating and the intensity of an earthquake.

Richter scale rating $= \log \dfrac{I}{I_0}$

Values of I include a factor of I_0, which will reduce with the denominator. Let x represent the quotient of $\dfrac{I}{I_0}$ and write the Richter scale as a function of x.

$y = \log x$

Enter function in a graphing calculator.

Set the Window for x-values 0–100,000,000 with a scale of 1,000 and for y-values 0–10 with a scale of 1.

Find the corresponding y-values for powers of 10 from 10^3 to 10^8. Use the CALC and the 1:VALUE features for each value of x.

x	y
1,000	3
10,000	4
100,000	5
1,000,000	6
10,000,000	7
100,000,000	8

The Richter scale rating for the intensity of an earthquake is equal to the exponent of the intensity expressed as a power of 10.

7 **Simplify Logarithmic Expressions by Using the Properties of Logarithms.**

Many applications of logarithms and exponential expressions require the understanding and use of the properties of logarithms. The laws of exponents are similar to the properties of logarithms because of the relationship between logarithms and exponents. Similar laws are appropriate for natural logarithms.

Properties of logarithms:

Logarithm of product	$\log_b mn = \log_b m + \log_b n$	The log of a product is the sum of the logs of the factors.
Logarithm of quotient	$\log_b \dfrac{m}{n} = \log_b m - \log_b n$	The log of a quotient is the difference of the logs of the numerator and denominator.
Logarithm of power	$\log_b m^n = n \log_b m$	The log of a quantity raised to an exponent is the exponent times the log of the quantity.
Logarithm of quantity with same base	$\log_b b = 1$	The log of a quantity with the same base as the quantity equals 1.

These laws can be illustrated with examples using a calculator.

EXAMPLE   Show that the statements are true by using a calculator.

(a) $\log 6 = \log 2 + \log 3$ (b) $\log 2 = \log 6 - \log 3$
(c) $\log 2^3 = 3 \log 2$ (d) $\log 20 = 1 + \log 2$

(a) $\log 6 = \log 2(3) = \log 2 + \log 3$

$\boxed{\log}\ 6\ \boxed{=}\ \Rightarrow \textbf{0.77815125,}$ $\boxed{\log}\ 2\ \boxed{+}\ \boxed{\log}\ 3\ \boxed{=}\ \Rightarrow \textbf{0.77815125}$

(b) $\log 2 = \log \dfrac{6}{3} = \log 6 - \log 3$

$\boxed{\log}\ 2\ \boxed{=}\ \Rightarrow \textbf{0.301029996,}$ $\boxed{\log}\ 6\ \boxed{-}\ \boxed{\log}\ 3\ \boxed{=}\ \Rightarrow \textbf{0.301029996}$

16–2 Logarithmic Expressions, Equations, and Formulas   643

(c) $\log 2^3 = 3 \log 2$

$\boxed{\log} \boxed{(} 2 \boxed{\wedge} 3 \boxed{)} \boxed{=} \Rightarrow \textbf{0.903089987}, \qquad 3 \boxed{\log} 2 \boxed{=} \Rightarrow \textbf{0.903089987}$

(d) $\log 20 = \log 10(2) = \log 10 + \log 2 = 1 + \log 2$

$\boxed{\log} 20 \boxed{=} \Rightarrow \textbf{1.301029996}, \qquad 1 + \boxed{\log} 2 \boxed{=} \Rightarrow \textbf{1.301029996}$

SECTION 16–2 SELF-STUDY EXERCISES

1 Write as logarithmic equations.

1. $3^2 = 9$

2. $2^5 = 32$

3. $9^{\frac{1}{2}} = 3$

4. $16^{\frac{1}{4}} = 2$

5. $4^{-2} = \dfrac{1}{16}$

6. $3^{-4} = \dfrac{1}{81}$

2 Write as exponential equations.

7. $\log_3 81 = 4$

8. $\log_{12} 144 = 2$

9. $\log_2 \dfrac{1}{8} = -3$

10. $\log_5 \dfrac{1}{25} = -2$

11. $\log_{25} \dfrac{1}{5} = -0.5$

12. $\log_4 \dfrac{1}{2} = -0.5$

Solve for x by using an equivalent exponential expression.

13. $\log_4 64 = x$

14. $\log_3 x = -4$

15. $\log_6 36 = x$

16. $\log_7 \dfrac{1}{49} = x$

17. $\log_5 x = 4$

18. $\log_4 \dfrac{1}{256} = x$

3 Evaluate with a calculator. Express the answer to the nearest ten-thousandth.

19. $\log 3$ **20.** $\log 6$ **21.** $\log 2.4$ **22.** $\log 4.2$

23. $\log 150$ **24.** $\log 0.0012$ **25.** $\ln 4$ **26.** $\ln 2.5$

27. $\ln 0.15$ **28.** $\ln 275$ **29.** $\ln 100$

4 Evaluate with a calculator.

30. $\log_5 125$

31. $\log_3 729$

32. $\log_{\frac{1}{2}} 0.03125$

33. $\log_7 49$

34. $\log_8 56$

5 The intensity of an earthquake is measured on the *Richter scale* by the formula

$$\text{Richter scale rating} = \log \frac{I}{I_0}$$

where I_0 is the measure of the intensity of a very small (faint) earthquake. Find the Richter scale rating of earthquakes having the following intensities:

35. $1,000 I_0$

36. $100,000 I_0$

37. $100,000,000 I_0$

38. How long does it take an investment of $20,000 to double if it is invested at 5% annual interest, compounded continuously? Round to tenths.

39. How long does it take for $48,000 to reach $52,000 when invested at 4%, compounded continuously? Round to tenths.

6 Graph the functions using the table-of-solutions method. Determine the domain and range.

40. $y = \ln 2x$ **41.** $y = -\ln 2x$ **42.** $y = \ln 5x$

43. $y = -\log 3x$ **44.** $y = -\log 6x$ **45.** $y = \log(-3x)$

46. HLTH/N The pH level of blood is the measure of the positive hydrogen ions present. A high pH indicates a high level of acidity and a low pH indicates the blood is alkaline. A pH of 7 is neutral (neither acidic nor alkaline). The change in pH level in blood can be modeled by the equation $pH = 6.1 + \log \dfrac{b}{c}$, where b is the concentration of bicarbonate and c is the concentration of bicarbonate acid. Graph the equation when $c = 5$ and use the TABLE function to determine what value of b produces a neutral pH. Report b to the nearest tenth.

47. HLTH/N Use the equation $pH = 6.1 + \log\left(\dfrac{b}{c}\right)$ for blood pH and $b = 35$. What value of c produces a neutral pH?

48. Graph the equation, Richter scale rating $= \log\left(\dfrac{I}{I_0}\right)$, where I represents the intensity of an earthquake and I_0 represents the intensity of a very small earthquake. Use the TABLE function to find the Richter scale rating of an earthquake that has an intensity of 19,500,000.

7 Solve for x by using an equivalent logarithmic expression and the facts that $\log_3 2 = 0.631$ and $\log_3 5 = 1.465$.

49. $\log_3 10 = x$ **50.** $\log_3 8 = x$ **51.** $\log_3 6 = x$ **52.** $\log_3 20 = x$

CHAPTER REVIEW OF KEY CONCEPTS

Learning Outcomes	What to Remember with Examples

Section 16–1

1 Evaluate formulas with at least one exponential term (pp. 620–626).

A scientific or graphing calculator can be used along with the order of operations to evaluate formulas with at least one exponential term.

> Use the formula for compound interest to find the compound amount for a loan of $5,000 for 3 years at an annual interest rate of 6% if the principal is compounded semiannually.
>
> $A = P\left(1 + \dfrac{r}{n}\right)^{nt}$ $P = \$5{,}000; r = 0.06; n = 2; t = 3$
>
> $A = 5{,}000\left(1 + \dfrac{0.06}{2}\right)^{2(3)}$ Perform operations inside grouping.
> Perform operation in exponent.
>
> $A = 5{,}000(1.03)^6$ Raise to power.
>
> $A = 5{,}000(1.194052297)$ Multiply.
>
> $A = \$5{,}970.26$ Compound amount (rounded)

2 Evaluate formulas that contain a power of the natural exponential, e (pp. 627–628).

Use a calculator to evaluate formulas containing the natural exponential, e.

> Use the formula $P = 760\,e^{-0.00013h}$ for atmospheric pressure to find P at 50 m above sea level (h).
>
> $P = 760e^{-0.00013(50)}$
> $P = 760e^{-0.0065}$
> $P = 760(0.9935210793)$
> $P = 755.08$ Rounded

3 Solve exponential equations in the form $b^x = b^y$, where $b > 0$ and $b \neq 1$ (pp. 629–630).

To solve an exponential equation in the form $b^x = b^y$, apply the property: When the bases are equal, the exponents are equal. If the bases are not equal, rewrite the bases as powers so that the bases are equal. The exponents will then be equal. This property applies only when b is positive and not equal to 1.

> Solve the equation $4^{3x+1} = 8$. Rewrite the bases: $4 = 2^2$ and $8 = 2^3$.
>
> $2^{2(3x+1)} = 2^3$ The bases are equal, so exponents are equal.
> $2(3x + 1) = 3$ Distribute.
> $6x + 2 = 3$ Sort terms and combine like terms.
> $6x = 1$ Divide.
>
> $x = \dfrac{1}{6}$

4 Graph an exponential function (pp. 630–632).

Use a table of values or the graphing calculator to graph exponential functions.

> The exponential equation for the breakdown of a particular drug in the human body is $D = D_0(1 + r)^t$, where D is the amount of drug in the bloodstream after a specified amount of time, D_0 is the initial amount of drug in the bloodstream, r is the rate of change as a decimal, and t is the amount of lapsed time. A certain drug breaks down at the rate of 25% per hour for the first 24 h. Find the amount of drug in the bloodstream after 6 h if the initial amount is 50 mg.
>
> $D = 50(1 - 0.25)^t$ Write as a function of x. Simplify grouping.
>
> $y = 50(0.75)^x$ Graph.
>
> $\boxed{y=}$ 50 $\boxed{(}$ .75 $\boxed{)}$ $\boxed{\wedge}$ $\boxed{X, \theta, T, n}$ $\boxed{\text{TBLSET}}$ TblStart $= \boxed{1}$ ΔTbl $= \boxed{1}$ $\boxed{\text{TABLE}}$
>
>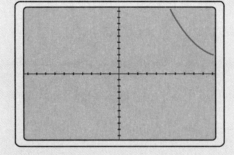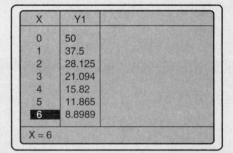
>
> **Approximately 8.9 mg of drug remains in the bloodstream after 6 hours.**

Section 16–2

1 Write exponential expressions as equivalent logarithmic expressions (pp. 636–637).

To write exponential expressions as equivalent logarithmic expressions, use the format: $x = b^y$ converts to $\log_b x = y$.

> Write $16 = 2^4$ in logarithmic form. $\log_2 16 = 4$

2 Write logarithmic expressions as equivalent exponential expressions (pp. 637–638).

To write logarithmic expressions as equivalent exponential expressions, use the format: $\log_b x = y$ converts to $x = b^y$.

Write $\log_5 125 = 3$ in exponential form. $5^3 = 125$

3 Evaluate common and natural logarithmic expressions using a calculator (pp. 638–639).

Use the $\boxed{\log}$ key to find common logarithms and the $\boxed{\ln}$ key to find natural logarithms. Some calculators require the number to be entered before the $\boxed{\log}$ or $\boxed{\ln}$ key. Most calculators require the $\boxed{\log}$ or $\boxed{\ln}$ key to be entered, followed by the number. If a parenthesis is opened, enter a close parenthesis after entering the argument of the logarithm.

Use a calculator to find log 25. Scientific calculator steps:

$\boxed{\log}$ 25 $\boxed{=}$ or   25 $\boxed{\log}$ $\Rightarrow$ 1.397940009

4 Evaluate logarithms with a base other than 10 or e (pp. 639–640).

To evaluate logarithms with bases other than 10 or e, divide the common or natural log of the number by the common or natural log of the base.

Evaluate $\log_4 64$:

$$\log_4 64 = \frac{\log 64}{\log 4} = 3 \qquad \boxed{\log}\ 64\ \boxed{)}\ \boxed{\div}\ \boxed{\log}\ 4\ \boxed{=}\ \Rightarrow 3$$

5 Evaluate formulas containing at least one logarithmic term (pp. 640–641).

To evaluate formulas containing at least one logarithmic term, use a calculator and follow the order of operations.

Use the formula $dB = 10 \log\left(\dfrac{I}{I_0}\right)$ to find the decibel rating for a sound that is 350 times the threshold sound ($350I_0$).

$$dB = 10 \log\left(\frac{350I_0}{I_0}\right)$$

$dB = 10 \log 350$ 10 $\boxed{\times}$ $\boxed{\log}$ 350 $\boxed{=}$

$dB = 25.44$ Rounded

6 Graph a logarithmic function (pp. 642–643).

A logarithmic function can be graphed using a table of values or a graphing calculator.

Graph the function, $y = \log(0.4x)$ and find the value of y when $x = 150$.

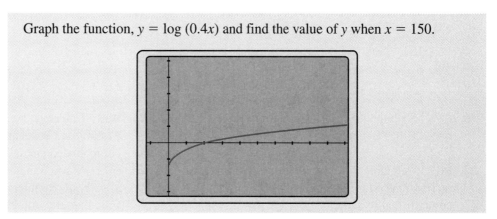

The preceding figure shows the graph for x between 0 and 10.

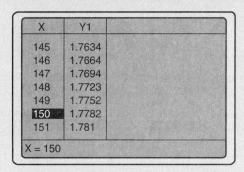

When $x = 150$, $y = 1.7782$.

7 Simplify logarithmic expressions by using the properties of logarithms (pp. 643–644).

The laws of exponents also apply to expressions with logarithms.

$$\log_b mn = \log_b m + \log_b n$$

$$\log_b \frac{m}{n} = \log_b m - \log_b n$$

$$\log_b m^n = n \log_b m$$

$$\log_b b = 1$$

Write $\log (3)(8)$ in another way. Use a calculator to verify that the equation is true.

$$\log (3)(8) = \log 3 + \log 8$$

$\boxed{\log}\ \boxed{(}\ 3\ \boxed{\times}\ 8\ \boxed{)}\ \boxed{=}\ \Rightarrow 1.380211242$

$\boxed{\log}\ 3\ \boxed{)}\ \boxed{+}\ \boxed{\log}\ 8\ \boxed{=}\ \Rightarrow 1.380211242$

CHAPTER REVIEW EXERCISES

Section 16–1

Solve for x.

1. $5^x = 5^8$

2. $2^x = 2^6$

3. $3^x = 3^{-2}$

4. $2^{x+3} = 2^7$

5. $4^{x-2} = 4^2$

6. $5^{2x-1} = 5^2$

7. $6^{3x+2} = 6^{-3}$

8. $2^x = 16$

9. $3^x = 81$

10. $3^x = \frac{1}{9}$

11. $2^x = \frac{1}{32}$

12. $4^{2x} = 64$

13. $5^{3x} = 125$

14. $3^{4-x} = \frac{1}{27}$

15. $6^{2-x} = \frac{1}{36}$

Evaluate.

16. e^3

17. e^{-4}

18. $e^{-0.12}$

19. e^{-10}

20. **BUS** Calculate the compound interest on a loan of $200 at 6%, compounded annually for 4 years.

21. **BUS** Calculate the compound interest on a loan of $1,600 for 3 years at 13% if the interest is compounded annually.

22. **BUS** Calculate the compound interest on a loan of $6,150 at $11\frac{1}{2}\%$ annual interest, compounded annually for 3 years.

23. **BUS** Maria Sanchez invested $2,000 for 2 years at 12% annual interest, compounded semiannually. Calculate the interest she earned on her investment.

24. **BUS** EZ Loan Company loaned $500 at 8% annual interest, compounded quarterly for 1 year. Calculate the amount the loan company will earn in interest.

25. **BUS** Find the future value on an investment of $3,000 made by Ling Lee for 5 years at 12% annual interest, compounded semiannually.

26. **BUS** Find the interest on $2,500 invested for 3 years at 3.75%, compounded quarterly.

27. **BUS** Find the interest on a certificate of deposit (CD) of $10,000 for 5 years at 4%, compounded semiannually.

28. **BUS** Find the interest on $10,000 for 2 years compounded monthly at a 6% annual interest rate.

29. **BUS** Find the future value on an investment of $8,000, compounded quarterly for 7 years at 8%.

30. **BUS** Find the compound interest on $5,000 for 2 years if the interest is compounded continuously at 12%.

31. **BUS** Find the compound interest on $5,000 for 2 years if the interest is compounded continuously at 12%.

32. **BUS** Find the effective interest rate for the loan described in Exercise 28.

33. **BUS** Find the effective interest rate for the loan described in Exercise 29.

34. **BUS** Find the accumulated amount on an investment of $8,000 invested for 5 years compounded quarterly at 5%.

35. **BUS** Find the amount of interest on $1,500 invested for 4 years at a 3% annual rate compounded monthly.

36. **BUS** Tommye Adams wishes to have $8,000 1 year from now to make a down payment on a lake house. How much should she invest at 9.5% annual interest to have her payment in 1 year?

37. **BUS** Billy Hill wishes to have $4,000 in 4 years to tour Europe. How much must he invest today at 8% annual interest, compounded quarterly, to have the $4,000 in 4 years?

38. **BUS** Kristin Ammons was offered $15,000 cash or $19,500 to be paid after 2 years for a resort cabin. If money can be invested in today's market for 3% annual interest, compounded quarterly, which offer should Kristin accept?

39. **BUS** An art dealer offered a collector $8,000 for a painting. The collector could sell the painting to an individual for $11,000 to be paid in 18 months. Currently investments yield 12% annual interest, compounded monthly. Which is the better deal for the collector?

40. **BUS** If you were offered $700 today or $800 in 2 years, which would you accept if money can be invested at 12% annual interest, compounded monthly?

41. How much should a family invest now at 10% compounded annually to have a $7,000 house down payment in 4 years.

42. **BUS** Dennis and Deb Walker had a baby in 2005. At the end of that year they began putting away $900 per year at 10% compounded annual interest for a college fund. When their child is 18 years old in 2023, the cost for 4 years of college is estimated to be about $20,000 per year.
    (a) How much money will be in the account when the child is 18 years old?
    (b) Will the Walkers have enough saved to send their child to college for 4 years?

43. Skip Quinn plans to deposit $2,000 at the end of every 6 months for the next 5 years to save for a boat. If the interest rate is 12% annually, compounded semiannually, how much money will Skip have in his boat fund after 5 years?

44. **BUS** A business deposits $4,500 at the end of each quarter in an account that earns 8% annual interest, compounded quarterly. What is the value of the annuity in 5 years?

45. **BUS** How much must be set aside at the end of each six months in a sinking fund by the Fabulous Toy Company to replace a $155,000 piece of equipment at the end of 8 years if the account pays 8% annual interest, compounded semiannually?

46. **BUS** Tasty Food Manufacturers, Inc., has a bond issue of $1,400,000 due in 30 years. If it wants to establish a sinking fund to meet this obligation, how much must be set aside at the end of each year if the annual interest rate is 6%?

47. **BUS** Lausanne Private School System needs to set aside funds for a new computer system. What monthly sinking fund payment would be required to amount to $45,000, the approximate cost of the computer system, in $1\frac{1}{2}$ years at 12% annual interest, compounded monthly?

48. BUS Zachary Alexander owns a limousine that will need to be replaced in 4 years at a cost of $65,000. How much must he put aside each year in a sinking fund at 8% annual interest to be able to afford the new limousine?

49. BUS Braddy's Department Store has a fleet of delivery trucks that needs to be replaced at a cost of $75,000. How much must it set aside every 3 months for 3 years in a sinking fund at 8% annual interest, compounded quarterly, to have enough money to replace the trucks?

50. BUS Danny Lawrence Properties, Inc., has a bond issue that matures in 25 years for $1 million. How much must the company set aside each year in a sinking fund at 12% annual interest to meet this future obligation?

51. BUS Brenda Pearson wants to save $25,000 for a new boat in 6 years. How much must be put aside in equal payments each year in an account earning 8% annual interest for Brenda to be able to purchase the boat?

52. BUS How much money needs to be set aside today at 10% annual interest, compounded semiannually, to have the same amount as a semiannual ordinary annuity of $500 for 5 years at 10% annual interest, compounded semiannually?

53. BUS Find the monthly payment on a mortgage of $238,000 if it is financed for 20 years at 7.5% annual interest.

54. BUS Martha Ann Dawson purchased a property with a mortgage of $528,260 at 6.8% annual interest. Find the monthly payment if the mortgage is amortized for 30 years.

A formula for electric current is $I = 1.50e^{-200t}$, where t is time in seconds. Calculate the current for each time. Express the answers in scientific notation.

55. ELEC 0.07 s

56. ELEC 0.2 s

57. ELEC 0.4 s

Graph the equations using the table-of-solutions method or a calculator.

58. INDTEC The number of grams of a chemical that will dissolve in a solution is given by the formula $C = 100e^{0.05t}$, where t = temperature in degrees Celsius. Evaluate when
  (a) $t = 10$ **(b)** $t = 20$
  (c) $t = 45$ **(d)** $t = 50$

59. $y = 5^x$
60. $y = -5^x$
61. $y = 5^{(x+4)}$
62. $y = 5^{(x-4)}$
63. $y = 5^{-x}$
64. $y = -5^{-x}$

65. HLTH/N In the equation $D = D_0(1 + r)^t$, D is the amount of drug in the bloodstream after a specified amount of time. D_0 is the initial amount of the drug in the bloodstream, r is the decimal equivalent of the rate of change, and t is the amount of lapsed time. A drug breaks down at the rate of 4% per hour for the first 24 h. Find the amount of drug after 5 h if the initial amount of drug in the bloodstream is 20 mg. (Round to tenths.)

66. HLTH/N Use the exponential model in Exercise 65 to find the approximate number of hours it will take to have less than 60 mg of a drug in the bloodstream if the drug breaks down at the rate of 25% per hour when the initial amount is 250 mg.

Section 16–2

Rewrite the following as logarithmic equations.

67. $2^3 = 8$

68. $5^2 = 25$

69. $3^4 = 81$

70. $81^{1/2} = 9$

71. $27^{1/3} = 3$

72. $5^{-3} = \frac{1}{125}$

73. $4^{-3} = \frac{1}{64}$

74. $8^{-1/3} = \frac{1}{2}$

75. $9^{-1/2} = \frac{1}{3}$

76. $121^{1/2} = 11$

77. $12^{-2} = \frac{1}{144}$

78. $49^{-1/2} = \frac{1}{7}$

Rewrite the following as exponential equations. Verify if the equation is true.

79. $\log_{11} 121 = 2$

80. $\log_3 81 = 4$

81. $\log_{15} 1 = 0$

82. $\log_{25} 5 = \frac{1}{2}$

83. $\log_7 7 = 1$

84. $\log_3 3 = 1$

85. $\log_4 \frac{1}{16} = -2$

86. $\log_2 \frac{1}{16} = -4$

87. $\log_9 \frac{1}{3} = -0.5$

88. $\log_{16} \frac{1}{4} = -0.5$

89. $\log_{10} 1,000 = 3$

90. $\log_{10} 100 = 2$

Evaluate the following with a calculator. Express the answers to the nearest ten-thousandth.

91. log 5   **92.** log 3.8   **93.** log 180   **94.** log 0.0015

95. log 0.4   **96.** ln 12   **97.** ln 270   **98.** ln 0.134

99. ln 0.8   **100.** ln 80   **101.** $\log_5 30$ **102.** $\log_7 120$

Solve for x by using an equivalent exponential expression.

103. $\log_4 16 = x$ **104.** $\log_7 49 = x$ **105.** $\log_7 x = 3$ **106.** $\log_5 x = -2$

107. $\log_6 \frac{1}{36} = x$ **108.** $\log_4 \frac{1}{64} = x$ **109.** $\log_2 x = 5$ **110.** $\log_x 125 = 3$

111. The intensity of an earthquake is measured on the Richter scale by the formula

$$\text{Richter scale rating} = \log \frac{I}{I_0}$$

where I_0 is the measure of the intensity of a very small (faint) earthquake. Find the Richter scale rating of the earthquakes having the following intensities:
(a) $100 I_0$ (b) $10{,}000 I_0$ (c) $150{,}000{,}000 I_0$

Solve for x by using an equivalent logarithmic expression and the facts that $\log_2 3 = 1.585$ and $\log_2 7 = 2.807$.

112. $\log_2 21$ **113.** $\log_2 9$ **114.** $\log_2 6$

115. BUS An investment of $100,000 is invested at 4.1%, compounded continuously. How long will it take for the investment to accumulate to $150,000?

116. BUS Quenesha McGee has $10,000 invested at 5.9%, compounded continuously. She needs to know how long it will take for the investment to accumulate to $18,000.

Graph the equations using a table of solutions or calculator.

117. $y = \ln 4x$ **118.** $y = -\ln 4x$ **119.** $y = \log 8x$ **120.** $y = \log (x - 8)$

121. HLTH/N Use the equation $\text{pH} = 6.1 + \log \left(\dfrac{50}{c} \right)$

to find the value of c that produces a neutral pH (7) to the nearest tenth.

TEAM PROBLEM-SOLVING EXERCISES

1. The Environmental Protection Agency (EPA) monitors atmosphere and soil contamination by dangerous chemicals. When possible, the chemical contamination is decomposed by using microorganisms that change the chemicals so they are no longer harmful. A particular microorganism can reduce the contamination level to about 65% of the existing level every 30 days. A soil test for a contaminated site shows 72,000,000 units per cubic meter of soil.

(a) Write a formula for determining the contamination level after x 30-day periods: the initial contamination times the percent reduction (65% or 0.65) raised to the xth power.

(b) What is the level of contamination after 60 days? After 150 days?

(c) A "safe level" is 60,000 units of contamination per cubic meter of soil. Estimate how long it would take for the soil to reach this safe level. Discuss your method of arriving at the estimate.

2. Complete the spreadsheet showing the Accumulated Amount and Interest for each entry (Fig. 16–7). The formula written for the spreadsheet is: $= P * (1+r) \wedge t$, where P is principal, r is rate per period, and t is number of periods.

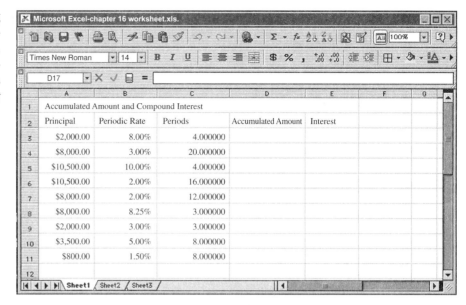

	A	B	C	D	E	F	G
1	Accumulated Amount and Compound Interest						
2	Principal	Periodic Rate	Periods	Accumulated Amount	Interest		
3	$2,000.00	8.00%	4.000000				
4	$8,000.00	3.00%	20.000000				
5	$10,500.00	10.00%	4.000000				
6	$10,500.00	2.00%	16.000000				
7	$8,000.00	2.00%	12.000000				
8	$8,000.00	8.25%	3.000000				
9	$2,000.00	3.00%	3.000000				
10	$3,500.00	5.00%	8.000000				
11	$800.00	1.50%	8.000000				
12							

Figure 16–7

CONCEPTS ANALYSIS

1. Give an example of an application that has a linear relationship and an application that has an exponential relationship.

2. Briefly describe how a linear relationship is different from an exponential relationship.

3. Describe how to recognize an equation as an exponential equation.

4. Explain why an equation of the form $b^x = b^y$ is only appropriate when $b > 0$ and $b \neq 1$.

5. Show symbolically the relationship between an exponential equation and a logarithmic equation. Give a numerical example illustrating this relationship.

6. Explain the difference between a common logarithm and a natural logarithm.

Find the mistakes in the examples. Explain the mistake and correct it.

7. $2^{x-3} = 4^2$
$x - 3 = 2$
$x = 2 + 3$
$x = 5$

8. Change $\log_5 125 = 3$ to an equivalent exponential equation.
$3^5 = 125$

PRACTICE TEST

Evaluate using a calculator.

1. 1.2^{45}

2. 5^{-8}

3. $15^{\frac{3}{2}}$

4. $e^{-0.25}$

5. 12^5

Solve for x.

6. $4^x = 4^{-3}$

7. $2^{x-4} = 2^5$

8. $3^x = \dfrac{1}{9}$

9. $2^{2x-1} = 8$

Write as logarithmic equations.

10. $2^8 = 256$

11. $4^{-\frac{1}{2}} = \dfrac{1}{2}$

Write as exponential equations.

12. $\log_5 625 = 4$

13. $\log_3 \dfrac{1}{27} = -3$

Evaluate with a calculator. Express the answer to the nearest ten-thousandth.

14. log 4.8

15. ln 32

Solve for *x* by using an equivalent exponential expression and a calculator.

16. $\log_4 x = -2$

17. $\log_6 216 = x$

Evaluate the following with a calculator. Round to ten-thousandths.

18. $\log_7 2$

19. $\log_8 21$

20. The formula for the population growth of a certain species of insect in a controlled research environment is

$$P = 1,000,000 \, e^{0.05t}$$

where *t* = time in weeks. Find the projected population after **(a)** 2 weeks and **(b)** 3 weeks.

21. The revenue in thousands of dollars from sales of a product is approximated by the formula

$$S = 125 + 83 \log (5t + 1)$$

where *t* is the number of years after a product was marketed. Find the projected revenue from sales of the product after 3 years.

22. The height in meters of the male members of a certain group is approximated by the formula

$$h = 0.4 + \log t$$

where *t* represents age in years for the first 20 years of life ($1 \le t \le 20$). Find the projected height of a 10-year-old male.

23. Find the accumulated amount on an investment of $5,000 at a rate of 5.8% per year, compounded annually for 2 years. Use the formula

$$A = P\left(1 + \frac{r}{n}\right)^{nt}$$

24. What is the future value of an ordinary annuity of $300 every 3 months for 4 years at 8% annual interest, compounded quarterly?

25. Find the future value of an ordinary annuity of $9,000 per year for 2 years at 15% annual interest.

26. What is the future value of an ordinary annuity of $985 every 6 months for 8 years at 8% annual interest, compounded semiannually?

27. What is the sinking fund payment required at the end of each year to accumulate to $125,000 in 16 years at 4% annual interest?

28. Find the compound interest on a loan of $3,000 for 1 year at 12% annual interest if the interest is compounded quarterly.

29. Find the effective interest rate for the loan described in Exercise 28.

30. Find the compound interest on an investment of $2,000 invested at 5.75% for 28 years compounded quarterly.

31. If you were offered $600 today or $680 in 1 year, which would you accept if money can be invested at 12% annual interest, compounded monthly?

32. Mabel Langston needs $12,000 in 10 years for her daughter's college education. How much must be invested today at 8% annual interest, compounded semiannually, to have the needed funds?

33. Which of the two options yields the greatest return on your investment of $2,000?
Option 1: 8% annual interest compounded quarterly for 4 years
Option 2: $8\frac{1}{4}$% annual interest compounded annually for 4 years

34. Harvey Barton plans to buy a house in 4 years. He will make an $8,000 down payment on the property. How much should he invest today at 6% annual interest, compounded quarterly, to have the required amount in 4 years?

35. If you invest $1,000 today at 4% annual interest compounded continuously, how much will you have after 20 years?

36. If you invest $2,000 today at 8% annual interest, compounded quarterly, how much will you have after 3 years?

37. How much money should Bryan Trailer Sales set aside today to have $15,000 in 1 year to purchase a forklift if the interest rate is 10.4%, compounded annually?

38. Ginger Canoy has purchased a home for $122,000. She plans to finance $100,000 for 15 years at $5\frac{1}{2}$% interest. Calculate the monthly payment and the total interest.

17

Inequalities and Absolute Values

Focus on Careers

Occupational therapists supervise the work of occupational therapist assistants who provide rehabilitative services to persons with mental, physical, emotional, or developmental impairments. Because more and more hands-on therapy work is being delegated to occupational therapist assistants and because the number of people with disabilities is increasing, persons choosing this career may expect excellent opportunities for a rewarding career of service to others.

An associate degree or certificate from an accredited community college or technical school is usually required for occupational therapist assistants. Students take courses in areas of basic medical terminology, anatomy, physiology, mental health, adult physical disabilities, gerontology, and pediatrics. Most states regulate occupational therapist assistants and require them to pass a national certification examination after graduation.

Employment for this career is expected to grow much faster than most other occupations through 2012. Job growth will result from the increasing number of individuals with disabilities and from an aging population who will need these services.

The highest 10% of occupational therapist assistants earned more than $48,480 in 2002 and the middle 50% earned between $31,090 and $43,030 in the same year.

Source: *Occupational Outlook Handbook,* 2004–2005 Edition. U.S. Department of Labor, Bureau of Labor Statistics.

17–1 | Inequalities and Sets

Learning Outcomes

1 Use set terminology.

2 Show inequalities on a number line and write inequalities in interval notation.

In Chapter 1 we defined an **inequality** as a mathematical statement that quantities *are not equal.* The symbol $\neq$ is read "is not equal to."

$$5 \neq 7 \qquad 5 \textit{ is not equal to } 7.$$

More specific inequality symbols are $<$ (is less than) and $>$ (is greater than).

$$5 < 7 \qquad 5 \textit{ is less than } 7. \qquad 7 > 5 \qquad 7 \textit{ is greater than } 5.$$

The symbol $\leq$ indicates *is less than or equal to,* and the symbol $\geq$ indicates *is greater than or equal to.* As in equations, we represent missing amounts in inequalities by letters.

$$x \leq 7 \qquad x \textit{ is less than or equal to } 7. \qquad x \geq 5 \qquad x \textit{ is greater than or equal to } 5.$$

To **solve an inequality,** we find the value or set of values of the unknown quantity that makes the statement true.

1 Use Set Terminology.

A **set** is a group or collection of items. For example, a set of days of the week that begin with the letter T includes Tuesday and Thursday. In this chapter, we examine sets of numbers. Numbers that belong to a set are called **members or elements** of a set. The description of a set clearly distinguishes between the elements that belong to the set and those that do not belong. This description can be given in words or by using **set notation.**

To illustrate set notation, we examine the set of whole numbers between 1 and 8. One notation is to make a list or **roster** of the elements of a set. These elements are enclosed in braces and separated with commas.

$$\text{Set of whole numbers between 1 and 8} = \{2, 3, 4, 5, 6, 7\}$$

TIP — Common Symbols for Sets

The following capital letters are used to denote the indicated set of numbers:

N = natural numbers W = whole numbers
Z = integers Q = rational numbers
I = irrational numbers R = real numbers
M = imaginary numbers C = complex numbers

Symbols that substitute for phrases that are often used in describing sets are

| is read "such that"
∈ is read "is an element of"

Another method of illustrating a set is **set-builder notation.** The elements of the set are written in the form of an inequality using a variable to represent all the elements of the set.

Set of whole numbers between 1 and 8 = $\{x \mid x \in W \text{ and } 1 < x < 8\}$

This statement is read "the set of values of *x such that* each *x is an element of* the set of whole numbers and *x* is between 1 and 8."

A special set is the empty set. The **empty set** is a set containing no elements. Symbolically, the empty set is identified as { } or ϕ (pronounced "fee"). An example of an empty set is the set of whole numbers between 1 and 2. The set of rational numbers between 1 and 2 includes numbers like $1\frac{1}{2}$ and 1.3, but there are no whole numbers between 1 and 2. Thus, the set of whole numbers between 1 and 2 is the empty set.

EXAMPLE Answer the statements as true or false.

(a) 5 is an element of the set of whole numbers.
(b) $\frac{3}{4}$ is an element of the set of Z.
(c) $-8 \in$ the set of real numbers with the property $\{x \mid x < -5\}$.
(d) 3.7 is an element of the set of rational numbers with the property $\{x \mid x > 3.7\}$.
(e) The set of prime numbers that are evenly divisible by 2 is an empty set.

(a) 5 is an element of the set of whole numbers. **True.**
(b) $\frac{3}{4}$ is an element of the set of integers. **False.** Integers include only whole numbers and their opposites.
(c) -8 is an element of the set of real numbers with the property $\{x \mid x < -5\}$. **True.** -8 is a real number and it is less than -5. Both conditions are satisfied.
(d) 3.7 is an element of the set of rational numbers with the property $\{x \mid x > 3.7\}$. **False.** 3.7 is a rational number. A number can equal itself, but a number cannot be greater than itself.
(e) The set of prime numbers evenly divisible by 2 is the empty set. **False.** 2 is a prime number, and 2 is divisible by 2. Therefore, the set contains the element 2.

2 Show Inequalities on a Number Line and Write Inequalities in Interval Notation.

The set of numbers that is represented by an inequality in one variable can be shown visually on a number line.

To graph an inequality in one variable:

1. Determine the boundaries of the inequality, if they exist.
2. Locate the boundaries on a number line.
   (a) If the inequality is $<$ or $>$, the boundary is *not* included. Represent this with a parenthesis.
   (b) If the inequality is $\leq$ or $\geq$, the boundary *is* included. Represent this with a bracket.
3. Shade the portion of the number line between the boundaries. If there is no boundary in one or either direction, the graph continues indefinitely in that direction.

Another type of notation used to represent inequalities is **interval notation.** The two *boundaries* are separated by a comma and enclosed with a symbol that indicates whether the boundary is included or not. If there is no boundary in one or both directions, an infinity symbol, ∞, is used. We can say the set is **unbounded.**

To write an inequality in interval notation:

1. Determine the boundaries of the interval, if they exist.
2. Write the left boundary or $-\infty$ first and write the right boundary or $+\infty$ second. Separate the two with a comma.
3. Enclose the boundaries with the appropriate grouping symbols. A parenthesis "(" or ")" indicates that the boundary is *not* included. A bracket "[" or "]" indicates that a boundary *is* included.

EXAMPLE Represent the sets of numbers on the number line and by using interval notation.

(a) $1 < x < 8$ (b) $1 \leq x \leq 8$ (c) $x < 3$
(d) $x \geq 3$ (e) all real numbers

(a) $1 < x < 8$ 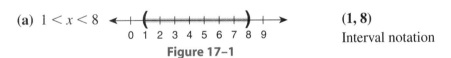   **(1, 8)**
Interval notation

Figure 17–1

Parentheses are used for both boundaries to indicate that 1 and 8 are *not* included in the solution set.

(b) $1 \leq x \leq 8$ **[1, 8]**
Interval notation

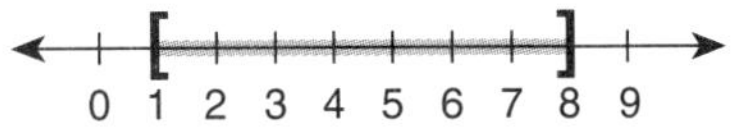

Figure 17–2

Brackets are used for both boundaries to indicate that 1 and 8 *are* included in the solution set.

(c) $x < 3$ **$(-\infty, 3)$**
Interval notation

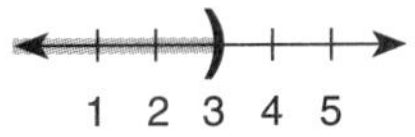

Figure 17–3

The left boundary is $-\infty$, and the right boundary is 3 but is not in the solution set so a parenthesis is used.

(d) $x \geq 3$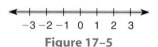

Figure 17–4

$[3, \infty)$
Interval notation

The left boundary is 3 and since 3 *is* included in the solution set, a bracket is used. The right boundary is ∞.

(e) All real numbers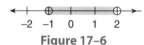

Figure 17–5

$(-\infty, \infty)$
Interval notation

All real numbers is an unbounded set of numbers, thus parentheses in the interval notation are used. The number line extends indefinitely in both directions.

TIP

Alternative Representation of an Inequality on a Number Line

Boundaries on a number line can also be indicated with open or shaded circles (Fig. 17–6).

$-1 \leq x < 2$

Figure 17–6

$[-1, 2)$

SECTION 17–1 SELF-STUDY EXERCISES

1 Answer the statements as true or false.

1. 0 is an element of the natural numbers.

2. $\frac{5}{8} \in N$

3. $-6 \in Z$

4. 5.3 is an element of the set of rational numbers with the property $\{x \mid x \leq 5.3\}$.

Write the sets as a roster.

5. The whole numbers between 5 and 12.

6. The even numbers between 3 and 8.

7. The whole numbers that are multiples of 5 and between 10 and 40.

8. The negative integers greater than -5.

Write the sets in set-builder notation.

9. $\{1, 3, 5, 7, 9\}$

10. $\{-7, -5, -3, -1\}$

2 Represent the sets on the number line and by using interval notation.

11. $5 < x < 9$

12. $-7 < x < -3$

13. $-5 \leq x \leq -3$

14. $8 \leq x \leq 12$

15. $x < 7$

16. $x < -2$

17. $x \geq 2$

18. $x \leq 3$

19. $x > 5$

20. $x \geq -3$

21. $x \leq -2$

22. All real numbers greater than -6.

23. University Trailer Company had sales of $843,000 for the previous year. The projected sales for the current year are more than the previous year, but less than $1,000,000, which is projected for the next year. Express sales for the current year using interval notation.

24. All real numbers with a minimum of 4 and a maximum of 18.

25. College classes generally must have a minimum number of students to avoid cancellation. If that number is 12, write and graph an inequality to represent the number of students that are in a cancelled class.

17–2 | Solving Linear Inequalities

Learning Outcomes
1 Solve a linear inequality with one variable.
2 Graph a linear inequality with two variables.
3 Solve a system of linear inequalities by graphing.

1 Solve a Linear Inequality with One Variable.

The procedures for solving inequalities are similar to the procedures for solving equations. The solution to an inequality is a *set* of numbers that satisfies the conditions of the statement.

The statement $x \leq 9$ means the *solution set* is 9 or any number less than 9.

Compare Figs. 17–7 and 17–8. In Fig. 17–7, the boundary 9 is *not* part of the solution set; thus, 9 is represented on the number line with a parenthesis. In Fig. 17–8, the boundary 9 *is* part of the solution set; thus, 9 is represented on the number line with a bracket.

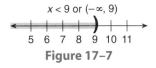

$x < 9$ or $(-\infty, 9)$

5 6 7 8 9 10 11

Figure 17–7

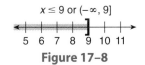

$x \leq 9$ or $(-\infty, 9]$

5 6 7 8 9 10 11

Figure 17–8

Look at the similarities in solving a linear equation and a linear inequality.

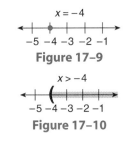

$x = -4$

−5 −4 −3 −2 −1

Figure 17–9

$x > -4$

−5 −4 −3 −2 −1

Figure 17–10

EXAMPLE Solve the equation $4x + 6 = 2x - 2$ and the inequality $4x + 6 > 2x - 2$.

$4x + 6 = 2x - 2$	$4x + 6 > 2x - 2$	Sort.
$4x - 2x = -6 - 2$	$4x - 2x > -6 - 2$	Combine.
$2x = -8$	$2x > -8$	Divide.
$\dfrac{2x}{2} = \dfrac{-8}{2}$	$\dfrac{2x}{2} > \dfrac{-8}{2}$	
$x = -4$	$x > -4$ or $(-4, \infty)$	Interval notation.

The solution is -4 (Fig. 17–9). The solution set is any number greater than, but not including, -4 (Fig. 17–10).

TIP   **Three Ways to Represent the Solution Set of Inequalities: Symbolically, Graphically, and Interval Notation**

In the preceding example, the solution to the inequality $4x + 6 > 2x - 2$ was written three ways.

- Symbolic representation: $x > -4$
- Graphical representation:
- Interval notation: $(-4, \infty)$

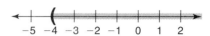

Figure 17–11

The symbolic representation evolves from the process of solving the inequality using the properties of inequalities and techniques for isolating the variable. Both the graphical representation and the interval notation help us interpret and visualize the solution (Fig. 17–11).

EXAMPLE   Solve the equation $3x + 5 = 7x + 13$ and the inequality $3x + 5 < 7x + 13$.

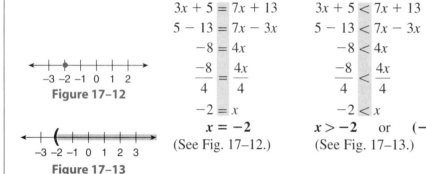

Figure 17–12

Figure 17–13

$3x + 5 = 7x + 13$	$3x + 5 < 7x + 13$	Sort.
$5 - 13 = 7x - 3x$	$5 - 13 < 7x - 3x$	Combine.
$-8 = 4x$	$-8 < 4x$	Divide.
$\dfrac{-8}{4} = \dfrac{4x}{4}$	$\dfrac{-8}{4} < \dfrac{4x}{4}$	
$-2 = x$	$-2 < x$	Interpret solution.
$x = -2$	$x > -2$ or $(-2, \infty)$	We prefer writing the variable on
(See Fig. 17–12.)	(See Fig. 17–13.)	the left. Saying -2 *is less than x* is the same as saying *x is greater than* -2.

The preceding example illustrates a very important difference in solving equations and solving inequalities. The sides of an equation can be interchanged without making any changes in the equal sign. This interchangeability is due to the *symmetric property of equality;* that is, if $a = b$, then $b = a$. The symmetric property does *not* apply to inequalities.

Interchanging the sides of an inequality:

When the sides of an inequality are interchanged, the sense of the inequality is reversed.

$$\text{If } a < b, \text{ then } b > a. \qquad \text{If } a > b, \text{ then } b < a.$$
$$\text{If } a \leq b, \text{ then } b \geq a. \qquad \text{If } a \geq b, \text{ then } b \leq a.$$

The **sense of an inequality** is the appropriate comparison symbol: less than, greater than, less than or equal to, and greater than or equal to.

To determine the effect of multiplying or dividing an inequality by a negative number, look at specific numbers and the number line (Fig. 17–14.). We start with a true statement.

$2 < 5$ 2 is to the left of 5 on the number line.

Then, multiply each side of the inequality by -1.

$-1(2) \overset{?}{<} -1(5)$ Is $-1(2) < -1(5)$?

$-2 \overset{?}{<} -5$ This is a false statement: -2 is to the right of -5 and is greater than -5.

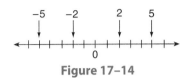

Figure 17–14

To make the statement true, the sense of the inequality must be reversed.

$$-2 > -5$$

Multiplying or dividing an inequality by a negative number:

If both sides of an inequality are multiplied or divided by a negative number, the sense of the inequality is reversed.

If $a < b$, then $-a > -b$.   If $a > b$, then $-a < -b$.
If $a \le b$, then $-a \ge -b$.   If $a \ge b$, then $-a \le -b$.

To solve linear inequalities:

1. Follow the same sequence of steps that is normally used to solve a similar linear equation.
2. The sense of the inequality remains the same unless one of the following situations occurs:
   (a) The sides of the inequality are interchanged.
   (b) The steps used in solving the inequality require that the entire inequality (both sides) be multiplied or divided by a negative number.
3. If either situation (a) or (b) in Step 2 occurs in solving an inequality, *reverse* the sense of the inequality; that is, less than ($<$) becomes greater than ($>$), and vice versa.

EXAMPLE   Solve the inequality, $4x - 2 \le 3(25 - x)$.

$4x - 2 \le 3(25 - x)$ Distribute.

$4x - 2 \le 75 - 3x$ Collect variable terms on the left and constants on the right (sort).

$4x + 3x \le 75 + 2$ Combine like terms.

$7x \le 77$ Divide by the coefficient of the variable. The coefficient is positive.

$\dfrac{7x}{7} \le \dfrac{77}{7}$

$x \le 11$ or $(-\infty, 11]$ Solution set. (See Fig. 17–15.)

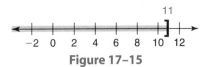

Figure 17–15

EXAMPLE Solve the inequality, $2x - 3(x + 2) > 5x - (x - 5)$.

Figure 17–16

$2x - 3(x + 2) > 5x - (x - 5)$	Distribute.
$2x - 3x - 6 > 5x - x + 5$	Combine like terms.
$-x - 6 > 4x + 5$	Sort.
$-x - 4x > 5 + 6$	Combine like terms.
$-5x > 11$	Divide by the coefficient of the variable. The coefficient is negative.
$\dfrac{-5x}{-5} < \dfrac{11}{-5}$	Both sides were divided by a negative number, so reverse the sense of the inequality.
$x < -\dfrac{11}{5}$ or $\left(-\infty, -\dfrac{11}{5}\right)$	

EXAMPLE An electrically controlled thermostat is set so the heating unit automatically comes on and continues to run when the temperature is equal to or below 72°F. At what Celsius temperatures will the heating unit come on? One formula relating Celsius and Fahrenheit temperatures is $°F = \frac{9}{5}°C + 32$.

Using the formula $°F = \frac{9}{5}°C + 32$, the heating unit will operate when the expression $\frac{9}{5}°C + 32$ is less than or equal to 72.

$$\frac{9}{5}°C + 32 \le 72$$

Estimation The Celsius value will be less than the Fahrenheit value.

$\dfrac{9}{5}°C + 32 \le 72$	Apply the addition axiom to isolate the variable term.
$\dfrac{9}{5}°C \le 72 - 32$	Combine like terms.
$\dfrac{9}{5}°C \le 40$	Multiply by the reciprocal of the coefficient of the variable term.
$\dfrac{5}{9}\left(\dfrac{9}{5}\right)°C \le \left(\dfrac{5}{9}\right)(40)$	
$°C \le \dfrac{200}{9}$	Interpret solution.
$°C \le 22.22$ or $(-\infty, 22.22]$	To nearest hundredth

Interpretation **The heating unit comes on and continues to run if the temperature is less than or equal to 22.22°C.**

2 Graph a Linear Inequality in Two Variables.

When we solved inequalities in one variable the solution was a set of numbers that we can represent graphically. The **solution of linear inequalities with two variables** is a set of coordinate points.

662   Chapter 17 / Inequalities and Absolute Values

The graphical solution for a linear inequality ($<$ or $>$) includes all points on one side of the line representing the graph of the similar equation. If either the "less than or equal" ($\leq$) or "greater than or equal" ($\geq$) symbol is used, the points on the boundary line are also included in the solution set.

To graph a linear inequality with two variables:

1. Find the boundary by graphing an equation that substitutes an equal sign for the inequality symbol. Make a solid line if the boundary is included ($\leq$ or $\geq$) or a dashed line if the boundary is not included ($<$ or $>$) in the solution set.
2. Test any point that is *not* on the boundary line.
3. If the test point makes a true statement with the original inequality, shade the side containing the test point.
4. If the test point makes a false statement with the original inequality, shade the side opposite the side containing the test point.

EXAMPLE   Graph the inequality $2x + y < 3$.

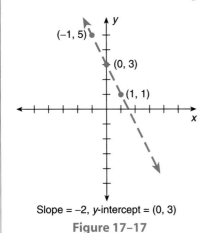

Slope = –2, y-intercept = (0, 3)

Figure 17–17

Graph the equation $2x + y = 3$ to establish the boundary. The boundary will not be included in the solution set and is represented with a dashed line (Fig. 17–17).

$2x + y = 3$ Solve the equation for y.

$\qquad y = -2x + 3$

y-intercept = 3 or (0, 3)   Constant term

slope = -2 or $\dfrac{-2}{1}$ or $\dfrac{2}{-1}$ Coefficient of x

Select one point, on either side of the boundary, to use as a test point. Suppose that we choose (2, 2) as the point above the boundary.

Is $2x + y < 3$ when $x = 2$ and $y = 2$?   Substitute values.
$\qquad 2(2) + 2 < 3$ Simplify.
$\qquad\quad 4 + 2 < 3$
$\qquad\qquad\quad 6 < 3$ False statement

Thus, the side of the line including the point (2, 2) is *not* in the solution set. Shade the opposite side. See Fig. 17–18.

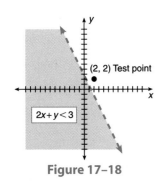

$2x + y < 3$

(2, 2) Test point

Figure 17–18

Solid or Dashed Boundary Lines

Examine the boundaries and shaded regions of the graphs in Figures 17–19, 17–20, and 17–21.

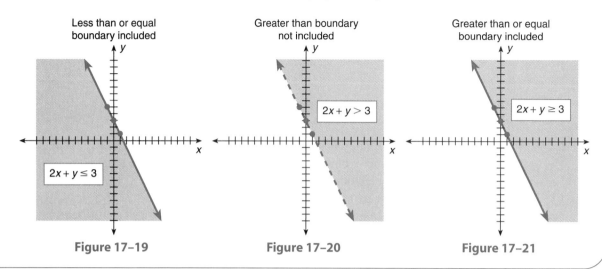

Less than or equal
boundary included

$2x + y \leq 3$

Figure 17–19

Greater than boundary
not included

$2x + y > 3$

Figure 17–20

Greater than or equal
boundary included

$2x + y \geq 3$

Figure 17–21

Selecting Test Points

Choose numbers that are easy to work with when selecting a test point. If the boundary line does not pass through the origin, a good point to use is $(0, 0)$.

EXAMPLE   Graph the inequality $y \geq 3x + 1$.

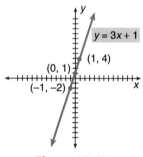

Figure 17–22

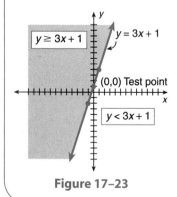

Figure 17–23

Graph the line $y = 3x + 1$.

y-intercept $= (0, 1)$ Constant term.

$\text{slope} = 3 \text{ or } \dfrac{3}{1} \text{ or } \dfrac{-3}{-1}$ Coefficient of x.

The boundary line is shown in Fig. 17–22 as a solid line because the inequality $y \geq 3x + 1$ *does* include the boundary.

The point $(0, 0)$ is not on the boundary line and it is on the right side of the boundary. We use $(0, 0)$ as our test point.

Is $y \geq 3x + 1$ when $x = 0$ and $y = 0$?   Substitute values.

$0 \geq 3(0) + 1$ Simplify.

$0 \geq 0 + 1$

$0 \geq 1$ False statement.

Thus, as shown in Fig. 17–23, the boundary and all points on the side of the boundary opposite the test point $(0, 0)$ are in the solution set.

General Observation About Inequalities

When the inequality is in the form $y < mx + b$ or $y \le mx + b$, the solution set includes the portion of the y-axis and all points *below* the boundary line. When an inequality is in the form $y > mx + b$ or $y \ge mx + b$, the solution set includes the portion of the y-axis and all points *above* the boundary line (Fig. 17–24).

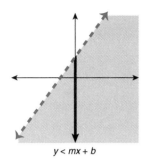

$y < mx + b$

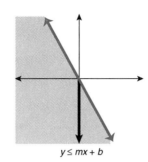

$y \le mx + b$

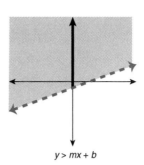

$y > mx + b$

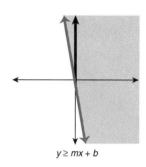

$y \ge mx + b$

Figure 17–24

3 Solve a System of Linear Inequalities by Graphing.

When two intersecting lines are plotted on the same set of axes, the rectangular coordinate system is separated into four regions (Fig. 17–25). The solution set of a system of linear inequalities will include all the points in one of the regions.

Figure 17–25

To solve a system of two linear inequalities with two variables by graphing:

1. Graph each inequality on the same pair of axes.
2. The solution set of the system will be the *overlapping* region of the solution sets of the two inequalities.

EXAMPLE Shade the portion on the graph that is represented by the following conditions: $y \le 3x + 5$ and $x + y > 7$.

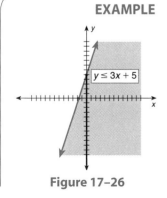

$y \le 3x + 5$

Figure 17–26

Graph the inequality $y \le 3x + 5$:

The boundary $y = 3x + 5$ has a slope of 3 and a y-intercept of 5. The boundary will be included in the solution set.

For the *less than* relationship, shade the region that includes the part of the y-axis that is *below* the boundary (Fig. 17–26).

Graph the inequality, $x + y > 7$:

The boundary $x + y = 7$ in slope-intercept form is $y = -x + 7$. The slope is -1 and the y-intercept is 7. The boundary will not be included.

17–2 Solving Linear Inequalities   665

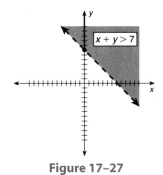

For the *greater than* relationship, shade the region that includes the part of the y-axis that is *above* the graph (Fig. 17–27).

Find the solution set common to both inequalities. Visualize both graphs on the same axes. The region that has overlapping shading represents the points that satisfy *both* conditions and forms the solution set for the system of inequalities (see Fig. 17–28).

Figure 17–27

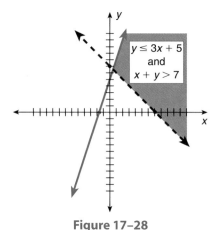

Figure 17–28

SECTION 17–2 SELF-STUDY EXERCISES

1 Solve the inequalities. Show the solution set on a number line and by using interval notation.

1. $y - 3 > 5$
2. $x + 7 < 8$
3. $x - 9 \leq -12$
4. $3x + 7x < 60$
5. $5a - 6a \leq 3$
6. $y + 3y \leq 32$
7. $b + 6 > 5$
8. $5t - 18 < 12$
9. $2y + 7 < 17$
10. $3a - 8 \geq 7a$
11. $4x + 7 \leq 8$
12. $2t + 6 \leq t + 13$
13. $4y - 8 > 2y + 14$
14. $3(7 + x) \geq 30$
15. $6(3x - 1) < 12$
16. $2x > 7 - (x + 6)$
17. $4a + 5 \leq 3(2 + a) - 4$
18. $15 - 3(2x + 2) > 6$

19. Kevin Presley sold $196 more than twice as much merchandise as Robyn Presley. If Kevin sold at least $52,800, how much did Robyn sell? Write an inequality to represent the facts and solve.

20. A supplier prices one brand of recordable CD at $3.60 and another brand of recordable CD at more than twice this price. Write an inequality to represent the facts and express the cost of the more expensive CD. Show the solution set on a number line.

2 Graph the following linear inequalities using test points.

21. $x - y < 8$
22. $x + y > 4$
23. $3x + y < 2$
24. $2x + y \leq 1$
25. $x + 2y < 3$
26. $2x + 2y \geq 3$
27. $y \geq 2x - 3$
28. $y \geq -3x + 1$
29. $y > \frac{1}{2}x - 3$
30. $y > -\frac{3}{2}x + \frac{1}{2}$

3 Graph each system of inequalities and shade the portion on the graph represented by the solution set.

31. $y < 2x + 1$ and $x + y > 5$
32. $x + y \geq 3$ and $x - y \geq 2$
33. $x - 2y \leq -1$ and $x + 2y \geq 3$
34. $x + y > 4$ and $x - y > -3$
35. $3x - 2y < 8$ and $2x + y \leq -4$
36. $x - 2y < -6$ and $2x + y \leq 5$

Learning Outcomes

1 Identify subsets of sets and perform set operations.
2 Solve compound inequalities with the conjunction condition.
3 Solve compound inequalities with the disjunction condition.

In many applications of inequalities, more than one condition is placed on the solution. As an example, when measurements are made, a range of acceptable values is specified. All measurements are approximations, and the range of acceptable values is generally stated as a tolerance. If an acceptable measurement is specified to be within ±0.005 in., then the range of acceptable values can be from 0.005 in. *less than* the ideal measurement to 0.005 in. *more than* the ideal measurement. This range can be stated symbolically as a compound inequality. A **compound inequality** is a mathematical statement that combines two statements of inequality.

The conditions placed on a compound inequality may use the connective **and** to indicate both conditions must be met simultaneously. Such compound inequalities may be written as a continuous statement. If an ideal measurement is 3 in. with a tolerance of ±0.005 in., the range of acceptable values would be

$$3 - 0.005 \leq x \leq 3 + 0.005$$

$$2.995 \leq x \leq 3.005$$

The conditions placed on a compound inequality may use the connective **or** to indicate that either condition may be met. Such compound inequalities *must* be written as two separate statements using the connective *or*.

1 Identify Subsets of Sets and Perform Set Operations.

Before solving compound inequalities, let's look at additional properties of sets. A concept related to sets is the concept of subsets. The symbol ⊂ is read "is a subset of." A set is a **subset** of a second set if every element of the first set is also an element of the second set. If $A = \{1, 2, 3\}$ and $B = \{2\}$, then B is a subset of A. This is written in symbols as $B \subset A$ and is illustrated in Fig. 17–29.

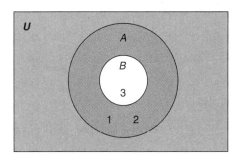

$B \subset A$

Figure 17–29

Two special sets are the universal set and the empty set. The **universal set** includes all the elements for a given description. The universal set is sometimes identified as U. The **empty set** is a set containing no elements. The empty set is identified as { } or φ.

EXAMPLE Using the given set definitions, answer the statements (a)–(h) as true or false.

$$U = \{0, 1, 2, 3, 4, 5, 6, 7, 8, 9\}; A = \{1, 3, 5, 7, 9\};$$

$$B = \{0, 2, 4, 6, 8\}; C = \{1, 2, 3\}; D = \{1\}; E = \{ \quad \}$$

(a) $A \subset U$ (b) $B \subset U$ (c) $A \subset B$ (d) $C \subset A$
(e) $C \subset U$ (f) $D \subset A$ (g) $E \subset U$ (h) $E \subset B$

(a) $A \subset U$ **True;** every element of A is also an element of U.
(b) $B \subset U$ **True;** every element of B is also an element of U.
(c) $A \subset B$ **False;** 1, 3, 5, 7, and 9 are not elements of B.
(d) $C \subset A$ **False;** 2 is not an element of A.
(e) $C \subset U$ **True;** every element of C is also an element of U.
(f) $D \subset A$ **True;** 1 is an element of A.
(g) $E \subset U$ **True;** the empty set is a subset of every set.
(h) $E \subset B$ **True;** the empty set is a subset of every set.

Two common set operations are union and intersection. The **union** of two sets is a set that includes all elements that appear in *either* of the two sets. Union is generally associated with the condition "or." The symbol for union is $\cup$. The colored portion in Fig. 17–30 represents $A \cup B$. The **intersection** of two sets is a set that includes all elements that appear in *both* of the two sets. Intersection is generally associated with the condition "and." The symbol for intersection is $\cap$. The colored portion of Fig. 17–31 represents $A \cap B$.

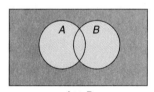

$A \cup B$

Figure 17–30

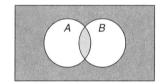

$A \cap B$

Figure 17–31

EXAMPLE If $A = \{1, 2, 3, 4, 5\}, B = \{1, 3, 5, 7, 9\}$, and $C = \{2, 4, 6, 8, 10\}$, list the elements in the following sets.

(a) $A \cup B$ (b) $A \cup C$ (c) $B \cup C$ (d) $A \cap B$ (e) $A \cap C$ (f) $B \cap C$

(a) $A \cup B = \{1, 2, 3, 4, 5, 7, 9\}$ All elements in either A or B
(b) $A \cup C = \{1, 2, 3, 4, 5, 6, 8, 10\}$ All elements in either A or C
(c) $B \cup C = \{1, 2, 3, 4, 5, 6, 7, 8, 9, 10\}$ All elements in either B or C
(d) $A \cap B = \{1, 3, 5\}$ Elements common to both A and B
(e) $A \cap C = \{2, 4\}$ Elements common to both A and C
(f) $B \cap C = \{ \quad \}$ **or** ϕ No elements common to both B and C

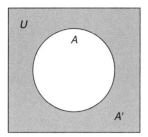

Figure 17–32

Another set operation is complement. The *complement* of a set is a set that includes every element of the universal set that is *not* an element of the given set. The symbol for the complement of a set is ′ and is read "prime." A' is represented by the colored portion of Fig. 17–32.

EXAMPLE If $U = \{0, 1, 2, 3, 4, 5\}$, $A = \{1, 3, 5\}$, $B = \{0, 2, 4\}$, and $C = \{1, 2, 3\}$, list the elements in the following sets.

(a) A' (b) B' (c) C' (d) $(A \cup B)'$ (e) $(A \cap C)'$

(a) $A' = \{0, 2, 4\}$ Elements of U not in A.
(b) $B' = \{1, 3, 5\}$ Elements of U not in B.
(c) $C' = \{0, 4, 5\}$ Elements of U not in C.
(d) $(A \cup B)'$
     $A \cup B = \{0, 1, 2, 3, 4, 5\}$ Elements in either A or B.
     $(A \cup B)' = \{\quad\}$ **or** ϕ Elements of U not in either A or B.
(e) $(A \cap C)'$
     $A \cap C = \{1, 3\}$ Elements common to both A and C.
     $(A \cap C)' = \{0, 2, 4, 5\}$ Elements of U not common to both A and C.

2   Solve Compound Inequalities with the Conjunction Condition.

As we said before, a compound inequality is a statement that places more than one condition on the variable of the inequality. The solution set is the set of values for the variable that meets all conditions of the problem. One type of compound inequality is *conjunction*. A **conjunction** is an intersection, or "and," set relationship. Both conditions must be met simultaneously in a conjunction. A conjunction can be written as a continuous statement.

Property for conjunctions:

When a, b, and x are real numbers:

$$\text{If } a < x \text{ and } x < b, \text{ then } a < x < b.$$

$$\text{If } a > x \text{ and } x > b, \text{ then } a > x > b.$$

Similar compound inequalities may also use $\leq$ and $\geq$.

$$\text{If } a \leq x \text{ and } x \leq b, \text{ then } a \leq x \leq b.$$

$$\text{If } a \geq x \text{ and } x \geq b, \text{ then } a \geq x \geq b.$$

To solve a compound inequality that is a conjunction:

1. Separate the compound inequality into two inequalities using the conditions of the conjunction.
2. Solve each inequality.
3. Determine the solution set that includes the *intersection* of the solution sets of the two inequalities.

EXAMPLE Find the solution set for each compound inequality. Graph the solution set on a number line and write the solution set in interval notation.

(a) $5 < x + 3 < 12$ (b) $-7 \le x - 7 \le 1$
(c) $-6 < 3x < -3$ (d) $17 \le 5x + 7 \le 32$

(a) $5 < \boxed{x + 3} < 12$	Separate into two inequalities.
$5 < \boxed{x + 3}$ and $\boxed{x + 3} < 12$	Solve each inequality.
$5 - 3 < x$ $x < 12 - 3$	
$2 < x$ $x < 9$	Find the overlap.

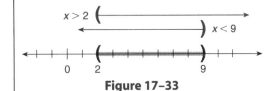

Figure 17–33

Figure 17–33 shows the solution set graphically. The solution set as a continuous statement is **$2 < x < 9$ and in interval notation, (2, 9)**.

(b) $-7 \le \boxed{x - 7} \le 1$	Separate into two inequalities.
$-7 \le \boxed{x - 7}$ and $\boxed{x - 7} \le 1$	Solve each inequality.
$-7 + 7 \le x$ $x \le 1 + 7$	
$0 \le x$ $x \le 8$	Find the overlap.

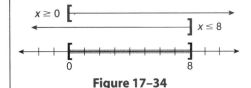

Figure 17–34

Figure 17–34 shows the solution set graphically. The solution set as a continuous statement is **$0 \le x \le 8$ and in interval notation, [0, 8]**.

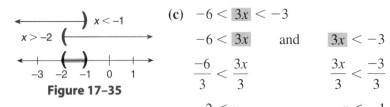

Figure 17–35

(c) $-6 < \boxed{3x} < -3$	Separate into two inequalities.
$-6 < \boxed{3x}$ and $\boxed{3x} < -3$	Solve each inequality.
$\dfrac{-6}{3} < \dfrac{3x}{3}$ $\dfrac{3x}{3} < \dfrac{-3}{3}$	
$-2 < x$ $x < -1$	Find the overlap.

Figure 17–35 shows the solution set graphically. The solution set is **$-2 < x < -1$ and in interval notation is $(-2, -1)$**.

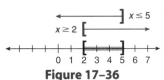

Figure 17–36

(d) $17 \le \boxed{5x + 7} \le 32$	Separate into two inequalities.
$17 \le \boxed{5x + 7}$ and $\boxed{5x + 7} \le 32$	Solve each inequality.
$17 - 7 \le 5x$ $5x \le 32 - 7$	
$10 \le 5x$ $5x \le 25$	
$2 \le x$ $x \le 5$	Find the overlap.

Figure 17–36 shows the solution set graphically. The solution set as a continuous statement is **$2 \le x \le 5$ and in interval notation, [2, 5]**.

Greater Than Versus Less Than

Even though continuous compound inequalities can be written using *greater than* or *less than* symbols, using *less than* symbols follows the natural positions of the boundaries on the number line. If $a > b > c$, then $c < b < a$. For instance, if $3 > 0 > -2$, then $-2 < 0 < 3$. Compound inequalities using the "less than" symbols are used most often. Do not mix symbols in a compound inequality.

As with equations and simple inequalities, a compound inequality may have no solution.

EXAMPLE   Find the solution set for the compound inequality, $7 < 2x - 1 < 5$.

Separate into two inequalities.

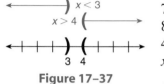

Figure 17–37

$$7 < 2x - 1 \quad \text{and} \quad 2x - 1 < 5$$
$$8 < 2x \qquad\qquad\qquad 2x < 6$$
$$4 < x \qquad\qquad\qquad\quad x < 3$$
$$x > 4 \qquad\qquad\qquad\qquad \text{Find the overlap.}$$

As shown in Fig. 17–37, there is no overlap, so it is impossible for both conditions to be met at the same time. **The compound inequality has no solution.**

The Solution of a Conjunction Relationship Is the Overlap

The importance of a graphical representation of the solution set of a conjunction is that you can visualize the overlap. Or, in the case of no solution to a conjunction, you can see there is no overlap.

A good way to emphasize the overlapped portion on a graph is to use two different colors of highlighters. Yellow and blue highlighters are good choices. The overlapped portion of the graph will be green.

3 Solve Compound Inequalities with the Disjunction Condition.

Disjunction is another type of compound inequality. A **disjunction** is a union, or "or," set relationship. Either condition is met in a disjunction. A disjunction uses the union symbol to indicate the intervals in the solution set.

To solve a compound inequality that is a disjunction:

1. Solve each simple inequality.
2. Determine the solution set that includes the *union* of the solution sets of the two simple inequalities.

EXAMPLE Find the solution set for each compound inequality. Graph the solution set on a number line and write the solution in symbolic and interval notation.

(a) $x + 3 < -2$ or $x + 3 > 2$ (b) $x - 5 \leq -3$ or $x - 5 \geq 3$

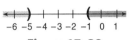

Figure 17–38

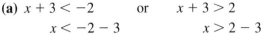

(a) $x + 3 < -2$ or $x + 3 > 2$ Solve each inequality.

$x < -2 - 3$ $x > 2 - 3$

$x < -5$ or $x > -1$ Symbolic notation. See Fig. 17–38.

$(-\infty, -5)$ ∪ $(-1, \infty)$ Interval notation

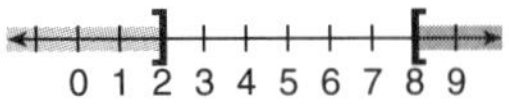

(b) $x - 5 \leq -3$ or $x - 5 \geq 3$ Solve each inequality.

$x \leq -3 + 5$ $x \geq 3 + 5$

$x \leq 2$ or $x \geq 8$ Symbolic notation. See Fig. 17–39.

$(-\infty, 2]$ ∪ $[8, \infty)$ Interval notation

Figure 17–39

TIP Let Your Graph Be Your Guide

The graphical representation of your solution set can be a great guide for writing the symbolic solution or the solution in interval notation.

Overlap on graph (Fig. 17–40)
The solution is one continuous statement.

Two intervals on graph (Fig. 17–41)
The solution is two separate statements.

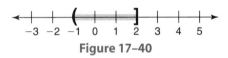

Figure 17–40

$-1 < x \leq 2$

$(-1, 2]$

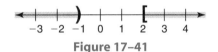

Figure 17–41

$x < 21$ or $x \geq 2$

$(-\infty, -1)$ ∪ $[2, \infty)$

SECTION 17–3 SELF-STUDY EXERCISES

1 Use the sets for Exercises 1–10.

$U = \{-5, -4, -3, -2, -1, 0, 1, 2, 3, 4, 5\}$, $A = \{-4, -2, 0\}$, $B = \{1, 2, 3, 4, 5\}$,
$C = \{2, -2\}$, $D = \{0\}$, $E = \{\ \}$

Answer each statement as true or false and justify your answer.

1. $B \subset U$ **2.** $C \subset A$ **3.** $E \subset B$ **4.** $A \subset B$

List the elements in each set.

5. $A \cap B$ **6.** $C \cup D$ **7.** $A \cup B$
8. A' **9.** $(A \cap D)'$ **10.** $(B \cup C)'$

2 Show the solution set of each inequality on a number line and by using symbolic and interval notation.

11. $6 \le 2x \le 8$ **12.** $15 \le 3x \le 21$ **13.** $-3 \le 3x \le 9$ **14.** $-3 \le 2x - 1 \le 3$
15. $-5 \le -x + 1 \le -7$ **16.** $1 < x - 2 < 5$ **17.** $0 < 5x < 15$ **18.** $4 \le 6 - x \le 8$
19. $6 < -2x < 12$ **20.** $1 < 6 - x < 3$ **21.** $-8 \le -3x - 2 \le 4$ **22.** $5 \le 5 + 7x \le 9$

3 Show the solution set of each inequality on a number line and by using symbolic and interval notation.

23. $x + 3 < 5$ or $x - 7 > 2$ **24.** $x - 1 \le 2$ or $x \ge 7$
25. $2x - 1 \le 7$ or $3x \ge 15$ **26.** $5 - x \le 2$ or $x + 1 \le 2$
27. $5x - 2 \le 3x + 1$ or $2x \ge 7$ **28.** $-2x > 8$ or $3x \ge 27$

29. The blueprint specifications for a part show it has a measure of 5.27 cm with a tolerance of ± 0.05 cm. Express the limit dimensions of the part with a compound inequality.

30. An automobile parts manufacturer makes a part that, according to the blueprint, measures 15.2 cm and has a tolerance of ± 0.5 cm. Write an inequality that expresses the range of measures an acceptable part must have.

17–4 | Solving Quadratic and Rational Inequalities in One Variable

Learning Outcomes

1 Solve quadratic inequalities in one variable.
2 Graph quadratic inequalities in two variables.
3 Solve rational inequalities in one variable.

Solving *quadratic* and *rational inequalities* is similar to solving quadratic and rational equations. The solution to these inequalities, like the solution to linear inequalities, is a *set* of numbers with specific boundaries.

1 **Solve Quadratic Inequalities in One Variable.**

Quadratic inequalities can be solved by all of the same methods we used to solve quadratic equations. You first treat the inequality as an equation to find the **critical values** or **boundaries.**

To solve quadratic inequalities:

1. Determine the critical values by solving an equation in which an equal sign is substituted for the inequality sign.
2. Plot the critical values on a number line to form the three regions.
3. Test *each* region of values by selecting any point within the region, substituting that value into the original inequality, solving the inequality, and deciding if the resulting inequality is a *true* statement.
4. The solution set for the quadratic inequality is the region or regions that produce a *true* statement in Step 3.
5. The critical values or boundary points are included in the solution set if the inequality is "inclusive" ($\le$ or $\ge$). The critical values or boundary points are not included in the solution set if the inequality is "exclusive" ($<$ or $>$).

EXAMPLE Solve the inequality $x^2 + 5x + 6 \leq 0$.

$x^2 + 5x + 6 = 0$ Write a similar equation and solve. Write in factored form.

$(x + 3)(x + 2) = 0$ Set each factor equal to 0.

$x + 3 = 0$ $x + 2 = 0$ Determine the critical values by solving each equation.

$x = -3$ $x = -2$ Critical values, or boundary points

Plot the critical values and label the corresponding regions (Fig. 17–42).

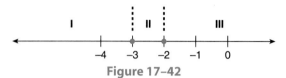

Figure 17–42

Region I: $x \leq -3$ To the left of the smaller critical value, -3

Region II: $-3 \leq x \leq -2$ Between the critical values, -3 and -2

Region III: $x \geq -2$ To the right of the larger critical value, -2

Test one point in each region.

Region I: $x \leq -3$	Region II: $-3 \leq x \leq -2$

Region I test point: $x = -4$ **Region II test point:** $x = -2.5$

$(x + 3)(x + 2) \leq 0$ $(-2.5 + 3)(-2.5 + 2) \leq 0$

$(-4 + 3)(-4 + 2) \leq 0$ $(0.5)(-0.5) \leq 0$

$(-1)(-2) \leq 0$ $-0.25 \leq 0$

$2 \leq 0$

The inequality is **false**, so Region I is *not* in the solution set.   The inequality is **true**, so Region II *is* in the solution set.

Region III: $x \geq -2$

Region III test point: $x = -1$

$(-1 + 3)(-1 + 2) \leq 0$

$(2)(1) \leq 0$

$2 \leq 0$

The inequality is **false**, so Region III is *not* in the solution set.

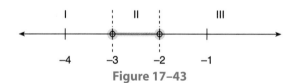

Figure 17–43

The solution set is $-3 \leq x \leq -2$ or $[-3, -2]$ (Fig. 17–43).

TIP

Selecting Test Points

Boundary points should not be used as test points. When possible, select integers as test points, and if zero is in a region, it makes an excellent test point.

    Chapter 17 / Inequalities and Absolute Values

EXAMPLE Solve the inequality $2x^2 + x - 6 > 0$.

$2x^2 + x - 6 = 0$	Write as an equation and solve.
$(2x - 3)(x + 2) = 0$	Write in factored form. Set each factor equal to 0.

$2x - 3 = 0 \qquad x + 2 = 0$ Determine the critical values by solving each equation.

$$2x = 3 \qquad\qquad x = -2$$

$$x = \frac{3}{2}$$

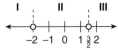

Figure 17–44

Plot the critical values and label the corresponding regions (see Fig. 17–44).

Region I: $x < -2$ Region II: $-2 < x < \frac{3}{2}$ Region III: $x > \frac{3}{2}$

Test each region.

Region I: $x < -2$	Region II: $-2 < x < \dfrac{3}{2}$
Region I test point: $x = -3$	**Region II test point**: $x = 0$
$(2x - 3)(x + 2) > 0$	$(2x - 3)(x + 2) > 0$
$[2(-3) - 3][-3 + 2] > 0$	$[2(0) - 3][0 + 2] > 0$
$(-6 - 3)(-1) > 0$	$(0 - 3)(2) > 0$
$(-9)(-1) > 0$	$(-3)(2) > 0$
$9 > 0$	$-6 > 0$

The inequality is true, so Region I *is* in the solution set.

The inequality is false, so Region II is *not* in the solution set.

Region III: $x > \dfrac{3}{2}$

Region III test point: $x = 2$

$$(2x - 3)(x + 2) > 0$$
$$[2(2) - 3][2 + 2] > 0$$
$$(4 - 3)(4) > 0$$
$$(1)(4) > 0$$
$$4 > 0$$

Figure 17–45

The inequality is true, so Region III *is* in the solution set.

The solution set is $x < -2$ or $x > \frac{3}{2}$. In interval notation the solution is $(-\infty, -2) \cup (\frac{3}{2}, \infty)$. See Fig. 17–45.

After examining several problems, you may notice a pattern for determining the solution set by inspection once the critical values are known. This pattern or generalization also can be applied to quadratic inequalities that do not factor.

Finding Solution Sets by Inspection for Quadratic Inequalities

Patterns and generalizations may allow short cuts, but they are appropriate only under special conditions. For quadratic inequalities, the inequality **must** be written in standard form ($ax^2 + bx + c < 0$ or $ax^2 + bx + c > 0$), where $a > 0$ (a is positive). The critical values are represented symbolically as s_1 and s_2, where $s_1 < s_2$.

Figures 17–46 and 17–47 illustrate the patterns for these conditions.

Figure 17–46

Figure 17–47

If you solve a quadratic equation by *any* method (square-root method, factoring, completing the square, or using the quadratic formula) you can determine the solution set of the associated quadratic inequality by inspection after you have found the critical values. The solution can be checked by testing one point in the projected solution set. Similar generalizations can be made for $ax^2 + bx + c \leq 0$ and $ax^2 + bx + c \geq 0$.

2 Graph Quadratic Inequalities in Two Variables.

The solution set for a quadratic inequality in two variables can be represented by a shaded region of a graph on a rectangular coordinate system.

To graph a quadratic inequality in two variables:

1. Write the inequality in function notation or solved for y.
2. Graph the boundary by substituting an equal sign for the inequality sign and solving the equation.
3. Select a test point that is *not* on the boundary.
4. Substitute the values from the test point into the original inequality and evaluate. The test point will be either *inside* or *outside* the boundary.
5. If the evaluated inequality makes a *true* statement, the solution set is the portion of the graph (inside or outside) that includes the test point.
6. If the evaluated inequality makes a *false* statement, the solution set is the portion of the graph (inside or outside) that does *not* include the test point.
7. The boundary *is not* included if the inequality is $<$ or $>$. The boundary *is* included if the inequality is $\leq$ or $\geq$.

EXAMPLE Graph the inequality $y \geq 2x^2 - 5x - 3$.
Graph the equation $y = 2x^2 - 5x - 3$ $\boxed{Y=}$ 2 $\boxed{X, \theta, T, n}$ $\boxed{\wedge}$ 2 $\boxed{-}$ 5 $\boxed{X, \theta, T, n}$ $\boxed{-}$ 3 (Fig. 17–48).

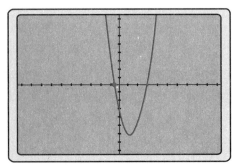

Figure 17–48

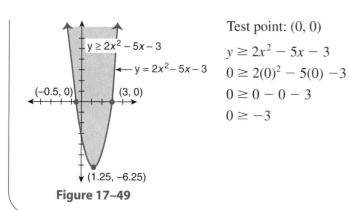

Test point: $(0, 0)$ Select a test point that is *not* on the boundary.

$y \geq 2x^2 - 5x - 3$ Substitute $x = 0$ and $y = 0$.

$0 \geq 2(0)^2 - 5(0) - 3$ Simplify.

$0 \geq 0 - 0 - 3$

$0 \geq -3$ True. Test point is included in the solution set.

Sketch the graph and shade inside the graph. The boundary is included (Fig. 17–49).

Figure 17–49

---

3   **Solve Rational Inequalities in One Variable.**

When is a rational expression greater than zero? For solving a rational inequality, we use strategies similar to those used in solving quadratic inequalities. To solve a **rational inequality** like $\dfrac{x + 2}{x - 5} > 0$, treat the numerator and denominator as factors and find the critical values by setting each factor equal to zero.

To solve a rational inequality like $\frac{x+a}{x+b} < 0$ *or* $\frac{x+a}{x+b} > 0$:

1. Find critical values by setting the numerator and denominator equal to zero.
2. Solve each equation from Step 1 and use the solutions (critical values) to divide the number line into three regions.
3. Evaluate the numerator and denominator for a test point in each region. Examine the signs of the results.
4. Regions that have like signs for both the numerator and denominator are in the solution set of $\frac{x+a}{x+b} > 0$ for $x \neq -b$ ($x = -b$ is an excluded value).
5. Regions that have unlike signs for the numerator and denominator are in the solution set of $\frac{x+a}{x+b} < 0$ for $x \neq -b$ ($x = -b$ is an excluded value).

EXAMPLE   Solve the following inequalities.

(a) $\dfrac{x + 2}{x - 5} < 0$ (b) $\dfrac{x + 2}{x - 5} \geq 0$

$x + 2 = 0$ $x - 5 = 0$ Set numerator and denominator equal to zero.

$x = -2$ $x = 5$ Critical values for inequalities in both (a)

Excluded value: $x = 5$ and (b).

Evaluate the numerator and denominator for one point in each region.

Numerator:

For $x < -2$, test $x = -3$.   For $-2 < x < 5$, test $x = 0$.   For $x > 5$, test $x = 6$.

$x + 2 = -3 + 2$ $x + 2 = 0 + 2$ $x + 2 = 6 + 2$

$= -1$ $= 2$ $= 8$

Critical values for numerator

Denominator:

For $x < -2$, test $x = -3$ For $-2 < x < 5$, test $x = 0$ For $x > 5$, test $x = 6$
$$x - 5 = -3 - 5$$
$$= -8$$
$$x - 5 = 0 - 5$$
$$= -5$$
$$x - 5 = 6 - 5$$
$$= 1$$

Critical values for denominator

Examine the signs of the results for the numerator and denominator in each region:

For $x < -2$, like signs.   For $-2 < x < 5$, unlike signs.   For $x > 5$, like signs.

See Fig. 17–50. The quotient in Region I is positive (like signs). The quotient in Region II is negative (unlike signs). The quotient in Region III is positive (like signs).

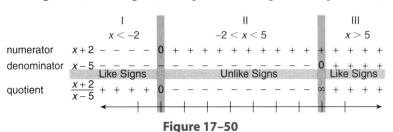

Figure 17–50

(a) Solution set (see Fig. 17–51):

For $\dfrac{x + 2}{x - 5} < 0$,

$$-2 < x < 5$$
$$(-2, 5)$$

(b) Solution set (see Fig. 17–52):

For $\dfrac{x + 2}{x - 5} \geq 0$,

$$x \leq -2 \quad \text{or} \quad x > 5$$
$$(-\infty, -2] \cup (5, \infty)$$

The boundary point, 5, is an excluded value. Therefore, it is not included in the solution set.

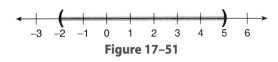

Figure 17–51

Figure 17–52

SECTION 17–4 SELF-STUDY EXERCISES

1 Graph the solution of the quadratic inequalities on a number line and write the solution in symbolic and interval notation.

1. $(x - 2)(x + 3) < 0$ **2.** $(2x - 1)(x + 3) > 0$ **3.** $(2x - 5)(3x - 2) < 0$
4. $(y + 6)(y - 2) \geq 0$ **5.** $(a + 4)(a - 6) < 0$ **6.** $x^2 - 7x + 10 \leq 0$
7. $x^2 - x - 6 \geq 0$ **8.** $x^2 - 4x + 3 \leq 0$ **9.** $4x^2 - 21x + 5 \leq 0$
10. $6x^2 + x > 1$ **11.** $6x^2 + 2 \geq 7x$ **12.** $2x^2 + 9x \leq 5$

2 Graph using a graphing calculator. Adjust the window if necessary to show the vertex of the parabola.

13. $y < 3x^2 + 5x - 2$ **14.** $y > (2x - 3)(x - 1)$ **15.** $y \leq 2x^2 - 9x - 5$ **16.** $y \geq -2x^2 + 9x + 5$

17. The revenue range for a leather wallet is based on the price of the item given by the formula $R \leq -4x^2 + 52x$, where R is the revenue and x is the price. Graph the revenue function and determine the greatest possible revenue.

18. The formula $d \geq 4.9t^2$ gives the distance d an object falls in t seconds. Graph the solution set to show the possible range of solutions.

     Chapter 17 / Inequalities and Absolute Values

3 Graph the solution of the rational inequalities on a number line and write the solution in symbolic and interval notation.

19. $\dfrac{x-3}{x+7} < 0$

20. $\dfrac{x+8}{x-3} > 0$

21. $\dfrac{2x-6}{3x-1} > 0$

22. $\dfrac{x-7}{x+7} \le 0$

23. $\dfrac{x}{x-1} < 0$

24. $\dfrac{5x-10}{6x-18} \ge 0$

25. $\dfrac{3x-9}{6x-12} \le 0$

26. $\dfrac{3x}{2x-8} \ge 0$

27. $\dfrac{5x-15}{4x+4} \ge 0$

17–5 │ Solving Equations and Inequalities Containing One Absolute-Value Term

Learning Outcomes
1 Solve equations containing one absolute-value term.
2 Solve absolute-value inequalities using the "less than" relationship.
3 Solve absolute-value inequalities using the "greater than" relationship.

1 Solve Equations Containing One Absolute-Value Term.

The equation $|x| = 5$ has two roots, $+5$ and -5. To solve an equation that contains an absolute-value term, we must examine each of two cases. One case is when the expression within the absolute-value symbol is positive. The other case is when the expression within the absolute-value symbol is negative. Recall that the symbolic definition of the absolute value of a number is $|a| = a$ for $a \ge 0$, and $|a| = -a$ for $a < 0$.

To solve an equation containing one absolute-value term:

1. Isolate the absolute-value term on one side of the equation.
2. Separate the equation into two cases. One case considers the expression within the absolute-value symbol to be positive. The other case considers it to be negative.
3. Solve each case to obtain the *two* roots of the equation.
4. $|x| = b$ has no solution if b is negative ($b < 0$).
5. $|x| = b$ has one root when $b = 0$.

EXAMPLE Find the roots for the equation $|x - 4| = 7$.

Case 1: $x - 4 = 7$	*Case 2:* $x - 4 = -7$
$x - 4 = 7$	$x - 4 = -7$
$x = 7 + 4$	$x - 4 = -7$
$\boxed{x = 11}$	$x = -7 + 4$
	$\boxed{x = -3}$

The roots of the equation are 11 and −3.

If $x = 11$,	If $x = -3$,	Check each solution.
$\lvert x - 4 \rvert = 7$	$\lvert x - 4 \rvert = 7$	
$\lvert 11 - 4 \rvert = 7$	$\lvert -3 - 4 \rvert = 7$	
$\lvert 7 \rvert = 7$	$\lvert -7 \rvert = 7$	
$7 = 7$	$7 = 7$	Each root checks.

EXAMPLE Find the roots of the equation $\lvert y + 3 \rvert - 5 = 6$.

$$\lvert y + 3 \rvert - 5 = 6 \qquad \text{Isolate the absolute-value term.}$$
$$\lvert y + 3 \rvert = 6 + 5 \qquad \text{Combine like terms.}$$
$$\lvert y + 3 \rvert = 11 \qquad \text{Separate into two cases.}$$

Case 1: $y + 3 = 11$	*Case 2:* $y + 3 = -11$
$y = 11 - 3$	$y + 3 = -11$
$y = 8$	$y = -11 - 3$
	$y = -14$

The roots of the equation are 8 and −14. Each root checks to be a true root.

TIP

The Importance of Isolating the Absolute-Value Term

The recommended procedure for finding the root for the case when the expression inside the absolute-value grouping is negative (case 2) is to isolate the absolute-value term first. Let's look at an alternative correct procedure and an incorrect procedure.

Find the root of $\lvert y + 3 \rvert - 5 = 6$ when $(y + 3)$ is negative (case 2).

Alternative Correct Procedure Without Isolating	**Incorrect Procedure Without Isolating**
$\lvert y + 3 \rvert - 5 = 6$	$\lvert y + 3 \rvert - 5 = 6$
Case 2: $-(y + 3) - 5 = 6$	Case 2: $y + 3 - 5 = -6$
$-y - 3 - 5 = 6$	$y - 2 = -6$
$-y - 8 = 6$	$y = -6 + 2$
$-y = 6 + 8$	$y = -4$
$-y = 14$	
$y = -14$	

Check: $y = -4$
$$\lvert y + 3 \rvert - 5 = 6$$
$$\lvert -4 + 3 \rvert - 5 = 6$$
$$\lvert -1 \rvert - 5 = 6$$
$$1 - 5 = 6$$
$$-4 = 6 \qquad \text{Incorrect.}$$

If we choose *not* to isolate the absolute-value term first, then we *must* take the opposite of the absolute-value grouping rather than the opposite of the value on the other side of the equation.

EXAMPLE Solve the equation $|2x - 5| + 14 = 7$.

$$|2x - 5| + 14 = 7 \qquad \text{Isolate the absolute-value term.}$$

$$|2x - 5| = 7 - 14 \qquad \text{Combine like terms.}$$

$$|2x - 5| = -7 \qquad \text{Cannot continue because the absolute value must be positive.}$$

$|2x - 5| + 14 = 7$ has no solution.

2 Solve Absolute-Value Inequalities Using the "Less Than" Relationship.

An inequality that contains an absolute-value term is interpreted by the sense of the inequality. If the inequality is a "less than" or "less than or equal to" relationship, the following property is used.

"Less than" relationships for absolute-value inequalities:

$$\text{If } |x| < b \text{ and } b > 0, \text{ then } -b < x < b.$$

or

$$\text{If } |x| \leq b \text{ and } b > 0, \text{ then } -b \leq x \leq b.$$

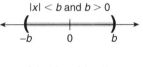

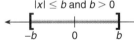

Figure 17–53

On a number line (Fig. 17–53), the solution set is a continuous set of values. The values in the solution set of $|x| < b$ are between b and $-b$ when b is positive. If $b = 0$, $|x| = 0$. Since zero is unsigned, we do not create an interval from -0 to 0. The solution is a single value, 0.

EXAMPLE Find the solution set for each inequality.

(a) $|x| < 3$ (b) $|x - 1| \leq -5$ (c) $|3x| + 2 \leq 14$

(a) $|x| < 3$ Apply the property of absolute-value inequalities having $<$ relationship.

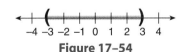

Figure 17–54

$-3 < x < 3$ **or** $(-3, 3)$. See Fig. 17–54.

(b) $|x - 1| \leq -5$ Apply the property of absolute-value inequalities having $\leq$ relationship.

An absolute-value term cannot be negative. **So $|x - 1| \leq -5$ has no solution.**

(c) $|3x| + 2 \leq 14$ Isolate the absolute-value term.

$|3x| \leq 14 - 2$ Combine like terms.

$|3x| \leq 12$ Apply the property of absolute-value inequalities having $\leq$ relationship.

$-12 \leq 3x \leq 12$ Separate into two inequalities.

$-12 \leq 3x$ and $3x \leq 12$ Divide.

$-4 \leq x$ $x \leq 4$ Find the overlap.

$-4 \leq x \leq 4$ **or** $[-4, 4]$ **is the solution.** See Fig. 17–55.

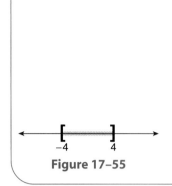

Figure 17–55

3   Solve Absolute-Value Inequalities Using the "Greater Than" Relationship.

The solution set for an absolute-value inequality with a "greater than" or "greater than or equal to" relationship is represented by the extreme values on the number line.

"Greater than" relationships for absolute-value inequalities:

If $|x| > b$ and $b > 0$, then $x < -b$ or $x > b$.

or

If $|x| \geq b$ and $b > 0$, then $x \leq -b$ or $x \geq b$.

The solution set for an absolute-value inequality with a $>$ or $\geq$ relationship is represented on the number lines in Figs. 17–56 and 17–57.

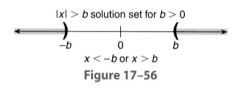

$x < -b$ or $x > b$

Figure 17–56

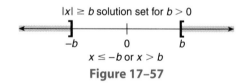

$x \leq -b$ or $x > b$

Figure 17–57

EXAMPLE   Find the solution set for each inequality.

(a) $|x| > 5$ (b) $|x| + 1 \geq 4$

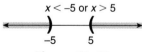

Figure 17–58

(a) $|x| > 5$ Apply the property of absolute-value inequalities having $>$ relationship.

$x < -5$ **or** $x > 5$; $(-\infty, -5) \cup (5, \infty)$. See Fig. 17–58.

(b) $|x| + 1 \geq 4$ Isolate the absolute-value term.

$|x| \geq 4 - 1$ Combine like terms.

$|x| \geq 3$ Apply the property of inequalities having $\geq$ relationship.

$x \leq -3$ or $x \geq 3$

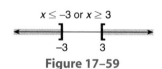

Figure 17–59

$x \leq -3$ **or** $x \geq 3$; $(-\infty, -3] \cup [3, \infty)$. See Fig. 17–59.

Develop Your Own Memory Flags

This chapter introduces many new concepts, yet most of these concepts build on previously learned similar concepts.

Develop your own list of "flags," or memory joggers, to get you started when trying to recall a process. Often that is all you need to get yourself on the right track during an exam or other situation that requires recalling the process from memory.

Some examples of memory flags are

Concepts	Memory Flags
union	join all
intersection	overlap
inequalities	solution set
conjunction	"and," overlap
disjunction	"or," separate pieces
quadratic inequalities	critical values or boundaries, test points in each region
rational inequalities	critical values or boundaries, test points in each region
absolute-value equation	two cases, isolate absolute-value term
absolute-value "less than" relationship	overlap
absolute-value "greater than" relationship	separate pieces

SECTION 17–5 SELF-STUDY EXERCISES

1 Solve the equations containing absolute values.

1. $|x| = 8$
2. $|x| = 15$
3. $|x - 3| = 5$
4. $|x - 7| = 2$
5. $|2x - 3| = 9$
6. $|3x - 7| = 2$
7. $|x| - 4 = 7$
8. $|x - 2| + 5 = 3$
9. $-5 + |x - 4| = -2$
10. $|2x - 1| - 3 = 4$
11. $|3x - 2| + 1 = -6$
12. $|3x - 5| + 12 = 18$
13. $|6 - 3x| + 2 = 8$
14. $|5 - 2x| = 15$
15. $|3x - 8| + 2 = 12$
16. $|7x - 9| - 1 = 8$
17. $|4x - 3| + 11 = 6$
18. $|5x - 3| - 7 = 8$
19. $|5x - 1| - 8 = 7$
20. $|2x - 3| - 8 = -5$
21. $|3x - 2| + 2 = 2$

2 Find the solution set for each inequality. Graph the solution set on a number line and write it in symbolic and interval notation.

22. $|x| < 5$
23. $|x| < 1$
24. $|x| < 2$
25. $|x| \leq 7$
26. $|x + 3| < 2$
27. $|x + 4| < 3$
28. $|x - 3| < 4$
29. $|x - 2| < 3$
30. $|x - 5| \leq 6$
31. $|x - 1| \leq 7$
32. $|3x - 5| < 7$
33. $|2x - 4| < 3$
34. $|5x + 2| \leq 3$
35. $|4x + 7| \leq 5$
36. $|3x - 1| + 2 < 5$

3 Find the solution set for each inequality. Graph the solution set on a number line and write it in symbolic and interval notation.

37. $|x| > 2$ **38.** $|x| > 3$ **39.** $|x| \geq 5$

40. $|x| \geq 6$ **41.** $|x - 5| > 2$ **42.** $|x - 4| > 6$

43. $|x + 3| \geq 5$ **44.** $|x + 4| \geq 6$ **45.** $|2x - 5| \geq 0$

46. $|3x - 2| \geq 4$ **47.** $|5x + 8| \geq 2$ **48.** $|4x - 3| \geq 9$

49. $|3x - 1| + 3 \geq 5$ **50.** $|2x - 3| + 1 \geq 5$ **51.** $|4x - 2| - 3 \geq 7$

52. $|3x - 2| + 4 \leq 5$ **53.** $|2x - 5| - 3 \leq 2$ **54.** $|5x - 3| - 6 \geq 1$

55. $|4x - 6| + 4 \geq 5$ **56.** $|3x - 8| - 2 \leq -6$ **57.** $|5x + 7| + 3 \leq -6$

Find the solution set for each inequality and write in symbolic notation.

58. $|x - 3| \geq 3$ **59.** $|4x + 1| > 1$ **60.** $|x - 4| \leq 3$

61. $|2x - 1| < 0$ **62.** $|x| < 7$ **63.** $|x + 8| < 3$

64. $|x| > 12$ **65.** $|x + 8| > 2$ **66.** $|2x - 7| < -5$

CHAPTER REVIEW OF KEY CONCEPTS

Learning Outcomes

Section 17–1
1 Use set terminology (pp. 655–656).

2 Show inequalities on a number line and write inequalities in interval notation (pp. 656–658).

Section 17–2
1 Solve a linear inequality in one variable (pp. 659–662).

What to Remember with Examples

Set-builder notation is used to show sets.

> Use set-builder notation to show the following set: the set of integers between −2 and 3, including 3.
>
> $$\{x \mid x \in Z \text{ and } -2 < x \leq 3\}$$

Solutions to inequalities can be shown on the number line or by using interval notation. Boundaries that are included are indicated with brackets. Boundaries that are not included are indicated with parentheses.

> The solution of $-2 < x \leq 3$ is shown on the number line in Fig. 17–60 and by using interval notation. $(-2, 3]$
>
>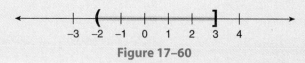
>
> **Figure 17–60**

To solve a linear inequality in one variable, isolate the variable as in solving equations with two exceptions. When interchanging the sides of an inequality, reverse the sense of the inequality, and when multiplying or dividing by a negative number, reverse the sense of the inequality. Show the solution set symbolically, graphically, and using interval notation.

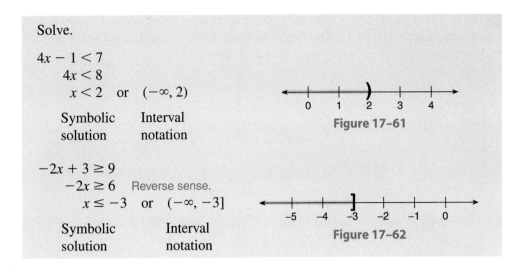

Solve.

$4x - 1 < 7$
$\quad 4x < 8$
$\qquad x < 2 \quad$ or $\quad (-\infty, 2)$

Symbolic   Interval
solution   notation

Figure 17–61

$-2x + 3 \geq 9$
$\quad -2x \geq 6 \quad$ Reverse sense.
$\qquad x \leq -3 \quad$ or $\quad (-\infty, -3]$

Symbolic   Interval
solution   notation

Figure 17–62

2 Graph a linear inequality in two variables (pp. 662–665).

To graph a linear inequality in two variables: (1) Graph the corresponding equation. (2) Select a point on one side of the boundary line and substitute each coordinate in place of the appropriate variable in the inequality. (3) If the statement is true, shade the side of the boundary that contains the point. (4) If false, shade the other side of the boundary. For $>$ or $<$ inequalities, make the boundary line dashed. For $\geq$ or $\leq$ inequalities, make the boundary line solid.

Graph $y \geq 2x - 1$.
First graph $y = 2x - 1$. Make the boundary line solid.
Test $(1, 3)$:

$3 \geq 2(1) - 1 \qquad$ Substitute $x = 1$ and $y = 3$.
$3 \geq 2 - 1 \qquad\quad$ Simplify.
$3 \geq 1 \qquad\qquad$ True.

Shade the side of the boundary line that contains the point since the coordinates of the test point made a true statement in the inequality (Fig. 17–63).

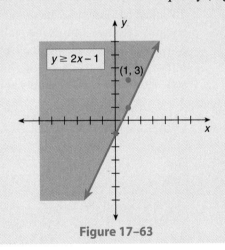

Figure 17–63

3 Solve a system of linear inequalities by graphing (pp. 665–666).

Rewrite each inequality as an equation. Graph each equation and shade the area that satisfies each inequality. Where shaded areas overlap is the set of solutions to the system of inequalities.

Graph $x + y \leq -2$ and $x - y > 3$.
For $x + y = -2$,

when $x = 0$, $y = -2$
when $y = 0$, $x = -2$
when $x = 1$, $y = -3$
shade *below* the line.

For $x - y = 3$,

when $x = 0$, $y = -3$
when $y = 0$, $x = 3$
when $x = 1$, $y = -2$
shade *below* the line.

The overlapping shaded area meets both conditions of the system of inequalities (Fig. 17–64).

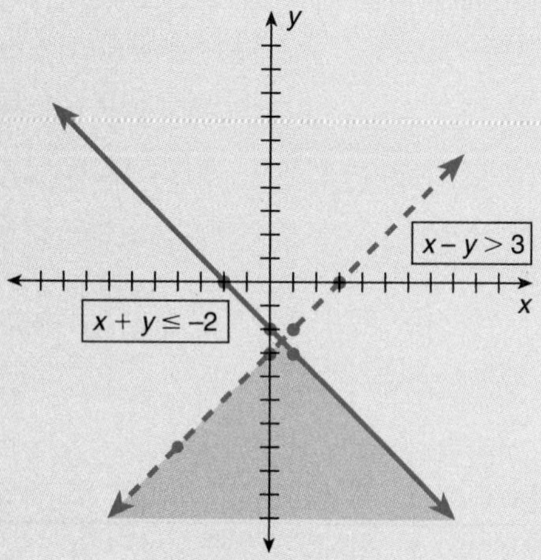

$x - y > 3$

$x + y \leq -2$

Figure 17–64

Section 17–3

1 Identify subsets of sets and perform set operations (pp. 666–669).

Every set has the set itself and the empty set (ϕ) as subsets. Set operations include union ($\cup$), intersection ($\cap$), and complements (A').

List the subsets of the set $\{3, 5, 8\}$:

ϕ, $\{3, 5, 8\}$ (the set itself), $\{3\}$, $\{5\}$, $\{8\}$, $\{3, 5\}$, $\{3, 8\}$, $\{5, 8\}$

If $U = \{1, 2, 3, 4, 5\}$, $A = \{1, 3\}$, and $B = \{2, 4\}$, find $A \cup B$, $A \cap B$, and A'.

$A \cup B = \{1, 2, 3, 4\}$ $A \cap B = \phi$ $A' = \{2, 4, 5\}$

2 Solve compound inequalities with the conjunction condition (pp. 669–671).

To solve a compound inequality that is a conjunction, separate the compound inequality into two simple inequalities using the conditions of the conjunction. Solve each simple inequality. Determine the solution set that includes the *intersection* of the solution sets of the two simple inequalities. *Note:* If there is no overlap in the sets, the solution is the empty set.

Indicate the solution using a number line and using interval notation.

$$-3 < x + 2 \leq 2$$

$-3 < x + 2$ and $x + 2 \leq 2$
$-3 - 2 < x$ $x \leq 2 - 2$
$-5 < x$ or $x > -5$ $x \leq 0$

Solution: $-5 < x \leq 0$ **Interval notation:** $(-5, 0]$

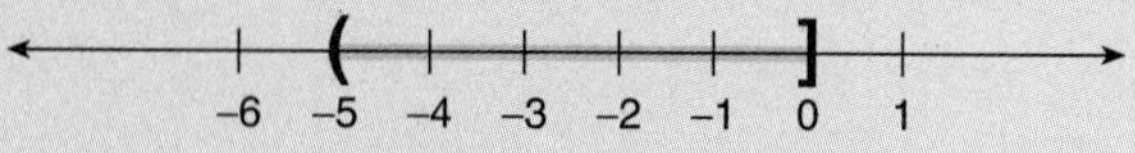

Figure 17–65

3 Solve compound inequalities with the disjunction condition (pp. 671–672).

To solve a compound inequality that is a disjunction: Solve each simple inequality. Determine the solution set that includes the union of the solution sets of the two simple inequalities.

Solve and indicate the solution set on the number line and by using interval notation.

$$x - 1 < 3 \quad \text{or} \quad x + 2 > 8$$
$$x < 3 + 1 \qquad\qquad x > 8 - 2$$
$$x < 4 \quad \text{or} \quad x > 6$$

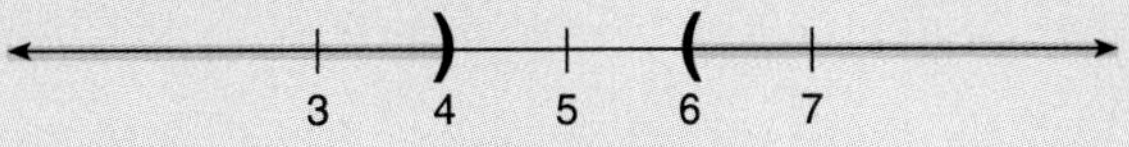

Figure 17–66

Interval notation: $(-\infty, 4) \cup (6, \infty)$

Section 17-4

1 Solve quadratic and rational inequalities in one variable (pp. 673–676).

To solve a quadratic inequality by factoring, rearrange the inequality so that the right side of the inequality is zero and the leading coefficient is positive. Write the left side of the inequality in factored form. Determine the critical values that make each factor equal to zero. Test each region of values. The solution set for the quadratic inequality is the region or regions that produce a true statement.

Solve $x^2 + 4x - 21 < 0$.

$(x + 7)(x - 3) = 0$ — Set each factor equal to zero.
$x + 7 = 0 \quad x - 3 = 0$ — Solve each equation.
$x = -7 \quad x = 3$ — Critical values.

Test Region I. Let $x = -8$.
$(-8)^2 + 4(-8) - 21 < 0?$
$64 - 32 - 21 < 0?$
$11 < 0$. *False.*

Test Region II. Let $x = 0$.
$0^2 + 4(0) - 21 < 0?$
$0 + 0 - 21 < 0?$
$-21 < 0$. *True.*

Test Region III. Let $x = 4$.
$4^2 + 4(4) - 21 < 0?$
$16 + 16 - 21 < 0?$
$11 < 0$. *False.*

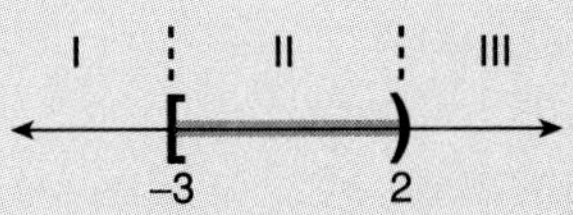

Figure 17–67

The solution set is Region II.
$-7 < x < 3$ or $(-7, 3)$.

2 Graph quadratic inequalities in two variables (pp. 676–677).

To graph quadratic inequalities. (1) Graph the boundary by substituting an equal sign for the inequality sign. (2) Use a test point to determine whether the solution set is inside or outside the boundary.

Graph $y \leq x^2 - 6x + 8$.

Graph the boundary $y = x^2 - 6x + 8$.

Clear the graphing screen.
Enter equation.
Graph.
Change viewing window if appropriate.

Select a test point such as $(0, 0)$.
$y \leq x^2 - 6x + 8$
$0 \leq 0^2 - 6(0) + 8$
$0 \leq 8$ True.

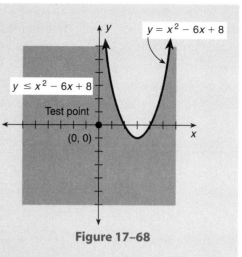

Figure 17–68

Shade the portion of the graph that includes the test point. The boundary is included in the solution set (Fig. 17–68).

3 Solve rational inequalities in one variable (pp. 677–678).

To solve a rational inequality, set both the numerator and denominator equal to zero and solve for the variable. These solutions are the boundaries for the solution regions of the inequality solution. The region or regions that have like signs for the numerator and denominator are solutions for the $>$ and $\geq$ inequalities. The region or regions that have unlike signs for the numerator and denominator are solutions for the $<$ and $\leq$ inequalities.

Solve the rational inequality $\dfrac{x + 3}{x - 2} \leq 0$.

$x = 2$ is an excluded value.	Set numerator and denominator equal to zero.
$x + 3 = 0 \qquad x - 2 = 0$	Solve each equation.
$x = -3 \qquad\qquad x = 2$	Critical values.

The critical value 2 is not included in the solution set because it is an excluded value.
Region I has like signs ($\frac{-}{-}$).

Region II has unlike signs ($\frac{+}{-}$).   Unlike signs indicate < 0.

Region III has like signs ($\frac{+}{+}$).

The solution set is Region II.
$-3 \leq x < 2; [-3, 2)$

Figure 17–69

Section 17–5

1 Solve equations containing one absolute-value term (pp. 679–681).

Isolate the absolute-value term, then form two equations. One equation considers the expression within the absolute-value symbol to be positive; the other equation considers the expression to be negative. Solve each case to obtain the two roots of the equation.

Solve $|x - 3| + 1 = 5$.

$|x - 3| + 1 = 5$ Isolate the absolute-value term.
$|x - 3| = 4$

Case 1: $x - 3 = 4$ *Case 2:* $x - 3 = -4$ or $-(x - 3) = 4$
$\qquad\qquad x = 7$ $\qquad\qquad\quad x = -4 + 3$ $\qquad\qquad -x + 3 = 4$
$\qquad\qquad\qquad\qquad\qquad\quad x = -1$ $\qquad\qquad\qquad -x = 1$
$\qquad\qquad\qquad\qquad\qquad\qquad\qquad\qquad\qquad\qquad x = -1$

Chapter 17 / Inequalities and Absolute Values

2 Solve absolute-value inequalities using the "less than" relationship (pp. 681–682).

To solve an absolute-value inequality using the "less than" relationship, use the property, if $|x| < b$ and $b > 0$, then $-b < x < b$.

Solve $|x - 3| < 4$.

$-4 < x - 3 < 4$

$-4 < x - 3 \quad$ and $\quad x - 3 < 4$

$-1 < x$ or $x > -1 \qquad x < 7$

$-1 < x < 7; (-1, 7)$

Show the solution set graphically, in symbolic and in interval notation.
Separate into two simple inequalities connected by *and*.
Solve each inequality.

Figure 17–70

3 Solve absolute-value inequalities using the "greater than" relationship (pp. 682–683).

To solve an absolute-value inequality using the "greater than" relationship, use the property, if $|x| > b$ and $b > 0$, then $x < -b$ or $x > b$.

Solve $|x + 7| > 1$.

$x + 7 < -1 \quad$ or $\quad x + 7 > 1$

$x < -8 \qquad\qquad x > -6; (-\infty, -8) \cup (-6, \infty)$

Show the solution set graphically, in symbolic and in interval notation.
Separate into two simple inequalities connected by *or*.

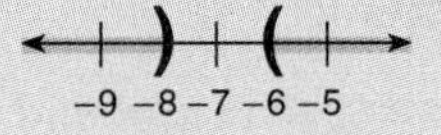

Figure 17–71

CHAPTER REVIEW EXERCISES

Section 17–1

1. Describe the empty set.
2. Write the symbols used to indicate the empty set.
3. Use symbols to write the statement: 5 is an element of the set of whole numbers.

Represent each set of numbers on the number line and with interval notation.

4. $x \le 3$
5. $x > -7$
6. $-3 < x < 5$
7. $-4 \le x < 2$
8. All real numbers
9. $-2 < x$
10. $-5 > x$

Section 17–2

Solve. Show the solution set on a number line and in symbolic and interval notation.

11. $42 > 8m - 2m$
12. $3m - 2m < 3$
13. $0 < 2x - x$
14. $1 \le x - 7$
15. $10 - 2x \ge 4$
16. $3x + 4 > x$
17. $10x + 18 > 8x$
18. $4x \le 5x + 8$
19. $12 + 5x > 6 - x$
20. $8 - 7y < y + 24$
21. $15 \ge 5(2 - y)$
22. $4t > 2(7 + 3t)$
23. $6x - 2(x - 3) \le 30$
24. $8x - (3x - 2) > 12$

25. A ream of 11×17 copy paper costs $5.60 more than a mechanical pencil. The total cost of two reams of paper and a dozen pencils must cost no more than $59.80. Write inequalities to find the cost range for a ream of paper and a single pencil.

26. A shirt costs $3 less than a certain tie. If the total cost for six ties and two shirts is less than $130, what is the most each shirt and tie can cost?

Use sets *U, A, B, C,* and *D* to give the elements in the sets in Exercises 27–31.

$U = \{-2, -1, 0, 1, 2, 3, 4, 5\}$, $A = \{-2, -1\}$, $B = \{0\}$, $C = \{0, 1, 2, 3, 4\}$, $D = \{-1, 0, 1\}$

27. $A \cap B$ **28.** $B \cup D$ **29.** $A \cap D$ **30.** A' **31.** $(A \cup C)'$

Solve. Show the solution on the number line and with interval notation.

32. $x - 7 < 2$ **33.** $x + 4 > 2$ **34.** $3x - 1 \le 8$ **35.** $3x - 2 \le 4x + 1$
36. $5 < x - 3 < 8$ **37.** $-3 < 2x - 4 < 5$ **38.** $-7 < x - 5 < 7$

Graph the linear inequalities using test points and verify with a graphing calculator.

39. $4x + y < 2$ **40.** $x + y > 6$ **41.** $3x + y \le 2$
42. $x + 3y > 4$ **43.** $x - 2y < 8$ **44.** $3x + 2y \ge 4$
45. $y \ge 3x - 2$ **46.** $y \le -2x + 1$ **47.** $y > \frac{2}{3}x - 2$

48. $y \ge x + 3$ and $x + y < 4$ **49.** $2x + y < 6$ and $x - y < 1$
50. $x + 2y < -1$ and $x + 2y > 3$ **51.** $2x + y > 3$ and $x - y \le 1$
52. $x + 2y > -4$ and $x - 2y > -1$

Section 17–3

Solve each compound inequality. Show the solution set on a number line and in symbolic and interval notation.

53. $x + 1 < 5 < 2x + 1$ **54.** $3x - 8 < x < 5x + 24$
55. $x + 2 < 7 < 2x - 15$ **56.** $x + 3 \le 7 \le 2x - 1$
57. $2x + 3 < 15 < 3x + 9$ **58.** $2 < 3x - 4 < 8$
59. $-5 \le -3x + 1 < 10$ **60.** $2x + 3 \le 5x + 6 < -3x - 7$
61. $-3 \le 4x + 5 \le 2$ **62.** $4x - 2 < x + 8 < 9x + 1$
63. $x + 3 < 5$ or $x > 8$ **64.** $x - 7 < 4$ or $x + 3 > 8$
65. $x - 3 < -12$ or $x + 1 > 9$ **66.** $3x < -4$ or $2x - 1 > 9$

67. The price of a particular brand of PC whiteboard is $1,365, and the stand costs $199. A college will spend between $15,000 and $30,000 for the equipment. Assuming an equal number of boards and stands are purchased, what is the range for the number of boards and stands to be purchased?

68. The blueprint for a modular desk shows the front edge to be 48 in. with a tolerance of 0.25 in. If the width of the desk is 8 in. less than the length and has the same tolerance, what is the allowable range for the width of the desk?

Section 17–4

Solve each inequality and graph the solution on a number line and in symbolic and interval notation.

69. $(x - 5)(x - 2) > 0$ **70.** $(3x + 2)(x - 2) < 0$ **71.** $(3x + 1)(2x - 3) < 0$
72. $(5x - 6)(x + 1) \ge 0$ **73.** $(x + 1)(x - 2) \le 0$ **74.** $x^2 + x - 12 < 0$
75. $2x^2 \le 5x + 3$ **76.** $2x^2 - 3x \ge -1$ **77.** $2x^2 + 7x - 15 < 0$
78. $x^2 - 2x - 8 \ge 0$ **79.** $\dfrac{x - 7}{x + 1} < 0$ **80.** $\dfrac{x + 1}{x - 3} > 0$

81. $\dfrac{x}{x + 8} > 0$ **82.** $\dfrac{x - 8}{x + 1} < 0$

Use a calculator to graph the following. Adjust the window to display the vertex.

83. $y \ge -2x^2$ **84.** $y < x^2 - 2x + 1$ **85.** $y \ge x^2 - 6x + 9$ **86.** $y \le \frac{1}{2}x^2 - 3$

Solve each equation containing an absolute value.

87. $|x| = 12$ **88.** $|x - 9| = 2$ **89.** $|x + 3| = 7$

90. $|x + 4| = 11$ **91.** $|x - 8| = 12$ **92.** $|x + 7| = 3$

93. $|4x - 7| = 17$ **94.** $|2x + 3| = 5$ **95.** $|7x + 8| = 15$

96. $|4x + 1| = 9$ **97.** $|7x - 4| = 17$ **98.** $|6x - 2| = 3$

99. $|3x - 9| = 2$ **100.** $|x| + 8 = 10$ **101.** $|x| + 12 = 19$

102. $|2x| - 1 = 9$ **103.** $|x| - 9 = 7$ **104.** $|x - 4| - 10 = 6$

105. $-5 + |x - 3| = 2$ **106.** $|4x - 2| + 1 = 5$ **107.** $|4x - 3| - 12 = -7$

108. $|3x + 4| - 5 = 17$

Graph the solution set for each inequality and write in symbolic and interval notation.

109. $|x - 3| < 4$ **110.** $|x + 7| > 3$ **111.** $|x - 4| - 3 < 5$

112. $|x - 1| < 9$ **113.** $|x - 3| < -4$

Write inequality statements.

114. You earn more than $35,000 annually.

115. The income for Riddle's market exceeds that of Smith's market and falls short of the income of Duke's market. Smith's income is $108,000, and Duke's income is $250,000. Write an inequality expressing Riddle's income compared to Smith's and Duke's.

TEAM PROBLEM-SOLVING EXERCISES

1. Your team has a travel agency and you have been asked to plan a trip for a high school math club. The math club has raised $8,700 for the trip. Your agency charges a one-time fee of $300 to set up the trip. The cost per person for the trip is $620.
   (a) Write an inequality that can be used to determine the maximum number of persons who can go on the trip with the money raised.
   (b) Solve the inequality created in part (a).
   (c) How much more money would need to be raised for one more person to be able to go on the trip?

2. The tip for finding the solution set for a quadratic inequality by inspection (pp. 675–676) is only appropriate when a, the coefficient of the leading coefficient, is positive.
   (a) How would the tip need to be modified if the leading coefficient were negative?
   (b) Illustrate your modified tip with a specific example for the "less than" relationship.
   (c) Illustrate your modified tip with a specific example for the "greater than" relationship.

CONCEPTS ANALYSIS

1. The symmetric property of equations states that if $a = b$ then $b = a$. Is an inequality symmetric? Illustrate your answer with an example.

2. In equations, if $a = b$, then $ac = bc$, if c equals any real number. Is the statement, if $a < b$ then $ac < bc$, true if c is any real number? For what values of c is the statement false? Illustrate your answer with an example.

3. Write in your own words two differences in the procedures for solving equations and for solving simple inequalities.

4. Explain the difference between the statements $x < 2$ and $x \leq 2$.

5. If $x < 2$ or $x > 10$, is it correct to write the statement as $2 > x > 10$? Why or why not? Explain your answer.

6. Explain the difference between an *and* relationship and an *or* relationship in inequalities.

7. In your own words, write a procedure for solving a compound inequality such as $5 < x + 2 < 9$.

8. Find the mistake in the inequality. Correct and briefly explain the mistake.

$$3x + 5 < 5x - 7$$
$$3x - 5x < -7 - 5$$
$$-2x < -12$$
$$x < 6$$

9. In the inequality $x^2 + 6x + 5 < 0$, what roles do the numbers -5 and -1 play? Explain the solution for the inequality in words.

10. If p_1 and p_2 are real numbers that solve the equation $ax^2 + bx + c = 0$ and $p_1 < p_2$, when will the solution of $ax^2 + bx + c < 0$ be $p_1 < x < p_2$? When will the solution be $x < p_1$ or $x > p_2$? Give an example to illustrate each answer.

PRACTICE TEST

Represent each set of numbers on the number line and by using interval notation.

1. $x \geq -12$

2. $-3 < x \leq -2$

Graph the solution on a number line and write the solution in symbolic and interval notation.

3. $3x - 1 > 8$
4. $2 - 3x \geq 14$
5. $10 < 2 + 4x$
6. $-5b > -30$
7. $\frac{1}{3}x + 5 \leq 3$
8. $2(1 - y) + 3(2y - 2) \geq 12$
9. $5 - 3x < 3 - (2x - 4)$
10. $7 + 4x \leq 2x - 1$
11. $-5 < x + 3 < 7$
12. $\frac{1}{4}x + 2 < 3 < \frac{1}{3}x + 9$
13. $3x - 1 \leq 5 \leq x - 5$
14. $(x + 4)(x - 2) < 0$
15. $(2x + 3)(x - 1) > 0$
16. $2y^2 + y < 15$
17. $2x - 3 < 1$ or $x + 1 > 7$
18. $2(x - 1) < 3$ or $x + 7 > 15$

Use the following sets to give the elements in the indicated sets in Exercises 19–22.

$U = \{1, 2, 3, 4, 5, 6, 7, 8, 9\}$, $\qquad A = \{5, 8, 9\}$, $\qquad B = \{1, 2, 3, 4, 5, 6, 7, 8\}$

19. $A \cup B$
20. $A \cap B$
21. $A \cap B'$
22. List all the subsets of set A.

Solve. Graph the solution set on a number line and write the solution set in interval notation.

23. $\dfrac{x - 2}{x + 5} < 0$

24. $\dfrac{x - 3}{x} > 0$

Solve.

25. $|x| = 15$
26. $|x + 8| = 7$
27. $|x| + 8 = 10$

Solve and graph the solution set on a number line and write the solution set in interval notation.

28. $|4x - 7| < 17$
29. $|x + 8| > 10$
30. $|x + 1| - 3 < 2$

31. An airplane manufacturer orders a part from a supplier that measures 96.8 cm and has a tolerance of ±0.01 cm according to the specifications. Write an inequality that shows the range of acceptable measures for the part.

32. An electronically controlled thermostat is set to cause the cooling unit to operate when the temperature is above 68°F. Write an inequality that expresses the temperature range in Celsius degrees. The formula $°F = \frac{9}{5}°C + 32$ may be used.

Shade the portion on the graph represented by the conditions.

33. $x + y < 4$ and $y > 3x + 2$

34. $y \le 2x + 1$ and $x + y \ge 5$

Graph the linear inequalities.

35. $2x - y \le 2$

36. $x + 2y < 4$

37. $x + y < 1$

38. $y \le 2x + 2$

Graph the quadratic inequalities.

39. $y \le x^2 - 6x + 8$

40. $y > 2x^2$

41. $y < -\dfrac{1}{2}x^2$

18

Geometry

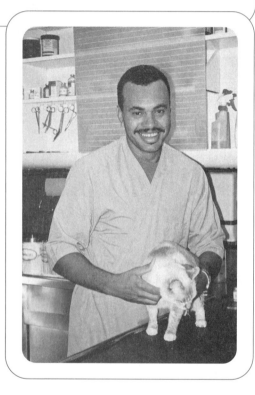

Focus on Careers

If you love animals, this may be the career for you! Veterinary technicians usually work in a private practice with a supervising veterinarian, in animal hospitals, or in research facilities. They perform laboratory tests, assist with dental prophylaxis, prepare tissue samples, take blood samples, and assist veterinarians with a variety of tests and analyses.

Most entry-level technicians have an associate degree from an accredited community college program in veterinary technology with clinical and laboratory settings using live animals. All states require veterinary technicians and technologists to pass a credentialing exam following their coursework.

The job outlook for veterinary technologists and technicians is expected to grow much faster than average through 2012. Although the top 10% of veterinary technologists and technicians earned more than $33,750 in 2002, the middle 50% earned between $19,210 and $27,890. Economic recession is less likely to cause layoffs than in some other occupations because animals continue to require medical care.

Working conditions in this career are usually indoors but may take place outdoors. The work can be physically and emotionally demanding and sometimes dangerous. Workers risk exposure to small-animal bites and scratches and other injuries when working with larger animals. When appropriate safety precautions are taken, injuries can be minimized.

Source: *Occupational Outlook Handbook,* 2004–2005 Edition. U.S. Department of Labor, Bureau of Labor Statistics.

Geometry is one of the oldest and most useful of the mathematical sciences. **Geometry** involves the study and measurement of shapes according to their sizes, volumes, and positions. A knowledge of geometry is necessary in many careers. Notice how the meanings of common words with which you already are familiar change when they are precisely defined for geometry.

Learning Outcomes

1 Use various notations to represent points, lines, line segments, rays, planes, and angles.

2 Classify angles according to size.

3 Determine the measure of an angle by using relationships among intersecting lines.

4 Convert angle measures between decimal degrees and degrees, minutes, and seconds.

1 Use Various Notations to Represent Points, Lines, Line Segments, Rays, Planes, and Angles.

Geometry is the study of size, shape, position, and other properties of the objects around us. The basic terms used in geometry are *point, line,* and *plane.* Generally, these terms are not defined. Instead, they are only described. Once described, they are used in definitions of other terms and concepts.

A **point** is a location or position that has no size or dimension. A dot is used to represent a point, and a capital letter may be used to label the point (Fig. 18–1).

A **line** extends indefinitely in both directions and contains an infinite number of points. It has length but no width. In our discussions, the word *line* always refers to a straight line unless otherwise specified. In Fig. 18–1, the line can be identified by naming any two points on the line (such as *A* and *B*).

A **plane** is a flat, smooth surface that extends indefinitely in all directions. A plane contains an infinite number of points and lines (Fig. 18–1).

Since a line extends indefinitely in both directions, most geometric applications deal with parts of lines. A part of a line is called a **line segment** or **segment.** A line segment starts and stops at distinct points that we call **endpoints.** A *line segment,* or *segment,* consists of all points on the line between and including the two *endpoints* (Fig. 18–2).

The notation for a line that extends through points *A* and *B* is $\overleftrightarrow{AB}$ (read "line *AB*"). The notation for the line segment including points *A* and *B* and all the points between is $\overline{AB}$ (read "line segment *AB*").

Another term used in connection with parts of a line is *ray.* Before we give the definition of a ray, consider the beam of light from a flashlight. The beam is like a ray. It seems to continue indefinitely in only one direction. A **ray** consists of a point on a line and all points of the line on one side of the point (Fig. 18–3).

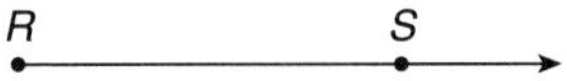

The point from which the ray originates is called the **endpoint,** and all other points on the ray are called **interior points** of the ray. A ray is named by its endpoint and any interior point on the ray. In Fig. 18–3, we use the notation $\overrightarrow{RS}$ to denote the ray whose endpoint is *R* and that passes through *S*.

Fig. 18–4 shows the contrast in the notation used for a line, line segment, and ray. To illustrate appropriate notations for lines, segments, and rays, consider a line with several points designated on the line (Fig. 18–5).

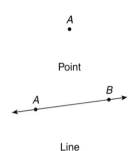

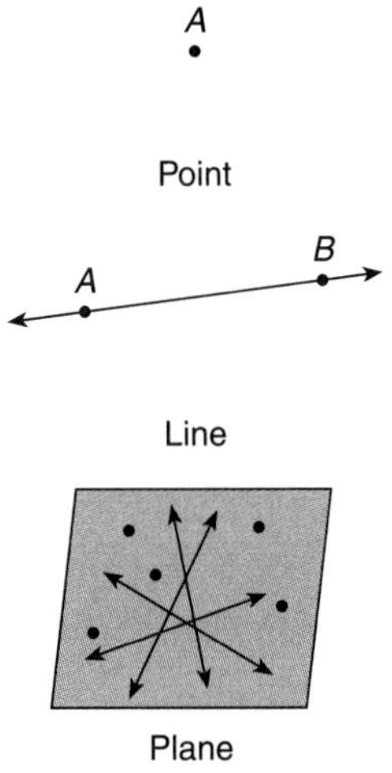

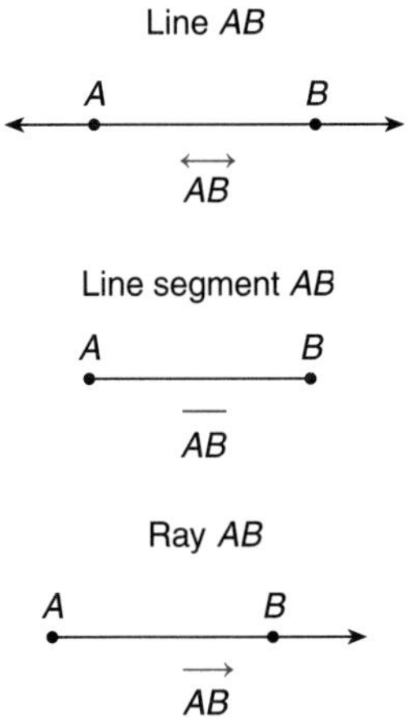

A
•

Point

A   *B*

Line

Plane

Figure 18–1

A   *B*

Figure 18–2

R   *S*

Figure 18–3

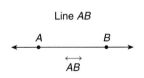

Line *AB*

A   *B*

$\overleftrightarrow{AB}$

Line segment *AB*

A   *B*

$\overline{AB}$

Ray *AB*

A   *B*

$\overrightarrow{AB}$

Figure 18–4

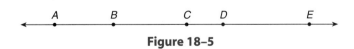

$A \qquad B \qquad C \quad D \qquad\qquad E$

Figure 18–5

Any two points can be used to name the line in Fig. 18–5. For example, $\overleftrightarrow{AB}$, $\overleftrightarrow{AC}$, $\overleftrightarrow{CE}$, $\overleftrightarrow{BD}$, and $\overleftrightarrow{BC}$ are some of the possible ways to name the line. However, a segment is named *only* by its endpoints. Thus, in Fig. 18–5, $\overline{AB}$ is not the same segment as $\overline{AC}$, but $\overleftrightarrow{AB}$ and $\overleftrightarrow{AC}$ represent the same line.

In Fig. 18–5, $\overrightarrow{BC}$ and $\overrightarrow{BD}$ represent the same ray, but $\overline{BC}$ and $\overline{BD}$ do not represent the same segment.

A line can be extended indefinitely in either direction. If two lines are drawn in the same plane, one of three situations may occur:

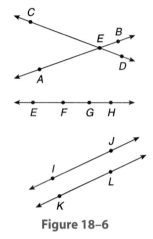

1. The two lines *intersect* in *one and only one point*. In Fig. 18–6, $\overleftrightarrow{AB}$ and $\overleftrightarrow{CD}$ intersect at point E.
2. The two lines *coincide;* that is, one line fits exactly on the other. In Fig. 18–6, $\overleftrightarrow{EF}$ and $\overleftrightarrow{GH}$ coincide.
3. The two lines never intersect. In Fig. 18–6, $\overleftrightarrow{IJ}$ and $\overleftrightarrow{KL}$ are the same distance from each other along their entire lengths and so never touch.

The relationship described in the third situation has a special name, **parallel lines.** The symbol ‖ is used for parallel lines. (Chapter 7, Section 4, Outcome 1.)

When two lines intersect in a point, four *angles* are formed, as shown in Fig. 18–7. An **angle** is a geometric figure formed by two rays that intersect in a point, and the point of intersection is the endpoint of each ray.

In Fig. 18–8, rays $\overrightarrow{AB}$ and $\overrightarrow{AC}$ intersect at point A. Point A is the endpoint of $\overrightarrow{AB}$ and $\overrightarrow{AC}$. $\overrightarrow{AB}$ and $\overrightarrow{AC}$ are called the **sides** or **legs** of the angle. Point A is called the **vertex** of the angle.

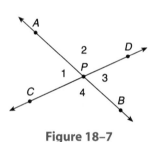

Angles can be named in several ways. An angle can be named by using a number or lowercase letter, by using the capital letter that names the vertex point, or by using three capital letters. If three capital letters are used, two of the letters name interior points of each of the two rays, and the middle letter names the vertex point of the angle. Using the symbol $\angle$ for angle, the angle in Fig. 18–9 can be named $\angle 1$, $\angle KLM$, $\angle MLK$, or $\angle L$. One capital letter is used only when it is perfectly clear which angle is designated by the letter. To name the angle in Fig. 18–10 with three letters, we write $\angle XZY$ or $\angle YZX$. This angle can also be named $\angle a$ or $\angle Z$.

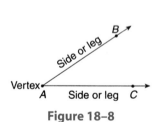

Figure 18–6

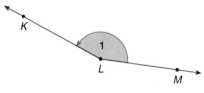

Figure 18–7

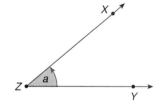

Figure 18–8

Figure 18–9

Figure 18–10

Fig. 18–11 illustrates how the intersection of two rays actually forms two angles. In this text, we refer to the smaller of the two angles formed by two rays unless the other angle is specifically indicated. Arcs (curved lines) and arrows are often used to clarify which angle we are considering.

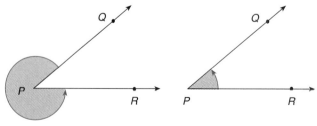

Figure 18–11

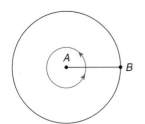

Figure 18–12

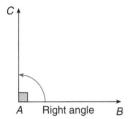

Figure 18–14

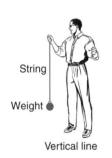

Figure 18–14 shows right angle A with B.

Figure 18–15

String

Weight

Vertical line

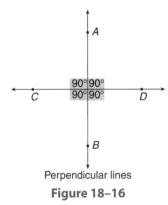

Figure 18–16

Perpendicular lines

2 Classify Angles According to Size.

The measure of an angle is determined by the amount of opening between the two sides of the angle. The length of the sides does not affect the angle measure. Two units are commonly used to measure angles, *degrees,* and *radians.* In this section, only degrees are used to measure angles. Radians are discussed in Section 18–3.

Consider the hands of a clock as the sides of an angle. When the two hands both point to the same number, the measure of the angle formed is 0 degrees (0°). An angle of 0 degrees is used in trigonometry but is seldom used in geometric applications. During 1 hour, the minute hand makes one complete revolution. Ignoring the movement of the hour hand, this revolution of the minute hand contains 360 degrees (360°). Fig. 18–12 shows a revolution or rotation of 360°. Note that *A* is kept as a fixed point and *B* rotates around point *A.* This rotation can be either clockwise or counterclockwise.

The customary rotation around a point progresses *opposite* to the normal rotation of the hands of a clock. The direction is called **counterclockwise** (Fig. 18–13).

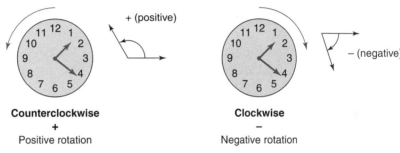

Counterclockwise
+
Positive rotation

Clockwise
−
Negative rotation

Figure 18–13

A **degree** is a unit for measuring angles. It represents $\frac{1}{360}$ of a complete rotation about the vertex. Suppose, in Fig. 18–14, that $\overrightarrow{AC}$ rotates from $\overrightarrow{AB}$ through one-fourth of a circle. Then $\overrightarrow{AC}$ and $\overrightarrow{AB}$ form a 90° angle ($\frac{1}{4}$ of 360 = 90). This angle is a **right angle.** The symbol for a right angle is ⌐.

If two lines intersect so that right angles (90° angles) are formed, the lines are **perpendicular** to each other. (See Section 9–4.) The symbol for "perpendicular" is ⊥. However, the right-angle symbol also implies the lines forming the angle are perpendicular.

If a string is suspended at one end and weighted at the other (Fig. 18–15), the line it forms is a **vertical line.** A line that is perpendicular to the vertical line is a **horizontal line.** In Fig. 18–16, $\overleftrightarrow{AB}$ is a vertical line, and $\overleftrightarrow{AB}$ and $\overleftrightarrow{CD}$ form right angles. Thus, $\overleftrightarrow{AB} \perp \overleftrightarrow{CD}$, and $\overleftrightarrow{CD}$ is a horizontal line.

Now look at Fig. 18–17. When $\overrightarrow{ML}$ rotates one-half a circle from $\overrightarrow{MN}$, an angle of 180° is formed ($\frac{1}{2}$ of 360 = 180). This angle is a **straight angle.**

An angle that is less than 90° but more than 0° is an **acute angle.** An angle that is more than 90° but less than 180° is an **obtuse angle** (see Fig. 18–18).

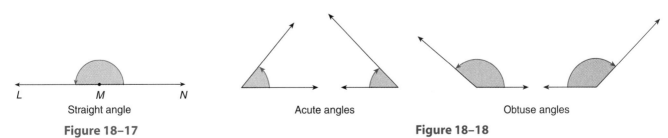

L   M   N
Straight angle

Figure 18–17

Acute angles

Obtuse angles

Figure 18–18

1. Examine the angle to determine the number of degrees it measures.
2. Classify the angle as a straight, right, obtuse, or acute angle based on its degree measure.

EXAMPLE   Classify the angles according to size (Fig. 18–19).

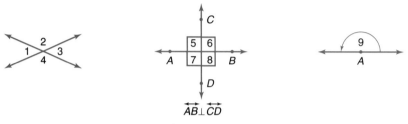

Figure 18–19

∠1 is acute.   ∠3 is acute.   ∠5, ∠6, ∠7, and ∠8 are all right angles.
∠2 is obtuse.   ∠4 is obtuse.   ∠9 is a straight angle.

The angle classifications used so far—right, straight, acute, and obtuse—deal with one angle at a time. If two angles together form a right angle or if their measures total 90°, they are **complementary angles.** If two angles together form a straight angle or if their measures total 180°, they are **supplementary angles** (Fig. 18–20).

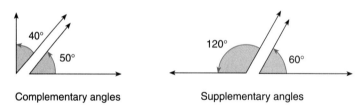

Complementary angles   Supplementary angles

Figure 18–20

To find the complement or supplement of an angle:

1. To find the complement of an angle, subtract its measure from 90°.
2. To find the supplement of an angle, subtract its measure from 180°.

EXAMPLE   Find the complement and supplement of an angle that measures 57°.

Complement	**Supplement**
90° − 57° = 33°	180° − 57° = 123°
The complement of 57° is 33°.	**The supplement of 57° is 123°.**

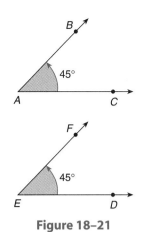

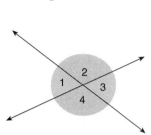

Figure 18–21

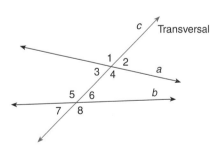

Figure 18–22

When the word *equal* is used to describe the relationship between two angles, it implies the measures of the angles are equal. Another word often used in geometry is *congruent*. When geometric figures are **congruent,** one figure can be placed on top of the other, and the two figures will match perfectly. If two angles are congruent, the measures of the angles are equal. Also, if the measures of two angles are equal, the angles are congruent. The symbol for congruence is ≅. For instance, in Fig. 18–21, the two angles have equal measures, both 45°. Because they have the same measures, they are congruent; that is, $\angle BAC \cong \angle FED$.

Traditionally, we indicate the measures of angles by placing the letter *m* before the angle symbol. Thus, $m\angle BAC = m\angle FED$ tells us that the measures of angles *BAC* and *FED* are equal. We interpret $\angle BAC = \angle FED$ and $m\angle BAC = m\angle FED$ as giving us the same information.

3 Determine the Measure of an Angle by Using Relationships Among Intersecting Lines.

Intersecting lines generate several angle relationships. When two lines intersect, two pairs of **vertical angles** are formed. In Fig. 18–22, angles 1 and 3 are vertical angles. Also, angles 2 and 4 are vertical angles. Two angles that share a common side as angles 1 and 2 are **adjacent angles.** Other adjacent angles are 2 and 3, 3 and 4, and 4 and 1.

Relationships among angles formed by intersecting lines can be used to find missing measures of angles (Fig. 18–22).		
Vertical angles are equal.	$\angle 1 = \angle 3$	$\angle 2 = \angle 4$
Adjacent angles are supplementary.	$\angle 1 + \angle 2 = 180°$	$\angle 2 + \angle 3 = 180°$
	$\angle 3 + \angle 4 = 180°$	$\angle 4 + \angle 1 = 180°$

A **transversal** is a line that intersects two or more lines in different points. When a transversal intersects (cuts) two lines, eight angles are formed (Fig. 18–23).

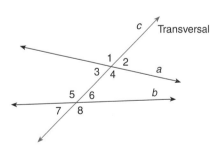

Figure 18–23

The four color-shaded angles *between* lines *a* and *b* ($\angle 3$, $\angle 4$, $\angle 5$, and $\angle 6$) are called **interior angles** (Fig. 18–24). The four gray-shaded angles *outside* lines *a* and *b* ($\angle 1$, $\angle 2$, $\angle 7$, and $\angle 8$) are called **exterior angles** (Fig. 18–24).

An exterior angle and an interior angle on the same side of the transversal are **corresponding angles.** Shaded angles $\angle 1$ and $\angle 5$ are corresponding angles (Fig. 18–25). Other corresponding angles are $\angle 2$ and $\angle 6$, $\angle 3$ and $\angle 7$, and $\angle 4$ and $\angle 8$. **Alternate**

angles are on opposite sides of the transversal. In Fig. 18–26, $\angle 1$ and $\angle 8$ are **alternate exterior angles** and $\angle 3$ and $\angle 6$ are **alternate interior angles.** There are other pairs of alternate angles.

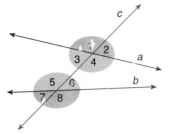

Figure 18–24 Interior and exterior angles.

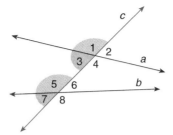

Figure 18–25 Corresponding angles.

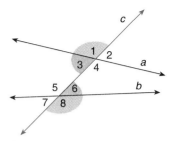

Figure 18–26 Alternate angles.

Relationships among angles on parallel lines cut by a transversal (Fig. 18–27) can be used to find missing measures of angles.

Corresponding angles are equal.	$\angle 1 = \angle 5$	$\angle 2 = \angle 6$
	$\angle 3 = \angle 7$	$\angle 4 = \angle 8$
Adjacent angles are supplementary.	$\angle 1 + \angle 2 = 180°$	$\angle 2 + \angle 4 = 180°$
	$\angle 3 + \angle 4 = 180°$	$\angle 3 + \angle 1 = 180°$
Alternate angles are equal.	$\angle 1 = \angle 8$	$\angle 2 = \angle 7$ (alternate exterior angles)
	$\angle 3 = \angle 6$	$\angle 4 = \angle 5$ (alternate interior angles)
Interior angles on the same side of the transversal are supplementary.	$\angle 3 + \angle 5 = 180°$	$\angle 4 + \angle 6 = 180°$
Exterior angles on the same side of the transversal are supplementary.	$\angle 1 + \angle 7 = 180°$	$\angle 2 + \angle 8 = 180°$

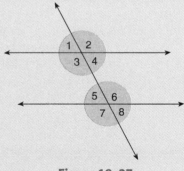

Figure 18–27

EXAMPLE In Fig. 18–27, $\angle 2$ is 125°. Find the measures of the other angles.

$\angle 1 = 180° - \angle 2$	$\angle 1$ and $\angle 2$ are adjacent angles
$\angle 1 = 180° - 125°$	and are supplementary.
$\angle 1 = 55°$	
$\angle 3 = \angle 2$	Vertical angles are equal.
$\angle 3 = 125°$	
$\angle 4 = \angle 1$	Vertical angles are equal.
$\angle 4 = 55°$	

700   Chapter 18 / Geometry

$\angle 5 = \angle 1$ Corresponding angles are equal.
$\angle 5 = 55°$

$\angle 6 = \angle 2$ Corresponding angles are equal.
$\angle 6 = 125°$

$\angle 7 = \angle 3$ Corresponding angles are equal.
$\angle 7 = 125°$

$\angle 8 = \angle 4$ Corresponding angles are equal.
$\angle 8 = 55°$

4 Convert Angle Measures Between Decimal Degrees and Degrees, Minutes, and Seconds.

A device used to measure angles is called a **protractor.** The most common protractor is a semicircle with two scales from 0° to 180°. An **index mark** is in the middle of the straight edge of the protractor (Fig. 18–28).

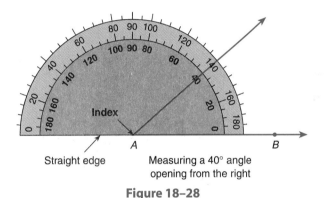

Straight edge   Measuring a 40° angle opening from the right

Figure 18–28

To measure an angle that opens from the right, read the degree measure from the lower scale. Notice in Fig. 18–28 the lower scale starts with 0 at the right. The index of the protractor is aligned with the vertex of the angle, and the straight edge lies along the lower side of the angle.

To measure an angle that opens from the left, read the degree measure from the upper scale. In Fig. 18–29 the upper scale starts with 0 at the left. The index of the protractor is aligned with the vertex of the angle and the straight edge lies along the lower side of the angle.

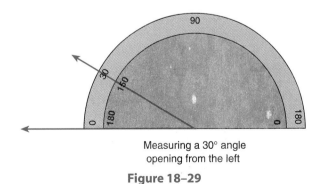

Measuring a 30° angle opening from the left

Figure 18–29

Degrees are divided into 60 equal parts. Each part is called a **minute.**

1 degree (1°) = 60 minutes (60′)

The symbol ′ is used for minutes. Similarly, 1 minute is divided into 60 equal parts called **seconds.**

$$1 \text{ minute } (1') = 60 \text{ seconds } (60'')$$

The symbol ″ is used for seconds. Thus,

$$1° = 60' = 3,600''$$

With the increased popularity of the calculator, it is sometimes desirable to change minutes or seconds to decimal equivalents. Minutes or seconds can be changed to a fractional part of a degree. Then the fraction is changed to its decimal equivalent by dividing the numerator by the denominator. Remember, $1' = \frac{1}{60}$ of a degree, and $1'' = \frac{1}{3,600}$ of a degree.

Scientific and graphing calculators have a key or menu choice that automatically converts between degrees, minutes, and seconds and decimal degrees. A key may be labeled $\boxed{°\ '\ ''}$ or $\boxed{\text{DMS}}$, or a menu choice on the angle menu may be DMS. This function key is also called a **sexagesimal function key** because of its relationship with the number 60. The same key is also used for hours, minutes, and seconds.

To Change the Format of Angle Measures Using a Calculator:

From degrees in decimal notation to degrees, minutes, and seconds (calculator steps will vary with different calculators):

1. Set the calculator to degree mode.
2. Enter the angle measure in decimal notation.
3. Access the sexagesimal function key or menu choice on the calculator (most often labeled $\boxed{°\ '\ ''}$ or $\boxed{\text{DMS}}$).
4. Press $\boxed{\text{ENTER}}$ or $\boxed{=}$.

From degrees, minutes, and seconds to decimal notation (calculator steps will vary with different calculators):

Option 1

1. Enter the series of calculations—degrees + minutes ÷ 60 + seconds ÷ 3,600.

Option 2

1. Enter the degrees followed by the degree symbol ($\boxed{2^{\text{nd}}}$, $\boxed{\text{ANGLE}}$, $\boxed{1:°}$)
2. Enter the minutes followed by the minute symbol ($\boxed{2^{\text{nd}}}$, $\boxed{\text{ANGLE}}$, $\boxed{2:'}$)
3. Enter the seconds followed by the second symbol ($\boxed{\text{ALPHA}}$, $\boxed{''}$)
4. Press $\boxed{\text{ENTER}}$ or $\boxed{=}$.

EXAMPLE Change 25.375° to degrees, minutes, and seconds.

25.375 $\boxed{2^{\text{nd}}}$ $\boxed{\text{ANGLE}}$ $\boxed{4:\blacktriangleright\text{DMS}}$ $\boxed{\text{ENTER}}$ Using a TI-83 or TI-84.

25.375° = **25°22′30″**

EXAMPLE Change 115°40′15″ to a decimal-degree equivalent. Round to the nearest thousandth.

115 + 40/60 + 15/3,600 = 115.6708333   Perform the series of calculations.
115°40′15″ = 115.671°   Round.

SECTION 18–1 SELF-STUDY EXERCISES

1 Use Fig. 18–30 for Exercises 1–3.

1. Name the line in three different ways.
2. Name in two different ways the ray with endpoint P and with interior points Q and R.
3. Name the segment with endpoints Q and R.

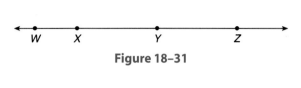

Figure 18–30

Use Fig. 18–31 for Exercises 4–10.

4. Does $\overleftrightarrow{XY}$ represent the same line as $\overleftrightarrow{YZ}$?
5. Does $\overrightarrow{XY}$ represent the same ray as $\overrightarrow{YZ}$?
6. Does $\overline{XY}$ represent the same segment as $\overline{YZ}$?
7. Is $\overleftrightarrow{WX}$ the same as $\overleftrightarrow{WY}$?

Figure 18–31

Use Fig. 18–32 for Exercises 8–12.

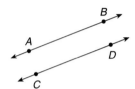

Figure 18–32

8. $\overleftrightarrow{AB}$ and $\overleftrightarrow{CD}$ are _____ lines.
9. $\overleftrightarrow{EF}$ and $\overleftrightarrow{GH}$ _____ at point O.
10. $\overleftrightarrow{IJ}$ and $\overleftrightarrow{KL}$ _____.
11. $\overleftrightarrow{AB}$ and $\overleftrightarrow{CD}$ will never _____.
12. Name two lines that intersect in exactly one point.

Use Fig. 18–33 for Exercises 13–15.

13. Name the angle in two different ways using three capital letters.
14. Name the angle using one capital letter.
15. Name the angle using a number.

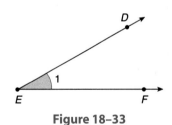

Figure 18–33

2 Fill in the blanks regarding angle rotation.

16. One complete rotation is _____°.
17. One-half a complete rotation is _____°.
18. One-fourth a complete rotation is _____°.
19. One-eighth of a complete rotation is _____°.

Classify the angle measures using the terms *right, straight, acute,* or *obtuse.*

20. 38°   21. 95°   22. 90°   23. 153°   24. 10°   25. 180°

Tell whether the angle pairs are complementary, supplementary, or neither.

26. 42°, 80°   27. 17°, 73°   28. 38°, 142°   29. 52°, 48°   30. 60°, 30°   31. 110°, 70°
32. 42°, 138°

3 Use Fig. 18–34 to answer Exercises 33–35.

33. Name two pairs of equal angles.
34. What are angles a and b called?
35. What is the sum of $\angle c$ and $\angle d$?

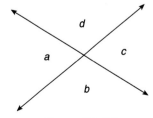

Figure 18–34

In Fig. 18–35 $l \parallel m$ and n is a transversal.

36. Name two pairs of alternate interior angles.
37. What is the sum of $\angle a$ and $\angle g$?
38. What name is given to $\angle d$ and $\angle h$?

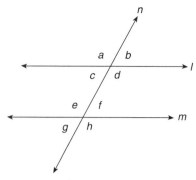

Figure 18–35

4 Change to decimal degree equivalents. Round to the nearest ten-thousandth if necessary.

39. 47′ **40.** 36″ **41.** 5′14″ **42.** 10′15″ **43.** 10°18′15″

Change to equivalent minutes and seconds. Round to the nearest second when necessary.

44. 0.35° **45.** 0.20° **46.** 0.12° **47.** 0.213° **48.** 0.3149°

18–2 | Polygons

Learning Outcomes **1** Find the perimeter of a polygon using the appropriate formula.
2 Find the area of a polygon using the appropriate formula.

1 Find the Perimeter of a Polygon Using the Appropriate Formula.

In Chapter 1, Section 3, Outcome 2, we introduced three polygons: the parallelogram, rectangle, and square. To review, a *polygon* is a plane or flat, closed figure described by straight-line segments and angles. Polygons have different numbers of sides and different properties. Some common polygons are the parallelogram, rectangle, square, rhombus, trapezoid, and triangle (Fig. 18–36).

The *base* of any polygon is the horizontal side or a side that would be horizontal if the polygon's orientation is modified.

The *adjacent side* of any polygon is the side that has an endpoint in common with the base.

A polygon that has four sides is a **quadrilateral.** Quadrilaterals with specific properties have other names.

A *parallelogram* is a quadrilateral with opposite sides that are parallel.

A *rectangle* is a parallelogram with angles that are all right angles.

A *square* is a parallelogram with sides all of equal length and with all right angles. A square can also be described as a rectangle with all sides of equal length.

A **rhombus** is a parallelogram with all sides of equal length.

A **trapezoid** is a quadrilateral having only two parallel sides.

A **triangle** is a polygon that has three sides.

The *perimeter* is the total length of the sides of a plane figure.

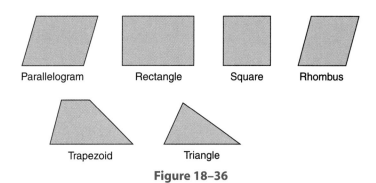

Parallelogram   Rectangle   Square   Rhombus

Trapezoid   Triangle

Figure 18–36

As we saw in Chapter 1, formulas provide a shortcut or guide for the calculations required to find the perimeter of a polygon. We will review the perimeter formulas for a parallelogram, a rectangle, and a square and introduce new formulas for the perimeter of a rhombus, a trapezoid, and a triangle.

Perimeter

Parallelogram	$P_{\text{parallelogram}} = 2b + 2s$ or $P_{\text{parallelogram}} = 2(b + s)$		b is the base s is the adjacent side
Rectangle	$P_{\text{rectangle}} = 2l + 2w$ or $P_{\text{rectangle}} = 2(l + w)$		l is length w is width
Square	$P_{\text{square}} = 4s$		s is length of each side
Rhombus	$P_{\text{rhombus}} = 4s$		s is length of each side
Trapezoid	$P_{\text{trapezoid}} = b_1 + b_2 + a + c$		b_1 and b_2 are bases, a and c are the other sides
Triangle	$P_{\text{triangle}} = a + b + c$		a, b, and c are sides

To find the perimeter of a polygon:

1. Select the appropriate formula.
2. Substitute the known values into the formula.
3. Evaluate the formula (perform the indicated operations).

A *trapezoid* is a four-sided polygon having only two parallel sides. Unlike a parallelogram, the trapezoid's four sides may all be of unequal size, as illustrated in Fig. 18–37.

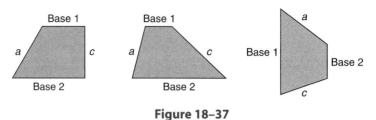

Figure 18–37

The *parallel* sides are called **bases** (b_1 and b_2) with the subscripts 1 and 2 used to distinguish them. The nonparallel sides are designated side a and side c.

EXAMPLE AUTO

The rear window of a pickup truck is a trapezoid whose shorter base is 40 in., whose longer base is 48 in., and whose nonparallel sides are each 16 in. How many inches of rubber gasket are needed to surround the window?

$P_{\text{trapezoid}} = b_1 + b_2 + a + c$ Select appropriate formula.

Estimation

By rounding, the perimeter is $40 + 50 + 20 + 20$ or 130 in.

$P_{\text{trapezoid}} = 40 + 48 + 16 + 16$ Substitute in formula.
$P_{\text{trapezoid}} = 120$ in.   Add.

Interpretation

The rubber gasket to surround the rear window must be 120 in. long.

Figure 18–38

A **triangle** is a three-sided polygon. The perimeter of a triangle is the sum of the lengths of the three sides (Fig. 18–38). The triangle, like the trapezoid, may have unequal lengths for all three sides.

EXAMPLE CON

The triangular gables of an apartment unit under construction are outlined in contrasting trim. If each gable is 32 ft wide and 18 ft on each side, how many feet of contrasting wood trim are needed for each gable? Disregard overlap at the corners (Fig. 18–39).

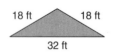

Figure 18–39

$P_{\text{triangle}} = a + b + c$ Select the appropriate formula and substitute in formula.
$P_{\text{triangle}} = 18 + 32 + 18$ Evaluate.
$P_{\text{triangle}} = 68$ ft

Each gable will require 68 linear feet of trim.

Shapes called **composite** figures are figures made up of two or more geometric figures.

To find a missing dimension of a composite shape:

1. Determine how the missing dimension is related to known dimensions.
2. Make a calculation of known dimensions according to the relationship found in Step 1.

EXAMPLE Find the missing dimensions *x* and *y* on the slab foundation (Fig. 18–40).

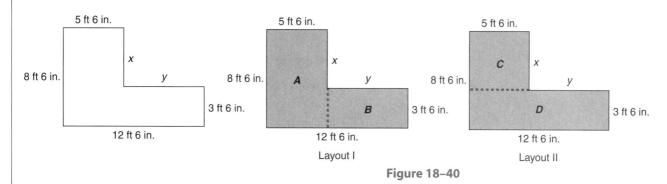

Figure 18–40

Separate the figure into parts.

In layout I (Fig. 18–40) the side of *B* opposite its 3′6″ side is also 3′6″ because opposite sides of a rectangle are equal. The side of *A* opposite its 8′6″ side is, for the same reason, 8′6″. Dimension *x* must therefore be the difference between 8′6″ and 3′6″.

$$x = 8'6'' - 3'6''$$

$$x = 5'$$

If we think of layout II as two horizontal rectangles, we can find dimension *y*. The side opposite the 5′6″ side of *C* must be 5′6″. The side of *D* opposite the 12′6″ side must also be 12′6″. Dimension *y* must therefore be the difference between 12′6″ and 5′6″.

$$y = 12'6'' - 5'6''$$

$$y = 7'$$

The missing dimensions are $x = 5'$ and $y = 7'$.

The *perimeter* of a composite figure is the sum of the lengths of the sides of a figure. The number and the length of the sides vary from one composite figure to the next, so no specific formula covers the entire variety of composite shapes that exist. However, we can use a general formula.

Perimeter of a composite figure:

$$P = a + b + c + \cdots$$

where $a, b, c, \ldots$ are lengths of all the sides.

EXAMPLE Find the number of feet of 4-in. stock needed for the base plates of a room that has the layout shown in Fig. 18–41. Make no allowances for openings when calculating the linear footage of the base plates.

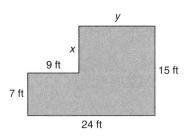

Figure 18–41

1. Find the missing dimensions.

$$x = 15 \text{ ft} - 7 \text{ ft} \qquad y = 24 \text{ ft} - 9 \text{ ft}$$
$$x = 8 \text{ ft} \qquad y = 15 \text{ ft}$$

2. Apply the general formula for the perimeter of a polygon.

$$P = a + b + c + \cdots$$
$$P = 7 \text{ ft} + 24 \text{ ft} + 15 \text{ ft} + 15 \text{ ft} + 8 \text{ ft} + 9 \text{ ft} = 78 \text{ ft}$$

3. Count the number of sides on the layout to make sure that each is substituted into the formula.

The room needs 78 ft of 4-in. stock for the base plates.

2 Find the Area of a Polygon Using the Appropriate Formula.

In Chapter 1, Section 3, Outcome 2, the concept of the area of a polygon was introduced and formulas were given for the area of a rectangle, a square, and a parallelogram. To review, the area of a polygon is the amount of surface of a plane figure. Area is expressed in square units. We will add formulas for the area of the rhombus, trapezoid, and triangle to our list of area formulas.

Area			
Rectangle	$A_{\text{rectangle}} = lw$		l is length w is width
Square	$A_{\text{square}} = s^2$		s is length of a side
Parallelogram	$A_{\text{parallelogram}} = bh$		b is base h is height
Rhombus	$A_{\text{rhombus}} = bh$		b is base h is height

Trapezoid	$A_{\text{trapezoid}} = \dfrac{1}{2}h(b_1 + b_2)$		b_1 and b_2 are bases
			h is height

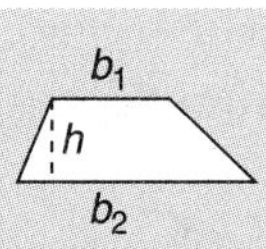

Triangle	$A_{\text{triangle}} = \dfrac{1}{2}bh$		b is base
			h is height

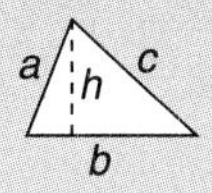

$$A_{\text{triangle}} = \sqrt{s(s - a)(s - b)(s - c)}$$

(Heron's formula)

$$s = \frac{1}{2}(a + b + c)$$

a, b, and c are sides

To find the area of a polygon:

1. Visualize the polygon.
2. Select the appropriate formula.
3. Substitute the known values into the formula.
4. Evaluate the formula.

EXAMPLE Find the area of the trapezoid in Fig. 18–42.

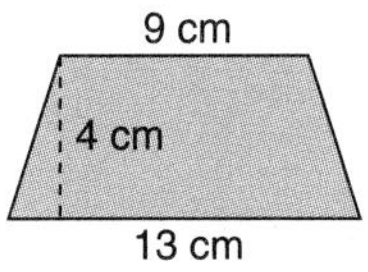

9 cm

4 cm

13 cm

Figure 18–42

$A_{\text{trapezoid}} = \dfrac{1}{2}h\,(b_1 + b_2)$ Select the appropriate formula and substitute the known values.

$A_{\text{trapezoid}} = \dfrac{1}{2}(4)(9 + 13)$ Add inside grouping.

$A_{\text{trapezoid}} = \dfrac{1}{2}(4)\,(22)$ Multiply.

$A_{\textbf{trapezoid}} = \textbf{44 cm}^2$ cm × cm = cm².

EXAMPLE Find the area of the triangle in Fig. 18–43.

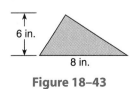

6 in.

8 in.

Figure 18–43

$A_{\text{triangle}} = \dfrac{1}{2}bh$ Select the appropriate formula and substitute the known values.

$A_{\text{triangle}} = \dfrac{1}{2}(8)(6)$ Multiply.

$A_{\textbf{triangle}} = \textbf{24 in}^2$ in. × in. = in²

18–2 Polygons

In some triangles, the height is measured along an imaginary line outside the triangle from the base to the highest point of the triangle (the *apex*) (Fig. 18–44).

EXAMPLE The base of ΔDEF in Fig. 18–44 is 15 cm and the height is 21 cm. Find the area.

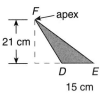

Figure 18–44

$$A_{\text{triangle}} = \frac{1}{2}bh \qquad \text{Substitute into the formula.}$$

$$A_{\text{triangle}} = \frac{1}{2}(15)(21) \qquad \text{Multiply.}$$

$$A_{\text{triangle}} = \frac{1}{2}(315)$$

$$A_{\text{triangle}} = 157.5 \text{ cm}^2$$

The area of ΔDEF is 157.5 cm².

When the three sides of a triangle are known and the height is not known, the area of a triangle can be found using *Heron's formula* (page 709).

EXAMPLE Find the area of the triangle in Fig. 18–45 by using Heron's formula. Round to four significant digits.

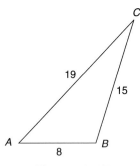

Figure 18–45

$$s = \frac{1}{2}(a + b + c) \qquad \text{Find } s. \text{ Substitute length of sides.}$$

$$s = \frac{1}{2}(19 + 15 + 8) \qquad \text{Add, then multiply.}$$

$$s = 21$$

$$\text{Area} = \sqrt{s(s - a)(s - b)(s - c)} \qquad \text{Heron's formula}$$

$$\text{Area} = \sqrt{21(21 - 19)(21 - 15)(21 - 8)} \qquad \begin{array}{l}\text{Substitute for } s, a, b, \text{ and } c, \text{ and} \\ \text{perform each subtraction.}\end{array}$$

$$\text{Area} = \sqrt{21(2)(6)(13)} \qquad \text{Multiply.}$$

$$\text{Area} = \sqrt{3{,}276} \qquad \text{Take the square root using a calculator.}$$

$$\text{Area} = 57.23635209 \qquad \text{Principal square root}$$

$$\textbf{Area} = \textbf{57.24 square units} \qquad \text{Round to four significant digits.}$$

As with the perimeter, the *area* of a composite figure is found by finding the sum of the areas of the parts of the figure.

Area of a composite figure:

$$A = A_1 + A_2 + A_3 + \cdots$$

where $A_1, A_2, A_3, \ldots$ are the areas of all the parts of the composite figure.

EXAMPLE Find the number of square yards of carpeting required for the room in Fig. 18–46.

1. Divide the composite shape into two polygons with areas we can compute (Fig. 18–47). In this case, A is a rectangle and B is a square.

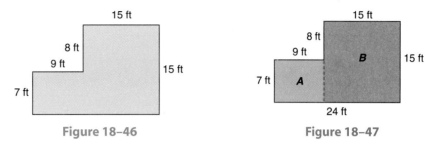

Figure 18–46 Figure 18–47

2. Find the areas of the smaller polygons and add them.

Rectangle A	**Square B**
$A_1 = lw$	$A_2 = s^2$
$A_1 = 9 \times 7$	$A_2 = 15^2$
$A_1 = 63 \text{ ft}^2$	$A_2 = 225 \text{ ft}^2$

$$A_1 + A_2 = \text{total area}$$
$$63 + 225 = 288 \text{ ft}^2$$

3. Convert square feet to square yards using a unit ratio.

$$\frac{\overset{32}{\cancel{288} \text{ ft}^2}}{1} \times \frac{1 \text{ yd}^2}{\underset{1}{\cancel{9} \text{ ft}^2}} = 32 \text{ yd}^2$$

The room requires 32 yd² of carpeting.

EXAMPLE Find the area of a gable end of a gambrel roof that has dimensions as shown in Fig. 18–48.

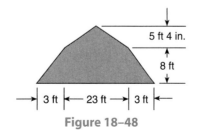

Figure 18–48

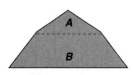

Figure 18–49

1. Divide the gable end of the gambrel roof into polygons with areas that can be calculated (Fig. 18–49).
2. Calculate the areas of the triangle and the trapezoid. Find the sum of the areas.

Triangle _A_	**Triangle _B_**
Base = 23′	Base$_1$ = 3′ + 23′ + 3′ = 29′
Height = 5′4″ = $5\frac{1}{3}$′	Base$_2$ = 23′
$A_{triangle} = \frac{1}{2}bh$	Height = 8′
$A_{triangle} = \frac{1}{2}(23)\left(5\frac{1}{3}\right)$	$A_{trapezoid} = \frac{1}{2}h(b_1 + b_2)$
$A_{triangle} = \frac{1}{2}(23)\left(\frac{\overset{8}{\cancel{16}}}{3}\right)$	$A_{trapezoid} = \frac{1}{2}(8)(29 + 23)$
$A_{triangle} = \frac{184}{3} = 61\frac{1}{3}$ ft^2	$A_{trapezoid} = 4(52)$
$A_{triangle} + A_{trapezoid} = A_{gable}$	$A_{trapezoid} = 208$ ft^2

$$61\frac{1}{3} \text{ ft}^2 + 208 \text{ ft}^2 = 269\frac{1}{3} \text{ ft}^2$$

The area of the gable end of the gambrel roof is 269$\frac{1}{3}$ ft^2.

Sometimes when we figure area, we find it more convenient to calculate an overall area and subtract a smaller area.

EXAMPLE   Find the area of the flat metal piece shown in Fig. 18–50.

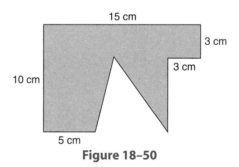

Figure 18–50

1. Divide the figure into polygons for which we can calculate areas (Fig. 18–51). Other strategies could be used.

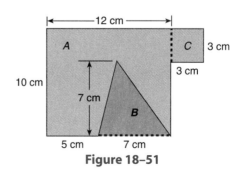

Figure 18–51

2. Find the missing dimensions.

3. Find the area of the square C (A_C), rectangle A (A_A), and triangle B (A_B). The area of the piece of metal is $A_C + A_A - A_B$.

Square C	Rectangle A	Triangle B
$A_C = s^2$	$A_A = lw$	$A_B = \dfrac{1}{2}bh$
$A_C = 3^2$	$A_A = 12(10)$	$A_B = \dfrac{1}{2}(7)(7)$
$A_C = 9 \text{ cm}^2$	$A_A = 120 \text{ cm}^2$	$A_B = \dfrac{1}{2}(49)$
		$A_B = 24.5 \text{ cm}^2$

Area of piece $= A_C + A_A - A_B = 9 + 120 - 24.5 = 104.5 \text{ cm}^2$

The area of the flat piece of metal is 104.5 cm².

SECTION 18–2 SELF-STUDY EXERCISES

1

1. Find the perimeter of the parallelogram in Fig. 18–52.

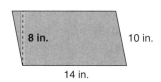

8 in. 10 in.

14 in.

Figure 18–52

2. Find the perimeter of the parallelogram in Fig. 18–53.

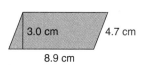

3.0 cm 4.7 cm

8.9 cm

Figure 18–53

Solve the problems involving perimeter.

3. An illuminated sign in the main entrance of a hospital is a parallelogram with a base of 56 in. and an adjacent side of 42 in. How many inches of aluminum molding are needed to frame the sign?

4. A customized van has a window cut in each side in the shape of a parallelogram with a base of 30 in. and an adjacent side of 14 in. How many inches of trim are needed to surround the two windows?

5. **CON** A contemporary building has a window in the shape of a parallelogram with a base of 80 in. and an adjacent side of 40 in. How many inches of trim are needed to surround the window?

6. **CON** A table for a mathematics lab has a top in the shape of a parallelogram with a base of 40 in. and an adjacent side of 24 in. How many inches of edge trim are needed to surround the tabletop?

7. Find the perimeter of the rectangle in Fig. 18–54.

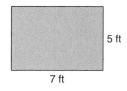

5 ft

7 ft

Figure 18–54

8. Find the perimeter of the rectangle in Fig. 18–55.

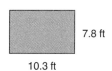

7.8 ft

10.3 ft

Figure 18–55

9. **CON** A rectangular parking lot is 200 ft by 145 ft. Find the perimeter of the parking lot.

11. **CON** How many feet of quarter-round molding are needed to finish around the baseboard after sheet vinyl flooring is installed if the room is 14 ft by 20 ft and there are three 3.5-ft-wide doorways?

10. **CON** A room is 18 ft by 15 ft. How many feet of chair rail are needed for the room? Disregard openings.

12. **CON** The swimming pool in Fig. 18–56 measures 36 ft by 20 ft. How much fencing is needed, including material for a gate, if the fence is to be built 9 ft from each side of the pool?

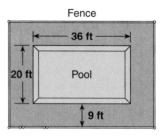

Figure 18–56

13. **CON** A Formica tabletop measures 40 in. by 62 in. How many feet of edge trim are needed (12 in. = 1 ft)?

15. Find the perimeter of Fig. 18–57.

Figure 18–57

14. **CON** A countertop requires rolled edging to be installed on all four sides. How much rolled edging material is needed if the countertop measures 25 in. by 40 in.?

16. Find the perimeter of Fig. 18–58.

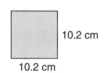

Figure 18–58

17. **CON** The square parking lot of a doctor's office is to have curbs built on all four sides. If the lot is 160 ft on each side, how many feet of curb are needed? Allow 12 ft for a driveway into the parking lot.

18. **CON** A border of 4-in. × 4-in. wall tiles surrounds the floor of a shower stall that is 52 in. × 52 in. How many tiles are needed for this border? Disregard spaces for grout (connecting material between the tiles).

Find the perimeter of Figs. 18–59 and 18–60.

19.

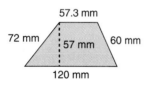

Figure 18–59

20.

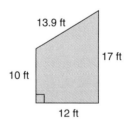

Figure 18–60

21. **CON** The six glass panes in a kitchen light fixture each measure $4\frac{1}{2}$ in. along the top and 10 in. along the bottom. The top and bottom are parallel. The two nonparallel sides of each pane are 10 in. What is the combined perimeter of the six trapezoidal panes?

23. A lot in an urban area has two nonparallel sides that are each 64 ft. The parallel sides are 120 ft and 154 ft. Find the perimeter of the trapezoidal property.

22. **CON** A section of a hip roof is a trapezoid measuring 38 ft at the bottom, 14 ft at the top, 10 ft high, and 11 and 12 ft, respectively, on each side. Find the perimeter of this section of the roof.

24. A swimming pool is fashioned in a trapezoidal design. The parallel sides are $18\frac{1}{2}$ ft and 31 ft. The other sides are $24\frac{1}{2}$ ft and 24 ft. What is the perimeter of the pool?

Find the perimeter of the triangles in Figs. 18–61 and 18–62.

25.

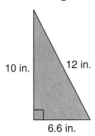

Figure 18–61

10 in. 12 in. 6.6 in.

26.

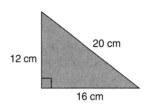

Figure 18–62

20 cm 12 cm 16 cm

27. CON Selena Henson is planning a patio that will adjoin the sides of her L-shaped home. One side of the patio is 24 ft and the other is 18 ft. The shape of the patio is triangular. Draw a representation of the patio and find the perimeter if the length of the third side is 30 ft.

28. Find the perimeter of a triangle that has sides measuring 2 ft 6 in., 1 yd 8 in., and 4 ft 6 in.

2

29. Find the area of the shape in Fig. 18–63.

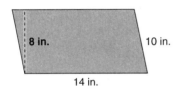

8 in. 10 in. 14 in.

Figure 18–63

30. Find the area of the parallelogram in Fig. 18–64.

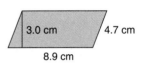

3.0 cm 4.7 cm 8.9 cm

Figure 18–64

31. Find the area of the shape in Fig. 18–65.

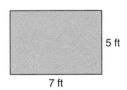

5 ft 7 ft

Figure 18–65

32. Find the area of the shape in Fig. 18–66.

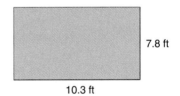

7.8 ft 10.3 ft

Figure 18–66

33. A rectangular parking lot is 340 ft by 125 ft. Find the number of square feet in the parking lot.

35. Find the area of Fig. 18–67.

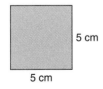

5 cm 5 cm

Figure 18–67

34. A room is 15 ft by 12 ft. How many square feet of flooring are needed for the room?

36. Find the area of the square in Fig. 18–68.

10.2 cm

Figure 18–68

37. CON Madison Duke is wallpapering a laundry room 8 ft by 8 ft by 8 ft high. How many square feet of paper will she need if there are 63 ft^2 of openings in the room?

38. AG/H Making no allowances for bases, the pitcher's mound, or the home plate area, how many square yards of artificial turf are needed to resurface an infield at an indoor baseball stadium? The infield is 90 ft on each side (9 ft^2 = 1 yd^2).

39. AG/H Ted Davis is a farmer who wants to apply fertilizer to a 40-acre field with dimensions $\frac{1}{4}$ mi $\times$ $\frac{1}{4}$ mi. Find the area in square miles.

41. CON Tiles that are 6 in. $\times$ 6 in. cover the floor of a shower. How many tiles are needed for the floor if the shower measures 4.5 ft by 6 ft?

43. Find the area of Fig. 18–69.

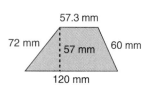

Figure 18–69

45. CON The six glass panes in a kitchen light fixture each measure $4\frac{1}{2}$ in. along the top and 10 in. along the bottom. The top and bottom are parallel. The height of each pane is 8 in. What is the combined area of the six trapezoidal panes?

47. Find the area of the triangle in Fig. 18–71.

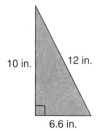

Figure 18–71

49. CON Carlee McAnally is planning a patio. The triangular-shaped patio has sides 24 ft, 18 ft, and 22 ft. Find the number of square feet of surface area to be covered with concrete. Round to the nearest square foot.

40. CON A 36-in. $\times$ 36-in. ceramic tile shower stall is being installed. How many 4-in. $\times$ 4-in. tiles are needed to cover the floor? Disregard the drain opening and grout spaces.

42. INDTR A 20-in. $\times$ 20-in. central heating and air conditioning return air vent is being installed in a wall. Find the area of the wall opening.

44. Find the area of Fig. 18–70.

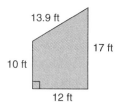

Figure 18–70

46. CON A section of a hip roof is a trapezoid measuring 38 ft at the bottom, 14 ft at the top, and 10 ft high. Find the area of this section of the roof in square feet.

48. Find the area of the triangle in Fig. 18–72.

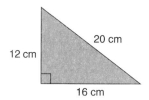

Figure 18–72

50. CON If aluminum siding costs $6.75 a square yard installed, how much does it cost to put the siding on the two triangular gable ends of a roof under construction? Each gable has a span (base) of 30 ft 6 in. and a rise (height) of 7 ft 6 in. Any portion of a square yard is rounded to the next highest square yard for each gable.

18–3 | Circles and Radians

Learning Outcomes

1 Find the circumference or area of a circle using the appropriate formula.
2 Convert angle measures between degrees and radians.
3 Find the arc length of a sector.
4 Find the area of a sector or segment.

A **circle** is a closed curved line with points that lie in a plane and are the same distance from the *center* of the figure (Fig. 18–73).

716   Chapter 18 / Geometry

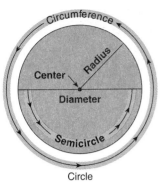

Figure 18–73

The **center** of a circle is the point that is the same distance from every point on the circle.

The **radius** (plural: *radii,* pronounced "ray · dē · ī") is a straight line segment from the center of a circle to a point on the circle. It is half the diameter.

The **diameter** of a circle is a straight line segment from a point on the circle through the center to another point on the circle.

The **circumference** of a circle is the perimeter or length of the closed curved line that forms the circle.

A **semicircle** is half a circle and is created by drawing a diameter.

1 **Find the Circumference or Area of a Circle Using the Appropriate Formula.**

The circle is a geometric form with a special relationship between its circumference and its diameter. If we divide the circumference of any circle by its diameter, the quotient is always the same number.

$$\pi = \frac{\text{circumference}}{\text{diameter}} = \frac{C}{d}$$

This number is a nonrepeating, nonterminating decimal approximately equal to 3.1415927 to seven decimal places. The Greek letter π (pronounced "pie") represents this value. Convenient approximations often used in calculations involving π are $3\frac{1}{7}$ and 3.14. Many calculators have a π key.

The formulas for the circumference and area of a circle are:

Circumference (C)	Area (A)			
$C = \pi d$ $C = 2\pi r$	$A = \pi r^2$			d is diameter ($d = 2r$) r is radius ($r = \frac{1}{2}d$)

To find the circumference or area of a circle:

1. Select the appropriate formula.
2. Substitute values for r or d as appropriate.
3. Evaluate the formula. Use the calculator value for π.

EXAMPLE Find the circumference (to the nearest tenth of a meter) of a circle that has a diameter of 1.3 m (Fig. 18–74).

1.3 m

Figure 18–74

$C = \pi d$	Select the circumference formula with diameter d.
$C = \pi(1.3)$	Use the π key on calculator.
$C = 4.08407045$ m	Evaluate.

The circumference is 4.1 m (rounded).　　Circumference is a linear measure.

18–3 Circles and Radians

TIP

Calculator Values of π

Calculations involving π are always approximations. Scientific or graphing calculators include a π function or menu option where π to seven or more decimal places is computed by pressing a single key. Other calculators require the use of more than one key to activate the $\boxed{\pi}$ key. However, for computations by hand or a calculator without the $\boxed{\pi}$ function, 3.14 is sometimes adequate. **We use the calculator value 3.141592654 for π in all examples and exercises unless stated otherwise.**

TIP

Fractions versus Decimals

When you use your calculator, you may find it convenient to convert mixed U.S. customary linear measurements to their decimal equivalents. For instance, if a diameter is 7 ft 6 in., convert it to 7.5 ft (from $7\frac{6}{12}$ ft, in which $\frac{6}{12} = 0.5$).

EXAMPLE Find to the nearest hundredth the circumference of a circle with a radius of 1 ft 9 in. (Fig. 18–75).

1 ft 9 in.

Figure 18–75

$C = 2\pi r$
$C = 2\pi(1.75)$
$C = 10.99557429$
$C = 11.00$ ft (rounded)

Substitute for π and r: $1\frac{9}{12}$ ft = 1.75 ft

Circumference is a linear measure.

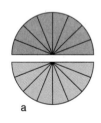

a

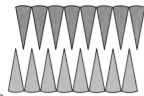

b

$\frac{1}{2}C$

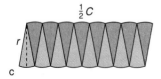

r

c

Figure 18–76

The area of a circle, like the circumference, is obtained from relationships within the circle. If we divide a circle into two semicircles and then subdivide each semicircle into pie-shaped pieces, we get something like Fig. 18–76a. If we then spread the upper and lower pie-shaped pieces, we get Fig. 18–76b. Now if we push the upper and lower pieces together, the result approximates the parallelogram in Fig. 18–76c, whose base is $\frac{1}{2}$ the circumference and whose height is the radius. Thus, the area of the circle is approximately the area of a parallelogram, that is, base times height. Since the base of the parallelogram is one-half the circumference and the height is the radius, the area of a circle equals one-half the circumference times the radius.

$A = \frac{1}{2}C \times r$

The formula for circumference is $C = 2\pi r$, so we substitute $2\pi r$ for C.

$A = \frac{1}{2}(2\pi r)(r)$

Multiply. Reduce where possible.

$A = \frac{1}{2}\overset{1}{(2\pi r)}(r)\over{1}$

Area is a square measure: $r \cdot r = r^2$

$A = \pi r^2$

EXAMPLE Find the area of a circle whose radius is 8.5 m (Fig. 18–77). Round to tenths.

Figure 18-77

$A = \pi r^2$ Select appropriate formula and substitute for r.

$A = \pi(8.5)^2$ Square the radius.

$A = \pi(72.25)$ Multiply by π. Use the π key on your calculator.

$A = 226.9800692 \text{ m}^2$ Area is a square measure.

The area of the circle is 227.0 m².

EXAMPLE Find the area to the nearest hundredth of the top of a circular tank with a diameter of 12 ft 8 in.

$A = \pi r^2$

$A = (\pi)\left(\dfrac{12 + \frac{8}{12}}{2}\right)^2$ Follow the order of operations. 8 in. $= \frac{8}{12}$ ft. The diameter, $12 + \frac{8}{12}$, divided by 2 is the radius.

$A = 126.012772 \text{ ft}^2$ Calculator result

$A = 126.01 \text{ ft}^2$ (rounded)   Area is a square measure.

The area of the top of the tank is 126.01 ft².

EXAMPLE A 15-in.-diameter wheel has a 3-in. hole in the center. Find the area of a side of the wheel to the nearest tenth (Fig. 18–78).

We are asked to find the area of the colored portion of the wheel in Fig. 18–78. To do so, we find A_{outside}, the area of the larger circle (diameter 15 in.) and *subtract* the area of A_{inside}, the smaller circle (diameter 3 in.). The colored portion is called a **ring.**

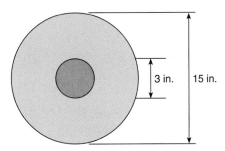

Figure 18-78

$A_{\text{outside}} = \pi r^2$ $\left(r = \dfrac{15}{2} = 7.5\right)$ $A_{\text{inside}} = \pi r^2$ $\left(r = \dfrac{3}{2} = 1.5\right)$

$A_{\text{outside}} = \pi(7.5)^2$ $A_{\text{inside}} = \pi(1.5)^2$

$A_{\text{outside}} = \pi(56.25)$ $A_{\text{inside}} = \pi(2.25)$

$A_{\text{outside}} = 176.7145868 \text{ in}^2$ $A_{\text{inside}} = 7.068583471 \text{ in}^2$

area of wheel (ring) $= A_{\text{outside}} - A_{\text{inside}}$

$A_{\text{wheel}} = 176.7145868 - 7.068583471 = 169.6460033$ or 169.6 in^2

The area of the wheel (ring) is 169.6 in².

EXAMPLE INDTR

A bandsaw has two 25-cm wheels spaced 90 cm between centers (Fig. 18–79). Find the length of the saw blade.

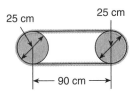

This layout is a composite figure which consists of a semicircle at each end and a rectangle in the middle. It is called a **semicircular-sided** figure. The two semicircles equal one whole circle, so we need to find the circumference of one circle (wheel) and add it to the lengths of the two sides of the rectangle.

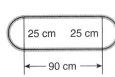

$C = \pi d$	total length of blade = $C + 2l$
$C = \pi(25)$	total length of blade = 78.53981634 + 2(90)
$C = 78.53981634$ cm	total length of blade = 258.5 cm (rounded)

Figure 18–79

The bandsaw blade is 258.5 cm in length.

Specific applications for a particular industry or career often use the formulas for area, perimeter, or circumference.

EXAMPLE INDTEC

Find the cutting speed of a lathe if a piece of work that has a 7-in. diameter turns on a lathe at 75 revolutions per minute (rpm) (Fig. 18–80).

The **cutting speed** is the speed of a tool that passes over the work, such as the speed of a lathe or a sander as it sands (passes over) a piece of wood. If the cutting speed is too fast or too slow, safety and quality are impaired. The formula for cutting speed is CS = C (in feet) $\times$ rpm.

$$CS = \text{cutting speed}$$
$$C = \text{circumference or one revolution (in feet)}$$
$$\text{rpm} = \text{revolutions per minute (r/min)}$$

Cutting speed is measured in *feet per minute* (ft/min)

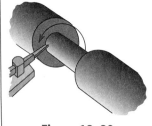

Figure 18–80

$C = \pi d$	Find the circumference.
$C = \pi(7)$	
$C = 21.99114858$ in.	Convert 21.99114858 in. into feet using a unit ratio.
	$\dfrac{21.99114858 \text{ in.}}{1} \times \dfrac{1 \text{ ft}}{12 \text{ in.}} = 1.832595715 \text{ ft}$
$C = 1.832595715$ ft	Circumference in feet or feet per revolution

$CS = C \times \text{rpm}$ = 1.832595715 ft/r $\times$ 75 r/min = 137 ft/min   Round.

The cutting speed of the lathe is approximately 137 ft/min.

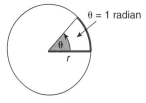

Figure 18–81

2 Convert Angle Measures Between Degrees and Radians.

In Section 18–1, we learned that angles can be measured in units called *degrees*. Angles can also be measured in *radians*. A **radian** is the measure of a central angle of a circle whose intercepted arc is equal in length to the radius of the circle (Fig. 18–81). The abbreviation for radian is **rad**. A **central angle** of a circle is an angle with its vertex at the center of the circle. The legs or sides of a central angle are radii of the circle.

The circumference of a circle is related to the radius by the formula $C = 2\pi r$. Thus, the ratio of the circumference to the radius of any circle is $\frac{C}{r} = 2\pi$; that is, the radius could be measured off 2π times (about 6.28 times) along the circumference. A complete rotation is 2π radians (see Fig. 18–82).

| 1 radian | 3 radians | 3.14 or π radians | 5 radians | 6.28 or 2π radians |

Figure 18–82

How are radians and degrees related? A central angle measuring 1 radian makes an arc length equal to the radius.

TIP

Degree and Radian Notation

Angles measured in degrees always require the word *degree* or the degree symbol to be written. No comparable symbol exists for the radian. The abbreviation *rad* or no unit at all indicates that radian is the measuring unit. Some calculators use a raised *r* to indicate a radian unit.

A complete rotation is $360°$ or 2π rad. To convert from one unit of angle measure to another, we multiply by a *unit ratio* that relates degrees and radians. Because $360° = 2\pi$ rad, we can simplify the relationship to $180° = \pi$ rad $\left(\dfrac{360°}{2} = \dfrac{2\pi}{2} \text{ rad} \right)$.

To convert degrees to radians:

1. Multiply degrees by the unit ratio $\dfrac{\pi \text{ rad}}{180°}$.
2. Write the product in simplest form.

Using a calculator (steps may vary):

1. Set angle MODE to radian.
2. Enter the degree measure followed by the degree symbol. Use $\boxed{° ' ''}$, $\boxed{\text{DMS}}$, or the angle menu, option $\boxed{1:°}$.
3. Enter the minute measure followed by the minute symbol. Use $\boxed{° ' ''}$, $\boxed{\text{DMS}}$, or the angle menu, option $\boxed{2:'}$.
4. Enter the second measure followed by the second symbol. Use $\boxed{° ' ''}$, $\boxed{\text{DMS}}$, or $\boxed{\text{ALPHA}}$ $\boxed{''}$.
5. Press $\boxed{\text{ENTER}}$.

EXAMPLE Convert the angle measures to radians. Use the calculator value for π to change to a decimal equivalent rounded to the nearest hundredth.

(a) 20°   **(b)** 175°

(a) $20° \times \dfrac{\pi \text{ rad}}{180°} = \dfrac{20(\pi)}{180} =$ **0.35 rad**   Multiply by $\dfrac{\pi \text{ rad}}{180°}$.

(b) 175° = 3.054326191 = **3.05 rad**   Set calculator in radian mode.

175 [2nd] [ANGLE] [1:°] [ENTER] [ENTER]

When converting from radians to degrees, we multiply by a unit ratio so that the radians cancel and are replaced by degrees.

To convert radians to degrees:

1. Multiply radians by the unit ratio $\dfrac{180°}{\pi \text{ rad}}$.
2. Write the product in simplest form.

Using a calculator (steps may vary):

1. Set angle MODE to degree.
2. Enter radian amount and radian symbol (angle menu, option [3:ʳ]).
3. If degrees, minutes, and seconds are desired, use DMS function (angle menu, option [4:▶DMS]).

EXAMPLE Convert to degrees. Round to the nearest ten-thousandth of a degree.

(a) 2 rad   **(b)** $\dfrac{\pi}{2}$ rad

(a) $2 \text{ rad} \times \dfrac{180°}{\pi \text{ rad}} = \dfrac{360°}{\pi} =$ **114.5916°**   Multiply by $\dfrac{180°}{\pi \text{ rad}}$. Use calculator value of π and round.

(b) $\dfrac{\pi}{2}$ rad = 90°

In degree mode:

[(] [π] [÷] [2] [)] [ANGLE] [3:ʳ] [ENTER].

EXAMPLE Convert to degrees, minutes, and seconds. Round to the nearest second.

(a) 1 rad   **(b)** 3.2 rad

(a) 1 rad = 57°17′ 44.806″   In degree mode:

Press 1 [ANGLE] [3:ʳ] [ANGLE] [4:▶DMS] [ENTER].

1 rad = **57°17′45″**   Round.

(b) 3.2 rad = 183°20′47.38″   In degree mode:

Press 3.2 [ANGLE] [3:ʳ] [ANGLE] [4:▶DMS] [ENTER].

3.2 rad = **183°20′47″**   Round.

Thus, 3.2 rad = 183°20′47″.

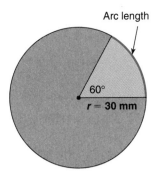

Arc length

60°
$r = 30$ mm

Figure 18-83

3 **Find the Arc Length of a Sector.**

We often work with figures that are less than a whole circle. For example, earlier we worked with the semicircle in composite figures. A **sector** of a circle is the portion of the area of a circle cut off by two radii. These two radii form a *central angle*. The sector is formed by a central angle of the circle and the arc connecting the sides of the angle (radii). See Fig. 18–83. We use the Greek letter **theta** (θ) to represent angle measures.

The **arc length** of a sector is the portion of the circumference intercepted by the sides of the sector (see Fig. 18–83).

To find the arc length of a sector:

Using degrees:

1. Substitute known values into the formula.

$$s = \frac{\theta}{360}(2\pi r) \qquad \text{or} \qquad s = \frac{\theta}{360}(\pi d)$$

where θ is a central angle measured in degrees and r is the radius of the circle.

2. Evaluate.

Using radians:

1. Substitute the given radian measure of the central angle and the radius in the formula

$$s = \theta r$$

where θ is the central angle measured in radians and r is the radius of the circle.

2. Simplify the expression.

EXAMPLE Find the arc length of the sector formed by a 60° central angle if the radius of the circle is 30 mm (Fig. 18–85).

$s = \dfrac{\theta}{360}(2\pi r)$ Select the formula for degrees. Substitute values.

$s = \dfrac{60}{360}(2)(\pi)(30)$ Evaluate.

$s = 31.41592654$ Round.

$s = \textbf{31.42 mm}$

Exact Solutions Involving π

There are times when exact solutions are desirable. An exact solution involving π leaves the factor of π in the solution. In the previous example, the exact solution is found by simplifying the expression

$$\frac{60}{360}(2)(\pi)(30) = \frac{60}{360}(2)(30)\pi = 10\pi$$

18-3 Circles and Radians

EXAMPLE Find the arc length intercepted on the circumference of the circle in Fig. 18–84 by

a central angle of $\frac{\pi}{3}$ radians (rad) if the radius of the circle is 10 cm.

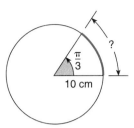

Figure 18–84

$$s = \theta r$$ Select the formula for radians. Substitute $\frac{\pi}{3}$ for θ and 10 cm for r.

$$s = \frac{\pi}{3}(10 \text{ cm})$$ Evaluate.

$$s = \frac{\pi(10 \text{ cm})}{3}$$ Use calculator value for π.

$$s = 10.47197551 \text{ cm}$$ Round to the nearest hundredth.

Thus, the arc length of the intercepted arc is 10.47 cm.

TIP Analyzing Arc Length Dimensions

In the preceding example, $\left(\frac{\pi}{3} \text{ rad}\right)(10 \text{ cm}) = 10.47 \text{ cm}$, what happened to the radians? In the definition of a radian, we relate the measure of an arc connecting the endpoints of a central angle to the measure of the radius of the circle. Therefore, the arc length will have the same measuring unit as the radius.

To find the central angle or radius of a sector given the arc length:

1. Substitute given values in the formula $s = \theta r$, where s is the arc length, θ is the central angle measured in radians, and r is the radius of the circle.
2. Solve for the missing value.

EXAMPLE Find the radian measure of an angle at the center of a circle of radius 5 m. The angle intercepts an arc length of 12.5 m (see Fig. 18–85).

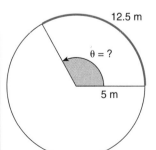

Figure 18–85

$$s = \theta r$$ Formula for arc length. Solve for θ.

$$\theta = \frac{s}{r}$$ Substitute 12.5 m for s and 5 m for r. Arc length and radius measuring units are compatible.

$$\theta = \frac{12.5 \text{ m}}{5 \text{ m}}$$

$$\theta = 2.5 \text{ rad}$$

The angle is 2.5 rad.

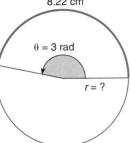

8.22 cm

θ = 3 rad

r = ?

Figure 18–86

EXAMPLE Find the radius of an arc if the length of the arc is 8.22 cm and the intercepted central angle is 3 rad (see Fig. 18–86).

$$s = \theta r$$ Formula for arc length. Solve for *r*.

$$r = \frac{s}{\theta}$$ Substitute 8.22 cm for *s* and 3 rad for θ.

$$r = \frac{8.22 \text{ cm}}{3}$$ Simplify.

$$r = 2.74 \text{ cm}$$ Same measuring unit as arc length

The radius of the arc is 2.74 cm.

---

4 **Find the Area of a Sector or Segment.**

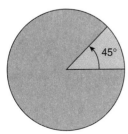

To find the area of a sector:

Using degrees:

1. Calculate the portion of the circle included in the sector. θ is a central angle measured in degrees.

$$\frac{\theta}{360} = \text{fractional part of circle}$$

2. Find the fractional part of the area of the circle.

$$A_{\text{sector}} = \frac{\theta}{360}\pi r^2 \text{ where } \pi r^2 = \text{area of circle}$$

Using radians:

1. Substitute known values into the formula $A_{\text{sector}} = \frac{1}{2}\theta r^2$, where θ is the central angle measured in radians and *r* is the radius of the circle.

2. Solve for the missing value.

EXAMPLE Find the area of a sector with a central angle of 45° in a circle with a radius of 10 in. Round to hundredths (see Fig. 18–87).

45°

Figure 18–87

$$A_{\text{sector}} = \frac{\theta}{360}\pi r^2$$ Substitute for θ and *r*.

$$A_{\text{sector}} = \frac{45}{360}(\pi)(10)^2$$ Perform the indicated operations.

$$A_{\text{sector}} = 0.125(\pi)(100)$$

$$A_{\text{sector}} = 12.5\pi$$ Exact solution

$$A_{\text{sector}} = 39.26990817 \text{ in}^2$$ Round.

The area of the sector is 39.27 in².

18–3 Circles and Radians

725

EXAMPLE
INDTR

A cone is made from sheet metal. To form a cone, a sector with a central angle of 40°20′ is cut from a metal circle whose diameter is 20 in. Find the area of the stretchout (portion of the circle) that is used to form the cone. Round to the nearest hundredth (Fig. 18–88).

Stretchout (flat)

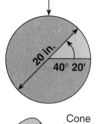

Figure 18–88

area used for cone = area of circle − area of sector

Circle	**Sector**	

$A_{circle} = \pi r^2$ $\qquad$ $A_{sector} = \dfrac{\theta}{360}\pi r^2$ $\qquad$ Substitute known values. Convert 40°20′ to 40.333333333°.

$A_{circle} = \pi(10)^2$ $\qquad$ $A_{sector} = \dfrac{40.3\overline{3}}{360}(314.1592654)$ $\qquad$ Substitute A_1 for πr^2.

$A_{circle} = \pi(100)$ $\qquad$ $A_{sector} = 0.112037037(314.1592654)$

$A_{circle} = 314.1592654 \text{ in}^2$ $\qquad$ $A_{sector} = 35.19747324 \text{ in}^2$

area used for cone $= A_{circle} - A_{sector}$

$A_{cone} = 314.1592654 - 35.19747324$

$A_{cone} = 278.9617922 \text{ in}^2$ $\qquad$ Round.

The area of the metal sector used to form the cone is 278.96 in².

EXAMPLE

Find the area of a sector that has a central angle of 5 rad and has a radius of 7.2 in.

$A_{sector} = \dfrac{1}{2}\theta r^2$ $\qquad$ Substitute $\theta = 5$ and $r = 7.2$.

$A_{sector} = \dfrac{1}{2}(5)(7.2 \text{ in.})^2$ $\qquad$ Simplify.

$A_{sector} = 129.6 \text{ in}^2$ $\qquad$ Area of sector

The area of the sector is 129.6 in².

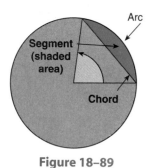

Figure 18–89

If a line segment (called a *chord*) joins the endpoints of the radii that form a sector, the sector is divided into two figures, a triangle and a *segment* (Fig. 18–89). A **chord** is a line segment joining two points on the circumference of a circle. The portion of the circumference cut off by a chord is an arc. A **segment** is the portion of the area of a circle bounded by a chord and an arc.

Because the chord divides the sector into a triangle and a segment, we can calculate the area of the segment by subtracting the area of the triangle from the area of the sector.

To find the area of a segment:

1. Substitute known values into the formula.

Using degrees:

$$A_{segment} = \frac{\theta}{360}\pi r^2 - \frac{1}{2}bh$$

Using radians:

$$A_{segment} = \frac{1}{2}\theta r^2 - \frac{1}{2}bh$$

where $\frac{\theta}{360}\pi r^2$ or $\frac{1}{2}\theta r^2$ is the area of the sector and $\frac{1}{2}bh$ is the area of the triangle.

2. Evaluate.

EXAMPLE Find the area to the nearest hundredth of the segment in Fig. 18–90.

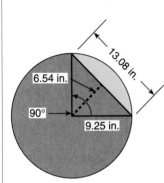

Figure 18–90

$$A_{segment} = \frac{\theta}{360}\pi r^2 - \frac{1}{2}bh \qquad \text{Substitute in formula.}$$

$$A_{segment} = \frac{90}{360}(\pi)(9.25)^2 - \frac{1}{2}(13.08)(6.54) \qquad \text{Evaluate.}$$

$$A_{segment} = 67.20063036 - 42.7716$$

$$A_{segment} = 24.42903036 \text{ in}^2 \qquad \text{Round.}$$

The area of the segment is 24.43 in².

EXAMPLE INDTR A segment in Fig. 18–91 is removed so that a template for a cam is made from the rest of the circle. What is the area of the template? Give the answer to the nearest hundredth.

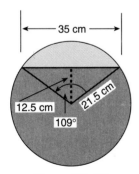

Figure 18–91

area of template = area of circle − area of segment

Circle

$$A_{circle} = \pi r^2$$

$$A_{circle} = \pi(21.5)^2$$

$$A_{circle} = 1{,}452.201204 \text{ cm}^2$$

Segment

$$A_{segment} = \frac{\theta}{360}\pi r^2 - \frac{1}{2}bh$$

$$A_{segment} = \frac{109}{360}(1{,}452.201204) - \frac{1}{2}(35)(12.5)$$

$$A_{segment} = 220.9442535 \text{ cm}^2$$

$$A_{template} = A_{circle} - A_{segment}$$

$$A_{template} = 1{,}452.201204 - 220.9442535$$

$$A_{template} = 1{,}231.256951 \text{ cm}^2 \qquad \text{Round.}$$

The area of the template is 1,231.26 cm².

SECTION 18–3 SELF-STUDY EXERCISES

1 Find the circumference of circles with the following dimensions. Round to tenths.

1. Diameter = 8 cm

2. Diameter = 15 m

3. Radius = 3 in.

4. Radius = 1.5 ft

5. Radius = $8\frac{1}{2}$ ft

6. Diameter = 5.5 m

Find the area of circles with the following dimensions. Round to tenths.

7. Diameter = 8 cm

8. Diameter = 15 m

9. Radius = 3 in.

10. Radius = 16 yd

11. Radius = 1.5 ft

12. Radius = $8\frac{1}{2}$ ft

13. Diameter = 5.5 m

14. Diameter = $5\frac{1}{4}$ in.

Find the color area of Figs. 18–92 through 18–95 to the nearest tenth.

15.

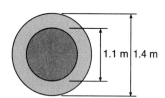

Figure 18–92

16.

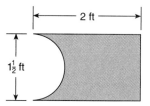

Figure 18–93

17.

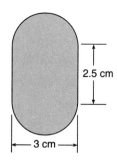

Figure 18–94

18.

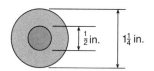

Figure 18–95

19. A swimming pool is in the form of a semicircular-sided figure. Its width is 20 ft and the parallel portions of the sides are each 20 ft (Fig. 18–96). What is the area of a 5-ft-wide walk surrounding the pool? Round to tenths.

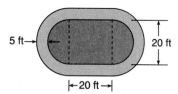

Figure 18–96

20. INDTR   A belt connecting two 9-in.-diameter drums on a conveyor system needs replacing. How many inches must the new belt be if the centers of the drums are 10 ft apart (Fig. 18–97)? Round to tenths.

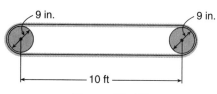

Figure 18–97

21. **INDTR** A 2-in.-inside-diameter pipe and a 4-in.-inside-diameter pipe empty into a third pipe whose inside diameter is 5 in. (Fig. 18–98). Is the third pipe large enough for the combined flow? (Justify your answer.)

22. **INDTR** A large pipe whose interior cross-sectional area is 20 in^2 empties into two smaller pipes that each have an interior diameter of 4 in. (Fig. 18–99). Are the smaller pipes together large enough to carry off the flow from the larger pipe? (Justify your answer.)

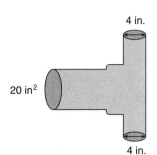

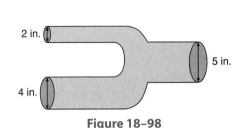

Figure 18–98   **Figure 18–99**

23. **INDTR** Cutting speed, when applied to a grinding wheel, is called *surface speed*. What is the surface speed in ft/min of a 9-in.-diameter grinding wheel revolving at 1,200 rpm? (Surface speed = circumference in feet × rpm.)

24. **INDTR** A 12-in.-diameter polishing wheel revolves at 500 rpm. What is the surface speed? (See Exercise 23 for formula.)

2 Convert the measures to radians rounded to the nearest hundredth.

25. 45°

26. 56°

27. 140°

Convert the measures to degrees rounded to the nearest ten-thousandth. Then convert to radians to the nearest hundredth.

28. 21°45′

29. 177°33′

30. 44°54′12″

Convert the measures to degrees. Round to the nearest ten-thousandth of a degree.

31. $\dfrac{\pi}{4}$ rad

32. $\dfrac{\pi}{6}$ rad

33. 2.5 rad

Convert the measures to degrees, minutes, and seconds. Round to the nearest second when necessary.

34. 0.5 rad

35. $\dfrac{\pi}{8}$ rad

36. 0.75 rad

3 Find the arc length of the sectors of a circle. Round to hundredths.

37. ∠ = 54°
    r = 30 mm

38. ∠ = 150°
    r = 5 in.

39. ∠ = 120°30′
    r = 1.52 ft

40. ∠ = 25°16′
    r = 16 cm

Solve the following problems. Round answers to hundredths if necessary.

41. Find the arc length intercepted on the circumference of a circle by a central angle of 2.15 rad if the radius of the circle is 3 in.

42. Find the arc length intercepted on the circumference of a circle by a central angle of 4 rad if the radius of the circle is 3.5 cm.

43. Use radians to find an angle at the center of a circle of radius 2 in. if the angle intercepts an arc length of 8.5 in.

44. Use radians to find an angle at the center of a circle of radius 4.3 cm if the angle intercepts an arc length of 15 cm.

45. Find the radius of an arc if the length of the arc is 14.7 cm and the intercepted central angle is 2.1 rad.

46. Find the radius of an arc if the length of the arc is 12.375 in. and the intercepting central angle is 2.75 rad.

4 Find the area of the sector of a circle using Fig. 18–100. Round to hundredths.

47. $\angle = 54°$
  $r = 16$ cm

48. $\angle = 25°16'$
  $r = 30$ mm

49. $\angle = 120°30'$
  $r = 1.52$ ft

50. $\angle = 65°$
  $r = 5$ in.

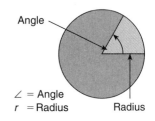

$\angle$ = Angle
r = Radius

Figure 18–100

51. **CON** The library of a contemporary elementary school is circular (Fig. 18–101). The floor plan includes sectors reserved for science materials, literary materials, reference materials, and so on. Find the area of the reference section excluding its storage area.

52. **CON** A mason lays a tile mosaic featuring a four-sector design (color portion of Fig. 18–102). What is the area of the design to the nearest hundredth?

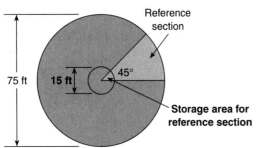

Figure 18–101

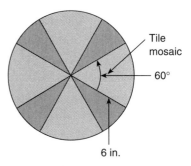

Figure 18–102

53. Find the area of a sector whose central angle is 2.14 rad and whose radius is 4 in.

54. Find the area of a sector that has a central angle of 6 rad and a radius of 1.2 cm.

55. Find the radius of a sector if the area of the sector is 7.5 cm² and the central angle is 3 rad.

56. How many radians does the central angle of a sector measure if its area is 1.7 in² and its radius is 2 in.?

57. Find the area of a sector whose central angle is 1.83 rad and whose radius is 7.2 cm. Round to hundredths.

58. Find the number of radians of a central angle of a sector whose area is 5.6 cm² and whose radius is 4 cm.

Find the area of the segments of a circle using Fig. 18–103. Round to hundredths.

59. $\angle = 60°$
  $r = 13.3$ cm
  $h = 11.52$ cm
  $b = 13.3$ cm

60. $\angle = 110°$
  $r = 10$ in.
  $h = 5.74$ in.
  $b = 16.38$ in.

61. $\angle = 105°$
  $r = 11.25$ in.
  $h = 6.85$ in.
  $b = 17.85$ in.

62. $\angle = 60°$
  $r = 24$ cm
  $h = 20.8$ cm
  $b = 24$ cm

$\angle$ = Angle
r = Radius
h = Height
b = Base

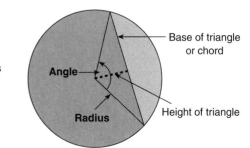

Figure 18–103

63. INDTEC A motor shaft has milled on it a flat for a setscrew to rest so that it can hold a pulley on the shaft (Fig. 18–104). What is the cross-sectional area of the shaft after being milled? Round to hundredths.

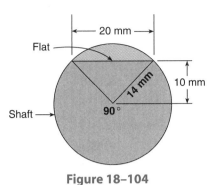

Figure 18–104

64. CON A contractor pours a concrete patio in the shape of a circle except where the patio touches the exterior wall of the house (Fig. 18–105). What is the area of the patio in square feet? Round to hundredths.

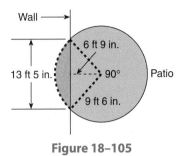

Figure 18–105

18–4 | Volume and Surface Area

Learning Outcomes

1 Find the volume of three-dimensional objects.
2 Find the surface area of three-dimensional objects.

Common household items like cardboard storage boxes and toy building blocks are examples of three-dimensional geometric figures classified generally as *prisms.* Cans and pipes are examples of *cylinders.*

1 Find the Volume of Three-Dimensional Objects.

A **prism** is a three-dimensional figure with polygonal bases (ends) that are parallel and faces (sides) that are parallelograms, rectangles, or squares (Fig. 18–106). In a **right prism,** the faces are perpendicular to the bases.

A **right circular cylinder** is a three-dimensional figure with a curved surface and two circular bases such that the height is perpendicular to the bases.

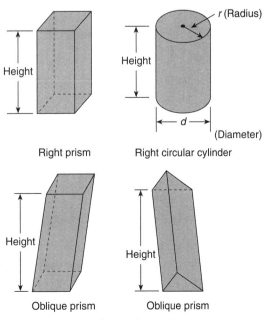

Figure 18–106

In an **oblique prism** or an **oblique cylinder** the faces are *not* perpendicular to the bases.

The **height** of a three-dimensional figure with two bases is the shortest distance between the two bases.

In right circular cylinders and in right prisms, the height is the same as the length of a side or face. However, in oblique prisms and cylinders the height is the perpendicular distance between the bases and is different from the length of a side or face.

As we saw in Chapter 1, Section 3, Outcome 2, the *volume* of an object, such as a container, is used to estimate how many containers can be loaded into a given-size storage area or shipped in a container of certain dimensions.

The **volume** of a three-dimensional geometric figure is the amount of space it occupies, measured in terms of three dimensions (length, width, and height). Volume is expressed as a **cubic measure.**

A general formula can be used for the volume of *any* right prism or right cylinder. If the base of the prism is a triangle, trapezoid, or other polygon, we use the appropriate formula for the area of the base.

Formula for the volume of right prism or right cylinder:

$$V = Bh$$

where B is the area of the base and h is the height of the prism or cylinder.

EXAMPLE Find the volume of the triangular prism in Fig. 18–107 if the height is 15 cm and the bases are triangles 3 cm on a side and 2.6 cm in height.

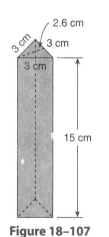

Figure 18–107

$V_{\text{prism}} = Bh$ Substitute the formula for the area of the triangular base for B.

$V_{\text{prism}} = \left(\dfrac{1}{2}bh_1\right)h_2$ $h_1 = 2.6$ cm (height of prism base), $h_2 = 15$ cm (height of prism).

$V_{\text{prism}} = \left[\dfrac{1}{2}(3)(2.6)\right]15$ Substitute values and evaluate.

$V_{\text{prism}} = 58.5$ cm^3

The volume of the prism is 58.5 cm³.

EXAMPLE What is the cubic-inch displacement (space occupied) of a cylinder (Fig. 18–108) whose diameter is 5 in. and whose height is 4 in.? Round to the nearest tenth.

Figure 18–108

$V = Bh$ Substitute the formula for the area of the circular base for B.

$V = \pi r^2 h$ Substitute values; $r = \frac{1}{2}$ diameter, or 2.5.

$V = \pi(2.5)^2(4)$ Evaluate.

$V = 25\pi$ Exact solution

$V = 78.5$ in^3 Round.

The cylinder displacement is 78.5 in³.

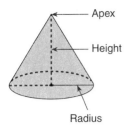

Figure 18–109   Circular cone.

A right **cone** is a three-dimensional figure whose base is a circle and whose side surface tapers to a point, called the **vertex** or **apex,** and whose height is the perpendicular line segment between the base and apex (Fig. 18–109). One example of a *cone* is the funnel or the circular rain cap placed on top of stove vent pipes extending through the roof of some homes.

A **pyramid** is a three-dimensional geometric shape that has a polygon for a **base** and **lateral faces** that are triangles with a common vertex (apex) (Fig. 18–110). The *height or altitude* of the pyramid is the perpendicular distance from the vertex to the base. If the base of the pyramid is a polygon that has all sides equal, such as a square, the height meets the base at the center of the base and is a **right** or **regular pyramid.** Fig. 18–110 shows right pyramids with 3-, 4-, and 5-sided bases.

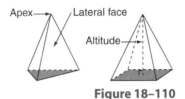

Figure 18–110

Volume of any cone or pyramid:

$$V = \frac{1}{3} Bh$$

where *V* is the volume, *B* is the area of the base, and *h* is the height of the cone or pyramid.

EXAMPLE   Find the volume of a pyramid with a square base that is 15 cm on a side and has a height of 28 cm (Fig. 18–111).

Figure 18–111

$V = \dfrac{1}{3} Bh$ Find the area of the base and substitute values for *B* and *h*.
$B = s^2 = 15^2 = 225$ cm^2

$V = \dfrac{1}{3}(225)(28)$ Multiply.

$V = \mathbf{2{,}100}$ **cm**^{3} Volume is cubic units.

A **frustum of a cone** is a part of a cone between the base and a plane passing through the cone parallel to the base (Fig. 18–112).

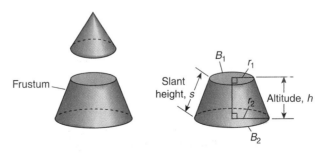

Figure 18–112

18–4 Volume and Surface Area

Similarly, a **frustum of a pyramid** is a part of a pyramid between the base and a plane passing through the pyramid that is parallel to the base (Fig. 18–113).

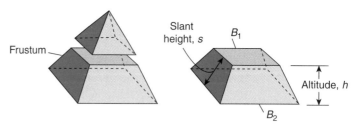

Figure 18–113

Volume of the frustum of a cone or pyramid:

$$V_{\text{frustum}} = \frac{1}{3} h \left(B_1 + B_2 + \sqrt{B_1 B_2} \right)$$

where V is the volume, B_1 is the area of the base, B_2 is the area of the top, and h is the height of the frustum.

EXAMPLE Find the volume of the frustum of a pyramid with a square base that is 4 in. on each side, a square top that is 2 in. on each side, and a height of 3 in. (Fig. 18–114).

Figure 18–114

$V_{\text{frustum}} = \dfrac{1}{3} h \left(B_1 + B_2 + \sqrt{B_1 B_2} \right)$

Substitute the values for B_1, B_2, and h. $B_1 = 4^2 = 16$ in^2; $B_2 = 2^2 = 4$ in^2

$V_{\text{frustum}} = \dfrac{1}{3} (3) \left(16 + 4 + \sqrt{16 \cdot 4} \right)$

Simplify the radicand.

$V_{\text{frustum}} = \dfrac{1}{3} (3) \left(16 + 4 + \sqrt{64} \right)$

Take the square root.

$V_{\text{frustum}} = \dfrac{1}{3} (3)(16 + 4 + 8)$

Simplify the grouping.

$V_{\text{frustum}} = \dfrac{1}{3} (3)(28)$

Multiply.

$V_{\text{frustum}} = \mathbf{28\ in^3}$

Volume is cubic units.

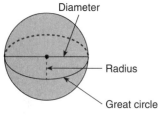

Figure 18–115 Sphere.

Soccer balls, golf balls, tennis balls, baseballs, and ball bearings are *spheres*. Spheres are also used as tanks to store gas and water because spheres hold the greatest volume for a specified amount of surface area. A **sphere** is a three-dimensional figure formed by a curved surface with points that are all equidistant from a point inside called the center (Fig. 18–115). A **great circle** divides the sphere in half and is formed by a plane passing through the center of the sphere.

A sphere does not have bases like prisms and cylinders. Because of the relationship of the sphere to the circle, the formula for the volume of a sphere includes elements of the formula for the area of a circle.

$$V_{sphere} = \frac{4\pi r^3}{3}$$

where r = radius.

Note that the radius is *cubed,* or raised to the power of 3, indicating volume.

EXAMPLE   Find the volume of a sphere that has a diameter of 90 cm.

$$V_{sphere} = \frac{4\pi r^3}{3}$$ Substitute values.

$$V_{sphere} = \frac{4(\pi)(45)^3}{3}$$ Cube 45, multiply, and divide.

$$V_{sphere} = \mathbf{381{,}704\ cm^3}$$ Round.

EXAMPLE   Find the weight of the cast-iron object shown in Fig. 18–116 if cast iron weighs
INDTR   0.26 lb per cubic inch. Round to the nearest whole pound.

The solution requires finding the volume of the cone that forms the top of the object, the volume of the cylinder that forms the middle portion of the object, and the volume of the **hemisphere** (half sphere) that forms the bottom of the object.

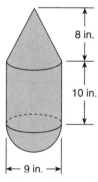

8 in.

10 in.

9 in.

Figure 18–116

$$V_{cone} = \frac{\pi r^2 h}{3}$$ $$V_{cylinder} = \pi r^2 h$$ $$V_{hemisphere} = \frac{1}{2}\left[\frac{4\pi r^3}{3}\right]$$

$$V_{cone} = \frac{(\pi)(4.5)^2(8)}{3}$$ $$V_{cylinder} = (\pi)(4.5)^2(10)$$ $$V_{hemisphere} = \frac{1}{2}\left[\frac{4(\pi)(4.5)^3}{3}\right]$$

$$V_{cone} = 169.6460033\ in^3$$ $$V_{cylinder} = 636.1725124\ in^3$$ $$V_{hemisphere} = 190.8517537\ in^3$$

total volume $= V_{cone} + V_{cylinder} + V_{hemisphere}$

total volume $= 996.6702694\ in^3$

Convert to pounds:

$$\frac{996.6702694\ in^3}{1} \times \frac{0.26\ lb}{1\ in^3} = 259\ lb$$ Rounded

To the nearest whole pound, the cast-iron object weighs 259 lb.

2 Find the Surface Area of Three-Dimensional Objects.

The surface area of a three-dimensional figure can refer to just the area of the *sides* of the figure. Or surface area can refer to the overall area, including the bases along with the sides.

The **lateral surface area (LSA)** of a three-dimensional figure is the area of its sides only.

The **total surface area (TSA)** of a three-dimensional figure is the area of the sides plus the area of its base or bases.

Lateral surface area of a right prism or a right circular cylinder:

$$LSA = ph$$

where p is the perimeter of the base and h is the height of the three-dimensional figure (Fig. 18–117).

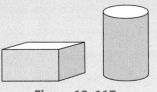

Figure 18–117

To get the total surface area (TSA) (Fig. 18–118), add the areas of the two bases to the lateral surface area.

Total surface area of a right prism or a right circular cylinder:

$$TSA = ph + 2B$$

where p is the perimeter of the base, h is the height of the three-dimensional figure, and B is the area of the base (Fig. 18–118).

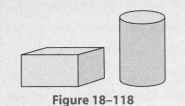

Figure 18–118

EXAMPLE
INDTR

Find the lateral surface area of a rectangular shipping carton measuring 24 in. in length, 12 in. in width, and 20 in. in height (Fig. 18–119).

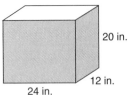

Figure 18–119

$LSA = \boxed{p}\,h$	Substitute the formula for the perimeter of a rectangle, $p = 2l + 2w$.
$LSA = (\boxed{2l + 2w})h$	Substitute numerical values.
$LSA = [2(24) + 2(12)]20$	Perform calculations inside grouping.
$LSA = [72]20$	Multiply.
$LSA = 1{,}440 \text{ in}^2$	Area requires square units.

The lateral surface area of the carton is 1,440 in².

EXAMPLE
INDTR

How many square centimeters of sheet metal are required to manufacture a can that has a radius of 4.5 cm and height of 9 cm? Assume no waste or overlap (Fig. 18–120).

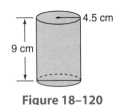

Figure 18–120

$TSA = \boxed{p}\,h + 2\boxed{B}$	Total surface area is needed. Substitute formulas $p = 2\pi r$, $B = \pi r^2$.
$TSA = \boxed{2\pi r}\,h + 2\,\boxed{\pi r^2}$	Substitute values.
$TSA = 2(\pi)(4.5)(9) + 2(\pi)(4.5)^2$	Square 4.5 and perform multiplications.
$TSA = 254.4690049 + 127.2345025$	Add.
$TSA = 381.70 \text{ cm}^2$	Round. Area requires square units.

The can requires 381.70 cm² of sheet metal.

Chapter 18 / Geometry

EXAMPLE Find the total surface area of the triangular prism shown in Fig. 18–121.

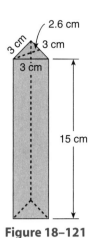

$$\text{TSA} = ph + 2B$$

Perimeter of triangular base is $a + b + c$.
Area of triangular base B is $\frac{1}{2}bh_1$, where h_1 is the height of the triangular base.

$$\text{TSA} = (a + b + c)h + 2\left(\frac{1}{2}bh_1\right)$$

Substitute values.

$$\text{TSA} = (3 + 3 + 3)(15) + 2\left(\frac{1}{2}\right)(3)(2.6)$$

Add inside grouping.

$$\text{TSA} = 9(15) + (2)\left(\frac{1}{2}\right)(3)(2.6)$$

Perform multiplications.

Figure 18–121

$$\text{TSA} = 135 + 7.8$$

Add.

$$\text{TSA} = 142.8 \text{ cm}^2$$

Area requires square units.

The total surface area of the triangular prism is 142.8 cm².

A sphere does not have bases like prisms and cylinders. The surface area of a sphere includes *all* the surface so there is only one formula. Because of the relationship of the sphere to the circle, the formula includes elements of the formula for the area of a circle. The total surface area of the sphere is 4 times the area of a circle with the same radius.

Total surface area of a sphere:

$$\text{TSA} = 4\pi r^2$$

where r = radius.

EXAMPLE Find the surface area of a sphere that has a diameter of 90 cm.

$$\text{TSA} = 4\pi r^2$$

Substitute values. $\frac{1}{2}d = r$, so $r = 45$ cm.

$$\text{TSA} = 4(\pi)(45)^2$$

Square 45 and multiply.

$$\textbf{TSA} = \textbf{25,447 cm}^2$$

Rounded

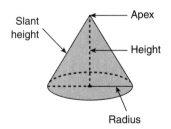

Figure 18–122
Circular cone.

The **slant height** of a cone is the distance along the side from the base to the apex. Fig. 18–122 shows the *perpendicular height* and the *slant height* of a cone.

The lateral surface area of a cone equals the circumference of the base times $\frac{1}{2}$ the slant height, or $\text{LSA} = 2\pi r\left(\frac{1}{2}s\right)$, which simplifies as $\pi r s$.

Lateral surface area of a cone:

$$\text{LSA} = \pi rs$$

where r is the radius and s is the slant height.

The total surface area, then, is the lateral surface area plus the area of the base.

Total surface area of a cone:

$$\text{TSA} = \pi rs + \pi r^2$$

where r is the radius of the circular base, s is the slant height, and πr^2 is the area of the base.

EXAMPLE Find the lateral surface area and total surface area of a cone that has a diameter of 8 cm, height of 6 cm, and slant height of 7 cm. Round to hundredths.

$\text{LSA} = \pi rs$	Substitute values: $r = \frac{1}{2}d$, or 4.
$\text{LSA} = (\pi)(4)(7)$	Multiply.
$\textbf{LSA} = \textbf{87.96 cm}^2$	Rounded from 87.9645943
$\text{TSA} = \pi rs + \pi r^2$	Substitute values. Use full calculator value for πrs.
$\text{TSA} = 87.9645943 + (\pi)(4)^2$	Perform operations using the proper order of operations.
$\textbf{TSA} = \textbf{138.23 cm}^2$	Round.

SECTION 18–4 SELF-STUDY EXERCISES

1 Find the volume in Figs. 18–123 and 18–124. Round to the nearest hundredth if necessary.

1.

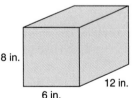

8 in.   12 in.   6 in.

Figure 18–123

2.

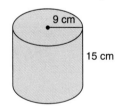

9 cm   15 cm

Figure 18–124

3. A right pentagonal (five-sided) prism is 10 cm high. If the area of each pentagonal base is 32 cm², what is the volume of the prism?

4. What is the volume of a triangular prism that has a height of 8 in., a triangular base that measures 4 in. on each side, and a height of 3.46 in.? Round to hundredths.

5. How many cubic inches are in an aluminum can with a $2\frac{1}{2}$ in. diameter and $4\frac{3}{4}$ in. height? Round to tenths.

6. What is the volume of a cylindrical oil storage tank that has a 40-ft diameter and 15-ft height? Round to the nearest whole number.

Solve. Round to tenths.

7. Find the volume of the cone in Fig. 18–125.

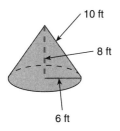

10 ft

8 ft

6 ft

Figure 18–125

8. CON How many cubic feet are in a conical pile of sand that is 30 ft in diameter and is 20 ft high?

9. BUS A cone-shaped storage container holds a photographic chemical. If the container is 80 cm wide and 30 cm high, how many liters of the chemical does it hold if $1 \text{ L} = 1,000 \text{ cm}^3$?

10. Find the volume of a pyramid that has a square base of 30 cm on a side and a height of 42 cm.

11. Find the volume of a pyramid that has a square base of 48 m on a side and a height of 100 m.

12. Find the volume of a pyramid that has an equilateral triangular base with altitude 10.39 m and a side of 12 m. The height is 20 m.

13. A frustum of a pyramid has a square base that is 18 in. on each side, a square top that is 10 in. on each side, and a height of 13 in. Find the volume of the frustum.

14. A frustum of a pyramid has a triangular base that has an area of 32 cm² and a triangular top that has a surface area of 28 cm². The height of the frustum is 81 cm. Find the volume of the frustum.

15. CON Cap blocks for a fence are molded in the shape of a frustum of a pyramid that has a square base and top. The base of the frustum is 30 in. on each side and the top is 24 in. on each side. The cap block is 5 in. thick (height of frustum). What is the volume of the frustum?

16. Find the volume of a sphere with a radius of 5 cm.

17. Find the volume of a sphere with a radius of 6 in.

18. If $1 \text{ ft}^3 = 7.48$ gal, how many gallons can a spherical water tank hold if its diameter is 45 ft?

19. INDTR If a spherical propane tank is filled to 90% of its capacity, how many gallons of propane does the tank hold if its diameter is 4 ft? ($1 \text{ ft}^3 = 7.48$ gal.)

2 Solve. Round to tenths unless otherwise indicated.

20. INDTEC How many square inches are in the lateral surface area and total surface area of an aluminum can with a $2\frac{1}{2}$ in. diameter and $4\frac{3}{4}$ in. height?

21. What are the lateral surface area and total surface area of a cylindrical oil storage tank that has a 40-ft diameter and 15-ft height?

22. A right pentagonal prism is 10 cm high. If the area of each pentagonal base is 32 cm² and the perimeter is 20 cm, what are the lateral and total surface areas of the prism?

23. What are the lateral and total surface areas of a triangular prism that has a height of 8 in. and a triangular base that measures 4 in. on each side with an altitude of 3.46 in.?

24. CON A cylindrical water well is 1,200 ft deep and 6 in. across. Find the lateral surface area of the well. Round to the nearest square foot. (*Hint:* Convert measures to a common unit.)

25. Find the lateral surface area and total surface area of the cone that has a radius of 6 ft, slant height of 10 ft, and height of 8 ft.

26. INDTR How many square centimeters of sheet metal are needed to form a conical rain cap 25 cm in diameter if the slant height is 15 cm?

27. Find the total surface area of a conical tank that has a radius of 15 ft and a slant height of 20 ft.

28. Find the surface area of a sphere with a radius of 5 cm.

30. INDTR A spherical propane tank has a diameter of 4 ft. How many square feet of surface area need to be painted?

29. INDTR How many square feet of steel are needed to manufacture a spherical water tank with a diameter of 45 ft?

31. INDTR A cylindrical water tower with a conical top and hemispheric bottom (see Fig. 18–126) needs to be painted. If the cost is $2.19 per square foot, how much does it cost (to the nearest dollar) to paint the tank?

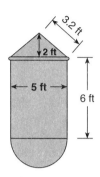

Figure 18–126

CHAPTER REVIEW OF KEY CONCEPTS

Learning Outcomes

What to Remember with Examples

Section 18–1

1 Use various notations to represent points, lines, line segments, rays, and planes and angles (pp. 695–696).

A dot represents a point and a capital letter names the point. A line extends in both directions and is named by any two points on the line (with double arrow above the letters, such as $\overleftrightarrow{AB}$). A line segment has a beginning point and an ending point, which are named by letters (with a bar above the letters, such as $\overline{CD}$). Line segments are sometimes named by letters only, such as CD. A ray extends from a point on a line and includes all points on one side of the point and is named by the endpoint and any other point on the line forming the ray (with an arrow above the letters, such as $\overrightarrow{AC}$). A plane contains an infinite number of points and lines on a flat surface.

> Use proper notation for the following:
>
> (a) Line GH (b) Line segment OP (c) Ray ST
>
> (a) $\overleftrightarrow{GH}$ (b) $\overline{OP}$ (c) $\overrightarrow{ST}$

Lines that intersect meet at only one point. Lines that coincide fit exactly on top of one another. Lines that are parallel are the same distance from one another and never intersect.

Classify the pairs of lines in Fig. 18–127.

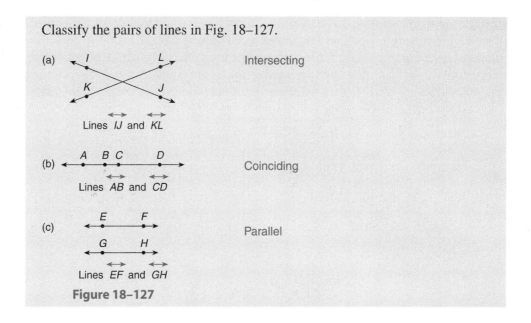

(a) Intersecting
Lines $\overleftrightarrow{IJ}$ and $\overleftrightarrow{KL}$

(b) Coinciding
Lines $\overleftrightarrow{AB}$ and $\overleftrightarrow{CD}$

(c) Parallel
Lines $\overleftrightarrow{EF}$ and $\overleftrightarrow{GH}$

Figure 18–127

When two rays intersect in a point (endpoint of rays), an angle is formed. Angles may be named with three capital letters (endpoint in middle and one point on each ray), such as $\angle ABC$. They may be named by only the middle letter (vertex of the angle), such as $\angle B$. They may be assigned a number or lowercase letter placed within the vertex, such as $\angle 2$ or $\angle d$.

Name the angle in Fig. 18–128 three ways.

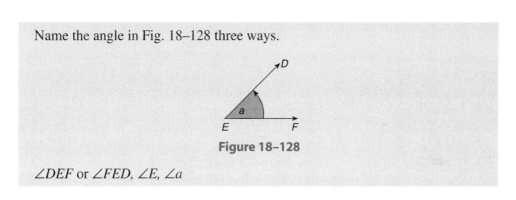

Figure 18–128

$\angle DEF$ or $\angle FED$, $\angle E$, $\angle a$

2 Classify angles according to size (pp. 697–699).

A right angle contains $90°$, or one-fourth a rotation. A straight angle contains $180°$, or one-half a rotation. An acute angle contains less than $90°$ but more than $0°$. An obtuse angle contains more than $90°$ but less than $180°$. The sum of the measures of complementary angles is $90°$. The sum of the measures of supplementary angles is $180°$.

Identify the following angles:

(a) $30°$ (b) $100°$ (c) $90°$ (d) $180°$

(a) acute   (b) obtuse   (c) right   (d) straight

3 Determine the measure of an angle by using relationships among intersecting lines (pp. 699–701).

Relationships of angles formed by intersecting lines: Vertical angles are equal. Adjacent angles are supplementary.

When two parallel lines are cut by a transversal: Corresponding angles are equal. Alternate interior angles are equal. Alternate exterior angles are equal. Exterior angles on the same side of the transversal are supplementary.

In Fig. 18–129 identify pairs of angles that are corresponding, alternate interior, and exterior on the same side of the transversal. Then state the relationship between each pair of angles.

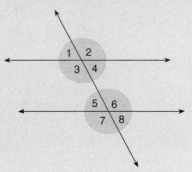

Figure 18–129

Corresponding angles:	$\angle 1$ and $\angle 5$	Corresponding angles are equal.
Alternate interior angles:	$\angle 3$ and $\angle 6$	Alternate interior angles are equal.
Exterior angles on the same side of the transversal:	$\angle 1$ and $\angle 7$	Exterior angles on the same side of the transversal are supplementary.

4 Convert angle measures between decimal degrees and degrees, minutes, and seconds (pp. 701–702).

To change the format of angle measures using a calculator:

From degrees in decimal notation to degrees, minutes, and seconds (calculator steps will vary with different calculators): **1.** Set the calculator to degree mode. **2.** Enter the angle measure in decimal notation. **3.** Access the sexagesimal function key or menu choice on the calculator (most often labeled $\boxed{°\,'\,''}$ or $\boxed{\text{DMS}}$). **4.** Press $\boxed{\text{ENTER}}$ or $\boxed{=}$.

From degrees, minutes, and seconds to decimal notation (calculator steps will vary with different calculators):

Option 1

1. Enter the series of calculations—degrees + minutes ÷ 60 + seconds ÷ 3,600.

Option 2

1. Enter the degrees followed by the degree symbol ($\boxed{2^{nd}}$, $\boxed{\text{ANGLE}}$, $\boxed{1:°}$) **2.** Enter the minutes followed by the minute symbol ($\boxed{2^{nd}}$, $\boxed{\text{ANGLE}}$, $\boxed{2:'}$) **3.** Enter the seconds followed by the second symbol ($\boxed{\text{ALPHA}}$, $\boxed{''}$) **4.** Press $\boxed{\text{ENTER}}$.

Change 110.625° to degrees, minutes, and seconds.

110.625 $\boxed{2^{nd}}$ $\boxed{\text{ANGLE}}$ $\boxed{4}$ $\boxed{\text{ENTER}}$

$110.625° = 110°37'30''$

Change 45°15′45″ to a decimal degree equivalent. Round to the nearest hundredth.

$$45 + 15/60 + 45/3600 = 45.2625$$
$$45°15'45'' = 45.2625°$$

Section 18–2

1 Find the perimeter of a polygon using the appropriate formula (pp. 704–708).

1. Select the appropriate formula. **2.** Substitute the known values into the formula. **3.** Evaluate the formula (perform the indicated operations). Formulas for perimeter:

$P_{parallelogram} = 2(b + s)$: b is the base, or the length of a side that is or can be rotated to a horizontal position; and s is the length of a side that joins with the base.
$P_{rectangle} = 2(l + w)$: l is length of the long side; w is width of the short side.
$P_{square} = 4s$: s is the length of a side.
$P_{rhombus} = 4s$, where s is the length of a side.
$P_{trapezoid} = b_1 + b_2 + a + c$, where b_1 and b_2 are the lengths of the bases or parallel sides and a and c are the lengths of the sides adjacent to the bases.
$P_{triangle} = a + b + c$, where a, b, and c are the lengths of the three sides.

What is the perimeter (P) of a triangular roof vent that has sides a, b, and c of 6 ft, 4 ft, and 4 ft, respectively?

$P_{triangle} = a + b + c$ Select the appropriate formula and substitute values.
$P_{triangle} = 6 + 4 + 4$ Add.
$P_{triangle} = 14$ ft

---

2 Find the area of a polygon using the appropriate formula (pp. 708–713).

1. Select the appropriate formula. **2.** Substitute the known values into the formula. **3.** Evaluate the formula (perform the indicated operations). Formulas for area:

$A_{rectangle} = lw$, where l is the length and w the width.
$A_{square} = s^2$, where s is the length of a side.
$A_{parallelogram} = bh$: b is the base; h is the height, or length of a perpendicular distance between the bases.
$A_{rhombus} = bh$, where b is the length of a side and h is the height.
$A_{trapezoid} = \frac{1}{2}h(b_1 + b_2)$, where h is the height between the parallel bases and b_1 and b_2 are the lengths of the bases or two parallel sides.
$A_{triangle} = \frac{1}{2}(bh)$, where b is the base and h is the height.
Heron's formula for the area of a triangle: $A = \sqrt{s(s - a)(s - b)(s - c)}$, where $s = \frac{1}{2}(a + b + c)$ and a, b, and c are sides.

Find the area of a trapezoid that has bases of 11 in. and 18 in. and height of 12 in.

$A_{trapezoid} = \frac{1}{2}h(b_1 + b_2)$ Select the appropriate formula and substitute values.

$A_{trapezoid} = \frac{1}{2}(12)(11 + 18)$ Perform operation in grouping.

$A_{trapezoid} = \frac{1}{2}(12)(29)$ Multiply (in. × in. = in²).

$A_{trapezoid} = 174$ in²

Section 18–3

1 Find the circumference or area of a circle using the appropriate formula (p. 716).

1. Select the appropriate formula. **2.** Substitute the known values into the formula. **3.** Evaluate the formula (perform the indicated operations). Formula for circumference: $C = \pi d$, or $C = 2\pi r$; d is the diameter or distance across the center of a circle; r is the radius, or half the diameter; π is approximated on a calculator as 3.141592654. Formula for area: $A_{circle} = \pi r^2$; r is the radius.

Chapter Review of Key Concepts

Find the area of a circle whose diameter is 3 m.

First, find the radius: $r = \dfrac{d}{2}$; 3 m ÷ 2 = 1.5 m.

$$A_{\text{circle}} = \pi r^2$$
$$A_{\text{circle}} = \pi(1.5 \text{ m})^2$$
$$A_{\text{circle}} = 7.07 \text{ m}^2 \text{ (rounded)}$$

2 Convert angle measures between degrees and radians (pp. 720–722).

Degrees to radians: Using a calculator (steps may vary): **1.** Set angle MODE to radian. **2.** Enter degree measure followed by the degree symbol. Use [° ′ ″], [DMS], or the angle menu, option 1. **3.** Enter minute measure followed by the minute symbol. Use [° ′ ″], [DMS], or the angle menu, option 1. **4.** Enter second measure followed by the second symbol. Use [° ′ ″], [DMS], or [ALPHA] [″]. **5.** [2nd] [ALPHA] [1] **6.** Press [ENTER].

Change 125°30′30″ to radians to the nearest hundredth.

125 [2nd] [ANGLE] [1:°] 30 [2nd] [ANGLE] [2:′] 30 [ALPHA] [″] [2nd] [ALPHA], [1:°] [ENTER] = 2.190533655 = 2.19 (rounded)

Radians to degrees: Using a calculator (steps may vary): **1.** Set angle MODE to degree. **2.** Enter radian amount and radian symbol (angle menu, option 3). **3.** If degrees, minutes, and seconds are desired, use DMS function (angle menu, option 4).

Change 1.245 rad to a decimal degree to the nearest thousandth.

1.245 [[2nd] [ANGLE] menu, option 3] [ENTER] = 71.33324549 = 71.333° (rounded)

3 Find the arc length of a sector (pp. 723–725).

Arc length is the portion of the circumference formed by the sides of a sector.

Using degrees: arc length $(s) = \dfrac{\theta}{360}(2\pi r)$ or $s = \dfrac{\theta}{360}(\pi d)$.

Using radians: $s = \theta r$.

Find the arc length of a sector formed by a 50° central angle if the radius is 20 cm (Fig. 18–130).

$s = \dfrac{\theta}{360}(2\pi r)$ Substitute θ = 50° and r = 20 cm.

$s = \dfrac{50}{360}(2)(\pi)(20)$ Evaluate.

$s = 17.45329252$ or   17.45 cm (rounded)

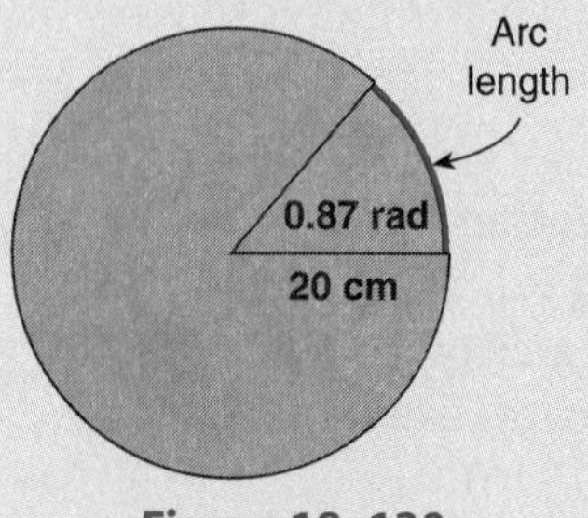

Arc length

0.87 rad

20 cm

Figure 18–130

4 Find the area of a sector or segment (pp. 725–727).

A sector is a portion of a circle cut off by two radii (r).

Using degrees: Area of a sector $= \dfrac{\theta}{360}\pi r^2$, where θ is the degree measure of the central angle formed by the radii.

Using radians: $A = \dfrac{1}{2}\theta r^2$, where θ is the radian measure of the central angle formed by the radii.

Find the area of a sector that has a central angle of 0.87 rad and a radius of 20 cm (Figure 18–131).

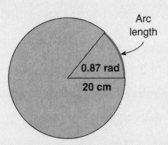

Figure 18–131

Area $= \dfrac{1}{2}\theta r^2$ Formula for area of a sector of a circle. Substitute $\theta = 0.87$ rad and $r = 20$ cm.

Area $= \dfrac{1}{2}(0.87)(20)^2$ Evaluate. (cm × cm = cm²)

Area $= 174$ cm²   Round.

A chord is a line segment joining two points on a circle. An arc is the portion of the circumference cut off by a chord. A segment is the portion of a circle bounded by a chord and an arc.

The area of a segment is the area of the sector minus the area of the triangle formed by the chord and the central angle of the sector.

Using degrees: $A = \dfrac{\theta}{360}\pi r^2 - \dfrac{1}{2}bh$. Using radians: $A = \dfrac{1}{2}\theta r^2 - \dfrac{1}{2}bh$.

Find the area of the segment formed by a 12-cm chord if the height of the triangle part of the sector is 6 cm and the radius is 8.5 cm. The central angle is 90° (Fig. 18–132).

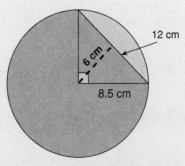

Figure 18–132

Chapter Review of Key Concepts

$$A_{\text{segment}} = \frac{\theta}{360}\,\pi r^2 - \frac{1}{2}\,bh \qquad \text{Substitute } \theta = 90°, r = 8.5,$$
$$b = 12, h = 6.$$

$$A_{\text{segment}} = \frac{90}{360}\,(\pi)(8.5)^2 - \frac{1}{2}(12)(6) \qquad \text{Evaluate.}$$

$$A_{\text{segment}} = 0.25(\pi)(72.25) - 0.5(72)$$

$$A_{\text{segment}} = 20.74501731 \quad \text{or} \quad 20.75 \text{ cm}^2 \qquad \text{Round.}$$

Section 18–4

1 Find the volume of three-dimensional objects (pp. 731–735).

Formulas:

Volume of a prism or cylinder: $V = Bh$, where B is the area of the base and h is the height.

Volume of a cone or pyramid: $V = \frac{1}{3}Bh$, where B is the area of the base and h is the height of the cone or pyramid.

Volume of a frustum of a cone or pyramid: $V = \frac{1}{3}h\left(B_1 + B_2 + \sqrt{B_1 B_2}\right)$, where h is the height of the frustum, B_1 is the area of the base, and B_2 is the area of the top.

Volume of a sphere: $V = \frac{4\pi r^3}{3}$, where r is the radius.

> Find the volume of a cylinder that has a diameter of 20 mm and a height of 80 mm.
>
> $V = Bh$ $\qquad\qquad\qquad\qquad B = \pi r^2.$
>
> $V = \pi r^2 h$ $\qquad\qquad\qquad\quad r = \frac{1}{2}d = 10$ mm.
>
> $V = \pi(10)^2(80)$
>
> $V = \pi(100)(80)$
>
> $V = 25{,}132.74123 \text{ mm}^3 \approx 25{,}132 \text{ mm}^3 \qquad \text{Round.}$

2 Find the surface area of three-dimensional objects (pp. 735–738).

Formulas:

Lateral surface area (area of sides) of a right prism or cylinder: $\text{LSA} = ph$, where p is the perimeter of the base and h is the height.

Total surface area (sides plus bases) of a right prism or cylinder: $\text{TSA} = ph + 2B$, where p is the perimeter of the base, h is the height, and B is the area of a base.

Total surface area of a sphere: $\text{TSA} = 4\pi r^2$, where r is the radius.

Lateral surface area of a cone: $\text{LSA} = \pi rs$, where r is the radius and s is the slant height.

Total surface area of a cone: $\text{TSA} = \pi rs + \pi r^2$, where r is the radius and s is the slant height. The area of the base is πr^2.

> A conical pile of gravel has a diameter of 30 ft and a slant height of 40 ft. What is the lateral surface area?
>
> $\text{LSA} = \pi rs$ $\qquad\qquad\qquad r = \frac{1}{2}d = \frac{1}{2}(30) = 15.$
>
> $\text{LSA} = \pi(15)(40)$
>
> $\text{LSA} = 1{,}884.96 \text{ ft}^2 \qquad \text{Rounded from } 1{,}884.955592$

Section 18–1

Use Fig. 18–133 for Exercises 1–3.

1. Which lines are parallel?
2. Which lines coincide?
3. Which lines intersect?

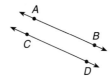
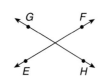
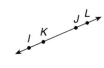

Figure 18–133

Use Fig. 18–134 for Exercises 4–5.

4. Name ∠*a* using three capital letters.
5. Name ∠*a* using one capital letter.

Figure 18–134

Classify the angle measures using the terms *right, straight, acute,* or *obtuse.*

6. 50°   7. 90°   8. 120°   9. 180°

State whether the angle pairs are complementary, supplementary, or neither.

10. 98°, 62°   11. 135°, 45°   12. 45°, 35°   13. 21°, 79°   14. 90°, 90°

Use Fig. 18–135 to answer Exercises 15–16.

15. Name two pairs of supplementary angles.
16. What are angles *a* and *c* called?

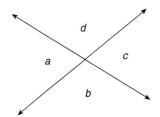

Figure 18–135

In Fig. 18–136, *l* ‖ *m* and *n* is a transversal.

17. Name two pairs of alternate exterior angles.
18. If ∠*a* = 147° what is the measure of ∠*e?*
19. If ∠*d* = 135° what is the measure of ∠*f?*

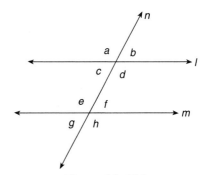

Figure 18–136

Change to decimal degree equivalents. Express the decimals to the nearest ten-thousandth.

20. 32°14′15″   21. 29′   22. 47″   23. 7′34″

Change to equivalent degrees, minutes, and seconds. Round to the nearest second when necessary.

24. 20.6°   25. 0.75°   26. 0.46°   27. 0.2176°

Section 18–2

Find the perimeter of Figs. 18–137 through 18–141.

28.

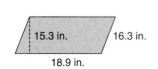

Figure 18–137

29.

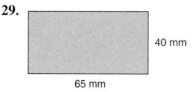

Figure 18–138

30.

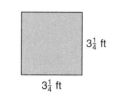

Figure 18–139

31.

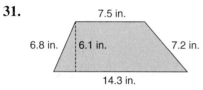

Figure 18–140

32.

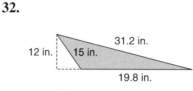

Figure 18–141

33. The Tennessee Highway Department has signs in the form of a parallelogram. One set of parallel sides each measures 15 ft and one set of parallel sides each measures 18 feet. Find the perimeter of the sign.

34. Antique tiles were often made in the form of a square that is 6 in. on each side. What is the perimeter of a tile?

35. A rectangular tablecloth measures 84 in. by 60 in. What length of lace is required to trim the edges of the cloth?

36. A classroom table in the form of a trapezoid has parallel sides that measure 26 in. and 48 in. The two non-parallel sides both measure 21.1 in. Find the length of trim needed to encase the edges of the table.

Find the area of Figs. 18–142 through 18–146.

37.

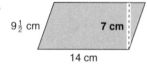

Figure 18–142

38.

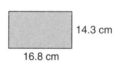

Figure 18–143

39.

Figure 18–144

40.

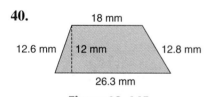

Figure 18–145

41.

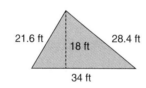

Figure 18–146

42. Find the area of a scalene triangle with sides of 15 in., 31.2 in., and 19.8 in. long.

43. **CON** If a parking lot for a new hospital in the shape of a parallelogram measures 300 ft by 175 ft and has a height of 120 ft, how many square feet need to be paved?

44. **CON** A hall wall with no windows or doors measures 30 ft long by 9 ft high. Find the number of square feet to be covered if paneling is installed on the two walls.

45. **CON** A den 21 ft by $18\frac{1}{2}$ ft is to be carpeted. How many square yards of carpeting are needed?

46. A square area of land is 10,000 m². If a baseball field must be at least 99.1 m along each foul line to the park fence, is the square adequate for regulation baseball?

47. CON   Debbie Murphy is building a contemporary home with four front windows, each in the form of a parallelogram. If each window has a 5-ft base and a height of 2 ft, how many square feet of the 25 ft × 11 ft wall will require stain?

48. CON   If the stain used on the wall in Exercise 47 is applied at a cost of $2.75 per square yard, find the cost to the nearest dollar of staining the front wall.

Section 18–3

Find the circumference (perimeter) and area of Figs. 18–147 through 18–150. Round to hundredths when necessary.

49.

4 m

Figure 18–147

50.

15.9 m

Figure 18–148

51.

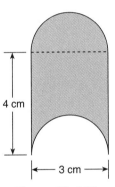

4 cm

3 cm

Figure 18–149

52.

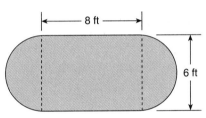

8 ft

6 ft

Figure 18–150

53. INDTR   A $\frac{1}{4}$-in. electric drill with variable speed control turns as slowly as 25 rpm. If an abrasive disk with a 5-in. diameter is attached to the drill driveshaft, what is the disk's slowest cutting speed in ft/min? Round to the nearest whole number. (Cutting speed = circumference measured in feet × rpm.)

54. CON   What is the cross-sectional area of the opening in a round flue tile whose inside diameter is 8 in.? Round any part of an inch to the next tenth of an inch.

Convert the degree measures to radians rounded to the nearest hundredth.

55. 60°

56. 212°

Convert the degree, minute, and second measures to degrees rounded to the nearest ten-thousandth. Then convert to radians to the nearest hundredth.

57. 99°45′

58. 120°20′40″

Convert the radian measures to degrees. Round to the nearest ten-thousandth of a degree when necessary.

59. $\frac{5\pi}{6}$ rad

60. 2.4 rad

Find the area of the sectors of a circle using Fig. 18–151. Round to hundredths.

61. ∠ = 45°9′
r = 2.58 cm

62. ∠ = 2.88 rad
r = 15 in.

63. ∠ = 40°
r = $2\frac{1}{2}$ ft

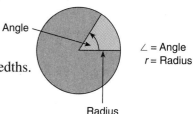

Find the arc length of the sectors of a circle using Fig. 18–151. Round to hundredths.

64. ∠ = 0.7 rad
r = 1.45 ft

65. ∠ = 180°
r = 10 in.

Figure 18–151

Find the area of the segments of a circle using Fig. 18–152. Round to hundredths.

66. ∠ = 55°
r = 10″
h = 8.9″
b = 9.2″

67. ∠ = 30°
r = 24 cm
h = 23.2 cm
b = 12.4 cm

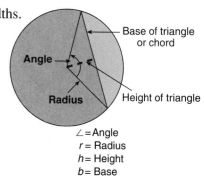

Figure 18–152

Solve.

68. INDTR A machine cuts a 12-in.-diameter frozen pizza into slices with sides that form 72° angles at the center of the pizza. What is the surface area of each slice to the nearest square inch?

69. CON A drain pipe with a 20-in. diameter has 4 in. of water in it (Fig. 18–153). What is the cross-sectional area of the water in the pipe to the nearest hundredth?

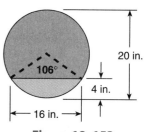

Figure 18–153

Section 18–4

70. Find the volume of the prism in Fig. 18–154.

71. Find the volume of the cylinder in Fig. 18–155.

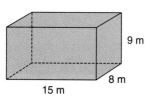

Figure 18–154

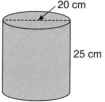

Figure 18–155

72. CON If concrete weighs 160 lb per cubic foot, what is the weight of a concrete circular slab 4 in. thick and 15 ft across? Round to the nearest pound.

73. CON How many cubic yards of topsoil are needed to cover an 85-ft by 65-ft area for landscaping if the topsoil is 6 in. deep? Round to the nearest whole number. (27 ft³ = 1 yd³)

74. **CON** An interstate highway is repaired in one section 48 ft across, 25 ft long, and 8 in. deep. If concrete costs $25.50 per cubic yard, what is the cost of the concrete needed to repair the highway rounded to the nearest dollar? ($27 \text{ ft}^3 = 1 \text{ yd}^3$)

75. Find the total surface area of the triangular prism in Fig. 18–156.

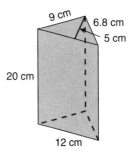

Figure 18–156

76. Find the lateral surface area of the cylinder in Fig. 18–157. Round to the nearest cm^2.

77. Find the total surface area of the cylinder in Fig. 18–157. Round to the nearest cm^2.

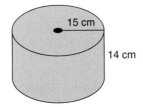

Figure 18–157

78. Find the total surface area of a cylinder that has a radius of 10 cm and a height of 30 cm. Round to the nearest cm^2.

79. **CON** A pipeline to carry oil between two towns 5 mi apart has an inside diameter of 18 in. If 1 mi = 5,280 ft, find the lateral surface area to the nearest ft^2.

80. How many barrels of oil will a conical tank hold if its height is $65\frac{1}{2}$ ft and its radius is 20 ft? Round to the nearest whole barrel. (31.5 gal = 1 barrel and $1 \text{ ft}^3 = 7.48$ gal.)

81. Find the volume of a pyramid that has a square base of 12 m on a side and a height of 36 m.

82. Find the volume of a frustum of a pyramid that has a square base of 15 yd on a side and a top of 12 yd on a side. The height is 12 yd.

Solve. Round the final answer to the nearest tenth unless otherwise specified.

83. Find the total surface area of a sphere with a radius of 9 m.

84. Find the volume of a sphere that has a diameter of 30 cm.

85. Find the lateral surface area of a cone with a radius of 6 cm and a slant height of 9 cm.

86. Find the total surface area of a cone that has a radius of 4 m and a slant height of 8 m.

87. **CON** The entire exterior surface of a conical tank with a slant height of 12 ft and a diameter of 18 ft is being painted. If the paint covers at a rate of 350 ft^2 per gallon, how many gallons of paint are needed for the job? Round any fraction of a gallon to the next whole gallon.

88. **AG/H** A hopper deposits grain in a cone-shaped pile with a diameter of 9'6″ and a height of 8'3″. To the nearest cubic foot, how much grain is deposited?

Identify the longest and shortest sides in Fig. 18–158.

89.

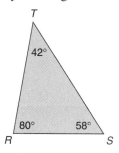

Figure 18–158

List the angles in order of size from largest to smallest in Fig. 18–159.

90.

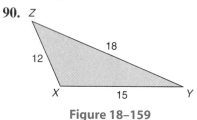

Figure 18–159

TEAM PROBLEM-SOLVING EXERCISES

1. Use an inductive (experimental) process to determine the sum of the angles of irregular triangles and quadrilaterals.

   (a) Draw three scalene triangles and cut them out. Taking one triangle at a time, number the angles and tear off each angle of the triangle. Arrange the vertices of each original triangle together at a common point. What type angle is formed in each case? Without measuring, what is the sum of the angles of each original triangle?

   (b) Draw three irregular quadrilaterals and cut them out. Taking one quadrilateral at a time, number the angles and tear off each angle of the quadrilateral. Arrange the vertices of the original quadrilateral together at a common point. What type angle is formed? Without measuring, what is the sum of the angles of each original quadrilateral?

2. Use an inductive (experimental) process to determine the sum of the angles of irregular polygons.

   (a) Draw an irregular pentagon. From one vertex draw lines connecting this vertex with each nonadjacent vertex. What types of polygons are formed? Use your knowledge of the sum of the angles of a triangle to determine the sum of the angles of the pentagon.

   (b) Draw an irregular hexagon. From one vertex draw lines connecting this vertex with each nonadjacent vertex. Use your knowledge of the sum of the angles of a triangle to determine the sum of the angles of the hexagon.

   (c) Use your findings from Exercise 1 and parts (a) and (b) of Exercise 2 to state the pattern that is developing. Test your proposed generalization (formula) with at least two other irregular polygons like a **septagon** (seven-sided polygon), **octagon** (eight-sided polygon), **nonagon** (nine-sided polygon), and so on.

1. Explain the process for finding the perimeter of a composite figure.

2. Explain the process for finding the area of a composite figure.

3. Draw a composite figure for which the area can be found by using either addition or subtraction and explain each approach.

4. Write in words the formula for finding the degrees in each angle of a regular polygon.

5. Describe a shortcut for finding the perimeter of an L-shaped figure. Illustrate your procedure with a problem.

6. Two pieces of property have the same area (acreage) and equal desirability for development. Both require expensive fencing. One piece is a square 600 ft on each side. The other piece is a rectangle 400 ft by 900 ft. Which piece of property requires the least amount of fencing and is thus more desirable?

7. Discuss the differences in a sector and a segment.

8. Why are two formulas needed to find the area of a sector? Create an example using each formula and solve your examples.

9. If the height of a cone is doubled, what effect does this have on the volume of the cone? Create several examples of cones with the same area of base but different heights to validate your answer.

10. If the height of a right circular cylinder is doubled, what effect does this have on the surface area of the cylinder? Create several examples of right circular cylinders that have the same area of base but different heights to validate your answer.

1. Change 0.3125° to minutes and seconds.

2. Change 15′32″ to a decimal degree to the nearest ten-thousandth.

Convert the degree measures to radians rounded to the nearest hundredth.

3. 35°

4. 122°

Convert the radian measures to degrees. Round to the nearest ten-thousandth of a degree when necessary.

5. $\dfrac{5\pi}{8}$ rad

6. 3.1 rad

Find the perimeter and area of Figs. 18–160 through 18–163. Round to the nearest tenth when necessary.

7.

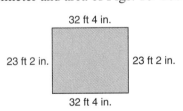

32 ft 4 in.

23 ft 2 in.   23 ft 2 in.

32 ft 4 in.

Figure 18–160

8.

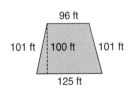

96 ft

101 ft | 100 ft | 101 ft

125 ft

Figure 18–161

9.

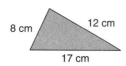

Figure 18–162

10.

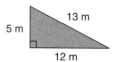

Figure 18–163

Find the circumference and area (Figs. 18–164 and 18–165).

11.

Figure 18–164

12.

Figure 18–165

Use the relationships of arc length s, area A, central angle (in radians) θ, and radius r to solve the problems relating to sectors. Round to hundredths if necessary.

13. Find s if $\theta = 0.5$ and $r = 2$ in.

14. Find θ if $s = 5.3$ m and $r = 7$ m.

15. Find r if $\theta = 1.7$ and $s = 2.9$ m.

16. Find A if $\theta = 0.6$ and $r = 7.3$ cm.

17. How many square feet of floor space are there in the plan shown in Fig. 18–166?

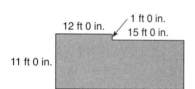

Figure 18–166

Solve and when necessary, round to hundredths.

18. Find the area of the colored portion of the tiled walk that surrounds a rectangular swimming pool (Fig. 18–167).

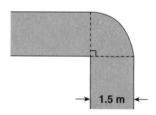

Figure 18–167

19. Find the area of the composite figure in Fig. 18–168.

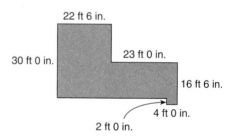

Figure 18–168

20. A section of a hip roof is a trapezoid measuring 35 ft at the bottom, 15 ft at the top, and 10 ft high. Find the area of this section of the roof in square feet.

21. Find the area of a sector of a circle that has a radius of 14 cm if the sector has an angle of 42°. Round to hundredths.

22. A segment is removed from a flat metal circle so that the piece rests on a horizontal base (Fig. 18–169). Find the area of the segment that is removed.

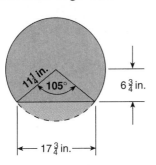

Figure 18–169

23. A pentagonal prism (five sides) measures 1 in. on each side of its base and has a height of 10 in. What is its lateral surface area?

24. Find the total surface area of the pentagonal prism in Exercise 23 if the area of the base is 2.17 in².

25. A spherical tank 12 ft in diameter can hold how many gallons of fluid if 1 ft³ = 7.48 gal? Answer to the nearest whole gallon.

26. The base of a brass pyramid is an equilateral triangle with sides of 3 in. and altitude of 2.6 in. If the pyramid's height is 8 in., what is the volume of the pyramid?

19

Triangles

Focus on Careers

When your education is completed, the next step is to search for a job. Job searches should be planned in a systematic and informed manner. Personal contacts—family, friends, and acquaintances—are often most effective in finding a job. These personal contacts may help you directly or put you in touch with someone who can help you.

Before you begin completing job applications, you should first prepare your job résumé. Lots of Internet sites provide excellent outlines for creating a job résumé. Next, you should develop a list of potential employers by searching the Internet, the library, or asking your teachers. Narrow the list to the companies that interest you most and learn as much as you can about each company. In particular, search the company's job announcements, which are often shown on the company website under the Human Resources office.

School career planning and placement offices help students and alumni find jobs. Placement offices have lists of jobs offered on campus, but they also have lists of companies and government organizations who have jobs to fill. Some colleges have "Career Day" and invite numerous companies to set up tables so students can talk with lots of prospective employers in a very short time. Investigate if your college has such service and take advantage of it.

Your state employment service office, sometimes called Job Service, operates in coordination with the U.S. Department of Labor's Employment and Training Administration to maintain a national database of job listings. This database can be accessed at http://www.CareerOneStop.org. You can get access to the Internet at any local public employment office or your local library. For more information on finding a job, see the U.S. Bureau of Labor Statistics website.

Source: U.S. Bureau of Labor Statistics, Office of Occupational Statistics and Employment Projections, http://www.bis.gov/OCO/. September, 2005.

19–1 | *Special Triangle Relationships*

Learning Outcomes

1 Classify triangles by the relationship of the sides or angles.

2 Determine if two triangles are congruent using inductive and deductive reasoning.

3 Solve problems that involve similar triangles.

We have examined various relationships among lines and angles in the previous chapter. In this chapter, we study one of the most useful mathematical figures for any technician who uses geometry—the *triangle*. The relationships among the sides and angles in the triangle enable us to obtain much information that is implied but not always expressed in certain applications.

Triangles can be classified according to the relationship of their sides. The symbol $\triangle$ is used to indicate a triangle.

1 Classify Triangles by the Relationship of the Sides or Angles.

Three relationships are possible among the three sides of a triangle and result in special names for each type of triangle.

An **equilateral triangle** is a triangle with three equal sides. The three angles of an equilateral triangle are also equal. Each angle measures 60° (Fig. 19–1, left).

An **isosceles triangle** is a triangle with *exactly* two equal sides. The angles opposite these equal sides are also equal (Fig. 19–1, middle).

A **scalene triangle** is a triangle with *all* three sides unequal (Fig. 19–1, right).

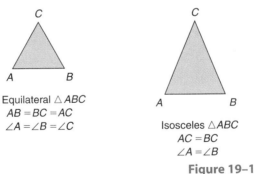

Equilateral $\triangle ABC$
$AB = BC = AC$
$\angle A = \angle B = \angle C$

Isosceles $\triangle ABC$
$AC = BC$
$\angle A = \angle B$

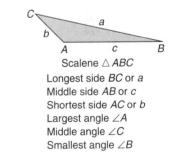

Scalene $\triangle ABC$
Longest side *BC* or *a*
Middle side *AB* or *c*
Shortest side *AC* or *b*
Largest angle $\angle A$
Middle angle $\angle C$
Smallest angle $\angle B$

Figure 19–1

In a scalene triangle, where no sides are equal, we can state an important relationship between the sides and their opposite angles (Fig. 19–1).

To determine the longest and shortest sides of a triangle:

If the three sides of a triangle are unequal, the *largest* angle is opposite the *longest* side and the *smallest* angle is opposite the *shortest* side.

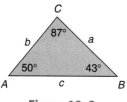

EXAMPLE   Identify the longest and shortest sides of the triangle in Fig. 19–2.

Figure 19–2

The longest side is *AB* or *c* (87° is the largest angle). **The shortest side is *AC* or *b*** (43° is the smallest angle).

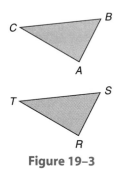

Figure 19–3

2 Determine if Two Triangles Are Congruent Using Inductive and Deductive Reasoning.

Triangles that have the same size and shape are **congruent** triangles. These triangles fit exactly on top of each other.

In Fig. 19–3, $\triangle ABC$ fits exactly over $\triangle RST$. They are congruent triangles. The symbol $\cong$ means congruent. That is, $\triangle ABC \cong \triangle RST$. Each angle in $\triangle ABC$ has an angle in $\triangle RST$ that is its equal. We say that these pairs of angles *correspond*. In Fig. 19–3, $\angle A$ corresponds to $\angle R$. Also, $\angle C$ corresponds to $\angle T$, and $\angle B$ corresponds to $\angle S$. The equal sides also correspond. $\overline{AB}$ corresponds to $\overline{RS}$, $\overline{CB}$ corresponds to $\overline{TS}$, and $\overline{AC}$ corresponds to $\overline{RT}$.

Congruent triangles are triangles in which the corresponding sides and angles are equal.

Use three sides to determine congruent triangles:

If the three sides of one triangle are equal to the corresponding three sides of another triangle, the triangles are congruent (side-side-side or SSS).

EXAMPLE   Use the side-side-side property to write the corresponding and equal sides of the two triangles in Fig. 19–4.

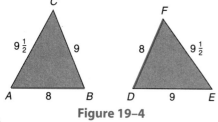

Figure 19–4

$\triangle ABC \cong \triangle FDE$

$\overline{AB}$ corresponds to $\overline{FD}$
$\overline{BC}$ corresponds to $\overline{DE}$
$\overline{AC}$ corresponds to $\overline{FE}$

Rotate $\triangle DEF$.

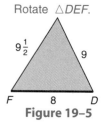

Figure 19–5

Use two sides and the included angle to determine congruent triangles:

If two sides and the *included angle* of one triangle are equal to two sides and the included angle of another triangle, the triangles are congruent (side-angle-side or SAS).

EXAMPLE Use side-angle-side property to list the corresponding and equal sides and angles of the two triangles in Fig. 19–6.

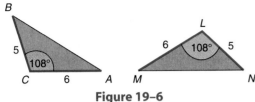

Figure 19–6

$$\overline{BC} = \overline{LN}$$
$$\overline{AC} = \overline{LM}$$
$$\angle C = \angle L$$

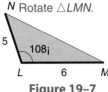

N Rotate △LMN.

Figure 19–7

The triangles are congruent because two sides and the included angle of one triangle are equal to two sides and the included angle of the other (SAS). Then, the other corresponding sides and angles are equal.

$\overline{AB} = \overline{MN}$ The sides are equal because they are opposite equal angles.
$\angle A = \angle M$ The angles are equal because they are opposite equal sides.
$\angle B = \angle N$ The angles are equal because they are opposite equal sides.

Use two angles and the included side to determine congruent triangles:

If two angles and the common side of one triangle are equal to two angles and the common side of another triangle, the triangles are congruent (angle-side-angle or ASA).

EXAMPLE Use the angle-side-angle property to list the corresponding and equal angles and sides of the two triangles in Fig. 19–8.

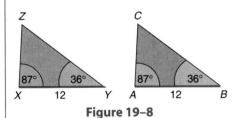

Figure 19–8

$$\angle X = \angle A$$
$$\angle Y = \angle B$$
$$\overline{XY} = \overline{AB}$$

The triangles are congruent because of the ASA relationship. Then,

$$\angle Z = \angle C$$
$$\overline{XZ} = \overline{AC}$$
$$\overline{YZ} = \overline{BC}$$

The study of geometry applies two types of reasoning, *inductive* and *deductive* reasoning. **Inductive reasoning** starts with investigation and experimentation. Early mathematicians discovered many properties of geometry through inductive reasoning. After extensive investigation, general conclusions are drawn from the results of specific cases.

Let's use inductive reasoning to examine the results of cutting a square into four parts by cutting along the two diagonals (Fig. 19–9). Compare the resulting four triangles. Are they congruent triangles? Make a square of a different size and repeat the exercise. Continued repetitions lead you to conclude that the diagonals of a square divide the square into four congruent triangles.

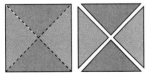

Figure 19–9

Deductive reasoning starts with accepted properties. Additional properties are concluded from these original properties. We accept the congruent triangle properties and an additional property that the diagonals of a square bisect (cut in half) the angles of the square. We can use deductive reasoning to show that the two diagonals of a square form four congruent triangles.

We will use a two-column format to show the flow of the deductive-reasoning process.

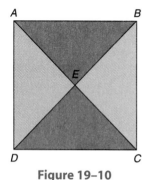

EXAMPLE *ABCD* is a square and the diagonals *AC* and *BD* are equals (Fig. 19–10). The diagonals also bisect each other. Are △*AED*, △*AEB*, △*BEC*, and △*DEC* congruent?

$AB = BC = CD = AD$	The sides of a square are equal.
$AE = BE = CE = DE$	The diagonals of a square are equal and bisect each other.
$\triangle AED \cong \triangle AEB$	SSS; $AB = AD$, $AE = AE$, $DE = BE$
$\triangle BEC \cong \triangle DEC$	SSS; $CD = BC$, $CE = CE$, $DE = BE$
$\triangle AED \cong \triangle BEC$	SSS; $AD = BC$, $DE = BE$, $AE = CE$

All other pairings of triangles can be shown to be congruent using a similar deductive-reasoning process.

Yes, △*AED*, △*AEB*, △*BEC*, and △*DEC* are congruent.

Figure 19–10

In a systematic or axiomatic study of geometric concepts, all properties are developed or proved from a limited number of basic properties. In our study of geometric concepts, we apply both inductive and deductive reasoning in our arguments for solving problems. We do not introduce all the concepts necessary to make formal proofs of all geometric properties.

3 Solve Problems That Involve Similar Triangles.

As we saw in the preceding learning outcome, *congruent triangles* have the same size and shape. **Similar triangles** have the same shape but not the same size (Fig. 19–11).

The corresponding angles in similar triangles are equal. The corresponding sides of similar triangles are directly proportional.

The symbol for showing similarity is ~ and is read "is similar to."

Properties of similar triangles:

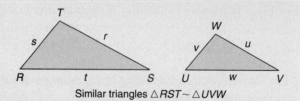

Similar triangles △*RST* ~ △*UVW*

Figure 19–11

Corresponding angles of similar triangles are equal in size.

$\angle R = \angle U$
$\angle S = \angle V$
$\angle T = \angle W$

Corresponding sides of similar triangles are directly proportional.

Side *r* corresponds to side *u*
Side *s* corresponds to side *v*
Side *t* corresponds to side *w*

The corresponding sides of similar triangles are directly proportional. That is, each pair of corresponding sides form a ratio, or fraction, and the ratios are equal (Fig. 19–12).

$$\frac{a}{m} = \frac{b}{n} = \frac{c}{q}$$

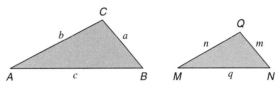

Figure 19–12

To find a missing side of a similar triangle:

1. Examine the given dimensions of the corresponding sides and find one pair for which both measures are known.
2. Pair the unknown side with its corresponding known side.
3. Set up a direct proportion with the two pairs.
4. Solve the proportion.

EXAMPLE   Find the unknown side in Fig. 19–13 if $\triangle ABC \sim \triangle DEF$.

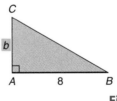 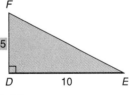

Figure 19–13

Write ratios of corresponding sides in a proportion. Use single lowercase letters to identify the sides. Side a is opposite $\angle A$, side b is opposite $\angle B$, and so on.

Pair 1: 8 from $\triangle ABC$ corresponds and is proportional to 10 from $\triangle DEF$.
Pair 2: b from $\triangle ABC$ corresponds and is proportional to 5 from $\triangle DEF$.

Estimation   Side b should be less than 5 since 8 is less than 10.

$$\frac{\text{Pair 1}}{\text{Pair 2}}\, \frac{8}{b} = \frac{10}{5} \qquad$$ Pair with both measures known

Pair with one measure known

$$10b = 8(5) \qquad$$ Cross multiply.

$$10b = 40 \qquad$$ Divide.

$$b = 4$$

Interpretation   **Side b is 4 units long.**

19–1 Special Triangle Relationships

761

EXAMPLE AG/H

A arborist must know the height of a tree to determine which way to fell it so that it does not endanger lives, traffic, or property. A 6-ft pole casts a 4-ft shadow when the tree casts a 20-ft shadow (Fig. 19–14). What is the height of the tree?

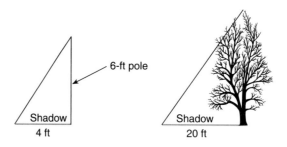

Figure 19–14

The triangles formed are similar. Let x = the height of the tree.

Pair 1: 6-ft pole casts a 4-ft shadow.　　Both measures known
Pair 2: x-ft tree casts a 20-ft shadow.　　One measure known

Estimation　The tree is more than 20 ft tall.　　Since the pole is taller than its shadow, the tree will be taller than its shadow of 20 ft.

$$\frac{6 \text{ (height of pole)}}{x \text{ (height of tree)}} = \frac{4 \text{ (shadow of pole)}}{20 \text{ (shadow of tree)}}$$　　Pair 1
　　　　　　　　　　　　　　　　　　　　　　Pair 2

$$6(20) = 4x$$　　Cross multiply.

$$120 = 4x$$　　Divide.

$$30 = x$$

Interpretation　**The tree is 30 ft tall.**

EXAMPLE CAD/ARC

A building lies between points A and B, so the distance between these points cannot be measured directly by a surveyor. Find the distance using the similar triangles shown in Fig. 19–15.

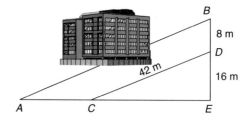

Figure 19–15

$\triangle ABE \sim \triangle CDE$. Note that CD must be made parallel to AB for the triangles to be similar. Parallel means the lines are the same distance apart from end to end.

Visualize the triangles as separate triangles (Fig. 19–16). Using the lowercase letter e for the missing AB measure is confusing because in $\triangle CDE$, side CD can also be thought of as side e.

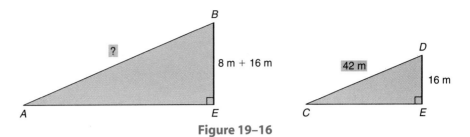

Figure 19–16

Pair 1: **BE** from △ABE corresponds to **DE** from △CDE.

Pair 2: **AB** from △ABE corresponds to **CD** from △CDE.

Estimation AB is more than 42 m.

$$\frac{BE}{AB} = \frac{DE}{CD}$$ Substitute values. CD = 42; DE = 16

  BE = BD + DE = 8 + 16 = 24

$$\frac{24}{AB} = \frac{16}{42}$$ Cross multiply.

$16AB = (24)42$ AB is interpreted as a single variable.

$16AB = 1,008$ Divide.

$AB = 63$ m

Interpretation **Thus, the distance from A to B is 63 m.**

SECTION 19–1 SELF-STUDY EXERCISES

1 Fill in the blanks.

1. A triangle with no equal sides is called a(n) _____ triangle.

2. A triangle with three equal sides is called a(n) _____ triangle.

3. A triangle with only two equal sides is called a(n) _____ triangle.

Identify the longest and shortest sides in Figs. 19–17 and 19–18.

4.

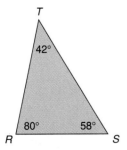

Figure 19–17

5.

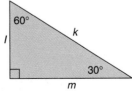

Figure 19–18

19–1 Special Triangle Relationships

List the angles in order of size from largest to smallest in Figs. 19–19 and 19–20.

6.

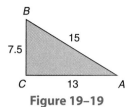

Figure 19–19

7.

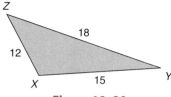

Figure 19–20

2 Write the corresponding parts not given for the congruent triangles in Figs. 19–21 to 19–23.

8.

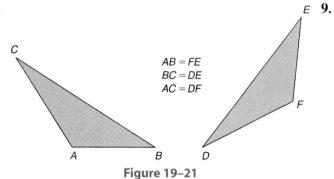

$AB = FE$
$BC = DE$
$AC = DF$

Figure 19–21

9.

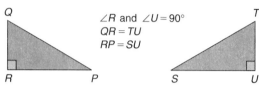

$\angle R$ and $\angle U = 90°$
$QR = TU$
$RP = SU$

Figure 19–22

10.

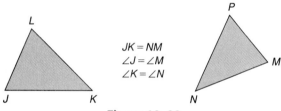

$JK = NM$
$\angle J = \angle M$
$\angle K = \angle N$

Figure 19–23

3 Find the indicated parts for the similar triangles in Exercises 11–14 (Figs. 19–24 to 19–27).

11. $AB = 12$ cm
$BC = 26$ cm
$DE = 20.8$ cm
Find *EF*.

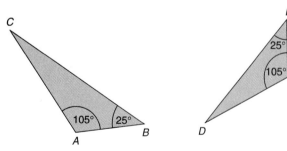

Figure 19–24

12. $\angle R$ and $\angle U = 90°$
$QR = 24$ in.
$RP = 28$ in.
$US = 22.4$ in.
Find *TU*.

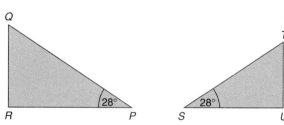

Figure 19–25

13. $JK = 8$ ft
$MN = 6$ ft
$JL = 5$ ft
$\angle J = \angle M = 50°$
$\angle K = \angle N = 18°$
Find PM.

14. $\triangle ABC \sim \triangle EDF$. $\angle A = \angle E$, $\angle C = \angle F$. Find a and d. Use Fig. 19–27.

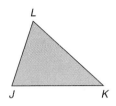
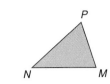

Figure 19–26

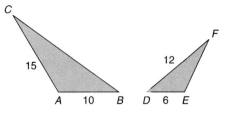

Figure 19–27

15. $\triangle ABC \sim \triangle DEC$. $\angle 1$ and $\angle 2$ have the same measure. Find DC and DE. (*Hint:* Let $DC = x$ and $AC = x + 3$. Use Fig. 19–28.) Round to tenths if necessary.

16. **AG/H** Find the height of a tree that casts a 30-ft shadow when a 6-ft 6-in. pole casts a shadow of 3 ft.

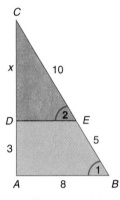

Figure 19–28

19–2 | **Pythagorean Theorem**

Learning Objectives   **1** Use the Pythagorean theorem to find the unknown side of a right triangle.

2 Use the properties of a 45°, 45°, 90° triangle to find unknown parts.

3 Use the properties of a 30°, 60°, 90° triangle to find unknown parts.

1 **Use the Pythagorean Theorem to Find the Unknown Side of a Right Triangle.**

One of the most famous and useful theorems in mathematics is the Pythagorean theorem. It is named for the Greek mathematician, Pythagoras. The theorem applies to a right triangle.

A **right triangle** has two sides that form a right angle (square corner). These two sides are called the **legs** of the triangle. The side opposite the right angle, or the third side of the triangle, is called the **hypotenuse:** The **Pythagorean theorem** states that the square of the hypotenuse of a right triangle is equal to the sum of the squares of the two legs of the triangle.

Formula for Pythagorean theorem:

$$c^2 = a^2 + b^2$$

where c is the hypotenuse of a right triangle, and a and b are legs (Fig. 19–29).

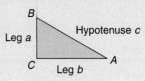

Figure 19–29

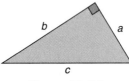

To find a leg or the hypotenuse of a right triangle:

1. Identify the two known sides of the triangle and the unknown side, and state the theorem symbolically.
2. Substitute known values in the Pythagorean theorem. $c^2 = a^2 + b^2$
3. Solve for the unknown value.
4. The solution is the principal square root.

EXAMPLE   Use the Pythagorean theorem to find b when $a = 8$ mm and $c = 17$ mm (Fig. 19–30).

$c^2 = a^2 + b^2$	State the theorem symbolically and substitute known values.
$17^2 = 8^2 + b^2$	Evaluate powers.
$289 = 64 + b^2$	Use the addition axiom to isolate b^2.
$289 - 64 = b^2$	Combine like terms.
$225 = b^2$	Take the square root of both sides.
$\sqrt{225} = b$	Solution is the principal square root.
$\mathbf{15 \ mm = b}$	

Figure 19–30

Many times we are given problems that contain "hidden" triangles. In these cases, we need to visualize the triangle or triangles in the problems. Drawing one or more of the sides of the "hidden" triangle helps solve the problem.

EXAMPLE
INDTR

Find the center-to-center distance between pulleys A and C (Fig. 19–31).

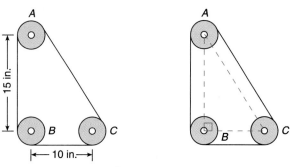

Figure 19–31

Connect the center points of the three pulleys to form a right triangle. The hypotenuse is the distance between the centers of pulleys A and C. Use the Pythagorean theorem and substitute the given values for the two known sides.

$(AC)^2 = (AB)^2 + (BC)^2$	State theorem symbolically and substitute values.
$(AC)^2 = 15^2 + 10^2$	Square both constants.
$(AC)^2 = 225 + 100$	Combine like terms.
$(AC)^2 = 325$	Take the square root of both sides.
$AC = \sqrt{325}$	Find the approximate principal square root.
$AC = 18.0277564$	

The distance from pulley C to pulley C is 18.0 in. (to the nearest tenth).

EXAMPLE
INDTR

The head of a bolt is a square 0.5 in. on each side (distance across flats). What is the distance from corner to corner (distance across corners)? (See Fig. 19–32.)

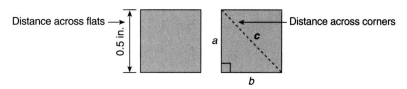

Figure 19–32

The sides of the bolt head form the legs of a triangle if a diagonal line is drawn from one corner to the opposite corner. This diagonal forms the hypotenuse of the right triangle. Because the legs of the triangle are known to be 0.5 in. each, we can substitute in the Pythagorean theorem to find the diagonal line, the hypotenuse, which is the distance across corners.

$c^2 = a^2 + b^2$	State theorem symbolically.
$c^2 = 0.5^2 + 0.5^2$	Substitute values and evaluate.
$c^2 = 0.25 + 0.25$	Combine like terms.
$c^2 = 0.5$	Take the square root of both sides.
$c = \sqrt{0.5}$	Evaluate.
$c = 0.707106781$	

The distance across corners is 0.71 in. (to the nearest hundredth).

Sometimes right triangles are used to represent certain relationships, such as forces acting on an object at right angles or electrical and electronic phenomena related in the way the three sides of a right triangle are related. Let's look at an example.

EXAMPLE   Forces A and B come together at a right angle to produce force C (Fig. 19–33). If force A is 74.8 lb and the resulting force C is 91.5 lb, what is force B?

Figure 19–33

Since the forces are related in the way the sides of a right triangle are related, we may use the Pythagorean theorem to find the unknown force B, a leg of the triangle.

$A^2 + B^2 = C^2$	State theorem symbolically and substitute values.
$74.8^2 + B^2 = 91.5^2$	Square both constants.
$5{,}595.04 + B^2 = 8{,}372.25$	Apply the addition axiom to isolate B^2.
$B^2 = 8{,}372.25 - 5{,}595.04$	Combine constants.
$B^2 = 2{,}777.21$	Take the square root of both sides.
$B = \sqrt{2{,}777.21}$	Find the approximate principal square root.
$B = 52.7$	Round to the nearest tenth.

Force B is 52.7 lb.

2 Use the Properties of a 45°, 45°, 90° Triangle to Find Unknown Parts.

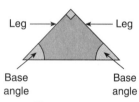

Figure 19–34

An isosceles triangle is a triangle with two equal sides. An **isosceles right triangle** is a right triangle that has equal legs. The angles opposite the equal legs are **base angles.** The two base angles are also equal because they are the angles opposite the equal sides (Fig. 19–34).

The sum of the angles of a triangle is 180°. The sum of the two base angles of a right triangle is $180° - 90°$ or 90°. If both angles are equal, as in the isosceles right triangle, then each angle is $\frac{1}{2}(90°)$ or 45°. These base angles are complementary.

Figure 19–35

A **45°, 45°, 90° triangle** is an isosceles right triangle. The two base angles are each 45°. This triangle is frequently used in applications of the Pythagorean theorem.

All isosceles right triangles (45°, 45°, 90°) have the same relationship between each leg of the triangle and the hypotenuse (Fig. 19–35).

Applying the Pythagorean theorem to find the hypotenuse when $x =$ length of each leg:

$$\text{hypotenuse}^2 = x^2 + x^2$$
$$\text{hypotenuse} = \sqrt{x^2 + x^2}$$
$$\text{hypotenuse} = \sqrt{2x^2}$$
$$\text{hypotenuse} = x\sqrt{2}$$

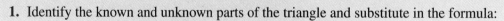

To find the hypotenuse or leg of a 45°, 45°, 90° triangle:

1. Identify the known and unknown parts of the triangle and substitute in the formula:

$$\text{Hypotenuse} = \text{Leg}\sqrt{2}$$

2. Solve for the unknown value.
3. Write as an exact solution or approximate solution, as desired.

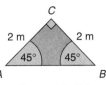

EXAMPLE Find the hypotenuse of an isosceles right triangle that has equal sides of 2 m (Fig. 19–36).

Figure 19–36

hypotenuse $= \text{leg}\sqrt{2}$ Substitute into appropriate formula.

hypotenuse $= \mathbf{2\sqrt{2}}$ Exact solution.

hypotenuse $= \mathbf{2.828}$ **m**   Approximate solution

EXAMPLE Find AC and BC if $AB = 5$ cm (Fig. 19–37).

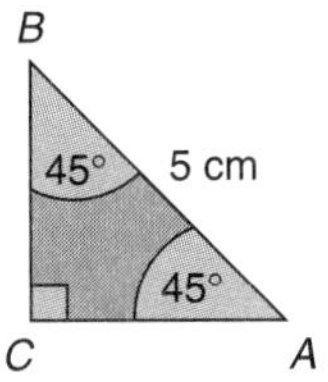

Figure 19–37

Hypotenuse $= \text{Leg }\sqrt{2}$ Substitute into appropriate formula.

$5 = AC\sqrt{2}$ Solve for AC, a leg. Divide both sides by $\sqrt{2}$.

$$\frac{5}{\sqrt{2}} = \frac{AC\sqrt{2}}{\sqrt{2}}$$

$$\frac{5}{\sqrt{2}} = AC$$ Rationalize the denominator. Multiply numerator and denominator by 1 in the form of $\frac{\sqrt{2}}{\sqrt{2}}$.

$$\frac{5}{\sqrt{2}} \cdot \frac{\sqrt{2}}{\sqrt{2}} = AC$$

$$\frac{5\sqrt{2}}{2} = \mathbf{AC}$$ Exact solution.

$$\mathbf{3.536 \text{ cm}} = \mathbf{AC}$$ Approximate solution

Since $AC = BC$, then $BC = 3.536$ cm.

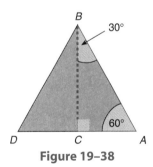

Figure 19–38

3 Use the Properties of a 30°, 60°, 90° Triangle to Find Unknown Parts.

Another special case of the Pythagorean theorem is the *30°, 60°, 90° triangle*, which arises often in applications. If we draw the altitude of an *equilateral* triangle, we form two 30°, 60°, 90° triangles (Fig. 19–38). The **altitude of an equilateral triangle** is a line drawn from the midpoint of the base to the opposite vertex, dividing the triangle into two congruent right triangles.

Because the altitude of an equilateral triangle bisects (divides in half) the vertex angle and the base, two 30° angles are formed at B, and $AC = CD$. We can also say that AC is $\frac{1}{2} AD$ or one-half any side of the original equilateral triangle. That means $AC = \frac{1}{2}AB$, or the side opposite the 30° angle is one-half the hypotenuse.

Apply the Pythagorean theorem to find the height of a 30°, 60°, 90° right triangle. When x is the length of the side opposite the 30° angle and $2x$ is the hypotenuse (Fig. 19–39):

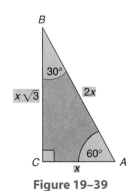

Figure 19–39

$$(2x)^2 = x^2 + \text{height}^2$$
$$4x^2 - x^2 = \text{height}^2$$
$$3x^2 = \text{height}^2$$
$$\text{height} = \sqrt{3x^2}$$
$$\text{height} = x\sqrt{3}$$

> **To find the unknown sides of a 30°, 60°, 90° triangle if one side is known:**
>
> 1. Identify the known side of the triangle.
> 2. If the side opposite the 30° angle is known, substitute in the formulas,
>
> $$\text{hypotenuse} = 2x$$
>
> $$\text{height} = x\sqrt{3}$$
>
> where x is the measure of the side opposite the 30° angle.
> 3. If either the hypotenuse or height is known, use the appropriate formula and solve for x. Use x and the remaining formula to find the remaining side.

EXAMPLE Find AB and BC if $AC = 7$ in. (Fig. 19–40).

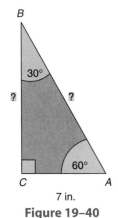

Figure 19–40

$\text{Hypotenuse} = 2x$	Substitute known values.
$AB = 2(7 \text{ in.}) = \textbf{14 in.}$	
$\text{Height} = x\sqrt{3}$	Substitute known values.
$BC = 7 \text{ in.}(\sqrt{3})$	
$BC = 7\sqrt{3} \text{ in.}$	Exact solution
$BC = 12.1$	Approximate solution to the nearest tenth

EXAMPLE Find AC and AB when $BC = 8$ cm (Fig. 19–41).

Find AC (side opposite the 30° angle):

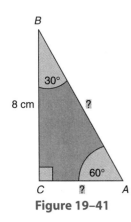

Figure 19–41

$\text{height} = x\sqrt{3}$	$x = AC$, $BC = \text{height} = 8$
$8 = x\sqrt{3}$	Solve for x. Divide both sides by $\sqrt{3}$.
$\dfrac{8}{\sqrt{3}} = \dfrac{x\sqrt{3}}{\sqrt{3}}$	
$\dfrac{8}{\sqrt{3}} = x$	Rationalize the denominator.
$x = \dfrac{8}{\sqrt{3}} \cdot \dfrac{\sqrt{3}}{\sqrt{3}}$	
$x = \dfrac{8\sqrt{3}}{3} \text{ cm}$	

$$AC = \frac{\mathbf{8\sqrt{3}}}{\mathbf{3}} \text{ cm (exact)} \quad \text{or} \quad \textbf{4.6 cm (approximate)}$$

Find AB (hypotenuse):

$$\text{hypotenuse} = 2x \qquad \text{hypotenuse} = AB, \; x = \frac{8\sqrt{3}}{3}$$

$$AB = 2\left(\frac{8\sqrt{3}}{3}\right)$$

$$AB = \frac{16\sqrt{3}}{3} \text{ cm (exact)} \qquad \text{or} \qquad \textbf{9.2 cm (approximate)}$$

TIP

Special Triangle Relationships

Visualize the relationships among the sides of these special triangles.

30°, 60°, 90° triangle (Fig. 19–42):   45°, 45°, 90° triangle (Fig. 19–43):

Figure 19–42

Figure 19–43

SECTION 19–2 SELF-STUDY EXERCISES

1 Find the unknown side of the right triangle. Round the final answers to the nearest thousandth. Use Fig. 19–44 for Exercises 1–2.

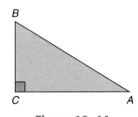

Figure 19–44

1. $AC = 24$ mm
$BC = ?$
$AB = 26$ mm

2. $AC = 15$ yd
$BC = 8$ yd
$AB = ?$

Use Fig. 19–45 for Exercises 3–5. Find the unknown dimensions. Round to the nearest thousandth where appropriate.

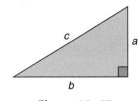

Figure 19–45

3. $a = 5$ cm
$b = 4$ cm
$c = ?$

4. $a = ?$
$b = 9$ m
$c = 11$ m

5. $a = 9$ ft
$b = ?$
$c = 15$ ft

Solve the problems. Round final answers to the nearest thousandth if necessary.

6. CON A light pole will be braced with a wire that is to be tied to a stake in the ground 18 ft from the base of the pole, which extends 26 ft above the ground. If the wire is attached to the pole 2 ft from the top, how much wire must be used to brace the pole? (See Fig. 19–46.)

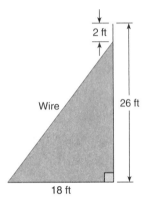

Figure 19–46

7. INDTR Find the center-to-center distance between holes *A* and *C* in a sheet-metal plate if the distance between the centers of *A* and *B* is 16.5 cm and the distance between the centers of *B* and *C* is 36.2 cm (Fig. 19–47).

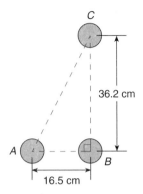

Figure 19–47

8. CON Find the length of a rafter that has a 10-in. overhang if the rise of the roof is 10 ft and the joists are 48 ft long (Fig. 19–48).

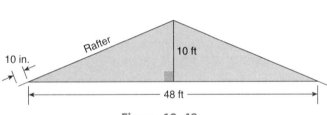

Figure 19–48

9. CON A stair stringer is 8 ft high and extends 10 ft from the wall (Fig. 19–49). How long will the stair stringer be?

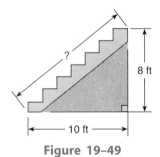

Figure 19–49

10. INDTR A machinist wishes to strengthen an L bracket that is 5 cm by 12 cm by welding a brace (hypotenuse of the triangle) to each end of the bracket. How much metal rod is needed for the brace?

12. Find the length of the side of the largest square nut that can be milled from a piece of round stock whose diameter is 15 mm (Fig. 19–50).

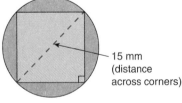

Figure 19–50

11. CON To make a rectangular table more stable, a diagonal brace is attached to the underside of the table surface. If the table is 27 dm by 36 dm, how long is the brace?

13. ELEC A rigid length of electrical conduit must be shaped as shown in Fig. 19–51 to clear an obstruction. What total length of the conduit is needed? (*Hint:* Don't forget to include *AB* and *CD* in the total length.)

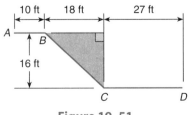

Figure 19–51

Chapter 19 / Triangles

2 Use Fig. 19–52 to solve. Round the final answers to the nearest thousandth if necessary.

14. $AC = 12$ cm; find BC and AB.

15. $AB = 10$ m; find AC and BC.

16. $BC = 7\sqrt{2}$ m; find AC and AB.

17. $AB = 8\sqrt{2}$ m; find AC and BC.

18. $AB = 12\sqrt{3}$ mm; find AC and BC.

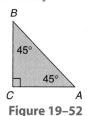

Figure 19–52

19. A rafter 19.5 ft long makes a 45° angle with a joist. If the rafter has an 18-in. overhang, find the length of the joist to the nearest inch (Fig. 19–53).

20. An elevated tank is connected to a pipe. The connecting pipe is 53.5 ft long and forms a 45° angle where it connects to the tank (Fig. 19–54). At what horizontal distance from the tank should the pipe stop to make the connection?

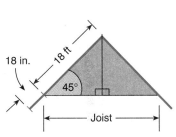

Figure 19–53

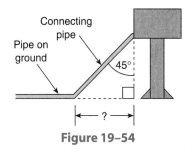

Figure 19–54

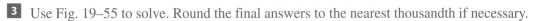

3 Use Fig. 19–55 to solve. Round the final answers to the nearest thousandth if necessary.

21. $AC = 6$ cm; find AB and BC.
22. $AB = 18$ mm; find AC and BC.
23. $BC = 8$ inches.; find AC and AB.
24. $BC = 7\sqrt{2}$ cm; find AC and AB.
25. $AC = 2$ ft 9 in.; find AB and BC to the nearest inch.

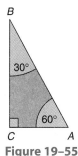

Figure 19–55

26. Find the length of the conduit $ABCD$ (Fig. 19–56) if $AB = 18$ ft, $CK = 6$ ft, $CD = 5$ ft, and $\angle KCB = 60°$. Round to the nearest foot.

27. A rafter makes a 30° angle with the horizontal. If the rise is 9 ft, find the rafter length and the run to the nearest inch (Fig. 19–57).

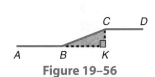

Figure 19–56

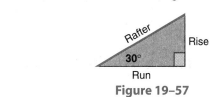

Figure 19–57

28. Find the depth of a V-slot in the form of an equilateral triangle if the cross-sectional opening is 5.2 cm across (Fig. 19–58). Round to the nearest tenth.

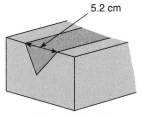

Figure 19–58

Learning Objectives

1 Find the number of degrees in each angle of a regular polygon.

2 Find the area of a regular polygon.

3 Use the properties of inscribed and circumscribed squares to find unknown amounts.

4 Use the properties of inscribed and circumscribed equilateral triangles to find unknown amounts.

5 Use the properties of inscribed and circumscribed regular hexagons to find unknown amounts.

1 **Find the Number of Degrees in Each Angle of a Regular Polygon.**

Polygons with equal sides and angles, such as squares, are called *regular polygons*. Several other regular polygons with definite shapes and specific names are treated as composite figures when calculating their areas. Let's examine regular polygons more closely.

A **regular polygon** is a polygon with equal sides and equal angles.

To find the number of degrees in each angle of a regular polygon:

1. Multiply the number of sides less 2 by 180°.
2. Divide by the number of sides.

$$\text{degrees per angle of a regular polygon} = \frac{180° (\text{number of sides} - 2)}{\text{number of sides}}$$

EXAMPLE Find the number of degrees in each angle of the regular polygons.

Triangle   (3 sides)   Pentagon   (5 sides)   Octagon   (8 sides)
Square   (4 sides)   Hexagon   (6 sides)

Equilateral Triangle: $\dfrac{180°(3-2)}{3} = \dfrac{180°(1)}{3} = 60°$ 3 sides

Square: $\dfrac{180°(4-2)}{4} = \dfrac{180°(2)}{4} = \dfrac{360°}{4} = 90°$ 4 sides

Regular Pentagon: $\dfrac{180°(5-2)}{5} = \dfrac{180°(3)}{5} = \dfrac{540°}{5} = 108°$ 5 sides

Regular Hexagon: $\dfrac{180°(6-2)}{6} = \dfrac{180°(4)}{6} = \dfrac{720°}{6} = 120°$ 6 sides

Regular Octagon: $\dfrac{180°(8-2)}{8} = \dfrac{180°(6)}{8} = \dfrac{1,080°}{8} = 135°$ 7 sides

2 **Find the Area of a Regular Polygon.**

A regular polygon can be divided into as many congruent triangles as there are sides of the polygon.

To form congruent triangles in a regular polygon:

1. Draw line segments from the center of the regular polygon to each vertex.
2. Congruent triangles are formed.

Figure 19–59

Fig. 19–59 shows how this property applies to regular polygons: the equilateral triangle, square, regular pentagon, and regular hexagon. To accept this property using inductive reasoning, construct various regular polygons, draw lines from each vertex to the center, and compare the resulting triangles.

To find the area of a regular polygon:

1. Find the area of one of the congruent triangles formed by connecting each vertex with the center of the polygon.
2. Multiply the result from Step 1 by the number of congruent triangles.

EXAMPLE Find the floor area of a recreational building at a park if it forms a regular hexagon and each side is 20 ft long. The perpendicular distance from one side to the center of the building is 17.3 ft (Fig. 19–60).

Divide the regular hexagon into congruent triangles by connecting the center of the hexagon with each vertex of the hexagon.

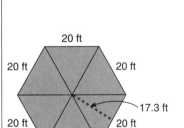

Figure 19–60

$$A = \frac{1}{2}bh$$

Find the area of one triangle. Substitute known values. Use the given distance from the side of the hexagon to the center for the height of the triangle.

$$A = \frac{1}{\cancel{2}}(\cancel{20}^{10})(17.3)$$

Simplify.

$$A = 173 \text{ ft}^2$$

$$173 \times 6 = 1{,}038 \text{ ft}^2$$

Multiply the area of the one triangle by 6 because the hexagon has been divided into six congruent triangles.

The floor area of the recreational building is 1,038 ft².

3 Use the Properties of Inscribed and Circumscribed Squares to Find Unknown Amounts.

Previously, we studied polygons and their areas and perimeters, and the area and circumference of a circle. In this section, we examine regular polygons *inscribed* in a circle and *circumscribed* about a circle.

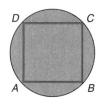

Inscribed square:
Vertices lie on circle
at points *A*, *B*, *C*, and *D*.

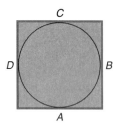

Circumscribed square:
Sides of square are tangent
at points *A*, *B*, *C*, and *D*.

Figure 19–61

A polygon is **inscribed in a circle** when it is inside the circle and all its *vertices* (points where sides of each angle meet) are on the circle. The circle is said to be **circumscribed about the polygon** (Fig. 19–61, top).

A polygon is **circumscribed about a circle** when it is outside the circle and all its sides are **tangent** to (intersecting in exactly one point) the circumference. The circle is said to be **inscribed in the polygon** (Fig. 19–61, bottom).

In this section, we examine the three most common polygons inscribed in or circumscribed about a circle: the triangle, the square, and the hexagon.

A square has four equal sides and four equal angles, each 90°. If a diagonal is drawn, the result is two congruent right triangles that have angles of 45°, 45°, 90°. A second diagonal divides the square into four 45°, 45°, 90° congruent right triangles, as shown in Fig. 19–62. Diagonals *AC* and *BD* are equal in length and bisect each other. Point *O,* where the diagonals intersect, is the center of the square and the center of any inscribed or circumscribed circle (Fig. 19–63).

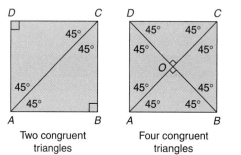

Two congruent
triangles

Four congruent
triangles

Figure 19–62

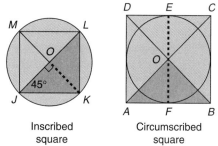

Inscribed
square

Circumscribed
square

Figure 19–63

Square–circle relationships

Square Inscribed in Circle

Radius of circle $= \dfrac{1}{2}$ Diagonal of square

Diameter of circle $= \dfrac{\text{Diagonal of square}}{\text{(across corners)}}$

$\dfrac{\text{Side of square}}{\text{(across flats)}} = \dfrac{\text{Diameter of circle}}{\sqrt{2}}$

Diameter of circle $= \sqrt{2}$(Side of square)

Square Circumscribed About a Circle

Side of square (across flats) = Diameter of circle

Radius of circle $= \dfrac{1}{2}$ Side of square

Diagonal of square (across corners) $= \sqrt{2}$ (Diameter of circle)

Diagonal of square (across corners) $= 2\sqrt{2}$ (Radius of circle)

EXAMPLE

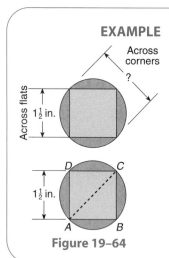

Figure 19–64

To mill a square bolt head $1\dfrac{1}{2}$ in. across flats, round stock of what diameter is needed? Answer to the nearest hundredth (see Fig. 19–64).

The distance across corners is the diameter of the circle and the diagonal of the square. The distance across flats is the side of a square inscribed in a circle.

Diameter $= \sqrt{2}$ (Side of square)   Substitute 1.5 for the side of the square.

$$1\dfrac{1}{2} \text{ in.} = 1.5 \text{ in}$$

Diameter $= \sqrt{2}\,(1.5)$

Diameter $= 2.12$ in.   Rounded

The diameter of the round stock is 2.12 in.

EXAMPLE Find the diagonal of an inscribed square if the area of the circle is 22 in². Round to hundredths (Fig. 19–65).

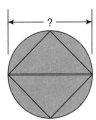

Figure 19–65

$$A = \pi r^2$$

Formula for the area of circle.
Substitute $A = 22$.

$$22 = \pi r^2$$

Solve for r.

$$\frac{22}{\pi} = r^2$$

$$\sqrt{\frac{22}{\pi}} = r$$

$$2.646283714 = r$$

Radius of the circle

Diagonal = Diameter = 2 × Radius

$$= 2(2.646283714) = 5.29 \text{ in.}$$

Rounded

The diagonal of the inscribed square is 5.29 in.

EXAMPLE What is the minimum depth of cut required to mill a circle on the end of a square piece of stock with a side measuring 4 cm? Answer to hundredths (Fig. 19–66).

The side of the square is 4 cm, so the diameter of the circle is 4 cm and the radius is 2 cm. If the radius is drawn perpendicular to the top side, a right triangle is formed as indicated (shading).

Let d = depth of milling, which equals the hypotenuse of the shaded triangle minus the radius of the circle.

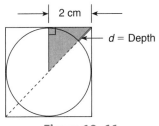

Figure 19–66

$$c^2 = a^2 + b^2$$

Find the hypotenuse using the Pythagorean theorem. Substitute $a = 2$ and $b = 2$.

$$c^2 = 2^2 + 2^2$$

Solve for c.

$$c^2 = 4 + 4$$
$$c^2 = 8$$
$$c = \sqrt{8}$$
$$c = 2.828427125 \text{ cm}$$

Hypotenuse

Depth of milling = Hypotenuse − Radius

$$d = c - r$$

Substitute $c = 2.828427125$ and $r = 2$.

$$d = 2.828427125 - 2$$
$$d = 0.83 \text{ cm}$$

Rounded

The minimum depth of milling is 0.83 cm.

---

4 Use the Properties of Inscribed and Circumscribed Equilateral Triangles to Find Unknown Amounts.

An equilateral triangle can be inscribed in a circle or circumscribed about a circle.

The height (or altitude) of the equilateral triangle has a special relationship with the radius of the circle. The height of an equilateral triangle divides the triangle into two

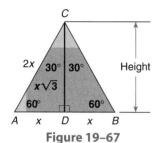

Figure 19–67

30°, 60°, 90° triangles (Fig. 19–67). By relating the radius of a circle to the height, we can then determine the length of the side of the equilateral triangle. Examine the relationships in Fig. 19–68.

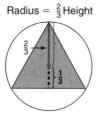

Radius = $\frac{2}{3}$ Height   Radius = $\frac{1}{3}$ Height

Inscribed triangle   Circumscribed triangle

Figure 19–68

Triangle–circle relationships

Equilateral Triangle Inscribed in a Circle

Radius of circle = $\frac{2}{3}$(Height of triangle)

Height of triangle = $\frac{3}{2}$(Radius of circle)

Side of triangle = $\dfrac{2(\text{Height of triangle})}{\sqrt{3}}$

Side of triangle = $\dfrac{3(\text{Radius of circle})}{\sqrt{3}}$

Radius of circle = $\dfrac{\sqrt{3}(\text{Side of triangle})}{3}$

Equilateral Triangle Circumscribed About a Circle

Radius of circle = $\frac{1}{3}$(Height of triangle)

Height of triangle = 3(Radius of circle)

Side of triangle = $\dfrac{2(\text{Height of triangle})}{\sqrt{3}}$

Side of triangle = $\dfrac{6(\text{Radius of circle})}{\sqrt{3}}$

Radius of circle = $\dfrac{\sqrt{3}(\text{Side of triangle})}{6}$

EXAMPLE Find the dimensions for the inscribed equilateral triangle in Fig. 19–69, where $CD = 13$ mm and $AO = 15$ mm.

(a) Height of $\triangle ABC$　　(b) BD　　(c) BC　　(d) $\angle CAD$

(e) $\angle ACD$　　(f) Area of $\triangle ABC$

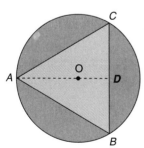

Figure 19–69

(a) Height of triangle = $\dfrac{3}{2}$ (radius of circle)

$AD = \dfrac{3}{2}(AO)$　　　Substitute $AO = 15$ mm.

$AD = \dfrac{3}{2}(15)$　　　Simplify.

$AD = \dfrac{45}{2}$

$AD = \mathbf{22.5}$ **mm**

(b) $BD = CD$　　　Corresponding sides of congruent triangles

$BD = \mathbf{13}$ **mm**　　　Substitute $CD = 13$ mm.

Chapter 19 / Triangles

(c) $BC = BD + CD$ Substitute 13 mm for BD and CD.

     $BC = 13\text{ mm} + 13\text{ mm}$ Simplify.

     $BC = 26$ mm

(d) $\angle CAD = 30°$ $\triangle CAD$ is a 30°, 60°, 90° $\triangle$, and $\angle CAD$ is opposite shortest side.

(e) $\angle ACD = 60°$ $\angle ACD$ is opposite height or other leg of $\triangle$.

(f) $A = \dfrac{1}{2}bh$ Formula for area of triangle. Substitute for base and height.

     $A = \dfrac{1}{2}(BC)(AD)$ $BC = 26$, $AD = 22.5$

     $A = \dfrac{1}{2}(26)(22.5)$ Simplify.

     $A = 292.5$ mm^2   Area of $\triangle ABC$

EXAMPLE   One end of a shaft 35 mm in diameter is milled as shown in Fig. 19–70. Find the radius of the smaller circle.

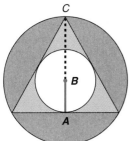

Figure 19–70

$BC = \text{Radius} = \dfrac{1}{2}\text{ Diameter of large circle}$

$BC = \dfrac{1}{2}(35)$ Substitute 35 for diameter.

$BC = 17.5$ mm   Radius of large circle

$AB = \dfrac{1}{2}BC$ From $AB = \dfrac{1}{3}AC$ and $BC = \dfrac{2}{3}AC$

$AB = \dfrac{1}{2}(17.5)$

$AB = 8.75$ mm   Radius of small circle

The radius of the smaller circle is 8.75 mm.

---

5   **Use the Properties of Inscribed and Circumscribed Regular Hexagons to Find Unknown Amounts.**

A regular hexagon is a figure with six equal sides and six equal angles of 120° each. Three diagonals joining pairs of opposite vertices divide the hexagon into six congruent equilateral triangles (all angles 60°) (see Fig. 19–71).

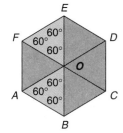

Figure 19–71

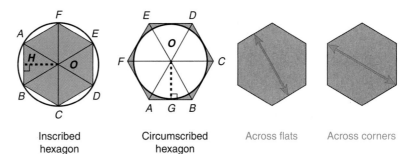

Inscribed hexagon   Circumscribed hexagon   Across flats   Across corners

Figure 19–72

Hexagon Inscribed in Circle	**Hexagon Circumscribed About a Circle**

diameter of circle $=\dfrac{\text{diagonal of hexagon}}{\text{(across corners)}}$

radius of circle = side of hexagon

across flats = (side of hexagon) $\left(\sqrt{3}\,\right)$

across flats = diameter of circle

across corners = 2 (side of hexagon)

radius of circle = height of one triangle formed by the three diagonals

$$=\dfrac{\text{side of hexagon}\left(\sqrt{3}\,\right)}{2}$$

EXAMPLE Find the indicated dimensions. When appropriate, round to hundredths. Use Fig. 19–73.

(a) ∠EOD (b) ∠COH (c) ∠BHO (d) Radius
(e) Distance across corners (f) FO

(a) ∠EOD = 60° This is an angle of an equilateral triangle.

(b) ∠COH = 30° Height OH bisects the 60° angle to form a 30°, 60°, 90° triangle.

(c) ∠BHO = 90° The height OH forms a right angle with the base of △BOC.

(d) Radius = 6 in. The radius is a side of an equilateral triangle or a side of the hexagon.

(e) Distance across corners = 12 in. The distance across corners is the diameter.

(f) FO = 6 in. FO is a radius or a side of an equilateral triangle.

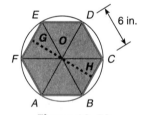

Figure 19–73

EXAMPLE Find the indicated dimensions. When appropriate, round to hundredths. Use Fig. 19–74.

(a) Distance across flats (b) EO (c) Diagonal (d) ∠FED

(a) Distance across flats = 8 cm This is the diameter of the circle.

(b) EO = 4.62 cm EO is a side of an equilateral triangle that equals the side of the hexagon.

(c) Diagonal = 9.24 cm The diagonal is twice the side of an equilateral triangle.

(d) ∠FED = 120° This is the angle at the vertex of a hexagon.

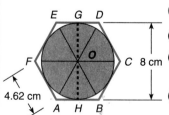

Figure 19–74

EXAMPLE If a regular hexagon is cut from a circle with a radius of 6.45 in., how much of the circle is not used? Round to hundredths (Fig. 19–75).

To solve, we find the area of the circle and the area of the hexagon and subtract to find the difference (portion not used).

$$A_{circle} = \pi r^2 \qquad \text{Substitute } r = 6.45 \text{ in.}$$

$$A_{circle} = \pi (6.45)^2 \qquad \text{Simplify.}$$

$$A_{circle} = \pi (41.6025)$$

$$A_{circle} = 130.6981084 \text{ in}^2 \qquad \text{Area of circle.}$$

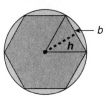

$$A_{hexagon} = 6\left(\frac{1}{2}bh\right)$$

$A_{hexagon}$ = area of six triangles.
The height of an equilateral triangle forms a 30°, 60°, 90° triangle,
$h = \dfrac{b}{2}\sqrt{3} = 3.225, \ \sqrt{3} = 5.585863854$ in.

Figure 19–75

$$A_{hexagon} = 6\left[\frac{1}{2}(6.45)(5.585863854)\right] \qquad \text{Substitute } b = 6.45 \text{ and } h = 5.585863854.$$

$$A_{hexagon} = 108.0864656 \text{ in}^2$$

$$\text{Waste} = A_{circle} - A_{hexagon} \qquad \text{Difference in areas}$$

$$\text{Waste} = 130.6981084 - 108.0864656$$

$$\text{Waste} = 22.61 \text{ in}^2 \qquad \text{Rounded}$$

There are 22.61 in² of waste from the circle.

EXAMPLE Use Fig. 19–76 to answer the questions. Round answers to hundredths.

(a) What is the smallest-diameter round stock from which a hex-bolt head $\frac{1}{4}$ in. on a side can be milled?
(b) What is the distance across the corners of the hex-bolt head?
(c) What is the distance across the flats?

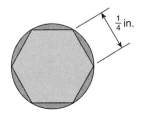

(a) Diameter = 2 × Side = Diagonal   Side = radius. $2 \times \dfrac{1}{4} = \dfrac{1}{2}$

Smallest-diameter round stock = $\dfrac{1}{2}$ **in.**

(b) Distance across corners = Diagonal

Distance across corners = $\dfrac{1}{2}$ **in.**

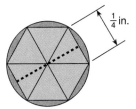

(c) Distance across flats = (side of hexagon)$\left(\sqrt{3}\right)$ Side of hexagon = $\dfrac{1}{4}$ in. = 0.25 in.

$$= 0.25\left(\sqrt{3}\right)$$

$$= 0.43 \qquad \text{To nearest hundredth}$$

The distance across the flats is 0.43 in.

Figure 19–76

1 Give the specific name of each regular polygon in Figs. 19–77 through 19–82 and the number of degrees in each angle.

1.

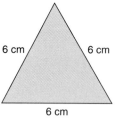

6 cm   6 cm

6 cm

Figure 19–77

2.

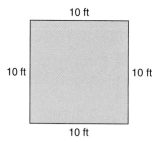

10 ft

10 ft   10 ft

10 ft

Figure 19–78

3.

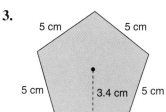

5 cm   5 cm

5 cm   3.4 cm   5 cm

5 cm

Figure 19–79

4.

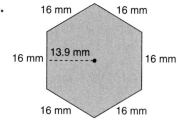

16 mm   16 mm

16 mm   13.9 mm   16 mm

16 mm   16 mm

Figure 19–80

5.

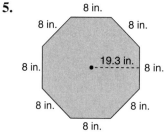

8 in.

8 in.   8 in.

8 in.   19.3 in.   8 in.

8 in.   8 in.

8 in.

Figure 19–81

6.

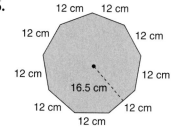

12 cm   12 cm

12 cm   12 cm

12 cm   12 cm

16.5 cm

12 cm   12 cm

12 cm

Figure 19–82   Nonagon

2

7. Find the area of Fig. 19–77.
9. Find the area of Fig. 19–79.
11. Find the area of Fig. 19–81.

8. Find the area of Fig. 19–78.
10. Find the area of Fig. 19–80.
12. Find the area of Fig. 19–82.

3 Find the measures of the inscribed square in Fig. 19–83.

13. ∠*DOC*
16. ∠*BOE*

14. ∠*BCO*
17. *CE*

15. *BO*

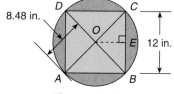

8.48 in.

D   *C*

O   *E*   12 in.

A   *B*

Figure 19–83

Find the measures of the circumscribed square in Fig. 19–84.

18. *DO*
20. ∠*AOB*

19. Radius of circle
21. *BE*

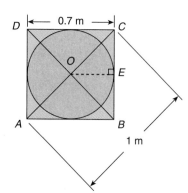

D   0.7 m   *C*

O   *E*

A   *B*

1 m

Figure 19–84

Solve the problems involving squares and circles. Round the answers to hundredths.

22. To what depth must a 2-in. shaft (Fig. 19–85) be milled on an end to form a square 1.414 in. on a side?

23. What is the smallest-diameter round stock (Fig. 19–86) needed to mill a square $\frac{3}{4}$ in. on a side on the end of the stock?

24. What is the area of the largest circle inscribed in a square with a perimeter of 36 dkm?

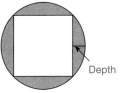

Figure 19–85

Figure 19–86

4 Use Fig. 19–87 to find the indicated dimensions for an equilateral triangle inscribed in a circle. Round to hundredths.

25. ∠CAB
26. ∠BCD
27. ∠ADC
28. CB
29. AC
30. Diameter
31. Area of △ABC
32. AD
33. Radius
34. Circumference

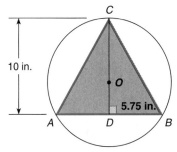

Figure 19–87

Solve the problems involving inscribed or circumscribed equilateral triangles. Round to hundredths.

35. The area of an equilateral triangle with a base of 10 cm is 43.5 cm². The triangle is circumscribed about a circle. What is the radius of the inscribed circle?

36. One end of a circular steel rod is milled into an equilateral triangle whose height is $\frac{3}{4}$ in. What is the diameter of the rod if the smallest diameter possible was used for the job?

5 Find the dimensions for Fig. 19–88 if $GO = 17.32$ mm and $FO = 20$ mm.

37. Distance across flats
38. ∠CDE
39. ∠EFO
40. Distance across corners
41. ∠FGO
42. GA
43. AB

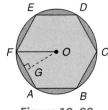

Figure 19–88

Find the dimension for Fig. 19–89 if $FO = 3$ in.

44. ∠ODC
45. ∠EOG
46. OC

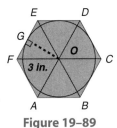

Figure 19–89

47. What is the distance across the flats of a hexagon cut from a circular blank with a circumference of 145 mm (Fig. 19–90)? Round to hundredths.

48. The end of a hexagonal rod is milled into the largest circle possible (Fig. 19–91). What is the distance across the corners of the hexagon if the area of the circle is 4.75 in²? Round to hundredths.

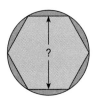

Figure 19–90

Figure 19–91

Learning Outcomes

1 Find the distance between two points on the rectangular coordinate system.

2 Find the coordinates of the midpoint of a line segment if given the coordinates of the end points.

1 **Find the Distance Between Two Points on the Rectangular Coordinate System.**

In Chapter 4 we found the distance between two points on a line. To expand this concept to the distance between two points on the rectangular coordinate system, we will apply the Pythagorean theorem.

$$c^2 = a^2 + b^2 \quad \text{or} \quad c = \sqrt{a^2 + b^2}$$

The distance between two points on the rectangular coordinate system is the shortest distance between the points.

To find the distance between two points on the rectangular coordinate system:

1. Use the formula

$$d = \sqrt{(\Delta x)^2 + (\Delta y)^2} \quad \text{or} \quad \sqrt{(x_2 - x_1)^2 + (y_2 - y_1)^2}$$

where P_1 and P_2 are two points with coordinates (x_1, y_1) and (x_2, y_2), respectively.

2. Substitute values for x_1, x_2, y_1, and y_2.
3. Evaluate to find d.

EXAMPLE Find the distance from $(4, 2)$ to $(7, 6)$.

These points are shown in Fig. 19–92. Point $(4, 2)$ is labeled P_1, and point $(7, 6)$ is labeled P_2.

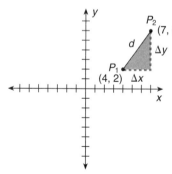

$d = \sqrt{(x_2 - x_1)^2 + (y_2 - y_1)^2}$	Substitute: $x_1 = 4$, $x_2 = 7$, $y_1 = 2$, and $y_2 = 6$
$d = \sqrt{(7 - 4)^2 + (6 - 2)^2}$	Combine in each grouping.
$d = \sqrt{3^2 + 4^2}$	Square each term in the radicand.
$d = \sqrt{9 + 16}$	Add terms in the radicand.
$d = \sqrt{25}$	Find the principal square root.
$d = 5$	

Figure 19–92 **The distance from (4, 2) to (7, 6) is 5 units.**

Point Selection Does Not Matter

TIP

The distance formula can be used to calculate the distance between two points, no matter which points we designate as P_1 and P_2. Let's rework the preceding example, letting $P_1 = (7, 6)$ and $P_2 = (4, 2)$. Thus, $x_1 = 7$, $x_2 = 4$, $y_1 = 6$, and $y_2 = 2$.

$$d = \sqrt{(4 - 7)^2 + (2 - 6)^2}$$ Substitute.

$$d = \sqrt{(-3)^2 + (-4)^2}$$ $(-3)^2 = +9; (-4)^2 = +16.$

$$d = \sqrt{9 + 16}$$

$$d = \sqrt{25}$$

$$d = 5$$

The change in vertical movement has the same *absolute value,* regardless of which point is used first. The result after squaring is positive whether the difference is positive or negative. For similar reasons, the change in horizontal movement squared is the same, regardless of which point is used first.

2 Find the Coordinates of the Midpoint of a Line Segment if Given the Coordinates of the Endpoints.

In Chapter 4 we found the midpoint between two points on a line. We will expand that concept to find the midpoint between two points on the rectangular coordinate system.

To find the coordinates of the midpoint of a line segment if given the coordinates of the endpoints:

1. Average the respective coordinates of the endpoints of the segment using the formula

$$\text{midpoint} = \left(\frac{x_1 + x_2}{2}, \frac{y_1 + y_2}{2} \right)$$

where P_1 and P_2 are endpoints of the segment, $P_1 = (x_1, y_1)$ and $P_2 = (x_2, y_2)$.
2. Substitute values for x_1, x_2, y_1, and y_2.
3. Evaluate to find the coordinates of the midpoint.

EXAMPLE Find the midpoint of each segment with the given endpoints.

(a) (2, 4) and (6, 10)
(b) (3, −5) and origin

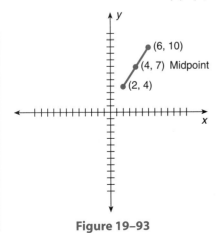

(a) (2, 4) and (6, 10). See Fig. 19–93.

$$\text{midpoint} = \left(\frac{x_1 + x_2}{2}, \frac{y_1 + y_2}{2} \right)$$ Substitute: $x_1 = 2, x_2 = 6,$ $y_1 = 4, y_2 = 10$

$$\text{midpoint} = \left(\frac{2 + 6}{2}, \frac{4 + 10}{2} \right)$$ Simplify.

$$\text{midpoint} = \left(\frac{8}{2}, \frac{14}{2} \right)$$ Reduce.

$$\textbf{midpoint} = \textbf{(4, 7)}$$

Figure 19–93

(b) $(3, -5)$ and origin. See Fig. 19–94.

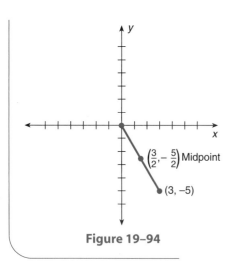

Figure 19–94

$$\text{midpoint} = \left(\frac{x_1 + x_2}{2}, \frac{y_1 + y_2}{2} \right)$$

Substitute: $x_1 = 3$, $x_2 = 0$, $y_1 = -5$, $y_2 = 0$

$$\text{midpoint} = \left(\frac{3 + 0}{2}, \frac{-5 + 0}{2} \right)$$

$$\textbf{midpoint} = \left(\frac{3}{2}, -\frac{5}{2} \right) \text{ or } \left(1\frac{1}{2}, -2\frac{1}{2} \right)$$

Midpoint Between the Origin and Another Point

To find the coordinates of the midpoint of a segment from the origin to any point, take one-half of each coordinate of the point that is not at the origin. Apply this tip to part (b) of the preceding example.

Visualize and Estimate Whenever Practical

When formulas are developed to find desired information, we often rely totally on our calculation skills and ability to manipulate formulas. We can strengthen our understanding of the concepts and avoid mistakes by visualizing the problem and by estimating.

In some of the examples in this section, we provided a visualization of the problem. Now, let's examine some estimates. To estimate the distance between points $(4, 2)$ and $(7, 6)$, we find the vertical and horizontal changes. The vertical change is 4 from $(6 - 2)$ and the horizontal change is 3 from $(7 - 4)$. Applying properties of the Pythagorean theorem, the distance between the points (hypotenuse) is more than the longest leg (4) and less than the sum of the legs, $3 + 4 = 7$. So, we estimate the distance to be between 4 and 7 units.

To estimate the coordinates of the midpoint between these points, find the range of values. The x-coordinate of the midpoint is halfway between 4 and 7. The y-coordinate of the midpoint is halfway between 2 and 6.

1 Graph the line segment determined by the given endpoints and find the distance between each of the pairs of points. Express the answer to the nearest thousandth when necessary.

  1. $(7, 10)$ and $(1, 2)$ **2.** $(7, -7)$ and $(2, 5)$ **3.** $(5, 0)$ and $(0, 5)$

  4. $(-2, -2)$ and $(3, -4)$ **5.** $(5, 7)$ and $(0, -3)$ **6.** $(8, -2)$ and $(-4, 3)$

  7. $(-4, 6)$ and $(2, -2)$ **8.** $(3, 5)$ and $(0, 1)$ **9.** $(5, 4)$ and $(-7, -5)$

10. $(7, 2)$ and $(-2, -3)$

2 Find the coordinates of the midpoints of the segments determined by the given endpoints.

11. $(7, 10)$ and $(1, 2)$ **12.** $(7, -7)$ and $(2, 5)$ **13.** $(5, 0)$ and $(0, 5)$

14. $(-2, -2)$ and $(3, -4)$ **15.** $(5, 7)$ and $(0, -3)$ **16.** $(8, -2)$ and $(-4, 3)$

17. $(-4, 6)$ and $(2, -2)$ **18.** $(3, 5)$ and $(0, 1)$ **19.** $(5, 4)$ and $(-7, -5)$

20. $(7, 2)$ and $(-2, -3)$

CHAPTER REVIEW OF KEY CONCEPTS

Learning Outcomes

What to Remember with Examples

Section 19–1

1 Classify triangles by the relationship of the sides or angles (pp. 757–758).

An equilateral triangle has three equal sides. An isosceles triangle has exactly two equal sides. A scalene triangle has three unequal sides.

> Identify the following triangles by the measures of their sides:
>
> (a) 6 cm, 6 cm, 6 cm   Equilateral
> (b) 12 in., 10 in., 12 in.   Isosceles
> (c) 10 cm, 12 cm, 15 cm   Scalene

An equilateral triangle has three equal angles. The equal sides of an isosceles triangle have opposite angles that are equal. If the three sides of a triangle are unequal, the largest angle is opposite the longest side, and the smallest angle is opposite the shortest side. In a right triangle, the hypotenuse is always the longest side.

> Identify the largest and the smallest angles in triangles with the following sides:
> 12 in., 10 in., 9 in.
> The largest angle is opposite the 12-in. side; the smallest angle is opposite the 9-in. side.

2 Determine if two triangles are congruent using inductive and deductive reasoning (pp. 758–760).

Triangles are congruent (one can fit exactly over the other) (a) if the three sides of one triangle equal the corresponding sides of the other triangle (SSS), (b) if two sides and the included angle of one triangle are equal to two sides and the included angle of the other triangle (SAS), or (c) if two angles and the common side of one triangle are equal to two angles and the common side of the other triangle (ASA).

Determine whether the pairs of triangles in Fig. 19–95 are congruent.

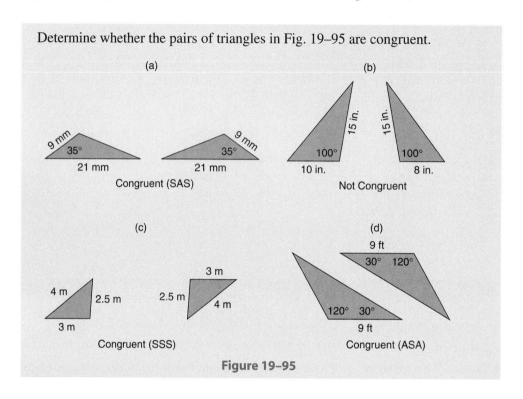

Figure 19–95

3 Solve problems that involve similar triangles (pp. 760–763).

Each angle of one triangle is equal to its corresponding angle in the other similar triangle. Each side of one triangle is proportional to its corresponding side in the other triangle.

Find the height of a building that makes a shadow of 25 m when a meter stick makes a shadow of 1.5 m.

Direct proportion: As shadow increases, height increases.

$$\frac{x\text{-m building}}{1\text{-m stick}} = \frac{25\text{-m building shadow}}{1.5\text{-m stick shadow}}$$ Pair 1. Estimation: meter stick.

Pair 2. Stick< shadow, so building <25 m.

$$\frac{x}{1} = \frac{25}{1.5}$$ Cross multiply.

$$1.5x = 25$$ Divide.

$$x = \frac{25}{1.5}$$

$$x = 16.67 \text{ m}$$

Section 19–2

1 Use the Pythagorean theorem to find the unknown side of a right triangle (pp. 765–768).

The square of the hypotenuse of a right triangle equals the sum of the squares of the other two sides. $c^2 = a^2 + b^2$

The hypotenuse (c) of a right triangle is 10 in. If side b is 6 in., find side a.

$$c^2 = a^2 + b^2 \qquad \text{Substitute values.}$$
$$10^2 = a^2 + 6^2 \qquad \text{Raise to powers.}$$
$$100 = a^2 + 36 \qquad \text{Sort.}$$
$$100 - 36 = a^2 \qquad \text{Combine.}$$
$$64 = a^2 \qquad \text{Take square root of both sides.}$$
$$\sqrt{64} = a \qquad \text{Find the principal square root.}$$
$$a = 8$$

Figure 19–96

2 Use the properties of a 45°, 45°, 90° triangle to find unknown parts (pp. 768–769).

To find the hypotenuse of a 45°, 45°, 90° triangle, multiply a leg by $\sqrt{2}$. To find a leg of a 45°, 45°, 90° triangle, divide the product of the hypotenuse and $\sqrt{2}$ by 2.

Given a 45°, 45°, 90° triangle, find

(a) the hypotenuse if a leg is 10 cm.
(b) a leg if the hypotenuse is 15 in.

(a) Hypotenuse $= \text{Leg}\left(\sqrt{2}\right)$

$\qquad = 10\sqrt{2}$

$\qquad = 14.14213562$ or 14.1 cm   Rounded

(b) Leg $= \dfrac{\text{Hypotenuse}\left(\sqrt{2}\right)}{2}$

$\qquad = \dfrac{15\sqrt{2}}{2}$

$\qquad = 10.60660172$ or 10.6 in.   Rounded

3 Use the properties of a 30°, 60°, 90° triangle to find unknown parts (pp. 769–771).

To find the hypotenuse of a 30°, 60°, 90° triangle, multiply the side opposite the 30° angle by 2. To find the side opposite the 30° angle in a 30°, 60°, 90° triangle, divide the hypotenuse by 2. To find the side opposite the 60° angle of a 30°, 60°, 90° triangle, multiply the side opposite the 30° angle by $\sqrt{3}$.

Given a 30°, 60°, 90° triangle, find

(a) the hypotenuse if the side opposite the 30° angle $= 4.5$ cm.
(b) the side opposite the 30° angle if the hypotenuse is 20 in.
(c) the side opposite the 60° angle if the side opposite the 30° angle is 35 mm.

(a) Hypotenuse $\quad = 2\,(\text{Side opposite } 30°)$

$\qquad\qquad\quad = 2(4.5)$

$\qquad\qquad\quad = 9$ cm

(b) Side opposite 30° $= \dfrac{\text{Hypotenuse}}{2}$

$\qquad\qquad\quad = \dfrac{20}{2}$

$\qquad\qquad\quad = 10$ in.

(c) Side opposite 60° $= \text{Side opposite } 30°\left(\sqrt{3}\right)$

$\qquad\qquad\quad = 35\left(\sqrt{3}\right)$

$\qquad\qquad\quad = 60.62177826$ or 60.6 mm.   Rounded

Section 19–3

1 Find the number of degrees in each angle of a regular polygon (p. 774).

To find the number of degrees in each angle of a regular polygon: Multiply the number of sides less 2 by 180° and divide the product by the number of sides.

> Find the number of degrees in each angle of a pentagon.
> A pentagon has five sides.
>
> $$\text{degrees per angle} = \frac{180(5 - 2)}{5} = \frac{180(3)}{5} = \frac{540}{5} = 108°$$

2 Find the area of a regular polygon (pp. 774–775).

A regular polygon has all sides equal. Regular polygons with 3, 4, 5, 6, and 8 sides are equilateral triangles, squares, pentagons, hexagons, and octagons, respectively. Lines drawn from the center of a regular polygon to each vertex will form as many congruent triangles as the polygon has sides.

To find the area of a regular polygon, find the area of one of the congruent triangles and multiply by the number of triangles.

> Find the area of the regular hexagon in Fig. 19–97.
> Each congruent triangle formed is an equilateral triangle (Fig. 19–98). Six congruent central angles make each central angle 60°: $\frac{360°}{6} = 60°$. Each base angle of the triangles equals one-half the angle at each vertex of the hexagon: $\frac{1}{2}(120°) = 60°$. The congruent triangles are 60°, 60°, 60° triangles, or equilateral.
> Find the area of one congruent triangle of the hexagon. 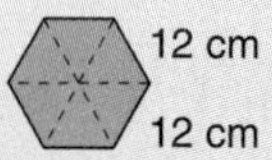
>
> **Figure 19–97** (12 cm, 12 cm)
>
> Height forms two 30°, 60°, 90° triangles. Hypotenuse = 12 cm, base (leg) = 6 cm.
>
> $$c^2 = a^2 + b^2$$
>
> Pythagorean theorem. Substitute $c = 12$ and $b = 6$.
>
> **Figure 19–98** (12 cm, 12 cm, 12 cm)
>
> $$12^2 = a^2 + 6^2$$ Solve for a.
> $$144 = a^2 + 36$$
> $$144 - 36 = a^2$$
> $$108 = a^2$$ Take the square root of both sides.
> $$10.4 \text{ cm} = a$$ Height of triangle. Rounded
>
> $$A_{\text{triangle}} = \frac{1}{2}bh$$ Substitute $b = 12$ and $h = 10.4$.
>
> $$A_{\text{triangle}} = \frac{1}{2}(12)(10.4)$$ Evaluate.
>
> $$A_{\text{triangle}} = 62.4 \text{ cm}^2$$ Area of one congruent triangle.
>
> $$A_{\text{hexagon}} = 6(A_{\text{triangle}})$$ Multiply by the number of congruent triangles. Substitute the area of one triangle.
>
> $$A_{\text{hexagon}} = 6(62.4)$$
> $$A_{\text{hexagon}} = 374.4 \text{ cm}^2$$

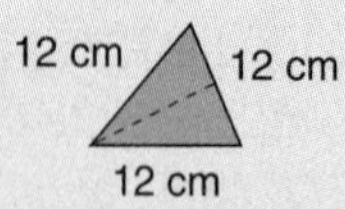

3 Use the properties of inscribed and circumscribed squares to find unknown amounts (pp. 775–777).

The diagonal of a square inscribed in or circumscribed about a circle forms congruent 45°, 45°, 90° triangles.

The diagonal of a square inscribed in a circle equals the diameter of the circle.

The side of a square circumscribed about a circle equals the radius of the circle.

The radius of a circle when a square is inscribed in the circle equals the height of a 45°, 45°, 90° triangle formed by one diagonal.

Chapter 19 / Triangles

The radius of a circle when a square is circumscribed about the circle equals the height of a 45°, 45°, 90° triangle formed by two diagonals or it equals one-half the length of a side of the square.

The end of a 20-mm diameter round rod is milled as shown in Fig. 19–99. Find the area of the square.

The radius of the circle equals the height of a right triangle formed by one diagonal. The diameter is the base. Find the area of the triangle and double it.

$$A_{\text{triangle}} = \frac{1}{2}bh$$

$$A_{\text{triangle}} = \frac{1}{2}(20)(10)$$

$$A_{\text{triangle}} = 100 \text{ mm}^2 \qquad \text{Area of one triangle}$$

Area of square end: $100 \times 2 = 200 \text{ mm}^2$

Figure 19–99

An alternative approach would be to find the side of the square using the proportions of a 45°, 45°, 90° triangle and then to find the area of the square.

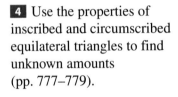 **4** Use the properties of inscribed and circumscribed equilateral triangles to find unknown amounts (pp. 777–779).

The height of an equilateral triangle forms two congruent 30°, 60°, 90° triangles.

The radius of a circle circumscribed about an equilateral triangle is two-thirds the height of the triangle or twice the distance from the center of the circle to the base of the triangle.

The radius of a circle inscribed in an equilateral triangle is one-third the height of the triangle or one-half the distance from the vertex of the triangle to the center of the circle.

One end of a shaft 30 mm in diameter is milled as shown in Fig. 19–100. Find the diameter of the smaller circle.

BC is the radius (half the diameter) of the large circle.

$$BC = \frac{1}{2}(30)$$

$$BC = 15 \text{ mm} \qquad \text{Radius of large circle.}$$

$$AB = \frac{1}{2}BC \qquad \text{Radius of small circle.}$$

$$AB = \frac{1}{2}(15)$$

$$AB = 7.5 \text{ mm}$$

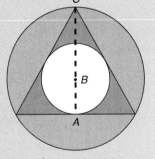

The diameter of the small circle $= 2(7.5) = 15$ mm. **Figure 19–100**

5 Use the properties of inscribed and circumscribed regular hexagons to find unknown amounts (pp. 779–781).

The three diagonals that join opposite vertices of a regular hexagon form six congruent equilateral triangles.

The height of each of the equilateral triangles forms two congruent 30°, 60°, 90° triangles. The diagonal of a regular hexagon inscribed in a circle is the diameter of the circle (distance across corners).

The radius of the circle when the regular hexagon is inscribed in the circle is a side of an equilateral triangle formed by the three diagonals and is one-half the distance across corners.

Chapter Review of Key Concepts

The radius of the circle when a regular hexagon is circumscribed about the circle is the height of an equilateral triangle formed by the three diagonals.

The diameter of the circle when a regular hexagon is circumscribed about the circle is the distance across flats or twice the height of an equilateral triangle formed by the three diagonals.

Find the following dimensions for Fig. 19–101 if the diameter is 10 cm.

(a) $\angle AOF$
(b) Radius
(c) Distance across corners

(a) $\angle AOF = 60°$ (angle of equilateral triangle)
(b) 5 cm (radius of a circle is one-half the diameter)
(c) 10 cm (diameter of a circle is the distance across corners of the hexagon)

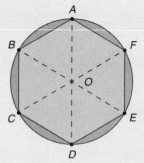

Figure 19–101

Section 19–4

1 Find the distance between two points on the rectangular coordinate system (pp. 784–785).

Use the distance formula: $d = \sqrt{(x_2 - x_1)^2 + (y_2 - y_1)^2}$.

Find the distance between the points $(3, 2)$ and $(-1, 5)$.

$d = \sqrt{(x_2 - x_1)^2 + (y_2 - y_1)^2}$	Substitute values.
$d = \sqrt{(-1 - 3)^2 + (5 - 2)^2}$	Simplify groupings in radicand.
$d = \sqrt{(-4)^2 + 3^2}$	Square terms in radicand.
$d = \sqrt{16 + 9}$	Combine terms in radicand.
$d = \sqrt{25}$	Find principal square root.
$d = 5$	

2 Find the coordinates of the midpoint of a line segment if given the coordinates of the endpoints (pp. 785–786).

Use the midpoint formula to find the coordinates of the midpoint of a line segment:

$$\left(\frac{x_1 + x_2}{2}, \frac{y_1 + y_2}{2} \right)$$

Find the midpoint of the segment joining the points $(3, 2)$ and $(-1, 5)$.

$\left(\dfrac{x_1 + x_2}{2}, \dfrac{y_1 + y_2}{2} \right)$	Substitute values.
$\left(\dfrac{3 + (-1)}{2}, \dfrac{2 + 5}{2} \right)$	Simplify numerators.
$\left(\dfrac{2}{2}, \dfrac{7}{2} \right)$	Reduce or simplify.
$\left(1, \dfrac{7}{2} \right)$ or $(1, 3.5)$	

Section 19–1

Identify the longest and shortest sides in Figs. 19–102 and 19–103.

1.

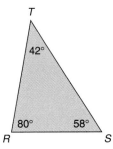

Figure 19–102

2.

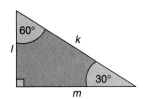

Figure 19–103

List the angles in order of size from largest to smallest in Figs. 19–104 and 19–105.

3.

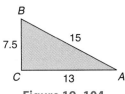

Figure 19–104

4.

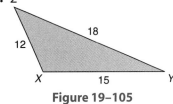

Figure 19–105

Write the corresponding parts that are not given for the congruent triangles in Figs. 19–106 through 19–108.

5.

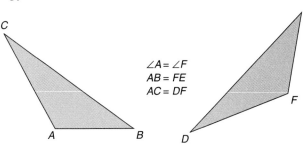

$\angle A = \angle F$
$AB = FE$
$AC = DF$

Figure 19–106

6.

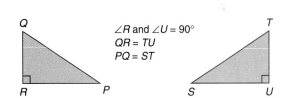

$\angle R$ and $\angle U = 90°$
$QR = TU$
$PQ = ST$

Figure 19–107

7.

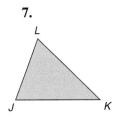

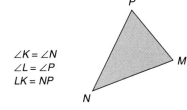

$\angle K = \angle N$
$\angle L = \angle P$
$LK = NP$

Figure 19–108

8. $\triangle LMN \sim \triangle XYZ$, $\angle L = \angle X$, $\angle M = \angle Y$. Find m and x (see Fig. 19–109).

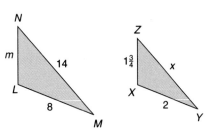

Figure 19–109

9. Write the proportions for the sides of the similar triangles in Fig. 19–110.

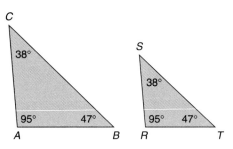

Figure 19–110

10. Find AB if $\triangle DEC \sim \triangle AEB$ (see Fig. 19–111).

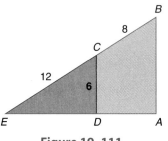

Figure 19–111

Section 19–2

Use Fig. 19–112 to solve the following exercises. Round the final answers to the nearest thousandth if necessary.

11. $a = 9$ in.
$b = 12$ in.
$c = ?$

12. $a = 8$ cm
$b = 15$ cm
$c = ?$

13. $a = 7$ ft
$b = ?$
$c = 10$ ft

14. $a = 8$ mm
$b = ?$
$c = 17$ mm

15. $a = ?$
$b = 15$ yd
$c = 17$ yd

16. $a = ?$
$b = 12$ km
$c = 15$ km

17. $a = 11$ mi
$b = 17$ mi
$c = ?$

18. $a = 10$ in.
$b = 24$ in.
$c = ?$

19. $a = ?$
$b = 40$ cm
$c = 50$ cm

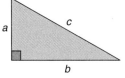

Figure 19–112

Solve the following problems. Round the final answers to the nearest thousandth if necessary.

20. CON If the base of a ladder is placed on the ground 4 ft from a house, how tall must the ladder be (to the nearest foot) to reach the chimney top that extends $18\frac{1}{2}$ ft above the ground?

21. AUTO In an automobile, three pulleys are connected by one belt. The center-to-center distance between the pulleys farthest apart cannot be measured conveniently. The other center-to-center distances are 12 in. and 18 in. (Fig. 19–113). Find the distance between the pulleys that are farthest apart.

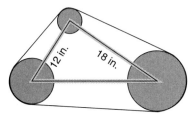

Figure 19–113

22. **CON** A central vacuum outlet is installed in one corner of a rectangular room that measures 9′ × 12′. How long must the nonelastic hose be to reach all parts of the room?

23. Find the diameter of a piece of round steel from which a 3-in. square nut can be milled.

24. **INDTR** Find the distance across the corners of a square nut that is 7.9 mm on a side.

Use Fig. 19–114 to solve the following exercises. Round your final answers to the nearest thousandth if necessary.

25. *RT* = 15 cm; find *RS* and *ST*.
26. *ST* = 7.2 ft; find *RS* and *RT*.
27. *RT* = 9√2 hm; find *RS* and *ST*.
28. *RT* = 8√5 dkm; find *RS* and *ST*.

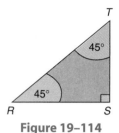

Figure 19–114

Use Fig. 19–115 to solve the following exercises. Round your final answers to the nearest thousandth if necessary.

29. *AC* = 12 dm; find *AB* and *BC*.
30. *AB* = 15 km; find *AC* and *BC*.
31. *BC* = 10 in.; find *AC* and *AB*.
32. *BC* = 8√2 hm; find *AC* and *AB*.
33. *AC* = 40 ft 7 in.; find *AB* and *BC* to the nearest inch.

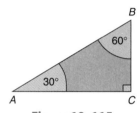

Figure 19–115

Solve. Round your final answers to the nearest thousandth if necessary.

34. The sides of an equilateral triangle are 4 dm in length. Find the altitude of the triangle. Round the answer to the nearest thousandth of a decimeter.

35. Find the total length to the nearest inch of the conduit *ABCD* if *XY* = 8 ft, *CE* = 14 in., and angle *CBE* = 30° (Fig. 19–116).

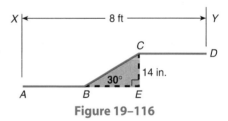

Figure 19–116

36. A piece of round steel is milled to a point at one end. The angle formed by the point is the angle of taper. If the steel has a 16-mm diameter and a 60° angle of taper, find the length *c* of the taper (Fig. 19–117).

37. A V-slot forms a triangle whose angle at the vertex is 60°. The depth of the slot is 17 mm (see Fig. 19–118). Find the width of the V-slot.

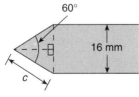

Figure 19–117

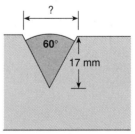

Figure 19–118

38. A manufacturer recommends attaching a guy wire to its 30-ft antenna at a 45° angle. If the antenna is installed on a flat surface, how long must the guy wire be (to the nearest foot) if it is attached to the antenna 4 ft from the top?

Find the area of Figs. 19–119 and 19–120.

39.

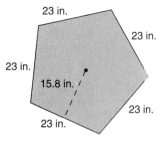

Figure 19–119

40.

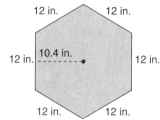

Figure 19–120

41. Identify the figure in Exercise 39 by giving its specific name and the number of degrees in each angle.

42. Identify the regular polygon in Exercise 40 by giving its specific name and the number of degrees in each angle.

For each exercise, supply the missing dimensions. For an inscribed triangle, see Fig. 19–121.

43. If $DO = 7$, $OB = $ _____

44. If $FB = 9$, $FO = $ _____

45. If $AB = 10$, $AE = $ _____

46. $\angle ABF = $ _____

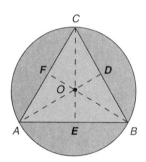

Figure 19–121

For a circumscribed square, see Fig. 19–122.

47. $\angle GJO = $ _____

48. If $GO = 7$, $HO = $ _____

49. If $KO = 10$, $IJ = $ _____

For an inscribed hexagon, see Fig. 19–123.

50. If $QO = 15$, $MN = $ _____

51. $\angle MOP = $ _____

52. $\angle MNO = $ _____

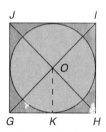

Figure 19–122

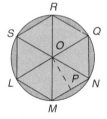

Figure 19–123

Solve the problems involving polygons and circles. Round the answers to hundredths.

53. A shaft with a 20-mm diameter is milled at one end, as illustrated in Fig. 19–124. What is the radius of the inscribed circle?

54. A hex nut 0.87 in. on a side is milled from the smallest-diameter round stock possible (Fig. 19–125). What is the diameter of the stock?

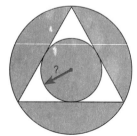

Figure 19–124

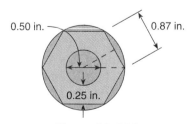

Figure 19–125

55. A square is milled on the end of a 5-cm shaft (Fig. 19–126). If the square is the largest that can be milled on the 5-cm shaft, what is the length of a side to hundredths?

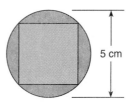

Figure 19–126

Section 19–4

Find the distance between each pair of points. Express the answer to the nearest thousandth when necessary.

56. (8, 3) and (3, −4)
57. (3, 6) and (−1, 4)
58. (4, 0) and (0, −1)
59. (3, −3) and (0, 7)
60. (2, 7) and (−3, −2)
61. (0, 0) and (−3, 5)
62. (−5, −5) and (4, 0)
63. (5, 2) and (−3, −3)
64. (1, −5) and (--2, −5)

Calculate the coordinates for the midpoint of each segment whose endpoints are given.

65. (3, 6) and (−1, 4)
66. (4, 0) and (0, −1)
67. (3, −3) and (0, 7)
68. (2, 7) and (−3, −2)
69. (0, 0) and (−3, 5)
70. (−5, −5) and (4, 0)
71. (5, 2) and (−3, −3)
72. (1, −5) and (−2, −5)
73. (−5, −4) and (2, −2)
74. (8, 3) and (3, −4)

TEAM PROBLEM-SOLVING EXERCISES

1. A *polygon is inscribed in a circle* if each vertex lies on the circumference of the circle. If a radius from the center of the circle is drawn to each vertex of the inscribed polygon, the central angles formed are equal.

The difference between the area of the circle and the area of the inscribed polygon can be found by calculating the areas of the circle and the polygon and subtracting. Another approach is to find the sum of the areas of the segments of the circle. There will be the same number of segments as sides of the polygon.

For a circle with a radius of 5 cm:
(a) Find the difference between the areas of the circle and an inscribed equilateral triangle and find the area of the triangle.
(b) Find the difference between the areas of the circle and an inscribed square and find the area of the square.
(c) Find the difference between the areas of the circle and the inscribed hexagon (six equal sides) in Fig. 19–127 and find the area of the hexagon.

2. Using the information found in Exercise 1:
(a) Estimate the area of the regular pentagon (five equal sides) in Fig. 19–128 by giving two values the area is between. Present a convincing argument for your estimate.
(b) Estimate the area of an inscribed regular octagon (eight equal sides) that has a radius of 5 cm and give an argument for your estimate.

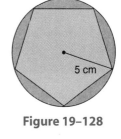

Figure 19–128

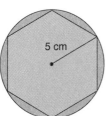

Figure 19–127

1. Can the measures 8 cm, 12 cm, and 25 cm represent the sides of a triangle? Illustrate and explain your answer.

2. The Pythagorean theorem, $a^2 + b^2 = c^2$, applies to right triangles where a and b are legs and c is the hypotenuse. For what type of triangle is the statement $a^2 + b^2 > c^2$ true? Illustrate your answer.

3. For what type of triangle is the statement $a^2 + b^2 < c^2$ true? Illustrate your answer.

4. A triangle has 6, 3, and $3\sqrt{3}$ as the measures of its sides. Sketch the triangle, place the measures on the appropriate sides, and show the measures of each of the three angles of the triangle.

5. Explain how the Pythagorean theorem is used to find the distance between two points on a coordinate grid.

6. Use the Internet to research Pythagorean triples and give 5 sets of Pythagorean triples that are natural numbers. Discuss the results of your research.

7. Find three sets of Pythagorean triples that are "primitive"—that is, they are not multiples of some other Pythagorean triple.

8. The formula for the radius of the inscribed circle in a triangle whose sides are length a, b, and c is:

$$r = \frac{\sqrt{(a + b + c)(a + b - c)(a - b + c)(-a + b + c)}}{2(a + b + c)}$$

What is the radius of the inscribed circle of a triangle that has sides 3, 4, and 5?

9. What is the radius of the inscribed circle of a triangle that has sides 5, 12, and 13?

10. Use the theorem to show that $p^2 - q^2$, $2pq$, and $p^2 + q^2$ is a Pythagorean triple if p and q are positive integers and q is less than p.

1. Identify the largest and the smallest angles in Fig. 19–129.

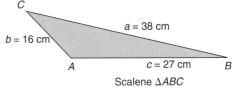

Scalene $\triangle ABC$

Figure 19–129

2. Find HI if $\triangle ABC \sim \triangle GHI$ (Fig. 19–130).

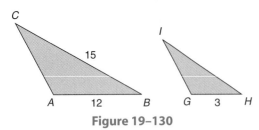

Figure 19–130

3. Find DB if $\triangle ABE \sim \triangle CDE$, $CD = 9$, $AB = 12$, $DE = 15$ (Fig. 19–131).

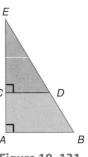

Figure 19–131

4. A model of a triangular part is shown in Fig. 19–132. Find side x if the part is to be similar to the model.

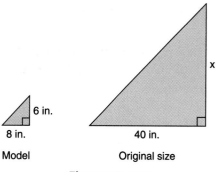

6 in.

8 in.

Model

40 in.

Original size

Figure 19–132

Use Figure 19–133 and the Pythagorean theorem to find the missing value. Round to the nearest whole number.

5. $a = 20$, $b = 48$, find c.

6. $b = 10$, $c = 18$, find a.

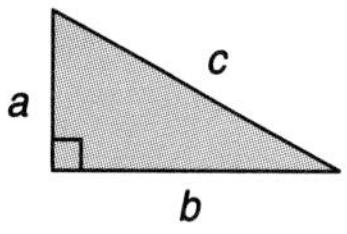

a

c

b

Figure 19–133

7. Three pulleys are designed so their centers form a right triangle when connecting lines are drawn. The distances between the centers forming the sides of the triangle are 8 in. and 15 in., respectively. What is the distance between the centers of the pulleys that form the hypotenuse?

Use Fig. 19–134 for Exercises 8–10. Round to the nearest hundredth if necessary.

8. $AC = 18$ ft; find AB and AC.

9. $AB = 62$ cm; find AC and BC.

10. $BC = 13.7$ mm; find AC and AB.

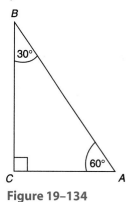

B

30°

60°

C

A

Figure 19–134

Use Fig. 19–135 for Exercises 11–12. Round to the nearest hundredth if necessary.

11. $AC = 5\frac{1}{2}$ cm; find AB and BC.

12. $AB = 15.3$ m; find AB and AC.

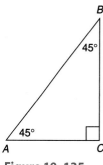

B

45°

45°

A

C

Figure 19–135

Practice Test

13. A conduit *ABCD* is made so that ∠*CBE* is 45° (Fig. 19–136). If *BE* = 4 cm and *AK* = 12 cm, find the length of the conduit to the nearest thousandth centimeter.

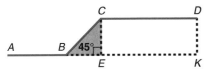

Figure 19–136

14. The rafters of a house make a 30° angle with the joists (Fig. 19–137). If the rafters have an 18-in. overhang and the center of the roof is 10 ft above the joists, how long must the rafters be cut?

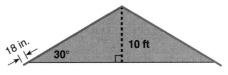

Figure 19–137

15. Find the circumference of a circle inscribed in an equilateral triangle if the height of the triangle is 12 in. (Fig. 19–138).

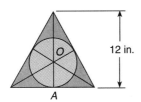

Figure 19–138

16. Ken Bennett milled a square metal rod so that a circle is formed at the end (see Fig. 19–139). If the square cross-section is 1.8 in. across the corners, what is the diameter of the circle?

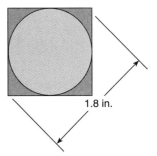

Figure 19–139

17. Laura Deskin milled a circle on the cross-sectional end of a hexagonal rod (Fig. 19–140). If the distance across the flats of the cross-section is $\frac{1}{2}$ in., what is the circumference of the circle?

Figure 19–140

Find the distance between each pair of points and find the coordinates of the midpoint of the line segment made by the pairs of points. Round to the tenths.

18. (5, 2) and (−7, −3)

19. (−2, 1) and (3, 3)

20. (3, 5) and (6, 9)

CHAPTERS 16–19 CUMULATIVE PRACTICE TEST

Write as logarithmic equations.

1. $5^3 = 125$

2. $36^{(-\frac{1}{2})} = \dfrac{1}{6}$

Write as exponential equations.

3. $\log_6 36 = 2$

4. $\log_2 \dfrac{1}{32} = -5$

5. Use the exponential model $D = D_0(1 + r)t$, where d is the amount of drug in the bloodstream after a specified amount of time and D_0 is the initial amount of drug in the bloodstream. The elapsed time is t, and r is the decimal equivalent of the breakdown rate. Find the amount of drug remaining in the bloodstream after 8 h if the initial amount of drug in the bloodstream is 300 mg and the drug breaks down by 12% each hour.

6. The height in meters of the female members of a population group is approximated by the formula $h = 0.3 + \log t$, where t represents age in years for the first 20 years of life. Find the projected height of a 12-year-old female.

7. If you invest $12,000 today at 4.5% annual interest compounded continuously, how much will you have after 15 years?

8. Find the perimeter and area of a trapezoid that has a height of 12 cm if the lower base is 20 cm and the upper base is 10 cm. The nonparallel sides are 13 cm each.

Write the solution set in symbolic form, graphical representation, and interval notation.

9. $5x - 3 < 7$

10. $12 - 3x \geq -33$

11. $-3 < x - 5 < 4$

12. $|x + 3| < 2$

Shade the portion of the graph represented by the conditions.

13. $x + y \leq 5$
    $x - y \geq 3$

14. $2x - y > 6$
    $x + y > 3$

15. $y \geq x^2 + x - 6$

16. $y < -x^2 + x + 6$

Classify the angle measures using the terms *right, straight, acute,* and *obtuse.*

17. $28°$

18. $150°$

19. $180°$

Find the measure of the complementary angle.

20. $46°$

21. $28°$

Find the measure of the supplementary angle.

22. $52°$

23. $143°$

Change to decimal degree equivalent rounded to the nearest ten-thousandth.

24. $42'15''$

25. $30°12'20''$

Change to minutes and seconds. Round to the nearest second when necessary.

26. $0.86°$

27. $0.352°$

Find the perimeter and area to the nearest whole unit.

28. A rectangle with length of 30 cm and width of 18 cm.

29. A triangle with sides that measure 42 m, 36 m, and 30 m.

30. Find the circumference and area of a circle that has a radius of 40 cm.

31. Find the circumference and area of a circle that has a diameter of 150 cm.

Convert to radians. Round to the nearest ten-thousandth radian.

32. $45°$

33. $15°22'$

Convert to degrees. Round to the nearest ten-thousandth of a degree.

34. $\dfrac{3\pi}{4}$ rad

35. 1.8 rad

36. Find the area of a sector of a circle that has a radius of 12 cm if the angle of the sector is $60°$.

37. Find the volume to the nearest centimeter of a cylinder that has a circular base with a radius of 12 cm if the cylinder has a height of 40 cm.

38. Find the lateral surface area of a cone with a radius of 72 cm and a slant height of 100 cm.

39. Use the Pythagorean theorem to find the hypotenuse to the nearest inch of a right triangle that has legs measuring 12 in. and 15 in.

Use Fig. 19–141 for Exercises 40–42. Round to tenths.

40. $AC = 28$ m; find AB and BC.

41. $AB = 32.5$ m; find AC and BC.

42. $BC = 17.32$ ft; find AC and AB.

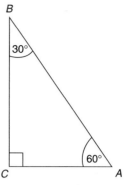

Figure 19–141

43. A square picture is 38 in. on each side. Find the length of the diagonal. Round to the nearest whole number.

44. The diagonal of a square is 48 in. Find the length of each side. Round to the nearest whole number.

45. Find the distance between two points described by (3, 14) and (9, 8).

46. Find the midpoint of the line segment described by $(-4, 5)$ and (5, 14). Round to tenths.

Selected Answers to Student Exercise Material

Answers to all Section Self-Study Exercises and answers to odd-numbered exercises in the Chapter Review Exercises and Practice Tests are included here. Solutions for the Chapter Review Exercises and Practice Tests are available separately in the Student Solutions Manual (odd-numbered) or the Instructor's Resource Manual (even-numbered).

Section 1–1 Self-Study Exercises

1 **1.** $6 < 8$; $8 > 6$ **2.** $32 < 42$; $42 > 32$ **3.** $148 < 196$; $196 > 148$ **4.** $2,517 < 2,802$; $2,802 > 2,517$ **5.** $7,809 < 8,902$; $8,902 > 7,809$ **6.** $42,999 < 44,000$; $44,000 > 42,999$ **7.** $\$183,500 < \$198,900$; $\$198,900 > \$183,500$ **8.** 786 flats $>$ 583 flats; 583 flats $<$ 786 flats **9.** 758 rooms $<$ 893 rooms; 893 rooms $>$ 758 rooms **10.** 5,982 e-mails $>$ 2,807 e-mails; 2,807 e-mails $<$ 5,982 e-mails

2 **11.** 0.5 **12.** 0.23 **13.** 0.07 **14.** 6.83 **15.** 0.079 **16.** 0.468

3 **17.** 3.72 **18.** 7.08 **19.** 0.3 **20.** 1.87, 1.9, 1.92 **21.** 72.07, 72.1, 73 **22.** 0.837 in. **23.** the micrometer reading **24.** yes **25.** 0.04 in. **26.** No. 10 wire **27.** 72.3 kg **28.** $4.2 > 3.8$; $3.8 < 4.2$ **29.** $1.68 > 1.6$; $1.6 < 1.68$ **30.** 0.026 kW $>$ 0.003 kW; 800-W toaster **31.** $0.394 < 0.621$; 0.621

4 **32.** 500 **33.** 430,000 **34.** 83,000,000,000 **35.** 300,000,000 **36.** 0.784 in. **37.** 2.8 A **38.** $3 **39.** Drivers A, B, and D

5 **40.** 97,614 **41.** 7,007 **42.** 30,133 **43.** 15.7 **44.** 34.18 **45.** 129.97 **46.** 3.077 in. **47.** 15.503 A **48.** $181.25 **49.** 391.4 ft **50.** 100.9° **51.** $28,334.01 **52.** 0; 56.365 **53.** 84,200; 84,213 **54.** 200; 183.405 **55.** 402,300; 402,199 **56.** 0; 81.401 **57.** $3,440; $3,443.60 **58.** $3,400; $3,365 **59.** 159 lb **60.** Yes, total capacity available is 115 gal. **61.** Yes, 479 pages are needed. **62.** 350 ft **63.** 1,020 **64.** 115 **65.** 53,036 **66.** 22 bags **67.** 341 boxes **68.** 291.82 **69.** 7.5 **70.** 310.8 **71.** 4.4 **72.** 5°F **73.** 15.1 lb **74.** 12.08 in., 12.10 in. **75.** 59.83 cm **76.** 4.189 in., 4.201 in. **77.** 186 bricks **78.** 213.8 in. **79.** 125.5 in. **80.** 45 m

6 **81.** 378 **82.** 630,000,000 **83.** 102,612 **84.** 4,096 washers **85.** 84 automobiles **86.** 700 tickets **87.** $288 **88.** 56.55 **89.** 3.2445 **90.** 0.05805 **91.** 0.08672 **92.** 170.12283 **93.** 0.38381871 **94.** 4.9386274 **95.** 30.66 **96.** 5.25 in. **97.** $27.48 **98.** 151.2 in. **99.** $64.20 **100.** 43 **101.** 47 **102.** 32 ft **103.** $1,245 **104.** 6 in. **105.** 10.9 **106.** 0.19 **107.** 25 **108.** 6 lb **109.** 23 rolls **110.** 600 revolutions **111.** 0.7 ft **112.** 0.16 **113.** 9 **114.** $575 **115.** $1.98 **116.** 1,070 bricks **117.** $10,500 **118.** $104.75 **119.** $1,950 **120.** 85 **121.** 454 lb **122.** $765.32 **123.** 1.69 in. **124.** 3.5 A **125.** 473 crates

Section 1–2 Self-Study Exercises

1   **1.** 4; 3   **2.** 9; 4   **3.** 2.7; 9   **4.** 15; 2   **5.** 1,000   **6.** 16   **7.** 11.56   **8.** 15   **9.** 8   **10.** 81   **11.** 8^1
12. 14.5^1 **13.** 12^1 **14.** 23^1 **15.** Leaves base unchanged   **16.** Use the base as a factor 3 times.

2   **17.** 64   **18.** 324   **19.** 1.96   **20.** 169   **21.** 10,000   **22.** 14,641   **23.** 12,996   **24.** 14.44   **25.** 64
26. 144   **27.** 324   **28.** 10,201   **29.** 5   **30.** 7   **31.** 9   **32.** 6   **33.** 14   **34.** 15   **35.** 11   **36.** 12
37. Use the number as a factor 2 times.   **38.** Find a number that is used as a factor 2 times to give the desired number.

3   **39.** 1,000   **40.** 1,200,000   **41.** 20,000   **42.** 10,200   **43.** 25   **44.** 21   **45.** 3   **46.** 250
47. Shift the decimal to the right in the number being multiplied by the power of 10 as indicated by the exponent.
48. Shift the decimal to the left in the number being divided by the power of 10 as indicated by the exponent.
49. 225   **50.** 343   **51.** 78,125   **52.** 20,736   **53.** 18   **54.** 28   **55.** 33   **56.** 14   **57.** 6   **58.** 8
59. 14   **60.** 33

Section 1–3 Self-Study Exercises

1   **1.** 26   **2.** 18   **3.** 9   **4.** 12.5   **5.** 24   **6.** 32   **7.** 18   **8.** 9   **9.** 10   **10.** 9   **11.** 30   **12.** 15
13. 156   **14.** 109   **15.** 23   **16.** 145   **17.** 7   **18.** 15   **19.** 160   **20.** 1,458   **21.** 80   **22.** 10
23. 24   **24.** 15.6   **25.** 25.04   **26.** 9   **27.** 5   **28.** 70   **29.** 13   **30.** 11   **31.** 29   **32.** 5   **33.** 39
34. 20   **35.** 30   **36.** 20.52   **37.** 483   **38.** 443   **39.** 79.16

2   **40.** $P = 40$ ft   **41.** $P = 61$ m   **42.** 123.54 ft^2 **43.** 1,164 m^2 **44.** 38 in.; 72 in^2 **45.** 21.2 cm;
13.8 cm^2 **46.** 156 in.   **47.** 124 in.   **48.** 160 in.   **49.** 108 in.   **50.** 10 ft; 6 ft^2 **51.** 43.8 ft; 116.84 ft^2
52. 930 ft   **53.** 54 ft   **54.** 59 ft   **55.** 156 ft   **56.** 12 cm; 9 cm^2 **57.** 35.6 cm; 79.21 cm^2 **58.** 590 ft
59. 48 tiles   **60.** 103.823 ft^3 **61.** 2,097.152 cm^3 **62.** 6,426 cm^3 **63.** 16.704 m^3

3   **64.** 29 boxes   **65.** $26,178.44   **66.** 5,660 people   **67.** 1,080 boxes of cards

Chapter Review Exercises, Chapter 1

1. (a) 0.3   (b) 0.15   (c) 0.04   **3.** (a) 500   (b) 50,000   (c) 41.4   (d) 6.90   (e) 23.4610
5. 4.79   **7.** 0.02; 0.021; 0.0216   **9.** $\dfrac{7}{8}$ **11.** (a) 28   (b) 25   **13.** 34.9 kW   **15.** (a) 4.61   (b) 3.127
(c) 204.899   (d) 12,140   **17.** 8.291 in.; 8.301 in.   **19.** 0.43 in.   **21.** 8.930 in.; 8.940 in.   **23.** 84
25. 13,725   **27.** 394,254,080   **29.** $1,407   **31.** $43,920   **33.** $14,800, $13,140   **35.** 1,140,000 ft^2,
1,204,010.28 ft^2 **37.** 0.12096 in.   **39.** 13   **41.** 2,008.4   **43.** 23 envelopes, 11 left over   **45.** 48.79 ft
47. (a) 7, 3, 343   (b) 2.3, 4, 27.9841   (c) 8, 4, 4,096   **49.** (a) 1   (b) 15,625   (c) 31.36   (d) 441
51. (a) 10^1 (b) 10^3 (c) 10^4 (d) 10^5 **53.** (a) 7   (b) 0.04056   (c) 0.605   (d) 2.3079   **55.** 75
57. 54   **59.** 15   **61.** 15   **63.** 13   **65.** 13.6   **67.** 113.608   **69.** 3   **71.** 18   **73.** 61
75. 49 boxes   **77.** $12,880   **79.** $53   **81.** 57 cm   **83.** 210 mm   **85.** 28.8 m   **87.** 66 ft
89. 288 in.   **91.** 162 cm^2 **93.** 2,450 mm^2 **95.** 51.84 m^2 **97.** 33,000 ft^2 **99.** 33 yd^2 **101.** $14.25

Practice Test, Chapter 1

1. 5.09   **3.** 48.3   **5.** 1,007   **7.** $9,271,314   **9.** 134   **11.** 106   **13.** $310, $310   **15.** $10, $14
17. 42,730   **19.** 11.6   **21.** 83   **23.** 0.6   **25.** $17,500   **27.** $P = 91$ ft; $A = 514.5$ ft^2 **29.** $P = 12.2$
in.; $A = 6.08$ in^2

Section 2–1 Self-Study Exercises

1 **1.** $5 = 5 \times 1, 10 = 5 \times 2, 15 = 5 \times 3, 20 = 5 \times 4, 25 = 5 \times 5, 30 = 5 \times 6$ **2.** $6 = 6 \times 1, 12 = 6 \times 2,$
$18 = 6 \times 3, 24 = 6 \times 4, 30 = 6 \times 5, 36 = 6 \times 6$ **3.** $8 = 8 \times 1, 16 = 8 \times 2, 24 = 8 \times 3, 32 = 8 \times 4,$
$40 = 8 \times 5, 48 = 8 \times 6$ **4.** $9 = 9 \times 1, 18 = 9 \times 2, 27 = 9 \times 3, 36 = 9 \times 4, 45 = 9 \times 5, 54 = 9 \times 6$
5. $10 = 10 \times 1, 20 = 10 \times 2, 30 = 10 \times 3, 40 = 10 \times 4, 50 = 10 \times 5, 60 = 10 \times 6$ **6.** $30 = 30 \times 1,$
$60 = 30 \times 2, 90 = 30 \times 3, 120 = 30 \times 4, 150 = 30 \times 5, 180 = 30 \times 6$ **7.** $6 \times 1 = 6, 6 \times 2 = 12, 6 \times 3 = 18,$
$6 \times 4 = 24, 6 \times 5 = 30$; Answers may vary. **8.** $12 \times 1 = 12, 12 \times 2 = 24, 12 \times 3 = 36, 12 \times 4 = 48,$
$12 \times 5 = 60$; Answers may vary. **9.** $13 \times 1 = 13, 13 \times 2 = 26, 13 \times 3 = 39, 13 \times 4 = 52, 13 \times 5 = 65,$
$13 \times 6 = 78$; Answers may vary. **10.** $3 \times 1 = 3, 3 \times 2 = 6, 3 \times 3 = 9, 3 \times 4 = 12, 3 \times 5 = 15$; Answers may
vary. **11.** $50 \times 1 = 50, 50 \times 2 = 100, 50 \times 3 = 150, 50 \times 4 = 200, 50 \times 5 = 250$; Answers may vary.
12. $4 \times 1 = 4, 4 \times 2 = 8, 4 \times 3 = 12, 4 \times 4 = 16, 4 \times 5 = 20$; Answers may vary. **13.** no; not divisible by 3
14. yes; ends in zero **15.** no, last two digits not divisible by 4 **16.** yes; sum of digits divisible by 3
17. yes; no remainder **18.** yes; the three digits divisible by 8 **19.** yes; sum of digits divisible by 3
20. yes; divisible by both 2 and 3 **21.** yes; last digit even number

2 **22.** $1 \cdot 4, 2 \cdot 2$ **23.** $1 \cdot 8, 2 \cdot 4$ **24.** $1 \cdot 12, 2 \cdot 6, 3 \cdot 4$ **25.** $1 \cdot 15, 3 \cdot 5$ **26.** $1 \cdot 16, 2 \cdot 8, 4 \cdot 4$
27. $1 \cdot 20, 2 \cdot 10, 4 \cdot 5$ **28.** $1 \cdot 24, 2 \cdot 12, 3 \cdot 8, 4 \cdot 6$ **29.** $1 \cdot 30, 2 \cdot 15, 3 \cdot 10, 5 \cdot 6$ **30.** $1 \cdot 36, 2 \cdot 18, 3 \cdot 12,$
$4 \cdot 9, 6 \cdot 6$ **31.** $1 \cdot 38, 2 \cdot 19$ **32.** $1, 2, 4, 5, 8, 10, 20, 40$ **33.** $1, 2, 23, 46$ **34.** $1, 2, 4, 13, 26, 52$
35. $1, 2, 4, 8, 16, 32, 64$ **36.** $1, 2, 3, 4, 6, 8, 9, 12, 18, 24, 36, 72$ **37.** $1, 3, 9, 27, 81$ **38.** $1, 5, 17, 85$
39. $1, 2, 4, 23, 46, 92$ **40.** $1, 2, 3, 4, 6, 8, 12, 16, 24, 32, 48, 96$ **41.** $1, 2, 7, 14, 49, 98$

3 **42.** prime **43.** composite **44.** composite **45.** prime **46.** composite **47.** composite
48. composite **49.** composite **50.** composite **51.** prime **52.** composite **53.** prime **54.** $2 \cdot 2 \cdot 3$
55. $2 \cdot 3 \cdot 3$ **56.** $2 \cdot 2 \cdot 5$ **57.** $2 \cdot 2 \cdot 2 \cdot 3$ **58.** $5 \cdot 5$ **59.** $3 \cdot 3 \cdot 3$ **60.** 29 **61.** $2 \cdot 3 \cdot 5$
62. $5 \cdot 7$ **63.** $2 \cdot 2 \cdot 2 \cdot 5$ **64.** 47 **65.** $7 \cdot 7$ **66.** $2 \cdot 5 \cdot 5$ **67.** $2 \cdot 2 \cdot 13$ **68.** $5 \cdot 13$ **69.** $3 \cdot 5 \cdot 5$
70. $2 \cdot 2 \cdot 5 \cdot 5$ **71.** $3 \cdot 5 \cdot 7$ **72.** $2 \cdot 2 \cdot 3 \cdot 3 \cdot 3$ **73.** $5 \cdot 23$ **74.** $11 \cdot 11$ **75.** $2 \cdot 2 \cdot 2 \cdot 2 \cdot 3 \cdot 3$
76. $2 \cdot 2 \cdot 3 \cdot 13$ **77.** 157 **78.** $2^3 \cdot 3^2$ **79.** $2^4 \cdot 7$ **80.** $2^2 \cdot 31$ **81.** $2^2 \cdot 41$ **82.** $2^3 \cdot 71$
83. $2^2 \cdot 3^2 \cdot 5^2$

4 **84.** 6 **85.** 30 **86.** 56 **87.** 12 **88.** 72 **89.** 60 **90.** 24 **91.** 18 **92.** 24 **93.** 700
94. 27 **95.** 16 **96.** 90 **97.** 120 **98.** 180 **99.** 30 **100.** 96 **101.** 72 **102.** 60 **103.** 300
104. 66 **105.** 312 **106.** 14 **107.** 144

5 **108.** 1 **109.** 1 **110.** 1 **111.** 6 **112.** 5 **113.** 12 **114.** 9 **115.** 5 **116.** 2 **117.** 6
118. 5 **119.** 3 **120.** 2 **121.** 2 **122.** 2

Section 2–2 Self-Study Exercises

1 **1.** $\dfrac{4}{5}, \dfrac{8}{10}, \dfrac{12}{15}, \dfrac{16}{20}, \dfrac{20}{25}, \dfrac{24}{30}$ **2.** $\dfrac{7}{10}, \dfrac{14}{20}, \dfrac{21}{30}, \dfrac{28}{40}, \dfrac{35}{50}, \dfrac{42}{60}$ **3.** $\dfrac{3}{4} = \dfrac{18}{24}$ **4.** 6 **5.** 12 **6.** 36 **7.** 5

8. 20 **9.** 21 **10.** 12 **11.** $\dfrac{1}{2}$ **12.** $\dfrac{3}{5}$ **13.** $\dfrac{3}{4}$ **14.** $\dfrac{5}{16}$ **15.** $\dfrac{1}{2}$ **16.** $\dfrac{7}{8}$ **17.** $\dfrac{5}{16}$ **18.** $\dfrac{1}{5}$

19. $\dfrac{1}{5}$ **20.** $\dfrac{6}{25}$ **21.** $\dfrac{5}{8}$ **22.** $\dfrac{1}{4}$ **23.** $\dfrac{3}{4}$ **24.** $\dfrac{3}{16}$ **25.** $\dfrac{7}{32}$ **26.** $4\dfrac{9}{16}$ in. **27.** $4\dfrac{1}{16}$ in.

28. $3\dfrac{13}{16}$ in. **29.** $3\dfrac{3}{8}$ in. **30.** $2\dfrac{1}{4}$ in. **31.** 2 in. **32.** $1\dfrac{3}{4}$ in. **33.** $1\dfrac{3}{16}$ in. **34.** $\dfrac{3}{4}$ in. **35.** $\dfrac{3}{8}$ in.

2 **36.** $2\frac{2}{5}$ **37.** $1\frac{3}{7}$ **38.** 1 **39.** $4\frac{4}{7}$ **40.** 4 **41.** $2\frac{1}{7}$ **42.** $2\frac{5}{9}$ **43.** $9\frac{2}{5}$ **44.** $9\frac{5}{9}$ **45.** $1\frac{17}{21}$

46. $3\frac{4}{5}$ **47.** 16 **48.** $7\frac{1}{5}$ **49.** $9\frac{1}{2}$ **50.** 9

3 **51.** $\frac{7}{3}$ **52.** $\frac{25}{8}$ **53.** $\frac{15}{8}$ **54.** $\frac{77}{12}$ **55.** $\frac{77}{8}$ **56.** $\frac{31}{8}$ **57.** $\frac{89}{12}$ **58.** $\frac{103}{16}$ **59.** $\frac{257}{32}$ **60.** $\frac{69}{64}$

61. $\frac{73}{10}$ **62.** $\frac{26}{3}$ **63.** $\frac{100}{3}$ **64.** $\frac{200}{3}$ **65.** $\frac{25}{2}$ **66.** 15 **67.** 18 **68.** 56 **69.** 32 **70.** 48

4 **71.** $\frac{1}{2}$ **72.** $\frac{1}{10}$ **73.** $\frac{1}{5}$ **74.** $\frac{7}{10}$ **75.** $\frac{1}{4}$ **76.** $\frac{1}{40}$ **77.** $3\frac{9}{10}$ **78.** $4\frac{4}{5}$ **79.** $\frac{189}{500}$ **80.** $\frac{7}{8}$

81. $\frac{3}{8}$ **82.** $\frac{5}{8}$ **83.** $\frac{3}{4}$ in. **84.** $\frac{3}{16}$ ft **85.** $2\frac{3}{8}$ in. **86.** $\frac{83}{100}$ lb **87.** $\frac{5}{16}$ in. **88.** $3\frac{1}{8}$ in. **89.** 0.6

90. 0.3 **91.** 0.875 **92.** 0.375 **93.** 0.45 **94.** 0.98 **95.** 0.21 **96.** 3.875 **97.** 1.4375 **98.** 4.5625
99. 0.17 **100.** 0.44 **101.** 1.43 **102.** 3.45 **103.** 0.67 **104.** 0.27 **105.** 0.78 **106.** 0.38
107. 0.83 **108.** 0.58 **109.** 2.38 **110.** 5.57 **111.** 2.046875 in. **112.** $0.045 **113.** 0.125 in.

114. $47\frac{3}{5}$ m

5 **115.** 72 **116.** 30 **117.** 50 **118.** 48 **119.** 24 **120.** $\frac{2}{3}$ **121.** $\frac{7}{16}$ **122.** $\frac{8}{9}$ **123.** $\frac{11}{16}$

124. $\frac{15}{32}$ **125.** $\frac{7}{12}$ **126.** $\frac{4}{5}$ **127.** $\frac{9}{10}$ **128.** $\frac{4}{15}$ **129.** $\frac{1}{2}$ **130.** 0.37 **131.** $\frac{4}{5}$ **132.** $\frac{5}{12}$ **133.** $\frac{1}{2}$

134. 0.34 **135.** No, $\frac{15}{64}$ is greater **136.** Yes, $\frac{3}{8}$ is greater **137.** Yes, $\frac{7}{16}$ is greater **138.** Yes

139. $\frac{5}{8}$ in. end **140.** No **141.** No **142.** Yes **143.** No **144.** too small

Section 2–3 Self-Study Exercises

1 **1.** $\frac{3}{8}$ **2.** $1\frac{3}{8}$ **3.** $\frac{5}{8}$ **4.** $\frac{25}{32}$ **5.** $\frac{9}{16}$ **6.** $1\frac{7}{16}$ **7.** $\frac{11}{64}$ **8** $1\frac{19}{40}$ **9.** $1\frac{23}{36}$ **10.** $1\frac{1}{12}$ **11.** $1\frac{2}{5}$

12. $\frac{19}{21}$ **13.** $\frac{15}{16}$ in. **14.** $1\frac{27}{32}$ in. **15.** $1\frac{15}{16}$ in. **16.** $1\frac{7}{8}$ in. **17.** $1\frac{1}{4}$ in. **18.** $6\frac{4}{5}$ **19.** $4\frac{1}{8}$

20. $16\frac{13}{16}$ **21.** $1\frac{11}{18}$ **22.** $4\frac{13}{16}$ **23.** $5\frac{17}{32}$ **24.** $4\frac{11}{16}$ **25.** $13\frac{1}{2}$ **26.** $1{,}001\frac{31}{36}$ **27.** $370\frac{37}{120}$

28. $1{,}115\frac{2}{45}$ **29.** $125\frac{21}{160}$ **30.** $8\frac{13}{16}$ in. **31.** $3\frac{15}{16}$ in. **32.** $11\frac{5}{8}$ gal **33.** $24\frac{5}{32}$ in.

34. $22\frac{7}{8}$ in. **35.** $5\frac{7}{8}$ c

2 **36.** $\frac{3}{16}$ **37.** $\frac{1}{16}$ **38.** $\frac{1}{8}$ **39.** $\frac{9}{64}$ **40.** $17\frac{3}{4}$ **41.** $4\frac{15}{16}$ **42.** $5\frac{21}{32}$ **43.** $1\frac{5}{7}$ **44.** $3\frac{19}{24}$

45. $18\frac{17}{48}$ **46.** $167\frac{9}{175}$ **47.** $759\frac{19}{56}$ **48.** $8\frac{13}{16}$ in. **49.** Yes; $60\frac{5}{16}$ in. < 60.4 in. **50.** $10\frac{1}{8}$ in.

51. $\frac{29}{32}$ in. **52.** $3\frac{1}{10}$ lb **53.** $\frac{3}{16}$ in.

Selected Answers to Student Exercise Material

Section 2–4 Self-Study Exercises

1 1. $\dfrac{3}{32}$ 2. $\dfrac{7}{32}$ 3. $\dfrac{7}{16}$ 4. $\dfrac{7}{12}$ 5. $\dfrac{1}{3}$ 6. $\dfrac{5}{32}$ 7. $21\dfrac{7}{8}$ 8. 75 9. $4\dfrac{1}{8}$ 10. $36\dfrac{1}{10}$

11. $1\dfrac{21}{40}$ 12. $2\dfrac{1}{6}$ 13. $18\dfrac{3}{4}$ L 14. 90 in. 15. 264 kg copper, 84 kg tin, 36 kg zinc

2 16. $\dfrac{1}{16}$ 17. $\dfrac{27}{125}$ 18. $\dfrac{1}{27}$ 19. $\dfrac{49}{81}$

3 20. $\dfrac{8}{5}$ 21. $\dfrac{5}{11}$ 22. $\dfrac{1}{8}$ 23. $\dfrac{10}{9}$ 24. $\dfrac{5}{9}$ 25. $\dfrac{6}{7}$ 26. $\dfrac{9}{10}$ 27. $\dfrac{11}{12}$ 28. 2 29. $4\dfrac{2}{3}$

30. $13\dfrac{1}{3}$ 31. $12\dfrac{1}{2}$ 32. $\dfrac{5}{8}$ 33. 4 34. $2\dfrac{1}{2}$ 35. 12 shovels 36. $23\dfrac{11}{12}$ in. 37. 20 ft × 15 ft

38. 4 whole pieces 39. 7 strips 40. 23 straws, $3\dfrac{3}{4}$ in. left

4 41. $17\dfrac{7}{40}$ 42. $82\dfrac{26}{245}$ 43. $32\dfrac{44}{45}$ 44. $5\dfrac{8}{9}$ 45. $5\dfrac{4}{9}$ 46. 3 47. $\dfrac{3}{4}$ 48. $4\dfrac{1}{6}$

49. $3\dfrac{37}{108}$ 50. $\dfrac{9}{14}$

Section 2–5 Self-Study Exercises

1 1. $\dfrac{2\text{ pt}}{1\text{ qt}}; \dfrac{1\text{ qt}}{2\text{ pt}}$ 2. $\dfrac{5{,}280\text{ ft}}{1\text{ mi}}; \dfrac{1\text{ mi}}{5{,}280\text{ ft}}$ 3. $\dfrac{12\text{ in.}}{1\text{ ft}}; \dfrac{1\text{ ft}}{12\text{ in.}}$ 4. $\dfrac{3\text{ ft}}{1\text{ yd}}; \dfrac{1\text{ yd}}{3\text{ ft}}$ 5. 48 in. 6. 21 ft

7. 4,400 yd 8. $9\dfrac{1}{3}$ yd 9. $2\dfrac{3}{10}$ lb or 2.3 lb 10. 732.8 oz 11. 20 qt 12. 13 pt

2 13. feet to inches = 12; inches to feet = $\dfrac{1}{12}$ or 0.0833333 14. quarts to gallons = $\dfrac{1}{4}$ or 0.25;
gallons to quarts = 4 15. 28.75 pt 16. 92 pt 17. 36.25 lb 18. 153 in. 19. 7,920 ft 20. 28 qt
21. 3 ft 8 in. 22. 2 mi 1,095 ft 23. 3 lb $3\dfrac{1}{2}$ oz 24. 2 gal 1 qt 25. 2 ft 10 in. 26. 6 lb 9 oz
27. 4 gal 2 qt 16 oz or 4 gal 2 qt 1 pt 28. 6 qt 20 oz or 6 qt 1 pt 4 oz

3 29. 2 lb 5 oz or 37 oz 30. 4 ft 7 in. or 55 in. 31. 15 lb 11 oz 32. 18 ft 3 in. 33. 8 qt $\dfrac{1}{2}$ pt
34. 14 gal 1 qt 35. 9 yd 1 ft 36. 7 yd 1 ft 5 in. 37. 2 ft 7 in. or 31 in. 38. 11 ft 39. 5 lb 15 oz
40. 12 lb 41. 6 in. 42. 1 pt 43. 2 ft 44. 4 lb 14 oz 45. 1 lb 6 oz 46. 6 lb 11 oz 47. 2 in.
48. 3 gal 3 qt $1\dfrac{1}{2}$ pt or 3 gal 3 qt 1 pt 1 c 49. 4 ft 2 in. 50. 2 ft 3 in. 51. 45 s 52. 39 s
53. 69 lb 7 oz 54. 16 lb 14 oz

4 55. 60 mi 56. 108 gal 57. 57 lb 8 oz 58. 58 ft 59. 36 lb 60. 7 qt 1 pt or 1 gal 3 qt 1 pt
61. 35 in^2 62. 108 ft^2 63. 180 yd^2 64. 108 mi^2 65. 378 tiles 66. 46 gal 18 oz 67. 7 qt 1 pt or
1 gal 3 qt 1 pt 68. 6 lb 4 oz 69. 1 day 15 h 70. 10 yd 1 ft 3 in. 71. 1 yd 1 ft 7 in. 72. 7 ft 6 in.
73. 3 gal 2 qt 74. 9 pieces 75. 5 ft 76. 3 gal 1 qt 5 oz 77. 24 lb 3 oz 78. 3 79. 18

80. 3   **81.** 8 pieces   **82.** 9 boxes   **83.** 24 cans   **84.** 9 tickets   **85.** $\dfrac{3}{4}\dfrac{\text{lb}}{\text{min}}$ **86.** $15,840\dfrac{\text{ft}}{\text{h}}$
87. $2,304\dfrac{\text{oz}}{\text{min}}$ **88.** $2\dfrac{\text{qt}}{\text{s}}$ **89.** $3\dfrac{\text{qt}}{\text{min}}$ **90.** $53\dfrac{1}{3}\dfrac{\text{lb}}{\text{min}}$

Chapter Review Exercises, Chapter 2

1. 24, 36, 48, 60, 72 Answers may vary.   **3.** 20, 30, 40, 50, 60 Answers may vary.   **5.** 42, 63, 84, 105, 126 Answers may vary.   **7.** 14, 21, 28, 35, 42 Answers may vary.   **9.** 16, 24, 32, 40, 48 Answers may vary.
11. Yes; sum of digits divisible by 3   **13.** No, divisible by 2 but not by 3   **15.** Yes; ends in zero   **17.** No; sum of digits not divisible by 9   **19.** $1 \times 48, 2 \times 24, 3 \times 16, 4 \times 12, 6 \times 8$; 1, 2, 3, 4, 6, 8, 12, 16, 24, 48
21. $1 \times 51, 3 \times 17$; 1, 3, 17, 51   **23.** $1 \times 74, 2 \times 37$; 1, 2, 37, 74   **25.** composite   **27.** composite
29. $2 \cdot 3 \cdot 7$ **31.** $2 \cdot 7 \cdot 7$ or $2 \cdot 7^2$ **33.** 360   **35.** 180   **37.** 2   **39.** 6   **41.** $\dfrac{15}{24}$ **43.** $\dfrac{25}{60}$ **45.** $\dfrac{10}{15}$
47. $\dfrac{24}{32}$ **49.** $\dfrac{11}{55}$ **51.** $\dfrac{16}{20}$ **53.** $\dfrac{1}{2}$ **55.** $\dfrac{1}{8}$ **57.** $\dfrac{1}{4}$ **59.** $\dfrac{17}{32}$ **61.** $\dfrac{3}{8}$ **63.** $\dfrac{3}{4}$ **65.** $5\dfrac{1}{4}$ in.   **67.** $4\dfrac{7}{16}$ in.
69. $3\dfrac{15}{16}$ in.   **71.** $3\dfrac{9}{16}$ in.   **73.** $2\dfrac{3}{4}$ in.   **75.** $1\dfrac{1}{2}$ in.   **77.** $2\dfrac{1}{8}$ in.   **79.** $\dfrac{7}{10}$ **81.** $\dfrac{19}{20}$ **83.** $\dfrac{109}{125}$ **85.** $\dfrac{1}{50}$
87. 0.2   **89.** 0.625   **91.** 0.818   **93.** $3\dfrac{3}{5}$ **95.** $4\dfrac{7}{8}$ **97.** $5\dfrac{3}{8}$ **99.** $87\dfrac{1}{2}$ **101.** $1\dfrac{1}{2}$ **103.** $\dfrac{8}{1}$ **105.** $\dfrac{57}{8}$
107. $\dfrac{147}{16}$ **109.** $\dfrac{23}{5}$ **111.** $\dfrac{12}{1}$ **113.** $\dfrac{16}{3}$ **115.** 20   **117.** 33   **119.** 60   **121.** 16   **123.** 60
125. larger   **127.** no   **129.** smaller   **131.** $\dfrac{3}{8}$ **133.** $\dfrac{3}{8}$ **135.** $\dfrac{3}{16}$ **137.** $\dfrac{27}{32}$ **139.** $\dfrac{9}{19}$ **141.** $\dfrac{3}{4}$
143. $\dfrac{3}{11}$ **145.** $\dfrac{21}{64}$ **147.** $1\dfrac{13}{30}$ **149.** $8\dfrac{19}{32}$ **151.** $10\dfrac{9}{32}$ **153.** $18\dfrac{1}{16}$ in.   **155.** $12\dfrac{19}{32}$ in.
157. $15\dfrac{25}{32}$ in.   **159.** $4\dfrac{3}{4}$ in.   **161.** $\dfrac{7}{8}$ in.   **163.** $\dfrac{1}{3}$ **165.** $1\dfrac{5}{8}$ **167.** $5\dfrac{31}{32}$ **169.** $7\dfrac{11}{16}$ **171.** $34\dfrac{3}{4}$
173. $\dfrac{7}{16}$ in.   **175.** $1\dfrac{29}{64}$ in.   **177.** $\dfrac{7}{24}$ **179.** $\dfrac{7}{24}$ **181.** $\dfrac{1}{2}$ **183.** $7\dfrac{7}{8}$ **185.** $1\dfrac{1}{5}$ **187.** 2   **189.** $100\dfrac{1}{2}$ in.
191. $2\dfrac{3}{4}$ cups   **193.** 150 cm   **195.** $\dfrac{9}{16}$ **197.** $\dfrac{16}{81}$ **199.** $\dfrac{1}{8}$ **201.** $\dfrac{81}{100}$ **203.** $\dfrac{1}{4}$ **205.** $\dfrac{10}{7}$ or $1\dfrac{3}{7}$
207. $1\dfrac{1}{6}$ **209.** $9\dfrac{1}{3}$ **211.** 24   **213.** 2   **215.** $\dfrac{3}{5}$ **217.** $2\dfrac{55}{64}$ in.   **219.** $\dfrac{3}{16}$ yd   **221.** 36 lengths
223. $\dfrac{1}{18}$ **225.** $5\dfrac{1}{3}$ **227.** $\dfrac{1}{4}$ **229.** $\dfrac{1}{8}$ **231.** $\dfrac{12 \text{ in.}}{1 \text{ ft}}; \dfrac{1 \text{ ft}}{12 \text{ in.}}$ **233.** $\dfrac{2,000 \text{ lb}}{1 \text{ T}}; \dfrac{1 \text{ T}}{2,000 \text{ lb}}$ **235.** 80 oz
237. $42\dfrac{1}{2}$ lb   **239.** 6,600 ft   **241.** 2 ft 7 in.   **243.** 13 lb $1\dfrac{1}{2}$ oz   **245.** 8 gal 2 qt   **247.** 14 oz
249. 17 in. or 1 ft 5 in.   **251.** 63 in^2 **253.** 10 yd 1 ft 3 in.   **255.** 4 lb 8 oz   **257.** 3.5 ft or 3 ft 6 in.
259. $300\dfrac{\text{mi}}{\text{h}}$ **261.** $1\dfrac{1}{2}$ qt or 1 qt 1 pt

Practice Test, Chapter 2

1. $\frac{3}{4}$　**3.** 3　**5.** $\frac{34}{7}$　**7.** $2^5 \cdot 3$　**9.** $2^5 \cdot 31$　**11.** 540　**13.** 6　**15.** $1\frac{1}{4}$ in.　**17.** $10\frac{1}{2}$　**19.** $\frac{7}{10}$

21. $7\frac{17}{30}$　**23.** $3\frac{8}{9}$　**25.** $3\frac{1}{2}$　**27.** $13\frac{1}{2}$　**29.** $\frac{2}{7}$　**31.** $3\frac{5}{6}$ c　**33.** 8 yd

Section 3–1 Self-Study Exercises

1　**1.** 40%　**2.** 70%　**3.** 62.5%　**4.** 77.8%　**5.** 0.7%　**6.** 0.3%　**7.** 20%　**8.** 14%　**9.** 0.7%
10. 1.25%　**11.** 500%　**12.** 800%　**13.** 133.3%　**14.** 350%　**15.** 430%　**16.** 220%　**17.** 305%
18. 720%　**19.** 1,510%　**20.** 3,625%

2　**21.** $\frac{9}{25}$, 0.36　**22.** $\frac{9}{20}$, 0.45　**23.** $\frac{1}{5}$, 0.2　**24.** $\frac{3}{4}$, 0.75　**25.** $\frac{1}{16}$, 0.0625　**26.** $\frac{5}{8}$, 0.625

27. $\frac{2}{3}$, 0.6667　**28.** $\frac{3}{500}$, 0.006　**29.** $\frac{1}{500}$, 0.002　**30.** $\frac{1}{2,000}$, 0.0005　**31.** $\frac{1}{12}$, 0.0833　**32.** $\frac{3}{16}$, 0.1875

33. 8　**34.** 4　**35.** $2\frac{1}{2}$, 2.5　**36.** $4\frac{1}{4}$, 4.25　**37.** $1\frac{19}{25}$, 1.76　**38.** $3\frac{4}{5}$, 3.8　**39.** $1\frac{3}{8}$, 1.375

40. $3\frac{7}{8}$, 3.875　**41.** $1\frac{2}{3}$　**42.** $3\frac{1}{6}$　**43.** 1.153　**44.** 2.125　**45.** 1.0625　**46.** $\frac{1}{10}$, 0.1　**47.** 25%, 0.25

48. 20%, $\frac{1}{5}$　**49.** $33\frac{1}{3}$%, $0.33\frac{1}{3}$　**50.** $\frac{1}{2}$, 0.5　**51.** 80%, 0.8　**52.** 75%, $\frac{3}{4}$　**53.** $\frac{2}{3}$, $0.66\frac{2}{3}$　**54.** 100%, $\frac{1}{1}$

55. 30%, 0.3　**56.** $\frac{2}{5}$, 0.4　**57.** 70%, $\frac{7}{10}$　**58.** 90%, 0.9　**59.** $\frac{3}{5}$, 0.6　**60.** 100%, 1　**61.** 25%, $\frac{1}{4}$

62. $66\frac{2}{3}$%, $0.66\frac{2}{3}$　**63.** $\frac{7}{10}$, 0.7　**64.** 50%, $\frac{1}{2}$　**65.** 60%, 0.6　**66.** 10%, $\frac{1}{10}$　**67.** $\frac{1}{5}$, 0.2　**68.** $33\frac{1}{3}$%, $\frac{1}{3}$

69. $66\frac{2}{3}$%, $\frac{2}{3}$　**70.** 75%, 0.75　**71.** $\frac{3}{10}$, 0.3　**72.** 90%, $\frac{9}{10}$　**73.** 20%, 0.2　**74.** 60%, $\frac{3}{5}$　**75.** $\frac{1}{1}$, 1

Section 3–2 Self-Study Exercises

1　**1.** $R = 40\%$; $B = 18$; P missing　**2.** $R = 66\%$; $B = \$35.99$; P missing　**3.** $R = 27\%$; $P = \$12.21$;
B missing　**4.** $R = 83\%$; $B = 12$ wh; P missing　**5.** R missing; $B = 10$; $P = 2$　**6.** $P = 2$; $R = 20\%$;
B missing　**7.** $P = 3$; R missing; $B = 4$　**8.** R missing; $B = 25$; $P = 5$　**9.** $P = 6$; $B = 15$; R missing
10. P missing; $R = 20\%$; $B = 15$　**11.** $R = 35\%$; B missing; $P = 70$　**12.** R missing; $B = \$45$; $P = \$3.15$

2　**13.** 5.4　**14.** 252　**15.** 1.8　**16.** 26　**17.** 50%　**18.** $333\frac{1}{3}$%　**19.** 87.5%　**20.** 150%　**21.** 64
22. 70,000

3　**23.** 75　**24.** 63　**25.** 206　**26.** 115.92　**27.** 0.675　**28.** 0.94　**29.** 154.1　**30.** 231　**31.** 924
32. 345　**33.** 25%　**34.** 30%　**35.** $33\frac{1}{3}$%　**36.** 32.9%　**37.** 15.75%　**38.** 16%　**39.** 0.8%

40. $0.66\frac{2}{3}$%　**41.** 500%　**42.** 111.25%　**43.** 72　**44.** 50　**45.** 344　**46.** 46　**47.** 360　**48.** 275

49. 75 **50.** 250 **51.** 18.4 **52.** 261 **53.** 3.75 **54.** 0.625 **55.** $66\frac{2}{3}\%$ **56.** $37\frac{1}{2}\%$ **57.** 14.25
58. 350 **59.** 9.375 **60.** 220 **61.** 20% **62.** 200

4 **63.** 1.0625 lb **64.** 3% **65.** 200 hp **66.** 0.021 lb **67.** 5% **68.** $575 **69.** 1,200 lb
70. 11% **71.** 9,625 swabs **72.** 100,880 welds **73.** $4.65 **74.** $6.45 **75.** 0.4 **76.** 0.4
77. $4,285.71 **78.** $1,364.16 **79.** $4.55, $80.38 **80.** $829.06 **81.** $23.30 **82.** 25% **83.** 23%
84. $41.93 **85.** $11,832 **86.** $17.52 **87.** $3,866.00 **88.** $944.50 **89.** $434.84 **90.** $3,205.75
91. $9.75 **92.** $150 **93.** 1.75% **94.** $171.50 **95.** $1,460 **96.** 2.5% **97.** 6.8% **98.** 30.1%
99. 111 lb **100.** $8\frac{3}{4}$ lb **101.** $32.91 **102.** 21,600 lb **103.** 170 lb **104.** $47.94

Section 3–3 Self-Study Exercises

1 **1.** 108 **2.** 109.2 **3.** 10.2 **4.** 33.75 **5.** 672,830; 2,878,830 **6.** 630,504; 2,576,504

2 **7.** 60% **8.** 82% **9.** 13.7% **10.** $66\frac{2}{3}\%$ **11.** 99.91% **12.** 99% **13.** 27 in.

14. 1,815 board feet **15.** 2,393 board feet **16.** 25,908 bricks **17.** $16,802.50 **18.** 1,366.4 yd^3
19. $53,350.20 **20.** $2,041.46 **21.** $3,961.82 **22.** $58.80

3 **23.** 7% **24.** 45% **25.** 20% **26.** 5.0% **27.** 350 hp **28.** 10 yd^3 **29.** 17% **30.** 25 lb
31. 40 in. **32.** 20% **33.** 150 hp **34.** 12%

Chapter Review Exercises, Chapter 3

1. 72% **3.** 23% **5.** 70% **7.** 83% **9.** 320% **11.** 12,500% **13.** 1,730% **15.** 0.72 or $\frac{18}{25}$
17. 0.125 or $\frac{1}{8}$ **19.** $0.0066\frac{2}{3}$ or $\frac{1}{150}$ **21.** 2.75 or $2\frac{3}{4}$ **23.** 1.125 or $1\frac{1}{8}$ **25.** 2.272 **27.** 3.4
29. 0.83 **31.** 0.625 **33.** $a = 4.5$ **35.** $y = 48$ **37.** $R = 5\%$; $B = 180$; P missing **39.** $R = 45\%$;
B missing; $P = \$36$ **41.** $P = 6$; R missing; $B = 25$ **43.** $R = 18\%$; $B = 150$; P missing **45.** 24
47. 0.4375 **49.** 60% **51.** 250% **53.** 83 **55.** 152 **57.** 15.3% **59.** 500% **61.** 84
63. 37.5% or $37\frac{1}{2}\%$ **65.** 1.14% **67.** $266 **69.** 4% **71.** 200 students **73.** $204.35 **75.** $439.84
77. $365.66 **79.** $1,707.60 **81.** 1.75% **83.** $3,935.63 **85.** $0.89 **87.** 8.5% **89.** $369.69
91. $355 **93.** 87.1% **95.** 78.5% **97.** $67\frac{3}{5}\%$ **99.** 8% **101.** 7% **103.** 14% **105.** 142.1 kg
107. 62 cm, 63 cm **109.** $1.69 **111.** 12.5% **113.** 57.5 lb **115.** $33\frac{1}{3}\%$ **117.** 4.8% **119.** 289 hp
121. 86% **123.** 26.6% **125.** 28.6%

Practice Test, Chapter 3

1. 80% **3.** 0.003 **5.** $R = 40\%$; $B = 10$; P missing **7.** $P = 9$; R missing; $B = 27$ **9.** $R = 12\%$; $B = 50$;
P missing **11.** 9 **13.** 305 **15.** 115 **17.** 67.10 **19.** 12% **21.** $525 **23.** $2,500 **25.** 21.14%
27. 13.17% **29.** 5% **31.** $104.87

Chapters 1–3 Cumulative Practice Test

1. 14   **3.** $P = 128$ cm   **5.** 42.82   **7.** Factored form $= (2)(2)(3)(5)(7)$; Exponential notation $= 2^2(3)(5)(7)$

9. 81.76   **11.** 454.11938   **13.** $19\frac{1}{8}$ **15.** 2   **17.** $1\frac{4}{17}$ **19.** 12.25   **21.** 1.4   **23.** 60

25. $10\frac{5}{12}$ ft or 10 ft 5 in.   **27.** 1,775 in.   **29.** $R = 83\%$ (rounded)   **31.** Percent of increase $= 7.14\%$

33. 86%   **35.** 44.4% protein

Section 4–1 Self-Study Exercises

1   **1.** (a) 1,000 m   (b) 10 L   (c) $\frac{1}{10}$ of a gram   (d) $\frac{1}{1,000}$ of a meter   (e) 100 g   (f) $\frac{1}{100}$ of a liter

2. b   **3.** b   **4.** c   **5.** b   **6.** a   **7.** c   **8.** b   **9.** c   **10.** a   **11.** b   **12.** c   **13.** b   **14.** a

15. b   **16.** b

2   **17.** 40   **18.** 70   **19.** 580   **20.** 80   **21.** 2.5   **22.** 210   **23.** 85   **24.** 142   **25.** 153 mL
26. 460 m   **27.** 75 dkg   **28.** 160 mm   **29.** 400   **30.** 8,000   **31.** 58,000   **32.** 800   **33.** 250
34. 2,100   **35.** 102,500   **36.** 8,330   **37.** 2,000,000   **38.** 70   **39.** 236 L   **40.** 467 cm
41. 38,000 dg   **42.** 13,000 cm   **43.** 2.8   **44.** 23.8   **45.** 10.1   **46.** 6   **47.** 2.9   **48.** 19.25
49. 1.7   **50.** 438.9 dm   **51.** 4.7 g   **52.** 0.225 dL   **53.** 2.743   **54.** 0.385   **55.** 0.15   **56.** 0.08
57. 2,964.84   **58.** 0.2983   **59.** 0.0003   **60.** 0.004   **61.** 0.002857   **62.** 15.285   **63.** 0.0297 hm
64. 0.00003 L

3   **65.** 11 m   **66.** 12 hL   **67.** 6 cg   **68.** 2.4 dm or 24 cm   **69.** 5.9 cL or 59 mL   **70.** cannot add
71. 10.1 kL or 101 hL   **72.** 0.55 g or 55 cg   **73.** cannot subtract   **74.** 7.002 km or 7,002 m   **75.** 1,000 mL
76. 1.47 kL or 147 dkL   **77.** 516 m   **78.** 40.8 m   **79.** 150.96 dm   **80.** 969.5 m   **81.** 4,680 mm
82. 16 g   **83.** 13 m   **84.** 9 cL   **85.** 163 g   **86.** 0.4 m or 4 dm   **87.** 5   **88.** 30   **89.** 16 prescriptions
90. 80 containers   **91.** 5.74 dL or 57.4 cL   **92.** 2.3 dkm or 23 m   **93.** 165.7 hm or 16.57 km
94. 9.1 kL or 91 hL   **95.** 1.25 cL   **96.** 70 mm   **97.** 7.5 dm   **98.** 50 mL   **99.** 21,250 containers
100. 19 vials

Section 4–2 Self-Study Exercises

1   **1.** 1.5 days   **2.** 9.67 min   **3.** 150 min   **4.** 318 s   **5.** 210 min   **6.** 3.03 min or 3 min 2 s
7. 432 min   **8.** 1.47 min or 1 min 28 s

2   **9.** 4,320 lb/h   **10.** 2,400 gal/h   **11.** 10,950 lb/yr   **12.** 1,680 vehicles/h   **13.** 0.0167 mi/s
14. 2.0833 gal/min   **15.** 60 lb/min   **16.** 1.6 gal/s   **17.** 50,000 species   **18.** 215,000 acres

3   **19.** 35   **20.** 0   **21.** 45   **22.** 5   **23.** 15   **24.** 10   **25.** 65   **26.** 50   **27.** 80   **28.** 120
29. 158   **30.** 59   **31.** 113   **32.** 122   **33.** 68   **34.** 419   **35.** 590   **36.** 770   **37.** 365   **38.** 32
39. 32.2 to 33.9°C   **40.** 35 to 37°C   **41.** 90°F   **42.** 68°F

4   **43.** 500,000,000 Hz   **44.** 0.42 H   **45.** 1,400,000 W   **46.** 20 images   **47.** 0.045 s   **48.** 0.805 s

Section 4–3 Self-Study Exercises

1 **1.** 354.33 in. **2.** 131.232 yd **3.** 26.0988 mi **4.** 6.3402 qt **5.** 9.463 L **6.** 59.5242 lb
7. 22.68 kg **8.** 17.78 cm **9.** 5.4864 m **10.** 1.3188 oz **11.** 21.7649 L **12.** 23.9498 ft
13. 5.0000 in. **14.** 0.4724 in. **15.** 146.029 mi **16.** 310.7 mi **17.** 711.2 mm **18.** 91.44 cm
19. 1.2192 m **20.** 1,609.344 m **21.** 91.44 m **22.** 321.86 km **23.** 52.835 qt **24.** 94.63 L
25. 0.8472 oz **26.** 44.092 lb **27.** 68.04 kg **28.** 20.412 kg **29.** 56.6572 g **30.** 14.1643 g
31. 6.6138 lb **32.** 30.48 m **33.** 27.216 kg **34.** 241.395 km **35.** 32.808 yd **36.** 236.132 mi
37. 7.4568 mi **38.** 11.3556 L **39.** 328.08 ft **40.** 22 **41.** 18

Section 4–4 Self-Study Exercises

1 **1.** 3 **2.** 4 **3.** 2 **4.** 4 **5.** 4 **6.** 2 **7.** 3 **8.** 5 **9.** 5 **10.** 3

2 **11.** $\frac{1}{32}$ in. **12.** 0.05 mm **13.** $\frac{1}{8}$ in. **14.** $\frac{1}{2}$ oz or 0.5 oz **15.** $\frac{1}{2}$ L or 0.5 L **16.** $\frac{1}{16}$ in.

17. $\frac{1}{64}$ in. **18.** $\frac{1}{16}$ oz **19.** 0.05 mi **20.** 0.05 cm **21.** 0.005 dg **22.** 0.05 cg **23.** 0.05 cL

24. 0.05 km **25.** 0.05 km **26.** 0.05 km

3 **27.** absolute error = 0.02 cm **28.** absolute error = 0.2 mm **29.** absolute error = 1.5 cm
relative error = 0.00038 relative error = 0.0038 relative error = 0.0308
percent error = 0.038% percent error = 0.38% percent error = 3.08%

30. absolute error = 0.29 in. **31.** absolute error = 0.2 L **32.** absolute error = 0.4 in.
relative error = 0.0204 relative error = 0.0040 relative error = 0.0267
percent error = 2.04% percent error = 0.40% percent error = 2.67%

4 **33.** 215.4 m **34.** 64.4 g **35.** 600 cm or 6 m **36.** 18.11 kg **37.** 900,000 m³ **38.** 17,000,000 m³
39. 43 **40.** 70

5 **41.** 115 mm or 11.5 cm **42.** 102 mm or 10.2 cm **43.** 96 mm or 9.6 cm **44.** 85 mm or 8.5 cm
45. 57 mm or 5.7 cm **46.** 50 mm or 5 cm **47.** 44 mm or 4.4 cm **48.** 30 mm, 30.5 mm, or 31 mm
(3.0 cm or 3.1 cm) **49.** 19 mm or 1.9 cm **50.** 10 mm or 1.0 cm **51.** 8.3 cm **52.** 5.1 cm **53.** 2.8 cm
54. 7.0 cm

6 **55.** 3 in. **56.** $2\frac{3}{4}$ in. **57.** 7.5 cm **58.** 13.6 cm **59.** $11\frac{1}{4}$ in. **60.** $10\frac{7}{8}$ in. **61.** 4.4 cm

62. 4.9 cm **63.** 7.9 cm **64.** 4.1 cm **65.** $11\frac{3}{8}$ in. **66.** $5\frac{7}{8}$ in.

Chapter Review Exercises, Chapter 4

1. kilo- **3.** milli- **5.** centi- **7.** 10 times **9.** $\frac{1}{1,000}$ of **11.** 1,000 times **13.** a **15.** a **17.** c

19. b **21.** 6.71 dkm **23.** 2,300 mm **25.** 12,300 mm **27.** 230,000 mm **29.** 413.27 km
31. 3.945 hg **33.** 30.00974 kg **35.** cannot add unlike measures **37.** 748 cg or 7.48 g **39.** 61.47 cg
41. 15 **43.** 8.5 hL **45.** 18.9 m **47.** 245 mL or 24.5 cL **49.** 6 m **51.** 100 servings **53.** 3 da
55. 2 h 38 min **57.** 4 da **59.** 3 yr 3 mo **61.** 59°F **63.** 203°F **65.** 104°F **67.** 185°C **69.** 100.2°F
71. 235.124 yd **73.** 15.8505 liq qt **75.** 70.5472 lb **77.** 22.86 cm **79.** 156.3916 qt **81.** 60.96 m

83. 281.6275 km   **85.** 21   **87.** 6   **89.** 4   **91.** $\frac{1}{4}$ **93.** $\frac{1}{32}$ **95.** 0.05 cm   **97.** 0.05 cm   **99.** 1%
101. 4 significant digits   **103.** 0.05 cg   **105.** 46.5 m or 4,650 cm   **107.** 117 mm or 118 mm   **109.** 99 mm

111. 60 mm   **113.** 45 mm   **115.** 20 mm   **117.** $5\frac{1}{4}$ in.   **119.** $4\frac{27}{32}$ in.   **121.** 2.7 cm   **123.** $20\frac{1}{4}$ in.
125. 5.6 cm   **127.** 7.4 cm

Practice Test, Chapter 4

1. 0.298 km   **3.** 9.48 L or 94.8 dL   **5.** 120.6975 km   **7.** 8.4536 pt   **9.** 9°C   **11.** 2,520 min
13. 235 h   **15.** 2   **17.** relative error = 0.0012; percent error = 0.12%   **19.** 99 mm or 9.9 cm
21. 746 m/s   **23.** 4.9 cm

Section 5–1 Self-Study Exercises

1 **1.** <   **2.** <   **3.** >   **4.** <   **5.** <   **6.** >   **7.** <   **8.** >   **9.** 7   **10.** 17   **11.** 8   **12.** 7
13. −42   **14.** 17   **15.** 78   **16.** −57

2 **17.** 23   **18.** −16   **19.** −13   **20.** −29   **21.** −55   **22.** −124   **23.** 29   **24.** −20   **25.** 99
26. −59   **27.** −48   **28.** −161   **29.** −53   **30.** 232   **31.** −1,005

3 **32.** 2   **33.** 2   **34.** 4   **35.** −4   **36.** 6   **37.** −6   **38.** −2   **39.** 4   **40.** −2   **41.** 2
42. −10   **43.** −40   **44.** 7   **45.** 15   **46.** −8   **47.** −33   **48.** 9   **49.** −11   **50.** 13   **51.** 9
52. 3   **53.** 0   **54.** 0   **55.** 0   **56.** 0   **57.** −3   **58.** −7   **59.** 0   **60.** −3   **61.** −12   **62.** −14
63. 13   **64.** −12   **65.** $24 million   **66.** −$98 million   **67.** $2 million   **68.** 10   **69.** $66   **70.** $36
71. −295°C   **72.** −23°F

Section 5–2 Self-Study Exercises

1 **1.** −12   **2.** 6   **3.** −6   **4.** 4   **5.** −25   **6.** −3   **7.** 8   **8.** −3   **9.** −9   **10.** 13   **11.** −1
12. −8   **13.** 62   **14.** −11   **15.** −58   **16.** −57   **17.** −15   **18.** −105   **19.** 18   **20.** −17
21. −167   **22.** 83   **23.** 1,413   **24.** −103   **25.** 15   **26.** −8   **27.** −12   **28.** 8   **29.** 7   **30.** 10
31. 56   **32.** −92   **33.** 14   **34.** −36   **35.** 0   **36.** 0

2 **37.** 4   **38.** 7   **39.** 6   **40.** −4   **41.** 1   **42.** 14   **43.** −3   **44.** −9   **45.** −5   **46.** 13
47. 2   **48.** −1   **49.** 1   **50.** −20   **51.** 26   **52.** −99   **53.** −45   **54.** −32   **55.** −60   **56.** −80
57. −35   **58.** 50   **59.** −21   **60.** 186   **61.** 107°F   **62.** 107°F   **63.** $115,054   **64.** Subtracting
zero from a number results in the same number with the same sign. Subtracting a number from zero results in the
opposite of the number.   **65.** −$8 million   **66.** $4.6 million   **67.** $5.5 million   **68.** $21.6 million

Section 5–3 Self-Study Exercises

1 **1.** 40   **2.** 12   **3.** 35   **4.** 21   **5.** 24   **6.** 6   **7.** −15   **8.** −10   **9.** −32   **10.** −12
11. −56   **12.** −24   **13.** 4(−$28) = −$112   **14.** (7)(−$40) = −$280   **15.** 336   **16.** −60   **17.** 36
18. 0   **19.** 90   **20.** 0   **21.** −42   **22.** 54   **23.** −210   **24.** 2,268   **25.** −20,160   **26.** 840
27. −240   **28.** 0   **29.** 0   **30.** 0   **31.** 0   **32.** 0   **33.** 0   **34.** 0   **35.** 0   **36.** 0   **37.** 0
38. −30   **39.** 168   **40.** 42   **41.** −8   **42.** 0   **43.** −16   **44.** The multiplicative inverse of a number is
the number that, when multiplied by the original number, results in 1, the multiplicative identity.   **45.** Answers
will vary, 5(−3) = −3(5) = −15.

2 **46.** 9 **47.** −8 **48.** 25 **49.** 0 **50.** −8 **51.** −512 **52.** 625 **53.** 81 **54.** 2,401
55. 1,764 **56.** 5 **57.** −28 **58.** 12 **59.** −8 **60.** −10 **61.** −8 **62.** −27 **63.** −121
64. 121 **65.** −125

3 **66.** −3 **67.** −4 **68.** −4 **69.** −4 **70.** 8 **71.** 5 **72.** −6 **73.** 8 **74.** 5

75. $-\$1,800; \dfrac{-1,800}{6} = -\$300$ per month **76.** −6°C per hour **77.** undefined **78.** undefined **79.** 0

80. 0 **81.** 0 **82.** undefined **83.** undefined **84.** undefined **85.** 0 **86.** undefined
87. multiplication and division **88.** −10°F per hour **89.** $3,472.50 per month **90.** The numerator must be zero and the denominator must be any number except zero.

Section 5–4 Self-Study Exercises

1 **1.** $-\dfrac{-5}{8}, -\dfrac{5}{-8}, \dfrac{-5}{-8}$ **2.** $-\dfrac{-3}{-4}, \dfrac{-3}{4}, \dfrac{3}{-4}$ **3.** $-\dfrac{2}{-5}, -\dfrac{-2}{5}, \dfrac{2}{5}$ **4.** $-\dfrac{7}{8}, \dfrac{-7}{8}, \dfrac{7}{-8}$ **5.** $-\dfrac{-7}{8}, -\dfrac{7}{-8}, \dfrac{-7}{-8}$

2 **6.** $-\dfrac{1}{4}$ **7.** $-1\dfrac{1}{10}$ or $-\dfrac{11}{10}$ **8.** $-\dfrac{1}{16}$ **9.** $\dfrac{3}{16}$ **10.** $-\dfrac{2}{9}$ **11.** $-10\dfrac{5}{8}$ or $-\dfrac{85}{8}$ **12.** $-2\dfrac{13}{24}$ or $-\dfrac{61}{24}$

13. $-3\dfrac{7}{8}$ or $-\dfrac{31}{8}$ **14.** $1\dfrac{1}{10}$ or $\dfrac{11}{10}$ **15.** $-1\dfrac{3}{8}$ or $-\dfrac{11}{8}$ **16.** $-\dfrac{1}{16}$ **17.** $2\dfrac{17}{24}$ or $\dfrac{65}{24}$ **18.** $-\dfrac{1}{2}$ **19.** $\dfrac{6}{11}$

20. $\dfrac{10}{27}$ **21.** 30 **22.** $-\dfrac{1}{32}$ **23.** −2 **24.** $-\dfrac{1}{10}$ **25.** $-\dfrac{1}{7}$ **26.** $-\dfrac{25}{32}$ **27.** −26.297 **28.** −1.11

29. −91.44 **30.** −110.72 **31.** −59.04 **32.** 340.71 **33.** −27.73 **34.** 0.41 **35.** 4.6 **36.** 0.03
37. −2.3 **38.** −5.5

3 **39.** −9 **40.** 20 **41.** 24 **42.** −1 **43.** −13 **44.** 21 **45.** −13 **46.** 13 **47.** 15 **48.** 8
49. 24 **50.** −5 **51.** −7 **52.** −3 **53.** −28 **54.** −12 **55.** −1,260 **56.** 27 **57.** −14

58. 28 **59.** −35 **60.** −29 **61.** −132 **62.** $-\dfrac{1}{5}$ **63.** $-5\dfrac{7}{8}$ **64.** −7 **65.** $-3\dfrac{3}{5}$ **66.** $\dfrac{37}{144}$

67. $-\dfrac{127}{245}$ **68.** −10.38 **69.** −3.992 **70.** −11.88 **71.** 18 **72.** 102 **73.** 128 **74.** 46

Section 5–5 Self-Study Exercises

1 **1.** 45,300 **2.** 27 **3.** 5,820 **4.** 0.897 **5.** 5.23 **6.** 0.00806 **7.** 0.573 **8.** 0.00293
9. 0.0457 **10.** 857.9 **11.** 4,370 **12.** 8,370 **13.** 37 **14.** 1,820 **15.** 0.56 **16.** 1.42
17. 780,000 **18.** 62 **19.** 0.00046 **20.** 0.61 **21.** 0.72 **22.** 42 **23.** 10^{12} **24.** 10 **25.** 10^2

26. 10^4 **27.** 1 **28.** 10^{-2} or $\dfrac{1}{10^2}$ **29.** 10^{-12} or $\dfrac{1}{10^{12}}$ **30.** 10^{-6} or $\dfrac{1}{10^6}$ **31.** 10^{10} **32.** 10^{-7} or $\dfrac{1}{10^7}$

33. 10^{-3} or $\dfrac{1}{10^3}$ **34.** 10 **35.** 10^3 **36.** 10^2 **37.** 10^{-3} or $\dfrac{1}{10^3}$ **38.** 1 **39.** 10^{-5} or $\dfrac{1}{10^5}$

40. 10^{-1} or $\dfrac{1}{10}$

2 **41.** 10^6 **42.** 10^{12} **43.** 10^{-4} or $\dfrac{1}{10^4}$ **44.** 10^{10} **45.** 10^{-8} or $\dfrac{1}{10^8}$

Section 5–6 Self-Study Exercises

1 **1.** 430 **2.** 0.0065 **3.** 2.2 **4.** 73 **5.** 0.093 **6.** 83,000 **7.** 0.0058 **8.** 80,000 **9.** 6.732
10. 0.00589 **11.** 78.3 **12.** 1,590 **13.** 397,000 **14.** 0.0004723 **15.** 0.00000991 **16.** 1,030,000

2 **17.** 3.92×10^2 **18.** 2×10^{-2} **19.** 7.03×10^0 **20.** 4.2×10^4 **21.** 8.1×10^{-2} **22.** 2.1×10^{-3}
23. 2.392×10^1 **24.** 1.01×10^{-1} **25.** 1.002×10^0 **26.** 7.21×10^2 **27.** 4.2×10^5 **28.** 3.26×10^4
29. 2.13×10^1 **30.** 6.2×10^{-6} **31.** 5.6×10^1 **32.** 1.97×10^{-6} **33.** 7.45×10^1 **34.** 1.8×10^4
35. 7.01×10^1 **36.** 7.25×10^{-1}

3 **37.** 2.144×10^7 **38.** 5.6×10^3 **39.** 2.36×10^{-2} **40.** 4.73×10^{-3} **41.** 7×10^2
42. 6.5×10^{-5} **43.** 7×10^2 **44.** 8×10^{-3} **45.** 2.45×10^1 **46.** 3.2285×10^{13} mi **47.** 4.2×10^2 Å

4 **48.** 5.7×10^8 **49.** 2×10^{-5} **50.** 3.72×10^{-8} **51.** 5.0×10^8 **52.** 3.6×10^{-19} **53.** 4×10^8
54. 5×10^{10} **55.** 7×10^{-1} **56.** 2.7×10^9 **57.** 5.9×10^7 **58.** 2.9×10^8 **59.** 1.9×10^{17}
60. 2×10^0 **61.** 2×10^4 **62.** 1×10^9 **63.** 2.9×10^8 **64.** 2.3×10^{-9} **65.** 3.8×10^6
66. 5.6×10^3 **67.** 78×10^0 **68.** 52×10^3 **69.** 80×10^6 **70.** 5.83×10^6 **71.** 1.7365×10^9
72. 41.98×10^6 **73.** 780×10^{-3} **74.** 330×10^{-3} **75.** 1.1×10^{-6} **76.** 8×10^{-9}
77. 983.2×10^{-6} **78.** 71.9×10^{-6} **79.** 12.0307×10^{-3} **80.** 675×10^{-6} **81.** 428 kΩ **82.** 5.7 MV
83. 3.52 GW **84.** 79 MHz **85.** 81 μs **86.** 97.3 mÅ **87.** 5.41 ns **88.** 890 ms **89.** 580 Ω
90. 770 ps **91.** 2.98 μs **92.** 7.81 kW **93.** 42.3 MV **94.** 1.572 GW **95.** 5.096 MΩ **96.** 8 nÅ
97. 182 kHz **98.** 1.6 kΩ **99.** 5.2 MV **100.** 97 MW

Chapter Review Exercises, Chapter 5

1. 5 **3.** 7 **5.** 12 **7.** 2 **9.** −87 **11.** −7 **13.** −6 **15.** gain of 8 yd; +8 yd **17.** $569
19. −13 **21.** 14 **23.** −15 **25.** 70°F **27.** −5.4°F **29.** 36°C **31.** −$298 million
33. −$133 million **35.** −14 **37.** 0 **39.** −168 **41.** 343 **43.** −16 **45.** 25 **47.** −10°
49. higher, 20, $-4 \times 5 = -20$ points **51.** 4 **53.** 4 **55.** 17 **57.** undefined **59.** −3
61. 56 **63.** 1 **65.** 73 **67.** $1\frac{5}{7}$ or $\frac{12}{7}$ **69.** −62 **71.** $-1\frac{7}{24}$ or $-\frac{31}{24}$ **73.** $-11\frac{13}{16}$ or $-\frac{189}{16}$
75. 23.76 **77.** 1,000,000,000,000 **79.** $\frac{1}{1,000}$ or 0.001 **81.** 8,730 **83.** 375,000 **85.** 0.0000387
87. 5.2×10^4 **89.** 1.7×10^{-4} **91.** 8×10^{-9} **93.** 3.4×10^{10} (rounded to tenths) **95.** 5.7×10^{-4}
97. 2.5×10^8 **99.** 920×10^{-9} **101.** 8.4×10^6 **103.** 41×10^0 **105.** 17×10^6
107. 3.084×10^9 **109.** 1.8×10^{-6} **111.** 7×10^{-3} **113.** 350×10^{-12} **115.** 490 μs
117. 588 mÅ **119.** 246.7 V **121.** 42 W **123.** 5.729 mW **125.** 4.8 THz

Practice Test, Chapter 5

1. < **3.** > **5.** −8 **7.** 4 **9.** −48 **11.** 2.1 **13.** 0 **15.** undefined **17.** −2 **19.** 4
21. 10^6 **23.** 42,000 **25.** 0.059 **27.** 2.1×10^{-2} **29.** 7.83×10^{-3} **31.** 3.5×10^2
33. 1,500,000 ohms **35.** 175°F **37.** 4.7 MΩ **39.** 23 ys

Section 6–1 Self-Study Exercises

1 **1.** 10% **2.** 36.8% **3.** 54.3% **4.** 20.8% **5.** 33.3% **6.** 75% **7.** 230,000,000 bushels
8. 1,150,000,000 bushels **9.** 34,776,000,000 lb **10.** 47 bushels/acre **11.** routine discharges
12. 17.5%

2 **13.** debt retirement **14.** misc. expenses and general government **15.** social projects and education costs **16.** 70,000,000 barrels **17.** approximately 10% increase **18.** approximately 20,000,000 gal **19.** approximately 20% increase **20.** 1975–1980 or 1995–2000 **21.** 1970 **22.** 7,355,000 barrels; 5,834,000 barrels **23.** 445.8% **24.** 2000

3 **25.** 5 A **26.** 50 V **27.** 35 V **28.** 25 Ω **29.** 1980 **30.** motor gasoline **31.** 1960–1970 **32.** 1960–1970 **33.** 20.3% **34.** beef cows and heifers **35.** feeder cattle **36.** 1,200,000 metric tons **37.** approximately 500% increase **38.** 1985–1990 **39.** dramatically increasing **40.** fluctuating with slight overall increase **41.** about 200% increase **42.** 1950–1960 and 1970 to 1980

Section 6–2 Self-Study Exercises

1 **1.** 16 **2.** 17 **3.** 66 **4.** 73.75 **5.** 33.7 **6.** 66.6 **7.** 42.33°F **8.** 12.67°C **9.** $34.80 **10.** $39.20 **11.** 14.67 in. **12.** 8 in. **13.** 20 **14.** 76 **15.** 17.3 runs **16.** 98 **17.** $24.6 \frac{\text{mi}}{\text{gal}}$ **18.** $10.2 \frac{\text{mi}}{\text{gal}}$ **19.** 737 thousand tons **20.** 1,300 million tons

2 **21.** 44 **22.** 43 **23.** 15.5 **24.** 26.5 **25.** $30 **26.** $66 **27.** $8.25 **28.** $8.85 **29.** 745 thousand **30.** 1,238 million **31.** 2 **32.** 5 **33.** no mode **34.** no mode **35.** $67 **36.** $32 **37.** 4 h **38.** $1.97 **39.** no mode **40.** no mode

3 **41.** 10 **42.** 2 **43.** $\frac{1}{3}$ **44.** $\frac{1}{7}$ **45.** 12% **46.** 28% **47.** 35–37 and 38–40 **48.** 20–22 and 23–25 **49.** 7 **50.** 16

	Midpoint	Tally	Class Frequency
51.	15	||	2
52.	12	||||	4
53.	9	ⅢⅢ ||	7
54.	6	ⅢⅢ ⅢⅢ ⅢⅢ ⅢⅢ	20

	Midpoint	Tally	Class Frequency
55.	93	||	2
56.	88	ⅢⅢ	5
57.	83	ⅢⅢ ||	7
58.	78	ⅢⅢ ⅢⅢ	10
59.	73	|||	3
60.	68	ⅢⅢ |	6
61.	63	||	2
62.	58	ⅢⅢ	5

	Class Interval	Midpoint	Tally	Class Frequency
63.	15–19	17	ⅢⅢ ||	7
	10–14	12	ⅢⅢ |||	8
	5–9	7	ⅢⅢ ||||	9
	0–4	2	ⅢⅢ |	6

	Class Interval	Midpoint	Tally	Class Frequency
64.	51–60	55.5	||	2
	41–50	45.5	||||	4
	31–40	35.5	||	2
	21–30	25.5	⫴| ||	7
	11–20	15.5	⫴|	5

	Class Interval	Midpoint	Tally	Class Frequency
65.	801–900	850.5	||	2
	701–800	750.5	⫴|	5
	601–700	650.5	|||	3

	Class Interval	Midpoint	Tally	Class Frequency
66.	1,751–2,000	1,875.5	||	1
	1,501–1,750	1,625.5	⫴|	5
	1,251–1,500	1,375.5	||||	4
	1,001–1,250	1,125.5	⫴| ||	7
	751–1,000	875.5	|||	3

67. 25 students   **68.** $8   **69.** 76 students   **70.** 10 h

Section 6–3 Self-Study Exercises

1 **1.** 24   **2.** 23   **3.** 13   **4.** 18   **5.** $25   **6.** $33   **7.** 48°F   **8.** 115,660 T   **9.** 132,600 T
10. 86,650 T   **11.** 66,000 T   **12.** 64,300 T   **13.** 1,155 T

2 **14.** 3.16   **15.** 8.29   **16.** 11.06°F   **17.** $17.36   **18.** 4   **19.** 8   **20.** 19,560   **21.** 18,480 T
22. 332 T   **23.** 15.87%   **24.** approximately 32 patients   **25.** 2.28%   **26.** approximately 2 pediatricians
27. 2.28%   **28.** approximately 11 patients

Section 6–4 Self-Study Exercises

1 **1.** Keaton Brienne Renee
Keaton Renee Brienne
Brienne Keaton Renee
Brienne Renee Keaton
Renee Keaton Brienne
Renee Brienne Keaton
$3 \cdot 2 \cdot 1 = 6$ ways

2. ABCD   BACD
ABDC   BADC
ACBD   BCAD
ACDB   BCDA
ADBC   BDAC
ADCB   BDCA
CABD   DABC
CADB   DACB
CBAD   DBAC
CBDA   DBCA
CDAB   DCAB
CDBA   DCBA
$4 \cdot 3 \cdot 2 \cdot 1 = 24$ ways

3. $4 \cdot 3 \cdot 2 \cdot 1 = 24$ ways   **4.** $5 \cdot 4 \cdot 3 \cdot 2 \cdot 1 = 120$ ways   **5.** $5 \cdot 4 \cdot 3 \cdot 2 \cdot 1 = 120$ ways   **6.** 60 ways

2 **7.** $\frac{1}{24}, \frac{1}{23}$ **8.** $\frac{1}{4}$ **9.** $\frac{11}{48}$ **10.** $\frac{2}{5}$ **11.** $\frac{1}{3}$ **12.** $\frac{1}{6}$

Chapter Review Exercises, Chapter 6

1. 2001, 2003, 2004 **3.** 2002, 2005, 2006 **5.** 12.9% **7.** 17.4% **9.** 7-10-2005 @ 4:00 P.M.
11. 6,575,000 barrels **13.** 27.4% **15.** 1997 **17.** 1994–1995 **19.** 12% **21.** 2003

23. 12.6 cars **25.** 13.8 $\frac{\text{mi}}{\text{gal}}$ **27.** $10.03 **29.** $1.85 **31.** 3.67 **33.** 83

	Midpoint	Tally	Class Frequency
35.	60.5	ҧҦ ҧҦ	10
37.	40.5	ҧҦ ҧҦ ‖	12
39.	20.5	ҧҦ ‖	7

41. 36–45 **43.** 15 **45.** $\frac{1}{2}$ **47.** $\frac{10}{54} = 18.5\%$

49.

Miles per Gallon	Midpoint	Tally	Frequency
20–24	22	ҧҦ ‖	7
25–29	27	ҧҦ ‖	6
30–34	32	‖‖	4

51. 26.1 **53.** range: 59 **55.** 21.47 **57.** approximately 841 men **59.** approximately 476 containers

61. $\frac{10}{13}; \frac{3}{4}$ **63.** 16 combinations

Practice Test, Chapter 6

1. bar **3.** line **5.** 2°C **7.** $65 **9.** 25% **11.** $\frac{15}{200} = \frac{3}{40}$ **13.** English dept., electronics dept.

15. approximately 104% **17.** $\frac{1}{2}$ **19.** range: 25 **21.** 8.49 **23.** 6 **25.** $\frac{3}{5}$ **27.** 34.13%
mean: 77.9
median: 78
mode: 81

Chapters 4–6 Cumulative Practice Test

1. 1.43 m **3.** 3.0002 kg **5.** $C = 30°$ **7.** $\frac{1}{8}$ cm **9.** 3 significant digits **11.** 30.21 cm rounds to 30.2 cm

13. -14 **15.** -6 **17.** $-1\frac{1}{4}$ **19.** 0.053 **21.** 3.5×10^{-3} **23.** 145.7×10^{-6} **25.** 2×10^{6}

27. 64,000 cwt **29.** Mississippi and Missouri **31.** China **33.** Mean, 489; median, 477; mode, 456

35. $p = \frac{4}{15}$

Section 7–1 Self-Study Exercises

1 **1.** $-15 = -15$ **2.** $11 = 11$ **3.** $2 = 2$ **4.** $-7 = -7$ **5.** $-35 = -35$ **6.** $-34 = -34$
7. $n = 11$ **8.** $m = 6$ **9.** $y = -4$ **10.** $x = 6$ **11.** $p = 9$ **12.** $b = -9$ **13.** $\boxed{7} + \boxed{c}$
14. $\boxed{4a} - \boxed{7}$ **15.** $\boxed{3x} - \boxed{2(x+3)}$ **16.** $\boxed{\dfrac{a}{3}}$ **17.** $\boxed{7xy} + \boxed{3x} - \boxed{4} + \boxed{2(x+y)}$ **18.** $\boxed{14x} + \boxed{3}$

19. $\boxed{\dfrac{7}{(a+5)}}$ **20.** $\boxed{\dfrac{4x}{7}} + \boxed{5}$ **21.** $\boxed{11x} - \boxed{5y} + \boxed{15xy}$ (Answers may vary.)   **22.** 5   **23.** -4 **24.** $\dfrac{1}{5}$

25. $\dfrac{2}{7}$ **26.** 6   **27.** $-\dfrac{4}{5}$ **28.** 7   **29.** -1 **30.** $-15c$ (Answers may vary.)

2 **31.** Four more than a number is 7. (Answers may vary.)   **32.** Five less than a number is 2. (Answers may vary.)   **33.** Three times a number is 15. (Answers may vary.)   **34.** One more than 3 times a number is 7. (Answers may vary.)

3 **35.** $x + 5 = 12$ **36.** $\dfrac{x}{6} = 9$ **37.** $4(x - 3) = 12$ **38.** $4x - 3 = 12$ **39.** $12 + 7 + x = 17$
40. $2x + 7 = 21$ **41.** $x + 15 = 48$ **42.** $x + 15$ mL $= 45$ mL   **43.** $x \cdot 5 = 45$
44. $\dfrac{18 - 3}{5} = x$ or $5x = 18 - 3$

4 **45.** $-3a$ **46.** $-8x - 2y$ **47.** $7x - 2y$ **48.** $-3a + 6$ **49.** 10   **50.** $3a + 6b + 8c + 1$
51. $10x - 20$ **52.** $6x + 13$ **53.** $-4x + 5$ **54.** $-12a + 3$ **55.** $-5x + 5y - 10z$ **56.** $-2 - 2a + 3b$
57. $-13x + 25y - 15$ **58.** $-6x - 20y + 20$

Section 7–2 Self-Study Exercises

1 **1.** $x = 8$ **2.** $x = 15$ **3.** $x = -3$ **4.** $x = -8$ **5.** $x = 4$ **6.** $x = -11$ **7.** $x = -26$ **8.** $x = 3$
9. $x = 5$ **10.** $x = -7$ **11.** $x = 16$ **12.** $x = 9$ **13.** $x = -2$ **14.** $x = 4$ **15.** $x = 6$ **16.** $x = 9$

2 **17.** $x = 8$ **18.** $x = 9$ **19.** $x = 5$ **20.** $n = 30$ **21.** $a = -9$ **22.** $b = -2$ **23.** $c = 7$
24. $x = -9$ **25.** $x = \dfrac{32}{5}$ **26.** $y = \dfrac{9}{2}$ **27.** $n = -\dfrac{5}{2}$ **28.** $b = \dfrac{28}{3}$ **29.** $x = -7$ **30.** $y = 2$

31. $x = 15$ **32.** $y = 8$ **33.** $x = \dfrac{10}{3}$ **34.** $y = -\dfrac{20}{3}$ **35.** $x = 12$ **36.** $n = 56$

3 **37.** $x = 6$ **38.** $m = 7$ **39.** $a = -3$ **40.** $m = -\dfrac{1}{2}$ **41.** $y = 8$ **42.** $x = 0$ **43.** $y = -7$
44. $y = -4$ **45.** $x = -2$

4 **46.** $x = -6$ **47.** $b = -1$ **48.** $x = 8$ **49.** $t = 6$ **50.** $x = 3$ **51.** $y = 5$ **52.** $x = -\dfrac{7}{2}$
53. $a = -2$ **54.** all real numbers   **55.** no solution   **56.** $x = -8$ **57.** $t = 7$ **58.** $x = -1$
59. $y = 11$ **60.** $y = -2$ **61.** $x = 27$ **62.** $y = 13$ **63.** $R = -68$ **64.** $P = \dfrac{5}{6}$ **65.** $x = 14$
66. $x = -\dfrac{4}{15}$ **67.** $m = \dfrac{49}{18}$ **68.** $s = \dfrac{8}{5}$ **69.** $m = \dfrac{8}{3}$ **70.** $T = \dfrac{68}{9}$ **71.** $x = 2$ **72.** $x = 2$
73. $R = 0.8$ **74.** $x = 6.03$ **75.** $x = 0.08$ **76.** $R = 0.34$

Section 7–3 Self-Study Exercises

1. $x = 7$ **2.** $y = 2$ **3.** $x = -5$ **4.** $x = -2$ **5.** $x = 5$ **6.** $x = -6$ **7.** $x = 1$ **8.** $x = 8$
9. $x = 1$ **10.** $x = 0$ **11.** no solution **12.** all real numbers **13.** no solution **14.** no solution

15. $x = 2$ **16.** $x = 5$ **17.** $x = \dfrac{1}{2}$ **18.** $x = -15$ **19.** $x = -9$ **20.** $x = 1$ **21.** $x = -2$

22. $x = -1$ **23.** $x = \dfrac{16}{5}$ **24.** $x = 5$ **25.** $x + 4 = 12; x = 8$ **26.** $2x - 4 = 6; 5$

27. $x + x + (x - 3) = 27$; two parts weigh 10 lb each; third part weighs 7 lb **28.** $24 + x = 60$; 36 gal
29. $4.03 - 3.97 = x$ or $4.03 - x = 3.97$; 0.06 kg **30.** $x + 3x = 400$; 100 ft of solid pipe; 300 ft of perforated pipe **31.** $x + 2x + 70 = 235$; Spanish text: \$55; calculator: \$110 **32.** $C = 14(4)(150) = 8,400$

33. $x + 2x = 325$; tank 1: $216\dfrac{2}{3}$ gal tank 2: $108\dfrac{1}{3}$ gal **34.** $A = 10w; w = 6$ ft

Section 7–4 Self-Study Exercises

1. 28 **2.** -11 **3.** $\dfrac{18}{7}$ **4.** 0 **5.** $-\dfrac{27}{5}$ **6.** $-\dfrac{5}{27}$ **7.** $\dfrac{8}{5}$ **8.** $\dfrac{11}{15}$ **9.** -350 **10.** 48

11. 27 **12.** $\dfrac{1}{3}; Q \neq 0$ **13.** 13 **14.** $\dfrac{25}{8}; p \neq 4$ **15.** 0 **16.** $\dfrac{1}{9}; P \neq 0$ **17.** 576 **18.** 4 **19.** 28

20. -68 **21.** $\dfrac{27}{35}$ **22.** $\dfrac{108}{5}$ **23.** $\dfrac{217}{24}$ **24.** 8 **25.** $-5; R \neq 0$ **26.** $\dfrac{3}{10}$ **27.** 70 **28.** 0 **29.** $\dfrac{32}{63}$

30. $\dfrac{32}{5}$ **31.** $-\dfrac{36}{11}$ **32.** $-\dfrac{1}{108}; x \neq 0$ **33.** 21 **34.** $\dfrac{4}{25}$ **35.** $\dfrac{2}{21}$ **36.** $\dfrac{9}{7}$

37. 80 vases **38.** $1\dfrac{1}{3}$ h **39.** $3\dfrac{1}{13}$ h **40.** $3\dfrac{3}{7}$ days **41.** $\dfrac{3}{5}$ h or 36 min **42.** 3 min **43.** 15 min

44. $2\dfrac{1}{3}$ min **45.** 8.57 Ω **46.** 5 Ω **47.** 7.38 Ω **48.** 3.6 Ω

49. 2 **50.** 0.8 **51.** 6.03 **52.** 16 **53.** 4.7 **54.** 0.08 **55.** 0.035 **56.** 0.34 **57.** -3.3
58. 38.8 **59.** 3.3 **60.** -0.2 **61.** 0.6 **62.** 1.128 **63.** 16 **64.** 87.5 lb **65.** 4.71 in. **66.** 12 h
67. 6.5 h **68.** 8.3 Ω **69.** \$11.30 **70.** 156.25 V **71.** \$264 **72.** \$136 **73.** \$3,325 **74.** 12.25%
75. 1.2%

Section 7–5 Self-Study Exercises

1. 280 mi **2.** \$120 **3.** 2,826 in^2 **4.** 117.6 m^2 **5.** 138 cm **6.** \$310 **7.** 14.25% **8.** 3 years
9. \$2,600 **10.** 90 lb **11.** 8% **12.** \$3,378.38 **13.** 13.6 **14.** 17% **15.** 66 in. **16.** 9,326.6 cm^2
17. \$10.50 **18.** 4 mi **19.** 153.9 in.^{2} **20.** 6.25 km^2 **21.** \$48.75 **22.** 4.8 Ω **23.** 61.6%
24. 35 mi/h **25.** 1.5 A **26.** 6 ft^3 **27.** 3 A **28.** 2 cylinders **29.** 1,600 rpm **30.** 71.71 in.
31. 17 Ω

32. $S = I - I_n$ **33.** $r = \dfrac{S}{2\pi h}$ **34.** $b = y - mx$ **35.** $h = \dfrac{V}{\pi r^2}$ **36.** $C = S - M$ **37.** $R = \dfrac{100\,P}{B}$

38. $b = \dfrac{P - 2s}{2}$ **39.** $r = \dfrac{C}{2\pi}$ **40.** $l = \dfrac{A}{w}$ **41.** $C = \dfrac{R + B}{A}$ **42.** $R = \dfrac{E}{I}$ **43.** $R = \dfrac{D}{T}$

44. $D = P - S$ **45.** $d = \dfrac{C}{\pi}$ **46.** $r = \dfrac{C}{2\pi}$ **47.** $I = A - P$ **48.** $T = \dfrac{I}{PR}$

49. B14 = B5 + B6 + B7 + B8 + B9 + B10 + B11 + B12

D14 = D5 + D6 + D7 + D8 + D9 + D10 + D11 + D12

C5 = B5 ÷ B14 × 100

C6 = B6 ÷ B14 × 100

C7 = B7 ÷ B14 × 100

$\vdots$

C12 = B12 ÷ B14 × 100

E5 = D5 ÷ D14 × 100

E6 = D6 ÷ D14 × 100

E7 = D7 ÷ D14 × 100

$\vdots$

E12 = D12 ÷ D14 × 100

F5 = (D5 − B5) ÷ B5 × 100

F6 = (D6 − B6) ÷ B6 × 100

F7 = (D7 − B7) ÷ B7 × 100

$\vdots$

F12 = (D12 − B12) ÷ B12 × 100

F14 = (D14 − B14) ÷ B14 × 100

50. See the figure.

	A	B	C	D	E	F
1		The 7th Inning: Budget Operating Expenses and Actual Expenses				
2						
3	Expense	Budget Amount	Percent of Total Budget	Actual Expenses	% of Actual Total Expense	% Difference from Budget
4						
5	Salaries	$45,000.00	17.93%	$42,000.00	16.57%	−6.7%
6	Rent	$37,000.00	14.74%	$36,000.00	14.20%	−2.7%
7	Depreciation	$12,000.00	4.78%	$14,000.00	5.52%	16.7%
8	Utilities and phone	$13,000.00	5.18%	$10,862.56	4.28%	−16.4%
9	Taxes and Insurance	$15,000.00	5.98%	$13,583.29	5.36%	−9.4%
10	Advertising	$2,000.00	0.80%	$2,847.83	1.12%	42.4%
11	Purchases	$125,000.00	49.80%	$132,894.64	52.41%	6.3%
12	Other	$2,000.00	0.80%	$1,356.35	0.53%	−32.2%
13						
14	Total	$251,000.00	100.01%	$253,544.67	99.99%	100.0%

Chapter Review Exercises, Chapter 7

1. $12 = 12$ **3.** $8 = 8$ **5.** $x = 14$ **7.** $y = -8$ **9.** $\boxed{15x} - \boxed{\dfrac{3a}{7}} + \boxed{\dfrac{(x-7)}{5}}$ **11.** A number increased by 5

equals 2. Answers will vary. **13.** The quotient of a number and 8 equals 7. Answers will vary.

15. $2x + 7 = 11$ **17.** $2(x + 8) = 40$ **19.** $5a$ **21.** $7y - 12$ **23.** $-3a - 11$ **25.** $-2x + 21$ or $21 - 2x$

27. $11 - 2x + 4y$ **29.** $30a - 21b + 12$ **31.** $x = 13$ **33.** $x = -2$ **35.** $x = 19$ **37.** $x = 3$

39. $x = -1$ **41.** $a = 5$ **43.** $x = 5$ **45.** $x = 7$ **47.** $x = 4$ **49.** $x = -3$ **51.** $x = 7$ **53.** $b = -\dfrac{15}{2}$

55. $x = 15$ **57.** $y = 7$ **59.** $x = 64$ **61.** $x = -49$ **63.** $x = 84$ **65.** $b = -2$ **67.** $x = 7$ **69.** $x = 5$

71. $x = -4$ **73.** $x = 2$ **75.** no real solution **77.** $y = -3$ **79.** $x = \dfrac{20}{13}$ **81.** $y = -12$ **83.** $y = 7$

85. $x = 9$ **87.** $x = \dfrac{7}{20}$ **89.** $c = -\dfrac{9}{32}$ **91.** $R = 0.61$ **93.** $y = -1$ **95.** $x = 1$ **97.** $x = 2$

99. $x = -9$ **101.** $x = 3$ **103.** $x = 2$ **105.** $x = -4$ **107.** $x = 0$ **109.** $x = 3$ **111.** $x = 3$

113. $x = -1$ **115.** $x = -\dfrac{1}{8}$ **117.** $x - 6 = 8; x = 14$ **119.** $5(x + 6) = x + 42; x = 3$

121. $x + (x - 3) = 51; 27\text{ h}, 24\text{ h}$ **123.** $720 = 2(2w + w); w = 120\text{ ft}; l = 2w = 240\text{ ft}$ **125.** $\dfrac{1}{2}$ **127.** $\dfrac{5}{6}$

129. $-\dfrac{4}{15}$ **131.** $\dfrac{49}{18}$ **133.** $\dfrac{8}{3}$ **135.** $6; P \neq 0$ **137.** $2\dfrac{1}{10}\text{ h}$ **139.** 25 fixtures **141.** $1.33\ \Omega$ **143.** 2

145. -11.8 **147.** $17\ \Omega$ **149.** 2 years **151.** 2.75 A **153.** 7 A **155.** $w = \dfrac{V}{lh}$ **157.** $r = s + d$

159. $t = \dfrac{v_0 - v}{32}$ **161.** $P = S + D$

Practice Test, Chapter 7

1. 14 **3.** 10 **5.** 3 **7.** 2 **9.** $-\dfrac{22}{7}$ **11.** $\dfrac{7}{25}$ **13.** $\dfrac{10}{21}$ **15.** 6.17 **17.** 254.24 **19.** -0.15

21. $I = 2.2$ A **23.** $L = \dfrac{AR}{P}$ **25.** $s = \dfrac{d}{\pi r^2 n}$ **27.** 12,000 cal

Section 8–1 Self-Study Exercises

1 **1.** $x = 3$ **2.** $x = 2$ **3.** $x = 29$ **4.** $x = \dfrac{5}{8}$ or 0.625 **5.** $x = \dfrac{34}{5}$ or 6.8 **6.** $x = \dfrac{17}{14}$ or 1.214

7. $x = 3$ **8.** $x = 21$ **9.** $x = 4$ **10.** $x = -\dfrac{4}{5}$ or -0.8 **11.** $x = 1$ **12.** $x = \dfrac{9}{5}$ or 1.8 **13.** $x = 30$

14. $x = 2.698$ **15.** $x = 836.9$ **16.** $x = -2$ **17.** $x = 2.835$ **18.** $x = 1.174$ **19.** $x = 2.769$
20. $x = 3.25$ **21.** $x = 214.7$ **22.** $x = 1.438$ **23.** $x = 29.18$ **24.** $x = 10^2$ or 100 **25.** $x = 10^{12}$
26. $x = \dfrac{15}{8}$ or 1.875 **27.** $x = \dfrac{5}{4}$ or 1.25 **28.** $x = \dfrac{23}{9}$ or 2.556 **29.** $x = 10$ in. **30.** $x = 22.5$ ft
31. $x = 9\text{ k}\Omega$ **32.** $x = 257.1$ W **33.** $x = 0.26$ V **34.** $x = 0.7317$ mA **35.** $x = 1.346$ **36.** $x = 1$
37. $x = \dfrac{5}{6}$ or 0.8333 **38.** $x = \dfrac{5}{24}$ or 0.2083 **39.** $x = \dfrac{10}{7}$ or $1\dfrac{3}{7}$ or 1.429 **40.** $x = \dfrac{28}{5}$ or $5\dfrac{3}{5}$ or 5.6
41. $x = 8$ h **42.** $x = 0.76$ h **43.** $x = \$98$ **44.** $x = \$95.19$ **45.** $x = \$0.21$ **46.** $x = 0.375$ gal
47. $x = 44.87$ mg **48.** $x = 5.833$ mg **49.** $x = 0.25$ gal **50.** $x = 132$ gal **51.** $x = 555.6$ mi
52. $x = 20$ mg **53.** $x = 8.333$ mg **54.** $x = 8.036$ mg

Section 8–2 Self-Study Exercises

1   **1.** $5.90   **2.** $74.13   **3.** 48 engines   **4.** 15 headpieces   **5.** 1,425 mi   **6.** 3,360 mi   **7.** 550 lb
8. 1.7 gal   **9.** $0.0019 per gram   **10.** $0.0022 per gram   **11.** The larger can has lower cost per gram.
12. 320 bottles per hour   **13.** 12 teeth   **14.** 32 teeth   **15.** 4 cm   **16.** 384 cm   **17.** 7 ft   **18.** 6 ft
19. 10 ft   **20.** 14 ft   **21.** 355 mg   **22.** 277.5 mg   **23.** 10 mg

2   **24.** 20 mL   **25.** 15 ft   **26.** $11\frac{2}{3}$ ft or 11 ft 8 in.   **27.** 30 in.   **28.** 40 in.   **29.** 0.02325 Ω

30. 0.576 Ω   **31.** 45 mi   **32.** 32 mi   **33.** 23 mi   **34.** 30 in.   **35.** 12.5 in.   **36.** 7.5 in.

37.
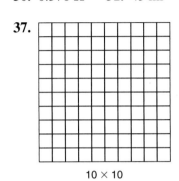
10 × 10

38.

4 × 3

39.

5 × 4

40.
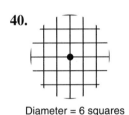
Diameter = 6 squares

41.
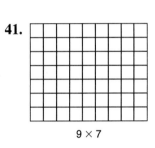
9 × 7

42. 1 mL   **43.** 250 mL   **44.** 0.4 mL   **45.** 15 oz   **46.** 12.5 V

3   **47.** 1,080   **48.** 750   **49.** 324   **50.** $2,400   **51.** $1,960   **52.** $P = 375$

Section 8–3 Self-Study Exercises

1   **1.** 900 rpm   **2.** 200 in³   **3.** 20 in.   **4.** 18 painters   **5.** 10 h   **6.** 50 rpm
7. 5 in.   **8.** 4 h   **9.** 5 machines   **10.** 120 rpm   **11.** 4 helpers (5 workers total)   **12.** 9 in.

2   **13.** 360   **14.** 960   **15.** 240 rpm   **16.** 225 rpm   **17.** 700 cm   **18.** 4 cm   **19.** 36 cm
20. 600 rpm   **21.** 400 rpm   **22.** 140 rpm   **23.** 300 teeth   **24.** 35 teeth   **25.** 37.5 rpm   **26.** 21 teeth
27. 10 rpm   **28.** 38 rpm   **29.** 180 teeth   **30.** 108 teeth   **31.** 9   **32.** 15   **33.** 6   **34.** 1,296

35. $A = \dfrac{kB^2}{C}; k = 25$ **36.** 5,000   **37.** 504 lb   **38.** 1,372 lb

Chapter Review Exercises, Chapter 8

1. $\dfrac{7}{6}$ **3.** $\dfrac{1}{2}$ **5.** $\dfrac{49}{12}$ **7.** $-\dfrac{21}{2}$ **9.** $\dfrac{3}{14}$ **11.** $-\dfrac{21}{23}$ **13.** 4,500 women   **15.** $4\frac{1}{5}$ ft   **17.** 129.7 gal

19. 5.0 h   **21.** 1,351 ft   **23.** $b = 240$ **25.** 700 in²   **27.** 23 machines   **29.** $2\frac{1}{2}$ h   **31.** 1,500 rpm
33. 3 days   **35.** 60 rpm   **37.** $p = 100$

Practice Test, Chapter 8

1. $\dfrac{15}{2}$ **3.** 9,600 **5.** $\dfrac{32}{5}$ **7.** 200 rpm **9.** 168.75 rpm **11.** $x = 88.74$ in^2 **13.** 54.7 L

15. $x = 8.8$ A **17.** 107 lb **19.** $450

Section 9–1 Self-Study Exercises

1 **1.** *R:* horizontal, 4; vertical, 2 **2.** *S:* horizontal, -4; vertical, 3 **3.** *T:* horizontal, 5; vertical, -3
4. *U:* horizontal, -2; vertical, -5 **5.** *V:* horizontal, 7; vertical, 0 **6.** *W:* horizontal, 0; vertical, 5
7. *X:* horizontal, 0; vertical, 0 **8.** *Y:* horizontal, -3; vertical, 0
9. $A = (4, 2)$ $B = (-3, 2)$ $C = (-2, -1)$ $D = (3, -2)$ **10.–15.**
16. *y*-value $= 0$ **17.** *x*-value $= 0$ **18.** $(-x, +y)$ **19.** $(+x, -y)$

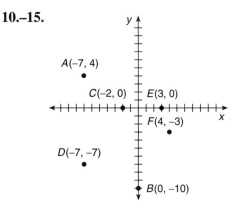

2 **20.** $f(x) = 5x - 7$ **21.** $f(x) = -3x + 2$ **22.** $f(x) = 4x - 5$ **23.** $f(x) = -\dfrac{7}{4}x - \dfrac{3}{4}$

24. $f(x) = -3x + 12$ **25.** $f(x) = 3x - 4$ **26.** $f(x) = -\dfrac{3}{2}x + \dfrac{9}{2}$ **27.** $f(x) = \dfrac{1}{7}x - 2$

28. $f(x) = \dfrac{5}{3}x - \dfrac{7}{3}$

3 Values chosen for table of values may vary.

29.
x	y
0	-7
1	-2
2	3

30.
x	y
-1	5
0	2
1	-1

31
x	y
0	-5
1	-1
2	3

32.
x	y
-1	1
0	$-\dfrac{3}{4}$
1	$\dfrac{5}{2}$

33.
x	y
1	9
2	6
3	3

34.
x	y
-1	-7
0	-4
1	-1

35.
x	y
0	-7
1	-4
2	-1

36.
x	y
-1	5
0	3
1	1
2	-1

37.
x	y
-1	9
0	5
1	1
2	-3
3	-7

38.
x	y
-6	0
-2	2
0	3
2	4
6	6

39.
x	y
-6	3
-3	1
0	-1
3	-3
6	-5

40.
x	y
-1	-3
0	2
1	7

41.
x	f(x)
-2	3
0	2
2	1
4	0

42.
x	f(x)
-6	0
-3	1
0	2
3	3
6	4

Selected Answers to Student Exercise Material

4

43.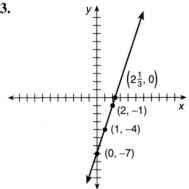

44.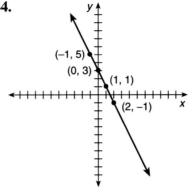

45.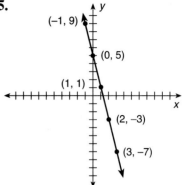

46.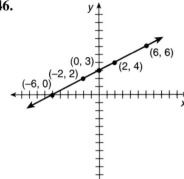

47.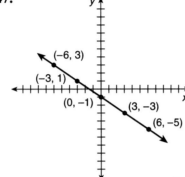

48.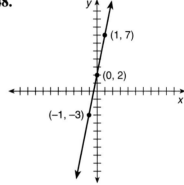

49.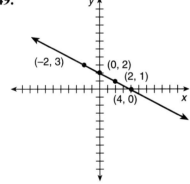

50.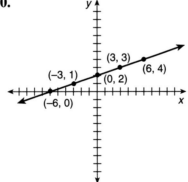

51. no **52.** yes **53.** yes **54.** no **55.** yes **56.** yes **57.** yes **58.** no **59.** yes **60.** no
61. yes **62.** yes **63.** (8, 5) **64.** (2, 5) **65.** (−6, 1) **66.** (21, 3) **67.** (5, 15) **68.** (2, 0)
69. (1, 3); (1, −2); (1, 0); (1, 1) **70.** (−1, 4); (3, 4); (0, 4); (2, 4) **71.** $12 **72.** $2 **73.** $34
74. $13,800

Section 9–2 Self-Study Exercises

1   **1.** x-intercept, $(5, 0)$
        y-intercept, $(0, 5)$

2. x-intercept, $(5, 0)$
   y-intercept, $\left(0, \dfrac{5}{3}\right)$

3. x-intercept, $(-4, 0)$
   y-intercept, $(0, 8)$

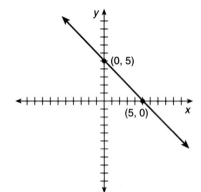

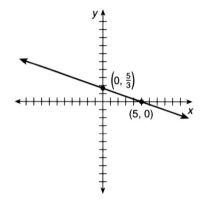

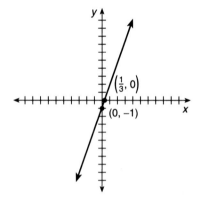

4. x-intercept, $\left(\dfrac{1}{3}, 0\right)$

   y-intercept, $(0, -1)$

5. x-intercept, $\left(\dfrac{2}{5}, 0\right)$

   y-intercept, $(0, -2)$

6. x-intercept, $(3, 0)$

   y-intercept, $\left(0, \dfrac{3}{2}\right)$

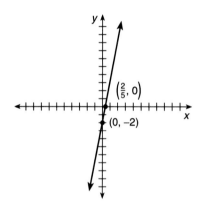

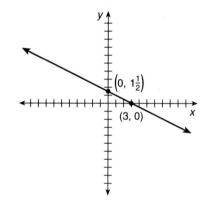

2   **7.** slope $= 4$; y-intercept $= 3$ **8.** slope $= -5$; y-intercept $= 6$ **9.** slope $= -\dfrac{7}{8}$; y-intercept $= -3$

10. slope $= 0$; y-intercept $= 3$ **11.** $y = 2x - 5$; slope $= 2$; y-intercept $= -5$

12. $y = \dfrac{5}{2}$; slope $= 0$; y-intercept $= \dfrac{5}{2}$ **13.** $y = -2x + 6$; slope $= -2$; y-intercept $= 6$

14. $y = 2x + 5$; slope $= 2$; y-intercept $= 5$ **15.** $x = 4$; slope not defined; no y-intercept

16. $x = 9$; slope not defined; no y-intercept

17. slope $= \dfrac{2}{1}$; y-intercept, $(0, -3)$ **18.** slope $= -\dfrac{1}{2}$; y-intercept, $(0, -2)$ **19.** slope $= -\dfrac{3}{5}$; y-intercept, $(0, 0)$

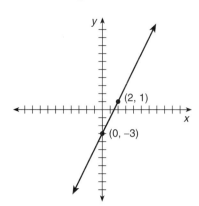

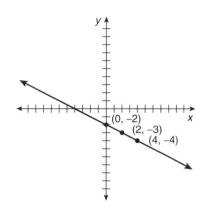

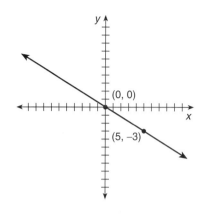

20. slope $= \dfrac{1}{2}$; y-intercept, $\left(0, -1\dfrac{1}{2}\right)$ **21.** slope $= -\dfrac{2}{1}$; y-intercept, $(0, 1)$ **22.** slope $= \dfrac{3}{4}$; y-intercept, $(0, 2)$

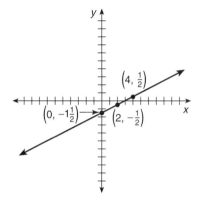

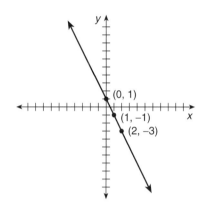

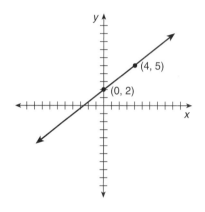

3

23.

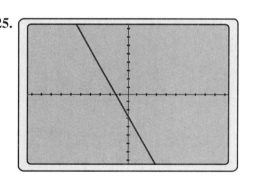

24. 0.38

25.

26. −5.08

27.

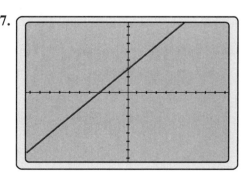

28. 1,817

29.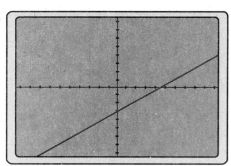

30. 136

31. $35,000; $95,000   **32.** $332; $17,264   **33.** $195 \dfrac{\text{beats}}{\text{min}}$; $155 \dfrac{\text{beats}}{\text{min}}$ **34.** 1,096 puzzles

4 **35.** $y = 3x - 5$
$x = \dfrac{5}{3}$ or 1.66667

36. $y = 5x - 6$
$x = \dfrac{6}{5}$ or 1.2

37. $y = 6x - 5$
$x = \dfrac{5}{6}$ or 0.83333

38. $y = x - 1$
$x = 1$

39. $y = x + 9$
$x = -9$

40. $y = 2x + 16$
$x = -8$

Section 9–3 Self-Study Exercises

1 **1.** $\dfrac{1}{2}$ **2.** $-\dfrac{5}{3}$ **3.** 3   **4.** 1   **5.** $\dfrac{3}{4}$ **6.** 0   **7.** $-\dfrac{5}{4}$ **8.** -1 **9.** undefined   **10.** -3 **11.** 0

12. undefined   **13.** $1,400   **14.** $10/backpak   **15.** 2,000 ft/min   **16.** $144.40 per year
17. $197.40 per year   **18.** $466.00 per year   **19.** $33.00 per year   **20.** $-$12.50 per year
21. The graph of the data in this table does not form a perfect straight line. If the slope (rate of change) remains the same between every two data points, the graph will be a straight line.   **22.** The rate of change is greater for public 4-year colleges ($172.50 per year) than it is for public 2-year colleges ($42.60 per year).

2

23.

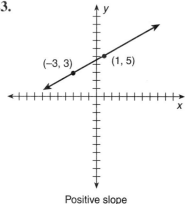

Positive slope

24.

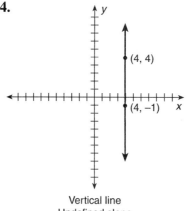

Vertical line
Undefined slope

25.

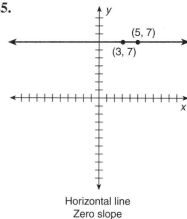

Horizontal line
Zero slope

26.

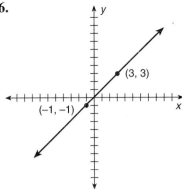

Positive slope

27.

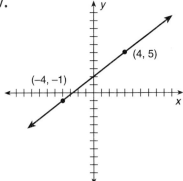

Positive slope

28.

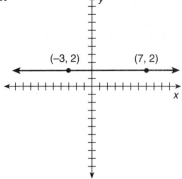

Zero slope

29.

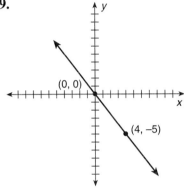

Negative slope

30.

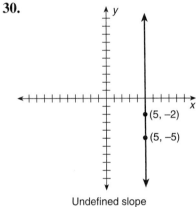

Undefined slope

31.

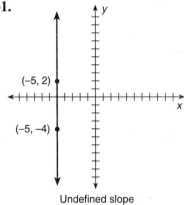

Undefined slope

32.

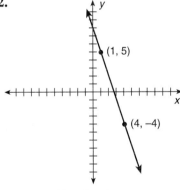

Negative slope

33.

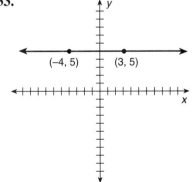

Zero slope

34.

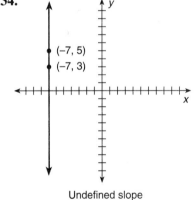

Undefined slope

Section 9–4 Self-Study Exercises

1   **1.** $y = \dfrac{2}{3}x + \dfrac{25}{3}$ **2.** $y = -\dfrac{1}{2}x + 3$ **3.** $y = 2x + 1$ **4.** $y = x - 1$

2   **5.** $y = -\dfrac{5}{3}x + \dfrac{38}{3}$ **6.** $x = -1$ **7.** $y = -3$ **8.** $x = -4$ **9.** $y = 2x$ **10.** $y = 0$

11. $y = 5x + 1{,}935$ **12.** $y = 22x + 50{,}000$

3 **13.** $y = \dfrac{1}{4}x + 7$ **14.** $y = -8x - 4$ **15.** $y = -2x + 3$ **16.** $y = \dfrac{3}{5}x - 2$ **17.** $y = x$

18. $y = 5x - \dfrac{1}{5}$ **19.** $y = 2x - 2$ **20.** $y = -\dfrac{3}{4}x$ **21.** $y = 0.2x + 3$; \$12   **22.** $y = 8x + 12{,}000$; \$76,000

23. $y = -3x + 1$ **24.** $y = \dfrac{2}{3}x - 2$ **25.** $y = 4$ **26.** $y = 2x + 3$ **27.** \$23

28. $y = -0.9x + 40$; 31 mi/h

4 **29.** $x + y = -1$ **30.** $2x + y = 9$ **31.** $x - 3y = 4$ **32.** $3x - y = -7$ **33.** $x + 3y = 7$
34. $3x - y = -3$ **35.** $2x + 3y = 5$ **36.** $3x + 2y = 6$ **37.** $2x - 5y = 11$ **38.** $6x - 8y = 3$
39. $y = 12x + 20{,}000$ **40.** $y = 0.5x + 45$

5 **41.** $x - y = -1$ **42.** $x - 2y = -13$ **43.** $3x + y = 12$ **44.** $x + 3y = -9$ **45.** $3x - y = 11$
46. $2x - y = -7$ **47.** $3x - 2y = 11$ **48.** $x - 4y = 0$ **49.** $2x - 2y = -3$ **50.** $x + 5y = 4$

Chapter Review Exercises, Chapter 9

1.–6. See figure to the right.

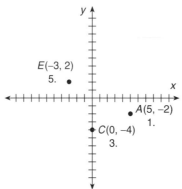

7. $(0, 0)$ **9.** $A(3, 0)$; $B(2, 2)$; $C(2, -5)$; $D(-4, -1)$; $E(-3, 1)$

Since plotted solutions will vary, check graphs by comparing x- and y-intercepts.

11.

x	y
-1	-5
1	-1
3	3

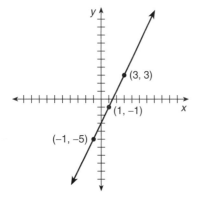

13.

x	y
-1	-3
0	0
1	3

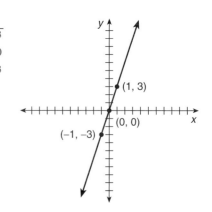

15.

x	y
−1	3
0	0
1	−3

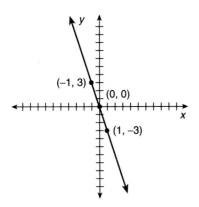

17.

x	y
−1	−1
0	1
1	3

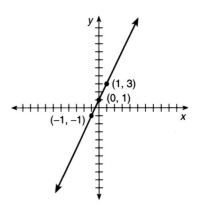

19.

x	y
−2	−10
−1	−6
0	−2
1	2
3	10

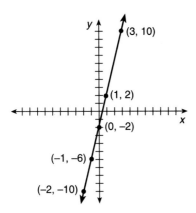

21.

x	y
−2	−3
0	−2
2	−1
4	0

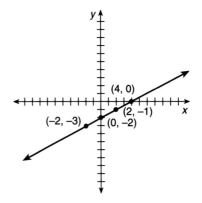

23. no   **25.** yes   **27.** yes   **29.** $y = -7$ **31.** $y = 1$

33. $x = -4y - 1$
x-intercept, $(-1, 0)$;
y-intercept, $\left(0, -\dfrac{1}{4}\right)$

35. $3x - y = 1$
x-intercept,
$\left(\dfrac{1}{3}, 0\right)$; y-intercept, $(0, -1)$

37. $\dfrac{1}{2}x + \dfrac{1}{3}y = 1$
x-intercept $(2, 0)$;
y-intercept $(0, 3)$

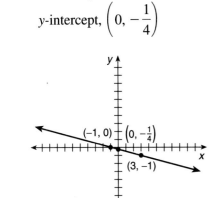

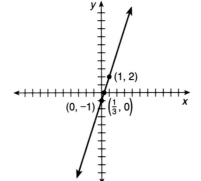

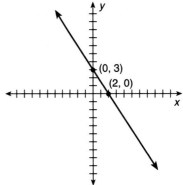

39. $y = 5x - 2$; $m = \dfrac{5}{1}$; $b = -2$

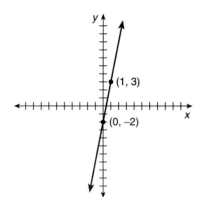

41. $y = -3x - 1$; $m = \dfrac{-3}{1}$; $b = -1$

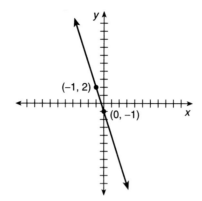

43. $y = x - 4$; $m = \dfrac{1}{1}$; $b = -4$

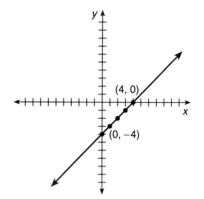

45.

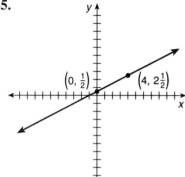

47.

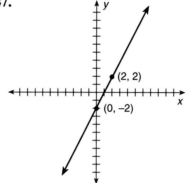

49.

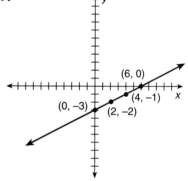

51.

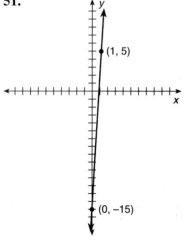

53. \$58,000

55. $y = x - 6$; $x = 6$

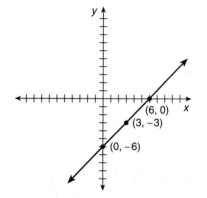

Selected Answers to Student Exercise Material

57. $y = 3x + 6; x = -2$

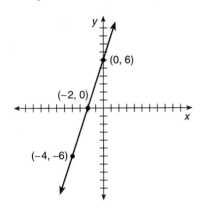

59. $y = 2x - 2; x = 1$

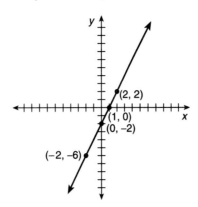

61. $m = 3; b = \dfrac{1}{4}$ **63.** $m = -5; b = 4$ **65.** undefined slope; no y-intercept   **67.** $m = \dfrac{1}{8}; b = -5$

69. $y = -2x + 8; m = -2; b = 8$ **71.** $y = \dfrac{3}{2}x - 3; m = \dfrac{3}{2}; b = -3$ **73.** $y = \dfrac{3}{5}x - 4; m = \dfrac{3}{5}; b = -4$

75. $y = \dfrac{5}{3}; m = 0; b = \dfrac{5}{3}$ **77.** $\dfrac{1}{3}$ **79.** 2   **81.** $\dfrac{5}{8}$ **83.** $-\dfrac{4}{3}$ **85.** undefined   **87.** $-\dfrac{4}{7}$ **89.** $\dfrac{7}{2}$

91. 0   **93.** undefined   **95.** $(-1, 2)$ $(3, 2)$; answers will vary.   **97.** \$56   **99.** \$37   **101.** The table of values is not a perfect linear function.   **103.** $y = \dfrac{1}{3}x + 4$ **105.** $y = \dfrac{3}{4}x - 3$ **107.** $y = 4x - 5$

109. $y = -\dfrac{1}{11}x + \dfrac{17}{11}$ **111.** $y = \dfrac{7}{4}x - \dfrac{5}{4}$ **113.** $y = \dfrac{9}{5}x + \dfrac{3}{5}$ **115.** $y = x - 3$ **117.** $y = x - 1$

119. $y = -2$ **121.** $S = 10x + 3{,}000$

123. $y = 3x - 2$ **125.** $y = 2x - 2$ **127.** $x + y = 7$ **129.** $x - 2y = 8$ **131.** $x - 3y = 20$
133. $x + 3y = -10$ **135.** $3x - 4y = -7$ **137.** $x - y = -4$ **139.** $2x - y = -4$ **141.** $x - 5y = -11$
143. $2x + 10y = 31$ **145.** $x + 4y = 2$

Practice Test, Chapter 9

Since plotted solutions will vary, check graphs by comparing x- and y-intercepts.

1.

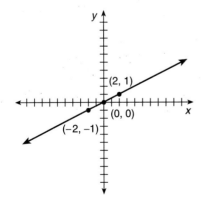

3.

x	y
-1	-6
0	-4
1	-2
2	0

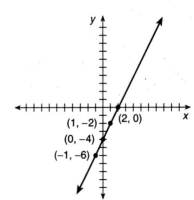

5. $y = 3x - 3$; $x = 1$

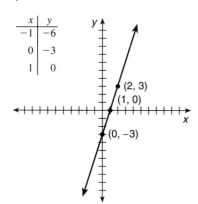

x	y
−1	−6
0	−3
1	0

7. $(2, 5)$

9. $23,200

11. x-intercept $(-5, 0)$
y-intercept $(0, -5)$

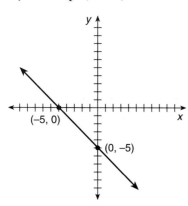

13. x-intercept $(8, 0)$
y-intercept $(0, 4)$

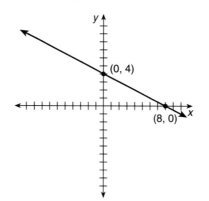

15. y-intercept $= (0, -3)$
slope $= \dfrac{-2}{1}$

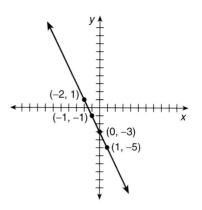

17. $-\dfrac{2}{3}$ **19.** $y = 2x + 34$; $m = 2$; $b = 34$ **21.** $y = \dfrac{1}{4}x$; $m = \dfrac{1}{4}$; $b = 0$ **23.** $y = \dfrac{2}{3}x - 7$

25. $y = \dfrac{2}{3}x + \dfrac{7}{3}$ **27.** $y = 2$ **29.** $y = \dfrac{3}{2}x + 3$ **31.** $2x + y = 5$ **33.** $x - 2y = 10$

Section 10–1 Self-Study Exercises

1

1. $x = 7$, $y = 5$

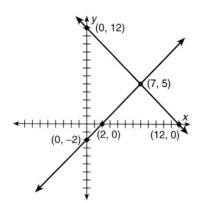

2. $x = 3$, $y = 3$

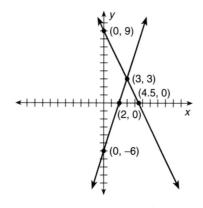

3. no solution; no intersection

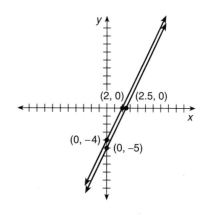

4. many solutions; lines coincide

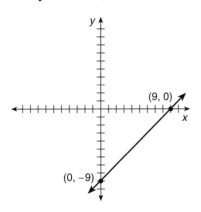

5. $x = 4, y = -2$

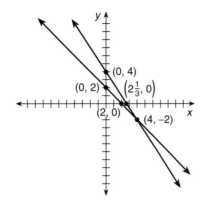

6. $x = 3, y = 1$

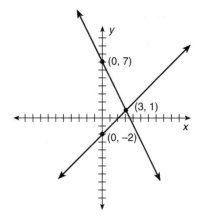

7. $x = 4, y = -3$

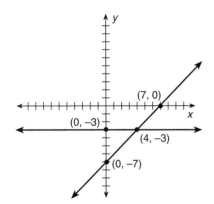

8. $x = -1, y = -5$

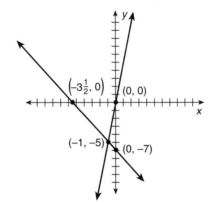

9. $x = 1, y = -2$

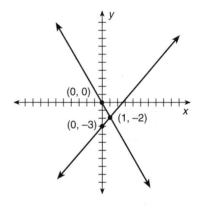

10. $x = 4, y = 1$

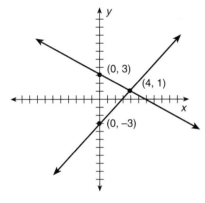

11. $x = 1, y = -7$

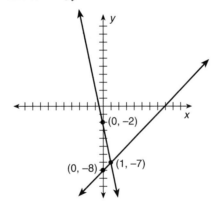

12. $x = 0, y = 3$

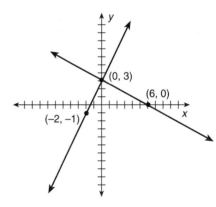

13. $x = 1, y = -5$

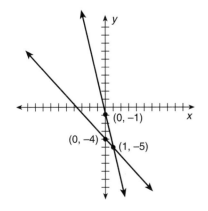

14. $x = 2, y = 1$

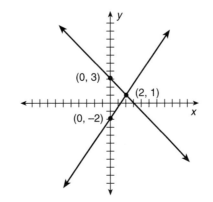

15. $x = 7, y = 9$

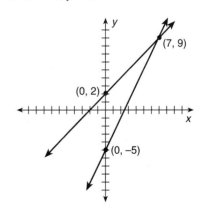

16. $x = -2, y = 4$

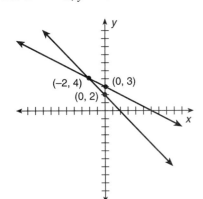

17. $x = 4, y = 1$

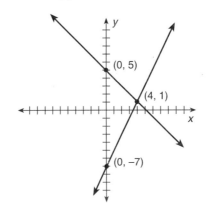

18. $x = 3, y = -4$

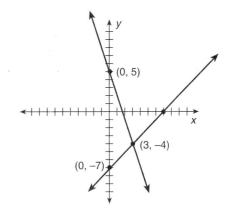

19. $x = 2, y = -2$

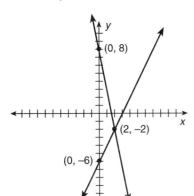

20. $x = 0, y = 0$

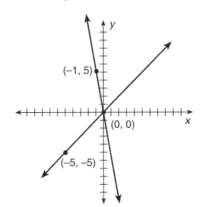

21. $x = 3, y = 1$

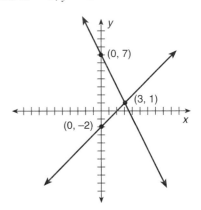

Section 10–2 Self-Study Exercises

1   **1.** $a = 5, b = -1$ **2.** $m = 4, n = -1$ **3.** $x = 1, y = -1$ **4.** $a = 3, b = -3$ **5.** $x = 2, y = \dfrac{3}{2}$

6. $x = 4, y = -\dfrac{3}{5}$ **7.** $x = -\dfrac{13}{5}, y = 5$ **8.** $x = -\dfrac{11}{4}, y = 5$ **9.** $x = -4, y = 3$

2   **10.** $x = 3, y = 0$ **11.** $x = 1, y = 5$ **12.** $a = 2, b = \dfrac{8}{3}$ **13.** $x = -1, y = 2$ **14.** $a = 6, y = 1$

15. $x = 1, y = 3$ **16.** $x = 2, y = -2$ **17.** $x = -4, y = 1$ **18.** $x = 5, y = -2$ **19.** $x = 0, y = 3$
20. $x = -1, y = 3$ **21.** $x = 13, y = 3$

3   **22.** inconsistent; no solution   **23.** inconsistent; no solution   **24.** dependent; many solutions
25. dependent; many solutions   **26.** inconsistent; no solution   **27.** dependent; many solutions
28. inconsistent; no solution   **29.** inconsistent; no solution   **30.** inconsistent; no solution

Section 10–3 Self-Study Exercises

1   **1.** $a = 20, b = 10$ **2.** $r = 5, c = 7$ **3.** $x = 13, y = 11$ **4.** $x = 7, y = 5$ **5.** $p = \dfrac{1}{2}, k = \dfrac{1}{3}$

6. $x = 3, y = 1$ **7.** $x = 13, y = 5$ **8.** $x = -2, a = -7$ **9.** $x = 1, y = 1$ **10.** $x = 1, y = 6$

11. $x = 8, y = 1$ **12.** $x = 4, y = 10$ **13.** $x = 3, y = 2$ **14.** $x = 3, y = 3$ **15.** $x = 1, y = \dfrac{3}{2}$

16. $x = \dfrac{1}{3}, y = \dfrac{1}{2}$ **17.** $x = \dfrac{1}{4}, y = \dfrac{2}{3}$ **18.** $x = -\dfrac{1}{2}, y = \dfrac{1}{2}$ **19.** $x = -2, y = 2$ **20.** $x = 2, y = -1$
21. $x = -8, y = 3$

Section 10–4 Self-Study Exercises

1 **1.** short 15.5 in.   **2.** $18,000 at 4%   **3.** $21 per shirt   **4.** seven 8-cylinder jobs
        long 32.5 in. $17,000 at 5% $16 per hat   three 4-cylinder jobs

5. resistor $0.25   **6.** rate of plane = 130 mi/h   **7.** $3,000 at 7%   **8.** rate of current = 1.67 mi/h
    capacitor $0.30   rate of wind = 10 mi/h $5,000 at 9%   rate of motorboat = 11.67 mi/h

9. dark roast, $4.10   **10.** 75% mixture, 123 lb   **11.** scientific $8.00   **12.** $4,000 at 10%
    with chicory, $3.75   10% mixture, 77 lb   graphing $72.00 $6,000 at 15%
                                Amounts rounded to
                                nearest pound.

13. reserved $20.00   **14.** 4 pt at 75%
    general $15.00   4 pt at 25%

15. $R_1 = 17\ \Omega, R_2 = 4\ \Omega$ **16.** potassium = 120 lb, nitrogen = 360 lb
17. holly = 40 shrubs, nandina = 20 shrubs   **18.** water = 2 ft^3, sand = 10 ft^3

Chapter Review Exercises, Chapter 10

1.

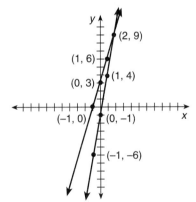

Solution: $x = 2, y = 9$

3.

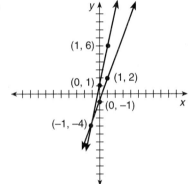

Solution: $x = -1, y = -4$

5.

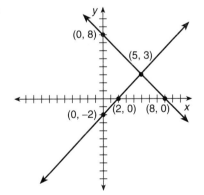

Solution: $x = 5; y = 3$

7.

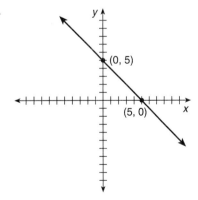

dependent; many solutions, lines coincide

9. $(3, 0)$ **11.** $x = 1; y = -1$ **13.** $(-2, 4)$ **15.** $(2, -2)$ **17.** $(6, 3)$ **19.** dependent, many solutions

21. $(5, 1)$ **23.** $\left(2, \dfrac{8}{3}\right)$ **25.** $(2, 0)$ **27.** $(3, 2)$ **29.** $(2, 1.5)$ **31.** $(4, 4)$ **33.** $(7, 5)$

35. $(4, -6)$ **37.** $(3, -4)$ **39.** $\left(\dfrac{4}{5}, \dfrac{2}{5}\right)$ **41.** $(28, 44)$ **43.** $\left(-3, \dfrac{5}{2}\right)$ **45.** $(4, -6)$ **47.** $(12, 2)$

49. $(1, -2)$

51. electrician = \$75
apprentice = \$35

53. shellac = \$2.50
thinner = \$3.50

55. $119°, 56°$

57. \$2,000 at 5%
\$3,000 at 6%

59. Colombian = \$5
blended = \$4

61. Ohio = \$1.95
Alaska = \$2.10

63. name brand = \$320
generic = \$175

65. telephone = \$15,000
showroom = \$25,000

Practice Test, Chapter 10

1. $(5, 0)$

3. $(14, -9)$

5. $(-6, -6)$

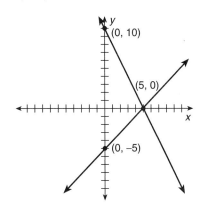

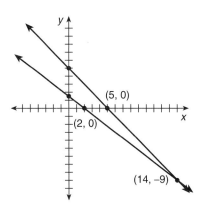

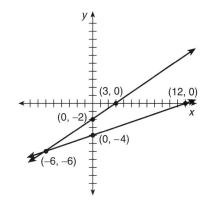

7. $(0, 0)$ **9.** $(1, 2)$ **11.** $(2, 2)$ **13.** $(-3, 15)$ **15.** $(-9, -11)$ **17.** 20 A, 15 A

19. $L = 51$ in.
$W = 34$ in.

21. \$20,000 at 3.5%
\$5,000 at 4%

23. 0.000215 F
0.000055 F

Chapters 7–10 Cumulative Practice Test

1. $-6x + 19$ **3.** $x = 1$ **5.** $x = \dfrac{31}{39}$ **7.** $Z = \sqrt{65}\ \Omega$ or $8.1\ \Omega$ **9.** $x = \dfrac{21}{10}$ **11.** 160 teeth

13.

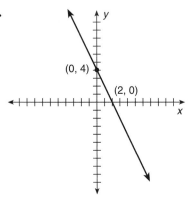

15.

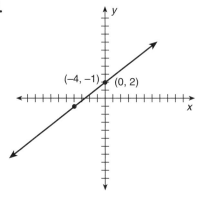

17. slope $= \dfrac{3}{2}$; y-intercept $= -5$ **19.** $y = -\dfrac{2}{3}x - \dfrac{2}{3}$ **21.** $y = 2x - 4$ **23.** $y = 4x + 3$ **25.** $x - y = 1$
27. $(1, 3)$ **29.** larger resistance $= 20\ \Omega$; smaller resistance $= 12\ \Omega$

Section 11–1 Self-Study Exercises

1   **1.** x^7 **2.** m^4 **3.** a^2 **4.** x^{10} **5.** y^6 **6.** a^2b **7.** $12a^5b^7$ **8.** $10x^{11}$ **9.** $-10x^3y^3z^4$ **10.** $24x^3y^4$
11. $\dfrac{2}{7}a^3b^5$ **12.** $\dfrac{3}{8}a^4b^3$ **13.** $-\dfrac{4.2n^{10}}{m^2}$ **14.** $-72a^6$ **15.** $24x^7$ **16.** $24ab^4$

2   **17.** y^5 **18.** x^4 **19.** $\dfrac{1}{a}$ **20.** b **21.** 1 **22.** $\dfrac{1}{x^2}$ **23.** y^4 **24.** n^7 **25.** $\dfrac{1}{x^{10}}$ **26.** $\dfrac{2}{n}$ **27.** $\dfrac{3x^7}{2}$
28. $\dfrac{1}{x^5}$ **29.** $\dfrac{x}{y^6}$ **30.** $\dfrac{a^3}{b^2}$ **31.** xy^2 **32.** $\dfrac{a}{b}$ **33.** $\dfrac{3a}{b^5}$ **34.** $\dfrac{3rs^{12}}{2}$ **35.** $\dfrac{2y^2}{3x^4}$ **36.** $\dfrac{3m^6n^3}{2}$

3   **37.** $4{,}096$ **38.** x^8 **39.** 1 **40.** x^{40} **41.** a^{21} **42.** $-\dfrac{1}{8}$ **43.** $\dfrac{4}{49}$ **44.** $\dfrac{a^4}{b^4}$ **45.** $8m^6n^3$ **46.** $\dfrac{x^6}{y^3}$
47. x^6y^{12} **48.** $4a^2$ **49.** x^6y^3 **50.** $-27a^3b^6$ **51.** $-16{,}807x^{10}$ **52.** $16x^4y^{16}$ **53.** $-\dfrac{7}{x^{17}}$ **54.** $-x^4$
55. x^4 **56.** $-6x^8$ **57.** $-x^{13}y$ **58.** $-3x^3$ **59.** $-\dfrac{x^2}{y^7}$ **60.** $-27a^6b^3c^3$

Section 11–2 Self-Study Exercises

1   **1.** binomial   **2.** binomial   **3.** monomial   **4.** monomial   **5.** monomial   **6.** monomial
7. trinomial   **8.** trinomial   **9.** monomial   **10.** binomial   **11.** binomial   **12.** binomial

2   **13.** 1 **14.** 2 **15.** $2, 1, 0$ **16.** $3, 1, 0$ **17.** $1, 0$ **18.** $2, 0$ **19.** 0 **20.** 0 **21.** $1, 0$ **22.** $2, 0$
23. $5, 2$ **24.** $5, 8$ **25.** 2 **26.** 3 **27.** 6 **28.** 5 **29.** 2 **30.** 1 **31.** 6 **32.** 6

3   **33.** $-3x^2 + 5x$; 2; $-3x^2$; -3 **34.** $-x^3 + 7$; 3; $-x^3$; -1 **35.** $9x^2 + 4x - 8$; 2; $9x^2$; 9
36. $5x^2 - 3x + 8$; 2; $5x^2$; 5 **37.** $7x^3 + 8x^2 - x - 12$; 3; $7x^3$; 7 **38.** $-15x^4 + 12x + 7$; 4; $-15x^4$; -15
39. $8x^6 - 7x^3 - 7x$; 6; $8x^6$; 8 **40.** $-14x^8 + x + 15$; 8; $-14x^8$; -14 **41.** $8x^4 + 15x^3 + 12x$; 4; $8x^4$; 8
42. $-7x^4 + 5x^3 + 3x - 8$; 4; $-7x^4$; -7 **43.** $-5x^4 + 2x^2 + 3x - 5$; 4; $-5x^4$; -5
44. $-8x^4 - x^3 + 7x - 3$; 4; $-8x^4$; -8

Section 11–3 Self-Study Exercises

1   **1.** $7a^2$ **2.** $3x^3$ **3.** $-2a^2 + 3b^2$ **4.** $2a - 2b$ **5.** $-5x^2 - 2x$ **6.** $5a^2$ **7.** $-x^2 - 2y$
8. $2m^2 + n^2$ **9.** $9a + 2b + 10c$ **10.** $-2x + 3y + 2z$ **11.** -4 **12.** $3x + 1$

2   **13.** $14x^3$ **14.** $2m^3$ **15.** $-21m^2$ **16.** $-2y^6$ **17.** $8x^3 - 28x^2$ **18.** $-6a^3b^2 + 15a^2b^3$
19. $35x^4 - 21x^3 - 49x$ **20.** $6x^3y^2 - 10x^2y^3$ **21.** $3x^2 - 18x$ **22.** $12x^3 - 28x^2 + 32x$ **23.** $-8x^2 + 12x$
24. $10x^2 + 4x^3$ **25.** $x^2 + 10x + 21$ **26.** $x^2 + 13x + 40$ **27.** $2x^2 + 3x - 2$ **28.** $3x^2 - 8x - 35$
29. $x^3 + 7x^2 + 9x - 5$ **30.** $x^3 + 4x^2 - 19x + 14$ **31.** $x^3 - 12x^2 + 37x - 14$ **32.** $x^3 - 11x^2 + 25x - 3$
33. $2x^3 - 7x^2 + 7x - 5$ **34.** $3x^3 + x^2 - 11x + 6$ **35.** $6x^3 + 11x^2 - 31x + 14$ **36.** $20x^3 - 22x^2 - 9x + 9$
37. $x^3 - 8$ **38.** $8x^3 - 27$ **39.** $x^3 + 27$ **40.** $27x^3 + 8$ **41.** $x^3 + 125$ **42.** $1{,}331x^3 - 27$
43. $a^2 + 11a + 24$ **44.** $x^2 + x - 20$ **45.** $y^2 - 11y + 18$ **46.** $y^2 - 10y + 21$ **47.** $2a^2 + 11a + 12$

48. $3a^2 - 2a - 5$ **49.** $3a^2 - 8ab + 4b^2$ **50.** $5x^2 - 26xy + 5y^2$ **51.** $6x^2 - 17x + 12$
52. $2a^2 - 7ab + 5b^2$ **53.** $21 - 52m + 7m^2$ **54.** $40 - 21x + 2x^2$ **55.** $x^2 + 11x + 28$
56. $y^2 - 12y + 35$ **57.** $m^2 - 4m - 21$ **58.** $3bx + 18b - 2x - 12$ **59.** $12r^2 - 7r - 10$
60. $35 - 22x + 3x^2$ **61.** $4 - 14m + 6m^2$ **62.** $6 + 13x + 6x^2$ **63.** $2x^2 + x - 15$ **64.** $20x^2 - 13xy - 21y^2$
65. $14a^2 + 19ab - 3b^2$ **66.** $30a^2 - 13ab - 10b^2$ **67.** $27x^2 + 30xy - 8y^2$ **68.** $20x^2 - 47xy + 24y^2$
69. $21m^2 + 29mn - 10n^2$

3 **70.** $a^2 - 9$ **71.** $4x^2 - 9$ **72.** $a^2 - y^2$ **73.** $16r^2 - 25$ **74.** $25x^2 - 4$ **75.** $49 - m^2$
76. $9y^2 - 25$ **77.** $64y^2 - 9$ **78.** $9a^2 - 121b^2$ **79.** $25y^2 - 9$ **80.** $x^2 - 49$ **81.** $x^2 - 121$
82. $4 - 12x + 9x^2$ **83.** $9x^2 + 24x + 16$ **84.** $Q^2 + 2QL + L^2$ **85.** $a^4 + 2a^2 + 1$ **86.** $4d^2 - 20d + 25$
87. $9a^2 + 12ax + 4x^2$ **88.** $9x^2 - 42x + 49$ **89.** $36 + 12Q + Q^2$ **90.** $y^2 + 10xy + 25x^2$
91. $16 - 24j + 9j^2$ **92.** $9m^2 - 12mp + 4p^2$ **93.** $m^4 + 2m^2p^2 + p^4$ **94.** $4a^2 - 28ac + 49c^2$
95. $81 - 234a + 169a^2$ **96.** $4x^2 - 20xy + 25y^2$ **97.** $x^3 + p^3$ **98.** $Q^3 + L^3$ **99.** $27 + a^3$
100. $8x^3 + 125p^3$ **101.** $27m^3 + 8$ **102.** $216 - p^3$ **103.** $125y^3 - p^3$ **104.** $x^3 - 8y^3$ **105.** $Q^3 - 216$
106. $27T^3 - 8$ **107.** $8x^3 + 27y^3$ **108.** $27x^3 - 64y^3$

4 **109.** $2x^2$ **110.** $-\dfrac{a}{2}$ **111.** $\dfrac{1}{2x^2}$ **112.** $-\dfrac{3x^3}{4}$ **113.** $3x - 2$ **114.** $4x^3 - 2x - 1$ **115.** $7x - \dfrac{1}{x}$

116. $4x^2 + 3x$ **117.** $\dfrac{x^2}{3} + \dfrac{4}{9}$ **118.** $5ab^2 - 3b - \dfrac{7}{ab}$ **119.** $\dfrac{5ab^2}{2} - \dfrac{b}{2}$ **120.** $\dfrac{3x}{y} - \dfrac{2y}{x} - 1$

121. $5x - 14x^4$ **122.** $24x^9y^6 + 4$ **123.** $-4 - 4x^2$ **124.** $\dfrac{x}{5y} - \dfrac{y}{3x} - 2x^2$ **125.** $x + 4$, yes

126. $x + 6$, yes **127.** $x^2 - 3x + 2$, yes **128.** $x^2 + x - 5$, yes **129.** $3x^2 - 2x + 3 - \dfrac{1}{x + 3}$, no

130. $2x^2 + 7x + 2 + \dfrac{18}{x - 5}$, no **131.** $x^2 + x - 6$, yes **132.** $x^2 - x - 2$, yes **133.** $x^2 + 3x + 9$, yes

134. $x^2 - 5x + 25$, yes

Chapter Review Exercises, Chapter 11

1. x^{10} **3.** $21x^9$ **5.** x^3 **7.** $7x^3$ **9.** $\dfrac{x}{y^3}$ **11.** x^{12} **13.** x^{15} **15.** $-27x^6$ **17.** $\dfrac{1}{y^2}$ **19.** $40x^2$ **21.** 4

23. 5 **25.** $3x^3 + x^2 + 5x - 8$; 3; $3x^3$; 3 **27.** $2x^4 + 5x^2 - 12x - 32$; 4; $2x^4$; 2 **29.** $-2x^4 - 2x^3 + 3x$
31. $-3x^2 + 5y^2$ **33.** $21x^6$ **35.** $2x^3 + 6x^2 - 10x$ **37.** $-2x^4 + 14x^3 - 30x$ **39.** $m^2 - 4m - 21$
41. $12r^2 - 7r - 10$ **43.** $4 - 14m + 6m^2$ **45.** $2x^2 + x - 15$ **47.** $14a^2 + 19ab - 3b^2$ **49.** $27x^2 + 30xy - 8y^2$
51. $21m^2 + 29mn - 10n^2$ **53.** $5x^3 - 12x^2 - 14x + 3$ **55.** $36x^2 - 25$ **57.** $49y^2 - 121$ **59.** $64a^2 - 25b^2$
61. 4 **63.** 10 **65.** $x^2 + 18x + 81$ **67.** $x^2 - 6x + 9$ **69.** $16x^2 - 120x + 225$ **71.** $64 + 112m + 49m^2$
73. $16x^2 - 88x + 121$ **75.** $g^3 - h^3$ **77.** $8H^3 - 27T^3$ **79.** $216 - i$ **81.** $z^3 + 8t^3$ **83.** $343T^3 + 8$

85. $-\dfrac{2x^3}{3}$ **87.** $-\dfrac{14}{5y^2}$ **89.** $2x^2 - 4x + 7$ **91.** $\dfrac{x^3}{2} - 6x^6$ **93.** $-4x - 3x^2 + 5x^3$ **95.** $x - 2$

97. $3x - 2$ **99.** $x^2 + x - 5$ **101.** $3x + 1$

Practice Test, Chapter 11

1. x^5 **3.** $\dfrac{16}{49}$ **5.** $\dfrac{1}{x^{10}}$ **7.** $\dfrac{x^4}{y^2}$ **9.** $12a^3 - 8a^2 + 20a$ **11.** $x^2 - 5x$ **13.** $56x^7$ **15.** $4xy - 3x + 6$

17. 1 **19.** $-3x^4 - 2x^3 + 6$ **21.** $m^2 - 49$ **23.** $a^2 + 6a + 9$ **25.** $2x^2 - 11x + 15$ **27.** $x^3 - 8$
29. $125a^3 - 27$ **31.** $2x + 7$

Section 12–1 Self-Study Exercises

1   **1.** $\sqrt{36}$; $36^{1/2}$ **2.** $\sqrt[3]{27}$; $27^{1/3}$ **3.** $\sqrt[4]{81}$; $81^{1/4}$ **4.** $\sqrt[5]{32}$; $32^{1/5}$ **5.** $\sqrt{81}$; $81^{1/2}$
6. $\sqrt[3]{1,331}$; $1,331^{1/3}$

2   **7.** 6 and 7   **8.** 8 and 9   **9.** 3 and 4   **10.** 10 and 11   **11.** 2 and 3   **12.** 3 and 4   **13.** 3 and 4
14. 4 and 5   **15.–22.** See number line and answers that follow. Numbers given for 16. −22. are approximate
square roots.   **15.** 5   **16.** 4.1   **17.** 4.5   **18.** 3.2   **19.** 4.2   **20.** 5.7   **21.** 2.6   **22.** 6.3

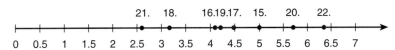

3   **23.** $x^{3/2}$; $\sqrt{x^3}$ **24.** $x^{2/3}$; $\sqrt[3]{x^2}$ **25.** $x^{4/3}$; $(\sqrt[3]{x})^4$ **26.** $x^{2/3}$; $(\sqrt[3]{x})^2$ **27.** $x^{3/4}$; $(\sqrt[4]{x})^3$
28. $x^{5/3}$; $(\sqrt[3]{x})^5$ **29.** $\sqrt[5]{x^3}$ **30.** $\sqrt[5]{x^2}$ **31.** $\sqrt[4]{x^3}$ **32.** $\sqrt[8]{x^5}$ **33.** $\sqrt[5]{y}$ **34.** $\sqrt[8]{y^7}$ **35.** $\sqrt[8]{y^3}$
36. $\sqrt[5]{y^4}$ **37.** $x^{2/3}$ or $x^{0.\overline{6}}$ **38.** $x^{1/2}$ or $x^{0.5}$ **39.** $x^{2/5}$ or $x^{0.4}$ **40.** $x^{3/5}$ or $x^{0.6}$ **41.** $x^{5/6}$ or $x^{0.8\overline{3}}$
42. $x^{1/2}$ or $x^{0.5}$ **43.** $x^{2/3}$ or $x^{0.\overline{6}}$ **44.** $x^{4/7}$ or $\approx x^{0.57}$

Section 12–2 Self-Study Exercises

1   **1.** x^5 **2.** x^3 **3.** $\pm x^8$ **4.** $\dfrac{x^7}{y^{12}}$ **5.** $a^3 b^5$ **6.** $-ab^2c^6$ **7.** $\dfrac{xy^2}{z^5}$ **8.** $\dfrac{a^2}{b^5c^6}$

2   **9.** $y^{1/3}$ **10.** $(5x)^{1/3}$ or $5^{1/3}x^{1/3}$ **11.** $(xy^{5/2})$ or $x^{5/2} y^{5/2}$ **12.** $r^{7/2}$ **13.** $3x^2$ **14.** $2r^3$ **15.** $16x^4y^8$
16. $32x^{5/4}y^{25/4}$ **17.** $x^{1/4}$ **18.** $(7x)^{1/5}$ or $7^{1/5}x^{1/5}$ **19.** $(xy)^{4/3}$ or $x^{4/3}y^{4/3}$ **20.** $3b^5$ **21.** $4x^{2/3}y^{14/3}$
22. x^4 **23.** y^3 **24.** $64ab^6$ **25.** $8x^2y^{5/2}$ **26.** $\dfrac{1}{x^{1/3}}$ **27.** $x^{3/5}$ **28.** $\dfrac{1}{y^{1/6}}$ **29.** $b^{1/6}$ **30.** $4x^{7/2}$
31. $5x^{7/3}$ **32.** 2   **33.** 192   **34.** $a^{5.2}$ **35.** $49a^8$ **36.** x^2 **37.** $3x^2$ **38.** $6x^6$

3   **39.** $2\sqrt{6}$ **40.** $7\sqrt{2}$ **41.** $4\sqrt{3}$ **42.** $x^4\sqrt{x}$ **43.** $y^7\sqrt{y}$ **44.** $2x\sqrt{3x}$ **45.** $2a^2\sqrt{14a}$
46. $6ax^2\sqrt{2a}$ **47.** $2x^2yz^3\sqrt{11xz}$ **48.** $\sqrt{8x^3} = 2x\sqrt{2x}$; Answers will vary.

Section 12–3 Self-Study Exercises

1   **1.** $12\sqrt{3}$ **2.** $-4\sqrt{5}$ **3.** $2\sqrt{7}$ **4.** $9\sqrt{3} - 8\sqrt{5}$ **5.** $-3\sqrt{11} + \sqrt{6}$ **6.** $43\sqrt{2} + 6\sqrt{3}$
7. $12\sqrt{3}$ **8.** $13\sqrt{5}$ **9.** $11\sqrt{7}$ **10.** $-3\sqrt{6}$ **11.** $8\sqrt{2} - 6\sqrt{7}$ **12.** $6\sqrt{3}$ **13.** $27\sqrt{5}$
14. $47\sqrt{2}$ **15.** $6\sqrt{10}$ **16.** $-\sqrt{3}$ **17.** $17\sqrt{7}$ **18.** $2\sqrt{2}$

2   **19.** $\sqrt{42}$ **20.** 6   **21.** $15\sqrt{10}$ **22.** 240   **23.** $20x\sqrt{15x}$ **24.** $28x^5\sqrt{6x}$ **25.** $7\sqrt{2} + 35$
26. $48 - 4\sqrt{3}$ **27.** $\sqrt{10} - 3\sqrt{5}$ **28.** $4\sqrt{7} + \sqrt{21}$ **29.** $\sqrt{15} + \sqrt{6}$
30. $\sqrt{22} + \sqrt{33} - 5\sqrt{2} - 5\sqrt{3}$ **31.** $2\sqrt{6} + 6\sqrt{3} - \sqrt{2} - 3$ **32.** $\sqrt{10} - 3\sqrt{5} - 4\sqrt{2} + 12$
33. $7 + 3\sqrt{35} + \sqrt{14} + 3\sqrt{10}$ **34.** -1 **35.** 5   **36.** $37 - 20\sqrt{3}$ **37.** $x\sqrt{3x}$
38. $\dfrac{2\sqrt{2}}{\sqrt{15y}}$ or $\dfrac{2\sqrt{30y}}{15y}$ **39.** $\dfrac{2\sqrt{2}}{x\sqrt{7}}$ or $\dfrac{2\sqrt{14}}{7x}$ **40.** $\dfrac{2}{3y\sqrt{2y}}$ or $\dfrac{\sqrt{2y}}{3y^2}$ **41.** $\dfrac{4x\sqrt{x}}{3}$ **42.** $x\sqrt{7}$ **43.** $4\sqrt{3}$
44. $\dfrac{x}{\sqrt{6}}$ or $\dfrac{x\sqrt{6}}{6}$ **45.** $\dfrac{y}{\sqrt{7}}$ or $\dfrac{y\sqrt{7}}{7}$

46. $2x\sqrt{2x}$ **47.** $3x\sqrt{6x}$ **48.** $6xy^2\sqrt{3x}$ **49.** 3 **50.** $\dfrac{\sqrt{10}}{2}$ **51.** $2\sqrt{2}$ **52.** $\dfrac{10\sqrt{3}}{3}$

53. $\dfrac{8}{5\sqrt{3}}$ or $\dfrac{8\sqrt{3}}{15}$ **54.** $\dfrac{15x^3\sqrt{x}}{4}$ **55.** $\dfrac{5\sqrt{3}}{3}$ **56.** $\dfrac{6\sqrt{5}}{5}$ **57.** $\dfrac{\sqrt{2}}{4}$ **58.** $\dfrac{\sqrt{55}}{11}$ **59.** $\dfrac{\sqrt{6}}{3}$

60. $\dfrac{5\sqrt{3x}}{3}$ **61.** $\dfrac{2x\sqrt{7}}{7}$ **62.** $\dfrac{2x^3\sqrt{3x}}{3}$

Section 12–4 Self-Study Exercises

1 **1.** $5i$ **2.** $6i$ **3.** $8xi$ **4.** $4y^2i\sqrt{2y}$

2 **5.** i **6.** i **7.** 1 **8.** -1 **9.** 1 **10.** i **11.** 1 **12.** $-i$

3 **13.** $15 + 0i$ **14.** $17 + 0i$ **15.** $0 + 7i$ **16.** $0 + 9i$ **17.** $0 + 33i$ **18.** $0 + 7i$ **19.** $-4 + 0i$
20. $5 + 2i$ **21.** $8 + 4i\sqrt{2}$ **22.** $7 - i\sqrt{3}$

4 **23.** $11 + 5i$ **24.** $(2\sqrt{3} + 2\sqrt{2}) - 8i\sqrt{3}$ **25.** $4 + i$ **26.** $12 + 20i$ **27.** $1 + 9i$ **28.** $12 - 8i$

5 **29.** 65 **30.** 26 **31.** -2 **32.** 17 **33.** -74 **34.** -7 **35.** $-9 + 7i$ **36.** $-16 - 3i$
37. $13 - 13i$ **38.** $8 - 6i$ **39.** $45 - 28i$ **40.** $7 + 24i$

Chapter Review Exercises, Chapter 12

1. $\sqrt{49}$; $49^{1/2}$ **3.** $\sqrt[4]{16}$; $16^{1/4}$ **5.** $\sqrt{121}$; $121^{1/2}$ **7.** 6 and 7 **9.** 11 and 12 **11.** 3 and 4
13.–15. See graph and answers that follow. **13.** approximately 3.9 **15.** approximately 2.2
17. $x^{7/2}$; $\sqrt{x^7}$ **19.** $x^{2/3}$; $(\sqrt[3]{x})^2$ **21.** $x^{1/2}$

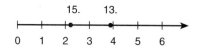

23. $x^{4/5}$ **25.** $x^{4/3}y^{4/3}$ **27.** $7^{1/2}$ **29.** $\sqrt[5]{y^3}$ **31.** y^6 **33.** $-b^9$ **35.** $x^{1/3}$ **37.** $(4y)^{1/5}$ or $2^{2/5}y^{1/5}$
39. $2b^4$ **41.** a^2 **43.** y **45.** $27x^{3/4}y^6$ **47.** $64a^3x^{3/2}$ **49.** $x^{1/2}$ **51.** $a^{7/6}$ **53.** $\dfrac{1}{x^{1/8}}$ **55.** $a^{8/3}$

57. $2a^{7/2}$; approximately 22.627 **59.** $\dfrac{3}{2a^{22/5}}$; approximately 0.071 **61.** $a^{6.3}$; approximately 78.793

63. $2a^3b^4$; 16 **65.** x **67.** $3p\sqrt{p}$ **69.** $3a\sqrt{2b}$ **71.** $4x^2y\sqrt{2x}$ **73.** $5x^5y^4\sqrt{3y}$ **75.** $7y^4\sqrt{3x}$
77. $-2\sqrt{3}$ **79.** $-\sqrt{7}$ **81.** $11\sqrt{6}$ **83.** $-28\sqrt{3}$ **85.** $-5\sqrt{2}$ **87.** $-17\sqrt{2}$ **89.** $24\sqrt{3}$
91. $40\sqrt{21}$ **93.** $-160\sqrt{6}$ **95.** $6 - 5\sqrt{3}$ **97.** $3\sqrt{2} - 3\sqrt{5}$ **99.** -59 **101.** $\dfrac{3}{4}$
103. $\dfrac{3\sqrt{3}}{4\sqrt{2}}$ or $\dfrac{3\sqrt{6}}{8}$ **105.** $\dfrac{\sqrt{3}}{2\sqrt{5}}$ or $\dfrac{\sqrt{15}}{10}$ **107.** $\sqrt{3}$ **109.** $\dfrac{9}{16}$ **111.** $\dfrac{6x^4}{9y^5}$ **113.** $\dfrac{\sqrt{21}}{6}$
115. $\dfrac{\sqrt{6}}{4}$ **117.** $\dfrac{5\sqrt{2}}{4}$ **119.** $\dfrac{\sqrt{10}}{3}$ **121.** $10i$ **123.** $\pm 2y^3i\sqrt{6y}$ **125.** -1 **127.** i **129.** $0 + 15i$
131. $0 - 12i$ **133.** $7 - 4i$ **135.** $11 + i$ **137.** -25

Practice Test, Chapter 12

1. $6\sqrt{14}$ **3.** $6\sqrt{3}$ **5.** $\dfrac{4\sqrt{6}}{3}$ **7.** $2\sqrt{6}$ **9.** $\dfrac{7\sqrt{15x}}{2x}$ **11.** $3x^5$ **13.** $x^2y^3z^6$ **15.** $5x^{\frac{1}{6}}y^2$ **17.** $2x$

19. $-i$ **21.** $-3 + 5i$ **23.** $-2\sqrt{3} + 8\sqrt{5}$ **25.** $\sqrt{35} - 4\sqrt{7}$ **27.** 41

Section 13–1 Self-Study Exercises

1   **1.** $7(a + b)$ **2.** $12(x + y)$ **3.** $m(m + 2)$ **4.** $y^2(5y + 8)$ **5.** $3x(2x + 1)$ **6.** $6y^3(2 + 3y)$
7. $6x^4(2x - 1)$ **8.** $5x(1 - 3y)$ **9.** prime   **10.** prime   **11.** $5(ab + 2a + 4b)$ **12.** $2ax(2x + 3a + 5ax)$
13. prime   **14.** $3(4a^2 - 5a + 2)$ **15.** $3x(x^2 - 3x - 2)$ **16.** $2ab(4a + 7b^2 + 14a^2b^2)$
17. $3m^2(1 - 2m + 4m^2)$ **18.** $6xy(2x - 3y^2 + 4xy)$ **19.** $3a^2bc(5 + 6abc^2 - 7a^2c^4)$
20. $5x^2y^2z(4y - 7x - 8)$ **21.** $4x^2y^2(2x^2 - 3y^2 - 1)$ **22.** $-(x + 7)$ **23.** $-(3x + 8)$ **24.** $-(5x - 2)$
25. $-(12x - 7)$ **26.** $-(x^2 - 3x + 8)$ **27.** $-(2x^2 + 7x + 11)$ **28.** $-2(x^2 - 3x + 4)$
29. $-3(x^2 + 3x - 5)$ **30.** $-7(x^2 + 3x - 2)$ **31.** $-6(2x^2 - 3x - 1)$ **32.** $(x + 3)(5x + 8y)$
33. $(2x - 1)(3x + 5)$ **34.** $(3y - 5)(4y + 7)$ **35.** $(a - b)(7a + 2b)$ **36.** $(2m - 3n)(5m - 7n)$
37. $(y - 2)(y - 3)$ **38.** $(2x - 7)(3x - 8)$ **39.** $(9y - 2)(7y - 5)$ **40.** $5\left(\sqrt{7} + 2\right)$
41. $4\left(2\sqrt{3} - 3\right)$ **42.** $3\left(\sqrt{2} - 3\sqrt{3}\right)$

Section 13–2 Self-Study Exercises

1   **1.** yes   **2.** no, not a binomial   **3.** yes   **4.** yes   **5.** no, not difference   **6.** yes
7. $(y + 7)(y - 7)$ **8.** $(4x + 1)(4x - 1)$ **9.** $(3a + 10)(3a - 10)$ **10.** $(2m + 9n)(2m - 9n)$
11. $(3x + 8y)(3x - 8y)$ **12.** $(5x + 8)(5x - 8)$ **13.** $(10 + 7x)(10 - 7x)$ **14.** $(2x + 7y)(2x - 7y)$
15. $(11m + 7n)(11m - 7n)$ **16.** $(9x + 13)(9x - 13)$ **17.** $(2a + 3)(2a - 3)$ **18.** $(5r + 4)(5r - 4)$
19. $(6x + 7y)(6x - 7y)$ **20.** $(7 - 12x)(7 + 12x)$ **21.** $(4x + 9y)(4x - 9y)$

2   **22.** no, not perfect square   **23.** yes   **24.** no, not perfect square   **25.** no, not perfect square
26. yes   **27.** yes   **28.** $(x + 3)^2$ **29.** $(x + 7)^2$ **30.** $(x - 6)^2$ **31.** $(x - 8)^2$ **32.** $(2a + 1)^2$
33. $(5x - 1)^2$ **34.** $(3m - 8)^2$ **35.** $(2x - 9)^2$ **36.** $(x - 6y)^2$ **37.** $(2a - 5b)^2$ **38.** $(y - 5)^2$
39. $(3x + 10y)^2$ **40.** $-(x + 6)^2$ **41.** $-(3x - 1)^2$ **42.** $-(x + 4)^2$

3   **43.** difference of two cubes   **44.** not a sum or difference of cubes   **45.** sum of two cubes
46. not a sum or difference of cubes   **47.** difference of two cubes   **48.** sum of two cubes
49. $(m - 2)(m^2 + 2m + 4)$ **50.** $(y - 5)(y^2 + 5y + 25)$ **51.** $(Q + 3)(Q^2 - 3Q + 9)$
52. $(c + 1)(c^2 - c + 1)$ **53.** $(5d - 2p)(25d^2 + 10dp + 4p^2)$ **54.** $(a + 4)(a^2 - 4a + 16)$
55. $(6a - b)(36a^2 + 6ab + b^2)$ **56.** $(x + Q)(x^2 - xQ + Q^2)$ **57.** $(2p - 5)(4p^2 + 10p + 25)$
58. $(3 - 2y)(9 + 6y + 4y^2)$ **59.** $-(a + 2)(a^2 - 2a + 4)$ **60.** $-(x + 3)(x^2 - 3x + 9)$

Section 13–3 Self-Study Exercises

1   **1.** $(x + 6)(x + 1)$ **2.** $(x - 6)(x - 1)$ **3.** $(x - 2)(x - 3)$ **4.** $(x + 3)(x + 2)$ **5.** $(x - 7)(x - 4)$
6. $(x + 6)(x + 2)$ **7.** $(x - 6)(x - 2)$ **8.** $(x + 12)(x + 1)$ **9.** $(x - 12)(x - 1)$ **10.** $(x + 4)(x + 3)$
11. $(x - 4)(x - 3)$ **12.** $(x - 3)(x - 1)$ **13.** $(x + 7)(x + 1)$ **14.** $(x + 5)(x + 2)$ **15.** $(x - 3)(x + 2)$
16. $(x + 3)(x - 2)$ **17.** $(x - 6)(x + 1)$ **18.** $(x + 6)(x - 1)$ **19.** $(x - 4)(x + 3)$ **20.** $(x + 4)(x - 3)$
21. $(x + 6)(x - 2)$ **22.** $(x - 6)(x + 2)$ **23.** $(x - 12)(x + 1)$ **24.** $(x + 12)(x - 1)$ **25.** $(y - 5)(y + 2)$
26. $(y - 5)(y + 4)$ **27.** $(b + 3)(b - 1)$ **28.** $(b - 7)(b + 2)$ **29.** $(4 - x)(3 - x)$ or $(x - 4)(x - 3)$
30. $-(5 + x)(6 - x)$ or $(x - 6)(x + 5)$ **31.** $(x + 9)(x + 2)$ **32.** $(x - 6)(x - 3)$ **33.** $(x - 9)(x + 2)$
34. $(x + 18)(x - 1)$ **35.** $(x + 5)(x + 4)$ **36.** $(x - 10)(x - 2)$ **37.** $(x - 8)(x - 2)$ **38.** $(x - 16)(x - 1)$
39. $(x - 14)(x + 1)$ **40.** $(x - 7)(x + 2)$

2 **41.** $(x + y)(x + 4)$ **42.** $(3x + 2)(2x − y)$ **43.** $(3x + 5)(m − 2n)$ **44.** $(6x − 7)(5y − 6)$
45. $(x − 2)(x + 8)$ **46.** $(3x − 1)(2x − 7)$ **47.** $(x − 4)(x + 1)$ **48.** $(2x − 1)(4x + 3)$
49. $(x − 5)(x + 4)$ **50.** $(x − 2)(3x + 5)$ **51.** $(x + 2)(4x − 3)$ **52.** $(x − 2)(4x + 3)$
53. $(x + 2)(4x + 3)$ **54.** $(x − 2)(4x − 3)$ **55.** $(2x + 1)(4x − 3)$

3 **56.** $(3x + 1)(x + 2)$ **57.** $(3x + 2)(x + 4)$ **58.** $(3x + 2)(2x + 3)$ **59.** $(4x + 3)(2x − 1)$
60. $(3x − 4)(2x − 3)$ **61.** $(2x − 5)(x − 2)$ **62.** $(3x − 5)(2x − 1)$ **63.** $(4x + 3)(2x + 1)$
64. $(6x − 5)(x − 1)$ **65.** $(4x + 3)(2x + 5)$ **66.** $(3x − 5)(5x + 1)$ **67.** $(2x − 1)(4x − 3)$
68. $(2x − 7)(x + 1)$ **69.** $(6x − 5)(2x + 3)$ **70.** $(5x + 3)(2x − 1)$ **71.** $(3x + 2)(4x − 1)$
72. $(3x − 2)(4x + 1)$ **73.** $(4x + 1)(3x + 2)$ **74.** $(4x − 1)(3x − 2)$ **75.** $(8x − 1)(3x + 1)$
76. $(3x − 1)(8x − 1)$ **77.** $(6x − 5y)(x + 2y)$ **78.** $(3a + 2b)(2a − 7b)$ **79.** $(6x − 5)(3x + 2)$
80. $(5x − 4y)(4x + 3y)$ **81.** $(7x + 8)(x − 1)$ **82.** $(2a + 19)(a + 1)$

4 **83.** $4(x − 1)$ **84.** $(x + 8)(x − 7)$ **85.** $(2x + 3)(x − 1)$ **86.** $(x + 3)(x − 3)$ **87.** $4(x + 2)(x − 2)$
88. $(m + 5)(m − 3)$ **89.** $2(a + 2)(a + 1)$ **90.** $(b + 3)^2$ **91.** $(4m − 1)^2$ **92.** $(x − 7)(x − 1)$
93. $(2m + 1)(m + 2)$ **94.** $(2m + 1)(m − 3)$ **95.** $(2a − 5)(a + 1)$ **96.** $(3x − 2)(x + 4)$
97. $(3x + 5)(2x − 3)$ **98.** $(4x−1)(2x+3)$ **99.** $−2(x − 2)(x − 1)$ **100.** $(x^2 + 4)(x + 2)(x − 2)$
101. $(x^3 + 9)(x^3 − 9)$ **102.** $(x^2 − 3)(x^4 + 3x^2 + 9)$ **103.** $3(x^2 + 4)(x + 2)(x − 2)$
104. $6(x − 2)(x^2 + 2x + 4)$ **105.** $3(x − 4)(x + 2)$ **106.** $(4a^3 + b)(16a^6 + 4a^3b + b^2)$

Chapter Review Exercises, Chapter 13

1. $5(x + y)$ **3.** $4(3m^2 − 2n^2)$ **5.** $2a(a^2 − 7a − 1)$ **7.** $5x(3x^2 − x − 4)$ **9.** $6a^2(3a − 2)$ **11.** $−2x(3x + 5)$
13. $5(\sqrt{3} + 3\sqrt{7}$ **15.** $2(\sqrt{3} − 5\sqrt{7}$ **17.** not difference **19.** difference **21.** difference
23. not perfect-square trinomial **25.** perfect-square trinomial **27.** perfect-square trinomial
29. difference of two cubes **31.** difference of two cubes **33.** sum of two cubes **35.** $(5y − 2)(5y + 2)$
37. *NSP*, this is a sum of two squares, not a difference. **39.** $(a + 1)^2$ **41.** $(4c − 3b)^2$ **43.** $(n − 13)^2$
45. $(6a + 7b)^2$ **47.** $(7 − x)^2$ **49.** *NSP*, this is a sum of two squares, not a difference. **51.** *NSP*, the
middle term needs a *y* factor. **53.** $(7 + 9y)(7 − 9y)$ **55.** $(3x + 10y)(3x − 10y)$ **57.** $(3x − y)^2$

59. $(3xy + z)(3xy − z)$ **61.** $(x + 2)^2$ **63.** $\left(\dfrac{2}{5}x + \dfrac{1}{4}y\right)\left(\dfrac{2}{5}x − \dfrac{1}{4}y\right)$ **65.** $(T − 2)(T^2 + 2T + 4)$

67. $(d + 9)(d^2 − 9d + 81)$ **69.** $(3k + 4)(9k^2 − 12k + 16)$ **71.** $(x + 3)(x + 8)$ **73.** $(x + 3)(x + 10)$
75. $(x − 1)(x − 8)$ **77.** $(x + 2)(x − 13)$ **79.** $(x + 8)(x − 3)$ **81.** $(6x + 1)(x + 4)$ **83.** $(5x − 4)(x − 6)$
85. $(3x + 7)(2x − 5)$ **87.** $(7x + 8)(x − 3)$ **89.** $(3a + 10)(3a − 10)$ **91.** $(2x + 1)(x − 2)$
93. $(a + 9)(a − 9)$ **95.** $(y − 7)^2$ **97.** $(b + 3)(b + 5)$ **99.** $(13 + m)(13 − m)$ **101.** $(x + 4)(x − 8)$
103. $(x + 20)(x − 1)$ **105.** $2(x + 2)(x − 4)$ **107.** $2x(x + 1)(x − 6)$

Practice Test, Chapter 13

1. $x(7x + 8)$ **3.** $7ab(a − 2)$ **5.** $(3x − 5)(3x + 5)$ **7.** $(x − 9)^2$ **9.** $(3r + 4s)(9r^2 − 12rs + 16s^2)$
11. $(3x + 2)(2x − 3)$ **13.** $(a + 8b)^2$ **15.** $(b + 2)(b − 5)$ **17.** $(3m − 2)(m − 1)$
19. $3(x + 2)(x − 2)$ **21.** $5(x + 2)(x − 2)$ **23.** $3(x + 2)^2$

Section 14–1 Self-Study Exercises

1 **1.** $\dfrac{4}{9}$ **2.** $\dfrac{3}{8}$ **3.** $2ab^2$ **4.** $\dfrac{3a}{2b^3}$ **5.** $\dfrac{x}{x + 2}$ **6.** $\dfrac{3(3x + 2)}{x(2x − 1)}$ or $\dfrac{9x + 6}{2x^2 − x}$ **7.** $\dfrac{x^2 − 3x − 10}{x^2 + 3x − 10}$

8. $\dfrac{1}{2}$ **9.** $\dfrac{1}{a − b}$ **10.** $\dfrac{x − 2}{x + 2}$ **11.** $\dfrac{2(x − 2)}{3}$ or $\dfrac{2x − 4}{3}$ **12.** $−1$ **13.** $−(x + 2)$ or $−x − 2$

14. $-\dfrac{1}{2}$ **15.** -3 **16.** $-\dfrac{1}{2}$ **17.** -5 **18.** $-\dfrac{2}{x-2}$ **19.** $-\dfrac{x-5}{x+1}$ **20.** -1 **21.** $\dfrac{x+3}{x+2}$

22. $\dfrac{x-3}{x+3}$ **23.** $\dfrac{x+1}{x+4}$ **24.** $\dfrac{2x-1}{3x-2}$

Section 14–2 Self-Study Exercises

1   **1.** $\dfrac{1}{2}$ **2.** $\dfrac{1}{10}$ **3.** $\dfrac{10x^3}{9y^2}$ **4.** $\dfrac{7}{15}$ **5.** $\dfrac{3}{x-2}$ **6.** $\dfrac{4}{x-2}$ **7.** $3(a-b)$ or $3a-3b$

8. 4 **9.** $\dfrac{a-7}{b+5}$ **10.** $\dfrac{3(x+1)}{x-1}$ or $\dfrac{3x+3}{x-1}$ **11.** $\dfrac{1}{12a}$ **12.** $\dfrac{3}{2}$ **13.** $\dfrac{2}{3}$ **14.** $\dfrac{3y^3}{2x}$ **15.** $\dfrac{3ab}{10}$

16. $\dfrac{1}{2}$ **17.** -2 **18.** 1 **19.** $\dfrac{x}{x-2}$ **20.** $\dfrac{50(a-b)}{ab}$ or $\dfrac{50a-50b}{ab}$ **21.** 1

2   **22.** $\dfrac{25}{27}$ **23.** $\dfrac{21}{20}$ or $1\dfrac{1}{20}$ **24.** $\dfrac{x-5}{12x}$ **25.** $\dfrac{2}{3x(x-2)}$ or $\dfrac{2}{3x^2-6x}$ **26.** $\dfrac{x+3}{2}$ **27.** $\dfrac{x-3}{3}$

28. $x+y$ **29.** $-\dfrac{x-1}{x+1}$ or $\dfrac{-x+1}{x+1}$ **30.** $3(x-6)$ or $3x-18$ **31.** $\dfrac{48x}{x-5}$

3   **32.** $-3(2+\sqrt{5})$ **33.** $5+2\sqrt{3}$ **34.** $\dfrac{22-9\sqrt{2}}{14}$ **35.** $\dfrac{14-\sqrt{7}}{9}$

36. $\dfrac{14+4\sqrt{3}+21\sqrt{5}+6\sqrt{15}}{37}$ **37.** $13+5\sqrt{6}$ **38.** $-\dfrac{1}{14+7\sqrt{5}}$ **39.** $\dfrac{14}{75+15\sqrt{11}}$

40. $\dfrac{1}{8-2\sqrt{6}}$ **41.** $\dfrac{1}{3-\sqrt{2}}$ **42.** $-\dfrac{2}{28-21\sqrt{2}}$ **43.** $\dfrac{13}{40-16\sqrt{3}}$

Section 14–3 Self-Study Exercises

1   **1.** $\dfrac{5}{7}$ **2.** $\dfrac{13}{16}$ **3.** $\dfrac{3x}{2}$ **4.** $\dfrac{23x}{24}$ **5.** $\dfrac{4+3x}{2x}$ **6.** $\dfrac{55}{6x}$ **7.** $\dfrac{9x-1}{(x+1)(x-1)}$ or $\dfrac{9x-1}{x^2-1}$

8. $\dfrac{10x+6}{(x+3)(x-1)}$ or $\dfrac{10x+6}{x^2+2x-3}$ **9.** $\dfrac{x-7}{(2x+1)(x-1)}$ or $\dfrac{x-7}{2x^2-x-1}$

10. $\dfrac{x+1}{(x-6)(x-5)}$ or $\dfrac{x+1}{x^2-11x+30}$

2   **11.** $\dfrac{5(2x+1)}{3x}$ or $\dfrac{10x+5}{3x}$ **12.** $\dfrac{8(3x+2)}{5x}$ or $\dfrac{24x+16}{5x}$ **13.** $-\dfrac{28}{9}$ or $-3\dfrac{1}{9}$ **14.** $-\dfrac{1}{7}$

15. $\dfrac{11}{3x^2}$ **16.** $\dfrac{10+9x}{3+10x}$ **17.** $\dfrac{-2(x-2)}{x+2}$ or $\dfrac{4-2x}{x+2}$ **18.** 9

Section 14–4 Self-Study Exercises

1 **1.** 0 **2.** 0 **3.** 5, 0 **4.** 0, −9 **5.** −8, 8 **6.** 2, −2 **7.** $\frac{3}{4}$, 0 **8.** 0, −3

2 **9.** 35 **10.** 5 **11.** 1 **12.** $\frac{1}{4}$ **13.** no solution **14.** no solution **15.** $1\frac{5}{7}$ days

16. 8 investors **17.** 25 mph going **18.** 3 min **19.** 50 lb dark-roast, 75 lb medium-roast, $2 per pound

20. 6 ohms

Chapter Review Exercises, Chapter 14

1. $\frac{3}{4}$ **3.** $\frac{b^2}{2ac}$ **5.** $\frac{x-3}{2(x+3)}$ **7.** −1 **9.** $\frac{m^2-n^2}{m^2+n^2}$ **11.** $\frac{1}{1+y}$ **13.** 5 **15.** $y+1$

17. $\frac{2}{x+6}$ **19.** 3 **21.** $\frac{5x^3}{4y^2}$ **23.** 15 **25.** $\frac{4(2y-1)}{y-1}$ **27.** 1 **29.** $\frac{x+3}{x+2}$ **31.** $\frac{1}{4}$

33. $\frac{(y-1)^3}{y}$ **35.** $\frac{y+3}{y+2}$ **37.** $\frac{3x^2}{2}$ **39.** $(y+4)(y-3)$ **41.** $\frac{5}{4x-12}$ **43.** $\frac{4x^2}{3}$

45. $\frac{72+12\sqrt{5}}{31}$ **47.** $\frac{26+7\sqrt{3}}{23}$ **49.** $\frac{35+15\sqrt{5}+7\sqrt{2}+3\sqrt{10}}{4}$ **51.** $\frac{26}{20-5\sqrt{3}}$

53. $\frac{1}{16+4\sqrt{13}}$ **55.** $\frac{9}{40+8\sqrt{7}}$ **57.** $\frac{2}{3}$ **59.** $\frac{4x}{7}$ **61.** $\frac{19x}{12}$ **63.** $\frac{15-7x}{3x}$ **65.** $\frac{13}{4x}$

67. $\frac{10x+5}{(x-3)(x+2)}$ **69.** $\frac{6x-38}{(x+3)(x-4)}$ or $\frac{2(3x-19)}{(x+3)(x-4)}$ **71.** $\frac{x+3}{x-5}$ **73.** $\frac{51}{28}$ **75.** $\frac{3x^2-30}{2x^2+12}$

77. 0, 2 **79.** $\frac{1}{2}$, 0 **81.** $-\frac{20}{3}$ **83.** $\frac{1}{7}$ **85.** 3 students **87.** $7\frac{1}{2}$ hr

Practice Test, Chapter 14

1. $\frac{1}{2}$ **3.** $\frac{3x-4}{x+3}$ **5.** $-\frac{x-2}{x+2}$ or $\frac{2-x}{x+2}$ **7.** $\frac{3a}{y}$ **9.** $\frac{1}{x^2(x+2y)}$ **11.** $\frac{2x+1}{x}$

13. $-\frac{5}{(x+2)(x-3)}$ **15.** $\frac{12+x}{4x}$ **17.** $-\frac{2}{3x-2}$ or $\frac{2}{2-3x}$ **19.** $\frac{1}{x+2y}$ **21.** $\frac{2x^2}{x-3}$

23. 0, −3 **25.** $\frac{11}{7}$ **27.** $\frac{1}{3}$ **29.** 5 persons

Section 15–1 Self-Study Exercises

1 **1.** $7x^2-4x+5=0$ **2.** $8x^2-6x-3=0$ **3.** $7x^2-5=0$ **4.** $x^2-6x+8=0$
5. $x^2-9x+8=0$ **6.** $x^2-4x-8=0$ **7.** $3x^2-6x+5=0$ **8.** $x^2-6x-5=0$ **9.** $x^2-16=0$
10. $8x^2-7x-8=0$ **11.** $8x^2+8x-10=0$ **12.** $0.3x^2-0.4x-3=0$

2 **13.** $x^2 - 5x = 0$
  $a = 1; b = -5; c = 0$

14. $3x^2 - 7x + 5 = 0$
  $a = 3; b = -7; c = 5$

15. $7x^2 - 4x = 0$
  $a = 7; b = -4; c = 0$

16. $3x^2 - 5x + 8 = 0$
  $a = 3, b = -5, c = 8$

17. $x^2 - 5x + 6 = 0$
  $a = 1, b = -5, c = 6$

18. $11x^2 - 8x = 0$
  $a = 11, b = -8, c = 0$

19. $x^2 - x = 0$
  $a = 1, b = -1, c = 0$

20. $9x^2 - 7x - 12 = 0$
  $a = 9, b = -7, c = -12$

21. $x^2 - 5 = 0$
  $a = 1, b = 0, c = -5$

22. $x^2 + 6x - 3 = 0$
  $a = 1, b = 6, c = -3$

23. $5x^2 - 0.2x + 1.4 = 0$
  $a = 5, b = -0.2, c = 1.4$

24. $\dfrac{2}{3}x^2 - \dfrac{5}{6}x - \dfrac{1}{2} = 0$

  $a = \dfrac{2}{3}, b = -\dfrac{5}{6}, c = -\dfrac{1}{2}$

25. $1.3x^2 - 8 = 0$
  $a = 1.3, b = 0, c = -8$

26. $\sqrt{3}x^2 + \sqrt{5}x - 2 = 0$
  $a = \sqrt{3}, b = \sqrt{5}, c = -2$

27. $15x^2 + 3x + 5 = 0$
  $a = 15, b = 3, c = 5$

28. $8x^2 - 2x - 3 = 0$

29. $x^2 + 3x = 0$

30. $5x^2 + 2x - 7 = 0$

31. $2.5x^2 - 0.8 = 0$

3 **32.** $x = \pm 3$ **33.** $x = \pm 7$ **34.** $x = \pm \dfrac{8}{3}$ **35.** $x = \pm \dfrac{7}{4}$ **36.** $y = \pm 3$ **37.** $x = \pm \dfrac{9}{2}$

38. $x = \pm \sqrt{5}$ or ± 2.236 **39.** $x = \pm 1$ **40.** $x = \pm 2$ **41.** $x = \pm 1.732$ **42.** $x = \pm \dfrac{3}{2}$ **43.** $x = \pm 2$

44. 23 ft **45.** 81.759 yd **46.** Isolate the quadratic term; then take the square root of both sides of the equation and simplify when necessary. **47.** opposites

Section 15–2 Self-Study Exercises

1 **1.** $x = 3$ or 0 **2.** $x = 0$ or 6 **3.** $x = 0$ or 2 **4.** $x = 0$ or $-\dfrac{1}{2}$ **5.** $x = 0$ or $\dfrac{1}{2}$ **6.** $x = 0$ or 3

7. $x = 0$ or $\dfrac{7}{3}$ **8.** $y = 0$ or -4 **9.** $x = 0$ or $-\dfrac{2}{3}$ **10.** $x = 0$ or $\dfrac{4}{3}$ **11.** $x = 0$ or 3 **12.** $x = 0$ or $\dfrac{2}{3}$

13. 8 or 0 **14.** 15 units **15.** An incomplete quadratic equation is missing the constant while a pure quadratic equation is missing the linear term. **16.** Yes, the common factor of x will be set equal to zero.

2 **17.** -3 or -2 **18.** 3 (double root) **19.** 7 or -2 **20.** 3 or -6 **21.** -4 or -3 **22.** 5 or 3
23. 14 or -1 **24.** 3 or 6 **25.** $\dfrac{1}{2}$ or 3 **26.** $-\dfrac{1}{3}$ or -4 **27.** $\dfrac{3}{5}$ or $-\dfrac{1}{2}$ **28.** $-\dfrac{1}{3}$ or $-\dfrac{3}{2}$ **29.** $-\dfrac{3}{2}$ or -5
30. $\dfrac{4}{3}$ or 2 **31.** $\dfrac{1}{6}$ or -3 **32.** -4 or $-\dfrac{2}{3}$ **33.** 5 or $\dfrac{3}{2}$ **34.** $\dfrac{1}{2}$ or $\dfrac{2}{3}$ **35.** $\dfrac{3}{4}$ or $\dfrac{1}{2}$ **36.** $\dfrac{2}{5}$ or -1
37. $\dfrac{5}{3}$ or $-\dfrac{3}{2}$ **38.** $-\dfrac{5}{3}$ or $-\dfrac{1}{3}$ **39.** $\dfrac{1}{3}$ or $\dfrac{3}{2}$ **40.** -3 or $\dfrac{2}{5}$ **41.** $w = 5$ ft $l = 11$ ft **42.** $l = 21$ in. $w = 18$ in.

Section 15–3 Self-Study Exercises

1   **1.** $-8, 6$ **2.** $-1, 9$ **3.** $1, 9$ **4.** $4, 6$ **5.** $-6, 4$ **6.** $-14, -2$ **7.** $\dfrac{3 \pm \sqrt{29}}{2}$ **8.** $\dfrac{5 \pm \sqrt{33}}{2}$

9. $\dfrac{5 \pm \sqrt{37}}{2}$ **10.** $\dfrac{3 \pm \sqrt{13}}{2}$ **11.** $\dfrac{2 \pm \sqrt{10}}{2}$ **12.** $\dfrac{1 \pm \sqrt{3}}{2}$ **13.** $\dfrac{3 \pm \sqrt{3}}{3}$ **14.** $\dfrac{-6 \pm 2\sqrt{3}}{3}$

15. $-1, 3$ **16.** $\dfrac{3 \pm \sqrt{205}}{14}$ **17.** $2, \dfrac{1}{5}$ **18.** $\dfrac{1 \pm \sqrt{85}}{6}$

2   **19.** $3, -\dfrac{2}{3}$ **20.** $3, -4$ **21.** $\dfrac{11}{5}, -1$ **22.** 3 (double root)   **23.** $3, -2$ **24.** $\dfrac{3}{4}, -\dfrac{1}{2}$ **25.** $1.84, -10.84$
26. $-0.18, -1.82$ **27.** $3.14, -0.64$ **28.** $2.39, 0.28$ **29.** $-0.75 \pm 0.97i$ or no real solution
30. $0.5 \pm 1.32i$ or no real solution   **31.** width = 5.55 cm, length = 8.55 cm   **32.** width = 4 in., length = 10 in.
33. 4 m   **34.** width = 11 ft, length = 16 ft   **35.** length = 110 ft, width = 70 ft (nearest ft); see figure below.
A solution that is larger than the orginal field must be disregarded.   **36.** 111.5 kg

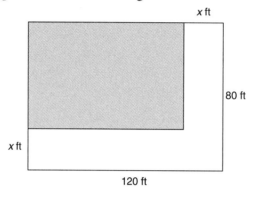

Section 15–4 Self-Study Exercises

1   **1.** The domain is the set of all real numbers. The range is the set of all real numbers greater than or equal to zero.

2. The domain is the set of all real numbers. The range is the set of all real numbers greater than or equal to zero.

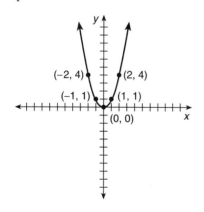

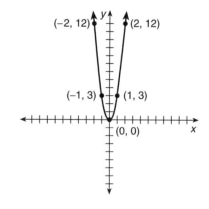

3. The domain is the set of all real numbers. The range is the set of all real numbers greater than or equal to zero.

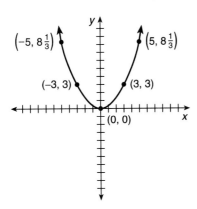

4. The domain is the set of all real numbers. The range is the set of all real numbers less than or equal to zero.

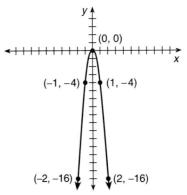

5. The domain is the set of all real numbers. The range is the set of all real numbers less than or equal to zero.

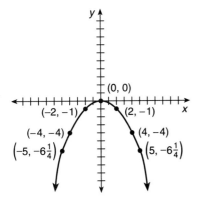

6. The domain is the set of all real numbers. The range is the set of all real numbers greater than or equal to -4.

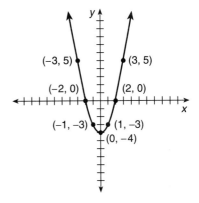

7. The domain is the set of all real numbers. The range is the set of all real numbers greater than or equal to 4.

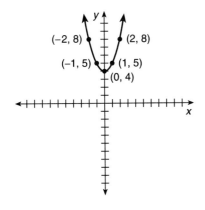

8. The domain is the set of all real numbers. The range is the set of all real numbers greater than or equal to zero.

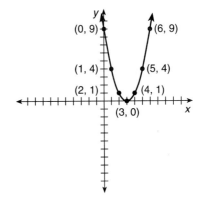

9. The domain is the set of all real numbers. The range is the set of all real numbers less than or equal to 0.

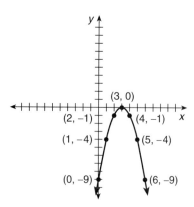

2 **10.** The domain is the set of all real numbers. The range is the set of all real numbers greater than or equal to 0.

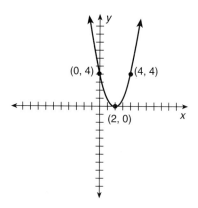

vertex: $(2, 0)$
x-intercept: $(2, 0)$

11. The domain is the set of all real numbers. The range is the set of all real numbers greater than or equal to -6.

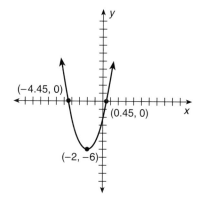

vertex: $(-2, -6)$
x-intercepts: $(-4.45, 0)$; $(0.45, 0)$

12. The domain is the set of all real numbers. The range is the set of all real numbers greater than or equal to -4.5.

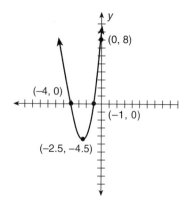

vertex: $(-2.5, -4.5)$
x-intercepts: $(-4, 0)$; $(-1, 0)$

13. The domain is the set of all real numbers. The range is the set of all real numbers less than or equal to 12.

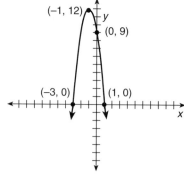

vertex: $(-1, 12)$
x-intercepts: $(-3, 0)$; $(1, 0)$

14. The domain is the set of all real numbers. The range is the set of all real numbers less than or equal to -3.

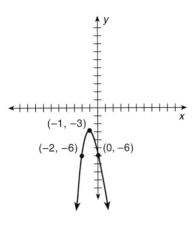

vertex: $(-1, -3)$
no x-intercepts

15. The domain is the set of all real numbers. The range is the set of all real numbers greater than or equal to -15.

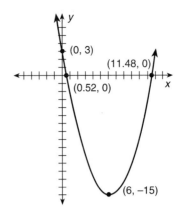

vertex: $(6, -15)$
x-intercepts: $(0.52, 0)$; $(11.48, 0)$

3 **16.** $x = 2, x = 3$ **17.** $x = 0$ **18.** no real solution **19.** \$0, \$9 **20.** $t = 0$ s

4 **21.** $y = 3x^2 + 5x - 2$

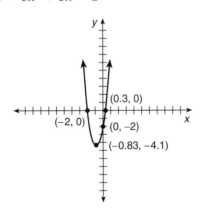

22. $y = (2x - 3)(x - 1)$

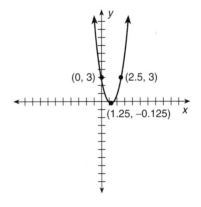

23. $y = 2x^2 - 9x - 5$

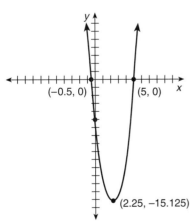

24. $y = -2x^2 + 9x + 5$

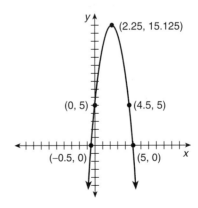

5 **25.** real, rational, $x = \dfrac{2}{3}, -1$ **26.** real, irrational, $x = \dfrac{3 \pm \sqrt{5}}{2}$ or 2.62, 0.38

27. real, irrational, $x = \dfrac{-1 \pm \sqrt{17}}{4}$ or 0.78, -1.28 **28.** no real solutions or $x = \dfrac{1 \pm i\sqrt{2}}{3}$ or $0.33 \pm 0.47i$

29. real, irrational, $x = \dfrac{3 \pm \sqrt{37}}{2}$ or 4.54, -1.54 **30.** real, irrational, $x = \dfrac{-5 \pm \sqrt{97}}{6}$ or 0.81 or -2.47

31. The discriminant must be greater than or equal to zero.   **32.** The discriminant must be greater than zero and a perfect square.

Section 15–5 Self-Study Exercises

1 **1.** 2   **2.** 1   **3.** 4   **4.** 1   **5.** 4   **6.** 7

2 **7.** $x = 0, 2, -3$ **8.** $x = 0, \dfrac{1}{2}, -3$ **9.** $x = 0, \dfrac{5}{2}, \dfrac{2}{3}$ **10.** $x = 0, 2, 5$ **11.** $x = 0, 3, -2$

12. $x = 0, -\dfrac{1}{2}, -2$ **13.** $x = 0, 3, -3$ **14.** $x = 0, \dfrac{1}{2}, -\dfrac{1}{2}$ **15.** $x = 0, \dfrac{3}{4}, -\dfrac{3}{4}$ **16.** $3; x = 0, 4, -4$

17. real root: $x = 0$; imaginary roots: $x = \pm i\sqrt{5}$ **18.** $w(w + 3)(w + 7) = 421$; No; The factored form of the equation does not equal to zero. When put into standard form a cubic equation is formed that cannot be factored using methods developed in this text.

3 **19.** $y = 2x^3$

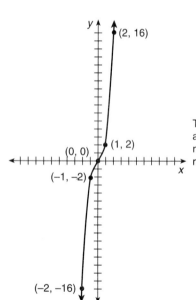

x	y
−2	−16
−1	−2
0	0
1	2
2	16

The domain is the set of all real numbers. The range is the set of all real numbers.

20. $y = \dfrac{1}{2}x^3$

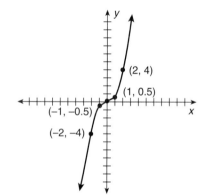

x	y
−2	−4
−1	−0.5
0	0
1	0.5
2	4

The domain is the set of all real numbers. The range is the set of all real numbers.

21. $y = x^3 - x^2 - 4x + 4$

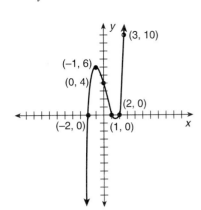

x	y
-3	-20
-2	0
-1	6
0	4
1	0
2	0
3	10

The domain is the set of all real numbers. The range is the set of all real numbers.

22. $y = x^3 + 2$

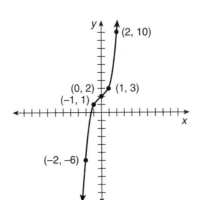

x	y
-2	-6
-1	1
0	2
1	3
2	10

The domain is the set of all real numbers. The range is the set of all real numbers.

23. $y = -x^3 + 4$

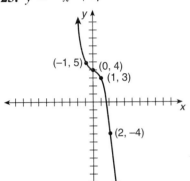

x	y
-2	12
-1	5
0	4
1	3
2	24
3	223

The domain is the set of all real numbers. The range is the set of all real numbers.

24. $y = -x^3 - 3x$

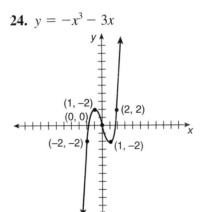

x	y
-3	-18
-2	-2
-1	2
0	0
1	-2
2	2

The domain is the set of all real numbers. The range is the set of all real numbers.

4 **25.** The graph is the graph of a function.

26. The graph is the graph of a function.

27. The graph is the graph of a relation but not a function.

28. The graph is the graph of a relation but not of a function.

29. The graph is the graph of a function.

30. The graph is the graph of a function.

5 **31.** domain—all real numbers; range—all real numbers greater than or equal to -2

32. domain—the real numbers between and including -5 and 5; range—the real numbers between and including -3 and 3

33. domain—the real numbers between and including -4 and 3; range—the real numbers between and including -6 and 3

34. domain—the real numbers greater than or equal to 0; range—the real numbers greater than or equal to 0

35. domain—the real numbers greater than or equal to -7 and less than or equal to 5; range—the real numbers greater than or equal to -2 and less than or equal to 5

36. domain—the real numbers greater than or equal to -3 and less than 3; range—the real numbers greater than or equal to 0

Chapter Review Exercises, Chapter 15

1. pure **3.** pure **5.** incomplete **7.** pure **9.** complete **11.** $2x^2 - 8x - 5 = 0$

13. $x^2 - 7x + 5 = 0$ **15.** $4x^2 + 3x - 1 = 0$ **17.** $x = \pm 10$ **19.** $x = \pm\dfrac{3}{2}$ or ± 1.5 **21.** $y = \pm 1.740$

23. $x = \pm 2.828$ **25.** $x = \pm 2.236$ **27.** $x = \pm 2$ **29.** $x = \pm 4.123$ **31.** $y = \pm 3.055$ **33.** $x = \pm 4$

35. $x = \pm 8$ **37.** 16 cm **39.** 0 or 5 **41.** 0 or 2 **43.** 0 or $-\dfrac{1}{2}$ **45.** 0 or 7 **47.** 0 or $-\dfrac{2}{3}$ **49.** 0 or -3

51. 0 or 9 **53.** 0 or -8 **55.** 0 or $\dfrac{5}{3}$ **57.** 0 or $\dfrac{1}{2}$ **59.** 0 or 4 **61.** 3 or 1 **63.** -5 or 2

65. -6 or -1 **67.** 2 or 4 **69.** $-\dfrac{2}{3}$ or $\dfrac{3}{2}$ **71.** $-\dfrac{2}{5}$ or $\dfrac{5}{2}$ **73.** $-\dfrac{3}{4}$ or -1 **75.** $\dfrac{3}{4}$ or $-\dfrac{1}{3}$ **77.** -21 or 2

79. $\dfrac{2}{3}$ or -1 **81.** 3 or 2 **83.** 6 or -3 **85.** -6 or -5 **87.** -9 or 2 **89.** width $= 12$ ft length $= 19$ ft

91. 2 (double root) **93.** 2 or 6 **95.** $4 - \sqrt{2}$ or $4 + \sqrt{2}$ **97.** $3 - i\sqrt{3}$ or $3 + i\sqrt{3}$ **99.** 1 or 4

101. $\dfrac{3 - \sqrt{37}}{2}$ or $\dfrac{3 + \sqrt{37}}{2}$ **103.** $a = 1$ $b = -2$ $c = -8$ **105.** $a = 1$ $b = 3$ $c = -4$ **107.** $a = 1$ $b = -3$ $c = 2$

109. 9 or -1 **111.** 2 or -4 **113.** 2 or $-\dfrac{1}{2}$ **115.** 1.78 or -0.28 **117.** -0.23 or -1.43

119. $w = 11$ ft, $l = 22$ ft **121.** $w = 14$ in., $l = 42$ in.

123.

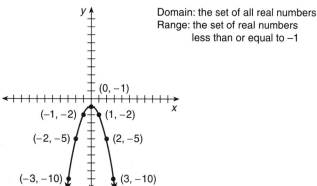

Domain: the set of all real numbers
Range: the set of real numbers less than or equal to –1

125.

Domain: the set of all real numbers
Range: the set of real numbers greater than –1

Selected Answers to Student Exercise Material

127.

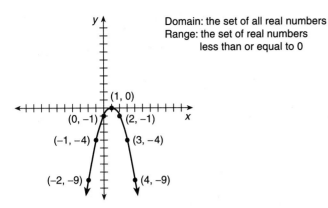

Domain: the set of all real numbers
Range: the set of real numbers
less than or equal to 0

129. vertex: $(1, -9)$
axis of symmetry: $x = 1$

131. vertex: $(4, 4)$
axis of symmetry: $x = 4$

133. $x = -2$ or 6 **135.** $x = -4$ (double root)

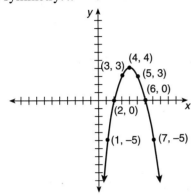

137. $y = -2x^2$

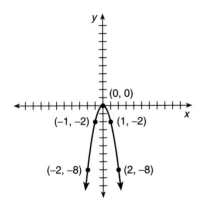

139. real, rational, double root   **141.** real, irrational, 2 unequal roots   **143.** real, rational, 2 unequal roots

145. 3   **147.** 1   **149.** 8   **151.** $0, 2, \dfrac{2}{3}$ **153.** $0, -2, -3$ **155.** $0, -5, \dfrac{1}{2}$ **157.** $0, 2$ **159.** $0, -2, -4$

161. $0, 5, -4$ **163.** $0, 3 \pm \sqrt{2}$ or $4.414, 1.586$ **165.** $0, -1, 4$ **167.** $-2, -1, 0$

169. $y = 5x^3$

x	y
−2	−40
−1	−5
0	0
1	5
2	40
3	135

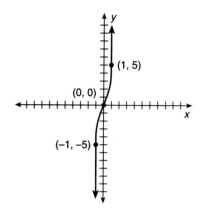

The domain is the set of all real numbers. The range is the set of all real numbers.

171. The graph is the graph of a relation but not a function since at least one vertical line can be drawn that intersects the graph in more than one point.

173. The graph is the graph of a relation but not a function. A vertical line intersects the graph in two points everywhere except at $(-5, 0)$ and $(5, 0)$.

175. domain—the set of all real numbers
range—the set of all real numbers less than or equal to 4

177. domain—the set of all real numbers from −2 to 2 inclusive
range—all real numbers from −3 to 9 inclusive

Practice Test, Chapter 15

1. pure   **3.** incomplete   **5.** ± 9 **7.** 0 or 2   **9.** 3 or 2   **11.** $\frac{3}{2}$ or 4   **13.** 1 or $-\frac{5}{2}$

15. $\dfrac{3 + \sqrt{29}}{2}$ or 4.19, $\dfrac{3 - \sqrt{29}}{2}$ or −1.19

17. real, rational, 2 unequal roots   **19.** $y = x^2 + 2x + 1$

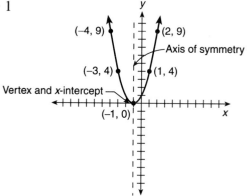

21. 169.79 mils   **23.** 1.98 amps   **25.** 6   **27.** 0 or $\frac{3}{2}$ or −5   **29.** 0 or 3

31. The graph is the graph of a relation but not a function. The domain is the set of all real numbers. The range is the set of all real numbers.

Selected Answers to Student Exercise Material

Chapters 11–15 Cumulative Practice Test

1. $40x^{10}$　**3.** x^{32}　**5.** $12x^4 + 5x^2$　**7.** $2x^2 - 4x^3 + 3x$　**9.** $8x(3x^2 - 2x - 1)$　**11.** $12x^2 - 23x + 10$
13. $x - 3$　**15.** $(x - 8)^2$　**17.** $(x + 7)(x - 6)$　**19.** $(2x + 3)(3x + 4)$　**21.** $x = \pm 3$　**23.** $x = 0; x = 8$
25. $x = \frac{1}{2}; x = 3$　**27.** $x \approx 1.18; x \approx -0.43$　**29.** $x = -2; x = 5; x = 0$

Section 16–1 Self-Study Exercises

1　**1.** $2,415.77　**2.** $2,050.40　**3.** (a) 122.14 (b) 149.18 (c) 164.87 (d) 182.21　**4.** (a) $1,197.22
(b) $1,576.91　**5.** 64　**6.** 0.0041　**7.** 9,765,625　**8.** 0.002　**9.** 243　**10.** 0.0032

2　**11.** 2.08×10^{-87}　**12.** 4.28×10^{-96}　**13.** 5.58×10^{-44}　**14.** 7.39　**15.** 0.05　**16.** 1.23
17. 0.03　**18.** $720.98　**19.** $14,448.89; $6,448.89　**20.** $15,373.05; $4,873.05　**21.** $14,414.25
22. $8\frac{1}{4}$% annually is the better deal.　**23.** $90,305.56　**24.** $9.26　**25.** $7.72　**26.** $27.99

27. $321.89　**28.** 8.24%　**29.** 10.25%　**30.** 6.14%　**31.** 12.55%　**32.** $2,252.25　**33.** $7,906.98
34. $19,462.47　**35.** $6,462.60　**36.** $1,587.66　**37.** $1,254.58　**38.** $6,736.25　**39.** $13,611.66
40. $6,270; $270　**41.** $8,280; $280　**42.** $16,232　**43.** $9,786.68　**44.** $127,391.11; $91,000
45. $60,743.42　**46.** $25,129.02　**47.** $7,243.28　**48.** $5,866.60

49.

Years	Total Investment	Total Interest
Ten-year	$5,000	$2,243.28
Five-year	$5,000	$866.60

50. $908.92　**51.** $1,155.89

The 10-year investment earned more interest even though half as much money was invested per year. At the same period interest rate, investing for twice as long gives a better yield on your investment than investing the same amount for half as long. Thus, the earlier you start saving, the better.

3　**52.** $x = 7$　**53.** $x = -3$　**54.** $x = 2$　**55.** $x = 10$　**56.** $x = 7$　**57.** $x = 3$　**58.** $x = 6$
59. $x = -4$　**60.** $x = -6$　**61.** $x = \frac{7}{6}$　**62.** $x = \frac{5}{2}$　**63.** $x = 5$　**64.** $x = 8$　**65.** $x = -3$　**66.** $x = 3$
67. $x = \frac{15}{11}$

68.

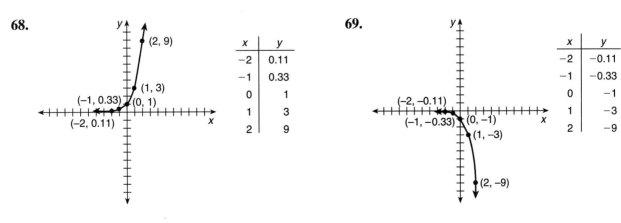

x	y
-2	0.11
-1	0.33
0	1
1	3
2	9

69.

x	y
-2	-0.11
-1	-0.33
0	-1
1	-3
2	-9

70.

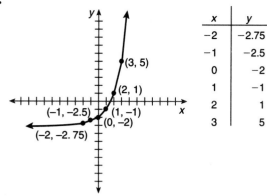

x	y
−2	−2.75
−1	−2.5
0	−2
1	−1
2	1
3	5

71.

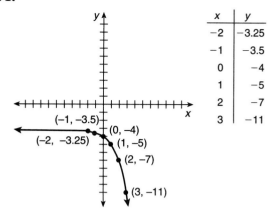

x	y
−2	−3.25
−1	−3.5
0	−4
1	−5
2	−7
3	−11

72.

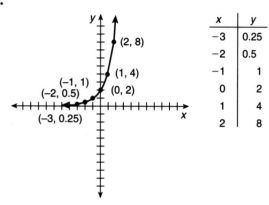

x	y
−3	0.25
−2	0.5
−1	1
0	2
1	4
2	8

73.

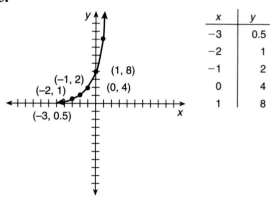

x	y
−3	0.5
−2	1
−1	2
0	4
1	8

74.

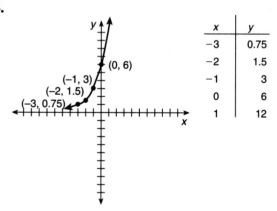

x	y
−3	0.75
−2	1.5
−1	3
0	6
1	12

75.

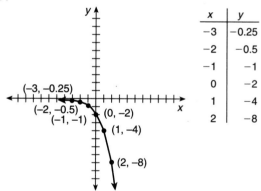

x	y
−3	−0.25
−2	−0.5
−1	−1
0	−2
1	−4
2	−8

76.

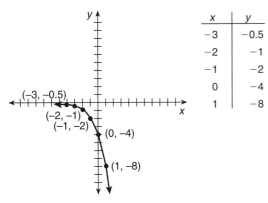

x	y
−3	−0.5
−2	−1
−1	−2
0	−4
1	−8

77. (a) 3.9721 units (b) between 5 and 6 min
78. between 13 and 14 min
79.

x	y
0	250
1	150
2	90
3	54
4	32.4
5	19.44
6	11.664

$x = 0$

54 mL remained in the bloodstream at 6:00 P.M.

80. between 1 and 2 h   **81.** after 1 year: \$1,050; after 2 years: \$1,102.50; after 10 years: \$1,628.89; after 30 years: \$4,321.94; after 50 years: \$11,467   **82.** 12.5; 2 cm

Section 16–2 Self-Study Exercises

1 **1.** $\log_3 9 = 2$ **2.** $\log_2 32 = 5$ **3.** $\log_9 3 = \dfrac{1}{2}$ **4.** $\log_{16} 2 = \dfrac{1}{4}$ **5.** $\log_4 \dfrac{1}{16} = -2$
6. $\log_3 \dfrac{1}{81} = -4$

2 **7.** $3^4 = 81$ **8.** $12^2 = 144$ **9.** $2^{-3} = \dfrac{1}{8}$ **10.** $5^{-2} = \dfrac{1}{25}$ **11.** $25^{-0.5} = \dfrac{1}{5}$ **12.** $4^{-0.5} = \dfrac{1}{2}$

13. $x = 3$ **14.** $x = \dfrac{1}{81}$ **15.** $x = 2$ **16.** $x = -2$ **17.** $x = 625$ **18.** $x = -4$

3 **19.** 0.4771   **20.** 0.7782   **21.** 0.3802   **22.** 0.6232   **23.** 2.1761   **24.** −2.9208   **25.** 1.3863
26. 0.9163   **27.** −1.8971   **28.** 5.6168   **29.** 4.6052

4 **30.** 3   **31.** 6   **32.** 5   **33.** 2   **34.** 1.9358

5 **35.** 3   **36.** 5   **37.** 8   **38.** 13.9 years   **39.** 2.0 years

40.

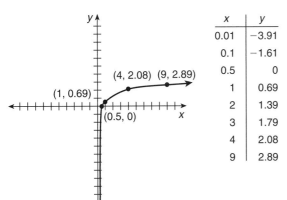

x	y
0.01	−3.91
0.1	−1.61
0.5	0
1	0.69
2	1.39
3	1.79
4	2.08
9	2.89

41.

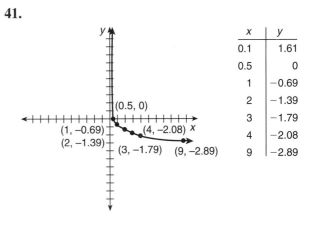

x	y
0.1	1.61
0.5	0
1	−0.69
2	−1.39
3	−1.79
4	−2.08
9	−2.89

42.

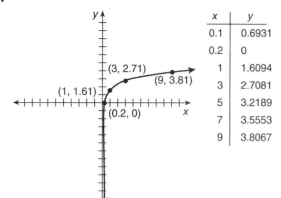

x	y
0.1	0.6931
0.2	0
1	1.6094
3	2.7081
5	3.2189
7	3.5553
9	3.8067

43.

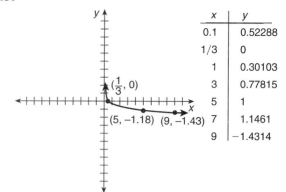

x	y
0.1	0.52288
1/3	0
1	0.30103
3	0.77815
5	1
7	1.1461
9	−1.4314

44.

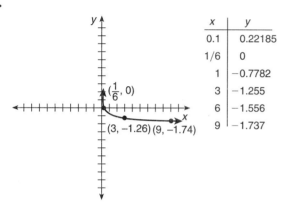

x	y
0.1	0.22185
1/6	0
1	−0.7782
3	−1.255
6	−1.556
9	−1.737

45.

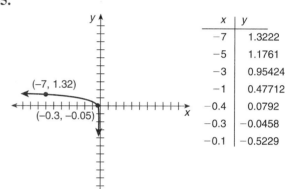

x	y
−7	1.3222
−5	1.1761
−3	0.95424
−1	0.47712
−0.4	0.0792
−0.3	−0.0458
−0.1	−0.5229

46. 39.8   **47.** 4   **48.** 7.29

7   **49.** 2.096   **50.** 1.893   **51.** 1.631   **52.** 2.727

Chapter Review Exercises, Chapter 16

1. $x = 8$ **3.** $x = -2$ **5.** $x = 4$ **7.** $x = -\dfrac{5}{3}$ **9.** $x = 4$ **11.** $x = -5$ **13.** $x = 1$ **15.** $x = 4$

17. 0.0183   **19.** 0.0000454   **21.** $\$708.64$ **23.** $\$524.95$ **25.** $\$5,372.54$ **27.** $\$2,189.94$
29. $\$13,928.19$ **31.** $\$1,356.25$ **33.** 8.24%   **35.** $\$190.99$ **37.** $\$2,913.78$ **39.** $\$11,000$ in
18 months is better.   **41.** $\$4,781.09$ **43.** $\$26,361.59$ **45.** $\$7,102.10$ **47.** $\$2,294.19$
49. $\$5,591.97$ **51.** $\$3,407.88$ **53.** $\$1,917.31$ **55.** 1.25×10^{-6} **57.** 2.71×10^{-35}

59.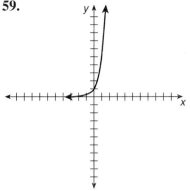

61.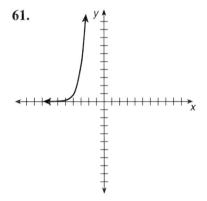

63.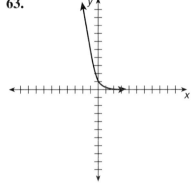

65. 16.3 mg **67.** $\log_2 8 = 3$ **69.** $\log_3 81 = 4$ **71.** $\log_{27} 3 = \dfrac{1}{3}$ **73.** $\log_4 \dfrac{1}{64} = -3$ **75.** $\log_9 \dfrac{1}{3} = -\dfrac{1}{2}$

77. $\log_{12} \dfrac{1}{144} = -2$ **79.** $11^2 = 121$ **81.** $15^0 = 1$ **83.** $7^1 = 7$ **85.** $4^{-2} = \dfrac{1}{16}$ **87.** $9^{-0.5} = \dfrac{1}{3}$

89. $10^3 = 1,000$ **91.** 0.6990 **93.** 2.2553 **95.** −0.3979 **97.** 5.5984 **99.** −0.2231 **101.** 2.1133

103. 2 **105.** 343 **107.** $x = -2$ **109.** $x = 32$ **111.** a. 2 b. 4 c. 8.1761 **113.** 3.17

115. 9.89 years

117.

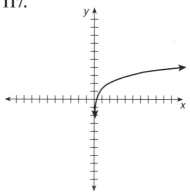

X	Y₁
1	1.3863
2	2.0794
3	2.4849
4	2.7726
5	2.9957
6	3.1781
7	3.3322

X = 1

119.

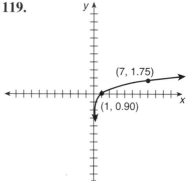

(7, 1.75)

(1, 0.90)

X	Y₁
1	.90309
2	1.2041
3	1.3802
4	1.5051
5	1.6021
6	1.6812
7	1.7482

X = 1

121. $c = 6.3$

Practice Test, Chapter 16

1. 3,657.26   **3.** 58.09   **5.** 248,832   **7.** $x = 9$ **9.** $x = 2$ **11.** $\log_4 \frac{1}{2} = -\frac{1}{2}$ **13.** $3^{-3} = \frac{1}{27}$

15. 3.4657   **17.** $x = 3$ **19.** 1.4641   **21.** $S = \$224.94$ thousands   **23.** \$5,596.82   **25.** \$19,350
27. \$5,727.50   **29.** 12.55%   **31.** \$680 in 1 year is better   **33.** Option 2 yields the greater return by \$0.69.
35. \$2,225.54   **37.** \$13,586.96

Section 17–1 Self-Study Exercises

1   **1.** F   **2.** F   **3.** T   **4.** T   **5.** $\{6, 7, 8, 9, 10, 11\}$ **6.** $\{4, 6\}$ **7.** $\{15, 20, 25, 30, 35\}$
8. $\{-4, -3, -2, -1\}$ **9.** $\{x \mid x < 10 \text{ and } x \text{ is odd natural number}\}$ **10.** $\{x \mid -7 \le x \le -1 \text{ and } x \text{ is odd integer}\}$

2   **11.** (5, 9)   **12.** (−7, −3)   **13.** [−5, −3]

14. [8, 12]   **15.** (−∞, 7)   **16.** (−∞, −2)

17. [2, ∞)   **18.** (−∞, 3]   **19.** (5, ∞)

20. 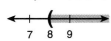 $[-3, \infty)$ **21.** 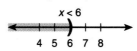 $(-\infty, -2]$ **22.** $(-6, \infty)$

23. $(843{,}000, 1{,}000{,}000)$ **24.** $[4, 18]$ **25.** $0 \le n < 12$ $[0, 12)$

Section 17–2 Self-Study Exercises

1 **1.** $y > 8; (8, \infty)$

2. $x < 1; (-\infty, 1)$

3. $x \le -3; (-\infty, 3]$

4. $x < 6; (-\infty, 6)$

5. $a \ge -3; [-3, \infty)$

6. $y \le 8; (-\infty, 8]$

7. $b > -1; (-1, \infty)$

8. $t < 6; (-\infty, 6)$

9. $y < 5; (-\infty, 5)$

10. $a \le -2; (-\infty, -2]$

11. $x \le \dfrac{1}{4}; \left(-\infty, \dfrac{1}{4}\right]$

12. $t \le 7; (-\infty, 7]$

13. $y > 11; (11, \infty)$

14. $x \ge 3; [3, \infty)$

15. $x < 1; (-\infty, 1)$

16. $x > \dfrac{1}{3}; \left(\dfrac{1}{3}, \infty\right)$

17. $a \le -3$; $(-\infty, -3]$

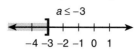

18. $x < \dfrac{1}{2}$; $\left(-\infty, \dfrac{1}{2}\right)$

19. $2x + 196 \ge 52{,}800$
$x \ge 26{,}302$

20. $x > \$3.60 \cdot 2$
$x > \$7.20$

2 **21.**

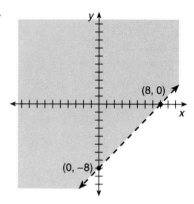

22.

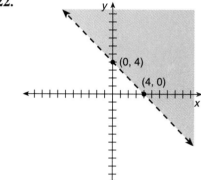

23.

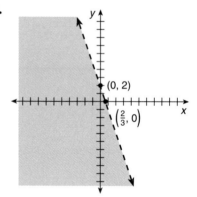

24.

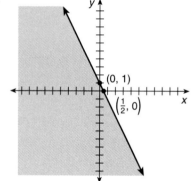

25.

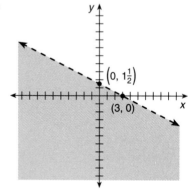

26.

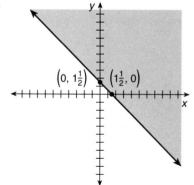

27.

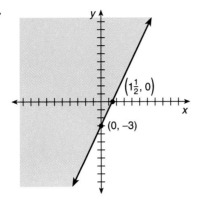

$\left(1\frac{1}{2}, 0\right)$

$(0, -3)$

28.

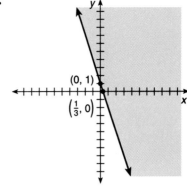

$(0, 1)$

$\left(\frac{1}{3}, 0\right)$

29.

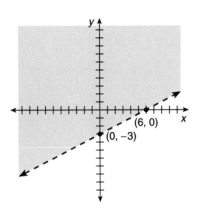

$(6, 0)$

$(0, -3)$

30.

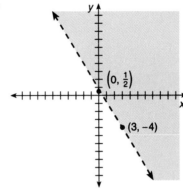

$\left(0, \frac{1}{2}\right)$

$(3, -4)$

31.

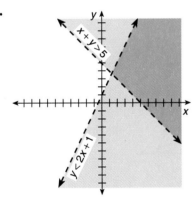

32.

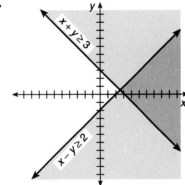

33. $x - 2y \leq -1$ $\qquad$ $x + 2y \geq 3$

$y \geq \frac{1}{2}x + \frac{1}{2}$ $\qquad$ $y \geq -\frac{1}{2}x + \frac{3}{2}$

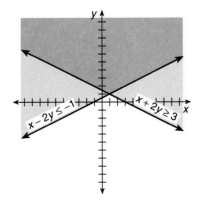

34. $x + y > 4$ $\qquad$ $x - y > -3$

$y > -x + 4$ $\qquad$ $y > x + 3$

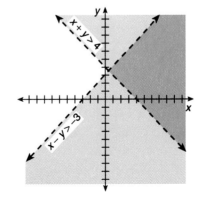

35. $3x - 2y < 8$ $\qquad$ $2x + y \le -4$

$y > \dfrac{3}{2}x - 4$ $\qquad$ $y \le -2x - 4$

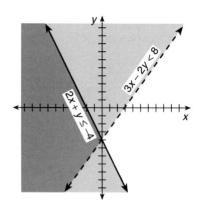

36. $y > \dfrac{1}{2}x + 3$

$y \le -2x + 5$

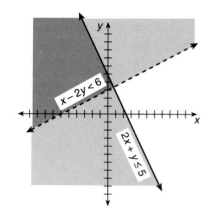

Section 17–3 Self-Study Exercises

1 **1.** *T;* all elements in B also in *U* $\quad$ **2.** *F;* 2 is not in *A* $\quad$ **3.** *T;* ϕ is a subset of all sets.
4. *F;* No element of *A* is in *B* $\quad$ **5.** { } or ϕ $\quad$ **6.** {2, 0, −2} $\quad$ **7.** {−4, −2, 0, 1, 2, 3, 4, 5}
8. {−5, −3, −1, 1, 2, 3, 4, 5} $\quad$ **9.** {−5, −4, −3, −2, −1, 1, 2, 3, 4, 5} $\quad$ **10.** {−5, −4, −3, −1, 0}

2 **11.** $3 \le x \le 4$; [3, 4]

12. $5 \le x \le 7$; [5, 7]

13. $-1 \le x \le 3$; [−1, 3]

14. $-1 \le x \le 2$; [−1, 2]

15. no solution; ϕ

16. $3 < x < 7$; (3, 7)

17. $0 < x < 3$; (0, 3)

18. no solution; ϕ

19. $-6 < x < -3$; (−6, −3)

20. $3 < x < 5$; (3, 5)

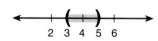

21. $-2 \le x \le 2$; [−2, 2]

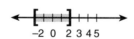

22. $0 \le x \le \dfrac{4}{7}$; $\left[0, \dfrac{4}{7}\right]$

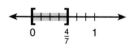

3 **23.** $x < 2$ or $x > 9$

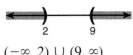

$(-\infty, 2) \cup (9, \infty)$

24. $x \leq 3$ or $x \geq 7$

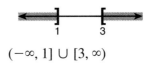

$(-\infty, 3] \cup [7, \infty)$

25. $x \leq 4$ or $x \geq 5$

$(-\infty, 4] \cup [5, \infty)$

26. $x \leq 1$ or $x \geq 3$

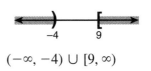

$(-\infty, 1] \cup [3, \infty)$

27. $x \leq 1\frac{1}{2}$ or $x \geq 3\frac{1}{2}$

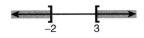

$\left(-\infty, 1\frac{1}{2}\right] \cup \left[3\frac{1}{2}, \infty\right)$

28. $x < -4$ or $x \geq 9$

$(-\infty, -4) \cup [9, \infty)$

29. $5.22 \leq 5.27 \leq 5.32;$

30. $14.7 \leq 15.2 \leq 15.7$

Section 17–4 Self-Study Exercises

1 **1.** $-3 < x < 2; (-3, 2)$

2. $x < -3$ or $x > \frac{1}{2}; (-\infty, -3) \cup \left(\frac{1}{2}, \infty\right)$

3. $\frac{2}{3} < x < \frac{5}{2}; \left(\frac{2}{3}, \frac{5}{2}\right)$

4. $y \leq -6$ or $y \geq 2; (-\infty, -6] \cup [2, \infty)$

5. $-4 < a < 6; (-4, 6)$

6. $2 \leq x \leq 5; [2, 5]$

7. $x \leq -2$ or $x \geq 3; (-\infty, -2] \cup [3, \infty)$

8. $1 \leq x \leq 3; [1, 3]$

9. $\frac{1}{4} \le x \le 5; \left[\frac{1}{4}, 5\right]$

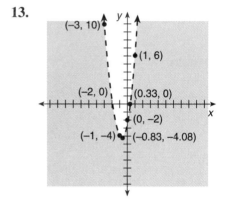

10. $x < -\frac{1}{2}$ or $x > \frac{1}{3}; \left(-\infty, -\frac{1}{2}\right) \cup \left(\frac{1}{3}, \infty\right)$

11. $x \le \frac{1}{2}$ or $x \ge \frac{2}{3}; \left(-\infty, \frac{1}{2}\right] \cup \left[\frac{2}{3}, \infty\right)$

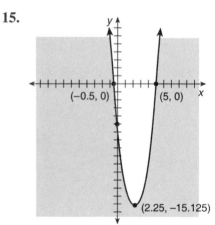

12. $-5 \le x \le \frac{1}{2}; \left[-5, \frac{1}{2}\right]$

13.

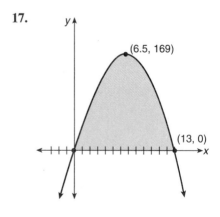

14.

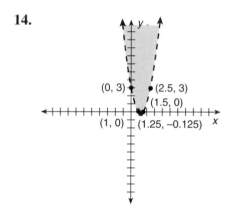

15.

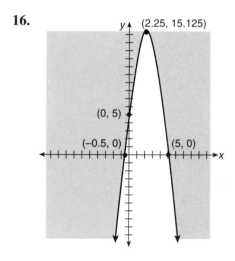

16.

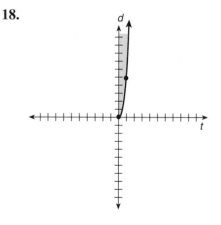

17.

18.

3 **19.** $-7 < x < 3;\ (-7, 3)$

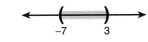

20. $x < -8$ or $x > 3;\ (-\infty, -8) \cup (3, \infty)$

21. $x < \dfrac{1}{3}$ or $x > 3;\ \left(-\infty, \dfrac{1}{3}\right) \cup (3, \infty)$

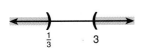

22. $-7 < x \le 7;\ (-7, 7]$

23. $0 < x < 1;\ (0, 1)$

24. $x \le 2$ or $x > 3;\ (-\infty, 2] \cup (3, \infty)$

25. $2 < x \le 3;\ (2, 3]$

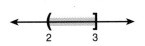

26. $x \le 0$ or $x > 4;\ (-\infty, 0] \cup (4, \infty)$

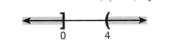

7. $x < -1$ or $x \ge 3;\ (-\infty, -1) \cup [3, \infty)$

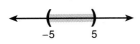

Section 17–5 Self-Study Exercises

1 **1.** ± 8 **2.** ± 15 **3.** $8, -2$ **4.** $9, 5$ **5.** $6, -3$ **6.** $3, \dfrac{5}{3}$ **7.** ± 11 **8.** no solution   **9.** $7, 1$

10. $4, -3$ **11.** no solution   **12.** $-\dfrac{1}{3}, +3\dfrac{2}{3}$ **13.** $0, 4$ **14.** $-5, 10$ **15.** $-\dfrac{2}{3}, 6$ **16.** $0, \dfrac{18}{7}$

17. no solution   **18.** $-\dfrac{12}{5}, \dfrac{18}{5}$ **19.** $-\dfrac{14}{5}, \dfrac{16}{5}$ **20.** $0, 3$ **21.** $\dfrac{2}{3}$

2 **22.** $-5 < x < 5;\ (-5, 5)$

23. $-1 < x < 1;\ (-1, 1)$

24. $-2 < x < 2;\ (-2, 2)$

25. $-7 \le x \le 7;\ [-7, 7]$

26. $-5 < x < -1;\ (-5, -1)$

27. $-7 < x < -1;\ (-7, -1)$

28. $-1 < x < 7; (-1, 7)$

29. $-1 < x < 5; (-1, 5)$

30. $-1 \leq x \leq 11; [-1, 11]$

31. $-6 \leq x \leq 8; [-6, 8]$

32. $-\dfrac{2}{3} < x < 4; \left(-\dfrac{2}{3}, 4\right)$

33. $\dfrac{1}{2} < x < \dfrac{7}{2}; \left(\dfrac{1}{2}, \dfrac{7}{2}\right)$

34. $-1 \leq x \leq \dfrac{1}{5}; \left[-1, \dfrac{1}{5}\right]$

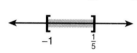

35. $-3 \leq x \leq -\dfrac{1}{2}; \left[-3, -\dfrac{1}{2}\right]$

36. $-\dfrac{2}{3} < x < \dfrac{4}{3}; \left(-\dfrac{2}{3}, \dfrac{4}{3}\right)$

3 **37.** $x < -2$ or $x > 2; (-\infty, -2) \cup (2, \infty)$

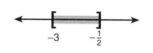

38. $x < -3$ or $x > 3; (-\infty, -3) \cup (3, \infty)$

39. $x \leq -5$ or $x \geq 5; (-\infty, -5] \cup [5, \infty)$

40. $x \leq -6$ or $x \geq 6; (-\infty, -6] \cup [6, \infty)$

41. $x < 3$ or $x > 7; (-\infty, 3) \cup (7, \infty)$

42. $x < -2$ or $x > 10; (-\infty, -2) \cup (10, \infty)$

43. $x \leq -8$ or $x \geq 2; (-\infty, -8] \cup [2, \infty)$

44. $x \leq -10$ or $x \geq 2; (-\infty, -10] \cup [2, \infty]$

45. $x \leq \dfrac{5}{2}$ or $x \geq \dfrac{5}{2}$ or all real numbers; $(-\infty, \infty)$

46. $x \leq -\dfrac{2}{3}$ or $x \geq 2$; $\left(-\infty, -\dfrac{2}{3}\right] \cup [2, \infty)$

47. $x \leq -2$ or $x \geq -\dfrac{6}{5}$; $\left(-\infty, -2\right] \cup \left[-\dfrac{6}{5}, \infty\right)$

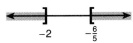

48. $x \leq -\dfrac{3}{2}$ or $x \geq 3$; $\left(-\infty, -\dfrac{3}{2}\right] \cup [3, \infty)$

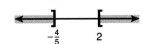

49. $x \leq -\dfrac{1}{3}$ or $x \geq 1$; $\left(-\infty, -\dfrac{1}{3}\right] \cup [1, \infty)$

50. $x \leq -\dfrac{1}{2}$ or $x \geq \dfrac{7}{2}$; $\left(-\infty, -\dfrac{1}{2}\right] \cup \left[\dfrac{7}{2}, \infty\right)$

51. $x \leq -2$ or $x \geq 3$; $(-\infty, -2] \cup [3, \infty)$

52. $\dfrac{1}{3} \leq x \leq 1$; $\left[\dfrac{1}{3}, 1\right]$

53. $0 \leq x \leq 5$

[0, 5]

54. $x \leq -\dfrac{4}{5}$ or $x \geq 2$

$\left(-\infty, -\dfrac{4}{5}\right] \cup [2, \infty)$

55. $x \leq 1\dfrac{1}{4}$ or $x \geq 1\dfrac{3}{4}$

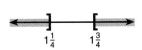

$\left(-\infty, 1\dfrac{1}{4}\right] \cup \left[1\dfrac{3}{4}, \infty\right)$

56. no solution; { } or ϕ

57. no solution; { } or ϕ **58.** $x \leq 0$ or $x \geq 6$ **59.** $x < -\dfrac{1}{2}$ or $x > 0$ **60.** $1 \leq x \leq 7$ **61.** no solution

62. $-7 < x < 7$ **63.** $-11 < x < -5$ **64.** $x < -12$ or $x > 12$ **65.** $x < -10$ or $x > -6$ **66.** no solution

1. a set with no elements { } or ϕ

3. $5 \in W$

5. $(-7, \infty)$

7. $[-4, 2)$

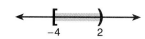

9. $(-2, \infty)$

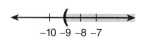

11. $m < 7; (-\infty, 7)$

13. $x > 0; (0, \infty)$

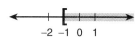

15. $x \le 3; (-\infty, 3]$

17. $x > -9; (-9, \infty)$

19. $x > -1; (-1, \infty)$

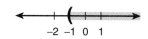

21. $y \ge -1; [-1, \infty)$

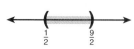

23. $x \le 6; (-\infty, 6]$

25. $D = P + \$5.60; 2D + 12P \le \$59.80$; therefore, $P \le \$3.47; D \le \$9.07$ **27.** { } or ϕ   **29.** $\{-1\}$
31. $\{5\}$

33. $x > -2; (-2, \infty)$

35. $x \ge -3; [-3, \infty)$

37.

39.

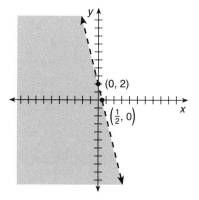

41.

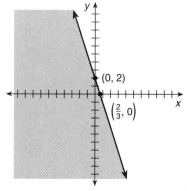

43.

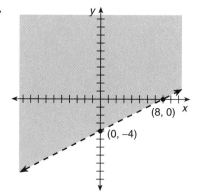

45.

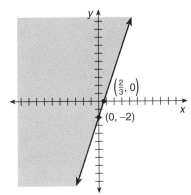

$\left(\frac{2}{3}, 0\right)$

$(0, -2)$

47.

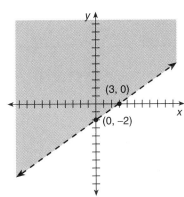

$(3, 0)$

$(0, -2)$

49.

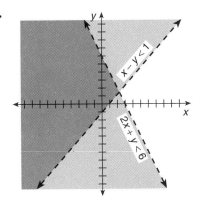

$x - y < 1$

$2x + y < 6$

51.

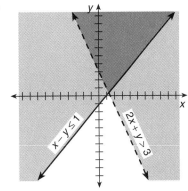

$x - y \leq 1$

$2x + y > 3$

53. $2 < x < 4;\ (2, 4)$

55. no solution; ϕ

57. $2 < x < 6;\ (2, 6)$

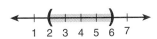

59. $-3 < x \leq 2;\ (-3, 2]$

61. $-2 \leq x \leq -\frac{3}{4};\ \left[-2, -\frac{3}{4}\right]$

63. $x < 2$ or $x > 8;\ (-\infty, 2) \cup (8, \infty)$

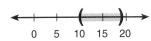

65. $x < -9$ or $x > 8;\ (-\infty, -9) \cup (8, \infty)$

67. $10 < x < 19$

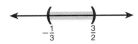

69. $x < 2$ or $x > 5;\ (-\infty, 2) \cup (5, \infty)$

71. $-\frac{1}{3} < x < \frac{3}{2};\ \left(\frac{1}{3}, \frac{3}{2}\right)$

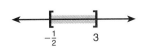

73. $-1 \leq x \leq 2;\ [-1, 2]$

75. $-\frac{1}{2} \leq x \leq 3;\ \left[-\frac{1}{2}, 3\right]$

77. $-5 < x < \frac{3}{2};\ (-5, \frac{3}{2})$

79. $-1 < x < 7;\ (-1, 7)$

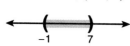

81. $x < -8 \cup x > 0;\ (-\infty, -8) \cup (0, \infty)$

83.

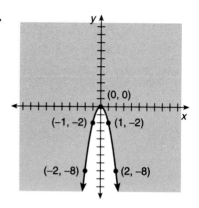

85.

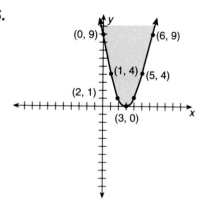

87. ± 12 **89.** $4, -10$ **91.** $20, -4$ **93.** $6, -\dfrac{5}{2}$ **95.** $1, -\dfrac{23}{7}$ **97.** $3, -\dfrac{13}{7}$ **99.** $\dfrac{11}{3}, \dfrac{7}{3}$ **101.** ± 7

103. ± 16 **105.** $10, -4$ **107.** $2, -\dfrac{1}{2}$ **109.** $-1 < x < 7;\ (-1, 7)$

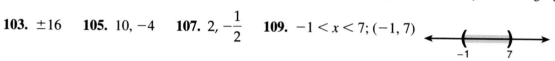

111. $-4 < x < 12;\ (-4, 12)$

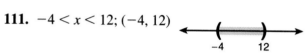

113. no solution   **115.** $\$108{,}000 < I < \$250{,}000$

Practice Test, Chapter 17

1. $x \geq -12;\ [-12, \infty)$

3. $x > 3;\ (3, \infty)$

5. $x > 2;\ (2, \infty)$

7. $x \leq -6;\ (-\infty, -6]$

9. $x > -2;\ (-2, \infty)$

11. $-8 < x < 4;\ (-8, 4)$

13. no solution

15. $x < -\dfrac{3}{2} \cup x > 1;\ \left(-\infty, -\dfrac{3}{2}\right) \cup (1, \infty)$

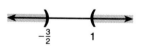

17. $x < 2 \cup x > 6;\ (-\infty, 2) \cup (6, \infty)$

19. $A \cup B = \{1, 2, 3, 4, 5, 6, 7, 8, 9\}$

Selected Answers to Student Exercise Material

21. $A \cap B' = \{9\}$

23. $-5 < x < 2; (-5, 2)$

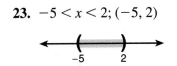

25. ± 15

27. ± 2

29. $x < -18 \cup x > 2; (-\infty, -18) \cup (2, \infty)$

31. $96.7 \leq x \leq 96.9$

33.

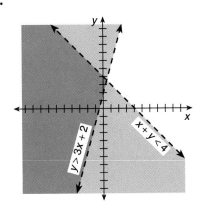

35.

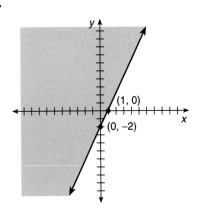

37.

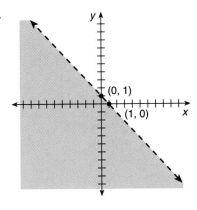

39.

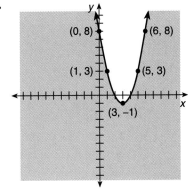

41.

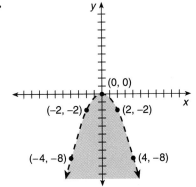

Section 18–1 Self-Study Exercises

1   **1.** $\overleftrightarrow{PQ}$, $\overleftrightarrow{QR}$, & $\overleftrightarrow{PR}$ or $\overleftrightarrow{QP}$, $\overleftrightarrow{RQ}$, and $\overleftrightarrow{RP}$ **2.** $\overrightarrow{PQ}$, $\overrightarrow{PR}$ **3.** $\overline{QR}$ or $\overline{RQ}$ **4.** Yes   **5.** No, endpoints are different.   **6.** no   **7.** yes   **8.** parallel   **9.** intersect   **10.** coincide   **11.** intersect   **12.** $\overleftrightarrow{EF}$ and $\overleftrightarrow{GH}$ **13.** $\angle DEF$, $\angle FED$ **14.** $\angle E$ **15.** $\angle 1$

2   **16.** 360°   **17.** 180°   **18.** 90°   **19.** 45°   **20.** acute   **21.** obtuse   **22.** right   **23.** obtuse   **24.** acute   **25.** straight   **26.** neither   **27.** complementary   **28.** supplementary   **29.** neither   **30.** complementary   **31.** supplementary   **32.** supplementary

3   **33.** $\angle a$ and $\angle c$; $\angle b$ and $\angle d$ **34.** supplementary angles   **35.** 180°   **36.** $\angle c$ and $\angle f$; $\angle d$ and $\angle e$ **37.** 180°   **38.** corresponding angles

4   **39.** 0.7833°   **40.** 0.01°   **41.** 0.0872°   **42.** 0.1708°   **43.** 10.3042°   **44.** 21′   **45.** 12′   **46.** 7′12″   **47.** 12′47″   **48.** 18′54″

Section 18–2 Self-Study Exercises

1   **1.** 48 in.   **2.** 27.2 cm   **3.** 196 in.   **4.** 176 in.   **5.** 240 in.   **6.** 128 in.   **7.** 24 ft   **8.** 36.2 ft   **9.** 690 ft   **10.** 66 ft   **11.** 57.5 ft   **12.** 184 ft   **13.** 17 ft   **14.** 130 in.   **15.** 20 cm   **16.** 40.8 cm   **17.** 628 ft   **18.** 52 tiles   **19.** 309.3 mm   **20.** 52.9 ft   **21.** 207 in.   **22.** 75 ft   **23.** 402 ft   **24.** 98 ft   **25.** 28.6 in.   **26.** 48 cm   **27.** 72 ft   **28.** 10 ft 8 in. or 3 yd 1 ft 8 in.

2   **29.** 112 in^2 **30.** 26.7 cm^2 **31.** 35 ft^2 **32.** 80.34 ft^2 **33.** 42,500 ft^2 **34.** 180 ft^2 **35.** 25 cm^2 **36.** 104.04 cm^2 **37.** 193 ft^2 **38.** 900 yd^2 **39.** $\frac{1}{16}$ mi^2 **40.** 81 tiles   **41.** 108 tiles   **42.** 400 in^2 **43.** 5,053.05 mm^2 **44.** 162 ft^2 **45.** 348 in^2 **46.** 260 ft^2 **47.** 33 in^2 **48.** 96 cm^2 **49.** 189 ft^2 **50.** \$175.50

Section 18–3 Self-Study Exercises

1   **1.** 25.1 cm   **2.** 47.1 m   **3.** 18.8 in.   **4.** 9.4 ft   **5.** 53.4 ft   **6.** 17.3 m   **7.** 50.3 cm^2 **8.** 176.7 m^2 **9.** 28.3 in^2 **10.** 804.2 yd^2 **11.** 7.1 ft^2 **12.** 227.0 ft^2 **13.** 23.8 m^2 **14.** 21.6 in^2 **15.** 0.6 m^2 **16.** 2.1 ft^2 **17.** 14.6 cm^2 **18.** 1.0 in^2 **19.** 592.7 ft^2 **20.** 268.3 in.   **21.** Yes, the cross-sectional area of the third pipe is larger than the combined area of the other two pipes.   **22.** Yes, the combined cross-sectional area of the two pipes is 25.1 in^2, which is greater than 20 in^2—the area of the large pipe.   **23.** 2,827 ft/min   **24.** 1,571 ft/min

2   **25.** 0.79 rad   **26.** 0.98 rad   **27.** 2.44 rad   **28.** 21.75°; 0.38 rad   **29.** 177.55°; 3.10 rad   **30.** 44.9033°; 0.78 rad   **31.** 45°   **32.** 30°   **33.** 143.2394°   **34.** 28°38′52″   **35.** 22°30′   **36.** 42°58′19″

3   **37.** 28.27 mm   **38.** 13.09 in.   **39.** 3.20 ft   **40.** 7.06 cm   **41.** 6.45 in.   **42.** 14 cm   **43.** 4.25 rad   **44.** 3.49 rad   **45.** 7 cm   **46.** 4.5 in.

4   **47.** 120.64 cm^2 **48.** 198.44 mm^2 **49.** 2.43 ft^2 **50.** 14.18 in^2 **51.** 530.14 ft^2 **52.** 75.40 in^2 **53.** 17.12 in^2 **54.** 4.32 cm^2 **55.** 2.24 cm   **56.** 0.85 rad   **57.** 47.43 cm^2 **58.** 0.7 rad   **59.** 16.01 cm^2 **60.** 48.98 in^2 **61.** 54.83 in^2 **62.** 51.99 cm^2 **63.** 561.81 mm^2 **64.** 257.93 ft^2

Section 18–4 Self-Study Exercises

1   **1.** 576 in^3 **2.** 3,817.04 cm^3 **3.** 320 cm^3 **4.** 55.36 in^2 **5.** 23.3 in^3 **6.** 18,850 ft^3 **7.** $V = 301.6$ ft^3 **8.** 4,712.4 ft^3 **9.** 50.3 L   **10.** 12,600 cm^3 **11.** 76,800 m^3 **12.** 415.6 m^3

13. 2,617.3 in^3 **14.** 2,428.2 cm^3 **15.** 3,660 in^3 **16.** 523.6 cm^3 **17.** 904.8 in^3 **18.** 356,892.8 gal
19. 225.6 gal **20.** $LSA = 37.3$ in^2; $TSA = 47.1$ in^2 **21.** $LSA = 1,885.0$ ft^2; $TSA = 4,398.2$ ft^2
22. $LSA = 200$ cm^2; $TSA = 264$ cm^2 **23.** $LSA = 96$ in^2; $TSA = 109.8$ in^2 **24.** 1,885 ft^2
25. $LSA = 188.5$ ft^2; $TSA = 301.6$ ft^2 **26.** 589.0 cm^2 **27.** 1,649.3 ft^2 **28.** 314.2 cm^2 **29.** 6,361.7 ft^2
30. 50.3 ft^2 **31.** $347

Chapter Review Exercises, Chapter 18

1. $\overleftrightarrow{AB}$ & $\overleftrightarrow{CD}$ **3.** $\overleftrightarrow{EF}$ & $\overleftrightarrow{GH}$ **5.** $\angle P$ **7.** right **9.** straight **11.** supplementary **13.** neither
15. $\angle a$ and $\angle b$; $\angle b$ and $\angle c$; $\angle c$ and $\angle d$; $\angle d$ and $\angle a$ **17.** $\angle a$ and $\angle h$; $\angle b$ and $\angle g$ **19.** 45° **21.** 0.4833°
23. 0.1261° **25.** 45′ **27.** 13′3″ **29.** 210 mm **31.** 35.8 in. **33.** 66 ft **35.** 288 in. **37.** 133 cm^2
39. 34.81 m^2 **41.** 306 ft^2 **43.** 36,000 ft^2 **45.** 44 yd^2 **47.** 235 ft^2 **49.** $A = 50.27$ m^2; $c = 25.13$ m^2
51. $A = 12$ cm^2; $p = 17.42$ cm **53.** 33 ft/min **55.** 1.05 rad **57.** 1.74 rad **59.** 150° **61.** 2.62 cm^2
63. 2.18 ft^2 **65.** 31.42 in. **67.** 6.96 cm^2 **69.** 44.50 in^2 **71.** 7,854 cm^3 **73.** 102 yd^3 **75.** 616 cm^2
77. 2,733 cm^2 **79.** 124,407 ft^2 **81.** 1,728 m^3 **83.** 1,017.9 m^2 **85.** 169.6 cm^2 **87.** 2 gal
89. longest TS, shortest RS

Practice Test, Chapter 18

1. 18′45″ **3.** 0.61 rad **5.** 112.5° **7.** $P = 111$ ft; $A = 749.1$ ft^2 **9.** $P = 37$ cm; $A = 43.5$ cm^2
11. $C = 144.5$ m; $A = 1,661.9$ m^2 **13.** 1 in. **15.** 1.71 m **17.** 282 ft^2 **19.** 1,016.5 ft^2
21. $A = 71.84$ cm^2 **23.** 50 in^2 **25.** 6,768 gal

Section 19–1 Self-Study Exercises

1 **1.** scalene **2.** equilateral **3.** isosceles **4.** longest TS, shortest RS **5.** longest k, shortest l
6. $\angle C$, $\angle B$, $\angle A$ **7.** $\angle X$, $\angle Z$, $\angle Y$

2 **8.** $\angle A = \angle F$, $\angle C = \angle D$, $\angle B = \angle E$ **9.** $\angle P = \angle S$, $\angle Q = \angle T$, $QP = TS$
10. $\angle L = \angle P$, $PM = LJ$, $LK = PN$ **11.** $EF = 9.6$ cm **12.** $TU = 19.2$ in. **13.** $PM = 3.75$ ft
14. $a = 20$; $d = 9$ **15.** $DC = 6$, $DE = 5.3$ **16.** 65 ft

Section 19–2 Self-Study Exercises

1 **1.** $BC = 10$ mm **2.** $AB = 17$ yd **3.** $c = 6.403$ cm **4.** $a = 6.325$ m **5.** $b = 12$ ft **6.** 30 ft
7. $AC = 39.783$ cm **8.** 26 ft 10 in. **9.** 12.806 ft **10.** 13 cm **11.** 45 dm **12.** 10.607 mm
13. 61.083 ft

2 **14.** $BC = 12$ cm, $AB = 16.971$ cm **15.** $AC = 7.071$ m, $BC = 7.071$ m **16.** $AC = 9.899$ m, $AB = 14.0$ m
17. $AC = 8$ m, $BC = 8$ m **18.** $AC = 14.697$ mm, $BC = 14.697$ mm **19.** 25′5″ **20.** 37.830 ft

3 **21.** $AB = 12$ cm, $BC = 10.392$ cm **22.** $AC = 9.0$ mm, $BC = 15.588$ mm
23. $AC = 4.619$ in., $AB = 9.238$ in. **24.** $AC = 5.715$ cm, $AB = 11.431$ cm
25. $AB = 5$ ft 6 in., $BC = 4$ ft 9 in. **26.** 35 ft **27.** rafter $= 18$ ft 0 in., run $= 15$ ft 7 in. **28.** 4.5 cm

Section 19–3 Self-Study Exercises

1 **1.** equilateral triangle; $\angle 60°$ **2.** square; $\angle 90°$ **3.** regular pentagon; $\angle 108°$ **4.** regular hexagon; $\angle 120°$
5. regular octagon; 135° **6.** regular nonagon; 140° **7.** 15.6 cm^2 **8.** 100 ft^2 **9.** 42.5 cm^2
10. 667.2 mm^2 **11.** 617.6 in^2 **12.** 891 cm^2

3   **13.** 90°   **14.** 45°   **15.** 8.49 in.   **16.** 45°   **17.** 6 in.   **18.** 0.5 m   **19.** 0.35 m   **20.** 90°
21. 0.35 m   **22.** 0.29 in.   **23.** 1.06 in.   **24.** 63.62 dkm²

4   **25.** 60°   **26.** 30°   **27.** 90°   **28.** 11.5 in.   **29.** 11.5 in.   **30.** 13.33 in.   **31.** 57.5 in²   **32.** 5.75 in.
33. 6.67 in.   **34.** 41.89 in.   **35.** 2.9 cm   **36.** 1 in.

5   **37.** 34.64 mm   **38.** 120°   **39.** 60°   **40.** 40 mm   **41.** 90°   **42.** 10 mm   **43.** 20 mm   **44.** 60°
45. 30°   **46.** 3 in.   **47.** 39.97 mm   **48.** 2.84 in.

Section 19–4 Self-Study Exercises

1   **1.** 10

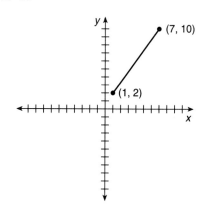

2. 13

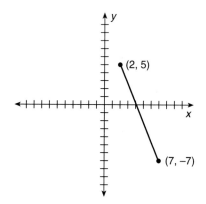

3. 7.071

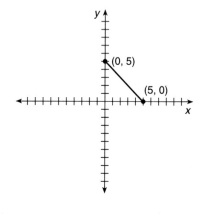

4. 5.385

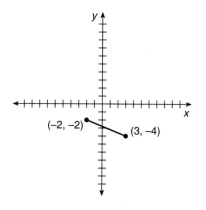

5. 11.180

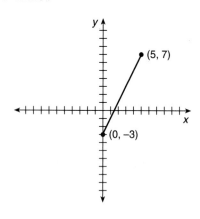

6. 13

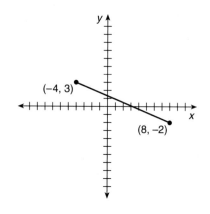

7. 10

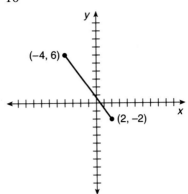

8. 5

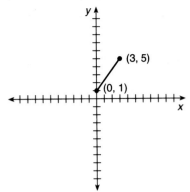

9. 15

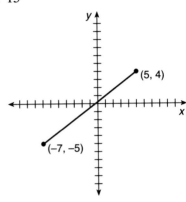

10. 10.296

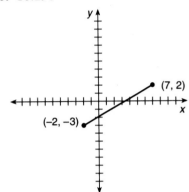

2 **11.** (4, 6) **12.** $\left(\dfrac{9}{2}, -1\right)$ **13.** $\left(\dfrac{5}{2}, \dfrac{5}{2}\right)$ **14.** $\left(\dfrac{1}{2}, -3\right)$ **15.** $\left(\dfrac{5}{2}, 2\right)$ **16.** $\left(2, \dfrac{1}{2}\right)$ **17.** (−1, 2)

18. $\left(\dfrac{3}{2}, 3\right)$ **19.** $\left(-1, -\dfrac{1}{2}\right)$ **20.** $\left(\dfrac{5}{2}, -\dfrac{1}{2}\right)$

Chapter Review Exercises, Chapter 19

1. longest *TS*, shortest *RS* **3.** ∠*C*, ∠*B*, ∠*A* **5.** ∠*C* = ∠*D*, ∠*B* = ∠*E*, *CB* = *DE*

7. ∠*M* = ∠*J*, *JL* = *PM*, *JK* = *MN* **9.** $\dfrac{AB}{RT} = \dfrac{BC}{ST} = \dfrac{AC}{RS}$ **11.** 15 in. **13.** 7.141 ft **15.** 8 yd

17. 20.248 mi **19.** 30 cm **21.** 21.633 in. **23.** 4.243 in. **25.** *RS* = 10.607 cm, *ST* = 10.607 cm
27. *RS* = 9 hm, *ST* = 9 hm **29.** *AB* = 13.856 dm, *BC* = 6.928 dm **31.** *AC* = 17.321 in., *AB* = 20.0 in.
33. *AB* = 46 ft 10 in., *BC* = 23 ft 5 in. **35.** 8 ft 4 in. **37.** 19.630 mm **39.** 908.5 in^2
41. regular pentagon, 108° **43.** 14 **45.** *AE* = 5 **47.** ∠*GJO* = 45° **49.** *IJ* = 20 **51.** ∠*MOP* = 30°
53. 5 mm **55.** 3.54 cm **57.** 4.472 **59.** 10.440 **61.** 5.831 **63.** 9.434 **65.** (1, 5)

67. $\left(1\dfrac{1}{2}, 2\right)$ **69.** $\left(-1\dfrac{1}{2}, 2\dfrac{1}{2}\right)$ **71.** $\left(1, -\dfrac{1}{2}\right)$ **73.** $\left(-1\dfrac{1}{2}, -3\right)$

Practice Test, Chapter 19

1. largest ∠*A*, smallest ∠*B* **3.** *DB* = 5 **5.** *c* = 52 **7.** 17 in. **9.** *AC* = 31 cm; *BC* = 53.69 cm
11. *AB* = 7.78 cm; *BC* = 5.5 cm **13.** 13.657 cm **15.** 25.13 in. **17.** 1.57 in. **19.** distance = 5.4;
midpoint = $\left(\dfrac{1}{2}, 2\right)$

Chapters 16–19 Cumulative Practice Test

1. $\log_5 125 = 3$ **3.** $6^2 = 36$ **5.** 107.9 mg **7.** \$23,568 **9.** $x < 2$ 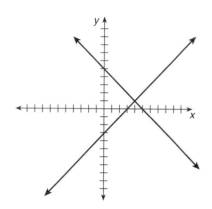 $(-\infty, 2)$

11. $2 < x < 9$ 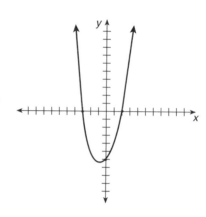 $(2, 9)$

13.

15.

17. acute **19.** straight **21.** 62° **23.** 37° **25.** 30.2056° **27.** 21′7″ **29.** $P = 108$ m; $A = 529$ m^2
31. $C = 471$ cm; $A = 17{,}671$ cm^2 **33.** 0.2682 rad **35.** 103.1324° **37.** 18,096 cm^3 **39.** 19 in.
41. $AC = 16.25$ m; $BC = 28.1$ m **43.** 53.7 in. **45.** 53.6

Glossary and Index

Deka-: the metric prefix used for a unit that is 10 times larger than the standard unit, 171

Delta (Δ), 404

Denominator: the denominator of a fraction is the number of parts one unit has been divided into. It is the bottom number of a fraction or the divisor of the indicated division, 3, 61

  common, 78

  decimal, 6

Dependent system: when solving a system of equations, if both variables are eliminated and the resulting statement is true, then the equations are dependent and have many solutions, 433, 438

Dependent variable: the variable that represents the results of calculations. In function notation, the dependent variable is y or $f(x)$, 357, 359

Descending order: arranging the terms of the polynomial beginning with a term with the highest degree, 463

Descriptive statistics, 265

Deviation, 275

Diameter: the straight line segment from a point on the circle through the center to another point on the circle, 717

  inside, 17

  outside, 17

Difference: the answer to a subtraction problem, 11

Difference of two perfect cubes, 472

  factoring, 525

Difference of two perfect squares, 470

Digits, 2

  nonzero, 10, 32

Dimension analysis, 43, 103, 358

  arc length, 724

  estimation, 106

Direct measurement

  metric rule, 195

  metric system (SI), 171

  U.S.-metric conversions, 188

  see also Metric system and U.S. customary system

Direct proportion: a proportion in which the quantities being compared are directly related so that as one quantity increases (or decreases), the other quantity also increases (or decreases). This relationship is also called direct variation, 357

Direct variation, 357

  constant of, 359

  distinguish from inverse, 370

Directional sign, 211

Discriminant: the radicand of the radical portion of the quadratic formula, $b^2 - 4ac$, 595

Disjunction, 671

Distance

  between two points on a line, 196

  rectangular coordinate system, 784

Distributive property of multiplication: the property stating that multiplying a sum or difference by a factor is equivalent to multiplying each term of the sum or difference by the factor, 17, 305, 315, 519

  applied to equations, 315

  common factors, 519

Dividend: the number being divided, 20

Divisibility tests, 62

Divisible: a number is said to be divisible by another number if the quotient has no remainder or if the dividend is a multiple of the divisor, 62

Division: the inverse operation for multiplication, 20

  check, 25

  commutativity or associativity, 20

  complex fractions, 97

  decimals, 23

  estimation, 24

  fractions, 94

  importance of first digit in quotient, 22

  inverse operation of multiplication, 20

  key words, 46, 304

  long, 23

  long division, 21

  metric system, 180

  mixed numbers, fractions, and decimals, 95

  multiplying by reciprocal, 93

  numerical average, 25

  polynomials, 474

  powers of 10, 235, 455

  rational expressions, 547

  rounding quotient, 24

  scientific notation, 241

  signed numbers, 222

  square-root radicals, 501

  symbols, 21

  U.S. customary system, 110

  whole numbers, 21

  with zero, 223

  zero and one, 22

Divisor: the number to divide by, 20

Domain, 385

  relation, 601

Double root, 575, 579, 591

Effective rate, 624

Efficiency, 148

Electronic spreadsheet, 337

Element: another name for the member of a set of data, 281

Elimination method, 434

Ellipsis, 2, 77

Empty set, 367, 656

Endpoint, 195, 695

Engine efficiency, 148

Engineering notation, 245

English rule

  see also U.S. customary rule

English system: a system of measurement that is currently called the U.S. customary system of measurement, 101

Equal

  not-equal symbol, 11

Equality

  addition property, 308

  multiplication property, 309

  principle, 307

  proportions, 354

  symmetric property, 301

Equation: a symbolic statement that two expressions or quantities are equal in value, 300

  applying distributive property, 315

  checking solutions, 311

  clearing decimals, 329

  clearing fractions, 319, 323

  cubic, 597

  excluded roots, 322

Equation (continued)
  exponential, 620
  horizontal line, 411
  linear, 307
  parallel lines, 413
  perpendicular lines, 415
  point-slope form, 409
  quadratic, 572
  second-degree, 597
  third-degree, 597
  verify, 300
  vertical line, 411
  with one absolute value term, 679
Equations in two variables
  see also Graph and Systems of equations
Equilateral triangle: a triangle with all three sides and angles equal, 757, 769, 774
Equilateral triangle
  altitude, 769
Equivalent fractions: fractions that represent the same value, 70
Equivalents
  approximate, 77
  exact, 77
Error
  absolute, 193
  greatest possible, 192
  percent, 193
  relative, 193
Estimation: finding a reasonable approximate answer to a problem, 10
  addition, 10
  dimension analysis, 106
  division, 24
  multiplication, 18
  new amount, 157
  percents, 142
  proportions, 367
  square root, 32
  subtraction, 14
Evaluation: to evaluate a formula is to substitute known numerical values for some variables and perform the indicated operations to find the value of the remaining variable in question, 39, 333
Even number: integers that are multiples of 2, 62
Exact equivalents, 77
Exact number, 8
Exact solutions
  with pi (π), 723
Exact values, 503
Excluded value: the value that will cause the denominator of any fraction to be zero. This situation produces an extraneous root, 322
Excluded value
  rational equations, 558
Expanded form, 519
Exponent: a natural number exponent indicates how many times a repeated factor (base) is used, 30
  base, 30
  decimal, 493
  exponential notation, 30, 236
  laws of, 454
  logarithm, 636
  natural number, 30

  negative integer exponents, 234, 456
  of 1, 31
  operations involving powers, 235
  powers, 30, 454
  powers of 10, 234
  rational, 488
  types of, 490
  variable, 620
  zero, 234
Exponential notation, 485
Exponential equation, 620
  solving equations in the form $b^x = b^y$, 629
  to logarithmic equation, 637
Exponential expression, 620
Exponential function, 620
  graph, 630
Exponential notation: a value written with a base and an exponent, 30
Exponential notation
  powers of 10, 236
Expressions, variable, 305
Exterior angle, 699
Extraneous root: solutions or roots that do not make a true statement in the original equation, 321
Extraneous root, 559

Factor: any of the numbers in a multiplication problem or an algebraic expression that are involved in multiplication, 15, 301
  algebraic expressions, 494, 519
  variable, 491
Factor an algebraic expression: to write the expression as the indicated product of two or more factors, that is, as a multiplication, 519
Factor pair: a factor pair of a natural number consists of two natural numbers whose product equals the given natural number, 63
Factored form, 519
Factoring
  algebraic expression, 519
  binomials or trinomials not prime, 534
  common factors, 519
  common factors from grouping, 528
  difference of two perfect cubes, 526
  difference of two perfect squares, 522
  general trinomials, 527
  general trinomials by grouping, 530
  perfect square trinomial, 524
  removing common factors, 519
  solving quadratic equations, 576
  sum of two perfect cubes, 525
Factorization, 66
Fahrenheit, 184
FOIL: a systematic method for multiplying two binomials, where F refers to the product of the first terms in each factor, O refers to the product of the two outer terms, I refers to the product of the two inner terms, and L refers to the product of the last terms in each factor, 469
Foot, 101
Formula: a symbolic statement that indicates a relationship among numbers, 39
Formula
  evaluation, 39

Formula rearrangement: the process of isolating a letter term other than the one already isolated (if any) in a formula, 335

Formulas
  accumulated amount for continous compounding, 628
  amortized payment, 626
  amount of work, 326
  arc length of sector, 723
  area of circle, 717
  area of composite figure, 710
  area of parallelogram, 42, 708
  area of rectangle, 19, 42, 708
  area of rhombus, 708
  area of sector, 725
  area of segment, 727
  area of square, 42, 708
  area of trapezoid, 709
  area of triangle, 709
  Celsius to Fahrenheit, 185
  circumference of circle, 717
  compound amount, 337, 621
  compound interest, 622
  cost, 335
  distance on rectangular coordinate system, 784
  effective rate, 624
  engine efficiency, 334
  evaluation, 333
  Fahrenheit to Celsius, 184
  finding time for investment, 641
  frustum of a cone, 734
  frustum of a pyramid, 734
  geometric, *see* Geometry
  Heron's formula, 709
  Kirchoff's law
  lateral surface area of cone, 738
  lateral surface area of prism or cylinder, 736
  markup, 336
  midpoint on rectangular coordinate system, 785
  ordinary annuity, 624
  parallelogram, 40
  percentage, 140
  perimeter of parallelogram, 705
  perimeter of rectangle, 18, 40, 705
  perimeter of rhombus, 705
  perimeter of square, 40, 705
  perimeter of trapezoid, 705
  perimeter of triangle, 705
  perimeter of composite figure, 707
  present value, 623
  pulley belt, 335
  Pythagorean theorem, 766
  quadratic formula, 583
  radius of circle inscribed in triangle, 798
  rearrangement, 335, 811
  resistance, 334
  Richter scale, 642
  simple interest, 329
  sinking fund payment, 625
  total resistance, 334
  total surface area of cone, 738
  total surface area of prism or cylinder, 736
  total surface area of sphere, 737
  volume of cone, 733

  volume of cube, 44
  volume of frustum of cone or pyramid, 734
  volume of prism, 44
  volume of prism or cylinder, 732
  volume of pyramid, 733
  volume of sphere, 735

Fraction: a number that is a part of a whole number; a value that can be expressed as the quotient of two integers, 3, 61
  addition, 82
  approximate equivalents, 77
  calculator, 98
  canceling, 90
  clearing denominators in equations, 319, 323
  common, 61
  common denominators, 78
  comparing, 78
  complex, 97
  decimal equivalents, 76
  decimal fraction, 4, 76
  denominator, 3, 61
  division, 94
  equivalent fractions, 70
  equivalent signed fractions, 225
  exact equivalents, 77
  fraction-decimal conversions, 76
  fundamental principle, 71
  improper fraction, 61
  kinds of fractions, 61
  least common denominator, 78
  like, 78
  mixed number, 61, 91
  multiplication, 89
  numerator, 3, 61
  percent equivalent, 134
  percents, 135
  power-of-10 denominator, 4
  powers, 93, 459
  proper, 61
  reduced to lowest terms, 71
  repeating decimal, 77
  subtraction, 85
  whole number equivalent, 74
  see also Rational expressions

Fraction line, 61
Fraction notation, 3

Frequency distribution: a compilation of class intervals, midpoints, tallies, and class frequencies, 269

Frustum
  cone, 733
  pyramid, 734
  volume, 734

Function, 385
  vertical line test, 600
  see also Equation

Fundamental principle of fractions: if the numerator and denominator of a fraction are multiplied by the same nonzero number, the value of the fraction remains unchanged, 71

Future value, 622
  continuous compounding, 628

Gallon, 102
GCF, greatest common factor, 68

General trinomial: a trinomial that is not a perfect square trinomial, 527

Geometry
    angle classification by size, 697
    angle notation, 696
    arc length of sector, 723
    area of sector, 725
    area of segment, 727
    composite figures, area, 710
    composite figures, perimeter, 707
    composite shape, 707
    degrees to radians, 721
    inscribed polygons, 776
    intersecting lines, 433, 696
    line, 695
    line segment, 695
    parallel lines, 696
    plane, 695
    point-slope form, 409
    ray, 695
    trapezoid, 706
    triangles, 704, 706, 757

Golden Ratio, 615

Golden Rectangle, 615

Gram: the standard unit for measuring weights in the metric system, 171, 174

Graph
    equations using a table of solutions, 386
    exponential function, 630
    higher-degree equation, 599
    linear equation, 386
    linear equations using calculator, 399
    linear equations using intercepts, 393
    linear equations using the slope-intercept method, 398
    linear inequalities in two variables, 663
    logarithmic function, 642
    one-dimensional, 382
    quadratic equation, 588
    quadratic inequalities in two variables, 676
    reading circle, bar, and line graphs, 258
    solutions of an equation, 401
    systems of equations, 432
    table of solutions, 387

Great Circle: a great circle divides a sphere in half at its greatest diameter and is formed by a plane through the center of the sphere, 734

Greater than (>), 3

Greatest common factor (GCF): the greatest common factor (GCF) of two or more numbers is the largest factor common to each number. Each number is divisible by the GCF, 68

Greatest possible error, 192

Grouped frequency, 269

Grouping, 36
    factoring common factors, 528
    see also Associative

Grouping symbols, 36

Hecto-: the prefix used for a metric unit that is 100 times larger than the standard unit, 171

Height: the perpendicular distance from the base to the highest point of the polygon above the base, 42, 44
    prism, 732

Height of a three-dimensional figure: the height of a three-dimensional figure with two bases is the shortest distance between the two bases, 732

Hemisphere, 735

Heron's formula, 709

Hexagon
    circumscribed, 780
    inscribed, 780
    regular, 774

Higher-degree equations, 597
    graph, 599
    solve by factoring, 598

Horizontal axis: number line that runs from left to right in a rectangular coordinate system. It is represented by the letter x, 382

Horizontal line, 697
    equation, 411
    slope, 406

Hour, 112

Hypotenuse: the side of a right triangle that is opposite the right angle, 765

i, 505

Identities, 313

Identity
    additive, 9, 214
    multiplicative, 93, 220

If . . . then, 629

Imaginary and complex numbers, 505

Imaginary number: an imaginary number has a factor of $\sqrt{-1}$, which is represented by i, 505
    powers, 507

Improper fraction: a common fraction that has a value equal to or greater than one unit; that is, the numerator is equal to or greater than the denominator, 61

Improper fraction, mixed-, and whole-number conversions, 74

Inch, 101

Incomplete quadratic equation: a quadratic equation that has a quadratic term and a linear term but no constant term, 573

Incomplete quadratic equation, 576

Inconsistent system: when solving a system of equations, if both variables are eliminated and the resulting statement is false, then the equations are inconsistent and have no solution. This system may also be called an independent system of equations, 433, 438

Increase
    amount, 155
    rate, 155

Independent system, 433

Independent variable: the variable that represents input values of a function, 357, 359

Indeterminate: the result of dividing zero by zero, 23, 223

Index mark on protractor, 701

Inductive reasoning, 759

Inequality: a mathematical statement showing quantities that are not equal. Inequality symbols are < and >, 3, 655
    absolute value greater than relationship, 682
    absolute value less than relationship, 681
    and relationship, 667
    compound, 667
    graph, 662
    graphical solution, 660
    graphing systems, 665

interchanging sides, 660
linear, 659
multiplying or dividing by a negative, 661
on number line, 658
one-dimensional graph, 659
one-variable, 659
or relationship, 668
quadratic, 673
rational, 677
sense, 660, 661
solution in interval notation, 660
solution set, 659
solving linear in two variables, 662
solving linear inequalities in one variable, 661
symbolic solution, 660
Infinity (∞), 2
Inscribed in a circle, 776
Inscribed in a polygon, 776
Inscribed polygon: a polygon is inscribed in a circle when it is inside the circle and all its vertices (points where sides of each angle meet) are on the circle. The circle is said to be circumscribed about the polygon, 776
Inside diameter, 17
Inspection, 62, 78
Integers: the set of whole numbers plus the opposite of each natural number, 209
see also Signed numbers
Intercept, 393
Interest: the amount charged for borrowing or loaning money, or the amount of money earned when money is saved or invested, 151, 329
amount, 151
compound, 621
percent, 151
principal, 151
rate, 151
simple, 151, 362
Interest rate per period, 622
Interior angle, 699
Interior point, 695
International System of Units (SI): an international system of measurement that uses standard units and power-of-10 prefixes to indicate other units of measure, 171
Intersect, 696
Intersection, 668
Interval notation, 657
Inverse
additive, 211, 214
multiplicative, 93
of addition, 11
of multiplication, 20
Inverse operations: a pair of operations that "do" and "undo" an operation addition/subtraction, and multiplication/division are pairs of inverse operations, 11
addition and subtraction, 11
multiplication and division, 20
Inverse proportion: a proportion in which the quantities being compared are inversely related so that as one quantity increases (or decreases), the other quantity decreases (or increases). This relationship is also called inverse variation, 366
constant of, 368
distinguish from direct, 370

Inverting: interchanging the numerator and denominator of a fraction, 93
Investments, 621
Irrational number: the root of a nonperfect power is an example of an irrational number. The decimal equivalent is nonterminating and nonrepeating, 486
Isolate, 307, 312
Isosceles right triangle, 768
Isosceles triangle: a triangle with exactly two equal sides. The angles opposite the equal sides are also equal, 757

j-factor, 506
Joint variation, 361
constant of, 361

Kelvin, 184
Key words
addition, 46, 304
division, 46, 304
equality, 46
multiplication, 46, 304
negative, 217
positive, 217
subtraction, 12, 46, 304
Kilo-, 171
Kilogram: the most often used unit for measuring weights in the metric system, 174
Kilometer: 100 meters; the kilometer is used to measure long distances, 173

Lateral face, 733
Lateral surface area: for a three-dimensional figure, the area of the sides only (bases excluded), 735
see also Surface area
Laws of exponents, 454
limitations, 460
rational exponents, 491
LCD, 78
see also Least common denominator
LCM, 67, 78
see also Least common multiple
Leading coefficient: the coefficient of the leading term of a polynomial, 464, 573
Leading term: the first term of a polynomial arranged in descending order, 464, 489, 573
Least common denominator (LCD): the smallest common denominator among two or more fractions, 78
Least common multiple: the least common multiple (LCM) of two or more natural numbers is the smallest number that is a multiple of each number. The LCM is divisible by each number, 67, 68
Leg: one of the two sides of a right triangle that is not the hypotenuse. Or, one of the two sides that forms the right angle, 765
Leg of an angle, 696
Length
metric, 172
U.S. customary, 101
Less than (<), 3
Letter term
see also Variable, 302
Like bases, 454
Like denominators, 9

multiplication, 90
subtraction, 85
Mode: the most frequently occurring quantity or quantities among the quantities considered in a data set, 268
Monomial: a polynomial containing one term. A term may have more than one factor, 461
Multiple: the product of a given number and any natural number, 61
Multiplicand: the first number in a multiplication problem, 15
Multiplication
    associative property, 15
    axiom, 309
    binomials (FOIL), 469
    check, 19
    commutative property, 15
    complex number, 509
    conjugate pairs, 500
    decimals, 16
    distributive property, 17
    ending zeros, 20
    estimation, 18
    fractions and mixed numbers, 89, 90
    identity, 220
    key words, 46, 304
    metric system, 180
    notations, 15, 301
    polynomials, 467, 469
    powers, 221, 454
    powers of 10, 235
    property of equality, 309
    radical expressions, 509, 547
    scientific notation, 241
    signed numbers, 219
    special products, 470
    square-root radicals, 498
    U.S. customary system, 109
    whole numbers, 16
    zero property, 15, 221
Multiplication axiom: both sides of an equation may be multiplied by the same nonzero quantity without changing the equality of the two sides. This axiom also applies to dividing both sides of an equation by the same nonzero quantity, 309
Multiplication axiom
    cross products, 355
Multiplication: repeated addition, 5
Multiplicative identity: one is the multiplicative identity because $a * 1 = 1 * a$ for all values of a, 93, 133
Multiplicative inverse, 93
Multiplier: the number to multiply by, 15

Natural exponential, e: an irrational number that is the limit that the value of the expression $\left(1 + \frac{1}{n}\right)^n$ approaches as n gets larger and larger without bound, 627
Natural logarithm, 638
Natural numbers: the set of counting numbers beginning with 1, 2, 3, and continuing indefinitely, 2, 63
Natural numbers
    exponent, 30
Negative
    key words, 217
Negative integer exponent: a notation for writing reciprocals $n^{-1} = \frac{1}{n}$ or $\frac{1}{n^{-1}} = n$, 234
Negative reciprocal, 415

New amount: when working with the amount of increase or decrease in percent problems, the original amount plus or minus the amount of change, 155
Nonnegative: positive or zero, $n > 0$, 210
Nonzero digit: a digit that is not a zero (1, 2, 3, 4, 5, . . .), 10, 32
Normal, 414
    see also Perpendicular
Normal distribution, 278
Notation
    angle, 696
    decimal, 4
    degree, 721
    degree and radian, 721
    exponential, 30, 236, 485
    fraction, 3
    interval, 657
    line, 695
    line segment, 695
    multiplication, 15, 18, 301
    ordinary, 238
    point, 382
    point, line, plane, 695
    point-slope form, 409
    powers and roots, 488
    radian, 721
    radical, 31, 485, 488
    rational exponents, 488
    ray, 695
    scientific notation, 238
    set, 655
    set-builder, 656
    standard, of measures, 106
    standard, of powers, 30
Number
    cardinal, 3
    ordinal, 3
Number line, 2
Number sense, 10
Number term or constant: a term that contains only numbers, 463
Numerator: the numerator of a fraction is the number of the parts being considered. It is the top number of a fraction, or the dividend of the indicated division, 3, 61
Numerical average: the sum of a list of values divided by the number of values, 25
    see also Mean
Numerical coefficient: the numerical factor of a term, 302
Numerical term or constant: a term that contains only numbers, 573

Oblique cylinder, 732
Oblique prism, 732
Obtuse angle: an angle that is more than 90° but less than 180°, 697
Octagon
    regular, 774
Odd numbers: integers that are not multiples of 2, 62
One-dimensional graph: a visual representation of the distance and direction that a value is from 0; for example a number line, 382
Opposite: a number that is the same number of units from 0 but in the opposite direction of the original number, 209
Or relationship of inequalities, 668

long division, 476
multiplication, 467
order, 463
prime, 520
special products, 470
Portion: in a problem involving percent, the number (P) that represents a portion of the base. Also called percentage, 139
identify in applied problems, 152
Positive
key words, 217
Pound, 102
Power: another term for the exponent or the result of raising a value to a power, 30
calculator, 34
divide, 455
fraction, 93, 459
imaginary number, 507
logarithm, 636
multiplication, 454
negative integer exponents, 234
perfect, 485
power of power, 458
product, 459
scientific notation, 244
signed numbers, 221
zero, 456
see also Powers of 10
Powers of 10: numbers whose only nonzero digit is one: 10, 100, 1,000 are examples of powers of 10, 3, 32, 233
engineering notation, 245
in expanded exponential, 236
multiplication and division, 32, 235
power, 236
scientific notation, 33, 238
Precision, 191, 192
Prefixes
common metric for large and small amounts, 172
metric, 171, 172, 244
Present value, 623
Prime, 64
polynomial, 520, 534
Prime factor: a factor that is divisible only by one and the number itself, 66
Prime factorization: writing a composite number as the product of only prime numbers, 66
Prime number: a whole number greater than 1 that has only one factor pair, the number itself and 1, 64
Prime polynomial, 534
Principal
simple interest, 329
Principle of equality, 307
Prism: a three-dimensional figure whose polygonal bases (ends) are parallel, congruent polygons and whose faces (sides) are parallelograms, rectangles, or squares. In a right prism the faces are perpendicular to the base or bases, 44, 731
oblique, 732
right, 731
right rectangular, 44
surface area, 736
volume, 44, 732
Probability: the chance of an event occurring. It is expressed as a ratio or percent of the number of possibilities for success to the total number of possibilities, 284

Problem solving
guess and check, 46
keywords, 46
linear equations, 317
six-step strategies, 45
stategies, 318
Product: the result of multiplication, 15
Product
cross, 354
Proper fraction: a common fraction whose value is less than one unit; that is, the numerator is less than the denominator, 61
Property of proportions, 354
Proportion: a mathematical statement that shows two fractions or ratios are equal; an equation in which each side is a fraction or ratio, 144, 145, 354
direct variation, 357
inverse variation, 366
problems involving similar triangles, 760
property of equality, 354
solving, 145
Protractor, 701
Pure quadratic equation: a quadratic equation that has a quadratic term and a constant term but no linear term, 573, 574
Pyramid, 733
frustum, 734
right or regular, 733
Pythagorean theorem: the theorem that states that the square of the hypotenuse of a right triangle equals the sum of the squares of the two legs of the triangle, 765

Quadrant, 384
Quadratic equation: an equation in which at least one letter term is raised to the second power and no letter terms have a power higher than 2 or less than 0. The standard form for a quadratic equation is $ax^2 + bx + c = 0$, where a, b, and c are real numbers and $a > 0$, 572
complete quadratic equations, 573, 578
graphing in two variables, 588
identifying coefficients, 573
incomplete quadratic equations, 573, 576
nature of roots, 595
pure quadratic equations, 573, 574
quadratic formula, 583
solve by completing the square, 580
solve by factoring, 576
solve by the square-root method, 574
standard form, 572
use of the discriminant in solving, 595
Quadratic formula, 583
Quadratic inequality, 673
graph, 676
Quadratic polynomial: a polynomial that has degree 2, 463
Quadratic term: a term that has degree 2, 463, 573
Quart, 102
Quotient: the result of a division problem, 20

Radian, rad: measure of a central angle of a circle with an intercepted arc that is equal to the radius of the circle. The abbreviation for radian is rad, 720
degrees to radians, 721
radians to degrees, 722
Radical
notation, 485

Radical expression: an expression including the radical sign and radicand that indicates a root such as square root or cube root, 31
 simplify, 504
Radical notation, 31, 488
Radical sign: an operational symbol indicating that a root is to be taken of the number under the bar portion of the radical sign, 31
Radicals
 like, 497
Radicand: the number or expression under the radical sign, 31
Radius: a straight line segment from the center of a circle to a point on the circle. It is half the diameter, 717
Range: the difference between the highest quantity and the lowest quantity in a set of data, 274, 385
Range
 relation, 601
Rankine, 184
Rate
 identify in applied problems, 152
Rate measure: the ratio of two different kinds of measures. It is often referred to as a rate. Some examples of rates are miles per hour (mi/h) and gallons per minute, 112, 325
Rate of work: the ratio of the amount of work completed to the time worked, 325
Ratio: a fraction comparing a quantity or measure in the numerator to a quantity or measure in the denominator, 102, 144, 354
Ratio, 354
Rational equation: an equation that contains one or more rational expressions, 558
 excluded values, 558
Rational exponent, 488
Rational expressions: an algebraic fraction in which the numerator or denominator or both are polynomials, 543, 544
 addition or subtraction, 553
 complex, 549
 multiply or divide, 547
 rationalize numerator or denominator, 551
 simplify, 545
Rational inequality, 677
Rational number, 225, 506
Rationalize, 502
 algebraic fractions, 551
Ray: a ray consists of a point on a line and all points of the line on one side of the point, 695
Real number, 8, 487
Real part of a complex number, 508
Real solutions, 595
Reciprocals: two numbers are reciprocals if their product is 1. For example, 1/2 and 2 are reciprocals because (1/2) (2) = 1; negative exponents are used for reciprocals, n and n^{-1} are reciprocals, n and $1/n$ are reciprocals, 93, 234, 456
Reciprocals
 negative, 415
Rectangle: a parallelogram whose angles are all right angles, 19, 704
Rectangle
 area, 19, 42, 708
 perimeter, 18, 40, 705
Rectangular coordinate system: a graphical representation of two-dimensional values. It is two number lines positioned to form a right angle or square corner, 382
 horizontal axis, 382
 origin, 382

plot a point, 383
 point notation, 382
 vertical axis, 382
 x-coordinate, 382
 y-coordinate, 382
Rectangular prism, 44
Reduce to lowest terms: to find an equivalent fraction that has smaller numbers and has no common factors in the numerator and denominator, 71, 90
Reference line, 259
Regrouping
 with addition, 9
 with subtraction, 12
Regular polygon: a polygon with equal sides and equal angles, 774
Regular polygons
 area, 775
Relation, 385, 600
 domain and range, 601
Relative error, 193
Remainder: the answer to a subtraction, 11
Remainder
 in division, 21
Repeating decimals, 77
Repetitions, 281
Result of exponentiation, 636
Rhombus, 704
 area, 708
 perimeter, 705
Richter scale, 642
Right angle: a right angle (90°) represents one-fourth of a circle or one-fourth of a complete rotation, 39, 697
Right cylinder, 731
Right prism, 731
Right rectangular prism, 44
Right triangle: a triangle that has one right (90°) angle. The sum of the other two angles is 90°, 765
Ring, 719
Rise: the change (difference) in the y-coordinates, 403
Root or solution: the value of the variable that makes the equation true, 307
 check or verify, 311
 double, 579, 595
 extraneous, 321
 nature of roots, 595
 real or complex, 595
Roots, 31
 calculator, 487
 square roots, 31
Roster, 655
Round: to express a number as an approximation, 6
 measurement, 194
 to a place value, 6
Rule
 metric, 195
 U.S. customary, 72
Run: the change (difference) in the x-coordinates, 403

Sampling with replacement, 222
Scalene triangle: a triangle with all three sides unequal, 757
Scientific notation: a number is expressed in scientific notation if it is the product of two factors. The absolute value of the first factor is a number greater than or equal to 1 but less than 10. The second factor is a power of 10, 238

solving by addition method, 434
solving by graphing, 432
solving by substitution method, 439
solving with a calculator, 433

Table of solutions or values, 386
    calculator, 387
    linear equation in two variables, 386
    linear function, 386
Tally: a process for determining the number of values in each class interval, 269
Tangent to: intersecting in exactly one point, 776
Taper, 359
Temperature conversions
    Celsius to Fahrenheit, 185
    Fahrenheit to Celsius, 184
Term: algebraic expressions that are single quantities (terms) or quantities that are added or subtracted. Algebraic expressions contain at least one variable term, 301
    binomial, 461
    combine, 466
    degree, 462
    leading term, 464
    like, 465
    monomial, 461
    trinomial, 461
Test points
    graphing inqualities, 664
    inequalities, 673
Theta (θ), 2
Thousands, 723
Threshold sound, 640
Time, 112, 183
    calculations, 183
    conversions, 183
    rate measures, 112
Toggle key: a calculator key that turns a feature on and off or that alternates between features, 234
Tolerance: the amount the part can vary from a blueprint specification, 13, 74, 84
Ton, 102
Total: the answer of an addition problem, also called sum, 8
Total surface area: for a three-dimensional figure the area of the sides plus the area of the base or bases, 736
Transversal, 699
Trapezoid: a four-sided polygon having only two parallel sides, 704, 706
Trapezoid
    area, 709
    perimeter, 705
Tree diagram: a method of counting the ways the elements in a set can be arranged; it allows each new set of possibilities to branch out from a previous possibility, 281
Triangle: a polygon that has three sides, 704, 706
    30°, 60°, 90°, 769
    45°, 45°, 90°, 768
    area, 709
    circumscribed, 778
    congruent, 758
    corresponding sides and angles, 760
    equilateral, 757, 769
    Heron's formula, 709
    inscribed, 778

    isosceles, 757
    isosceles right, 768
    perimeter, 705
    Pythagorean theorem, 765
    right triangles, 765
    scalene, 757
    sides and angles, 757
    similar, 760
    special relationships, 771
Trinomial: a polynomial containing three terms. Examples include $a + b + c, x^2 + 2x - 1$, 461
    factoring by grouping, 530
    general, 527
    perfect square, 471
Two-dimensional graph, 382

U.S. customary system: the system of measurement commonly used in the United States. Formerly known as the English system, 101
    addition, 107
    capacity, 102
    conversions within system, 103
    division, 110
    length, 101
    metric conversion factors, 188
    mixed measures, 108
    multiplication, 109
    rate measures, 112
    rule, 72
    subtraction, 108
    using conversion factors, 105
    volume, 102
    weight or mass, 101
Unbounded, 657
Undefined, 23, 223
Undefined slope, 407
Union, 668
Unit cost, 360
Unit ratio: a ratio of measures whose value is 1, 102
Units, 2
Universal set, 667
Unknown, 300
Unlike measures: measures with different units. Before any mathematical process can be done the measures need to be changed so that all measures have the same units, 107
Unlike signs
    addition, 212, 213
    multiplication, 219

Variable or unknown: a letter that represents an unknown or missing number in an equation, 300
    dependent, 357
    exponent, 620
    independent, 357
Variable term: a term that contains one letter, several letters used as factors, or a combination of letters and numbers used as factors, 302
Variance, 276
Variation
    combined, 370
    constant of combined, 370
    constant of direct, 359
    constant of inverse, 368

Variation *(continued)*
    constant of joint, 361
    direct, 357
    inverse, 366
    joint, 361
Verify root or solution, 311
Vertex of a cone or pyramid, 733
Vertex of an angle, 696
Vertex of a parabola: the point of the graph of a parabola that crosses the axis of symmetry, 589
Vertical angle, 699
Vertical axis: the number line that runs from top to bottom on a rectangular coordinate system. It is represented by the letter *y*, 382
Vertical line, 697
    equation, 411
Vertical line test, 600
Volume: the amount of space a three-dimensional geometric figure occupies, measured in terms of three dimensions (length, width, and height). Measures of volume will always be cubic measures, 44, 732
    cone, 733
    cube, 44
    cylinder, 732
    frustum, 734
    prism, 44
    pyramid, 733
    right cylinder, 732
    right prism, 732
    sphere, 735
    see also Capacity

Watt, 205
Week, 112
Weight
    metric, 174
    U.S. customary, 101
Whole number: a number made up of one or more digits; the set of natural numbers and 0, 2
    addition, 9
    compare, 2

decimal number system, 2
division, 21
exponents, 30
improper fraction conversions, 74
multiplication, 16
percent equivalents, 134
place values, 2
place-value chart, 2
powers of 10, 4, 32
rounding, 6
standard notation, 30
subtraction, 12
Without repetition, 281
Without replacement, 281
Work: the result of the rate of work times the time worked, 325

x-coordinate: the value of the horizontal movement of a point on a rectangular coordinate system, 382
x-intercept: the point on the *x*-axis through which the line of the equation passes; that is, the *y*-value is zero $(x, 0)$, 393

yard, 101
y-coordinate: the value of the vertical movement of a point on a rectangular coordinate system, 382
y-intercept: the point on the *y*-axis through which the line of the equation passes; that is, the *x*-value is zero $(0, y)$, 393

Zero
    as exponent, 234
    in addition, 9
    in division, 223
    in multiplication, 20, 221
    in subtraction, 11, 12
Zero function
    calculator, 590
Zero property of addition: the sum of zero and any number is the number itself, 9
Zero property of multiplication: the product of zero and any number is 0, 15, 221
Zero slope, 407
Zero-product property, 576

Geometric Formulas

Perimeter and Area of Two-Dimensional Figures

Square

Perimeter $= 4s$, where $s =$ length of a side of the square
Area $= s^2$, where $s =$ length of the side of the square

Rectangle

Perimeter $= 2l + 2w$, where $l =$ length of the rectangle and $w =$ width of the rectangle
Area $= lw$, where $l =$ length of the rectangle and $w =$ width of the rectangle

Parallelogram

Perimeter $= 2b + 2s$, where $b =$ the base and $s =$ length of the adjacent side
Area $= bh$, where $b =$ length of the base and $h =$ height of the parallelogram

Triangle

Perimeter $= a + b + c$, where a, b, and c are the lengths of the sides of the triangle
Area $= \frac{1}{2}bh$, where $b =$ base of triangle and $h =$ height of triangle
Area of a triangle if the *height* is not known (Heron's formula):

Area $= \sqrt{s(s - a)(s - b)(s - c)}$, where $s = \frac{1}{2}(a + b + c)$ and $a =$ side 1, $b =$ side 2, and $c =$ side 3 of the triangle

Trapezoid

Perimeter $= a + b + c + d$, where a, b, c, and d are the lengths of the sides

Area $= \frac{1}{2}h(b_1 + b_2)$, where $h =$ height of trapezoid, $b_1 =$ length of short base, and $b_2 =$ length of long base

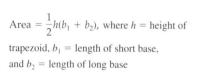

Circle

Circumference $= 2\pi r$ or πd, where $r =$ radius and $d =$ diameter
Area $= \pi r^2$, where $r =$ radius

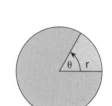

Sector

Area of a sector $= \frac{\theta}{360°}(\pi r^2)$, where $\theta =$ angle (measured in degrees) that forms the sector and $r =$ radius of circle

Area of a sector $= \frac{\theta r^2}{2}$, where $\theta =$ angle (measured in radians) that forms the sector and $r =$ radius of circle

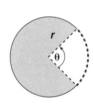

Arc Length

Length of an arc $= \frac{\theta}{360°}(2\pi r)$, where $\theta =$ angle (measured in degrees) that forms the arc and $r =$ radius of circle
Length of an arc $= r\theta$, where $\theta =$ angle (measured in radians) that forms the sector and $r =$ radius of circle

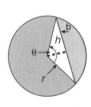

Segment

Area of a segment $= \frac{\theta}{360°}(\pi r^2) - \frac{1}{2}bh$,

$\theta =$ angle (measured in degrees), $r =$ radius of the circle, $b =$ base of triangle formed by two radii and the chord and $h =$ height of the triangle.

Area and Volume of Three-Dimensional Objects

Cube

Lateral Surface Area $= 4s^2$, where $s =$ length of one side of cube
Total Surface Area $= 6s^2$, where $s =$ length of one side of cube
Volume $= s^3$, where $s =$ length of one side of cube

Right Rectangular Prism or Right Circular Cylinder

Lateral Surface Area (LSA) $= ph$ where $p =$ perimeter of the base and $h =$ height of the three-dimensional figure

Total Surface Area (TSA)$= ph + 2B$, where $p =$ perimeter of the base and $h =$ height of the three-dimensional figure and $B =$ area of the base.